Věra Kůrková
Nigel C. Steele
Roman Neruda
Miroslav Kárný (eds.)

Artificial Neural Nets and Genetic Algorithms

Proceedings of the International Conference in Prague, Czech Republic, 2001

Springer-Verlag Wien GmbH

Dr. Věra Kůrková
Institute of Computer Science

Dr. Roman Neruda
Institute of Computer Science

Dr. Miroslav Kárný
Institute of Information Theory and Automation

Academy of Sciences of the Czech Republic
Prague, Czech Republic

Dr. Nigel C. Steele
Division of Mathematics
School of Mathematical and Information Sciences
Coventry University, Coventry, U.K.

Originally published by Springer- Verlag Wien New York in 2001

Camera-ready copies provided by authors and editors
Printed by Novographic Druck G.m.b.H., A-1230 Wien
Printed on acid-free and chlorine-free bleached paper
SPIN 10832776

With 375 Figures

ISBN 978-3-211-83651-4 ISBN 978-3-7091-6230-9 (eBook)
DOI 10.1007/978-3-7091-6230-9

Preface

The first *ICANNGA* conference, devoted to biologically inspired computational paradigms, *Neural Networks* and *Genetic Algorithms*, was held in Innsbruck, Austria, in 1993. The meeting attracted researchers from all over Europe and further afield, who decided that this particular blend of topics should form a theme for a series of biennial conferences. The second meeting, held in Ales, France, in 1995, carried on the tradition set in Innsbruck of a relaxed and stimulating environment for the exchange of ideas. The series has continued in Norwich, UK, in 1997, and Portoroz, Slovenia, in 1999.

The Institute of Computer Science, Czech Academy of Sciences, is pleased to host the fifth conference in Prague. We have chosen the Liechtenstein palace under the Prague Castle as the conference site to enhance the traditionally good atmosphere of the meeting. There is an inspirational *genius loci* of the historical center of the city, where four hundred years ago a fruitful combination of theoretical and empirical method, through the collaboration of Johannes Kepler and Tycho de Brahe, led to the discovery of the laws of planetary orbits.

In this conference, theoretical insights and reports on successful applications are both strongly represented. It is especially pleasing to observe the dual affinity with biology. There are papers dealing with cognition, neurocontrol, and biologically inspired brain models, while there are also descriptions of successful applications of computational methods to biology and environmental science. Theoretical contributions cover a variety of topics including nonlinear approximation by feedforward networks, representation of spiking perceptrons by classical ones, recursive networks and associative memories, learning and generalization, population attractors, and proposals and analyses of new genetic operators or measures. These theoretical studies are augmented by a wide selection of application-oriented papers on topics ranging from signal processing, control, pattern recognition and time series prediction to routing tasks.

To keep track of the rapid development of the field of computational intelligence and to further extend the original focus of ICANNGA on blending various computational methods and approaches to solve hard real-world problems, we have extended the scope of the conference by adding some special sessions.

The first one, *Hybrid Methods and Tools*, concentrates on theoretical, algorithmic and applied aspects of combinations of methods from the soft computing area, neural networks, evolutionary algorithms, fuzzy logic and probabilistic computing and symbolic artificial intelligence.

The second one, *Computer-Intensive Methods in Control and Data Processing*, aims to attack the "curse of dimensionality". It specializes in methods for handling the problem of implementation of optimal mathematical procedures of approximation, inference and decision-making in a variety of fields such as control, data processing, image reconstruction, pattern recognition and nonparametric estimation. This session is a continuation of a series of workshops organized by the Institute of Information Theory and Automation that have been held in Prague every two years since 1992.

To emphasize the importance of applications of combined computational methods to environmental issues, we have also included a special session devoted to *Data Mining in Meteorology and Air Pollution.* The session focuses on data mining and statistical methods applied to modelling and forecasting of highly complex time series and spatio-temporal data from air pollution and meteorology. The methods include neural networks, genetic algorithms, statistical data processing, wavelet functions, adaptive filters and rule-based approaches.

It is our pleasure to express our gratitude to everyone who contributed in any way to the success of the event and the completion of this volume. In particular, we thank ICANNGA's founding members, who, via the Advisory Committee, keep the tradition of the series and help with the organization. We thank all the members of the Programm Committee for careful and prompt reviewing of the submissions. With deep gratitude we thank the members of the Organizing Committee, whose efforts made the vision of the 5th ICANNGA a practical reality. Our excellent ICANNGA secretary Hana Bílková deserves special thanks. We thank Action M Agency for perfect local arrangements. Last but not least, we thank all authors who contributed to this volume to share their new ideas and results with the community of researchers in this rapidly developing field of biologically motivated computer science.

We hope that you enjoy reading and find inspiration for your future work in the papers contained in this volume.

Věra Kůrková, Roman Neruda, Miroslav Kárný

Contents

Part IV: Genetic Algorithms – Theory

Part V: Genetic Algorithms – Applications

Part VI: Soft Computing – Applications

Part VII: Hybrid Methods and Tools

Part VIII: Probabilistic Models and Clustering

Part IX: Data Mining in Meteorology and Air Pollution

ICANNGA 2001

International Conference on Artificial Neural Networks and Genetic Algorithms
Prague, Czech Republic, April 22-25, 2001

International Advisory Board

Programme Committee

What Can a Neural Network Do without Scaling Activation Function?

Yoshifusa Ito*

*Aichi-Gakuin University, Nisshin-shi, Aichi-ken, Japan, E-mail: ito@psis.aichi-gakuin.ac.jp

Abstract

It has been proved that a three layered feedforward neural network having a sigmoid function as its activation function can 1) realize any mapping of arbitrary n points in $\mathbf{R}^d$ into $\mathbf{R}$, 2) approximate any continuous function defined on any compact subset of $\mathbf{R}^d$, and 3) approximate any continuous function defined on the compactification of $\mathbf{R}^d$, without scaling the activation function [3],[4],[6],[11]. In this paper, these results are extended doubly in the sense that the activation function defined on $\mathbf{R}$ is not restricted to sigmoid functions and the concept of activation function is extended to functions defined on higher dimensional spaces $\mathbf{R}^c$ ($1 \leq c \leq d$). In this way sigmoid functions and radial basis functions are treated on a common basis. The conditions on the activation function for 1) and 2) are exactly the same, but more restricted for 3).

1 Introduction

Many authors have proved that any continuous function defined on any compact set of the Euclidean space $\mathbf{R}^d$ can be well approximated by a three layered neural network having hidden layer units equipped with a nonlinear activation function. Though they supposed that the activation functions can be scaled, it was later shown that the approximation of continuous functions on compact sets and on the compactification $\bar{\mathbf{R}}^d$ of $\mathbf{R}^d$ can be realized without scaling activation functions [3], [4], [6]. Moreover, the necessary and sufficient condition on the sigmoid function for the approximation on $\bar{\mathbf{R}}^d$ was obtained [6]. It is not difficult to show that a three layered neural network having n hidden layer units can implement a mapping of n points in $\mathbf{R}^d$ into $\mathbf{R}$ [5] if the activation function can be scaled, but this mapping can also be realized without scaling activation functions [11].

The standard three layered neural network has d linear input layer units, a finite number, say n, of nonlinear hidden layer units and one linear output layer unit. The number d coincides with the dimension of the space where the input pattern is defined, as the inputs to the individual input units are the components of $x \in \mathbf{R}^d$. The number n is rather flexible. It is usually supposed that any number of hidden layer units are available depending on necessity. The activation function defined on $\mathbf{R}$ was originally restricted to sigmoid functions in many literatures, but soon extended to bounded functions [2]. Nowadays the activation function is not restricted to those defined on $\mathbf{R}$. Higher dimensional functions, such as a radial basis function on $\mathbf{R}^d$, are commonly used.

This paper focuses on the use of activation functions without scaling. We extend the activation function to higher dimensional slowly increasing functions, so that various kinds of functions defined on $\mathbf{R}^c$, $1 \leq c \leq d$, are treated on a common basis. The topics are restricted to the realization of three kinds of functions. They are the mapping of a finite number of points in $\mathbf{R}^d$ into $\mathbf{R}$, the approximate realization of continuous functions defined on compact sets $\mathbf{K} \in \mathbf{R}^d$ and that of continuous functions defined on $\bar{\mathbf{R}}^d$. It is shown that if the support of the Fourier transform of a slowly increasing continuous function on $\mathbf{R}^c$ includes a dense subset of a nonempty open set, it can be an activation function for both the mapping of points and the approximation on compact sets without scaling. If a continuous function converges to 0 as $x \to \infty$ and the support of its Fourier transform is dense in $\mathbf{R}^c$, it can be an activation function for the approximation on $\bar{\mathbf{R}}^d$ without scaling. In these results the continuity condition of activation functions can be weakened to some extent. The results extend the theorems described in [3], [4], [6] and [11]. Moreover, some of the proofs are improved, which leads on to the treatment of various kinds of activation functions in the same framework.

This paper refrains from some details, which are described in [8], [9], [10].

2 Preliminaries

We denote by $\mathbf{S}^{d,c}$, $1 \leq c \leq d$, a set of $c \times d$ matrices having orthonormal row vectors. Then, $\mathbf{S}^{d,1}(= \mathbf{S}^{d-1})$ is the unit sphere in $\mathbf{R}^d$ and $\mathbf{S}^{d,d}(= O(d))$ is the orthogonal transformation group of order d.

For $W \in \mathbf{S}^{d,c}$, set

$$\mathbf{H}_W = \{x \in \mathbf{R}^d | Wx = 0\},$$

$$\mathbf{L}_W = \{{}^tWz | z \in \mathbf{R}^c\}.$$

The spaces $\mathbf{H}_W$ and $\mathbf{L}_W$ are $d-c$ dimensional and c dimensional spaces respectively, and $\mathbf{R}^d = \mathbf{H}_W \oplus \mathbf{L}_W$ holds. Let $W^\perp$ be a $(d-c)\times d$ matrix such that $\begin{pmatrix} W \\ W^\perp \end{pmatrix} \in O(d)$. Then, ${}^tWWx$ and ${}^tW^\perp W^\perp x = x - {}^tWWx$ are the projections of $x \in \mathbf{R}^d$ onto $\mathbf{L}_W$ and $\mathbf{H}_W$ respectively. Let $W \in \mathbf{S}^{d,c}$ and let $z \in \mathbf{R}^c$. For a function g defined on $\mathbf{R}^c$, we define three functions g_z on $\mathbf{R}^c$, g^W and g_z^W on $\mathbf{R}^d$ by $g_z(x) = g(x-z)$, $g^W(x) = g(Wx)$, and $g_z^W(x) = g(Wx - z)$ respectively.

If g is locally integrable and there is n for which $g(z)(1+|z|^2)^{-n}$ is bounded, g is called a slowly increasing function. The Fourier transform $\mathcal{F}_c g$ of an integrable function g on $\mathbf{R}^c$ is defined by $\mathcal{F}_c g(y) = \int_{\mathbf{R}^c} g(z) e^{-\sqrt{-1}y\cdot z} dz$. This transform can be extended in the sense of distribution. Though we do not mention each time, measures are signed Borel measures in this paper. Let g be a slowly increasing function defined on $\mathbf{R}^c$. If $c < d$, the Fourier transform of g_z^W, $W \in \mathbf{S}^{d,c}$, $z \in \mathbf{R}^c$, is a tensor product of the Fourier transform of g on $\mathbf{L}_W$ and the delta function on $\mathbf{H}_W$ [9]:

$$\mathcal{F}_d g_z^W(y) = \delta_{\mathbf{H}_W}(W^\perp y) \mathcal{F}_c g(Wy) e^{-\sqrt{-1}z\cdot Wy}. \quad (1)$$

Note that this expression doesn't depend on the choice of $W^\perp$. In the case where g is a sigmoid function, the Fourier transform $\mathcal{F}_d g^W$, $W \in \mathbf{S}^{d-1}$, is supported by a line $\mathbf{L}_W = \{Wt | -\infty < t < \infty\}$ [4], [6]. If g is a slowly increasing function and ν is a measure with compact support, or g is a bounded function and ν is a measure of bounded variation, the convolution $g*\nu$ is well defined and we have that $\mathcal{F}_d(g^W * \nu) = \mathcal{F}_d g^W \mathcal{F}_d \nu$.

3 Mapping of finite points

We may omit the proofs of the two lemmas below.

Lemma 1. Let $x_1, \cdots x_n$ be distinct points in $\mathbf{R}^d$. Then, there is $W \in \mathbf{S}^{d,c}$ for which $Wx_1, \cdots, Wx_n$ are distinct points in $\mathbf{R}^c$.

We say that functions $g_1, \cdots, g_n$ are linearly independent on a set $\mathbf{G}$, if $\sum_{i=1}^n a_i g(x) = 0$ for all $x \in \mathbf{G}$ implies $a_1, \cdots, a_n = 0$.

Lemma 2. Let $z_1, \cdots z_n$ be distinct points in $\mathbf{R}^c$. Then, $e^{-\sqrt{-1}z_i\cdot z}$, $i = 1, \cdots, n$, are linearly independent on any dense subset of a nonempty open set.

Lemma 3. Let g be a slowly increasing function on $\mathbf{R}^c$ such that the support of $\mathcal{F}_c g$ is dense in a nonempty open set, and let z_i, $i = 1, \cdots, n$, be distinct points in $\mathbf{R}^c$. Then, the g_{z_i} are linearly independent in $\mathbf{R}^c$.

Proof. Let $\mathbf{G}$ be a dense subset of a nonempty open set included in the support of $\mathcal{F}_c g$. Suppose that $\sum_{i=1}^n a_i g_{z_i} = 0$. Then,

$$\mathcal{F}_c \sum_{i=1}^n a_i g_{z_i}(x) = \mathcal{F}_c g(x) \sum_{i=1}^n a_i e^{-\sqrt{-1}z_i\cdot z} = 0$$

on $\mathbf{R}^c$. Hence,

$$\sum_{i=1}^n a_i e^{-\sqrt{-1}z_i\cdot z} = 0 \qquad \text{on} \qquad \mathbf{G},$$

Accordingly, $a_1 = \cdots = a_n = 0$ by Lemma 2. Thus, the lemma follows.

Let $\mathbf{F} = \{x_1, \cdots, x_n\}$ be a set of distinct points in $\mathbf{R}^d$. A mapping of $\mathbf{F}$ into $\mathbf{R}$ by the neural network is denoted by

$$\bar{f}(x_i) = \sum_{j=1}^n a_j g(W_j x_i - z_j), \quad i = 1, \cdots, n. \quad (2)$$

We can regard (2) as simultaneous equations with respect to a_j. If the coefficient matrix $(g(W_j x_i - z_j))_{i,j=1}^n$ is nonsingular we can realize any finite mapping from $\mathbf{F}$ into $\mathbf{R}$, determining the coefficients a_j by Cramér's formula [5], [11].

Theorem 4. Let g be a slowly increasing function defined on $\mathbf{R}^c$, $c \le d$, such that the support of $\mathcal{F}_c g$ is dense in a nonempty open set. Then, for any distinct points $x_i \in \mathbf{R}^d$ and any constants b_i, $i = 1, \cdots, n$, there exist coefficients a_j, matrices $W_j \in \mathbf{S}^{d,c}$ and points $z_j \in \mathbf{R}^c$, $j = 1, \cdots, n$, for which

$$\sum_{j=1}^n a_j g(W_j x_i - z_j) = b_i, \quad i = 1, \cdots, n. \quad (3)$$

Proof. By Lemma 1, there is a matrix $W_j \in \mathbf{S}^{d,c}$ for which $W_j x_i$, $i = 1, \cdots, j$ are distinct. Then, by Lemma 3, $g(W_j x_i - z)$, $i = 1, \cdots, j$, are linearly independent. Hence, there is $z_1 \in \mathbf{R}^c$ for which $g(W_1 x_1 - z_1) \ne 0$. If the determinant $|g(W_j x_i - z_j)|_{i,j=1}^k$ is nonzero, we can obtain $c_1, \cdots, c_k$ for which $\Sigma_{i=1}^k c_i g(W_j x_i - z_j) + g(W_j x_{k+1} - z_j) = 0$ for $j = 1, \cdots, k$, applying Cramér's formula. Hence, we obtain that

$$|g(W_j x_i - z_j)|_{i,j=1}^{k+1}$$

$$= \begin{vmatrix} g(W_1x_1 - z_1) & \cdots & g(W_{k+1}x_1 - z_{k+1}) \\ \cdots & \cdots & \cdots \\ g(W_1x_k - z_1) & \cdots & g(W_{k+1}x_k - z_{k+1}) \\ 0 & \cdots & 0 \quad \sum_{i=1}^{k+1} c_i g(W_{k+1}x_i - z_{k+1}) \end{vmatrix}$$

$$= \sum_{i=1}^{k+1} c_i g(W_{k+1}x_i - z_{k+1}) |g(W_j x_i - z_j)|_{i,j=1}^{k},$$

where $c_{k+1} = 1$. Since $g(W_{k+1}x_i - z)$ are linearly independent and $c_{k+1} = 1 \neq 0$, there is a point $z_{k+1} \in \mathbf{R}^c$ for which $\sum_{i=1}^{k+1} c_i g(W_{k+1}x_i - z_{k+1}) \neq 0$, implying that $|g(W_j x_i - z_j)|_{i,j=1}^{k+1} \neq 0$. Thus, we obtain $|g(W_j x_i - z_j)|_{i,j=1}^{n} \neq 0$. Applying Cramér's formula again, we can finally obtain the coefficients a_j, $j = 1, \cdots, n$, for which (3) holds.

This theorem implies that any mapping of any distinct n points in $\mathbf{R}^d$ into $\mathbf{R}$ can be realized by a three layered neural network without scaling the activation function. The matrices W_j can be the same. The number of the hidden layer units n is the upper bound which cannot be decreased. However, if the mapping does not need to be exact and/or the activation function can be scaled, the number of the hidden layer units may be considerably decreased (See [12]). The linear independence of linear transforms of a function on $\mathbf{R}$ is discussed in [7] and [15].

4 Approximation on compact sets

In this section and the next, the usual functional analytic argument is applied to prove the respective theorems.

Definition. Let ν be a measure defined on $\mathbf{R}^d$ with compact support. A function g defined on $\mathbf{R}^c$, $c \leq d$, is strongly discriminatory if

$$\int g(Wx - z)d\nu(x) = 0 \tag{4}$$

for all $W \in \mathbf{S}^{d,c}$ and $z \in \mathbf{R}^c$ implies that $\nu = 0$.

Though an integral $\int g(Wx - z)d\nu(x)$ is not a simple convolution of g and ν, we still have a lemma below. We write $\check{g}(x) = g(-x)$ in the proof. Note that W^tW is the unit matrix of order c.

Lemma 5. Let g be a slowly increasing function defined on $\mathbf{R}^c$, $c \leq d$, and let ν be a measure on $\mathbf{R}^d$ with compact support. Then, if and only if (4) holds,

$$\mathcal{F}_d g^W(y) \mathcal{F}_d \nu(y) = 0 \tag{5}$$

for all $W \in \mathbf{S}^{d,c}$ and $y \in \mathbf{R}^d$.

Proof. Suppose that (4) holds. Then, for any $x_{W^\perp} \in \mathbf{H}_W$, we have that

$$\int g(Wx - z)d\nu(x) = \int g(W(x - x_{W^\perp}) - W^tWz)d\nu(x)$$

$$= \int g^W(x - x_{W^\perp} - {}^tWz)d\nu(x) = \check{g}^W * \nu(x_{W^\perp} + {}^tWz).$$

Since $x_{W^\perp}$ is an arbitrary point of $\mathbf{H}_W$ and $\mathbf{L}_W = {}^tW\mathbf{R}^c$, $x_{W^\perp} + {}^tWz$ can be any point of $\mathbf{R}^d$. Hence, $\check{g}^W * \nu = 0$. Since the support of $\mathcal{F}_d g^W$ is symmetric, we have (5). The "only if" part can be proved by tracing the proof in the opposite direction.

Lemma 6. Let g be a slowly increasing function defined on $\mathbf{R}^c$. If the support of the Fourier transform of g is dense in a nonempty open set, g is strongly discriminatory.

Proof. By the equality (1), $\mathrm{supp}\mathcal{F}_d g^W$ is dense in a nonempty open subset of $\mathbf{L}_W$, where "supp" stands for the support. Hence, $\bigcup_{W \in \mathbf{S}^{d,c}} \mathrm{supp}(\mathcal{F}_d g^W)$ is dense in a nonempty open subset $\mathbf{G} \in \mathbf{R}^d$. Let ν be a measure on $\mathbf{R}^d$ with compact support, for which (4) holds. Then, by Lemma 5, $\mathcal{F}_d\nu = 0$ in $\mathbf{G}$. Since ν is with compact support, $\mathcal{F}_d\nu$ is analytic on $\mathbf{R}^d$, implying that the equality $\mathcal{F}_d\nu = 0$ can be extended to $\mathbf{R}^d$. Hence, $\nu = 0$. This concludes the proof.

Any sigmoid function is strongly discriminatory [4]. If a Gauss kernel is not degenerated, it is strongly discriminatory. These are typical strongly discriminatory functions defined on $\mathbf{R}$ and $\mathbf{R}^d$ respectively.

Theorem 7. Let g be a slowly increasing continuous function defined on $\mathbf{R}^c$, $c \leq d$, and let f be an arbitrary continuous function defined on a compact set $\mathbf{K} \subset \mathbf{R}^d$. If the support of $\mathcal{F}_c g$ is dense in a nonempty open set, then, for any $\varepsilon > 0$, there are coefficients a_i, matrices $W_i \in \mathbf{S}^{d,c}$, and points $z_i \in \mathbf{R}^c$, $i = 1, \cdots, n$, for which a finite sum

$$\bar{f}(x) = \sum_{i=1}^{n} a_i g(W_i x - z_i) \tag{6}$$

satisfies

$$|\bar{f}(x) - f(x)| < \varepsilon \qquad \text{on} \qquad \mathbf{K}.$$

Proof. Let $C(\mathbf{K})$ be the space of continuous functions defined on $\mathbf{K}$ with the uniform topology. Let $\mathbf{A}$ be a set of functions of the form $\sum_{i=1}^{n} a_i g(W_i x - z_i)$, $a_i \in \mathbf{R}$, $W_i \in \mathbf{S}^{d,c}$, $z_i \in \mathbf{R}^c$. It is sufficient to prove that the closure $\bar{\mathbf{A}}$ of $\mathbf{A}$ in the uniform topology includes $C(\mathbf{K})$. Suppose that there is a function

$f_0 \in C(\mathbf{K})$ which does not belong to $\bar{\mathbf{A}}$. Then, by the Hahn-Banach theorem (see [14]), there exists a linear functional Λ such that $\Lambda(f_0) = 1$ and $\Lambda(f) = 0$ for all $f \in \bar{\mathbf{A}}$. Accordingly, by the Riesz representation theorem (see [13]), there exists a measure ν on $\mathbf{K}$ such that $\Lambda(f) = \int f(x)d\nu(x)$ for all $f \in C(\mathbf{K})$. Since $g(Wx - z)$ belongs to $\bar{\mathbf{A}}$, $\int g(Wx - z)d\nu(z) = 0$ for all $W \in \mathbf{S}^{d,c}$ and all $z \in \mathbf{R}^c$. By Lemma 6, g is strongly discriminatory. Hence, $\nu = 0$. This contradicts to $\Lambda(f_0) = \int f_0(x)d\nu(x) = 1$. Thus, the theorem is proved.

This result can be extended to discontinuous activation functions. To do so, two lemmas below are useful.

Lemma 8. Let g be a slowly increasing function defined on $\mathbf{R}^c$ and let φ be an integrable nonnegative nonzero function defined on $\mathbf{R}^c$ with compact support. If the support of $\mathcal{F}_c g$ is dense in a nonempty open set, so is the support of $\mathcal{F}_c(g * \varphi)$.

Proof. Since suppφ is compact, $\mathcal{F}_c\varphi$ is an analytic nonzero function. Hence, supp$\mathcal{F}_c\varphi$ coincides with $\mathbf{R}^c$. Accordingly, this lemma is obvious.

If g is a monotone increasing function defined on $\mathbf{R}$ and φ is a continuous function with compact support, $g * \varphi$ can be locally uniformly approximated by a finite sum of the form $\Sigma a_i g(t - t_i)$ [3]. Accordingly, a function of bounded variation can be approximated by a linear sum of this form.

We may call $g(x_1, \cdots, x_c)$ a monotone increasing function if $x_i \le x'_i, i = 1, \cdots, c$, implies that $g(x_1, \cdots, x_c) \le g(x'_1, \cdots, x'_c)$, and its orthogonal transformations, $g(Wx)$, $W \in \mathbf{S}^{c,c}$, a monotone function. The proof of the lemma below is described in [9].

Lemma 9. Let g be a monotone increasing function on $\mathbf{R}^c$ and let φ be a nonnegative continuous function with compact support. Then, the convolution $g * \varphi$ can be locally uniformly approximated by a finite sum of the form $\Sigma a_i g(z - z_i)$.

We can extend Theorem 7 to discontinuous activation functions, if they are expressed as linear sums of monotone functions. In fact, let g be a linear sum of monotone functions such that the support of $\mathcal{F}_c g$ is dense in a nonempty open set. Then, $g * \varphi$ satisfies the condition of Theorem 7 by Lemma 8. Hence, $\bar{f}(x) = \sum_{i=1}^n a_i g * \varphi(W_i x - z_i)$ approximates the function f for an appropriate choice of a_i, W_i and z_i. By Lemma 9, $g * \varphi$ can be approximated by a linear sum of shifts of g. Hence, a linear sum of the form (6) approximates f uniformly on $\mathbf{K}$.

5 Approximation on $\bar{\mathbf{R}}^d$

We denote by $C_0(\mathbf{R}^c)$ the space of continuous functions defined on $\mathbf{R}^c$ such that $\lim_{x\to\infty} f(x) = 0$, and by $C(\bar{\mathbf{R}}^c)$ the space of continuous functions defined on $\bar{\mathbf{R}}^c$. Of course, $C_0(\mathbf{R}^c) \subset C(\bar{\mathbf{R}}^c)$.

Definition. Let ν be a measure of bounded variation on $\mathbf{R}^d$. We call a function g defined on $\mathbf{R}^c$, $c \le d$, strongly completely discriminatory, if

$$\int g(Wx - z)d\nu(x) = 0 \tag{7}$$

for all $W \in \mathbf{S}^{d,c}$ and $z \in \mathbf{R}^c$ implies that $\nu = 0$.

The lemma below can be proved similarly to Lemma 5.

Lemma 10. Let g be a bounded function defined on $\mathbf{R}^c$, $c \le d$, and let ν be a measure of bounded variation on $\mathbf{R}^d$. Then, if and only if (7) holds,

$$\mathcal{F}_d g^W(y)\mathcal{F}_d\nu(y) = 0 \tag{8}$$

for all $W \in \mathbf{S}^{d,c}$ and $y \in \mathbf{R}^d$.

We remark that if and only if $\bigcup_{W\in\mathbf{S}^{c,c}}$ supp $\mathcal{F}_c g^W$ is dense in $\mathbf{R}^c$, $\bigcup_{W\in\mathbf{S}^{d,c}}$ supp$\mathcal{F}_d g^W$ is dense in $\mathbf{R}^d$

Lemma 11. Let g be a bounded function defined on $\mathbf{R}^c$. Then, if and only if $\bigcup_{W\in\mathbf{S}^{c,c}}$ supp $\mathcal{F}_c g^W$ is dense in $\mathbf{R}^d$, g is strongly completely discriminatory.

Proof. First, suppose that $\bigcup_{W\in\mathbf{S}^{c,c}}$ supp$\mathcal{F}_c g^W$ is dense in $\mathbf{R}^c$. Then, as remarked above, $\bigcup_{W\in\mathbf{S}^{d,c}}$ supp$\mathcal{F}_d g^W$ is dense in $\mathbf{R}^d$. Let ν be a measure of bounded variation on $\mathbf{R}^d$ such that (7) holds for any $W \in \mathbf{S}^{d,c}$ and $z \in \mathbf{R}^d$. Then, by Lemma 10, we have (8) for any $W \in \mathbf{S}^{d,c}$. Hence, $\mathcal{F}_d\nu = 0$ on a dense subset of $\mathbf{R}^d$. Since $\mathcal{F}_d\nu$ is a continuous function, this implies that $\mathcal{F}_d\nu = 0$ as well as $\nu = 0$.

Conversely, suppose that $\bigcup_{W\in\mathbf{S}^{c,c}}$ supp $\mathcal{F}_c g^W$ is not dense in $\mathbf{R}^c$. Then, by the remark below Lemma 10, $\bigcup_{W\in\mathbf{S}^{d,c}}$ supp $\mathcal{F}_d g^W$ is not dense in $\mathbf{R}^d$. Since this set is spherically symmetric, there is a spherically symmetric open set $\mathbf{G}$ such that $\mathbf{G} \cap \bigcup_{W\in\mathbf{S}^{d,c}}$ supp $\mathcal{F}_d g^W = \phi$. Hence, there is a spherically symmetric, infinitely many times differentiable, nonzero function φ with support included in $\mathbf{G}$. The inverse Fourier transform $\mathcal{F}_d^{-1}\varphi$ is a rapidly decreasing function. Hence, there is a nonzero measure ν_φ of bounded variation with density $\mathcal{F}_d^{-1}\varphi$. Since the support of $\mathcal{F}_d\nu_\varphi (= \varphi)$ and that of $\mathcal{F}_d g_z^W$ are disjoint for any $W \in \mathbf{S}^{d,c}$ and $z \in \mathbf{R}^d$, we have that $\mathcal{F}_d \check{g}_z^W \mathcal{F}_d\nu_\varphi = \mathcal{F}_d g^W \mathcal{F}_d\nu_\varphi = 0$. Hence, by Lemma 10, $\int g(Wx - z)d\nu_\varphi(z) = 0$, which implies that g is not strongly discriminatory.

Lemma 12. Let g be a function defined on $\mathbf{R}^c$, $c \le d$, such that $\bigcup_{W \in \mathbf{S}^{c,c}} \mathrm{supp}\mathcal{F}_c g^W$ is dense in $\mathbf{R}^c$. Then, there is a measure μ on $\mathbf{S}^{d,c}$ for which

$$g_\mu(x) = \int_{\mathbf{S}^{d,c}} g(Wx)d\mu(W) \tag{9}$$

belongs to $C_0(\mathbf{R}^d)$ and $\bigcup_{W \in \mathbf{S}^{d,c}} \mathrm{supp}\mathcal{F}_d g_\mu^W$ is dense in $\mathbf{R}^d$.

Proof. It is sufficient to show that, given a function $g \in C_0(\mathbf{R}^c)$ satisfying the condition, we can construct a function $g_1 \in C_0(\mathbf{R}^{c+1})$ satisfying the condition. We describe the proof concretely for $c = 2$. Let $g \in C_0(\mathbf{R}^2)$ be a function defined on $\mathbf{R}_x^2 = \{(x_1, x_2)\}$ such that $\bigcup_{W \in \mathbf{S}^{c,c}} \mathrm{supp}\mathcal{F}_c g^W$ is dense in $\mathbf{R}_y^2 = \{(y_1, y_2)\}$. Let

$$E = \begin{pmatrix} 1 & 0 & 0 \\ 0 & 1 & 0 \end{pmatrix} \in \mathbf{S}^{3,2}.$$

Then, g^E is a continuous function on $\mathbf{R}^3$ but does not belong to $C_0(\mathbf{R}^3)$. Let

$$R_\theta = \begin{pmatrix} 1 & 0 & 0 \\ 0 & \cos\theta & -\sin\theta \\ 0 & \sin\theta & \cos\theta \end{pmatrix} \in \mathbf{S}^{3,3},$$

and let μ be an absolutely continuous, nonzero measure fully supported by $[0, \pi]$ having a continuous density converging to 0 as $\theta \to 0$ and $\theta \to \pi$. Then, μ can be regarded as a measure on $\mathbf{S}^{3,2}$ supported by $\{ER_\theta | \theta \le \theta \le \pi\} \subset \mathbf{S}^{3,2}$. Set

$$g_1(x) = \int_{\mathbf{S}^{3,2}} g(Wx)d\mu(W).$$

Then, $g_1 \in C_0(\mathbf{R}^3)$. Let $\mathbf{P}_\theta = \{R_\theta x | x = (x_1, x_2, 0)\}$. If $\theta_1 \ne \theta_2$, $\mathbf{P}_{\theta_1} \ne \mathbf{P}_{\theta_2}$. The rotation R_θ in $\mathbf{R}_x^3$ corresponds to a rotation $R_{-\theta}$ in $\mathbf{R}_y^3$. Hence, $\mathrm{supp}\ \mathcal{F}_3 g_1 = \bigcup_{0 \le \theta \le \pi} \mathrm{supp}\mathcal{F}_3 g^{ER_{-\theta}}$, implying that $\bigcup_{W \in \mathbf{S}^{3,2}} \mathrm{supp}\mathcal{F}_3 g_1^W = \bigcup_{W \in \mathbf{S}^{3,2}} \mathrm{supp}\mathcal{F}_3 g^W$ is dense in $\mathbf{R}^3$.

We can repeat this procedure, starting from $g \in C_0(\mathbf{R}^c)$ and define a sequence of rotations $R_{\theta_1}, \cdots, R_{\theta_{d-c}}$. When the rotation R_{θ_i} is defined, a measure μ_i is simultaneously defined and later it is rotated. In this way, we finally construct a measure μ on $\mathbf{S}^{d,c}$ which is supported by a subset $\{ER_{\theta_1} \cdots R_{\theta_{d-c}} | 0 \le \theta_i \le \pi\} \in \mathbf{S}^{d,c}$. Then, the function g_μ defined by (9) for the measure μ belongs to $C_0(\mathbf{R}^d)$ and the set $\bigcup_{W \in \mathbf{S}^{d,c}} \mathrm{supp}\ \mathcal{F}_d g_\mu^W$ is dense in $\mathbf{R}^d$.

Theorem 13. Let $g \in C_0(\mathbf{R}^c)$, $c \le$ d. If the set $\bigcup_{W \in \mathbf{S}^{c,c}} \mathrm{supp}\ \mathcal{F}_c g^W$ is dense in $\mathbf{R}^d$, then, for any $f \in C_0(\mathbf{R}^d)$ and any $\varepsilon > 0$, there are coefficients a_i, matrices $W_i \in \mathbf{S}^{d,c}$, and points $z_i \in \mathbf{R}^c$, $i = 1, \cdots, n$, such that a finite sum

$$\bar{f}(x) = \sum_{i=1}^{n} a_i g(W_i x - z_i) \tag{10}$$

satisfies

$$|\bar{f}(x) - f(x)| < \varepsilon \qquad \text{on} \qquad \mathbf{R}^d.$$

Conversely, if $\bigcup_{W \in \mathbf{S}^{c,c}} \mathrm{supp}\ \mathcal{F}_c g^W$ is not dense in $\mathbf{R}^d$, there is a function $f \in C_0(\mathbf{R}^d)$ that cannot be approximated by a linear sum of the form (10).

Proof. First, suppose that $\bigcup_{W \in \mathbf{S}^{c,c}} \mathrm{supp}\mathcal{F}_c g^W$ is dense in $\mathbf{R}^c$. Then, by Lemma 12, there is a measure μ on $\mathbf{S}^{d,c}$ for which g_μ defined by (9) belongs to $C_0(\mathbf{R}^d)$ and $\bigcup_{W \in \mathbf{S}^{d,c}} \mathrm{supp}\mathcal{F}_c g_\mu^W$ is dense in $\mathbf{R}^d$. Hence, g_μ is strongly completely discriminatory. Let $\mathbf{A}$ be a set of linear sums of the form $\sum_{i=1}^n a_i g_\mu(W_i x - z_i)$, $a_i \in \mathbf{R}$, $W_i \in \mathbf{S}^{d,c}$, $z_i \in \mathbf{R}^c$. We prove that the closure $\bar{\mathbf{A}}$ of $\mathbf{A}$ in the uniform topology coincides with $C_0(\mathbf{R}^d)$. Suppose that there is a function $f_0 \in C(\mathbf{R}^d)$ which does not belong to $\bar{\mathbf{A}}$. Then, by the Hahn-Banach theorem (see [14]), there exists a linear functional Λ such that $\Lambda(f_0) = 1$ and $\Lambda(f) = 0$ for all $f \in \bar{\mathbf{A}}$. Accordingly, by the Riesz representation theorem (see [13]), there exists a measure ν on $\mathbf{R}^d$ of bounded variation such that $\Lambda(f) = \int f(x)d\nu(x)$ for all $f \in C_0(\mathbf{R}^d)$. Since g is strongly completely discriminatory, this is a contradiction as in the case of Theorem 7.

If $\bigcup_{W \in \mathbf{S}^{c,c}} \mathrm{supp}\ \mathcal{F}_c g^W$ is not dense in $\mathbf{R}^c$, set $f_0 = \mathcal{F}_d^{-1}\varphi$ for the φ defined in Lemma 11. If there is a sequence $f_n \in \mathbf{A}$ for which $f_n - f_0$ converges uniformly to 0, $\mathcal{F}_d(f_n - f_0)$ converges to 0. This is impossible because the support of $\mathcal{F}_d f_0$ and that of $\mathcal{F}_d f_n$ are disjoint.

Even if g is not continuous, the statement of Theorem 13 may hold similarly to the case of Theorem 7. But we omit details because of the similarity of the discussion.

6 Concluding Remark

The discriminatory function was introduced in [1] for approximation of continuous functions on compact sets. The strongly discriminatory and strongly completely discriminatory functions were defined in [4] and [6] respectively and named after [1]. The completely discriminatory function was also introduced in [6]. Among them the strongly and strongly completely discriminatory functions are relevant to the mapping of points and approximations of functions without scaling.

Though the activation function on $\mathbf{R}$ was restricted to sigmoid functions in [3], [4] and [6] like many other literatures at that time, the methods used in [4] and [6] can be applied to non-sigmoidal functions in its original form. The two hold extension of activation functions in Section 4, not only to slowly increasing functions but also to higher dimensional functions, was realized by applying the method. On this occasion we remark that if they can be scaled, the approximation capability of nonpolynomial slowly increasing functions on compact sets can be proved similarly to but more easily than Theorem 7. In fact, the support of the Fourier transform of a nonpolynomial function, say g, has a point other than the origin. Hence, $\mathcal{F}_d g^{rW} \mathcal{F}_d \nu = 0$ for any $W \in \mathbf{S}^{d,c}$ and $r \in \mathbf{R}$ is equivalent to $\nu = 0$, which implies that g is discriminatory.

It is obvious that an arbitrary function $f \in C(\mathbf{R}^d)$ cannot be uniformly approximated by a linear combination of linear transforms of a given function. Accordingly, the target of approximation was restricted to $C(\bar{\mathbf{R}}^d)$ in [6]. Furthermore, since $g(Wx - z)$, $g \in C_0(\mathbf{R}^c)$, $c < d$, is not a function of $C(\bar{\mathbf{R}}^d)$, the usual functional analytic argument cannot be directly applied. Consequently, the strongly completely discriminatory function was defined on $\bar{\mathbf{R}}^d$ in [6] and constructed from a sigmoid function by combining rotation, integration and convolution. This method works well as long as the activation function is restricted to sigmoid functions. However, to extend activation functions consistently to higher dimensional functions, we had to overcome the discrepancy among the definitions of distinct kinds of discriminatory functions. To do so, Lemma 12 played a key role. As a result, the distinct kinds of discriminatory functions can be treated in the same framework. In the case where $g \notin C_0(\mathbf{R}^c)$ but a linear sum of linear transforms of g belongs to $C_0(\mathbf{R}^c)$, the uniform approximation in $C(\bar{\mathbf{R}}^d)$ might be realized (see [6]).

Finally, we illustrate three kinds of discriminatory functions for comparison. Set

$g_1(x) = \cos x$,

$g_2(x) = \int_{-1}^{1} \cos \omega x d\omega = \frac{2}{x} \sin x$,

$g_3(x) = \int_{-\infty}^{\infty} e^{-w^2/2} \cos \omega x d\omega = \sqrt{2\pi} e^{x^2/2}$.

Then,

$\mathcal{F}_1 g_1(y) = \pi(\delta_1(y) + \delta_{-1}(y))$, $\mathrm{supp}\mathcal{F}_1 g_1 = \{-1,1\}$,

$\mathcal{F}_1 g_2(y) = 2\pi I_{[-1,1]}(y)$, $\mathrm{supp}\mathcal{F}_1 g_2 = [-1,1]$,

$\mathcal{F}_1 g_3(y) = 2\pi e^{-y^2/2}$, $\mathrm{supp}\mathcal{F}_1 g_3 = \mathbf{R}$.

These results on the supports may be easily foreseen from the definition of the functions. Since the support of $\mathcal{F}_1 g_1$ has a point other than the origin, g_1 is discriminatory; since the support of $\mathcal{F}_1 g_2$ includes an open set, g_2 is strongly discriminatory; and, since the support of $\mathcal{F}_1 g_3$ is full in $\mathbf{R}$, g_3 is strongly completely discriminatory.

References

[1] Cybenko, G., Approximation by superpositions of sigmoidal function. Math. Control Signals Systems, 2, 303-314.(1989)

[2] Hornik, K., Multilayer feedforward networks are universal approximators. Neural Networks 2, 359-366. (1989)

[3] Ito, Y., Representation of functions by superpositions of a step or sigmoid function and their applications to neural network theory. Neural Networks 4, 385-394.(1991a)

[4] Ito, Y., Approximation of functions on a compact set by finite sums of a sigmoid function without scaling. Neural Networks 4, 817-826.(1991b)

[5] Ito, Y., Finite mapping by neural networks and truth functions. Math. Scien. 17, 69-77.(1992a)

[6] Ito, Y., Approximation of continuous functions on $\mathbf{R}^d$ by linear combinations of shifted rotations of a sigmoid function with and without scaling. Neural Networks 5, 105-115.(1992b)

[7] Ito, Y., Nonlinearity creates linear independence. Adv. Compt. Math. 5, 189-203.(1996)

[8] Ito, Y., Independence of unscaled basis functions and finite mappings by neural networks, to appear in Math. Scien. 26 (2001).

[9] Ito, Y., Approximations with unscaled basis functions on compact sets, to appear.

[10] Ito, Y., Approximations with unscaled basis functions on $\bar{\mathbf{R}}^d$, to appear.

[11] Ito, Y. and K. Saito, Superposition of linearly independent functions and finite mapping by neural networks. Math. Scientists 21, 27-33.(1996)

[12] Kůrková, V., P. Savický and K. Hlaváčková, Representations and rates of approximation of real-valued Boolean functions by neural networks, Neural Networks 11, 651-659, (1998).

[13] Rudin, W., Real and complex analysis. New York: McGraw-Hill Book Company.(1966)

[14] Rudin, W., Functional analysis. New York: McGraw-Hill Book Company. (1973)

[15] Sussmann, H.J., Uniqueness of the weights for minimal feedforward nets with a given input-output map. Neural Networks 5, 0589-593. (1992)

Ratio Scales are Critical for Modeling Neural Synthesis in the Brain

Thomas L. Saaty
University of Pittsburgh, Pittsburgh, PA. 15260, USA, e-mail: saaty@katz.pitt.edu

Abstract

The brain generally miniaturizes its perceptions into what may be regarded as a model of what happens outside. We experience the world according to the capacity of our nervous system to register the stimuli we receive. In order to understand and control the environment there needs to be proportionality between the measurements represented in the miniaturized model that arise from the firings of our neurons, and the actual measurements in the real world. Thus our response to stimuli must satisfy the fundamental functional equation $F(ax) = bF(x)$. In other words, our interpretation of a stimulus as registered by the firing of our neurons is proportional to what it would be if it were not filtered through the brain. This equation is the homogeneous part of the inhomogeneous equation $F(ax) - bF(x) = G(x)$ with the forcing function $G(x)$. What interests us here is the mode of operation of the (firing) system that needs to always satisfy the homogeneous part.

1. Introduction

The brain is a network with several hierarchies as sub-networks. Connections in a hierarchy imply a special kind of integration and synthesis of signals. Integration means the addition and sequencing of signals in their temporal order. The signals are neural firings of a form that can be represented by Dirac delta functions.

The observation that the brain is a network means that it allows for feedback and cycling which also requires a synthesis of signals. Signals can have different strengths that in the brain are reflected in their frequency rather than in their amplitudes as the brain is a frequency modulating system. To combine different frequencies and obtain an outcome that has an underlying order requires a method of synthesis that maintains ratio and proportionality among these signals. We have used multicriteria decision-making methods to synthesize the many and varied signals in the brain. This paper is divided into four parts: 1) How the subconscious brain surrenders its hidden mode of operation through conscious discrete comparisons and synthesis in decision making; 2) How the discrete process generalizes to the continuous case embedded in which as a necessary condition for a solution is a proportionality functional equation whose solution is a damped periodic oscillation solution whose space-time Fourier transform takes the form of Dirac type distributions (firings of neurons that respond to stimuli); 3) How the synthesis of such distributions can serve as the basis for forming images and creating sound oscillations for example; and finally, 4) How the neural representation is made See my book [5].

2. The Secrets of the Brain as Revealed by Decision-Making Give Rise to Normalized Ratio Scales

The brain reveals its essential secrets through the acts of a conscious mind. At the conscious level, decision-making gives us an excellent idea how ratios arise in the workings of the brain. We all make comparison judgments to create scales of relative values of the qualities we experience. Because there is no single scale built into the brain to measure things, we need to derive scales by making comparisons in relative terms so that all the information can be accurately integrated on a single ratio scale of priorities. To see how this is done and that its results are credible and valid, we first illustrate the approach with a measurable attribute, the area of several geometric figures. Figure 1 gives five geometric figures that we wish to compare according to area.

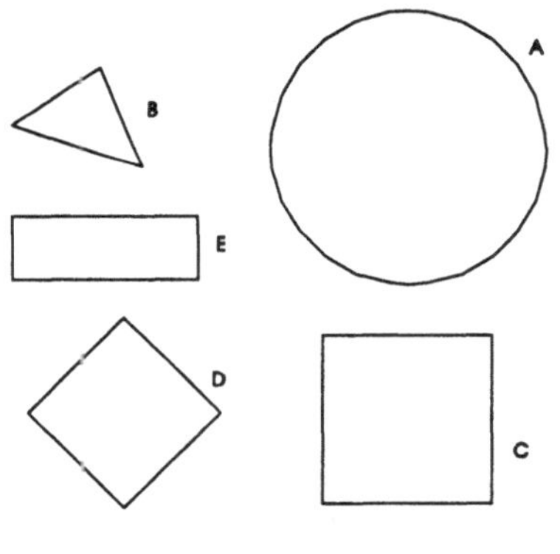

Fig. 1. Comparing the Areas of Five Forms.

To compare two forms according to the relative sizes of their corresponding areas, one determines which of the two is the smaller, and then estimates how many

comparisons is arranged in a matrix $A=\{a_{ij}\}$. If the area of form i is a_{ij} times larger than the area of form j, then the area of form j is $a_{ji}=1/a_{ij}$ times larger than the area of form i, a reciprocal relationship. The matrix for these comparisons, the derived scale of relative values (priorities), and the relative values obtained from actual measurement are shown in Table I. Later we show how the priorities are derived from the comparisons. Each area represented on the left of the matrix is compared with each area at the top of the matrix to determine how many times it is larger or smaller than the one on top.

A more abstract form of comparisons would involve elements with tangible properties that one must think about but cannot be perceived through the senses. See the judgments in Table II for estimating the relative amount of protein in seven food items.

Both examples show that an experienced person can provide informed numerical judgments from which relative good estimates are derived. The same process can be applied to intangibles. My theories of decision making, the Analytic Hierarchy Process (AHP) and the Analytic Network Process (ANP) are based on the comparisons of both tangibles and intangibles. Before we proceed to deal with intangibles we give a sketch of the theoretical concepts underlying the way scales are derived from paired comparisons.

Consider n stocks, $A_1, \ldots, A_n$, with known worth $w_1, \ldots, w_n$, respectively, and suppose that a matrix of pairwise ratios is formed whose rows give the ratios of the worth of each stock with respect to all others as follows:

$$\begin{bmatrix} \frac{w_1}{w_1} & \cdots & \frac{w_1}{w_n} \\ \vdots & \ddots & \vdots \\ \frac{w_n}{w_1} & \cdots & \frac{w_n}{w_n} \end{bmatrix} \begin{bmatrix} w_1 \\ \vdots \\ w_n \end{bmatrix} = n \begin{bmatrix} w_1 \\ \vdots \\ w_n \end{bmatrix}$$

A more abstract form of comparisons would involve elements with tangible properties that one must think about but cannot be perceived through the senses. See the judgments in Table II for estimating the relative amount of protein in seven food items.

Both examples show that an experienced person can provide informed numerical judgments from which relative good estimates are derived. The same process can be applied to intangibles. My theories of decision making, the Analytic Hierarchy Process (AHP) and the Analytic Network Process (ANP) are based on the comparisons of both tangibles and intangibles. Before we proceed to deal with intangibles we give a sketch of the theoretical concepts underlying the way scales are derived from paired comparisons.

We can recover the scale w using the following equation:

$$Aw = nw$$

$$\begin{array}{c} \\ A_1 \\ \vdots \\ A_n \end{array} \begin{array}{c} \begin{array}{ccc} A_1 & \cdots & A_n \end{array} \\ \begin{bmatrix} \frac{w_1}{w_1} & \cdots & \frac{w_1}{w_n} \\ \vdots & \ddots & \vdots \\ \frac{w_n}{w_1} & \cdots & \frac{w_n}{w_n} \end{bmatrix} \end{array}$$

Table I. Judgments for Estimating Relative Areas of Five Forms

What food has more protein?

Protein in Food	A	B	C	D	E	F	G	Estimated Values	Actual Values
A: Steak	1	9	9	6	4	5	1	0.345	0.370
B: Potatoes	1/9	1	1	1/2	1/4	1/3	1/4	0.031	0.040
C: Apples	1/9	1	1	1/3	1/3	1/5	1/9	0.030	0.000
D: Soybean	1/6	2	3	1	1/2	1	1/6	0.065	0.070
E: Whole Wheat Bread	1/4	4	3	2	1	3	1/3	0.124	0.110
F: Tasty Cake	1/5	3	5	1	1/3	1	1/5	0.078	0.090
G: Fish	1	4	9	6	3	5	1	0.328	0.320

Table II. Judgments for Estimating the Relative Amount of Protein in Seven Foods.

FIVE FIGURES

	A	B	C	D	E	Estimated Relative Areas	Actual Relative Areas
A	1	9	2.5	3.5	5	0.490	0.471
B	1/9	1	1/5	1/2.5	1/2	0.050	0.050
C	1/2.5	5	1	2	2.5	0.235	0.234
D	1/3.5	2.5	1/2	1	1.5	0.131	0149
E	1/5	2	1/2.5	1/1.5	1	0.094	0.096

where A has been multiplied on the right by the vector of weights w. The result of this multiplication is nw. Thus, to recover the scale w from the matrix of ratios A, one must solve the eigen- value problem $Aw = nw$ or $(A - n\mathbf{I})w = 0$. This is a system of homogeneous linear equations. It has a nontrivial solution if and only if the determinant of A-$n\mathbf{I}$ vanishes, that is, n is an eigenvalue of A. Now A has unit rank since every row is a constant multiple of the first row. Thus all its eigenvalues except one are zero. The sum of the eigenvalues of a matrix is equal to its trace, the sum of its diagonal elements, and in this case the trace of A is equal to n. Thus n is the principal eigenvalue of A, and one has a nonzero solution w. The solution consists of positive entries and is unique to within a multiplicative constant. To make w unique, one can normalize its entries by dividing by their sum. Note that if one divides two readings from a ratio scale one obtains an absolute number. Normalization transforms a ratio scale to an absolute scale. Thus, given the comparison matrix, one can recover the original scale in relative terms. In this case, the solution is any column of A normalized. The matrix A is not only reciprocal, but also consistent. Its entries satisfy the condition $a_{jk} = a_{ik}/a_{ij}$. It follows that the entire matrix can be constructed from a set of n elements that form a spanning tree across the rows and columns. We have shown that if values from a standard scale are used to make the comparisons, the principal eigenvector recovers these values in normalized form.

In the general case when only judgment but not the numbers themselves are available, the precise value of w_i/w_j is not known, but instead only an estimate of it can be given as a numerical judgment. For the moment, consider an estimate of these values by an expert who is assumed to make small perturbations of the ratio w_i/w_j. This implies small perturbations of the eigenvalues. The problem now becomes $A'w' = \lambda_{max}w'$ where λ_{max} is the largest eigenvalue of A'. To determine how good the derived estimate w is we note that if w is obtained by solving $A'w' = \lambda_{max}w'$, the matrix A whose entries are w'_i/w'_j is a consistent matrix. It is a consistent estimate of the matrix A' which need not be consistent. In fact, the entries of A' need not even be transitive; that is, A_1 may be preferred to A_2 and A_2 to A_3 but A_3 may be preferred to A_1. What we would like is a measure of the error due to inconsistency. It turns out that A' is consistent if and only if $\lambda_{max} = n$ and that we always have $\lambda_{max} \geq n$. The existence of the vector w' with positive components and its uniqueness to within a multiplicative constant in the inconsistent case, is proven by using the concept of dominance and the limiting matrix of powers of A' that yields the vector w' perturbation theory applied to the consistent case. Thus w' belongs to a ratio scale.

Note that while in the first example the eye perceives different size areas, in the second example the mind, through wide experience, derives a feeling for the amount of protein different foods contain. Feelings are ususally distinguished qualitatively and associated with numerical values. In practice, the entries a_{ij} are estimated using the scale given in Table III. The values in this scale are derived from the stimulus-response psychophysical law of Weber-Fechner, and the scale has been validated in numerous applications.

Since small changes in a_{ij} imply a small change in λ_{max}, the deviation of the latter from n is a deviation from consistency and can be represented by $(\lambda_{max} - n)/(n-1)$, which is called the *consistency index (C.I.)*. If we write

$$a_{ij} = \frac{w_i}{w_j}\varepsilon_{ij}$$

then

$$\mu \equiv \frac{\lambda_{max} - n}{n-1} = -1 + \frac{1}{n(n-1)} \sum_{1 \le i < j \le n} \left[\varepsilon_{ij} + \frac{1}{\varepsilon_{ij}} \right]$$

which represents the average cumulative inconsistency of the matrix. When the consistency has been calculated, the result is compared with those of the same index of a reciprocal matrix whose entries are randomly selected from the scale in Table III. This index called the *random index (R.I.)*, is obtained from an ensemble of reciprocal matrices whose entries are randomly selected from the scale 1 to 9 and their reciprocals. Thus for a matrix of order n, a number of matrices of the same order are randomly constructed and the consistency index $(\lambda_{max} - n)/(n-1)$ computed. Table IV gives the order of the matrix n (first row) and the average *R.I.* (second row).

Instead of assigning two numbers wi and wj and forming the ratio w_i/w_j we assign a single number drawn from the fundamental 1-9 scale of absolute numbers to represent the ratio $(w_i/w_j)/1$. It is a nearest integer approximation to the ratio w_i/w_j. The derived scale will reveal what the w_i and w_j are. This is a central fact about the relative measurement approach and the need for a fundamental scale.

The ratio of C.I. to the average R.I. for the same order matrix is called the *consistency ratio(C.R.)*. The consistency ratio needs to be kept "small", e.g., less than 10 percent, indicating deviations from non-random entries (informed judgments) of less than an order of magnitude.

Factors that contribute to the consistency of a judgment are: (1) the homogeneity of the elements in a group, that is, not comparing directly a cherry tomato with a watermelon; (2) the sparseness of the elements in the group, and (3) the knowledge of the individual making the comparisons about the elements being compared to ensure validity of the outcome.

Table III. Numerical Ratings Associated with Pairwise Comparisons

Intensity of Importance	Definition	Explanation
1	Equal Importance	Two activities contribute equally to objective
2	Weak	
3	Moderate importance	Experience and judgment slightly fav- or one activity
4	Moderate plus	
5	Strong importance	Experience and judgment strongly fav- or one activity
6	Strong plus	
7	Very strong or demonstrated importance	An activity is favored very strongly over another.
8	Very, very strong	
9	Extreme importance	The evidence favoring one activity is of highest possible order of affirmation
Reciprocals of above	$a_{ji}=1/a_{ij}$	If A=5B, then B=(1/5)A

In situations where the scale 1-9 is inadequate to cover the spectrum of comparisons needed, that is, the elements compared are inhomogeneous, as for example in comparing a cherry tomato with a watermelon according to size, one uses a process of clustering with a pivot from one cluster to an adjacent cluster that is one order of magnitude larger or smaller than the given cluster, and continues to use the 1-9 scale within each cluster, and in doing that, the scale is extended as far out as desired. What determines the clusters is the relative size of the priorities of the elements in each one. If a priority differs by an order of magnitude or more, it is moved to the appropriate cluster. Hypothetical elements may have to be introduced to make the transition from cluster to cluster a well-designed operation.

Table IV. Average Consistency Ratios (*C.R.*)

1	2	3	4	5	6	7	8	9	10
0	0	.52	.89	1.1 1	1.2 5	1.3 5	1.4 0	1.4 5	1.4 9

3. Generalization to the Continuous Case

The foregoing discrete formulation (see my book: The Brain),

$$\sum_{j=1}^{n} a_{ij}\, w_j = \lambda_{\max}\, w_i \tag{1}$$

$$\sum_{i=1}^{n} w_i = 1 \tag{2}$$

with $a_{ji}=1/a_{ij}$ or $a_{ij}\, a_{ji}=1$ (the reciprocal property), $a_{ij} > 0$ generalizes to the continuous case through Fredholm's integral equation of the second kind (which we can also derive directly from first principles to describe the response of a neuron to stimuli):

$$\int_a^b K(s,t)\, w(t)\, dt = \lambda_{\max}\, w(s) \tag{3}$$

where instead of the matrix A we have a positive kernel, $K(s,t) > 0$. A solution $w(s)$ of this equation is a right eigenfunction.

The standard way in which (3) is written is to move the eigenvalue to the left hand side which gives it the reciprocal form

$$\lambda \int_a^b K(s,t) w(t) dt = w(s) \tag{4}$$

with the normalization condition:

$$\int_a^b w(s) ds = 1 \tag{5}$$

An eigenfunction of this equation is determined to within a multiplicative constant. If $w(t)$ is an eigenfunction corresponding to the charateristic value 8 and if C is an arbitrary constant, we see by substituting in the equation that $Cw(t)$ is also an eigenfunction corresponding to the same 8. The value 8=0 is not a characteristic value because we have the corresponding solution $w(t)=0$ for every value of t, which is the trivial case, excluded in our discussion. Here also, we have the reciprocal property

$$K(s,t)\,K(t,s) = 1 \qquad (6)$$

so that $K(s,t)$ is not only positive, but also reciprocal. An example of this type of kernel is $K(s,t) = e^{s-t} = e^s/e^t$. As in the finite case, the kernel $K(s,t)$ is consistent if it satisfies the relation

$$K(s,t)\,K(t,u)= K(s,u), \text{ for all s, t, and u} \qquad (7)$$

It follows by putting $s = t = u$, that $K(s,s)=1$ for all s which is analogous to having ones down the diagonal of the matrix in the discrete case.

The most important part of what follows is the derivation of the fundamental equation, a functional equation whose solution is an eigenfunction of our basic Fredholm equation.

Theorem 1 K(s,t) is consistent if and only if it is separable of the form:

$$K(s,t)=k(s)/k(t) \qquad (8)$$

Theorem 2 If K(s,t) is consistent, the solution of (4) is given by

$$w(s) = \frac{k(s)}{\int_S k(s)ds} \qquad (9)$$

We note that this formulation is general and applies to all situations where a continuous ratio scale is needed. It applies equally to the derivation or justification of ratio scales in the study of scientific phenomena. We now determine the form of $k(s)$ and also of $w(s)$.

In the discrete case, the normalized eigenvector was independent of whether all the elements of the pairwise comparison matrix A are multiplied by the same constant a or not, and thus we can replace A by aA and obtain the same eigenvector. Generalizing this result we have:

$$K(as, at)=aK(s,t)=k(as)/k(at)=a\,k(s)/k(t) \qquad (10)$$

which means that K is a homogeneous function of order one.

Theorem 3 A necessary and sufficient condition for w(s) to be an eigenfunction solution of Fredholm=s equation of the second kind, with a consistent kernel that is homogeneous of order one is that it satisfy the functional equation

$$w(as)=bw(s) \qquad (11)$$

where b=∀a.

It is clear that whatever aspect of the real world we consider, sight, sound, touch, taste, smell, heat and cold, at each instant, their corresponding stimuli impact our senses numerous times. A stimulus S of magnitude s, is received as a similarity transformation as, $a > 0$ referred to as a dilation of s. It is a *stretching* if $a > 1$, and a *contraction* if $a < 1$. When relating response to a dilated stimulus of magnitude as to response to an unaltered stimulus whose magnitude is s, we have the proportionality relation we just wrote down:

$$\frac{w(as)}{w(s)} = b$$

We refer to equation (11) as: The Functional Equation of Ratio Scales. Because of its wider implications in science, we may call it: *The Fundamental Equation of Proportionality and Order.*

If we substitute $s=a^u$ in (11) we have (see Aczél and Kuczma [1]):

$$w(a^{u-1}) - bw(a^u) = 0$$

Again if we write,

$$w(a^u) = b^u\,p(u)$$

we get:

$$p(u+1) - p(u)=0$$

which is a periodic function of period one in the variable u (such as cos u/2B). Note that if a and s are real, then so is u which may be negative even if a and s are both assumed to be positive.

If in the last equation $p(0)$ is not equal to 0, we can introduce $C=p(0)$ and $P(u)=p(u)/C$, we have for the general response function $w(s)$:

$$w(s) = Ce^{\log b \frac{\log s}{\log a}} P\left(\frac{\log s}{\log a}\right) \qquad (12)$$

where P is also periodic of period 1 and $P(0)=1$. Note that $C > 0$ only if $p(0)$ is positive. Otherwise, if $p(0) < 0$, $C < 0$.

Near zero, the exponential factor which is equal to $s^{\log b/\log a}$, "slims" $w(s)$ if $\log b/\log a > 0$ and "spreads" $w(s)$ if $\log b/\log a < 0$. Because s is the magnitude of a stimulus and cannot be negative, we do not have a problem with complex variables here so long as both a and b are real and both positive. Our solution in the complex domain has the form:

$$w(z)=z^{\ln b/\ln a}\, P(\ln z/\ln a) \qquad (13)$$

Here $P(u)$ with $u = \ln z/\ln a$, is an arbitrary multivalued periodic function in u of period 1. Even without the multivaluedness of P, the function $w(z)$ could be multivalued because ln b/ln a is generally a complex number. If P is single-valued and ln b/ln a turns out to be an integer or a rational number, then $w(z)$ is a single-valued or finitely multivalued function, respectively. This generally multivalued solution is obtained in a way analogous to the real case.

This Solution Leads to the Weber-Fechner Law

Note in (12) that the periodic function P(u) is bounded and the negative exponential leads to an alternating series. Thus, to a first order approximation one obtains the Weber-Fechner law for response to a stimulus s:

A log s + B. We assume that B=0, and hence the response belongs to a ratio scale.

In 1846 Weber found, for example, that people while holding in their hand different weights, could distinguish between a weight of 20 g and a weight of 21 g, but could not if the second weight is only 20.5 g. On the other hand, while they could not distinguish between 40 g and 41 g, they could between 40g and 42g, and so on at higher levels. We need to increase a stimulus s by a minimum amount Δs to reach a point where our senses can first discriminate between s and $s + \Delta s$. Δs is called the just noticeable difference (jnd). The ratio $r = \Delta s/s$ does not depend on s. Weber's law states that change in sensation is noticed when the stimulus is increased by a constant percentage of the stimulus itself. This law holds in ranges where Δs is small when compared with s, and hence in practice it fails to hold when s is either too small or too large. Aggregating or decomposing stimuli as needed into clusters or hierarchy levels is an effective way for extending the uses of this law.

In 1860 Fechner considered a sequence of just noticeable increasing stimuli. He denotes the first one

$$s_1 = s_1 + \Delta s_0 = s_0 + \frac{\Delta s_0}{s_0} s_0 = s_0(1+r)$$

by s_0. The next just noticeable stimulus is given by Weber's law.

Similarly

$$s_2 = s_1 + \Delta s_1 = s_1(1+r) = s_0(1+r)^2 \equiv s_0\alpha^2$$

In general

$$s_n = s_{n-1}\alpha = s_0\alpha^n \ (n = 0,1,2,...)$$

Thus stimuli of noticeable differences follow sequentially in a geometric progression. Fechner noted that the corresponding sensations should follow each other in an arithmetic sequence at the discrete points at which just noticeable differences occur. But the latter are obtained when we solve for n. We have

$$n = \frac{(\log s_n - \log s_0)}{\log \alpha}$$

and sensation is a linear function of the logarithm of the stimulus. Thus if M denotes the sensation and s the stimulus, the psychophysical law of Weber-Fechner is given by

$$M = a\log s + b, \quad a \neq 0$$

We assume that the stimuli arise in making pairwise comparisons of relatively comparable activities. We are interested in responses whose numerical values are in the form of ratios. Thus $b = 0$, from which we must have $\log s_0 = 0$ or $s_0 = 1$, which is possible by calibrating a unit stimulus. Here the unit stimulus is s_0. The next noticeable stimulus is $s_1 = s_0\alpha = \alpha$ which yields the second noticeable response $a \log \alpha$. The third noticeable stimulus is $s_2 = s_0\alpha^2$ which yields a response of $2a \log \alpha$. Thus we have for the different responses:

$M_0 = a \log s_0$, $M_1 = a \log \alpha$, $M_2 = 2a \log \alpha, \dots$, $M_n = na \log \alpha$.

While the noticeable ratio stimulus increases geometrically, the response to that stimulus increases arithmetically. Note that $M_0 = 0$ and there is no response. By dividing each M_i by M_1 we obtain the sequence of absolute numbers 1, 2, 3, ... of the fundamental 1-9 scale. Paired comparisons are made by identifying the less dominant of two elements and using it as the unit of measurement. One then determines, using the scale 1-9 or its verbal equivalent, how many times more the dominant member of the pair is than this unit. In making paired comparisons, we use the nearest integer approximation from the scale, relying on the insensitivity of the eigenvector to small perturbations (discussed below). The reciprocal value is then automatically used for the comparison of the less dominant element with the more dominant one. Despite the foregoing derivation of the scale in the form of integers, someone might think that other scale values would be better, for example using 1.3 in the place of 2. Imagine comparing the magnitude of two people with respect to the magnitude of one person and using 1.3 for how many there are instead of 2.

Stimuli received by the brain from nature are transformed to chemical and electrical neural activities that result in summation and synthesis. This is transformed to awareness of nature by converting the electrical synthesis (vibrations caused by a pattern) to a space-time representation. The way the brain goes back and forth from a pattern of stimuli to its electro-chemical synthesis and then to a representation of its response to that spacio-temporal pattern is by applying the Fourier transform to the stimulus and the inverse Fourier transform to form its response. What we have been doing so far is concerned with the inverse Fourier transform. We now need to take its inverse to develop expressions for the response.

We now show that the space-time Fourier transform of (13) is a combination of Dirac distributions. Our solution of Fredholm's equation here is given as the Fourier transform,

$$f(\omega) = \int_{-\infty}^{+\infty} F(x)\, e^{-2\pi i \omega x}\, dx = Ce^{\beta\omega}\, P(\omega) \quad (14)$$

whose inverse transform is given by:

$$(1/2\pi)\log a \sum_{-\infty}^{\infty} a'_n \left[\frac{(2\pi n+\theta(b)-x)}{(\log a\,|b| + (2\pi n+\theta(b)-x)}\, i \right] \cdot \delta(2\pi n+\theta(b)-x) \quad (15)$$

where *(2B*n*+2(*b*)-*x*) is the Dirac delta function. This is supporting evidence in favor of our ratio scale model.

4. The Formation of Images and Sounds with Dirac Distributions [2,3,4]

Complex valued functions cannot be drawn as one does ordinary functions of three real variables. The reason is that complex functions contain an imaginary part. Nevertheless, one can make a plot of the modulus or absolute value of such a function. The basic assumption we made to represent the response to a sequence of individual stimuli is that all the layers in a network of neurons are identical, and each stimulus value is represented by the firing of a neuron in each layer. A shortcoming of this representation is that it is not invariant with respect to the order in which the stimuli are fed into the network. It is known in the case of vision that the eyes do not scan pictures symmetrically if they are not symmetric, and hence our representation must satisfy some order invariant principle. Taking into account this principle would allow us to represent images independently of the form in which stimuli are input into the network. For example, we recognize an image even if it is subjected to a rotation, or to some sort of deformation. Thus, the invariance principle must include affine and similarity transformations. This invariance would allow the network to recognize images even when they are not identical to the ones from which it recorded a given concept, e.g., a bird. The next step would be to use the network representation given here with additional conditions to uniquely represent patterns from images, sounds and perhaps other sources of stimuli such as smell. Our representation focuses on the real part of the magnitude rather than the phase of the Fourier transform. Tests have been made to see the effect of phase and of magnitude on the outcome of a representation of a complex valued function. There is much more blurring due to change in magnitude than there is to change in phase. Thus we focus on representing responses in terms of Dirac functions, sums of such functions, and on approximations to them without regard to the coefficients in (15).

$$\{t^{\alpha} e^{-\beta t},\ \alpha,\ \beta \ge 0\}$$

The functions result from modeling the neural firing as a pairwise comparison process in time. It is assumed that a neuron compares neurotransmitter-generated charges in increments of time. This leads to the continuous counterpart of a reciprocal matrix known as a reciprocal kernel. A reciprocal kernel $\boldsymbol{K}$ is an integral operator that satisfies the condition $\boldsymbol{K}(s,t)\boldsymbol{K}(t,s) = 1$, for all s and t. The response function $w(s)$ of the neuron in spontaneous activity results from solving the homogeneous equation (4). If

$$\lim_{\xi \to 0} K(\xi s, \xi t)$$

exists, where $\boldsymbol{K}$ is a compact integral operator defined on the space $L_2[0,b]$ of Lebesgue square integrable functions. If the reciprocal kernel $K(s,t)\exists 0$, on o#s, t#b, is Lebesgue square integrable and continuously differentiable, then

$$w(t) = t^{\alpha} e^{g(t)} / \int_0^b t^{\alpha} e^{g(t)} dt$$

satisfies (4) for some choice of $g(t)$. Because finite linear combinations of the functions $\{t^{\alpha} e^{-\beta t}, \alpha, \beta \geq 0\}$ are dense in the space of bounded continuous functions $\mathbf{C}[0,b]$ we can approximate $t^{\alpha} e^{g(t)}$ by linear combinations of $t^{\alpha} e^{-\beta t}$ and hence we substitute $g(t) = -\beta t, \beta \geq 0$, in the eigenfunction $w(t)$. The density of neural firing is not completely analogous to the density of the rational numbers in the real number system. The rationals are countably infinite, the number of neurons is finite but large. In speaking of density here we may think of making a sufficiently close approximation (within some prescribed bound rather than arbitrarily close).
We use the functions:

$$\{t^{\alpha} e^{-\beta t}, \alpha, \beta \geq 0\}$$

to represent images and sounds.
Before we describe how the network can be used to represent images and sound, we summarize the mathematical model on which the neural density representation is based.

Neural responses are impulsive and hence the brain is a discrete firing system. It follows that the spontaneous activity of a neuron during a very short period of time in which the neuron fires is given by:

$$w(t) = \sum_{k=1}^{R} \gamma_k (t - \tau_k)^{\alpha} e^{-\beta(t-\tau_k)} \quad (2)$$

if the neuron fires at the random times τ_k, k=1, 2 ,..., R. The empirical findings of Poggio and Mountcastle (1980) support the assumption that R and the times τ_k, k=1, 2 ,..., R are probabilistic. However, as observed by Brinley (1980), the parameters α and β vary from neuron to neuron, but are constant for the firings of each neuron. Non-spontaneous activity can be characterized as a perturbation of background activity. To derive the response function when neurons are stimulated from external sources, we consider an inhomogeneous equation to represent stimuli acting on the neuron in addition to existing spontaneous activity. Thus, we solve the inhomogenous Fredholm equation of the 2nd kind given by:

$$w(s) - \lambda_0 \int_0^b K(s,t) w(t) dt = f(s)$$

This equation has a solution in the Sobolev space

$$W_p^k(\Omega)$$

of distributions (in the sense of Schwartz) in $\mathbf{L}_p(W)$ whose derivatives of order k also belong to the space $\mathbf{L}_p(W)$, where W is an open subset of $\mathbf{R}^n$.

5. Neural Representation [6,7]

We created a 2-dimensional network of neurons consisting of layers. For illustrative purposes, we assume that there is one layer of neurons corresponding to each of the stimulus values. Thus, if the list of stimuli consists of n numerical values, we created n layers with a specific number of neurons in each layer. Under the assumption that each numerical stimulus is represented by the firing of one and only one neuron, each layer of the network must also consist of n neurons with thresholds varying between the largest and the smallest values of the list of stimuli. We also assumed that the firing threshold of each neuron had the same width. Thus, if the perceptual range of a stimulus varies between two values θ_1 and θ_2, and each layer of the network has n neurons, then a neuron in the ith position of the layer will fire if the stimulus value falls between

$$\theta_1 + (i-1)\frac{\theta_2 - \theta_1}{n-1} \quad \text{and} \quad \theta_1 + i\frac{\theta_2 - \theta_1}{n-1}$$

Picture Experiment
In the graphics experiment the bird and rose pictures required 124 and 248 data points, respectively, whereas the sound experiment required 1000 times more data points. Once the (x,y) coordinates of the points were obtained, the x-coordinate was used to represent time and the y-coordinate to represent response to a stimulus. The numerical values associated with the drawings in Figures 2 and were tabulated and the numbers provided the input to the neurons in the networks built to represent the bird and the rose.

Sound Experiment
In the sound experiment we first recorded with the aid of Mathematica the first few seconds of Haydn's symphony no.102 in B-flat major and Mozart's symphony no. 40 in G minor. The result is a set of numerical amplitudes between -1 and 1. Each of these amplitudes was used to make neurons fire when the amplitude falls within a prescribed threshold range. Under the assumption that each neuron fires in response to one stimulus, we would need the same number of neurons as the sample size, i.e., 117,247 in Haydn's symphony and 144,532 in Mozart's symphony.

Our objective was to approximate the amplitude using one neuron for each amplitude value, and then use the resulting values in Mathematica to play back the music. A small sample of the numerical data for Mozart's symphony is displayed in Figure 4.

This task is computationally demanding even for such simple geometric figures as the bird and the flower shown in Figures 2 and 3. For example, for the bird picture, the stimuli list consists of 124 values, and we would need 124^2=15376 neurons, arranged in 124 layers of 124 neurons each The network and the data sampled to form the picture given in Figure 2, were used to create a 124H124 network of neurons consisting of 124 layers with 124 neurons in each layer. Each dot in the figures is generated by the firing of a neuron in response to a stimulus falling within the neuron's lower and upper thresholds.

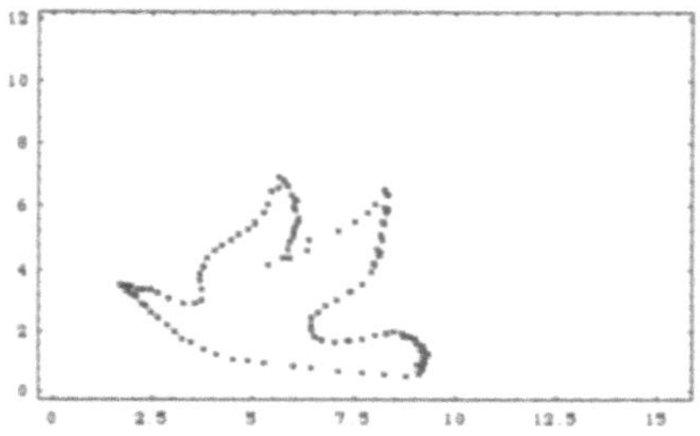

Fig. 2. Bird

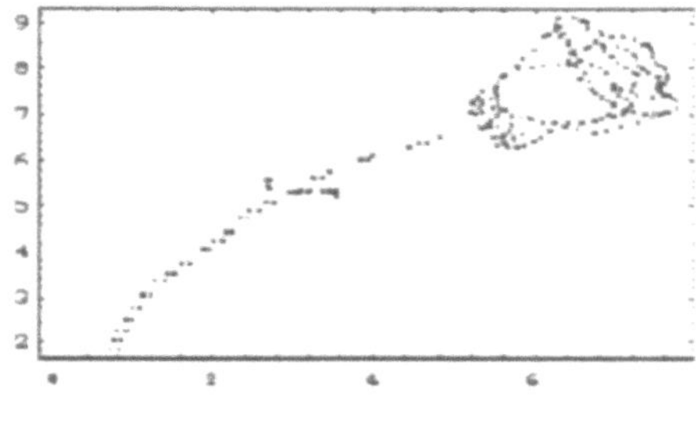

Fig. 3. Rose

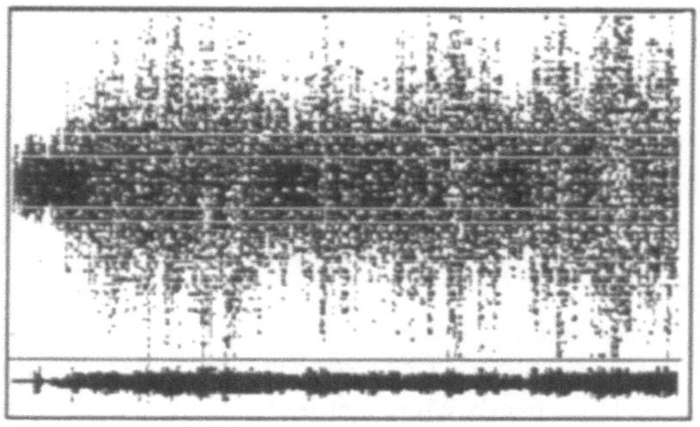

Fig. 4. Mozart's Symphony No. 40

References

[1] Aczél, J. D. and M. Kuczma: Generalizations of a Folk-Theorem, in Vol. 19, Results in Mathematics *5-21*. Basel: Birkhäuser Verlag (1991).

[2] Brinley Jr., F.J.: Excitation and Conduction in Nerve Fibers. Chapter 2 in Medical Physiology, V.B. Mountcastle (Ed.). St. Louis: C.V. Mosby Co. 1980.

[3] Hodgkin, A.L. and A.F. Huxley: A Quantitative Description of Membrane Current and its Applications to Conduction and Excitation in Nerves. J. of Physiol. *500-544,* 117 (1952).

[4] Poggio, G.F. and V.B. Mountcastle: Functional Organization of Thalamus and Cortex. Chapter 9 in Medical Physiology, V.B. Mountcastle (Ed.). St. Louis: C.V. Mosby Co. 1980.

[5] Saaty, T.L.: The Brain, Unraveling the Mystery of How it Works, the Neural Network Process, Pittsburgh, Pennsylvania: RWS Publications, 4922 Ellsworth Avenue, Pittsburgh, PA 15213 2000.

[6] Saaty, T.L. and L.G. Vargas: A Model of Neural Impulse Firing and Synthesis. J. of Math. Psych. *200-219*, 2 (1993).

[7] Saaty, T.L., and L.G. Vargas: Representation of Visual Response to Neural Firing. Math. and Comp. Mod. *17-23*, 18/7 (1993).

Hypermatrix of the Brain

Thomas L. Saaty*

*University of Pittsburgh, Pittsburgh, PA 15260, USA e-mail: saaty@katz.pitt.edu

Abstract

Decision-making, a natural and fundamental process of the brain, involves the use of pairwise comparisons. They are represented by a matrix whose entries belong to a fundamental scale, and from which an eigenvector of priorities that belongs to a ratio scale is derived. A simple decision is represented by a hierarchic structure, more complex ones by a feedback network. The alternatives from which the choice is made belong to the bottom level of the hierarchy whose upper levels contain the criteria and objectives of the decision. The derived eigenvectors are successively used to synthesize the outcome priorities by weighting and adding. A simple example of choosing the best school for the author's son is used to illustrate this process. When there is dependence and feedback in a decision, synthesis requires the use of a stochastic supermatrix whose entries are block matrices of column normalized eigenvectors derived from paired comparisons. Stochasticity is ensured by also comparing the influence of the components that give rise to the blocks.

The synthesis of ratio scale signals in different parts of the brain leading to an overall state of awareness and feeling can be represented by a hypermatrix. The entries of this hypermatrix are block supermatrices each representing synthesis in one organ of the brain. The entries of a supermatrix are in turn blocks of matrices representing the contribution of a suborgan or part of an organ, whose columns are eigenfunctions of the general form [2]

$$w(z)=z^{\ln b/\ln a}\, P(\ln z/\ln a)$$

with a, b, and z complex. Here $P(u)$ with $u = \ln z/\ln a$, is an arbitrary multivalued periodic function in u of period 1. Even without the multivaluedness of P, the function $w(z)$ could be multivalued because $\ln b/\ln a$ is generally a complex number. If P is single-valued and $\ln b/\ln a$ turns out to be an integer or a rational number, then $w(z)$ is a single-valued or finitely multivalued function, respectively. This generally multivalued solution is obtained in a way analogous to the real case. The solution represents the relative contribution of each member neuron of that suborgan to the synthesis of signals in the parent organ. The resulting hypermatrix is raised to powers (cycles signals) until stability is reached. Within each organ we have a different complex variable to represent synthesis within that organ. Feedback from control organs like the hypothalamus occurs in such a way as not to interfere with the variable itself but acts on muscles that reorient the activity of that organ. In the *sense* organs there must be a single valued analytic function, whereas in the *feeling* and perhaps also the *perceiving* organs we have a multivalued analytic function.

1. Priorities in Hierarchies and Networks

In decision making, hierarchies and networks take the forms illustrated in Figures 1 and 2 respectively. To illustrate an elementary form of synthesis in decision making, we give a simple example of a decision to choose the best of three schools in Philadephia for the author's son in terms of six criteria shown in Figure 3. The pairwise comparison matrix of the criteria in terms of the goal is given in Table I. It is followed in Table II by the six comparison matrices of the schools in terms of each criterion and then the synthesis of the priorities of the schools, shown in Table III. (For the scale see [3]).

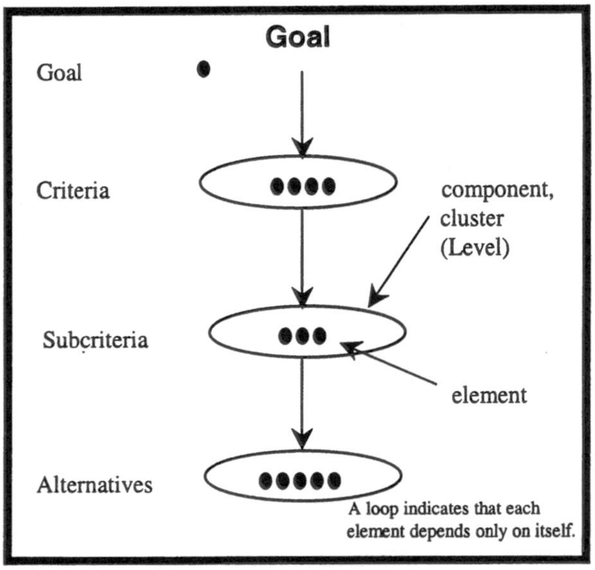

Figure 1. Hierarchy

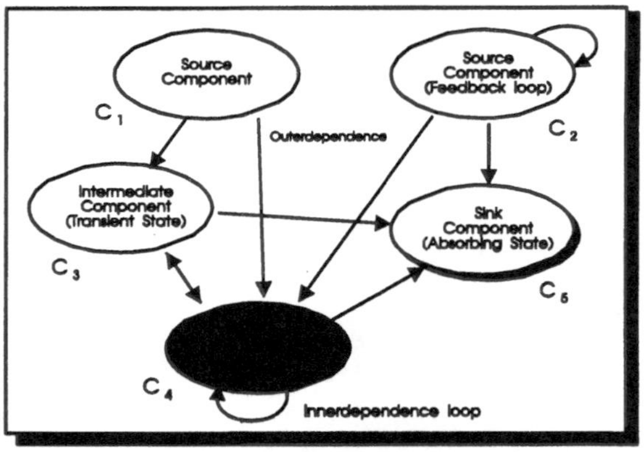

Figure 2. Network

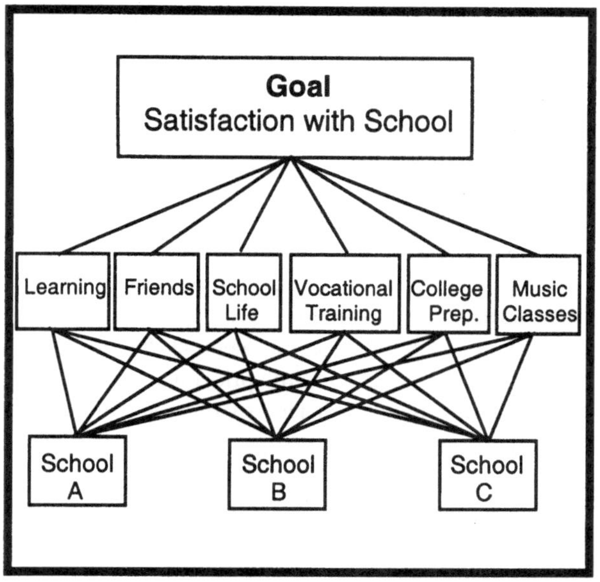

Figure 3. School Hierarchy

Table I. Pairwise Comparisons of Criteria

	L	F	SL	VT	CP	MC	Weights
Learning	1	4	3	1	3	4	.32
Friends	1/4	1	7	3	1/5	1	.14
School Life	1/3	1/7	1	1/5	1/5	1/6	.03
Vocational Trng.	1	1/3	5	1	1	1/3	.13
College Prep.	1/3	5	5	1	1	3	.24
Music Classes	1/4	1	6	3	1/3	1	.14

Table II. Comparisons of Schools with respect to the Six Criteria

Learning	A	B	C	Priorities
A	1	1/3	1/2	.16
B	3	1	3	.59
C	2	1/3	1	.25

Friends	A	B	C	Priorities
A	1	1	1	.33
B	1	1	1	.33
C	1	1	1	.33

School Life	A	B	C	Priorities
A	1	5	1	.45
B	1/5	1	1/5	.09
C	1	5	1	.46

Vocational Trng.	A	B	C	Priorities
A	1	9	7	.77
B	1/9	1	1/5	.05
C	1/7	5	1	.17

College Prep.	A	B	C	Priorities
A	1	1/2	1	.25
B	2	1	2	.50
C	1	1/2	1	.25

Music Classes	A	B	C	Priorities
A	1	6	4	.69
B	1/6	1	1/3	.09
C	1/4	3	1	.22

Table III. Synthesis of Schools

	.32 L	.14 F	.03 SL	.13 VT	.24 CP	.14 MC	Composite Impact of Schools
A	.16	.33	.45	.77	.25	.69	.37
B	.59	.33	.09	.05	.50	.09	.38
C	.25	.33	.46	.17	.25	.22	.25

Three facts attest to the general validity of the AHP/ANP:

1) The *priority* vector of the matrix of paired comparisons gives back the relative values of actual measurements;
2) The *composite vector* gives back the relative values of the composition of actual measurements on multicriteria and is thus naturally generalizable to the measurement of intangibles;
3) The *supermatrix composition* in the ANP includes the AHP as a special case. The ANP also gives back relative values of actual measurements used and yields outcomes that can be very close to known numerical data when used in prediction.

2. The Supermatrix of a Network

Assume that we have a system of N components whereby the elements in each component interact or have an impact on or are influenced by some or all of the elements of another component with respect to a property governing the interactions of the entire system, such as energy or capital or political influence. Assume that component h, denoted by C_h, $h = 1, \ldots, N$, has n_h elements, which we denote by $e_{h1}, e_{h2}, \ldots, e_{hn_h}$. The impact of a given set of elements in a component on another element in the system is represented by a ratio scale priority eigenvector derived from paired comparisons in the usual way.

In Figure 2 no arrow feeds into a source component, no arrow leaves a sink component, and arrows both leave and feed into a transient component. A priority vector belongs to the appropriate column in the supermatrix of Table IV.

Table IV. The Supermatrix

$$
W \ni
\begin{array}{cc}
 & \begin{array}{cccc} C_1 & C_2 & J & C_N \\ e_{11}e_{12}J\,e_{1n_1} & e_{21}e_{22}J\,e_{2n_2} & & e_{N1}e_{N2}J\,e_{Nn_N} \end{array} \\
\begin{array}{cc} & e_{11} \\ & e_{12} \\ C_1 & ! \\ & e_{1n_1} \\ & e_{21} \\ C_2 & e_{22} \\ & ! \\ & e_{2n_2} \\ ! & ! \\ & e_{N1} \\ & e_{N2} \\ C_N & ! \\ & e_{Nn_N} \end{array} &
\begin{bmatrix} W_{11} & W_{12} & J & W_{1N} \\ W_{21} & W_{22} & J & W_{2N} \\ ! & ! & !!! & ! \\ W_{N1} & W_{N2} & J & W_{NN} \end{bmatrix}
\end{array}
$$

where the i,j block of this matrix is given by:

$$
W_{ij} \ni \begin{bmatrix} w_{i_1 j_1} & w_{i_1 j_2} & J & w_{i_1 j_{n_j}} \\ w_{i_2 j_1} & w_{i_2 j_2} & J & w_{i_2 j_{n_j}} \\ ! & ! & !!! & ! \\ w_{i_{n_i} j_1} & w_{i_{n_i} j_2} & J & w_{i_{n_i} j_{n_j}} \end{bmatrix}
$$

Each of whose columns is a principal eigenvector that represents the impact of all the elements in the ith component on each of the elements in the jth component. The discussion of this section will focus on deriving limit priorities for the supermatrix. It must first be reduced to a matrix, each of whose columns sums to unity. In each block of the supermatrix, a column is either a normalized eigenvector with possibly some zero entries, or all of its elements are equal to zero. In either case it is weighted by the priority of the corresponding component represented on the left of the supermatrix. If it is zero, that column of the supermatrix must be normalized after weighting by the components' weights.

The outcome is a *column stochastic* or simply a *stochastic* matrix. *If the matrix is stochastic, the limit priorities depend on the reducibility, primitivity, and cyclicity of that matrix.* Interaction in the supermatrix may be measured according to one of several different criteria. To display and relate the criteria, we need a separate *control hierarchy* (see section 3) that includes these criteria with their priorities. For each criterion, a different supermatrix of impacts is developed.

A component is a collection of elements whose function derives from the synergy of their interaction and hence has a higher-order specialized function not found in any single element. A component is like the audio or visual component of a television set or likes an arm or a leg, consisting of muscle and bone, in the human body. The components of the system should generally be synergistically different from the elements themselves.

The components are compared in terms of a control criterion according to their relative impact (or absence of impact) on each other component at the top of the supermatrix, thus developing priorities to weight the corresponding block matrices of eigenvector columns under that component in the supermatrix. In other words, each component of the brain itself has a priority that contributes to the outcome of synthesis. The resulting supermatrix would then be a stochastic matrix, and its powers would converge.

Figures 4 and 5 and their accompanying supermatrices represent a hierarchy and a holarchy whose bottom level is connected to its top level of criteria and has no single element goal as in a hierarchy.

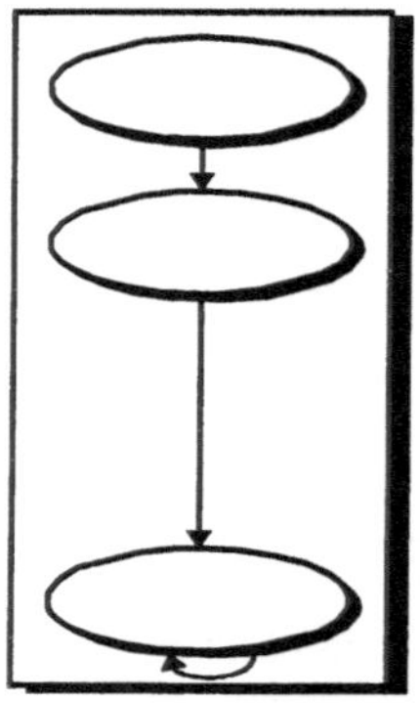

$$
W = \begin{bmatrix} 0 & 0 & 0 & \ldots & \bullet & 0 & 0 \\ W_{21} & 0 & 0 & \ldots & \bullet & 0 & 0 \\ 0 & W_{32} & 0 & \ldots & \bullet & 0 & 0 \\ \vdots & \vdots & \vdots & \vdots & \vdots & \vdots & \vdots \\ \bullet & \bullet & \bullet & \ldots & W_{n-1,n-2} & \bullet & \bullet \\ 0 & 0 & 0 & \ldots & \bullet & W_{n,n-1} & I \end{bmatrix}
$$

Figure 4. The Structure and Supermatrix of a Hierarchy

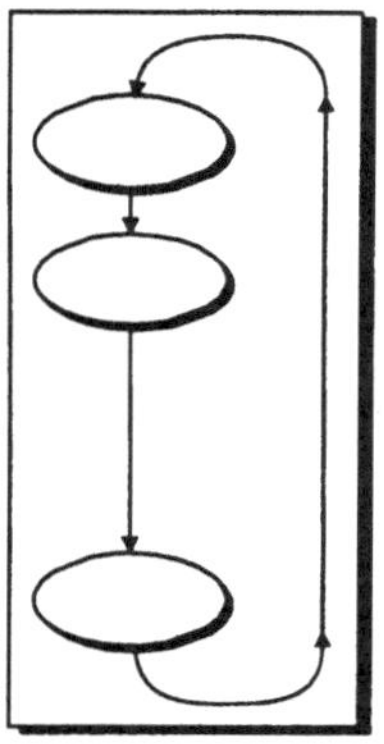

$$W=\begin{bmatrix} 0 & 0 & 0 & \bullet & \bullet & \bullet & 0 & W_{1,n} \\ W_{21} & 0 & 0 & \bullet & \bullet & \bullet & 0 & 0 \\ 0 & W_{32} & 0 & \bullet & \bullet & \bullet & 0 & 0 \\ \bullet & \bullet & \bullet & \bullet & \bullet & \bullet & \bullet & \bullet \\ \bullet & \bullet & \bullet & \bullet & \bullet & \bullet & \bullet & \bullet \\ \bullet & \bullet & \bullet & \bullet & \bullet & W_{n-1,n-2} & \bullet & \bullet \\ 0 & 0 & 0 & \bullet & \bullet & \bullet & W_{n,n-1} & 0 \end{bmatrix}$$

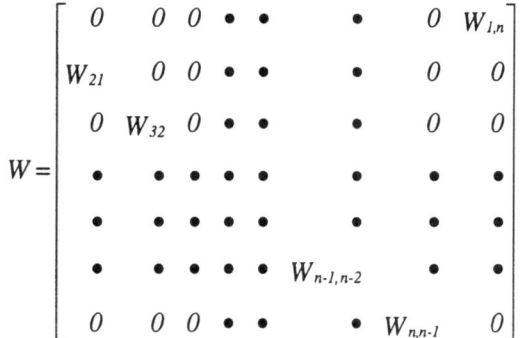

Figure 5. The Structure and Supermatrix of a Simple Cycle (a Holarchy)

3. The Control Hierarchy

Analysis of priorities in a system can be thought of in terms of a control hierarchy with dependence among its bottom-level subsystem arranged as a network as in Figure 4. Dependence can occur within the components and between them. A control hierarchy at the top may be replaced by a control network with dependence among its components. More generally, one can have a cascading set of control networks, the outcome of one used to synthesize the outcomes of what it controls. For obvious reasons relating to the complexity of exposition, apart from a control hierarchy, we will not discuss such complex control structures here. A control hierarchy can also be involved in the network itself with feedback from the criteria to the elements of the network and back to the criteria to modify their influence. This kind of closed-circuit interaction between the operating parts and the criteria that drive the parts is likely to be prevalent in the brain.

The criteria in the control hierarchy that guide the workings of the components are usually the major parent criteria whose subcriteria guide elements in the component.

4. The Hypermatrix of the Brain

How do we maintain an ongoing record of the signals transmitted in the brain that is updated, revised, and synthesized, to capture the transient information that is communicated through neural firings?

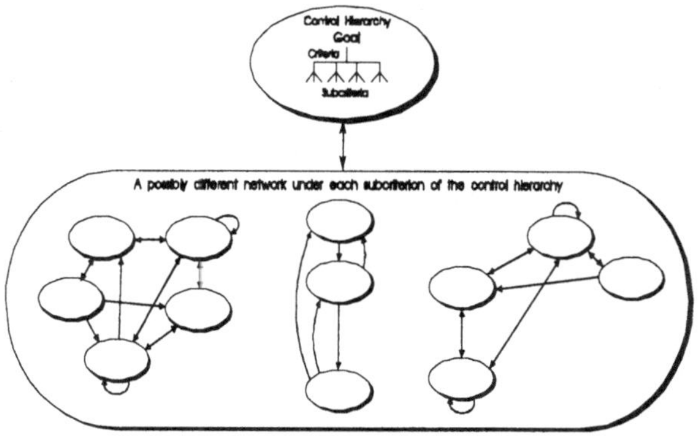

Figure 6. Control System

The approach we follow to represent interactions in the brain is a result of representing the network of neurons by a graph, whose nodes are the neurons themselves, and whose synaptic connections with other neurons are arcs or line segments that connect the nodes. Electric potentials are measured across the arcs. We assign the potentials direction so that they are read in a positive direction. In the opposite direction we assign them a zero value.

Thus we represent the neurons of the brain conceptually by listing them in a column on the left of the hypermatrix (Table V and Figure 7) and listing them again in a row above the columns of the matrix. We assume that for convenience, we have arranged the neurons into components, which correspond to subcomponents or layers and these in turn according to the components or modules to which they belong. One can then enter a zero or a one in each column of the matrix to indicate whether a neuron in a row synapses with a neuron in a column. In fact, instead of the number one, a positive integer can be used to indicate the number of synapses that a neuron on the left has with a neuron at the top of the matrix. In that case each column of the matrix would represent the number of synapses of all neurons with a given neuron represented at the top of the matrix. That would be the most elementary way to represent the connections of the brain. It all hangs together in one piece, because every element is connected to some other element. Such a representation in a matrix can be modified by multiplying its nonnegative integer entries by the damped periodic oscillations of period one corresponding to the neuron with which the synapses are associated. In neural terminology, summing the elements in a column corresponds to spatial summation at an instant of time.

The different components of the hypermatrix are

represented as block supermatrices (Table V). The control subsystems are connected to the supermatrices, which they control, and among themselves, and also to a higher-level control components. We shall see in the next section that the outcome obtained from the hypermatrix is somewhat different from that of the supermatrix.

$$
W^{mn} = \begin{array}{cc} & \\ c_1^m & \begin{array}{c} e_{11} \\ e_{12} \\ \vdots \\ e_{1m_1} \end{array} \\ c_2^m & \begin{array}{c} e_{21} \\ e_{22} \\ \vdots \\ e_{2m_2} \end{array} \\ \vdots & \vdots \\ c_M^m & \begin{array}{c} e_{m1} \\ e_{m2} \\ \vdots \\ e_{Mm_M} \end{array} \end{array}
\begin{array}{c} \begin{array}{cccc} c_1^n & c_2^n & \cdots & c_M^n \\ e_{11}e_{12}\cdots e_{1n_1} & e_{21}e_{22}\cdots e_{2n_2} & & e_{N1}e_{N2}\cdots e_{Nn_N} \end{array} \\ \begin{bmatrix} W_{11} & W_{12} & \cdots & W_{1N} \\ W_{21} & W_{22} & \cdots & W_{2N} \\ \vdots & \vdots & \vdots\vdots\vdots & \vdots \\ W_{M1} & W_{M2} & \cdots & W_{MN} \end{bmatrix} \end{array}
$$

Table V. The Hypermatrix

The *i,j* block of this matrix is a supermatrix as defined above under the supermatrix.

We offer the reader in Figures 7 and 8 a hard-worked-on but very rudimentary application of the hypermatrix to modules and submodules of the brain to illustrate what we have in mind. We warn the reader that we are simply using imagination to brave the complexity. The size of the hypermatrix to describe the brain would be of the order of 100 billion by 100 billion (we have not consulted the Cray Research people about whether the development of their supercomputers comes close to satisfying the dicta of brain computing). It is far beyond our capability to handle the size of such a matrix, or know enough about the physical reality of the brain and its synapses to create the entries in the matrix. Figure 8 shows the supermatrix of vision as part of the hypermatrix.

5. Synthesis

The most significant observation about the brain, which consists of many individual neurons, is that it is primarily a synthesizer of the firings of individual neurons into clusters of information and these in turn into larger clusters and so on, leading to an integrated whole.

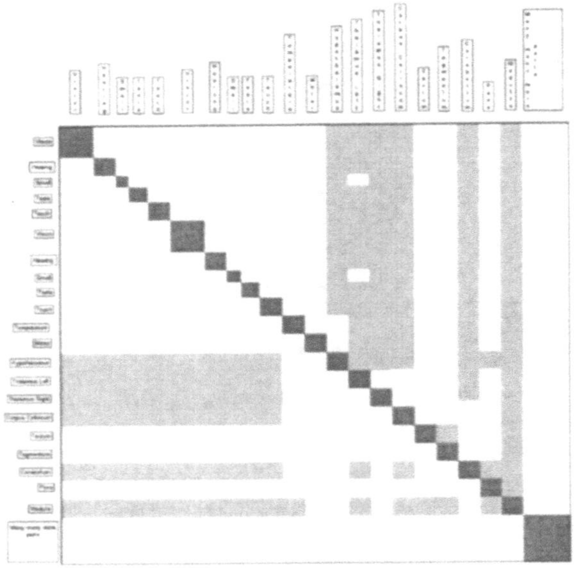

Figure 7. The Hypermatrix of the Brain

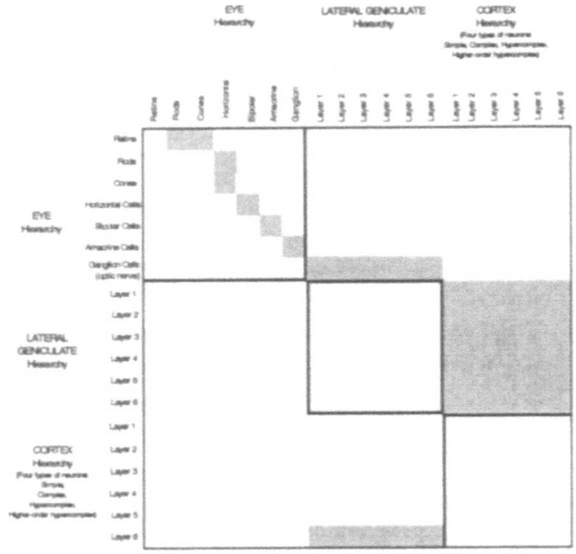

Figure 8. The Supermatrix of Vision

Due to their sequential nature, the firings of a neuron that precede other neurons would be lost unless there is something like a field in which all the firings fit together to form a cohesive entity which carries information. Is there a field in the brain and where is it? We believe that the process of analytic continuation in the theory of functions of a complex variable provides insight into how neurons seem to know one another. On page 373, Kuffler and Nicholls in their often cited book [3] say, ***"The nervous system appears constructed as if each neuron***

had built into it an awareness of its proper place in the system.*"** That is what analytic continuation does. It conditions neurons to fall on a unique path to continue information that connects with information processed by adjacent neurons with which it is connected. *The uniqueness of analytic continuation has the striking consequence that something happening on a very small piece of a connected open set* ***completely determines *what is happening in the entire set, at great distances from the small piece.*

By raising the hypermatrix to powers one obtains transitive interactions. This means that a neuron influences another neuron through intermediate neurons. All such two step interactions are obtained by squaring the matrix. Three step interactions are obtained by cubing the matrix and so on. By raising the matrix to sufficiently large powers, the influence of each neuron on all the neurons with which one can trace a connection, yields the transient influence of neurons in the original hypermatrix Multiplying the hypermatrix by itself allows for combining the functions that represent the influence from pre-to post- synaptic neurons to accumulate all the transitive influences from one neuron to another and allow for feedback. The Fourier transform that takes place as a result of firing and the density of the resulting firings give us the desired synthesis. Depending on what parts of the brain are operational and participating in the synthesis, different physical and behavioral attributes are observed to take place, including consciousness related to the Fourier transform of the single valued sensory functions.

References

[1] Kuffler, S. and J.G. Nichols: *From Neuron to rain.* Sunderland, MA, USA: Sinauer Associates (1976).

[2] Saaty, T.L.: *The Brain: Unraveling the Mystery of How it Works.* 4922 Ellsworth Avenue, Pittsburgh, PA 15213, USA: RWS Publications (2000).

[3] Saaty, T.L.: *In the Brain Ratio Scales are Critical for Synthesis. Proceedings of the 5th ICANNGA.* Prague, The Czech Republic: Academy of Sciences (2001).

The Computational Capabilities of Neural Networks[1]

Jiří Šíma*

*Institute of Computer Science, Academy of Sciences of the Czech Republic, P.O. Box 5, 182 07 Prague 8, Czech Republic, Institute for Theoretical Computer Science (ITI), Charles University, sima@cs.cas.cz

1 Introduction

The (artificial) neural networks represent a widely applied computational paradigm that is an alternative to the conventional computers in many areas of artificial intelligence. By analogy with classical models of computation such as Turing machines which are useful for understanding the computational potential and limits of conventional computers, the capability of neural networks to realize general computations have been studied for more than decade and many relevant results have been achieved [1, 2, 3, 5]. The neural networks are classified into a computational taxonomy according to the restrictions that are imposed on their parameters. Thus, various models are obtained which have different computational capabilities and enrich the traditional repertoire of computational means. In particular, the computational power of neural networks have been investigated by comparing their variants with each other and with more traditional computational tools including finite automata, Turing machines, Boolean circuits, etc. The aim of this approach is to find out what is, in principle, or efficiently, computable by particular neural networks, and how to optimally implement required functions.

In the present paper a taxonomy of neural network models is reviewed from the computational point of view. The *digital* computations with binary inputs and outputs are mostly assumed here although their values may be encoded by analog states. The respective theoretical results concerning the computational power of neural networks are briefly surveyed. The main focus is on the *perceptron networks* in Section 2 whose computational properties are now fairly well understood. The computational taxonomy is extended in Section 3 with other neural network models whose computational capabilities have only recently been analyzed including the *RBF*, *winner-take-all*, and *spiking* networks. The full version of this paper appeared as technical report [4] which contains relevant references.

[1]Research partially supported by project LN00A056 of The Ministry of Education of the Czech Republic.

2 Perceptron Networks

The *perceptron network* consists of s computational *units (neurons, perceptrons)* $V = \{1, \ldots, s\}$ where s is its *size*. Assume that n *input* and m *output* units serve as an interface while the remaining ones are called *hidden* neurons. The units are connected into an oriented graph—*architecture* with each edge from i to j labeled with a *weight* $w_{ji} \in \Re$ ($w_{ji} = 0$ iff there is no edge). The *computational dynamics* determines for each $j \in V$ the evolution of its *state (output)* $y_j^{(t)} \in \Re$ in time $t \geq 0$ which establishes the *network state* $\mathbf{y}^{(t)} = (y_1^{(t)}, \ldots, y_s^{(t)}) \in \Re^s$. At $t = 0$, the network is placed in an *initial state* $\mathbf{y}^{(0)}$ (e.g. including an external input). An *excitation* $\xi_j^{(t)} = \sum_{i=0}^{s} w_{ji} y_i^{(t)}$ is assigned to each $j \in V$ at time $t \geq 0$ as the weighted sum of outputs from incident units including a *bias* $w_{j0} \in \Re$ which can be viewed as the weight of formal input $y_0^{(t)} \equiv 1$. In the **discrete-time** dynamics, the network state is updated at time instants $t = 1, 2, \ldots$ so that units $j \in \alpha_{t+1}$ from a certain subset $\alpha_{t+1} \subseteq V$ compute their new outputs $y_j^{(t+1)}$ by applying an *activation function* $\sigma : \Re \longrightarrow \Re$ to $\xi_j^{(t)}$, i.e. $y_j^{(t+1)} = \sigma(\xi_j^{(t)})$ while $y_j^{(t+1)} = y_j^{(t)}$ for $j \notin \alpha_{t+1}$. The *binary* perceptron networks with $y_j \in \{0, 1\}$ usually employ the *hard limiter* activation function: $\sigma(\xi) = 1$ for $\xi \geq 0$ and $\sigma(\xi) = 0$ for $\xi < 0$; while the *analog* networks use continuous *sigmoid* functions, e.g. the *saturated-linear* function: $\sigma(\xi) = 1$ for $\xi \geq 1$, $\sigma(\xi) = \xi$ for $0 < \xi < 1$, and $\sigma(\xi) = 0$ for $\xi < 0$; or the *standard sigmoid* $\sigma(\xi) = 1/(1+e^{-\xi})$. For the purpose of computing Boolean functions the analog states y_j of output units j are interpreted as binary outputs 0 or 1 with *separation* $\varepsilon > 0$ if $y_j \leq h - \varepsilon$ or $y_j \geq h + \varepsilon$, respectively, for some fixed threshold $h \in \Re$.

A **single perceptron** computes an n-variable *linearly separable* Boolean function $f : \{0,1\}^n \longrightarrow \{0,1\}$. The class of such functions contains only a small fraction $2^{\Theta(n^2)}$ of all the 2^{2^n} Boolean functions, including AND, OR, NOT, and is closed under the negation. Any linearly separable Boolean function can be realized by using only *integer* weights with $\Theta(n \log n)$ bits in their binary representations. The parity (XOR), on the other hand, represents the most prominent example among the functions not computable by a single perceptron. In general, the issue of whether a given Boolean function is linearly separable, is co-NP-complete problem.

The architecture of *feedforward* perceptron networks is an acyclic graph which is split into $d+1$ pairwise disjoint *layers* $\alpha_0, \ldots, \alpha_d \subseteq V$ where d is called its *depth*, so that units in any layer α_t are connected only to neurons in subsequent layers α_u, $u > t$. Usually, α_0 is the input layer (not counted in the number of layers) and α_d represents the output layer. The *network function* $\mathbf{f} : \{0,1\}^n \longrightarrow \{0,1\}^m$ is evaluated, layer after layer, in parallel time d. *Any* Boolean input-output mapping can be implemented with a **binary feedforward network** of $\Theta(\sqrt{m2^n/(n - \log m)})$ perceptrons in depth 4 (constant time) which, in the worst case, cannot be reduced even for unbounded depth (this size is optimal also for $m = 1$). The integer weights in feedforward networks are usually bounded by a polynomial in terms of input length n which clearly decreases their computational power. However, unbounded weights can still be replaced with polynomial ones by adding one more layer while only a polynomial overhead in the network size is needed. Also the exponential size in the previous universal networks is far from being practically realizable for larger input lengths although there is no specific function known so far that would require superlinear number of units. In fact, many important functions can be computed by the feedforward perceptron networks of *polynomial size* and *constant depth* d which, for *polynomial weights*, correspond to the complexity classes TC_d^0, $d \geq 1$. Thus, TC_1^0 contains all the functions computable by single perceptrons with polynomial weights, and $TC_1^0 \subsetneqq TC_2^0$ since $XOR \in TC_2^0 \setminus TC_1^0$. A less known result is a deeper separation $TC_2^0 \subsetneqq TC_3^0$. The witnessing function from $TC_3^0 \setminus TC_2^0$ is the *Boolean inner product* $IP : \{0,1\}^{2k} \longrightarrow \{0,1\}$ ($k \geq 1$) defined as $IP(x_1, \ldots, x_k, x_1', \ldots, x_k') = \bigoplus_{i=1}^k AND(x_i, x_i')$ where $\bigoplus$ stands for the parity. This means that polynomial-size and polynomial-weight three-layered perceptron networks are computationally more powerful than two-layered ones. The separation above depth 3 is unknown and it is even conceivable that NP-complete problems can be computed in three layers with a linear number of perceptrons. Another important case is the class of *symmetric Boolean functions* (including the parity) whose values depend only on the number of 1s within the input, which can be implemented by polynomial-weight three-layered networks with $O(\sqrt{n})$ perceptrons. This size cannot generally be reduced below $\Omega(\sqrt{n/\log n})$ even if arbitrary depth and unbounded weights are allowed. Hence, the feedforward perceptron networks are computationally more powerful than the *AND-OR circuits* of logical gates AND, OR, NOT which cannot compute the parity within polynomial size and constant depth. In addition, the feedforward perceptron networks of small constant depths compute efficiently basic *arithmetic functions* and thus, implement analytic functions with a large precision. For example, the two-layered perceptron networks of polynomial size and weights can be constructed for comparing two n-bit binary numbers, or for computing the multiple sum of n such numbers. Further, the product and division of two n-bit binary numbers, powering to n, and sorting of n such numbers can be achieved with three-layered perceptron networks whose depth (except for powering) cannot be reduced when polynomial size and weights are required. In some cases, one can build smaller networks if the optimal depth is increased by few layers (e.g. quadratic-size four-layred perceptron multiplier). Even more general *trade-off* results are known among different complexity measures such as size, depth, and connectivity, e.g. for every $d > 0$, the n-variable parity can be computed by a depth-$O(d \log n / \log f)$ feedforward network of size $O(dn/f^{1-1/d})$ with $O(dnf^{1/d})$ polynomial-weight edges, and *fan-in* (the maximum number of inputs to a single unit) bounded by f.

The computational capabilities of **analog feedforward networks** with e.g. the standard sigmoid which is the most widely applied model in practical neurocomputing (back-propagation), can be compared to that of binary perceptrons. Obviously, binary units can be replaced with analog ones within the same architecture. The real states, on the other hand, may bring more efficiency in analog networks of *constant size*: the *unary squaring* functions $SQ_k : \{0,1\}^{k^2+k} \longrightarrow \{0,1\}$ ($k \geq 1$) defined as $SQ_k(x_1, \ldots, x_k, z_1, \ldots, z_{k^2}) = 1$ iff $(\sum_{i=1}^k x_i)^2 \geq \sum_{i=1}^{k^2} z_i$, can be computed by only two analog units with polynomial weights and separation $\varepsilon = \Omega(1)$ while any binary feedforward network computing SQ_k requires $\Omega(\log k)$ units even for unbounded depth and weights. Hence, the size of feedforward networks can sometimes be reduced by logarithmic factor when binary units are replaced by analog ones. However, the analog versions $TC_d^0(\sigma)$ with separation $\varepsilon = \Omega(1)$ equal TC_d^0 for every $d \geq 1$.

The architecture of *recurrent* perceptron networks is, in general, a cyclic graph, and hence, their computations may not terminate. *Sequential* updates satisfy $|\alpha_t| \leq 1$ for every $t \geq 1$ while *synchronous (fully) parallel* computations with $\alpha_t = V$ for all $t \geq 1$ will mostly be assumed here. Thus, a network *converges* or reaches a *stable state* $\mathbf{y}^{(t^\star)}$ at time $t^\star \geq 0$ if $\mathbf{y}^{(t^\star)} = \mathbf{y}^{(t^\star+k)}$ (within some precision for analog states) for all $k \geq 1$. To decide whether a given binary recurrent network has a stable state is NP-complete whereas the *halting problem* which is the issue of whether a network converges for a given input, is computationally even harder: PSPACE-complete for binary networks and algorithmically undecidable for analog nets with 25 units and rational weights. For the purpose of universal computations over binary inputs of arbitrary lengths, different input protocols have been introduced. In *finite* networks the input is either presented *online*, each bit every *period* of $p \geq 1$ steps, or in analog nets the input string can be encoded into a real input state. The computational power of **finite recurrent perceptron networks** with the saturated-linear function depends on the descriptive complexity of weights. For *integer weights* such (binary) networks are computationally equivalent to finite automata and are shortly called *neuromata.* In particular, a neuromaton of size $\Theta(\sqrt{q})$ can be constructed which, for period $p = 4$, simulates a given deterministic finite automaton with q states, and in the worst case, this size cannot be reduced if either $p = O(\log q)$ or polynomial weights are assumed. Also an optimal-size neuromaton with $\Theta(\ell)$ units can be build recognizing a regular language described by a given regular expression of length ℓ. Furthermore, with *rational weights* arbitrary Turing machines can be simulated step per step. Thus, any function computable by a Turing machine in time $T(n)$ can be computed by a fixed *universal* analog recurrent network with only 886 units in time $O(T(n))$ whose size can be reduced to 25 neurons at the cost of $O(n^2T(n))$ time overhead. The Turing universality is also proved for the standard sigmoid although the simulation requires exponential time per each step. Finally, with *arbitrary real weights* the finite analog nets have even super-Turing computational capabilities: polynomial time corresponds to the nonuniform complexity class P/poly and within exponential time *any* input-output mapping can be computed. Moreover, a proper hierarchy of complexity classes between P and P/poly has been shown for polynomial-time computations of analog nets with increasing Kolmogorov complexity (information contents) of real weights. However, any amount of *analog noise* reduces their power to that of neuromata or even more (definite languages).

Special attention has been paid to symmetric **Hopfield nets** with undirected architectures whose weights satisfy $w(i,j) = w(j,i)$ for all $i, j \in V$. The fundamental property of binary Hopfield nets with $w(j,j) \geq 0$ for every $j \in V$ is that, under sequential update, they always converge towards some stable state corresponding to a local minimum of bounded *Liapunov (energy)* function E defined on the state space which decreases along any nonconstant computation, e.g. $E(\mathbf{y}) = -1/2 \sum_{j=1}^{s} \sum_{i=0}^{s} w_{ji} y_i y_j$. The parallel binary Hopfield nets either reach a stable state (e.g. when E is negative definite) or eventually alternate between two different states. These convergence results hold also for analog symmetric networks which, under mild hypotheses, converge to a fixed point or to a limit cycle of length at most two for parallel updates. The *convergence time* in binary Hopfield nets of size s is trivially upper bounded by 2^s and may indeed be exponential, e.g. $\Omega(2^{s/3})$ parallel steps. In the average case, on the other hand, a very fast convergence $O(\log\log s)$ is guaranteed under reasonable assumptions. The previous bounds can be expressed in terms of the network *weight* $W = \sum_{j,i \in V} |w_{ji}|$. An upper bound $O(W)$ on the convergence time of binary Hopfield nets follows from the characteristics of the energy which yields polynomial-time convergence for networks with polynomial weights. Moreover, an analog Hopfield net can be constructed whose computation terminates later than that of any other binary symmetric network of the same representation size (the number of bits in the binary representation of weights). Further, the *capacity* of the Hopfield memory is upper bounded by the number of stable states. In average, there are asymptotically $1.05 \times 2^{0.2874s}$ many stable states in binary Hopfield nets of size s, with $w_{jj} = w_{j0} = 0$ ($j \in V$) whose other weights are independent identically zero-mean Gaussian random variables. For a particular binary symmetric network, however, the issues of deciding whether there are at least one, two, or three stable states, are NP-complete. Also the problem of *attraction radius*, i.e. how many bits may be flipped in a given stable state so that the Hopfield net still converges to it, is NP-hard and probably cannot even be efficiently approximated. Hopfield nets have alternatively been applied to the fast approximate *combinatorial optimization*. Hence, the issue of finding a state with *minimal energy* for a given Hopfield net is of special interest. However, this problem is in general NP-complete although it can be solved in polynomial time for binary nets with planar architectures.

In addition, there is a polynomial-time *approximate* algorithm that solves the minimum energy problem within absolute error less than $0.243W$ in binary Hopfield nets with weight W. Finally, the *computational power* of finite Hopfield nets will be reviewed. For binary states, the Hopfield nets are computationally equivalent with *convergent* asymmetric networks within only a linear overhead in time and size. This means that not only do all binary symmetric networks converge, but all convergent computations can be realized efficiently with symmetric weights (for feedforward networks even within the same architecture). However, the finite binary Hopfield nets recognize a strict subclass of the regular languages (*Hopfield languages*). Only if the analog symmetric networks with rational weights are augmented with an external oscillator of a certain type then such devices are Turing universal.

In an alternative input protocol, **infinite families of binary recurrent networks** $\{N_n\}$, each N_n with n input units for one input length $n \geq 0$, are exploited for universal computations. The computational power of such families of polynomial-size networks correspond to the nonuniform complexity class PSPACE/poly even if one restricts only to Hopfield nets. However, their power reduces to P/poly for polynomial symmetric weights.

The computational properties of various stochastic versions of perceptron networks have also been analyzed for which the reference model of **probabilistic networks** is considered. In particular, the corresponding deterministic network is augmented with additional *random binary* input units i so that for all $t \geq 0$ the probability that $y_i^{(t)} = 1$, is given by a fixed $p_i \in [0,1]$ while $y_i^{(t)} = 0$ with probability $1 - p_i$. This probabilistic model is polynomially (in parameters) related to neural networks with other stochastic behavior including Boltzmann machines, networks with unreliable states and connections, etc. Thus, required Boolean functions are computed by such networks possibly with an *error* whose probability is bounded by $0 < \varepsilon < 1/2$. This error probability can be reduced in *probabilistic binary feedforward networks* of size s and depth d to arbitrarily small $0 < \lambda < \varepsilon$ at the cost of size $\lceil 2 \log_{4\varepsilon(1-\varepsilon)} \lambda \rceil s + 1$ and depth $d+1$. Hence, for n binary (deterministic) inputs, $\lceil 8\varepsilon \ln 2/(1-2\varepsilon)^2 + 1 \rceil ns + 1$ units should in principle suffice to make such networks deterministic. Also, the complexity classes TC_d^0 $(d \geq 1)$ are generalized to their stochastic versions RTC_d^0 with error probabilities $\varepsilon(n) = 1/2 - 1/n^{O(1)}$ depending on input length $n \geq 0$. Thus, $RTC_d^0 \subseteq TC_{d+1}^0$ for every $d \geq 1$ (nonuniform networks) and the probabilistic feedforward networks may indeed be more efficient since $IP \in RTC_2^0 \setminus TC_2^0$. The computational analysis has been extended to *probabilistic recurrent networks* with the saturated-linear function. With integer weights such networks recognize regular languages with a bounded error probability. For rational weights, polynomial time corresponds to the nonuniform complexity class Pref-BPP/log where a weak super-Turing capability originates from arbitrary reals p_i while rational values p_i reduce the power to the recursive class BPP. For arbitrary real weights, polynomial time corresponds to $P/poly$.

The computational dynamics of **continuous-time perceptron networks** determines analog states $\mathbf{y}(t) \in \Re^s$ for every real time $t > 0$, e.g. by a system of s differential equations $dy_j/dt(t) = -y_j(t) + \sigma(\xi_j(t))$ $(j \in V)$ with boundary conditions given by $\mathbf{y}(0)$ and the saturated-linear function σ. By Liapunov function argument any continuous-time symmetric network converges to some stable state with $dy_j/dt = 0$ for all $j \in V$ which can take exponential time in terms of s. Moreover, a continuous-time asymmetric network can be constructed that simulates a given finite binary discrete-time recurrent network within only a linear-size overhead. For convergent computations such a simulation can even be achieved with the continuous-time Hopfield nets. Hence, the polynomial-size infinite families of continuous-time (symmetric) networks compute at least PSPACE/poly.

3 Other Neural Network Models

The computational capabilities of neural network models that are alternative to perceptron networks have recently been analyzed. For example, unit $j \in V$ in the **RBF networks** (*radial basis functions*) computes its "excitation" as $\xi_j^{(t)} = \|\mathbf{x}_j^{(t)} - \mathbf{w}_j\|/w_{j0}$ where $\mathbf{x}_j^{(t)} = (y_{ji_1}^{(t)}, \ldots, y_{ji_{n_j}}^{(t)}) \in \Re^{n_j}$ consists of states of units $i_1, \ldots, i_{n_j} \in V$ incident to j, and the "weight" vector $\mathbf{w}_j = (w_{ji_1}, \ldots, w_{ji_{n_j}}) \in \Re^{n_j}$ is called its *center* while the positive "bias" $w_{j0} > 0$ determines its *width*. Also, the shape of activation function differs from that of sigmoid functions, e.g. the *Gaussian function* $\sigma(\xi) = e^{-\xi^2}$ is employed. For the maximum norm and representation of binary values by two different analog states, the RBF unit can robustly implement the universal Boolean NAND gate over multiple literals for a large class of activation functions. Thus, the deterministic finite automata with q states can be implemented by recurrent networks with $O(\sqrt{q \log q})$ RBF units.

The **winner-take-all networks** represent another neural network model that employs powerful k-WTA_n gates computing a mapping k-WTA_n : $\Re^n \longrightarrow \{0,1\}^n$ defined as k-$WTA_n(x_1,\ldots,x_n) = (y_1,\ldots,y_n)$ where $y_i = 1$ $(1 \leq i \leq n)$ iff $|\{j;\, x_j > x_i,\, 1 \leq j \leq n\}| \leq k-1$. In particular, even a simple $1-WTA_n$ $(n \geq 3)$ cannot be implemented by any feedforward perceptron network of size less than $\binom{n}{2}+n$. Moreover, any Boolean function from TC_0^2 can be computed by a *single* k-WTA_r gate applied to polynomially many r weighted sums of inputs with positive polynomial weights.

The most prominent position among alternative models is occupied by **networks of spiking neurons** which are supposed to be more biologically plausible units than perceptrons. The states in spiking networks (including interface units) are encoded by temporal differences between so-called *spikes (firing times)* of neurons. Thus, a sequence of firing times $0 \leq y_j^{(1)} < y_j^{(2)} < \cdots < y_j^{(\tau)} < \ldots$ is associated with each $j \in V$, and for continuous time $t \geq 0$ denote $Y_j(t) = \{y_j^{(\tau)} < t;\, \tau \geq 1\}$ and $y_j(t) = \max Y_j(t)$ for $Y_j(t) \neq \emptyset$. Formally choose $y_j^{(0)} < 0$ and define $y_j(t) = t$ for all $0 \leq t < y_j^{(1)}$ $(j \in V)$. In addition, for every $j, i \in V$ a *response function* $\varepsilon_{ji} : \Re_0^+ \longrightarrow \Re$ of j to spikes from i in continuous time $t \geq 0$ is either excitatory (EPSP) $\varepsilon_{ji} \geq 0$ or inhibitory (IPSP) $\varepsilon_{ji} \leq 0$ $(\varepsilon_{ji}(t) \equiv 0$ for $w_{ji} = 0)$. The examples of the EPSP functions include e.g. $\varepsilon_{ji}(t) = 0$ for $0 \leq t \leq \Delta_{ji}$ or $t \geq \Delta_{ji}+2$, and $\varepsilon_{ji}(t) = 1-|t-\Delta_{ji}-1|$ for $\Delta_{ji} < t < \Delta_{ji}+2$ where $0 < \Delta_{\min} \leq \Delta_{ji} \leq \Delta_{\max}$ is a *delay* of connection from i to j while the IPSP can be chosen as $-\varepsilon_{ji}$. Furthermore, for each non-input unit j a function $w_{j0} : \Re_0^+ \longrightarrow \Re_0^- \cup \{-\infty\}$ determines its nonpositive bias at time $t \geq 0$ while $w_{ji} \geq 0$ for all $i \in V$, e.g. $w_{j0}(t) = w_{j0}(0) < 0$ for $t \geq t_{ref}$, and $w_{j0}(t) = -\infty$ for a *refrectory period* $0 < t < t_{ref}$. The spikes $y_j^{(\tau)}$, $\tau \geq 1$ for each non-input j are computed recursively as $y_j^{(\tau)} = \inf\{t \geq 0;\, t > y_j^{(\tau-1)} \,\&\, \xi_j(t) \geq 0\}$ according to its excitation $\xi_j(t) = w_{j0}(t-y_j(t)) + \sum_{i=1}^{s} \sum_{y \in Y_i(t)} w_{ji} \cdot \varepsilon_{ji}(t-y)$. First the *lower bounds* on the computational power of spiking networks are reviewed. Any binary feedforward perceptron network of size s and depth d can be simulated by a network of $O(s)$ spiking units within time $O(d)$. Or any deterministic finite automaton with q states can be implemented with $O(\sqrt{q})$ spiking neurons. Finally, one can build a finite spiking network with rational weights from $[0,1]$ that simulates any Turing machine in linear time, and any input-output mapping can be computed with arbitrary real weights. The computational power of spiking networks can completely be characterized by introducing the *upper bounds*. The spiking networks with any piecewise linear response- and bias-functions are computationally equivalent to discrete-time analog recurrent perceptron networks with any piecewise linear activation functions (e.g. the hard limiter and saturated-linear functions together are universal in this context). Such networks are further equivalent to so called *N-RAMs*, which are random access machines with $O(1)$ registers working with arbitrary reals of bounded absolute values. This equivalence is shown by linear-time pairwise simulations and is still valid when all the numerical parameters are rational numbers. The Turing universality holds even for *piecewise constant* response functions but with exponential-time overhead. Indeed, the small temporal differences in the firing times (e.g. different Δ_{ji} in the piecewise constant response functions) may bring additional power: the Boolean *coincidence-detection* function $CD_k : \{0,1\}^{2k} \longrightarrow \{0,1\}$ $(k \geq 1)$ that formalizes a simple pattern-matching task as $CD_k(x_1,\ldots,x_k,x_1',\ldots,x_k') = 1$ iff there is $1 \leq i \leq k$ such that $x_i = x_i' = 1$ can easily be implemented by a single spiking neuron whose inputs are suitably delayed. Any feedforward perceptron network computing CD_k, on the other hand, requires at least $k/log(k+1)$ or $\Omega(\sqrt{k})$ or $\Omega(k^{1/4})$ units with the hard limiter or saturated-linear or standard sigmoid activation function, respectively.

In more realistic **noisy spiking networks** the excitation of unit j just governs the probability that j fires. Any Boolean function can be implemented by a sufficiently large network of noisy spiking neurons with arbitrarily high probability of correctness.

References

[1] I. Parberry: *Circuit Complexity and Neural Networks.* The MIT Press, Cambridge, MA, 1994.

[2] V.P. Roychowdhury, K.-Y. Siu, A. Orlitsky (eds.): *Theoretical Advances in Neural Computation and Learning.* Kluwer Academic Publishers, 1994.

[3] H.T. Siegelmann: *Neural Networks and Analog Computation: Beyond the Turing Limit.* Birkhäuser, Boston, 1999.

[4] J. Šíma: The computational theory of neural networks. TR V–823, ICS, AS CR, Prague, 2000.

[5] K.-Y. Siu, V.P. Roychowdhury, T. Kailath: *Discrete Neural Computation: A Theoretical Foundation.* Prentice Hall, Englewood Cliffs, NJ, 1995.

Activation Functions

N.C. Steele, C.R. Reeves, E.I. Gaura *

*School of MIS, Coventry University, Priory Street, Coventry CV1 5FB, UK

Abstract

This paper considers an alternative activation function for use with MLP networks. The performance on parity problems is considered and it has been found that only $n-1$ hidden units were needed to resolve the n-bit problem. Also, insight has been gained into the families of network parameters generated. Use as the kernel of a support vector machine for particular problems is anticipated.

1 Introduction

It is probably true to say that the vast majority of applications of artificial neural networks make use of multi-layer perceptron (MLP) networks made up of hidden units with logistic (sigmoid) activation functions and which are trained using a variant of the classical back-propagation algorithm. Recently, the radial basis function (RBF) network has found favour in some quarters, particularly when the network is designed to implement a fuzzy inference system. In this latter application, the shape of the graph of the node activation function as a function of a single variable has influenced the choice of network paradigm. By contrast, despite the broad classes of functions known to possess the universal approximation property, there does not seem to have been much work done on alternative activation functions for MLP networks. The reason for this is probably two-fold, first, the 'standard' sigmoid function has a 'natural' shape, and second, the availability of packaged software often prevents adventurous choices for node activation functions. Recently, Reeves and Johnston [3] were considering the problem of fitting distributions to data sets using MLP type networks, and this gave rise to thoughts on the appropriateness of the sigmoid activation function for this type of problem. This and succesful experience gained using the RBF paradigm with Gaussian activation functions led to considering the possibility of using the derivative of the usual sigmoid as the node activation function. The output of the Gaussian activation function is, of course, localised in the input space, whilst the sigmoid derivative, with a biased dot-product as its argument, is directionally sensitive rather than localised. We return to this latter point later in the paper. Considerable attention has been paid to the sigmoid function and its derivative in connection with artificial neural networks, for example by Menon *et al* [1] and Minai and Williams [2]. However, we are not aware of any studies in the literature on the use of the derivative as the activation function itself.

Statistically, RBF networks with Gaussian activations are very similar to Normal mixture distributions. The work by Reeves and Johnston [3] has suggested an interesting link between mixture distributions and the derivative of sigmoidal functions. The successful experience we have had with Gaussian RBFs thus led us to consider the possibility of using the derivative of the usual sigmoid as an activation function. The remainder of this paper discusses the results achieved and some of the insight gained.

2 The sigmoid function and its derivative

The sigmoid or logistic activation function is defined as

$$y = \frac{1}{1+e^{-x}} \quad \forall\, x \tag{1}$$

with $y \in (0,1)$. It is easy to show that

$$\frac{\mathrm{d}y}{\mathrm{d}x} = y(1-y)$$

and since $0 \le \frac{\mathrm{d}y}{\mathrm{d}x} \le 0.25$, it is natural when thinking about possible use as an activation function, to consider a normalised function $f(x)$ defined by

$$f(x) = 4\frac{\mathrm{d}y}{\mathrm{d}x} = 4y(1-y) \tag{2}$$

with y as defined above. In much the same way as $\frac{\mathrm{d}y}{\mathrm{d}x}$ can be written in terms of y, we find that

$$\begin{aligned} \frac{\mathrm{d}f}{\mathrm{d}x} &= 4y(1-y)(1-2y) \\ &= f(x)(1-2y) \\ &= -\mathrm{sgn}(x)f(x)\sqrt{1-f(x)} \end{aligned}$$

Figure 1 shows the sigmoid function $y(x)$ and the normalised derivative $f(x)$.

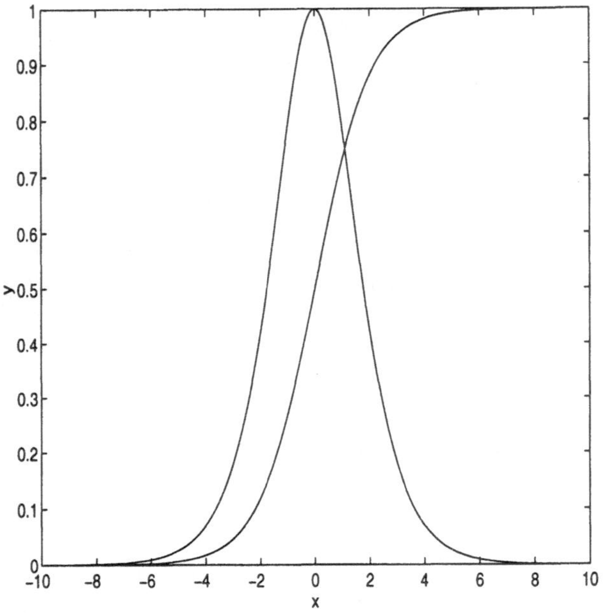

Fig. 1. The sigmoid function and its (normalised) derivative.

In the next section we consider some results obtained when using the 'bump' function $f(x)$ as hidden unit activation function.

3 Some results

The normal starting point in attempting to assess the functionality of a new network is to consider the parity problems. These problems are classed as hard learning problems and the ability to learn to resolve these is a good indicator of the likely utility of the design. Experiments with the 2, 3 and 4 bit problems were carried out using the 'standard' architecture of n hidden units for the n-bit problem. It soon became clear that networks with 'bump' activation functions learned to resolve these much more rapidly, were less troubled by local mininima and were less sensitive to choice of training heuristics, than the corresponding networks equipped with sigmoid activation functions. Table I gives a summary of the results for the 2 bit problem. The limit on the number of training cycles was set at 10^4, and random starting points were chosen in weight-space. A variety of training heuristics were applied, and the learning rate was varied between 0.1 and 1.

Trials were conducted with the 4 bit problem. They were carried out in a similar way but with a limit of 2000 training cycles imposed and, in this case, 10/14 training attempts converged when the 'bump' activation function was used, while only 1/10 did so with the sigmoid function.

The ease with which the parity problems were resolved prompted the question as to whether n hidden units were actually required to resolve the n-bit problem. In fact, networks containing just $n-1$ hidden units (and a simple summation at the output) were able to learn to resolve the problem in all cases tested, that is $n = 2, 3, 4, 5$. It is our conjecture that this result is general, but as yet, it is not established as such. Figure 2 shows the surface generated by the mapping discovered in one experiment for the case $n = 2$. The overall mapping in this case was found as

$$u = 1.686x_1 + 1.692x_2 - 1.648 \quad (3)$$

$$z = 1.828f(u) - 0.890 \quad (4)$$

where u is the input to the hidden unit, z the network output and $f(.)$ is as defined in equation (2). Notice that $u = 0$ approximately when

$$x_1 + x_2 - 1 = 0$$

and thus the activation function achieves its maximum along this line. The effect of this is clear in Figure 2.

Given the simplicity of the network which learns to resolve the 2-bit problem, it is possible to investigate exactly what happens, and to gain some insight into how 'bump' networks work. The 2-bit parity problem (XOR) is specified as in Table II

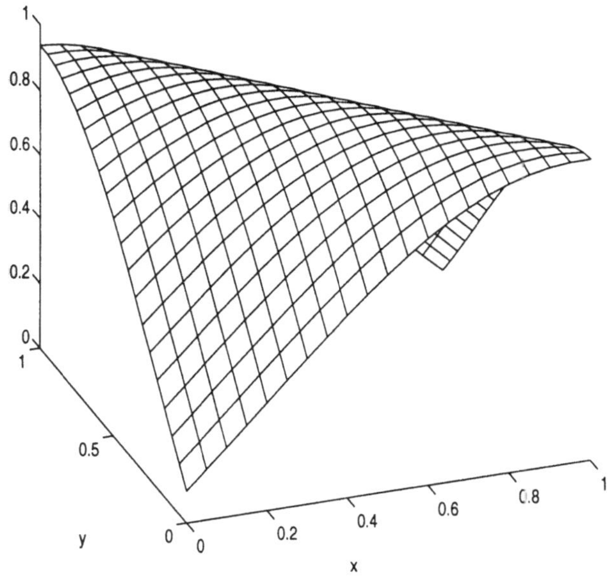

Fig. 2. The surface generated for the XOR-problem using one 'bump' node

Activation Function	Number Converging	Average Number of Cycles
Sigmoid	15	3475.6
Bump	20	287.45

Table I. Results for 20 training runs for the 2 bit parity problem, with a limit of 10^4 cycles

With the notation in Figure 3, and with $z_A = 0$ and $z_B = 1$ denoting the two desired output values, we have

$$z_A = 0 = w_3 f_A + b_2 \tag{5}$$

$$z_B = 1 = w_3 f_B + b_2 \tag{6}$$

Here f_A and f_B denote the values of the activation functions which, if equations (5) and (6) are to have a solution, must be equal for both members of class A and for both members of class B. If we examine class A we find that in the two cases the argument of the activation function is given by

1. b_1 or

2. $w_1 + w_2 + b_1$

Since $f(.)$ is an even function, we must then have

$$b_1 = \pm(w_1 + w_2 + b_1) \tag{7}$$

A similar argument for class B gives

$$w_1 + b_1 = \pm(w_2 + b_1) \tag{8}$$

It follows from (7) and (8) that

$$\begin{aligned} w_1 &= w_2 = -b_1 \quad \text{or} \\ w_1 &= -w_2,\ b_1 = 0 \end{aligned} \tag{9}$$

x_1, x_2	z
0,0	0
0,1	1
1,0	1
1,1	0

Table II. The 2-bit parity problem.

The mapping described above in (3) and (4) is an (approximate) example of that described by (9), and members of this 'family' were most often, but not exclusively, observed. In the case of the three-bit problem, similar families of mappings were observed, but it has not yet been possible to provide the underlying analysis.

'Bump' networks were tested on a number of classification tasks. In summary, the performance in training and also in testing on unseen data, was comparable to that of a network made up of sigmoid units. In some cases training performance was

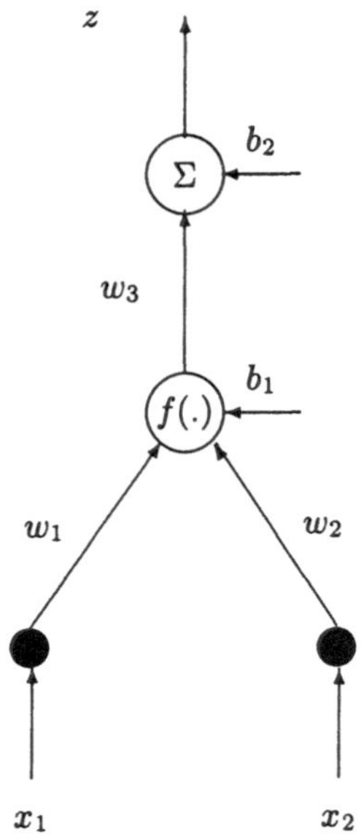

Fig. 3. The architecture for the 2-bit problem

noticeably more robust in the choice of training parameters.

4 Conclusions and Further Work

An alternative activation function has been proposed for MLP networks, and some of its properties have been explored in the context of parity and other classification problems. This has led us to believe that there may be advantages in using networks with this type of activation function for at least some of its nodes.

It should also be pointed out that Eq. (2) is not the only type of sigmoidal function that could be employed. In principle, any smooth probability density function could be used, and this opens up the possibility of using 'bumps' with marked asymmetries for example, such as the Weibull or Gumbel functions.

As mentioned above the argument of the proposed activation function consists of a biased dot-product. This opens the possibility that a related inner product Kernel might be found on which to base the design of a support vector machine. We have found parameters which provide such a Kernel for particular classification problems, and this aspect of the study will be further developed.

References

[1] Menon A, Mehrotra K, Mohan C, and Ranka S: Characterization of a Class of Sigmoid Functions with Applications to Neural Networks. Neural Networks *9*, No 5, pp 818-835, (1996).

[2] Minai A, and Williams R: On the Derivatives of the Sigmoid. Neural Networks *6*, pp 845-853, (1993).

[3] Reeves C R, and Johnston C: Fitting densities and hazard functions with neural networks. Accepted by ICANNGA 2001.

A Simulation of Spiking Neurons by Sigmoid Neurons

Vladimír Kvasnička*

*Dept. Mathematics, Slovak Technical University, 812 37 Bratislava, Slovakia, email: kvasnic@cvt.stuba.sk

Abstract

A simple feed-forward neural networks with sigmoid neurons are studied as a potential effective simulation device of neural networks with spiking neurons. A back-propagation method of calculation of partial derivatives could not be immediately used for spiking neurons. In particular, an adaptation process of the studied neural networks explicitly distinguishes a "time" of activities. We suggested a simple generalization of the back-propagation method such that instead of an original acyclic neural network we consider its unfolded tree form with a root corresponding to the output neuron. We have formulated a conjecture that the presented type of feed-forward neural networks is a universal approximator of any deterministic training pattern specified by binary spiking vectors.

1 Introduction

The theory of spiking neurons [1,2] belongs to most popular recent topics of modern artificial neural networks, that are developed to be more biologically plausible than previous older neural models. For instance, current theories of cognitive activities (e.g. reflexive reasoning [3]) are built up almost entirely on spiking neurons, these theories are biologically more plausible because they use spiking neurons endowed with simple computational activities and their mutual connections are able to transfer only simple unstructured signals of spikes. Unfortunately, there does not exist a simple, well established theory of learning of neural networks composed of spiking neurons. It means that the current theories of cognitive activities with spiking neurons are based on "invariant" neural substructures that are "prewired", and we only "a-posteriori" study their properties, particularly whether they are able to simulate the respective cognitive activity. The purpose of this short communication is to demonstrate that there exists an alternative approach to spiking neurons, we show that it is still possible to study some of their properties within the framework of analog sigmoid neurons. We suggest a special architecture of feed-forward neural networks together with its learning process. The produced networks are able to simulate discrete activities of spiking neurons as well as to carry out their systematic adaptation (learning) process towards a gradual increase of fulfillment of the required output spiking activities.

2 Theory

Let us consider a *feed-forward neural network* $N=(G,W,\Theta)$, where $G=(V,E)$ is an oriented, connected and acyclic graph, and $W=(w_{ij})$ and $\Theta=(\vartheta_i)$ are weight and threshold coefficients assigned to connections and neurons, respectively. We say that the graph G specifies an *architecture* of the neural network N. According to our assumption that the graph G is acyclic, neurons may be always indexed in such a way that input neurons are indexed at first, then hidden neurons are indexed, and finally the highest indices are used for output neurons. Moreover, for any edge $e=(v,v')\in E$ that starts at vertex v and terminates at v', both respective vertices are may be indexed such that index(v)<index(v'). An architecture of the network is displayed in Fig. 1.

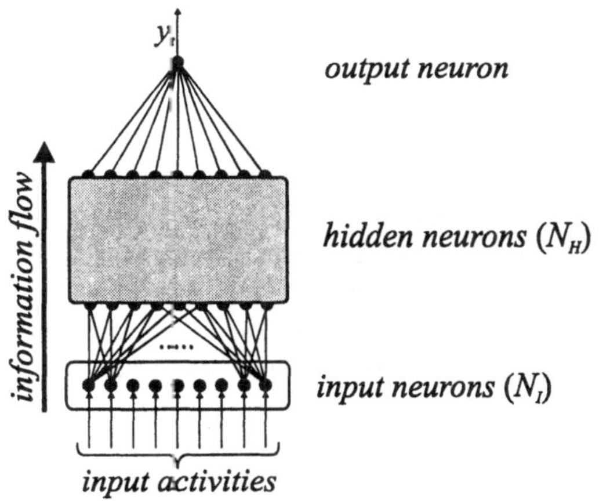

Fig. 1: An architecture of the used feed-forward neural networks. It is composed of N_I input neurons, N_H hidden neurons, and one output neuron. Hidden neurons are mutually interconnected in such a way that a produced graph should be acyclic.

Activities of hidden and output neurons are determined by

$$x_i^{(t)}=s(\xi_i)=s\left(\sum_{(j,i)\in E} w_{ij}x_j^{(t-1)}+\vartheta_i\right) \tag{1}$$

where $s(\xi)$ is a sigmoid activation function. In our forthcoming considerations this sigmoid function will be explicitly expressed by $s(\xi)=1/(1+\exp(-a\xi))$, where

a is a positive slope coefficient, for $a\to\infty$ the exponential sigmoid function tends to behave as a step function, $s(\xi)=1$, for $\xi>0$, and $s(\xi)=0$, for $\xi<0$.

Activities of input neurons are determined by the so-called input spiking binary vectors composed of t_{max} entries

$$\omega_i = \left(\omega_1^{(i)}, \omega_2^{(i)}, \ldots, \omega_{t_{max}}^{(i)}\right) \in \{0,1\}^{t_{max}} \quad (2a)$$

$$\omega_j^{(i)} = \begin{cases} 0 \Rightarrow \left(i^{\text{th}} \text{ neuron at time } j \text{ is not firing}\right) \\ 1 \Rightarrow \left(i^{\text{th}} \text{ neuron at time } j \text{ is firing}\right) \end{cases} \quad (2b)$$

Neural network may be formally considered as a parametric mapping F (its actual form depends on weight and threshold coefficients) of input spike binary vectors onto output spiking binary vectors

$$y = F\left(\omega_1, \omega_2, \ldots, \omega_{N_I}\right) \quad (3)$$

The output vector $y = \{0,1\}^{t_{max}}$ is a response of the respective neural network to input binary vectors $\omega_1, \omega_2, \ldots, \omega_{N_I}$. For better understanding of our ideas we shall study a simple example of neural network displayed in Fig. 2.

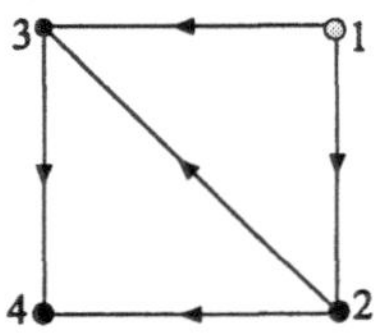

Fig. 2: Simple acyclic graph composed of one input neuron (1), two hidden neurons (2 and 3), and one output neuron (4). All its weight and threshold coefficients are set to 1. The input spiking binary vector is ω_1=(1,0,0,1, 0,0,0,1,0), where t_{max}=9. Activities of hidden and output neurons are calculated by eq. (1).

We postulate, in this illustrative example, that activation functions of hidden and output neurons are simulated by the hard nonlinearity (step function). The resulting activities of all neurons are listed in Table I. The first row in Table corresponds to the input neuron whereas the last fourth row corresponds to the output neuron, for time steps $t=1,2,\ldots,t_{max}$. The activities of hidden neurons (considered as intermediate results) are listed in the second and third rows. Formally, the used illustrative network is considered as a parametric mapping

$$\underbrace{(001101100)}_{output} = F\left(\underbrace{100100010}_{input}; \underbrace{w, \vartheta = 1}_{parameters}\right) \quad (4)$$

It maps input spiking binary vectors ω_1=(1,0,0,1,0,0,0,1,0) onto output binary vector y=(0,0,1,1,0,1,1,0,0), and algorithm of this mapping is hidden in a sequential way of calculation of neuron activities, where the input activities are determined by $x_1^{(t)} = \omega_1^{(t)}$, for $t=1,2,\ldots,t_{max}$ (see also the first row in Table I)

Table I. Activities of neurons for different t

v	t: 1	2	3	4	5	6	7	8	9
1	1	0	0	1	0	0	0	1	0
2	0	1	0	0	1	0	0	0	1
3	0	1	1	0	1	1	0	0	1
4	0	0	1	1	0	1	1	0	0

A *training pattern* is specified by a pair of an assemble of spiking binary vectors $\left(\omega_1, \omega_2, \ldots, \omega_{N_I}\right)$ and the so-called *required* output spiking binary vector $\omega^{(req)}$

$$\left(\omega_1, \omega_2, \ldots, \omega_{N_I}\right) / \omega^{(req)} \quad (5)$$

In order to formulate the present theory in an appropriate form for a simple specification of an adaptation process, the original input binary vectors from (5) are rewritten in a sequence of new input binary vectors

$$\left(\sigma_1, \sigma_2, \ldots, \sigma_{t_{max}}\right) \quad (6)$$

such that σ_i is composed of ith entries of all input binary vectors $\left(\omega_1, \omega_2, \ldots, \omega_{N_I}\right)$, then $\sigma_i \in \{0,1\}^{N_I}$.

Activities of hidden and output neurons are formally determined as follows

$$\begin{aligned} y_1 &= F\left(\sigma_1; w, \vartheta\right) \\ y_2 &= F\left(\sigma_1, \sigma_2; w, \vartheta\right) \\ &\ldots\ldots\ldots\ldots\ldots \\ y_{t_{max}} &= F\left(\sigma_1, \sigma_2, \ldots, \sigma_{t_{max}}; w, \vartheta\right) \end{aligned} \quad (7)$$

It means that the ith output entry is fully specified by the first i new input binary vectors. From output activities we construct t_{max}-dimensional real vector

$$y = \left(y_1, y_2, \ldots, y_{t_{max}}\right) \in (0,1)^{t_{max}} \quad (8)$$

We may say that this output binary vector is a response of the neural network onto new input binary vectors $\left(\sigma_1, \sigma_2, \ldots, \sigma_{t_{max}}\right)$.

An *objective function* with respect to the training pattern (5) is defined as follows

$$E(w) = \frac{1}{2}\left(\omega^{(req)} - y\right)^2 \geq 0 \quad (9)$$

If $y=\omega^{(req)}$, then the objective function is zero, on the other hand, if $y\neq\omega^{(req)}$, then the objective function is positive. Optimal values of parameters of the respective neural network are achieved by a minimization of the objective function $E(w)$ in a space Ω of all feasible weight coefficients

$$w_{opt} = \arg \min_{w \in \Omega} E(w) \tag{10}$$

Providing that a sigmoid activation function is used, then the objective function $E(w)$ is *differentiable* with respect to weight and threshold coefficients. An adaptation process is numerically realized by the simplest steepest-descent gradient method applied to a minimization of the objective function $E(w)$, an *updating of weight parameters* of the neural network is then specified as follows

$$\begin{aligned} w &:= w - \alpha \cdot grad_w E(w) \\ w &:= w + \sum_{t=1}^{t_{max}} \Delta w_t \\ \Delta w_t &= \alpha\left(\omega_t^{(req)} - y_t\right) grad_w y_t \end{aligned} \tag{11}$$

where α is a positive learning rate. Standard back-propagation method [4] could not be used immediately for the calculation of gradients of output activities y_t, for $t=1,2,\ldots,t_{max}$. This is caused by the fact that the same neurons have different activities at different time t (see Table I). In order to overcome this problem, we unfold the graph G into a tree structure, where the output neuron plays a role of a root, see Fig. 3 and Algorithm 1.

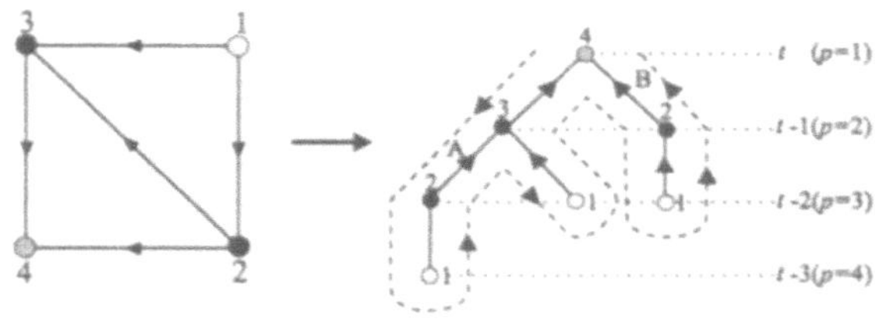

Fig. 3: An unfolding of the acyclic graph G into a tree with the output neuron as the root. The back-propagation method for the calculation of partial derivatives of the output activities y_t is easily performed over the unfolded diagram by a backtrack (see Algorithm 1), the created path is represented by an oriented dashed line

```
p:=1; U₁:=[N]; count:=1;
grad(0,N,t,1,1);
while p>0 do
if Uₚ≠[] then
begin path[p]:=Uₚ; Uₚ:=Uₚ-[path[p]];
      if p>1 then
      begin count:=count+1;
            i_upper:=path[p];
            i_lower:=path[p-1];
            grad(i_upper,i_lower,t,p,count);
      end;
      if (Γ[path[p]]≠[]) and (p<t) then
      begin p:=p+1; Uₚ:=Γ[path[p-1]];
      end;
end else p:=p-1;
```

Algorithm 1. A combination of back-propagation method with a back-track move on the unfolded graph G, in the course of construction of paths the respective partial derivatives are calculated. Procedure `grad` calculates a partial derivative for actual indices `i_upper` and `i_lower` and the integer variable `count` is used for sequential store of calculated partial derivatives. Set variable $\Gamma(i)$ represents predecessors of i in the tree.

The partial derivatives of an output activity y_t with respect to weight and threshold coefficients are easily constructed by the back-propagation method over the unfolded graph, we get the following four partial derivatives with respect to threshold coefficients

$$\frac{\partial y^{(t)}}{\partial \vartheta_4} = y^{(t)}\left(1 - y^{(t)}\right) \tag{12a}$$

$$\frac{\partial y^{(t)}}{\partial \vartheta_3} = x_3^{(t-1)}\left(1 - x_3^{(t-1)}\right) \cdot \frac{\partial y^{(t)}}{\partial \vartheta_4} w_{43} \tag{12b}$$

$$\left(\frac{\partial y^{(t)}}{\partial \vartheta_2}\right)_A = x_2^{(t-2)}\left(1 - x_2^{(t-2)}\right) \cdot \frac{\partial y^{(t)}}{\partial \vartheta_3} w_{32} \tag{12c}$$

$$\left(\frac{\partial y^{(t)}}{\partial \vartheta_2}\right)_B = x_2^{(t-1)}\left(1 - x_2^{(t-1)}\right) \cdot \frac{\partial y^{(t)}}{\partial \vartheta_4} w_{42} \tag{12d}$$

The partial derivative (12a) is calculated initially, there is necessary to know only a derivative of the output activity y_t. Knowing this derivative, we may continue a back propagation process in a calculation of partial derivatives that are assigned to neurons in a forthcoming lower positions, see (12b-c). Finally, we may calculate the last partial derivative (12d). In this process some partial derivatives may be calculated more than one times (see eqs. (12c-d)), then the resulting partial derivative is a sum of both of them. The partial derivatives with respect to weight coefficients are easily calculated from (12) by applying the general back-propagation formula

$$\frac{\partial y_t}{\partial w_{ij}} = \frac{\partial y_t}{\partial \vartheta_i} x_j \tag{13}$$

An activity x_j is taken at a respective time $t\text{-}p$, where p is a depth of the lower neuron v_j from the edge (v_j, v_i) on the unfolded graph (see Fig. 3).

Theoretical formulae for calculation of partial derivatives were checked numerically, i.e. the partial derivatives were approximately calculated by $\partial E/\partial w \approx (E(w+\delta)-E(w))/\delta$, for $\delta=10^{-3}$. We have obtained a perfect agreement between theoretically and numerically predicted derivatives up to three figures after decimal point.

In order to numerically verify the present theory we used training pattern $(\omega_1,\omega_2,\omega_3)/\omega^{(req)}$ for $t_{max}=12$, its single binary vectors are

$$\begin{aligned}\omega_1 &= (010100100100), \omega_2 = (010001100000) \\ \omega_3 &= (000000101000), \omega^{(req)} = (000101011000)\end{aligned} \tag{14}$$

The used feed-forward neural network is composed of five hidden neurons with connections randomly generated with probability $P_{connect}$=0.9. The adaptation process was realized by (11) with learning rate α=0.1. The obtained numerical results are presented in Figs 4. We see that the present theory of feed-forward networks composed of sigmoid neurons perfectly reproduces the required behavior of spiking neural networks with respect to the training pattern (15). The similar results were also achieved for different numbers of hidden neurons with randomly generated feed-forward topology and also for different types of training patterns.

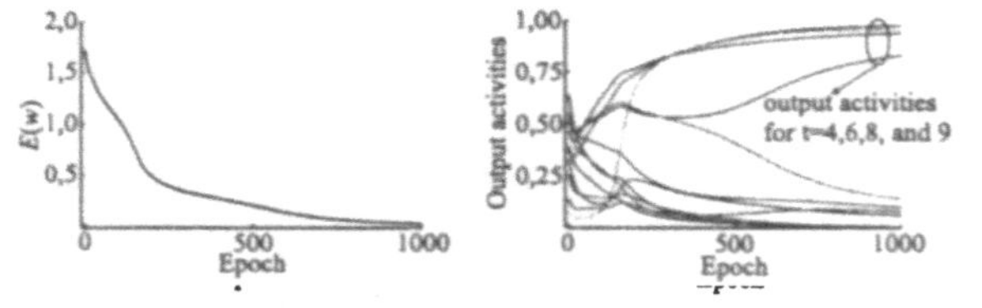

Fig. 4: The left hand side figure shows a plot of objective function with respect to epochs of adaptation process. We see that the objective function monotonously decreases asymptotically to zero, i.e. the adaptation process is successful. The right hand side figure displays plots of different output activities y_t (for t=1,2,...,12), we see that four of them tend to unit activities while other ones tend to zero activities (which was to be demonstrated).

3 Conclusions

We have demonstrated that the present feed-forward neural networks with simple sigmoid neurons *are able to simulate* the required output spiking activities as a response to the input spiking activities. They are *adapted by a simple modification of the back-propagation method* such that activities of neurons for different time steps are distinguished. This approach manifests many similarities with the well-known "back propagation through time" method [4], often used in the theory of recurrent neural networks.

Let us consider a training pattern

$$(\omega_1, \omega_2, \ldots, \omega_{N_I}) / \omega^{(req)} \tag{16}$$

We say that the pattern is *deterministic* if any entry of the required output spike-train vector at time t may be interpreted as a consequence of past entries at time $t' < t$ of the input spike-train vectors.

Our numerical results allow us to formulate a conjecture that the present type of feed-forward neural networks is a universal *approximator* for any deterministic training pattern of binary spiking vectors.

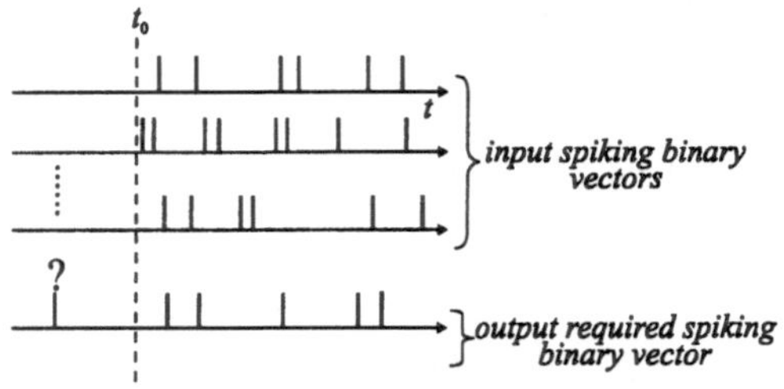

Fig. 5: An illustrative counterexample of training pattern that is not deterministic. The required output spiking activity contains an entry that could not be interpreted as a response of input spiking activities. In particular, the output vector contains an entry that appears earlier than first spikes in all input activities.

Conjecture. *For a given deterministic training pattern there exists a feed-forward network composed of sigmoid neurons, such that it reproduces the required output spike binary vector as a response to the respective input spike binary vectors.*

Let a neural network from the above theorem be represented by a parametric mapping F, then the required output spike-train vector is determined by

$$\omega^{(req)} = y = F(\omega_1, \omega_2, \ldots, \omega_{N_I}) \tag{15}$$

After the above conjecture, the present type of neural networks manifest interesting spiking properties that are usually proposed only for spiking neural networks. Now, we see that a property "to be spiking" may be alternatively well achieved and simulated by making use of the standard feed-forward neural network with sigmoid neurons.

Acknowledgement. This work was supported by the grants # 1/4209/97 and # 1/5229/98 of the Scientific Grant Agency of Slovak Republic.

References

[1] W. Maas, C. M. Bishop (Editors), *Pulsed Neural Networks*. The MIT Press, Cambridge, MA, 1999.

[2] F. Rieke, D. Warland, and W. Bialek (Editors), *Spikes: Exploring the Neural Code (Computational Neuroscience)*. The MIT Press, Cambridge, MA, 1999.

[3] S. Wermter, R. Sun (Editors): *Hybrid Neural Systems*. Springer Verlag, Berlin, 2000.

[4] S. Haykin: Neural Networks. *A Comprehensive Foundation*. Prentice Hall, Upper Saddle River, New Jersey, 1999.

Tightness of Upper Bounds on Rates of Neural-Network Approximation

Věra Kůrková *, Marcello Sanguineti † 1

* Institute of Computer Science, Academy of Sciences of the Czech Republic, P.O. Box 5, 182 07, Prague 8, Czech Republic, e-mail: vera@cs.cas.cz † DIST, University of Genoa, Via Opera Pia 13, 16145 Genoa, Italy, e-mail: marcello@dist.unige.it

Abstract

Tightness of upper bounds on neural network approximation is investigated in the framework of variable-basis approximation. Conditions are given on a variable basis that do not allow a possibility of improving such bounds beyond $\mathcal{O}(n^{-(\frac{1}{2}+\frac{1}{d})})$, where d is the number of variables of the functions to be approximated. Such conditions are satisfied by Lipschitz sigmoidal perceptrons.

1 Introduction

Feedforward networks are mostly simulated on classical computers; for such simulations, one of the limiting factors is the *number n of hidden units*. Jones [7] has obtained insight into the reason that some high-dimensional tasks can be performed efficiently by neural networks with a moderate number of hidden units. He constructed incremental approximants with rates of convergence of the order of $\mathcal{O}(n^{-\frac{1}{2}})$. The same estimates had earlier been proved by Maurey using a probabilistic argument (see Pisier [17] and also Barron [2]). Barron [2] improved Jones's [7] upper bound and applied it to neural networks. Using a weighted Fourier transform, he described sets of multivariable functions that can be approximated by perceptron networks having n hidden units within an accuracy of the order of $\mathcal{O}(n^{-\frac{1}{2}})$. Such bounds are sometimes called "dimension-independent" as they do not depend on the number of variables. However, such a term can be misleading, as sets of multivariable functions to which such estimates apply become more and more constrained as the number of variables increases.

The Maurey-Jones-Barron upper bound is quite general, as it applies to *nonlinear approximation of the variable-basis type*, i.e., approximation by linear combinations of n-tuples of elements of a given set of basis functions. This approximation scheme has been widely investigated (see, e.g., DeVore and Temlyakov [4] and the references therein): it includes splines with free nodes, trigonometric polynomials with free frequencies, sums of wavelets and feedforward neural networks.

Several authors have further improved or extended these dimension-independent bounds. An extension to $\mathcal{L}_p$-spaces, with $p \in (1, \infty)$, has been derived by Darken et al. [3] and an extension to $\mathcal{L}_\infty$-spaces has been obtained by Barron [1], Girosi [5], Gurvits and Koiran [6], Makovoz [15] and Kůrková, Savický and Hlaváčková [13].

Makovoz [14] improved Maurey's probabilistic argument by combining it with a concept from metric entropy theory, which he also used to show that in the case of Lipschitz sigmoidal perceptron networks, the upper bound cannot be improved to $\mathcal{O}(n^{-\alpha})$ for $\alpha > \frac{1}{2} + \frac{1}{d}$, where d is the number of variables of the functions to be approximated. A similar tightness result for perceptron networks was earlier obtained by Barron [1], who used a more complicated proof technique. For the special case of orthonormal variable-basis, Mhaskar and Micchelli [16], Kůrková, Savický and Hlaváčková [13] and Kůrková and Sanguineti [11] have derived tight improvements of Maurey-Jones-Barron's bound.

In this paper, we extend tightness results derived by Barron [1] and Makovoz [14] for approximation by convex combinations of functions computable by sigmoidal perceptrons to combinations of more general basis functions satisfying certain conditions, that are fulfilled by standard neural-network hidden units. These conditions are defined in terms of (i) polynomial growth of the number of sets of a given diameter needed to cover such basis and (ii) sufficient "capacity" of the basis, in the sense that its convex hull has an orthogonal subset that for each positive integer k contains at least k^d functions with norms greater or equal to $\frac{1}{k}$. The proofs of our results, which are only sketched here, are given in [12].

[1]Both authors were partially supported by NATO Grant PST.CLG.976870. V. Kůrková was partially supported by grant GA ČR 201/00/1489. M. Sanguineti was partially supported by the Italian Ministry for the University and Research (MURST) and by grant D.R.42 of the Univ. of Genoa.

2 Approximation by Neural Networks and by Variable-Basis Functions

Approximation by feedforward neural networks can be studied in a more general context of *approximation by variable-basis functions.* In this approximation scheme, elements of a real normed linear space $(X, \|.\|)$ are approximated by linear combinations of at most n elements of a given subset G. The set of such combinations is denoted by $span_n G = \{\sum_{i=1}^n w_i g_i; w_i \in \mathcal{R}, g_i \in G\}$; it is equal to the union of n-dimensional subspaces generated by all n-tuples of elements of G. G can represent the set of functions computable by *hidden units* in neural networks. Such units compute functions of the form $\phi : \mathcal{R}^p \times \mathcal{R}^d \to \mathcal{R}$, where $\mathcal{R}$ denotes the set of real numbers, ϕ corresponds to the type of unit, and p and d to the dimension of a *parameter space* and an *input space*, resp.. The set of input/output functions of a network with a single linear output unit and n hidden units computing the function ϕ is equal to $span_n G_\phi$, where $G_\phi = \{\phi(\mathbf{a}, \cdot); \mathbf{a} \in \mathcal{R}^p\}$. Also multilayer networks with a single linear output unit and n units in the last hidden layer belong to this approximation scheme; they compute functions from $span_n G$ with G depending on the number of units in the previous hidden layers.

Recall that a *perceptron* with an activation function $\psi : \mathcal{R} \to \mathcal{R}$ computes functions of the form $\phi((\mathbf{v}, b), \mathbf{x}) = \psi(\mathbf{v} \cdot \mathbf{x} + b) : \mathcal{R}^{d+1} \times \mathcal{R}^d \to \mathcal{R}$, where $\mathbf{v} \in \mathcal{R}^d$ is an *input weight* vector and $b \in \mathcal{R}$ is a *bias.* By $P_d(\psi) = \{f : [0,1]^d \to \mathcal{R}; f(\mathbf{x}) = \psi(\mathbf{v}\cdot\mathbf{x}+b), \mathbf{v} \in \mathcal{R}^d, b \in \mathcal{R}\}$ we denote the set of functions on $[0,1]^d$ computable by ψ-perceptrons. The most common activation functions are *sigmoidals*, i.e., functions $\sigma : \mathcal{R} \to [0,1]$ such that $\lim_{t\to-\infty} \sigma(t) = 0$ and $\lim_{t\to\infty} \sigma(t) = 1$; the discontinuous sigmoidal defined as $\vartheta(t) = 0$ for $t < 0$ and $\vartheta(t) = 1$ for $t \geq 0$ is called *the Heaviside function.* A function $\sigma : \mathcal{R} \to \mathcal{R}$ is Lipschitz if there exists $M > 0$ such that $|\sigma(t) - \sigma(t')| \leq M|t - t'|$ for all $t, t' \in \mathcal{R}$.

Rates of approximation of functions from a set Y by functions from a set M can be studied in terms of the *worst-case error* formalized by the concept of *deviation of Y from M* defined as $\delta(Y, M) = \delta(Y, M, (X, \|.\|)) = \sup_{f\in Y} \|f - M\| = \sup_{f\in Y} \inf_{g\in M} \|f - g\|$. To formulate estimates of deviation from $span_n G$ we need to introduce a few more concepts and notations. If G is a subset of $(X, \|.\|)$ and $c \in \mathcal{R}$, then we define $cG = \{cg; g \in G\}$ and $G(c) = \{wg; g \in G, w \in \mathcal{R} \,\&\, |w| \leq c\}$. The *closure* of G is denoted by $cl\, G$ and defined as $cl\, G = \{f \in X; (\forall\varepsilon > 0)(\exists g \in G)(\|f - g\| < \varepsilon)\}$. G is *dense* in $(X, \|.\|)$ if $cl\, G = X$. The *convex hull* of G, denoted by $conv\, G$, is the set of all convex combinations of its elements, i.e., $conv\, G = \{\sum_{i=1}^n a_i g_i; a_i \in [0,1], \sum_{i=1}^n a_i = 1, g_i \in G, n \in \mathcal{N}_+\}$. $conv_n\, G$ denotes the set of all convex combinations of n elements of G, i.e., $conv_n\, G = \{\sum_{i=1}^n a_i g_i;\ a_i \in [0,1], \sum_{i=1}^n a_i = 1, g_i \in G\}$. $B_r(x, \|.\|)$ denotes the ball of radius r with respect to the norm $\|.\|$ centered at $x \in X$, i.e., $B_r(x, \|.\|) = \{y \in X; \|y - x\| \leq r\}$. We write shortly $B_r(\|.\|)$ instead of $B_r(0, \|.\|)$.

The following estimate is a version of Jones' result as improved by Barron [2] and also of earlier result of Maurey. Recall that a Hilbert space is a normed linear space with the norm induced by an inner product.

Theorem 2.1 *Let $(X, \|.\|)$ be a Hilbert space, b a positive real number, G a subset of X such that for every $g \in G$ $\|g\| \leq b$, and let $f \in cl\, conv\, G$. Then, for every positive integer n, $\|f - conv_n G\| \leq \sqrt{\frac{b^2 - \|f\|^2}{n}}$.*

In the following, we shall sometimes refer to Theorem 2.1 and to its bound as Maurey-Jones-Barron's theorem and bound, resp. As $conv_n G \subseteq span_n G$, the upper bound from Theorem 2.1 also applies to rates of approximation by $span_n G$. However, when G is not closed up to multiplication by scalars, $conv\, G$ is a proper subset of $span\, G$, and hence also $cl\, conv\, G$ is a proper subset of $cl\, span\, G$. Thus density of $span\, G$ in $(X, \|.\|)$ does not guarantee that Theorem 2.1 can be applied to all elements of X. As $conv_n G(c) \subset span_n G(c) = span_n G$ for any $c \in \mathcal{R}$, by replacing the set G by $G(c) = \{wg; w \in \mathcal{R}, |w| \leq c, g \in G\}$ we can apply Theorem 2.1 to all elements of $\cup_{c\in\mathcal{R}_+} cl\, conv\, G(c)$. This approach can be mathematically formulated in terms of a norm tailored to a set G (in particular, to sets G_ϕ corresponding to various computational units ϕ in neural networks). Let $(X, \|.\|)$ be a normed linear space and G be its subset, then *G-variation* (variation with respect to G) denoted by $\|.\|_G$ is defined as the Minkowski functional of the set $cl\, conv\, G(1) = cl\, conv(G \cup -G)$, i.e.,

$$\|f\|_G = \inf\{c \in \mathcal{R}_+; f \in cl\, conv\, G(c)\}.$$

G-variation has been introduced by Kůrková [9] as an extension of Barron's [1] concept of variation with respect to half-spaces (more precisely, variation with respect to characteristic functions of half-spaces) corresponding to perceptrons with Heaviside activation function. For functions of one variable, variation with respect to half-spaces coincides, up to a constant, with the notion of total variation studied in integration theory; for G orthonormal, it is equal to the l_1-norm with respect to G (see [11]).

The following theorem is a corollary of Theorem 2.1 formulated in terms of G-variation (see [9]). Recall that for any G, the unit ball in G-variation is equal to $cl\, conv(G \cup -G)$.

Theorem 2.2 *Let $(X, \|.\|)$ be a Hilbert space and G be its subset. Then, for every $f \in X$ and every positive integer n, $\delta(B_1(\|.\|_G), span_n G) \leq \frac{s_G}{\sqrt{n}}$, where $s_G = \sup_{g \in G} \|g\|$.*

Thus all functions from the unit ball in G_ϕ-variation can be approximated within $\frac{s_{G_\phi}}{\sqrt{n}}$ by ϕ-networks with n hidden units independently on the number d of variables. However, with increasing number of variables, the condition of being in the unit ball in G_ϕ-variation becomes more and more constraining (see [13] for examples of functions with variations depending exponentially on d).

3 Tightness of the Bound $\mathcal{O}(n^{-\frac{1}{2}})$ on Variable-Basis Approximation

To disprove for certain sets G the possibility of an improvement of Maurey-Jones-Barron's upper bound beyond $\mathcal{O}(n^{-(\frac{1}{2}+\frac{1}{d})})$, we shall assume that such an improvement is possible and derive a contradiction by considering its consequences on the growth of certain covering numbers of the unit ball in G-variation. Recall that for $\varepsilon > 0$, the ε-*covering number* of a subset K of a normed linear space $(X, \|.\|)$ is defined as $cov_\varepsilon K = cov_\varepsilon(K, \|.\|) = \min\{n \in \mathcal{N}_+ ; K \subseteq \cup_{i=1}^n B_\varepsilon(x_i, \|.\|), x_i \in K\}$ if the set over which the minimum is taken is nonempty, otherwise $cov_\varepsilon(K) = +\infty$. The following lemma gives a lower bound on the ε-covering number of balls in G-variation containing a subset of orthogonal elements, for ε defined in terms of the cardinality of such a subset and the minimum of norms of its elements. The proof of the lemma is based on the exponentional growth of quasiorthogonal dimension studied in [8]. $H(p) = p \log_2(p) + (1-p)\log_2(1-p)$, $p \in [0,1]$, denotes the binary entropy function.

Lemma 3.1 *Let $(X, \|.\|)$ be a Hilbert space, G, A be its subsets such that $A \subseteq B_1(\|.\|_G)$, A is a set of m orthogonal elements and $\min_{h \in A} \|h\| = a$. Then $cov_{\frac{a}{2\sqrt{m}}} B_1(\|.\|_G) \geq 2^{bm}$, where $b = H(\frac{1}{4})$.*

We shall apply this lemma to a ball containing a sequence of subsets with increasing cardinality, that contain orthogonal elements with norms that do not vanish "too quickly". More precisely, for a positive integer d (corresponding, in the following, to the number of variables of functions in X), we call a subset A of a normed linear space $(X, \|.\|)$ *not quickly vanishing with respect to d* if $A = \cup_{k \in \mathcal{N}_+} A_k$, where, for each $k \in \mathcal{N}_+$, $card\, A_k \geq k^d$ and for each $h \in A_k$, $\|h\| \geq \frac{1}{k}$ (see [10]). We shall also need the following relationships between covering numbers.

Lemma 3.2 *Let $(X, \|.\|)$ be a normed linear space, G be its bounded subset and $s_G = \sup_{g \in G} \|g\|$. Then, for every $\varepsilon > 0$,*
(i) $cov_{\varepsilon(1+s_G)} conv_n G \leq \left(\frac{2}{\varepsilon} cov_\varepsilon G\right)^n$;
(ii) $cov_\varepsilon(G \cup -G) \leq 2 cov_\varepsilon G$.

Recall that for $f, g : \mathcal{N}_+ \to \mathcal{N}_+$, $g(n) \leq \mathcal{O}(f(n))$ if there exists $c \in \mathcal{R}_+$ such that for all but finitely many $n \in \mathcal{N}_+$, $g(n) \leq c f(n)$. Makovoz [14] proved that when σ is a Lipschitz sigmoidal, then the rate of the order of $\mathcal{O}(n^{-\frac{1}{2}})$ in approximation of elements of the unit ball in $P_d(\sigma)$-variation by $conv_n(P_d(\sigma) \cup -P_d(\sigma))$, that is guaranteed by Maurey-Jones-Barron's theorem, cannot be improved to $\mathcal{O}(n^{-\alpha})$ for $\alpha > \frac{1}{2} + \frac{1}{d}$. Our main theorem extends this Makovoz's result to sets G of functions of d variables that have covering numbers depending only polynomially on the number of variables d and for which the unit ball in G-variation contains an orthogonal subset that is not quickly vanishing with respect to d.

Theorem 3.3 *Let $(X, \|.\|)$ be a Hilbert space of functions of d variables and G be its bounded subset satisfying the following conditions:*
(i) there exists a polynomial $p(d)$ and $a \in \mathcal{R}_+$ such that, for every $\varepsilon > 0$, $cov_\varepsilon(G) \leq a \left(\frac{1}{\varepsilon}\right)^{p(d)}$;
(ii) there exists $r \in \mathcal{R}_+$ for which $B_r(\|.\|_G)$ contains a set of orthogonal elements which is not quickly vanishing with respect to d.
Then $\delta(B_1(\|.\|_G), conv_n(G \cup -G)) \leq \mathcal{O}(n^{-\alpha})$ implies $\alpha \leq \frac{1}{2} + \frac{1}{d}$.

Sketch of the proof. Assume that there exists $\alpha > \frac{1}{2} + \frac{1}{d}$ such that, for all but finitely many $n \in \mathcal{N}_+$, $\delta(B_1(\|.\|_G), conv_n(G \cup -G)) \leq \frac{c}{n^\alpha}$. Set $\delta = \frac{2c}{n^\alpha}$. We shall derive a contradiction by comparing an upper bound on $cov_\delta B_1(\|.\|_G)$ (obtained from the assumption (i) and this hypothetical upper bound) with a lower bound on the same covering number (obtained from the assumption (ii) and Lemma 3.1). Without loss of generality assume $s_G = 1$. By the triangle inequality, Lemma 3.2 and the assumption (i), we get $cov_\delta B_1(\|.\|_G) \leq cov_{\delta/2} conv_n(G \cup -G) \leq (2 cov_{\delta/4} G)^n (\frac{8}{\delta})^n \leq a^n 4^{n(2+p(d))} \delta^{-n(1+p(d))} = a(n,d) n^{\alpha n(1+p(d))}$, where $a(n,d) = a^n 4^{n(2+p(d))} (2c)^{-n(1+p(d))}$. On the other hand, using the assumption (ii) set for each positive integer k, $A_{r,k} = \frac{1}{r} A_k$. We have $A_{r,k} \subset B_1(\|.\|_G)$ and by Lemma 3.1, $cov_{\varepsilon_k} B_1(\|.\|_G) \geq cov_{\varepsilon_k} B_1(\|.\|_{A_{r,k}}) \geq 2^{bk^d}$, where $b = H(\frac{1}{4})$ and $\varepsilon_k =$

$\frac{1}{2rk^{d/2+1}}$. If $k \leq \bar{k} = \frac{n^{\alpha}}{4cr}^{\frac{2}{d+2}}$, then $\delta \leq \varepsilon_k$. So for $\bar{k}$ an integer, set $k = \bar{k}$. Then we get $cov_{\delta} B_1(\|.\|_G) \geq cov_{\varepsilon_k} B_1(\|.\|_G) \geq 2^{bk^d} \geq 2^{c_d n^{\frac{\alpha}{1/2+1/d}}}$, where $c_d = b\left(\frac{1}{4cr}\right)^{\frac{1}{1/2+1/d}}$, which gives for large n a contradiction. If $\bar{k}$ is not integer, set $k = \lfloor \bar{k} \rfloor \geq \bar{k} - 1 \geq \frac{\bar{k}}{2}$ for $\bar{k} \geq 2$, and get a contradiction in a similar way.

As both assumptions of Theorem 3.3 are verified in the case of perceptrons with Lipschitz sigmoidal activation, we get the following corollary.

Corollary 3.4 *Let d, n be positive integers and let $\sigma : \mathcal{R} \to \mathcal{R}$ be a Lipschitz sigmoidal function. Then in $(\mathcal{L}^2([0,1]^d), \|.\|_2)$, $\delta(B_1(\|.\|_{P_d(\sigma)}), span_n(P_d(\sigma) \cup -P_d(\sigma)) \leq \mathcal{O}(n^{-\alpha})$ implies $\alpha \leq \frac{1}{2} + \frac{1}{d}$.*

Sketch of the proof. Both conditions (i) and (ii) from Theorem 3.3 are satisfied by $P_d(\sigma)$. For (i), see [14, Lemma 2]. The condition (ii) is guaranteed by the following construction from [10]: set $A_d = \cup_{k \in \mathcal{N}_+} A_{d,k}$, where $A_{d,k} = \{h_{\mathbf{v}}; \mathbf{v} \in \{1, \ldots, k\}^d\} \subset (\mathcal{L}_2([0,1]^d), \|.\|_2)$, with $h_{\mathbf{v}}(\mathbf{x}) = c_{\mathbf{v}} \cos(2\pi \mathbf{v} \cdot \mathbf{x}) : [0,1]^d \to \mathcal{R}$, $c_{\mathbf{v}} = d/(\sqrt{2}\lceil \sum_{j=1}^{d} v_k \rceil)$, and $\mathbf{v} = (v_1, \ldots, v_d)$. It is shown in [10] that for any positive integer d, $A_d \subset B_{d/\sqrt{8}}(\|.\|_{P_d(\sigma)})$ and that $A = \cup_{d \in \mathcal{N}_+} A_d$ is orthogonal not quickly vanishing with respect to d.

4 Discussion

We have stated conditions that prevent an improvement of Maurey-Jones-Barron's upper bound to $\mathcal{O}(n^{-\alpha})$ for $\alpha > \frac{1}{2} + \frac{1}{d}$. As sets of functions computable by Lipschitz sigmoidal perceptrons satisfy these conditions, it follows that one cannot improve the upper bound on the approximation rate for one-hidden-layer networks with such perceptrons when the sum of the absolute values of the output weights is kept below a fixed bound. It is an open problem whether Theorem 3.3 can be generalized to approximation by linear instead of convex combinations (a special case of this problem concerning one-hidden layer networks with a Lipschitz sigmoidal activation function and unconstrained output weights was stated by Makovoz [14]).

Better rates than $\mathcal{O}(n^{-(\frac{1}{2}+\frac{1}{d})})$ might be achievable using networks with more than one hidden layer: for some of such networks, sets of basis functions might be large enough not to satisfy condition (i) on polynomial growth of covering numbers.

References

[1] Barron, A.R.: Neural net approximation. *Proc. 7th Yale Work. on Adaptive and Learning Systems* (K. Narendra, Ed.), pp. 69-72. Yale Univ. Press, 1992.

[2] Barron, A.R.: Universal approximation bounds for superpositions of a sigmoidal function. *IEEE Trans. on Information Th.* 39, pp. 930-945, 1993.

[3] Darken, C., Donahue, M., Gurvits, L., and Sontag, E.: Rate of approximation results motivated by robust neural network learning. *Proc. Sixth Annual ACM Conf. on Computational Learning Th.*. The Association for Computing Machinery, New York, N.Y., pp. 303-309, 1993.

[4] DeVore, R.A., and Temlyakov, V.N.: Nonlinear approximation by trigonometric sums. *The J. of Fourier Analysis and Appl.* 2, pp. 29-48, 1995.

[5] Girosi, F.: Approximation error bounds that use VC-bounds. *Proc. Int. Conf. on Artificial Neural Networks ICANN'95.* Paris: EC2 & Cie, pp. 295–302, 1995.

[6] Gurvits, L., and Koiran, P.: Approximation and learning of convex superpositions. *J. of Computer and System Sciences* 55, pp. 161–170, 1997.

[7] Jones, L.K.: A simple lemma on greedy approximation in Hilbert space and convergence rates for projection pursuit regression and neural network training. *Annals of Statistics* 20, pp. 608–613, 1992.

[8] Kainen, P.C., and Kůrková, V.: Quasiorthogonal dimension of Euclidean spaces. *Appl. Math. Lett.* 6, pp. 7–10, 1993.

[9] Kůrková, V.: Dimension-independent rates of approximation by neural networks. In *Computer-Intensive Methods in Control and Signal Processing. The Curse of Dimensionality* (K. Warwick, M. Kárný, Eds.). Birkhauser, Boston, pp. 261-270, 1997.

[10] Kůrková, V., and Sanguineti, M.: Tools for comparing neural network and linear approximation. *Submitted to IEEE Trans. on Information Th.*

[11] Kůrková, V., and Sanguineti, M.: Bounds on rates of variable-basis and neural-network approximation. To appear in *IEEE Trans. on Information Th.*

[12] Kůrková, V., and Sanguineti, M.: Covering numbers and rates of neural-network approximation. Research Report ICS-00-830, 2000.

[13] Kůrková, V., Savický, P., and Hlaváčková, K.: Representations and rates of approximation of real–valued Boolean functions by neural networks. *Neural Networks* 11, pp. 651-659, 1998.

[14] Makovoz, Y.: Random approximants and neural networks. *J. of Approx. Th.* 85, pp. 98–109, 1996.

[15] Makovoz, Y.: Uniform approximation by neural networks. *J. of Approx. Th.* 95, pp. 215–228, 1998.

[16] Mhaskar, H.N. and Micchelli, C.A.: Dimension-independent bounds on the degree of approximation by neural networks. *IBM J. of Research and Development* 38, n. 3, pp. 277–283, 1994.

[17] Pisier, G.: Remarques sur un resultat non publié de B. Maurey. *Seminaire d'Analyse Fonctionelle*, vol. I, no. 12. École Polytechnique, Centre de Mathématiques, Palaiseau, 1980-81.

Incremental Function Approximation Based on Gram-Schmidt Orthonormalisation Process

Bartlomiej Beliczynski *[1]

*Warsaw University of Technology,Institute of Control and Industrial Electronics, ul. Koszykowa 75, 00-662 Warszawa, Poland, b.beliczynski@isep.pw.edu.pl

Abstract

In this paper we present an incremental function approximation in Hilbert space, based on Gram-Schmidt orthonormalisation process. Two bases of approximation space are determined and mantained during approximation process. The first one is used for neural network implementation, the second - orthonormal one, is treated as an intermediate step of calculations. Only after terminating all iterations the output weights are calculated (once).

1 Introduction

Among many different neural network function approximation schemes one might be specially distinguished being, in various opinion, effective and controversial, simple and natural or wasteful. There, contrary to traditional fixed architecture where initially neural network structure is chosen and then its parameters are tuned, the network architecture is not fixed. It is selected to be one-hidden-layer architecture, but the number of hidden layer units changes during incremental learning process. In every incremental step one unit is added to the network and its parameters are determined. All previously selected units' parameters are frozen. This is symbolically shown in Figure 1. Dashed lines denote connections calculated in every iteration.

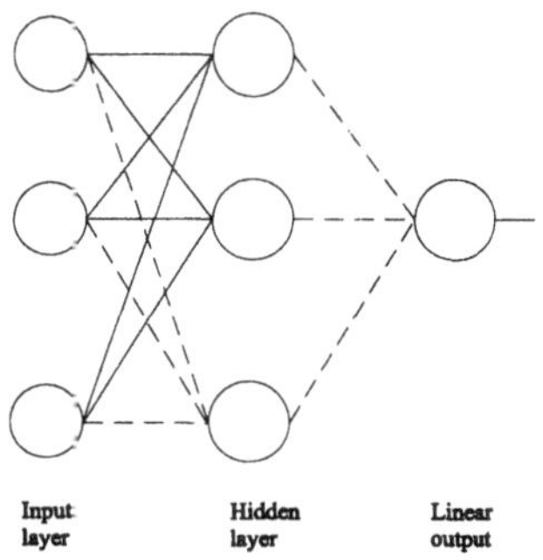

Fig. 1. Incremental aproximation scheme. In principle, in every iteration one-hidden-unit is added to the network and all output weights are calculated.

This kind of decomposed approximation, when only one node is optimised at a time, makes optimisation task easier. However it generates also questions about convergence and doubts about efficiency of the optimisation. For many engineers well experienced with optimisation techniques this scheme is far away from being intuitively accepted.

From theoretical side the incremental approximation might be variously originated. In 1992 Jones [5] introduced a recursive construction of approximation converging to elements in convex closures of bounded subsets of a Hilbert space. Then Barron proved convergence of incremental algorithms satisfying certain conditions imposed in each iteration on optimisation of parameters of a new hidden unit as well as of all output weights [1]. Various other versions (e.g. Fahlman and Lebiere [3], Fritzke [4]) were demonstrated.

Different origins were referred by Kwok and Yeung [7], [8] whose papers were very much practise oriented. For similar concepts as "incremental approximation" they used term "constructive approximation". The constructive approximation was discussed in line with pruning and regularisation techniques and treated as a method for matching the network complexity to the problem.

For many practitioners such properties of incremental approximation as lack of local minima in the error surface and relatively easy complexity control are great advantages over traditional fixed architecture.

In this paper we demonstrate that the incremental approximation concept arrises naturally from XIX century Gram-Schmidt basis orthonormalisation process (see for instance [6]) and if described in this way its properties are more understandable. Some of them become transparent, others easy provable. In fact, a general condition to achieve decrease of error in every step is almost obvious and the for-

[1]This work was supported in part by Warsaw University of.Technology grant.

mula to obtain maximum speed of approximation in every step comes easily. In various cases of practical schemes it was observed that if many hidden units were used and approximation error was small enough, some numerical difficulties occurred. This again is an observable and understandable phenomena in our scheme. In the incremental approximation presented here two bases of the approximation space are calculated: a basis for implementation and an orthonormal basis which is used for intermediate calculations and examining stopping conditions etc. We select iteratively a sequence of hidden units represented by functions $g_1, ..., g_i$ named implementation basis and a related to it sequence of orthonormal functions $o_1, ..., o_i$ named orthonormal basis. In each step the output weight of newly selected orthonormal element is calculated. Stopping conditions are tested and if the iterations are terminated the output weights of the implementation scheme are calculated (once only).

This paper is organised as follows. In Section 2 best approximation property in a space with fixed basis is discussed. In Section 3 an adaptive selection of basis for the approximation space which utilises Gram-Schmidt orthonormalisation process is proposed and incremental approximation algorithm presented. Then in Section 4 some properties of the approximation are formulated and proved. Section 5 contains an example and finally conclusions are drawn in Section 6.

2 Best Approximation and Orthonormal Basis

Let us consider the following function

$$f_n = \sum_{i=1}^{n} g_i w_i, \tag{1}$$

where $g_i \in \mathcal{G} \subset \mathcal{H}$, and $\mathcal{H}$ is a Hilbert space $\mathcal{H} = (\mathcal{H}, ||.||)$, $i = 1, ..., n$, and $w_i \in \Re$, $i = 1, \ldots, n$.

The formula (1) describes one-hidden-layer neural network. We do not assume that the functions representing hidden layer $g_1, ..., g_n$ are of the special form neither they are sigmoidal nor radial. If the functions g_i are linearly independent, then the set of functions $\{g_1, ..., g_n\}$ is a basis of n-dimensional space. The functions of the form (1) belong to $span\{g_1, ..., g_n\}$.

For any function f from a Hilbert space $\mathcal{H}$ and a closed (finite dimensional) subspace $\mathcal{G} \subset \mathcal{H}$ with basis $\{g_1, ..., g_n\}$ there exists unique best approximation of f by elements of $\mathcal{G}$. Let denote it by g_0. Quite naturally one may search for it. The coefficients w_i can be determined in the way described further. Because error of the best approximation is orthogonal to all elements of the approximation space $f - g_0 \bot \mathcal{G}$, the coefficients w_i may be calculated from the set of linear equations

$$\langle g_i, f - g_0 \rangle = 0 \text{ for } i = 1, ..., n \tag{2}$$

where $\langle ., . \rangle$ denotes inner product.

The formula (2) can also be written as

$$\left\langle g_i, f - \sum_{k=1}^{n} w_k g_k \right\rangle = \langle g_i, f \rangle - \sum_{k=1}^{n} w_k \langle g_i, g_k \rangle = 0 \tag{3}$$

for $i = 1, ..., n$ or in matrix form

$$\Gamma w = G_f \tag{4}$$

where $\Gamma = [\langle g_i, g_j \rangle]$, $i, j = 1, ..., n$, $w = [w_1, ..., w_n]^T$, $G_f = [\langle g_1, f \rangle, ..., \langle g_n, f \rangle]^T$ and "T" denotes transposition.

Because there exists an unique best approximation of f in an n-dimensional space $\mathcal{G}$ with basis $\{g_1, ..., g_n\}$, the matrix Γ is nonsingular and $w_0 = \Gamma^{-1} G_f$.

For any basis $\{g_1, ..., g_n\}$ one can find such orthonormal basis $\{o_1, ..., o_n\}$, $\langle o_i, o_j \rangle = 1$ when $i = j$ and $\langle o_i, o_j \rangle = 0$ when $i \neq j$ that $span\{g_1, ..., g_n\} = span\{o_1, ..., o_n\}$. In such case Γ is an unit matrix and $w_0 = \begin{bmatrix} \langle o_1, f \rangle & \langle o_2, f \rangle & \ldots & \langle o_n, f \rangle \end{bmatrix}^T$.

When additional element o_{n+1} is added to the basis than only one output weight has to be calculated $\langle o_{n+1}, f \rangle$. All previously determined output weights remain unchanged.

The idea of this approximation is to calculate and store simultaneously two bases of the approximation space. One $\{g_1, ..., g_n\}$ is used for implementation and another orthonormal basis for analysis, as an intermediate step in calculations and testing stopping conditions. Conversion between $\{g_1, ..., g_n\}$ and $\{o_1, ..., o_n\}$ can be done by a simple and elegant Gram-Schmidt orthonormalisation process.

3 Gram-Schmidt Process and the Incremental Approximation Algorithm

The original Gram-Schmidt algorithm of basis orthonormalisation is an incremental process demonstrated below:

Given: $n, g_1, ..., g_n$, **Searched for:** $o_1, ..., o_n$

1.$o_1 = g_1/||g_1||$, **2.**$for\ i = 2 : n$, **3.**$v_{i-1}(g_i) = v_{i-1} = g_i - \sum_{k=1}^{i-1} \langle g_i, o_k \rangle o_k$, **4.**$o_i = v_{i-1}/||v_{i-1}||$, **5.***end*.

In step 3 of the algorithm v_{i-1} is calculated as a difference between g_i and its projection to space $span\{o_1,...,o_{i-1}\} = span\{g_1,...,g_{i-1}\}$. It is easy to demonstrate that $\left\langle v_{i-1}, \sum_{k=1}^{i-1} \langle g_i, o_k\rangle o_k \right\rangle = 0$ i.e. the projection is orthogonal. The only numerically sensitive point in this algorithm is $||v_{i-1}||$ which appears at the denominator in Step 4. Note however that $v_{i-1} \neq 0$ unless $g_i \in span\{g_1,...,g_{i-1}\}$.

In an incremental approximation process one has to select a finite length sequence of functions $g_1,...,g_n$, $g_i \in \mathcal{G} \subset \mathcal{H}$ describing hidden nodes and the output weights $w_1,...,w_n, w_i \in \Re$. Well chosen output weights ensure the best approximation of $f \in \mathcal{H}$ in the approximation space $\mathcal{G}$ with basis $g_1,...,g_n$ i.e. the orthogonal projection of f function into space $span\{g_1,...,g_n\}$.

Here the algorithm modification is provided in such a way that $g_2, g_3, ...$ functions are selected within a loop, n is not necessary given or fixed, but the stopping conditions trap is tested. The modified algorithm is given next. Capital letters denote blocs described elsewhere or not discussed at all. Before we proceed with The Algorithm, a useful definition is given.

Definition 1 *Let $\mathcal{A}_n$ be a set of orthonormal elements of Hilbert space $\mathcal{H}$, $\mathcal{A}_n = \{o_1,...,o_n\} \subset \mathcal{H}$. Function $v_i : \mathcal{H} \to \mathcal{H}$, $i = 1,...,n$,*

$$v_i(f) = f - \sum_{k=1}^{i} \langle f, o_k\rangle o_k \tag{5}$$

will be named the function of accuracy of approximation.

The $v_i(f)$ describes error function between f and its best approximation in i-dimensional approximation space.

Now The Algorithm can be presented.

The Algorithm
Given: $n_{\max}$, f, **Searched for:** $g_1,...,g_i \in \mathcal{G}$, $i \leq n_{\max}$

1. FIND g_1
2. $o_1 = g_1/||g_1||$
3. *for* $i = 2 : n_{\max}$
4. FIND g_i
5. $v_{i-1} = v_{i-1}(g_i) = g_i - \sum_{k=1}^{i-1} \langle g_i, o_k\rangle o_k$
6. $o_i = v_{i-1}/||v_{i-1}||$
7. *if* STOP. COND. *then go to* 9 *end*
8. *end*
9. CALCULATE OUTPUT WEIGHTS
10. *stop*

4 Approximation Properties

The Algorithm presented in Section 3 is correct, if $g_1,...,g_i$ are linearly independent thus the value $||v_{i-1}||$ in step 6 is different from zero. The condition of linear independence is fulfilled when $g_1,...,g_i$ possesses so called universal approximation property (see [9]) and then The Algorithm ensures an improvement of approximation in every incremental step. Basic properties of such approximation are formulated in Theorem 1.

Theorem 1 *Let $\mathcal{H}$ be a Hilbert space and let $f \in \mathcal{H}$, $\mathcal{G} \subset \mathcal{H}$ and $g_1,...,g_n \in \mathcal{G}$ be a set of linearly independent elements $||g_i|| < b$ for some $b \in \Re$, let $o_1,...,o_n \in span\mathcal{G}$ be an orthonormal set such that for every $i = 1,...,n$, $span\{o_1,...,o_i\} = span\{g_1,...,g_i\}$, then $||v_{i-1}(f)||^2 - ||v_i(f)||^2 \geq 0$ for $i = 2,...,n$ and if $||v_{i-1}(f)||^2 - ||v_i(f)||^2 = 0$ then $v_{i-1}(f) = 0$. Morever if $span\mathcal{G}$ is dense in $\mathcal{H}$, then $\lim_{n\to\infty}||v_{n-1}(f)|| = 0$ and if $\mathcal{G}$ is complete then also $\lim_{n\to\infty}||v_{n-1}(g_n)|| = 0$.*

Proof: $||v_{i-1}(f)||^2 - ||v_i(f)||^2 = ||f||^2 - \sum_{k=1}^{i-1} \langle f, o_k\rangle^2 - ||f||^2 + \sum_{k=1}^{i} \langle f, o_k\rangle^2 = \langle f, o_i\rangle^2 \geq 0$. Because The Algorithm Step 6, one may write $\langle f, o_i\rangle^2 = \left\langle f, \frac{v_{i-1}(g_i)}{||v_{i-1}(g_i)||} \right\rangle^2$. Note that by a simple manipulation we obtain $\langle f, v_{i-1}(g_i)\rangle = \langle v_{i-1}(f), g_i\rangle$, then $\langle f, o_i\rangle^2 = \frac{1}{||v_{i-1}(g_i)||^2} \langle v_{i-1}(f), g_i\rangle^2$. Now as g_i is linearly independent from $\{g_1,...,g_{i-1}\}$ then $||v_{i-1}(g_i)|| > 0$ and because $||v_{i-1}(g_i)|| \leq ||g_i|| < b$, then $\langle f, o_i\rangle = 0 \iff \langle v_{i-1}(f), g_i\rangle = 0$ for every g_i thus $v_{i-1}(f) = 0$. If $span\mathcal{G}$ is dense in $\mathcal{H}$, then the sequence $(||v_n(f)||)$ is decreasing and bounded from below and because $\mathcal{H}$ is metric and complete $\lim_{n\to\infty}(||v_{n-1}(f)||^2 - ||v_n(f)||^2) = 0$ i.e. $\lim_{n\to\infty}||v_{n-1}(f)|| = 0$ for every $f \in \mathcal{H}$. Similarly for $g \in \mathcal{H}$. If $\mathcal{G}$ is complete then also $\lim_{n\to\infty}||v_{n-1}(g_n)|| = 0$. ∎

Examining Theorem 1 one has to admit that fulfiling rather weak conditions of the incremental approximation, the error will decrease in every iteration. There is no problem of local minima at all.

However for large i of hidden units, $||v_{i-1}(g_i)||$ may naturally decrease, touching accuracy of calculation level. Thus for practical approximation schemes one should wish to decrease $||v_{i-1}(f)||$ as fast as it is possible ensuring however that $||v_{i-1}(g_i)|| > \varepsilon > 0$.

By inspecting proof of Theorem 1 it is easy to notice that if g_i is added to the basis, the decrease of the squared error will be $|< f, o_i >|^2 = \left|< f, \frac{v_{i-1}(g_i)}{||v_{i-1}(g_i)||} >\right|^2$. So maximum decrease of the approximation error will take place if g_i is chosen according to the following condition

$$g_i = \arg\sup_{g \in \mathcal{G}} \left\langle f, \frac{v_{i-1}(g)}{||v_{i-1}(g)||} \right\rangle^2. \tag{6}$$

As proved in [2] there is an exchange rule between rate of approximation and $||v_{i-1}(g)||$ value. So in order to avoid numerical problems one may increase $||v_{i-1}(g)||$ in the expense of error. Special smoothing functions were defined [2] and the formula (6) was modified to

$$g_i = \arg\sup_{g \in \mathcal{G}} \left\langle f, \frac{v_{i-1}(g)}{||v_{i-1}(g)||} \varphi(||v_{i-1}(g)||) \right\rangle^2 \tag{7}$$

where $\varphi : \Re^+ \to [0,1]$ is a nondecreasing function fulfiling the three conditions $\lim_{t\to 0}\varphi(t) = 0$, $\lim_{t\to\infty}\varphi(t) = 1, \lim_{t\to 0}\frac{d}{dt}\varphi(t) = 1$.

5 Example

Consider the function $f : \Re^2 \to \Re$ given by $z = 3(1-x)^2 e^{-x^2-(y+1)^2} - 10(\frac{x}{2} - x^3 - y^4)e^{-x^2-y^2} - \frac{e^{-(x+1)^2-y^2}}{3}$. We wish to approximate f over the set $\Omega = \{\{x,y\} \in \Re;\ -3 \le x, y \le 3\}$. The function in this range has six extrema and was represented by 441 triples uniformly sampling Ω . Hidden nodes were selected to be sigmoidal and their parameters chosen according to (7). The smoothing function was $\varphi_3(t) = 1 - e^{-t}$. In Fig. 2 average squared error over 100 units range is presented.

Some parameters of the network with 100 units are the following: maximum absolute value of the parameters set in the hidden layer is 5800, maximum absolute value of the output weights is 27 and the main parameter influencing reliability of numerics and smoothing of the solution is $\frac{\min}{i} ||v_{i-1}(g_i)|| = 1.3$, $i = 1, ..., 100$. The given numbers seems to be in a good range as far as successful applications are concerned.

6 Conclusions

We presented the incremental function approximation based on modified Gram-Schmidt process.

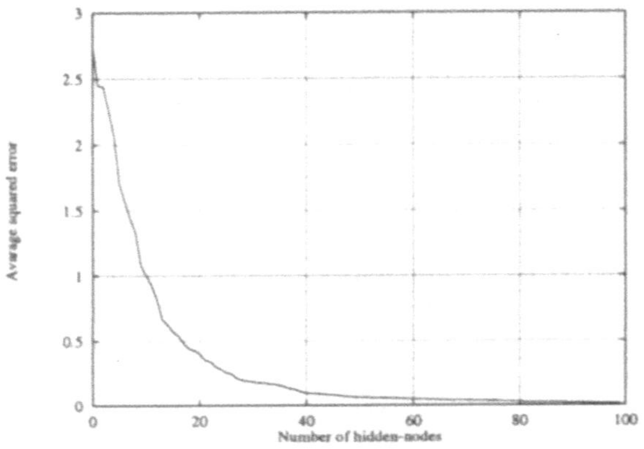

Fig. 2. The learning curve of the approximated function

Its main features are the following. Two bases of the approximation spaces are maintained. Fulfilling very weak conditions, the approximation error decreases in every iteration, so there is no local minima problem. The output weights are calculated only once when the iterative process is terminated. The approximation is fast and efficient for highly complicated practical problems.

References

[1] A. R. Barron. Universal approximation bounds for superpositions of a sigmoidal function. *IEEE Transactions of Information Theory*, 39(3):930–945, 1993.

[2] B. Beliczynski. *Incremental Function Approximation by Using Neural Networks*, volume 112 of *Electrical Engineering Series*. Warsaw University of Technology Press, 2000 (in Polish).

[3] S. E. Falhman and C. Lebiere. The cascade correlation learning architecture. Technical report, CMU-CS-90-100, 1991.

[4] B. Fritzke. Fast learning with incremental rbf networks. *Neural Processing Letters*, 1(2), 1994.

[5] L. K. Jones. A simple lemma on greedy approximation in hilbert space and convergence rates for projection pursuit regression and neural network training. *Annals of Statistics*, 20:608–613, 1992.

[6] E. Kreyszig. *Introductory Functional Analysis with Applications*. J.Wiley, 1978.

[7] T. Y. Kwok and D. Y. Yeung. Constructive algorithms for structure learning in feedforward neural networks for regression problems. *IEEE Trans. Neural networks*, 7:1168–1183, 1996.

[8] T. Y. Kwok and D. Y. Yeung. Objective functions for training new hidden units in constructive neural networks. *IEEE Trans. Neural Networks*, 8(5):1131–1148, 1997.

[9] M. Leshno, V. Lin, A. Pinkus, and S. Schocken. Multilayer feedforward networks with a nonpolynomial activation function can approximate any function. *Neural Networks*, 13:350–373, 1993.

Wavelet Based Smoothing in Time Series Prediction with Neural Networks

Uroš Lotrič, Andrej Dobnikar *

*University of Ljubljana, Faculty of Computer and Information Science, Slovenia, uros.lotric@fri.uni-lj.si

Abstract

To reduce the influence of noise in time series prediction, a neural network, the multilayered perceptron, is combined with smoothing units based on the wavelet multiresolution analysis. Two approaches are compared: smoothing based on the statistical criterion and smoothing which uses the prediction error as the criterion. For the latter an algorithm for simultaneous setting of free parameters of the smoothing unit and the multilayered perceptron is derived. Prediction of noisy time series is shown to be better with the model based on the prediction error.

1 Introduction

Noise, inherently present in the majority of the real time series, makes modeling and prediction difficult. To simplify the neural network model, we have introduced wavelet based smoothing of neural network inputs. Namely, we assume that a neural network combined with wavelet based smoothing is better suited for noise reduction than a general neural network. Since the smoothing mainly removes the noise from input time series the task of the neural network is simplified.

Two approaches towards time series smoothing based on the wavelet multiresolution analysis are presented: the first uses the statistical smoothing criterion [1] while the second minimizes the prediction error [2]. A brief background of wavelet based smoothing is given in the next section. In the third section, a multilayered perceptron is combined with smoothing units and equations for parameter settings are derived. Both types of smoothing are compared on several time series in the fourth section. The main conclusions are drawn in the last section.

2 Smoothing Based on the Wavelet Analysis

In the wavelet multiresolution analysis [3] a time series is observed on different time scales or at different resolutions. On each time scale 2^j the time series is decomposed into an approximation which is described with scaling functions $\phi_{j,k}(t)$ and the remaining details, described with wavelets $\psi_{j,k}(t)$, $j,k \in \mathbf{Z}$. Scaling functions and wavelets are obtained by scaling and translating the father wavelet $\phi(t)$, $\phi_{j,k}(t) = 2^{-j/2}\phi(2^{-j}t - k)$, and the mother wavelet $\psi(t)$, $\psi_{j,k}(t) = 2^{-j/2}\psi(2^{-j}t - k)$. The largest scale 2^J to which the time series is decomposed is usually limited. In this case the time series $x(t)$ can be written as

$$x(t) = \sum_{k\in\mathbf{Z}} a_{J,k}\phi_{J,k}(t) + \sum_{j\leq J}\sum_{k\in\mathbf{Z}} d_{j,k}\psi_{j,k}(t) \quad , \qquad (1)$$

where the sum with coefficients $a_{J,k}$ presents the approximation on scale 2^J, while the sums with coefficients $d_{j,k}$ present details on all scales.

In the wavelet multiresolution analysis the coefficients $a_{j,k}$ and $d_{j,k}$ on adjacent scales are connected by the decomposition

$$a_{j,k} = \sum_{n\in\mathbf{Z}} h^*_{n-2k} a_{j-1,n} \ , \ d_{j,k} = \sum_{n\in\mathbf{Z}} g^*_{n-2k} a_{j-1,n} \ , \qquad (2)$$

as well as by the reconstruction

$$a_{j-1,k} = \sum_{n\in\mathbf{Z}} (h_{k-2n} a_{j,k} + g_{k-2n} d_{j,k}) \quad , \qquad (3)$$

where $h_n = \int_{-\infty}^{+\infty} \phi(t)\phi_{-1,n}(t)dt$, $g_n = (-1)^n h^*_{1-n}$, and * denotes complex conjugates. The coefficients h_n and g_n of most popular wavelets are given in [3]. To simplify the calculation on a discrete time series with N values, $a_{0,k} = x_k$ is set. In this case the number of decompositions J is limited to the largest integer, smaller than $\log_2 N$ [3]. The decomposition and the reconstruction of a time series are schematically presented in Fig. 1, while an example is shown in Fig. 2.

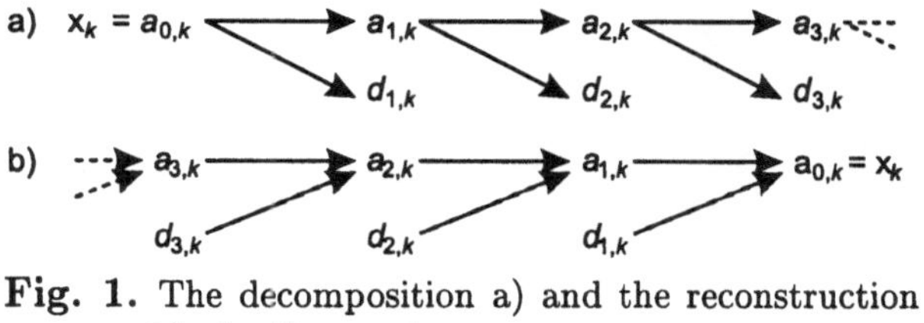

Fig. 1. The decomposition a) and the reconstruction b) of a time series.

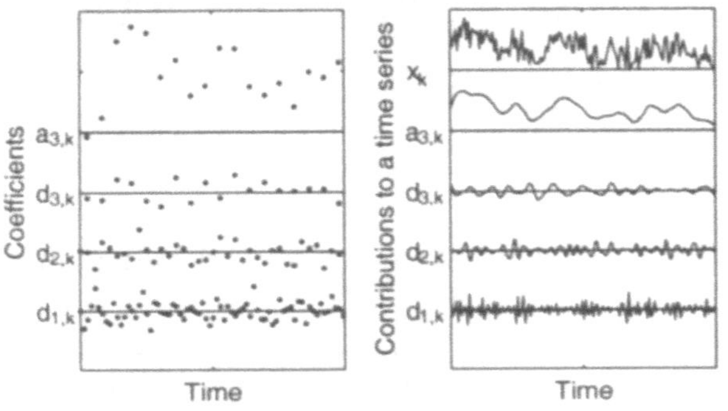

Fig. 2. Coefficients $a_{3,k}$ and $d_{j,k}$, $j = 1, \ldots, 3$, and their contributions to a time series x_k, $k \in \mathbf{Z}$.

Usually the decomposition results in many small coefficients $d_{j,k}$ which mainly contribute to the noise [1]. They can be removed with the threshold function

$$T(d_{j,k}, \tau_j) = \begin{cases} 0, & |d_{j,k}| \le \tau_j \\ d_{j,k} - \operatorname{sign}(d_{j,k})\tau_j, & \text{otherwise} \end{cases} , \quad (4)$$

which at the same time reduces coefficients $d_{j,k}$ larger than a threshold. If the modified coefficients $\tilde{d}_{j,k} = T(d_{j,k}, \tau_j)$ are used instead of the original ones in equation (1) the smoothed time series $\tilde{x}_k$ is obtained. In this case the reconstruction equation (3) becomes

$$\tilde{a}_{j-1,k} = \sum_{n \in \mathbf{Z}} (h_{k-2n}\tilde{a}_{j,k} + g_{k-2n}\tilde{d}_{j,k}) \quad , \quad (5)$$

with $\tilde{a}_{J,k} = a_{J,k}$. Smoothing methods differ in the way thresholds τ_j are calculated. Two ways of determining these thresholds are presented in the following section.

3 Models

Effects of neural network inputs smoothing on modeling and prediction were studied in combination with a multilayered perceptron. Neurons in the multilayered perceptron are arranged in layers

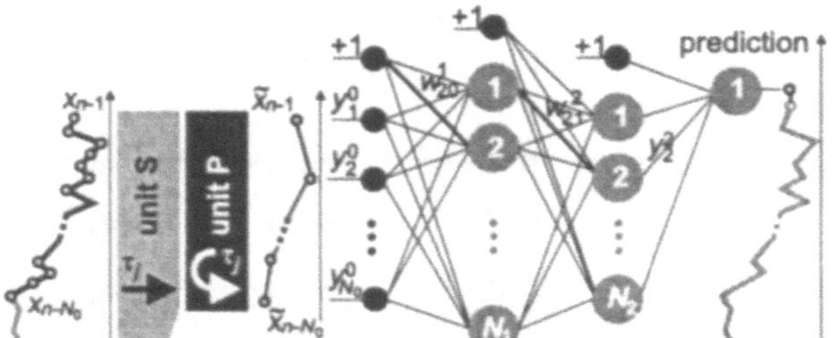

Fig. 3. A multilayered perceptron N_0-N_1-N_2-1 with smoothing units based on the statistical criterion (S) and the prediction criterion (P).

(Fig. 3). Each neuron is connected to all neurons on adjoining layers. Output of the n-th neuron on the l-th layer depends on the sum of weighted outputs from a previous layer and the activation function φ, $y_n^l = \varphi(\sum_{i=1}^{N_{l-1}} w_{n,i}^l y_i^{l-1} + w_{n,0}^l)$.

In time series modeling the multilayered perceptron usually connects a value of a time series x_n at time t_n with N_0 previous values. If the inputs of the multilayered perceptron are set as $y_i^0 = x_{n-i}$, $i = 1, \ldots, N_0$, then the output of the multilayered perceptron y_1^L is an estimate of the correct value at time t_n. To establish such a connection the multilayered perceptron is trained on a set of known input–output samples. Weights $w_{n,i}^l$ are set iteratively and the squared difference between the target value x_n and the calculated value y_1^L, $\mathcal{E} = \frac{1}{2}(x_n - y_1^L)^2$ is used as a criterion. Following the back propagation algorithm [4] weights are changed proportionally to the negative gradient of the squared error,

$$w_{n,i}^l \leftarrow w_{n,i}^l - \eta\, \partial\mathcal{E}/\partial w_{n,i}^l \quad . \quad (6)$$

3.1 The statistical smoothing criterion

Prevailing smoothing techniques based on the wavelet analysis are those developed by Donoho and Johnstone [1]. The thresholds in their basic method are set to

$$\tau_j = \sigma_j \sqrt{2 \ln n_j} \quad , \quad (7)$$

where σ_j denotes the estimate of the standard deviation and n_j the number of coefficients $d_{j,k}$ on the scale 2^j. Thus, the defined threshold is the maximal absolute value of Gaussian white noise in the limit $n_j \to \infty$ [2]. In order to remove those coefficients $d_{j,k}$ which mainly contribute to the noise, the threshold value should therefore be calculated from long time series. Hence, the smoothing unit based on the statistical criterion (unit S in Fig. 3) uses all known values in the time series, x_0 to x_{n-1}, to obtain a smoothed time series $\tilde{x}(t)$, but only smoothed values $\tilde{x}_{n-N_0}, \ldots, \tilde{x}_{n-1}$ are used as inputs to the multilayered perceptron.

3.2 Criterion based on the prediction error

Smoothing with the statistical criterion may not be optimal for prediction. Therefore, we have combined the multilayered perceptron and the smoothing unit into a uniform model. Both, the weights of the multilayered perceptron and thresholds of the smoothing unit are iteratively set, based on the prediction error [2]. Since the prediction error is used as the smoothing criterion smoothing is no longer treated separately from prediction.

When extending the back propagation algorithm (equation (6)) to the thresholds τ_j,

$$\tau_j \leftarrow \tau_j - \eta\, \partial\mathcal{E}/\partial\tau_j \ , \tag{8}$$

we need to calculate partial derivatives $\partial\mathcal{E}/\partial\tau_j$, $j = 1,\ldots,J$. Following the chain rule we obtain

$$\frac{\partial\mathcal{E}}{\partial\tau_j} = \sum_{i=1}^{N_0} \frac{\partial\mathcal{E}}{\partial y_i^0}\frac{\partial y_i^0}{\partial\tau_j} \ . \tag{9}$$

The first factor of the product is derived from the multilayered perceptron back propagation algorithm [4]. Considering the relations $y_i^0 = \tilde{x}_{n-i} = \tilde{a}_{0,i}$, the second factor is obtained recursively from equation (5),

$$\frac{\partial\tilde{a}_{l,k}}{\partial\tau_j} = \sum_{i\in\mathbf{Z}} \frac{\partial\tilde{a}_{l,k}}{\partial\tilde{a}_{l+1,i}}\frac{\partial\tilde{a}_{l+1,i}}{\partial\tau_j} = \sum_{i\in\mathbf{Z}} h_{k-2i}\frac{\partial\tilde{a}_{l+1,i}}{\partial\tau_j} \tag{10}$$

on the scales $l = 1,\ldots,j-2$ and

$$\frac{\partial\tilde{a}_{j-1,k}}{\partial\tau_j} = \sum_{i\in\mathbf{Z}} \frac{\partial\tilde{a}_{j-1,k}}{\partial\tilde{d}_{j,i}}\frac{\partial\tilde{d}_{j,i}}{\partial\tau_j} = \sum_{i\in\mathbf{Z}} g_{k-2i}\frac{\partial\tilde{d}_{j,i}}{\partial\tau_j} \tag{11}$$

on the scale $l = j-1$. Finally, the partial derivative in equation (11) is obtained from equation (4). The back propagation algorithm does not take into account that the thresholds τ_j are nonnegative and bounded to a finite interval. Namely, the threshold τ_j larger than maximal absolute value of coefficients $d_{j,k}$ on the scale 2^j taken over all input samples, $d_{j,\max}$, gives the same result as the threshold $\tau_j = d_{j,\max}$. To apply the algorithm we introduce unbounded thresholds τ_j^∞,

$$\tau_j = d_{j,\max}/(1 + e^{-\tau_j^\infty}) \ . \tag{12}$$

The partial derivatives are calculated as

$$\frac{\partial\tilde{a}_{l,k}}{\partial\tau_j^\infty} = \frac{\partial\tilde{a}_{l,k}}{\partial\tau_j}\frac{\partial\tau_j}{\partial\tau_j^\infty} = \frac{\partial\tilde{a}_{l,k}}{\partial\tau_j}\tau_j(1 - \tau_j/d_{j,\max}) \ . \tag{13}$$

In this case smoothing can be performed only on a sample of the same length as used further in the multilayered perceptron (unit P in Fig. 3).

4 Results

The performance of the classical multilayered perceptron and perceptrons with both types of smoothing is compared on several time series: a second order process, Feigenbaum sequence and the real time series, obtained from the quality control of compounds from a national producer. From each time series input–output samples are prepared. First 85% of the samples are used to set free parameters of the models (training set) and the last 15% are used to compare the models (test set). The number of model inputs is limited to 20 and the number of free parameters to 30% of the number of samples in training set. The models are designed to predict only the next value in a time series.

4.1 Second order process

We have considered a second order process given by the equation $d^2x(t)/dt^2 + \omega^2 x(t) = u(t)$, where $\omega^2 = 5\,\mathrm{s}^{-2}$ and $u(t)$ denotes Gaussian white noise with the standard deviation $\sigma_p = 0.1\,\mathrm{s}^{-2}$. Simulated measurements with the standard error $\sigma_m = 0.3$ were taken 250 times, once every 0.2 s.

Root mean squared errors between the evaluated and the target values on the training and the test set, normalized to the time series standard deviation $\sigma = 0.66$, are presented in Table I. The best predictions are obtained with the multilayered perceptron with the smoothing unit based on the prediction error. As can be seen in Fig. 4, the smoothing unit based on the prediction error has transformed its inputs (gray circles) in such a way that its outputs (the thick black line) are close to the real process without the measurement noise (the gray line).

Table I. Normalized root mean squared errors for the prediction of the second order process.

Model	Training set	Test set
MLP, 10-3-1	0.47	0.56
MLP+unit S, 19-3-1	0.45	0.56
MLP+unit P, 14-5-1	0.47	0.51

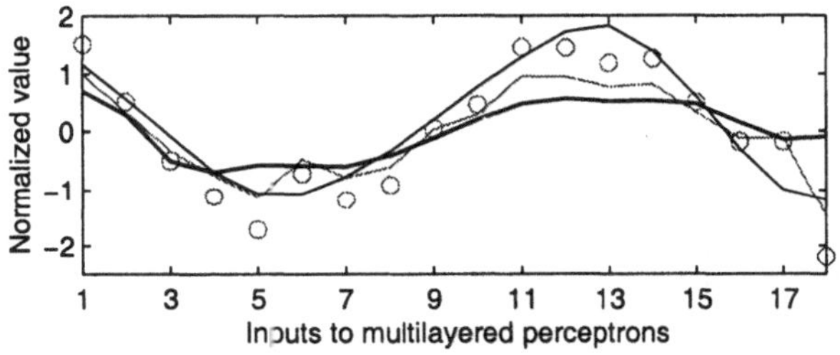

Fig. 4. The impact of smoothing on an input sample of the second order process.

When these values are used as the inputs to the multilayered perceptron, its job is easier than in the case of the classical perceptron and the perceptron with inputs smoothed with the statistical criterion (the thin black line).

4.2 Feigenbaum sequence

Feigenbaum sequence is generated by the recursive relation $x_t = r\,x_{t-1}(1 - x_{t-1})$. For $r = 4$ the obtained time series is chaotic. We have used 250 values, calculated with 15-digit precision from the initial value $x_0 = 0.01$.

Table II gives the prediction errors, normalized to the time series standard deviation $\sigma = 0.35$. It is

Table II. Normalized root mean squared errors for the prediction of the Feigenbaum sequence.

Model	Training set	Test set
MLP, 4-2-1	$3.1 \cdot 10^{-5}$	$3.1 \cdot 10^{-5}$
MLP+unit S, 16-2-1	0.98	0.95
MLP+unit P, 4-2-1	$3.4 \cdot 10^{-5}$	$3.4 \cdot 10^{-5}$

obvious from the large prediction error of the model with the statistical smoothing criterion that such a smoothing is not appropriate. Namely, the statistical smoothing can not distinguish between a chaotic time series and a pure noise. On the other hand, both, the classical multilayered perceptron and the perceptron with the smoothing unit based on the prediction error, managed to learn the relationship between the consecutive values. The error is slightly higher for the multilayered perceptron with smoothing based on the prediction error. This is expected since, due to the nature of the learning procedure, the thresholds can never reach zero [2].

4.3 Quality control

An important characteristic which determines the quality of rubber compounds is its hardness. It is measured in Shore units on the scale reading from 0 to 100. Hardness of a compound, used for bicycle and motorcycle tubes, is shown in Fig. 5. The time series consists of 199 measurements, taken after successive mixings.

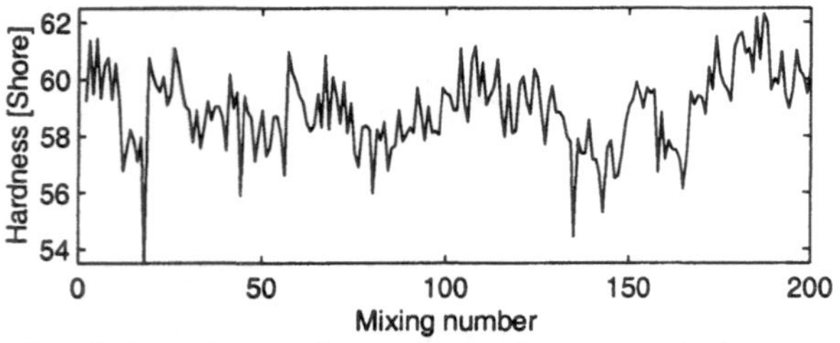

Fig. 5. Hardness of a compound measured after successive mixings.

Prediction errors, normalized to the time series standard deviation $\sigma = 1.40$ Shore, are presented in Table III. Compared to the classical multilayered perceptron, the smoothing unit based on the prediction error improves prediction, while the statistical smoothing makes it worse. In this case we can see how important the choice of smoothing level is. If the statistical smoothing criterion is used, the level of smoothing is too high, as it can be observed

Table III. Normalized root mean squared errors for the prediction of the quality control time series.

Model	Training set	Test set
MLP, 15-1-1	0.82	0.68
MLP+unit S, 8-3-1	0.80	0.71
MLP+unit P, 10-6-1	0.84	0.65

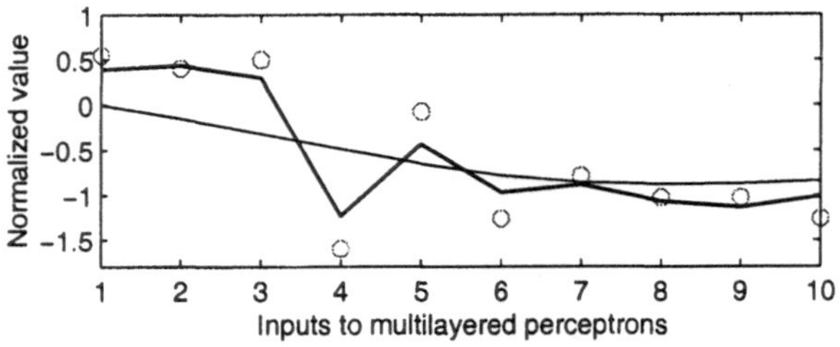

Fig. 6. The impact of smoothing on an input sample of the quality control time series.

in Fig. 6. In this figure, the time series is presented with gray circles, output of the smoothing unit based on the statistical criterion with the thin black line and output of the smoothing unit based on the prediction error with the thick black line.

5 Conclusions

We have proposed a smoothing method which uses the prediction error as the smoothing criterion. Prediction of noisy time series is more successful with the smoothing unit which uses prediction error as the criterion than with statistical smoothing and the classical multilayered perceptron without smoothing. The multilayered perceptron with smoothing based on the prediction error can also detect if smoothing is not needed and in such cases it performs equally well as the classical multilayered perceptron.

References

[1] D. L. Donoho and I. M. Johnstone, "Adapting to unknown smoothness via wavelet shrinkage," *J. Am. Stat. Assoc.*, vol. 90, no. 432, pp. 1200–1224, 1995.

[2] U. Lotrič, *Using Wavelet Analysis and Neural Networks for Time Series Prediction.* PhD thesis, University of Ljubljana, Faculty of Computer and Information Science, Ljubljana, Slovenia, 2000.

[3] I. Daubechies, *Ten Lectures on Wavelets.* Philadelphia: SIAM, 1992.

[4] S. Haykin, *Neural Networks: A Comprehensive Foundation.* New Jersey: Prentice-Hall, 2 ed., 1999.

Fitting Densities and Hazard Functions with Neural Networks

Colin Reeves, Charles Johnston *

*School of Mathematical and Information Sciences, Coventry University, UK

Abstract

In this paper, we show that Artificial neural networks (ANNs)—in particular, the Multi-Layer Perceptron (MLP)—can be used to estimate probability densities and hazard functions in survival analysis. An interpretation of the mathematical function fitted is given, and the work is illustrated with some experimental results.

1 Introduction

Artificial neural networks (ANNs) have attracted a lot of attention over the past decade; partly this must be due to a certain fascination with the idea of imitating the human brain, but while the biological background is interesting, it is sufficient for practical purposes to consider ANNs as a means of solving non-linear regression or classification problems, which comprise most of their practical applications (see [1] for a comprehensive survey). In this paper, we show that they can also be used for the problem of estimation of probability densities and hazard functions in survival analysis.

The most popular and pervasive ANN model is probably still the *Multi-Layer Perceptron* (MLP), which has produced impressive results in regression and classification problems.

2 Mathematical Formulation

The mathematical model that maps inputs to output for the MLP with several hidden layers is constructed as follows: We define the input to the jth unit in the kth layer to be

$$s_j^k = \sum_{i=0}^{N_{k-1}} w_{ij}^{(k-1)} x_i^{(k-1)}.$$

Here, $x_i^{(k-1)}$ is the output of the ith unit in layer $(k-1)$, $w_{ij}^{(k-1)}$ is the weight associated with a connection from unit i in layer $(k-1)$ to unit j in layer k, and N_k is the number of units in layer k. Note that a variable $x_0^{(k-1)} \stackrel{\text{def}}{=} 1$ is assumed at each level, to allow for a bias or constant term. The labelling given in this formulation implies that the input layer is labelled as layer 0, and it provides a consistent picture of the relationship between network layers, the associated net functions, and the sets of weights. The transmission section of each unit computes the 'activation' a_j^k of the unit, defined as

$$a_j^k = g(s_j^k)$$

which becomes the input x_j^k to the next layer. The 'squashing' function $g(\cdot)$ is most commonly the logistic function

$$\sigma(x) = \frac{1}{(1+e^{-x})}, \tag{1}$$

which provides a continuous approximation to a threshold function by mapping the input to each unit onto the range (0,1).

In the simplest case, we have a MLP with a single hidden layer of p units and activation function $g(\cdot)$, and the mapping formed simplifies to the following composition of functions:

$$\varphi(\mathbf{x}) = g\left(\beta_0 + \sum_{i=1}^{p} \beta_i g(\alpha_{i0} + \sum_{j=1}^{n} \alpha_{ij} x_j)\right) \tag{2}$$

where, for clarity, we use α to denote the input-to-hidden weights, and β for the hidden-to-output weights. The biases are denoted by a subscript 0. This is a common formulation; in most cases the activation functions are the same for both the input-hidden and hidden-output connections. However, in some applications, a non-linearity at the output may not be necessary, and a simple sum is sufficient. In the latter case, Eq. (2) becomes

$$\varphi(\mathbf{x}) = \beta_0 + \sum_{i=1}^{p} \beta_i g(\alpha_{i0} + \sum_{j=1}^{n} \alpha_{ij} x_j) \tag{3}$$

3 Fitting Densities and Hazard Functions

A common problem in statistics is to use a sample of m observations to estimate the probability distribution of the population from which the sample values were drawn. This is particularly difficult in the case of a continuous variable: simple methods such as the construction of histograms can be heavily influenced by the choice of group intervals and starting points. In this paper, we shall focus on the continuous univariate case. More sophisticated methods of density estimation rely on smoothing the histogram, perhaps by using kernel methods [2]. Fitting densities by these means is relatively demanding in terms of computational effort and understanding, although these is considerably alleviated by the development of libraries such as those available with modern statistical software.

A simple alternative is to use ANNs to fit the empirical distribution function (EDF), and given the rapidly spreading availability of software for fitting ANNs, this may be an attractive option. Thus the output values are given as $(i-0.5)/m$, where i runs from 1 to m—approximating the cumulative probabilities from 0 to 1, while the corresponding input values are the order statistics $x_{[i]}$.

Another area where such a technique can be applied is in the estimation of hazard functions in survival analysis. A common approach in addressing survival data is to construct the Kaplan-Meier step-function as a nonparametric estimate of the survival curve, representing graphically the way the hazard changes over the course of time. Again, it seems obvious that this step-function could be smoothed by using a neural net.

3.1 A Neural Net Approach

As has been made clear above, the result of fitting a neural net is a mathematical function. Thus, if we fit the EDF, we obtain a model for $F(x)$ which can be differentiated in order to estimate the density $f(x)$. Similarly, the Kaplan-Meier function is an empirical estimate of the survival function $S(t) = 1 - F(t)$. In fitting this with a neural net, we can find its derivative $-f(t)$, which in turn enables the estimation of the *hazard* function

$$h(t) = f(t)/S(t).$$

From this the *cumulative hazard*

$$H(t) = \int h(u)du = -\log S(t)$$

can also be estimated.

In the case of the logistic function of Eq. (1), this approach is particularly simple: the derivative of $\sigma(x)$ is

$$\sigma'(x) = \frac{e^{-x}}{(1+e^{-x})^2} = \sigma(x)(1-\sigma(x)).$$

This is in fact a special case of the *logistic density*, which is in general a 2-parameter function

$$\frac{e^{-\frac{x-\mu}{\lambda}}}{\lambda\left(1+e^{-\frac{x-\mu}{\lambda}}\right)^2}$$

whose mean is μ and variance $(\pi\lambda)^2/3$. This is in fact a good approximation to a Normal distribution—the maximum difference between standard Normal and logistic distribution functions is only 0.0228 [3]. In the univariate case of interest, the input to hidden layer unit i is of the form

$$\alpha_{i0} + \alpha_{i1}x$$

(cf. Eq. (2)), from which it is easily seen that the unit represents a logistic density with mean $-\alpha_{i0}/\alpha_{i1}$ and variance $\pi^2/(3\alpha_{i1}^2)$.

3.2 What does it mean?

For the case of Eq. (3), the function $\varphi(x)$ that is fitted to $F(x)$ will on differentiation become

$$\varphi'(x) = \sum_{i=1}^{p} \beta_i \alpha_{i1} g'(\alpha_{i0} + \alpha_{i1}x) \qquad (4)$$

Since in the case of a logistic activation function,

$$\alpha_{i1}\sigma'(\alpha_{i0} + \alpha_{i1}x)$$

is the logistic density, Eq. (4) will also yield a density, provided that

$$\sum_{i=1}^{p} \beta_i = 1.$$

This is, of course, not guaranteed by the neural net fitting procedure, although it would be possible to modify the algorithm by including a Lagrange multiplier λ and adding (say)

$$\lambda(\sum_{i=1}^{p} \beta_i - 1)^2$$

to the mean squared error term. Whether this is worth doing will probably depend on the application: if all that is required is to obtain a reliable estimate of the (composite) distribution function, the individual components and their weights are of minor importance. However, if it is also hoped to identify the distinct components that make up the composite density, something like this approach would seem to be necessary.

We should note also that if Eq. (2) is used instead, with an additional non-linearity at the output, i.e.,

$$\varphi(\mathbf{x}) = g\left(\beta_0 + \sum_{i=1}^{p} \beta_i g(\alpha_{i0} + \alpha_{i1}x)\right),$$

we are guaranteed to have a density. However, how this relates to individual components that are assumed to exist is not clear, since the derivative of $\varphi(x)$ is no longer a simple linear combination of p logistic densities:

$$\varphi'(x) = g'(v) \sum_{i=1}^{p} \beta_i \alpha_{i1} g'(\alpha_{i0} + \alpha_{i1}x) \qquad (5)$$

where we define

$$v = \beta_0 + \sum_{i=1}^{p} \beta_i g(\alpha_{i0} + \alpha_{i1}x)$$

for clarity. This is still a sum of p components, and if we assume a logistic activation,

$$\alpha_{i1}\sigma'(\alpha_{i0} + \alpha_{i1}x)$$

is still a density. However, we cannot assume a simple sum of logistic densities, since the weights would be the values

$$\beta_i \sigma'(v),$$

which of course are a function of x. Alternatively, we could take the $\sigma'(v)$ term 'inside' by treating

$$\sigma'(v)\alpha_{i1}\sigma'(\alpha_{i0} + \alpha_{i1}x)$$

as a density, and force $\sum_{i=1}^{p} \beta_i = 1$ as before. However, we could equally well assume that the proportions are defined by the α weights, and force $\sum_{i=1}^{p} \alpha_{i1} = 1$.

4 Experimental Work

To test these ideas out further, we generated 100 simulations, each consisting of 100 data points generated by a mixture of two Normal distributions with means 0 and 4 respectively, and a common variance of 1. The mixture was in the ration 1 : 2. In each case, the methodology developed above was used to fit the EDF using a single hidden layer ANN with 2 hidden nodes and a logistic activation function. The fit obtained was extremely close, as measured by the χ^2 criterion, but more interesting were the values of the logistic parameters. We would like to know if the above interpretation is practically useful. As we did not force the condition $\sum_{i=1}^{p} \beta_i = 1$, the implied estimates of the proportions were not informative. (These are usually the most difficult to estimate using conventional methods anyway.) The most useful parameters would be the values of the means obtained from the logistic mixture that is implicit in the ANN. These were −0.98 and 5.10, with a mean squared error (MSE) of 1.22 and 1.61.

Now these values seem to be rather poor; similar experiments in [4] using maximum likelihood (ML) techniques found MSE values two orders of magnitude better. However, on closer examination, it appears that the experiments in [4] were somewhat compromised by the fact that the *true* values of the means were used as initial values for the ML iteration. If you tell it what the answer is, it's not surprising if it works! In [4], the point was to evaluate the effectiveness of iterative ML techniques—and it is known that 'good' initial estimates are very important. At least our ANN technique, making no assumptions about the location of the means, and using a logistic approximation to the Normal in any case, has obtained very respectable values, which could be refined by ML methods if that is required.

As a second experiment, a pilot study was carried out using a data set of patients suffering from malignant melanoma. The Kaplan-Meier (KM) survival probabilities for members of all subsets of the patient data can be used as a training set for the neural net. Patient data were categorised and the subsets were defined by a hierarchical tree diagram. For each subset, a non-parametric survival analysis was performed. The output from this is a set of event and censoring times (i.e. the times when each patient leaves the study) together with values of the estimated KM probability of survival for each patient. The input to the neural net was then the patient factors (categorised covariates) and the times when they left the study.

While this should not be taken as a report on the survival of such patients, since the data have been superseded, it is nonetheless useful as a vehicle for comparing the type of fits available by Cox proportional hazards (PH) and the neural net approach. The explanatory variables for these data consisted of Age (greater or less than 60 years), Gender, Clarke level (an indicator of involvement of skin layers, which had four levels in this study), Pathological depth of the cancer (log-transformed and categorised into three levels for this study).

The most popular parametric (or semi-

Fig. 1. Survival curves for malignant melanoma patients

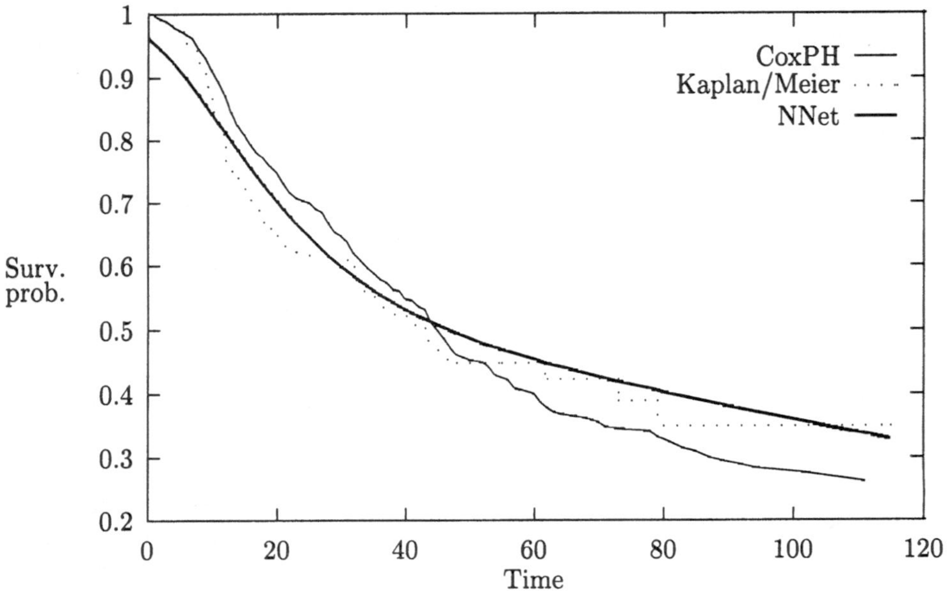

parametric) model for survival analysis is Cox's proportional hazards (PH) model. Using the same data, a Cox PH model was produced using S-Plus that required an interaction term between age and Clarke level. The results are illustrated in Figure 1 for the subset of older men with fairly severe skin cancer. The fit to the Kaplan-Meier survival plot obtained from the neural net was at least as good as that given by the Cox PH model for nearly all subsets, and in many cases (as here) actually better.

References

[1] S.Haykin (1999) *Neural Networks : a Comprehensive Foundation*, 2nd edition, Prentice Hall, London.

[2] M.P.Wand and M.C.Jones (1995) *Kernel Smoothing*, Chapman and Hall, London.

[3] N.L.Johnson, S. Kotz and N.Balakrishnan (1995) *Continuous Univariate Distributions*, Vol.II, Wiley, Chichester.

[4] B.S.Everitt and D.J.Hand (1981) *Finite Mixture Distributions*, Chapman and Hall, London.

Optimization of Expensive Functions by Surrogates Created from Neural Network Ensembles

Jiří Pospíchal* [1]

*Dept. Mathematics, Slovak Tech. Univ., 812 37 Bratislava, Slovakia, e-mail: pospich@cvt.stuba.sk

Abstract. The goal of this paper is to model hypothesis testing. A "real situation" is given in the form of a response surface, which is defined by a derivative–free continuous expensive objective function. An ideal hypothesis should correspond to a global minimum of this function. Thus, hypothesis testing is converted into optimization of a response surface. First, an objective function is evaluated at a few points. Then, the hypothetical (surrogate) surface landscape is created from an ensemble of approximations of the objective function. Approximations result from neural networks, which use already evaluated samples as the training set. The hypothesis landscape adapted by a merit function estimates a possibility of getting at a given point a better value, than is the currently achieved value from already evaluated points. The most promising point (a minimum of the adapted function) is used as the next sample point for the true expensive objective function. Its value is then used to adapt neural networks, creating a new hypothesis landscape. The results suggest that (1) in order to get a global minimum, it may be useful to have an estimation of the whole response surface, and therefore to explore also those points, where maxima are predicted, and (2) an assembly of modules predicting the next sample point from the same set of sample points can be more advantageous than a single neural network predictor.

1 Introduction

In psychology [1], the mind modules take representational form of input and relay information to a central system, which carries out inference and belief fixation. The degree of confirmation of a hypothesis depends not only on its intrinsic features but also on its relation to all other beliefs (so called Quinean system [1]). Modules are believed to be specialized for different types of tasks. Similarly a function approximation can be carried out by an ensemble of neural networks, with each network responsible mainly for one part of the approximated function [2]. A weighted average of predictions of neural networks from ensembles is used to get more reliable approximation, where variance can also guide the next sample selection during the so called active learning [3]. A similar approach is used for regression models in general [4-6]. Different hypotheses can be derived from the same data available and for the same part of the search space. Tyler and Marslen-Wilson [7] argue, that the module's job is now redefined as that of making candidate suggestions, while selection occurs at a later, post-modular stage of processing. In order to optimize the subject's behavior, which translates into fitness, a subject has to test the hypotheses, while each test is expensive for time and energy resources. Therefore, one has to create an optimization procedure, consisting from (1) creation of hypotheses about his environment (2) finding out the best possible behavior for each hypothesis, and (3) selecting or combining the outcome behavior from the hypotheses predictions. When the outcome does not correspond to hypotheses predictions, the hypotheses must be adjusted to accommodate new facts.

There have been various attempts to use a single predictor (neural network [8,9] or other regression models [10-12]) as the hypothesis generator or function approximator for optimization, and there have been many combinations of ensembles for classification or approximation [6], but neural network ensemble was not thoroughly tested as a surrogate for optimization with a large sampling cost. Evolutionary optimization for ensembles was used only for their selection [2] or for an optimization of a single predictor [8], not for optimization of a function created by their combination.

The present work models the process of optimization of behavior by creating hypotheses combined with their testing. A simplified computational process uses neural network ensemble followed by a postmodular stage of processing, which does not use neural networks. The modules are of a connectionist nature, but the model does not pretend to capture any low level features of cognitive processes. The "classical" feedforward backpropagation neural networks are used and the search for their optimal in-

[1]The work was supported by grants # 1/7336/20 and # 1/5229/98 of the Slovak Republic Scientific Grant Agency.

put is carried out by an evolutionary algorithm. The "environmental" objective function is modeled by a simple mathematical function. Nevertheless, even such a simplified model may bring a fresh insight to a higher-level processing in a brain.

2 Formalization

The "environment" objective function is represented by a simple function $f(x)=e^{-2x}sin(10\pi x)$ for $x\in[-0.5,0.5]$ (see Fig. 1). It models a problem with a very expensive function with no derivative.

The compromise between a local search and a global search must be reached in every optimization algorithm. In the presented approach, the new search points are chosen not at the optimum of the approximation itself, but at the optimum of the approximation adapted by another function, called the merit function [12], that predicts the uncertainty of prediction. This merit function is $f_m(x) = \min \sqrt{|x - x_i|}$, where the minimum is computed over all points x_i, at which we know the value of the domain objective function. The function $f_{adapted}(x, sig) = f_{prediction}(x, sig) - \rho \cdot f_m(x)$ is then minimized, where $sig \in\{1,-1\}$. When sig=1, the method searches for a minimum, when sig=−1, the method searches for a maximum. For a single neural network, the prediction was used directly as a result of this neural network. For an ensemble of independent neural networks, the prediction uses the bounds of a confidence interval computed from the predictions of neural networks at a given x. When the method searches for a maximum, the predictions of neural networks are multiplied by a value sig=−1: $f_{prediction}(x, sig) = lbc(sig \cdot f_{ith\mathrm{NN}}(x) \,|i = 1, ..., n)$. The function lbc gives the lower bound of a confidence interval calculated from several values, e.g. predictions $f_{ith\mathrm{NN}}(x)$ of n neural networks (indexed i) at a given point x, for the gaussian distribution with the confidence parameter alpha=0.001. The parameter ρ is defined as $\rho = \max\limits_{x\in[a,b]} (|f_{prediction}(x, sig)|) \Big/ \max\limits_{x\in[a,b]} (f_m(x))$, so that the uncertainty caused by the distance from the closest true known value is as important as the prediction by the surrogate(s).

The feed-forward neural networks used for the given problem had 1 input neuron, 2 hidden layers with 10 and 5 neurons, respectively, and 1 output neuron. The hidden neurons had sigmoid transfer functions with sum of inputs and bias, and the output neuron transfer function was linear with bias. The Levenberg-Marquardt training method was used with a learning rate 0.01. The networks were trained from the scratch after each addition of a new sample point. This ad-hoc designed neural network architecture proved to be satisfactory for the given problem, therefore neither architecture nor parameters were further optimized. The training of each network was stopped, when a least square performance error value of 0.001 was reached.

Outline of the algorithm:

1. Choose an initial grid of sample points over a search domain as a set $X = \{ x_i \}$ (in our case with one variable we choose two initial sample points, at 1/3 and at 2/3 of the interval $x\in[a,b]$). $sig := 1$
2. Evaluate the domain objective function $f(x)$ at the given sample points $x_i \in X$
3. Create a predictor (e.g. train n neural networks, in our case n=10) to predict values of the objective function at all sample points (a training set $\{x_i, f(x_i)\} \,|x_i \in X$).
4. Use an optimization algorithm (e.g. evolution strategy in our case) to find out a minimum (or a maximum) of the adapted objective function created by a combination of a merit function preferring unexplored regions with a function generated by a predictor (e.g. lower (or upper) bounds of confidence interval calculated from predictions made by neural networks) $x_{new}: = \arg\min\limits_{x\in[a,b]} (f_{adapted}(x, sig))$ $= \arg\min\limits_{x\in[a,b]} (f_{prediction}(x, sig) - \rho \cdot f_m(x))$
5. $X: = X \cup x_{new}$
6. Evaluate the domain objective function $f(x_{new})$ at the new search point
7. $sig := sig * (-1)$ (switch between seeking a minimum and a maximum)
8. If the number of allowed evaluations is exhausted, return $x_{result}: = \arg\min\limits_{x_i\in X} (f(x_i))$, else go to step 3.

3 Results

Average values (from 10 runs) of the achieved minima of the true objective function were compared for various optimization approaches. The averages were calculated at each iteration, when a new sample point was evaluated. Optimization started with 2 sample points and then 20 other points were predicted and sampled. Surprisingly, optimization with a single neural network predictor combined with a merit function promoting greater dispersion of sample points had even worse results than the random choice of sample points (see Fig. 2). Even though the merit function was multiplied by some weight to become as important as the prediction by a neural network, it was not still enough to prevent

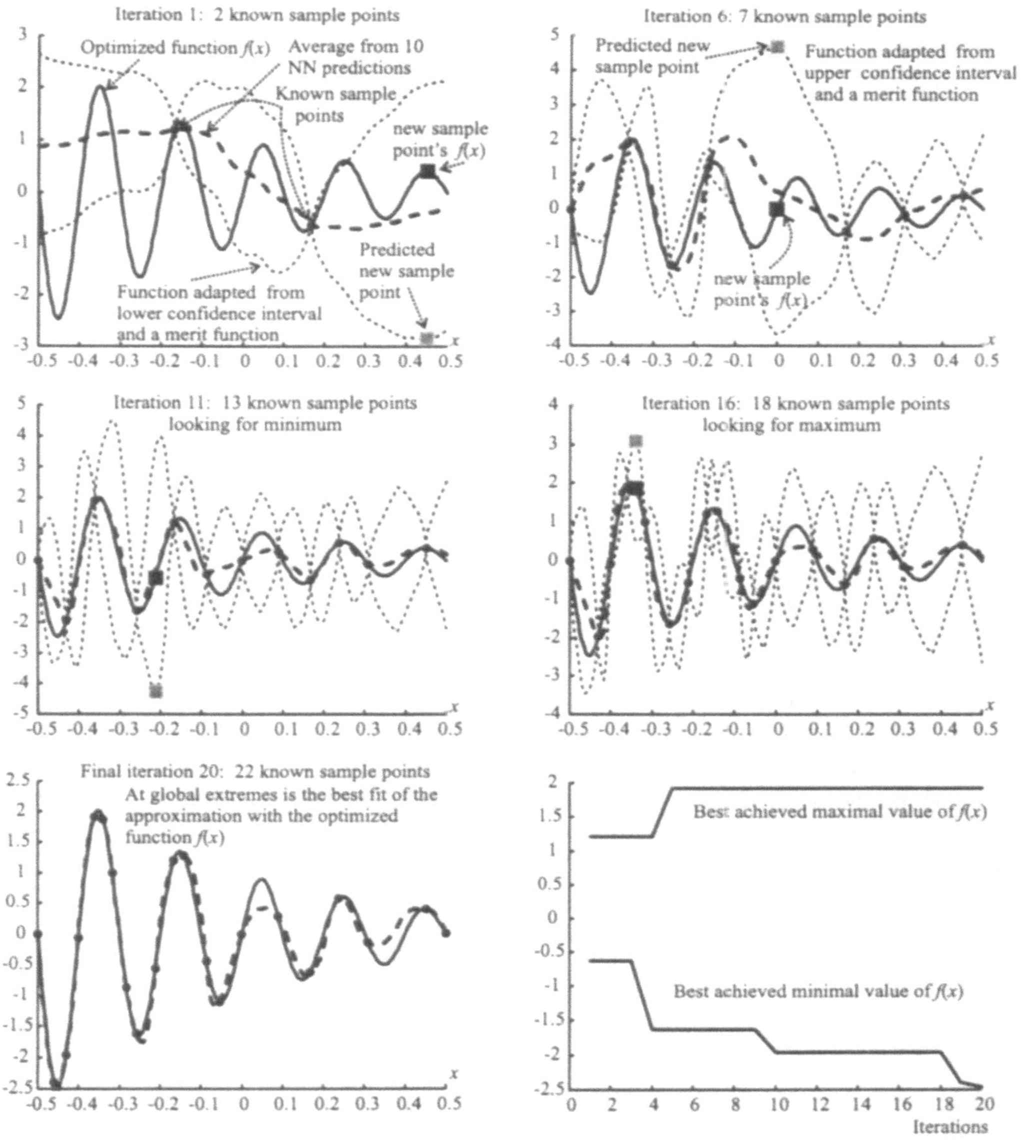

Fig. 1. Results of one run of an algorithm with ensemble of 10 neural networks, which are used to predict lower and upper bounds of confidence intervals, adapted further by a merit function to predict next suitable sample points. The algorithm searches alternatively for minima and maxima. This approach approximates the whole optimized function, which also helps to predict the minimum.

the algorithm to fall into a local minimum, while random sampling was able to find out better functional values.

The single neural network using a merit function, but searching alternatively for a minimum and a maximum, better explored the whole search space, learning the whole landscape. Most of the time, this prevented the algorithm to fall into a local minimum and enabled it to find out a global minimum. The merit function is necessary; without it, the method would fall into a local optimum, since the neural network prediction for the used architecture and learning method tends to give very similar results for a small training set.

The best results were achieved for an ensemble of neural networks, using not only the merit function, but also the lower and upper bounds of confidence intervals, when the algorithm searched alternatively for minima and maxima. This version of the algorithm also predicted the minimum with the best reliability; all the runs achieved a value close to the global minimum after evaluation of 20 sample points

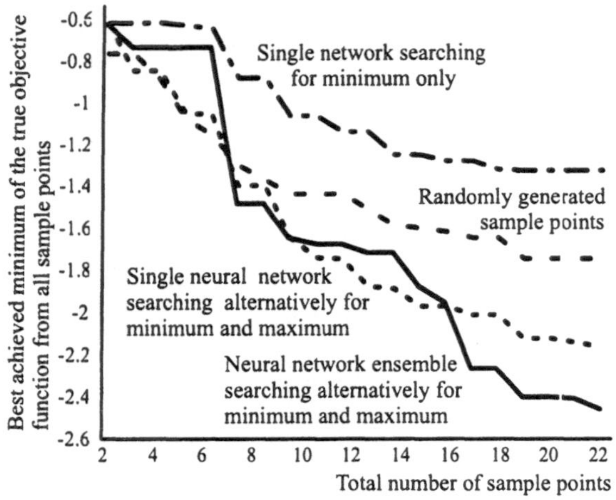

Fig. 2. Comparison of various algorithms used for minimization of the function $f(x)=e^{-2x}sin(10\pi x)$ for $x\in[-0.5,0.5]$. The averages of best results from 10 runs are displayed against the number of sample points, where the next sample point was always calculated from a function approximator.

(Fig. 2), even though with fewer sample points the algorithm did not perform better than a single neural network predictor with a merit function. Minimum of the used objective function would be achieved even faster (in 12 iterations) with a spline interpolation [12] as a surrogate.

If the described approach was used as an optimization algorithm, still many questions would be left open, for instance, when the optimization should be stopped. The answer however depends on the number of predicted future evaluations, the cost of one evaluation and the decrease of cost achieved by optimization, type of the objective function and so on. The real predictive power of the approximator could be evaluated by an absolute difference between the predicted values and true optimized function values of new sample points. This error can be used to adjust the weight value of the merit function and the confidence parameter alpha. Another possible extension can use the ensemble of functions predicted by the genetic programming. Also, the alternative search for minima and maxima can be replaced by a more rigorous approach balancing between learning of a function landscape and a search for the optimum.

These problems however go beyond the scope and intentions of this paper. Achieved results showed:

(1) The neural network ensemble prediction of the next search point on the basis of confidence interval combined with a merit function may be a viable approach for testing hypotheses, where the number of sampling points is highly restricted. The results might suggest that, apart from different modules for different tasks in the human brain, several modules that seem to do the same job, might be of some value.

(2) The alternative search for minima and maxima can lead to a landscape prediction, which helps to achieve a global optimum with much greater probability than the utilitarian direct search for the desired optimum. It might indicate, that some apparently disadvantageous actions may be in fact useful in a long term for adjustment of our hypotheses or mental images of our environment.

References

[1] Fodor, J.A.: *The Modularity of Mind.* Cambridge, MA: MIT Press 1983.

[2] Opitz, D., Shavlik, J.: A Genetic Algorithm Approach for Creating Neural Network Ensembles. In: Sharkey A. (ed.): *Combining Artificial Neural Nets*, pp. 79-97. London: Springer 1999.

[3] Krogh, A., Vedelsby J.: Neural Network Ensembles, Cross Validation and Active Learning. In: Touretzky, D.S., Tesauro, G., Leen, T.K., (eds.): *Adv. Neural Inf. Process. Syst.*, *7*, 231–238 (1995).

[4] Wolpert, D.H.: Stacked Generalization. *Neural Networks*, *5*, 241–259 (1992).

[5] Perrone, M.: *Improving regression estimation: averaging methods for variance reduction with extensions to general convex measure optimization.* PhD thesis, Brown Univ., Phys. Dept. 1993.

[6] Alpaydin, E.: Techniques for Combining Multiple Learners. In: Alpaydin, E. (ed.): *Proc. Eng. Intell. Syst. '98*, vol. 2, pp. 6–12. ICSC Press 1998.

[7] Tyler, L.K., Marslen-Wilson, W.D.: Conjectures and refutations: a reply to Norris. *Cognition*, *11*, 103–107 (1982).

[8] Bull, L.: On Model-Based Evolutionary Computation. *Soft Comp.*, *3*, 76-82 (1999).

[9] Jin, Y., Olhofer, M., Sendhoff, B.: On Evolutionary Optimization with Approximate Fitness Function. In: Whitley, D., Goldberg, D., Cantú-Paz, E., Spector, L., Parmee, I., Beyer, H.-G. (eds.): *GECCO-2000 Proc.*, pp. 786-793. M. Kaufmann Pub. 2000.

[10] Jones, D., Schonlau, M., Welch, W.: Efficient Global Optimization of Expensive Black-Box Functions. *J. Global Optim.*, *14*, 455-492 (1998).

[11] Booker, A.J., Dennis, J.E. Jr., Frank, P.D., Serafini, D.B., Torczon, V., Trosset, M.W.: A rigorous framework for optimization of expensive functions by surrogates. *Struct. Optim.*, *17*, 1-13 (1999).

[12] Torczon, V., Trosset M.W.: Using approximations to accelerate engineering design optimization. In: *Proc. 7th Symp. Multidisc. Anal. Optim.*, St. Louis, MO, Sept. 2-4, 1998.

Local-Global Neural Networks for Interpolation

Carlos E. Pedreira*, Luiz Carlos Pedroza†, Mayte Fariñas*

*Catholic University, PUC-RIO; C.P. 38063; CEP 22452-970, Rio de Janeiro, Brazil, {pedreira or mayte}@ele.puc-rio.br, †CEFET-RJ, Av. maracanã 229; CEP 20271-110, Rio de Janeiro, Brazil, pedroza@cefet-rj.br;

Abstract
In this paper a new connectionist model is proposed. The proposed architecture is trained by a scheme based on partition of the function domain, approximating the generator function by a set of very simple supporting functions. This method has an interesting ability concerning interpolation. A synthetic experiment and a real data missing data application are presented.

1 Introduction

Many interesting problems, concerning real world applications are related to interpolation. Among those, missing value filling is of particularly importance. In this kind of problems the main goal is to emulate a function in a sector of the domain where only a fraction of points is known. A new algorithm to reconstruct a generator function, based on local estimates, is proposed. Prediction, i.e. estimations outside the pre-established domain, is not a goal here.

The proposed architecture is trained by a scheme based on partition of the function domain. The main idea is to approximate the original function by a set of very simple supporting functions. Although, there are no theoretical limitations concerning these functions complexity, the supporting function are in general linear. The input-output mapping is expressed by a piecewise structure. The network output is constituted by a combination of several pairs, each of those, composed by a supporting function and by a membership function. The membership functions define the role of an associated supporting function, for each subset of the domain. Partial superposition of membership functions is allowed. In this way, the problem of approximation functions is approached by the specialisation of neurons in each of the sectors of the domain. In other word, the neurons are formed by pairs of membership and supporting functions, that emulate the generator function in different parts of the domain. The level of specialisation in a given sector is proportional to the value of the membership function.

2 The Proposed Architecture

Let us consider a network with m nodes or neurons. Let $\{x_i\}_1^n$ be the subset of the available data that is used for training. For algebrical and notational simplicity we will restrict ourselves to the case where $x \in \Re$ (x subscript is omitted). Generalisation for $x \in \Re^n$ is straightforward. Let us define, for each point x, m membership functions:

$$B_j(x) \equiv -C_j\left[\frac{1}{1+\exp(d_j(x-h_j^{(1)}))} - \frac{1}{1+\exp(d_j(x-h_j^{(2)}))}\right] \qquad j=1,\ldots,m$$

where C_j, d_j, $h_j^{(1)}$ e $h_j^{(2)}$ are parameters to be adjusted. Note that parameter C_j reflects the membership functions level, while d_j is related to this function declivity. Parameters $h_j^{(1)}$ and $h_j^{(2)}$ delimit the domain sector where the associated support function is more active (see figure 1).

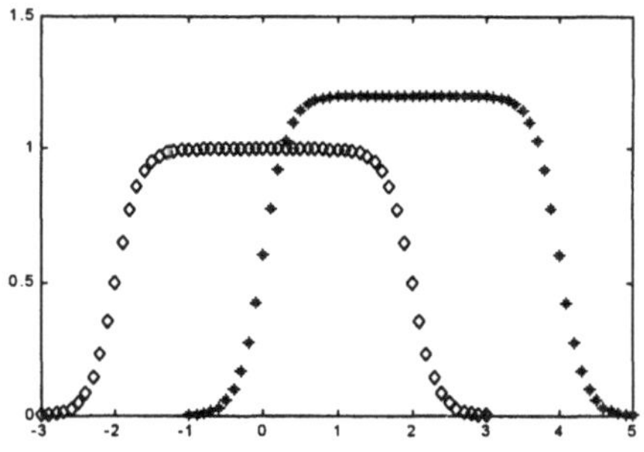

Fig. 1. Example of activation functions
◊ $C=1, d=6; h_j^{(1)} = -2; h_j^{(2)} =2$,
* $C=1.2, d=6; h_j^{(1)} = 0; h_j^{(2)} =4$

The supporting functions are typically linear or quadratic. Although more complex functions may be used, it seams that this additional complexity does not bring a correspondent refinement of the model. Let us consider linear supporting functions:

$$\kappa_j(x) = a_j x + b_j \qquad j=1,\ldots,m$$

where a_j and b_j are the parameters to be estimated. Each node or neuron is constituted by a pair {membership function; support function} (see figure 2). Then, for each node one need to estimate 6 parameters (7 for the quadratic function case). As usual, the model complexity may be inferred by the number of nodes.

The input is connected to the nodes producing as its output the membership and supporting function, $B_j(x)$ and $\kappa_j(x)$ product. Note that there no weight links the nodes outputs to the network output (see figure 2). The output of the jth neuron is $B_j(x)\ \kappa_j(x)$, and the network output is given by:

$$g^m(x) = \sum_{j=1}^{m} B_j(x)\kappa_j(x) \qquad (2.1)$$

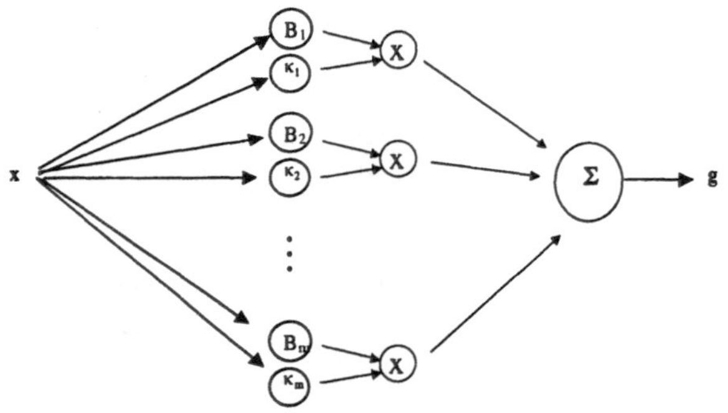

Fig. 2. The proposed architecture

The central goal is to design a network that is able to approximate as well as possible a target function f(x). With this objective in mind we define an error function as a convex combination of two quadratic error measures E_1 e E_2:

$$E \equiv \alpha \sum_{i=1}^{k} E_1^2(x_i) + (1-\alpha)\, E_2^2 \qquad (2.2)$$

where

$$E_1(x_i) \equiv g^m(x_i) - y(x_i) \quad e \quad E_2 \equiv 1 - \sum_{j=1}^{m} C_j \qquad (2.3)$$

Parcel E_1 is associated to the quality of the obtained approximation while E_2 has the target to keep the membership functions bounded. By using this E_2 one is penalising the solutions where the sum of parameters C_j exceed 1. The unit-valued limit is not mandatory although it may confer interpretably to the results.

If one defines, for each neuron, a vector of parameters $\Im^j \equiv (C_j, d_j, h_j^{(1)}, h_j^{(2)}, a_j, b_j)$, the main goal will be to find $\Im^j$ that minimizes the error function E.

3 Theoretical Results

Theorem T-1 gives theoretical consistence to the proposed methodology. It is shown that any L_2-integrable function may be approximated by functions $g^m(x)$. The following auxiliary results are need to prove the theorem:

Auxiliary result AR-1: The simple functions as

$S(x) = \sum_{j=1}^{m} \alpha_i X_{Ai}(x)$ with constant $\alpha_i \in \Re$ and where $X_{Ai}(x)$ is an indicator function of the set Ai, are dense in L_2.

Auxiliary result AR-2: There exits a sequence of functions $g_n(x) \in \{g^m(x)\}$ such that $g_n(x) \rightarrow S(x)$ (L^2 convergence), $\forall$ S(x).

Proof: Let $S(x) = X_{[0,\ 1]}$. Let us consider $\kappa_n(x) = 1$, then, $g_n(x) = B_n(x)$ with $h_1=0$ $h_2=1$.

$$B_n(x) = -C_n\left[\frac{1}{(1+e^{d_n x})} - \frac{1}{(1+e^{d_n(x-1)})}\right]$$

where $C_n = \left(e^{d_n/2}+1\right)/\left(e^{d_n/2}-1\right)$ and $d_n \rightarrow \infty$. It is not difficult to show that $B_n(x) \xrightarrow[n]{} S(x)$ punctually, and in L^2 by using the Lebesgue convergence theorem [3]. Since punctual convergence is already shown, it will be enough to prove that $B_n(x)$ is dominated by an L^2- integrable function. It is easy to verify that an L^2-integrable function g(x) defined as:

$$g(x) = \begin{cases} 1+e^{x} & x<0 \\ 2 & x \in [0,1] \\ 1+e^{-x} & x>1 \end{cases}$$

limits function $B_n(x)$ for all x $\forall$n, as we desire to prove. It is still needed to generalize from interval [0 1] for an arbitrary one. For that, one has to find a sequence Bn(x) that converges for functions of the type X_A $A=[\delta_1, \delta_2]$. With this goal in mind one may consider functions Bn(x) with $h^1=\delta_1$, $h_2=\delta_2$. Extension of this result for any simple function S(x) follows by considering that L^2 convergence is preserved with adding and constant multiplication operations. Since simple functions are expressed as finite linear combinations of X_{Ai} functions, one obtains that the sequence Bn(x) that converge to S(x) in L^2.

∎

Note that by choosing $g_n(x)$ one considered $\kappa_n(x) = 1$. So, The theorem is valid inclusive in this case. The choice of the linear form $\kappa(x) = ax + b$ (sub-indices are omitted) has the purpose of increase the approximation for a limited number of neurons and to accelerate convergence.

Theorem T-1: Let $g^m(x) = \sum_{j=1}^{m} B_j(x)\kappa_j(x)$ where $\kappa_j(x) = a_j x + b_j$, $B_j(x) \equiv$

$$-C_j\left[\frac{1}{1+\exp(d_j(x-h_j^{(1)}))} - \frac{1}{1+\exp(d_j(x-h_j^{(2)}))}\right]$$

$j=1,\ldots,m$

any L^2-integrable function may be approximated by functions of the form $g^m(x)$.

Proof: One wants to prove that the set of functions $\{g^m(x)\}$ with the norm $\|\cdot\|_{L_2}$ is a dense set in L_2.

It is needed to shown that for any function in L2, there exists a sequence of functions $g_n(x) \in \{g^m(x)\}$ that converge for f(x) in L^2. From **AR-1** we have that simple functions form a dense set in L_2. So, there exists a sequence of simple functions, $S_k(x)$ that converge for f(x) in L^2. From **AR-2**, we have that the simple function may be approximated by functions of the type $g^m(x)$. For each simple function $S_k(x)$, there exists a sequence $g_{kn}(x)$ that converges to $S_k(x)$ in L^2. To construct $f_n(x)$ we use the following result (Cantor diagonal [3]): Let $g_n(x)=g_{nn}(x)$, then $g_n(x) \to f(x)$ in L^2.

4 On the Initial Choice of Parameters

The relationship between the network input and output is learned by estimating the parameters that define the membership and supporting functions. The membership functions may overlap in part of the domain allowing that a given point is estimated by a balanced combination of more than one supporting function.

The initial choice of the parameters $h_1^{(1)}$ and $h_m^{(2)}$ may reflect an a priori knowledge on the function domain. with the goal of accelerating convergence, one may use the following initialization heuristic. The central idea is to divide the domain obtaining intervals where the function is approximately monotonic. The starting point is to adjust on data a polinomium of the degree equal to the number of supporting functions one decided to use. By calculating the maximum and minimum of this polinomium one determine the regions of the domain where the function remain monotonic. To define the values for a and b associated to a linear supporting function one should adjust, for each interval, a straight line by using linear regression.

5 Numerical Results

In the first numerical experiment (see figure 3) 100 points were generated by using the function sen(x) +2 + noise. Noisy signal was obtained by adding a gaussian signal with zero mean and 3 different standard deviations: 0.1; 0.4 and 0.7. Table I resumes these experiment results. In all simulations 3 pairs of membership functions were used. The initialization heuristic was used in experiments 1, 2 and 3, but on the fourth.

Noise level	Number of epochs	Noise MAPE	In sample MAPE	Out of sample MAPE
0	111	0	0.14	0.157
0.1	137	4.79	4.64	1.04
0.4	78	16.13	16.80	4.91
0.7	37	30.88	52.36	7.87

Table I - Numerical results for different noise levels

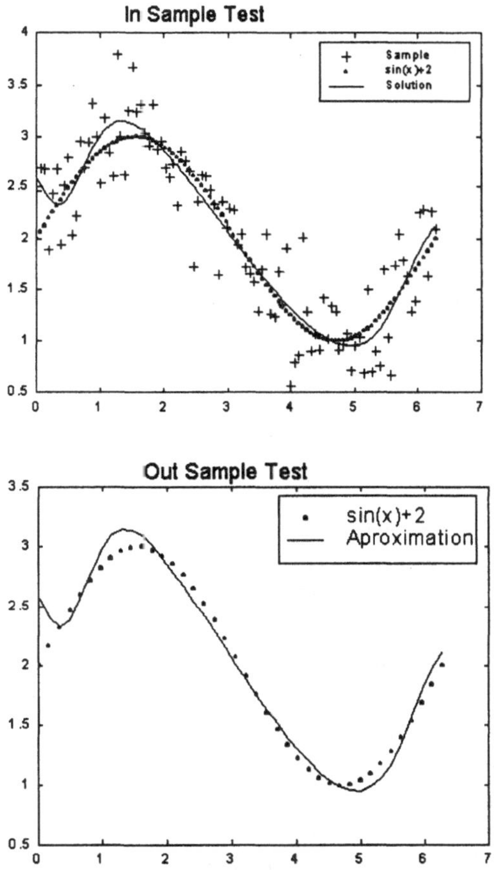

Fig. 3. noise level = 0.4

We used the same data of experiment 4, but deliberated inicializated the algorithm with a 'bad initial condition'. After 355 iterations we obtained the following errors: In-sample MAPE = 17.12 and out-of-sample MAPE = 3.20. The convergence may be observed in figure 4.

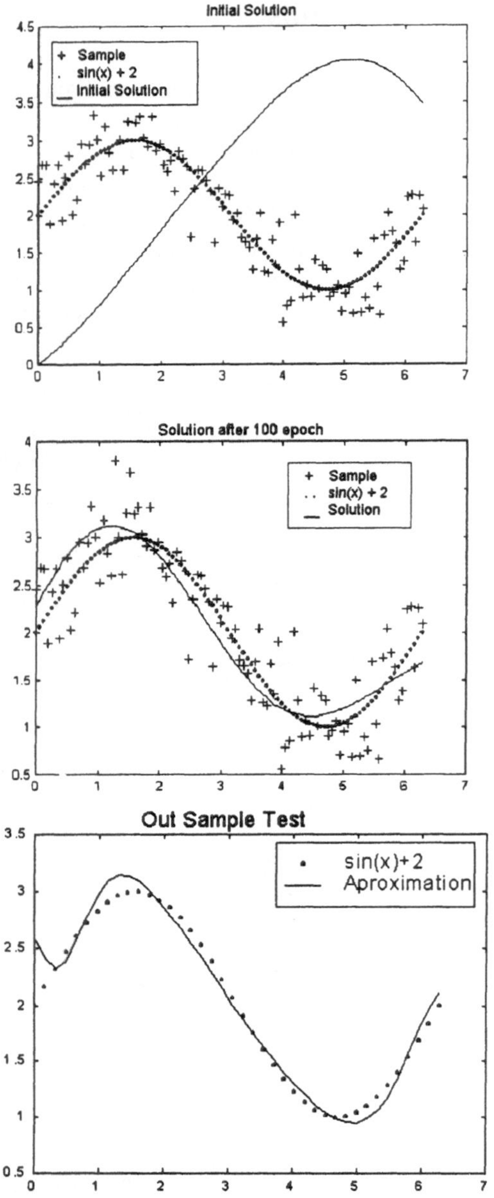

Fig. 4. Convergence evolution without the use of the initialization heuristic

6 An Application for Missing Data

In this section we present an application of the proposed methodology for the problem of missing data. Real electricity load data was used. The series concerns minute measures of electricity load in Brazil for 1st July 1999. Please note that missing data problem is quite frequent in minute measures of electricity load.

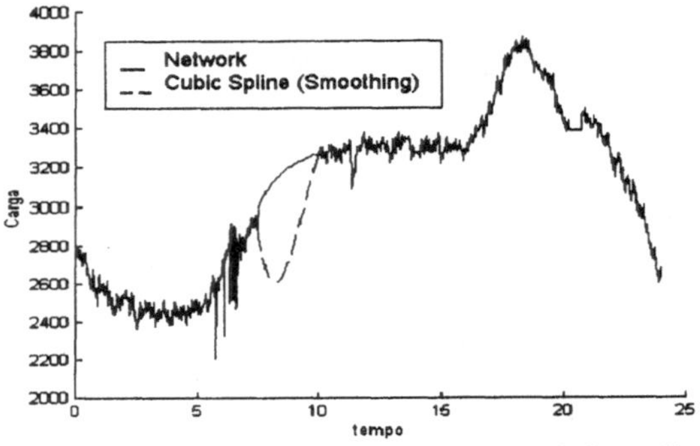

Fig. 5. Proposed algorithm versus Cubic spline

In order to simulate a missing data problem we withdrew a fixed quantity of points from the original serie. This serie is then recomposed by using the proposed algorithm. The points were withdrawn randomly in different percentuals: 5%; 10%; 20%; 30% and 40 %. The results can be found in table II.

Withdrawn points (%)	**Number of epochs**	**In sample Mape**	**Mape (Out of sample)**
5%	75	0.99	1.21
10%	333	1.01	1.15
20%	106	1.06	1.116
30%	379	1.00	1.119
40%	105	1.07	1.117

Table II – Missing data simulation - real electricity load data

We compared our method with smooth cubic spline. One can notice by observing figure 5 that the spline algorithm did not produce good results when one has a considerable number of consecutive missings.

7 Final Remarks

In this paper a new connectionist architecture is proposed. This method has an interesting ability concerning interpolation. It has an intrinsic ability to produce regular solutions. A synthetic experiment and a real data missing data application were presented. Simulated results for noisy environment were particularly encouraging, producing very nice generalization results.

References

[1] Haykin S. "Neural Networks – A Comprehensive Foundation", Prentice Hall, second edition, 1999.

[2] Pedroza L.C e Pedreira C.E. "Multilayer Neural Networks and Function Reconstruction by Using a priori Knowledge" International Journal of Neural Systems, Volume 9, number 3, pp 251-256, 1999.

[3] Bartle, R.G. Elements of integration. Wiley. New York, 1966

Improving the Competency of Classifiers through Data Generation

Herna L. Viktor * [1], Iryna Skrypnik †

* Department of Informatics, School of Information Technology, University of Pretoria, South Africa
† Department of Computer Science and Information System, University of Jyväskylä, Finland

Abstract

This paper describes a hybrid approach in which subsymbolic neural networks and symbolic machine learning algorithms are grouped into an ensemble of classifiers. Initially, each classifier determines which portion of the data it is most competent in. The competency information is used to generated new data that are used for further training and prediction. The application of this approach in a difficult to learn domain shows an increase in the predictive power, in terms of the accuracy and level of competency of both the ensemble and the component classifiers.

1 Introduction

When acquiring knowledge from data, it has been observed that, in complex domains, many individual classifiers fail to find adequate concept descriptions for all of the concepts contained in the data. Rather, the techniques usually find complementary results with respect to these concepts. This paper discusses an approach that builds on this observation through the combination of classifiers into an ensemble. In this approach, the subspace of the problem domain in which a classifier is most competent, is identified. This knowledge is subsequently used to improve the classification ability of the ensemble and its components.

The paper is organized as follows. Section 2 introduces the ensemble approach. This is followed, in Section 3, with a description of the incremental ensemble training method. The section also discusses how data are generated to facilitate retraining. Section 4 illustrates the method by means of the well-known Monk2 data set. Finally, Section 5 concludes the paper.

2 Ensembles of Classifiers

The ensemble machine learning approach refers to a method in which a set of modules are generated from separately trained inductive machine learning algorithms [5]. Experimental results indicate that ensembles can greatly improve the generalization accuracy when compared to the individual algorithms [2, 7, 9]. An ensemble of classifiers is a set of $l > 1$ classifiers, denoted by $h_1, \ldots, h_l$ where the individual decisions are combined when classifying new instances. An effective ensemble consists of a set of models that are not only highly correct, but with errors on different parts of the input space [5].

The general idea is as follows. Assume that a learning algorithm C_i is presented with a total of m training instances of the form $(\mathbf{x_1}, y_1), \ldots, (\mathbf{x_m}, y_m)$ for some unknown function $y = f(\mathbf{x})$. Here $\mathbf{x}_i$ denotes a vector of the form $(x_{i1}, \ldots, x_{il})$, where each x_{ij} is called a feature of $\mathbf{x}_i$. Given a set S of training instances, a learning algorithm outputs a classifier h_i, which represents a hypothesis that estimates the function f and, when presented with a new instance, predicts the corresponding output y.

Figure 1 illustrates the basic framework of ensemble classification. Initially, each classifier in the ensemble uses a set of training instances to construct a hypothesis. Then, for each instance, the predicted output of each of these classifiers is combined to produce the output of the ensemble. It follows that the combination of outputs of several classifiers is useful only if there is disagreement on some inputs, i.e. the component classifiers make uncorrelated errors.

A large number of ensemble combination methods have been developed. Majority voting is one of the basic methods of combining classifiers. Other methods include weighed average, bagging, boosting and its variants [1, 3], Approaches designed particularly for neural networks are described by [2, 5], amongst others.

The rather simplistic majority voting scheme has shown to yield good results in several domains [1, 3] and will thus be used here to illustrate the effectiveness of data generation and retraining on the ensemble's accuracy and competence. The general idea is as follows. Consider an ensemble that consists of n classifiers, namely $h_1, h_2 \ldots h_n$. For a particular

[1] Currently on sabbatical leave at the University of Jyväskylä, Finland, email: hlviktor@hakuna.up.ac.za

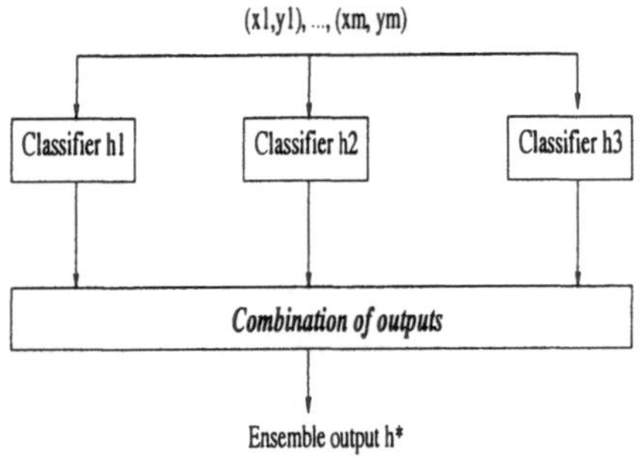

Fig. 1. An ensembles consisting of three classifiers

training instance, each classifier outputs a prediced class values i, where $i \in \{1, \ldots, L\}$. Here, L denotes the number of classes. In the majority voting scheme, the final class is the class selected by the majority of the classifiers.

3 Incremental Ensemble Training

This section describes the program that is followed by each component classifier that forms part of the ensemble. Note that, prior to training, the data set is divided into a training, test and validation set. The test set is used to measure the predictive power of the initial ensemble and its components. At the completion of the incremental ensemble training process, the accuracy, the final predictions and the ensemble's competency are evaluated against the validation set.

Step 1. Initial training and prediction. Firstly, each classifier used the training data to construct an initial description of the problem domain. The ensemble is constructed and the initial quality there-of are evaluated by using the majority voting approach as introduced above. Also, the areas of competence are measured by considering the percentage of the input space covered by the ensemble and its components. In this way, the initial predictive power of the component classifiers and the ensemble are determined.

Step 2. Data set partitioning. Next, the results of training are used to determine the subspaces of the problem domain in which a particular classifier produces good results. In the current implementation, this portion of the data set is determined through the extraction of a concept description that is expressed as disjunctive normal form (DNF) rules of the form $(A_1\ Op\ V_1) \cap \ldots \cap (A_n\ Op\ V_n) \rightarrow C_i$. Here $(A_i\ Op\ V_i)$ refers to a feature value test, where A_i denotes the feature identifier, Op the relational operator and V_i the threshold value for A_i, e.g. temperature > 20. C_i denotes the concept class. A rule thus describes the decision boundaries in terms of one or more input feature.

The accuracy of each rule R_i is evaluated against the test set, as follows. Each rule with an accuracy that is higher than a pre-determined accuracy threshold Acc_R is used as an indicator of a portion of the data that the classifier excels in. That is, the rules act as "competency maps" that describe the classifier's current knowledge [6].

Step 3. Data generation. The data generation step uses the results of Step 2 to focus the training process. For each classifier h_i, the portion of the data in which it is considered to be competent is further emphasized through the automatic generation of new instances that fall in this range. Note that two classifiers h_1 and h_2 may produce rules that overlap. In this case, both h_1 and h_2 use these descriptions to generate new data and retrain. Any conflicts will be resolved during the final evaluation and prediction step (Step 5). The data generation process is discussed in Section 3.1.

Step 4. Retraining. Each classifier uses the new training set, as created during Step 3, to reiterate the training phase.

Step 5. Ensemble evaluation and prediction. Finally, the classifiers are combined into the final ensemble, which again uses the majority voting scheme to predict the classes against the validation set. The final accuracy and competency values are used as a measure to evaluate the success of the incremental ensemble training process.

3.1 Data generation approach

The data generator generates a new training set that is based on the areas of competence, as described by one or more high quality rules R_i. The data generation process uses the number of instances describing a particular concept, as found in the original training set, when producing a new set of instances. For example, consider a subset of the data space covered by rule R_1 that concerns class C_1. Assume that the training set contains E_1 training instances describing concept C_1. The training set contains m concepts, $C_1, \ldots, C_m$, where $m \geq 1$. Consider the number of instances $[n_1, n_j]$ covered by R_1. Assume that rule R_1 covered $n_1 \subseteq E_1$ training instances for class C_1 and n_j training instances that were not of class C_1. These n_j instances describe one or more of the other classes C_i, where $i \neq 1$, contained in the data set.

The data generator produces new training instances that contain n_1 occurrences for class C_1 and $n_k \subseteq n_j$ occurrences for each of the other classes C_i, where rule R_1 covers x_k instances of class C_i. The instances are added to the original training set.

Monk2 Robot if **EXACTLY** two choices of (Head-shape = round), (Body-shape = round), (Smiling = yes), (Holding = sword), (Jacket-colour = red), (Has-tie = yes)

Table I. The Monk2 problem

The data generation process is constrained by generating new data without changing the class distribution of the original training instances. Also, the original distributions of feature values in the original training set are used when constructing the new training set. For Boolean and nominal valued features, the number of occurrences of the feature with respect to each class is determined. When considering continuous-valued features, the mean values, variances and the original range of feature values are determined for each class. The new training set combines this distribution information with the knowledge as contained in the feature-value tests of the rule.

For the newly generated data, the values of the features that are not included in the rule reflect the original distribution of the training data. Also, the data generation process generates data that reflect the original feature distribution of those concepts that are not covered by the rule. This results in a training set that is biased towards the feature values as contained in the rule, while maintaining the knowledge of the original training set. In this way, a learner benefits from the knowledge of another [12].

Section 4 illustrates the incremental ensemble training approach by means of the Monk2 problem.

4 An Illustrative Example

The Monk2 problem is one of the three Monk's problems that have been widely used to test machine learning algorithms [8, 10]. The problems are based on an artificial robot domain, in which robots are described by six different features, i.e. the Head-shape (*round, square* or *octagon*), Body-shape (*round, square* or *octagon*), Smiling (*yes* or *no*), Holding (*sword, balloon* or *flag*), Jacket-color (*red, yellow, green* or *blue*) and Has-tie (*yes* or *no*). The three Monk's problems are described by means of a population of 432 instances defined over these six features.

For each problem, the training instances are used to distinguish between two classes, namely Class 0 (Not-Monk) and Class 1 (Monk). For the Monk2 problem, the individual rule set accuracy of the machine learning algorithms are generally low. For example, [10] reported that the average accuracy of 25 machine learning algorithms was only 69.6% when applied to this problem. The Monk2 problem is therefore especially suited to a domain where the individual classifiers do not perform well in the entire problem space. The rules describing the Monk2 robots are depicted in Table 1.

	C4.5	CN2	NN-1	NN-2	Ensem.
Acc (I)	68.0%	69.0%	66.8%	68.9%	65.8%
Comp (I)	65.0%	67.0%	74.0%	64.5%	64.1%
Acc (F)	68.0%	76.4%	69.0%	71.8%	70.8%
Comp (F)	65.0%	67.0%	64.7%	68.0%	75.9%

Table II. Accuracy and competence values (I)-Init, (F)-Final

The Monk2 training set contained 169 training instances without noise [10], the number of instances in Class 0 was 105 and the number of instances for Class 1 was 64. For the purpose of the set of experiments presented here, the full test data set was randomly divided into a 50% (216 instances) test set and a 50% (216 instances) validation set.

Prior to learning, the accuracy threshold value Acc_R was set to 69.6%. That is, the quality of a competent classification should be at least as good as the average of the 25 algorithms as tested by [10]. The ensemble consisted of four classifiers, namely two 3-layers neural networks, with topologies 6-4-2 (NN-1) and 6-5-2 (NN-2) respectively, with monotonicly increasing differentiable functions, trained using gradient descent optimization. The other two classifiers used the C4.5 decision tree algorithm [8] and the CN2 rule extraction approach [4] to construct their initial representations of the problem domain.

The first two rows of Table 2 shows the results of the initial training step. The decision tree algorithm produced a 68.0% accurate classification and the accuracy of the results produced by the CN2 algorithm was 69.0%. Neural networks NN-1 and NN-2 converged after 72 and 152 epochs respectively, with accuracies of 71.8% and 65.8%. The ANNSER rule extraction approach as described in [11] was used to extract rules from the two trained neural networks producing 66.8% and 68.9% accurate rule sets.

The initial results of the four classifiers were transformed to uniform DNF format [11] and subsequently evaluated using the CN2 evaluation function [4]. The table shows that the ensemble had an accuracy of 65.8% and was considered to be competent in 64.1% of the problem domain.

During the second step of the program, the areas of competencies were identified using the rule based approach introduced in Section 3. Each classifier completed the data generation process and used this data to retrain.

The following observation is noteworthy. The symbolic induction programs used in this paper do not learn incrementally. When supplied with additional data, the methods induce new rules, without considering the knowledge contained in prior rule sets. Incremental learning was facilitated by adding the newly generated data of the symbolic classifiers to the original training set.

The final results, as obtained during Step 5 of the ensemble training process, is shown in the third and fourth rows of Table 2. The table shows, that the accuracy and the competency of the ensemble increased to 70.8% and 75.9% respectively. Also, the results of the ensemble were higher than that of its components. The accuracies of the individual classifiers increased with an average value of 3.1%. Importantly, the percentage of the problem domain covered by the individual classifiers showed a small average decrease of 1.4%. This indicated that the component classifiers learned to excel in those areas of the problem space they are most competent in. That is, the component classifiers specialized within the areas of expertise, thus producing more informative classifications.

These results indicate that, for this particular domain, the approach described here improved the predictive capabilities of the ensemble and the component classifiers by focussing the areas of competence.

5 Conclusion

This paper introduced an approach in which an ensemble of classifiers utilized the results of data generation and retraining to improve their predictive power. Each member of the ensemble was trained to excel in that portion of the problem domain of which it had the best description. The method was illustrated by means of the Monk2 problem domain, where the initial results indicated that the incremental ensemble training method increases the predictive power of an ensemble and its component classifiers.

It follows that our approach should be thoroughly tested against a number of diverse data sets, using a variety of classifiers and ensemble combination schemes. Current research include the evaluation of our approach against a real world Human Resources data set. The use of other data generation methods such as the method described by [6], which uses decision boundaries to generate new instances, as well as the use of feature selection to reduce the input space, need to be further investigated.

References

[1] E Bauer and R Kohavi, 1999. An Empirical Comparison of Voting Classification Algorithms: Bagging, Boosting and Variants. Machine Learning, 36, pp.105-139.

[2] L Breiman, 1996. Bagging Predictors. Machine Learning, 24 (2), pp.123-140.

[3] S Cost and Salzberg, 1993. A Weighted Nearest Neighbor Algorithm for Learning with Symbolic Features. Machine Learning, 10(1), pp. 57-78.

[4] P Clark and T Niblett, 1989. The CN2 rule induction program, Machine learning, pp.261-283.

[5] T Dietterich, 1997. Machine Learning Research: Four Current Directions. Artificial Intelligence, 18 (4), pp.97-136.

[6] J-N Hwang, *et al*, 1991. Query-based Learning applied to Partially Trained Multilayer Perceptrons, IEEE Transactions on Neural Networks, 2(1), pp.131-136.

[7] R Maclin and D Opitz, 1997. An Empirical Evaluation of Bagging and Boosting. Proc. of 14th National Conf. on Artificial Intelligence, AAAI/MIT Press, 1997, pp. 546-551.

[8] JR Quinlan, 1994. *C4.5: Programs for Machine Learning*, Morgan Kaufmann, California: USA.

[9] R Shapiro, Y Freud, P Bartlett and W Lee, 1997. Boosting the margin: A new explanation of the effectiveness of the voting methods. Proc. of 14th Intern. Conf. on Machine Learning, Morgan Kaufmann, pp. 322-330.

[10] SB Thrun *et al*, 1991. The Monk's problems: A Performance Comparison of Different Learning Algorithms. Technical Report CMU-CS-91-17. Computer Science Department, Carnegie Mellon University, Pittsburgh: USA.

[11] HL Viktor, AP Engelbrecht and I Cloete, 1995. Reduction of Symbolic Rules from Artificial Neural Networks using Sensitivity Analysis, IEEE International Conference on Neural Networks (ICNN'95), Perth: Australia, pp.1788-1793.

[12] HL Viktor. 2000. Generating new patterns for information gain and improved neural network learning, The International Joint Conference on Neural Networks (IJCNN-00), Como: Italy.

Graded Rescaling in Hopfield Networks

Xinchuan Zeng, Tony R. Martinez *

*Computer Science Department, Brigham Young University, Provo, Utah 84602, e-mail: zengx@axon.cs.byu.edu, martinez@cs.byu.edu

Abstract

In this work we propose a method with the capability of improving the performance of the *Hopfield network* for solving optimization problems by using a *graded rescaling* scheme on the distance matrix of the energy function. This method controls the magnitude of rescaling by adjusting a parameter (scaling factor) in order to explore the optimal range for performance. We have evaluated different scaling factors through 20,000 simulations, based on 200 randomly generated city distributions of the 10-city *traveling salesman problem*. The results show that the graded rescaling can improve the performance significantly for a wide range of scaling factors. It increases the percentage of valid tours by 72.2%, reduces the error rate of tour length by 10.2%, and increases the chance of finding optimal tours by 39.0%, as compared to the original Hopfield network without rescaling.

1 Introduction

Hopfield and Tank [1] proposed a neural network approach to find approximate solutions to combinatorial optimization problems, such as the *traveling salesman problem* (*TSP*). Much research work has focused on analyzing the model and improving its performance.

Wilson and Pawley [2] showed that the Hopfield network often failed to converge to valid solutions, and many solutions were far from optimal. Brandt et al. [3], Aiyer et al. [4], and Protzel et al. [5] showed that they could improve the convergence of the network by modifying constraints in the energy function. Li [6] combined the Hopfield network with the "augmented Lagrange multipliers" algorithm from optimization theory. Catania et al. [7] applied a fuzzy scheme to tune the parameters in the Hopfield network. Although these approaches have demonstrated some improvement, most were tested on only a single or a small number of TSP city distributions, and general performance on a large number of city distributions were not reported.

Despite the improvement of its performance over the past decade, the Hopfield network still has some basic problems [8, 9] that remain unsolved. One problem is the inconsistent performance, which is usually better for simple city distributions, but poor for those with complex topologies and multiple clusters. Another problem is that the performance is sensitive to the choice of parameters in the energy function. In previous work, we proposed a new activation function to reduce effects of noise [10], a new relaxation procedure to reach a smoother relaxation process [11], and a simple rescaling scheme to reduce the effects of clustering [12]. These approaches have been shown capable of significantly improving performance based on a large number of simulations.

In this work we propose an approach which uses *graded rescaling* of the distance matrix in the energy function to reduce the negative impacts of distribution clusterings on the performance. The proposed graded rescaling scheme has the capability to improve the balance of the distance matrix. This approach has been tested with a large number (20,000) of simulations based on 200 randomly generated city distributions of the 10-city *TSP*. The results indicate that it is capable of improving the performance significantly: 72.2% increase in the percentage of valid tours, 10.2% decrease in the error rate, and 39.0% increase in the chance of finding optimal tours.

2 Graded Rescaling of Distance Matrix

Hopfield's original energy function for an N-city *TSP* is given by [1]:

$$E = \frac{A}{2}\sum_{X=1}^{N}\sum_{i=1}^{N}\sum_{j=1,j\neq i}^{N} V_{Xi}V_{Xj}$$
$$+\frac{B}{2}\sum_{i=1}^{N}\sum_{X=1}^{N}\sum_{Y=1,Y\neq X}^{N} V_{Xi}V_{Yi} + \frac{C}{2}\Big(\sum_{X=1}^{N}\sum_{i=1}^{N} V_{Xi} - N_0\Big)^2$$
$$+\frac{D}{2}\sum_{X=1}^{N}\sum_{i=1}^{N}\sum_{Y=1,Y\neq X}^{N} d_{XY}V_{Xi}(V_{Y,i+1} + V_{Y,i-1}) \qquad (1)$$

where X, Y are row indices, and i, j are column indices, V_{Xi} is the activation for neuron (X, i), and d_{XY} is the distance between cities X and Y. The first three terms enforce the constraints for a valid tour, and the last term represents the cost function for obtaining a short tour. The value of each parameter (A, B, C, and D) measures the importance of the corresponding term. The connecting weight between neuron (X,i) and (Y,j) is set according to:

$$W_{Xi,Yj} = -A\delta_{XY}(1-\delta_{ij}) - B\delta_{ij}(1-\delta_{XY}) - C - Dd_{XY}(\delta_{j,i+1} + \delta_{j,i-1}) \quad (2)$$

During relaxation, the input value $U_{Xi}^{(n+1)}$ for neuron (X,i) at iteration step *(n+1)* is the sum of its value $U_{Xi}^{(n)}$ at step n and ΔU_{Xi}. The value of ΔU_{Xi} is given by the following equation:

$$\Delta U_{Xi} = (-\frac{U_{Xi}}{\tau} + \sum_{Y=1}^{N}\sum_{j=1}^{N} W_{Xi,Yj}V_{Yj} + I_{Xi})^{(n)}\Delta t \quad (3)$$

where $\tau(= RC)$ is the time constant of an RC circuit and $I_{Xi}(= CN_0)$ is an external input current connected to each neuron (X,i).

The distance matrix d_{XY} in the energy function has the function of controlling the quality of solutions. From Eq. (2) and (3), we can see that the second term in Eq. (3) for the activation change ΔU_{Xi} is proportional to the value: $-D\sum_{Y=1}^{N} d_{XY}V_{XY}$. Since each V_{XY} has a similar average level, that term is approximately proportional to the value $-D\sum_{Y=1}^{N} d_{XY}$.

Now consider a case of two clusters C_1 and C_2, where N_1 and N_2 are the number of cities in C_1 and C_2 respectively. Assume that cluster C_1 is much larger than C_2, i.e. $N_1 >> N_2$. For a city X_1 in C_1 and X_2 in C_2, d_{X_1Y} would be smaller than d_{X_2Y} for $(N_1 - 1)$ cities in C_1 while larger for N_2 cities in C_2. Since $N_1 >> N_2$, the value $\sum_{Y=1}^{N} d_{X_1Y}$ would be smaller than $\sum_{Y=1}^{N} d_{X_2Y}$. The net effect is that the activations for the cities in cluster C_1 have overall higher values than those in C_2 (a smaller $\sum_{Y=1}^{N} d_{X_2Y}$ leads to a larger $-D\sum_{Y=1}^{N} d_{XY}$). A potential consequence is the formation of an invalid tour – a closed loop inside cluster C_1 without inclusion of those cities in C_2. Even when a valid tour is formed, its quality may be poor since each cluster constructs its own local configuration with less consideration of the other cluster because of the imbalance in their overall activations. We have observed such phenomenon for some city distributions where one or two cities are isolated from the main cluster.

Based on this analysis we propose the following approach, called *graded rescaling*, to rescale the distance matrix so that activations in a smaller cluster become closer to those in a larger cluster. For each row X $(X = 1, 2, ..., N)$, we first calculate the summation $S_X = \sum_{Y=1}^{N} d_{XY}$, and then calculate the average of S_X over all N rows using $\overline{S} = (\sum_{X=1}^{N} S_X)/N$. Next we rescale each original matrix element d_{XY} to be a new element d'_{XY} using the formula:

$$d'_{XY} = \frac{((1-\alpha)S_X + \alpha\overline{S})d_{XY}}{S_X} \quad (4)$$

where α is a parameter, called *scaling factor*, which controls the amount of rescaling and is in the range of $0 \leq \alpha \leq 1$. For the limit $\alpha = 0$, there is no rescaling and the matrix keeps its original form: $d'_{XY} = d_{XY}$. For the limit $\alpha = 1$, we have $d'_{XY} = (\overline{S}/S_X)d_{XY}$. Summation S'_X over row X gives $S'_X = \sum_{Y=1}^{N} d'_{XY} = (\overline{S}/S_X)\sum_{Y=1}^{N} d_{XY} = (\overline{S}/S_X)S_X = \overline{S}$, i.e. all rows are normalized to the same constant $\overline{S}$. In this case, the two terms $\sum_{Y=1}^{N} d'_{X_1Y}V_{XY}$ and $\sum_{Y=1}^{N} d'_{X_2Y}V_{XY}$ would have a similar overall level even though they are located in two unbalanced clusters. When $0 < \alpha < 1$, however, it is easy to show that the range for S'_X is between S_X and $\overline{S}$ ($S_X < S'_X < \overline{S}$ if $S_X < \overline{S}$, and $\overline{S} < S'_X < S_X$ if $S_X > \overline{S}$). The choice of α is important for the performance. A small α value has little influence on balancing activations. However, too large an α value can also have a negative effect since it causes too much distortion of the distance matrix, which can reduce the chance of finding optimal tours, and also degrades the percentage and quality of valid solutions. We have found empirically that an α value of 0.5 is both robust and balances the above concerns.

3 Simulation Results

The performance of the proposed graded rescaling approach has been evaluated through a large number of simulations based on 200 randomly generated 10-city *TSP* city distributions (including wide varieties of topologies). Many previous studies evaluated the performance of their algorithms based on only one (e.g., [1]) or a small number of city distributions (e.g., 10 distributions in [2]). In this simulation, 100 runs are conducted for each of 200 city distributions. Different random noise is added to the initial neuron values for each of the 100 runs of a given city distribution. The quantities that measure the performance are evaluated by first averaging over 100 runs for each city distribution, and then averaging over the entire 200 city distributions. Thus 20,000 runs are used to calculate each simulation data point shown in the following figures. We have experimented with different sets of 20,000 runs for a fixed set of parameters. The results show that the estimated quantities are fairly stable and their values vary within a range of about 1% among different sets of 20,000 runs. This demonstrates the robustness of the evaluation for the scaling factor α. The original energy function of the Hopfield network was used in the simulation, and the parameters in the energy function are those used by Hopfield and Tank [1].

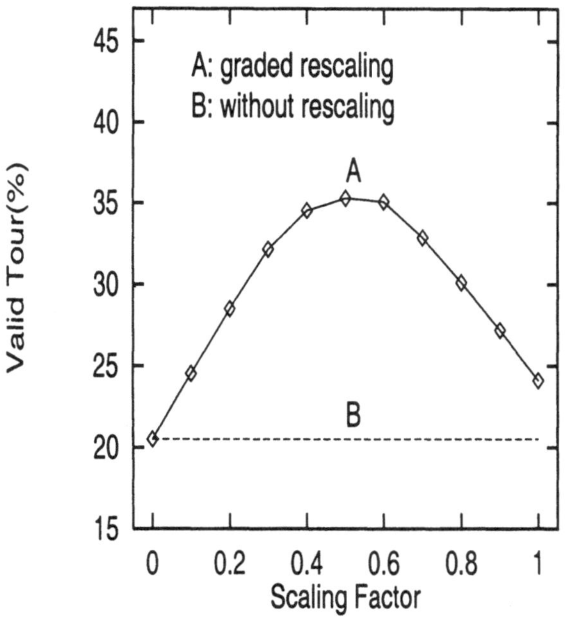

Fig. 1. The percentage of valid tours of different scaling factors using the graded rescaling procedure compared to the original Hopfield network without rescaling.

Fig. 1 shows the percentage of valid tours with different scaling factors using the proposed graded rescaling procedure and compares them to that using the original procedure without rescaling. For each city distribution, valid tour percentage is evaluated based on number of valid tours among 100 runs. Each data point shown in the figure is the average of valid tour percentages over 200 city distributions. The result shows that the rescaled distance matrix significantly improves the percentage of valid tours over the original one. The largest improvement is at the middle range for α from 0.4 to 0.6. At $\alpha = 0.5$, the percentage of valid tours (35.3%) with rescaling is increased by 72.2% compared to that (20.5%) without rescaling.

Fig. 2 shows the error rate of different scaling factors using the proposed graded rescaling procedure. For city distribution i, let $d_{i,optimal}$ be its optimal (shortest) tour length, and let $d_{i,j}$ be tour length of a valid tour j. Then the error rate of valid tour j is defined as $(d_{i,j} - d_{i,optimal})/d_{i,optimal}$. The error for each city distribution is averaged over all valid tours for that distribution. Each data point shown

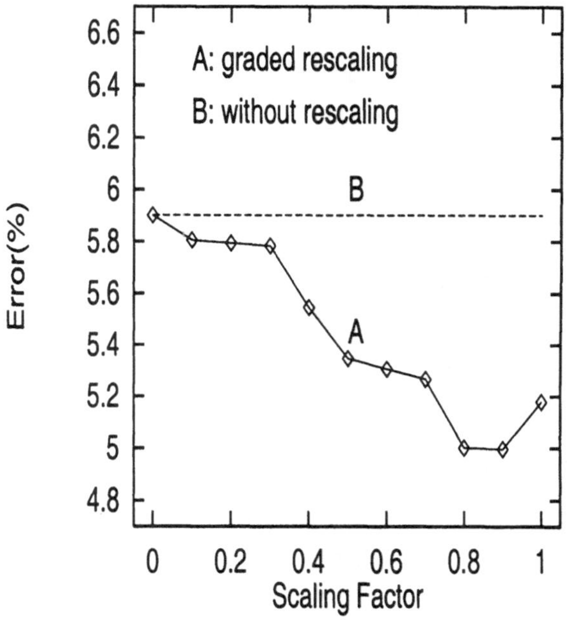

Fig. 2. The errors (%) of different scaling factors using the graded rescaling procedure compared to the original Hopfield network without rescaling.

in the figure is the average error of valid tours in 200 city distributions and is weighted by the percentage of valid tours for each city distribution. From the result we can see that all the errors with graded scaling are smaller than that without rescaling. The best improvement is at $\alpha = 0.9$, with a decrease of 15.3% in error (5.0% vs 5.9%). At $\alpha = 0.5$, the decrease in error is 10.2% (5.3% vs 5.9%).

Fig. 3 compares the graded rescaling and original scheme in terms of the percentage of city distributions for which the optimal tour is by at least one of the 100 runs, which measures the chance of finding optimal tours by the network. The result shows that all scaling factors have better chances to find optimal tours. At $\alpha = 0.5$, the improvement is 39.0% (82.0% vs 59.0%).

The above results demonstrate the advantage of using an adjustable graded scaling factor to achieve an optimal range of rescaling. A proper choice of scaling factor can lead to significant improvement in performance over that without rescaling ($\alpha = 0$), and also has an overall better performance than simple rescaling ($\alpha = 1$) [12].

4 Summary

In this paper, we have proposed a graded rescaling approach to rescale the distance matrix in the Hopfield network for solving *TSP*. A parameter (scaling factor α) is applied to adjust the amount of rescaling in order to achieve better performance. The pro-

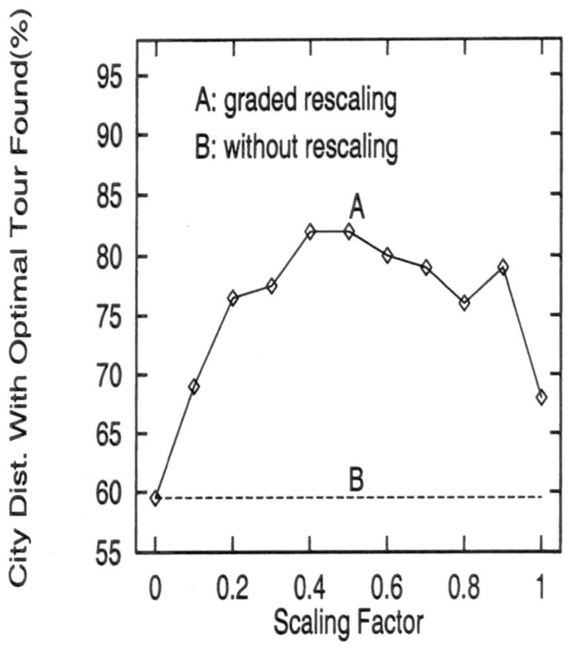

Fig. 3. The percentage of city distributions which optimal tour is found by at least one of 100 runs, showing results of different scaling factors using the graded rescaling procedure compared to the original Hopfield network without rescaling.

posed approach has capability of reducing activation imbalances in the network caused by distribution clustering so that more valid tours with higher qualities can be obtained. We have evaluated its performance for different scaling factors based on a large number (20,000) of simulations using 200 randomly generated city distributions of the 10-city TSP. The results show that a significant improvement can be achieved by this approach. With scaling factor of 0.5, the network obtains a 72.2% increase in the percentage of valid tours, a 10.2% decrease in the error rate, and a 39.0% increase in the chance of finding optimal tours, compared to those without rescaling.

5 Acknowledgments

This research is funded in part by a grant from *fonix* Corp.

References

[1] J. J. Hopfield and D. W. Tank, "Neural Computations of Decisions in Optimization Problems", *Biological Cybernetics*, vol. 52, pp. 141–152, 1985.

[2] G. V. Wilson and G. S. Pawley, "On the Stability of the Traveling Salesman Problem Algorithm of Hopfield and Tank", *Biological Cybernetics*, vol. 58, pp. 63–70, 1988.

[3] R. D. Brandt, Y. Wang, A. J. Laub and S. K. Mitra, "Alternative Networks for Solving the Traveling Salesman Problem and the List-Matching Problem", *Proceedings of IEEE International Conference on Neural Networks*, San Diego, CA, vol. 2: pp. 333–340, 1988.

[4] S. V. B. Aiyer, M. Niranjan and F. Fallside, "A Theoretical Investigation into the Performance of the Hopfield Model", *IEEE Transactions on Neural Networks*, vol. 1, no. 2, pp. 204-215, 1990.

[5] P. W. Protzel, D. L. Palumbo and M. K. Arras, "Performance and Fault-Tolerance of Neural Networks for Optimization", *IEEE Transactions on Neural Networks*, Vol. 4, pp. 600-614, 1993.

[6] S. Z. Li, "Improving Convergence and Solution Quality of Hopfield-Type Neural Networks with Augmented Lagrange Multipliers", *IEEE Transactions On Neural Networks*, vol. 7, no. 6, pp. 1507-1516, 1996.

[7] V. Catania, S. Cavalieri and M. Russo, "Tuning Hopfield Neural Network by a Fuzzy Approach", *Proceedings of IEEE International Conference on Neural Networks*, pp. 1067-1072, 1996.

[8] B. S. Cooper, "Higher Order Neural Networks-Can they help us Optimise?", *Proceedings of the Sixth Australian Conference on Neural Networks (ACNN'95)*, pp. 29-32, 1995.

[9] D. E. Van den Bout and T. K. Miller, "Improving the Performance of the Hopfield-Tank Neural Network Through Normalization and Annealing", *Biological Cybernetics*, vol. 62, pp. 129-139, 1989.

[10] X. Zeng and T. R. Martinez, "A New Activation Function in the Hopfield Network for Solving Optimization Problems", *Fourth International Conference on Artificial Neural Networks and Genetic Algorithms*, pp. 67-72, 1999.

[11] X. Zeng and T. R. Martinez, "A New Relaxation Procedure in the Hopfield Network for Solving Optimization Problems", *Neural Processing Letters*, vol. 10, pp. 1-12, 1999.

[12] X. Zeng and T. R. Martinez, "Rescaling the Energy Function in Hopfield Networks", *Proceedings of the IEEE International Joint Conference on Neural Networks* vol. 6, pp. 498-504, 2000.

Principles of Associative Computation

Andrzej Wichert, Birgit Lonsinger-Miller *

*Department of Neural Information Processing, University of Ulm, D-89069 Ulm, Germany, email: {wichert,birgit}@neuro.informatik.uni-ulm.de

Abstract

Currently neural networks are used in many different domains. But are neural networks also suitable for modeling problem solving, a domain which is traditionally reserved for the symbolic approach? This central question of cognitive science is answered in this paper. It is affirmed by a corresponding neural network model. The model has the same behavior as a symbolic model. However, also additional properties resulting from the distributed representation emerge. It is shown by comparison of those additional abilities with the basic behavior of the model, that the additional properties lead to a significant algorithmic improvement. This is verified by statistical hypothesis testing.

1 Introduction

Production systems are composed of if-then rules which are also called productions. The complete set of productions constitute long term memory. Productions are triggered by specific combinations of symbols which describe items. These items represent a state and are stored in short term memory. A computation is performed with the aid of productions by the transformation from an initial state in the short term memory to a desired state. In systems which model human behavior backtracking to a previous state of short term memory is allowed. By allowing backtracking and the exclusion of loops, a search from the initial state to the desired state is executed. The search defines a problem space, problems are solved by searching in a problem space whose state includes the initial situation and the desired situation [3, 2, 1].

Production system by itself do however not explain satisfactory many aspects of the human behavior, like learning, intuition and the treatment of noise.

It is an old idea to describe the process of human thinking by the formation of a chain of associations. The problem is however, that no satisfactory computational model exists until today.

Associations should describe links between different pictures. Pictures, like two dimensional sketches which are often used as examples in works concerning the symbolical representation, to ease the comprehension for the human reader. The intuitive idea is that pattern representation could give us an hint to the answer concerning learning, intuition and the treatment of noise by the principles of the similarity resulting from the distributed representation.

Equivalent to the production system the associative computer is composed out of long term memory in which the associations are stored, and in the short term memory in which the states are represented by structured pictures. Structuring of the pictures is needed to allow their manipulation. The computational task concerning problem solving corresponds to the manipulation of pictures.

2 Associative Memory for State Transitions

Not structured pictures can be represented by binary vectors and stored in an usual associative memory. The associative memory [4, 5] is composed of a cluster of units which represent a simple model of a real biological neuron. Structured pictures which represent states requires a new concept of associative memory model. A vector is structured by the division in parts. An association describes a correlation between different vector parts, which corresponds to different objects. Those parts are called cognitive entities. Each cognitive entity should contain a plain description of an object. An example is the representation of objects with their corresponding coordinates in a picture by different cognitive entities. From a state different states follow dependent on the executed associations. The associative memory which recognizes those different associations is called the permutation associative memory. The permeation associative memory represents the long term memory of the associative computer. A state is represented by Δ cognitive entities. Associations represent transitions between the states representing pictures. The premise of an association is represented by δ cognitive entities which describe a correlation of objects which should be present (see fig. 1). If present, they are replaced by δ cognitive entities of the conclusion.

Generally the premise is describe by fewer cognitive entities then the state, $\delta \leq \Delta$. In the recogni-

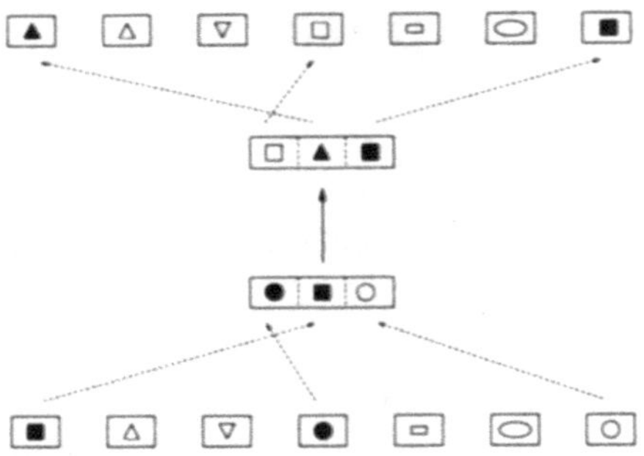

Fig. 1. A copy of the state representation is formed and the corresponding cognitive entities are replaced by the conclusion pattern.

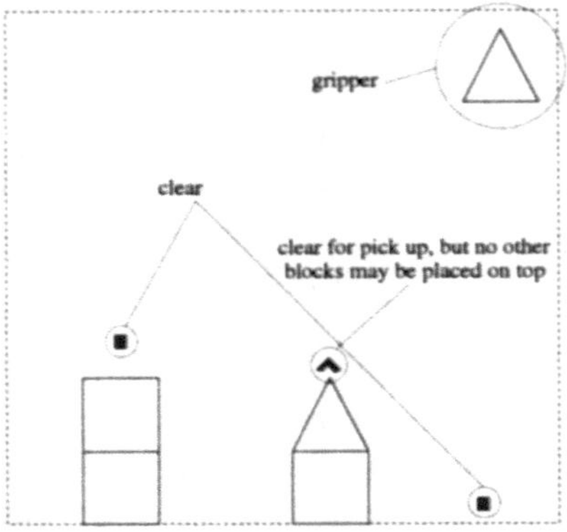

Fig. 2. A state in the blocks world.

tion phase, all posible δ-permutation of Δ cognitive entities should be composed to test if the premise of an association is valid.

$$\Xi := P(\Delta, \delta) = \frac{\Delta!}{(\Delta - \delta)!}$$

This is done because the premise can describe any correlation between the cognitive entities. In the retrieval phase after learning Ξ permutations are formed. Each permutation represents a question vector. To each question vectoran answer vector is determined. A copy of the state representation is formed and the corresponding cognitive entities are replaced by the conclusion pattern (see fig. 1).

3 Representation

Similarity between states represented as binary pictures can be measured. Beside the representation of a state by a binary vector a structural reprtesentation is needed. This is done with the aid of cognitive entities. Each cognitive entity represents the identity of the object and its position by the coordinates. The identity of an object is represented by a

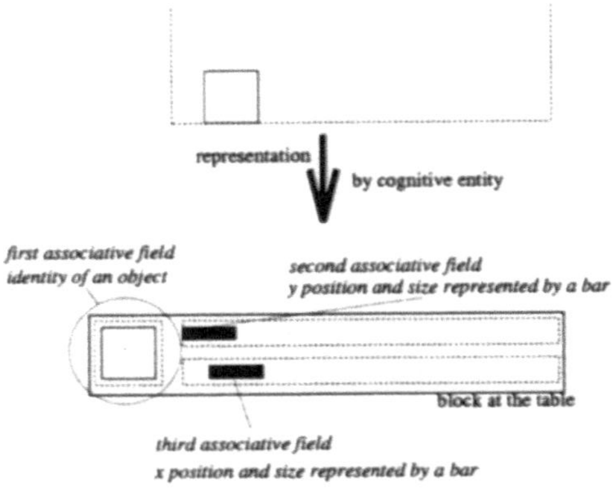

Fig. 3. A cognitive entity.

binary pattern which is normalized for size and orientation. Its location corresponding to the abscissa is represented by a binary vector of the dimension of the abscissa of the picture representing the state. The location corresponding to the ordinate is likewise represented by a binary vector of the dimension of the ordinate of the picture representing the state. A binary bar of the size and position of the object in the picture of the state represents in each of those vectors the location and size (see fig. 2, fig. 3).

3.1 Associations

Cognitive entities can represent associations which represent transitions between states. The first pattern represented by the cognitive entities describes the state which should be present before the transition (the premise). The second pattern describes the world state after the transition (the conclusion). In order to preserve the equality of cognitive entities in the premise and in the conclusion pattern, a notation for an empty cognitive entity is used (see fig 4). The associations are defined with respect to the frame problem. In fig. 4 an example from the block world is shown. Both, an empty roboter arm, which is represented by the right corner, or a "clear" position are represented by a dot.

The cognitive entities of the premise pattern are replaced by the conclusion pattern in case the similarity between the condition pattern and the corresponding part of the state picture is sufficient.

4 Problem Solving

4.1 Search

The problem space is represented by a search chain. After the temporary parallel execution of a chosen state, s new states emerge. From the s^t states, one state is chosen and the new s^{t+1} states

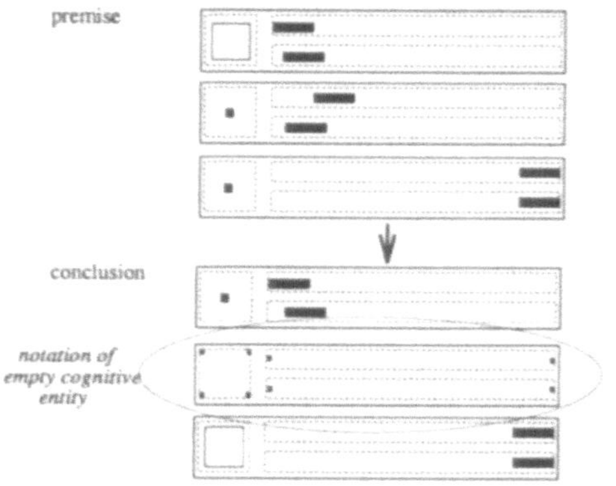

Fig. 4. Representation of the association: If a block is at a certain position and above it it is clear and the gripper is empty then the block is grasped by the gripper. Because of the frame problem, the old position of the block is represented as clear. One cognitive entity of the conclusion pattern is not used. In the inverse association, the premise pattern is interchanged with the conclusion pattern.

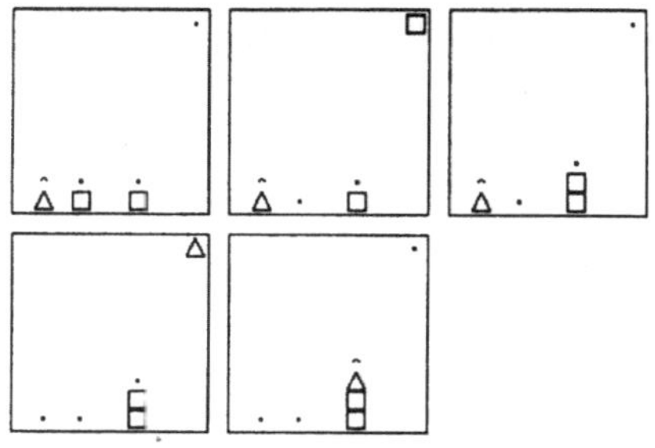

Fig. 5. Planning with pattern heuristic function, 4 steps are needed. Without paterrern heuristic function 24 steps were needed.

task	examples	blind	heuristic
blocks	30	14.4	27.29%

Table I. For a task for which a number of examples was presented; the blind search strategyneeded on average the number of steps shown in column; the pattern heuristic brought a significant improvement in %.

are determined. A state can cause an impasse when no valid transition to a succeeding state exists. In this case, backtracking to the previous state is performed. Another state can be chosen, if possible, or backtracking is repeated. The resulting search strategy is the "deep search" strategy. The resulting search strategy is either a blind search strategy or the hill climbing search strategy. If the biological essential local spreading activation, in which values are propagated by links in the search chain, is ignored, a best-first search which does not get stuck in local maxima can be easily implemented.

4.2 Pattern heuristics

From the s^t states, the most similar state to the desired state is chosen, and the new s^{t+1} states are determined. This is possible, because the similarity between states represented as binary pictures can be measured. This kind of heuristic is called the pattern heuristic.

The pattern heuristic speeds up the search. The real world give us insights on how to solve the problem. Such an insight is used by the pictorial representation and the resulting similarity criterion. This information is not always correct, but it is generally better to obey it than to ignore it.

5 Conclusion

The computational task concerning problem solving corresponds to the manipulation of pictures. A problem is described by the associations in the long term memory, by the initial state, and by the desired state. The solution to the problem is represented by a chain of the associations which successively change the state from the initial state to the desired state. The human ability to process images and understand what they mean in order to solve a problem holds an important clue to how the human thought process works and give hints about the structure of a new computer architectures. This clue was examined by empirical experiments with the associative computer. One general conclusion from the experiments is the claim that it is possible to use systematically associative structures to perform reasoning by forming chains of associations. In addition, beside symbolical problem solving, pictorial problem solving is possible.

References

[1] Anderson, J. R. (1995). *Cognitive Psyhology and its Implications.* W. H. Freeman and Company, fourth edition.

[2] Newell, A. (1990). *Unified Theories of Cognition.* Harvard University Press.

[3] Newell, A. and Simon, H. (1972). *Human Problem Solving.* Prentice-Hall.

[4] Steinbuch, K. (1961). Die Lernmatrix. *Kybernetik,* 1:36–45.

[5] Steinbuch, K. (1971). *Automat und Mensch.* Springer-Verlag, fourth edition.

An Experimental Assessment of the Performance of Several Associative Memory Models

S. P. Turvey, S.P. Hunt, N. Davey, R.J. Frank *

*Department of Computer Science, University of Hertfordshire, Hatfield, Herts, AL10 9AB. United Kingdom
e-mail: {s.p.turvey, s.p.hunt, n.davey, r.j.frank}@herts.ac.uk

Abstract
The performance characteristics of four different associative memory models are examined. The models differ in the training algorithm employed, although all four employ algorithms that are iterative, and use local information. They are classified using the method of Abbott [1], their attractor performance is examined, and the time taken to train them is measured.

1 Introduction

The dynamics and performance of the Hopfield associative memory model have been thoroughly investigated and are well understood. Several alternative training algorithms have been proposed, each of which leads to an increase in capacity over the original Hopfield model, and an improvement in attractor performance, usually at the expense of an increase in training time.

This paper compares the performance of a number of such high capacity models, with respect to training time, attractor performance and the stability of stored patterns. The work has two goals. First, to classify the models in question using the method developed by Abbott [1]. Second, to evaluate the models against a consistent set of criteria in order to ascertain which of them gives the best balance of performance.

2 Models Examined

2.1 Common properties

Each model employs a fully connected network of N bipolar (+1/-1) processing elements, as used in the Hopfield model. Each network is trained using a set, Π, of N-ary, bipolar pattern vectors, $\{\xi^p\}$. The N by N weight matrix which results will be denoted by $\mathbf{W}$, and the state (output) of the i'th unit by S_i.

The *local field* (input) of the i'th unit is h_i, where

$$h_i = \sum_{j \neq i} w_{ij} S_j \quad (1)$$

The *aligned local field* of the i'th unit for ξ^p is

$$h_i \xi_i^p \quad (2)$$

If all aligned local fields for a ξ^p are non-negative it is guaranteed to be stable.

The temporal evolution of unit states during recall is governed by:

$$S_i' = \begin{Bmatrix} 1 & \text{if } h_i > 0 \\ -1 & \text{if } h_i < 0 \\ S_i & \text{if } h_i = 0 \end{Bmatrix} \quad (3)$$

Unit states may be updated either synchronously or asynchronously. All models investigated here employ asynchronous, random-order updates, and updating continues until the network reaches a stable state. These dynamics, coupled with a symmetric weight matrix, guarantee simple point attractors [2].

Each ξ^p that is a stable state of the trained network is known as a *fundamental memory*. The *capacity* of a network, C, is the maximum number of fundamental memories it can hold. The *loading* of a network, α, is a measure of the size of the training set relative to the number of processing elements in the network, giving

$$\alpha = \frac{\#(\Pi)}{N} \quad \text{and} \quad \alpha_{\max} = \frac{C}{N} \quad (4)$$

2.2 The Iterative Local Learning rule (ILL)

This learning rule, devised by Diederich and Opper [3], attempts to push the values of all units' aligned local fields to be greater than or equal to the training threshold, T, for all ξ^p, as follows:

Beginning with a zero weight matrix
Repeat until all aligned local fields are correct
For each training pattern, ξ^p, in turn
Clamp the pattern onto the network
For each processing element in turn
If $h_i^p \xi_i^p < T$, change the weights on connections into unit i according to:

$$\Delta w_{ij} = \frac{\xi_i^p \xi_j^p}{N} \quad (i \neq j)$$

Note that the resulting **W** will have a zero diagonal, but is unlikely to be symmetric. A variant of this rule exists that enforces $\Delta w_{ij} = \Delta w_{ji}$ for each weight change, guaranteeing symmetry and, hence, simple point attractors. We have chosen not to examine it here because its attractor performance is similar to that of the ILL rule (see [4] for details).

2.3 The Iterative Local Learning with Equal Fields rule (ILL-Eq)

Diederich and Opper [3] proposed a modification to ILL in which weights are changed so that the aligned local fields of all units asymptotically approach 1 for every pattern. In the implementation employed here, training continues until the value of every aligned local field falls within the range 0.998 .. 1.002:

Beginning with a zero weight matrix
Repeat until all aligned local fields are correct
For each training pattern, ξ^p, in turn
Clamp the pattern onto the network
For each processing element in turn
Update incoming weights according to:

$$\Delta w_{ij} = \frac{\left(1 - h_i^p S_i^p\right) S_i^p S_j^p}{N} \quad (i \neq j)$$

Performance may be varied by changing the acceptable range of values for aligned local fields. The effect this has on attractor performance and training time is the subject of work to be published.

2.4 The Krauth - Mezard learning rule (KM)

Another modification to ILL, this rule was proposed by Krauth and Mezard [5]. It attempts to present each training pattern an optimal number of times. At each unit, the pattern with the smallest aligned local field is chosen for presentation. Weights are changed until all aligned local fields are greater than or equal to the training threshold, T:

Beginning with a zero weight matrix
Repeat until all aligned local fields are correct
For each unit, i, in turn
Select the pattern, ξ^p, with the smallest aligned local field for this unit
Update the incoming weights according to:

$$\Delta w_{ij} = \frac{\xi_i^p \xi_j^p}{N} \quad (i \neq j)$$

This rule has been shown produce optimal γ values, γ is described in section 3.1.

2.5 The Blatt - Vergini learning rule (BV)

Blatt and Vergini [6] propose a training algorithm that is guaranteed to find an appropriate weight matrix within a finite number of presentations of each pattern.

The minimum number of presentations to perform, P, is calculated as being the smallest integer conforming to:

$$P \geq \log_k\left(\frac{N}{(1-T)^2}\right) \tag{5}$$

where k and T are real valued constants such that $1 < k \leq 4$ and $0 \leq T < 1$, and N is the number of units in the network. k is known as the *memory coefficient* of the network, because the larger it is, the fewer steps are required to train the network. In this work, k=4 and T=0.5 for all networks trained by this rule.

Beginning with a zero weight matrix
For each pattern in turn
Clamp the pattern onto the network
For m := 1 to P
For each processing element in turn
Update incoming weights according to:

$$\Delta w_{ij} = \left(\frac{k^{m-1}}{N}\right)\left(\xi_i^p - h_i\right)\left(\xi_j^p - h_j\right)$$

Remove all self-connections

Note that patterns are added incrementally without interfering with patterns learnt previously.

2.6 Relationship to the pseudo-inverse rule

The algorithms employed in the ILL-Eq and BV models are both designed to generate weight matrices that are approximations of the weight matrix generated by the *pseudo-inverse* rule of Personnaz et al [7]. According to this rule $\mathbf{W}=\Xi\Xi^{I}$, where Ξ is the matrix whose columns are the ξ^p, and Ξ^{I} is its pseudo-inverse.

ILL-Eq and BV both employ iterative learning algorithms that use local information to generate a weight matrix, $\mathbf{W} \approx \Xi\Xi^{I}$, with its diagonal set to zero.

Whilst the BV rule guarantees a solution to the problem within a finite, and calculable, number of iterations through the training set, there is no upper bound on the number of iterations that may be required for the ILL-Eq to satisfy its stopping criterion. Blatt and Vergini also state there to be no restrictions on the training set with respect to correlation or linear dependency.

3 Experimental Procedure

All networks were of size N=100. Each model was tested by training networks with sets containing 50 random training patterns (α=0.5). The γ distribution analyses were performed on single networks. R is averaged over 50 networks, and training time is averaged over 100 networks in each case.

3.1 γ distribution analysis

Abbott [1] identifies 3 classes of associative memory, based upon the distribution of γ values for a trained network. The γ value for unit i for a pattern, ξ^p, is obtained by dividing the aligned local field by the magnitude of the incoming weight vector:

$$\gamma_i^p = \frac{h_i \xi_i^p}{|W_i|} \qquad (6)$$

For each network we obtain a set, Γ, containing the γ values for all the network's fundamental memories.

Abbott's classification system is based upon the distribution of γ values in Γ. The three classes are:

1) Networks with a Gaussian distribution of γ values
2) Networks with all γ values the same ($\forall \gamma_i \bullet \gamma_i = \gamma_0$)
3) Networks with a clipped Gaussian distribution of γ values, where $\forall \gamma_i \bullet \gamma_i \geq \gamma_{min}$

Class 1 includes networks trained using Hopfield's algorithm, so they are referred to as Hopfield-type networks. Interestingly, Abbott calculates that the upper bound for α_{max} in Class 1 is 1.4, which is much higher than α_{max} for the 'vanilla' Hopfield network. Class 2 is made up of networks trained using the *pseudo-inverse* rule, or derivatives thereof. Networks trained using this rule can hold N linearly independent patterns (giving $\alpha_{max} \approx 1$). Class 3 is known as the Gardner class, after the work of Gardner [9], whose training algorithm gives networks with α_{max}=2.

3.2 Performance

Kanter and Sompolinsky's [10] R value was used as a measure of attractor performance:

$$R = \left\langle \left\langle \left[\frac{1 - m_0}{1 - m_1} \right] \right\rangle \right\rangle \qquad (7)$$

A series of sample starting states are chosen, each of which is a corrupted fundamental memory, acting as the target final state for the network. m_0 is the proportion of each sample pattern which must be the same as its target state in order that all sample patterns will converge upon their targets. m_1 is the greatest overlap of each sample state with the fundamental memories other than its target. Details of the method used for calculating this value are presented in [11].

We also measured both the number of passes through the training set required to train a network and the time taken to complete training, so as to allow for the differing complexity of the algorithms used. From this we calculated the mean time for a single iteration.

Measurements were conducted on a PC with a 600MHz AMD Athlon CPU and 128MB of RAM, running Windows 98SE. All simulations were written and run in Java, using the Sun JDK 1.3.

4 Results

4.1 γ distribution analysis

The distributions of measured γ_i values confirm the classes to which the models belong. The pseudo-inverse approximators have very similar distributions:

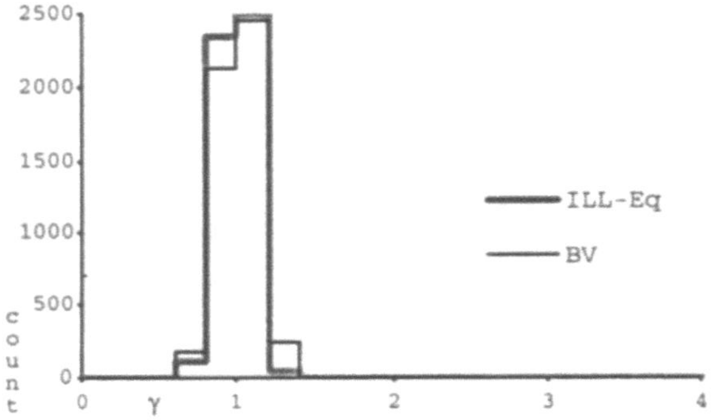

Fig. 1 Distribution of γ_i values in ILL-Eq and BV networks.

Figure 1 shows very tight distributions for both network types . This is not surprising: both models are designed to find weight matrices that *approximate* $\Xi\Xi^I$. Thus, we place them in Class 2. The distribution of the γ_i values for the ILL and KM models shown in figure 2 confirms that both models fall into Class 3:

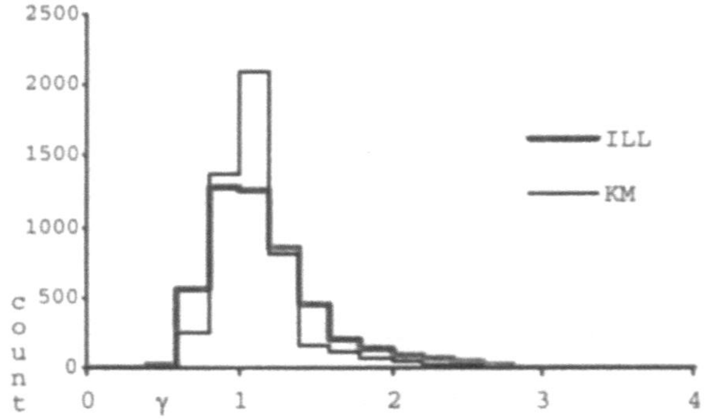

Fig. 2. Distribution of γ_i values for ILL and KM networks.

4.2 Performance

Table I shows *R* values for the different models, (including ILL and KM at several different training thresholds). In terms of attractor performance, the model with the highest *R* and therefore 'best' attractor performance would appear to be KM with *T*=100.

Table I R Values for a range of models and parameters

Model	*R*
ILL (T=1)	0.196
(T=10)	0.246
(T=100)	0.262
KM (T=1)	0.253
(T=10)	0.254
(T=100)	0.270
ILL-Eq	0.215
BV	0.214

Table II shows the mean number of times a training set needs to be presented to complete training, the mean total training time for each model, and the mean time taken for a single pass through the training set.

Table II Training times for a range of models/parameters

Model	Ave. no. of epochs	Ave. time to train (secs)	Approx. time per epoch (secs)
ILL (T=1)	16.5	1.6	0.1
(T=10)	95.6	9. 8	0.1
(T=100)	895.4	91.4	0.1
KM (T=1)	4.6	18.7	4.1
(T=10)	33.4	144.7	4.3
(T=100)	320.5	1416.5	4.4
ILL-Eq	52.9	18.6	0.4
BV	4.0	2.2	0.6

It should be noted that the KM model does not run through the entire training set in the same way as the other models, so the KM figures represent the time taken to present 50 patterns.

5 Discussion

A number of interesting observations may be made. Firstly, for the models where a training threshold is used it is clear that *R* increases with *T* indicating that this is one means of improving performance, though at the expense of training time. Secondly, the KM rule performs better than the ILL rule, though only marginally so at the higher thresholds. Thirdly, both KM and ILL generally perform better than the pseudo-inverse rules with respect to *R*, and have a higher maximum capacity (2*N* vs *N*).

The KM (*T*=100) rule seems best if the sole gauge of performance is to be *R*. However, the time taken to train this network is around 24 minutes (Table II). Whilst this is not a prohibitively long time, it has to be acknowledged that the increase in *R* that results when *T* is changed from 1 to 100 is relatively small, and the trade-off is probably not worthwhile.

The ability of the BV rule to learn new fundamental memories without re-training with the whole training set makes it a better choice for on-line applications. A further investigation of the behaviour of these models is warranted, taking into account such issues as the nature and number of spurious attractor states, in order to determine just how important these relatively small differences in *R* values are.

References

[1] L.F.Abbott, "Learning in Neural Network Memories" *Network: Comp. Neural Sys.* vol.1, pp.105-122, 1990

[2] D.J.Amit, *Modelling brain function: the world of attractor neural networks*. Cambridge, UK: Cambridge University Press, 1989

[3] S.Diederich and M.Opper, "Learning of Correlated Patterns in Spin-Glass Networks by Local Learning Rules" *Physical Review Letters* vol.58, pp.949-952, 1987

[4] N.Davey, R.G.Adams, S.P.Hunt, "High performance associative memory models and symmetric connections", *Proceedings of ISA 2000*, to be published December 2000

[5] W.Krauth and M.Mezard, "Learning algorithms with optimal stability for neural networks" *J. Phys.* vol.A20, pp.L745-L752, 1987

[6] M.G.Blatt and E. G. Vergini (1991). "Neural Networks: A Local Learning Prescription for Arbitrary Correlated Patterns", *Physical Review Letters*, vol.66, pp.1793-1797

[7] L.Personnaz, I. Guyon, and G. Dreyfus, "Collective Computational Properties of Neural Networks: New Learning Mechanisms" *Physical Review A* vol.34, pp.4217-4228, 1986

[8] E.Gardner, "The space of interactions in neural network models" *Journal of Physics* vol.A21, pp.257-270, 1988

[10] I. Kanter and H. Sompolinsky, "Associative Recall of Memory without Errors", *Physical Review A* vol.35, pp.380-392, 1987

[11] N.Davey and S.P.Hunt, "The Capacity and Attractor Basins of Associative Memory Models", *Proceedings of IWANN99*, 1999

Recall Time in Densely Encoded Hopfield Network: Results from KFS Theory and Computer Simulation

A.A. Frolov*, D. Husek†, P. Combe‡, V. Snášel§ [1]

*Institute of Higher Nervous Activity and Neurophysiology Russian Academy of Sciences, Butlerova 5, Moscow, Russia, E-mail: aafrolov@mail.ru †Institute of Computer Science, Academy of Science of the Czech Republic, Pod Vodárenskou věží 2, Prague 8, Czech Republic, E-mail: dusan@cs.cas.cz ‡University de Provance CNRS PR 7061 France CPT, CNRS-Luminy case 907, Marseille cedex 09, France, E-mail: combe@cpt.univ-mrs.fr §Silesian University Ostrava, Ostrava, Czech Republic, E-mail: vsnasel@si.univ.cz

Abstract

Recall time in Hopfield attractor neural network with parallel dynamics is investigated analytically and by computer simulation. The method of the recall time estimation is based on calculation of overlaps between successive patterns of network dynamics. Recall time is estimated as the time when the overlap reaches the value $1 - \Delta m$ where Δm is the minimal increment of overlap for the network of a given size. It is shown, first, that this time actually gives rather accurate estimation for the recall time and, second, that the overlap between successive patterns of network dynamics can be rather accurately estimated by the theory recently developed by [10]. It is shown that recall process has three very different phases: the search of the recalled prototype by large steps with low convergence rate, fast convergence to the attractor in the vicinity of the recalled prototype and again slow convergence to the attractor when it is almost reached. If recall process ends at two first phases then point attractors dominate. If it ends at the third phase then cyclic attractors of the length 2 dominate. Transition to the third phase can be revealed by computer simulation of networks of extremely large size (up to the number of neurons in the order 10^5). Special algorithm is used to avoid storing in the computer memory both connection matrix and the set of stored prototypes.

1 Introduction

Hopfield-like neural network is a fully connected network which acts as an autoassociative memory on the base of correlational Hebbian learning rule. Many analytical methods have been developed to reveal its informational capacity [2] and the size of attraction basins [1], [11]. However, there is no method which allows to estimate its convergence time. Theoretical estimations have shown that generally the search of the global maximum of the Lyapunov function of the network dynamics is NP-problem [6]. However, these estimations are very far from the results obtained by computer simulation [9] which revealed very week dependence of the convergence time on the network size. The main goal of the present paper is to develop the method for estimation of the convergence time in Hopfield network and to prove its validity by computer simulation. The method is based on the new approach to the analysis of parallel dynamics in Hopfield network of infinite size elaborated by Koyama, Fujie and Seyama (KFS) [10]. This method is based in turn on the method developed in [4] the exact general theory of such dynamics was suggested but computable formulae only for two first steps of the recall process were obtained. Koyama et al. [10], suggested some reasonable approximations in this theory and derived computable form of neurodynamic equations for the whole recall process.

In the contrary to all other analytical approaches [11], these two methods operate not only with overlaps between current patterns of network dynamics and the recalled prototype but also with overlaps between successive patterns. Since for the network of finite size there exists a minimal overlap increment Δm, one can expect that the time when the overlap between successive patterns of network dynamics reaches the value $1 - \Delta m$, provides sufficiently accurate estimation of the convergence time. This expectation is reasonable because at the next time step the overlap between successive patterns must reach 1, i.e. the recall process must converge. To substantiate the validity of our approach we, first, checked whether the time when overlap between successive patterns of network dynamics

[1]This work was supported by grants from Grant Agency of the Czech Republic, No.201/01/1192 and 201/00/1031 and by grant BARRANDE No.99010-2

reaches the value $1 - \Delta m$, actually provides accurate estimation of the convergence time, and second, checked the accuracy of KFS in prediction of this overlap.

Due to the study [9], it is known that finite size effects which are the most important for estimation of the convergence time in Hopfield network can be revealed only by computer simulations of the networks of extremely large size (number of neurons N is in the order of 10^5). To perform the simulation of such large network special algorithm which allows to avoid storing connection matrix and the set of the stored patterns in the computer memory was developed. In our computer simulations performed with the use of PC Pentium, the size of the network reached $1.5 \cdot 10^5$ neurons.

We used the special case of Hopfield network described in [3]. This network has three peculiarities comparing to the original Hopfield network [7]. First, activity of neurons is given by values 0 or 1 instead of -1 and 1. Second, the number of active neurons in each stored prototype is fixed and equal to $N/2$ instead of their random number in the original network. Third, the threshold of the neuron activation is variable and is chosen at each time step in a way that exactly $N/2$ winners are active instead of fixed zero threshold and random number of active neurons in original network. These three peculiarities altogether make the model completely identical by informational and dynamical properties to the original Hopfield model [3]. We prefer to use this model only because of technical reason: the algorithm of computer simulation for this model happened to be faster than for the original one.

As the measure of the relative informational loading we use parameter $\alpha = L/N$ where L is the number of stored prototypes. The closeness between two patterns of the network activity $\mathbf{X}^l$ and $\mathbf{X}(t)$ is measured by their overlap which is defined as

$$m(\mathbf{X}^l, \mathbf{X}(t)) = \sum_{i=1,N} (X_i^l - 0.5)X_i(t)/(N/4)). \quad (1)$$

As pointed above the main idea of our method is to estimate the convergence time on the base of overlaps between adjacent and each second patterns of the network dynamics $m_1(t) = m(\mathbf{X}(t), \mathbf{X}(t-1))$ and $m_2(t) = m(\mathbf{X}(t), \mathbf{X}(t-2))$ calculated by KFS for the case $N \to \infty$. For the network of finite size the recall process stops when $m_1(t)$ or $m_2(t)$ reach 1. Overlap $m_1(t)$ becomes 1 for point attractors and $m_2(t)$ becomes 1 for cyclic attractors of the length 2. The convergence time S could be estimated as the time step just before m_1 or m_2 reach 1. Since for the used model the minimal increment of overlap amounts to $\Delta m = 4/N$, then S is suggested to estimate from the equation

$$m_i(S) = 1 - 4/N \quad (2)$$

where $i = 1$ for point and $i = 2$ for cyclic attractors. If for given N $m_1(t)$ reaches the value $1-4/N$ before $m_2(t)$ it is reasonable to suggest that for this N point attractors dominate. And opposite if $m_2(t)$ reaches this value before $m_1(t)$ then it is reasonable to suggest that for this N cyclic attractors dominate.

2 Results

Fig. 1 demonstrates the calculated values of $m_1(t)$ and $m_2(t)$ for $\alpha = 0.1$ and initial overlap $m_{in} = 0.4$ and $m_{in} = 0.5$. The results of computer simulation are compared with those obtained by KFS. It is shown in Fig. 1 that recall process has three very different phases. At the first phase overlaps are relatively small and change slowly. This means that at each time step during this phase the network activity changes strongly but mainly not in the direction of the recalled prototype. At the second phase, after the network activity reaches the vicinity of the recalled prototype, it tends to the attractor very quickly and the distance between successive patterns of the network activity decreases exponentially. Particularly the overlap m_1 can be approximated in this phase by formula

$$m_1(t)) \simeq 1 - \exp(-A(t - t_0)) \quad (3)$$

The corresponding tangents to curves $\ln(1 - m_1(t))$ are shown in Fig. 1. At the second phase all four presented families of curves vary in parallel. It means that the network dynamics in the vicinity of the recalled prototype is completely determined by informational loading α and only the time t_0 when this vicinity is reached depends on the initial overlap.

In two first phases m_2 is smaller than m_1, i.e. the pattern of the network activity changes during two recall steps more than during one step. This situation paradoxically inverses at the third phase when m_1 becomes smaller than m_2. In this phase m_1 is almost constant and m_2 continues to increase, i.e. the network dynamics tends to the cycle of the length 2. However, the rate of the recall convergence in the third phase is much less than in the second one.

Since in phases 1 and 2 $m_1 > m_2$ then m_1 reaches critical value $1 - 4/N$ earlier than m_2 and thus one can expect that for relatively small N (when this critical value is reached in phases 1 and 2) point

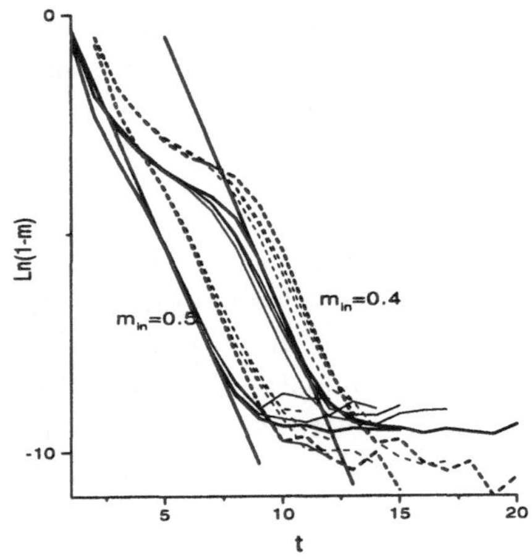

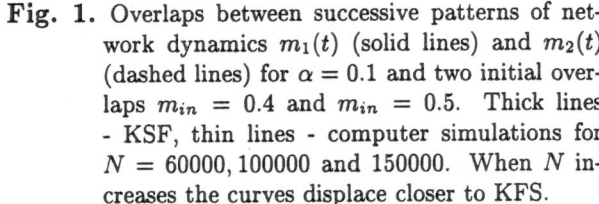

Fig. 1. Overlaps between successive patterns of network dynamics $m_1(t)$ (solid lines) and $m_2(t)$ (dashed lines) for $\alpha = 0.1$ and two initial overlaps $m_{in} = 0.4$ and $m_{in} = 0.5$. Thick lines - KSF, thin lines - computer simulations for $N = 60000, 100000$ and 150000. When N increases the curves displace closer to KFS.

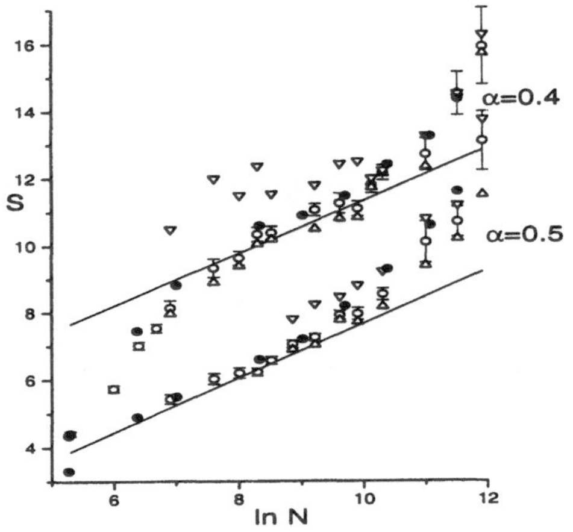

Fig. 2. Dependence of the number of recall steps S on N for the same values of α and m_{in} as in Fig. 1. $\triangle$ - point attractors, ∇ - cyclic attractors of the length 2, $\circ$ - averaged (with errors bars) over point and cyclic attractors, $\bullet$ - the data taken from [9]. Straight lines correspond to straight lines in Fig. 1.

attractors dominate and the dependence of the convergence time S on N is determined by the dependence of m_1 on t in two first phases. When S is determined by this dependence in the phase 2 then according to (2) and (3)

$$S \simeq t_0 + \gamma \ln(N/4) \tag{4}$$

where $\gamma = 1/A$.

The recall time S for point attractors, cyclic attractors and averaged over both types of attractors is shown in Fig. 2. The values of averaged S are compared with those presented in [9] for original Hopfield network. Both sets of data completely coincide what demonstrates once more the identity of the models. It is shown in Fig. 2 that formula (4) actually gives rather accurate estimation of the recall time. The difference between this estimation and experimental results are large for small N when the recall time is determined by the first phase of the recall process and for large N when it is determined by the third phase. For intermediate N the deviation of experimental data from formula (4) slightly increases when N increases. It is explained by the increase of the portion of cyclic attractors. As shown in Fig. 2 and predicted earlier, the recall time for attractors of the length 2 is greater than for point ones.

As shown in Fig. 3 the portion R of cyclic attractors increases when N increases. It does not depend on the initial overlap: the points obtained for different m_{in} are randomly mixed. In Fig. 3 the values of R are transformed into the values $F = -\ln(1/R-1)$. Then all the data can be rather accurately approximated by the unique simple relation

$$F = 0.6/\alpha + 0.5 \ln N.$$

Straight lines corresponding to this relation are also shown in Fig. 3. The transition from point to cyclic attractors occurs at $\ln N_{tr} \simeq 1.2/\alpha$, that is the smaller α the later this transition occurs. Thus for smaller α the estimation of S by (4) is valid for larger range of N. The obtained result for N_{tr} is compared in Fig. 4 with that predicted by KFS. For each α the transition from point to cyclic attractors was estimated by the point of intersection of $m_1(t)$ and $m_2(t)$ for $m_{in} = 0.5$. It is shown in Fig. 4 that KFS gives rather accurate estimation of N_{tr}.

Fig. 4 demonstrates also the dependence of γ on α obtained both by KFS and experimentally. For KFS the data were obtained for $m_{in} = 0.5$. Experimental data were obtained for $m_{in} = 0.5$ and $N = 1.5 \cdot 10^5$. Fig. 4 shows that KFS rather accurately estimates this parameter of neurodynamics

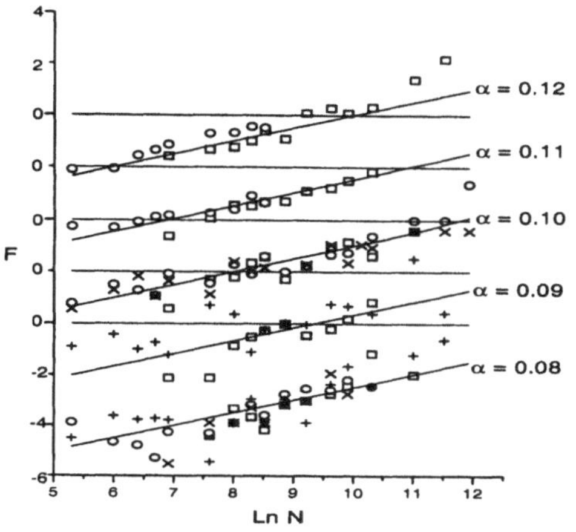

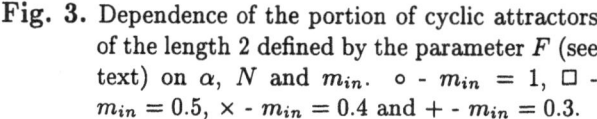

Fig. 3. Dependence of the portion of cyclic attractors of the length 2 defined by the parameter F (see text) on α, N and m_{in}. $\circ$ - $m_{in} = 1$, $\square$ - $m_{in} = 0.5$, $\times$ - $m_{in} = 0.4$ and $+$ - $m_{in} = 0.3$.

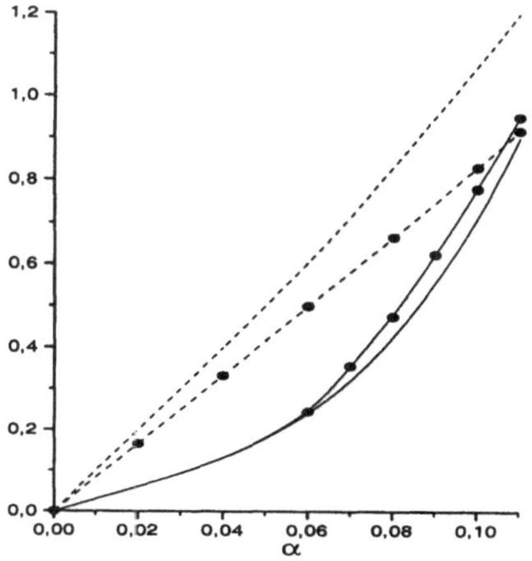

Fig. 4. Coefficient γ in the formula $S \simeq t_0 + \gamma \ln N$ (solid lines) and network size N_{tr} of transition from point to cyclic attractors given in the form $10/\ln N_{tr}$ (dashed lines) in dependence on α. Unmarked curves - KFS, curves with $\bullet$ - computer simulation.

almost for the whole range of informational loading before saturation.

References

[1] Amari, S., & Maginu, K. (1988). Statistical neurodynamics of associative memory. *Neural Networks*, **1**, 63-73.

[2] Amit, D. J., Gutfreund, H., & Sompolinsky, H. (1987). Statistical mechanics of neural networks near saturation. *Annal of Physics*, **173**, 30-67.

[3] Frolov, A. A., Husek, D., & Muraviev, I. P. (1997). Informational capacity and recall quality in sparsely encoded Hopfield-like neural network: Analytical approaches and computer simulation. *Neural Networks*, **10**, 845-855.

[4] Gardner, E., Derrida, B., & Mottishaw, P. (1987). Zero temperature parallel dynamics for infinite range spin glasses and neural networks. *J. Physique*, **48**, 741-755.

[5] Goles-Chacc, E., Fogelman-Soulie, F., & Pellegrin, D. (1985). Decreasing energy functions as a tool for studying threshold networks. *Discrete mathematics*, **12**, 261-277.

[6] Godbeer, G. H. (1988). The computantional complexity of the stable configuration problem for connectionist models. In *On the Computational Complexity of Finding Stable State Vector in Connectionist Models (Hopfield Nets)*. Technical Report 208/88. University of Toronto, Department of Computer Science.

[7] Hopfield, J.J. (1982). Neural network and physical systems with emergent collective computational ability. *Proceeding of the National Academy of Science USA*, **79**,2544-2548.

[8] Kohring, G.A. (1990a). A high-precision study of the Hopfield model in the phase of broken replica symmetry. *Journal of Statistical Physics*, **59**, 1077-1086.

[9] Kohring, G.A. (1990b). Convergence time and finite size effects in neural networks. *Journal of Physics A: Math. Gen.*, **23**, 2237-2241.

[10] Koyama, H., Fujie, N., & Seyama, H. (1999). Results from the Gardner-Derrida-Mottishaw theory of associative memory. *Neural Networks*, **12**, 247-257.

[11] Okada, M. (1996). Notions of associative memory and sparse coding.*Neural Networks*, **9**, 1429-1458.

On-Line Identification and Rule Extraction of Finite State Automata with Recurrent Neural Networks

Ivan Gabrijel, Andrej Dobnikar*

*University of Ljubljana, Faculty of Computer and Information Science, Tržaška 25, SI-1001 Ljubljana, Slovenia
E-mail: *Ivan.Gabrijel@fri.uni-lj.si*

Abstract

The on-line identification of an unknown finite state automaton with a generalized recurrent neural network and an on-line learning scheme, together with an on-line rule extraction algorithm is presented. Several tests were made on different, strongly connected automata with structures ranging between 2 and 32 states and the results of both training and extraction processes are very promising.

1 Introduction

Identification of an unknown plant and, in general, its sequential function is an important step in a system control process. There already exist several ideas and applications based on neural networks [4], [5]. In this paper we describe a new on-line approach to the identification of a finite state automaton with a generalized recurrent neural network and its special learning algorithm. An algorithm is further proposed for extracting a structure of the observed finite state automaton from the recurrent neural network learnt.

The identification task is to build a model of an unknown plant by observing its input and output sequences. In this research work an unknown system is assumed to be a finite state automaton, that we treat as a causal, time-invariant, observable, controllable, stable, discrete dynamic system [4], [9].

The paper is set out as follows. Sections 2 and 3 outline the generalized architecture of recurrent neural networks and its learning scheme. The on-line rule extraction algorithm is proposed in Section 4. Section 5 describes the experimental work and gives the results with comments. The conclusion summarizes the paper.

2 The Generalized Architecture of Recurrent Neural Networks (GARNN)

We propose the architecture of recurrent neural networks, shown in Fig. 1, whose learning is based on a gradient descent. It consists of three layers of neurons: the input layer, one hidden layer and the output layer. The number of neurons in the output layer is equal to $m+n$, where m is the number of outputs from the neural network and n is the number of neurons in the hidden layer.

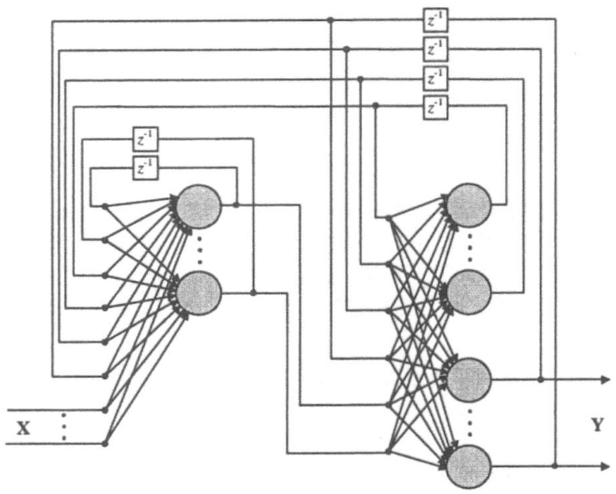

Fig. 1. The generalized architecture of the recurrent neural networks.

There are some reasons for establishing the generalized architecture. It is well known [9] that at least two feedforward layers of neurons can approximate an arbitrary function. As a consequence, the function of the next state from the state-space model of the recurrent computation in GARNN is computed over the two layers, thus ensuring its descriptive universality. There are no special output layers, separated from the state layers, in order that the teacher forcing technique, used in the learning scheme, may be applied. This technique helps learning considerably.

3 Learning Procedure for the GARNN

The proposed learning algorithm for the GARNN is based on the gradient descent, where the gradients are computed using a generalized version of a real time recurrent learning (RTRL) algorithm [2]. All the network weights are updated with the decoupled extended Kalman filtering process (DEKF) [9], together with the teacher forcing technique [2].

We are familiar with other learning approaches for recurrent neural networks which are not based on the gradient descent (like Alopex [3], learning automata, genetic algorithms, etc.). Although such learning methods are much easier to implement and easier to adapt to different neural network architectures, we found them far less effective in our experiments. However, it is important to note that there are problems to which gradient-based methods cannot be applied.

Real time recurrent learning (RTRL) was originally designed by Williams and Zipser [2] for learning fully connected recurrent neural networks. We generalized it to calculate the gradients of an output error for all weights in the proposed architecture (GARNN).

Extended Kalman filtering is adopted for updating the weights during the learning procedure, as it is well suited for on-line training. It tracks the whole history of the observed input-output mapping, which is essential in the temporal computation. The decoupled version (DEKF) is used because it performs calculations on the lower dimensional matrices.

The use of the teacher forcing technique is also important for a successful on-line learning. The proposed generalized architecture is designed in such a way that the teacher forced signals are fed back to each neuron at each step, which makes the learning much easier. Alpha-projected teacher forcing is used, where instead of using the desired output vector **d** as pure teacher forcing signals, the projection

$$\alpha \mathbf{d} + (1\text{-}\alpha)\mathbf{y} \tag{1}$$

is applied, where **y** is the actual output vector and the value of projection parameter α is linearly adapted from value 1 at the beginning of the learning to the value 0 at the end of the learning.

4 Extraction of Finite State Automaton from the Recurrent Neural Network

The on-line extraction process is a natural addition to the on-line learning of the deterministic finite state automaton behavior. Normally, it starts after the learning is finished. However, because of its on-line nature, it can be used along with the learning process, after the learning becomes stable and its performance is satisfactory. Extraction consists of three stages: on-line clustering, on-line extraction and state minimization.

On-Line Clustering: In the first stage, we applied the attraction force on-line clustering algorithm [6], [7]. It originates from the attraction force between the two mass points in the Newtonian planetary system. For the two point objects with the masses m_a and m_b separated by the distance d, the attraction force is given by

$$F = \mathrm{G}\frac{m_a m_b}{d^2}\,, \tag{2}$$

where G is the gravitational constant. The procedure dynamically determines the clusters of the hidden state activities of the recurrent neural network. The state vector at time t in the proposed architecture consists of the output values of all the neurons.

On-line clustering proceeds as follows. The first state vector is taken as the center of the first cluster. The mass of this cluster is set to 1. For every next state vector $\mathbf{q}_p$ the attraction force (2) to all existing clusters is calculated. The cluster with the maximal value is selected. If it is higher than some threshold value F^*, we update its position and mass by the following equations:

$$\mathbf{c}_i = \frac{m_i \mathbf{c}_i + \mathbf{q}_p}{m_i + 1} \tag{3}$$

$$m_i = m_i + 1 \tag{4}$$

Otherwise, we form a new cluster and set its mass to the value 1. The clustering is supposed to be stable if there are no new clusters generated for a certain amount of time, or if enough patterns have been processed.

On-Line Extraction: This starts with an empty automaton. A new state of the automaton is formed, labeled with a cluster index of the current state vector. The output value of the first state of the automaton is set to the current desired vector. For every new input the recurrent neural network is processed and a new state vector is obtained. The clustering step is followed, yielding the index of the updated group. The automaton is updated as follows:

1. A new state is added, labeled with the current cluster index, if no state exists with such a label. The output value of the new state of the automaton is then set to the current desired vector.
2. A new transition is made, designated with the current input value, from the current state of the automaton to the state, labeled with the current cluster index, if no transition yet exists with such a designation.
3. The automaton is processed one step further.

This procedure is repeated until the end of the extraction interval or until we find that the extracting automaton is strongly connected. The resulting automaton usually has more states than required.

State Minimization: After the on-line extraction phase is completed, a well established state minimization of the extracted automaton is performed.

5 Experimental Work

5.1 Problem Domains

The first problem domain is derived from the well known Tomita automata [1]. Most of the seven original automata are modified so that they become strongly connected. They are shown in Fig. 2.

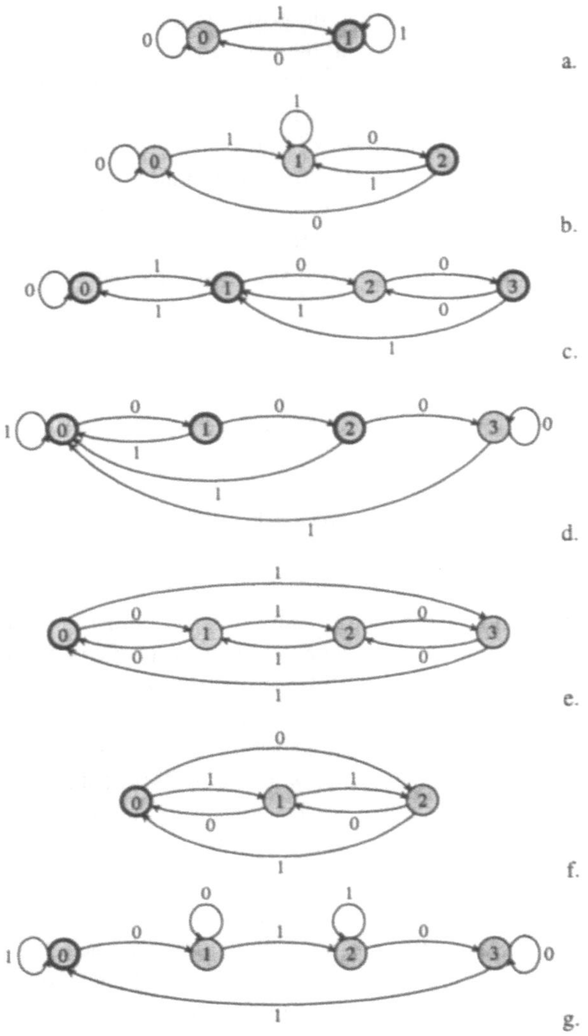

Fig. 2. The seven on-line Tomita automata: a. 1', b. 2', c. 3', d. 4', e. 5, f. 6, g. 7'.

The second problem domain covers the temporal XOR problems with different values of delay d [3]. The desired value at time t is the XOR function of the inputs at times t - d and t - d - 1. The values 0, 1, 2 and 3 are used for the delay d.

The goal is to identify the underlying automata (Fig. 2, 3) in an on-line manner by applying the randomly generated inputs and calculated corresponding desired outputs. The on-line requirement of the solution means that it is not allowed to reset the observed system, nor to explicitly buffer the input or desired output sequences or any part of them in order to cycle through the collected data.

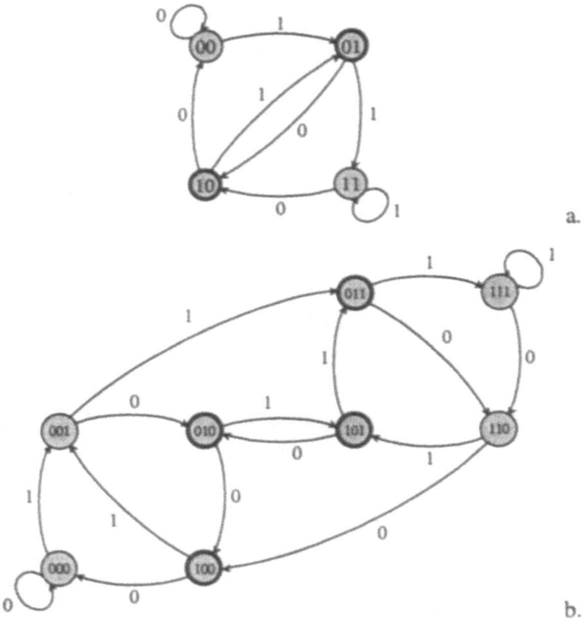

Fig. 3. The two temporal XOR problems automata: a. $d = 0$, b. $d = 1$.

5.2 Description of Experiments

A randomly generated sequence of inputs with a corresponding output sequence is used for each experiment. The length of the sequences is 31000. The first 20000 patterns are used for learning, followed by the next 2000 patterns which are intended to test the learnt neural network. The last 9000 patterns are used for the extraction process and are further segmented as follows. The first 70% are intended for clustering, followed by 25% of extraction patterns. The last 5% of patterns are used to test the extracted automaton.

All neural networks used in the experiments have one input and two outputs. Input values are coded as 0.1 for value 0 and 0.9 for value 1. The desired output values are coded as vector $[0.1, 0.9]^T$ for value 0 and vector $[0.9, 0.1]^T$ for value 1. The mean squared error and the winner-takes-all strategy is used for evaluating the success of training.

The initial weights of the neural networks are set to the random values, uniformly distributed on the real interval (-0.5, 0.5).

There are 10 runs for each experiment for each particular set of parameters with different initial settings of the weights. Several experiments are made for each problem, starting with the smallest GARNN network and followed by progressively larger structures until the networks have successfully learnt in all of the ten runs and 100% hit ratio is obtained on the test set.

5.3 Results and Comments

The results of the first successful experiment for each problem of the two problem domains are given in Table I. The numbers of neurons in the state layers of the successful neural networks are shown in the second column. The average, minimum, maximum and variance of mean square errors for ten runs of each experiment are given in the next columns of the table.

Table I. The results of on-line identification.

Problem	NN Size	MSE Ave	MSE Min	MSE Max	MSE StDev
Tomita 1'	1-3	1.60E-6	2.85E-8	5.07E-6	1.65E-6
Tomita 2'	2-4	5.55E-4	1.07E-6	2.31E-3	8.71E-4
Tomita 3'	3-5	2.32E-4	1.58E-5	9.89E-4	2.99E-4
Tomita 4'	1-3	1.46E-3	9.39E-4	2.51E-3	5.41E-4
Tomita 5	9-11	5.24E-6	2.16E-6	1.02E-5	2.37E-6
Tomita 6	7-9	2.60E-5	3.39E-7	9.77E-5	3.33E-5
Tomita 7'	4-6	1.66E-4	3.88E-6	1.13E-3	3.30E-4
T-XOR 0	5-7	5.70E-6	3.52E-7	3.19E-5	9.31E-6
T-XOR 1	7-9	1.01E-5	2.50E-6	2.58E-5	7.81E-6
T-XOR 2	11-13	4.29E-5	1.83E-6	2.55E-4	7.17E-5
T-XOR 3	16-18	4.58E-5	1.85E-5	8.37E-5	1.97E-5

Table II. The results of the on-line rule extraction.

Problem	F^*/G Lower	F^*/G Upper	N States Min
Tomita 1'	3	3E15	2
Tomita 2'	13	1E7	3
Tomita 3'	9	25000	4
Tomita 4'	1135	1E6	4
Tomita 5	3	85000	4
Tomita 6	4	25000	3
Tomita 7'	25	10000	4
T-XOR 0	3	9000	4
T-XOR 1	6	20000	8
T-XOR 2	7	250	16
T-XOR 3	27	400	32

The results of the extraction process are shown in Table II. Approximate lower and upper boundaries of an interval of good values for selection of the threshold force parameter of the clustering process are given in the second and third columns. The numbers of states after the minimization are shown in the last column.

The lower boundary of the architecture of size 2-4 for the Tomita 4' problem is approximately 46, and for the network of size 3-5 is 12, which is practically indistinguishable from the lower boundaries of the other problems.

6 Conclusion

We have shown how an unknown, strongly connected finite state automaton can be identified on-line by using the generalized recurrent neural network with the proposed learning scheme. It was shown also how the structure of the observed automaton can be extracted on-line from the learnt neural network. The results of our experimental work clearly show the real value of the approach.

References

[1] M. Tomita. Dynamic Construction of Finite Automata from Examples Using Hill-Climbing. *Proceedings of the 4th Annual Cognitive Science Conference*, pp. 105-108, Ann Arbor, MI 1982.

[2] R. J. Williams, D. Zipser. A Learning Algorithm for Continually Running Fully Recurrent Neural Networks. *Neural Computation*, Vol. 1, pp. 270-280, 1989.

[3] K. P. Unnikrishnan, K. P. Venugopal. Alopex: A Correlation-Based Algorithm for Feedforward and Recurrent Neural Networks. *Neural Computation*, Vol. 6, pp. 469-490, 1994.

[4] A. U. Levin, K. S. Narendra. Identification of Nonlinear Dynamical Systems Using Neural Networks. *Neural Systems for Control*, Edited by O. Omidvar, D. L. Elliott. Academic Press, pp. 129-160, San Diego, 1997.

[5] H. T. Su, T. Samad. Neuro-Control Design: Optimization Aspects. *Neural Systems for Control*, Edited by O. Omidvar, D. L. Elliott. Academic Press, pp. 259-288, San Diego, 1997.

[6] I. Gabrijel, A. Dobnikar. Adaptive RBF Neural Network. *Proceedings of the SOCO'97 Conference*, Nimes, France, pp. 164-170, September 17-19, 1997.

[7] I. Gabrijel, A. Dobnikar, N. Steele. RBF Neural Networks and Fuzzy Logic Based Control - a Case Study. *Proceedings of the 2nd IMACS International Multiconference CESA'98*, Hammamet, Tunisia, vol. 2, pp. 284-289, April 1-4, 1998.

[8] S. Das, M. Mozer. Dynamic On-Line Clustering and State Extraction: An Approach to Symbolic Learning. *Neural Networks*, Vol. 11, No. 1, pp. 53-64, 1998.

[9] S. Haykin. *Neural Networks - A Comprehensive Foundation - Second Edition*. Prentice-Hall, Inc., New Jersey, 1999.

Using Decision Surface Mapping in the Automatic Recognition of Images

Colin Reeves, Gurdeep Singh Billan *

*School of Mathematical and Information Sciences, Coventry University, UK

Abstract

The decision surface mapping (DSM) version of Kohonen's LVQ technique seems to have been neglected in comparison to other neural network paradigms. In this paper, we explore its use for the swift learning and recognition of objects flowing along a conveyor line in a small-scale manufacturing plant. The experimental work reported suggests that DSM has considerable potential in image recognition problems.

1 Introduction

The project concerned the use of artificial neural networks (ANNs) for learning and recognizing objects flowing along a conveyor line in a small-scale manufacturing plant. The objective was to recognise different objects (some of which may be very similar to each other) such that the objects could be sorted into the correct boxes.

Requirements in terms of training and recognition times dictated that a fast supervised learning algorithm was needed. The one chosen for development was the decision surface mapping (DSM) variant of Kohonen's learning vector quantization (LVQ) approach. This paper is a case study that reports on a successful implementation of DSM in Matlab which met the stipulated targets for a series of test images.

2 The Problem

Objects flow along a conveyor belt and it is required to pack them into boxes of homogeneous objects. Currently this is carried out manually. The first essential step in automating the process is a reliable means of recognizing the objects and distinguishing perhaps quite similar objects from each other.

In the learning phase, an image of the object is captured using a simple webcam, which uses its accompanying software to store a black-and-white image in the memory of a PC. The image is then loaded into MATLAB and digitized using a built-in function. The digitised image is then fed through a supervised learning algorithm, which learns an internal representation of the image and stores it in a file. This has to be completed in under 3 minutes.

In the recognition phase, an image of the object is again captured, digitised, and passed to a Matlab program that compares it with all the representation of the stored images. The recognition algorithm determines the best match and the result is sent to the monitor screen and to a set of speakers. The system requirement was for this to take place in under 10 seconds. It was also required that the system should have a recognition rate of 98%.

In principle, this could be done by storing the digitised images themselves and applying a 'best match' algorithm directly. In practice, recognition would take far too long. It was for this reason that it is necessary to form a small-scale representation of each image. While this could be done in a number of ways, the method chosen for investigation here was a neural net approach.

3 LVQ and DSM

Kohonen [1] originally introduced the idea of learning vector quantization, and it has subsequently been refined in various ways. Haykin [2], for example, gives a review of these. The DSM approach was introduced by Geva and Sitte [3], who claimed it was more efficient and effective than other LVQ approaches. However, it does not seem to have been tested very frequently, although it seemed to be a good candidate for this application. Geva and Sitte report that DSM is particularly suited to the mapping of piecewise linear boundaries, which was the case for the application envisaged for these experiments. (However, some curved boundaries were also included later to test its performance in these cases.)

The basic idea is to use a specified number of prototype vectors or *exemplars* in order to cover the input space, which here is simply a 2-dimensional image coded as 0s and 1s. The objects used have fairly well-defined areas, so they can be well characterised by learning the boundaries between black (1) and white (0) space. Effectively, points in the black and white areas are treated as two classes. DSM is claimed [3] to be very good at defining such boundaries closely.

Firstly, n_e exemplars are chosen at random points in the space and randomly assigned class membership values. As the sizes of the black and white areas differ, it is sensible (as suggested in [3]) to assign the numbers in each class in a roughly proportional way.

Secondly, the positions of the exemplars are changed in a learning phase. The training data consists of the set of all pixel positions along with their 'class membership', presented to the learning algorithm in a random order. For each point in turn, the closest exemplar is found, and if this is of a different class, the closest exemplar of the same class is also found. If the exemplar and the training case x are of the same class, no action is necessary. Otherwise, the exemplars are 'rewarded' or 'punished' appropriately.

$$e_c \leftarrow e_c + \alpha(x - e_c)$$

is the reward for the closest correct exemplar, while

$$e_i \leftarrow e_i - \alpha(x - e_i)$$

is the punishment for the incorrect (but closest) exemplar. The term α is a value in $[0, 1)$ which adjusts the exemplar positions in the right direction.

Initially, α is set at 0.3, as recommended in [3], which allows fairly large adjustments in exemplar positions; after all training patterns have been used, α is reduced in value and the procedure repeated. In all, 10 training phases were used, with a linear reduction of α to 0.03 for the last phase.

4 Application

In order to test the ability of DSM to perform the task outlined above, 6 items were used: two circles, one being 15mm larger in diameter than the other; two rectangles with circles in the centers (one circle is 6mm larger in diameter), and a socket and a socket with a switch. The first two were used to test the general recognition capability of the system, the relative sizes being such that it should be an easy task. The next two were rather similar, so that detection was expected to be much harder, giving an indication of the limits of the process. The last two were examples of the objects that were to be used in the real system, of which this was the prototype.

The image capture device was placed at a uniform height above the conveyor during both learning and recognition phases, chosen to ensure a reasonable-sized border around the object. (Clearly there had to be plenty of contrast between the object and the conveyor belt.) The images obtained from the screen had $176 \times 144 = 25344$ pixels

The DSM algorithm was implemented using Matlab, with MEX functions in C used to perform some of the more computationally expensive parts. Initial experiments on identification of alphabetical characters on a coarse grid were used to validate the program, and then experiments were carried out on the real system. It was found that around 50 exemplars were needed in order to guarantee good coverage of the space, and the number of presentations for each phase (i.e., at each α value) needed to be about one-third of the number of pixels. The whole process took less than 20 seconds on a Pentium II PC (350MHz). Learning the circular objects was less effective than the ones with mostly straight lines. (This is expected, as DSM essentially uses piecewise linear decision boundaries, so it will tend to find curved objects less easy to distinguish.) Having learned the boundaries, both internal and external, of the image, the exemplar positions for each image were saved as the 'stored image' of the object.

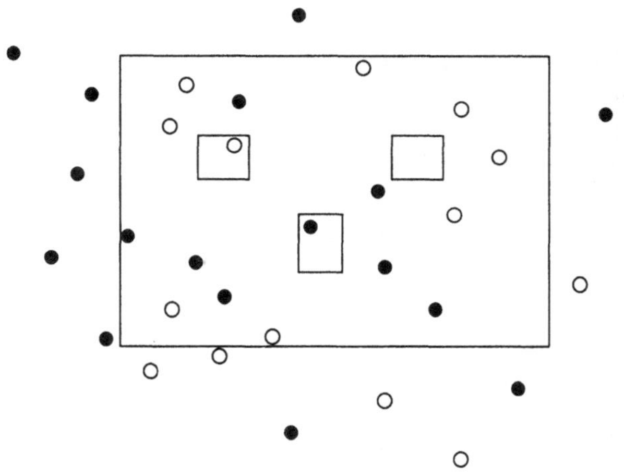

Fig. 1. Socket exemplars before training. Filled and unfilled circles represent different classes (black/white).

Figure 1 shows an example with 30 exemplars randomly placed around an image of a socket. The holes and the surrounding area of the socket would appear black in the actual image, but are left blank here for clarity. The stored image at this point, as implied by the exemplar positions, would barely resemble a socket at all. In Figure 2, we see the positions after several cycles of DSM training have taken place. The shape of the socket is becoming more clearly defined by the exemplar positions.

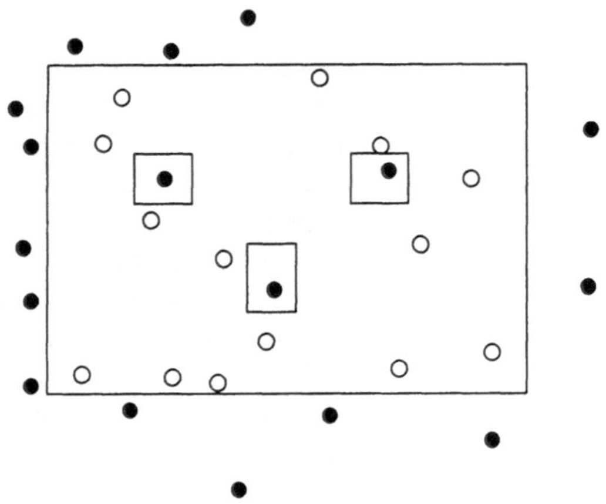

Fig. 2. Socket exemplars after a few cycles of training. Notice how filled circles have migrated to socket holes and exemplars are starting to 'line up' around the socket's outer boundary.

4.1 Rotational effects

Two of the sets of objects were not rotationally invariant, and it was realized at an early stage that this would create problems. This was dealt with by learning each image in these sets in 4 different orientations—turning the image through 90, 180 and 270 degrees respectively. This created 4 templates for each image, and although this meant 4 times as much learning was needed, the total time taken was still well below the system requirements.

In the recognition phase the first step was to identify a corner of the image, then the angle of one of the edges incident at that corner, and finally to rotate the image using simple trigonometry so that its edges fell along one of the templates. Once this problem has been solved, the recognition algorithm proceeds one by finding for each pixel in the image, the nearest exemplar in each 'stored image'. A count is made of the number of incorrect classifications arising, and the one with the smallest error is declared the winner. For the set of 6 objects used in these experiments, this required about 1 second per stored image, well within the initial specification of 10 seconds. Recognition rates of 98% and 99% were found for the socket and circle images, so this also met the specification. However, distinguishing the rectangle-containing circles was (as expected) harder, with 80% correct. However, this was a difficult task, given that the images were very similar, and 'only' 20% error was considered to be quite an achievement.

The prototype system was also enhanced by a user interface, with both spoken and displayed messages concerning the system's determination of each object.

5 Potential Improvements

DSM was successfully applied to the problem of automatic recognition of objects on a conveyor, and the pilot study has shown it to be capable of meeting the system specifications. The pilot study also revealed some of the proposed system's limitations. To use it in an industrial-scale application, it should be realized that detection of curves is likely to be less reliable than straight lines. The speed of recognition will obviously scale linearly with the number of objects to recognize, so faster decisions would be needed to cope with more than about 10 objects. It is clear that further research is needed before a full prototype can be constructed.

Using a more powerful PC, and perhaps a lower-resolution camera (to reduce the number of pixels), would obviously speed up the process. However, there are also some potential algorithmic speed-ups that are to be investigated. Better edge detection techniques and image clean up after rotation could also improve the recognition software.

It is also conceptually possible to reduce the number of exemplars by dispersing them more efficiently, which would further speed up the recognition phase. (Alternatively, the same number could be put to better use, especially in the case of curved edges.) It was noticed that some exemplars were 'stimulated' by relatively few points. In some cases this is necessary, as when there is an island of one class in the middle of a sea of the other. However, in some cases, it is believed that removing them would not cause any problems, and a method of detecting them and modifying the exemplar pattern is currently under investigation. Another topic to be considered is the use of gray scale or even colour, which should make recognition more effective, although whether this has costs in efficiency (e.g., more exemplars) will need further research.

6 Conclusion

This project achieved its initial aims, which were to implement a NN-based image recognition system that could operate within the specified parameters. It has also demonstrated the effectiveness of DSM as a tool for such applications, and suggests that it deserves further development. It is anticipated that some of the improvements under investigation could further increase recognition reliability, while speeding up the process so that up to 100 objects could be distinguished within the stipulated time limit.

References

[1] T.Kohonen (1989) *Self-Organization and Associative Memory*, 2nd edition, Springer-Verlag, Berlin.

[2] S.Haykin (1999) *Neural Networks : a Comprehensive Foundation*, 2nd edition, Prentice Hall, London.

[3] S.Geva and J.Sitte (1991) Adaptive nearest neighbour pattern classification. *IEEE Trans. Neural Networks*, **2**, 318-322.

Competitive Learning and Its Convergence Analysis

Ya-Jun Zhang *, Zhi-Qiang Liu*†

*Department of Computer Science and Software Engineering, The University of Melbourne, Australia
†School of Creative Media, City University of Hong Kong, Hong Kong, P.R.China

Abstract

Competitive learning in the neural network literature can be viewed as the adaptive versions of Kohonen's Self-Organizing Feature Map (SOFM). The common method to guarantee the convergence property is to use the so-called step-size schedule which decays from one to zero during the learning process. To define the step-size schedule, several factors and conditions must be considered. In this paper we review the convergence conditions and investigate the effectiveness of different step-size schedules available in the literature as well as some new schedules in the recently proposed competitive learning algorithms.

1 Introduction

In Competitive Learning (CL) there exists a rather large number of models that have similar goals but differ considerably in the ways they work. A common goal of the CL algorithms is to distribute a certain number of prototype vectors in a possibly high-dimensional space. The distortion of these prototype vectors should reflect the probabilty distribution of the input data set which in general is not known *a priori* but only through sample vectors. So far many CL algorithms have been proposed in the literature (see [12] for a survey). Although they are implemented in different neural models, e.g., fixed-dimensionality *vs* non-fixed dimensionality (incremental or decremental) network, batch learning *vs* online learning, *winner-take-all* (WTM) paradigm *vs winner-take-most* (WTM) paradigm etc., for each single prototype vector the learning process can be formularized as an adaptive version of the classical K-means or Kohonen's self-organizing feature map (SOFM). Theoretically it is a stochastic gradient descent process given a proper energy function, which incorporates the assignment of the data to the ptototypes and at the same time measures the quality of the representation.

In this paper, we give a review on the convergence property for competitive learning in the neural network literature. We start with an introduction to the SOFM algorithm since many other CL algorithms can be regarded as its adaptive versions. Kohonen's SOFM is an application of the least mean squares (LMS) to a nonquadratic cost function [11]. The convergence property for multiple prototypes is very difficult to analyze since it is a dynamic, stochastic process, and in fact there is no rigious convergence proofs for nonquadratic distortion functions available. However, the analysis for a single prototype is valuable for assessing the convergence properties of the multiple prototype case.

2 Fundamentals and Notations

Given a finite subset of prototypes $P = \{\vec{P}_1, \ldots, \vec{P}_K\}$, the performance of classical competitive learning is measured by the average distortion, $D(P) = E[d(\vec{X}, Q(\vec{X}))]$, between the source vectors and their associated prototypes, where $d(\bullet)$ is some distortion function, usually taken as the squared-error distortion, and $E[\bullet]$ denotes the expectation operator. The statistics of the source are represented by a finite training set of vectors. The objective of the design procedure is to minimize the average distortion [11]. The prototype set P is said to be optimal if no other set has a smaller average distortion, for the same source, vector dimension, and the number of prototypes. There are two well-known necessary conditions to achieve the optimal learning: the *Nearest Neighbor* Condition and the *Centroid* Condition [12].

Many CL algorithms, (see [1, 5, 9, 10, 11], etc.), can be regarded as the adaptive versions of SOFM. They have similar formulations on estimating prototypes. A general learning formula is

$$\vec{P}_i(n) = \vec{P}_i(n-1) + \alpha_i(n) h_i(j^*, n) \beta_i(n) [\vec{X}(n) - \vec{P}_i(n-1)],$$

$$j^* = \arg\min_j \|\vec{X}(n) - \vec{P}_j(n-1)\|, \tag{1}$$

where $\vec{P}_{j^*}$ is the prototype vector closest to $\vec{X}(n)$ in the Euclidean space. The determination of $\vec{P}_{j^*}$ can be thought of as a competition between the prototypes. The function $h_i(j^*, n)$ is a neighborhood function that at a time instance, n is used to alter the step size of the ith prototype as a function of the physical distance of its associated node on the network grid from that of the winner j^*. Typically $h_i(j^*, n)$ is non-zero if prototype $\vec{P}_i$ is close to the winner $\vec{P}_{j^*}$ (inside the "neighborhood"), and is zero outside the "neighborhood". Hence, only those prototypes inside the neighborhood of the winner j^* get a chance to update their weights upon the presentation of $\vec{X}(n)$. The size of the neighborhood is reduced to unity in a predefined manner as the learning progresses. For CL algorithms based on WTA paradigm, the size of the neighborhood function is fixed as unity so that only a single prototype is the winner each time when a training vector is presented. As for WTM-based CL algorithm, however, the size of the neighborhood is always larger than or equal to unity, and is less than or equal to the number of the total prototypes. Thus, there is no ultimate "winner"; each prototype is updated to a certain degree.

$\beta_i(n)$ may be regarded as a part of the learning rate. In general, for the WTM paradigms, $\beta_i(n)$ is a fraction which denotes the probability of winning for the ith prototype. In the WTA paradigm, $\beta_i(n) = 1$ since one and only one prototype wins the chance to be updated at each step. $\beta_i(n)$ usually satisfies the following conditions,

$$0 \leq \beta_i(n) \leq 1 \text{ for } i = 0..K, \tag{2}$$

$$\sum_{i=1}^{K} \beta_i(n) = 1 \text{ for all } n \geq 1, \tag{3}$$

where K is the number of the winning prototypes. Liu *et al.* suggested that K may be set as the number of prototypes initially and reduced to unity as the learning progresses [5].

$\alpha_i(n) \in [0,1]$ is the learning rate of the ith prototype at time instant n. Together with the $\beta_i(n)$ mentioned above, they form the step-size schedule which decays to zero as a function of time in a predefined manner. Hereafter, for simplicity, we use $\alpha_i(n)$ to denote the final step-size value for ith prototype at time instant n.

In SOFM learning scheme and some of its variants and applications [1, 7, 10], the step-size $\alpha(n)$ was independent of i. This, however, could result in "insufficient" learning since some slowly-learning prototypes have to be updated at the same speed as those frequently winning prototypes. In SCS and Fuzzy Competitive Learning (FCL) [2], this problem is solved to some extent by adding each prototype a winning counter n_i so that each prototype may have its own step-size $\alpha_i(n_i)$. n_i is incremented only after the ith prototype is updated. With this notation, the update formula of (1) for the ith prototype can be written as

$$\vec{P}_i(n_i) = \vec{P}_i(n_i - 1) + \alpha_i(n_i)[\vec{X}^{(i)}(n_i) - \vec{P}_i(n_i - 1)], \tag{4}$$

where $\vec{X}^{(i)}(n_i)$ is the n_ith traning vector for which the ith prototype wins. Since we will focus on the convergence property for a single prototype, we simplify the formula (4) as the following

$$\vec{P}(n) = \vec{P}(n-1) + \alpha(n)[\vec{X}(n) - \vec{P}(n-1)]. \tag{5}$$

where n is the value of the winning counter for this specific prototype, and $\vec{X}(n)$ is the nth traning vector for which this prototype wins. It has been proved that to guarantee that both the bias and variance of each prototype will converge to zero, the step-size schedule $\alpha(n)$ must be monotonically nonincreasing, decaying to zero as $n \to \infty$, and the sum of $\alpha(n)$ must be infinity as $n \to \infty$. Moreover, as another necessary condition, the schedule should satisfy $\sum \alpha^2(n) < \infty$. However, if the step-size schedule decreases too quickly, the variance will converge to a nonzero value. To solve this problem, a sufficient condition is suggested as,

$$\alpha(n+1) \geq \frac{\alpha(n)}{1+\alpha(n)} \text{ for all } n \geq 1. \tag{6}$$

A key problem with the competitive learning is how to reduce the step-size values to zero so that the prototype set is produced which satisfies both the *nearest neighbour* condition and the *centroid* condition. Several schedules in practice have been available in the literature. Different schedules may lead to different results and have different influence on the learning performance. In the next section, we investigate these schedules and their influences on the learning process respectively.

3 Convergence Analysis

In the following discussions, six classes of schedules are examined. For analysis, we rewrite (5) as follows,

$$\vec{P}(n) = (1-\alpha(n))\vec{P}(n-1) + \alpha(n)\vec{X}(n), \tag{7}$$

whose solution is given by

$$\vec{P}(n) = \prod_{j=1}^{n}(1-\alpha(j))\vec{P}(0) + \sum_{i=1}^{n}\prod_{j=i+1}^{n}(1-\alpha(j))\alpha(i)\vec{X}(i). \tag{8}$$

3.1 Constant schedule

If the step-size schedule is predefined as a constant, the learning will not converge. The value of each prototype represents an *exponentially decaying average* of those input samples for which the prototype has been the winner. It can be proved that the influence of past input training vectors decays exponentially fast with the number of further input training vectors for which this prototype is the winner. The most recent input training vectors, however, always play the principle role on determining the learning process. This is not desirable since the learning process stays non-stationary and there is no convergence. Even after a large number of training vectors, the current input vector can cause a considerable change of the prototype vector of the winner. Thus, the constant step-size schedule is only suitable for non-stationary distributions.

3.2 Hyperbolic schedule

Here we consider a hyperbolic schedule $\alpha(n) = n^{-r}$ with $r > 0$. For a hyperbolic schedule, to satisfy the requirements for the step-size schedule discussed in Section 2, a necessary and sufficient condition is $r \leq 1$.

The schedule with $r = 1$ is very offen and frequently used in competitive learning algorithms. A famous example is K-Means [6], which is a rather appropriate name, because each prototype vector is always the exact arithmatic mean of the input training vectors for which it has been the winner so far. However, a big disadvantage with $r = 1$ is that some of the input vectors that contribute to the update of the prototype at some time are actually in the regions belonging to other prototypes. This happens especially in the early learning stage since all the prototypes are randomly initialized in the input space. For those CL algorithms that use the frequency distance (see, [1]) to determine the winner, this schedule is extremely unsuitable.

While for $r < 1$, the hyperbolic schedule implies that the weighting sequence for the input training vectors is an increasing sequence with the learning algorithm, thus giving more weight to the latest data presented. It can be seen that when r equals to 0, the hyperbolic schedule reduces to a special case of the constant schedule ($\alpha(n) = 1$). While $r = \frac{1}{2}$, it becomes the case proposed by Darken and Moody in [3].

3.3 Exponential schedule

Exponential schedule can be written as $\alpha(n) = \lambda\rho^n$, whereby $\rho < 1$ and $\lambda \in (0, 1]$. This schedule is actually inadequate for SOFM and its variants. It has been proved by Yair *et al.* [11] that there is a freezing point with the exponential schedule after which the presentation of new data will have a diminishing effect on the prototype locations. Informally, this means that for some values of n the learning algorithm "freezes". Moreover, it is unfortunately that the freezing point, say n_c, is independant of the learning instant n. It can be solved by,

$$n_c = \frac{\ln(1-\rho) - \ln\lambda}{\ln\rho}. \tag{9}$$

Thus, n_c becomes larger if ρ increases. When ρ equals to unity, the exponential schedule reduces to the constant schedule. Thus, according to the equation (9), n_c is in the infinity, the effect of the input data is always increasing with n. Therefore the most recent data presented is weighted more than older data. This observation is in complete agreement with the analysis in the constant schedule. The larger ρ is, the higher n_c will be. Also λ has a significant effect on n_c. n_c decreases with descreasing λ. For any $\lambda \leq (1-\rho)$, the freezing point becomes zero, inducing the algorithm to start freezing at the very first input data.

3.4 Adaptive exponential schedule

One adaptive version of the exponentially decaying schedule has been proposed by Ritter *et al.* [8] in the context of self-organizing maps. They proposed an exponential decay according to

$$\epsilon(t) = \epsilon_i(\epsilon_f/\epsilon_i)^{t/t_{max}}. \tag{10}$$

Normally the ϵ_i is much larger than ϵ_f, consequently, the freezing instant is zero. Thus, the learning process start freezing from the very begining. However, in this modified schedule, it can be seen that the exponential term is not t but t/t_{max}. t_{max} is a very large number. Thus, t/t_{max} is a relatively steady value close to the freezing point. Even t increases to t_{max}, this exponential term just has a small change as 1. Therefore, all the input data presented have almost the same effect on the learning process.

3.5 Linear schedule

The linear schedule is considered in this subsection, which can be useful to design when the data presented may have the most effect on the learning process. The form of this schedule is that $\alpha(n) = \lambda(1 - n/N)$ for $0 \leq n \leq N$ and $\lambda \in (0, 1]$. Similar to the exponential schedule, linear schedule has a critical point, called n_r, which is given by,

$$n_r = N - \sqrt{\frac{N}{\lambda}}. \tag{11}$$

Note that n_r decreases as λ decreases. Unlike the exponential case in which n_c is a fixed point independent of N, the critical point n_r is relatively close to N. Hence, while a learning algorithm with the exponential schedule tends to favor data presented at the begining of the learning and ignores the most recent data, the linear schedule favors recent data samples

3.6 Gradient descent schedule

Gradient descent scheme is an effective method to generate the step-size schedule. It tends to minimize the system distortion or other cost functions by iteratively updating the prototypes in an attempt to satisfy both the centroid and nearest neighbour rules. For batch competitive learning such as generalized Lloyd [4], it may have the problem of local minima. This problem can be solved by online gradient competitive learning which combines the gradient descent scheme into the deterministic Kohonen's SOFM. Kohonen's learning scheme ensures that the distortion will be minimized when the learning converges. However, it does not ensure that the distortion function decrease monotonically during learning process; it thus may avoid being trapped in poor local minima [11]. To get the step-size schedule for online gradient learning, we need an objective function. Suppose it is $E = f(X, P)$ where $X = \{\vec{X}_1, \ldots, \vec{X}_N\}$ is the data set, and $P = \{\vec{P}_1, \ldots, \vec{P}_K\}$ is the prototype set. Gernerally, for online gradient learning, E can be rewritten as

$$E = \sum_{n=1}^{N} E_n, \tag{12}$$

where E_n is a function that $E_n = g(\vec{X}(n), P)$. For each element $\vec{P}_i$ in P with the input data $\vec{X}(n)$, it can be updated by the gradient descent method,

$$\Delta\vec{P}_i(n) = -\eta\frac{\partial E_n}{\partial\vec{P}_i(n)}, \tag{13}$$

where η is a constant fraction. $\Delta\vec{P}_i(n)$ can be further written by

$$\Delta\vec{P}_i(n) = \alpha_i(n)(\vec{X}(n) - \vec{P}_i(n-1)) + \vec{R}_i(n), \tag{14}$$

where $\alpha_i(n)$ is the step-size schedule for the ith prototype. $\vec{R}_i(n)$ is a penalization term reflecting the influence from other prototypes, it does not have the effect on the convergence property of $\vec{P}_i$. Gradient descent schedule can be used in the tasks for which the objective function is easy to obtain.

3.7 OPTOC schedule

OPTOC (*one prototype takes one cluster*) schedule is a new method to guarantee the convergence property. It assigns each prototype an asymptotic property vector which guides the learning of its associated prototype during the learning process(see [12] for details).

4 Conclusions

The convergence property of competitive learning usually relies on the step-size schedule. In this paper we have investigated several step-size schedules and their influences on the learning process. Since we usually do not have the adequate prior knowledge in the input data, we suggest that the parameters of these schedules be defined adaptively from the analysis of data distribution and the task at hand.

References

[1] S.C. Ahalt, A.K. Krishnamurty, P. Chen, D.E. Melton, "Competitive learning algorithms for vector quantization," *Neural Networks*, vol.3, no.3, pp.277-291, 1990.

[2] F.L. Chung, T. Lee, "Fuzzy competitive learning," *Neural Networks*, vol.7, no.3, pp.539-551, 1994.

[3] C. Darken, J. Moody, "Fast adaptive K-Means clustering: Some empirical results," *Proc. IJCNN*, vol.2, pp.233-238, IEEE Neural Council, 1990.

[4] Y. Linde, A. Buzo, R.M. Gray, "An algorithm for vector quantizer design," *IEEE Trans. Commun.*, vol.COM-28, pp.84-95, Jan., 1980.

[5] Z.-Q. Liu, M. Glickman, Y.-J. Zhang, "Soft-Competitive Learning Paradigms," in *Soft Computing and Human-Centered Machines*, Z.-Q. Liu and S. Miyamoto (eds), pp.131-161, Tokyo: Springer-Verlag, 2000.

[6] J. MacQueen, "Some methods for classification and analysis of multivariate observations," *Proc. of the 5th Berkeley Symposium on Mathematical Statistics and Probability*, vol.1, pp.281-297, 1967, Univeristy of California Press.

[7] N.M. Nasrabadi, Y. Feng, "Vector quantization of images based upon the Kohonen self-organization feature maps," *Proc. 2nd ICNN Conf.*, vol.1, pp.101-105, 1988.

[8] H.J. Ritter, T.M. Martinetz, K. J. Schulten, *Neuronale Netze*, Addison-Wesley, München, 1991.

[9] D.E. Rumelhart, D. Zipser, "Feature discovery by competitive learning," *Cognitive Science*, vol.9, pp.75-112, 1985.

[10] L. Xu, A. Krzyzak, E. Oja, "Rival penalized competitive learning for clustering analysis, RBF Net, and curve detection," *IEEE Trans. Neural Networks*, vol.4, pp.636-649, July 1993.

[11] E. Yair, K. Zeger, A. Gersho, "Competitive learning and soft competition for vector quantizer design," *IEEE Trans. Signal Processing.*, vol.40, no.2, Feb., pp. 294-309, 1992.

[12] Y.-J. Zhang, Z.-Q. Liu, "Self-Splitting Competitive Learning: a new algorithm," Submitted to *IEEE Trans. Neural Networks*, Mar., 2000

Structural Information Control for Flexible Competitive Learning

Ryotaro Kamimura*, Taeko Kamimura†, Thomas R. Shultz‡

*Information Science Laboratory, Tokai University, 1117 Kitakaname, Hiratsuka, Kanagawa, 259-1292, Japan, ryo@cc.u-tokai.ac.jp †Department of English, Senshu University, 2-1-1 Higashimita, Tama-ku, Kawasaki, Kanagawa, 214-8580, Japan, taekok@isc.senshu-u.ac.jp ‡Department of Psychology, McGill University, Montreal, Quebec, H3A 1B1, Canada, shultz@psych.mcgill.ca

Abstract

In this paper, we propose a new information theoretic method called *structural information* to overcome fundamental problems inherent in conventional competitive learning such as dead neurons and deciding on the appropriate number of neurons in the competitive layer. Our method is based on defining and controlling several kinds of information, thus generating particular neuron firing patterns. For one firing pattern, some neurons are completely inactive, meaning that some dead neurons are generated. For another firing pattern, all neurons are active, that is, there is no dead neurons. This means that we can control the number of dead neurons and choose the appropriate number of neurons by controlling information content. We applied this method to simple pattern classification to show that information can be controlled, and that different neuron firing patterns can be generated.

1 Introduction

The majority of unsupervised learning methods have been based upon competitive learning [1]. In competitive learning, neurons compete with each other and finally one neuron (winner-take-all) wins the competition for given input patterns. Though competitive learning is a simple and powerful method for unsupervised learning, several problems such as dead neurons and deciding on the appropriate number of neurons have been pointed out. For overcoming these problems, a number of heuristic methods have been proposed: for example, leaky learning [2], a conscience method [3], frequency-sensitive learning [4], rival penalized competitive learning [5] and lotto-type competitive learning [6]. Though each of these heuristic methods may be effective in eliminating one specific problem, none of them can solve the dead neurons problems and the appropriate number of neurons in a unified way.

In this context, we propose a novel approach called *structural information* to control the number of dead neurons. Our method is based upon information theoretic approaches to competitive learning. Information theoretic methods applied to neural computing have so far given promising results [7], [8]. One thing that these approaches have in common is that they have been exclusively concerned with the quantity of information obtained by learning. However, it is easily pointed out that the same amount of information can produce a number of unpredictable network final states or information representations. Consequently, current information theoretic methods are not as powerful as they might be. Structural information has been introduced to overcome these problems. In the structural information method, we can define several different kinds of information content, corresponding to different neurons activation patterns. By controlling this structural information, we can control competitive unit activation patterns. For example, by controlling structural information, we can create an activation pattern in which some neurons are always off, and thus not used for classification. This activation pattern can be used to deal with a situation where the number of neurons is redundantly larger than the expected classes. Thus, structural information control is used to control the number of dead neurons and to choose the appropriate number of competitive units.

2 Structural Information

In this paper, we are concerned not with information to be transmitted through information channels but with stored information in systems [9]. Thus, *information* can be defined as a decrease in uncertainty. Structural information is introduced to see this information content in a more detailed way.

We can distinguish different kinds of structural information. First, we consider simple structural information with two random variables to be immediately extended to n random variables. This structural information is defined by the uncertainty decrease from maximum to actually observed uncertainty. The maximum uncertainty H_0 is computed by $\log M$, where M is the number of elements in a system and actual uncertainty H_1 is described by first order entropy or uncertainty

$$H_1 = -\sum_j p(j) \log p(j), \tag{1}$$

where $p(j)$ denotes that the jth unit in a system occurs with probability $p(j)$. Thus, information independent of input patterns, that is, first order information is defined by

$$\begin{aligned} D_1 &= H_0 - H_1 \\ &= \log M + \sum_j p(j) \log p(j). \end{aligned} \tag{2}$$

First order entropy or uncertainty H_1 may be further decreased to second order uncertainty H_2, that is, the decrease in uncertainty after receiving input signals:

$$H_2 = -\sum_s \sum_j p(s)p(j \mid s) \log p(j \mid s), \tag{3}$$

where $p(s)$ represent the probability of input signals s. This uncertainty decrease, that is, second order information is defined by

$$\begin{aligned} D_2 &= H_1 - H_2 \\ &= -\sum_j p(j) \log p(j) \\ &\quad + \sum_s \sum_j p(s)p(j \mid s) \log p(j \mid s). \end{aligned} \tag{4}$$

Using the structural parameter α, second order structural information $SI^{(2)}$ is defined by

$$SI^{(2)} = \alpha D_1 + (1-\alpha) D_2, \tag{5}$$

where α is the structural parameter, ranging between 0 and 1. When the structural parameter α is 0, structural information is equivalent to second order information. On the other hand, when the structural parameter is 1, structural information is equal to first order information.

Finally, we should note that the structural information just explained can easily be extended to a more general n case as follows.

$$SI^{(n)} = \sum_n \alpha D_n, \tag{6}$$

where D_n denotes the nth order structural information, and $\sum_n \alpha_n = 1$.

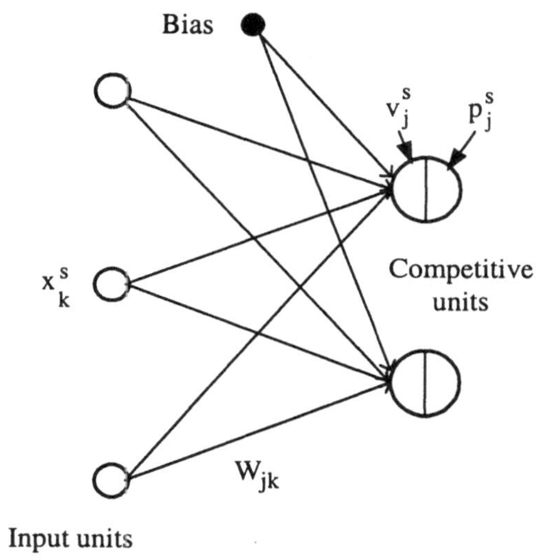

Fig. 1. Network architecture for defining structural information.

3 Application to Neural Learning

In this section, we attempt to apply the structural information method to a neural network architecture. As shown in Figure 1, the network architecture is composed of input units x_k^s and competitive units v_j^s.

Let us define information for competitive units and try to control competitive unit activation patterns. The jth competitive unit receives a net input from input units, and an output from the jth competitive unit can be computed by

$$v_j^s = f\Big(\sum_k w_{jk} x_k^s\Big), \tag{7}$$

where w_{jk} denotes a connection from the kth input unit to the jth competitive unit. In modeling competition among units, one of the easiest ways is to normalize the outputs from the competitive units as follows:

$$p_j^s = \frac{v_j^s}{\sum_m v_m^s}. \tag{8}$$

The conditional probability $p(j \mid s)$ is approximated by this normalized competitive unit output, that is,

$$p(j \mid s) \approx p_j^s. \tag{9}$$

Since input patterns are supposed to be uniformly given to networks, the probability of the jth competitive unit is approximated by

$$\begin{aligned} p(j) &= \sum_s p(s)p(j \mid s) \\ &\approx \frac{1}{S} \sum_s p_j^s \\ &= p_j. \end{aligned} \tag{10}$$

Using these approximated probabilities, first order information is approximated by

$$D_1 = \log M + \sum p(j) \log p(j) \approx \log M + \sum_j p_j \log p_j. \quad (11)$$

Second order information D_2 is approximated by

$$\begin{aligned} D_2 &= -\sum_j p(j) \log p(j) \\ &\quad + \sum_s \sum_j p(s) p(j \mid s) \log p(j \mid s) \quad (12) \\ &\approx -\sum_j p_j \log p_j + \frac{1}{S} \sum_s \sum_j p_j^s \log p_j^s. \end{aligned}$$

As second order information is larger, specific pairs of input patterns and competitive units are strongly correlated. Second order structural information is approximated by

$$\begin{aligned} SI^{(2)} &= \alpha D_1 + (1-\alpha) D_2 \\ &\approx \alpha \log M + (2\alpha - 1) \sum_j p_j \log p_j \\ &\quad + (1-\alpha) \sum_s \frac{1}{S} \sum_j p_j^s \log p_j^s. \quad (13) \end{aligned}$$

Differentiating structural information with respect to input-competitive connections w_{jk}, we have the update rule:

$$\begin{aligned} \Delta w_{jk} &= \beta(2\alpha - 1) \sum_s \left(\log p_j - \sum_m p_m^s \log p_m \right) Q_{jk} \\ &\quad + \beta(1-\alpha) \sum_s \left(\log p_j^s - \sum_m p_m^s \log p_m^s \right) \\ &\quad \times Q_{jk}, \quad (14) \end{aligned}$$

where

$$Q_{jk} = \frac{1}{S} p_j^s (1 - v_j^s) \xi_k^s, \quad (15)$$

and where β is the learning rate parameter.

4 Application to Simple Pattern Detection

In this experiment, we used artificial data to show how the structural information method can produce different types of activation patterns, depending upon the chosen structural parameter α. Figure 4(a) shows four input patterns used for this experiment. Because the number of input patterns is four, the number of competitive units is restricted to four. Even if the number of competitive units is larger than four, the results obtained will not change. The number of input units was 16, as shown in Figure 2. The learning parameter β was always set to 1. The learning epochs were set to 1,000. No further adjustments were made to the learning process.

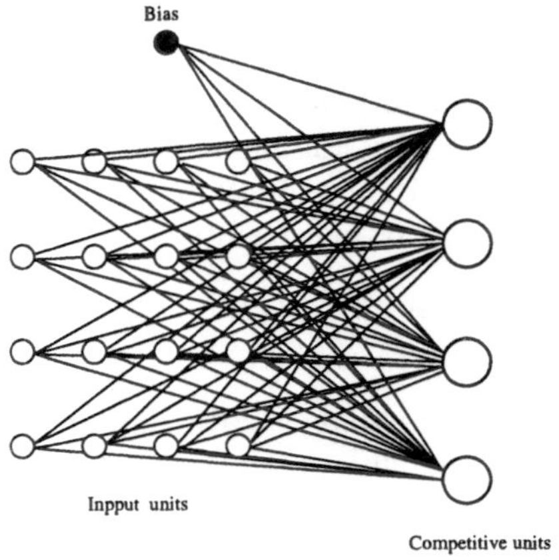

Fig. 2. A network architecture for artificial data in which the numbers of input and hidden units are 16 and 4, respectively. The bias is exclusively used to break the symmetry of competitive unit activation patterns.

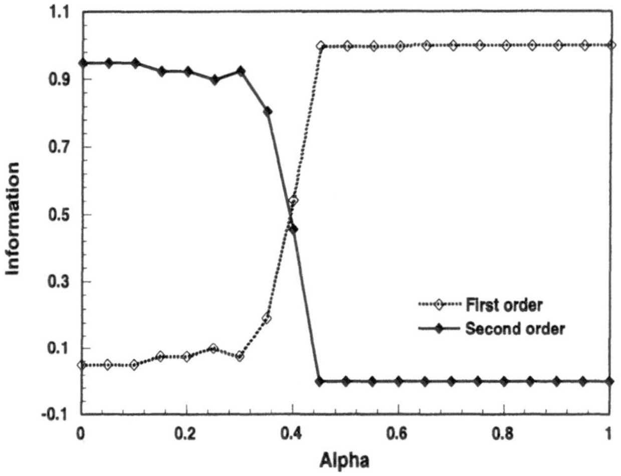

Fig. 3. Information as a function of the structural parameter α. First order and second information represent normalized information, that is, information divided by the maximum information. First and second order information are represented by a dotted and a solid line respectively.

Figure 3 shows information as a function of the structural parameter α. The information was averaged over ten different runs, and it was normalized for its values to range between 0 and 1. As can be seen in the figure, second order information represented by a solid line is close to the maximum point when the structural parameter α is 0. As the structural parameter is increased, second order information is decreased. Actually second order information reaches almost the zero level after the structural parameter α becomes 0.45. On the other hand, first order information represented by the dotted line is close to 0 when the structural parameter is 0. As the structural parameter is increased, first order information is increased, and it reaches approximately the maximum point after the structural parameter α becomes 0.45. When the structural parameter α is 0.4, first and second order information become almost equal to each other.

Figure 4 shows four input patterns (a) and different kinds of competitive unit activation patterns for three different parameter values (b), (c) and (d). In (b), (c) and (d), black and white squares represent competitive unit activation levels p_j^s close to 1 and close to 0, respectively.

As can be seen in (b), when the structural parameter is 0, the four different competitive units respond to different input patterns respectively. This means that we have completely specialized competitive units. In this case, we there are no dead neurons, and all neurons are equally used. In principle, conventional competitive learning can produce only this final state [2]. When the structural parameter is increased to 0.4(c), we have many different kinds of internal representations, depending upon different initial conditions. The activation pattern (c1) shows that the third competitive unit responds to three input patterns and that the fourth competitive unit responds to just one input pattern. The representation (c2) shows that only one competitive unit responds to all the input patterns. The representation (c4) illustrates that the network can classify four input patterns into two groups. The numbers of dead neurons in (c1), (c2), (c3) and (c4) are 2, 3, 2 and 2, respectively. Finally, as shown in Figure (d), when the structural parameter is 1, that is, when first order information is only used, just one competitive unit always responds to all four input patterns. The number of dead neurons here is always three. These results have shown that structural information can control the number of dead neurons and choose the appropriate number of competitive units by adjusting the structural parameter α.

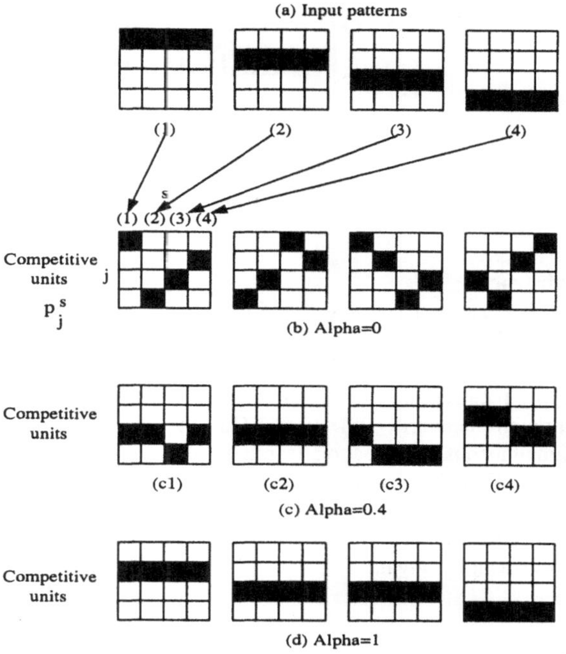

Fig. 4. Input patterns and final competitive unit activation patterns p_j^s for three different structural parameter values. Black and white squares in (b), (c) and (d) represent the activation values p_j^s close to 1 and 0, respectively. No intermediate activation levels could be seen in the experiments.

References

[1] S. Grossberg, "Competitive learning: from interactive activation to adaptive resonance," *Cognitive Science*, vol. 11, pp. 23–63, 1987.

[2] D. E. Rumelhart and D. Zipser, "Feature discovery by competitive learning," in *Parallel Distributed Processing* (D. E. Rumelhart, G. E. Hinton, and R. J. Williams, eds.), vol. 1, pp. 151–193, Cambrige: MIT Press, 1986.

[3] D. Diesieno, "Adding a conscience to competitive learning," in *Proceedings of IEEE International Conference on Neural Networks*, pp. 117–124, IEEE, 1988.

[4] S. C. Ahalt, A. K. K. andd P. Chen, and D. E. Melton, "Competitive learning algorithms for vector quantization," *Neural Networks*, vol. 3, pp. 277–290, 1990.

[5] L. Xu, "Rival penalized competitive learning for clustering analysis, rbf net, and curve detection," *IEEE Transaction on Neural Networks*, vol. 4, no. 4, pp. 636–649, 1993.

[6] A. Luk and S. Lien, "Properties of the generalized lotto-type competitive learning," in *Proceedings of International conference on neural information processing*, pp. 1180–1185, 2000.

[7] R. Linsker, "Self-organization in a perceptual network," *Computer*, vol. 21, pp. 105–117, 1988.

[8] R. Kamimura and S. Nakanishi, "Hidden information maximization for feature detection and rule discovery," *Network*, vol. 6, pp. 577–622, 1995.

[9] L. L. Gatlin, *Information Theory and Living Systems.* Columbia University Press, 1972.

Recursive Learning Scheme for Recurrent Neural Networks Using Simultaneous Perturbation and Its Application

Yutaka Maeda*

*Department of Electrical Engineering, Faculty of Engineering, Kansai University, JAPAN

Abstract

In this paper, a learning rule for recurrent neural networks using the simultaneous perturbation optimization method is described. The back-propagation method is complicated to apply recurrent networks, since the error have to propagated through network and time. On the other hand, the simultaneous perturbation learning rule requires only error value. Therefore it is easily applicable to recurrent networks. Examples for the Hopfield networks and the bi-directional associative memories are detailed. Application of vehicle movement is shown.

1 Introduction

Recurrent type of neural networks (NNs) have complicated and interesting properties compared with the ordinary mutilayered NNs. However, it is difficult to set up values of weights in a recurrent network for a specific purpose.

Hopfield neural network (HNN) is a typical example of recurrent neural networks with symmetrical fully- connected weights[1]. HNNs are used to store patterns or to solve combinatorial optimization problems like the traveling salesman problem. Moreover, bidirectional associative memory (BAM) is an extended network of HNN with two layered structure[2]. Adaptive BAM, in which weights are updated based on Hebb's learning rule, has also been proposed [2]. However, in these networks, weights in the network are ordinarily set by patterns to be memorized or problems themselves to be solved.

The simultaneous perturbation optimization method was introduced by J. C. Spall[3],[4]. J. Alespector *et al.*[5] and G. Cauwenberghs[6]. Y. Maeda also independently proposed a learning rule using the simultaneous perturbation and reported a feasibility of the learning rule[7-9]. At the same time, the merit of the learning rule was demonstrated in VLSI implementation of analog NNs using this rule[10-12].

The advantage of the simultaneous perturbation method is its simplicity. The simultaneous perturbation can estimate the gradient of a function using only the values of the function. Therefore, it is relatively easy to implement as a learning rule of NNs, compared with the other learning rules such as the back-propagation learning rule. At the same time, this learning rule is applicable to recurrent types of NNs very easily, since final error values are used to estimate the gradient of the error function.

2 Learning Scheme Using the Simultaneous Perturbation

When we think about BAMs, the weights in BAMs are determined using patterns to be memorized or problems to be solved ordinarily. Then, these determinations are based on energy functions. Therefore, it is important to find proper energy functions describing the corresponding problems.

However, it is generally hard to find such energy functions for arbitrary problems. If we can implement problems without knowing such energy functions, we will be able to find broader range of applications of these types of the networks.

In many applications, we know an ideal situation for the network to be learnt. Using this information, we can evaluate how much a network works well. For example, when a recurrent network stores patterns, we know the patterns for the network to be stored. Therefore, if a stable state i.e. a result of the network is different from the pattern to be stored, we can measure the difference. Such an evaluation function gives a clue to optimize weights in the network.

Unfortunately, in order to use the back-propagation, the error quantity must propagate through time from a stable state to an initial state. It seems difficult to use such a method directly, because it will take a long time to compute the modifying quantities corresponding to all weights.

On the other hand, the simultaneous perturbation optimization method requires only a value of an evaluation function. If we know the evaluation of a stable state, we can obtain modifying quantities of the weights of the network without complicated error propagation through time.

$$\Delta w_t^i = \frac{J(w_t + cs_t) - J(w_t)}{cs_t^i} \tag{1}$$

$$w_t^i = \begin{cases} w_{\max} & \text{if } \left(w_t^i - \alpha \Delta w_t^i\right) > w_{\max} \\ -w_{\max} & \text{if } \left(w_t^i - \alpha \Delta w_t^i\right) < -w_{\max} \\ w_t^i - \alpha \Delta w_t^i & \text{if otherwise} \end{cases} \quad (2)$$

The above recursion shows the simultaneous perturbation learning rule for recurrent neural networks. $\boldsymbol{w}$ is the weight vector in a network, α is a positive constant. $\Delta \boldsymbol{w}$ is a modifying vector and Δw^i represents the i-th element of the vector $\Delta \boldsymbol{w}$. s_t and s_t^i denote the sign vector and its i-th element that is 1 or –1, respectively. The sign of s^i_t is randomly determined.

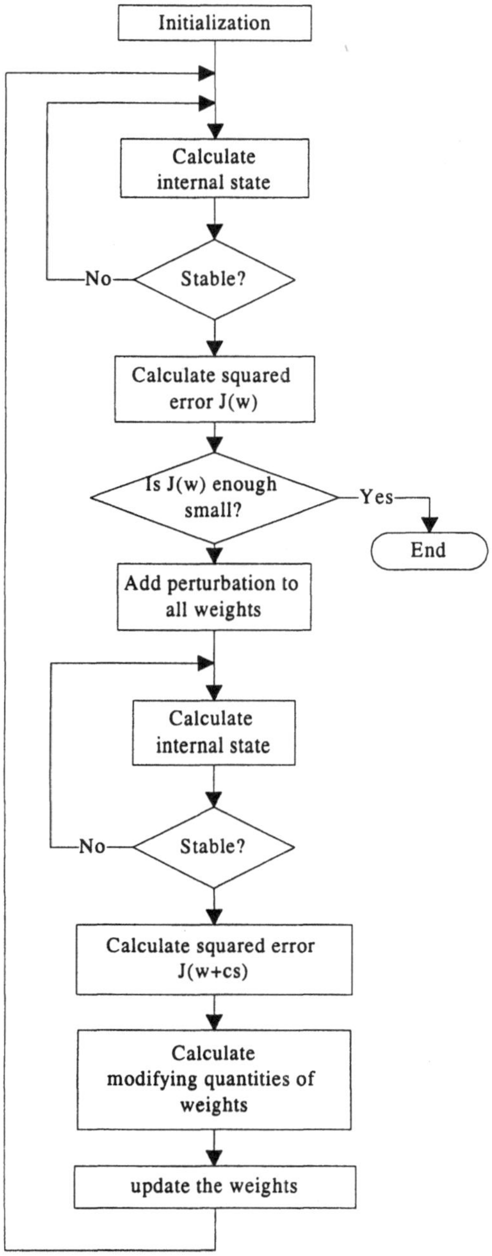

Fig.1 Flowchart

Moreover, the sign of s^i_t is independent of the sign of the j-th element s^j_t of the sign vector. That is, $E(s^i_t)=0$, $E(s^i_t, s^j_t)=0$ $(i \neq j)$. E denotes the expectation, c is a magnitude of the perturbation. $J(\boldsymbol{w})$ denotes an error or an evaluation function defined by outputs of neurons in stable state and a pattern to be embedded.

When we expand the right-hand side of Eq.(2) at the point $\boldsymbol{w}_t$, there exist $\boldsymbol{w}_{s1}$ such that

$$\Delta w_t^i = s_t^i s_t^T \frac{\partial J(w_t)}{\partial w} + \frac{c s_t^i}{2} s_t^T \frac{\partial^2 J(u(w_{S1}))}{\partial w^2} s_t \quad (3)$$

We take an expectation of the above quantity. From the conditions of the sign vector s_t, for sufficiently small perturbation c, we have

$$E\left(\Delta w_t^i\right) = \frac{\partial J(w_t)}{\partial w_t^i} \quad (4)$$

That is, Δw^i_t approximates $\partial J_p(w_t)/\partial w^i_t$. Since the right-hand side of Eq.(2) is an estimated value of the first-differential coefficient of the error function, the learning rule is a type of a stochastic gradient method[9],[10].

An important point is that this learning rule requires only two values of an error function. Therefore, we can apply this learning scheme to not only so-called feedforward type of NNs but also recurrent type of NNs like HNNs and BAMs.

2.1 Learning scheme for HNNs

We explain the procedure of this learning scheme for HNNs.

1. Set an initial state of a HNN with random weights.
2. The HNN arrived at a stable state. Then, we calculate a value of the error which is defined as follows;

$$J(\boldsymbol{w}) = \sum_i (o_i - d_i)^2 \quad (5)$$

   where, o_i and d_i denote a stable state of the HNN and its corresponding ideal output, respectively. This error is a difference between an ideal pattern to be memorized and a present final state of the network.
3. Add the perturbation to all weights in the network. Then, we have the HNN to operate and obtain the error.(Obtain $J(\boldsymbol{w}_t+c\boldsymbol{s}_t)$.)
4. Using Eqs.(1) and (2), we update the weights of the network.
5. The error is not small enough to end, then go to procedure 2.

2.2 Learning scheme for BAMs

Next, we explain the procedure of this learning scheme for BAMs.

1. Set an initial state for two layers of a BAM with random weights.
2. The BAM arrived at a stable state. Then, we calculate a value of the error which is defined as follows;

$$J(w)=\sum_{i}(o_i-d_i)^2+\sum_{j}(o_j-d_j)^2 \quad (6)$$

where, o_i and d_i denote a stable state of a layer and its corresponding ideal output, respectively. o_j and d_j denote a stable state of the other layer and its corresponding ideal output, respectively. This error is a difference in the two layers between an ideal pattern to be memorized and a present final state of the network.
3. Add the perturbation to all weights in the network. Then, we have the BAM to operate and obtain the error.(Obtain $J(w_t+cs_t)$.)
4. Using Eqs.(1) and (2), we update the weights of the network.
5. The error is not small enough to end, then go to procedure 2.

In these procedures, if we know only a stable state of the network, we can update the weights of the network. A flowchart of these procedures for BAMs is shown in Figure 1.

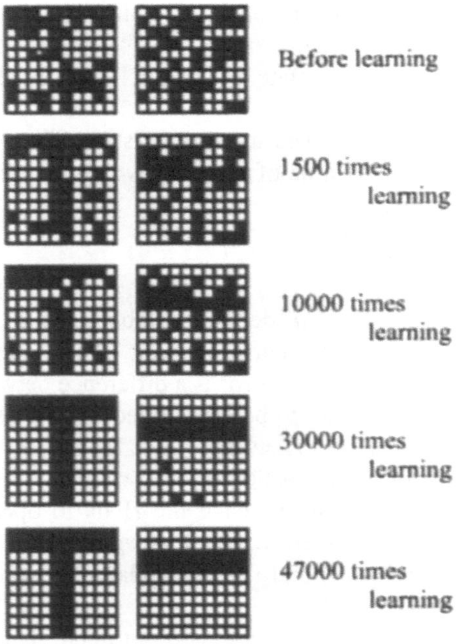

Fig. 2 Result for one pair of patterns.

3. Simulation Results

In this section, pairs of patterns are used for BAM learning. The BAM consists of two layers with 100 (10 by 10) neurons respectively. The BAM must memorize pairs of the patterns. Initial values of the BAM are determined randomly.

Figure 2 shows a simulation result by our scheme for one pair of patterns; T and -. The more learning proceeds, the more stable state closes to the patterns T and -. This result shows a feasibility of the scheme proposed here. After 47000 times learning, the stable patterns in Fig. 2 are as same as the patterns to be memorized.

Next, we use two pairs of patterns; T and -, H and -. These two pairs of patterns are embedded by the learning rule proposed here.

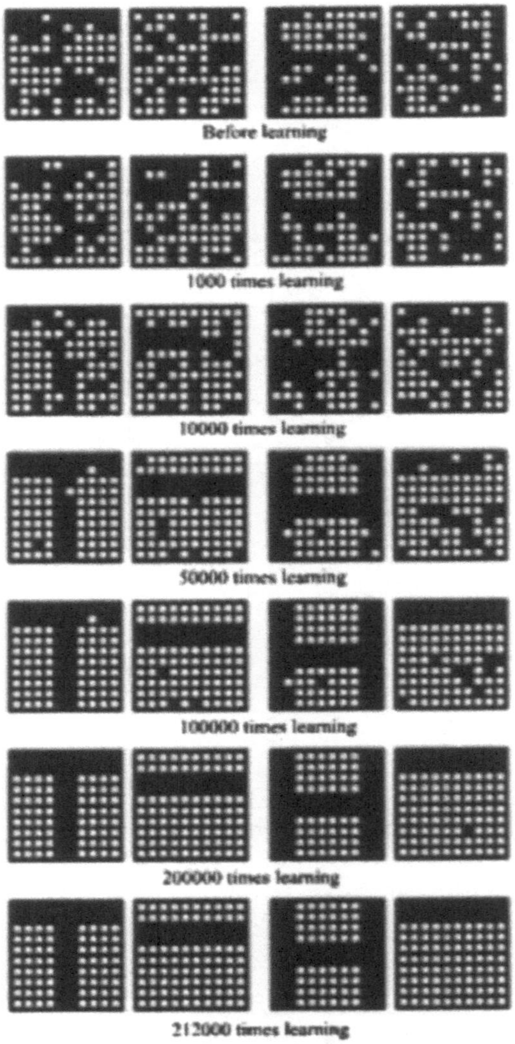

Fig. 3 Result for two pairs of pattern

Fig. 3 shows a result. After 212000 times learning, the stable four patterns in Fig. 3 are as same as the patterns to be embedded.

4. Application of HNN via Simultaneous Perturbation

We consider a control problem for vehicles with four 180 degree revolving wheels. A kind of electric car are designed to move aside using 180 degree revolving four wheels.

We would like to move the vehicle from a certain initial position to a certain destination position. Outputs of HNN correspond to two angles of wheels and its speed. An initial state of the HNN corresponds to an initial state of the vehicle. The HNN moves to a stable state; the vehicle is controlled by the HNN to a destination position.

In this problem, we assume that we know the initial position and the destination. Using the error between a final position of the vehicle controlled by a network and its destination, we can apply the simultaneous perturbation learning rule. The HNN learns a final state as a final destination of the vehicle by the learning rule.

Figure 4 shows a locus of a vehicle controlled by a HNN through the learning using simultaneous perturbation. Using the learning scheme, we can obtain a HNN that can control that vehicle to a certain destination.

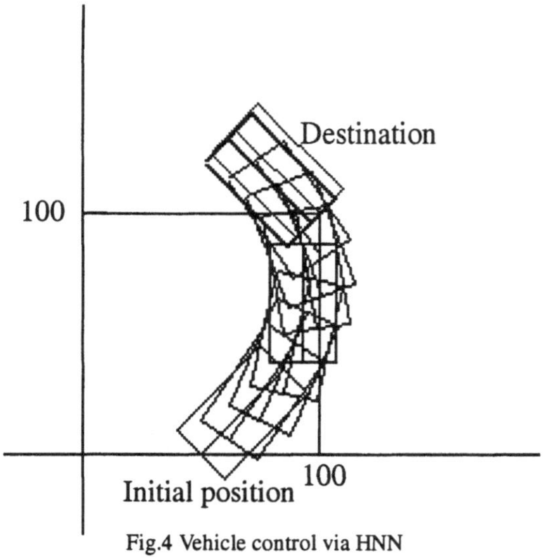

Fig.4 Vehicle control via HNN

5. Conclusion

This paper proposed a recursive learning scheme for recurrent NNs. This scheme requires only values of an error function. Therefore, it is relatively easy to implement. The learning scheme is applicable to learning of oscillatory solution. Some examples and its application were also shown.

Acknowledgement

This work is partly supported by Kansai University Frontier Sciences Center and High Technology Research Center.

References

[1] J.J.Hopfield : Neurons with graded response have collective computational properties like those of two-state neuron, *Proc. of the National Academy of Sciences*, **81**, 3088-3092, 1984

[2] B.Kosko : Neural networks and fuzzy systems, Prentice-Hall Inc., 1992

[3] J.C.Spall : A stochastic approximation technique for generating maximum likelihood parameter estimates, *Proc. of the 1987 American Control Conference*, 1987, pp.1161-1167

[4] J.C.Spall : Multivariable stochastic approximation using a simultaneous perturbation gradient approximation, *IEEE Trans.*, **AC-37**, 1992, pp.332-341

[5] J.Alespector, R.Meir, B.Yuhas, A.Jayakumar and D.Lippe, A parallel gradient descent method for learning in analog VLSI neural networks, in S.J.Hanson, J.D.Cowan and C.Lee(eds.), *Advances in neural information processing systems 5*, Morgan Kaufmann Publisher, 1993, pp.836-844

[6] G.Cauwenberghs : A fast stochastic error-descent algorithm for supervised learning and optimization," in S.J.Hanson, J.D.Cowan and C.Lee(eds.), *Advances in neural information processing systems 5*, Morgan Kaufmann Publisher, 1993, pp.244-251

[7] Y.Maeda and Y.Kanata : Learning rules for recurrent neural networks using perturbation and their application to neuro-control, *Trans. of the Institute of Electrical Engineers of Japan*, **113-C**, 1993, pp.402-408 (in Japanese)

[8] Y.Maeda and Y.Kanata : A learning rule of neural networks for neuro-controller, *Proc. of the 1995 World Congress of Neural Networks*, **2**, 1995, pp.II-402-II-405

[9] Y.Maeda and R.J.P. de Figueiredo : Learning rules for neuro-controller via simultaneous perturbation, *IEEE Trans. on Neural Networks*, **8**, 1997, pp.1119-1130

[10] Y.Maeda, H.Hirano and Y.Kanata : A learning rule of neural networks via simultaneous perturbation and its hardware implementation, *Neural Networks*, **8**, 1995, pp.251-259

[11] G.Cauwenberghs : An analog VLSI recurrent neural network learning a continuous-time trajectory, *IEEE Trans. on Neural Networks*, **7**, 2, 1996, pp.346-361

[12] Y.Maeda, Nakazawa and Y.Kanata : Hardware Implementation of a Pulse Density Neural Network Using Simultaneous Perturbation Learning Rule, *Analog Integrated Circuits and Signal Processing*, **18**, 1999, pp.153-162

Reinforcement Learning for Rule Generation

D. Vogiatzis*, A. Stafylopatis†

*Department of Electrical and Computer Engineering, National Technical University of Athens, 15773 Zographou Athens, Greece. dimitrv@central.ntua.gr †Department of Electrical and Computer Engineering, National Technical University of Athens, 15773 Zographou Athens, Greece. andreas@cs.ntua.gr

Abstract

The algorithm extracts propositional rules from a labeled data set. The constituent parts of a rule are the features of the labeled data-set, each accompanied by an appropriate interval of activation and a label denoting the class. Initially, the input space is partitioned using tiles. The algorithm tries to compose the largest possible orthogonal intervals out of tiles. After the creation of intervals for each feature the rule receives credit for its classification ability. This credit will be used to improve the rule. We have obtained encouraging results on 5 different classification problems: the iris data set, the concentric data, the four gaussians, the pima-indians set and the image segmentation data set.

1 Introduction

There have been many contributions (for instance see [1]) in the direction of rendering the inner workings of a multilayered perceptron (MLP) in a symbolic form. Our contribution is based on the *Temporal-Difference Learning* (TD) method for reinforcement learning (RL) [2] and it extracts rules directly from a data set.

In Section 2 we present a formal description of the problem; then Section 3 elaborates on our method with the appropriate equations. Finally, Section 4 presents experimental results and conclusions.

2 Definition of the Problem

Suppose there are M vectors of n features each. Each of these vectors receives a category label from $L = \{c_1, c_2, \ldots, c_m\}$. We are seeking a solution which provides propositional rules for each label — the rules being of the form

$$if \quad j_1 \in I_1 \quad and \quad \ldots \quad j_k \in I_k \rightarrow c_i \qquad (1)$$

where $\{j_1, \ldots, j_k\}$ is a subset of the set of features, as it may happen that not all features are significant in a rule. $I_1 \ldots, I_k$ are intervals, that is $I_i = [l_i, h_i]$. The constituent parts of the rules are *intervals* and *features* (the left part) for a certain label (the right part of the rule).

We make discrete the continuous input space with the aid of tiling. We define each tile to be a hypercube whose dimensions correspond to the features/dimensions of the input space. Furthermore, all tile edges are equal in length, this implies that the number of tiles per feature is different.

Each rule prescribes a set of intervals corresponding to features. Each interval is composed of a number of consecutive tile edges (along the respective dimension). Thus, the rule defines a hyperbox-shaped subspace within the n-dimensional input space.

It might be the case that the initial partitioning is not fine enough, in that case we apply a finer partioning and try again to find rules. The edges of the tiles are parallel to the axes of the orthonormal system of coordinates. However this might not be optimal in the sense that there might be tiles that correspond to empty areas of the input space. This phenomenon can be partly alleviated by applying *principal component analysis (PCA)* [3] to the data before anything else. In that case, the axes of the orthonormal system are in the direction of the highest variance of the input data.

3 Implementation Details

The extraction of a single rule for a single label is a repetition of a $3n$ step process; n steps to determine which of the n features of the data will be included and then another $2n$ steps to determine the lower and the upper limits of the features. In the end we receive credit (see Eq. 5) for the quality of the rule. This problem fits into the framework of RL if we think of these $3n$ steps as time steps from an initial state to a final state. The final state represents a completely formed rule, whereas the intermediate states represent partially formed rules.

Normally, at each state there are many next states and we must select the best one, which is defined in terms of the credit that will be received at the end. Recall, that the only feedback we receive from the environment comes at the end of the $3n$ time steps. In the meantime we can only make an estimate of the quality of each state which amounts to the learning of the *value function*.

3.1 Learing the Value Function

We train a MLP to approximate the value function, Thus the MLP input is a set of candidate next states and the MLP output is a value for each of the candidate states. The choice of the next best state value is performed according to the Boltzmann-Gibbs distribution. The encoding of a state follows an 1-out-of-n scheme. Hence, there are 2 bits for each of the n features; the first denotes presence whereas the second denotes absence of the corresponding dimension. There are also l bits for the lower limit and l bits for the upper limit of each feature. Actually, each of the l bits is an index in a table of tiles. Thus, the total number of required bits is $2n + 2nl$. At the beginning of an episode a state is a string of zeroes. During the episode certain states are selected, and certain bits of the state vector are turned on so as to denote the selections. Due to this scheme, at each step the state uniquely encodes the history of selections. The result is that the Markov property holds true.

Supposing that s states are evaluated by the MLP and their values are: $y_1, y_2, \ldots, y_s$, then the probability of choosing a specific action m is

$$p_m = \frac{e^{y_m/\tau}}{\sum_{i=1}^{s} e^{y_i/\tau}} \tag{2}$$

where τ is a "temperature" parameter. High values of τ make all actions equally probable and low values favor the action with the greatest y_m.

We shall be using the $TD(\lambda)$ algorithm and we provided a sketch of the equations for training the hidden to output connections of the MLP (loosely based on [4]).

$$\vec{w}_{t+1} = \vec{w}_t + \alpha(\gamma\, y_{t+1} - y_t) \sum_{k=1}^{t} (\gamma\lambda)^{t-k} \nabla_{\vec{w}} y_k \tag{3}$$

$$\vec{w}_{t+1} = \vec{w}_t + \alpha(r - y_t) \sum_{k=1}^{t} (\gamma\lambda)^{t-k} \nabla_{\vec{w}} y_k \tag{4}$$

where t denotes the time step, vector $\vec{w}$ represents the weights from the hidden units to the output unit, y_t is the network's output and r is the reinforcement signal. For the parameters we have $\lambda \in (0,1)$, $\gamma \in (0,1)$ and $\alpha \in (0,1)$. λ denotes a decaying dependance upon all previous gradients of y, γ is a discount of the next evaluation y_{t+1} and α is the learning rate, Eq. 3 is valid for all time steps apart from the last one, while Eq. 4 is used only for the last time step, when the reinforcement signal is available. Finally, $\nabla_{\vec{w}}$ is the gradient of the output with respect to the weights. The aforementioned equations tell that weight changes occur according to the difference between the current and the next estimate of the MLP coupled with the gradients of the past choices.

Provided we are looking for a rule for cat_k, we will adopt the following form for the reinforcement signal which should be maximised:

$$r = \frac{a_1 \cdot \frac{data(cat_k)}{total(cat_k)} - a_2 \cdot \frac{data(other)}{total(other)}}{a_1 + a_2} \tag{5}$$

where $data(cat_k)$ is the number of data from cat_k that fall within the current rule and $total(cat_k)$ is the total number of data from cat_k. The rest, i.e. $data(other)/total(other)$ represents the percentage of the misclassified data, which is a form of penalty. The factors a_1 and a_2 tune the penalty of misclassified data items.

Eq. 3 and 4 assume a recursive form if we set:

$$\vec{e}_{t+1} = \sum_{k=1}^{t+1} (\gamma\lambda)^{t+1-k} \nabla_{\vec{w}} y_k \tag{6}$$

Thus, using Eq. 6 and adding a momentum term with parameter μ, Eq. 3 and Eq. 4 assume the following form:

$$\vec{w}_{t+1} = \vec{w}_t + \alpha \cdot err_t \cdot \vec{e}_t + \mu \cdot (\vec{w}_t - \vec{w}_{t-1}) \tag{7}$$

where,

$$err_t = \begin{cases} \gamma\, y_{t+1} - y_t \\ r - y_t \end{cases} \tag{8}$$

We set $\vec{e}_0 = 0$, the quantity $\vec{e}_t$ is called ***eligibility trace*** and carries the previous "behaviour" of the MLP. We should also observe that the size of $\vec{e}_t$ equals the size of $\vec{w}$.

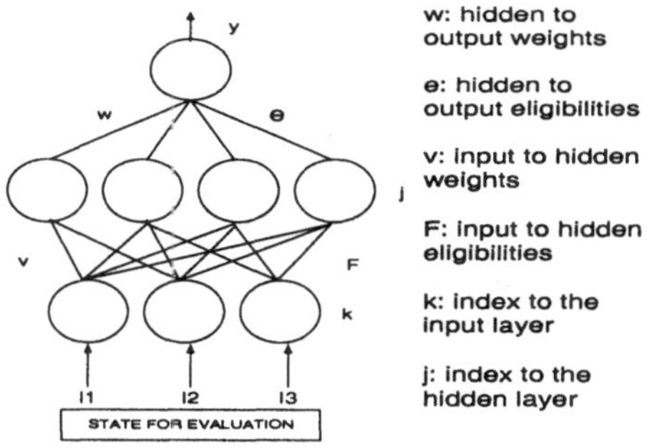

Fig. 1. The MLP used for the estimation of value functions.

The weights from the input to the hidden layer form a two dimensional matrix $V = [\vec{v}_1, \vec{v}_2, \ldots]^{\top}$, each row of this matrix is the weight vector from the input units to a hidden unit. The eligibility traces for the weights of the hidden layer are represented by the matrix $F = [\vec{f_1}, \vec{f_2}, \ldots]^\top$, which has the same size as V since there is an eligibility trace for every weight. The eligibility traces of the weights from the input to the hidden layer are determined as follows: Let I^k be the kth external input to the MLP and $\vec{I} = [I^1, I^2, \ldots]$. We set

$$h^j = \sigma(\sum_k v^{jk} I^k) \quad (9)$$

where $\vec{h}$ is the output of the hidden layer and $\sigma(x) = \frac{1}{1+e^{-x}}$. The superscripts denote connection to a target unit from a source unit (left to right). (see Fig. 1). Similarly to the $\vec{w}$ weights, the equations for the V weight changes are as follows (setting $f_0^j = 0$ for all j):

$$v^{jk}_{t+1} = v^{jk}_t + \alpha\, err_t\, f^{jk}_t + \mu(v^{jk}_{t+1} - v^{jk}_t) \quad (10)$$

$$\vec{f}^j_{t+1} = y'_{t+1} w^j_{t+1} (h^j_{t+1})' \vec{I}_{t+1} + \gamma\lambda \vec{f}^j_t \quad (11)$$

We can stop the training of the MLP when the error is below a threshold, where the error is defined in Eq. 8. Another stopping criterion is when the probability of selecting a state outweights the probabilities of the alternative states and this has been stabilised. It is not necessary for a rule to contain all the features of the data set, as a matter of fact it is desirable that it contains as few as possible. Initially, there is a pool of data along with their labels. During the rule generation process, the data vectors that are classified by the previously generated rules are removed from the pool. The rule extraction algorithm is summarised below (see also Fig. 2).

1. Perform Principal Component Analysis.
2. Partition the input space.
3. Search for rules (Eq. 1) for each label:
   (a) The symbols that constitute a rule are: *features* and *intervals*.
   (b) For a number of episodes or until a criterion is met
      i. For t=1..3*(number of features):
         A. The MLP suggests a value for the quality of each state.
         B. The best state is selected according to the Boltzmann-Gibbs distribution.
         C. Weight changes are performed.
      ii. A reinforcement signal is produced depending on the classification capability of the newly created rule, which is tested on the given data.
      iii. The data vectors that are classified by the previous rule are removed from the pool.
4. If there are still unclassified points go to step 2, but apply a finer partitioning. A rule of thumb value for a finer partitioning is in powers of 2, i.e. in the finer partitioning the tile edge is two times smaller.

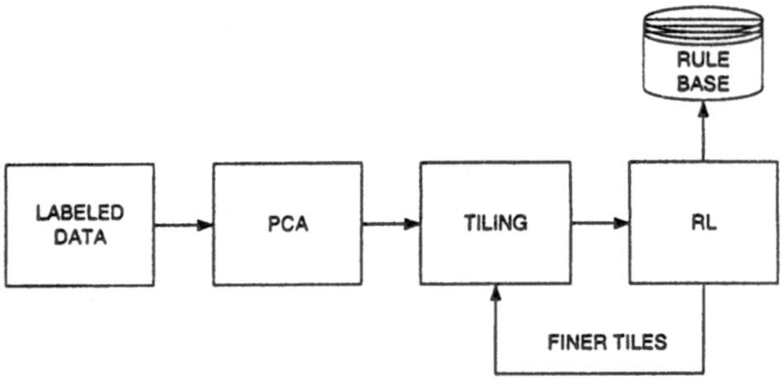

Fig. 2. The rule extraction process.

4 Experiments and Conclusions

In all experiments we set $\lambda = 0.9$ and $\gamma = 1$ (Eq. 3, 4, 10, 11). In Eq. 2 we set a variable τ which decreases monotonically as episodes pass. This seems to improve results compared to a constant value of τ, since we allow a heavy amount of exploration to take place in the beginning, and turn to exploitation of the best solutions near the end.

We have evaluated our method on the following data sets [5]

Iris: 150 vectors, 4-features, 3-labels. *Concentric*: 2500 vectors, 2-features, 2-labels. *4-Gaussians*: 1200 vectors, 2-features, vectors, 4-labels. *Pima Indians*: 768 vectors, 8-features, vectors, 2-labels. *Image Segmentation*: 2310 vectors, 19-features, 7-labels. We performed two experiments for each of the Indians and the Image Segmentation data sets; the first without data preprocessing, the second with PCA. For the classification ability we obtained similar results for both experiments but there is substantial difference in the number of rules. We believe that this is due to the tiling which in the first one does not take into account the density of the data points.

For each experiment, we provide the classification ability on the training set and the testing set, the number of rules that were created, the average number of data points that are classified by a rule and the different tilings that were used. These tilings refer to the population of tiles along the feature with the largest range. Furthermore, we report on the average number of antecedents per rule as well as on their range (min and max value). The numbers appearing are averages over 10 times, where half of the data set was used for training and the other half for testing.

The equations for the TD(λ) presuppose discrete time, states and actions. There have been many attempts to pass from the continuous to the discrete [6]. Furthermore many authors have suggested the use of a *skewing function* which projects the state space onto the skewed state space, so that less important regions are contracted and important ones are expanded [7]. We could say that our skewing function is the PCA. However PCA is not universally applicable thus *Independent Component Analysis* or the discovery of *non linear principal curves* might be good suggestions [3]. In our method, we remove the data that are classified by a rule, thus the next rule has fewer data to classify —this renders the task of rule extraction somewhat different from control problems, where the aforementioned partitioning techniques are very effective.

A crucial point in RL is the value function, which in our case has been approximated by an MLP. For this case there is no proof of convergence which might affect the rule extraction process. However, we could implement the approximation function as a weighted set of gaussian functions which is based on *on-line EM*. In this case there exists a proof of convergence, an efficient implementation as well as a possibility to escape from local maxima [8].

The generalisation ability of the rules could be improved if we applied more sophisticated techniques, for instance, a reinforcement signal could be produced based on a validation data set. Finally, a clustering of the input space could facilitate a parallel implementation, whereby a separate rule extraction would be activated for each cluster.

	Iris	Con	4-Gauss
Tr	96%	97%	100%
Te	95%	96%	96%
Num Rules	4	25	32
Data per Rule	18.75	41.8	17.4
Tiling pop.	8-16	8-16	8-16
Num.Antec.	3	1.72	1.96
Ant. range	2,4	1,2	1,2

	Pima	Pima PCA	Seg	Seg PCA
Tr	82%	84%	70%	80%
Te	78%	75%	60%	71%
Num Rules	95	41	90	85
Data per Rule	4	9.24	12.8	13.5
Tiling pop.	8-128	8-128	64-128	32-128
Num.Antec.	2.8	3.73	10.7	5
Ant. range	1,5	2.5	1-15	1-8

References

[1] A. Tickle, R. Andrews, M. Golea, and J. Dietrich, "The truth will come to light: Directions and challenges in extracting the knowledge embedded withing trained artificial neural networks," *IEEE Transactions on Neural Networks*, vol. 9, pp. 1057–1068, 1998.

[2] R. S. Sutton and A. G. Barto, *Reinforcement Learning*. The MIT Press, 1998.

[3] V. Cherkassky, *Learning from Data*, ch. 6. John Wiley & Sons, INC., 1998.

[4] R. S. Sutton, "Implementation details of the TD(λ) procedure for the case of vector predictors and backpropagation," Tech. Rep. 87-509.1, GTE Laboratories Incorporated, Aug 1989.

[5] "ftp.ics.uci.edu/pub/machine-learning-data bases, ftp.dice.ucl.ac.be/pub/neural-nets/elena/databases/artificial/concentric/."

[6] R. Munos and A. Moore, "Barycentric interpolators for continous space & time reinforcement learning," *Neural Information Processing Systems*, 1998.

[7] J. Santamaria, R. Sutton, and A. Ram, "Experiments with reinforcement learning in problems with continuous state and action spaces," *Adaptive behavior*, vol. 6, no. 2, pp. 163–217, 1997.

[8] M. Sato and S. Ishii, "Reinforcement learning based on on-line em algorithm," *Advances in Neural Information Processing Systems*, vol. 11, 1999.

Learning Feed-Forward Multi-Nets

R.S. Venema, L. Spaanenburg*

*Rijksuniversiteit Groningen, Dept. of Mathematics and Computing Science, P.O. Box 800, NL-9700 AV Groningen, The Netherlands

Abstract

Multi-nets promise an improved performance over monolithic neural networks by virtue of their distributed implementation. This potential lacks popularity as, without precautions, the learning rate has to drop considerably to eliminate the occurrence of unlearning. This paper introduces extensions of the Error Back-Propagation algorithm to enable function-preserving merging of neural modules at full learning rate.

1 Introduction

Multi-nets are combinations of several neural networks [1]. They are of growing interest, as the implied feature redundancy is believed to make the overall net more accurate than the parts. Moreover, multi-nets can be easier to understand and to modify. The current practice in ensemble networks confirms this expectation. Here, the various solutions to the problem are trained independently, then frozen and subsequently combined in a voting arrangement. For modular networks in general, the use of frozen parts is less acceptable and the assembly becomes problematic.

The monolithic neural network is a combination of neurons with a characteristic transfer function. This function can be explicitly imposed or locally created from a small sub-network. For instance, a sub-network of neurons with linear transfer can behave as a single neuron with a sigmoid transfer. In this sense, the monolithic neural network is already a modular one in disguise and one may therefore expect that the learning problems will be the same. From the observation that, with growing problem size, the monolithic network has increasing difficulty to learn with sufficient quality [2], one may expect not better from a multi-net.

In both cases, the learning process suffers from the entropy in the example set. This can only be resolved (a) by data preprocessing, (b) by inclusion of pre-knowledge or (c) by domain structuring. The claim that more than 80% of the development time for monolithic networks is spent on the data preprocessing underlines this observation [3].

R.S. Venema is currently with the Mathematics Department of Boise State University, Boise (ID).

Modular neural networks address this learning problem by structuring the multi-net into modules, of which some will merely perform functions that were previously part of the data preprocessing. Training such parts individually is not guaranteed to provide for fully matching inputs and outputs. The merging of the parts into a modular neural network will then have errors on the module boundaries, which are directly propagated at the continuation of learning and may wipe out all previously learned information. This effect has been studied in a companion paper [4].

First we discuss some topologies in the construction of a neural network from modules. Subsequently the learning problem is illustrated in the monolithic and modular implementation of the ill-famed Exclusive-OR [5]. Then we focus on the initial merging conditions. We propose the use of a default global learning rate, while each individual module will train by local adaptation to the specific needs, and see that a small delay in the moment on which each module starts to be influenced by the overall learning can be very helpful. This can be achieved by a slightly extended version of the basic Error Back Propagation algorithm.

2 Unlearning Effects

It has been found in a companion paper [4] that unlearning appears in multi-nets in more ways than in monolithic structures. The basic phenomenon is *data interference*: presented data may comprise conflicting targets to the input and/or hidden features and thereby cause non-determinism during error back-propagation.

In multi-nets, a new problem appears because modules are in different phases of knowledge development and thereby suggest unwanted shifts in target. For instance, "empty" modules generate initially large errors that will subsequently propagate through other modules and wipe out the already stored information. To limit the effect of back propagation, the learning rate can be decreased by typically a factor of 25 to 100, leading to a large-scale slowing down of the learning process.

In the following we will discuss 4 different concatenations of network parts. Suppose that a neural network is created from the serial composition of two modules A and B, where A receives the primary inputs and passes intermediate results to B who in turn creates the output signals. Depending on the degree of previous training, there are four situations to be distinguished: none is pre-trained {A,B;}, A is pre-trained {B;A}, B is pre-trained {A;B}, and both A and B are pre-trained {;A,B}. In the following we will look at each of them separately.

{A,B;}. The two modules are empty. The overall network will behave like a single monolithic one when supplying the input/output pairs. The only difference is the possibly incomplete wiring between the modules. Practice shows that this makes the network even more vulnerable to unlearning effects.

{A;B}. Module B is pre-trained, and A is not. Assume A is initialized with small random numbers, then the intermediate results will be small and randomized too. When the network is trained from the overall input/output pairs, B will actually receive small, randomized inputs. Error back-propagation will result in large errors that tend to wipe away the information as initially installed. Keeping the weights in module B fixed will have little effect, as it does not help module A to receive input/output pairs from which it can learn.

{B;A}. Module A is pre-trained, and B is not. The paper will initially focus on this situation. In our XOR example, the trained AND- and OR-modules A are merged with an empty module B. This will cause a large mismatch between the outputs of A and the inputs of B, that shows as a clear error disruption. Preferably, we will try to reach a more fully trained initial situation, as described below.

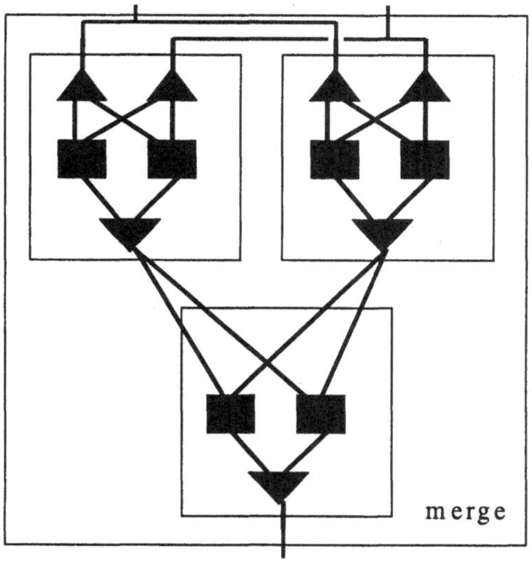

Figure 1 Merging modules into a single neural system

{;A,B}. Both module A and module B are pre-trained. For instance, in the modular XOR example, the B part may have been trained from and(x,y) and nor(x,y) inputs rather than (x,y). The overall network will behave like a single monolithic one when supplying the input/output pairs. However, as training will not be totally perfect, there remains an error disruption on the interface between the modules. Further it requires complex operations on the data file to produce the partial examples fit for the merged network. The paper describes how to handle this situation.

3 Basics of a Modular Learning

As an example of a modular multi-net, we start from a trained AND- and OR-function and combine them by feeding their outputs into a new 2-level Perceptron, see Figure 1. Basically this is a multi-layer Perceptron with some partially trained parts.

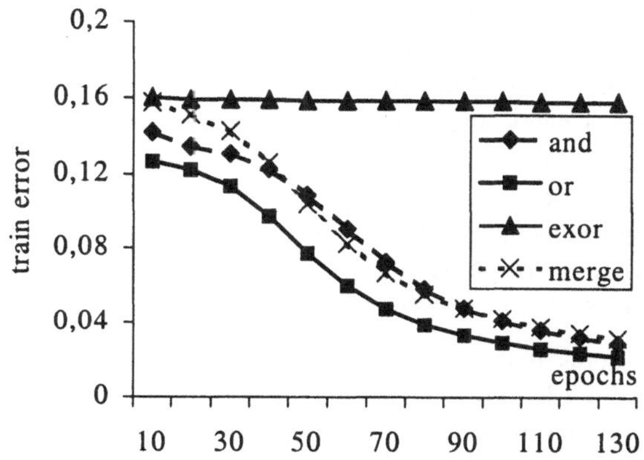

Figure 2 Monolithic versus modular training.

The major benefit of such a modular design shows from Figure 2, where the XOR function is trained directly on a feed-forward network. Though the function mapping is feasible in principle, the actual training is extremely slow and not guaranteed to terminate. On the other hand, training the AND and the OR function is relatively easy. Then, merging the AND- and OR-modules into a multi-net, followed by training for the EXOR is also relatively easy.

This combination may be interpreted as a single monolithic network with 4 hidden neurons to explain the better performance. Taking a closer look at the behavior of monolithic XOR implementations falsifies this suggestion. Though 4 hidden neurons seem optimal for a monolithic network (Figure 3), training will still take much longer than for the smallest modular network. Apparently, the XOR multi-net implementation has a constructional advantage.

The evident success of this approach for real-life problems (for instance [6]) leads to the question whether this can be parallelized, i.e. whether the modules can be trained simultaneously? This may seem a question that is almost too trivial to answer. When the modules can be trained independent of one another, they can clearly be trained in parallel also.

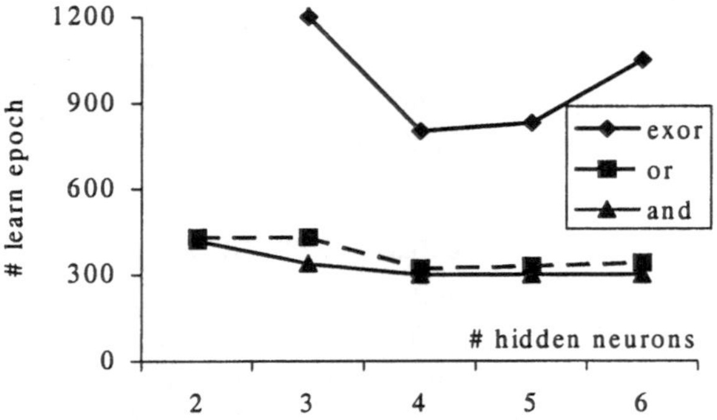

Figure 3 Dependence of learning on hidden neurons.

The results of such an experiment are shown in Figure 4. When the structure is treated as a single monolithic block, training is still bad. However, when the modules are first trained independently and only in combination after a delay, the learning time improves considerably. With the increase in delay, the subsequent improvements get smaller. Finally, there is hardly any improvement visible. What remains is a small switching response, the size of which is related to the error size at the interface between the modules.

It seems that a delay of 60 epochs is sufficient for a minimal learning time. This illustrates that the effect of error disruption in modular networks with pre-trained parts can be effectively remedied with a minor adaptation of the classical Error Back-Propagation algorithm, that handles a flexible switch-over from local to global adaptation. In this paper we focus on the means to detect the onset of unlearning.

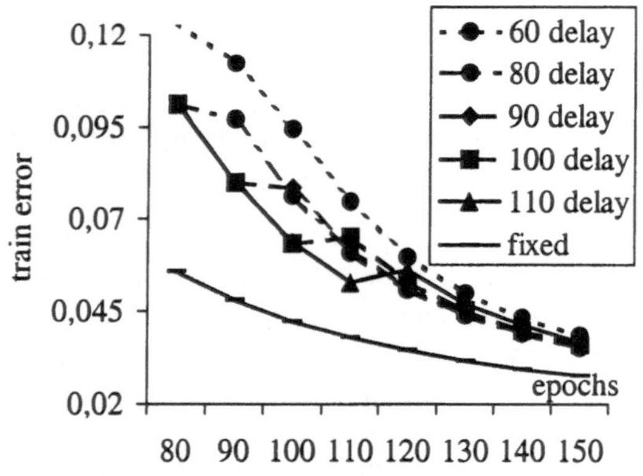

Figure 4 Impact of large scheduling delay.

4 Onset of Unlearning

Where the merging of neural modules can be scheduled such as to preserve the pre-trained functionality, the same mechanism can also be used to detect the onset of unlearning. This is of special relevance in applications, where the neural network is involved in autonomous on-line monitoring and diagnosis [7].

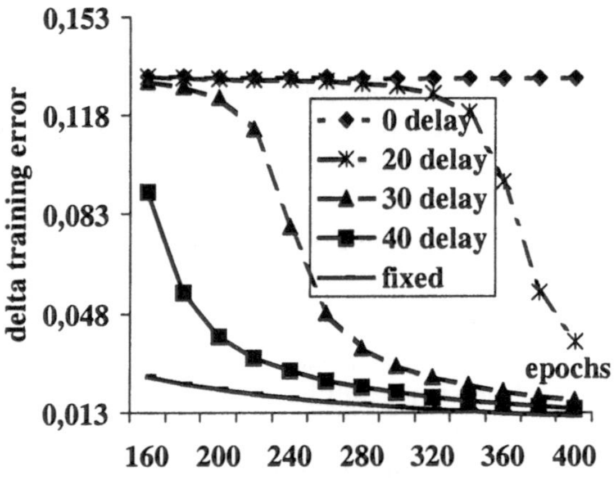

Figure 5 Impact of small scheduling delay.

Within a monolithic feed-forward structure, unlearning is directly connected to a re-orientation of the hidden layer. The neural outputs are combinations of the hidden features and some adaptation of the weights in the synapses between the hidden and the output layer will be sufficient to handle small changes in the required output function. In other words, the existing hidden features define the functional range of the neural network. No unlearning will start, when the hidden features can comfortably produce the output value.

When the examples become intrinsically different, it will become necessary to change the definition of the hidden features too. Then the weights between the input and the hidden layer will be drastically changed. Such changes will clearly involve unlearning of the old features, followed by re-learning of the new features. From this statement it may be clear that unlearning is easily characterized by a prolonged learning time. This is the same characteristic that can be found in Figure 2, where any remaining training problem on the interface between the modules is directly reflected in a noticeable increase in learning time.

A typical example is shown in Figure 5, where a trained XOR network is suddenly receiving samples from an AND function. The difference is small as the AND is already available as an internal module, but the output module must suddenly be changed to an identity relation. We can see two effects. First, it takes some time before the internal modification becomes visible on the output. Second, the re-learning may look in the first instance as a chaotic internal movement.

The first observation is of direct importance here, as this means that though large errors are propagated this does not become immediately visible. Hence the outputs may be false for some time, which may have a direct consequence to the reliability of the overall system. This is similar in effect, as when incomplete learning is pursued to investigate the dynamical frontiers of training [8]. Therefore, an immediate detection of large internal learning efforts is only possible after a long enough period of training.

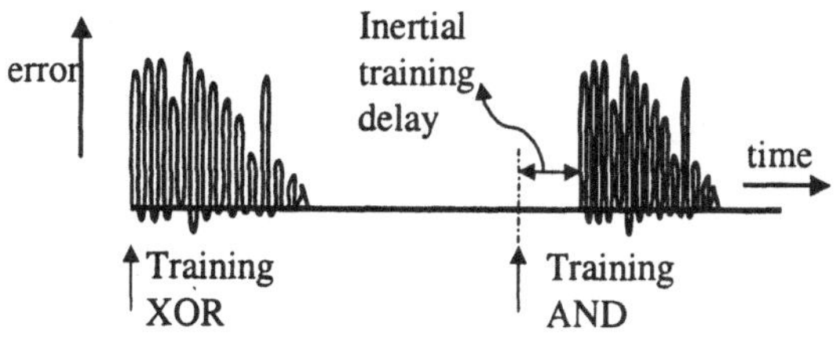

Figure 6 System output during re-training.

Based on these observations, we propose here a means to apply the Error Back-Propagation to the training of feed-forward multi-nets.

5 Error Back Propagation

Apparently there is an easy test on whether a module can be merged into the overall learning process. Barakova has already indicated such a test when analyzing the effect of data interference on monolithic neural networks [9]. She assumes that all modules should first be submitted to a short training cycle to find their effective initialization rate. This sets the value for the delay in switching from local to global learning. We take the switching delay as the time elapse till the output error has dropped 90%.

Here we conjecture that we can also handle this propensity to unlearn on-line, by trial-and-error. On the basis of this insight, we propose the following algorithm for the EBP training of modular networks:

Step 1 [Initialization] Let the modular neural network be modeled by the graph N=(M,E) with the nodes M for the modules and the edges E for the inter-module connections. The global train list GTL starts with the nodes M without outgoing edges; all remaining modules are part of the local train list LTL.

Step 2 [Classical EBP] Present a new pattern to the network inputs and evaluate the network outputs. Perform the standard error back-propagation on the GTL modules with the global learning rate. Perform the standard error back-propagation for the modules in the LTL on an individual basis.

Step 3 [Connection check] Check the errors on the edges between GTL and LTL modules. For any module with all output errors within a 10% margin, time stamp the modules if so far unstamped. For any module with some output errors outside a 10% margin, remove the existing time stamp.

Step 4 [Conformity Check] Move all LTL modules with a time stamp older than one epoch to the GT list. Move all GTL modules with a time stamp younger than one epoch to the LT list. If LTL is empty and no module in GTL is younger than one epoch, stop Otherwise goto Step 2.

This procedure can be used for the initial multi-net, but also for the detection of novel data. Here the criterion is based on the duration of the training effort. If the learning is performed in just a couple of epochs, this is evidence of a mere adaptation. When the training takes considerably longer, unlearning is in effect, which signals the presence of abnormalities.

The difference is that for normalities, the training will be performed in only a small number of samples. The interpretation of Figure 4 and Figure 5 is then that where a degree of abnormality is amplified in the increase of the required scheduling delay. This makes the neural model a sensitive device for learning aberrations. More on this will be discussed in [4].

References

[1] Sharkey, A.J.C. (ed.): Combining artificial neural nets. Heidelberg: Springer 1999.

[2] Macready, W.G., Siapas, A.G, and Kauffman, S.A.: Criticality and parallelism in combinatorial optimization, Science *271*, pp. 56-59 (1996).

[3] Schuermann, B.: Applications and Perspectives of Artificial Neural Networks, VDI Berichte *1526*, pp. 1-14 (2000).

[4] Spaanenburg, L.: Unlearning in feed-forward nets, this proceeding.

[5] Sprinkhuizen-Kuyper, I.G., and Boers, E.J.W.: A local minimum for the 2-3-1 XOR network, IEEE Trans. on Neural Networks *10*, pp. 968-971 (1999).

[6] TerBrugge, M.H., Nijhuis, J.A.G., and Spaanenburg, L.: License-Plate Recognition, pp. 263-296, in: Jain, L.C., and Lazzarini, B.: Intelligent Techniques in Character Recognition: Practical Applications: CRC Press 1999.

[7] VanVeelen, M., Nijhuis, J.A.G. and Spaanenburg, L.: Process fault detection through quantitative analysis of learning in neural networks, Proceedings ProRISC'2000, pp. 557-565 (2000).

[8] Nijhuis, J.A.G., Siggelkow, A., and Spaanenburg, L.: Delay-sensitive learning of neural networks, Digest INNC pp. I-697 (1990).

[9] Barakova, E.I., Learning Reliability: a study on indecisiveness in sample selection, Ph.D. thesis Groningen University: Groningen 1999.

Unlearning in Feed-Forward Multi-Nets

L. Spaanenburg*
*Rijksuniversiteit Groningen, Dept. of Mathematics and Computing Science, P.O. Box 800, NL-9700 AV Groningen The Netherlands

Abstract
Multi-nets promise an improved performance over monolithic neural networks by virtue of their distributed implementation. Modular neural networks are multi-nets based on an judicious assembly of functionally different parts. This can be viewed as again a monolithic network, but with more complex neurons (the neural modules). Therefore they will share the same learning problems, notably the unlearning effect. In this paper we will look more closely into the reasons for unlearning and discuss how this can be applied to detect novelties.

1 Introduction

Multi-nets are combinations of neural networks [1]. Different names are used depending on the nature of the combination. *Ensemble* networks are based on a redundant collection. Each part gives more or less the solution; together they give the best. In *hierarchical* networks, partial solutions are structurally combined in a "part-of" fashion. The partial solutions are not necessary redundant but a degree of overlap to ensure continuity is well advised. When the combination is of a different nature (child/parent, co-operation, supervision), the notion of *modular* network is used [2].

Multi-nets are of growing interest, as the impact of redundancy is believed to make them more accurate than single-nets. Moreover, multi-nets can be easier to understand and to modify. The current practice in ensemble networks confirms this expectation. Here, the various solutions to the problem are trained independently, then frozen and subsequently combined in a *voting* arrangement. For modular networks in general, the use of frozen parts is less acceptable and the assembly becomes more of a problem.

The monolithic neural network is a combination of neurons with a characteristic transfer function. This function can be explicitly imposed or locally created from a small sub-network. For instance, a sub-network of neurons with linear transfer can behave as a single neuron with a sigmoid transfer. In this sense, the monolithic neural network is already a modular one in disguise and one may therefore expect that the learning problems will be the same. As it has been noted that with growing problem size the monolithic network has increasing difficulty to learn with sufficient quality [3], one may expect no better from a multi-net.

In both cases, the learning process suffers from the entropy in the example set, which in individual cases can lead to catastrophic forgetting [4] (also called interference or cancellation). This can only be resolved by giving suitable general directions, as (a) by data preprocessing, (b) by inclusion of pre-knowledge or (c) by domain structuring. The claim that for monolithic networks more than 80% of the development time is spent on data preprocessing supports this [5].

Modular neural networks address this learning problem by structuring the multi-net into modules, of which some will merely perform functions that are previously part of the data preprocessing. Such modules can often have a strict mathematical content; in this case the multi-net is also called *heterogeneous*. With suitable initialization, the multi-net shows the structure of the domain problem. The major benefit is to the designer, who wants to prototype his solution directly from the raw input data. This enables an immediate view on the problem and its solution by recalling the multi-net.

Apparently modular neural networks may have an improved performance, if the data interference can be solved by a proper transformation of the input data. The further difference with monolithic networks is the facility to segment the domain solution, again reducing the chance on the appearance of learning problems. In a companion paper we discuss how a modular network can be suitably trained. Here, the focus is rather on how unlearning can be utilized as a detection mechanism for novelties.

After an introduction to the nature of data interference, we will illustrate its operation by analyzing the performance of the ill-famed Exclusive-OR [6]. This small and seemingly trivial network displays clearly all the negative effects of data interference without cluttering by other phenomena. Subsequently we move on to techniques to make the network explicitly vulnerable to external aberrations and demonstrate the use of unlearning in the detection of novelties.

2 Catastrophic Forgetting

Symmetries in the operation of neural learning are usually beneficial as the randomized selection of weights together with the random presentation of examples provide enough leeway to eliminate the detrimental side-effects. However, in function approximation, example randomization brings in the longest-ruin effect: in the long run enough symmetry is available in the presentation set that unlearning can not be excluded beforehand.

First of all, symmetry can already be present in the example set. When for instance the sine-wave y=sin(x) is trained, one example may force the result y_1 to be caused from x_1 in the first quadrant, while another sees y_1 to be caused from x_2 in the second quadrant. The mixed presentation of such signals will cause internal interference as $\sin(x) = \sin(\pi\text{-}x)$, leading mostly to increased learning time or even erroneous results.

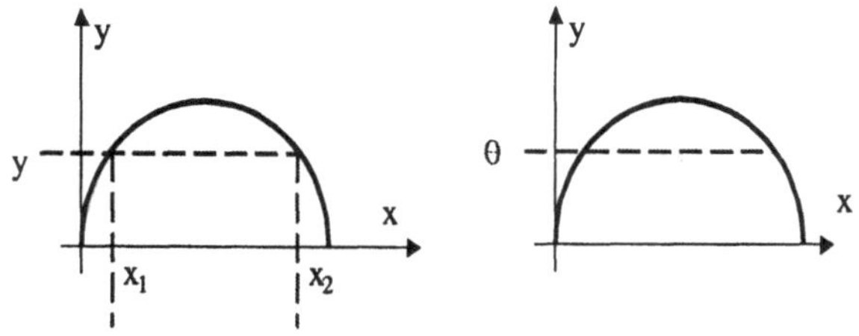

Figure 1 Training a half sine wave.

A second symmetry may come from the architecture of the neural net, more specifically from the discriminatory function that creates the axon value. A number of discriminatory functions have been posed in the past, of which especially the zero-centered sigmoid $f(x) = (1-e^{-x})/(1+e^{-x})$ appears to make the network very vulnerable for symmetry effects. In [4], this discriminatory function has been exploited to make the detrimental occurrence of data interference due to symmetry effects most visible.

The architecture of the fully connected feed-forward net brings a third symmetry as any hidden neuron can be trained to a specific feature. Assuming all required features to be present, the trained function is invariant to the binding of features to hidden neurons. In other words, whereas $y = f(\sum_{k=0}^{n} w_{kj}(f(\sum_{j=0}^{m} w_{ji}x_j)))$, any structural ordering of the m hidden neurons will do.

Connected to the above hidden invariance is the observation that the function of the feed-forward network is based on building solutions from the hidden features. Such hidden features are created in the first layer(s) of the network and do not have to be unique as long as the composition in the output layer provides the correct result. It has been indicated, that from the architecture of the feed-forward net a number of symmetries in the error landscape result that each will provide the desired response on the output [8]. On the other hand, the network may migrate during learning from the one set of hidden features to the other.

The data interference seems caused by indeterminacy and therefore repairable by adding input features. However, the conventional random initialization may often lead to internal indeterminacy which is harder to diagnose and may be facilitated by the many alternatives offered by structural symmetry. The impact of structural alternatives may be offset by realization choices. In [1] it has been shown that the three symmetries will in combination have a severe deleterious effect on the outcome of neural training.

Modules can be used to disperse symmetry in a local and manageable structure. Such symmetries can occur on any level of packaging and will therefore complicate the discussion on modular networks. For the earlier mentioned sine-wave problem, there are a number of suitable, goniometric solutions. The simplest one is to model one quadrant only and to combine this with a phase indicator to model the full signal. However, the potential for unlearning can still be set off by the error disruption caused by module assembly.

3 The XOR Multi-Net

To illustrate the way that data interference leads to unlearning in a modular setting we discuss here a simple experiment, wherein the XOR function is trained directly on a feed-forward network. Though the training is feasible, it is also extremely slow and not guaranteed to terminate. On the other hand, training the AND and the OR function is relatively easy. Then, *connecting* the AND- and OR-modules over a single additional neuron, followed by training for the EXOR, is also relatively easy.

In a way, the difference between the monolithic and the modular approach can be viewed as an initialization problem: by separating in front the two symmetrical parts of the function, the data interference is taken out of the learning. The way we have trained the connected network has been straightforward, as we are able to correctly insert the knowledge in the right place. We start from a trained AND- and OR-function and *merge* them into an overall modular Perceptron by feeding their outputs into a new 2-level module.

When we perform learning on this structure, there appears to go a disruption through the network in the early phases of learning for reason of the sudden contact between pre-trained and "empty" parts. This makes that initially the function output reflects the empty part, while the back-propagated error is intensified when it returns at the interface with the trained parts (Figure 2). Fortunately this disruption soon dies out. We see no difference in the error curve whether the interconnections between the pre-trained functions and the merger are fixed or trainable.

As reference, we have also plotted the learning curve for the monolithic XOR with 4 hidden neurons. Though a 2-2-1 solution is feasible, a 2-3-1 solution is more comfortable to reach by training while 2-4-1 seems optimal. One may assume that the modular construction of a 2-hidden OR and a 2-hidden AND may be equivalent to a 4-hidden monolithic construction, as found earlier to give the optimal response time. Apparently, this is not the case. The learning curve for the 2-4-1 XOR comes only down at around 430 epochs.

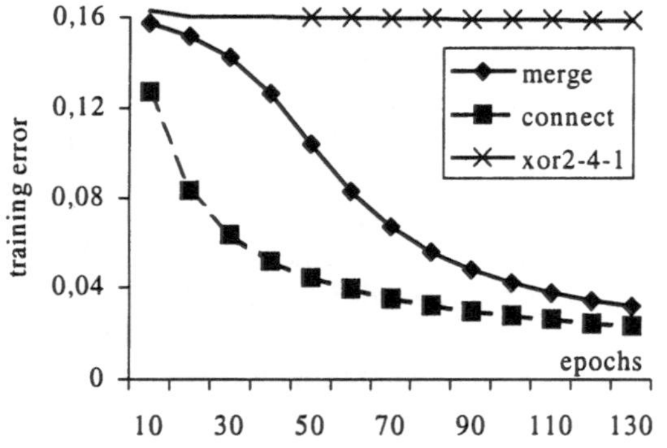

Figure 2 Training with modular inserted knowledge

On closer look, the weight settings have shifted when learning the assembly with trainable interconnect (Figure 3). Though the overall performance does not seem to be affected, the modules are not functionally optimal anymore. For the individual input responses, the change is sometimes for the better, but regretfully the opposite may also be true. It appears that the adaptation is consistent in one direction for those vectors (1='10' and 2='01'), that are redundant for the overall function, and disruptive for the ones (0='00' and especially 3='11'), that are internally conflicting.

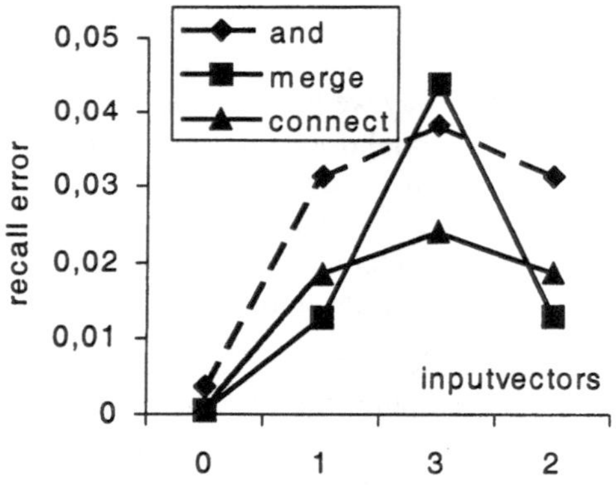

Figure 3 How training of the overall XOR influences the modular parts.

The above example has shown that assembling pre-trained neural modules into a more complex function has two effects. The first is the error disruption caused by the contact between the learned and the unlearned part; the second is the personalization of the modules to the benefit of the overall function. Both effects are potentially harmful to the integrity of the assembly network and the stored knowledge may disappear [9]. Usually this is attacked by considerably lowering the learning rate, leading to lengthy training [10].

The easy alternative seems to assemble only trained networks. In our example, this involves the trained AND- and OR-modules to combine with an XOR-module that was already learned from and(x,y) and nor(x,y) inputs rather than (x,y). However, as training will not be totally perfect, there will remain an error disruption on the interface of the modules. Further it requires complex operations on the data file to produce the partial examples fit for the merged network. The evident success of this approach for real-life problems (for instance [11]) leads to the question whether the modules can be trained simultaneously?

4 Unlearning in Novelty Detection

The desired handling of modular assembly is that the new, unlearned part is first brought into unison with the included knowledge and later on allowed to personalize the module content in such a way that the initial knowledge is always retained and never overruled. In other words, the back-propagation of an error needs to have a different effect on inserted knowledge as compared to fresh ignorance.

The proposition is therefore to have a single learning rate, which is globally effective, and an error-dependent derivation, effective per module. This is a different adaptive behavior than published in [12] where the rate is gradually lowered at the end of learning to improve convergence.

To have a quantitative feeling for the impact of such local learning rates we have first experimented with degrees of error back-propagation on the modular interfaces for the modular EXOR solution. As stated in the discussion of Figure 2, the overall performance will not be effected but the functional correctness of the parts is. Hence in Figure 4 we show the functional correctness of the AND-module after using it in a modular EXOR.

It appears that the amount by which the error is passed back to the learned modules leaves a permanent effect on the initialized knowledge. Even a minimal back propagation sets the destruction process in motion. This can be explained from the observation, that the output reference is lacking for the AND module and therefore even a small error will be used for weight adaptation.

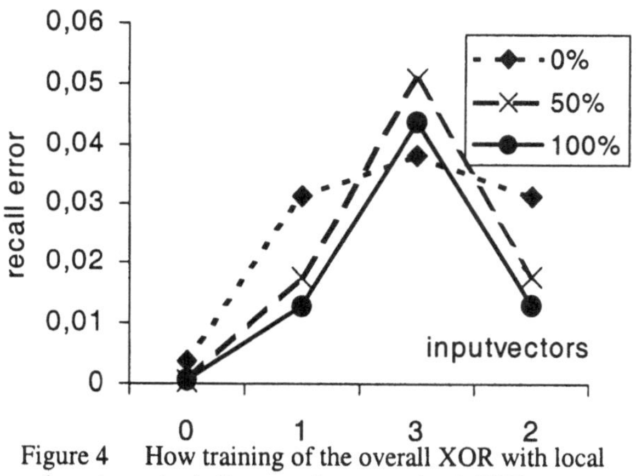

Figure 4 How training of the overall XOR with local learning rates influences the pre-trained AND part.

Where in a companion paper [7] we have continued at this point to find the proper way to learn, the interest lies here in the proper way to unlearn. Apparently, the knowledge within the AND module is affected during the training in a degree, that may cause permanent harm. In Figure 4 there is still room to recover, but at a potentially prolonged learning.

Barakova has already concluded for a range of 9 complex signals [4], that increasing the amount of data interference in the training set has at first little effect. However, learning time increases steeply when the amount crosses a critical value: the *unlearning threshold*. This has been largely explained from the indecisiveness of the learning process.

We conclude here, that even a correct assembly will be vulnerable during post-training. Due to the existing initialization of the parts, the unlearning threshold will have shifted but will still be there. As a consequence, novelties and abnormalities that are large enough to bring the network (or one of its parts) across the threshold will still cause unlearning.

This makes the neural network very suitable for abnormality detection. A neural network trained on existing examples will react on new but different examples by a noticeable prolonged period of post-training. Exceeding the expected training time can be easily measured.

In [13], a comparison is made of several classical and neural fault detectors for a number of signals. It appears that the classical techniques handle random disturbances well, but have a problem with chronic disturbances. On the other hand, a neural network has a degree of robustness for random disturbances, while they react quickly on structural changes: the chronic disturbances.

Some examples are reproduced in Figure 5. They show the learning time in case of five typical noise disturbance: (a) small uniform, (b) medium uniform, (c) block wave, and (d) saw-tooth.

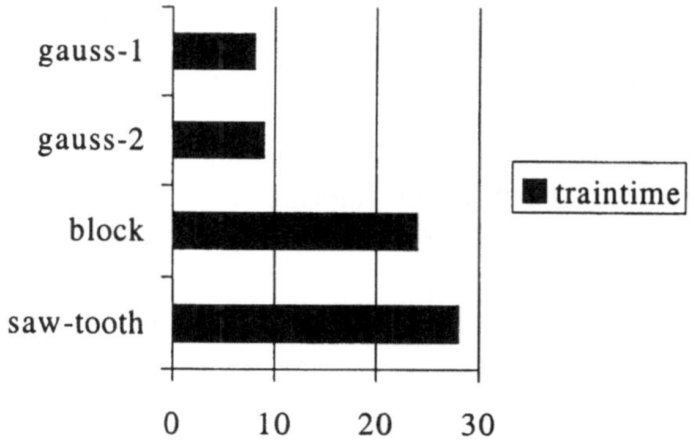

Figure 5 Prolonged training due to disturbance

References

[1] Sharkey, A.J.C. (ed.): Combining artificial neural nets, Heidelberg: Springer 1999.
[2] Caelli, T., Guan, L., and Wan, W.: Modularity in Neural Computing, Proc. of the IEEE *87*, pp. 1497-1518 (1999.
[3] Macready, W.G., Siapas, A.G. and Kauffman, S.A.: Criticality and parallelism in combinatorial optimization, Science *271*, pp. 56-59 (1996).
[4] Barakova, E.I.: Learning Reliability: a study on indecisiveness in sample selection, Ph.D. thesis (Groningen University, Groningen) 1999.
[5] Schuermann, B.: Applications and Perspectives of Artificial Neural Networks, VDI Berichte *1526*, pp. 1-14 (2000).
[6] Sprinkhuizen-Kuyper, I.G., and Boers, E.J.W.: A local minimum for the 2-3-1 XOR network, IEEE Tr. on Neural Networks *10*, pp. 968-971, (1999).
[7] Venema, R.S., and Spaanenburg, L.: Learning feed-forward networks, this proceeding.
[8] Saad, D.: Explicit symmetries and the capacity of multi-layer neural networks, Journal Physics A *27*, pp. 2719-2734 (1994).
[9] Spaanenburg, L., Jansen, W.J., and Nijhuis, J.A.G.: Over multiple rule-blocks to modular nets, Proceedings EUROMICRO'97, pp. 698-705 (1997).
[10] Yamauchi, K., Yamaguchi, Y., and Ishii, N.: Incremental learning methods with retrieving of interfered patterns, IEEE Tr. on Neural Networks *10*, pp. 1351-1365 (1999).
[11] TerBrugge, M.H., Nijhuis, J.A.G., and Spaanenburg, L.: License-Plate Recognition, pp. 263-296, in: Jain, L.C., and Lazzarini, B.: Intelligent Techniques in Character Recognition: Practical Applications, CRC Press 1999.
[12] Yu, X.L., Chen, G.K.C., and Cheng, S.: Dynamic learning rate optimization of the back-propagation Algorithm, IEEE Tr. on Neural Networks *6*, pp. 669-677 (1995).
[13] VanVeelen, M., Nijhuis, J.A.G. and Spaanenburg, L.: Process fault detection through quantitative analysis of learning in neural networks, Proceedings ProRISC'2000, pp. 557-565 (2000).

Context Neural Network for Temporal Correlation and Prediction

Mingo L.F.*, Aslanyan L.†, Castellanos J.‡, Riazanov V.§, Díaz M.A.¶

*Dpto. Organización y Estructura de la Información. Escuela Universitaria de Informática. UPM. Crta. de Valencia Km. 7. 28031 Madrid - Spain †Institute for Informatics and Automation Problems. Laboratory of Discrete Analysis and Modelling Technologies. AM - 375014 Yerevan. Armenia ‡Dpto. Inteligencia Artificial. Facultad de Informática. UPM. Campus de Montegancedo. Boadilla del Monte 28660 Madrid - Spain §Computing Centre. Dept. of Recognition Problems and Combinatorial Analysis. RU - 141700 Moscow. Russia ¶Dpto. Organización y Estructura de la Información. Escuela Universitaria de Informática. UPM. Crta. de Valencia Km. 7. 28031 Madrid - Spain

Abstract

This paper is focused on the application of *Enhanced Neural Networks* to the load demand forecasting. These nets can be considered as *Context Networks* since they are able to process the pattern set in order to obtain a valid context according to the input presented to the net, the context data are expressed in the weights of a neural network. Concerning the load demand forecasting, classical methods require two stages, first stage is a classification net to organize data, and second stage is a forecasting net to output desired response. With *Context Neural Networks*, only one stage is required. The net will perform a classifation and a forecasting process at the same time. Results of *Context Networks* against classical neural methods are compared, showing the improvement of proposed networks.

1 Introduction

To obtain different configurations depending on the input to the net, several architectures could be stated, but surely they will not have any connection with the biologist conception of the human brain.

According to Barbizet [1], different classes of connection among biological neurons could be presented in human brain. These connections propagate the input signal to other neurons increasing or decreasing the power of the signal.

The most usual connection type is the axo-dendritic connection. This connection is based on the fact that the axon of an afferent neuron is connected to another neuron via a synapse on a dendrite, and modelized in *ANN* model by a weighted activation transfer function. But, there exists many other connection types as: axo-somatic, axo-axonic and axo-synaptic [2].

This paper is focused on the second kind of connection type *axo-axonic*. Merely, the structure of the axo-axonic connection can be sketched by three neurons with a classical axo-dendritic connection and the synaptic axonal termination of N_3 connected to the synapse S_{12}. The principle consists on propagating the action of neuron N_3 as synapse S_{12}, see figure 1.

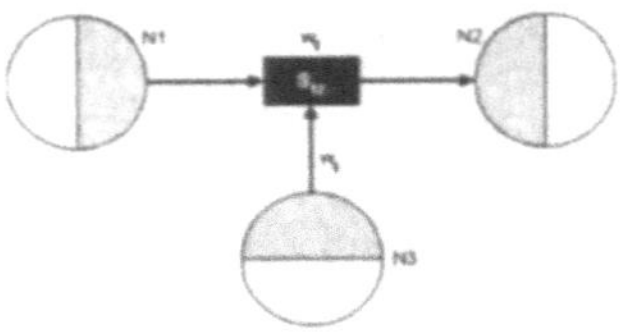

Fig. 1. Modelized connection type.

In order to model previous connection type, two neural networks are required [3]. The first one (assistant net) will compute the weight matrix of the second one (principal net). And, the second network will output a response, using the previously computed weight matrix, this architecture is named Enhanced Neural Networks *ENN* [3, 4].

Figure 2 shows a possible *ENN* architecture. This is the simplest one, since it has no hidden layer and it has only one input and one output neuron, being all of them linear Processing Elements *PEs*. Right net computes weights $W = (w, b)$ of left network in order to output a valid or desired response o. Taking into account previous equations concerning the behaviour of the net, it can be said that the output of the net is ruled by (assuming linear units):

$$\begin{aligned} o &= wx + b \\ &= (w_1x + b_1)x + w_2x + b_2 \\ &= w_1x^2 + (b_1 + w_2)x + b_2 \end{aligned} \tag{1}$$

Equation (1) is a quadratic polynomial, therefore the separation surface generated by this network

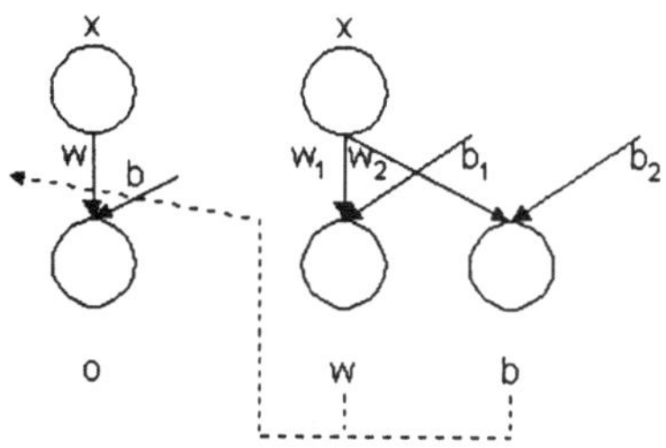

Fig. 2. Enhanced Neural Network Architecture.

(figure 2) is a parabola, instead a plane (in case a linear *MLP* is used). This is a the basic idea of this architecture: *more free parameters* and *a high degree output equation.*

This way, the search in the weight space has more possible solutions (local or non local minima) than *MLP* search, since the space is divide into complex regions in order to fit data, and the number of free parameters is highly increased. As an example, in a classical $5-3-1$ full connected *MLP*, there are 22 weights but with *ENN* there will exist, at least, 132 free parameters.

In order to improve the performance of previous net, polynomial output, activation functions could be changed in such a way that the output equation will be a non polynomial expression. That is, if the net uses sinusoidal activation functions the output will be a cuasi *Fourier* expression that approximates the pattern set. This way, the activation functions can be *Wavelets*, *Ridge* and *Sigmoidals* functions to obtain different separation hyper spaces.

2 Context Neural Networks Overview

Hilberts's 13th problem, although not formulated in the following terms, was interpreted by some as conjecturing that not all functions of 3 variables could be represented as *superpositions* (compositions) and *sums* of functions of two variables. Surprinsingly it turned that all functions could be so represented, and even more was true. *Kolmogorov* and his student *Arnold* proved in a series of papers in the late 50's that there exist fixed coninuous one variable functions h_{ij} such that every continous funcion f of n variable on $[0,1]^n$ could be represented in the form:

$$f(x_1,\cdots,x_n) = \sum_{i=1}^{2n+1} g_i \left(\sum_{j=1}^{n} h_{ij} \right) \tag{2}$$

where one chooses the continuos one variable functions $g_i(x)$.

Function approximation is a finite application of Kolmogorov's theorem, it tries to find a function close enough to given data, but it not necessary to obtain the exact value at these points. Most approximation methods are based on a norm or distance minimization between the real function (or data set) and its approximation.

Observing figure 2, there is an special topic to deal with. Main network output is a non lineal combination of assistant network output (a non lineal combination of inputs). This fact makes this architecture a bit more powerful than classical *MLPs* since they are only a non lineal combination of inputs. This way another free dimension is opened when learning.

Linear architecture behaves like *Taylor Series* when approximating a given function or a data set. An enhanced neural network can be used to approximate a data set with a n-degree polynomial provided that the network had $n-2$ hidden layers. This way, if the data source is a polynomial one then the approximation will be exact, that is, with no error. The only required condition is that the data set had a large amount of data, if it hadn't the approximation could be achieved with many different polynomials.

Figure 3 shows the output of the network corresponding to the learning of different polynomials in the 2D and 3D space. This response has been obtaining by means of the network output corresponding to a grid (projection of the surfaces into a plane).

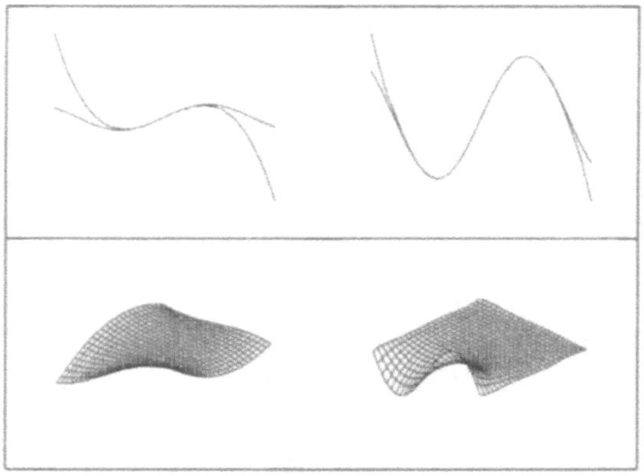

Fig. 3. Approximation with lineal activation functions.

According to previous ideas, *linear ENNs* are better than linear *MLPs*, or at least, they are able to generate complex regions in order to divide the output space. When working with a *MLP*, only hyperplanes can be obtained. And moreover, the degree of the output equation increases according to the number of hidden layers.

In order to obtain a functional basis, one constraint must be made. It consists on implementing the network architecture with lineal *PEs* except the output neurons of assistant network. These neurons must have an activation function $g(x)$ which is used to computed the functional basis as the application of $g(x)$ to a non lineal combination of inputs.

Depending on the activation function of output neurons belonging to assistant network, the main network will output an approximation function based on non lineal combination of elements belonging to the basis. That is if a sinusoidal activation function is implemented, then a cuasi-Fourier approximation is computed by the network; is a Ridge activation function is implemented, then a cuasi-Ridge approximation is computed and so on.

Main advantage of this new approximation method is that is absolutely easy to implement. And moreover, a global approximation to all the pattern set is perform. This way, if there are enough input patterns, then the generalization error will be minimized if there are enough learning iterations.

3 Load Demand Forecasting

Solution using a classical approach involves two modules, the former is an auto associative map that deals with the classification of input patterns into several prototypes[8, 6, 7]. The latter is a set of *MLPs* that deals with the short term load demand problem. Each prototype has an *MLP*, so the patterns that have been classified in the same prototype use same *MLP* to forecast next load sample [5].

Training process is independent in each subsystem. First, the classification is performed by a *SOM* using the *Kohonen* algorithm; and then all the patterns that belong to the same prototype make up the training set for one *MLP* network. In the test process you must identify the prototype of input test pattern and use the *MLP* network together with the prototype to obtain the next sample of load demand.

Figure 4 shows some obtained prototypes after the *Kohonen* algorithm was applied. This means that there exist nine classes, therefore, all days belonging to one class have similar behaviour concerning the load demand forecasting and then, each class is associated to one *MLP* in order to output the desired response.

Obtained results are very encouraging since the performance of the network is 92% in the training and 92% in the test, with a mean square error of 0.02. This error is equivalent to a 600 units deviation of the output of the network (the maximum magnitude of the samples is 30000 units). This section presented a method for the short term load demand prediction. Two stages have been considered : the former performs a classification into several day prototypes and it is based on an auto associative map; and the latter performs the prediction of a sample that approach the real value of load demand of the next hour considered in the input to the net.

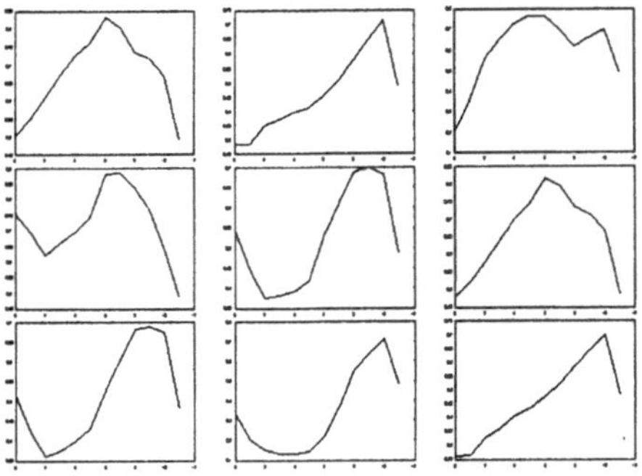

Fig. 4. Prototypes Features using *SOM* networks.

The error obtained by the network is 0.02 units, that is lower enough to assure a reliable behaviour of the system. Same way the performance of both networks is about 92result if the abrupt changes, that the load demand could show, are considered. The obtained results show that this mixed system could be applied in other places or different social situations than the considered in this paper due to the general nature of the proposed solution.

This system improves other methods employed in the time series prediction (weather forecast, economic simulation, etc.) based on statistical models or even expert systems. The principal advantage is the simplicity and speed in the learning of the base behaviour of the patterns that conform the training set.

3.1 Context networks

According to the proposed architecture in section 1 and section 2, this net can approximate any pattern set using a functional basis, lineal or non lineal. In case a lineal basis is employed the net will approximate using a n degree polynomial, being $n-2$ the number of hidden layers; but is a non lineal basis is employed, the net will compute the best approximation using a non lineal combination os the basis, obtaining a cuasi-*Fourier* expression is activation functions are sinusoidal.

First, a lineal context network is proposed to solve the load demand problem. A polynomial was obtained in order to approximate the pattern set, but the *Mean Squared Error* was relatively high. Also, the generalization of this net was very poor. Due to these results a non lineal context network was employed in order to improve previous results.

Patterns are made up of 13 elements, corresponding to the load demand samples every each hour along a day together with the mean temperature in that day. All these patterns are overlapped in order to input the net with the maximum amount of information. The network output must be the next hour load demand.

Table I shows the performance of classical nets against Context Networks, showing both training and test performances.

Context Network		Classical Approach	
Overall Perf.	95%	Overall Perf.	92%
Training Perf.	97%	Training Perf.	93%
Test Perf.	93%	Test Perf.	91%

Table I. Context Networks vs. Classical Approach

Best solution, see table I, has been obtained with a sinusoidal context network with 5 hidden layers with only 1 processing element in each hidden layer. This architecture obtained a 0.01 *Mean Squared Error*, bettern than classical methods, and moreover, the generalization once the net has been trained gets a lower error.

Previous results confirm the power of *Context Networks* against classical network models. The architecture does not introduce computational complexity, it only modifies ,in a light sense, the concept of varying weights from biology point of view. This improvement permits to approximation a function $f(x)$ using a functional basis represented by the net, and therefore, the application of this net to a given problem, load demand forecasting, obtains good results.

4 Conclusions

Taking into account the properties of the proposed neural network architecture in this paper, the function of the assistant network is to preprocess the input data in order to extract enough contextual information to approximate the input to output mapping. Such contextual information is not an input to the net, it modifies the weights of the connections, an important difference with respec to to the classical Multi-Layer Perceptrons.

Another advantage is that the addition of more hidden layers increases the degree of the output equation corresponding to an output neuron in such a way that, if the activation function is a lineal one, then the output of the net corresponds to the polinomial $P(x)$ used in the mathematical approximation based on Taylor series, or based on other method depending on the chosen activacion function. The mathematical definition of the network behavior makes possible to approximate with no error any n-degree polinomial.

Load demand forecasting using *Enhanced Neural Networks* improve obtained results when classical methods are applied. This is mainly due to the fact that *ENN* could be treated as *Context Networks* since they obtain, via the assistant network, usefull information about the pattern configuration in order to use it as weights in the main net.

References

[1] Barbizet, J. & Duizabo, Ph.: *Manual de Neuropsicología*. Barcelona: Toray-Masson S.A. Translate from *Abrege de Neuropsychologie*. Paris: Masson S.A. Pp. 17-21. (1978).

[2] Delacour, J.: *Apprentissage et Memoire: Une Approache Neurobiologique*. Masson(Ed.) September. (1987).

[3] Mingo L.F., Giménez V. & Castellanos J.: *A New kind of Neural Networks and its Learning Algorithm*. 7th Intenational Conference on Information Processing and Management of Uncertainty in Knowledge-based Systems. IPMU'98. France. July (1998). Pp: 1913-1914.

[4] Mingo L.F., Arroyo F., Luengo C. & Castellanos J.: *Learning HyperSurfaces with Neural Networks*. 11th Scandinavian Conference on Image Analysis. SCIA'99. Greenland. June (1999).

[5] Carpintero A., Castellanos J., Leiva S., Mingo L.F., Rios J.; *Short-Term Load Demand with a Mixed Neural Network Systems*. International Conference on Intelligence and Cognitive Systems, ICICS 96. Tehran - Iran. Sept. Pp:60-63. 1996.

[6] Baumann T. & Germond A.J., *Application Of The Kohonen Network To Short-Term Load Forecasting* in Proceedings of the ANNPS International Forum, 1993.

[7] Ligomenides P., Daw-Tung L., *Adaptive Time-Delay Nn For Temporal Correlation And Prediction*. Proceedings of SPIE'92. Boston. 1992.

[8] Cosculluela M.J., Dominguez M.J., Montes R. and Garca Tejedor A.J., *Day Type Identification For Electric Hourly Load Demand Forecasting Using Selforganized Maps*. In Proceedings of NeuroNimes 1993.

A Pruning Self-Organizing Algorithm to Select Centers of Radial Basis Function Neural Networks

Leandro Nunes de Castro, Fernando J. Von Zuben *

*School of Electrical and Computer Engineering, State University of Campinas/SP, Brazil, e-mail: {lnunes,vonzuben}@dca.fee.unicamp.br

Abstract
The appropriate operation of a radial basis function (RBF) neural network depends mainly upon an adequate choice of the number and position of its basis function centers. The simplest approach to train an RBF network is to assume fixed radial basis functions defining the activation of the hidden units, followed by the application of a linear regression procedure to determine the output weights. The main drawback of this strategy is the lack of an efficient algorithm to determine the amount and position of the RBF centers. In this paper, a pruning self-organizing algorithm makes use of the training data in order to initialize the radial basis functions. The method starts with a large amount of centers and prunes the least meaningful ones while positioning them in representative regions of the input space. Simulation results are reported concerning a regression and a classification problem.

1 Introduction

RBF neural networks represent a powerful approximation tool to generate multivariate nonlinear mappings. Their architecture is quite simple and the learning algorithm corresponds to the solution of a linear regression problem, resulting in a fast training process. The performance of the RBF network strongly depends upon the number and positions of the basis functions composing its hidden layer. Traditional methods to determine the centers are: randomly choose input vectors from the training data set [1]; vectors obtained from unsupervised clustering algorithms, such as *k*-means, applied to the input data [2]; or vectors obtained through a supervised learning scheme [3].

A self-organizing feature map (SOM) [4] has been proposed to produce mappings from high dimensional spaces to spaces whose topological dimension is usually inferior to the original one. These mappings are capable of preserving neighborhood relationships among the input data, a property to be explored in a wide range of applications. The SOM can be used to implement the unsupervised part of the RBF neural network [5], with the difficulty that the number of output units, corresponding to the centers of the RBFs, is unknown a priori. An efficient method to determine the number of output units of a SOM can be readily applied to the problem of defining an appropriate set of basis functions for RBF neural networks.

2 RBF Neural Networks

An RBF neural network can be regarded as a feedforward network composed of three layers of neurons with entirely different roles (see Figure 1) [6]. The input layer is made up of sensory units that connect the network to its environment. The second layer, the single hidden layer, applies a nonlinear transformation from the input space into the hidden space. The activation of the nonlinear hidden units is the output of radial basis functions. The output layer supplies each network response with a linear combination of the hidden responses [5, 7].

For a *p*-dimensional input vector $\mathbf{x} = (x_1, x_2, \ldots , x_p)$, where $\mathbf{x} \in \mathbf{X} \subset \Re^p$, the RBF network outputs can be computed by the following expression:

$$y_i = \mathbf{w}_i^T \mathbf{g} = \sum_{j=1}^{m} w_{ij} g_j , \quad i = 1,\ldots,o, \tag{1}$$

where $\mathbf{w}_i = [w_{i1}, \ldots , w_{im}]^T$, $i = 1,\ldots,o$, are the weight vectors for each output neuron *i*, $\mathbf{g} = [g_1, g_2, \ldots , g_m]^T$ is the vector of basis function activations at the hidden layer, and *o* is the number of network output units. Given a set of prototype vectors $\mathbf{c}_j \in \Re^p, j = 1,\ldots,m$, the output of each radial basis function is

$$g_j = h_j(\| \mathbf{x} - \mathbf{c}_j \|), \quad j = 1,\ldots,m , \tag{2}$$

where $h_j(\cdot)$ is the *basis function* and $\| \cdot \|$ is a norm, usually the Euclidean norm, defined on the input space.

In the context of an interpolation problem, a mapping function $y: \Re^p \rightarrow \Re$ satisfying Equation (1), for $o = 1$, has to be determined. Consider a set of *N* data points $\{\mathbf{x}_i \in \Re^p | \, i = 1,\ldots,N\}$. If the desired values are known for all these *N* data points, i.e. $\{d_i \in \Re | \, i = 1,\ldots,N\}$, then each basis function $h_j(\cdot)$ may be centered on one of these data points. So, there are as many centers (prototype vectors) $\mathbf{c}_j$ as data points: $m = N$ [8]. In matrix notation,

$$\mathbf{H}\,\mathbf{w} = \mathbf{d}, \tag{3}$$

where the *N*-by-1 vectors **d** and **w** represent the desired response vector and linear weight vector, respectively. **H** is an *N*-by-*N* matrix, called interpolation matrix. The solution to the problem stated in Equation (3) is given by Equation (4).

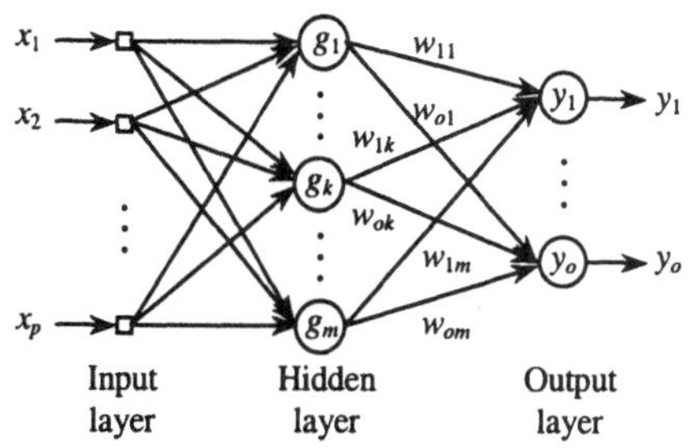

Fig. 1. Radial Basis Function (RBF) network.

$$\mathbf{w} = \mathbf{H}^{-1}\mathbf{d} \tag{4}$$

Micchelli's theorem [9] stated that the only pre-requisite for the existence of $\mathbf{H}^{-1}$, is that the N data points are *distinct*, regardless the values of N and p.

Broomhead & Lowe [6] argued that, the strict interpolation problem described above may not be a good strategy for RBF network training because of poor generalization to previously unseen data.

In addition, if N is very large, and/or if there is a great amount of redundant data (linearly dependent vectors), the likelihood of obtaining an ill-conditioned matrix $\mathbf{H}$ will be much higher. The constraint of having as many RBFs as data points makes the problem overdetermined. To overcome these computational difficulties, the complexity of the network has to be reduced, requiring an approximation to a regularized solution [3]. This approach involves the search for a suboptimal solution in a lower dimensional space. A new set of basis functions $\{h_j(\cdot), j = 1,...,m_1, m_1 < N\}$, assumed to be linearly independent, has to be defined. The set of centers $\{\mathbf{c}_j \mid j = 1,...,m_1\}$ must be determined, such that in the case of $m_1 = N$, $\mathbf{d}_i = \mathbf{x}_i$, $\forall\ i$. If we disregard the regularization parameter, the solution to the overdetermined least-squares data-fitting problem, for $m_1 < N$, is simply given by

$$\mathbf{w}^* = \mathbf{H}^+\mathbf{d} = (\mathbf{H}^T\mathbf{H})^{-1}\mathbf{H}^T\mathbf{d}, \tag{5}$$

where $\mathbf{H}^+$ is the pseudo-inverse of matrix $\mathbf{H}$ [6]. This solution obeys the following equation:

$$\mathbf{w}^* = \arg\min_{\mathbf{w}} \| \mathbf{d} - \mathbf{H}\mathbf{w} \|. \tag{6}$$

Haykin [5] argued that experience with this method showed it is relatively insensitive to the use of regularization, as far as an appropriate choice of the radial basis functions centers is performed.

The simplest approach to train an RBF network is to assume fixed basis functions as the activation of the hidden units. The locations of the centers might be chosen somehow, usually at random or from the training data set. For the RBFs, we will employ a Gaussian function whose standard deviation is fixed according to the spread of the centers

$$h_j(\| \mathbf{x} - \mathbf{c}_j \|) = \exp\left(-\frac{m_1}{d_{\max}^2} \| \mathbf{x} - \mathbf{c}_j \|^2 \right), \tag{7}$$

where $j = 1,...,m_1$ is the number of centers (hidden units), $d_{\max}$ is the maximum distance between the chosen centers, $\mathbf{x}$ is the input vector, and $\mathbf{c}_j$ is the j-th center location. The width, or standard deviation, of the Gaussian radial basis functions is fixed at

$$\sigma = \frac{d_{\max}}{\sqrt{2m_1}}. \tag{8}$$

This formula guarantees that the individual radial basis functions are not too peak or too flat; conditions that should be avoided [5].

So, the only parameters to be learnt are the weights in the network output. A straightforward procedure is to use the pseudo-inverse method presented in Equation (5).

3 The Kohonen Self-Organizing Map

The main goal of the SOM is to group similar input data into clusters. Each output unit may represent a *cluster*, limiting the number of clusters to the number of output units. During the training process, the net determines the output unit that is "closer" to the input vector (called "winner"), based on a given distance metric; the weight vector associated with this unit is adjusted according to a learning algorithm.

In the SOM learning algorithm, some units within a certain neighborhood Nc of the winner have their weight vectors proportionally adjusted, and the learning parameters (like the learning rate, α, and neighborhood radius, Nc) are tuned throughout iterations. The number of input units depends on the training data set, but the output *grid*, in principle, can have a varied number of nodes, arranged in a well-defined topological structure, like the one-dimensional grid presented in Figure 2.

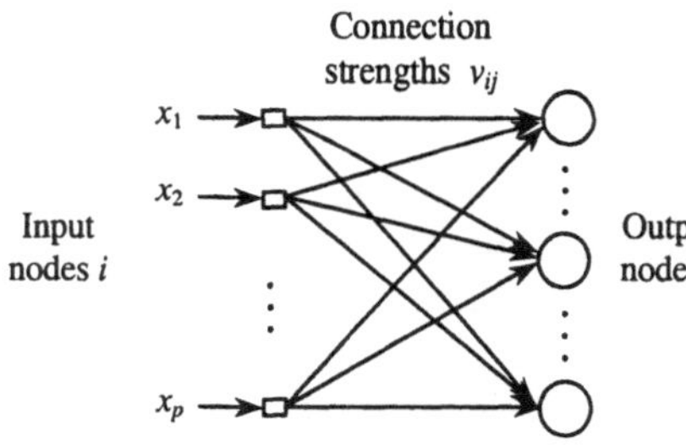

Fig. 2. Typical one-dimensional SOM.

4 RBF Center Selection with Pruning

A great number of variations of the original SOM [4], aiming at determining the most adequate architecture given a certain problem, have been proposed in the literature [10, 11]. De Castro & Von Zuben [12] presented a formal pruning technique, for one-dimensional SOM, that can be directly applied to the problem of selecting RBF network centers.

If the m_1 output units of a SOM represent a mapping from the input space into a space of different dimension, usually smaller, the weight vectors, v_j, $j = 1,...,m_1$, connecting the output units to the network inputs can be used as the prototypes for the RBF centers, $\{\mathbf{c}_j \mid j = 1,...,m_1\}$. The questions that still remain are: (1) How many prototypes m_1 (output units) might we take for the SOM? (2) How can we efficiently determine the values (positions) of these prototype vectors? The algorithm presented by de Castro & Von Zuben [12], called PSOM (for Pruning SOM), can suitably answer to these questions and be directly applicable to the RBF centers selection problem. In this context, this algorithm will be named CSP, for center selection with pruning.

The goal of the CSP algorithm is to find and prune prototype vectors that do not represent a pre-specified minimum number of input patterns. A clustering measure (CM) is defined to evaluate the representative degree of each prototype vector. Those vectors whose CM < ξ (a pre-specified threshold) are candidates to be pruned.

Usually, the stopping criterion for the SOM is a limited number of iterations, or a minimum value for the learning rate (α), assuming that α decreases during the adaptive process. In order to obtain the CM, the pruning procedure must be delayed, i.e., the network must be allowed to run a few iterations, until a stable configuration is reached. Due to the dynamically decreasing learning rate (α) and neighborhood radius (Nc), a restarting procedure for the algorithm is necessary to re-initialize these parameters. As a consequence, every time a unit is pruned, it follows the restoration of the training parameters. The initial state (prototype vectors) of the pruned network corresponds to the final state of the previous network before the pruning procedure.

Making use of the final state of the previous network increases the chance that the new network presents superior results, once we guarantee that the remaining net is capable of appropriately representing the data set. In this study, the suggested stopping criterion is a minimum value (α_{min}) of the learning rate (α), together with a fixed number of iterations (epochs). The parameter ξ is a percent value, i.e., if a prototype vector does not represent at least ξ% of the data, then it is a candidate to be pruned. The idea here is to appropriately define a set of prototype vectors $\{\mathbf{c}_j \mid j = 1,...,m_1\}$ in a number much smaller than the total number of samples ($m_1 << N$). With the pruning phase, the final number of prototype vectors, m_1, is usually smaller than the initial number of prototype vectors, m.

For small data sets, with no more than some dozens of samples, the initial number of prototype vectors (output units) may be taken as the number of available patterns, i.e., $m = N$. This choice can be explained by the fact that in a data set with N samples, there can be no more than N different groups of data (each sample representing a different group). If the data set is very large, this strategy is not recommended due to the high computational cost at the very beginning of the training process, and smaller values are suggested. The pseudocode for the CSP algorithm can be described as:

1. Initialize the parameters:
   - v_{ij}, $Nc_0 \leftarrow floor(m/2)$, ξ, i $\leftarrow$ 1, α_0, m_{min};
   - I: after each I iterations the pruning procedure will be called; and
   - *delay*: number of iterations before the pruning starts.
2. While the stopping criterion is false, do:
   2.1. For each pattern l determine:
      2.1.1. $J = \arg\min_j \{\|v_j - x_l\|\}$
      2.1.2. Increase CM(J)
      2.1.3. $\forall j \in Nc(J)$:
      $$\mathbf{v}_j(new) = \mathbf{v}_j(old) + \alpha\,[\mathbf{x}_l - \mathbf{v}_j(old)]$$
   2.2. If ($m > m_{min}$) and (epochs > *delay*) and (epochs mod I == 0)
      2.2.1. Prune the unit that groups less than ξ% of the samples, $\alpha = \alpha_0$ and $Nc = floor(m/2)$
      2.2.2. Else, keep all the units
   2.3. Decrease Nc and α, and update i$\leftarrow$i+1
3. Test the stopping criterion

The resulting output units correspond to the prototype vectors c_j, $j = 1,...,m_1$ to be used as the RBF centers.

5 Performance Evaluation

In order to illustrate the performance of the proposed center selection method, this algorithm was applied to a regression (REG) and a classification (CLASS) problem. Its results were compared to those of a random and a k-means initialization technique [5]. Although these examples seem to be quite simple, they serve to provide a graphical explanation of the capabilities of the proposed approach.

The proposed algorithm, CSP, automatically generates a set of radial basis function centers satisfying Micchelli's theorem (see Section 2). As can be seen from Figure 3, this algorithm naturally accounts for a regularized solution. The result depicted in Figure 4, shows that CSP determines the RBF centers uniformly distributed according to the input data configuration. The parameters used to train the CSP method were ($delay = 20$):

1. CLASS: $\xi = 0.1$; $m = 50$; $\alpha_0 = 0.9$; $m_{\min} = 5$.
2. REG: $\xi = 0.015$; $m = 100$; $\alpha_0 = 0.9$; $m_{\min} = 5$.

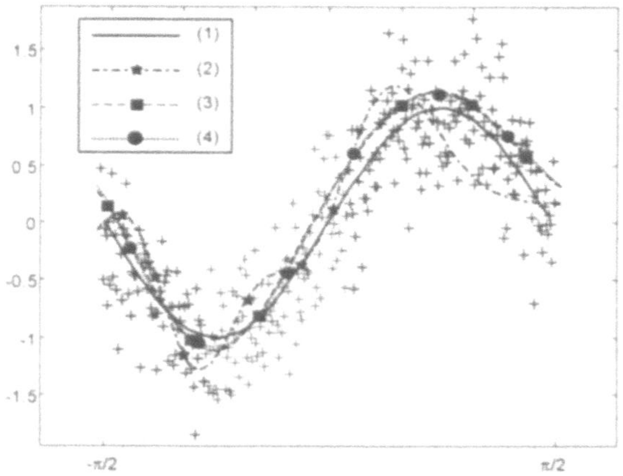

Fig. 3. Regression problem, $m_1 = 6$. *Crosses*: training data; (1): ideal function to be approximated; (2) approximation obtained by the random initialization of centers; (3) *k*-means selection of centers; (4) CSP method.

Fig. 4. Classification problem, $m_1 = 5$ (dark stars). Decision surface determined by the CSP method.

6 Concluding Remarks

This paper presented the development and evaluation of a new paradigm for defining the number and position of radial basis function neural network centers. The proposed learning scheme makes use of the training set for the creation of the prototype vectors based upon an unsupervised algorithm with pruning.

The strategy is plastic in nature, positioning the remaining prototype vectors on locations of the input space which are crucial for the implementation of the input-output mapping. It allows the representation of the input space with different resolution levels by distributing the prototypes according to the density distribution of the data set in the input space.

The performance of the proposed technique was compared to that of the random and *k*-means self-organized initialization procedures. Experiments demonstrated that a random initialization of centers might lead to misclassification and biased approximation. The *k*-means unsupervised selection can waste network resources by creating prototypes in insignificant regions of the input space while ignoring regions that are important for the input-output mapping.

The method presented has the advantage that it allows the construction of a reduced set of radial basis function centers in which linearly dependency among the centers is avoided, satisfying the Micchelli's condition for the application of the simplest training algorithm for RBF output learning: the pseudo-inverse method. This way, the algorithm is capable of generating sets of prototype vectors such that the RBF network training becomes a simpler task than when using supervised training, for example. The main drawback of the proposed strategy is the existence of parameters to be tuned by the user.

Acknowledgements

Leandro Nunes de Castro would like to thank FAPESP (Proc. n. 98/11333-9) and Fernando Von Zuben would like to thank FAPESP (Proc. n. 98/09939-6) and CNPq (Proc. n. 300910/96-7) for their financial support.

References

[1] Lippmann, R. P.: Pattern Classification Using Neural Networks, IEEE Comm. Mag., 47-63, 1989.
[2] Moody, J. & Darken, C.: Fast Learning in Networks of Locally-Tuned Processing Units, NEUROCOM, *1*, 281-294, 1989.
[3] Poggio, T. & Girosi, F.: Networks for Approximation and Learning, Proc. of IEEE, *78*(9), 1481-1497, 1990.
[4] Kohonen T.: Self-Organized Formation of Topologically Correct Feature Maps, Biol. Cyb., *43*, 59-69, 1982.
[5] Haykin S.: Neural Networks – A Comprehensive Foundation, 2nd Ed. Prentice Hall, 1999.
[6] Broomhead, D. S. & Lowe, D.: Multivariable Functional Interpolation and Adaptive Networks, Complex Systems, *2*, 321-355, 1988.
[7] Karayannis, N. B. & Mi, G. W.: Growing Radial Basis Neural Networks: Merging Supervised and Unsupervised Learning with Network Growth Techniques, IEEE TNN, *8*(6), 1492-1506, 1997.
[8] Powell, M. J. D.: Radial Basis Functions for Multivariable Interpolation: A Review, in IMA CAAFD, J. C. Mason & M. G. Cox (eds.), Oxford, U.K.: Oxford Univ. Press, 143-167, 1987.
[9] Micchelli, C. A.: Interpolation of Scattered Data: Distance Matrices and Conditionally Positive Definite Functions, Const. Approx., *2*, 11-22, 1986.
[10] Cho S.-B.: Self-Organizing Map with Dynamical Node Splitting: Application to Handwritten Digit Recognition, NEUROCOM, *9*, 1345-1355, 1997.
[11] Fritzke, B.: Growing Cell Structures – A Self-Organizing Network for Unsupervised and Supervised Learning, Neural Networks, *7*(9), 1441-1460, 1994.
[12] De Castro, L. N. & Von Zuben, F. J.: An Improving Pruning Technique with Restart for the Kohonen Self-Organizing Feature Map, In Proc. of IJCNN'99, *3*, 1916-1919, 1999.

Modular Clustering by Radial Basis Function Network for Complexity Reduction in System Modeling

Ö. Ciftcioglu, S. Sariyildiz *

*Delft University of Technology, Faculty of Architecture, Department of Building Technology

Abstract.

Clustering is one of the dominant techniques of exploratory data analysis and data driven dynamic system modeling. In the context of model complexity, one of the powerful clustering methods is the method of orthogonal least squares (OLS) applied to radial basis functions (RBF) network. However, conventional way of utilization of OLS learning for a set of complex data is not desirable even though the learning process might be feasible for the amount of data at hand. This is due to the fact that some singular associations in the data can easily obscure many interrelations of interest among the data and also because of this, the rest of the associations for identification can heavily be limited. Therefore, as novel RBF clustering for system modeling, a set of time-series data is divided into several subsets as modules so that each subset is subjected to clustering separately. The dominant clusters in each subset are accumulated throughout the modular processing of total data set. The newly formed reduced data set which comprises the patterns from the clusters of the subsets, is subjected to final RBF network clustering by OLS for hierarchical cluster gradation from the subsets to identify the input-output model being searched for. In this way RBF clustering is accomplished substantially fast and at the same time effective reduction in complexity is obtained. The paper deals with the details of the novel RBF clustering.

1 Introduction

Clustering is one of the dominant techniques of exploratory data analysis and data driven dynamic system modeling. There are a number of ways for clustering [1]. They can broadly be divided into two categories as clustering with unsupervised and with supervised learning. In the unsupervised case, the clustering is performed where given a data set and the distance function, one starts with the number of clusters assigning each pattern to a separate cluster and proceeds to merge the clusters that are the closest. In the supervised case, we are concerned with a collection of labeled data that come in the form of ordered pairs. This can be seen as a feature vector describing the data and its class assignment. Eventually, the associations between the feature vectors and the class assignments are established. In the context of complexity reduction in system modelling perhaps the major disadvantage of unsupervised clustering is due to that the clusters are established from the input data and the final associations with some output functions are performed afterwards. In this case the associations may not be optimal since the established clusters may not be the optimal representations with respect to associations being looked for. To improve this unfavourable situation in the unsupervised clustering, not only the input space but also the output space in the clustering process is proposed [2]. In contrast with this, in the supervised case the clustering and associations may be performed in a constructive way so that the most effective representation of associations is obtained. In this respect, one of the powerful supervised clustering method is the method of orthogonal least squares (OLS) [3] applied to radial basis functions (RBF) network where the clusters are hierarchically ordered according to their effectiveness on the model error reduction. As a variant of feed-forward neural networks, RBF networks are widely treated in the literature in the last decade so that their potential in various applications spanning a wide range of sciences from Engineering to knowledge modeling in Architecture. By means of clustering, the data modeling is simplified and in the case of modeling by RBF network, a complex system is represented by means of limited number of adequate basis functions so that the model complexity is reduced. That is, if the basis functions are considered to be fuzzy membership functions in a fuzzy system represented in an RBF structure [4,5], an adequate number of fuzzy If-Then rules are also kept limited. In this research, a novel RBF clustering for effective complexity reduction in dynamic system modeling is presented.

2 RBF Network

The RBF architecture consists of an input layer, a hidden layer and an output layer. The hidden layer consists of a set of radial basis functions as nodes. Each node has a parameter vector **c** defining a cluster centre whose dimension is equal to the input vector. The hidden layer node calculates the Euclidean distance between the centre and the network's input vector. The distance calculated is used to determine the radial basis function output. Conventionally, all the radial basis functions in the hidden layer nodes are the same type and usually gaussian. The response of the output layer node(s) can be seen as a map f: $R^n \rightarrow R^m$ (n and m are the dimensions of the input and output spaces respectively), of the form

$$\mathbf{f(x)} = \Sigma\, \mathbf{w}_i\, \Phi(\|\mathbf{x} - \mathbf{c}_i\|)^2$$

Here the summation is over the number of training data N. $\mathbf{c}_i$ (i=1,2,....,N) is the i-th centre which may be equal to the input vector $\mathbf{x}_i$ or may be determined in some other way. Once the basis function outputs are determined, the connection weights from hidden layer to the RBF net output are determined from a linear set of equations. As a result, accurate functional approximation is obtained. The complexity increases exponentially as the size of the data increases. For a large data set this may even become unpractical. Therefore it is desirable to use limited number of hidden layer nodes in place of having a number equal to N. In the present description for sake of simplicity in representation and description, a single function is considered so that neural network has one output for each multivariable input. For this case the output is given by

$$f(x) = w_o + \sum_{j=1}^{N} w_i \phi(\| x - c_j \|)$$

where $\mathbf{x} \in R^n$ is the input vector . There are several methods to determine the w_i and $\mathbf{c}_i$ parameters. Khonen's self organising feature map (SOFM) can be used for c_is and algebraic computation for w_is afterwards or neural network oriented error back-propagation methods, for instance. In the latter case, the method is referred to as *network training*. However, in this research, orthogonal least-squares method is used in a special form as a training process due to its outstanding features for this study.

3 Modular Clustering

Conventionally, the main drawback of training RBF networks with OLS method with large volume of data is that the learning process becomes formidably cumbersome since each pattern forms a cluster center subject to a hierarchical gradation. Therefore, the dimension of a square matrix used in the OLS algorithm becomes formidably large for computation. Since the complexity increases exponentially with the number of nodes, next to significant round of errors and computational instability problems, excessive computing time may be necessary. However, added to the training inconvenience of large volume of data described above, OLS training in the presence of large volume of data is not desirable. This is due to the fact that some singular associations in the data can easily obscure many interrelations of interest among the input and output data at hand since such input pattern does not belong to any particular cluster. In this case, OLS algorithm may still select this pattern as a cluster centre next to the other ones selected, although the case is not a cluster centre in a real sense. Such a situation may occur simply as result of a mathematical treatment where cost function as a squared error is minimised. With the increase in severity of such singularity, the possibility of being selected as cluster centre increases. This is clearly illustrated by the outcomes shown in *figure1* and *figure2*.

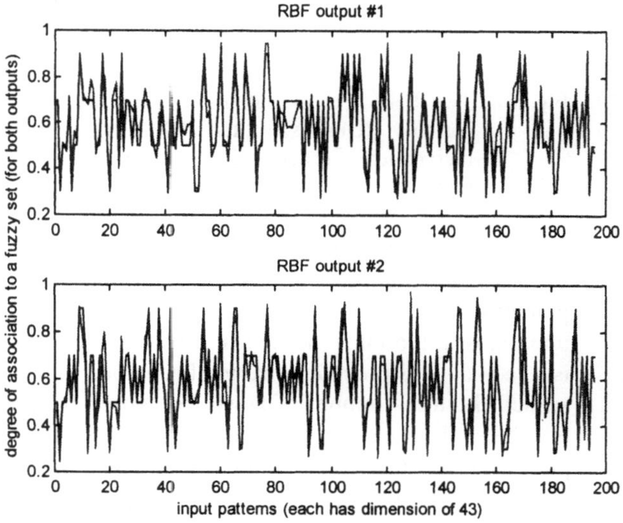

Fig.1. Two RBF network outputs with the sequence of patterns (clusters) after OLS training, in the training order. Each output variation represents the data and the estimates.

In figure 1, the OLS training results, that is both data and the estimates, are presented for an architectural data set where the majority of the output data are between 0.5 and 0.7, the extremities being 0.3 and 0.9. The input space is 43 dimensional and number of patterns for training is 196. The same results are represented once more but differently with respect to the hierarchical cluster sequence obtained from the OLS training. As it is seen in figure 2, the extremities took the precedence although these are relatively in minority; namely, each set of extremity patterns comprised about 18% of the total patterns. The major drawback of this situation is due to the degradation effect on the generalisation capability of the model, which may represent a system dynamics or a knowledge model, for instance.

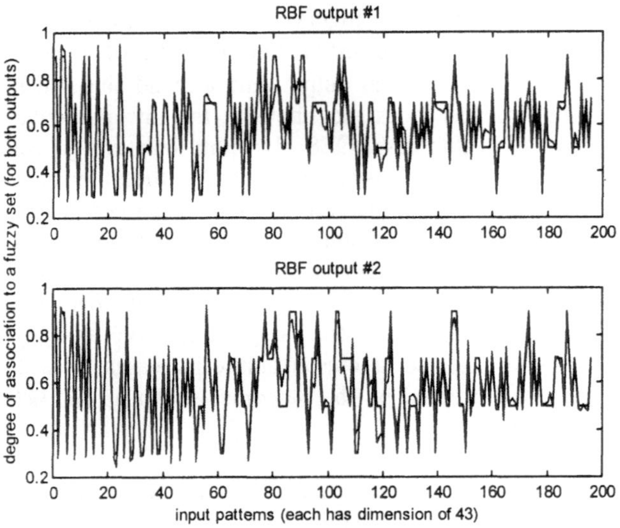

Fig.2. Two RBF network outputs with the sequence of patterns (clusters) after OLS training, in hierarchical order. Each output variation represents the data and the estimates.

To circumvent this, the large data set is divided into several subsets and each subset is subjected to OLS gradation and limited number of patterns from the uppermost importance in the sense of error-reduction is selected. This procedure having been done for each subset, the total number of newly composed patterns is subjected altogether again to a final OLS training procedure. In this way, the number of nodes that are finally selected are the representatives of total number of patterns, while total number of selected nodes (M) can be so arranged that it is much less then the total number of patterns (N), i.e., $M << N$. For instance, let us assume N=1000. Each subset can be selected as $N_i=100$ and $M_i=10$, (i=1,..,10) so that the eventually modularly clustered RBF network is formed where $M = \Sigma M_i = 100$. That is final OLS training is made based on these final hundred nodes. Referring to preceding hierarchical process described, a data set is divided into several subsets so that each subset is subjected to OLS clustering separately. The M_i dominant clusters in each subset are accumulated throughout the total data set. The newly formed reduced data set with M nodes, $M = \Sigma M_i$, is subjected to final RBF clustering by OLS for hierarchical cluster gradation to identify the input-output model being searched for. In this way the input-output associations, which might be time-dependent as well as time-independent, are discovered in a perspective without probable obscurity that may be caused by some strong but local associations in the given data set of large volume of data. Conspicuously, the clustering virtually requires no limitation for the number of patterns in use, for all practical purpose. In the terminology of neural networks, a clustering is said to be modular if the clustering performed by the network can be decomposed into two or more clustering processes that each clustering operates independently and the clustering outcomes are appropriately combined afterwards. One of the applications for modular clustering is the efficient (fast) and effective (accurate) wavelet transform by RBF network. The accurate wavelet transform by RBF networks has already been reported earlier [6]. Very briefly, discrete wavelet transform is similar to discrete Fourier transform in the sense that a bock of time series data is transformed into another block of data of the same size. In the case Fourier transform we obtain frequency spectrum of the data. In the case of wavelet transform [7,8] we obtain coefficients that are called wavelet coefficients and they are used for time-frequency representation of the time series data. Although, RBF approach can provide accurate wavelet transform, it uses relatively high number of basis functions. In real time applications, the real added value of soft computing approach for wavelet transform is due to the speed of computation. However, for high number of basis functions the speed of computation with RBF network might be comparable with actual computation using a fast transform algorithm. Therefore, parsimony in the use of number of basis functions is important. For wavelet transform by soft computing, to train the RBF network originally the number of patterns used is 100 and the number of basis functions employed was relatively high to achieve high accuracy. In this research, for the same transform task the training of RBF network is carried out using larger number of data. In this way more representative centres of radial basis functions are identified from a larger set of patterns relative to a smaller data set used earlier. By doing so, the anticipated result is that the number of centres subject to transform computation for the same accuracy in multivariable functional approximation can be reduced.

4 Experimental research

To illustrate the effect of the modular clustering total 500 blocks of time series pump data are considered. Each block contained 32 samples and it is referred to as pattern here. The network contained 32 inputs for each pattern and 32 outputs as the corresponding wavelet transforms. The number of patterns used was first 400 blocks out of 500 total and the number of basis functions was selected to be as low as 20.

Following the OLS training, as a test, the trained network is used to estimate the transform of the last 100 consecutive blocks of time series data. The same experiment was repeated with the same experimental conditions, i.e., 20 basis functions with OLS training using modular clustering described in this research. Explicitly, four times a set of 100 patterns with a length of 32 was used for the first phase of modular clustering.; each time hierarchically uppermost 25 patterns are selected out of each set of 100. In the second phase of the clustering the corresponding 4x25=100 patterns are used for final phase. Following the clustering, the trained network is used to estimate the wavelet transform of the last 100 consecutive blocks of the time series data set as it was used for the test before. The results from modular OLS approach are shown in *figure 3*.

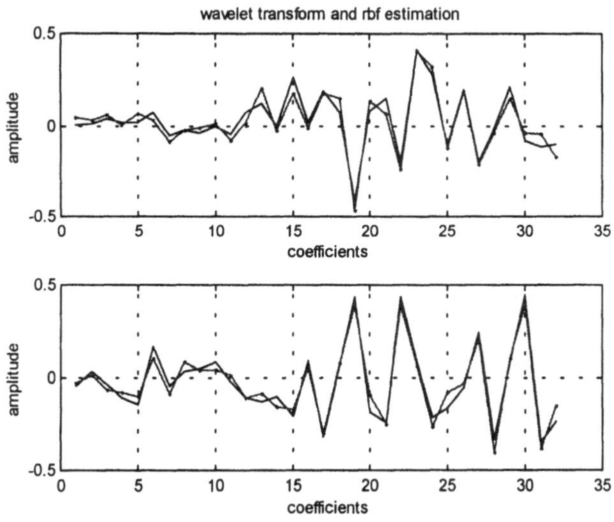

Fig.3. Two wavelet tranformed blocks of data and the estimation counterparts obtained from a modularly clustered RBF network

Comparison of the results from the RBF network and the actual results from the computational wavelet transform revealed that the proposed modular clustering method is substantially faster relative to standard clustering methods in use. At the same time the RBF network so designed is more accurate relative to its non-modular counterparts, i.e., straightforward OLS training with total number of patterns at once. This is due to selective feature for cluster centres from a large repertory of input patterns, in the case of modular OLS clustering.

Two enhancement features of modular clustering pointed out above are of particular importance in the case of complex data while for a small size of data, the enhancement features do not prevail.

5 Conclusions

The presented novel modular clustering method eliminates the unfavourable role of the excessive number of patterns in RBF network training, for all practical purposes. Next to this, basically, the selection of the clusters by this method is an important step for the enhancement of the generalization capability of RBF networks used for modeling at large.

References

[1] Bezdek, J.C., Pattern Recognition with Fuzzy Objective Function Algorithms, Plenum, New York (1981)

[2] Uykan Z. et al. :Analysis of Input-Output Clustering for Determining Centers of RBFN, IEEE Trans. on Neural Networks, Vol.11 No.4 (2000)

[3] Chen S, C.F.N. Cowan and P.M. Grant, Orthogonal Least Squares Algorithm for Radial Basis Function Networks, IEEE Trans. on Neural Networks, Vol.2, No.2 (1991)

[4] Jang J.-S R. and Sun T.C.: Functional Equivalence Between Radial Base Function Networks and Fuzzy Inference Systems, IEEE Trans. on Neural Networks, Vol.4, No.1, (1993)

[5] Hunt K.J., Haas R. and Murray-Smith R.M.: Extending the Functional Equivalence of Radial basis Function Networks and Fuzzy Inference Systems, IEEE Trans. on Neural networks, Vol.7, No.3 (1996)

[6] Ciftcioglu Ö., Wavelet Transform by Soft Computing in 'Advances in Soft Computing' by R. John and R. Birkenhead (Eds.), Physica-Verlag,, Heidelberg, New York, (2000)

[7] Mallat S. Wavelet Tour of Signal Processing, Academic Press, San Diego, London (1999)

[8] Daubechies I, Ten Lectures on Wavelets, CBMS-NSF regional conference series in applied mathematics Vol.61, SIAM (1992)

A Fuzzy Approach to Sociodynamical Interactions

David William Pearson*, Gérard Dray†

*University of Saint-Etienne, I.U.T. de Roanne, 20, avenue de Paris, 42334 Roanne, France, e-mail: david.pearson@univ-st-etienne.fr †LGI2P, EMA Site-EERIE, Parc Scientifique G. Besse, 30319 Nimes, France, e-mail: gerard.dray@site-eerie.ema.fr

Abstract

In this paper we present a sociodynamical model of a network of people and their attitudes towards each other. The model is based on concepts from fuzzy logic. The dynamic element of the model allows us to simulate the temporal evolution of the attitudes and to see how this evolution might converge to a stable state.

1 Introduction

Sociodynamics is a relatively new field of research which combines elements mainly from psychology, sociology, mathematics, informatics and physics [2, 3, 4, 5, 6, 8]. The main idea is to develop mathematical models of social phenomena that can be used in computer simulations and that will ultimately lead to the possibility of controlling and optimising certain situations where these phenomena play an important rôle.

In our work, we are particularly interested in modelling groups of companies (enterprises) [1]. There are obvious economic indicators which enable us to model the dynamical evolution of a company or a group of companies. But, economical factors are not the only ones governing the evolution and the state of health of a company. Social factors come into play and, in particular, attitudes should be taken into consideration. By attitudes we mean someone displaying a positive or negative attitude towards his neighbour.

Attitudes can be based on a number of different criteria. We are interested in attitude changes caused by interactions between individuals. We have looked at the problem of detecting pair interactions [7], for this we used a probability based model. In the present article we look at the network structure in more detail and we adopt a fuzzy logic based approach.

2 Fuzzy Model

Our model is based on a graph representing a network of individuals. In our particular case these individuals are heads of companies, although the model could be adapted for other types of situation.

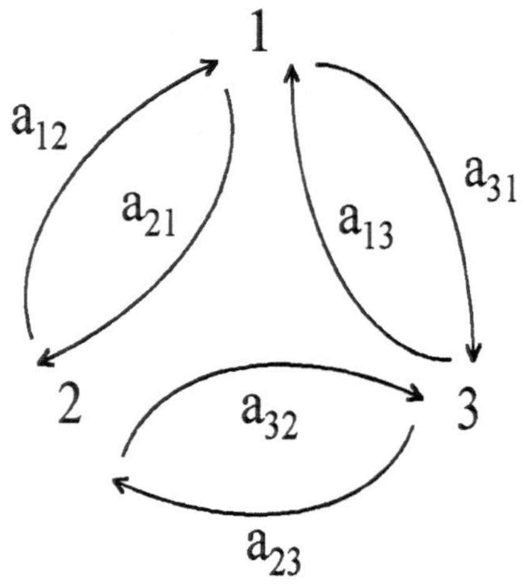

Fig. 1. A typical network

We place the individuals at points on the network and mark them simply with a number. Then we indicate by directional arcs when a relationship exists between two individuals, these arcs represent the attitudes of individuals towards others. Where a relationship exists between two individuals there are always two directional arcs because attitudes are not always reciprocal. A typical network, with three individuals, or companies, involved, is illustrated in figure 1. Here we illustrate a system where each individual is in contact with all the others. We denote by a_{ij} the attitude that individual j has towards individual i. The attitudes have values $a_{ij} \in [-1, 1]$, a value of -1 indicating a totally negative or no confidence attitude and $+1$ a totally positive or absolute confidence attitude.

From a fuzzy logic point of view, if two social attitudes are totally opposed then we would like the corresponding membership function to return a value of 0. At the other end of the scale, if two attitudes are in total agreement then we would like the membership function to return a value of 1. With these constraints in mind, we chose the following for a membership function

$$\mu(a_{ij}, a_{kl}) = e^{-k(a_{ij}-a_{kl})^2} \qquad (1)$$

which can be interpreted as attitude a_{ij} is similar to attitude a_{kl}.

The fuzzy rule for updating the social attitudes is based on all the possible chains from a person i to a person j. For each chain c_γ of $n+1$ links such that

$$a_{ki} \to a_{k+1k} \to a_{k+2k+1} \to \cdots \to a_{jk+n}$$

we make use of the membership function (1) to build a fuzzy rule of the form

$$\begin{aligned} &\text{if } \mu(a_{ki}, a_{k+1k}) \wedge \mu(a_{k+1k}, a_{k+2k+1}) \wedge \\ &\quad \cdots \wedge \mu(a_{k+nk+n-1}, a_{jk+n}) \\ &\qquad \text{then attitude } a_{ji} \text{ is reinforced} \qquad (2) \end{aligned}$$

The rule (2) is meant to model the situation where an individual will compare his/her attitude towards another individual with all the attitudes towards the same individual that he/she knows of, ie within the network. If all the attitudes towards the particular individual are similar, then his/her own attitude towards this individual will most likely be reinforced ie increased if the attitude is positive and decreased if negative. If, however, the individual finds that there is a general dissagreement with his/her attitude towards another individual within the network, then he/she is likely to modify his/her attitude towards a more positive value if originally negative or a more negative value if originally positive.

The conjunction of all the rules like (2) for all the chains c_γ provides us with a new fuzzy set, that we simply call *agreement*, that can be used to update the attitudes in an iterative manner as follows

$$\begin{aligned} &\text{if } \textit{agreement} \geq T_{\text{high}} \\ &\quad \text{then } a_{ij}(t) \geq 0 \\ &\Rightarrow a_{ij}(t+1) = a_{ij}(t) + \alpha \times \textit{agreement} \\ &\quad \text{or } a_{ij}(t) < 0 \\ &\Rightarrow a_{ij}(t+1) = a_{ij}(t) - \alpha \times \textit{agreement} \\ &\quad \text{elseif } \textit{agreement} \leq T_{\text{low}} \\ &\quad \text{then } a_{ij}(t) \geq 0 \\ &\Rightarrow a_{ij}(t+1) = a_{ij}(t) - \alpha \times \textit{agreement} \\ &\quad \text{or } a_{ij}(t) < 0 \\ &\Rightarrow a_{ij}(t+1) = a_{ij}(t) + \alpha \times \textit{agreement} \\ &\quad \text{else } a_{ij}(t+1) = a_{ij}(t) \qquad (3) \end{aligned}$$

where $a_{ij}(t)$ denotes attitude a_{ij} at time instant t.

In (3), if *agreement* is greater than a threshold T_{high} then the attitude is reinforced (increased or decreased by an amount dependent upon a parameter α), if *agreement* is lower than a threshold T_{low} then the attitude is modified (increased or decreased by an amount dependent upon a parameter α) and if neither of these conditions are satisfied then the attitude remains at the same value. The model parameters T_{high}, T_{low} and α can be the same for all the attitude iterations (3) or can vary from one attitude a_{ij} to another a_{kl}.

Clearly, (3) could invlove a lot of chains even for a relatively small number of individuals. However, we could expect that a real network would not be completely connected and so the number of chains would be reduced. Our particular objective is to model groups of small companies and it is usually the case that there are few direct interactions between them.

3 Example

We present an example based on figure 1 with, in the first instance, the following initial values for the attitudes

$$\begin{aligned} a_{12} &= -0.5 \\ a_{21} &= 0.5 \\ a_{13} &= 0.6 \\ a_{31} &= 0.7 \\ a_{23} &= 0.4 \\ a_{32} &= 0.5 \end{aligned}$$

We can interpret this as individual 1 overestimating his/her attitudes with respect to the others in the network and the only negative attitude present is from individual 2 towards individual 1.

We chose the following values for the parameters in the model. All the increments $\alpha = 0.1$, all of the upper thresholds where attitudes are reinforced we set to $T_{\text{high}} = 0.5$ and all of the lower thresholds where attitudes are modified we set to $T_{\text{low}} = 0.1$. We let the system run for 300 iterations.

In figure 2 we present the trajectories for the variables a_{12}, a_{21} and a_{13} and the trajectories for the variables a_{31}, a_{23} and a_{32} can be seen in figure 3.

As a brief interpretation of these results we can make the following comments.

- 1 increases his/her confidence in 3
- 1 increases his/her confidence in 2 even though 2 eventually has less confidence in 1 than his/her original confidence

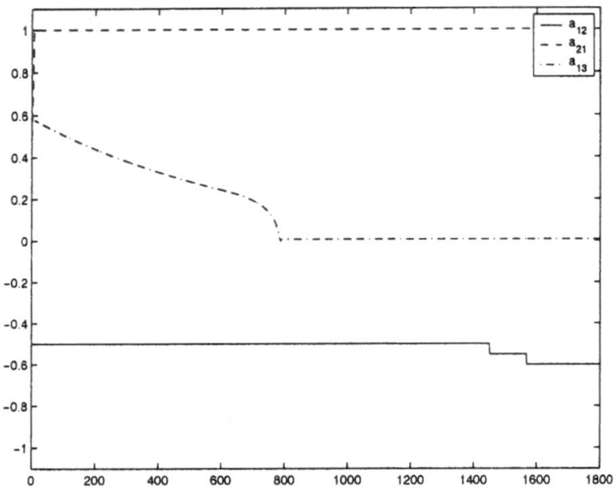

Fig. 2. Attitude trajectories

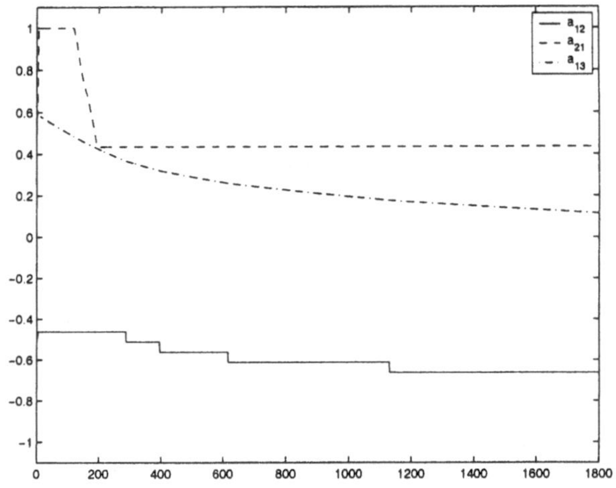

Fig. 4. Attitude trajectories

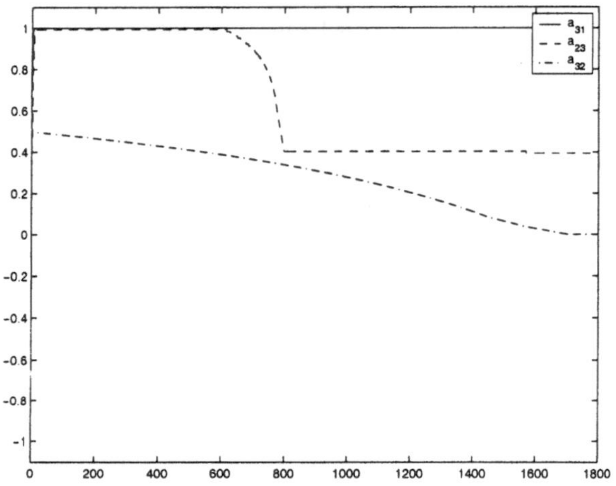

Fig. 3. Attitude trajectories

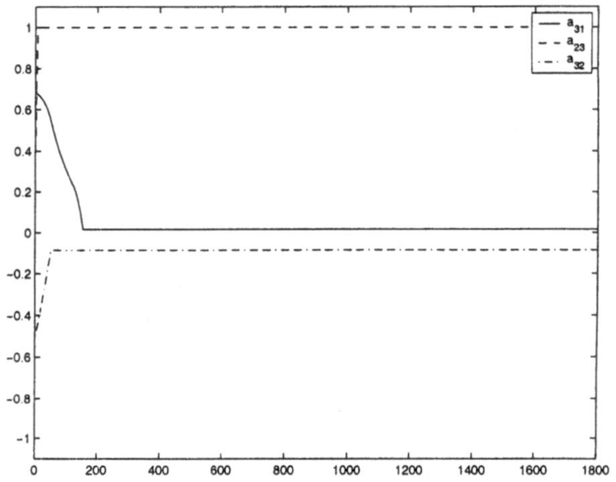

Fig. 5. Attitude trajectories

- 2 really doesn't like 1 and his/her confidence in 3 is reduced to a neutral value

- 3 has his/her confidence in 1 reduced but keeps confidence in 2

For a second simulation we use the same values for all the parameters and initial values except for the initial value for a_{32} which was originally set to 0.5 and now we set it to -0.5. In other words individual 2 now doesn't like anybody in the network. The results from this simulation can be seen in figures 4 and 5.

We see that the totally negative attitude of individual 2 has resulted in an overall decrease of confidence in the network. Only individual 3 has increased his/her confidence and this increase in confidence is towards individual 2.

4 Conclusion

We have presented a model, based on fuzzy logic concepts, that can simulate certain sociodynamical phenomena. Quite obviously, the model is in its early stages and there remains a lot of research work to be done.

In particular we can think of two main areas of future work.

- modelling advances
- analysis of the model

By modelling advances we mean that we will develop and improve the model to make it as realistic as possible. Also, we will tackle the problem of dynamically fitting the model to an actual network.

By analysis of the model we refer to the underlying dynamical structure that may be susceptible to analysis. Although this is quite a task because the model is essentially nonlinear.

References

[1] B. Besombes and G. Mnemoi, "Performance Indicators and Cooperation Network in SME: An Approach Based on the Ecograi Concept", *Proceedings Management and Control of Production and Logistics (MCPL2000)*, Grenoble, France, 2000.

[2] R. Conte, R. Hegselmann and P. Terna (Eds.), *Simulating Social Phenomena*, Springer-Verlag, 1997.

[3] N. Gilbert and J. Doran (Eds.), *Simulating Societies: the Computer Simulation of Social Phenomena*, UCL Press, London, 1994.

[4] R. Hegselmann, U. Mueller and K.G. Troitzsch (Eds.), *Modelling and Simulation in the Social Sciences from a Philosophy of Science Point of View*, Dordrecht, Boston (Kluwer), 1996.

[5] D. Helbing, *Quantitative Sociodynamics*, Kluwer Academic Publishers, 1995.

[6] W.B.G. Liebrand, A. Nowak and R. Hegselmann (Eds.), *Computer Modelling of Social Processes*, Sage, 1998.

[7] D.W. Pearson, "Detection of Pair Interactions in Group Decision Making", submitted to *Mathematical and Computer Modelling*, 2000.

[8] W. Weidlich and G. Haag, *Concepts and Models of a Quantitative Sociology*, Springer-Verlag, 1983.

An Immunological Approach to Initialize Feedforward Neural Network Weights

Leandro Nunes de Castro, Fernando J. Von Zuben *

*School of Electrical and Computer Engineering, State University of Campinas/SP, Brazil, e-mail: {lnunes,vonzuben}@dca.fee.unicamp.br

Abstract

The initial weight vector to be used in supervised learning for multilayer feedforward neural networks has a strong influence in the learning speed and in the quality of the solution obtained after convergence. An inadequate initial choice may cause the training process to get stuck in a poor local minimum, or to face abnormal numerical problems. In this paper, we propose a biologically inspired method based on artificial immune systems. This new strategy is applied to several benchmark and real-world problems, and its performance is compared to that produced by other approaches already suggested in the literature.

1 Introduction

The importance of a proper choice for the initial set of weights (weight vector) is stressed by Kolen and Pollak [1]. They showed that it is not feasible to perform a global search to obtain the optimal set of weights. So, for practical purposes, the learning rule should be based on optimization techniques that employ local search to find the solution [2]. As an important outcome of their procedure, there is the fact that a local search process results in a solution strongly related to the initial configuration of the weight vector. It happens because each initial condition belongs to the basin of attraction of a particular local optimum in the weight space, to which the solution will converge [3]. Consequently, only a local optimum can be produced as the result of a well-succeeded training process. If such a solution happens to be the global or a good local optimum, the result is a properly trained neural network. Otherwise, an inferior result will be achieved, so that the poorer the local optimum, the worse the performance of the trained neural network.

This correlation between the initial set of weights and the quality of the solution resembles the existing correlation between the initial antibody repertoire and the quality of the response of natural immune systems, that can be seen as a complex pattern recognition device with the main goal of protecting our body from malefic external invaders, called antigens. Antibodies are the primary immune elements that bind to antigens for their posterior destruction by other cells [4]. The number of antibodies contained in our immune system is known to be much inferior to the number of possible antigens, making the diversity and individual binding capability the most important properties to be exhibited by the antibody repertoire. In this paper, we present a simulated annealing approach, called SAND (Simulated ANnealing for Diversity), that aims at generating a dedicated set of weights that best covers the weight space, to be searched in order to minimize the error surface. The strategy assumes no a priori knowledge about the problem, except for the assumption that the error surface has multiple local optima. In this case, a good sampling exploration of the error surface is necessary to improve the chance of finding a promising region to search for the solution. The algorithm induces diversity in a population by maximizing an energy function that takes into account the inverse of the affinity among the antibodies. The weights of the neural network will be associated with antibodies in a way to be further elucidated.

2 The Simulated Annealing Algorithm

The simulated annealing algorithm makes a connection between statistical mechanics and combinatorial optimization [5,6]. The origin of the method is associated with aggregate properties of a large number of atoms found in samples of liquids or solid matters. The behavior of the system in thermal equilibrium, at a given temperature, can be characterized experimentally by small fluctuations around the average behavior. Each atomic position is weighted by a probability factor

$$P(\Delta E) = \exp(-\Delta E / T), \tag{1}$$

where E is the energy of the configuration, T the temperature and ΔE a small deviation in the energy measured. At each step of this algorithm, an atom is given a small random displacement and the resulting change, ΔE, in the energy of the system is computed. If $\Delta E \leq 0$, the displacement is accepted, and the configuration with the displaced atom is used as the starting point of the next step. The case $\Delta E > 0$ is treated probabilistically: the probability of accepting the new configuration is given by Equation (1).

The temperature is simply a control parameter in the same unit as the cost (energy) function. The simulated annealing process consists of first "melting" the system being optimized at a high effective temperature, then lowering the temperature by slow stages until the system "freezes" and no further change occurs (steps of increasing temperature can also be incorporated). At each temperature, the simulation must proceed long enough for the system to reach a steady state. Notice that, transitions out of a local optimum are always possible at nonzero temperatures. The sequence of temperatures and the size of the ΔE variation are considered an annealing schedule.

3 An Immunological Approach

The immune system model used in this work is a simplification of the biological one. Real-valued vectors represent the antibodies (Ab). In our problem, the antigen (Ag) population (training set) will be disregarded, so the energy measure of the population of antibodies (set of weights) will be determined based solely on the individuals of the antibody repertoire. The binding Ag–Ab represents a complementarity measure between an antigen and an antibody, an idea that can be adapted to determine the complementarity among members of the antibody repertoire. The goal is to maximize the distance among antibodies (Ab–Ab), with the purpose of reducing the amount of similarities within the population.

An abstract model to describe Ag–Ab interactions was introduced by Perelson & Oster [7]. In this model, it is assumed that the features of an antibody receptor (combining region) relevant to antigen binding can be described by specifying a total of L shape parameters. It is also assumed that the same L parameters can be used to describe an antigen. Combining these L parameters into a vector, the antibody receptors and the antigen determinants can be described as points Ab and $\overline{\text{Ag}}$, respectively, in an L-dimensional Euclidean vector space, called shape-space S. Here we use $\overline{\text{Ag}}$ (the complement of Ag) because the affinity is directly proportional to complementarity, and affinity will be associated with the proximity of Ab to $\overline{\text{Ag}}$. By defining a metric on S, the proximity between Ab and $\overline{\text{Ag}}$ is a measure of their affinity. The antibodies and the complement of the antigens were represented by a set of real-valued coordinates. Thus, mathematically, each molecule could be regarded as a point in an L-dimensional real-valued space, and the affinity Ag–Ab was related to the inverse of the Euclidean distance between them.

Any given antibody is assumed to recognize some set of antigens and therefore covers some portion of the space.

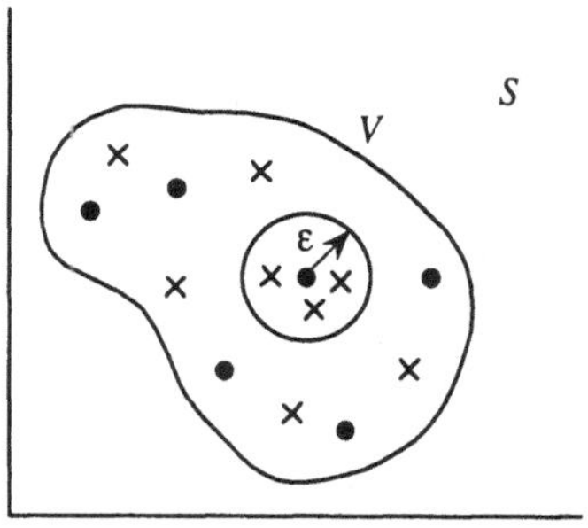

Fig. 1. Within shape-space S, there is a volume V in which the antibodies (•) and the complement of antigens (x) are located. An antibody is assumed to be able to bind any complement of antigen within a distance ε (region of stimulation).

In the natural immune system, the binding Ag–Ab might not be fully complementary for the lymphocytes to become activated, since a partial matching might suffice. It implies that, if no error were allowed, the immune cells would become activated only when a perfect match occurred. Based on this argument, it is defined an acceptable matching distance (ε), that determines the coverage provided by the antibodies. Some authors called ε a ball (or region) of stimulation [8], because it represents the group of antigens that can stimulate the antibody contained in its center. Figure 1 depicts the shape-space model and the region of stimulation, where the dots and crosses denote the location of antibodies and the complement of antigens, respectively. The circle of radius ε around one of the antibodies shows its coverage. This is an illustrative abstraction, once Ag-Ab recognition happens through shape complementarity, instead of similarity as in $\overline{\text{Ag}}$-Ab recognition.

As we are dealing with real-valued vectors, the inverse of the Euclidean distance can be used as the measure of affinity between the molecules:

$$Aff(x_i, x_j) = \frac{1}{ED(x_i, x_j) + \delta} \tag{2}$$

where x_i and x_j represent independent vectors of length L, δ is a small positive constant, and

$$ED(x_i, x_j) = \sqrt{\sum_{k=1}^{l} (x_{ik} - x_{jk})^2}\ , \tag{3}$$

Assuming an Euclidean search-space, the energy measure to be optimized can be simply defined as the sum of the Euclidean distances among all vectors that represent the antibody population

$$E = \sum_{i=1}^{N} \sum_{j=i+1}^{N} ED(x_i, x_j)\,. \tag{4}$$

A stopping criterion for the simulated annealing algorithm, which takes into account the diversity among the vectors, has to be defined. The approach to be proposed here, among other possibilities, involves the analysis of directional data.

Given the vectors x_i, $i = 1,..., N$, it is initially necessary to transform them into unit vectors, resulting in a set $\{\bar{x}_i, i = 1,..., N\}$ of unit vectors of length L. The average directional vector is

$$\bar{x} = \frac{1}{N}\sum_{i=1}^{N}\bar{x}_i \, . \quad (5)$$

A metric to estimate the diversity, or equivalently the uniformity, of the distribution of the unit vectors in a hypersphere can be simply given by

$$\|\bar{x}\| = \left(\bar{x}^T \bar{x}\right)^{1/2}, \quad (6)$$

where $\|\cdot\|$ represents the Euclidean norm.
The stopping criterion (*SC*) is then based on the index

$$I_{SC} = 100\times(1-\|\bar{x}\|) \, . \quad (7)$$

Equation (7) is the percentile norm of the resultant vector $\bar{x}$, and is equal to 100% when this norm is zero. In practical terms, a value of I_{SC} close to 100% for the stopping criterion (*SC*) is a reasonable choice, indicating a distribution close to uniform.

3.1. Neural network weights and antibodies

Each antibody corresponds here to a vector that contains the weights of a given neuron in a layer of a multilayer neural network. Thus, generating the most diverse population of antibodies in $\Re^L$ corresponds to producing a set of neurons with well-distributed weight vectors. This way, the SAND approach will have to be applied separately to each layer of the network. Another important aspect of the strategy is that, as we are generating vectors with unitary norms, these vectors can be normalized to force the activation of each neuron to occur near to the linear part of the activation functions, in order to avoid saturation at the initialization.

4 Algorithms and Benchmarks

We compare the performance of SAND with five other methods to generate the initial set of weights: BOERS [9], WIDROW [10], KIM [11], OLS [12], and INIT [13]. All methods are applied to seven benchmark problems.

To specify the benchmark problems used, let N be the number of samples, SSE the desired sum squared-error (stopping criterion) and *net* the net architecture represented by $[n_i\text{-}n_h\text{-}n_o]$. Where n_i is the number of inputs, n_h is the number of hidden units and n_o is the number of outputs of the network.
The benchmarks used for comparison were:

- parity 2 (XOR): $N = 4$, *net*: [2-2-1], SSE = 0.01;
- parity 3: $N = 8$, *net*: [3-3-1], SSE = 0.01;
- sin(x).cos($2x$): $N = 25$, *net*:[1-10-1], SSE = 0.01;
- ESP: real-world problem used by [14]; $N = 75$, *net*: [3-10-5], SSE = 0.1;
- SOYA: another real-world problem used by [15], $N = 116$, *net*: [36-10-1], SSE = 0.1;
- IRIS: this benchmark is part of the machine learning database and is available in [16]; $N = 150$, *net*: [4-10-3], SSE = 0.15; and
- ENC/DEC: the family of encoder/decoder problem is very popular and is described in [17]. $N = 10$, *net*: [10-7-10].

The training algorithm used in all cases was the Moller scaled conjugate gradient [18], with the exact calculation of the second order information [19].

For each method and each benchmark problem we performed 10 runs. The results presented in Figure 2 correspond to the percentage of times each method produced the best performance in terms of maximum, minimum, mean and standard deviation of the number of epochs necessary for convergence to high quality solutions for all 7 problems and 10 runs evaluated. This picture shows that SAND, INIT and OLS are superior to the others (BOERS, WIDROW, and KIM). If the name of a method do not appear in the comparison, it is because the approach does not converge given the maximum number of epochs, or converges to a poor local optimum. The advantage of SAND is that it does not make use of the training data to estimate the initial set of weights, like INIT and OLS.

5 Discussion

The proposed strategy (SAND) is inspired by the diversity generation and maintenance characteristics of the immune system. The SAND performance shows that neurons with well-distributed weight vectors lead to faster convergence rates. The necessity to re-scale the weight vectors reinforces the theory proposed by de Castro & Von Zuben [13]: initializing the weights in the approximately linear part of the neurons' activation function reduces numerical instabilities and results in improved convergence rates.

The performance of the proposed algorithm leads to two important conclusions concerning feedforward neural network initialization:

- It is necessary to avoid an initial set of weights that guides to the saturation of the neuron's response, what can be achieved by properly setting the weights' initial interval; and
- The generation of weight vectors mostly spread over the search-space results in smaller training times.

The method proposed also shows that there are still many biological phenomena in which to search for mechanisms and inspiration for solving computational intelligence problems, like ANN initialization, architecture optimization, learning, among others.

It is also important to stress that the proposed strategy does not take into account the training data, preparing the initial set of weights to deal appropriately with any input data, a process similar to the definition of the initial antibody repertoire of immune systems. In addition, the other approaches that were competitive, OLS and INIT, require the determination of inverse or pseudo-inverse matrices, being subject to numerical instabilities in cases the training samples contain linearly dependent vectors, i.e., redundancy, what is usually the case while training ANN.

Although the SAND algorithm was biologically inspired, all the tools employed in its development (e.g., the simulated annealing algorithm) have a strong mathematical formulation and are being successfully used in other areas of research, including multimodal optimization.

Acknowledgements

Leandro N. de Castro thanks FAPESP (Proc. 98/11333-9) and Fernando Von Zuben thanks FAPESP (Proc. 98/09939-6) and CNPq (Proc. 300910/96-7) for their financial supports.

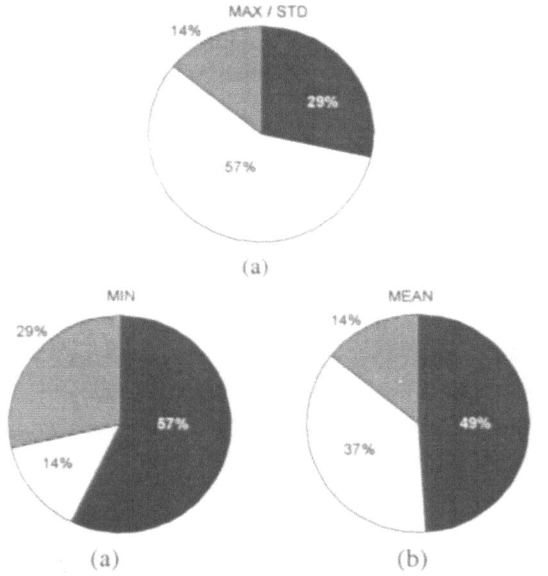

Fig. 2. Performance comparison of the methods (considering 10 runs of 7 benchmark tasks). OLS: dashed; INIT: dark gray; SAND: white. (a) Percentage of times a method required the smallest maximum number of epochs for convergence and smallest STD. (b) Percentage of times a method required the smallest MIN number of epochs for convergence. (c) Percentage of times a method presented the smallest MEAN number of epochs for convergence.

References

[1] Kolen, J. F. & Pollack, J. B.: Back Propagation is Sensitive to Initial Conditions, Technical Report TR 90-JK-BPSIC, 1990.

[2] Shepherd, A. J.: Second-Order Methods for Neural Networks – Fast and Reliable Methods for Multi-Layer Perceptrons, Springer, 1997.

[3] Hertz, J., Krogh, A. & Palmer, R.G.: Introduction to the Theory of Neural Computation. Addison-Wesley Publishing Company, 1991.

[4] Janeway Jr., C. A. & P. Travers: Immunobiology The Immune System in Health and Disease, garland Publishing Inc., N.Y., 2nd ed., 1994.

[5] Haykin S.: Neural Networks – A Comprehensive Foundation, Prentice Hall, 2nd ed., 1999.

[6] Kirkpatrick, S., Gelatt Jr., C. D. & Vecchi, M. P.: Optimization by Simulated Annealing, Science, *220*(4598), 671-680, (1987).

[7] Perelson, A. S. & Oster, G. F.: Theoretical Studies of Clonal Selection: Minimal Antibody Repertoire Size and Reliability of Self-Nonself Discrimination, J. theor. Biol., *81*, 645-670, (1979).

[8] Smith, D. J., Forrest, S., Hightower, R. R. & Perelson, S. A.: Deriving Shape Space Parameters from Immunological Data, J. theor. Biol., *189*, 141-150, (1997).

[9] Boers, E. G. W. & Kuiper, H.: Biological Metaphors and the Design of Modular Artificial Neural Networks, Master Thesis, Leiden University, Netherlands, (1992).

[10] Nguyen, D. & Widrow, B.: Improving the Learning Speed of Two-Layer Neural Networks by Choosing Initial Values of the Adaptive Weights, Proc. IJCNN'90, *3*, 21-26, (1990).

[11] Kim, Y. K. & Ra, J. B.: Weight Value Initialization for Improving Training Speed in the Backpropagation Network, Proc. of IJCNN'91, *3*, 2396-2401, (1991).

[12] Lehtokangas, M., Saarinen, J., Kaski, K. & Huuhtanen, P.: Initializing Weights of a Multilayer Perceptron by Using the Orthogonal Least Squares Algorithm, NEUROCOM, *7*, 982-999, (1995).

[13] De Castro, L. N. & Von Zuben F. J.: A Hybrid Paradigm for Weight Initialization in Supervised Feedforward Neural Network Learning, Proc. ICS'98, Workshop on Artificial Intelligence, 30-37, (1998).

[14] Barreiros, J. A. L., Ribeiro, R. R. P., Affonso, C. M. & Santos, E. P.: Estabilizador de Sistemas de Potência Adaptativo com Pré-Programação de Parâmetros e Rede Neural Artificial, LAC: EGT, 538-542, (1997).

[15] De Castro, L. N., Von Zuben, F. J. & Martins, W.: Hybrid and Constructive Neural Networks Applied to a Prediction Problem in Agriculture. Proc. of IJCNN'98, *3*, 1932-1936, (1998).

[16] ftp://ftp.ics.uci.edu/pub/machine-learning-databases

[17] Fahlman, S. E.: An Empirical Study of Learning Speed in Back-Propagation Networks, Tech. Rep., CMU-CS-88-162, Carnegie Mellon University, Pittsburg, (1988).

[18] Moller, M. F.: A Scaled Conjugate Gradient Algorithm for Fast Supervised Learning, Neural Networks, *6*, 525-533, (1993).

[19] Pearlmutter, B. A.: Fast Exact Calculation by the Hessian, NEUROCOM, *6*, 147-160, (1994).

How Certain Characteristics of Cortical Frequency Representation May Influence our Perception of Sounds

Lubica Beňušková*

*Department of Computer Science and Engineering, Slovak Technical University, 81219 Bratislava, Slovakia
e-mail: benus@elf.stuba.sk

Abstract

We present two attractor neural network (ANN) [1] models introduced in [2, 3] to show how two particular characteristics of the sound frequency representation in the auditory cortex may influence the way in which we process sounds and sound sequences. In particular, we consider neurophysiologically recognized isofrequency stripes and different amount of cortical surface area devoted to low versus high frequencies. Although we apply these models to explain several phenomena in the perception of music, they can be generalized to other sounds as well.

1 Introduction

The primary auditory cortex (AI) has been neuroscientifically identified in all mammalian species studied so far including humans [4, 5, 6]. In all of them there is a tonotopic representation of sound frequencies. That is, neurons with similar *best frequencies* (the frequencies to which the neurons are best tuned) within AI define "isofrequency" stripes that are oriented orthogonal to the low-to-high best-frequency gradient. The cortical surface area in AI devoted to low versus high frequencies varies among species, and this distribution reflects their natural experience. For instance in macaque monkey, low frequencies occupy more space than high frequencies. In the case of musical and natural sounds, there is a general rule, that on average the intensity decreases when the frequency increases [7]. Moreover, it is known from psychophysical experiments that accuracy of frequency (pitch) discrimination for frequencies $\geq$ 200 Hz decreases with frequency [7]. Taken together this leads us to the assumption that the number of neurons representing higher frequencies in humans may be lower than the number of neurons representing lower frequencies (figure 1). Tone templates are spatial maps of the result of Fourier transform that is performed in the inner ear. We assume that these templates are created by experience.

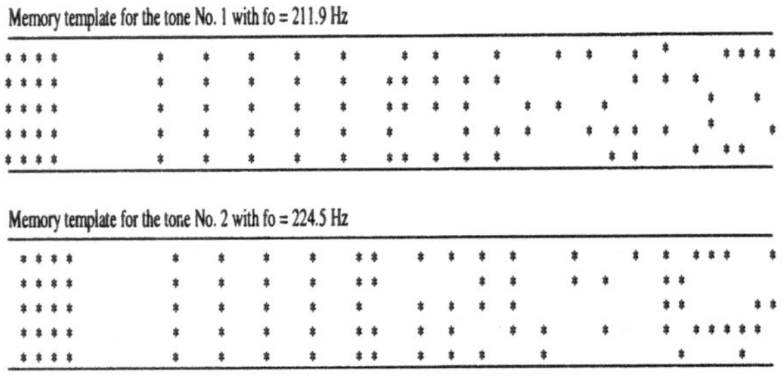

Fig. 1. Two examples of the ANN complex tone representation that is a stripe-like pattern of activity with more neurons representing low frequencies than representing high frequencies. Isofrequency stripes are orthogonal to the low-to-high frequency gradient. An asterisk means that the neuron in the isofrequency stripe is active while a blank means that the neuron is not active. The four lines of asterisks on the left represent the fundamental frequency 4-stripe.

First, we show which role a different amount of cortical surface area devoted to low versus high frequencies may play in the phenomenon of the missing fundamental [2]. Second, we propose a mechanism of transpositionally-invariant melody recognition which is founded on the existence of isofrequency stripes [3].

2 Modelling the Effect of the Missing Fundamental

In music theory, the lowest frequency of the complex tone is called the *fundamental frequency*, f_0. The higher frequencies, that are all integral multiples of f_0, are called the *harmonics*. The 2nd harmonic has a frequency $f_0 \times 2$, the 3rd harmonic has a frequency $f_0 \times 3$, and so on. A tone's height, *pitch*, is determined by the tone's fundamental frequency, and expressed as a note in musical notation. Sounds like those produced by speech, slamming a door, etc., are musically unharmonic, i.e. the higher frequencies are not integral multiples of the sound lowest frequency. *Pitch perception* can be considered to be a *labeling process* by which the auditory system collects certain frequencies and tags them with a pitch identifier, and simultaneously collects another set of frequencies and tags them with another identifier, and so on [8]. Thus, it is curious that humans are able to recognize the sameness of pitches of complex tones with the same fundamental frequencies but with different occurances and intensities of the tone harmonics. However, the most striking instance of spectrum-invariant pitch recognition is *the effect of the missing fundamental* [8]. The effect of the missing fundamental means that our auditory system is able to perceive pitches that correspond to the fundamental frequencies of complex tones while those fundamentals are physically totally absent.

Over the decades there has been a debate whether the neural mechanisms responsible for the effect of the missing fundamental in particular, and pitch perception in general, are peripheral or central. However, there are several experimental findings that strongly undermine the idea about the peripheral mechanisms being *always* responsible for the effect of the missing fundamental. One of the strongest arguments comes from the psychophysical experiments [9] which showed that the corresponding musical pitch can be generated only by two upper harmonics (2nd and 3rd) presented dichotically, i.e. delivered simultaneously through separate ears. However, each of them alone leads to the corresponding pure tone frequency percept. It is clear that with the dichotic stimulus there is no possible way for the two components to fall into the same peripheral channel. Important findings in several pitch identification psychophysical experiments [10, 11], showed that the salience of the missing fundamental pitch percept degrades when harmonic order $n(h)$ increases (figure 2).

Our neural network model is the stochastic ANN [2]. We deal with non-orthogonal memory patterns for which we use the dynamics proposed by Amit *et al.* [12] to extend the original Hopfield dynamics [13]. A global control on the dynamics of the network that prevents too high or too low activity patterns is achieved by imposing a finite energy cost on fluctuations away from the optimal mean level of activity. The network dynamics is asynchronous, i.e. at every time instant only one randomly chosen neuron updates its state according to a probabilistic rule [14]. When the time evolution of a stochastic network ends up close enough to one memory state, it is interpreted as a recall of the given pattern from memory. An observable variable is the overlap of the current state with individual memory patterns.

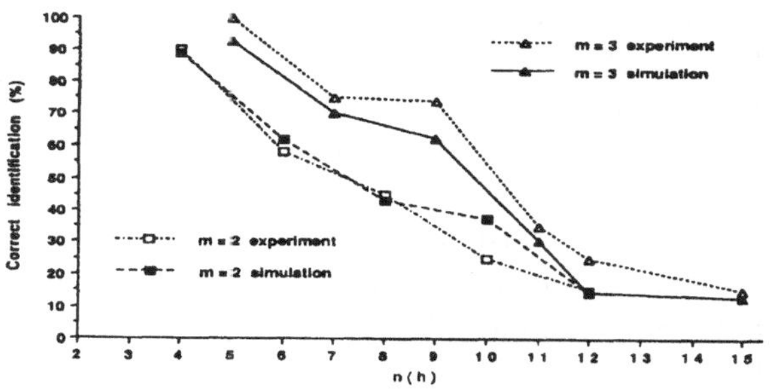

Fig. 2. Identification of the pitch of the missing fundamental as a function of the lowest harmonic number $n(h)$ for the two overall numbers of harmonics, $m = 2$ and $m = 3$. Each point of the results of psychophysical *experiment* [10] and our computer *simulation* [2] is the average from 400 trials, i.e. 50 trials for each of the eight tones. The ANN has $N = 455$ neurons.

In psychophysical experiments, humans were presented with the sounds lacking the fundamental frequencies and consisting of small numbers of higher harmonics, for instance $m = 2$ and $m = 3$, starting with different lowest harmonic number $n(h)$ [10, 11]. They were asked to identify the pitch they heard. In our computer simulations, the initial configuration of the ANN was the activity present only in the two or three corresponding isofrequency stripes. Neuronal relaxation led the ANN close to one of the tone patterns. In the criterion for the correct retrieval of the tone template three conditions had to be met simultaneously: (1) The overlap between the final network state and the desired tone template was the largest compared with the overlaps with all other tone templates. (2) The partial overlap between the activity retrieved at the place of the missing fundamental 4-stripe was the largest compared with overlaps with other missing fundamental 4-stripes. (3) The final partial overlap with the desired fundamental 4-stripe was > 0.25.

Houtsma and Smurzynski [11] interpret their results in such a way that they reflect the existence of either two distinct and separate pitch mechanisms in the auditory system, one for lower harmonics and the second for higher harmonics, or there is a single neural mechanism that behaves differently for these two kinds of frequencies.

An explanation offered by our ANN model is that there is a single cortical mechanism for pitch perception, and differences in pitch salience arise as a consequence of the characteristics of the tone cortical representation. Namely, *because of the larger area of representation for the lower harmonics than for the higher harmonics.* The quality of pitch retrieval in the ANN, in terms of the magnitude of the final overlap, is proportional to the initial overlap between the incomplete input and the corresponding perceptual template. Thus, the bigger part of the representation devoted to lower harmonics conveys the pitch stronger. Simultaneously, the more harmonics is present in the spectrum, the pitch is more salient (figure 2, the curves for $m = 3$ are higher than the curves for $m = 2$).

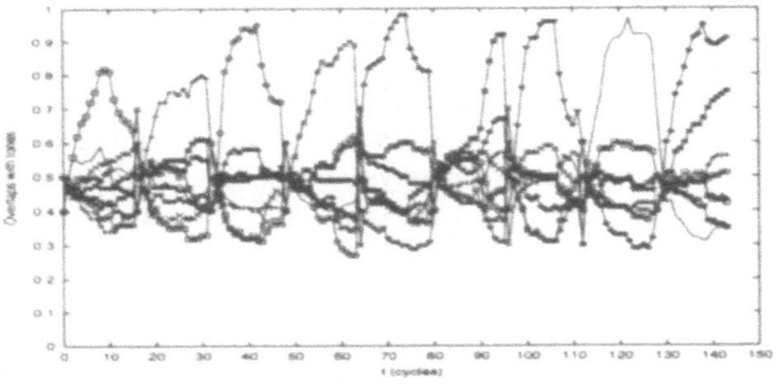

Fig. 3. Computer simulation of recognition of the tone sequence under transpositional invariancy, after medium initial transposition. The ANN has $N = 400$ neurons and stores 9 patterns as a sequence. We plot the overlaps with the memorized, i.e. untranslated tone memory patterns. Translation-invariant pattern retrieval of pattern μ is followed by a retrieval of the $(\mu+1)$-th pattern in the tone sequence via slow synapses. Evolution is averaged over 50 trials.

3 Modelling Transpositional Invariancy of Melody Recognition

A common but nonetheless remarkable human faculty is the ability to recognize a melody even if it is presented in a key different from that in which it was originally heard [15, 16]. Melodies transposed to different keys are recognized even though the absolute frequencies of the tones have changed. Moreover, transpositionally invariant melody recognition does not require musical training. A melody can be considered to be a temporal sequence of complex tones (with some timbre, loudness, rhythm, etc.). A particular tonal melody is characterized by a melodic contour – the pattern of ups and downs – applied to the musical scale. In the transposed melody, there is a constant frequency ratio between all the original and transposed tones, while the melody contour is preserved. Interestingly, recognition of tone sequences under transposition invariancy displays the key distance effect [15, 16].

In our model [3], the recognition of a tone sequence is performed by an ANN with fast and slow synapses designed for storage and recognition of sequences of patterns [17, 18]. Here, recognition is defined as a completed set of transitions from one former attractor to another, in response to a particular input. An external stimulus, e.g. a transposed tone, first initiates a process of transposition-invariant recall of the original tone pattern. If this transposition-invariant recall was succesfull, the recalled state leads to an associative retrieval of the next pattern in a predetermined sequence. The network proceeds from one pattern to the next in a predetermined sequence only in the case when the transposition-invariant recall of individual original tones is succesfully completed (figure 3).

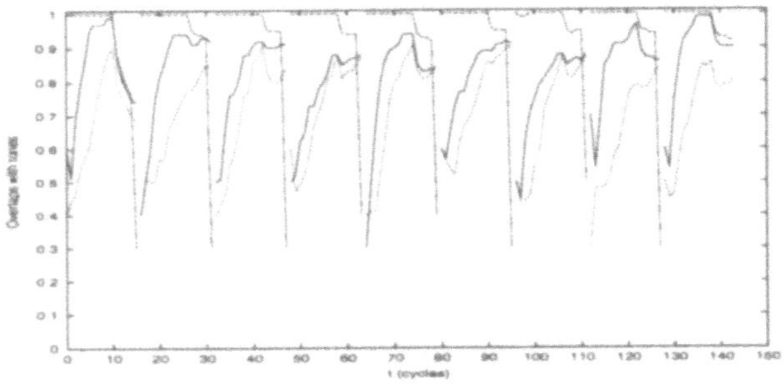

Fig. 4. Illustration of the effect of melody transposition distance on the quality of transposition-invariant melody recognition (an analogy to the key distance effect). We present the results of computer simulations of the retrieval for the 3 different initial shifts of all tone patterns: small (dashed line) medium (solid line), and large (dotted line). All overlaps are averaged over 50 trials.

Thus, there are two processes taking place in the ANN: (1) transposition-invariant recall of tones comprising the original untransposed melody, and (2) recognition whether these patterns belong to the original sequence of tones. As for the tone patterns, each complex tone was comprised of a particular combination of stripes (figure 1). *Such a representation allows for treating the problem of transposition-invariant recognition of tones as a translation-invariant retrieval of their stripe representations.* For translation-invariant retrieval of tone patterns, we used the Dotsenko algorithm for translation-, rotation- and scale-invariant recognition of patterns [19]. This algorithm uses firing threshold dynamics to guide the network evolution in the state space towards the desired attractor, and neuronal dynamics performs the error correction. We extended the original proposal first by modification of the neuronal dynamics to allow for the storage and retrieval of biased patterns [12], and second, by replacing the gradient optimization of threshold dynamics with simulated annealing [20]. In the computer simulations of the proposed ANN model, we found regular effects of transformation distance. That is, the retrieval quality, in terms of magnitude of the overlaps with desired tone patterns and the time spent in the corresponding quasi-attractors, increases with decreasing the initial transposition distance of a melody (figure 4).

4 Conclusion

The proposed ANN models for spectrum-invariant pitch recognition and for transposition-invariant melody recognition can be applied to invariant recognition of unharmonic sounds, for instance speech. Recently, areas in human auditory cortex have been demostrated that are selective only to human voices [21]. In speech, each phoneme is an unharmonic frequency complex that activates a particular combination of frequency stripes in the primary auditory cortex in the same way as the complex tone (figure 1). *Phoneme identification* can be considered to be a *labeling process* by which the auditory system collects certain frequencies and tags them with a phoneme identifier, and simultaneously collects another set of frequencies and tags them with another identifier, and so on. Spectrum-invariant recognition of phonemes can be modeled as an auto-associative recall of the phoneme template. Sequencies of phonemes correspond to words, and thus their transposition-invariant recognition can be modeled by the same mechanism as the transposition-invariant recognition of melodies.

Supported by the VEGA grants 2/6018/99 and 1/7611/20.

References

[1] D.J.Amit, *Modeling Brain Function. The World of Attractor Neural Networks*, Cambridge: Cambridge University Press, 1989.

[2] L. Beňušková, "Modelling the effect of the missing fundamental with an attractor neural network", *Network*, vol. 5, pp. 333–349, 1994.

[3] L. Beňušková, "Modelling transpositional invariancy of melody recognition with an attractor neural network", *Network*, vol. 6, pp. 313–331, 1995.

[4] R.A. Reale and T.J. Imig, "Tonotopic organization in auditory cortex of the cat", *J. Comp. Neurol.*, vol. 192, pp. 265–292, 1980.

[5] A. Morel and J.H. Kaas, "Subdivisions and connections of auditory cortex in owl monkeys", *J. Comp. Neurol.*, vol. 318, pp. 27–63, 1992.

[6] J.L. Lauter, P. Herscovitch, C. Formby and M.E. Raichle, "Tonotopic organization in human auditory cortex revelaed by positron emission tomography", *Hearing Res.*, vol. 20, pp. 199–205, 1985.

[7] A. Špelda, *Úvod do akustiky pro hudebníky*, Praha: Státní nakladatelství krásné literatury, hudby a umění , 1958.

[8] W.M. Hartmann W M, "Pitch perception and the segregation and the integration of auditory entities", *Auditory Function. Neurobiological Bases of Hearing*, G.M. Edelman, W. Einar Gall and W.M. Cowan (eds), New York: John Wiley and Sons, 1988.

[9] A.J.M. Houtsma and J.L. Goldstein, "The central pitch origin of the pitch of complex tones: evidence from musical interval recognition", *J. Acoust. Soc. Amer.*, vol. 51, pp.520–529, 1972.

[10] J.L. Goldstein, A. Gerson, P. Srulovicz and M. Furst, "Verification of the optimal probabilistic basis of aural processing in pitch of complex tones", *J. Acoust. Soc. Amer.*, vol. 63, pp. 486–497, 1978.

[11] A.J.M. Houtsma and J. Smurzynski, "Pitch identification and discrimination for complex tones with many harmonics", *J. Acoust. Soc. Amer.*, vol 87, pp. 304–310, 1990.

[12] D.J. Amit, H. Gutfreund and H. Sompolinsky, "Information storage in neural networks with low levels of activity", *Phys.Rev. A*, vol. 35, pp. 2293–2303, 1987.

[13] J.J. Hopfield, "Neural networks and physical systems with emergent collective computational abilities", *Proc. Natl. Acad. Sci. USA*, vol. 79, pp. 2554–2558, 1982.

[14] P. Peretto and J.-J. Niez J.-J., "Stochastic dynamics of neural networks", *IEEE Trans. Syst. Man Cybern.*, vol. 16, pp. 73–83, 1986.

[15] J.C. Bartlett and W.J. Dowling, "Recognition of transposed melodies: a key-distance effect in developmental perspective", *J. Exp. Psychol.: Hum. Percept. Perform.*, vol. 6, pp. 501–515, 1980.

[16] E.C. Carterrete, D.V. Kohl and M.A. Pitt, "Similarities among transformed melodies: the abstraction of invariants", *Music Perception*, vol. 3, 393–410, 1986.

[17] H. Sompolinsky and I. Kanter, "Temporal association in asymetric neural networks", *Phys. Rev. Lett.*, vol. 57, pp. 2861–2864, 1986.

[18] H. Gutfreund and M. Mézard, "Processing temporal sequences in neural networks", *Phys. Rev. Lett.*, vol. 61, pp. 235–238, 1988.

[19] V.S. Dotsenko, "Neural networks: translation-, rotation- and scale-invariant patterns recognition", *J. Phys. A: Math. Gen.*, vol. 21, pp. L783–L787, 1988.

[20] V. Černý, "Thermodynamical approach to the traveling salesman problem: an efficient simulation algorithm", *J. Optim. Appl.*, vol. 45, pp. 41–51, 1985.

[21] P. Belin, R.J. Zatorre, P. Lafaille, P. Ahad and B. Pike, "Voice-selective areas in human auditory cortex", *Nature*, vol. 403, pp. 309–312, 2000.

Long Short-Term Memory Learns Context Free and Context Sensitive Languages

Felix A. Gers, Jürgen Schmidhuber *

*IDSIA, Galleria 2, 6928 Manno, Switzerland, www.idsia.ch

Abstract

Previous work on learning regular languages from exemplary training sequences showed that Long Short-Term Memory (LSTM) outperforms traditional recurrent neural networks (RNNs). Here we demonstrate LSTM's superior performance on context free language (CFL) benchmarks, and show that it works even better than previous hardwired or highly specialized architectures. To the best of our knowledge, LSTM variants are also the first RNNs to learn a context *sensitive* language (CSL), namely, $a^n b^n c^n$.

1 Introduction

Until recently standard recurrent neural networks RNNs (see survey by Pearlmutter [4]) have been plagued by a major practical problem: the gradient of the total output error with respect to previous inputs quickly vanishes as the time lags between relevant inputs and errors increase. Hence standard RNNs fail to learn in the presence of time lags exceeding as few as 5-10 discrete time steps between relevant input events and target signals.

The recent *"Long Short-Term Memory"* (LSTM) method [3], however, is not affected by this problem. LSTM can learn to bridge minimal time lags in excess of 1000 discrete time steps by enforcing *constant* error flow through "constant error carousels" (CECs) within special units, without loss of short time lag capabilities. Multiplicative gate units learn to open and close access to the constant error flow. Moreover, LSTM's learning algorithm is more efficient than previous RNN algorithms such as real time recurrent learning (RTRL) and back propagation through time (BPTT): it is local in space and time, with computational complexity $O(1)$ per time step and weight.

Previous work showed that LSTM outperforms traditional RNN algorithms on numerous tasks involving real-valued or discrete inputs and targets [2], [3], including tasks that require to learn the rules of regular languages (RLs) describable by deterministic finite state automata (DFA). Until now, however, it has remained unclear whether LSTM's superiority carries over to tasks involving context free languages (CFLs), such as those discussed in the RNN literature [7], [9], [8], [5], [6]. Their recognition requires the functional equivalent of a stack. It is conceivable that LSTM has just the right bias for RLs but might fail on CFLs. Here we will focus on the most common CFL benchmarks: $a^n b^n$ and $a^n b^m B^m A^n$. Finally we will apply LSTM to a context *sensitive* language (CSL). The CSLs include the CFLs, which include the RLs. We will focus on the classic example $a^n b^n c^n$, which is a CSL but not a CFL. In general, CSL recognition requires a linear-bounded automaton, a special Turing machine whose tape length is at most linear in the input size. To our knowledge no RNN has been able to learn a CSL.

2 LSTM

We are using LSTM with forget gates and the recently introduced peephole connections [1]. The basic unit of an LSTM network is the *memory block* containing one or more *memory cells* and three adaptive, multiplicative gating units shared by all cells in the block. Each memory cell has at its core a recurrently self-connected linear unit called the "Constant Error Carousel" (CEC) whose activation is called the cell *state*. The CECs enforce *constant* error flow and overcome a fundamental problem plaguing previous RNNs: they prevent error signals from decaying quickly as they "back in time". The adaptive gates control input and output to the cells (*input and output gate*) and learn to reset the cell's state once its contents are out of date (*forget gate*). Peephole connections connect the CEC to the gates. All errors are cut off once they leak out of a memory cell or gate, although they do serve to change the incoming weights. The effect is that the CECs are the only part of the system through which errors can flow back forever, while gates etc. learn the nonlinear aspects of sequence processing. This makes LSTM's updates efficient without significantly affecting learning power: LSTM's learning algorithm is local in space and time; its computational complexity per time step and weight is $O(1)$. The CECs permit LSTM to bridge huge time lags (1000 discrete time steps and more) between relevant events, while traditional RNNs already fail to learn in the presence of 10 step time lags, despite requiring more complex update algorithms.

Forward Pass. See [1] for a detailed description of LSTM's forward pass with forget gates and peephole connections. Essentially, the cell output, y^c, is calculated based on the current cell state s_c and four sources of input: to the cell itself, to the input gate, to the forget gate and input to output gate. All gates have a sigmoid squashing functions with range $[0, 1]$. The state of memory cell $s_c(t)$ is calculated by adding the squashed (the squashing function g is a sigmoid with range $[-1, 1]$), gated input to the state at the previous time step $s_c(t-1)$, which is multiplied by the forget gate activation. The cell output y^c is calculated by multiplying (gating) $s_c(t)$ by the output gate activation.

Gradient-Based Backward Pass. Essentially, LSTM's backward pass (for details see Hochreiter and Schmidhuber [3] and Gers et. al. [2]) is an efficient fusion of slightly modified, truncated BPTT and a customized version of RTRL. We are using iterative gradient descent, minimizing an objective function E, here the usual mean squared error function. Unlike BPTT and RTRL, LSTM's learning algorithm is local in space and time.

3 Experiments

The network sequentially observes exemplary symbol strings of a given language, presented one input symbol at a time.

Following the traditional approach in the RNN literature we formulate the task as a prediction task. At any given time step the target is to predict the possible next symbols, including the terminal symbol T. When more than one symbol can occur in the next step *all* possible symbols have to be predicted, and none of the others.

Every input sequence begins with the start symbol S. The empty string, consisting of ST only, is considered part of each language. A string is accepted when all predictions have been correct. Otherwise it is rejected.

This prediction task is equivalent to a classification task with the two classes "accept" and "reject", because the system will make prediction errors for all strings outside the language. A system has learned a given language up to string size n once it is able to correctly predict all strings with size $\leq n$.

Symbols are encoded locally in d-dimensional vectors, where d is equal to the number of symbols of the given language plus one for either the start symbol in the input or the terminal symbol in the output (d input units, d output units, each standing for one of the symbols). +1 signifies that a symbol is set and −1 that it is not set; the decision boundary for the network output is 0.0.

CFL a^nb^n [9], [8], [7], [6]. Here the strings in the input sequences are of the form a^nb^n, input and output vectors are 3-dimensional. Before the first occurrence of b either a or b, or a or T at sequence beginings, are possible in the next step. Thus, e.g., for $n=4$:

Input:	S	a	a	a	a	b	b	b	b
Target:	a/T	a/b	a/b	a/b	a/b	b	b	b	T

CFL $a^nb^mB^mA^n$ [5]. The second half of a string from this palindrome or mirror language is completely predictable from the first half. This task involves an intermediate time lag of length $2m$. Input and output vectors are 5-dimensional. Before the first occurrence of B two symbols are possible in the next step. Thus, e.g., for $n = 2, m = 2$:

Input:	S	a	a	b	b	B	B	A	A
Target:	a/T	a/b	a/b	b/B	b/B	B	A	A	T

CSL $a^nb^nc^n$. Input and output vectors are 4-dimensional. Before the first occurrence of b two symbols are possible in the next step. Thus, e.g., for $n = 3$:

Input:	S	a	a	a	b	b	b	c	c	c
Target:	a/T	a/b	a/b	a/b	b	b	c	c	c	T

3.1 Training and testing

Learning and testing alternate: after 1000 training sequences we freeze the weights and run a test. Training and test sets incorporate all legal strings up to a given length: $2n$ for a^nb^n, $3n$ for $a^nb^nc^n$ and $2(n+m)$ for $a^nb^mB^mA^n$. Only positive exemplars are presented. Training is stopped once all training sequences have been accepted, or after at most 10^7 training sequences. The *generalization set* is the largest accepted test set.

Weight changes are made after each sequence. We apply the momentum algorithm with learning rate α is 10^{-5} and momentum parameter 0.99. All results are averages over 10 independently trained networks with different weight initializations (the same for each experiment).

CFL a^nb^n. We study training sets with $n \in \{1,..,N\}$. We test all sets with $n \in \{1,..,M\}$ and $M \in \{N,..,1000\}$ (sequences of length ≤ 2000).

CFL $a^nb^mB^mA^n$. We use two training sets: a) The same set as used by Rodriguez and Wiles (1999) [5] $n \in \{1,..,11\}$, $m \in \{1,..,11\}$ with $n+m \leq 12$ (sequences of length ≤ 24). b) The set given by $n \in \{1,..,11\}$, $m \in \{1,..,11\}$ (sequences of length ≤ 48). We test all sets with $n \in \{1,..,M\}$, $m \in \{1,..,M\}$ and $M \in \{11,..,50\}$ (sequences of length ≤ 200).

CSL $a^nb^nc^n$. We study training sets with the same parameters as for the CFL a^nb^n. We test all sets with $n \in \{1,..,M\}$ and $M \in \{N,..,500\}$ (sequences of length ≤ 1500).

3.2 Topology and experimental parameters

The input units are fully connected to a hidden layer consisting of memory blocks with 1 cell each. The cell outputs are fully connected to the cell inputs, to all gates, and to the output units, which also have direct "shortcut" connections from the input units. All gates, the cell itself and the output unit are biased. The bias weights to input gate, forget gate and output gate are initialized with -1.0, $+2.0$ and -2.0, respectively (these are standard values, which we use for all our experiments; precise initialization is not critical here). All other weights are initialized randomly in the range $[-0.1, 0.1]$. The cell's input squashing function g is the identity function. The squashing function of the output units is a sigmoid function with the range $[-2, 2]$.

CFL $a^n b^n$. We use one memory block (with one cell). Without the 3 peephole connections there are 37 adjustable weights (28 unit-to-unit and 7 bias connections).

CFL $a^n b^m B^m A^n$. We use two blocks (with one cell each), resulting 114 adjustable weights (91 unit-to-unit and 13 bias connections).

CSL $a^n b^n c^n$. We use the same topology as for the $a^n b^m B^m A^n$ language, but with 4 input and output units instead of 5, giving 84 adjustable weights (72 unit-to-unit and 12 bias connections).

3.3 Previous results

CFL $a^n b^n$. Published results on the $a^n b^n$ language are summarized in Table I. RNNs trained with plain BPTT tend to learn to just reproduce the input [9], [8], [6]. Sun et al. [7] used a highly specialized architecture, the "neural pushdown automaton", which also did not generalize well.

CFL $a^n b^m B^m A^n$. Rodriguez and Wiles (1998) [5] used BPTT-RNNs with 5 idden nodes. After training with $n + m \leq 12$ (sequences of length ≤ 24), the best network generalized to sequences up to length 36 ($n = 9, m = 9$). But their networks did not learn the complete training set, and generalization was restricted to few strings with $n, m \in \{1,.., 9\}$.

CSL $a^n b^n c^n$. To our knowledge no previous RNN ever learned a CSL.

3.4 LSTM results

CFL $a^n b^n$. 100% solved for all training sets (lhs of Table II). Small training sets ($n \in \{1,.., 10\}$) were already sufficient for perfect generalization up to the tested maximum: $n \in \{1,.., 1000\}$. Note that long sequences of this kind require very stable, finely tuned control of the network's internal counters.

This performance is much better than in previous approaches, where the largest set was learned by the specially designed neural push-down automaton [7] $n \in \{1,.., 160\}$. The latter, however, required training sequences of the same length as the test sequences. From the training set with $n \in \{1,.., 10\}$ LSTM generalized to $n \in \{1,.., 1000\}$, whereas the best previous result (see Table I) generalized only to $n \in \{1,.., 18\}$ (even with a slightly larger training set: $n \in \{1,.., 11\}$).

In contrast to Tonkes and Wiles [8], we did not observe our networks forgetting solutions as training progresses. So unlike all previous approaches, LSTM reliably finds solutions that generalize well.

CFL $a^n b^m B^m A^n$. Training set a): 100% solved; after $29 \cdot 10^3$ training sequences the best network of 10 generalized to at least $n, m \in \{1,.., 22\}$ (all strings until a length of 88 symbols processed correctly); the average generalization set was the one with $n, m \in \{1,.., 16\}$ (all strings until a length of 64 symbols processed correctly), learned after $25 \cdot 10^3$ training sequences on average.

Training set b): 100% solved; after $26 \cdot 10^3$ training sequences the best network generalized to at least $n, m \in \{1,.., 23\}$ (all strings until a length of 92 symbols processed correctly). The average generalization set was the one with $n, m \in \{1,.., 17\}$ (all strings until a length of 68 symbols processed correctly), learned after $82 \cdot 10^3$ training sequences on average.

Unlike the previous approach of Rodriguez and Wiles [5], LSTM easily learns the complete training set and reliably finds solutions that generalize well.

CSL $a^n b^n c^n$. LSTM learns 4 of the 5 training sets in 10 out of 10 trials (only 9 out of 10 for the training set with $n \in \{1,.., 40\}$) and generalizes well (rhs Table II). Small training sets ($n \in \{1,.., 40\}$) were already sufficient for perfect generalization up to the tested maximum: $n \in \{1,.., 500\}$, that is, sequences of length up to 1500.

3.5 Analysis

How do the solutions discovered by LSTM work?

CFL $a^n b^n$. The cell state s_c increases while a symbols are fed into the network, then decreases (with the same step size) while b symbols are fed in. At sequence beginnings (when the first a symbols are observed), however, the step size is smaller due to the closed input gate, which is triggered by s_c itself. This results in "overshooting" the initial value of s_c at the end of a sequence and leads to the prediction of the sequence termination.

CFL $a^n b^m B^m A^n$. The network learned to establish and control two counters, the two symbol pairs (a, A) and (b, B) are treated separately by two different cells, c_2 and c_1, respectively. Cell c_2 tracks the difference between the number of observed a and A symbols. It opens only at the end of a string, where it predicts the final T. Cell c_1 treats the embedded $b^m B^m$ substring in a similar way. While values are stored and manipulated within a cell, the output gate remains closed. This prevents the cell from disturbing the rest of the network and also protects its CEC against incoming errors.

Table I. Previous results for the CFL $a^n b^n$, showing (from left to right) the number of hidden units or state units, the values of n used during training, the number of training sequences, the number of found solutions/trials and the largest accepted test set.

Reference	Hidden Units	Train. Set [n]	Train. Str. [10^3]	Sol./Tri.	Best Test [n]
[7][1]	5	1,.., 160	13.5	1/1	1,.., 160
[9]	2	1,.., 11	2000	4/20	1,.., 18
[8]	2	1,.., 10	10	13/100	1,.., 12
[6][2]	2	1,.., 11	267	8/50	1,.., 16

Table II. Results for the CFL $a^n b^n$ (lhs) and for the CSL $a^n b^n c^n$ (rhs), showing (from left to right) the values for n used during training, the average number of training sequences until best generalization was achieved (average over all networks given in parenthesis), the percentage of correct solutions and the best generalization (average over all networks given in parenthesis).

	CFL $a^n b^n$		CSL $a^n b^n c^n$	
Train. Set [n]	Train. Str. [10^3]	Generalization Set [n]	Train. Str. [10^3]	Generalization Set [n]
1,.., 10	22 (19)	1,.., 1000 (1,.., 118)	54 (62)	1,.., 52 (1,.., 28)
1,.., 20	18 (19)	1,.., 587 (1,.., 148)	28 (43)	1,.., 160 (1,.., 66)
1,.., 30	16 (19)	1,.., 1000 (1,.., 408)	37 (43)	1,.., 228 (1,.., 91)
1,.., 40	25 (28)	1,.., 1000 (1,.., 628)	51 (48)	1,.., 500 (1,.., 120)
1,.., 50	42 (40)	1,.., 767 (1,.., 430)	60 (94)	1,.., 500 (1,.., 409)

CSL $a^n b^n c^n$. The network solutions use a combination of two counters, instantiated separately in the two memory blocks. Here the second cell counts up, given an a input symbol. It counts down, given a b. The second memory block does the same for b, c, and a, respectively.

4 Conclusion

We found that Long Short-Term Memory (LSTM) clearly outperforms previous RNNs not only on regular language benchmarks (according to previous research) but also on context free language (CFL) benchmarks. It learns faster and generalizes better. LSTM also is the first RNN to learn a context sensitive language.

Acknowledgment. This work was supported by SNF grant 2100-49'144.96 "Long Short-Term Memory."

References

[1] F. A. Gers and J. Schmidhuber. Recurrent nets that time and count. In Proc. IJCNN'2000, Int. Joint Conf. on Neural Networks, Como, Italy, 2000.

[2] F. A. Gers, J. Schmidhuber, and F. Cummins. Learning to forget: Continual prediction with LSTM. Neural Computation, 12(10):2451-2471, 2000.

[3] S. Hochreiter and J. Schmidhuber. Long short-term memory. Neural Computation, 9(8):1735-1780, 1997.

[4] B. A. Pearlmutter. Gradient calculations for dynamic recurrent neural networks: A survey. IEEE Transactions on Neural Networks, 6(5):1212-1228, 1995.

[5] P. Rodriguez. J. Wiles, and J Elman. A recurrent neural network that learns to count. Connection Science, 11(1):5-40, 1999.

[6] Paul Rodriguez and Janet Wiles. Recurrent neural networks can learn to implement symbol-sensitive counting. In Advances in Neural Information Processing Systems, volume 10, pages 87-93. The MIT Press, 1998.

[7] G. Z. Sun, C. Lee Giles, H. H. Chen, and Y. C. Lee. The neural network pushdown automaton: Model, stack and learning simulations. Technical Report CS-TR-3118, University of Maryland, College Park, August 1993.

[8] B. Tonkes and J. Wiles. Learning a context-free task with a recurrent neural network: An analysis of stability. In Proceedings of the Fourth Biennial Conference of the Australasian Cognitive Science Society, 1997.

[9] J. Wiles and J. Elman. Learning to count without a counter: A case study of dynamics and activation landscapes in recurrent networks. In In Proceedings of the Seventeenth Annual Conference of the Cognitive Science Society, pages pages 482-487, Cambridge, MA, 1995. MIT Press.

[1] Sun's training set was augmented stepwise by sequences misclassified during testing, and in the final accepted set n was in {1,.., 20} except for 20 random sequences up to length $n = 160$ (the exact generalization performance was unclear).

[2] Applying brute force search to the weights of the best network of Rodriguez et al. [6] further improves performance to acceptance up to $n = 28$.

An Artificial Neural Network Model Based on Neuroscience: Looking Closely at the Brain

João Luís Garcia Rosa *

*Instituto de Informática - PUC-Campinas, Rodovia D. Pedro I, km. 136, Caixa Postal 317, 13086-900, Campinas, SP, Brasil - joaol@ii.puc-campinas.br

"Almost all aspects of life are engineered at the molecular level, and without understanding molecules we can only have a sketchy understanding of life itself."
Francis Crick, *What Mad Pursuit*, 1988

Abstract
Classical connectionist models [3, 8, 11] are based upon a simple description of the neuron taking into account the presence of pre-synaptic cells and their synaptic potentials, the activation threshold, and the propagation of an action potential. Certainly, this is an impoverished explanation of human brain characteristics [1, 9, 12]. In this paper, a mechanism to generate a biologically plausible artificial neural network model is presented [10], which is taken to be closer to some of the human brain features. In such a mechanism, the classical framework is redesigned in order to encompass not only the "traditional" features but also labels that model the binding affinities between transmitters and receptors. This is accomplished by a restricted data set, which explains the neural network behavior. In addition to feed-forward networks, the present model also contemplates recurrence in its architecture, which allows the system to have re-entrant connections [2].

1 Introduction

The presented paper [10] departs from a classical connectionist model and proposes a biologically plausible neural network model. Such model is defined by a restricted data set, which explains the neural network behavior. Unlike other models, this one introduces transmitter, receptor, and controller variables in order to account for the binding affinities between neurons. The following feature set thus defines the neurons:

$$N = \{\{w\}, \theta, g, T, R, C\}$$

where w represents the connection weights, θ is the neuron activation threshold, g stands for the activation function, T symbolizes the transmitter, R the receptor, and C the controller. θ, g, T, R, and C are part of the genetic information. T, R, and C are the labels, absent in other models.

As stated in Ramón y Cajal's *principle of connectional specificity*, "nerve cells do not communicate indiscriminately with one another" [6]. In the presented model, each neuron is connected to another neuron not only in relation to its connection weight, activation threshold, and activation function, but also in relation to its labels. Neuron i is only connected to neuron j if there is binding affinity between the transmitter of i and the receptor of j. Binding affinity means compatible types, enough amount of substrate, and compatible genes.

In addition, the coupling result of a transmitter T with a receptor R generates a controller C, which can act over other neuron connections.

2 The Biological Support

The ordinary biological neuron has many dendrites usually branched, which receive information from other neurons, and an axon which transmits the processed information, usually by propagation of an action potential [1, 5]. The axon is divided into several branches, which make synapses onto the dendrites and cell bodies of other neurons. The nervous cells influence others by (a) excitation, that is, they contribute to produce impulses on other cells, and (b) inhibition, that is, they prevent the releasing of impulses on other cells.

The predominant type of synapse in the mammalian brain is chemical, and operates through the releasing of a transmitter substance from the pre-synaptic to the post-synaptic terminal [5-7]. This release occurs in active zones, inside pre-synaptic terminals. Certain chemical synapses lack active zones, so synaptic actions between these cells are slower and more diffuse. The coupling result of a neurotransmitter with a receptor makes the post-synaptic cell releases a protein.

The synaptic contacts can be morphologically classified in two basic types: type I and type II synapses [1, 6]. Type I synapses seem to be excitatory because they have larger membrane thickness on the post-synaptic side, and the pre-synaptic process has rounded synaptic vesicles, presumably containing packets of neurotransmitter. Type II synapses seem to be inhibitory because they have smaller and flattened synaptic vesicles and the contact zone is usually smaller than that of type I synapses.

This picture, however, can be much more complicated than implied above. In the first place, the action of a transmitter in the post-synaptic cell does not depend on the chemical nature of the neurotransmitter, but instead on the properties of the receptors with which the transmitter binds. In some cases, it is the receptor that determines whether a synapse is excitatory or inhibitory, and whether an ion channel will be activated directly by the transmitter or indirectly through a second messenger [5, 6].

Secondly, instead of propagating an action potential, an axon can produce a graded potential [1]. Because of attenuation, one should expect that this form of information signaling does not occur over long distances. These graded potentials can occur in another level. For instance, an axon terminal that makes synapse in a given cell can receive a synapse. The pre-synaptic synapse can produce only a local potential change, which is then restricted to that axon terminal.

In view of these biological facts, it was decided to model two features. On the one hand, the binding affinities between transmitters and receptors were modeled through labels T and R. On the other hand, the role of the "second messenger," the effects of graded potential, and the protein released by the coupling of transmitter and receptor were all modeled under only one label, the controller C.

3 The Roles of the Controller

Within the model, the controller can modify the binding affinities between neurons, through three main functions. Firstly, it can modify the degrees of affinity of receptors. Secondly, it can modify the amount of substrate (that is, the amount of transmitters and receptors). Finally, it can modify the gene expression, in the case of mutation. Let's consider the biological motivation for each of these functions in detail.

Degrees of affinity, at chemical synapses, are related to the way receptors gate ion channels, through which transmitter material enters the post-synaptic cell: in direct gating, receptors produce relatively fast synaptic actions, while in indirect gating, receptors produce slow synaptic actions. These slower actions often serve to *modulate* behavior [5] because they modify the degrees of affinity of receptors.

In addition, modulation can be related to the action of peptides. There are many distinct peptides, of several types and shapes, that can act as neurotransmitters [4]. There are, however, reasons to suspect that peptides are different from many conventional transmitters [1]: peptides appear to "modulate" the synaptic function instead of activating it; the action of peptides usually appears to spread slowly and persist for some time, much more than conventional transmitters; and in some cases, peptides do not act where they were released, but at some distant site.

As transmitters, peptides act at very restrict places, display a slow rate of conduction, and do not sustain the high frequencies of impulses. As neuromodulators of the synaptic function, its activity is more intense. The excitatory effects of substance P (a peptide) are very slow in the beginning but longer in duration (more than one minute) and cannot cause, per se, enough depolarization to excite the cells. The effect, however, is to make neurons more readily excited by other excitatory inputs – a clear example of "neuromodulation". Controllers, in the model presented, explain this function by modifying the degrees of affinity of receptors.

An additional function of the controller is to account for variation in the amount of substrate. In biological systems, the acetylcholine (a neurotransmitter) is spread over a short distance toward the post-synaptic membrane and acts at the specific receptor molecules in that membrane. Then, the acetylcholine is enzymatically divided and part of it is taken up again for synthesis of a new transmitter, causing an increase in the amount of substrate. In this model, the controller represents substrate increase by a variable acting over the initial substrate amount.

The final function of the controller concerns gene expression. It was shown that peptides are a second, slower, means of communication between neurons – which is more economical than using extra neurons for this purpose. This second messenger, besides altering the affinities between transmitters and receptors, can regulate gene expression thereby endowing synaptic transmission with long-lasting consequences [5]. In the model, this is achieved by the modification of the variable that represents gene expression. Consequently, mutation can be accounted for in this model.

4 The Labels and Their Dynamic Behaviors

The aim of this paper is to present a more sophisticated mathematical model of the neuron, through the definition of a restrict data set, thus explaining the behavior of a biologically plausible artificial neural network. In this sense, it is important to define the labels (T, R, and C) and their dynamic behaviors in the following way, as stated in [10]:

A. For the network genesis:

1. the specification of the number of layers;
2. the specification of the number of neurons in each layer;
3. the definition of the initial amount of substrate (transmitters and receptors) in each layer; and
4. the definition of the genetics of each layer (type of transmitter and its degree of affinity, type of receptor and its degree of affinity, and genes (name and gene expression)).

B. For the evaluation of the controllers and how they act:

1. the controllers can modify the degree of affinity of receptors;
2. the controllers can modify the initial substrate storage; and
3. the controllers can modify the gene expression value (mutation).

It is expected that these specifications lead to an artificial neural network displaying some distinctive characteristics. In the first place, each neuron has a genetic code (a set of genes plus a gene expression controller). The controller can cause *mutation*, because it can regulate gene expression.

Second, the substrate (amount of transmitter and receptor) is defined by layer. Because substrate amounts are limited, there is a chance that some post-synaptic neurons, to which a certain pre-synaptic neuron should be connected, will not be activated. Such a network, then, can be seen as favoring *clustering*.

Third, the substrate increase is related to the gene specified in the controller, because the synthesis of a new transmitter occurs in the pre-synaptic terminal (origin gene). The modification of the genetic code, that is, mutation, as well as the modification of the degree of affinity of receptors, however, is related to the target gene. The reason is that the modulation function of controller is better explained at some distance of the emission of neurotransmitter, therefore at the target.

5 A Network Simulation

In table 1, a data set for a five-layer network simulation is presented. For the specifications displayed in table 1, the network architecture and its activated connections are shown in figure 1. For the sake of simplicity, all degrees of affinity are set at 1 (the degree of affinity is represented by a real number in the range [0..1]; so that the greater the degree of affinity is the stronger the synaptic connection will be).

In figure 1, one can notice that every unit in layer 1 (the input layer) is linked to the first nine units in layer 2 (first hidden layer). The reason why not every unit in layer 2 is connected to layer 1, although the receptor of layer 2 has the same type of the transmitter of layer 1, is that the amount of substrate in layer 1 is eight units. This means that, in principle, each layer-1 unit is able to connect to at most eight units. But controller 1, from layer 1 to 2, incremented by 1 the amount of substrate of the origin layer (layer 1). The result is that each layer 1 unit can link to nine units in layer 2. Observe that from layer 2 to layer 3 (the second hidden layer) only four layer-2 units are connected to layer 3, because also of the amount of substrate of layer 3, which is 4.

Table I. The data set for a five-layer network

layer	*1*	2	3	4	5
number of neurons	10	10	5	5	1
amount of substrate	8	10	4	5	2
type of transmitter	1	2	1	2	1
degree of affinity of transmitter	1	1	1	1	1
type of receptor	2	1	2	1	2
degree of affinity of receptor	1	1	1	1	1
genes (name/gene expression)	abc/1	abc/1	abc/1 def/2	abc/1 def/2	def/ 2

Controllers: 1/1-2: abc/s/abc/1
1/1-4: abc/e/abc/2
2/2-3: abc/a/def/0.5

(Controller syntax: *number/origin layer-target layer*: *og/t/tg/res*, where *og* = origin gene (name); *t* = type of synaptic function modulation: a = degree of affinity, s = substrate, e = gene expression; *tg* = target gene (name); *res* = control result: for t = a → *res* = new degree of affinity of receptor (target), for t = s → *res* = substrate increasing (origin), for t = e → *res* = new gene expression controller (target). The controllers from layer 2 to 5, from layer 3 to 4, and from layer 4 to 5 are absent in this simulation.)

As a result of the compatibility of layer-2 transmitter and layer-5 receptor, and the existence of remaining unused substrate of layer 2, one could expect that the first two units in layer 2 should connect to the only unit in layer 5 (the output unit). However, this does not occur because their genes are not compatible. Although gene compatibility exists, in principle, between layers 1 and 4, their units do not connect to each other because there is no remaining substrate in layer 1 and because controller 1 between layers 1 and 4 modified the gene expression of layer 4, making them incompatible. The remaining controller has the effect of modifying the degrees of affinity of receptors in layer 3 (target). Consequently, the connections between layers

2 and 3 became weakened (represented by dotted lines). Notice that, in order to allow connections, in addition to the existence of enough amount of substrate, the genes and the types of transmitters and receptors of each layer must be compatible.

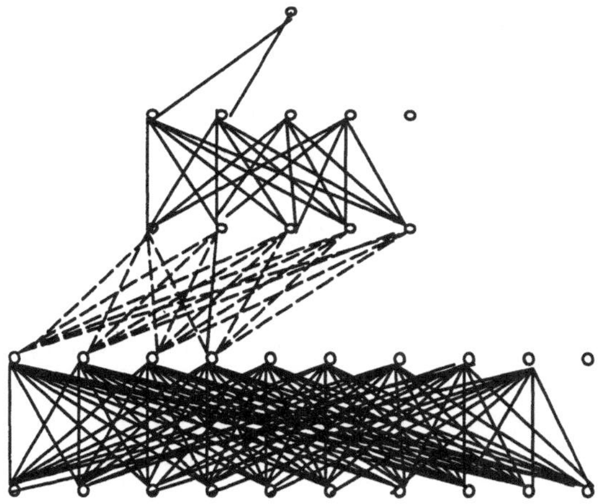

Fig. 1. A five-layer neural network for the data set in table 1. In the bottom of the figure is the layer 1 (input layer) and in the top is the layer 5 (output layer). Between them, there are three hidden layers (layers 2 to 4)

Although the architecture shown in figure 1 is feed-forward, recurrence, or re-entrance, is permitted in this model. This kind of feedback goes along with Edelman and Tononi's "dynamic core" notion [2]. This up-to-date hypothesis suggests that there are neuronal groups underlying conscious experience, the dynamic core, which is highly distributed and integrated through a network of reentrant connections.

6 Conclusion

Nowadays, models of artificial neural networks are in debt with human brain physiology. That is, for mathematical simplicity reasons mainly, conventional neural network models are too simple and thus lack several biological features of the cerebral cortex. The aim here is to present a biologically plausible artificial neural network model [10], which seeks to be closer to the human brain capacity, although only a few brain features are considered. In this model, the possibility of connections between neurons is related not only to synaptic weights, activation threshold, and activation function, but also to labels that embody the binding affinities between transmitters and receptors. This type of neural network would be closer to human evolutionary capacity, since it purports to be a genetically well-suited model of the brain. The recent hypothesis of the "dynamic core" [2] is also contemplated because this model allows reentrancy in its architecture connections.

Acknowledgements: I would like to thank Armando Freitas da Rocha, Márcio Luiz de Andrade Netto, and Edson Françozo (of Unicamp) for their extensive comments on previous drafts of this paper. In addition, I am in debt with ICANNGA 2001 anonymous reviewers for their valuable suggestions.

References

[1] Crick, F. and Asanuma, C.: Certain Aspects of the Anatomy and Physiology of the Cerebral Cortex, in J. L. McClelland and D. E. Rumelhart (eds.), Parallel Distributed Processing, Vol. 2, Cambridge, Massachusetts - London, England, The MIT Press, 1986.

[2] Edelman, G. M. and Tononi, G.: A Universe of Consciousness – How Matter Becomes Imagination, Basic Books, 2000.

[3] Haykin, S.: Neural Networks - A Comprehensive Foundation, 2nd ed. Prentice Hall, Upper Saddle River, New Jersey, 1999.

[4] Iversen, L. L.: Amino Acids and Peptides: Fast and Slow Chemical Signals in the Nervous System?, Proceedings of the Royal Society of London, B, 221 (1224), 22 may, pp. 245-260, (1984).

[5] Kandel, E. R., Schwartz, J. H., and Jessell, T. M. (eds.): Principles of Neural Science. 4th ed., McGraw-Hill, 2000.

[6] Kandel, E. R., Schwartz, J. H., and Jessell, T. M. (eds.): Essentials of Neural Science and Behavior. Appleton & Lange, Stamford, Connecticut, 1995.

[7] Kuffler, S. W., Nicholls, J. G., and Martin, A. R.: From Neuron to Brain - A Cellular Approach to the Function of the Nervous System, 2nd ed., Sinauer Associates Inc. Publ. Sunderland, Mass., 1984.

[8] McCulloch, W. S. and Pitts, W.: A Logical Calculus of the Ideas Immanent in Nervous Activity, Bulletin of Mathematical Biophysics, 5, pp. 115-137, (1943).

[9] Rocha, A. F.: Neural Nets - A Theory for Brains and Machines, Berlin, Heidelberg: Springer-Verlag, 1992.

[10] Rosa, J. L. G. and Françozo, E.: A Biologically Fine-grained Artificial Neural Network: Towards a Hybrid Model, in Proceedings of the IASTED Intl. Conference on Artificial Intelligence and Soft Computing. Cancun, Mexico. May 27-30, pp. 302-305, (1998).

[11] Rosenblatt, F.: The Perceptron: A Perceiving and Recognizing Automaton, Rep. 85-460-1, Proj. PARA, Cornell Aeronautical Lab., Ithaca, New York, 1957.

[12] Rumelhart, D. E. and McClelland, J. L.: PDP Models and General Issues in Cognitive Science, in D. E. Rumelhart and J. L. McClelland (eds.), Parallel Distributed Processing, Vol. 1, Cambridge, Massachusetts - London, England, The MIT Press, 1986.

STARBRAIN and EUROBRAIN Starlab's and Europe's Artificial Brain Projects An Overview

Hugo de Garis*

* STARLAB, Blvd. St.Michel 47, B-1040, Brussels, Belgium, E-mail: degaris@starlab.net, http://foobar.starlab.net/~degaris, http://www.cs.usu.edu/~degaris

Abstract

This paper gives an overview of two research projects which aim to build artificial brains. The first is called STARBRAIN (an abbreviation of the words "Starlab", the author's research lab, and "Artificial Brain"). The STARBRAIN Project aims to use the newly constructed CAM-Brain Machine (CBM) to build an artificial brain of nearly 100 million artificial neurons to control the hundreds of behaviors of a cute life-sized robot kitten "Robokitty". The second is EUROBRAIN, a larger scale more ambitious project at European level, which aims to conceive, design and build a 2nd generation brain building machine called BBM2 within 4 years which will be used to build an artificial brain of a billion neurons, based more closely on biological brain principles. A possible third project is also discsussed.

1 Introduction

This paper gives an overview of two projects to build artificial brains, and briefly discusses a possible third project of a more commercial nature which will also use artificial brains. The first is called STARBRAIN, which is a Brussels based local project, funded by a 1,000,000 dollar grant from the Brussels government, which should start late 2000. Its aim is to use the recently constructed and tested CAM-Brain Machine (CBM) [1] that Starlab has bought (500,000 dollars) (one of four in the world) to build an artificial brain of nearly 100 million artificial neurons. This artificial brain will be used to control several hundred behaviors of a cute life-sized kitten robot "Robokitty". This kitten robot will be used to show off what an artificial brain can do, and hence "prove the concept" that brain building is possible. Section 2 of this paper will be devoted to a description of this project. Section 3 will discuss the second such brain building project which is a lot more ambitious, called EUROBRAIN, which at the time of writing (late September 2000) has not yet been funded. The EUROBRAIN project is of European scale (including some Americans acting as advisors). The EUROBRAIN consists of 3 parts, a) showing what the CBM can do, which is similar to the STARBRAIN project, b) conceiving, designing and building a prototype of a second generation brain building machine called BBM2 that will handle an artificial brain of a billion neurons, which must automate to a much higher degree the evolutionary engineering process, and c) using the billion neuron artificial brain to implement a range of biological brain models, thus making the second brain far more real-brain like. Section 4 discusses a possible industrial R and D project that hopes to use artificial brain (as far as possible, and other) technologies to control thousands of robots as a component in a multi billion dollar application. This contract has not yet been signed (as of Sept 2000) so remains still tentative. If it is signed it will have a major impact on the field of brain building, since some 100 researchers will be involved. Section 5 concludes.

2 The STARBRAIN Project

The STARBRAIN project is essentially a "proof of concept" project that aims to show that building artificial brains is possible. It will use a CAM-Brain Machine (CBM) [1,2,3,4] (Fig. 1) to build an artificial brain of nearly 100 million neurons to control a cute life-sized robot kitten "Robokitty"(Fig. 2) to give it several hundred behaviors.

The "CAM-Brain Machine"(CBM) is a Xilinx XC6264 FPGA based piece of hardware that is used to evolve 3D cellular automata based neural network circuit modules (Fig. 3) at electronic speeds, i.e. in about a second per module. 64,000 of these modules can then be assembled into a gigabyte of RAM according to humanly specified artificial brain architectures. This RAM is updated by the CBM fast enough (130 billion CA cell updates/sec) for real time control of robots. The CBM was built and delivered to the author's previous lab (ATR, Kyoto,

Japan) in March 1999 and another to his present lab (STARLAB, Brussels, Belgium) in July 2000. A third CBM can be found at Lernout and Hauspie (L and H, a speech processing company) in Ypres (WW1 German gas attack site), Belgium, and a fourth at Genobyte Inc. (the designer and builder of the CBMs) in Boulder, Colorado, USA.

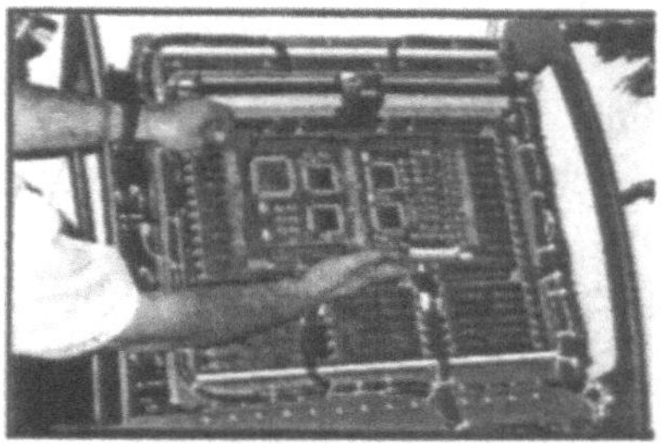

Fig. 1. The CBM, the first generation brain building machine

Fig. 2. Robokitty, a life-sized kitten robot to be controlled by the CBM

The CBM is intended to make practical the creation of artificial brains, which are defined to be assemblages of tens of thousands (and higher magnitudes) of evolved neural net modules into humanly defined artificial brain architectures. An artificial brain consists of a large RAM memory space, into which individual CA modules are downloaded once they have been evolved, and interconnected according to the architectures of human BAs (brain architects). The CA cells in this RAM are updated by the CBM fast enough for real time control of a robot kitten "Robokitty".

3 The EUROBRAIN Project

The EUROBRAIN Project is much bigger than the STARBRAIN project and a lot more ambitious. Its basic aim is to build a billion neuron artificial brain, using principles more powerful than those used in the CBM based STARBRAIN project. Common sense says that it will be quite impractical to specify and evolve a million modules individually. The evolution of the modules and their interconnections will need to be automated in some way. One of the fundamental conceptual challenges of the EUROBRAIN project will be to find a way to do this. The EUROBRAIN project consists of 3 main parts whose descriptions given below are taken from the project's proposal document (with a little editing).

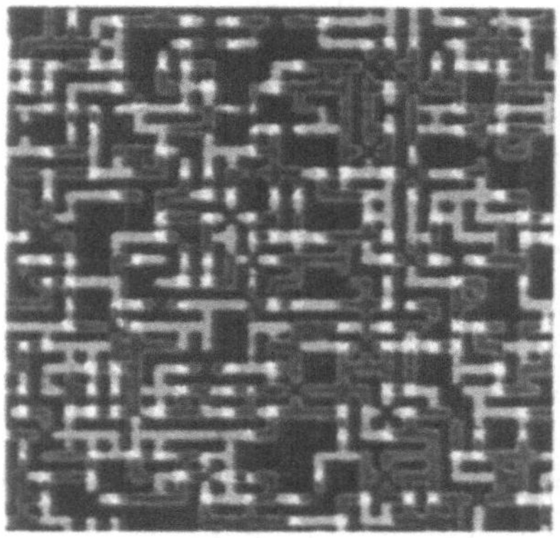

Fig. 3. A cellular automata based neural net circuit evolved by the CBM

A) The first aim is to use the existing CAM-Brain Machine (CBM), an FPGA based hardware device that evolves a 3D cellular automata based neural network circuit module of some 1000 neurons in about 1 second each, to grow/evolve/design an artificial brain of nearly 100 million neurons. This artificial brain will consist of 64000 separately evolved and humanly interconnected modules, with the CBM executing the neural signaling of the whole brain at 130 billion CA cell updates a second, which is fast enough for real time control of the hundreds of behaviors of a robot kitten "Robokitty".

B) The second aim is to conceive, design and build a prototype of a second-generation brain building machine (BBM2) that will handle a billion neurons. The hardware implementation may use the self-configuring circuit ideas of Nick Macias [5] (an adviser to the team). A Xilinx Virtex like device with 64 megagates may be the basis chip or we may try our own ASIC.

C) The third aim is to use the existing CBM (to a limited extent) and mainly the BBM2 to implement Prof. T's (a world top neural net researcher member of the Eurobrain team) models of brain function (e.g. layering, hyper-columnarisation, neurochemical modulation, spiking neurons, retina/LGN, thalamus and V1, early associate visual cortex, temporal lobe/hypothalamus, etc)

If even a subset of part C) can be implemented on the BBM2, this second generation machine will be much more biologically brain like than the CBM, which was only indirectly inspired by neuroscience.

4 A Much Larger Possible Third Project

I had felt until about the spring of 2000, that it would be another 2-3 years before industry would get interested in building artificial brains. I was wrong. Dr. Michael Korkin and I received an email from a European professional project manager who wants to use our brain building technology (as far as possible, and combined with other technologies) to control thousands of robots in a major industrial application, with a budget for Korkin and me of several hundred million dollars. In fact, the artificial brain and robotics component is only a small fraction of the total project cost, which will be several billion dollars (billion with a "b"). I cannot give more specific details at the moment because the contract is not yet signed. At the present time (late Sept 2000) the financiers of this project have asked the organizer to proceed to the next step, which is to approach a particular government to get its permission to proceed. If the government agrees (hopefully by Xmas 2000), it is highly likely that the project will go ahead. Korkin and I will then have the money to hire some 100 researchers and developers to execute the main task of the project. Interestingly and in parallel with the execution of this task, will be the development of more intelligent artificial brains to control a later generation of robots to be used in similar tasks in a second (probably 5 year) project. If this project comes through, then its announcement at this conference could be one of its highlights and have a strong impact on the future of brain building in general.

5 Conclusions

If the larger project contract is signed (probably about the spring of 2001, if it is to be signed at all), then my hardware colleague Dr. Michael Korkin (who designed and built all the CBMs) and I will form a spinoff R and D company from Starlab. The CBMs will have remote accessibility, so that brain building teams around the world (consisting of EEs ("Evolutionary Engineers"), who specify and evolve individual circuit modules with the CBM, and BAs ("Brain Architects") who architect the artificial brain) will be able to benefit from the incredible speed of the CBMs. Such teams already exist in Belgium, Japan, Poland, China and the US. Since there are 64000 modules to be individually specified and evolved, there will be a ton of work. Many researchers will be needed to undertake this task. However, as stated in an earlier section, individually designing and evolving a million modules for a billion neuron Eurobrain will be quite impractical. So the number of EEs and BAs needed for the Eurobrain project is more uncertain at this time. Much of the individual module and interconnection evolution will have to be automated.

References

[1] Michael Korkin, Hugo de Garis, Felix Gers, Hitoshi Hemmi, CAM-Brain Machine (CBM) : A Hardware Tool which Evolves a Neural Net Module in a Fraction of a Second and Runs a Million Neuron Artificial Brain in Real Time, Genetic Programming Conf., July 1997, Koza John R., Deb Kalyanmoy, Dorigo Marco, Fogel David B., Garzon Max, Iba Hitoshi, Riolo Rick L. (eds.), Stanford University, San Francisco, CA, USA.

[2] Hugo de Garis, Felix Gers, Michael Korkin, Arvin Agah, Norberto Eiji Nawa, CAM-Brain, ATR's billion neuron artificial brain project : A three year progress report, Artificial Life and Robotics Journal, Vol.2, 1998, pp56-61

[3] Hugo de Garis, Michael Korkin, Felix Gers, Eiji Nawa, Michael Hough, Building an Artificial Brain Using an FPGA Based CAM-Brain Machine, Applied Mathematics and Computation Journal, Special Issue on Artificial Life and Robotics, Artificial Brain, Brain Computing and Brainware, North Holland, to appear 2000.

[4] Hugo de Garis, Michael Korkin, The CAM-Brain Machine (CBM) : Real Time Evolution and Update of a 75 Million Neuron FPGA-Based Artificial Brain, Journal of VLSI Signal Processing Systems (JVSPS), Special Issue on Custom Computing Technology, to appear 2000.

[5] Hugo de Garis, "Review of Proceedings of the First NASA/DoD Workshop on Evolvable Hardware", IEEE Transactions on Evolutionary Computation (IEEE-TEC), Nov 1999, Vol. 3 No. 4

Application of Feature Extraction in Text-to-Speech Processing[1]

Václav Šebesta[*], Jana Tučková[†]

[*]Institute of Computer Science, Academy of Sciences of the Czech Republic, E-mail: vasek @ cs.cas.cz, [†]Faculty of Electrical Engineering, Czech Technical University, E-mail: tuckova @ feld.cvut.cz

Abstract

A speech signal synthesis in real time, with an unlimited vocabulary is very complicated task for all languages. The synthesizers usually work in the frequency domain and the fundamental frequency F_0 and duration must be determined for all phonemes or diphones by conventional equipment based on linguistic rules [4]. Our effort is to minimize the difference between the synthetic speech of the synthesizer, which is usually more monotonous, and the natural speech of people. Because of this, a special functional block is included into the synthesizer for prosody control. A multilayer artificial neural network (ANN) is used for prosody control in our case. In this part of the synthesizer the fundamental frequency is "a little bit" modified in such a way that speech can sound as natural as possible.

The number of input training parameters for ANN training must be generally kept as small as possible because of the optimal generalization ability of the network. An original method for the determination of the most important features (input parameters) for the training of ANN for prosody control is described in this paper. This method is based on the "data mining" from the database of the training patterns by the GUHA method described in [2].

1 Text-to-Speech (TTS) System for Czech Language

In conventional synthesizer the prosodic parameters correspond to the grammatical rules of the national language. The background of our TTS system is in the fundamentals of phonetics, in linguistic research of the spoken Czech language, in the digital signal processing and in the research of neural network applications. The Czech system is based upon the concatenation of elementary speech units - diphones and phonemes of natural speech.

The corpus for a speech synthesis consists of 441 elementary speech units (diphones and subphonemic segments) of natural speech read by a professional male speaker. The sampling frequency is 8 kHz, the segment length is 24 ms with 12 ms overlapping.

The Czech TTS system includes five parts: linguistic preprocessing, phonetic transcription of the text, transformation of the transcribed text into a sequence of speech units, prosody control and synthesis of the speech signal. The system runs in real time and can be used as a reading machine for blind people [5].

2 Modeling of Prosody Parameters

Small corpus of natural speech data for this research have been created by a careful choice of available sentences. The preliminary database consists of only several tens of sentences divided to the training and testing set. The success of the prosody control is surely dependent on the labeling of the natural speech signal in the database. By labeling we understand the determination of all speech units, i.e. the decision where diphones or phonemes begin and end. The speech signal was labeled by hand in our case, but an automatic approach is also under construction.

The processed sentences are split into phonemes and each phoneme creates separate training or test patterns. The phonemes are represented by 23 parameters (at the beginning of a data mining), which represent information about phonemes, words and sentences, speech signal information, and stress and syllable boundaries (see Tab. I).

One of the most important parameters for the fundamental frequency and duration control is the accent alias the stress describing the prominence of a current uttered syllable in relation with other syllables (different degrees of accent are marked by different pitch changes).

A great part of the acoustic information about the consonants is located on the boundary between the vowels and the consonants. Because the syllables are important for coarticulation (the influence of pronunciation on the context) the syllabification of the phonetic transcription is a very important step in speech processing. The rules for the assignment of a syllable boundary does not exist for the Czech language and therefore the syllabification procedure has to be made by hand.

[1] This research was supported by grant GA AS CR No. A2030801 and by grant GA CR No. 102/96/K087.

3 Feature extraction

The multilayer neural network in our application is arranged into one input, one hidden and one output layer. The number of neurons in the input layer is given by the important language parameters, which are needed for characterization of the Czech language. To improve a generalization ability of neural network the standard pruning method [3] was used for the minimization of the number of neurons in the hidden layer. Two neurons in the output layer produce the values of fundamental frequency and phoneme duration.

The fast back-propagation learning algorithm with a moment and an adaptive learning rate and with the feed-forward recall was used. The transfer functions were a sigmoid function in the hidden layer and a linear function in output layer. The optimal number of the training iterations (epochs in Matlab) was determined in the training process by the increment of sum square errors of test patterns.

It is generally known that a simpler ANN usually has a better generalization ability. Therefore the optimization of ANN topology has been performed in several steps. We have used the "data mining" procedure based on the GUHA method described in [2] for markers determination.

CODE	CHARACTERIZATION OF PARAMETERS	MODE
P_1	End of vowel	Binary
P_2	Beginning of the stress unit	Binary
P_3	Word stress position	Binary
P_4	Intonation pattern type	5 categories
P_5	Number of syllables in a stress unit	integer 0 - 6
P_6	Number of stress units in the sentence	integer 0 - 6
P_7	Type of punctual mark in sentence	integer 1- 6
P_8	Type of sentence	5 categories
P_9	Type of phoneme on the first left place from the focus place	6 categories[2]
P_{10}	Type of phoneme on the focus place	6 categories[3]
P_{11}	Type of phoneme on the first right place from the focus place	6 categories[3]
P_{12}	Type of phoneme on the second right place from the focus place	6 categories[3]
P_{13}	Type of phoneme on the first left place from the focus place	6 categories[3]
P_{14}	Type of phoneme on the focus place	6 categories[4]
P_{15}	Type of phoneme on the first right place from the focus place	6 categories[4]
P_{16}	Type of phoneme on the second right place from the focus place	6 categories[4]
P_{17}	Duration of phoneme in ms divided by the number of syllables in the stress unit	Real
P_{18}	Number of segments in phoneme	Real
P_{19}	Type of accent	5 categories
P_{20}	Position of the phoneme in the syllable	3 categories
P_{21}	Word boundaries	3 categories
P_{22}	Phrase boundaries	3 categories
P_{23}	Number of segments in the syllable	Real

Tab. I.: Features for prosody control at the beginning of extraction process

[2] fricative, nasal, plosive, semiplosive, vowel, diphtong

[3] vowel, diphtong, sonar consonant, voiced consonant, unvoiced consonant, pause

The GUHA method (General Unary Hypotheses Automaton) can be used for the determination of relations in experimental data. The processed data form a rectangular matrix where the rows correspond to the different objects (in our case diphones) and the columns correspond to the different input parameters from P_1 to P_{23} (features) or output parameters investigated (in our case frequency F_{PROS} and duration of diphones). The type of parameter values can be either binary or categorical or real numbers. In the case of real values of data a corresponding interval of the parameter must be divided into several subintervals and the value of a corresponding attribute is equal to one in only one relevant subinterval. The value is equal to zero in all other subintervals. The real number parameters are transformed in such a way to the categorical ones. Parameters are split into antecedents (e.g. input parameters) and succedents (e.g. output parameters).

The program generates and evaluates the hypotheses of an association in the form A → S, where A is an elementary conjunction of antecedents, S is an elementary conjunction of succedents and → is a quantifier of implication. The implication quantifier estimates, in a sense, the conditional probability P (S| A). The user must specify the number of elementary conjunctions in antecedent and succedent and the program in sequence generates all possible hypotheses about the relation between antecedents and succedents in the form of a four-fold-table:

VARIABLE	S = 1	S = 0	TOTAL
A = 1	a	b	a + b
A = 0	c	d	c + d
Total	a + c	b + d	n = a + b + c +d

where parameters a, b, c and d are frequencies of the "1"-value occurence.

All hypotheses can be evaluated by several different quantifiers [2], e.g. LIMPLE or FISCHER:

Quantifier LIMPLE (Lower critical implication quantifier) is valid iff $a \geq a_{min}$ and

$$\sum_{i=a}^{a+b} \binom{a+b}{i} p^i \cdot (1-p)^{a+b-i} \leq LBOUND \quad .$$

Quantifier FISCHER (Fischer test) is valid iff $a \geq a_{min}$ and

$$\sum_{i=a}^{\min(a+b,a+c)} \frac{\binom{a+c}{i}\binom{b+d}{a+b-i}}{\binom{n}{a+b}} \leq \alpha_{FISCH}$$

It is possible to determine different levels of hypotheses, e.g.

- "weak hypotheses" when p = 0,8, LBOUND = 0,05 and $\alpha_{FISCH} = 10^{-20}$,
- "stronger hypotheses" when p = 0,9, LBOUND = 0,01 and $\alpha_{FISCH} = 10^{-25}$,
- "strong hypotheses" when p = 0,95, LBOUND = 0,001 and $\alpha_{FISCH} = 10^{-30}$.

At first all 23 input parameters were declared as antecedents, and outputs, i.e. the fundamental frequency F_{PROS} and the duration was taken as a succedent. The maximal number of elements in the antecedents conjunction was 3. The total number of verified hypotheses was 1615. The numbers of verified hypotheses are shown in II. It can be seen that parameters P_2, P_3, P_9 and P_{13} have a minimal influence on the output parameters.

The verification by listening of this elimination of four input parameters have been successful. Therefore we tried to determine some more input parameters which could be deleted.

Besides the parameters which have only a limited influence on the output parameters, also some input parameters having serious mutual correlation can be omitted. Their influences on the output parameter can be very similar. Such parameters were looked for in the following step. In this case we took 19 input parameters as antecedents and the same 19 parameters as succedents. The maximal numbers of elements in antecedent and succedent conjunctions were 2. The total number of verified hypotheses was 3182. The resulting numbers of hypotheses are shown in Tab. III. The greatest values show the greatest mutual correlation between parameters P_{10} and P_{14}, P_{11} and P_{15} and between P_{12} and P_{16}. We decided to omit P_{10}, P_{11} and P_{12} because of a higher total number of hypotheses. This omission caused a further improvement of practical results, proved by listening to synthetic speech. After each reduction of input parameters a new ANN were trained and our mathematical approach were verified by the listening to several training and testing sentences.

Both described steps were repeated once more and parameters P_1, P_3 were also found redundant. It was also verified by listening. The hidden layer starting from 30 neurons was pruned in a way described in [3] for all variants of inputs. The optimal resulting numbers are 25 - 27 (see Fig.1).

Parameter	P_1	P_2	P_3	P_4	P_5	P_6	P_7	P_8	P_9	P_{10}	P_{11}	P_{12}
about F_{PROS}	29	0	3	168	168	169	39	3	44	76	219	242
about duration	89	9	45	224	260	164	158	50	120	169	237	209
Total number	92	9	45	293	327	231	178	51	120	196	331	315
Parameter	P_{13}	P_{14}	P_{15}	P_{16}	P_{17}	P_{18}	P_{19}	P_{20}	P_{21}	P_{22}	P_{23}	**total**
about F_{PROS}	1	194	172	176	208	344	242	209	216	333	130	500
about duration	52	233	200	192	353	1002	225	252	285	764	684	1443
Total number	52	311	265	254	447	1100	334	333	371	872	688	1615

Tab. II: Numbers of hypotheses that the resulting values F_{PROS} and duration depends on the features P_I.

	P_6	P_7	P_9	P_{10}	P_{11}	P_{12}	P_{14}	P_{15}	P_{16}	P_{17}	P_{18}	P_{19}	P_{20}
P_9	...	8	0										
P_{10}	...	44	42	0									
P_{11}	...	45	21	73	0								
P_{12}	...	51	17	54	78	0							
P_{14}	...	43	33	176	56	45	0						
P_{15}	...	42	15	57	210	65	54	0					
P_{16}	...	33	17	44	68	215	46	62	0				
P_{17}	...	31	6	39	29	20	37	24	18	0			
P_{18}	...	23	10	41	39	24	39	32	23	90	0		
P_{19}	...	12	12	23	16	14	17	16	14	10	11	0	
Total	424	452	288	716	723	647	645	645	586	411	505	240	386

Tab. III: An interesting part of matrix with the numbers of hypotheses about mutual dependencies of input parameters P_I

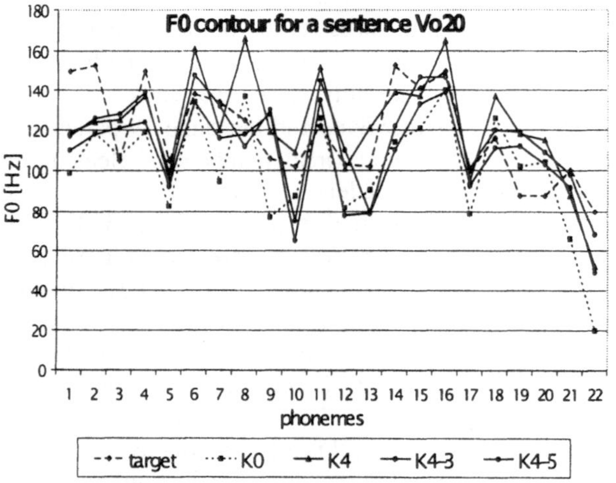

Fig.1: Fundamental frequency contour for a selected testing sentence containing 22 phonemes. Target means values extracted from the natural speech, K0 means values from the ANN with all 23 features, K4 means values from ANN after pruning of input layer (14 features) and K4-3 means values after pruning of hidden layer.

References

[1] Tučková J., Šebesta V.: Prosody Modeling for a Text-to-Speech System by Artificial Neural Networks. Proc. IASTED Int. Conf. "Signal and Image Processing 2000", IASTED/ACTA Press, November 2000, Las Vegas, USA, pp 312 - 317.

[2] Hájek P., Sochorová A., Zvárová J.: GUHA for personal computers., Computational Statistics and Data Analysis, Vol.19, 1995, North Holland, pp.149-153.

[3] Šebesta V.: Pruning of Neural Networks by Statistical Optimization. Proc. of the 6th School of Neural Networks, Theory and Applications. Micro-computer '94. Sedmihorky, Czech Rep., September 1994, pp.209-214., ISBN 80-2140564-3.

[4] Tučková J.,Vích R.: Fundamental Frequency Control in Czech Text-to-Speech Synthesis. Proc. IASTED Int. Conference SIP'97, ISBN 0-88986-247-7, New Orleans, Louisiana, USA, December 1997, pp.85-87.

[5] Vích R.: Pitch Synchronous Linear Predictive Czech and Slovak Text-to-Speech Synthesis. Proc. of the 15th Internat. Congress on Acoustics, ICA'95, Trondheim, Norway, June 1995.

[6] Sejnovski T. J. Rosenberg C. R.: NETtalk: a parallel network that learns to read aloud. The Johns Hopkins University Electrical Engineering and Computer Science, Technical Report JHU/EECS-86/01, 32p.

Gaussian Synapse Networks for Handwritten Character Recognition

J. L. Crespo, R. J. Duro *

*Grupo de Sistemas Autónomos, Universidade da Coruña, Spain, crespo@cdf.udc.es, richard@udc.es

Abstract

In the context of improved higher order neural architectures for pattern recognition we have made use of a new type of higher order network containing gaussian synapses and developed the Gaussian Synapses Backpropagation Algorithm (GSBP) for the implementation of different types of pattern detectors. This paper concentrates on the presentation of the algorithm and its comparison to other structures in a typical benchmark problem i.e. the recognition of handwritten characters. The inclusion of gaussian functions in the synapses of the network allows the training algorithm to select the appropriate spatial information and filter out all that is irrelevant according to the training it has received. With this strategy it is possible to obtain very good recognition results with hardly any preprocessing and independently of backgrounds and slant using small networks that are quite easy to train.

1 Introduction

Most neural network based solutions for addressing pattern recognition problems have been based on multilayer feedforward architectures or recurrent networks [1][2]. In these structures the neurons in each layer are completely connected to those of the next, being a synaptic weight the element that determines the importance of each connection. In general, this weight presents a fixed value whatever the input to the synapse. These types of networks have been proven capable of great plasticity and adaptation to different classes of problems. However, due to the limited processing capacity of their nodes and connections, their scalability to complex problems has usually implied a significant increase in their sizes and thus made their training more time consuming and liable to local minima.

To prevent these drawbacks, one approach of great interest is to seek structures that correspond in a more direct manner to the context in which they are employed [3]. Obviously, if we increase the processing order of nodes and/or connections appropriately, we may obtain a larger processing capacity with an architecturally simpler structure. If this structure is clearly adapted to the problem, the generalization and learning capacities of the system will not be affected, and will even improve. In fact, what we discuss is going from general-purpose networks to networks that allow for an easy procedure to concentrate their attention on whatever is relevant to the problem in hand, and ignore whatever is not, through an appropriate architecture and training process.

Thus, one possible strategy is to employ activation functions that differ from the usual step or sigmoid. Another option is to increase the processing power of synaptic weights. In typical MLP and recurrent architectures, the weight is a numerical value, which once the network has been trained, will have the same effect on any value that circulates through this synapse. On the other hand, the use of non-constant functions, converts the synapses into active elements, whose outputs depend not only on the higher or lower intrinsic importance of the connection, but also on the value that circulates through it. In the case in hand, we consider the problem of establishing spatial relationships in detection and classification tasks, and consequently, we must seek structures that adapt as closely as possible to the problem. We have chosen an architecture that involves gaussian synapses which, in the case of recognizing patterns, allow for the establishment of relative intensity spatial relationships.

The next sections are devoted to describing the architecture of the network and the training algorithm we have developed to train it. After this we provide an example of its use in the task of recognizing handwritten numerals.

2 Structure of the Network and GSBP

The architecture employed in this type of networks is very similar to the classical Multiple Layer Perceptron. In fact, the activation functions of the nodes are simple sigmoids. The only difference, as displayed in figure 1, is that each synaptic connection implements a gaussian function determined by three parameters: the center of the gaussian, its amplitude and its variance:

$$g(x) = A * e^{B(x-C)^2}$$

To train this structure we have developed an extension of the backpropagation algorithm, called Gaussian Synapses Backpropagation (GSBP) [4]. In what follows we will provide a brief overview of it.

First, as in any other backpropagation algorithm, we must determine what the outputs of the different layers are. We must also define the error with respect to the target values we desire and backpropagate it to the parameters determining the synaptic connections, in this case the three parameters that correspond to the gaussian function. In order to do this, we must obtain the gradients of the error with respect to each one of the parameters for each synapse. Consequently, if we define the error as the classical sum of the squares of the differences between what we desire and what we obtain:

$$E_{tot} = \sum_k \frac{1}{2}(T_k - O_k)^2$$

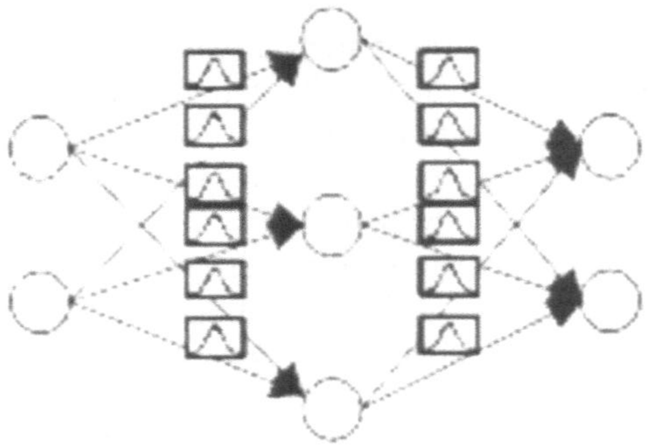

Fig 1: Structure of the Network

And taking into account that the outputs of the neurons in the hidden and output layers are:

Output: $$O_k = F\left(\sum_j h_j A_{jk} e^{B_{jk}(h_j - C_{jk})^2}\right) = F\left(O_{Net_k}\right)$$

Hidden: $$h_j = F\left(\sum_i I_i A_{ij} e^{B_{ij}(I_i - C_{ij})^2}\right) = F\left(h_{Net_j}\right)$$

If we now calculate the gradients of the error with respect to each one of the parameters of the gaussians in each layer we obtain the following equations that will be used for the modification of the gaussian corresponding to each synapses every iteration.

1. Output Layer

The gradient of the error with respect to $\mathbf{A_{jk}}$ is:

$$\frac{\partial E_{tot}}{\partial A_{jk}} = h_j (O_k - T_k) F'(O_{Netk}) \, e^{B_{jk}(h_j - C_{jk})^2}$$

In the case of $\mathbf{B_{jk}}$, and $\mathbf{C_{jk}}$ we obtain:

$$\frac{\partial E_{tot}}{\partial B_{jk}} = h_j (O_k - T_k) F'(O_{Netk}) \, A_{jk} (h_j - C_{jk})^2 \, e^{B_{jk}(h_j - C_{jk})^2}$$

$$\frac{\partial E_{tot}}{\partial C_{jk}} = -2 h_j A_{jk} B_{jk} (O_k - T_k) F'(O_{Netk}) (h_j - C_{jk}) \, e^{B_{jk}(h_j - C_{jk})^2}$$

2. Hidden Layer

This result was calculated for the output layer and now we call:

$$\Theta_j = \frac{\partial E_{tot}}{\partial h_{Netj}} = \frac{\partial E_{tot}}{\partial h_j} \frac{\partial h_j}{\partial h_{Netj}} = \frac{\partial E_{tot}}{\partial h_j} F'(h_{Netj})$$

and the variation of the error with respect to $\mathbf{A_{ij}}$ is:

$$\frac{\partial E_{tot}}{\partial A_{ij}} = I_i \Theta_j \, e^{B_{ij}(I_i - C_{ij})^2}$$

For $\mathbf{B_{ij}}$ and $\mathbf{C_{ij}}$ the procedure is similar and we obtain:

$$\frac{\partial E_{tot}}{\partial C_{ij}} = -2 \Theta_j I_i A_{ij} B_{ij} (I_i - C_{ij}) \quad e^{B_{ij}(I_i - C_{ij})^2}$$

$$\frac{\partial E_{tot}}{\partial B_{ij}} = \Theta_j I_i A_{ij} (I_i - C_{ij})^2 \, e^{B_{ij}(I_i - C_{ij})^2}$$

3 Handwriting Recognition

Much effort has been devoted to the recognition of handwritten characters in an unconstrained setting [1][5]. In fact, this problem has become one of the benchmarks for testing pattern recognition and classification algorithms. Over the years, the trend for more complex structures, involving multiple experts and comprising all kinds of preprocessing and post processing stages such as multi-resolution analysis [6] have been implemented, obtaining recognition rates of more than 95% in many cases. As indicated in [1], most current systems include *thresholding* stages to separate text from background, *noise removal* stages to clean up the image before processing, line and word *segmentation* and then the *character recognition* phase. This last phase usually involves *slant correction*, *size correction* and, in many cases, *feature extraction* from the segmented image with a correctly positioned detector.

In this paper, we will make use of this problem as an example application for our network and algorithm and will show that with a very simple structure and with a single preprocessing step, which is just a size normalization of the characters, the results obtained are comparable to those of much more complex architectures. The method presents the advantage that these detectors operate correctly by sweeping them over the image. Consequently, all of the the segmentation stages [7] and positioning mechanisms (such as in [8] for the Kodak Imagelink software) added to other architectures are unnecessary. Additionally, as shown elsewhere [4], thresholding operations and noise removal stages are also avoided, except for very extreme cases, due to the background independent recognition offered by the filtering operations the gaussian synapses are able to intrinsically carry out.

4 Experimental Results

In the handwritten character recognition experiments presented here, we have made use of the ETL character database (from the Electrotechnical Laboratory in Tsukuba. Japan) containing totally unconstrained handwritten characters. The character subset employed was ETL1. This subset consisted of 18547 characters including numbers, western characters and Japanese characters. The original size of each image was 64x63 pixels, but in the experiments we carried out we reduced this resolution to 16x16 pixels so that the comparisons with other authors could be made more accurately.

Because of the fact that most papers we have found in the literature regarding this subject contained more precise numerical results for the case of numerals than for letters, and as comparison is the objective of these experiments, we will use this part of the database in this paper. In figure 2 we display some randomly selected characters from the database. Notice the large shape variability.

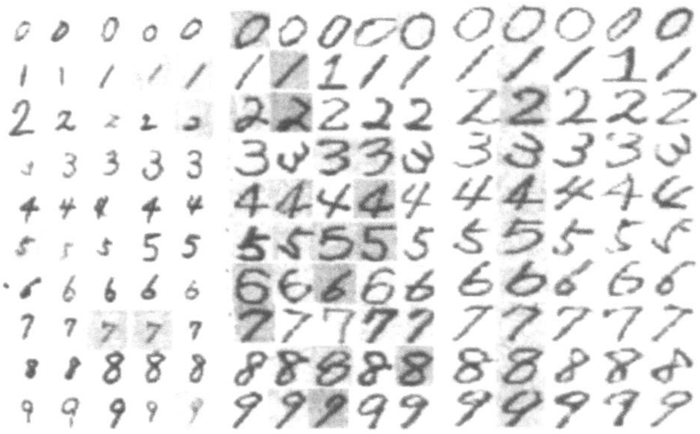

Fig 2: Three sets of samples from the database

For training, we have randomly selected 800 instances of numerals from the database, normalized them in size, generated a strip by concatenating them and swept the detector from left to right over the strip once per epoch. The networks were then tested with randomly selected test sets of around 1100 numerals which were also concatenated and swept.

Figure 3 displays the learning curve of a GSBP trained network as compared to a Feedforward neural

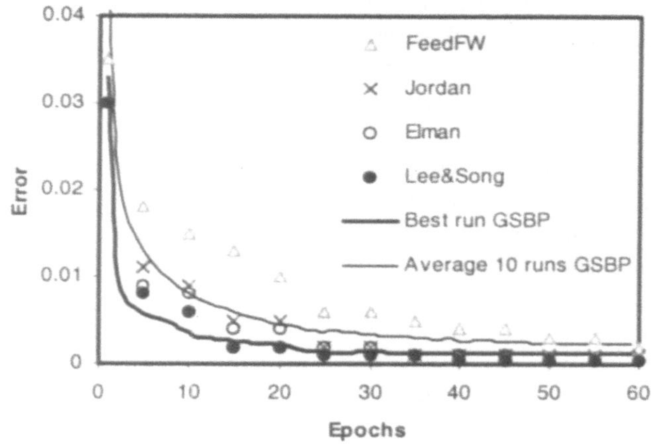

Fig 3: Comparison of learning curves

network and three recurrent networks [2]. In the case of our network, we display the best learning curve and an average one for 10 training runs of different numerals. The training sets for the learning curve data provided by [2] contained 4000 numerals, as compared to 800 in our case and, thus, each of their epochs presented 5 times more numbers to the networks than ours. In fact these authors indicate that if only 800 numerals are used in their training, the error rate goes up to 22%. Despite all of this, as shown in the figure, the results provided by the GSBP on average are comparable to the best results obtained by other authors and were they to be corrected for the factors we indicate above, the learning curve would show that these networks learn much faster. All the networks we have used have between 14 and 17 neurons in the hidden layer.

Numeral	Correctly recognized	False positives
0	100.0%	2.1%
1	99.0%	4.5%
2	91.0%	3.2%
3	95.0%	3.5%
4	96.2%	2.5%
5	92.9%	1.3%
6	98.0%	3.6%
7	96.0%	2.4%
8	94.1%	1.3%
9	94.0%	6.4%
Average	95.6%	3.1%

The table displays the results of testing the networks with 1100 randomly selected numerals from the database. We present the percentage of correctly recognized characters as well as the percentage of false positives for each numeral. These results compare quite well with the reference values we are using from [2] of between 95.5% and 98.0% recognized and 2.0% and 3.5% substituted, and [6] of between 94.3% and 97.9% accuracy on test data. Obviously, they are quite superior to those provided by MLPs using backpropagation, especially using the same number of neurons.

5 Conclusions

We have presented a pattern recognition system based on Artificial Neural Networks with High-Order Gaussian Synapses and a new algorithm for training them, Gaussian Synapses Backpropagation (GSBP). The inclusion of gaussian functions in the synapses of the network allows the network to select the appropriate spatial information and filter out all that is irrelevant according to the training it has received. The networks that result, as they are clearly adapted to the problem of spatial pattern recognition, are much smaller than if other network paradigms were used. The pattern recognizers only require a very small training set due to their great generalization capabilities. We have tested these networks on a benchmark pattern recogniton problem: handwritten numeral recognition. The results are comparable to the best results we have found in the literature and much better than those of MLPs and other similar architectures. In fact, if we take into account the number of exemplars used for training and the fact that our networks worked on numerals without any type of segmentation, positioning or noise reduction preprocessing or any background elimination or slant correction, the results indicate that these networks and the new training algorithm we propose (GSBP) are quite suited for real problems where spatial patterns must be recognized.

References

[1] Plamondon, R. and Srihari, S. N.: On-Line and Off-Line Handwriting Recognition: A Comprehensive Survey IEEE Trans on PAMI, V22, N1, (2000), 63-84.

[2] Seong-Whan Lee and Hee-Heon Song: A New Recurrent Neural Network Architecture for Visual Pattern Recognition, IEEE Trans on Neural Networks, V8, N2, (1997), 331-339.

[3] Gori, M., Scarselli, F.: Are Multilayer Perceptrons Adequate for Pattern Recognition and Verification?, *IEEE Trans on PAMI,* Vol. 20. No. 11 (1998) 1121-1132.

[4] Duro, R.J., Crespo, J.L., and Santos, J.: Training Higher Order Gaussian Synapses. Lecture Notes in Computer Science, Vol. 1606 Springer-Verlag, Berlín (1999) 537-545.

[5] Suen C.Y., Nadal, C., Legault, R, May, T. A. and Lam, L.: Computer Recognition of Handwritten Numerals, Proc IEEE, V80, (1992), 1162-1180.

[6] Morns, I.P. and Dlay, S. S.: The DSFPN, a New Neural Network for Optical Character Recognition, IEEE Trans on Neural Networks, V10, N6, (1999), 1465-1473.

[7] Kim, G., Govindaraju, V. and Srihari, S. N.: An Architecture for Handwritten Text Recognition Systems, Int'l J. Document Analysis and Recognition, 2 (1999),37-44 .

[8] Shustorovich, A. and Thrasher, C. W.:Neural Network Positioning and Classification of Handwritten Characters, Neural Networks, V9, N4, (1996), 685-693.

A Generic Pretreatment for Spiking Neuron Application on Lipreading with STANN (Spatio-Temporal Artificial Neural Networks)

Renaud Séguier, David Mercier *

*Suplec - quipe Traitement du Signal et Neuromimetisme, Avenue de la Boulaie - BP28 35511 Cesson Svign, France, E-mail: Renaud.Seguier@supelec.fr, David.Mercier@supelec.fr

Abstract

Spiking neurons treat sequences of impulses. However the signals to which we have access in the majority of the applications evolve generally continuously with time and are not of impulse nature. If one wants to use spiking neurons, a pretreatment should then be found adapted to the application to convert the raw signals into sequences of impulses. We propose here a simple generic pretreatment which carries out this conversion. We illustrate then this proposal within the framework of the lipreading by STANN (Spatio-Temporal Artificial Neural Networks) and show that this pretreatment is simpler and more effective than that which had been used in [1] for this same application.

1 STANN and Lipreading

1.1 STANN

The STAN (Spatio-Temporal Artificial Neuron) is an artificial neuron which codes discrete events (spikes, impulses) using complex numbers [12]. The impulse of amplitude η_1 emitted at time t_1 is coded at current time t by the complex number:

$$\eta_1 e^{-\mu_S \tau_1} e^{i \arctan \mu_T \tau_1}$$

$$i = \sqrt{-1},\ \tau_1 = t - t_1 \text{ et } \mu_S = \mu_T = 1/TW$$

TW depends on the application and represents the size of the temporal window inside which one wishes to identify sequences of impulses (see Figure 1). The component j of the vector of entry X is thus represented by:

$$x_j(t) = \sum_{p=1}^{P} \eta_p e^{-\mu_S \tau_p} e^{i \arctan \mu_T \tau_p}$$

Each neuron has a vector weight W with N complex components. To compare the entry X with the vector W, one can use the scalar product:

$$V = \sum_{j=1}^{N} \overline{w_j}.x_j \text{ with}$$

$\overline{w_j}$ the complex conjugate of w_j

or the hermitian distance:

$$D(X,W) = \sqrt{\sum_{j=1}^{N} (x_j - w_j)\overline{(x_j - w_j)}}$$

The function of activation F applied to V or D determines the exit y.

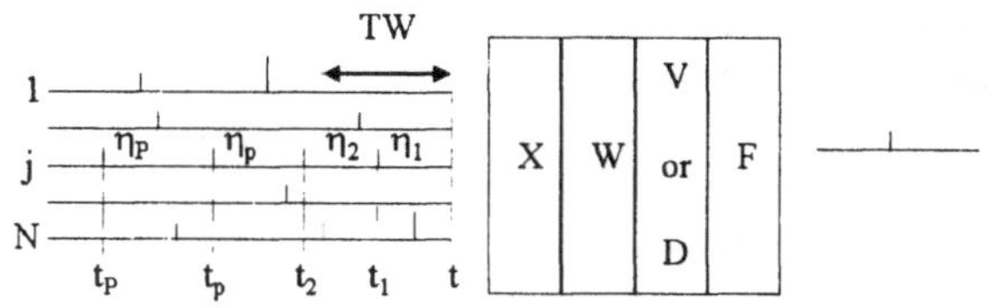

Fig. 1. The STAN (Spatio-Temporal Artificial Neuron)

The STANN (Spatio-Temporal Artificial Neural Networks) are the neuronal architecture built around the STAN. They are able to analyze signals whose space and temporal forms are significant (gestures in a sequence of images for example). A general procedure of use of the STANN was recently proposed [1].

The complete system making it possible to carry out a classification of space-time signals is indicated Figure 2. The purpose of the module of pretreatment is to generate sequences of impulses on the basis of input signal. The module of Vector Quantization emits impulses when it identifies sub-sequences (of a size TW1 temporal units) which it will have learned how to recognize (using ST-Kmeans or ST-Kohonen). The last module takes care with classification and generates impulses which are characteristic of the class of the sequences of impulses (of a size TW0 temporal units) that one will have learned to identify (with a ST-RCE or a ST-MLP).

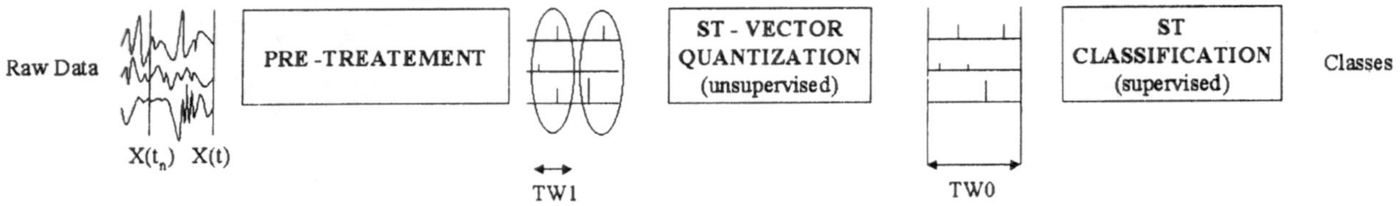

Fig. 2. General classification system with STANN

1.2 Lipreading

Within the framework of the voice recognition, the audio method is not always enough when the ambient conditions are disturbed. For example in a car, the voice recognition can prove to be difficult because of the presence of the noise resulting from the road traffic or the vocal messages from the radio. It is then useful to exploit the video method and thus of reading on the lips to work out a robust system [3]. Three tools are often used: the HMM (Hidden Markov Models) [11], the TDNN (Time Delay Neural Network) [10] and the STANN [2].

In [2], the movement of four points on the lips (corner of the lips, middle of the lips top and bottom) had been judged relevant to recognize the words pronounced by a person. From this movement, impulses had been generated.

This particular pretreatment which consisted in passing from the raw signal to the impulses thus did not take into account the position of the tongue and the teeth, even if these features could contain relevant information for classification.

It is a more general problem having reference to the type of pretreatment used in spiking neurons domain.

2 Pretreatment and Spiking Neurons

2.1 Problematical

In the real world, the signals which we must treat are often of space-time nature (sequences of images, multisensor signals etc). In the major part of the cases (Figure 3 a, b and d), they evolve continuously in time (level of brightness of a pixel in a sequence of image, sound level for voice recognition etc)..

Spiking neurons (especially PCNN-Pulsates Coupled Neural Network, I&F-Integrated and Fire and STANN) accept in entry only sequences of impulses (Figure 3d).

An operation of pretreatment is thus necessary when one uses spiking neurons to convert the continuous signals into impulse signals.

This operation is seldom generic. The PCNN works directly on temporal series [4]. When one wants to use them on natural signals like images to make segmentation for example [6], one then converts the level of gray of a pixel into temporal information to be able to generate impulses spaced in time. In handwriting recognition, [7] bases itself mainly on the movement of the pen in four directions (high, low, right-hand and left-hand side) to generate impulses. It is the moment when the movement changes direction which defines temporal localization of the impulse; its amplitude being equal to the movement quantity cumulated in the considered direction . The I&F in [8] use the increases in contrast of a pixel in an image in the course of time to generate a spike.

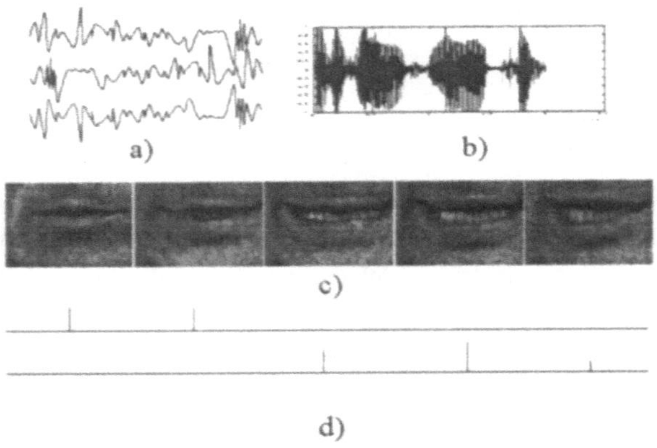

Fig. 3. raw signals: a) multisensor signal, b) speech signal, c) image sequence. Impulse signals: d)

Unfortunately, this stage of pretreatment being completely dependent on the application, one is often forced to introduce much information a priori. The disadvantage of taking into account so early information a priori is that we can lose relevant information not well identified.

2.2 A generic pretreatment

To generate impulses in a simple way starting from signals which evolve continuously in time, we propose to carry out a static Vector Quantization (VQ) on the basis of signal taken at every moment. This VQ makes it possible to associate with the static form (define at one moment given by the whole of the sensors) a prototype of form. To illustrate our matter let us consider for example the sequence of images of Figure 3. Our objective is to generate impulses. We proceed in four stages:

Training:

- Definition of the M static prototypes. K-means [5] is carried out on a meaningful basis of multisensor signals taken every moment (for example: an image in a sequence is an element of the base). *From the sequence of Figure 3, two prototypes will thus be extract: P_1 which will characterize the first two images, P_2 which will be associated with the three last.*
- Measure of a ray r_j for each prototype: evaluation of the set E_j containing the examples of the training base which are associated with the prototype P_j, r_j is the distance between P_j, and the example most distant from E_j. *In Figure 3c), it is the last of the three images associated with P_2 which is most distant from P_2. The ray of P_2 is thus the distance between P_2 and the fifth image of the training base.*

Relaxation:

- Identification at every time of the prototype P_j nearest to the static signal $X(t)$ in entry. *If the sequence of Figure 3 is analysed: P_1 will be identified at t_1 and t_2, P_2 at t_3, t_4 and t_5.*
- Emission of an impulse. The M outputs of the pretreatment module are null except that which corresponds to the prototype j on which an impulse is generated whose height $y(t)$ depends on the distance between the input signal $X(t)$ and P_j:

$$y(t) = e^{-\frac{D^2(X(t),P_j)}{2r_j^2}}$$

In Figure 3c), the last impulse at t_5 will be thus lower than that of t_3 or t_4.

3 Results

To illustrate it, we applied this new pretreatment to lipreading by STANN. The raw input signal consists of a sequence of images of mouth moving, of which it will be necessary to extract from the prototypes of the characteristic shape of mouth. Since this new pretreatment considers the whole of the pixels of the image, it can take account of characteristics a priori non-obvious like the presence of the tongue or the asymmetrical shapes of the mouth.

There is very little of public database making it possible to carry out benchmark in lipreading. Moreover among those which are available none is single-speaker. For example Tulips1 [1] propose 12 subjects saying the first 4 digits in English.

Our objective is to recognize the ten digits in French from 0 to 9, each one of these numbers being expressed the ones following the others during a sequence. The bases of training and test were carried out according to the same protocol as in [1]: the sequences were acquired during three sessions, at fifteen days of interval, under different brightness conditions, the size and the localization of the mouth in the image not being rigorously identical (see figure 4).

Insofar as we treat video images of big size and that the operation of Vector Quantization relates to the whole of the pixels, we subsample initially factor eight in line and column each image in order to accelerate the treatments. We do not need indeed the space resolution which had been necessary (192 X 288 pixels) to identify each of the four points on the lips; very small images of the mouth are now sufficient (24 X 36 pixels). In order to make the system robust to the change of brightness, we carry out an operation of histogram equalization. The pretreatment suggested in 2 is then applied: 20 prototypes characterize the set of the possible shapes of mouths. The figure 5 summarizes the whole of the procedures installation for this application.

Let us recall that Kmeans is sensitive to the initialization which is random, so that for the same number of prototype, the results can be different. Those are thus evaluated on average.

In [1], when a ST-RCE is used as ST classifier, without exploiting the module of ST-VQ (see Figure 2), the results are 58% of good classification. When the module of ST-VQ is used, the results are on average 72% and 77% in the best of the cases.

With the new pretreatment, that we use or not a module of ST-Vector Quantization (ST-Kmeans, see Figure 5), the percentage of correct classification is 85%. With the module of ST-VQ, it reaches even the 90% in best case.

This generic pretreatment thus improves to a significant degree (13 points) the performances of the complete system.

4 Conclusion

Until now, the use of the spiking neurons required a specific pretreatment on the raw data to pass from signals evolving continuously in time to sequences of impulses. The pretreatment that we proposed in this article makes it possible to carry out this

[1] ftp://markov.ucsd.edu/pub/tulips1/

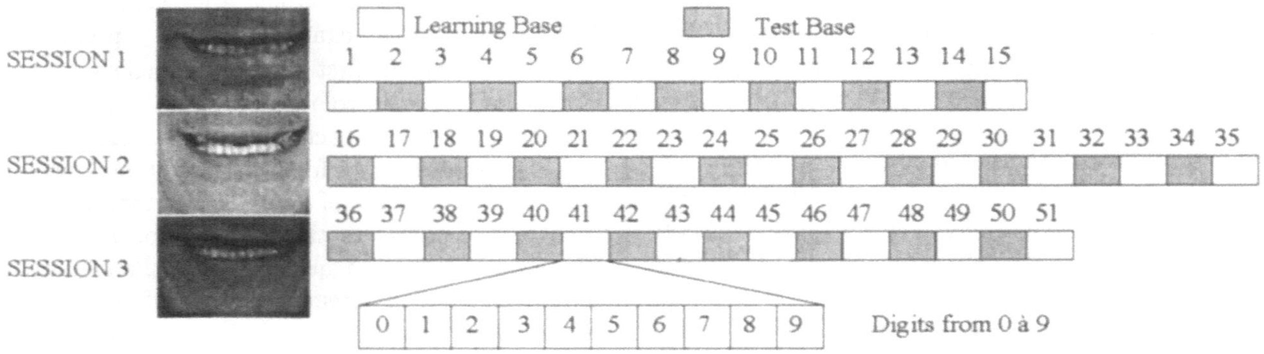

Fig. 4. The base built under three sessions different

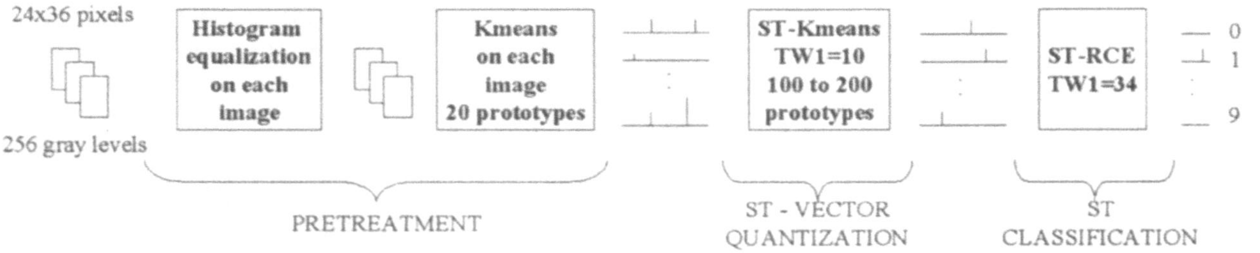

Fig. 5. The complete system of lipreading

passage in an automatic way. It is moreover more extremely simple to implement and does not require any information a priori. In certain cases (lipreading with STANN) it makes it possible to improve the results of treatment of the complete system of classification.

We currently work within the framework of the lipreading, on a "one shoot" training. The idea is to extract sufficient information on the first sequence of the first session (see figure 4) to be able to generalize classification on the whole of the three sessions.

References

[1] A.R. Baig "Une approche mthodologique de l'utilisation des STAN applique la reconnaissance visuelle de la parole" *Universit Rennes I*, PhD Report, 2000.

[2] A.R. Baig and R. Séguier and G. Vaucher "A Spatio-temporal Neural Network applied to visual speech recognition" *ICANN*, 1999.

[3] C.C. Chibelushi and F. Deravi and J.S.D. Mason "Audio-Visual Integration in Multimodal Communication" *Proceedings IEEE*, May 1998.

[4] J.L. Johnson and M.L. Padgett "PCNNModels and Applications" *IEEE Transactions on NeuralNetworks*, vol. 10, no 3, May 1999.

[5] L. Lebart and A. Morineau and M. Piron "Statistique exploratoire multidimensionelle" *Dunod*, 1997

[6] T. Lindblad and J.M. Kinser "Image Processing using Pulse-Coupled Neural Networks" *Springer-Verlag*, 1998

[7] N. Mozayyani and A.R. Baig and G. Vaucher "A Fully Neural Solution for on-Line Handwritten Character Recognition" *IJCNN*, 1998.

[8] W. Paquier and A. Delorme and R. Vanrullen and S. Thorpe "Detection de mouvements apparents par codage asynchrone" *NSI - Neurosciences et Sciences de l'ingnieur*, 2000.

[9] H.S. Ranganath and G. Kuntimad "Object Detection Using Pulse Coupled Neural Networks" *IEEE Transactions on NeuralNetworks*, vol. 10, no 3, May 1999.

[10] R. Stiefelhagen and U. Meier and J. Yang "Real-Time Lip-Tracking for Lipreading" *Proc. of Eurospeech*, 1997.

[11] T. Wark and S. Sridharan and V. Chandran "The use of temporal speech and lip information for multi-modal speaker identification via multi-stream HMM's" *ICASSP*, PhD Report, 2000.

[12] G. Vaucher "A la recherche d'une algèbre neuronale spatio-temporelle" *Universit Rennes I*, PhD Report, 1996.

A Comparison of a Neural Network and an Observer Approach for Detecting Faults in a Benchmark System

D. N. Shields, S. Du, E. Gaura *

*Maths-MIS, Coventry University, Priory St, Coventry, CV15FB, U.K., E-mail: d.n.shields@coventry.ac.uk

Abstract

The detection of faults is considered for a class of nonlinear systems. A fault detection observer approach is compared to a neural network approach. Both approaches are applied to a an experimental (benchmark) three-tank system.

1 Introduction

Fault detection and isolation for hydraulic systems has been investigated by using several methods including: observers [1], parity equation with neuro-fuzzy identification [2] and neural networks [3]. A comparison of observer and neural network methods has been lacking, and this motivates the work here.

2 Design of Observer and Residuals

A system is assumed modelled in the polynomial form

$$\begin{aligned}\dot{x}(t) &= Ax(t) + E_a d(t) + Kf(t) + Bu(t) \\ &+ \sum_{i=1}^{m} u^i(t) A^i_{ux} x(t) \\ &+ \sum_{i=1}^{k} x^i(t)[A^i x(t) + E^i d(t) + K^i f(t)] \\ &+ \sum_{i_1=1}^{k}\sum_{i_2=1}^{k} x^{i_1}(t) x^{i_2}(t)[A^{i_1 i_2} x(t) \\ &+ E^{i_1 i_2} d(t) + K^{i_1 i_2} f(t)] + \cdots \\ &+ \sum_{i_1=1}^{k} \cdots \sum_{i_N-1=1}^{k} x^{i_1}(t)\cdots x^{i_N-1}(t) \cdot \\ &\quad [A^{i_1\cdots i_N-1} x(t) + E^{i_1\cdots i_N-1} d(t) + \\ &\quad K^{i_1\cdots i_N-1} f(t)], \qquad (1)\\ y(t) &= Cx(t) + Qf(t) \qquad (2)\end{aligned}$$

where $x(t) \in I\!R^k, u(t) \in I\!R^m, y(t) \in I\!R^p, d(t) \in I\!R^q$ and $f(t) \in I\!R^v$ are the state, input, output, disturbance and fault vectors respectively.

It is assumed all matrices in (1)-(2) are constant. For the system described by (1) and (2) a nonlinear observer is propcsed. The observer, linear in $z(t)$ and involving bilinear and polynomial nonlinearities up to degree N in $y(t)$ and $u(t)$is given by:

$$\begin{aligned}\dot{z}(t) &= Fz(t) + Ju(t) + Hy(t) \\ &+ \sum_{i=1}^{m} u^i(t)[H^i_{ux} y(t) + F^i_{ux} z(t)] \\ &+ \sum_{i=1}^{p} y^i(t)[H^i y(t) + F^i z(t)] \\ &+ \sum_{i_1=1}^{p}\sum_{i_2=1}^{p} y^{i_1}(t) y^{i_2}(t)[H^{i_1 i_2} y(t) + F^{i_1 i_2} z(t)] \\ &+ \cdots \\ &+ \sum_{i_1=1}^{p} \cdots \sum_{i_N-1=1}^{p} y^{i_1}(t)\cdots y^{i_N-1}(t) \cdot \\ &\quad [H^{i_1\cdots i_N-1} y(t) + F^{i_1\cdots i_N-1} z(t)], \qquad (3)\end{aligned}$$

where $z(t) \in I\!R^d$, is a linear estimate of $Tx(t)$. A fault residual (detection signal) is defined as

$$\epsilon(t) = L_1 z(t) + L_2 y(t), \qquad (4)$$

where $\epsilon(t) \in I\!R^{d_o} (1 \leq d_o \leq d)$, $k \geq 0$.

Design Solution: If T ($T \neq 0$), J, H, F, L_1 ($L_1 \neq 0$), L_2, H^i_{ux}, F^i_{ux}, $H^{i_1\cdots i_{N-1}}$, F^i, and $F^{i_1\cdots i_{N-1}}$ can be found such that the following conditions are satisfied (for some $d \geq 1$, $d_0 \geq 1$)

$$\begin{aligned} 0 &> \Re e(\lambda_i(F))\,; i = 1, \cdots, d & (5)\\ H &= [TA - FT]\,\Phi & (6)\\ J &= TB & (7)\\ L_2 &= -L_1 T\Phi & (8)\\ 0_{d,(k-p)} &= [TA - FT]\,\Psi & (9)\\ 0_{d,q} &= TE_a & (10)\\ 0_{d,(k-p)} &= L_1 T\Psi & (11)\\ H^i_{ux} &= [TA^i_{ux} - F^i_{ux} T]\Phi; & \\ & \quad i = 1, \cdots, m & (12)\\ 0_{d,(k-p)} &= [TA^i_{ux} - F^i_{ux} T]\Phi; & \end{aligned}$$

$$
\begin{aligned}
& i = 1, \cdots, m && (13)\\
H^{i_1 \cdots i_r} &= [TA^{i_1 \cdots i_r} - F^{i_1 \cdots i_r} T]\Phi;\\
& i_1, \cdots, i_r = 1, \cdots, p;\\
& r = 1, \cdots, N-1 && (14)\\
0_{d,(k-p)} &= [F^{i_1 \cdots i_r} T - (r+1) TA^{i_1 \cdots i_r}]\Psi;\\
& i_1, \cdots, i_r = 1, \cdots, p;\\
& r = 1, \cdots, N-1 && (15)\\
0_{d,k} &= TA^{i_1 + p i_2 \cdots i_r}; i_1 = 1, \cdots, k-p;\\
& i_2, \cdots, i_r = 1, \cdots, k;\\
& r = 1, \cdots, N-1 && (16)\\
0_{d,v} &= TK^{i_1 + p i_2 \cdots i_r}; i_1 = 1, \cdots, k-p;\\
& i_2, \cdots, i_r = 1, \cdots, k;\\
& r = 1, \cdots, N-1 && (17)\\
0_{d,q} &= TE^{i_1 \cdots i_r}; i_1, \cdots, i_r = 1, \cdots, k;\\
& r = 1, \cdots, N-1 && (18)
\end{aligned}
$$

where

$$\Phi = \begin{bmatrix} I_p \\ 0_{(k-p)\times p} \end{bmatrix}, \; \Psi = \begin{bmatrix} 0_{p\times(k-p)} \\ I_{k-p} \end{bmatrix},$$

then $e(t)$ and $\epsilon(t)$ are implicitly decoupled from $d(t)$ and have the forms, respectively,

$$\dot{e}(t) = W^e(t)e(t) + W^*(t)f(t), \quad (19)$$

and

$$\epsilon(t) = L_1 \left[e(t) - T\Phi Q f(t)\right]. \quad (20)$$

Using the algorithms developed in [5], F, T, L_1 and L_2 (and the other gains) can be calculated using SVD decompositions. Detectability conditons are used and developed from work in [4].

3 The Application System (Benchmark)

A three-tank system has been set up in a laboratory and consisting of three cylinders tank1, tank3 and tank2 with cross section A. These three tanks are connected serially with each other by cylindrical pipes with the cross section S_n. Tank2 has an outflow valve which also has a circular cross section S_n. The water levels, $x(t)_i$, satisfy the nominal differencial equations:

$$
\begin{aligned}
\dot{x}_1 &= \frac{Q_1}{A} + d_1\\
&- \frac{a_z S_n sgn(x_1 - x_3)\sqrt{2g|x_1 - x_3|}}{A} && (21)\\
\dot{x}_2 &= \frac{Q_2}{A} - \frac{a_z S_n \sqrt{2gx_2}}{A} - \frac{f_1}{A}\\
&+ \frac{a_z S_n sgn(x_3 - x_2)\sqrt{2g|x_3 - x_2|}}{A} && (22)\\
\dot{x}_3 &= \frac{a_z S_n sgn(x_1 - x_3)\sqrt{2g|x_1 - x_3|}}{A}\\
&- \frac{a_z S_n sgn(x_3 - x_2)\sqrt{2g|x_3 - x_2|}}{A}\\
&- \frac{f_2}{A} && (23)\\
y &= Cx && (24)
\end{aligned}
$$

where $f_1 = a_z S_l \sqrt{2gx_2}$ represents leak from tank 2, $f_2 = a_z S_l \sqrt{2gx_3}$ represents leak from tank 3, $d_1 = 10^{-5} sin(0.01t)$ is a disturbance, $C = I_3$, $Q_i (i = 1, 2)$ is the inflow through pump$i$, a_z is the flow correction term and g is the gravity constant.
Parameters: $a_z = 1$, $A = 0.0154m^2$, $g = 9.81m/s^2$, $S_n = 5 * 10^{-5}m^2$, $S_l = 2.7 * 10^{-5}m^2$, $Q_{1max} = 1 * 10^{-4}m^3/sec$, $Q_{2max} = 1 * 10^{-4}m^3/sec$.
For the observer method approximate models of order $N = 1$ (linear, Model 3), $N = 2$ (cubic, Model 1) and $N = 5$ (fifth order, Model 2) were derived of the form (1)-(2).
The following testing input was chosen, $u = [u_1, u_2]'$,
where $u_1 = 2 * 10^{-5}, 0 \leq t < 1000$;

$$u_2 = \begin{cases} 3 * 10^{-5} & 0 \leq t < 100 \\ 0 & 100 \leq t < 800 \\ 3 * 10^{-5} & 800 \leq t < 900 \\ 0 & 900 \leq t < 1000 \end{cases}$$

Here, u_2 is chosen as a pulse function with 0 values in intervals $[100, 800]$ and $[900, 1000]$ for the purposes of producing a different interaction between input and residual performance.
Output: $y_1 = x_1, y_2 = x_2, y_3 = x_3$.

4 Performance Using Observer-Residuals

For the system (21)-(24), three observers were designed. Observer 1 for Model 1, with observer state z_1 and residual ϵ. Observer 2 for Model 2, with observer state z_2 and residual r. Observer 3 for Model 3. Results are poor for this Model and are not displayed. Figure 7 shows the performance of ϵ (shown as ϵ_1) and r (shown as r_1) when a simulated fault f_1, in the form of a leak from tank2, is imposed on the system (see (22)). We observe that both observer residuals contain modelling errors, especially during time histories from 0 to 150 and from 800 to 950. However, the maximum error in residual r_1 is smaller than 0.134×10^{-3}, while the maximum error in residual ϵ_1 is greater than 3.1×10^{-3}. Therefore, the threshold for residual r_1 is chosen as $\theta_{r_1} (= 0.134 \times 10^{-3})$ and $\theta_{\epsilon_1} (= 3.2 \times 10^{-3})$ is for residual ϵ_1. Clearly the residual r_1 based on

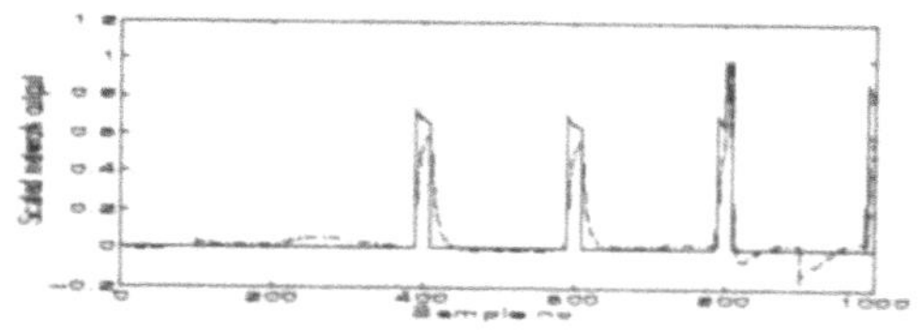

Fig. 1.

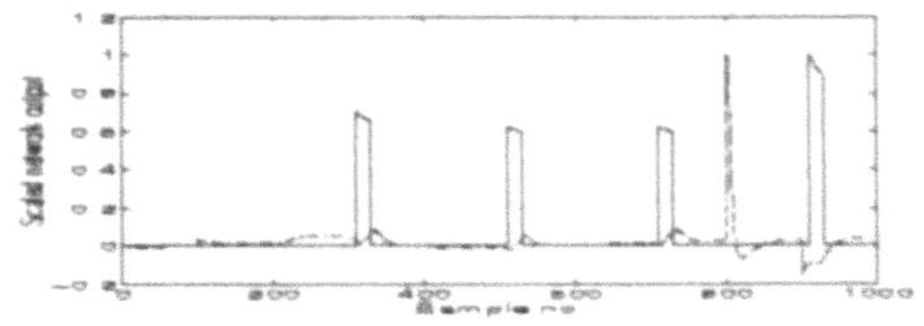

Fig. 2.

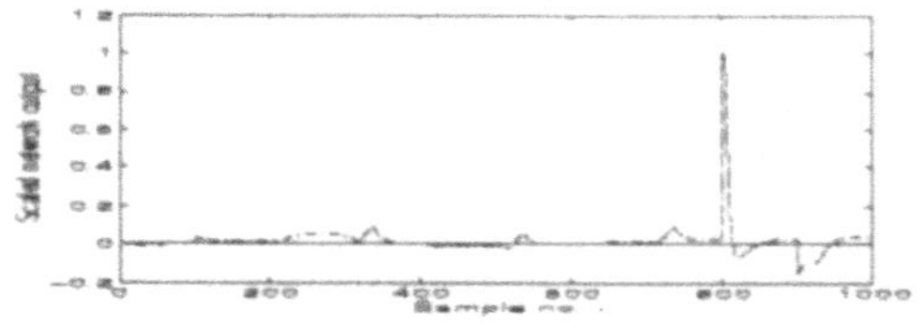

Fig. 3.

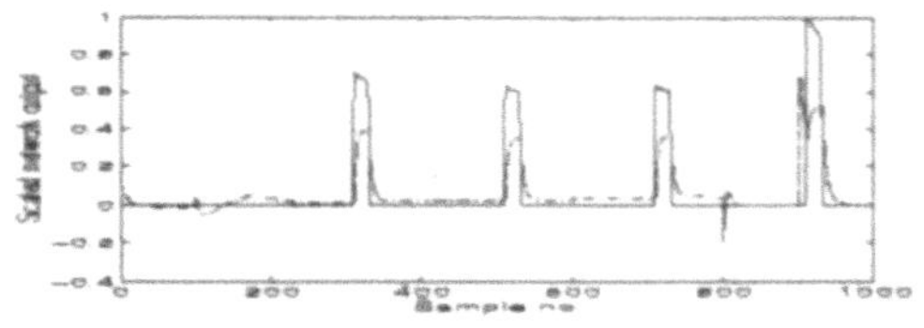

Fig. 4.

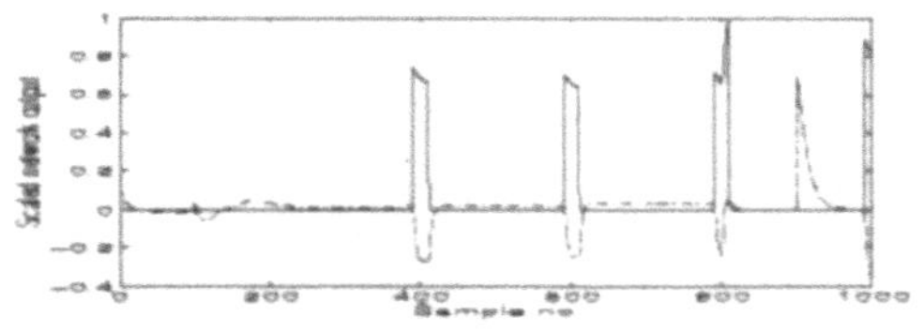

Fig. 5.

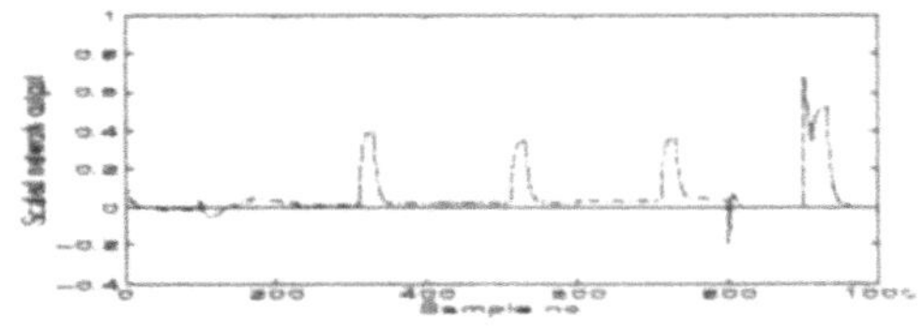

Fig. 6.

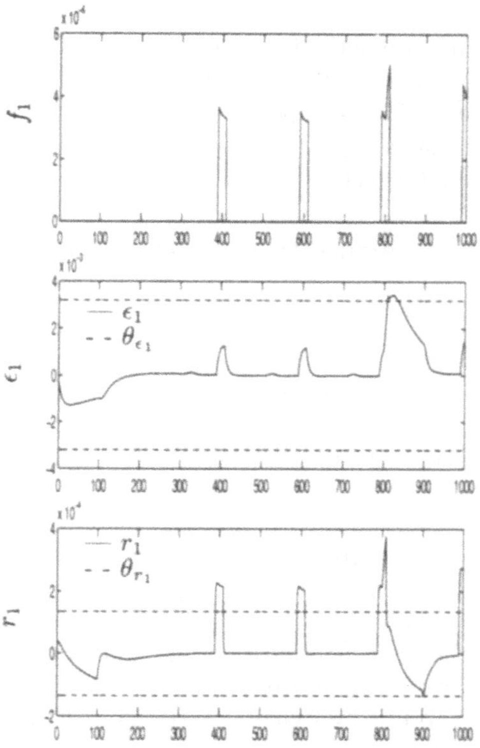

Fig. 7

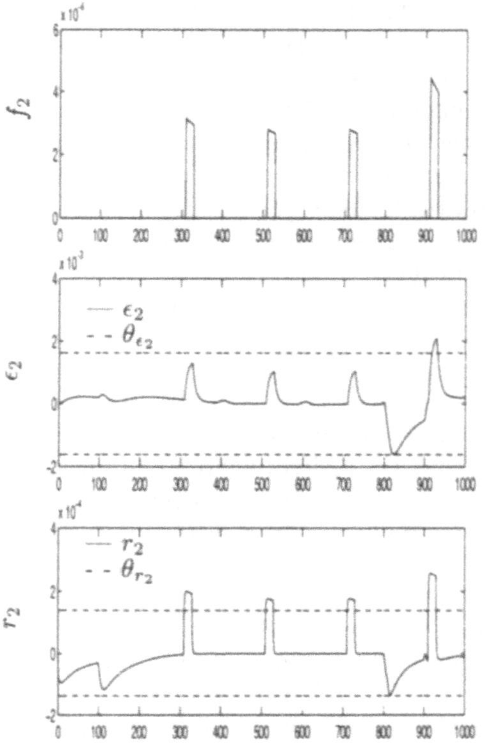

Fig. 8

Model 2 (fifth order) is better has it picks up all the faults (with respect to the theshold). Figure 8 shows the performance of ϵ (shown as ϵ_2) and r (shown as r_2) when a simulated fault f_2, in the form of a leak from tank3, is imposed on the system (see (23)). The residual r_2 corresponding to Model 2 is the only residual detecting f_2.

5 Performance Using Neural Networks

Two neural networks are used here for the detection of f_1 and f_2 (same as for observer case). Each network is trained to identify one fault only. Unlike the observer method the neural network strategy is not based on the system model. Information available to each network consists of: the input signal to the system, the three measurable states of the system for the fault/no fault cases and the two faults. Three data sets were generated in Matlab, as follows: set 1 (u, y, f_1); set 2 (u, y, f_2); set 3 (u, y). Feedforward networks were chosen for solving the fault identification task. The two networks, network 1 and network 2, were trained, using Matlab, and using a dynamic version of BKP, including variable learning rate and a momentum term. The output for network 1 was f_1, trained on set 1. That for network 2 was f_2 trained on set 2. Each of the networks had two hidden layers, with 14 neurons in the first hidden layer and 9 neurons in the second one. The performance of the networks on the two training sets (set 1 and set 2, respectively) after an average of 30000 epochs is shown in figures 1 and 4. Both network 1 and network 2 have outputs which follow the faults to be identified in terms of position, duration and magnitude, although the training error remains fairly large.

The networks performance was then tested on the remaining data sets: sets 2 and 3 for network 1 and sets 1 and 3 for network 2 (figures 2 & 3 and 5 & 6 respectively). It can be noticed in the tests that, while network 1, for example, does not react to fault 2 (as desired, figure 2) and has an average of zero output for no faults (figure 3), it seems to preserve the information learned during training on fault 1 and react accordingly at the position where this fault had a maximum magnitude, even in the absence of the fault. Similar behaviour is exhibited by network 2 as far as fault 2 is concerned. It is inferred therefore, that the networks mapped strongly the relationship between input signal and each fault rather then states and faults. It can be concluded that, by supplying the networks with no information on the order/structure of the system, the fault detection was poor. Better results are hoped to be achieved by modifying the networks structure as to include the past history of the stated up to an order equal to the system's order.

Figure 1: output of network1 (training set,set 1)-dotted line; desired output f_1-full line. Figure 2: output of network1 (on test set, set 2)-dotted line; desired output,zero; f_2-full line. Figure 3: output of network1 (on the test set, set 3)-dotted line; desired output, zero. Figure 4: output of network2 (training set, set 2)-dotted line; desired output f_2-full line. Figure 5: output of network 2 (on test set, set 1)-dotted line; desired output zero; f_1-full line. Figure 6: output of network 2 (on test set, set 3)-dotted line; desired output zero (no faults).

6 Conclusion

Three residuals are designed using nonlinear fault detection observers of different degree nonlinearities. Full system knowledge is assumed and disturbances are present. Two neural networks are designed using only three data sets from a three-tank system. Comparisons of the performance of the two methods shows that care is needed for the network approach when considering highly nonlinear systems with interactions between inputs and faults and with disturbances.

References

[1] Frank, P.M. (1994). On-line Fault Detection in Uncertain Nonlinear Systems Using Diagnostic Observers : A Survey. *Int.J.Systems Sci*, **Vol. 25**, pp. 2129-2154.

[2] Garcia, F. J., V. Izquierdo, L. de Miguel, J. Peran (1997). Fuzzy Identification of Systems and its Applications to Fault Diagnosis Systems. *IFAC Symposium on Fault Detection, Supervision and Safety for Technical Processes "SAFEPROCESS'97", Kingston Upon Hull*, **Vol.2**, 705-712.

[3] Han, Z., P. M. Frank(1997). Physical Parameter Estimation Based FDI with Neural Networks. *IFAC Symposium on Fault Detection, Supervision and Safety for Technical Processes "SAFEPROCESS'97", Kingston Upon Hull*, **Vol.1**, 294-299.

[4] Shields, D.N. and S. Daley (1998). A Quantitative Fault Detection Method for a Class of Nonlinear Systems. *Trans. Inst. MC 20(3)*, 125-133.

[5] Yu, D. and D.N. Shields (1996). Bilinear Fault Detection Observer and its Application to a Hydraulic System. *Int. Jnl. of Control*, **Vol. 64**, pp. 1023-1047.

A Neuro-Fuzzy Image Analysis System for Biomedical Imaging Applications

L. Patino, M. Razaz *

*Royal Society Wolfson Bioinformatics Laboratory, University of East Anglia, Norwich England, e-mail: luis@sys.uea.ac.uk or mr@sys.uea.ac.uk

Abstract

Intelligent image analysis is becoming increasingly important in biological and medical imaging applications. We present here an adaptable and intelligent image analysis system based on a combination of neural networks and fuzzy logic. The system has been applied successfully for diagnostic application such as recognition of the left ventricle in blood pool myocardial Single Photon Emission Computed Tomography (SPECT) images. We briefly discuss this system and present typical results from the application to SPECT images.

1 Introduction

During the last decade, biomedical applications have benefited from the development of new acquisition techniques that enable clinicians to achieve non-invasive diagnostics through image analysis and visualization with increasingly more efficiency and accuracy than before.

To automate the process of diagnostics from an image in a typical biomedical imaging application, we need first to enhance the information content of the image, and then segment, recognise and analyse the objects in the image. The key to this automation is the provision of an adaptable segmentation and recognition strategy that can deal intelligently with variations in the acquisition conditions (such as noise, distortion, signal attenuation etc.) and the wide variety of possible medical cases encountered in practice. The problem of adaptability in this filed has been partially addressed in the literature by using procedures that can learn from a training set. Most of these procedures are neural networks-based [1, 2, 3, 4], but more recently fuzzy logic-based approaches are being used [5]. The neural network (NN) based procedures have learning capability but are slow, and it is hard to transfer experience gained from one application to another. The Fuzzy Logic (FL) based procedures on the other hand cannot learn and the choice of the membership functions is tricky but they are fast, easily trainable and the experience gained from one application is easily transferable to a different one. The latter is due to the clear global decision making process of fuzzy logic. However in biological as well as medical image processing there is still a lack of intelligent systems that are capable of self-tuning and self-adapting to a wide variety of imaging situations, see for example [6]. Combining fuzzy logic and neural nets is capable of overcoming this problem.

In the past we had developed an image segmentation and recognition system that has been applied to gated blood pool emission tomographic images in order to automatically calculate the left ventricular blood volume ejected in a cardiac cycle [7, 8, 9]. We suggest now a new segmentation procedure that uses a competitive neural network. The structure of the network and its weights are devised by implementing a technique similar to the "follow the leader" algorithm [10]. In section 2 we discuss this procedure. In order to recognize the left ventricle among the different "objects" present in the image, our system then executes a neuro-fuzzy recognition procedure, which has the important features of self-tuning, automatic creation of rules and membership functions. The recognition procedure is presented in section 3. Section 4 shows some of our results and section 5 presents our conclusions.

2 Segmentation Procedure

Our segmentation procedure uses a clustering technique similar to the "follow the leader" algorithm introduced by Carpenter and Grossberg in their Adaptive Resonance Theory (ART) [10]. For this classification task, the number of classes does not have to be specified before launching the process. The system itself can produce new neurons to increase the number of classes. An element to be classified is arbitrarily chosen as the representative or the "leader" of a first class defined by its characteristics. Then, a measure of similarity is calculated to compare the retained leader with the remaining elements. This measure of similarity is simply the euclidian distance between a feature vector and the leader of a class. If the result of the comparison indicates that an element is similar to the leader, then it is retained in the corresponding class. This element can also be used to define a new class in which it will be the leader. At the end of the process, the totality of the initial data is partitioned into a certain number of classes, each represented by a leader element belonging also to the initial set.

Our segmentation is capable of creating a feature vector including all the neighbours of a pixel at the center of a hypercube with a connectivity and window size set up by the user. The process uses a 3-D segmentation to take more accurately the space continuity into account as established in [12]. As a result we do not need to pre-establish several leaders to initialize the "follow the leader" algorithm. This is a significant improvement over the algorithms in [7, 8, 9], where the segmentation was solely based on the gray level of the pixel in a similar way to the thresholding-based semiautomatic methods [11].

In our actual implementation the feature vectors are handled in a parallel rather than sequential process. This allows us, at each iteration, to create a new class for the most poorly classified feature vector than for the first significantly different input as stated in [10]. As a result the segmented regions are better contrasted and separated than in [8, 9].

The new segmentation algorithm would be as follows:

Step1. Select the first input vector as the leader or representative of the first cluster.

Step2. Classify all the input vectors using the present neural network.

Step3. Search for the most poorly classified input vector.

Step4. IF the selected input vector has a similarity degree lower than a threshold, THEN create a new cluster with the selected input vector as the leader or representative element and go to step 2, ELSE stop the procedure.

3 Recognition Procedure

One of the most difficult steps in biomedical image evaluation systems is trying to recognize automatically organs, such as ventricles in G-SPECT images, from the other organs represented in the image such as vessels, etc. This is in particular due to changes, in both position and form of the ventricle, from one patient to the other or even from one image slice to the next for a given patient. It is possible to overcome this difficulty using an adaptable neuro-fuzzy recognition procedure. This has the important features of self-tuning and automatic creation

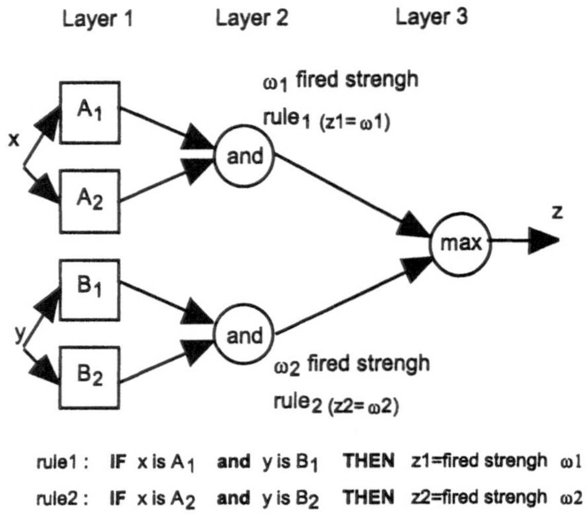

Fig. 1. Neuro-Fuzzy recognition system.

of rules and membership functions. The self-tuning is achieved by an optimization of the parameters defining the membership functions in the fuzzy rules using a steepest descent algorithm. We then make use of an algorithm called ALGORAM (Automatic Leader Generation Of Rules And Membership functions) [9]. This involves applying the already mentioned follow the leader procedure directly to the existing rules in order to modify the membership functions and the rule base. New membership functions and rules can also be created.

Briefly explained, the system is constructed by a set of rules, R, of the form:

R_i: IF x_1 is A_{i1} and and x_j is A_{ij} and and x_N is A_{iN} THEN z_i= *firing strengh of* R_i

The input parameters x_j are associated with the spatio-temporal features measured on each segmented image such as *area of the region*, *circularity of the region*, and *volume relatioship between different regions*. A typical membership function A is a triangular function defined by its centre and support. The firing strengh z, is calculated according to the neural implementation shown in Figure 1.

We create and/or update the knowledge system by activating ALGORAM. The training of the system is done in a supervised way knowing a priori the Leaders or representative regions of the left ventricle in each slice of a 3-D set of myocardial images. The mechanism is as follows:

Step1. Rule construction. Select a leader among the regions in the first slice of the image sequence. Construct the rule and membership functions matching the input parameters of the leader.

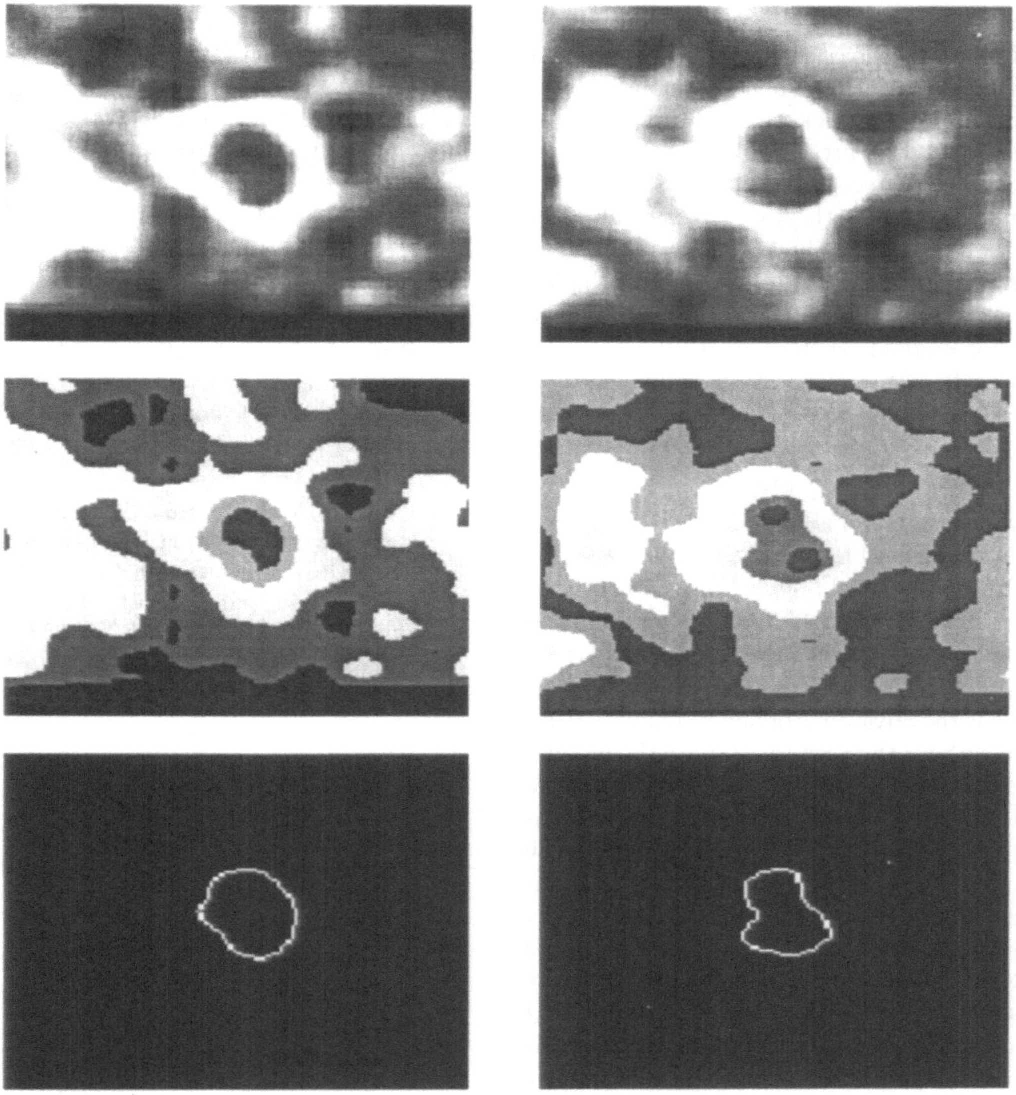

Fig. 2. a) Two slices of original 3-D gated-SPECT myocardial image (top row). b) Result of segmentation with our new improved algorithm (middle row). c) Left ventricle recognized contour (bottom row).

Step2. Verification of leader. This step checks if the leader in the next slice of the image sequence leads to firing.
Step3. Training membership functions. Addition of rules and membership functions for the parameters, which have not led to firing.
Step4. Verification of classification quality. During this step, we verify that the leader returned by the recognition system is the same as the previously selected leader.
Step5. IF the leader do not correspond THEN add rules and possibly membership functions to avoid misclassification.
Step6. Achieve fine-tuning of the system by optimising all membership functions.
Step7. IF last slice image has not been reached THEN go to step2. ELSE stop.

In the above algorithm optimisation of the membership function parameters is achieved by using a gradient-based approach similar to that proposed by Bersini and Gorrini [13].

4 Results

The segmentation and recognition procedures have been successfully tested on synthetic and real images. For example the combined segmentation and recognition system has been tested on gated-SPECT cardiac images related to a set of 30 patients. Here we present some typical results. Figure 2a represents two slices of a 3-D myocardial SPECT image. Figure 2b shows the segmentation of the images in Figure 2a using our neural network classification method. We took a [3x3x3] window with 4-connectivity and a threshold of 0.2. The resulting left ventricular recognized contours are shown in Figure 2c. To assess the validity of results produced by our system, we have compared the computed correlation coefficient (of the sum of the left ventricular areas) by our system with another medical evaluation modality such as Magnetic Resonance Imaging (MRI). Comparison of the results shows a good linear correlation, typically 0.94.

5 Conclusions

We have presented an adaptable neuro-fuzzy system which can be used for intelligent image analysis in biological and medical applications. The process consists mainly of two procedures, segmentation and recognition. The segmentation is achieved by an improved "Follow the leader" algorithm. This new technique has the advantage of being completely unsupervised without the need to establish the number of clusters in the image in advance. Also the weights in the network do not need any particular initialization to be able to separate the two ventricles in the image as in [8, 9]. The recognition exploit a combination of neural networks and fuzzy logic. The system has been tested on SPECT images, and has proven to produce consistent results. We are currently applying the system to segmentation of 3-D confocal microscopy images. We are also investigating the development of morphological techniques such as marker-based and weighted watershed segmentation [14] as a complimentary or alternative to NN-based segmentation to provide further flexibility and robustness in our system.

References

[1] Porenta G. et al., "Evaluation of a neural network classifier for PET scans of normal and Alzheimer s disease subjects", *Journal of Nuclear Medicine*, vol. 33, pp. 1459–1467, 1992.

[2] Kippenhan J. S. et al., "Neural-network classification of normal and Alzheimer s disease subjects using high-resolution and low resolution PET cameras", *Journal of Nuclear Medicine*, vol. 35, pp. 7–15, 1993.

[3] Tourassi G. D., Floyd C. E. Jr., "Lesion size quantification in SPECT using an artificial neural network classification approach", *Computers and Biomedical Research*, vol. 28, pp. 257-270, 1995.

[4] Yan Z., Hong Y., "Computerized tumor boundary detection using a Hopfield neural network", *IEEE Transactions on Medical Imaging*, vol. 16, pp. 55–67, 1997.

[5] Teodorescu H. N., Kandel A., Jain L. C., "Fuzzy and Neuro-Fuzzy Systems in Medicine", *CRC Press*, pp. 3–16, 1999.

[6] Aurengo A., "L'intelligence artificielle en imagerie médicale : pourquoi le mythe tarde-t-il à dévenir réalité ?", *Médecine Nucléaire Imagérie Fonctionnelle et Métabolique*, pp. 53–55, 1997.

[7] Patino L., Mertz L., Hirsh E., Constantinesco A., "Segmentation and contouring of blood pool myocardial SPECT images with wavelet-fuzzy constraints", *Proceedings of EUFIT*, pp. 2086–2090, 1996.

[8] Patino L., Mertz L., Hirsh E., Dumitrescu B., Constantinesco A., "Contouring blood pool myocardial gated SPECT images with a neural network leader segmentation and a decision-based fuzzy logic", *Proceedings of FUZZ-IEEE*, pp. 969–974, 1997.

[9] Patino L., Constantinesco A., Hirsh E., "Contouring blood pool myocardial gated SPECT images with a sequence of three techniques based on wavelets, neural networks, and fuzzy logic", *CRC Press*, pp. 95–136, 1999.

[10] Carpenter G. A., Grossberg S., "The ART of adaptive pattern recognition by a self-organizing neural-network", *Computer*, vol. 21, pp. 77–88, 1988.

[11] Chin B. B., "Right and left ventricular volume and ejection fraction by tomographic gated blood pool scintigraphy", *Journal of Nuclear Medecine*, vol. 38, pp. 942–948, 1997.

[12] Razaz M., Hagyard D.M.P., Lee R.A., "A Segmentation Methodology for Real 3D Images", *Nonlinear Model Based Image Analysis*, pp. 269–276, 1998.

[13] Bersini H., Gorrini V., "FUNNY (FUzzY or Neural Net) methods for adaptive process control", *Proceedings of EUFIT*, pp. 55–61, 1993.

[14] Razaz M., Hagyard D.M.P., "Morphological Segmentation of Multidimensional Images", *Image Processing*, vol. 2, pp. 294–312, 2000.

Real Time Identification and Control of a DC Motor Using Recurrent Neural Networks

Ieroham Baruch*, José Martín Flores*[1], Ruben Garrido*, Boyka Nenkova†

*CINVESTAV-IPN, México †IIT-BAS, Bulgaria

Abstract

The paper proposes to use three Recurrent Trainable Neural Network RTNN models for real time DC motor system identification and state feedback-feedforward control. The proposed RTNN model have a Jordan canonical structure which permits to use the generated vector of states directly for DC motor feedback control. A Backpropagation through-time type learning algorithm for RTNN model training, is also described. The experimental results, confirms the applicability of the described identification and control methodology in practice. The given results of nonlinear mechanical system identification and control by means of three RTNN models show a good convergence and confirm RTNN qualities.

1 Introduction

Recent developments in science and technology provide a wide scope of applications of high performance electric motor drives in various industrial processes. In high-performance motor drive applications involving mechatronics, such as robotics, rolling mills, machine tools, etc., an accurate speed or position control is of critical importance and there DC-motors are still widely used to accomplish this task. There is an increasing number of applications in high precision motion control systems in manufacturing, i.e., ultra precision machining, assembly of small components and micro drives. It is very difficult to assure high positioning accuracy and high trajectory tracking ability due to many factors affecting the precision of motion, such as load-torque variations, friction, backlash and stiffness in the drive system, [5]. Friction is a natural resistance to relative motion between two contacting bodies. The friction model has been widely studied by numerous researchers. Extensive work can be found in [1], [4]. It is commonly modelled as a linear combination of Coulomb friction, stiction, viscous friction, and Stribeck effect, [1], [4]. The presence of nonlinear friction forces is unavoidable in high performance motion control system. In servo systems, if the controller is designed without consideration of the friction, the closed-loop system may show steady-state tracking error and/or oscillations. In addition, the friction characteristics may change easily due to the environment's changes like load variations, temperature and humidity changes, and some dynamic effects could be observed, [1]. So, the standard PID type servo control algorithm is not capable of delivering the desire precision under the influence of friction. To compensate the friction effects, adaptive schemes were developed. Adaptive friction compensation for DC-motor drives and robot control systems are given in [5]. Some advanced works also are done on Neural Networks (NN) application for adaptive friction compensation. The cited in [4], works applied CAMAC based NN for robust control of systems with friction. Kim and Lewis, [4], applied a reinforcement adaptive learning, based on Fuctional Link NN for friction compensation of high speed precise mechanical system. Weerasooriya and El-Sharkawi, [5]; applied a Feedforward NNs for identification and control of DC-motor drives. As it can be seen, the proposed schemes in the literature of NN learning control systems possesses higher complexity and higher dimensionality, which makes them hardly applicable. To avoid this complexity, it is appropriate to use the recurrent NN approach. Some works in this field, allowing to identify nonlinear dynamic objects by means of recurrent neural network, has been done by Baruch et al., [2], [3]. So, the purpose of this paper is to apply the Recurrent Trainable Neural Network (RTNN) approach, [2], [3], for real-time identification and control of mechanical system with friction and unknown load characteristics, driven by a DC-motor.

2 Brief Description of the Recurrent Neural Network Topology and Learning

In [2], [3], a discrete-time model of RTNN, and the dynamic Backpropagation (BP) weight updating rule, are given. The RTNN model is a parametric one. It gives information about the state, control

[1]Sponsored by CONACYT-MÉXICO

and output matrices, and the state vector as well. The RTNN topology is described by the following vector-matrix equations:

$$X(k+1)=JX(k)+BU(k) \quad (1)$$
$$Z(k)=S\left[X(k)\right] \quad (2)$$
$$Y(k)=S\left[CZ(k)\right] \quad (3)$$
$$|J|>1 \quad (4)$$

where $X(.)$ is a n - state vector of the system; $U(.)$ is a M- input vector; $Y(.)$ is a $l-$ output vector; $Z(.)$ is an auxiliary vector variable with dimension l , $S(.)$ is a vector-valued sigmoid function with appropriate dimension; J is a weight-state block-diagonal matrix with (1×1) blocks; B and C are weight input and output matrices with appropriate dimensions and block structure, corresponding to the block structure of J. As it can be seen, the given $RTNN$ model is a completely parallel parametric one, so it is useful for identification and control purposes. The general BP learning algorithm is given in the form:

$$W_{ii}(k+1) = W_{ii}(k)+\eta\Delta W_{ii}(k)+\alpha\Delta W_{ii}(k-1) \quad (5)$$

where: $W_{ij}(C, J, B)$ is the ij-th weight element of each weight matrix (given in parenthesis) of the RTNN model to be updated; ΔW_{ij} is the weight correction of Wij; η, α are learning rate parameters. The updates ΔC_{ij} , ΔJ_{ij}, ΔB_{ij} of model weights C_{ij} , Jij, Bij are given by:

$$\Delta C_{ij}(k)=[T_j(k)-Y_j(k)]\,Y_j(k)\,[1-Y_j(k)]\,Z_i(k) \quad (6)$$
$$\Delta J_{ij}(k)=RX_i(k-1) \quad (7)$$
$$R=C_i(k)\,[T(k)-Y(k)]\,Z_i(k)\,[1-Z_i(k)] \quad (8)$$
$$\Delta B_{ij}(k)=RU_i(k) \quad (9)$$

where: T is a target vector with dimension l and $[T-Y]$ is an output error vector also with the same dimension; R is an auxiliary variable.

As it could be seen from equations, the proposed canonical RNN architecture is a two-layer hybrid one, with one Feedforward (FFNN) output layer and one Recurrent (RNN) hidden layer. The linearization of the sigmoid function allows study the dynamic properties of this RTNN architecture, such as stability, observability and controllability. To preserve the RTNN stability during the learning, the model poles must remain inside the unity circle. To do this, some restrictions of the weights in the feedback loop J of the hidden layer, are imposed.

The aim of this paper is to apply this model for a real-time identification and control of nonlinear mechanical system. It is expected that the application of a learning adaptive model like the RTNN model will be well suited for identification and control of such nonlinear process with unknown variable parameters and dynamic effects.

3 General Model of the DC-Motor Driven System

The general equation of a 1-DOF mass system with friction is given in the form, [1], [4]:

$$m\ddot{q}(t) + fr(\dot{q}, t) + d(t) = k_o u(t) \quad (10)$$

where m is the mass, $q(t)$ is the relative displacement, $v(t) = \dot{q}(t)$ is the velocity, $fr(v, t)$ is the friction force, $u(t)$ is the control force, k_o is the system gain, and $d(t)$ is a bounded external disturbances due to measurement noises or other load forces. It is assumed that the external disturbance is bounded by an unknown upper bound d as follows:

$$|d(t)| \leq d; t > 0 \quad (11)$$

The stick-slip friction force $fr(v, t)$ is described in [1], [5]. The friction force is modelled as a summation of the Coulomb friction, viscous friction, and the Stribeck effect. The Stribeck effect models the fact that friction force is decreasing with increasing fluid lubrication. Some models, [1], consider dynamic lag effect of the friction force with respect to the velocity, which effect can be neglected. The DC-motor model is also described by Weerasooriya and El-Sharkawi, [5].

4 Experimental Results

In this section, we illustrate the effectiveness of the RTNN algorithm by a real-time DC-motor sys-

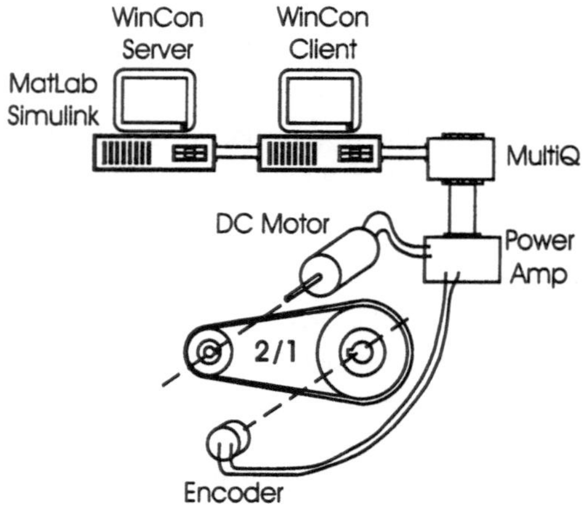

Fig. 1. Real-time system configuration for a DC Motor System identification and control.

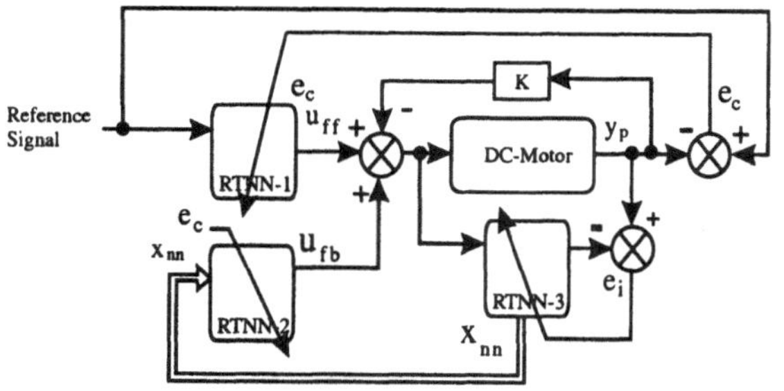

Fig. 2. DC motor identification and control. The RTNN topologies are (1,4,1), (1,4,1) and (1,4,1). $\alpha, \eta = 0.005$. The sampling rate $Ts = 0.001sec$.

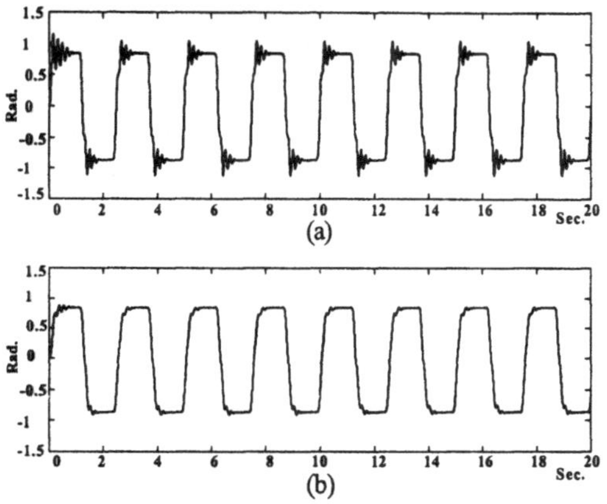

Fig. 3. Real-time DC motor drive identification by a RTNN model, (a) Position of the DC-motor. (b) Output of the RTNN.

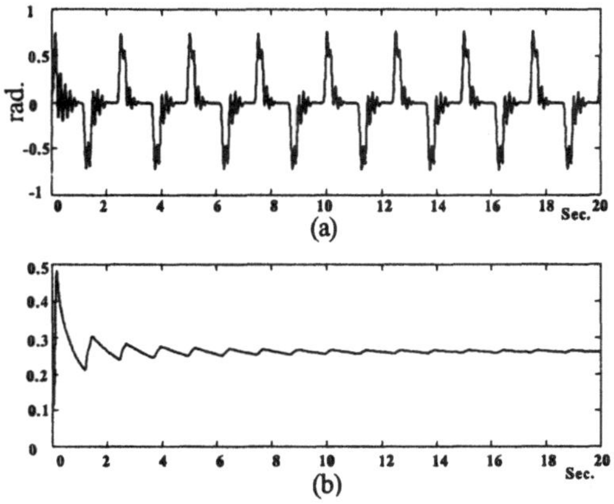

Fig. 4. (a) Identification error. (b) Mean Square Error of identification.

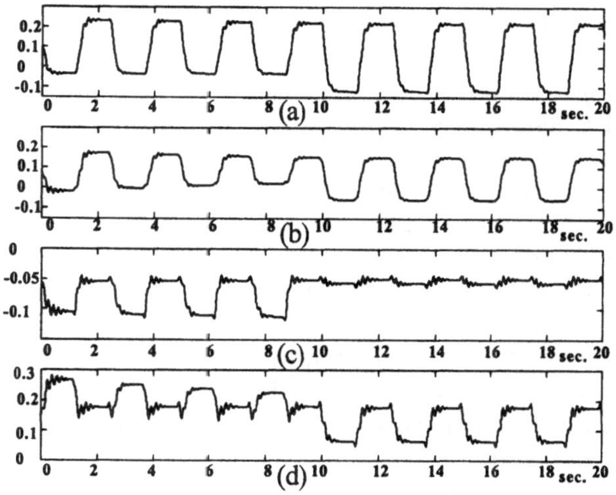

Fig. 5. Systems states estimated by a RTNN model. (a) $x1(k+1)$; (b) $x2(k+1)$; (c) $x3(k+1)$; (d) $x4(k+1)$

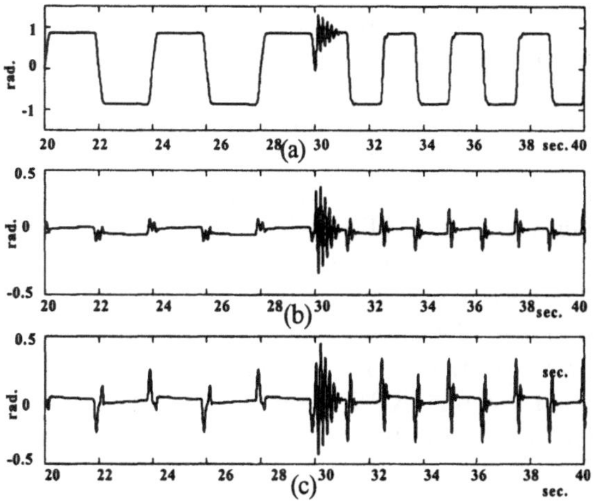

Fig. 6. Real-time state-feedback/feedforward control by two RTNN models and states, estimated by an identification. The reference signal changes from the given eqn. (12) to $u(k) = 0.277sat[\pi sin(0.5\pi k)]$. (a) Position of the DC-motor; (b) Control signal; (c) Instantaneous control error.

tem identification and control. The objective is to apply a the real time identification and control scheme of DC motor using one RTNN for system identification and two RTNNs for system control. The block diagram of this control scheme is given in the Fig. 2. The feedback part of the control uses the state vector given from the state estimation/identification RTNN. The identification RTNN is learned by the error between the DC-motor output (position) and the RTNN output. The two control RTNNs are learned by the error between the DC-motor output and the reference signal. The Fig.1 shows the configuration of the experimental DC-motor mechanical system together with the control and measurement components.

We use a 24 Volts, 8 Amperes DC motor driven by a power amplifier and connected by a data acquisition and control board. Multi-Q with the PC. The RTNN was programmed in Matlab-Simulink and WinCon that is a realtime Windows 95 application that runs Simulink generated code using Real-time Workshop to achieve digital real-time control on a PC equipped with a data acquisition and control board.

The first experiment (Fig. 3, 4, 5) corresponds to an open-loop input/position real-time identification of the DC-Motor. The RTNN model has the following topology: one input neurone, four hidden neurone nodes and one output neurone. The learning rate parameters are: $\eta = 0.005$, and $\alpha = 0.005$. The sampling period is $T_o = 0.001$. The DC- Motor input is a saturated sinusoid given by:

$$u(k) = 0.277 sat\left[\pi \sin(0.8\pi k)\right] \qquad (12)$$

Fig. 3, 4, 5 shows the output, error and state signals generated by the RTNN. The output position of the DC-motor is also given.

The second experiment is the real-time closed loop DC-motor identification and state feedback/ feedforward control. The system identification is performed by one RTNN and the control - by two RTNNs, separately for the state feedback and the feedforward parts. Fig. 6. shows the control results (DC-motor position output, control signal and control error). In this experiment the reference signal changes its frequency from 0.8 to 0.5 and a good control system reaction could be observed.

5 Conclusions

A comparative study of various mechanical control systems with friction compensation is done. The paper proposes to use three RTNN models for real-time DC-motor driven mechanical system identification and state feedback/feedforward control. The proposed RTNN model is a Jordan canonical model permitting to use the generated vector of states directly for DC-motor feedback control. A dynamic Backpropagation-type learning algorithm for RTNN model training, is also described. The experimental results, confirms the applicability of the described identification and control methodology in practice. The given results of nonlinear mechanical system identification and control by means of three RTNN models show a good convergence and confirm RTNN qualities.

References

[1] 1. B.Amstrong-Helouvry, P. Dupont and C. Canudas De Wit: A survey of models, analysis tools and compensation methods for the control of machines with friction. Automatica *30* , 1083 (1994)

[2] I.Baruch, T.Arsenov, S.Koynov: Recurrent Neural Network Models for Systems Identification and Control. Proc. of the 2-nd IFAC Workshop on New Trends in Design of Control Systems, Sept. 7-10, 360 (1997)

[3] Baruch I ,Stoyanov I. and T.Arsenov: An Improved RTNN Model for Dynamic Systems Identification and Time-Series Prediction, Proc. of the 4th International Symposium on "Methods and Models in Automation and Robotics", 26-29 August, Miedzyzdroje, Poland, Vol *2*, 711 (1997)

[4] Y.H. Kim, and F. L. Lewis: High-level Feedback Control with Neural Networks, World Scientific Publ. Co. (1998)

[5] S.Weerasooriya and M.A. El-Sharkawi: Identification and control of a DC-motor using back-propagation neural networks. IEEE TEC *6*, 663 (1991)

Recurrent Neural Networks in a Mobile Robot Navigation Task

Branko Šter*

*Faculty of Computer and Information Science, University of Ljubljana, Slovenia, e-mail: Branko.Ster@fri.uni-lj.si

Abstract

Recurrent neural networks are applied to the forward modeling of the sensory-motor flow of a miniature mobile robot. It is shown that the robot is able to predict the sensory flow a few steps ahead, which suffices for simple environments. The proposed method requires mainly topological information (little geometrical information is used), simplifying the problem considerably.

1 Introduction

To be useful, a mobile robot should be able to navigate at least in indoor environments. Basic skills include obstacle avoidance and approaching a goal. While the obstacle avoidance skill is a reactive-type behaviour, i.e., it requires no memory, approaching a goal usually requires a certain amount of knowledge about the environment. Unless the goal is observed at a given moment, the robot must have some knowledge (model) of the environment. Classic approaches to environment modeling consist of geometrical 2-D space reconstruction using measurements from robot sensors. This approach could be error-prone due to noise and inaccuracy of the sensors. The alternative approach includes topological modeling of the environment. It leads to a simpler description of the space, preserving essential information about the environment. Provided that certain low-level behaviour exists, a topological map of the space (in the form of a graph or a finite-state automaton) suffices to the robot to distinguish among qualitative distinct behaviours. This type of description is much simpler, but has a drawback. Whenever an error occurs (due to sensor noise, for example), this symbolic representation can break down.

It was argued [1] that for truly intelligent agents the type of representation must not be a designer's choice, because the abstraction is the key part of the intelligence. Human designers tend to decompose the problem to the blocks-world, i.e., they do all the abstraction and leave an "intelligent" program merely to search in this simplified world. In fact, we did here exactly the same thing. The approach taken here still simplifies the world to a large degree. More than that, the world is relatively constraint in the very beginning. Although we also consider this as a kind of deception, it is still useful to simplify a problem in the beginning, but having in mind this is not the world an agent is supposed to act in. The approach, taken here, has connections to [2] and [3].

2 Recurrent Neural Networks

One of the classic algorithms for training recurrent neural networks (RNN) is described in [4]. The algorithm is gradient following and is applicable for fully connected RNNs, i.e., each unit (neuron) has a feedback connection to each unit in the network, see Figure 1. The outputs of certain units represent outputs of the network, while the other units are called context-units, because they provide information relevant to sequence-processing problems.

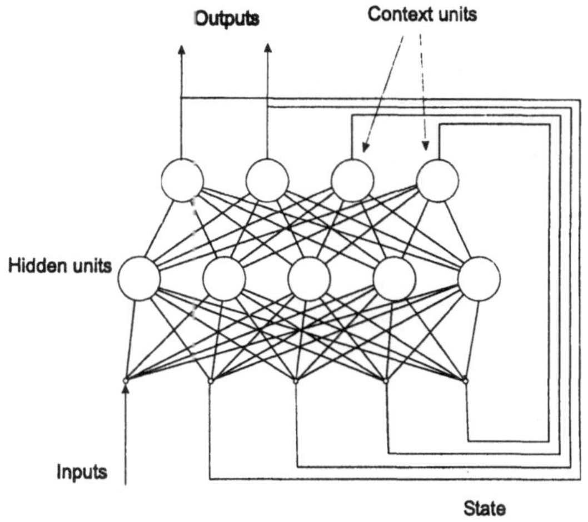

Fig. 1. Recurrent neural network.

Inputs to the network are fed to each unit. Thus RNN can be trained to simulate any dynamical system, at least conceptually. The batch variant of the algorithm requires a sequence of input vectors $x(t)$ and output vectors $y(t)$ for $t = 1, ..., N$, where N is the length of the sequence. RNN is trained to produce the sequence of the desired outputs, if fed with the input sequence. A caution must be exercised while training the network; namely, when RNN is trained to produce desired outputs at each cycle, the functional ability of the network is restricted to a single perceptron layer, i.e., a linear transformation plus sigmoid activation function. To increase the functional ability of the network, inputs and outputs may change after two or more cycles of the network processing. We applied a recurrent neural network with a hidden layer, which increases the functional ability of the network.

3 Building the Internal Map of the Environment

Sensory information consists of infrared proximity sensors, wheel-encoders (not used at this stage) and processed image from a video camera. The miniature robot Khepera (K-team from Lausanne, see http://diwww.epfl.ch/Khepera/) with an additional video color camera was applied. The task was considerably simplified using the camera solely for detection of colored objects (red, green and blue in this case). A video image was thresholded to extract the intensities of red, green and blue colour, which subsequently form a three-dimensional sensory vector. Two images at resolution 80x60, seen by the robot before the thresholding operation, are shown in Figures 2 and 3. They correspond to the sensory outputs "Green" and "Red observed" in Figure 4, respectively. It should be mentioned here that these labels do not actually represent states, but rather the outputs of a finite-state machine, which describes the environmental dynamics. Actual states are really not directly accessible via sensors.

Fig. 2. Sensory output "Green" close to object G, as seen by the robot.

The low-level or reactive behaviour of the robot consists of following the right wall using information from proximity sensors. The robot is equipped with eight proximity sensors. The low-level behaviour is

Fig. 3. Sensory output "Red observed", where object R is observed, as seen by the robot.

deterministic and rather short to program manually, but a little longer to find experimentally. It turned out that the following portion of the low-level program (written in C language) leads to very efficient right-wall following:

```
if ( proxSen[2]>100 || proxSen[3]>100 ||
        proxSen[4]>100 )   //
   SetSpeed(-h/2,h/2);
else if ( proxSen[5] > 800) // too close
   SetSpeed(-h,h);          // left
else if ( proxSen[5] < 300) // too far away
   SetSpeed(2*h,h);         // right and forward
else
   SetSpeed(2*h,2*h);       // forward
```

The h represents the speed and proxSen[0..7] is the vector of proximity sensors. The robot advances following the wall until it observes a colored object.

There is a certain, experimentally determined threshold for each colour to signify the observation of an object (about 100 pixels), as well as vicinity of the object (about 600 pixels). When the robot observes a colored object, it has to decide whether to still follow the wall or to approach the object. This situation is called branching point. In the vicinity of a colored object, the decision is to approach another colored object or to find the wall. Looking for another object, the robot turns left until another object is observed with a specified threshold. It subsequently approaches the object. Looking for the wall, the robot turns right for approximately 90 degrees, and from then on follows Breitenberg's algorithm for low-level behaviour, which is basically forward-motion, avoiding obstacles. The wheels' velocities are linear functions of the proximity sensors plus an offset term.

Thus the whole environment could be represented as a graph or as a finite-state automaton (finite-state machine, FSM). One possible approach could be simply to induce the FSM and to use it for navigation.

It was shown that finite-state automata can be represented by recurrent neural networks. There has been a lot of effort to induce FSMs for regular languages with RNNs on the basis of shown examples, mainly successful. On the other hand, on-line learning of RNN to a FSM is not so well understood.

FSMs, corresponding to typical indoor environments, are usually relatively simple (providing that there are not many branching points). Therefore, in practical cases RNNs should not have problems learning the corresponding FSMs. Of course, a complex maze would produce a complex FSM.

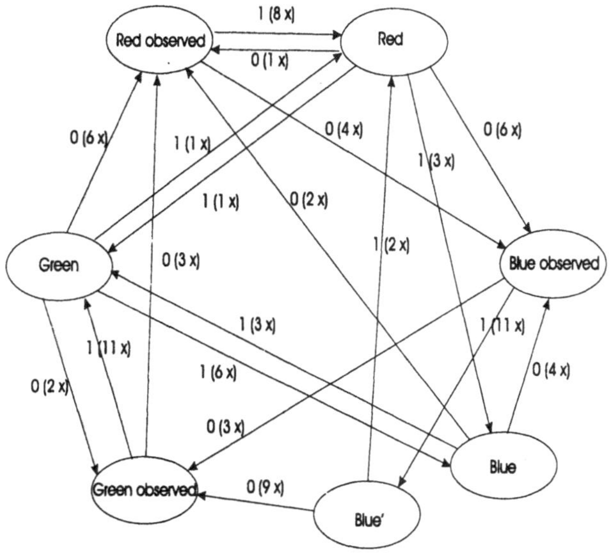

Fig. 4. A description of the environment in a similar form to a finite-state machine. The difference is that sensory outputs (instead of states) are represented with bubbles. This difference is essential although in our simple environment it is not big. Labels on the arcs represent type of action (0: wall-following and wall-seeking, 1: object-approaching and another-object-seeking) and its multiplicity.

There is a question whether a RNN is able to induce a finite-state automaton only on the basis of a limited number of training examples. It is obvious that the network can be trained (fitted) to a particular sequence, but it is not clear whether the complete structure of an automaton can be induced. This is not a question of functional ability, since a RNN can simulate any FSM, in principle. It is probably a question of an efficient training algorithm. Recurrent neural nets are harder to train than feedforward nets.

4 Experiments

During training the robot traveled around using random decisions at branching points. The whole path included 101 branching points. An automatic preprocessing, which dealt with situations, where the robot observed the same object again after a short period of time, was done. Thus the path with 87 branching points remained. A small part of the path is shown in Figure 5. At the first branching point from the start the robot observes object R (red). Therefore, the starting sensory output is "Red observed" in the FSM, see Figure 4. The training trajectory can be represented as a motor-program describing the action sequence. The vectors, extracted from images, represent the outputs of the FSM. The outputs do not exactly correspond to the states, so there is not a one-to-one mapping from the states to the outputs. Images at distinct places may look the same, so there is an obvious need to have internal states, i.e., memory.

RNN was trained off-line many times on the sensory-motor sequence of the first 80 samples for 5000 iterations (recurrent nets typically require large number of iterations) and tested on the remaining 6 samples. It takes about two minutes on a 433MHz PC computer. In each iteration the weights were updated following the gradient of the mean-squared error (MSE) on the whole training set ("batch" variant). MSE amounted on average from 0.01 to 0.02 on the training set and from 0.01 to 0.25 (RMSE from 0.1 to 0.5) on the testing set. Sometimes the network actually did very poor predictions. We were interested in cases where RNN was able to predict good enough in a closed-loop on the testing set. The five-dimensional input vector consists of the current sensory vector (red, green and blue intensities), the distance from the previous to the current branching point, and the current decision (action).

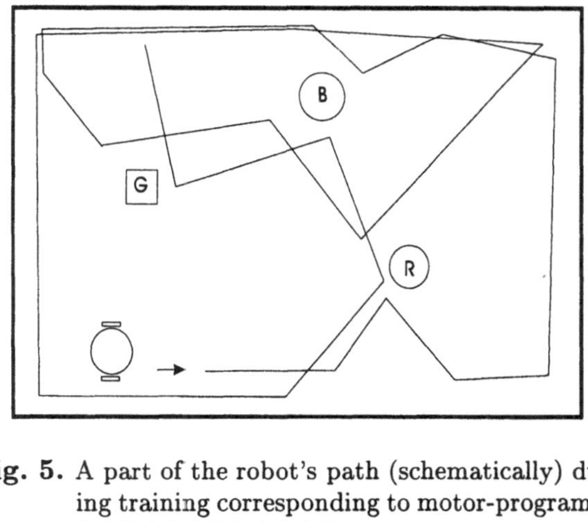

Fig. 5. A part of the robot's path (schematically) during training corresponding to motor-program 1-0-1-0-1-1-0-0-0-1-1-1-0.

Given an input vector, a trained RNN predicts the next sensory vector and distance to the next branching point, which can be used in turn as the input at the next branching point, and so on. Given a motor sequence, RNN should be able, at least in principle, to predict the corresponding sensory flow.

There are two types of overtraining or overfitting here. Firstly, due to too many hidden neurons (standard overfitting in static neural networks), and secondly, due to too many context neurons. The latter might be called "dynamic overfitting", namely the RNN use its context units to "invent" a kind of context to lower the prediction error on the particular training sequence, especially when the latter is short. This problem would possibly be avoided with an on-line training, long enough to resolve uncertainties. For example, when turning to the left near the red object and looking for another object, the robot may overlook the blue object. The cause would be probably the video camera. In the training sequence there were four occurences of the red object followed by the action "1", i.e., looking for another object (after turning to the left.) Three times the robot observed B and once G. In the latter case it overlooked B, unfortunately. In case of larger sensor errors a training sequence should be much longer in order to reliably "collect statistics".

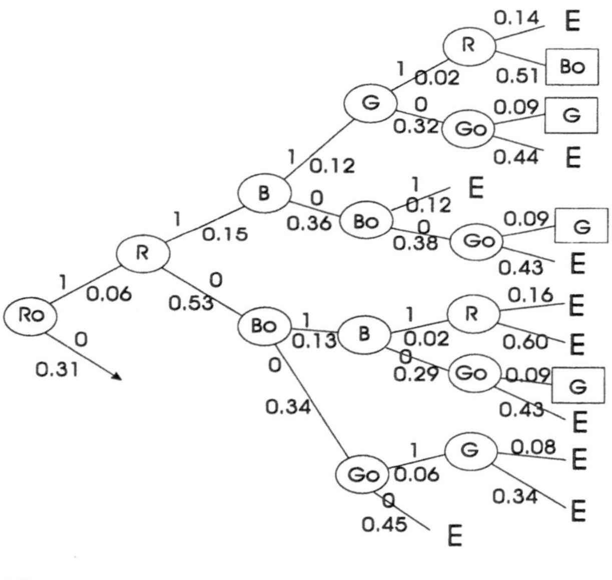

Fig. 6. A half of the tree, representing motor programs of the length five. Only the programs, beginning with 1, are shown. At each arc the motor command (0 or 1) and the distance are written. Unclear predictions are labeled with E.

Prediction of the RNN was tested applying all possible motor programs five steps (branching points) ahead, i.e., $2^5 = 32$ programs: 00000, 00001, 00010,...,11111.

The sense of forward-modeling of the environment is applying it in a "mental" planning. The term "navigation" stands for finding the (shortest) path to a specified point. For example, we would like to find the shortest path from beginning to the green object. It is obvious from the figure 6, that using the program 1-1-1 is the best way to do it. All we have to do is closed-loop simulation of all the motor programs, searching for the matching object, specified as the goal, and finding the least costly path to the goal. One possibility is proposed in [2].

5 Conclusion

It was shown that recurrent neural networks can be applied to build the internal map of the environment in a simplified robot navigation task. The robot is able to induce the FSM of a simple environment in the form of RNN and to use it for further planning. Because of the simplicity of the environment in the given task, the robot has to predict sensory flow for very short motor-programs, and it does it successfully approximately a half of the time. Further work should reveal how this behaviour scales up to larger problems and more realistic sensors, i.e., where a lower level of abstraction of sensory information is provided. Also more reliable and possibly on-line training is required. The ultimate goal would be no human-designed representation, which seems to be still a very complex task.

References

[1] R. A. Brooks, "Intelligence without representation", *Artificial Intelligence*, vol. 47, pp. 262–276, 1991.

[2] J. Tani, "Model-Based Learning for Mobile Robot Navigation from the Dynamical Systems Perspective", *IEEE Transactions on Systems, Man, and Cybernetics - Part B:Cybernetics*, vol. 26, no. 3, pp. 421–436, 1996.

[3] M. J. Mataric, "Integration of representation Into Goal-Driven Behavior-Based Robots", *IEEE Transactions on Robotics and Automation*, vol. 8, no. 3, pp. 304–312, 1992.

[4] R. J. Williams and D. Zipser. "A Learning Algorithm for Continually Running Fully Recurrent Neural Networks", *Neural Computation*, vol. 1, pp. 270–280, 1989.

Angular Memory and Supervisory Modules in a Neural Architecture for Navigating NOMAD

Catarina Silva*†[1], Manuel Crisóstomo, Bernardete Ribeiro†

*Escola Superior de Tecnologia e Gestão de Leiria †Departamento de Engenharia Informática, Centro de Informática e Sistemas, Universidade de Coimbra, 3030 Coimbra, Portugal, email:{catarina,bribeiro}@dei.uc.pt

Abstract

This paper presents a neural modular architecture for navigating mobile robots. The proposed architecture, based on functional task division, has been shown to be efficient in previous work [1, 2].

The traditional difficulties associated with monolithic neural networks are circumvent by introducing a neural modular architecture, developed for the NOMAD mobile robot. The modularity introduced retrieves all the available information, minimizing the incoherence in the training sets, that arises from conflicting training patterns.

After presenting the modular architecture, two additional modules are introduced: Supervisory and Angular Memory. The combination of all defined modules was first tested in simulation and then with the real robot, providing a solution to the problem of navigating NOMAD mobile robot to an objective in an unknown environment.

1 Introduction

The navigation task proposed consists in guiding a mobile robot to a goal position, while avoiding all obstacles in the path.

Real environments are usually dynamic and unstructured, giving rise to serious difficulties when applications rely on *a priori* knowledge of obstacle localisation. Thus, the construction of a map can present important drawbacks when the environment is too much changeable.

Consequently, if the presence of obstacles is unknown and there is no map of the environment, flexibility in robot's response is required.

Neural networks can provide the mentioned flexibility through its adaptive and learning abilities. Nevertheless, monolithic neural networks, i.e., systems where there is only one, usually large, network that combines all the information and delivers the desired result, are often fallible. When the task involves a large input space, as occurs in mobile robot navigation, the performance decreases [3], the training time increases and generalisation capabilities impoverish.

Section 2 describes the mobile robot system used as a testbed for this work and section 3 introduces the concept of modular neural networks.

The proposed architecture is explained in section 4 and section 5 refers to the Supervisory and Angular Memory modules.

Section 7 shows some results and, finally, in the last section the conclusion is presented.

2 Nomad Mobile Robot

NOMAD is a wheeled cylindrical robot whose diameter is about 46 centimetres, and is shown in figure 1. The robot can move forward or backward with varying speed and turn right or left with varying degrees/s. The robot has also a dead-reckoning system for keeping track of the robot's orientation and position. It has 16 ultrasonic sensors and 16 infrared sensors for observing the surrounding environment. Ultrasonic and infrared sensors are evenly distributed around the robot, yielding a 22.5 degrees angle between any two adjacent sensors, distributed according to figure 2.

Fig. 1. NOMAD mobile robot.

[1]This work was partially supported by the portuguese Ministério da Ciência e Tecnologia and the European Union through the R&D Unit 326/94 (CISUC).

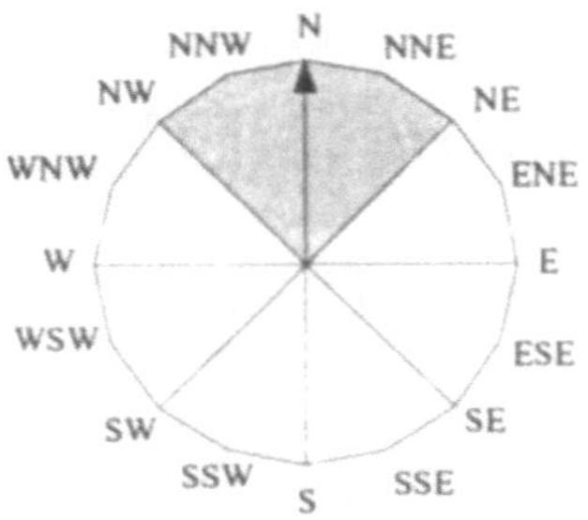

Fig. 2. Localisation of NOMAD sensors.

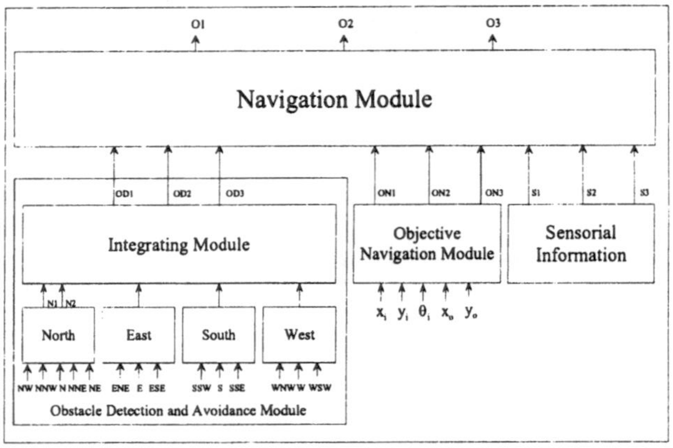

Fig. 3. Modular architecture.

3 Modular Neural Networks

The building block of any neural network is the neuron. The organisation of neurons in layers and interconnection of layers are the next steps in the neural architecture. However, this organisation can go one step higher. Specifically, a neural network can consist of a multiplicity of networks in a modular structure [6].

Modularity can be viewed as a manifestation of the *principle of divide and conquer*, which allows us to solve complex problems, by diving them into smaller sub problems, easier to conquer, combining their individual solutions to achieve the final solution.

One of the greater improvements achieved by modular neural networks is better generalisation. The convention in neural networks is to use an architecture as small as possible to obtain better generalisation, because the ability of feed-forward neural networks to generalise is inversely proportional to the number of weights involved [7].

Moreover, by partitioning a complex task, modular neural architectures achieve representations that are more easily understood than those generated by monolithic architectures. Finally, modularity can enable a learning economy, since some modules can be reused or modified without altering the other modules. This economy can also become crucial when hardware implementations are an objective, since modular neural networks can facilitate hardware execution of neural networks [4].

4 Modular Architecture

The navigation task was initially divided in two main modules: obstacle avoidance and objective navigation, described in the next two subsections. A third one - the navigation module - was developed to combine the obstacle avoidance module and the objective navigation module. Figure 3 shows the modular architecture proposed in [2].

4.1 Obstacle detection and avoidance module

The task of obstacle detection and avoidance was further divided into 4 sub-tasks (one for each quadrant), which cooperate in deciding the direction the robot should take to fulfil the task of obstacle detection and avoidance. This module will play a central role in the resolution of the problem, since it will receive all the sensorial information available.

As referred in section 2, there are two types of sensors aligned in each sensorial direction: an infrared sensor and an ultrasonic sensor. The difference between their ranges can be used to optimise the information in each of the 16 possible sensorial directions:are ultrasonic sensors are used to give global information about the area, but when the robot is close to obstacles, it uses the infrared sensors. A system of cardinal points has been assigned to the sensorial directions available (see figure 2). As mentioned, the robot always faces North in the pseudo-system. Thus, the North direction presents the higher probability of collision. According to this analysis, all the sensors in the North quadrant (NW, NNW, N, NNE and NE) were used as inputs to the North module and two outputs were considered.

The training set of the North module was derived from actual trajectories driven by the supervisor in simulated environments. It consists of 120 sets of sensorial information with the desired direction associated with it. The other three modules are smaller and identical. They all watch one of the other quadrants and have three sensorial inputs, that correspond to three sensorial readings of the respective quadrant, as illustrated in figure 3. There is only one output neuron that indicates the trust in the presence of an obstacle in that quadrant. The training set, composed by 30 patterns, was directly

OD1	OD2	OD3	Direction
0	0	0	N
0	0	1	NW
0	1	1	NE
0	1	0	E
1	1	0	S
1	0	0	W

Table I. Obstacle detection and avoidance module codification.

derived from the robot's simulator by defining environments in both cases of existence or not of obstacles.

To integrate the four modules (North, East, South and West), an integrating module was defined (see figure 3). The integrating module is a neural network with 5 inputs: two from the North module and one from each of the other three modules. It has three output neurons, which codify the direction to be followed according to the table I. The training set for the integrating module was generated by the supervisor and consisted of 32 training examples, that constitute the possible combinations of the 5 binary inputs.

4.2 Objective navigation module

The Objective Navigation Module computes the heading direction to the goal, considering the goal position and the actual robot's position (x_i, y_i) and orientation, θ_i. The orientation is computed according to (1):

$$\theta_o = atan\frac{y_o - y_i}{x_o - x_i} \qquad (1)$$

where (x_i, y_i) is the actual robot's position, θ_i is the actual robots's orientation, (x_o, y_o) is the objectives's position and θ_o represents the angle to the objective. Having the angle that represents the direction, a mapping is made to the cardinal directions and codified according to table I.

4.3 Navigation module

Using the two modules, in which the main task was divided, the navigation module was built to determine the final direction to follow. It also receives three sensorial inputs corresponding to the sensorial values for the heading direction calculated, yielding 9 inputs and 3 outputs, that codify the output direction according to table I.

5 Supervisory and Angular Memory Modules

Two additional modules were defined to cope with situations were the proposed architecture could be improved. The supervisory module detects obstacles whose dimension is much larger that the robot's dimension and determines the functioning mode: obstacle contour or navigation. The navigation mode is basically the architecture defined in the last section, i.e., the direction is determined by the outputs of the navigation module. The obstacle contour mode uses the outputs of the integrating module to establish the direction to follow.

The Angular Memory module keeps track of the angular directions followed when there is a change in the functioning mode, to detect cycles in the trajectory. Using a circular memory, the transition points are memorised and when a repetition is found an alternative direction is followed, as a way to break the cycle.

6 Modules Construction

The topology and parameters of all the modules defined so far were constructed and not determined *ad-hoc*. A *Dynamic Node Creation* [8] method was used with a variant of the back propagation algorithm. Construction algorithms avoid problems usually bound to trial and error solutions. These algoritms search, not only the correct set of weights for a topology, but also the most appropriate topology for a given problem.

Table II resumes the results obtained, presenting for each module the number of training patterns, the final configuration achieved and the Mean Square Error (MSE) obtained.

Module	Test Set	Configuration	MSE
North	120	$5 \to 5 \to 2$	0.028
East	30	$3 \to 1 \to 1$	0.001
South	30	$3 \to 1 \to 1$	0.001
West	30	$3 \to 1 \to 1$	0.001
Integrating	32	$5 \to 4 \to 3$	0.001
Navigation	396	$9 \to 4 \to 3$	0.002
Supervisory	25	$2 \to 2 \to 1$	0.002

Table II. Results obtained with the Dynamic Node Creation method.

7 Results

The purpose of the simulation was to test the behaviour of the mobile robot using the trained modular neural network system. The simulation was carried in environments where the modular neural network had to control the robot to a pre defined goal

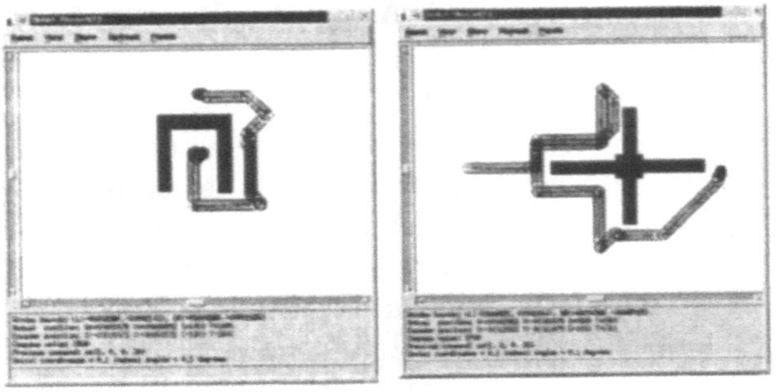

Fig. 4. Results in environments with unknown obstacles.

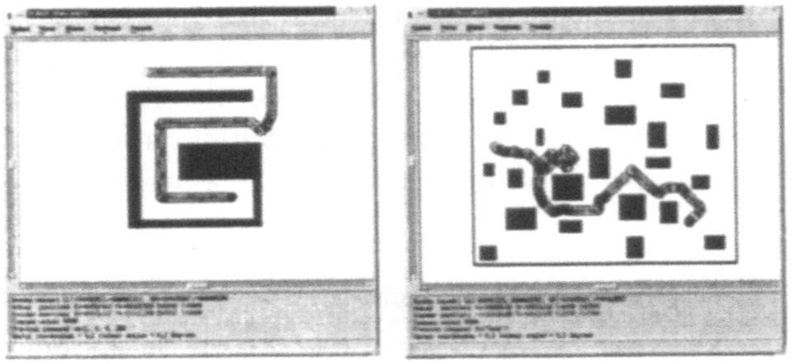

Fig. 5. More results in environments with unknown obstacles.

position avoiding collisions with obstacles without *a priori* knowledge of the environment. Tests with the real robot were also carried out, strenghtening the confidence in the proposed architecture.

Figures 4 and 5 present some of the results obtained.

The modular structure presented proved to be efficient. Neural networks inherent adaptive capabilities enable a good performance as can be seen in the examples presented. The modular architecture avoids robot collisions and produces smooth trajectories, thus showing generalisation after learning.

8 Conclusion

We showed how modular neural networks can be used to navigate a mobile robot among obstacles, without knowledge of the environment. The results presented show the strategy to be robust and adaptive.

Future work will include the study of on-line learning as a way to improve the system performance.

References

[1] C. Silva, M. Crisóstomo, B. Ribeiro, "MONODA: A Neural Modular Architecture for Obstacle Avoidance Without Knowledge of the Environment", in *Proceedings of the 2000 International Joint Conference on Neural Networks*, 24-27 July 2000.

[2] C. Silva, M. Crisóstomo, B. Ribeiro, "A Modular Learning Architecture for Navigating NOMAD Mobile Robot", 8th International Conference on Information Processing and Manageament of Uncertainty in Knowledge Based Systems, 3-7 July 2000.

[3] A. Schmidt, Z. Bandar, "A Modular Neural Network Architecture with Additional Generalisation Abilities for Large Input Vectors", in *Proceedings of the ICANNGA97*, Springer-Verlag, pp 40-43, April 1997.

[4] G. Auda, M. Kamel, "Modular Neural Networks: A Survey", in *International Journal of Neural Systems*, Vol. 9, No. 2, pp. 129-151, 1999.

[5] A. Sharkey (Ed.), "Combining Artificial Neural Networks", Springer-Verlag, 1999.

[6] S. Haykin, "Neural Networks - a comprehensive foundation", Prentice Hall, 1999.

[7] E. Baum, D. Haussler, "What Size Net Gives Valid Generalization?", in *Neural Computation*, vol. 1, pp. 151-160, 1989.

[8] T. Kwok, D. Yeung, "Constructive Algorithms for Structure Learning in Feed Forward Neural Networks for Regression Problems", in *IEEE Transactions on Neural Networks*, Vol. 8(3), pp. 630-645, 1997.

A Recurrent Neural Network for Controlling a Fed-Batch Fermentation of B. t. [1]

Barrera Cortés J., Baruch I., Valdez Castro L., Vázquez Cervantes V.*

*CINVESTAV-IPN, Av. IPN No. 2508. A. P. 14470 México D. F., México, E-mail: jbarrera@mail.cinvestav.mx

Abstract

The paper proposed to use a new Recurrent Neural Network Model (RNNM) to stabilize fermentation process of Bacillus thuringiensis from fermentation kinetic data. The multi-input multi-output RNNM proposed, have ten inputs, six outputs, sixteen neurones in the hidden layer, and also global and local feedbacks. The weight update learning algorithm designed, is a version of the well known backpropagation through time algorithm, directed to the RNNM learning. The approximation error for the last epoch of learning is about 2% and the total time of learning is 201 epochs, where the size of epoch is 115 iterations.

1 Introduction

The lack of information related to the metabolic mechanisms of many microorganisms had conduced to find an alternative modeling techniques for control systems design [1]. The control schemes commonly applied to the biological process are based over the operational conditions, considering that if the microorganism is maintained under the adequate operational conditions, they will have a good growth. Unfortunately, this strategy of control is not applicable in the case of metabolic mechanisms manipulation. So, in this paper we propose to use Neural Networks (NN) for a predictive control systems design, applied to a fermentation process of Bacillus thuringiensis (B.t) [2]. The advantages of the NN are in the field of non linear function approximation [3], [4], as the fermentation process is. For this aim, a Recurrent Neural Network (RNN) has been chosen, because of its lower training time required.

As a object of interest, we study the Bacillus thuringiensis, variety kurstaky strain HD-73. The importance of this microorganism is its property to produce a toxic protein with selective toxic characteristic for certain kind of insects in its larval phase [5]. The problem of B.t. fermentation is the high variation detected in the yield and toxicity quality of Cry 1A(c) protein. The factors detected as responsible of such variation are the operational conditions (temperature, pH, oxygen dissolved and agitation speed), the culture characteristics, the viability of the strain as well as the specific growth rate (m) of B.t. From all these variables, we are interested to study the last one.

To study the effect of m in the production of Cry1A(c), we start to implement a method of feeding that let us to manipulate this variable. According to the literature [5], the specific growth rate of this microorganism is a function of the nutrient concentration available in the culture. The method implemented, is based on a continuous feeding where the variables to manipulate are the initial time of the feeding and the duration of the feeding with a constant feeding flux.

For control purposes we develop a predictive recurrent neural network model, whose inputs are: concentration of bacteria's, glucose, spores, temperature, pH, specific growth rate, nutrient flux, acid and base pump states, as well as the air flux at a initial concentration of total solids of 40 g/l. The NN outputs are: count of bacteria and spores (variable directly related to the protein production), glucose concentration, temperature, pH and the nutrient flux.

The training of the RNN was carried out with five fermentation data sets which difference is in the specific growth rate of B.t.

2 Description of the Fermentation Process

Fed-batch fermentation of B.t. consists of microorganisms cultivation into a sterile reactor, maintained under the operational conditions, adequate for microorganisms growing and with a nutrient flux according to the specific growth rate desired for that microorganisms. The nutrients comprehend mineral salts, a source of carbon (glucose) and nitrogen (corn solids and soya flour) are in a 7:1 rate. The more adequate operational conditions are: temperature of

[1] This paper was supported by CONACYT, project No. 31511-B

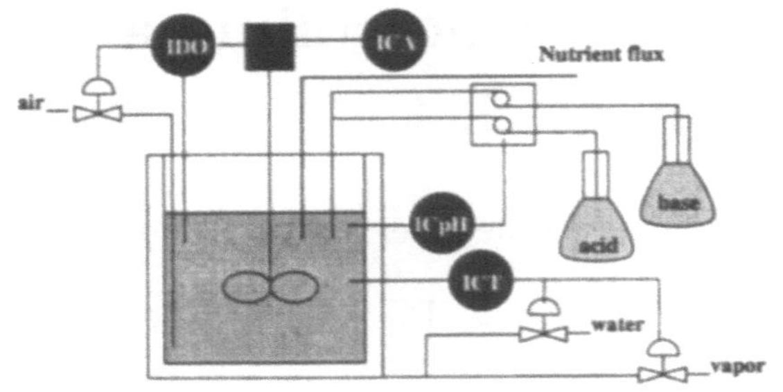

Fig. 1. Fed batch Fermentation Process of B.t. Indicator of dissolved oxygen (IDO), indicator and controller of agitation (ICA), indicator and controller of potential of hydrogen (IcpH) and indicator and controller of temperature (ICT).

30°C, pH of 7.1 and the available oxygen of concentration, greater than 40%. For sake of sterilization, the pressure is maintained higher than the atmospheric one. The frequently used pressure value is of 1.3 Kg/m^2.

The purpose of the RNN is to predict the specific growth rate of cells from the nutrient flux fed to the fermentor.

3 Description of the RTNN Topology and Its Lerning

In the present paper, it is proposed to use the Recurrent Trainable Neural Network (RTNN), described in [6], [7], as a neural predictive model of the fermentation process variables. The RTNN two-layer architecture contains hidden and output layers. The output layer is a Backpropagation (BP) one and the hidden layer is a recurrent one.

To improve the prediction abilities of this RTNN model, it is necessary to introduce some local and global output feedbacks. So, the RTNN model, [6], [7], obtains the form:

$$X(k+1)=S\left[A_1X(k)+BU(k)+DY(k)\right]; \quad (1)$$

$$Y(k+1)=S\left[CX(k)+A_2(k)Y(k)\right]; \quad (2)$$

$$A_1<1; \; A_2<1 \quad (3)$$

Where, X, U are output, state and input vectors with dimensions l, n, m, respectively; A_1 =block-diag(A_{1i}), $A2$ = block-diag(A_{2i}) are $(n \times n)$ and $(l \times l)$- local feedback block-diagonal weight matrices respectively; A_{1i} , A_{2i} are blocks of A_1, A_2 with (1×1) dimensions, and the equation (3) represents the local stability conditions, imposed on all blocks A_{1i} , A_{2i} of A_1, A_2; B and C are $(n \times m)$ and $(l \times n)$-weight matrices; D is a $(n \times l)$- global output closed loop matrix; $S(x)$ is a vector-valued sigmoid activation function. The saturation function could be used as approximation of the sigmoid function to improve the RTNN architecture, [7]. The stability of the RTNN model is assured by the activation functions S and by the local stability conditions (3).

Simultaneously with the RTNN topology improvements, some advanced researches have been done on the methods of RTNN learning so to adapt the given in [6], [7], learning algorithm to the extended RTNN model, given by the equations (1), (2). The most common used BP updating rule, is the following:

$$W_{ij}(k+1) = W_{ij}(k) + \eta\Delta W_{ij}(k) + \alpha\Delta W_{ij}(k-1) \quad (4)$$

where: Wij is a general weight, denoting the ij-th weight element of each weight matrix (C, D, A_1, A_2, B) in the RTNN model, to be updated; ΔW_{ij}, $(\Delta C_{ij}, \Delta D_{ij}, \Delta A_{1,ij}, \Delta J_{2,ij}, \Delta B_{ij})$, is the weight correction of W_{ij}; η, α are learning rate parameters.

The weight corrections of the updated matrices in the discrete-time RTNN model, described by equations (1), (2), are given as follows:

- For the output layer:

$$R_1=[T_j(k)-Y_i(k)]\,Y_i(k)\,[1-Y_i(k)] \quad (5)$$

$$\Delta C_{ij}=R_lX_i(k) \quad (6)$$

$$\Delta A_{2,i}=R_lY_i(k) \quad (7)$$

Where: ΔC_{ij}, $\Delta A_{2,i}$ are weight corrections of the $ij-th$ elements of the $(l \times n)$ learned matrix C and an $i-th$ element of the $(l \times l)$ learned matrix A_2; T_j is an $j-th$ element of the target vector; Y_j is an $j-th$ element of the output vector; Xi is an $i-th$ element of the output vector of the hidden layer, R_1 is an auxiliary variable.

- For the hidden layer:

$$\Delta B_{ij}(k)=RU_i(k) \quad (8)$$

$$\Delta D_{ij}(k)=RY_i(k) \quad (9)$$

$$\Delta A_{ij}(k)=RX_i(k-1) \quad (10)$$

$$R=C_i(k)\,[T(k)-Y(k)]\,X_i(k)\,[1-X_i(k)] \quad (11)$$

Where: ΔB_{ij}, ΔD_{ij}, are weight corrections of the $ij-th$ elements of the $(n \times m)$ learned matrix B and the $(n \times l)$ learned matrix D; C_i is a row vector of dimension $(1 \times l)$, taken from the transposed matrix C'; $[T-Y]$ is a $(l \times 1)$ output error vector, through which the error is back-propagated to the hidden layer; U_i is an $i - th$ element of the input vector U; X_i is an $i - th$ element of the vector X; ΔA_{1i} is the weight correction of the $i - th$ elements of the $(n \times n)$ diagonal matrix A_1 under learning; R is an auxiliary variable.

4 Experimental Results

The multi-input multi-output RTNN model, used for process identification and prediction has ten inputs (BAC, ESP, GLU, TEM, pH, m, F, PA, PB, FA), six outputs (BAC, ESP, GLU, TEM, pH, F), sixteen neurons in the hidden layer, and also main and local feedbacks. The applied weight update learning algorithm is a version of the backpropagation through time one, specially designed for this RTNN topology. The described above learning algorithm is applied simultaneously for 5 normalized between 0 and 1 fermentation kinetics data, containing 23 points each. The fermentation kinetics variables are: BAC (concentration of bacteria's), GLU (glucose), ESP (concentration of spores), TEM (temperature), pH, m (specific growth rate), F (nutrient flux), PA, PB (acid and base pump states) and FA (air flux) at a initial concentration of total solids of $40g/l$. The mean square error obtained for the last epoch of learning is below of 2% and the total time of learning is 202 epoch size, corresponding to the number of data, is 115 iterations. After each epoch of training, the 5 sets of 23 points fermentation kinetics data are interchanged in an arbitrary manner from one epoch to another. An unknown fermentation of kinetic data, containing 23 points and repeated 5 times, is used as a validation (generalization) set. The graphical results of RNNM training and generalization, given in Fig. 1, 2, 3, 4. of the Appendix for last epoch of training, together with the mean square errors (MSE) of whole learning and generalization process shows a good convergence.

5 Conclusions

The paper proposed to use a new Recurrent Trainable Neural Network Model to identify and predict a fermentation process of Bacillus thuringiensis from fermentation kinetic data (concentration of bacterias-BAC, glucose-GLU, spores-ESP, temperature-T, pH, specific growth rate-m, nutrient flux-F, acid and base pump states, PA and PB, as well as the air flux-FA). The multi-input multi-output RTNN proposed, have ten inputs, six outputs, sixteen neurons in the hidden layer, and also global and local feedbacks. The weight update learning algorithm designed, is a version of the well known backpropagation through time algorithm, directed to the RTNN learning. The learning and generalization errors for the last epoch of learning are less than 2% and the total time of learning is 201 epochs, where the size of epoch is 115 iterations. The learning process is applied simultaneously for five fermentation kinetics data of different m (0.18, 0.23, 0.3 and 0.38 g/l) and good learning and generalization results has been obtained.

References

[1] J.Glassey, G.A.Montague, A.C.Ward and B.V.Kara. "Artificial neural network based experimental design procedures for enhancing fermentation development". Biotechnology and Bioengineering, pp. 397-405, Vol. 44, 1994.

[2] G.Muralikrishnan and M.Chidambaram. "Control of bioreactors using a neural network model. Bioprocess Engineering", pp. 35-39, Vol. 12, 1995.

[3] J.Thibault, V.Van Breusegem and A.Cheruy. "Online prediction of fermentation variables using neural networks". Biotechnology and Bioengineering, pp. 1041-1048, Vol. 36, 1990.

[4] Q.Zhang, J.F.Reid, J.B.Litchfield, J.Ren and S. Wu Chang. "A prototype neural network supervised control system for bacillus thuringiensis fermentation". Biotechnology and Bioengineering, pp. 483-489, Vol. 43, 1994.

[5] R.R. Farrera , "Carbon:Nitrogen ratio interacts with initial concentration of total solids on insecticidal crystal protein CryIA(c) and spore production in Bacillus thuringiensis HD-73". Appl. Microbiol Biotechnol. pp. 758-765, 49, 1998.

[6] I.Baruch, I.Stoyanov, E.Gortcheva. "Topology and Learning of a Class Recurrent Neural Networks", ELEKTRIK, Turkish Journal of Electrical Engineering and Computer Sciences, pp. 35-43, Vol. 4, 1996.

[7] I.Baruch, I. Stoyanov and T.Arsenov. "An improved RTNN model for dynamic systems identification and time series prediction". In: Proc. of the 4-th Int. Symp. MMAR'97, Aug. 26-29, 1997, Miedzyzdroje, Poland, (S. Domek, Z. Emirsajlov, R.Kaszynski, Eds.), pp. 711-716, TU of Szczecin, Poland, Vol. 2, 1997.

Appendix

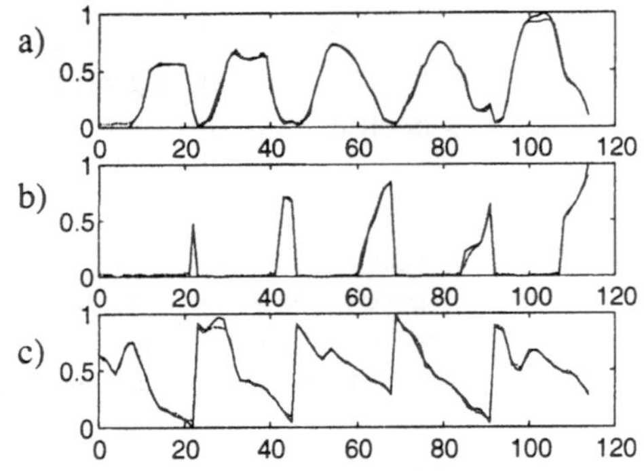

Fig. 2. B.t. fermentation process learning. Plant outputs (continuous line) and RTNN outputs (dashed line) for last epoch of learning. The epoch size is n=115, a) BAC, b) GLU, c) ESP.

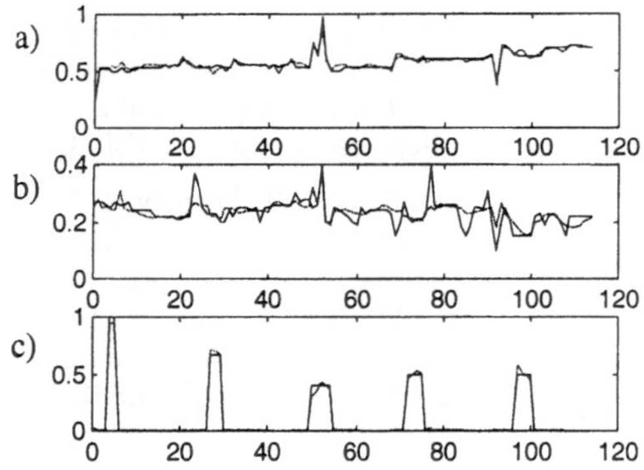

Fig. 3. B.t. fermentation process learning. Plant outputs (continuous line) and RTNN outputs (dashed line) for last epoch of learning. The epoch size is n=115, a)TEM, b) pH, c) F.

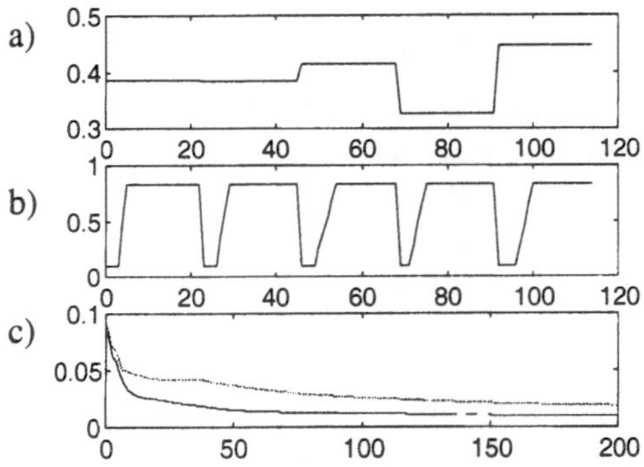

Fig. 4. B.t. fermentation process learning. Two plant inputs, the generalization and learning MSE The epoch size is n=115, a) MU, b) PB, c) generalization and learning MSE for all trainning time of N=201 epochs.

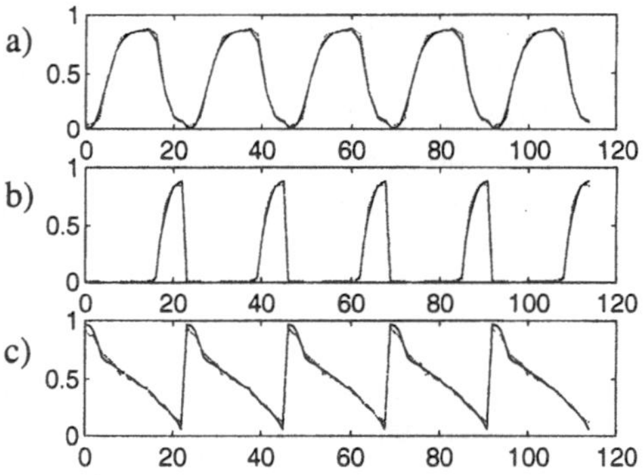

Fig. 5. B.t. fermentation process generalization. Plants outputs (continuous line) and RTNN outputs (dashed line) for last epoch of generalization; The epoch size is n=115 iterations; a) BAC, b) GLU, c) ESP.

A Draft on Modelling The Behaviour of Bioreactor Landfill Using Neural Nets

Lenka Landryova*, John P. Robinson†

*Faculty of Mechanical Engineering, VSB-Technical University of Ostrava, Czech Republic lenka.landryova@vsb.cz †Department of Environmental Health and Safety, QMW, University of London, UK j.p.robinson@qmw.ac.uk

Abstract

This draft reviews biological and technical aspects driving the development of the sustainable large-scale bioreactor concept of landfilling. At the present time, landfilling is still the most widely used disposal route in many European countries and it seems likely that it will continue to be an important waste management option for many years. Another consideration is the need to provide appropriate upgrading of waste management practices in those Central European countries wishing to join the EU.

1 Introduction: Making the Choice of Method

Neural networks are one of a group of intelligence technologies for data analysis that differ from other classical analysis techniques by learning about our chosen subject from the data we provide them, rather than being programmed by the user in a traditional sense. Neural networks gather their knowledge by detecting the patterns and relationships in our data, learning from relationships and adapting to change.

The range of functions that neural nets can perform is very large. In working out an application, the standard method is to break the overall function into blocks of code. These blocks, often called an algorithm, receive input data from outside the system (user input) or from other algorithms, perform some processing depending on the code written into it and then output data to outside the system (system output).

At the Department of Control System and Instrumentation, VSB-TUO, we have access to many standard algorithms which can be incorporated into programs for such tasks, analyzing a system's behavior, modeling it as a mathematical model, observing it in such way. However, sometimes there is no algorithm available for a certain type of system or its behavior.

A key benefit of neural networks is that we can use them to build a model of the system or subject we are interested from just the data we provide them. We know the inputs and outputs that are important but may not know what happens internally. Then the neural network will model this system for us from the data.

In engineering we are especially interested in these tasks, which neural nets can work out for us:

- machinery defect diagnosis, signal processing, character recognition process
- supervision, process fault analysis, speech recognition machine vision,
- speech recognition, radar signal classification

In addition to these, we are looking for tasks and applications which are connected to environment we live in. There has been interest in applying neural nets to gene recognition, botanical taxonomy and classification and bacterial identification. Since neural nets are used in engineering for such applications as machine and robot control and navigation as well as process control, there may also be possible applications in prediction and control of chemical and biological processes.

This draft reviews biological and technical aspects driving the development of the sustainable bioreactor concept of landfilling. At the present time, landfilling is still the most widely used disposal route in many European countries and it seems likely that it will continue to be an important waste management option for many years. Another consideration is the need to provide appropriate, upgrading of waste management practices in those former Eastern Bloc countries wishing to join the EU.

Given this reality a sustainable approach to landfill management is urgently required in order to minimise the environmental impact, particularly of landfill gases and leachates. The traditional model of a landfill as a permanent waste deposit, in which environmental impact is minimised by restriction of biological decomposition processes, has given way to the concept of landfill as a process managed as a large-scale bioreactor. The fundamental aim of the sustainable landfill is to optimise and control the natural degradation processes in the waste and to contain the products of degradation to prevent environmental pollution.

This controlled bioreactor landfill is seen as a flexible, cost effective, and sustainable waste management option which, when combined with on-site material recovery and post decomposition "landfill mining", represents a large-scale waste processing system rather than a waste deposit.

2 Modelling the Behaviour of Bioreactor Landfill

It is generally accepted that finding a satisfactory mathematical model for such processes as microbial decomposition is much more difficult compared with technical and mechanical devices. Although the general elements and basic biochemistry of the process are fairly well characterised, there are many parameters, some uncharacterised, whose effects on the process are not known. This makes classical numerical modelling difficult and uncertain and the number of reliable models in this area is very limited. Furthermore selection of appropriate sensors for on-line measurements in landfill systems is a problem due both to the large-scale and to the heterogeneity of the environment. Thus we have to work with limited information about both the process status and culture composition, both of which vary with time and location in the reactor.

Neural nets are an appropriate modelling tool when we have to work with or analyse data in different form, and the problems are either complex, laborious, 'fuzzy' or simply un-resolvable with present classical methods. Neural networks outperform current methods of analysis because they can successfully:

- deal with the non-linearities typical of the environment,
- be developed from data without an initial system model,
- handle noisy or irregular data from the real world,
- quickly provide answers to complex issues,
- be easily and quickly updated.

The feature we want to enforce is supervised learning. It is a process of training a neural network by giving it examples of the task we want it to learn. That is learning with a teacher. The way this is done is by providing a set of pairs of vectors (patterns), where the first pattern of each pair is an example of an input pattern that the network might have to process and the second pattern is the output pattern that the network should produce for that input which is known as a target output pattern for whatever input pattern. This technique is mostly applied to feed forward type of neural networks.

An algorithm that requires no code generation and little maintenance is a powerful and useful tool. The functions that neural nets perform overlap many performed by expert systems, but the development and maintenance times can be reduced tenfold with neural nets.

There are, however, some tasks we have to deal with, the two main issues are:

- Data pre-processing
- Network architecture design.

Some experimentation was required in order to prepare data suitable for our problem. Too much data is rarely a problem, but too little or ambiguous data may prevent the network from finding a result. The issue of data pre-processing is a big topic in its own right.

3 The Sustainable Landfill

The fundamental aim of the sustainable bioreactor landfill is to optimise and control the natural degradation processes in the deposited waste and to prevent pollution of the environment by containing and treating the products of degradation. The process is represented schematically in Figure 1.

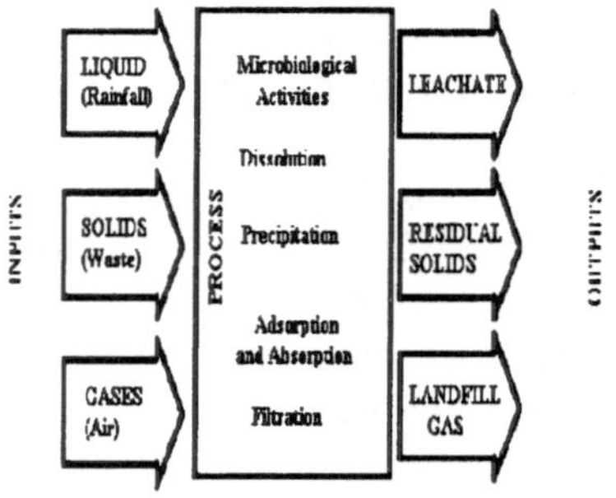

Fig. 1. Natural decomposition process

Using current landfill management methods the time required to achieve stabilisation is variable and is usually of the order of decades. The aim of a sustainable landfill management regime is to initiate methane formation and achieve maximum rates of waste degradation as soon as possible after deposition. This optimises the economics of methane recovery and stabilises the landfill quickly so that the land can be returned to amenity use or the residual material extracted, referred to as landfill mining, and the void space can be re-used for waste disposal.

The decomposition processes in landfilled waste depend on the type of waste, particularly the proportion of degradable organic compounds, the size of the particles and on the water content. In general a high water content combined with small, homogenous particle size and a high proportion of biodegradable organic compounds, leads to rapid microbial decomposition.

The anaerobic degradation of organic matter under anoxic conditions to methane and carbon dioxide can be represented by the following equation:

$$CaHbOcNdSe + [a - \tfrac{b}{4} - \tfrac{c}{2} + \tfrac{3d}{4} + \tfrac{e}{2}]H_2O$$

$$\Downarrow$$

$$[\tfrac{a}{2} + \tfrac{b}{8} - \tfrac{c}{4} - \tfrac{3d}{8} - \tfrac{e}{4}]CH_4 +$$

$$[\tfrac{a}{2} - \tfrac{b}{8} + \tfrac{c}{4} + \tfrac{3d}{8} + \tfrac{e}{4}]CO_2 +$$

$$dNH_3 + eH_2S$$

Some average elemental compositions of waste components are:

- Municipal solid waste $C_{99}\ H_{27}\ O_{59}\ N$
- Paper $C_{203}\ H_{334}\ O_{138}\ N$
- Food waste $C_6\ H_{27}\ O_8\ N$

The moisture content of waste has a major effect on the rate of gas formation. Between 25 and 70% moisture the rate of gas formation increases approximately 1000-fold (Rees et al). This suggests that water content of the waste should be maximised to achieve rapid, complete decomposition, gas formation and consequent stabilisation of the waste. The requirement for a high moisture content in the waste means that leachate containment and management systems are a particularly important aspect of sustainable landfill design and operation.

4 Microbial Processes in Landfill

Effective management of a sustainable bioreactor landfill requires an appreciation of both the engineering imperatives and the biochemical processes that occur in the waste.

In the initial stages of decomposition the fraction of organic matter that dissolves readily in the leachate will be metabolised to carbon dioxide in oxygen-dependant respiration. This is rapid and causes the oxygen concentration in the waste gas phase to fall quickly to zero and the temperature to increase. After the oxygen is used, decomposition is carried out by fermentative bacteria which form a range of fermentation products which are reduced organic compounds, such as alkenic acids and alcohols in addition to hydrogen and carbon dioxide. During this stage of decomposition the leachate becomes acidic due to the high concentration of volatile fatty acids in the leachate. The conversion of these acids into methane and carbon dioxide is a key step in the process.

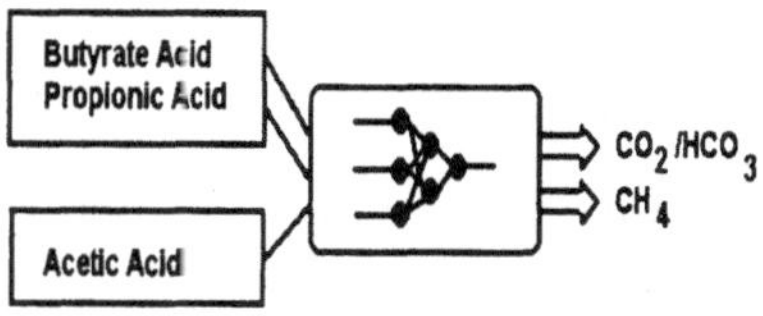

Fig. 2. The idea behind implementing the neural network

A simplified schematic representation of the microbiological breakdown of the fermentation acids under anaerobic conditions can also be described as a table of inputs and outputs of the process and fed into a neural network.

Acetic Acid	Propionic Acid	Butyric Acid	Bicarbonate
1700	2553	2136	1724
1277	2590	599	1164
996	2234	356	5287
1117.8	1791.4	179.5	2027
1234.4	1895.7	100.5	2101
631.3	1516.6	35.3	2994
⋮	⋮	⋮	⋮

Table I. Format of inputs and outputs of the process as they were measured and stored in Excel spreadsheet

The design of a network can be done in Matlab, neural net toolbox, or other suitable neural networks tools used at our department. Properly trained backpropagation networks tend to give reasonable answers when presented with inputs that they have never seen. Typically, a new input will lead to an output similar to the correct output for input vectors used in training that are similar to the new input being presented. This generalisation property makes it possible to train a network on a representative set of inputs target pairs and get good results without training the network on all possible input/output pairs. The number of hidden units and the number of layers can be set by the user. The number of inputs and outputs is extracted directly from our data. The transfer function type of each layer can also be set by the user. For example: the following command will create a two layer network. There will be one input vector with two elements, three neurons in the first layer and one neuron in the output layer. The transfer function in the first layer will be tan-sigmoid, and the output layer transfer function will be linear. The training function will be traingd [Demuth, Beale].

```
net=newff([-1 2;0 5],[3,1], \
      {'tansig', 'purelin'}, 'traingd');
```

The environment of MATLAB is user friendly and allows creating many different types of the network:

- Matlab Editor : configuring the neural net with its parameters given according the type of the net
- Matlab Command Window: looking through the
- results, if the trained and actual outputs are the same

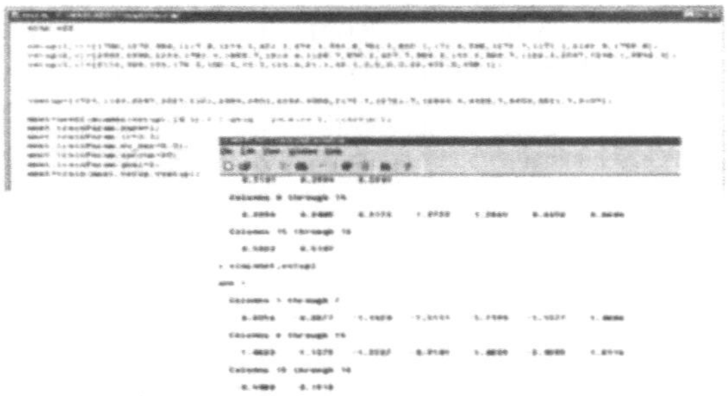

Fig. 3. Working with Neural Network Toolbox

5 Conclusion

In this draft, we have described some aspects of waste management with its essential problems considering modeling the process and data acquiring. The cost of alternatives to landfilling is high and has technical and economic constraints. For instance in the UK incineration plants with energy recovery handling less than 100,000 tonne.annum-1 are unlikely to be economic. Another important strategic consideration is the effect that increased recycling, reuse and pre-treatment will have on the composition of the waste stream. The composition of waste will change at rates which are difficult to predict and waste management authorities will have to take account of changes in planning disposal facilities. In this context landfill bioreactors using artificial neural net modeling have the potential to provide a flexible, environmentally acceptable approach to and a basis for decision making about waste disposal for some decades.

The fields of applications of neural networks are limitless. In many situations, artificial neural networks are best solution, especially when it comes to dealing with non-linear and complex problems. The contribution was written in the framework of grant No. 105/01/0311 of GA CZ

References

[1] Rees, J.F & Grainger, J.M,, "Rubbish dump or fermenter? Prospects for the control of refuse fermentation to methane in landfills", *Process Biochemistry*, vol. 17, pp. 41–44, 1982.

[2] Landryova, L. & Robinson, J.P., "Landfill as a Sustainable Bioreactor: A Waste Management Option For the Twenty-First Century.", *In: Proceedings of International Carpathian Control Conference Podbanske, Slovakia: : Faculty BERG Technical University of Kosice, 23–26 May, 2000*, pp. 705–709, ISBN 80-7099-510-6.

[3] Landryova, L. & Zolotova, I., "Integrating Methods of Artificial Intelligence Into Control.", *In: Proceedings of International Carpathian Control Conference Podbanske, Slovakia: : Faculty BERG Technical University of Kosice, 23–26 May, 2000*, pp. 447–451, ISBN 80-7099-510-6.

[4] Hunt K.J., Irwin G.R. & Warwick, K. "Neural Network Engineering in Dynamic Control Systems. Advances in Industrial Control.", *Springer Verlag London.*, ISBN 3-540-19960-8. 1995.

[5] Demuth, H., Beale, M. "Neural Network Toolbox User's Guide." For use with MATLAB.

Neural Network Combustion Optimisation in Naantali Power Plant

Jussi Mäkilä* [1], Juha-Pekka Jalkanen†

*Fortum Engineering Ltd, POB 10, 00048 Fortum, Finland, e-mail: jussi makila@fortum.com †Fortum Engineering Ltd, POB 10, 00048 Fortum, Finland, e-mail: juha-pekka.jalkanen@fortum.com

Abstract

An optimisation system giving advice to the operators is installed to unit 2 of a coal-fired power plant located in the city of Naantali, Finland.

The basis of the optimisation is a model of the steady state behaviour of the boiler. Due to the complexity and severe non-linearity of the process, a neural network approach was chosen.

Control inputs to the boiler model include oxygen in the flue gas, the percentage of burning air fed as over fire air, the coal/air -ratio and the tilting angle of the burners. Outputs of the model are the boiler efficiency, the level of NO_xemissions, the unburned carbon in the fly ash and the maximum material temperature in the reheater tubes of the boiler.

The optimiser operates in an advisory mode allowing the operators to acquire optimal setpoints for control variables. It can either optimise any of the outputs while specifying limits for the other outputs and allowable values for the control variables. Or with the what-if simulation, the operator can study the effect of changes in the control variables in advance.

1 Introduction

1.1 Naantali Power plant

The project was carried out at the Unit 2, a pulverized coal-fired boiler at Naantali power plant located in the city of Naantali, Finland. The boiler is of tangential firing type and has a capacity of producing 420 tons of steam per hour. It has three levels of tilting burners each fed by a separate coal mill. The combustion air is staged to reduce the nitrogen oxide emissions.

1.2 Optimisation goals

The purpose of the optimisation is to maximise the boiler efficiency while simultaneously keeping the plant emissions as low as possible, assuring the quality of the fly ash and preventing too high material temperatures in the boiler tubes.

The optimiser is designed to operate in an advisory mode allowing the operators to acquire optimal setpoints for control variables. It can either optimise any of the outputs while specifying limits for the other outputs and allowable values for the control variables. This optimisation is performed so that the resulting control recommendations are always in the operating region well known to the boiler model. Or with the what-if simulation, the operator can study the effect of changes in the control variables in advance.

2 Modelling

The basis of the optimisation is a model of the steady state behaviour of the boiler. The process to be modelled is both complex and severely non-linear.

2.1 Modelled variables

Nitrogen oxides: The NO_x-emission level at the boiler exit is measured directly as ppm (parts per million), but emission limits in Finland are defined in mg per MJ energy produced. A conversion from ppm to mg/MJ taking into account the oxygen content in flue gases has to be done, as for the same net power an increase in O_2-level also increases the flue gas flow. This conversion is done according to a standard formula given in Energia-aapinen by Energia-Ekono [3].

Efficiency: The efficiency of the boiler is modelled through two of the most important losses i.e. flue gas loss and loss due unburned carbon. Of these the flue gas loss is the dominating one and can be calculated and modelled based on the flue gas temperature and flow.

Unburned carbon in fly ash: Unburned carbon in fly ash or LOI (loss of ignition) is the amount of carbon that does not burn in the boiler. A high level of LOI thus lowers the efficiency of the boiler

[1]Fortum Service, Fortum Engineering, Fortum Corporation and TEKES funded the project. Project team acknowledges personnel at Naantali power plant from co-operation and help. Operators and other personnel at the plant have been very helpful.

but more importantly it affects the possibility of selling the ash to be used for example in cement industry. So far the attempts to measure LOI directly on-line have not given satisfactory results. To measure the LOI content for modelling, samples were taken from the dust silo of the electric precipitator.

Maximum material temperature: To avoid too high material temperatures in reheater tubes as a result of optimisation the effect of control variables to the maximum material temperatures had to be modelled.

2.2 Control variables

These are the variables that can be adjusted by the plant operators within a operating window to achieve improvement in the boiler outputs described in the previous section. In addition to these, there are other factors (like boiler power and mill combination) that affect the outputs and must therefore be included to the model as inputs.

Excess oxygen in flue gas: The burning air flow is controlled by the plant DCS to keep the O_2-level in the flue gas at a specified setpoint.

Burner tilt: All burners can be tilted vertically from -25 to +25 degrees. The tilt angle is same for all burners.

Over fire air / total burning air -ratio: Over fire air (OFA) percentage is the ratio of the air led to a level above the burners to the total air amount. Use of over fire air significantly reduces NO_x-emissions.

Primary air amount: Primary air is the airflow that carries the pulverised coal form the pulverisers to the burners. Primary air flow is controlled according to mill load by the DCS, but the operator can make limited adjustments to it.

2.3 A priori knowledge

Table I indicates the most significant effects of control variables to the modelled boiler outputs, known from general power plant experience. This knowledge was utilised during the processes of data collection and model contruction.

	O_2	Tilt	OFA	Pr.a.
Eff.	-			-
NO_x	+	+	-	+
LOI	-			
Temp.		+		

Table I. Dominant effects of control variables to modelled variables, known *a priori*.

2.4 Collection of the data

The data required for constructing the model was gathered with test runs as normal plant operation does not contain enough variation for all the dependencies to be estimated. The tests had to be performed without disturbing the normal operation of the plant. This was done during spring season when there is a lot of natural variation in power demand.

Test plan was a modified 3^k-method because of the non-linear process. On the other hand the number of time-consuming tests had to be limited and the plant production requirements had to be taken into account. Therefore only the most important variables were tested in a 3^k approach. The total amount of individual tests made at the plant still rose up to over 100.

2.5 Neural networks

Neural networks were chosen as the modelling tool because of the complexity and non-linearity of the process.

To avoid typical pitfalls, process knowledge was already applied when carefully selecting the model inputs and outputs and gathering the data used to train and test the models. During the actual model construction extra care was taken to assure the quality of model predictions outside the clusters made up by the test points.

In the partitioning of data into training and testing sets whole clusters were excluded from the training set to get a better estimate of the real performance of the neural network model in capturing the boiler behaviour.

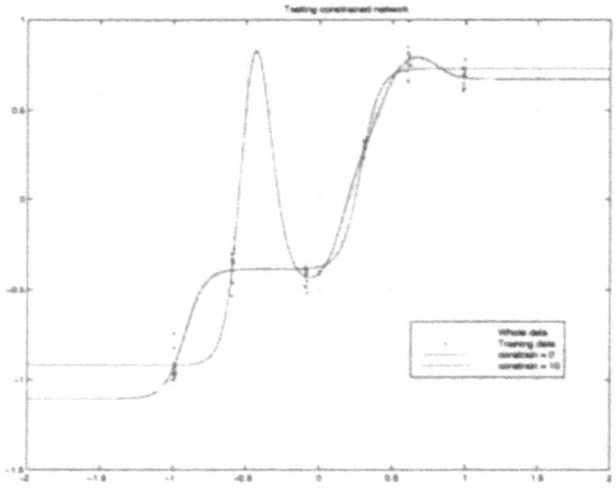

Fig. 1. Due the modified learning method neural network performs smoothly and does not overlearn teaching data so easily.

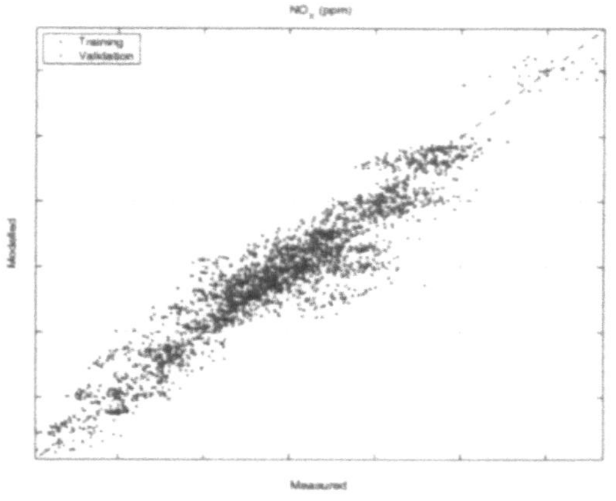

Fig. 2. Results obtained when modelling the NO_xlevel, represented as actual vs. modelled xy-scatter.

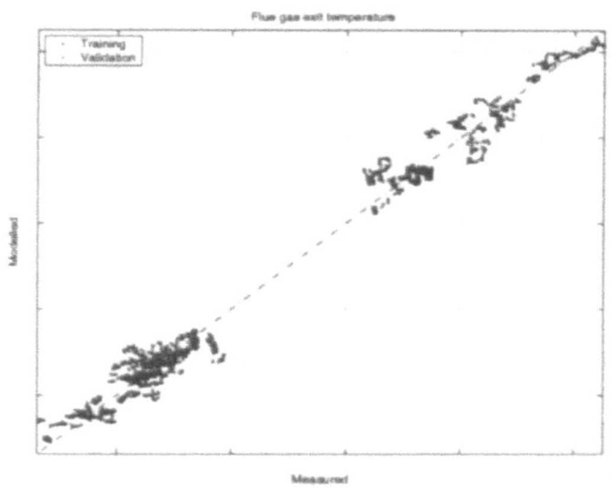

Fig. 3. Modelled flue gas exit temperature versus measured. Model extrapolates well at higher temperatures.

To additionally limit the inherent tendency of NN models to generate undesired nonlinearity (data overfitting), a training method with regularisation [4, 2] was used. In this case the training method was further tailored to suit the relatively sparse test sceme used to acquire the modelling data. This nonsymmetric regularisation method allows also to some extent *a priori* knowledge about the behaviour of the system to be incorporated to the network during the training.

With the training method the non-linearity of the neural network can clearly be settled down, as shown in figure 1. As a result, the neural network models aquired are still non-linear but perform more smoothly and their interpolation properties are improved.

2.6 Modelling results

The performance and behaviour of models was examined both with error measures (for training and testing sets) and several visual examination methods. The model accuracy was considered to be adequate for the coming use.

Figures 2 and 3 show as an example modelling results two boiler outputs as scatter plots. Due to constraining of the non-linearity the behaviour of the neural network model is smooth, which was also confirmed by graphical inspections. From figure 3 it can be seen that under suitable conditions the models can even extrapolate.

3 Optimisation Engine

The standard notation for an optimisation problem [1] shall be used:

$$\min_{x} f(x) \quad \text{when } G(x) \leq 0 \text{ and } H(x) = 0 \tag{1}$$

The target function f maps the vector x to a scalar (cost) value. This value is minimised subject to the inequality and equality constraints expressed by the vector valued functions G and H. If all these functions are allowed to be non-linear then a general constrained non-linear programming problem is to be solved.

In the boiler optimisation problem the vector x consists of the four controlled variables. The boiler models (based on neural networks, described above) are included in both functions f and G. These functions are parameterised with parameter vectors a and b, respectively, resulting in $f(x;a)$ and $G(x;b)$ to allow the same functions to be used in different situations. Equality constraints (H) are not used.

This problem is solved with an iterative algorithm, that uses derivative information, but the gradient of f or the Jacobians of G or H need not be calculated analytically. This is required for the utilisation of neural network models inside the functions.

In order to make the operator's task easy, the user inputs required for parameter vectors a and b have been minimised. Pre-set parameter combinations for a are used based on the optimisation target choices that are available to the user. Vector b for the inequality constraints contains the control and output limitations. Only a limited set of user defined values for output limitations are needed and these are specified by the user in the form of allowable changes in the boiler outputs. All of the control limitations are automatically calculated based on the current boiler operation point.

One of the elements of G is used to limit the feasible area of the optimisation problem to cover only the area where the models are considered to be reliable. This acts as a safeguard preventing possible problems due to extrapolation at points too far away from the region that was covered with process experiments.

4 Implementation

The user interface of the optimizer is integrated to the plant information system while the actual processing is performed on a separate Windows NT-workstation. Information is shared trough the plant LAN.

4.1 Graphical user interface

The graphical user interface consists of two different displays in the plant information system; one display for optimising purposes and one for what-if (simulation) analyses.

The optimising display enables operators to choose among different optimisation goals that are minimisation of NO_x-emissions, maximisation of boiler efficiency and minimisation of unburned carbon in fly ash.

In addition to choosing the optimisation task, operator can introduce constrains to optimisation engine i.e. limits for allowable changes in NO_x-emissions, unburned carbon in fly ash, boiler efficiency and maximum material temperature. Any control variable can also be excluded from optimisation, thus freezing it to the current value.

Operators are able to inspect the effects of the changes in control variables with the help of simulation display. In this simulation display the operators set values for the control variables. These values are fed to the boiler model to calculate the corresponding outputs.

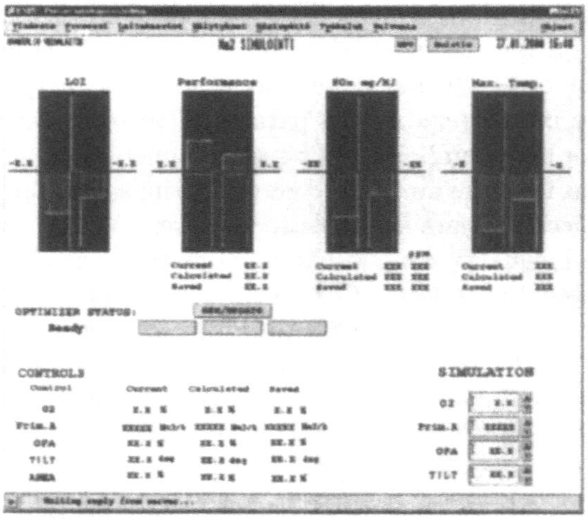

Fig. 4. The simulation display of the optimiser.

4.2 Operation experiences

The optimisation system was under field tests and a trial period during winter and spring 2000. In long-term monitoring the performance of the boiler model was found to be good, especially concerning the predictions for NO_x and boiler efficiency. Modelling of unburned carbon in fly ash and material temperatures was more difficult, but these can in all cases be used at least as limiting factors during the optimisation.

The tests have not yet been finished as the use of optimisation at larger boiler loads is yet to be verified. Final results from the optimisation benefits will be ready after another test period that is to be performed during winter 2001. So far it seems to be possible to minimize NO_x-emissions without significant loss on efficiency and vice versa.

5 Conclusions

This introduced system uses neural network modelling to capture the steady state behaviour of a boiler. An optimisation engine working based on the model can be used to calculate optimal setpoints regarding the goals specified by the operator.

During practical testing the model has so far proven to be of adequate accuracy. For example concerning the NO_x-emissions the model is so accurate that with the described optimising system emissions could be controlled as any other controllable variable in the plant.

References

[1] Mokhtar S. Bazaraa, Hanif D. Sherali, and C. M. Shetty. *Nonlinear Programming; Theory and Algorithms.* John Wiley & Sons, second edition, 1993.

[2] Howard Demuth and Mark Beale. *Neural Network Toolbox User's Guide.* The MathWorks Inc., 1998.

[3] Energia-Ekono Oy. Ekonon energia-aapinen.

[4] Simon Haykin. *Neural Networks; A Comprehensive Foundation.* Prentice Hall, 1994.

Control Sensitivity SVM for Imbalanced Data A Case Study on Automotive Material

K.K. Lee, C.J. Harris, S.R. Gunn*, P.A.S. Reed†

*Department of Electronics and Computer Science, University of Southampton, Highfield Road, Southampton SO17 1BJ, Email: [kkl98r;CJH;SRG]@ecs.soton.ac.uk †Material Research Groups, School of Engineering Sciences, University of Southampton, Highfield Road, Southampton SO17 1BJ, Email: PASReed@soton.ac.uk

Abstract

In many classification problems the data is imbalanced, that is the class priors are different. Here we consider the classification problem of fatigue crack initiation in automotive camshafts, where this imbalance is significant. The standard averaging technique used to access the performance of a model is inappropriate for imbalanced data and therefore the geometric mean, was used to evaluate the performance of the model. It has been shown elsewhere that the original SVM estimate concurs with that of the Bayes optimal decision rule. As such, a comparison was investigated using Support Vector Machine (SVM) and Controlled Sensitivity (CS) SVM using two different training sets, with different class ratios (1:8 and 1:1) between the "crack" and "no crack" respectively. Result show that the obtained balanced training set gave improved performance for the SVM. Alternatively, using imbalanced training data the CS SVM outperformed the SVM. Although, the computation speed for balanced data is faster, however, the emphasis in this application is for model performance, as such, the CS SVM with imbalanced produced an average estimated generalisation performance of over 71%.

1 Introduction

With appropriate heat treatment, Austempered Ductile Iron (ADI) provides good resistance to rolling fatigue, high strength and good wear resistance. This makes it a suitable candidate for camshafts used in automotive industries. However, there is a tradeoff between high strength and fatigue cracks. As such, it is important to investigate why the crack was initiated from the graphite nodule. Clearly, the number of "no crack" nodules exceeds those of "crack" nodules and consequently the data is imbalanced. The graphite nodule size and/or distribution morphology can be obtained from Finite Body Tessellation (FBT). These are used as the features for the classifier to investigate upon causing the fatigue cracks initiation. The ultimate goal is to understand their effect on the existence of fatigue crack.

The best classification rule is obtained from Bayes rule when the posterior probabilities of the classes are equal. In most real world applications, such as medical diagnosis, the data set is often limited within one class which is under-represented as compared to the other class. In the diagnosis of cancer, less data can be obtained from those with cancer than those without cancer. Usually, the cost of misclassification associated with the under-represented class are usually more severe than the heavily-represented class (i.e. diagnosing a patient as not having cancer when actually they do). The use of the standard averaging technique used for measuring the performance is not applicable in imbalanced data. An unequal cost can be incorporated to each misclassification of the diagnosis in order to distinguish more appropriately between false negative and false positive results. This addresses the problem caused by imbalanced data in predicting fatigue cracks in camshaft. The structure of this paper is as follows: Sections 2 describes how the performance can be evaluated. Section 3 describes the SVM; Section 4, provides the model specification and section 5 the simulation results.

2 Performance Criteria for Imbalanced Data

A confusion matrix visualises the performance of most classification problems. It consists of the number of points in the data set corresponding to four categories: False Positive (FP), False Negative (FN), True Positive (TP) and True Negative (TN); TP and TN are the correct prediction. Due to the curse of the imbalance of data, the standard performance criteria such as the average accuracy of the test set is not applicable in this case. For example the system with an average accuracy of 70% may be dominated by the performance of the class with largest prior. In imbalanced data applications, the prediction from the minority class is usually more important. As such, the above system cannot be used and an alternative based upon the imbalance and domain knowledge must be used.

Receiver Operating Characteristic (ROC) analysis measures the classifier performance over the whole range of thresholds from 0 to 1. This performance is based on measuring the sensitivity(Se) and specificity (Sp). Sensitivity and specificity define the performance in terms of predicted classifications within each of the true classes (FP and FN):

$$Sensitivity, Se = \frac{TP}{TP + FN}; \qquad Specificity, Sp = \frac{TN}{TN + FP} \tag{1}$$

The average accuracy of the test set is then the summation of Se and Sp. ROC curve describes the trade-off between sensitivity and specificity with the plot of TP against FP. The ROC curve then allows us to represent simultaneously the classifier performance by two degrees of freedom for a range of possible classification thresholds. Within the ROC curve, a point that has a high value when the TP and TN of the classifier is balanced is known as geometric mean [2]. This essentially reduces the two degrees of freedom in ROC curve to a scalar and it is define as:

$$GMean = \sqrt{TN.TP} \tag{2}$$

The motivation for using the geometric mean as our measure of performance is that we would want the performance of the crack and no crack class to have similar results.

3 Support Vector Machine (SVM)

SVM was developed based on the idea of *"Structural Risk Minimisation"* (SRM) [3]. Given a training set $D = \{(\mathbf{x}_i, y_i)\}_{i=1}^{N}$ with input $\mathbf{x}_i$ and output $y_i \in \{\pm 1\}$, the optimal hyperplane can be obtained by maximising a margin which is defined as the minimum distance from the hyperplane to the closest point. The optimal hyperplane weight ($\mathbf{w}$) is given by:

$$\mathbf{w} = \sum_{i=1}^{n} \alpha_i y_i \mathbf{x}_i \tag{3}$$

where α_i are known as the *Support Vectors* which can be obtained by solving a quadratic program (QP) by maximising:

$$Q(\alpha) = \sum_{i}^{n} \alpha_i - \frac{1}{2} \sum_{i,j}^{n} \alpha_i \alpha_j y_i y_j (\mathbf{x}_i, \mathbf{x}_j) \tag{4}$$

subject to the constraint $\alpha_i \geq 0$, $\sum_i^n \alpha_i y_i = 0$. For the case where a linear boundary is inappropriate, a *Kernel*, can be used to transform the input vector to a higher dimensional space via non-linear transformation. Kernel methods calculate the nonlinear transformation from the input space explicitly, only the dot product of the input vector is required. For the case where the data are non-separable in the feature space, a slack variable, ξ, is introduced to allow some misclassification error and a capacity control, C, for controlling the tradeoff between complexity of decision boundary and the number of errors allowed. The solution for the QP is similar to that of separable case except that it has a upper limit on α (i.e. $0 \leq \alpha \leq C$).

Splitting the C according to the respective classes implies that the misclassification cost associated with each class is different [4]. This is known as CS SVM. This leads to a minimisation problem,

$$\phi(\mathbf{w}, \xi, \xi^*) = \frac{1}{2}||W||^2 + C^+ \sum_{i|y_i=+1}^{n} \xi_i + C^- \sum_{i|y_i=-1}^{n} \xi_i^* \tag{5}$$

subject to the constraint:

$$\begin{aligned} y_i(\mathbf{w}^T\mathbf{x} + b) &\geq 1 - \xi_i - \xi_i^* \\ \xi_i \, , \, \xi_i^* &\geq 0 \end{aligned} \tag{6}$$

where C^+ and C^- are the misclassification costs associated with positive and negative classes. This is suitable for our imbalanced data for fatigue crack prediction in camshafts as the misclassification costs of each class are different. Furthermore, there is a relationship between misclassification cost and class size. Eq. 5 and 6 can be solved with a simple modification to Eq. 4.

4 Model Specification

The ADI materials data set for the automotive camshaft application contains a total of 2923 examples of which 116 samples are crack initiation sites ("Crack" class) while 2807 samples did not act as crack initiation sites ("No Crack" class). These data were obtained from a FBT of ADI [5]. A set of nine measurements relating to the spatial distributions and measures of the object (graphite nodules) were obtained from the tessellation. This set of nine features describe the prior domain knowledge of the microstructural distributions e.g. morphology of secondary particles. FBT involves three

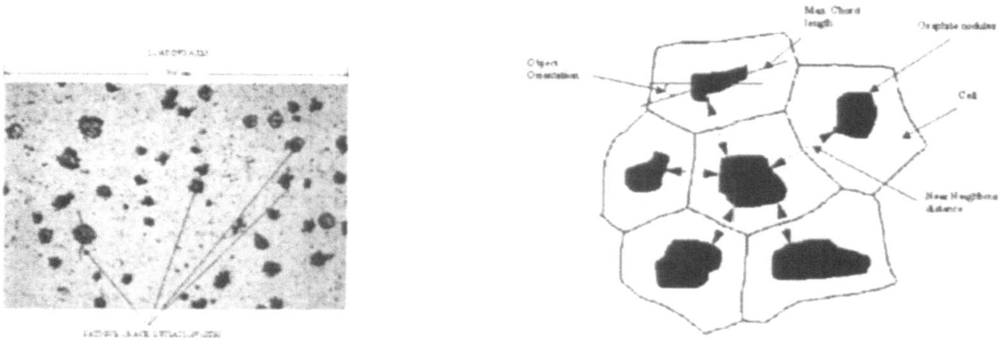

(a) (b)

Fig. 1. (a) Original microstructural Image (b) FBT Image.

Classification Performance		Data Set 1			Data Set 2		
Approaches	Kernels	TP	TN	Gmean	TP	TN	Gmean
SVM	Spline	0.36 $C^+ = 1000$	0.94 $C^- = 1000$	0.58	0.68 $C^+ = 1000$	0.76 $C^- = 1000$	0.72
	RBF (σ=0.5)	0.39 $C^+ = 1000$	0.90 $C^- = 1000$	0.59	0.56 $C^+ = 1$	0.83 $C^- = 1$	0.68
CS SVM	Spline	0.71 $C^+ = 1$	0.78 $C^- = 0.1$	0.74	0.67 $C^+ = 1000$	0.77 $C^- = 10000$	0.72
	RBF(σ=0.5)	0.76 $C^+ = 1$	0.75 $C^- = 0.1$	0.75	0.56 $C^+ = 1$	0.83 $C^- = 1$	0.68

Table I. Summary of the test result using two different data sets for training on the SVM and CS SVM. TP and TN are the true classification rate for "crack" and "no crack" class calculated by averaging the five randomly selected data set in order to provide a good generalisation.

stages, binarisation of images, a distance transformation and a watershed transformation [6]. During these stages, noise in the background is attenuated and holes within bodies are filed. Then the features for each nodule for learning are generated from the following measurements: 1. Nodule area; 2. Nodule aspect ratio, 3. Nodule angle, 4. Cell area surrounding the nodule, 5. Local cell area fraction, 6. Number of near neighbours, 7. Nearest neighbour distance, 8. Mean near neighbour distance and Nearest neighbour angle. Note: that the near neighbour cells are defined as the cells that are sharing the same boundaries. Fig. 1 a & b show the microstructural image before and after FBT.

Prior to using the different approaches to classify the graphic nodule, the input features is normalised. This will ensure that the input feature is restricted to a unit domain and it provide no bias on the significant for each feature. Upon normalising, the data needs to be partitioned for training and testing. Due to the extensive computation time required for large data, we consider a reduced data set (data set 1) which consist of 700("no crack") and 90("crack") for training and the rest of the data for testing. This is repeated five times with random selection of the data each time to provide good generalisation.

5 Results

The spline and the radial basis function (RBF) kernels were used, and the capacity control values were sample logarithmically on $[0.01, 10000]^2$. Better results may be possible with a greater and more intelligent sampling methods. Several widths of RBF was used and the width of 0.5 were selected as it provide the best result. The Geometric mean was used to select a point on the ROC curve where the TN and TP are balanced.

Results obtained using Data set 1 show that the SVM has high classification rate on TN class. When the CS SVM was used, the TN was reduced by about 10% but the TP sensitivity increased by 50%. This increased the overall performance of our model based on geometric mean by 15% with both TN and TP about 71%. C has a larger value in the SVM than in the CS SVM. This indicates that while using the SVM, the minimisation emphasises on the misclassification rather than maximising the margin (see table I). Notice that the use of different kernels has not changed the performance significantly. The misclassification cost between the crack and no crack were 0.1 and 1 respectively. This suggests that the ratio between the misclassification cost may be due to the imbalanced data that we have used for training. As such, a second data set, Data set 2, which only reduced the "no crack" class training size to 120 (similar to that of the "crack" class) was investigated.

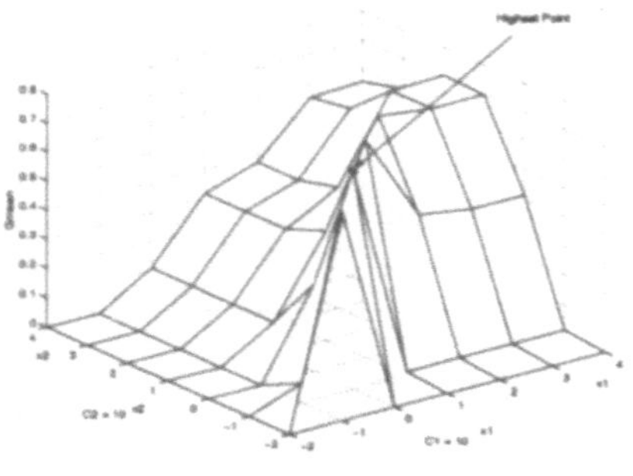

(a) Data set 1

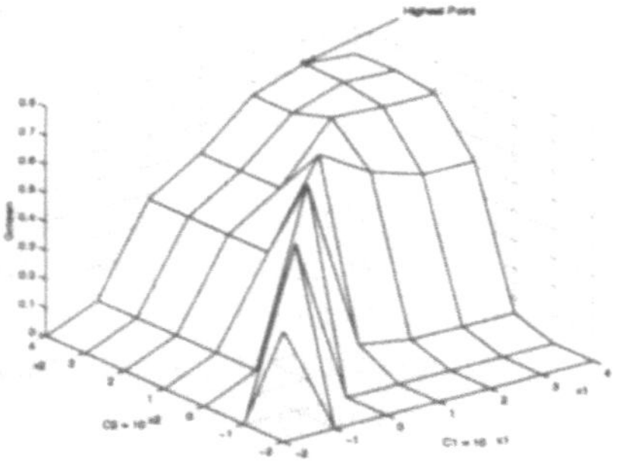

(b) Data set 2

Fig. 2. Surface plot of the geometric mean result for Spline Kernel using different sizes of training data with L1-norm error. C1 & C2 are the "crack" and "no crack" misclassification cost and its scale are in logarithmic increment. The plot of the RBF (σ=0.5) is very similar to this.

Using a more balanced ratio between the classes for training increased the performance of the SVM. The significant increase in performance of the CS SVM as highlighted for data set 1 is no longer present. The best results lie around C's which have much higher values than the previous data set (see figure 2) . This is reasonable because we would expect an increase in C to compensate for a reduced number of misclassifications. The use of spline kernel showed approximately 5% better performance than that of RBF.

6 Conclusion

Transforming the ratio of each class for training to be almost equal increases the performance of the SVM. This is advantageous for computation speed. The kernels used show consistent result in both training sets. The use of the balanced data shows that the result is consistent with that of [1], showing that the SVM estimate coincides with that of the Bayes optimal decision rule. In this application the emphasis is on the model performance, as such, the CS SVM of the imbalanced data set is preferred as it provides a well balanced classification, with an average estimated generalisation performance of over 71% for each class.

Acknowledgments

I would like to acknowledge the financial support from Federal Mogul Technology. I would like to thank Tony Dodd for valuable discussions.

References

[1] M. Kubat, R. Holte, and S. Matwin, "Learning when negative examples abound," Machine Learning, vol. 30, pp. 195-215, 1998.

[2] C. Burges, "A tutorial on support vector machines for pattern recognition," Data Mining and Knowledge Discovery, vol. 2, pp. 121-167, 1998.

[3] K. Veropoulos, C. Campbell, and N. Cristianini, "Controlling the sensitivity of support machines," Proceedings of the Int. Joint Conf. on Artifical Intelligence (IJCAI99), Sweden, 1999.

[4] R. Hockley, D. Thakar, J. Boselli, I. Sinclair, and P. Reed, "Effect of graphite nodule distribution on 'crack' initiation and early growth in austempered ductile iron," Small Cracks Mechanics and Mechanisms, 1999.

[5] J. Boselli, P. Pitcher, P. Gregson, and I. Sinclair, "Secondary phase distribution analysis via finite body tessellation," Journal of Microscopy, vol. TM 140, 1998.

[6] Y. Lin, Y. Lee, and G. Wahba, "Support vector machines for classification in nonstandard situations," Tech. Rep. 1016, University of Wisconsin, March 2000. 4

Analysis of Defectoscopy Data to Be Used by Neural Classifier

Jan Grman, Rudolf Ravas, Livia Syrova *

*Department of Measurement, Slovak University of Technology, Ilkovicova 3, 812 19 Bratislava, Slovakia; Phone:++421 7 654 29 600; E-mail:janog@pluto.elf.stuba.sk

Abstract

At present a very perspective solution of indications classification in defectoscopy is neural network application. One of the fields is classification of indications into classes that are characterized by the signal shape, or by the signatures relating to the signal shape. Nondestructive defectoscopy of steam generator tubes of nuclear power plants by multifrequency eddy current method is the field in which the use of classifiers based on neural network architecture is very perspective.

The contribution concentrates on the choice of a suitable representation of indications for neural classifier represented by probabilistic neural network. Selected representations are compared using real records of steam generator tubes and also using artificial defects and imitations of construction elements.

1 Introduction

Eddy current testing is one of the methods of nondestructive testing. In our case it is testing of heat-exchanger tubing using a differential probe. Tubes are made of a non-magnetic material. The shape of output signal from the probe reflects properties of tested material. The fundamental problem is to determine, according to the signal shape, whether there is some defect, structural element, roughness or impurity in the tube. Potential locations of defect in the signal are called indications (figure 1).

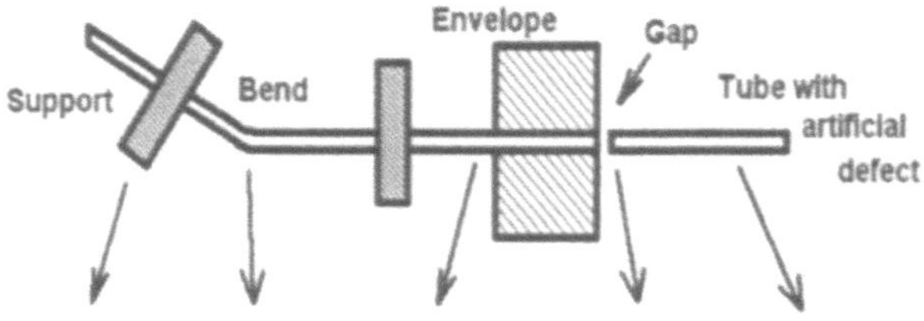

Fig. 1. Testing environment

Indication (figure 2) can be processed into different representations and output signal from the probe includes measurement data for different frequencies. Signals of different frequency describe tube changes in the different depth of the tested material. Higher frequencies can better map the changes of internal structure (e.g. internal scratch), lower frequencies can better describe the changes of the outside cover (e.g. presence of support).

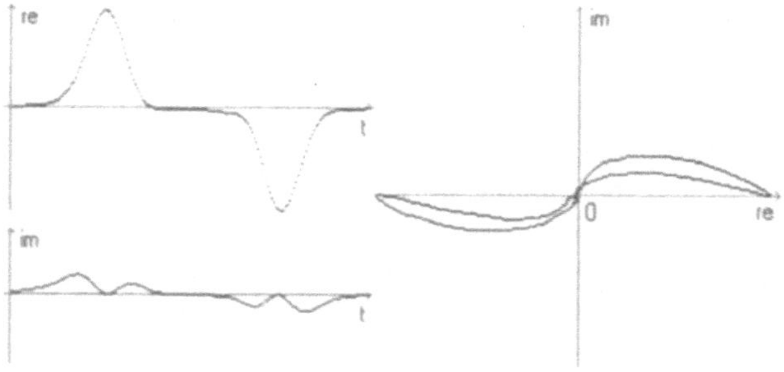

Fig. 2. Support plate indication

2 Probabilistic Neural Network

Network topology depends on the choice of input data representation. The dimension of space of indications data is equal to the size of network input layer. The size of output network layer depends on the number of tested classes [1].

Neural networks have the ability of generalization and universal approximation. This is a result of the general approximation theorem. They were successfully used in this type of defectoscopy methods [2]. For classification usually feed-forward supervised NN are used. The most popular is the multilayer perceptron (MP). MP containing one hidden layer is adequate for approximation of any arbitrary continuous function [3]. The input space of signature vectors of indications must be separable into disjunctive subspaces (clusters). Every cluster then contains indications of the same class.

The data of indications are a bit specific. Training of typical feed-forward MP is quite difficult because it is difficult to obtain a big amount of real data of indications. We are forced to use the data of artificial indications.

Probabilistic neural networks (PNN) can be used for classification problems [4]. Their design is straightforward and does not depend on training. A PNN is guaranteed to converge to a Bayesian classifier providing it is given enough training data. These networks generalize well. When an input is presented, the first layer computes distances from the input vector to the training input vectors, and produces a vector whose elements indicate how close the input is to a training input. The second layer sums these contributions for each class of inputs to produce as its net output a vector of probabilities. Finally, a compete transfer function on the output of the second layer picks the maximum of these probabilities, and produces a one for that class and a zero for the other classes.

Our experiments show that probabilistic neural network is able to divide the input space into disjunctive subspaces which correspond to the set of required types of indications (figures 3,4). Moreover, PNN gives stabler results than MP even if small trainnig set was used.

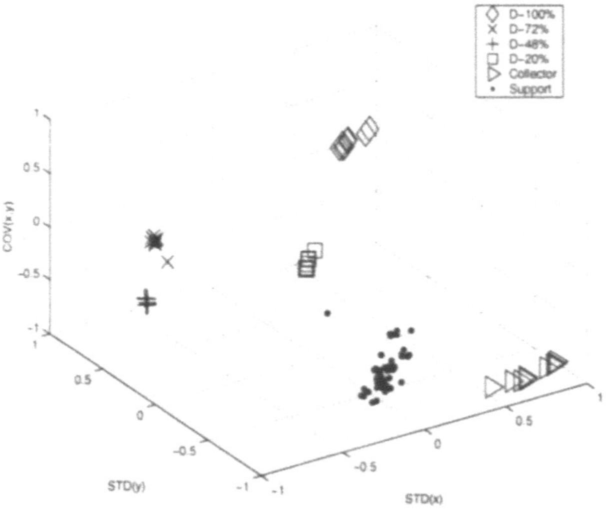

Fig. 3. Clusters of indications

3 Data Representations

The choice of suitable representation of indication data is very important. It is more important than choice of data classifier. The contribution concentrates on a comparison of data representations to be used as input of neural classifier implemented by PNN. Representation convenience is compared in terms of data input space separability into disjunctive clusters.

One of the most popular methods used in signal processing is the well known Fourier transformation [5]. By this transformation we are able to achieve the frequency domain of the tested indication. The

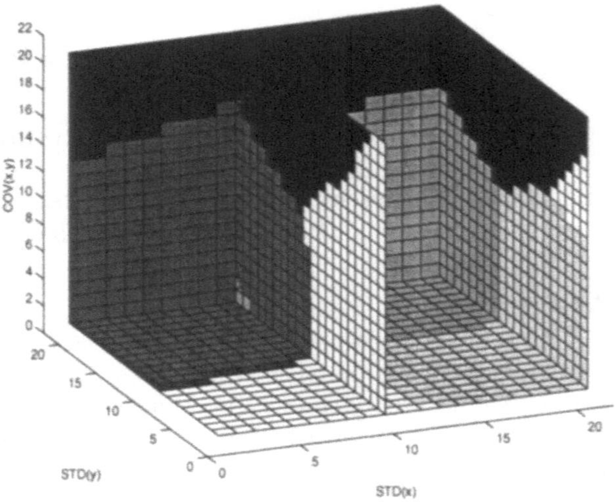

Fig. 4. Subspaces defined by PNN

indication can be then represented by vector of the most significant Fourier coefficients. This transformation can reduce the dimension of indication data and it is possible to reconstruct original data.

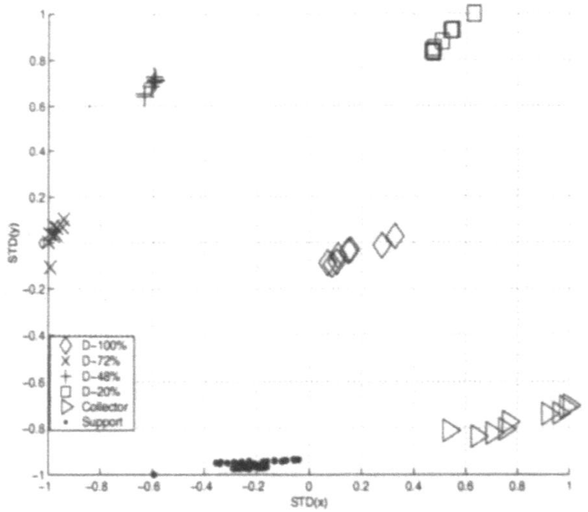

Fig. 5. Dependency of standard deviations from the database of indications

Of course, we can also use different types of 2D-curve natural parameters [6]. The output signal from the probe consists of two orthogonal parts (also called real and imaginary parts). The following list contains representations used in our experiments with PNN:

- R1 standard deviation of real and imaginary parts of indication signal and their covariance coefficient (3 numbers)(figures 3, 5)

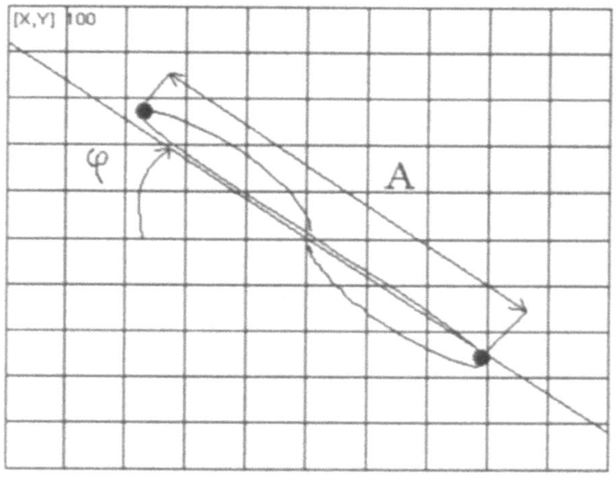

Fig. 6. Indication of 100% defect, peaks distance & angle

R2 peak-angle-peak representation (peaks angle a their distances in X-axis and Y-axis direction (3 numbers)(figure 6)

R3 Fourier transformation based representation defined as the vector of the most significant coefficients (15 complex numbers = 30 numbers)[7]

R4 Invariant Fourier based representation (6 complex numbers = 12 numbers)[7]

4 Results

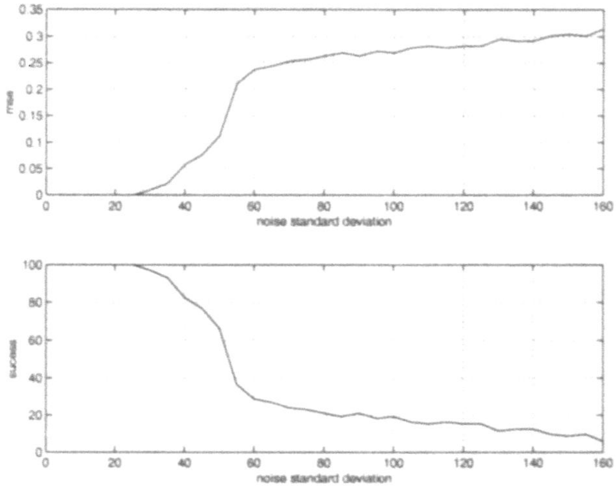

Fig. 7. PNN success ratio for representation R1

PNN was created by artificial indications without noise and then tested by the same set of indications with added Gaussian white noise in different levels. All representations were tested using indi-

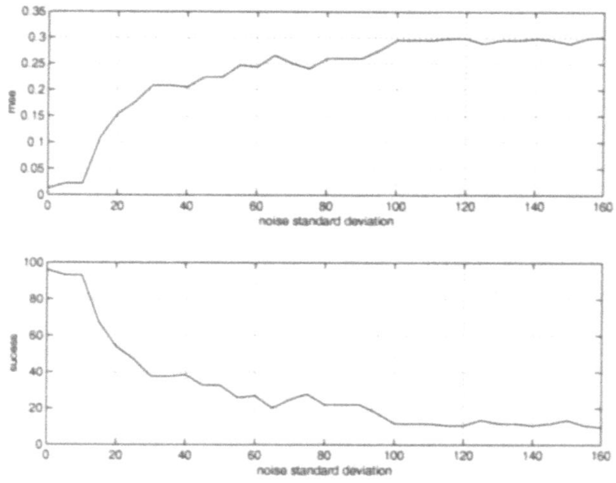

Fig. 8. PNN success ratio for representation R2

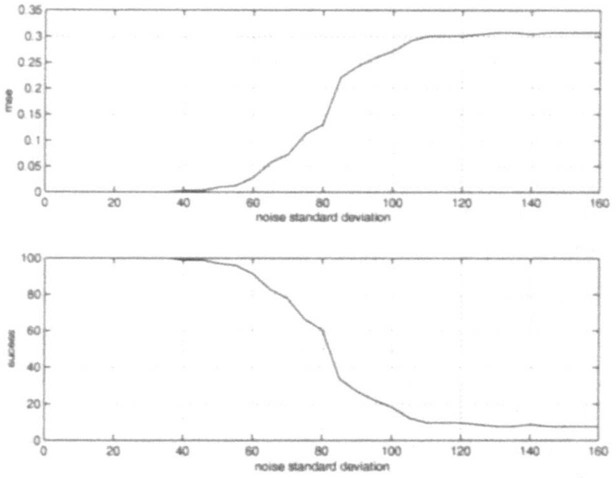

Fig. 9. PNN success ratio for representation R3 (15 coefficients)

cations with added noise whose standard deviation was from the interval (0, 160).

The presented results make us believe that representations R1 (figure 7) and R3 (figure 9) are the best of all tested representations. Representation R1 gives very compact class clusters (figures 3, 5) and gives also good classification results for noisy data of indications. Stability of representation R2 was very influenced by noise (figure 8), but this result was expected. The best results in this test were obtained for representation R3. There was a mistake in our first experiments. It is not true that Fourier based representations are sensitive to noise. Successfulness of final result for Fourier based coefficients depends on a number of used coefficients (figures 9,10). Representation R4 is calculated from representation R3 and this is the main reason why it

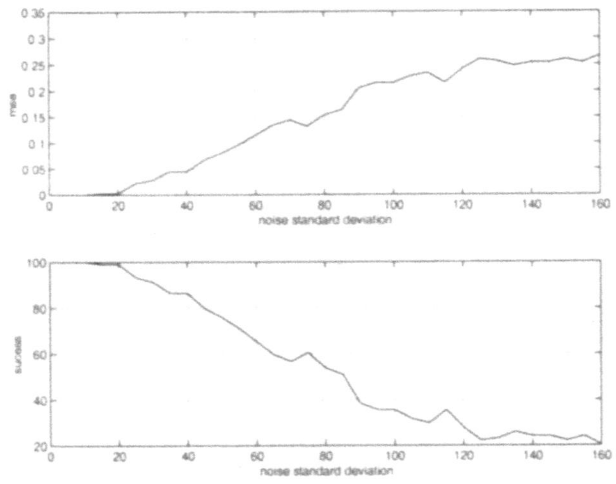

Fig. 10. PNN success ratio for representation R4 (6 coefficients)

losts some information and then gives worse results.

It is interesting that PNN using representation R1 is able to give good results (good answer in more than 80% cases) even if noise with standard deviation less than 45 was added. Remember that standard deviation of real [imaginary] parts of indications data are from the interval (28, 137) [(22, 413)].

5 Future Steps

The choice of suitable representation of indication data is very important. Every representation has its own possitive and negative features.

Finally, let us define the main topics of possible experimental steps:

- representation based on representation R1 invariant to change of measurement probe (there is dependency)
- classification of indication using her representations based on data in different frequences (to increase classification stability)
- classification of indication using representation based on wavelet data analysis

6 Conclusion

As for tested representations, we suppose that representation R1 and Fourier based representations R3 and R4 give good results. Benefit of representation R4 is that it is independent of position, scale and angle (except it's first coefficient). Representation R2 is very sensitive to noise and we believe that can be fully substituted by representation R1. Presented results indicate that PNN gives better results than MP and is easy to use. Training of MP has a lot of difficulties: learning parameters, learning time and esspecially overfiting. PNN has many advantages but it suffers from one major disadvantage. PNN is slower to operate because it uses more computation than other kinds of networks to do their function approximation or classification. Nevertheless, the speed is not so important. Important is that PNN does not depend on training. Every measurement probe is a unique device and its output may be a bit different that the output of other probe. The probe time evolution was notified too. So, it is necessary to calibrate the probe before every measurement and because PNN does not need training, there is possibility to save a lot of time. Strong classification algorithms can be used as a modules in decision-making systems in nuclear plant diagnostic centres.

References

[1] C. Rajagopalan, Baldev Raj, P. Kalyanasundaram, "The Role of Articial Intelligence in Non-destructive Testing and Evaluation", *INSIGHT*, Vol 38, No 2, February 1996.

[2] C. Charlton, "Investigation into the Suitability of a Neural Network Classifier for use in an Automated Tube Inspection System", *British Journal of NDT*, vol. 35, No 8, August 1993.

[3] Fu. LiMin, "Neural Networks in Computer Inteligence", *McGraw-Hill Companies*, March 1994.

[4] P.D. Wasserman, "Advanced Methods in Neural Computing", *New York: Van Nostrand Reinhold*, 1993, pp. 35-55, 155-61

[5] Ch.T. Yahn, R.Z. Roskies, "Fourier Descriptors for Plane Closed Curves", *IEEE Transactions on computers*, March 1972.

[6] R. Palanisamy, W. Loyd, "Finite Element Simulation of Support Plate and Tube Defect Eddy Current Signals in Steam Generator NDT", *Materials Evaluation*, Vol 39, June 1981.

[7] J. Grman, "Neural Network Application in the Defectoscopy", *IMEKO 2000 - XVI IMEKO World Congress*, Wienna, Austria, 25.-28.September 2000.

NeuroHough: A Neural Network for Computing the Hough Transform

M.Köppen, A. Soria-Frisch, R. Vicente-García *

*Fraunhofer IPK, Dept. Pattern Recognition, Pascalstr. 8-9, 10587 Berlin, Germany

Abstract. A new paradigm for the implementation of the Hough Transform (HT) is presented in this paper. The paradigm makes use of the neural networks' properties as function approximators in order to avoid some problems of the standard HT implementation. Some encouraging results are presented.

1 Introduction

Originally designed as a procedure to detect patterns on binary images [8] the Hough Transform (HT) is nowadays a methodology used for the resolution of a wide variety of problems in image processing and understanding. Beyond the classical application for linking contours on edge maps [5] the HT is applied in object recognition [6], shape parametrization [11][12], shape detection [13][17], and movement analysis on image sequences [4].

The performance of the original HT has been improved with the apparition of numerous modifications, e.g. generalized HT [2], adaptive HT [9], fast HT [14], randomized HT [16], fuzzy HT [7], whose abundance can be taken as a sign of its regard as processing tool. Moreover the research on the subject has been encouraged by the uncertain classification of the HT from a theoretical point of view. The HT has been considered as a paradigm of a more general connectionist model for low- and intermediate-level visions [3], as a product of Bayes theorem [18], as an evidence gathering procedure in the context of a computational evolutionary strategy [15], and as a particular case of the mathematical transformation called Radon transform [19].

The here presented paradigm does not want to bring this fruitful tradition to an end but to widen it into the theoretical framework of neural networks. This point was scarcely considered in [3] and [18]. In this case the paradigm makes use of a neural network as function approximator, a new terrain for the implementation of the HT.

A brief review on the HT is presented in Section 2. In Section 3 the neural architecture is discussed. Finally some results, the conclusions and the projective work can be found in Section 4.

2 The Hough Transform on Review

The HT considers the transformation of the image space to a multidimensional parameter space, where a set of image points (x,y) belonging to a determined geometrical element in the image space is represented by a combination of its characterizing parameters. This parameter space consists of a set of discrete accumulator cells, which are incremented when a point in the image space fulfills the analytical expression of the geometrical element being searched for. In this so-called accumulation process the fulfillment of the analytical expression acts as a piece of evidence being accumulated in the parameter space. The parameter space is finally analyzed to detect the cells where the evidence is mostly accumulated. Therefore the geometrical element can be characterized as a function of the parameters related with the most voted accumulator cell.

In the most basic application of the HT a straight line is for instance characterized through the length (ρ) and orientation (θ) of its normal vector:

$$f((x,y),(\rho,\theta)) = \rho - x\cos\theta - y\sin\theta = 0 \qquad (1)$$

In this case the parameter space is two-dimensional and the straight line is eventually parameterized through (ρ,θ) of the accumulator cell with a greatest value (characterization that includes the error produced by the discretization of the parameter space).

The generalization of the HT for the detection of arbitrary shapes was introduced by Ballard [2]. The generalizing strategy is to increase the dimensionality of the parameter space in order to include not only changes in the geometrical element to be detected, but also in its translation, scale and rotation.

2.1 Properties of the HT for shape detection

The HT has demonstrated its suitability for the detection of shapes on edge maps [13][17]. This suitability is based on the reduction of the complex problem of line analysis to a more tractable one of peak detection in the parameter space. This is true even for the detection of arbitrary shapes, thanks to the already mentioned generalized HT. The robustness of the HT in front of noisy images, light deformations of the searched shapes respect to its model, and discontinuities of some parts of the edges are very appreciated features of this kind of analysis.

Another interesting property is the parallelism of the computations undertaken in the calculation of the HT. The analysis of a complex form can be carried out after a decomposition in simpler geometrical elements of the model shape. Moreover the structure of the accumulator cell allows also the parallelization of this computation for each line. This property has been exploited in numerous parallel architectures implementing the polar parameters finding approach [10], and even in a general connectionist framework for modeling low and intermediate human vision [3].

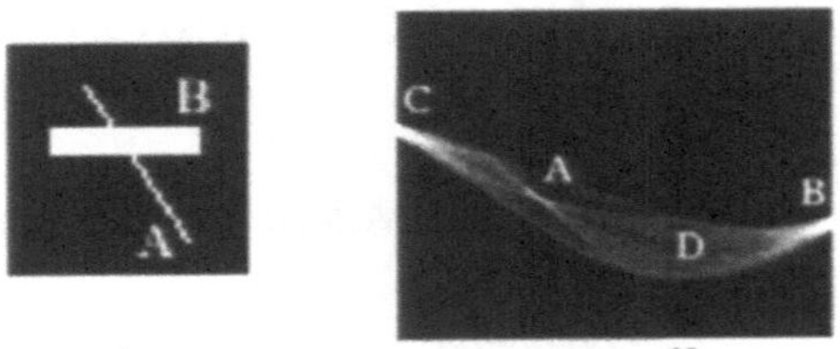

1a 1b

Fig. 1: Hough Transform pair that show different not-desired effects in the parameter space (1b). Lines A and B in the image space (1a) should appear as unique points A and B in the parameter space (1b). B appears not as a point but as a surface, due to the thickness of the line in image space (1b). The same as C, which is a point induced by a reflective adjacency relationship with B. The cloud of points D, caused by alignments of different points in lines A and B (1a), avoids a clearer appearance of point A in the parameter space (1b).

2.2 HT drawbacks

The extensive memory usage and the computational cost, both proportional to the dimensionality of the parameter space, are the most known disadvantages of the HT for shape detection [10]. The problem of computational cost becomes more evident when trying to detect complex forms or to implement the methodology in applications where real-time response is needed. Beside this, quantization errors appear when applying the methodology in real applications due to the consideration of a discrete parameter space. In order to successfully implement the methodology a trade-off is needed [1]. Parameter accuracy on the one hand, and computational time and tractability of 'thick' lines (see figure 1) on the other have to be considered.

The apparition of false peeks is a minor problem when considering the HT from a mathematical point of view, but plays an important role in the resolution of real problems. They appear in front of spurious edges and of aligned points of objects that are separately analyzed, and due to reflective adjacency relationships of lines occupying extreme positions in polar accumulator spaces [5] (see figure 1).

2.3 Neural Networks Solution

In the following a neural network paradigm for the computation of the HT, which will be called NeuroHough Transform, will be presented. The purpose of this implementation is the avoidance of some drawbacks of the classical implementation taking advantage of the approximation capability of neural networks.

One of the general goals to be attained through the neural implementation of the HT is the constancy in terms of computation time independently from the complexity of the analyzed shape and the quantization step of the parameter space. This is a significant point when trying to implement the HT in real-time. The neural implementation should be also easier to generalize, what will allow the analysis of more complex scenes than edge maps, i.e. the analysis of surfaces and textures. Finally the NeuroHough Transform is thought to achieve the elimination of some false peeks, those caused by adjacency reflection and by the presence of spurious contours, through usage of pre-processed training data. Being these goals quite ambitious the main objective of the here presented work was succeeding in implementing the HT through a neural network.

3 NeuroHough Transform

In this section, a method is proposed to represent the computations of the HT by a neural network. The proposed architecture can be observed in figure 2. Some previous aspects have to be taken under consideration before realizing the neural architecture. Assume the HT constructed from a mapping H of the four-tupel (x,y,ρ,θ) into the interval [0, 1] , with x and y coordinates in the image space and ρ and θ the coordinates in the accumulator space. Thereby, the assignment $H(x,y,\rho,\theta)=1$ means that the presence of the pixel (x,y) in the image foreground domain induces the accumulator cell with the coordinates (ρ,θ) to be incremented by 1. For $H(x,y,\rho,\theta)=0$, the accumulator cell remains unchanged. Thus, the HT is realized by going for each (x,y) in the image foreground over all (ρ,θ) according to the chosen quantization of the accumulator space, and adding H(x,y,ρ,θ) to the corresponding cells:

$$A(\rho,\theta) = \sum_{(x,y)\ in\ I} H(x,y,\rho,\theta) \qquad (2)$$

This will give the same result as for the standard Hough transform algorithm, but can be computed cell-wise. Now, the task for the neural network is to approximate H(x,y,ρ,θ).

For the function approximation no special neural architecture is necessary and thus a 3-layer Backpropagation Network was chosen for the sake of simplicity. However a special representation of the given input and output data is considered.

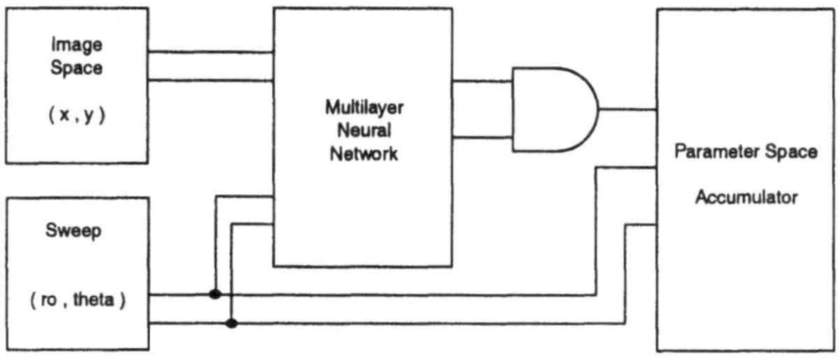

Fig. 2. Proposed neural architecture for the computation of the HT, NeuroHough.

The neural network is fed with the following input data: x, y, x*sinθ, y*cosθ, ρ, sinθ. This is regarded to the fact that neural networks may have problem to internally approximate products of input data. Furthermore the inclusion of redundant inputs help to empirically adjust the relevance of the input parameters. So, for the standard Hough transform, the task is presented as a linear separation problem to the network.

For the output, two neurons are used. Considering the Hough equation, as in (1), it could be assumed that the neural network should compute the value 0 for "correct" (x,y,ρ,θ) tupels (and "non-zero" otherwise) just using one output neuron. However, this approach is not practicable, since the training data become ambiguous. It is simpler to train a network on having either output value larger or smaller than a given threshold (due to the sigmoidal transfer functions used). So, for the "neural computer", a=0 is taken as (a>=0) AND (a<=0), which gives the right motivation to decide considering two output neurons. Then, the training data for output neurons O1 and O2 are given by:

I. ρ-x sinθ-y cosθ > 0: O1=1, O2=0,
II. ρ-x sinθ-y cosθ < 0: O1=0, O2=1,
III. ρ-x sinθ-y cosθ = 0: O1=1, O2=1
(the case O1=O2=0 never happens).

As a consequence the NeuroHough network is trained by randomly selecting (x,y) positions and (ρ,θ) values, checking for case 1, 2 or 3 and setting accordingly the training data. The training set itself should be planned in a manner so that the loading of values 0 or 1 into the output neurons is balanced.

This procedure will give a neural network, which is able to perform the computations of the standard HT and can be directly trained from the analytical expression of the transform. The same training procedure can be used for slight modifications of the transform, since it is based on a more general interpretation of the Hough transform not considered so far (as a special case of an arbitrary mapping of R^4 into the interval [0,1]).

For a superposed training regime, the NeuroHough procedure can be used for the initial configuration. Then, the error backpropagation procedure has to be modified according to the given recognition task. This is mostly addicted to further works on this approach. But before doing so, it is essential to prove the validity of the approach by representing the standard HT in its neural form.

4 Preliminary Results, Conclusions and Future Work

In order to approximate the desired function using a Multilayer Backpropagation Network the empirical performance of the paradigm was researched. The training and test data sets do not include any noise added and the amount is usually about 7000 examples.

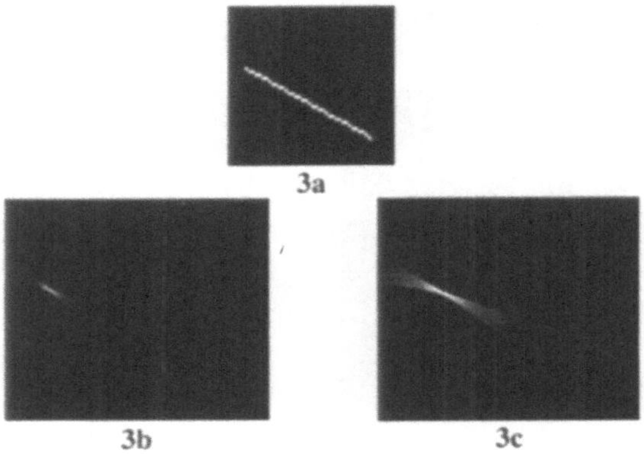

Fig. 3: Hough Transformation of an image (3a) with the classical procedure (3b) and through the here presented neural network paradigm, NeuroHough (3c).

Till now the best results were obtained with a 3-layer structure, 7 neurons in the input layer, 2 in the output one, and between 18 and 20 neurons in the hidden layer. Larger checked network sizes did not improve the results significantly and needed very large training time, while smaller sizes underfit the targets.

The gain is set to 3 in order to sharpen decision areas of the network, the momentum factor, the learning rate, and the initial weights were and to achieve a good generalization in a reasonable training time.

The results obtained till now are excellent in terms of approximation capability. The Hough Transformation could be approximated through the NeuroHough, what can be observed in the case of the transformation of a straight line (see figure 3). The NeuroHough shows a good generalization capability, as the result (3c) were obtained with a neural network trained for the transformation of a line different from the one shown (3a).

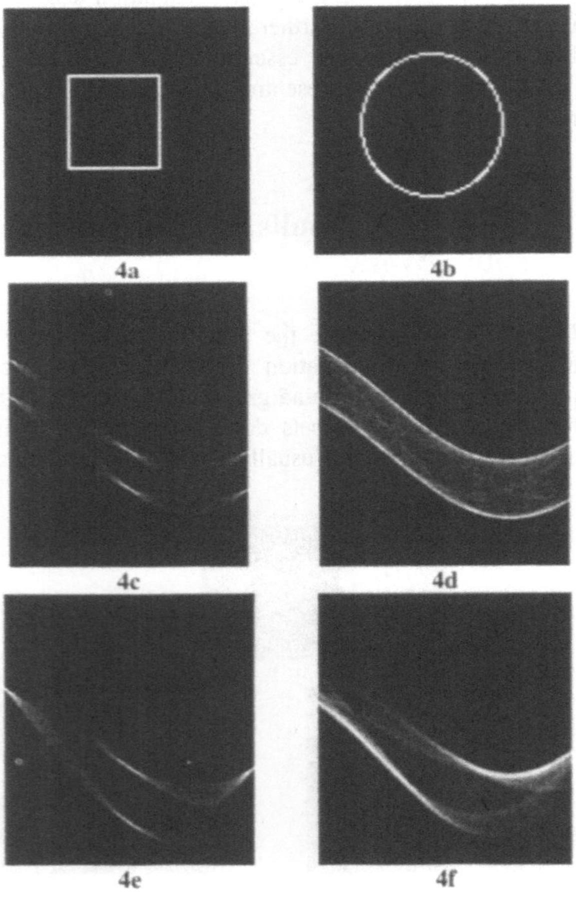

Fig. 4. Transformation of more complex forms through the here presented NeuroHough architecture. Original images of a square (4a) and a circumference (4b) were transformed with the classical HT (respectively, 4c and 4d) and the NeuroHough Transform (respectively, 4e and 4f). The suppression of points due to reflective adjacency relationship (4c) can be avoided using this methodology (4e).

Also the suppression of points due to reflective adjacency relationship in the analysis of more complex shapes, which was one of the aims to be attained by the neural implementation of the HT, was reached through the implementation of the NeuroHough architecture (see figure 4c and 4e). Although having only been trained with straight lines the transformation of a circumference was encouraging (4f).

The complete results of the here presented work, which is on progress at the present, will be presented in future communications.

References

[1] R.C. Agrawal, R.K. Shevgaonkar, S.C. Sahasrabudhe (1996). *A fresh look at the Hough transform*, Pattern Rec. Letters 17 (10) 1065-1068.

[2] D.H. Ballard (1981). *Generalizing the Hough Transform to Detect Arbitrary Shapes*, Pattern Recognition 13 (2) 111-122.

[3] D.H. Ballard (1984). *Parameter Nets*, Artificial Intelligence 22, 235-267.

[4] I. Fermin, A. Imiya, A. Ichikawa (1996). *Randomized polygon search for planar motion detection*, Pattern Rec. Letters 17 (10) 1109-1115.

[5] R.C. Gonzalez, R.E. Woods (1993). Digital Image Processing, Addison-Wesley Pub. Co.

[6] W.E.L Grimson, D.P. Huttenlocher (1990). *On the Sensitivity of the Hough Transform for Object Recognition*, IEEE Trans. PAMI 12 (3) 255-273.

[7] J.H. Han, L.T. Kóczy, T. Poston (1994). *Fuzzy Hough Transform*, Pat. Rec. Letters 15 (7) 649-658.

[8] P.V.C. Hough (1962). *A method and means for recognizing complex patterns*, U.S. Patent 3,069,654.

[9] J. Illingworth, J. Kittler (1987). *The Adaptive Hough Transform*, IEEE Trans. PAMI 9 (5) 690-697.

[10] J. Illingworth, J. Kittler (1987). *A Survey of the Hough Transform*, Comp. Vision, Graphics, and Image Processing 44, 87-116.

[11] D. Ioannou, E.T. Dugan, A.F. Laine (1996). *On the uniqueness of the representation of a convex polygon by its Hough transform*, Pat. Rec. Letters 17 (12) 1259-1264.

[12] V.F. Leavers, J.F. Boyce (1987). *The Radon transform and its application to shape parametrization in machine vision*, Image and Vision Computing 5 (2) 161-166.

[13] V.F. Leavers (1992) *Use of the Radon transform as a method of extracting information about shape in two dimensions*, Image and Vision Comp. 10 (2) 99-107.

[14] H. Li, M.A. Lavin, R.J. LeMaster (1986). *Fast Hough Transfor: A Hierarchical Approach*, Comp. Vision, Graphics, and Image Processing 36, 139-161.

[15] J. Louchet (2000). *From Hough to Darwin: An Individual Evolutionary Strategy Applied to Artificial Vision*, Lecture Notes in Comp. Sci. (1829): Artificial Evolution, Springer.

[16] R.A. McLaughlin (1998). *Randomized Hough Transform: Improved ellipse detection with comparison*, Pattern Rec. Letters 19 (3-4) 299-305.

[17] D.C.W. Pao, H.F. Li, R. Jayakumar (1992). *Shapes Recognition Using Straight Line Hough Transform: Theory and Generalization*, IEEE Trans. PAMI 14 (11) 1076-1089.

[18] A.S. Rojer, E.L. Scwartz (1992). *A Quotient Space Hough Transform for Space-Variant Visual Attention*, Neural Networks for Vision and Image Processing, MIT Press.

[19] P. Toft (1996). The Radon Transform, Theory and Implementation, PhD Thesis, TU of Denmark.

Fast Iris Detection Using Cooperative Modular Neural Nets

H. M. El-Bakry*

*Faculty of Computer Science &Information Systems, Mansoura University – Egypt,
helbakry1@hotmail.com

Abstract

In this paper, a combination of fast and cooperative modular neural nets to enhance the performance of the detection process is introduced. We have applied such approach successfully to detect human faces in cluttered scenes [10]. Here, this technique is used to identify human irises automatically in a given image. In the detection phase, neural nets are used to test whether a window of 20x20 pixels contains an iris or not. The major difficulty in the learning process comes from the large database required for iris / non-iris images. A simple design for cooperative modular neural nets is presented to solve this problem by dividing these data into three groups. Such division results in reduction of computational complexity and thus decreasing the time and memory needed during the test of an image. Simulation results for the proposed algorithm show a good performance.

1. Introduction

To have a more practical biometric system, here we concentrate on personal identification through human iris recognition, a relatively new biometric technology that has some key advantages in terms of speed, simplicity, accuracy, and applicability [9]. Of course, face recognition does not need any voluntary action and can be used for covert surveillance. But, the problem of classification of twins makes the identification of persons based on face recognition unreliable. Moreover, for a given person, the patterns of right and left irises are different from each other, although they are genetically identical as the four irises of two identical twins. Furthermore, the spatial patterns which apparent in the human iris are highly distinctive to an individual as shown in Fig. 1 [1, 9]. Also, the size of iris is stable throughout life. We may have noticed how large the eyes of a baby seem to be when their heads are so small. This is because even from birth, the human iris is about 10 millimeters in diameter. More importantly, the iris pattern itself is also stable throughout life, except that the color may change in the first months. As the pupil size is constantly changed, neural networks are used to compensate for this distortion in the iris patterns.

The goal of this paper is to solve the problem of requiring large database to build an automatic system in order to detect the location of irises in scenes. This paper explores the use of modular neural network (MNN) classifiers. Non-modular classifiers tend to introduce high internal interference because of the strong coupling among their hidden layer weights [2]. As a result of this, slow learning or over fitting can occur during the learning process. Sometimes, the network could not be learned for complex tasks. Such tasks tend to introduce a wide range of overlap which, in turn, causes a wide range of deviations from efficient learning in the different regions of input space [3]. High coupling among hidden nodes will then, result in over and under learning at different regions [8]. Enlarging the network, increasing the number and quality of training samples, and techniques for avoiding local minina, will not stretch the learning capabilities of the NN classifier beyond a certain limit as long as hidden nodes are tightly coupled, and hence cross talking during learning [7]. A MNN classifier attempts to reduce the effect of these problems via a divide and conquer approach. It generally, decomposes the large size / high complexity task into several sub-tasks, each one is handled by a simple, fast, and efficient module. Then, sub-solutions are integrated via a multi-module decision-making strategy. Hence, MNN classifiers, generally, proved to be more efficient than non-modular alternatives [5,6]. In section 2, a method for detection of human irises in photo images is presented. Also, an algorithm during the searching procedure is described. A fast searching algorithm for iris detection which reduces the computational complexity of neural nets is presented in section 3.

2. Human Iris Detection Based on Neural Nets

Image acquisition of the iris can not be expected to yield an image containing only the iris. Rather, image acquisition will capture the iris as part of a larger image that also contains data derived from the immediately surrounding eye region. Therefore, prior to iris pattern matching, it is important to localize precisely that portion of the acquired image corresponding to an iris. In particular, it is necessary to localize that portion of the image derived from inside the limbus (the border between the sclera and the iris) and outside the pupil. Further if the eyelids are occluding part of the iris, then only that portion of the image below the upper eyelid and above the lower eyelid should be included. Moreover, the detection algorithm must be invariant to changes in pupillary constriction and overall iris image size, and hence also invariant to camera zoom factor and distance to the eye. Previous work in iris detection is accomplished by using 2D Gabor wavletes. Then a doubly dimensionaless coordinate system is used to map the tissue [1]. But, these methods are complex and require large number of computations and consume more time. Here neural networks are used to detect the iris with all of these variations and compensate automatically for the stretching of the iris tissue as the pupil dilates.

2.1 A Proposed algorithm for iris detection using MNNs

First, in an attempt to equalize the intensity values of the iris image, the image histogram is equalized. This not only reduce the variability of generated by illumination conditions, and enhance the image contrast but also increases the number of correct pixels that can be actually encountered [10]. The next component of the proposed system is a classifier that receives an input of 20x20 pixel region of gray scale image and generates an output region ranging from 1 to -1, signifying the presence or absence of a iris, respectively. This classification must have some invariance to position, rotation, and scale. To detect irises anywhere in the input, the classifier is applied at every location in the image. To detect irises larger than the window size, the input image is repeatedly reduced in size. The classifier is applied at every pixel position in the image and scale the image down by a factor of 1.2 for each step as shown in Fig. 2. So, the classifier is invariant to translation and scaling. To have rotation invariant, the neural network is trained for images rotated from 0° to 355° by a step of 5°. In order to train neural networks used in this stage, a large number of iris and non-iris images are needed. A sample of non-iris images, which are collected from the world wide web, is shown in Fig. 3. So, conventional neural nets are not capable of realizing such a searching problem. As a result of this, MNNs are used for detecting the presence or absence of human irises for a given image. Images (iris and non-iris) in the database are divided into three groups which result in three neural networks. More divisions can occur without any restrictions in case of adding more samples to the database. Each group consists of 400 patterns (200 for iris and 200 for non-iris). Each group is used to train one neural network. Each network consists of hidden layer containing 13 neurons, and an output layer which contains only one neuron. Here, two models of MNNs are used. The first is the ensemble majority voting which gives a result of 80% detection rate. The other is the average voting which gives a better result of 84% detection rate.

2.2 Enhancement of detection performance

To enhance the detection d`ecision, the detection results of neighboring windows to confirm the decision at a given location can be used. This will reduce false detection as neighboring windows may reveal the non-iris characteristics of the data. For each location, the number of detections within a specified neighborhood of that location can be counted. If the number is above a threshold, then that location is classified as an iris. Among a number of windows, the location with the higher number of detections is preserved in range of one pixel, and locations with fewer detections are eliminated. In our case, a threshold of 4 is chosen. Such strategy increases the detection rate to 96% (average voting), as a result of reducing the false detections. It is clear that, the use of MNNs with this enhancement improves the results over those obtained in the previous section. As shown in Fig. 4, when the light beam is focused more to inside the pupil, the pupil itself is compressed. The iris contained in the eye of Fig. 4(a) is fed to the neural net during the learning process while the other (which found in the eye of Fig. 4(b)) could be detected correctly during the test phase.

3. Fast Neural Nets for Iris Detection

In subsection 2.1, modular neural network for object detection is presented using a sliding window to test a given input image. In this section, a fast algorithm for object detection (used with each of the neural nets presented in section 2.1) based on two dimensional cross correlations that take place between the tested image and the sliding window. Such window is represented by the neural net weights situated between the input unit and the hidden layer. The convolution theorem in mathematical analysis says that a convolution of f with h is identical to the result of the following steps: let F and H be the results of the Fourier transformation of f and h in the frequency domain. Multiply F and H in the frequency domain point by point and then transform this product into spatial domain via the inverse Fourier transform [4]. As a result of this, these cross correlations can be represented by a product in frequency domain. So, by using cross correlation in frequency domain, speed up in an order of magnitude can be achieved during the detection process.

In the detection phase, a sub image I of size mxn (sliding window) is extracted from the tested image which has a size PxT and fed to the neural network. Let X_i be the vector of weights between the input sub image and the hidden layer. This vector has a size of mxn and can be represented as mxn matrix. The output of hidden neurons h(i) can be calculated as follows:

$$h_i = g\left(\sum_{j=1}^{m}\sum_{k=1}^{n} X_i(j,k)I(j,k) + b_i\right) \quad (1)$$

where g is the activation function and b(i) is the bias of each hidden neuron (i). Equ. 1 represents the output of each hidden neuron for a particular sub-image I. It can be obtained to the whole image Z as follows:

$$h_i(u,v)=g\left(\sum_{j=-m/2}^{m/2}\sum_{k=-n/2}^{n/2} X_i(j,k)\, Z(u+j,v+k)+b_i\right) \quad (2)$$

Equ.2 represents a cross correlation operation. Given any two functions f and d, their cross correlation can be obtained by:

$$f(x,y)\otimes d(x,y)=\left(\sum_{m=-\infty}^{\infty}\sum_{n=-\infty}^{\infty} f(m,n)d(x+m,y+n)\right) \quad (3)$$

Therefore, equ. 2 may be written as follows:

$$h_i = g\left(X_i \otimes Z + b_i\right) \quad (4)$$

where h_i is the output of the hidden neuron (i) and h_i (u,v) is the activity of the hidden unit (i) when the sliding window is located at position (u,v) and (u,v) $\in$[P-m+1,T-n+1].

Now, the above given cross correlation can be expressed in terms of Fourier Transform:

$$Z\otimes X_i = F^{-1}\left(F(Z)\bullet F^*\left(X_i\right)\right) \quad (5)$$

Hence, by evaluating this cross correlation, a speed up ratio can be obtained compared to conventional neural networks. Also, the final output of the neural network can be evaluated as follows:

$$O(u,v) = g\left(\sum_{i=1}^{q} w_o(i)\, h_i(u,v) + b_o\right) \quad (6)$$

O(u,v) is the output of the neural network when the sliding window located at the position (u,v) in the input image Z.

For a tested image of NxN pixels, the 2D FFT requires $O(N^2\log_2 N^2)$ computation steps. For the weight matrix X_i , the 2D FFT can be computed off line since these are constant parameters of the network independent of the tested image. The 2D FFT of the tested image must be computed. As a result, q backward and one forward transforms have to be computed. Therefore, for a tested image, the total number of the 2DFFT to compute is $(q+1)N^2\log_2 N^2$. Moreover, the input image and the weights should be multiplied in the frequency domain. Therefore, computation steps of (qN^2) should be added. Finally, a total of $O((q+1)N^2\log_2 N^2+qN^2)$ computation steps must be evaluated for fast the neural algorithm.

Using sliding window of size nxn, for the same image of NxN pixels, $O((N-n+1)^2n^2q)$ computation steps are required when using traditional neural networks for the face detection process. The theoretical speed up factor K can be evaluated as follows:

$$K = \frac{q(N-n+1)^2n^2}{(q+1)N^2\log_2 N^2 + qN^2} \quad (7)$$

In our case, for N=200, n=20, q=13, a speed up factor of 18.7595 may be achieved. The speed up factor introduced in [11] for object detection which is given by:

$$K = \frac{qn^2}{(q+1)\log^2 N} \quad (8)$$

is not correct since the number of computation steps required for the 2D FFT is $O(N^2\log_2 N^2)$ and not $O(N^2\log^2 N)$. Also, this is not a typing error as the curve in Fig. 2 in [11] realizes equ.8. Moreover, the speed up ratio presented in [11] is not only contains an error but also not precise. This is because for fast neural nets, the term (qN^2) must be added. Such term has a great effect on the speed up ratio. Furthermore, for conventional neural nets, the number of operations is $(q(N-n+1)^2n^2)$ and not (qn^2).

4. Conclusion

A fast modular neural network approach has been introduced to identify human irises. Such approach can be applied to detect other objects in an image. Gray scale images of resolution 20x20 up to 500x500 pixels have been manipulated. The technical problem associated with large database (iris/non-iris) required for training neural networks has been solved using MNNs. A simple algorithm for fast iris detection based on cross correlation in frequency domain has been presented in order to speed up the execution time. Simulation results have shown that the proposed algorithm is an efficient method for finding locations of irises when the size of the iris is unknown as well as rotated, occluded, noised, and mirrored irises are detected correctly.

References

[1] Wildes, R. P.: Iris Recognition : An Emerging Biometric Technology. Proceedings of IEEE, Vol. 85, No.9, pp.1347-1363, September, (1997).

[2] Jacobs, R., Jordan, M., and Barto, A.: Task Decomposition Through Competition in a Modular Connectionist Architecture: The what and where vision tasks. Neural Computation 3, pp. 79-87, (1991).

[3] Auda, G., Kamel, M., Raafat, H.: Voting Schemes for cooperative neural network classifiers. IEEE International Conference on Neural Networks ICNN95, Vol. 3, Perth, Australia, pp. 1240-1243, November, (1995).

[4] Klette, R., and Zamperon.: Handbook of image processing operators. John Wiley & Sonsltd, (1996).

[5] Auda, G. and Kamel, M.: CMNN: Cooperative Modular Neural Networks for Pattern Recognition. Pattern Recognition Letters, Vol. 18, pp. 1391-1398, (1997).

[6] Alpaydin, E.: Multiple Networks for Function Learning. Int. Conf. on Neural Networks, Vol.1 CA, USA, pp. 9-14, (1993).

[7] Waibel, A.: "Modular Construction of Time Delay Neural Networks for Speech Recognition. Neural Computing 1, pp. 39-46, (1989).

[8] K. Joe, Y. Mori, and S. Miyake.: Construction of a large scale neural network: Simulation of handwritten Japanese Character Recognition. on NCUBE Concurrency 2 (2), pp. 79-107, (1990).

[9] Jain, A., Bolle, R., and Pankanti, S.: BIOMETRICS: Personal Identification in Networked Society. Chap5, pp. 103-122, Kluwer Academic Publishers, (1998).

[10] El-Bakry, H. M., Abo-Elsoud, M. A. , and Kamel, M. S.: Fast Modular Neural Networks for Human Face Detection. Proc. of IJCNN International Joint Conference on Neural Networks 2000, Como, Italy, Vol. III, pp. 320-323, 24-27 July, (2000).

[11] Ben-Yacoub, S.: Fast Object Detection using MLP and FFT. IDIAP-RR 11, IDIAP, (1997).

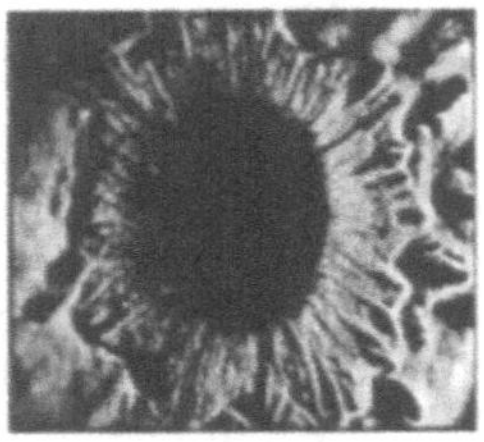

Fig 1. The distinctiveness of the human iris.

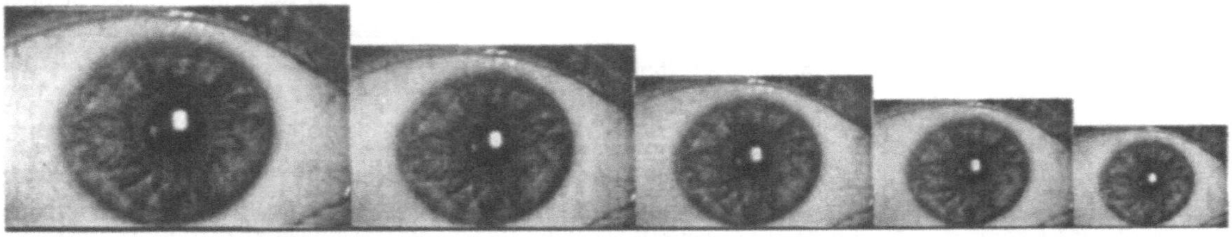

Fig 2. Image resizing by a factor of 1.2 during iris detection.

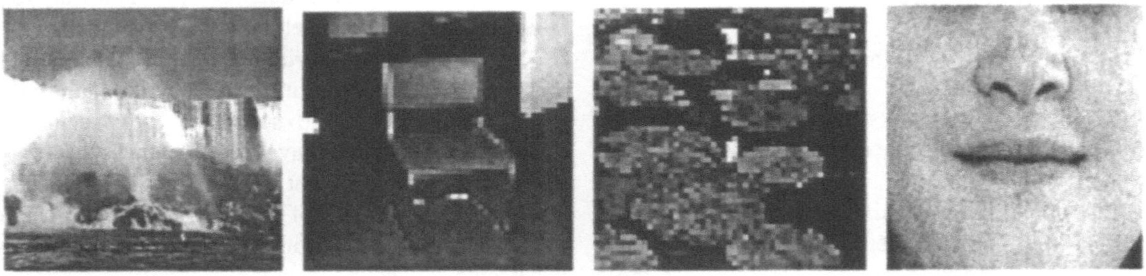

Fig 3. Examples of non-iris images.

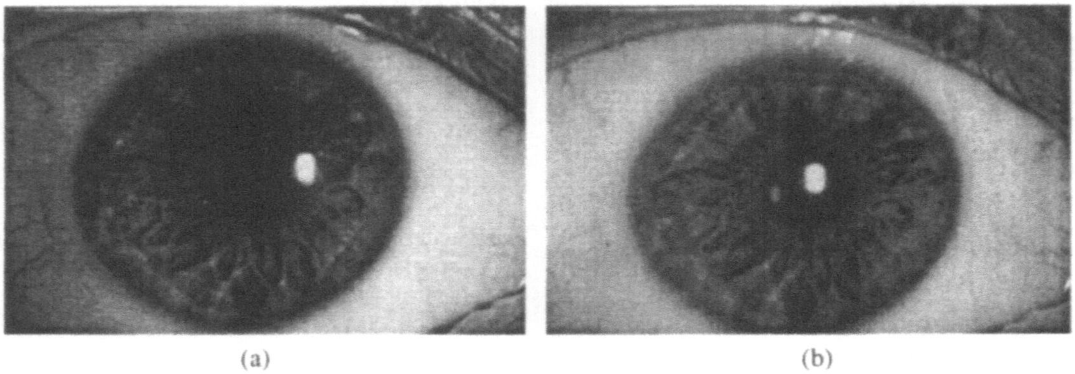

(a) (b)

Fig 4.Variation in pupil size when the light beam is more focused into the eye.

Fig 4.Variation in pupil size when the light beam is more focused into the eye.

Symbolic Representation of a Multi-Layer Perceptron

Fériel MOURIA-BEJI (*Membre IEEE*)(1), (2)
(1) *ENSI/LIA. Artificial Intelligence Group.*
BP. 275, Cité Mahrajène, 1082 Tunis, Tunisia
E-mail : feriel.beji@ensi.rnu.tn
(2) *INRIA/LORIA. Villers-lès-Nancy, France*

Abstract

We propose a Top-Down Inferring algorithm TD-INFER for artificial neural network rule extraction. These rules formalize the decision process of a standard multi-layer network and make its prediction explicit and understandable. They do not involve any weight values and no restrictions are made on the activation values. The algorithm is applied to a speech and character recognition problems.

1 Introduction

Artificial neural networks (ANN's) are often used to predict or classify a given output on the basis of one or several inputs. Although we can compare neural net prediction or classification success rates with standard techniques such as multiple regression or discriminant analysis [1, 2], no method has been accepted that assesses the relative importance of the inputs used by the network to arrive at its conclusions [3]. Instead, ANNs have been presented to users as a sort of "black box" whose unimaginably complex inner workings somehow magically transform inputs into predicted outputs [4, 5, 6, 7, 8]. ANN's are relatively straightforward: The user enters a training set of input output examples and the ANN heuristics attempt to model the process by which the inputs become outputs [13, 11]. This process is reminiscent of inductive expert systems, which also infer rules from training examples. However, the iterative passing of data among the processing elements of the neural net can be much more complex than inductive expert-system heuristics [14, 15]. Information is stored in a perfectly reasonable form in both a computer program and an expert system. We can list all the rules, variables, or instructions and see what they mean. But in a neural network, though we can easily obtain a list of all the weights and connections, it is far from obvious what those values have to do with the problem of deciding the credit worthiness of an applicant, or which phoneme corresponds to the next letter in a paragraph of text. Unlike the easily understood information resident in an expert system or computer program, knowledge of a neural network is unintelligible to a person [7, 8]. Unfortunately, knowing the connection strength between the fourth neuron in the input layer and the seventh neuron in the middle layer is totally uninformative if what we really want to know is how the network gives a response or a solution. To make sense of the weights and connections, we need to explore exactly how a neural network performs its tasks. Users nearly always want to know why a system comes up with a particular answer so that they can improve their own understanding of the problem [7, 8]. Building an explanation facility into an average neural network is difficult or impossible. Some problems lend themselves naturally to a collection of rules. All or part of these problems may be easily solved by a set of *if-then* conditions which are not easy to build into a neural network [14, 15, 17].

We believe that this is so only because we do not have the techniques to understand how a neural network makes a prediction. If we could extract rules from a neural network, we would be able to understand better its prediction process. Rules are a form of knowledge that human experts can easily verify, pass on, and expand.

Recent works [9, 5, 10, 11, 12] show that rules can be extracted from ANNs by search-based heuristics with exponential complexity. These heuristics search subsets of weights that exceed the bias on a unit, then rewrite such subsets as rules. To simplify the search process, the algorithms make some assumptions. One of these assumptions is that the activation value of a node is either very close to 1 or very close to 0. But this can restrict the network capability, when using the sigmoid transfer function, since this value can be anywhere between 0 and 1. This article presents a very simple Top-Down Inferring algorithm for understanding an ANN TD-INFER. In this algorithm the connection weights from the input layer to the hidden nodes to the output layer are used to partition the relative share of the output associated with each input.

The tests we conducted showed that there is a strong relationship between the activation values of the hidden units and the values of the input and output units. Furthermore, integration of these activation values in the rule generation process is very important for the accuracy of these rules.

Section 2 presents the principle of the TD-INFER. Section 3 concerns the description of an application of the algorithm to a speech and a character recognition problems. Some concluding remarks are given in section 4.

2 TD-INFER algorithm

Most neural networks used in real applications have three layers or more. It is conventionally understood that the input layer distributes the pattern through the network, the output layer generates an appropriate response and the middle layer acts as a collection of feature detectors. Each feature detector node in the middle layer looks for a key feature or features in the input pattern and reacts strongly when one is found. The output layer can then construct an appropriate output pattern based on the combination of features the middle layer has detected.

How are these feature detectors established, given that the network designer has not specified them? The answer to this question is through the training procedure. Neural network training can be viewed as nothing more then a process establishing an appropriate set of feature detectors in the middle layer and the proper response in the output layer. Once the network is trained, we would like to be able to look at the knowledge it contains and confirm it has learned what it was supposed to. How can we examine the knowledge embodied by a network?

There is no magic technique to decipher the network's recognition strategy; we have to infer what it is doing based on the weighted connections established during training.

Networks generate their own set of key features, and those features may not correspond to those people find obvious.

From our analysis, however, it should be clear that the nodes are not the keys to the operation of the network. Instead, the starring roles belong to the connections between nodes and especially to the weights on those connections. The nodes themselves are little more than waystations on the path through the network; all the actual information processing is performed in the way the pattern is modified as it passes along the connections from the input layer to the output layer.

2.1 Output Layer

For neuron k in the output layer L, we select those neurons in layer $L-1$ most influencing on its activation value z_k^L. That is we select neuron $j \in \mathcal{J}_{L-1}$ if $| w_{jk}^{L-1} z_j^{L-1} | > a_{L-1}$, where w_{jk}^{L-1} is the weight of the connection from the jth neuron in layer $L-1$ to the kth neuron in layer L , z_j^{L-1} is the activation value of the jth neuron in layer $L-1$ and a_L is a chosen constant in such a way that $\alpha\%$ say of the nodes are selected. Let $\mathcal{J}_{L-1}^*(k) \subseteq \mathcal{J}_{L-1}$ be the set of neurons so selected. For the output layer and $j \in \mathcal{J}_{L-1}^*(k)$, we set $W_{jk}^{L-1} = | w_{jk}^{L-1} |$.

2.2 Hidden Layers

For the remaining hidden layers, we obtain the paths from the nodes in $\mathcal{J}_{L-1}^*(k)$ to the input layer that have a large impact on z_k^L in the following way.

We select neuron j in layer l, $1 \leq l < L-1$ if for some $j' \in \mathcal{J}_{l+1}^*(k)$:

$$| w_{jj'}^l W_{j'k}^{l+1} z_j^l | > a_l \qquad (1)$$

where a_l is a chosen constant.

Let $\mathcal{J}_l^*(k) \subseteq \mathcal{J}_l$ be the set of nodes so selected. For each j in $\mathcal{J}_l^*(k)$, we set $W_{jk}^l = w_{js_j} W_{s_j k}^{l+1}$ where:

$$s_j = \arg\max_{j' \in \mathcal{J}_{l+1}^*(k)} | w_{jj'}^l W_{j'k}^{l+1} | \qquad (2)$$

is the node in layer $l+1$ which succedes to node j in layer l. In this way, each node in $\mathcal{J}_l^*(k)$ has a significant output z_j^l and lies along one of the highly weighted paths from the input layer to the output node k.

2.3 Input Layer

Let $\mathcal{J}_1^*(k)$ be the set of selected nodes of the input layer according to section 2.2 and $\{W_{jk}^1 : j \in \mathcal{J}_1^*(k)\}$ their corresponding path weights to node k of the output layer.

These nodes indicate inputs that have contributed to the conclusion at node k of the output layer L. It may happen that $\mathcal{J}_1^*(k) = \emptyset$, indicating that no clear justification may be provided for a particular input output case. This means that no suitable path can be selected and the process terminates. Otherwise, we sort these nodes in decreasing order according to their impact $\mid W_{jk}^1 z_j^1 \mid$. Then we generate clauses for a symbolic rule from this ordered list until the following inequality holds:

$$\sum_{j_s \in \mathcal{J}_s^*(k)} \mid W_{j_s k}^1 z_{j_s}^1 \mid > \lambda \sum_{j_s \in \mathcal{J}_r^*(k)} \mid W_{j_r k}^1 z_{j_r}^1 \mid \qquad (3)$$

where λ is a chosen constant, $\mathcal{J}_1^*(k) = \mathcal{J}_s^*(k) \cup \mathcal{J}_r^*(k)$, and $\mathcal{J}_s^*(k)$ is the set of currently active input nodes contributing the most to the final conclusion (among those lying along the maximum weighted paths to the output node k) (see figure 1). We use all the nodes in $\mathcal{J}_s^*(k)$ to generate a set of conjonctive antecedent clauses for the rule regarding inference at output node k. The consequent part of the rule can be stated in quantitative form as membership value z_k^L to class k. In principle, it should be possible to examine a connectionist network and produce every such *if-then* rule.

3 Experimentations and results

3.1 Speech recognition

The TD-INFER algorithm was applied to a set of 400 french vowels uttered in ongoing speech by five male speakers. The data set has three features F_1, F_2 and F_3 corresponding to the first, second and third vowel formant frequencies obtained through spectrum analysis of the speech corpus. Thus, the dimension of the input vector for the proposed model is 3. We used an ANN with one hidden layer of 7 nodes. The training was done on a body of specific phrases where only the vowels have been labeled (around 200 examples). This training corpus has the complete set of input features in the 3-dimensional form while the desired output gives the membership to the vowel classes (/a/ , /u/ , /i/ , /e/, ...). The test set uses complete partial sets of inputs and the appropriate classification is inferred by the trained ANN model.

Some rules representing the sample test set patterns are:

- *Rule*$_1$: If F_1 is about 350 and F_2 is greater than 2300 then unlikely class /e/
- *Rule*$_2$: If F_1 is about 550 and F_2 is about 1600 then likely class /e/
- *Rule*$_3$: If F_1 is about 700 and F_2 is about 1000 then likely class /a/

3.2 Character recognition

This experiment concerns character recognition. The characters' used are A, C, E, Q, S, T and V. Each character is represented by a 7×5 matrix of 0's and 1's. The output is an 7 dimensional vector of O's and 1's. The error sum of squares decreases considerably with the number of hidden units. For 9 and 10 hidden units this error reaches a reasonable value (0.01) and stabilizes at 0.009. For this example we retained 9 hidden units. The network recognized all the letters. This result enables us to choose the optimal weights generated with 9 hidden units. In the final architecture of the network the matrix of the connections between the input and the hidden units is a 9×35 matrix and that of the connections between the hidden unit and the output unit is a 9×8 matrix.

Some of the generated rules are:

- *Rule*$_1$: If $X_{11} = 1$ and $X_{12} = 0$ and $X_{21} = 1$ and $X_{22} = 0$ and $X_{24} = 0$ and $X_{26} = 0$ and $X_{27} = 0$ then $SORT_5 = 0$.
- *Rule*$_2$: If $X_7 = 0$ and $X_{10} = 0$ and $X_{13} = 0$ and $X_{19} = 0$ then $SORT_1 = 1$
- *Rule*$_3$: If $X_6 = 0$ and $X_{14} = 0$ and $X_{29} = 0$ then $SORT_3 = 1$

where X_i is the ith input attribute.

For comparison, the algorithm presented in [13] is based on the principle of causal index and gives as a results an influence degree of the input values on the output values of the network. These results can be interesting in a probabilistic context. With regard to the heuristic proposed in [5] it differs with our algorithm by the fuzzy nature of the data corpus used by Mitra whereas our algorithm uses exact data corpus. The treatment is done from every output unit to the input units following an optimal path. Thus, our treatment is based on the values of the connection weights as well as on the activation values of the hidden units.

4 Conclusion

ANNs can be trained to provide for solutions in some domains where clear rules based on symbolic solutions are not available. However, they suffer from a major disadvantage since there is no explanation for why a particular solution is given by the network. TD-INFER algorithm extracts *if-then* rules from the network to represent its decision process. The magnitudes of the connection weights of a trained ANN are used in every stage of the algorithm inference. Furthermore, the tests we conducted showed that there is a strong relationship

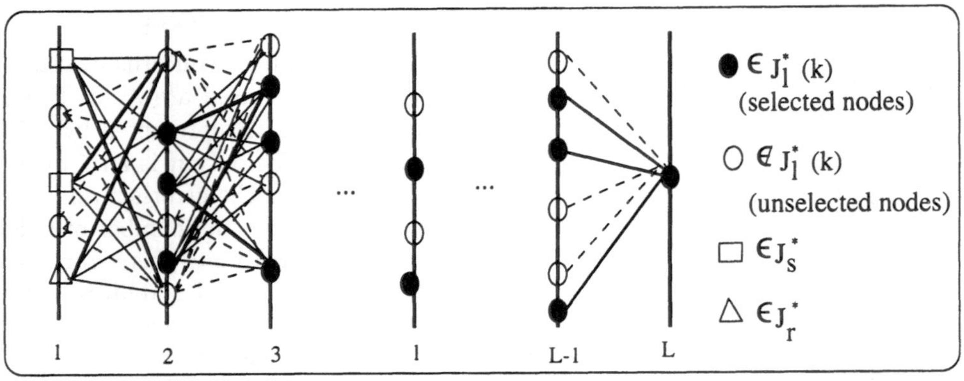

Figure 1: An example to demonstrate the rule generation scheme.

between the activation values of the hidden units and the values of the input and output units. Integration of these activation values in the rule generation process is very important for the accuracy of these rules.

References

[1] Mouria-Beji, F.: CODEPHON-NN: A context-dependent phonemic model based on neural networks. In Proc. IEEE Int. multiconference on Computational Engineering in Systems Applications. IEEE-SMC, April (1998).

[2] Mouria-Beji, F.: Neural network use in a non-linear vectorial interpolation technique for speaker recognition. In IEEE World Congress on Computational Intelligence, IEEE/IJCNN. vol. 2, pp. 1200-1205, Anchorage, Alaska. May (1998).

[3] Peifer, H., Gutknecht, M. and Stolze, M.: Cooperative hybrid systems. In Proceedings of 11th IJCAI, pages 824–829, (1991).

[4] Choi, E. C. Y. and Gedeon, T. D.: Comparison of extracted rules from multiple networks. IEEE Transactions on Neural Networks, 6, January (1995).

[5] Mitrad, S.: Fuzzy multi-layer perceptron, inferencing and rule generation. IEEE Transactions on Neural Networks, 6(l), January (1995).

[6] Terrace, H. M. K. and Ridge, K.: Integating rules and neural computation. IEEE Transactions on Neural Networks, 6, January (1995).

[7] Quinlan, J. R.: Comparing connectionist and symbolic learning methods. In Drastall, G. A., Hanson, S. J. and Rivest, R. I., editors, Computational Learning Theory and Natural Learning Systems. MIT Press, Cambrige, Mass., (1994).

[8] Mooney, R. J., Shavilk, J. W. and Towell, G. G.: Symbolic and neural learning algorithms: An experimental comparison. Machine Learning, 6(2):111-143, Mar. (1991).

[9] Yamashita, K., Hirose, Y. and Hijiya, S.: Back propagation algorithm wich varies the number of hidden unitsintegating rules and neural computation. Transactions on Neural Networks, 4:61-66, January (1991).

[10] Towell, G. G. and Shavlik, J. W., Extracting refined rules from knowledge-based neural networks. Machine Learning, 13(l):71-101, Oct. (1993).

[11] Fu, K. S.: Error-correcting parsing for syntactic pattern recognition. In Klinger, A., Fu, K. S. and Kunii, T. L., editors, Data Structures, Computer Graphics, and Pattern Recognition, pages 449-492. Academic Press, Inc., (1977).

[12] Mouria-Beji, F. and Boulahia, J.: ANNREX: An algorithm for neural network rule extraction. In Proc. IEEE Int. multiconference on Computational Engineering in Systems Applications. IEEE-SMC, April (1998).

[13] Sima, J.: Neural expert systems. Transactions on Neural Networks, 8(2):261-271, January (1995).

[14] Hayes, S., Ciesielsk, U. and Kelly, B.: Comparaison of an expert system and a hybrid neural network. In AAAI-92 Workshop on Integrating Neural and Symbolic Process, the Cognitive Dimension, San Jose, California, (1992).

[15] Pellegrini, C., Hilario, M. and Alexandre, F.: Modular integration of connexioniste and symbolic processing in knowledge based systems. In Proceedings of Int. Symposium on Integrating Knowledge and Neural Heuristics, pages 824-829, Pensacola Beach, Florida, (1994).

[16] Hinton, G. E., Rummelhart, D. E. and Williams, R. J.: Learning representations by error propagation. In Mc Clelland, J. L., Rummelhart, D. E. and the PDP Research Group, editors, Parallel Distributed Processing. Mc Graw-Hill Book Company, Cambrige, MA, MIT Press.

[17] Mouria-Beji, F. and Boulahia, J.: Extraction and insertion rules during the training process of a neural network. In the International Conference on Artificial and Computational Intelligence for Decision, Control and Automation in Engineering and Industrial Applications, March (2000).

Software Generation of Random Numbers by Using Neural Network

Choi-Kuen Chan, Chi-Kwong Chan and L.M. Cheng*

*Department of Electronic Engineering, City University of Hong Kong, 83 Tat Chee Avenue, Kowloon Tong, Hong Kong

Abstract. A new scalable random number generator is proposed. The proposed generator is based on the chaotic behavior of a clipped Hopfield neural network. By executing the neural network in a parallel architecture and performing simple permutation operations on the neural synaptic weight matrix in the proposed generator, random outputs of variable length can be obtained. Sequences produced by the proposed generator have passed standard statistical tests, exhibiting a good random behavior. The parallel structure of the clipped Hopfield neural network results in fast operation and can be implemented in software easily. A practical use of the proposed generator is the generation of secret keys for data encryption and keystreams in one-time pad for password authentication.

1 Introduction

Software cryptography is coming into wider use. Cryptography is an essential requirement for communication privacy or concealment of data. To secure information exchanges via the public communication channel, data encryption and password authentication are commonly used. The security of the cryptographic systems like DES, RSA, PGP, etc., depends upon the generation of unpredictable keys that must be of sufficient size and random. Here, a new scalable random number generator based on the chaotic property of a clipped Hopfiled neural network is introduced. The proposed generator is used to generate secret keys for data encryption and keystreams in one-time pad for password authentication.

Neural networks have recently been attracting much interest because of their nonlinear mapping property, such as the Hopfield Neural Network (HNN) [1]. Due to the highly nonlinear properties of HNN, it is extremely favorable for cryptography. HNN can be viewed as associative memory. Memorized patterns are stored in the stable points of the network in which they are auto-recalled in minimum hamming distance (MHD). The stable points are referred to as the attractors of the network. Each attractor has a basin of attraction in which state vectors surrounding the attractor are attracted to it in MHD. The basin of attraction is called the convergent domain of the corresponding attractor. In the case a of clipped HNN (CHNN) [2,3], the number of stored patterns will be larger than that of a HNN and the system initial state will converge to one of the attractors chaotically rather than in a predictable form as in a HNN [4]. This property can be used to randomize any outcomes of an event. A CHNN also has a property that for a few small changes in the neural synaptic weight matrix parameters, the distribution of system energy can be modified chaotically [5]. Thus a set of attractors with a large variation of chaotic domains of attraction can be obtained. This chaotic property of CHNN is extremely suitable for designing new cryptographic schemes.

In the proposed generator, by cascading blocks of CHNN in parallel, sequences of any length can be obtained according to the number of CHNN used. By passing through the CHNN, the input is nonlinearly mapped into one of the chaotic attractors of the network. This property randomizes the outputs from the generator. Simulated results show that the resulting sequences obtain a good statistical property. Due to its simplicity, the proposed generator is suitable to be implemented efficiently in software.

2 The Clipped Hopfield Neural Network

The clipped Hopfield neural network (CHNN) used here is a modified version of the general HNN, in which the synaptic weight matrix of the network is clipped to three values {-1, 0, 1}. For a network of n neurons, it stores a set of $2n + 1$ n-tuple chaotic attractors which are denoted as $\boldsymbol{A}$, $\boldsymbol{B}$ and $\boldsymbol{C}$, where $\boldsymbol{A} = \{A_i \mid i = 0, 1, \ldots, n-1\}$, $\boldsymbol{B} = \{B_i \mid i = 0, 1, \ldots, n-1\}$ and $\boldsymbol{C} = \{1^n\}$, with $A_0 = \{1^{n/2}\, 0^{n/2}\}$, $B_0 = \{1^{n/2+1}\, 0^{n/2-1}\}$ and $A_{i+1} = \Theta(A_i)$ for $i = 0, 1, \ldots, n-1$, $B_{i+1} = \Theta(B_i)$ for $i = 0, 1, \ldots, n-1$. Here,
and Θ (.) is a cyclic shift function. That is, if $x = \{x_0,$

$$\{a^n\} = \{\underbrace{aa......a}_{n}\} \tag{1}$$

$x_1, \ldots, x_{n-1}\}$, $\Theta(x) = \{x_1, x_2, \ldots, x_{n-1}, x_0\}$. Every state vectors in the n-dimension vector space converge to one of the attractors.

Let $\xi = \{\xi_\mu \mid \mu = 1, 2,\ldots, 2n + 1\}$ and $\xi = \{A, B, C\}$. Let $\xi_{\mu i}$ denotes the ith element of ξ_μ. The synaptic weight matrix W from neuron i to neuron j of the network is formed as

$$W_{ji} = \sigma\left(\left(\sum_{\mu=0}^{2n}(2\xi_{\mu,j}-1)(2\xi_{\mu,i}-1)\right)-1\right) \qquad (2)$$

Let S denotes the state vector of the network, the next state of each neuron $S_i(t+1)$ depends on the current states of other neurons in the following way:

$$S_i(t+1) = f\left(\sum_{j=0}^{n-1} W_{ij}S_j(t)\right), \quad i = 0,1,\ldots,n-1 \qquad (3)$$

where $f(.)$ is a non-linear function with

$$f(x) = \begin{cases} 1 & x \geq 0 \\ 0 & x < 0 \end{cases} \qquad (4)$$

Let Λ_i denote the convergent domain of attractor ξ_i. For all $i = 0,1,\ldots,2n$, we have

$$\Lambda_i \cap \Lambda_j = \phi \qquad \forall i \neq \qquad (5)$$

If a simple permutation is performed on the attractors ξ_i and the synaptic weight matrix W, i.e.,

$$\xi'_i = \xi_i P \qquad \text{for } i = 0,1,\ldots,2n \qquad (6)$$

and

$$W' = PW\tilde{P} \qquad (7)$$

where P is a permutation matrix and $\tilde{P}$ is the transpose of P. For example, consider $P = (0, 1, 2, 3, 4, 5, 6, 7) \rightarrow (3, 1, 0, 5, 7, 6, 4, 2)$, which can be written as

$$P = \begin{pmatrix} 0&0&0&1&0&0&0&0 \\ 0&1&0&0&0&0&0&0 \\ 1&0&0&0&0&0&0&0 \\ 0&0&0&0&0&1&0&0 \\ 0&0&0&0&0&0&0&1 \\ 0&0&0&0&0&0&1&0 \\ 0&0&0&0&1&0&0&0 \\ 0&0&1&0&0&0&0&0 \end{pmatrix} \qquad (8)$$

The network formed with W' and ξ'_i has the same properties as that formed with W and ξ_i, while $\Lambda'_i \neq \Lambda_i$, for $i = 0,1,\ldots,2n$. As a result, for the same input imposed on the network, it will converge to different attractor for different P. By changing P, different sequences are obtained.

3 The Proposed Generator

The proposed generator is constructed as shown in Fig. 1. In this generator, a random and secret 64-bit seed is firstly divided into 8 sub-seed. Each sub-seed is then xored with the time T generated from the computer timer. The 8 sub-seed, from seed0 to seed7, is further divided into 8 groups according to their bit position. They become the inputs to the CHNN. The 8 CHNN each with 8 neurons (n = 8) and have different neural synaptic weight matrix W (by randomly changing P) are cascaded together in parallel. Every inputs imposed on the CHNN will eventually converge to one of the chaotic attractors. The final 64-bit random output will become the new seed of the next generation. By simply cascading more CHNN in parallel, random sequences with more than 64 bits can be generated, e.g. 128 bits and 512 bits. Thus the proposed generator is scalable.

4 Tests and Results

The criterion used for random number generators is whether they were adequately uniform. To test for the distribution of the bits generated and the randomness, the output sequences of the generator has to go through standard statistical tests as specified in FIPS 140-1 [6]. This standard specifies security requirements for the design and implementation of random number generator. The following properties of the bit stream have been tested: distribution of single bit (monobit test), the number of occurrence of each of the 16 possible 4-bit combinations (poker test), the consecutive occurrence of 1's or 0's (runs test), and the maximum consecutive occurrence (long run test). If any of the above tests fail, the generator is rejected from being considered as a random number generator. The results from Table I show that the sequence has a good statistical property. (The output remained within acceptable limits of the tests for most of the sequences, generated with different seeds.)

5 Discussions

For a random number generator, the most important issue that we are concerned is the distribution of the bits generated, i.e. the randomness of the output sequences. The proposed generator and many other random number generators [7,8] also have good statistical properties and posses satisfactory degree of randomness. However, the proposed generator has an advantage that random outputs of variable length can be produced by cascading several CHNN in parallel, which is simple in architecture and can be implemented easily.

In the proposed design, eight CHNN each with 8 neurons ($n = 8$) are cascaded together in parallel which is same as a single neural network with size $n = 64$. This arrangement will not scarify the security as $n = 64$ but on the other hand the efficiency is enhanced by a basic network of $n = 8$, since CHNN can be implemented and executed in a parallel architecture.

6 Conclusions

In this paper, a new approach for constructing a scalable random number generator is proposed. By simply cascading several CHNN in parallel and changing the synaptic weight matrix in the neural network, different sequences with sufficient size are obtained. The resulting sequences possess a satisfactory degree of randomness. Since the proposed generator is simple in architecture in which only simple exclusive or functions and neural synaptic weight matrixes are used, it is suitable for software implementation.

References

[1] J.J.Hopfield: Neural Networks and Physical Systems with Emergent Collective Computational Abilities, Proc. Natl. Acad. Sci. USA *79*, 2554-2558 (1982).

[2] Donghui Guo, Zhengxiang Chen, Ruitang Liu and Boxi Wu: A modified Hopfield Model of Neural Network, Journal of Xiamen University (Natural) *32(1)*, 33-40 (1993).

[3] Chi-Kwong Chan and L.M.Cheng: Configurable Nonlinear Filter Generator, Electronic Letters, *34(4)*, 349-350 (1998).

[4] R.J.McEliece, E.C.Posner, E.R.Rodemich and S.S.Vankatesh: The Capacity of the Hopfield Associative Memory, IEEE Trans. Inform. Theory *IT-33(4)*, 461-482 (1987).

[5] Guo, L.M.Cheng and L.L.Cheng: A New Symmetric Probabilistic Encryption Scheme Based on Chaotic Attractors of Neural networks, Applied Intelligence *10(1)*, (1999).

[6] FIPS 140-1, Security Requirements for Cryptographic Modules, Federal Information Processing Standard Publications 140-1, U.S. Department of Commerce / N.I.S.T., National Technical Information Services, Springfield, Virginia, (1994).

[7] ANSI X9.17, Financial Institution Key Management(Wholesale), American National Standards Institute, (1995).

[8] National Institute for Standards and Technology, Digital Signature Standard, NIST FIPS PUB 186, U.S. Department of Commerce, (1993).

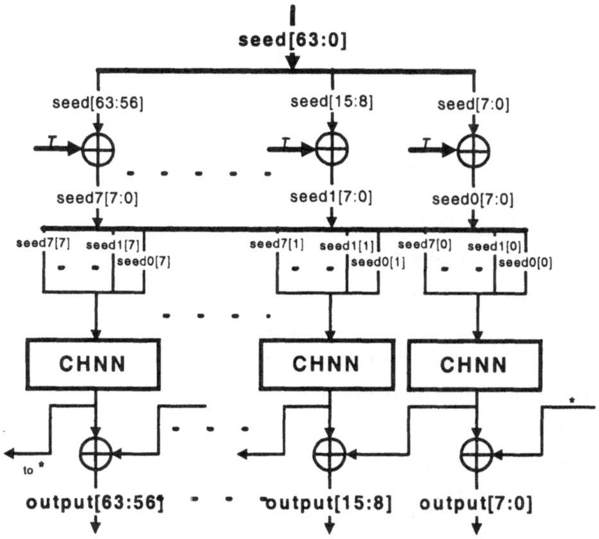

X[z, y] denotes the bit position from y to z of label X.
X[y] denotes the bit position y of label X.

Fig.1. The proposed generator

Table I. Simulated result of the output sequence

Statistical test	**Required interval**	**Output X**	**Result**
Monobit test	$9654 < X < 10346$	10036	Pass
Poker test	$1.03 < X < 57.4$	6.1576	Pass
Runs test	Run = 1, $2267 \le X \le 2733$	2505	Pass
	Run = 2, $1079 \le X \le 1421$	1227	Pass
	Run = 3, $502 \le X \le 748$	616	Pass
	Run = 4, $223 \le X \le 402$	303	Pass
	Run = 5, $90 \le X \le 223$	176	Pass
	Run $\ge$ 6, $90 \le X \le 223$	177	Pass
Long run test	Run $\ge$ 34, $X = 0$	0	Pass

FPGA implementation of a Spike-Based Sound Localization System

Marek Ponca *, Carsten Schauer †

*Dept. of Electornics, TU Ilmenau, Germany e-mail: marek.ponca@et.stud.tu-ilmenau.de †Dept. of Neuroinformatics, TU Ilmenau, Germany e-mail: carsten.schauer@informatic.tu-ilmenau.de

Abstract

In this paper we describe an implementation of a part of binaural sound localization system, which uses Interaural Time Differences (ITDs). All neurons are simulated by a spike response model, which includes postsynaptic potentials (PSPs) and refractory period. A winner-take-all (WTA) network selects the dominant source from the representation of the sound's angles of incidence. At the beginning, we explain the principles of the localization system, and then details of implementation of its input part: Inner Hair Cells (IHCs), responsible for transforming input sound signals into spikes.

1 Introduction

Computational neuroscience has established and evolved as a rich source for neuroinformatics and robotics, such that one can appreciate the subject of perceptional models as a significant example of their synergy.

It is obvious, that artificial systems which are more and more inspired by the biological model will benefit from a brain–like processing of their sensory input. The drawback of a simulation of the massively parallel processing in the visual and auditory system is the enormous computational expense. Hence there is an increasing interest in saving computational power, e.g. by implementing the algorithms in special hardware solutions. A multitude of approaches ranges from programmable analog and digital hardware to special VLSI chip designs – the technique of neuromorphic engineering.

In our interdisciplinary project it is the part of neuroinformatics to investigate models and methods of spatial hearing on the basis of real world experiments on mobile robots and, thus providing an objective for the work on hardware implementations of artificial neural systems.

2 Concept of the Sound Localization System

Our computational model of spatial hearing is designed to support the camera–based visual system of a mobile service–robot, to focus its attention in a certain direction or to localize a user just by his speech signal. Thus we are interested in an orientation in the horizontal plane. Further constraints for the model are robustness to noise and reverberations and the use of two–channel stereo signals instead of a larger microphone array. Consequently the basis of the system is a two–microphone configuration that enables the processing of spatial information, hidden in temporal and spectral characteristics of its stereo signal. Firstly the interaural time delay (ITD), corresponding to a different distance between the sound source and the two microphones, is used to built a 180°–map of angles from the left–most to the right–most direction. In addition, the sound color is analyzed to evaluate, wether a sound arrived from front or behind, so that finally a 360° enclosing detection is performed. The combination of the auditory and and visual space representation is simplified by the usage of omnidirectional cameras which obviates problems of coordinate transformation.

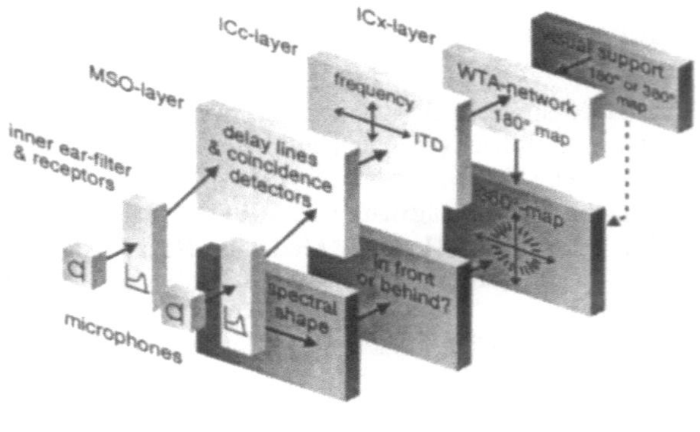

Fig. 1. System architecture, composed of the biologically inspired model of ITD detection (light blocks, MSO & ICc/x term the corresponding brain stem nuclei), a spectrum–based in front/behind discrimination (dark blocks) and the possibility to integrate spatial cues from the visual system.

Figure 1 sketches the model architecture. With certain restrictions, its essential part, the ITD detection in an artificial neural network (compare [7]), is inspired by the mammalian auditory brainstem.

Like the majority of the known ITD–detectors, it is based on Jeffress' coincidence model [5] [1], whose basic idea is the cross-correlation of corresponding nerve fibers from the left and the right hemisphere in a highly specialized neural structure. The time-window, necessary for the correlation function, is realized by counterpropagating neural delay lines. Coincidence cells integrate correlated inputs from the left and the right fibers and transform the time coded ITD information into the space code of the 180°–map. Since the coincidence detection is similar to the calculation of the cross-correlation of periodical signals its result is just as periodical. Also noise, echos and interferences with other sources often disarrange the spatial representation, and so we need to simulate a focusing mechanism to select reliably a dominant direction. Our model uses a winner-take-all (WTA) structure containing lateral and self excitation and recurrent inhibition via an interneuron, whereby only a single region of dominant feature representation can maintain activity [6].

The simulation of the system is based on spike–coded signals and a uniform spike–response neuron model. First the spike code is generated by a cochlear model which performs a frequency analysis in an all–pole gammatone filter [8] and than uses a hair–cell receptor to transform the analog signals into spike trains. Compared to a simple cross-correlation of the microphone signals, the presented system provides significant advantages:

- The spike-pattern mainly codes temporal information like phasing - amplitude is coded indirectly by the spike rate - the effect is a sharp peak as the result of the coincidence detection instead of a smooth sine–response.
- The broadband response of the cochlea filter together with a distributed processing in parallel frequency bands, effectively prevents phase–ambiguities.
- Because competing sound sources, noise or the disturbance by interferences need energy and time to shift the focus in the WTA–layer, the system shows a hysteresis property and thus prefers the signals onsets where the spatial cues are more reliable.

A further motivation to use spikes is, that an implementation of the whole network, either in a neuromorphic hardware or in programmable logic, will be simplified by an uniform description of the entire system. However, the utilization of silicon cochleas and the design of an analog VLSI-implementation as described in [4] turned out to be complicated and could not yet be realized.

As an alternative approach we are working on a demonstrator system in which the cochlear filter runs on a DSP and the artificial neural network is to be realized on a FPGA. To discuss basic aspects of the FPGA implementation, the realization of the hair–cell receptor is explained in this paper.

2.1 Neuron model

The neuron model (figure 2), a spike response model inspired by Gerstner's work [2] [3], takes up fundamental properties of biological cells: the spatial and temporal integration of stimuli via postsynaptic potentials (PSP) in the dendritic tree, the generation of an action potential when reaching a threshold, and the effect of diminished sensitivity during a period of refraction. An absolute refractory period and axonal delays are not modeled. To

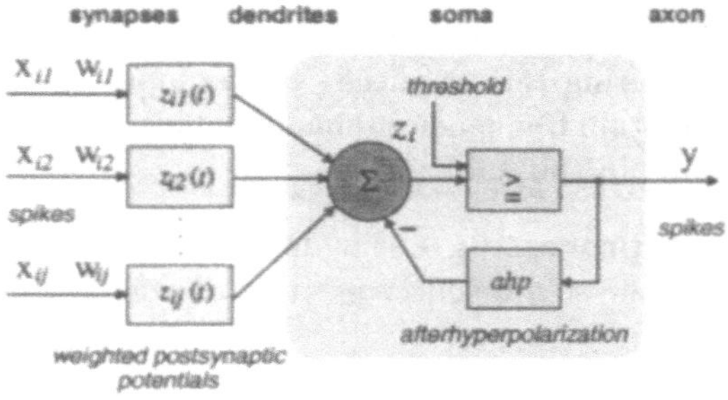

Fig. 2. Neuron model

describe the impulse response of a synapse, we chose the so called α–function $f_\alpha(t) = \frac{t}{\tau}e^{1-\frac{t}{\tau}}$, the afterhyperpolarization (AHP) follows a simple exponential fading function. The combination of these potentials results in a biological plausible behavior, which is more complex than the performance of leaky integrate-and-fire models.

The receptor model is similar to the spike response model with the exception, that the output of the all-pole gammatone filter is used as a generator potential instead of PSPs at the dendrites.

3 Implementation of the System for Sound Source Localization

3.1 Architectural overview

When we look at the input stage of the system described above, we can see the details of the spikes generation. We can divide the input–sound processing into three stages, as it can be seen on the figure 3.

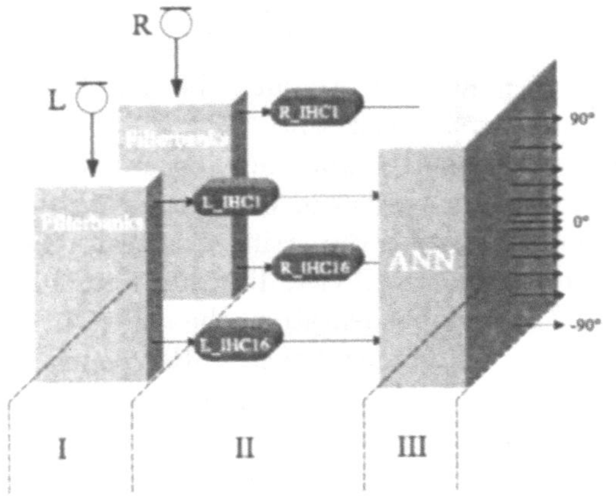

Fig. 3. Architectural overview of the system for sound source localization. I. Filtering, II. Spike generation, III. Neural processing

Filtering - dividing of a digital wideband signals from L and R microphone into 16 frequency subbands (in each sound channel) using the model of the biological cochlea [8]

Spike generation - generation of the spikes, considering the amplitude of the input signal in certain frequency subband and history of spiking in this subband

Neural processing - using I&F neurons and WTA is the sound source position decided

3.2 Filterbanks

The filterbanks act as silicon cochleas. They are implemented in Analog Devices - ADSP2101 Starter Kit, utilizing its built-in support for digital filtering.

3.3 Inner hair cells

To achieve the temporal resolution, which is required for ITD detection, we chose the sampling frequency to be 44.1kHz (65 resulting directions from +90° to -90°). Considering this fact, it is advantageous to implement only a core of the IHC (figure 4) and to process 16 subband's values (filterbank's outputs) of one sound channel sequentially on this common hardware. To do this, we use internal RAM (figure 5), where the actual AHP and TH values of every IHC are stored, to process it in the next cycle. The function of the IHC–bank can be explained as follows:

- DATA and ADDR are loaded into IN_REG, in ADDR(4:0) is the RAM address of appropriate IHC selected

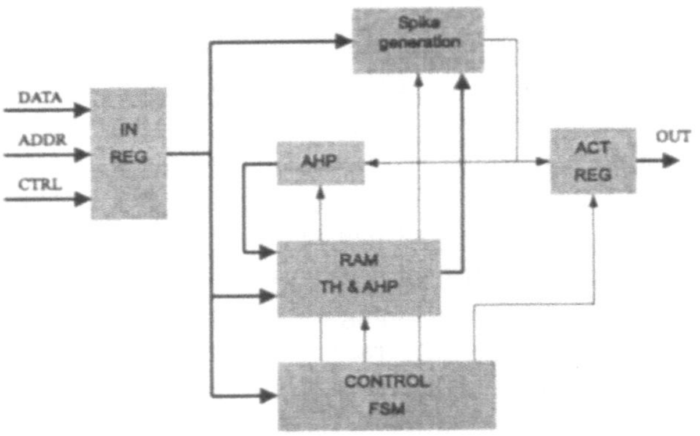

Fig. 4. Inner Structure of the IHC-bank

- RAM output (TH value) is in ACT_GEN (Spike generation block) stored, with inversion of RAM_address-MSB is the AHP space selected
- ACT_GEN elaborates the next-cycle RAM output (AHP) together with stored TH and actual DATA. If DATA-AHP>TH, then a spike is produced on an output signal, the output activities are registered in ACT_REG.
- Steps above are repeated for all IHCs.
- Content of the ACT_REG is propagated into the ANN, next sound sample is converted and filtered.

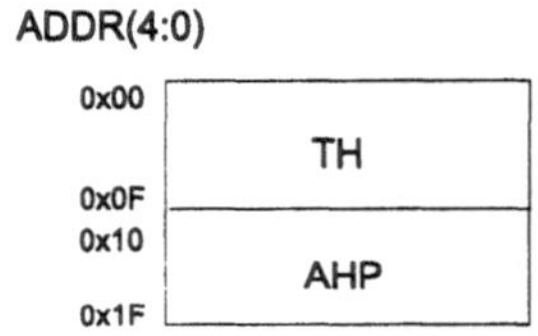

Fig. 5. Mapping of the TH and AHP values in internal RAM. The subbspaces differ in value of MSB

To realize the exponential decay of AHP, three different algorithms were approached:

Filter algorithm – works as a 1st order recursive filter, the decay is modelled with an impulse response to a spike

Sum of Shifts (SOS) – instead of constant $\frac{1}{e}$ for ($\tau = 1$) we can multiply the input value by a constant $\frac{3}{8}$. This simplifies arithmetic operation, we only need to shift input value 2, respectively 3 bits to the right, and to add this shifted values. For other τ it is possible to find another rational number, approximating the $e^{-\frac{1}{\tau}}$. For example $e^{-\frac{1}{10}}$ can be approximated as $\frac{7}{8}$.

Difference of Shifts (DOS) – the same principle as above, *Signed Digit Number* format is used [11], what makes it more suitable for approximation of decays with bigger τ. For example $A * e^{-\frac{1}{10}}$ can be approximated as $A * (1 - \frac{A}{16} - \frac{A}{32})$.

Results and characteristics of the algorithms descibed above are to be seen on figure 6. The filtering approach is the highest precision (depending on filter, that is implemented), the most area-effective is the SOS algorithm. Operating mode of the IHC-bank is within ADDR specified. After reset, the content of internal RAM must be initialized - proper TH and AHP starting values are loaded (INI mode). Than it is possible to test the whole function of a system: to connect all internal signals to the output or to step internal clock signal (TEST mode). In this mode the IHC-bank acts as an in-circuit simulator. In the most often used mode (NORMAL) the inputs are processed in real time. Here it is also possible to monitor the internal signals.

Method	Max. Error	No. Gates
Filter	6,49E-06	1156
DOS	8,77E-06	612
SOS	2,17E-05	795

Fig. 6. Simulation and synthesis results using different algorithms for modelling of an exponential decay for τ=10

4 I&F Neural Network

The output activities of IHCs are correlated, and processed separately for every subband in ANN. The WTA network chooses the dominant direction, where the sound seems to come from. To increase the capabilities of this stage, an implementation of I&F neurons operating with floating-point numbers will be investigated.

5 Conclusion

At the present we are implementing the IHC-bank into Altera FLEX10k FPGA. The hardware realization of I&F ANN will be done in the near future. In this stage the use of prototyping hardware is prefered (DSP Starter Kits, FPGA universal board). After testing of the system, a special-purpose PCI board will be developed to interface the sound environment to an experimental mobile robot. Further work will also concern the problem in which way the precision of the arithmetic affects the behaviour of the recurrent WTA structure.

References

[1] Carr, C.E., Konishi, M.: A circuit for detection of interaural time differences in the brainstem of the barn owl, *Journal of Neuroscience 10*, 3227–3246 (1990)

[2] Gerstner, W.: Kodierung und Signalübertragung in Neuronalen Systemen, *Reihe Physik*, Verlag Harri Deutsch, Thun - Frankfurt am Main (1993)

[3] Gerstner, W., Kempter, R., Leo van Hemmen, J., Wagner, H.: A neuronal learning rule for sub-millisecond temporal coding, *Nature 383*, 76–78 (1996)

[4] Ižák, R., Scarbata, G., Paschke, P.: Sound source localization with an integrate-and-fire neural system, *Proceedings of 7th International Conference on Microelectronics for Neural, Fuzzy, and Bio-Inspired Systems MicroNeuro'99*, 103–109, Granada, Spain (1999)

[5] Jeffress, L.A.: A place theory of sound localization, *J. Comp. Physiol. Psychol. 41*, 35–39 (1948)

[6] Kaski, S., Kohonen, T.: Winner-Take-All Networks for Physiological Models of Competitive Learning, *Neural Network 7*, 973–984 (1994)

[7] Lazzaro, J., Mead, C.: A silicon model of auditory localization, *Neural Computation 1(1)*, 41–70 (1989)

[8] Slaney, M.: Lyon's cochlea model, *Technical Report 13*, Apple, Advanced Technology Group (1988)

[9] Schonauer, T., Jahnke, A., Roth, U., Klar, H.: Digital Neurohardware: Principles and Perspectives, *Proceedings of Neuronale Netze in der Anwendung NN'98*, 101–106,Magdeburg, Germany (1998)

[10] Koren, I.: *Computer arithmetic Algorithms*, Prentice Hall (1994)

[11] Avizienis, A.: Signed-Digit Number Representation for Fast Parallel Arithmetic, *IRE Trans. on Elec. Computers EC-10*, 389–400 (1961)

A Simulator to Parallelise Large Biologically-Inspired Artificial Neural Networks

Yann Boniface *

*Cortex Team, LORIA - INRIA Lorraine, Campus scientifique, B.P. 239, F-54506 Vandœuvre-lès-Nancy Cedex. E-mail : Yann.Boniface@loria.fr

Abstract

This paper presents parallel simulator for artificial neural networks development. This simulator uses the intrinsic neuron parallelism of the connectionists models to map the neural networks onto shared memory MIMD general purpose parallel computers. Due to this method, the simulator is more especially dedicated to large biologically inspired neural networks.

1 Introduction

Nowadays artificial neural networks are commonly applied to numerous technical problems. Many connectionist models are proposed according to the different problems to solve. However, these models are well known to be very time consuming, especially during the training phase. Moreover, the practical problems to be solved are always larger.

At present, biologically inspired neural networks are one of the more studied fields of connectionism. Using these models, researchers try to map the brain, which is often described like a parallel model composed with many simple units, the neurons. In the same way, biologically inspired networks are, in theory, totally distributed models. However the elementary unit of these model are often more complex than the formal neuron, used in classical artificial neural networks.

To help researchers to develop theirs own models, we propose in this paper a new artificial neural networks simulator. This simulator allows to implement any kind of neural networks, and to use general purpose parallel computers for their executions. In order to make the implementations easier, our simulator uses the intrinsic parallelism of neural networks both for the development of the network and for the mapping onto parallel machines. Due to the differences between the two parallelisms, our simulator is rather dedicated to large biologically inspired neural networks than small classical artificial neural networks.

2 Overview of the Simulator

To propose a user friendly tool, we have developed a 'C' library. This library is restricted to a few set of functions necessary to define the simulator paradigm. With this library, an artificial neural network is considered as a sum of independent neurons, synchronised by cycles [1]. To implement it, one needs to describe each neurons of this network[1] as an object like the one is depicted in figure 2.

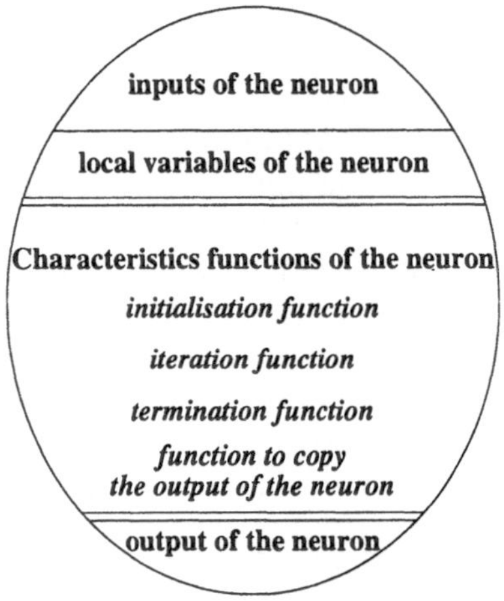

Fig. 1. Description of the object *neuron* within our simulator.

To describe a neuron, it is necessary to declare the inputs of the neuron, its unique output and its *characteristic functions*. The output of the neuron represents the only data that the other neurons of the networks can read. The inputs are the connections of the neuron to some others neurons of the network. Through one input, the neuron can read the output declared by the connected neuron. The characteristic functions describe the method used by the neuron to, for each cycle of the execution, compute his output function of his inputs and his locals variables. The *initialisation function* is dedicated to the first cycle of the neuron. The *termination*

[1]In fact each different types of neurons present into the network.

function is dedicated to the last cycle of the neuron. And the *iteration function* refers to all the others cycle of life of the neuron.

In order to use and study the famous intrinsic parallelism of neural models, the networks implemented with the simulator use the *neuron parallelism* of connectionist models [2]. Each neuron of the network works continuously, in parallel. In the same way, with our simulator, we consider that all the neurons could be executed in parallel (see figure 2) i.e the result of the network computation is independent of the neural processing order.

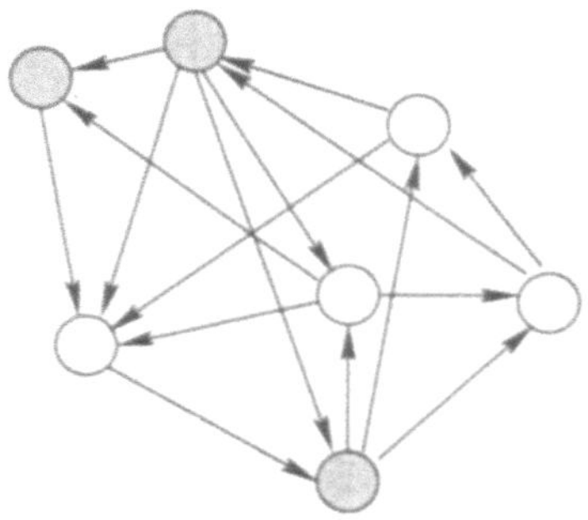

Fig. 2. Neuron parallelism of connectionism : each neuron could be executed in parallel. Here grey neurons are executed at the same time during the network execution.

3 A parallel Simulator

Our simulator allows parallel executions of artificial neural networks. When a network is implemented with our library, the user can execute it on both sequential or parallel computers. In term of parallel machines, our simulator works onto *shared memory MIMD*[2] general purpose parallel computers. To allow the use of parallel computers by connectionists, the parallel implementation of neural networks is totally transparent. The code is the same for both sequential and parallel compilation, and the user needs no specific parallel knowledge to develop these programs. The main difficulty is due to the fact that both artificial neural networks and computer parallelisms are notably different. Neuron parallelism is a fine grain parallelism whereas MIMD parallelism is a coarse grain parallelism [3]. To resume, neural networks parallelism is defined by networks with many units (neurons), a great connectivity and a lot of small messages exchanged between the different units. On the contrary, an MIMD parallel computer has few processors, and the communication power is weak compared to the computation capabilities of these machines. So, an efficient program onto an MIMD parallel computer needs to have long computing periods for, proportionally, a few communication phases. In order to make transparent the use of parallel machines, the simulator uses the neuron parallelism of neural models to map connectionists networks onto the parallel machine (see figure 3).

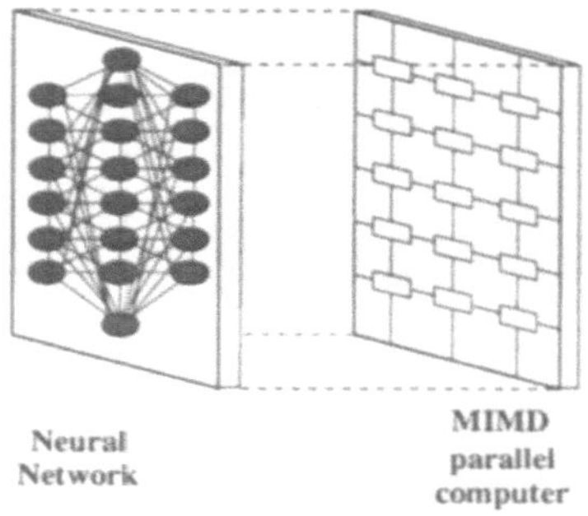

Fig. 3. The simulator map the neural network onto the processor network of parallel machines in a transparent way for the user.

The generic aspect of our parallel simulator is due to the new large spread of shared memory parallel computers. Before the shared memory, parallelisation of neural networks was often restricted to specific neural topologies[4, 5], and hard to use [6, 7]. With the use of shared memory, the parallelisation of neural networks has to be extracted from the neural network parallelism. The developer uses the classical algorithms to implement these models, without any modification. It is only the library which maps the implementation onto the parallel machine. This method can yield good results when the neural network has the good properties, that are present in large biologically inspired artificial neural networks.

4 A Simulation Model Close to the Biological Model

Due to the use of neuron parallelism for the mapping onto parallel machines, the simulator needs correspondent models to gain parallel performance. The simulator uses neuron parallelism to parallelise networks. The statistical networks, that do not have this property cannot allow parallel performance. For instance, back-propagation learning networks do not match the neuron parallelism property. In this class of models, neurons are computed layer by layer : all the layers are not executed simultaneously. When one layer is being computed, all the neurons from other layers do not work. So the parallel perfor-

[2]Multiple Instructions Multiple Data

mance of the simulator are poor with this kind of networks.

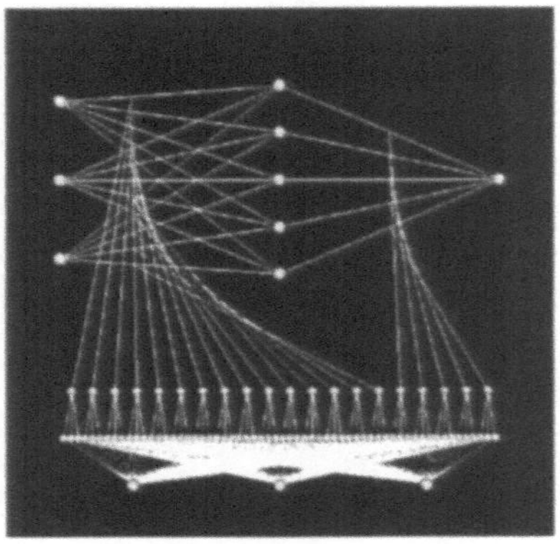

Fig. 4. Example of an OWE neural network with 3 × 5 × 1 neurons.

We have implemented a large model of back-propagation network, the OWE (see figure 4). This network is a contextual MLP : each weight of the MLP is compute by a another specific MLP [8]. We have implemented a 32 × 10 × 1 neurons MLP, and each weight is compute by a specific 32 × 10 × 1 neuron MLP. Figures 5 and 6 show the parallel performance obtained with this network.

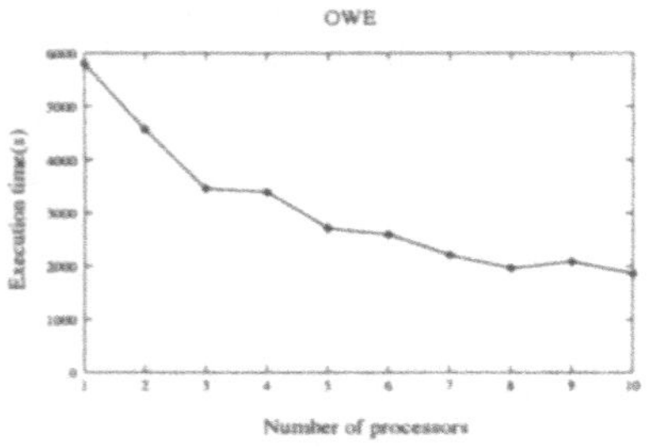

Fig. 5. Execution time for an OWE neural network

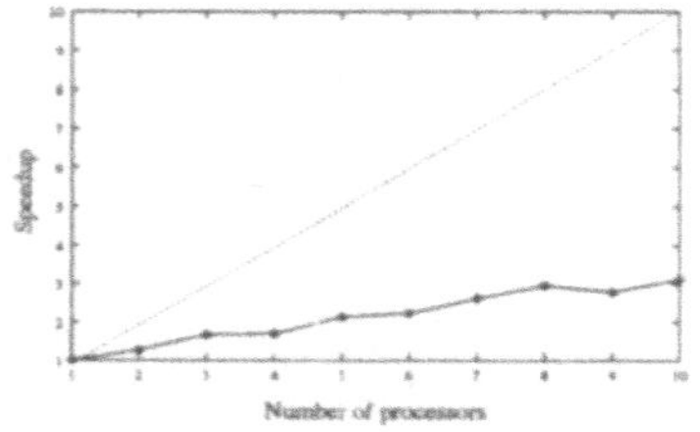

Fig. 6. Speedup for a Growing Neural Gas network with 2000 neurons.

Despite the large number of neurons present in the network, the simulator obtains a poor speedup of the executions. With 10 processors, the execution is just 3 times faster.

On the other hand, the simulator obtains good parallel performance for networks with neuron a parallelism property. Figure 7 shows the speedup obtains with a Growing Neural Gas network. With this dynamic model, each neuron is computed for each example. So all the neurons can be computed at the same time, and the simulator can map efficiently this kind of network onto a parallel computer. The execution is more than 30 times faster with 32 processors.

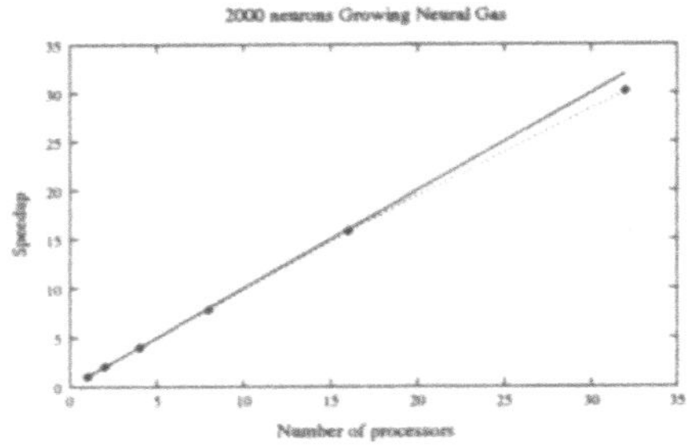

Fig. 7. Execution time for an 2000 neurons Growing neural gas.

5 Efficient Parallelisations Need Large Networks

When an artificial neural network has neuron parallelism, the parallel simulator allows good performance. But, due to the differences between both neuron and MIMD parallelisms, only large networks allow for an efficient parallelisation. An efficient implementation onto MIMD computer needs large computation phases and few communication or synchronisation phases between the different processors whereas neural networks are known to have small computation phases and many communication phases between the different neurons. Consequently, the simulator needs large networks to increase the number of computing phases and to compensate for the number of communications between processors.

As shown in figure 8, with few neurons (100 here), results of parallelisation are not satisfying. After 8 processors, due to the increase of the number of communications, the computation time increases with the number of processors.

On the other hand, with large networks, the simulator can keep performance with more processors. Figure 9 shows that simulator have good parallel performance until 32 processors, with a 2000 neurons Growing Neural Gas.

These good parallel performance have two dif-

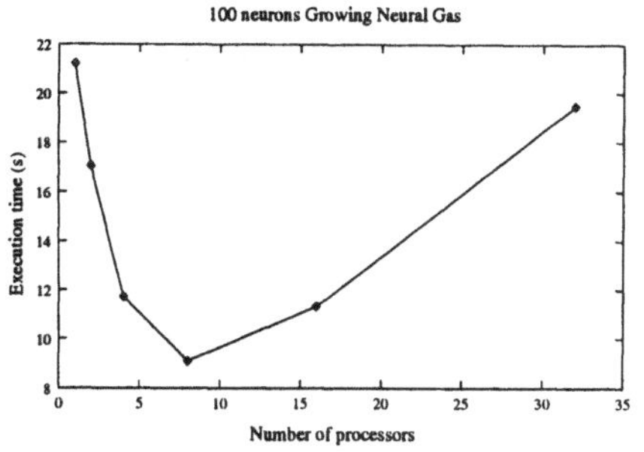

Fig. 8. Computation time as a function of number of processors for a 100 neurons Growing Neural Gas.

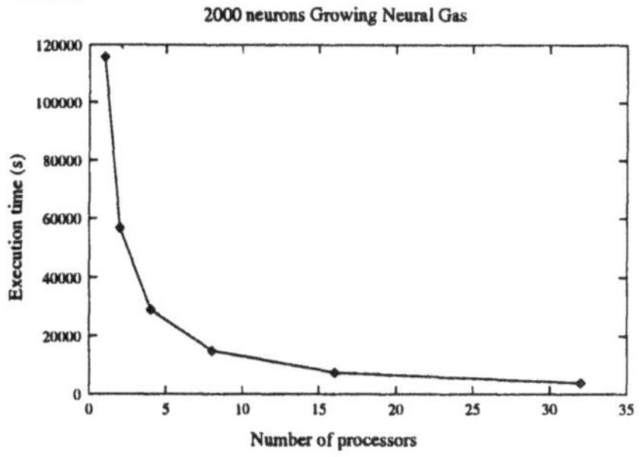

Fig. 9. Computation time as a function of number of processors, for a 2000 neurons Growing Neural Gas.

ferent causes. First, the computation phases are larger, and they mask better the communication phases. Second, artificial neural networks are known for their memory cost. In a parallel machine, the size of cache memory available increases with the number of processors. Gains get by the cache are better if the network has a great cost in memory : if the network has a large number of neurons.

Figure 10 shows the speedup obtained with different sizes of networks. The efficiency of the simulator parallelisation increases with the size of the networks. For 50 or 100 neurons networks, the parallelisation is quickly inefficient, whereas the parallelisation is still efficient with 32 processors for 1500 and more neurons.

6 Conclusion

In conclusion, the parallel applications of our artificial neural networks simulator library are interesting for large networks with neuron parallelism property. These two characteristics allows to obtain efficient performance in terms of speedup of the networks execution time. By their properties and this implementation type, implementation of a network is based on the implementation of the neuron of this network, the model of efficient networks

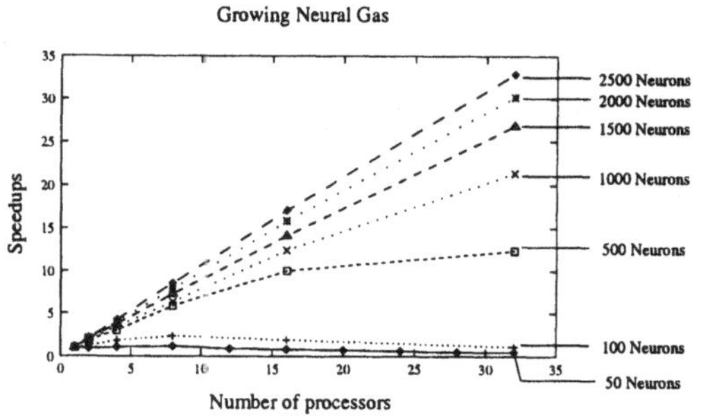

Fig. 10. Speedup comparisons between different size of Growing Neural Gas networks.

is close to the biologically-inspired artificial neural networks. Commonly these networks are based on independent neurons, and they are required to have larger and larger number of neurons. Thus, the simulator presented in this paper is especially designed to implemented large biological models.

References

[1] Y. Boniface, F. Alexandre, and S. Vialle, "A bridge between two paradigms for parallelism: Neural networks and general purpose MIMD computers.," in *IJCNN*, 1999.

[2] T. Nordström and B. Svensson, "Using and designing massively parallel computers for artificial neural networks," *Journal of Parallel and Distributed Computing*, vol. 14, no. 3, pp. 260–285, 1992.

[3] Y. Boniface, F. Alexandre, and S. Vialle, "A library to implement neural networks on MIMD machines," in *EuroPar'99*, 1999.

[4] T. Reski, *Mapping and Parallel, Distributed Simulation of Neural Networks on Message Passing Multiprocessors.* PhD thesis, Universität Paderborn, 1999.

[5] N. Sundararajan and P. Saratchandran, *Parallel architectures for artificial neural network, Paradigms and Implementation.* IEEE computer society, 1998.

[6] H. Paugam-Moisy, "Multiprocessor simulation of neural networks.," in *The Handbook of Brain Theory and Neural Network.*, pp. 605–608, The MIT Press., 1995.

[7] S. Shams and J. Gaudiot, "Parallel implementations of neural networks," *International Journal on Artificial Intelligence Tools*, vol. 2, no. 4, pp. 557–581, 1993.

[8] L. Bougrain and Y. Boniface, "Détermination de l'architecture de modèles neuromimétiques par implémentation parallèle," in *Journées Scientifiques du CCH*, 2000.

Time Perception and Reinforcement Learning — a Neural Network Model of Animal Experiments

J. L. Shapiro, John Wearden *

*Manchester University, Manchester M13 9PL, U.K.

Abstract

Animal data reveals that animals can time intervals associated with rewards. The data shows a striking property — the data for different time intervals collapses into a single curve when the data is scaled by the time interval. This is called the scalar property of interval timing. Here a simple model of a neural clock is presented and shown to give rise to the scalar property. The model is an accumulator consisting of noisy, linear spiking neurons. It is analytically tractable and contains only three parameters. When coupled with simple reinforcement learning, it simulates experiments on estimation of time duration in animals.

1 Introduction

Often animals need to carry out actions in order to receive rewards later. As examples, an animal may travel to a foraging site and expend time and effort looking for food before any food is found; two animals may engage in combat with the expectation of winning a mate; conference attendees may walk through an unfamiliar city with the expectation of finding their way to the conference. In these examples, the animal must carry out a series of actions before any reward is found.

How does the animal know when the action is not working and when it is time change to another behavior? One answer could be that the animal has a sense of the length of time that it has been carrying out the current strategy and has knowledge of the expected time to find a reward. If the action has been tried for a time longer than the expected time to reward, it could be time to abandon this approach and try something else. In the examples above: if no food has been found after a given length of time, the foraging patch could be depleted; the animal may move to another patch. If the combat does not result in the other animal backing down after a given time, the animal may look for an area defended by a less aggressive male. If the conference goer wonders around the city without finding the conference site for too long, he might go and buy a map.

It is clear that the ability wait the *appropriate* length of time before changing behaviors would be advantageous. In the case of foraging, an animal which left too soon, always exploring, would never remain in one place long enough to find food. Likewise, the animal who remained in spent locations would also go hungry. In order to have this ability, an animal needs to be able to perceive time intervals — it needs an internal clock. In addition, if the waiting time is to be learned from experience, it needs to solve the delayed-reward reinforcement learning problem. These issues have been discussed by Grossberg and Merrill [4] and Moore et. al. [7].

It has been shown in a number of psychological experiments that animals can learn to time intervals and wait for rewards, so they do have an internal clock. In addition, the data shows a special property, called the scalar property. When the responses are scaled with respect to the time interval being learned, the data collapse onto a single curve. This scalar property could be revealing something about the mechanisms of animals' internal clock, but this has not be understood up to now. This paper describes a neural network model of a clock used which can be used to perceive time intervals. This model reproduces the scalar property found in the psychological data. We show that this can be coupled to reinforcement learning algorithms to produce a model which reproduces behavior experiments on time perception in animals.

1.1 The scalar property of the internal clock

Numerous experiments have investigated the ability of animals to estimate time intervals and wait appropriately for delayed rewards. A typical type of experiment is the *peak procedure* experiments. The animal is rewarded for the first response after a given time interval has elapsed. For example, a light may be turned on at the start of the experiment. The animal must push a lever after given time interval has passed to receive food. There is no penalty for pushing the lever too soon, but only after the time interval has passed does the lever push result in food. Once the reward is received, the light goes off and the trial ends.

On some trials, however, no reward is given even when the animal responds appropriately. This is done to see when the animal stops responding. What happens in non-reward trials is that the animal will respond for a period, and then stop responding. Responses averaged over many trials give a smooth curve. The highest response is at the time interval, but there is variation around this. The variation increases as the time interval gets longer. Thus, on average the animal measures the interval correctly, but the inaccuracy in this increases with the length of the timed interval.

A striking feature of the data on peak procedure and other experiments on time intervals is the so called "scalar property". The inaccuracy in the measured time interval increases in proportion to the time interval. Thus, in its weak form, the scalar property states that the ratio of the standard deviation to the mean response time (the *coefficient of variation*) is a constant independent of the time interval.

A stronger form of the scalar property is also found. When the response probability is multiplied by the time interval and is plotted against the relative time (time divided by the time interval), the data from different time intervals collapse on to one curve. This has been found not only in peak procedure experiments, but in other experiments on interval timing. It is found in many species, including rats, pigeons, turtles; humans will show similar results if the time intervals are short or if they are prevented from counting through distracting tasks (see below). For reviews of interval timing phenomena, see [5] and [3].

This strong form of the scalar property can be expressed mathematically as follows. Let T be the actual time since the start of the trial and $\tilde{T}$ be subjective time. Subjective time is the time duration which the animal perceives to have occurred (or appears to perceive judging from its behavior). The variation of $\tilde{T}$ for given T can be expressed as the conditional probability, $P(\tilde{T}|T)$. The fact that the data collapses implies this probability depends on T and $\tilde{T}$ in a special way,

$$P(\tilde{T}|T) \approx \frac{1}{T} P_{\text{inv}} \left(\frac{\tilde{T}}{T} \right). \tag{1}$$

Here P_{inv} is the function onto which the data collapses. Thus, time acts as a scale factor. This is a strong and striking result.

A key question which remains unanswered is: what is the origin of the scalar property. Since the scalar property is ubiquitous across experiments and species, it may be revealing something fundamental about the nature of animals' internal clock. Up to now, it has been very difficult to devise a mechanism to produce this property. One mechanism which can be ruled out is a noisy clock. For example, one could imagine that the variation times from the use of a clock with variable pulse-rate, or which miscounts pulses. It is well known that such a model, or any model based on the accumulation of independent errors, would not produce the scalar property. In such a model it would be the ratio of the *variance* to the mean response time which would be independent of the time interval. This is what is found in humans if counting is not suppressed. Thus, it would appear that humans achieve higher accuracy in measuring time intervals by counting interval clock ticks. When counting is suppressed, humans show the scalar property in analogous experiments to the animal experiments.

In this paper, a simple neural network model of an internal clock is described which gives rise to the scalar property. When coupled with reinforcement learning algorithms [10], like those applied to classical conditioning experiments (e.g. Sutton and Barto [9], and Moore et. al. [6]), it reproduces the results of peak procedure and other experiments.

2 Previous Work

Most models of interval timing assume a scalar clock, and use that to address other questions. One question which has been studied is, given a scalar measure of time, how does the organism convert that into behavior. This is a primary part of Scalar Expectancy Theory) [1].

Connectionists models of scalar timing were proposed by Church and Broadbent [1] and Grossberg and Merrill [4]. Both used a number of clocks tuned to different frequencies. The former put the scalar property ab initio by couplying the clocks to a multiplicative noise source. The latter did not test the scalar property quantitatively, although the tuning curves of each clock appeared to have increasing width for increasing delay. This originated from the first order differential equations which described the neural dynamics, but required tuning of a large number of parameters.

3 A Connectionist Model of the Clock

We now consider a very simple connectionist model which reproduces the scalar property. It consists of a network of spiking neurons. The neurons are noisy and have linear response. The network encodes time as the quantity of activity, i.e. the number of spikes in the network. This grows over time.

The network consist of N identical neurons. Any neuron can potentially be connected to any other; the actual connectivity pattern is defined by a connection matrix $\mathbf{C}$, where C_{ij} is 1 if neuron j receives input from neuron i and 0 otherwise. Each node is connected to C other nodes on average. Time is divided into discrete units of size τ which are assumed short (10 mS, say). During a time interval, a neuron can emit one or more spikes. There is no limit to the spiking rate of these neurons, so saturation effects are ignored.

The neurons are noisy linear neurons. Let $x_i(t)$ denote the number of spikes emitted by neuron i between times t and $t + \tau$. The number of spikes arriving at that neuron (pre-synaptically) $h_i(t)$ is

$$h_i(t) = \sum_j C_{ji} x_j(t) + I_i(t), \tag{2}$$

where I_i is external input. The assumption is that the number spikes emitted by this neuron by time $t+\tau$ is drawn from a probability distribution with mean $\gamma h_i(t)$, with gain $\gamma < 1$. For example, if there is an independent probability γ of each spike crossing the synapse, then the distribution of spikes emitted by neuron i is binomial.

At each time-step, the number of spikes will grow due to the fan-out of the neurons; each neuron can excite several other neurons. At the same time, the number of spikes will shrink due to the fact that a spike invokes another spike with a probability less than 1. An essential assumption of this work is that these two processes balance each other.

$$C\gamma = 1. \tag{3}$$

This insures that the number of spikes internally excited remains constant over time on average.

Finally, in order for this network to act as an accumulator, it receives internal input during the time interval which is being perceived. The input is assumed statistically the same at each time-step.

This simple network gives rise to the scalar property in timing. Let $P(n|t)$ be the probability of network having total activity n at time t. What is proved is,

$$\lim_{t \to \infty} P(n|t) = \frac{1}{t} P_{\text{inv}}(n/t), \tag{4}$$

with corrections of order $1/t$. Thus, if an interval of time is measured through the activity of this network, the perception of time will obey the scalar property equation (1). The invariant distribution is a gamma-function,

$$P_{\text{inv}}(x) = \frac{\exp(-x/b) x^{a-1}}{b^a \Gamma(a)}; \ a = \frac{2m_I}{C\sigma_\eta^2}; \ b = \frac{C\sigma_\eta^2}{2\tau}, \tag{5}$$

where m_I is the mean external excitation per unit time and σ_η is the standard deviation of the neuron noise.

4 Learning Time Intervals

The above model represents a way of representing and measuring time. How would this be used to produce behavior? We have considered two models for learning time intervals. In the first, time is converted into a spatial code which are associated with time intervals. In the second, the accumulator output is used by the system directly. The spatially encoded model with be considered here. The details will be presented elsewhere [8].

Learning is done with a form of temporal difference learning. In this model, the accumulator network feeds into a set of nodes which encode the activity of the accumulator. Each of the spatial code nodes feed into response nodes which stimulate the responses. The connection between the the ith spatial node and the jth response node learns via a temporal difference learning rule.

This model has some similarities to that of Desmond and Moore [2]. The spatial nodes play the role of the tapped delay-line nodes in that model. However, here they are stimulated by the accumulator rather than each other, and they will follow a stochastic trajectory due to the fluctuating nature of the accumulator. Learning uses an eligibility trace for the spatial nodes and learning is via a simple TD rule. The model also as some similarities to the Grossberg and Merrill model [4]. The spatial nodes respond to different times and the response is associated with time. However, the spatial nodes require no special tuning curves.

The model has been used to simulate peak procedure. In the simulations, the model is forced to respond for the first set of trials (50 trials in the simulations); otherwise the model would never respond. This could represent shaping in real experiments. After that the model learns using reward trials for an additional number of trials (100 trials in these simulations). The model is then tested on a number of non-reward trials. Figure 1 shows the average responses for non-reward trials, averaged over 100 runs. The model clearly exhibits the scalar property.

5 Discussion

Previous models of interval timing fail to explain its most striking feature — the collapse of the data when scaled by the time interval. We have presented a simple model of an accumulator clock based on spiking, noisy, linear neurons which produces this effect. The model is analytically tractable and has only three adjustable parameters. The major weakness of this model is that it requires fine-tuning of a pair of parameters, equation (3).

Once a scalar clock is produced, simple reinforcement learning can be used to associate the clock signal with appropriate responses. Simple temporal difference association between the intermediate nodes at reinforcement and an eligibility trace simulates peak procedure. Another set of experiments, bisection experiments, have also been modeled using this approach.

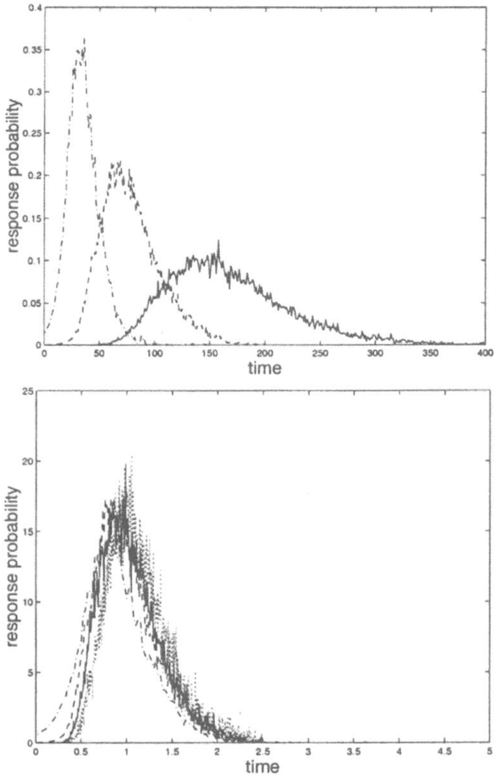

Fig. 1. Top: Average response of the network for non-reward trials. 100 non-reward trials have been averaged. Bottom: The response probabilities scaled with the time interval and plotted against relative time. The strong scalar property is clearly observed. The time intervals are 40τ (dot-dashed line), 80τ (dashed line), 160τ (solid line), and 320τ (bottom figure only).

References

[1] R. Church and H. Broadbent. Alternative representations of time, number, and rate. *Cognition*, 37:55–81, 1990.

[2] John E. Desmond. Temporally adaptive responses in neural models: The stimulus trace. In Michael Gabriel and John Moore, editors, *Learning and Computational Neuroscience: Foundations of Adaptive Networks*, A Bradford Book, pages 421–456. The MIT Press, 1990.

[3] John Gibbon and Russell M. Church. Representation of time. *Cognition*, 37:23–54, 1990.

[4] Stephen Grossberg and John W. L. Merrill. A neural network model of adaptively timed reinforcement learning and hippocampal dynamics. *Cognitive Brain Research*, 1:3–38, 1992.

[5] S. C. Hinton and W. H. Meck. How time flies: Functional and neural mechansims of interval timing. In C. M. Bradshaw and E. Szadabi, editors, *Time and Behaviour: Psychological and Neurobehavioural Analyses.* Amsterdam: Elsevier Science, 1997.

[6] J. W. Moore, J. E. Desmond, and N. E. Berthier. Adaptively timed conditioned responses and the cerebellum: A neural network approach. *Biological Cybernetics*, 62:17–28, 1989.

[7] John W. Moore, Neil D. Berthier, and Diana E. J. Blazis. Classical eye-blink conditioning: Brain systems and implementation of a computational model. In Michael Gabriel and John Moore, editors, *Learning and Computational Neuroscience: Foundations of Adaptive Networks*, A Bradford Book, pages 359–387. The MIT Press, 1990.

[8] Jonathan Shapiro, John Wearden, and Rossano Barrone. Simple models of scalar timing. In preparation.

[9] Richard S. Sutton and Andrew G. Barto. Time-derivative models of pavlovian reinforcement. In Michael Gabriel and John Moore, editors, *Learning and Computational Neuroscience: Foundations of Adaptive Networks*, A Bradford Book, pages 497–537. The MIT Press, 1990.

[10] Richard S. Sutton and Andrew G. Barto. *Reinforcment Learning: An Introduction.* A Bradford Book. The MIT Press, 1998.

ADFs: An Evolutionary Approach to Predicate Invention

C. Giraud-Carrier*, C.J. Kennedy†1

*Department of Computer Science, University of Bristol, Bristol, UK †Defence Evaluation and Research Agency, EX Building, St Andrew's Road, Malvern, UK

Abstract

This review paper makes explicit, for the first time, the conceptual similarity between Genetic Programming's ADFs and Machine Learning's invented features/predicates, and shows how the evolutionary nature of ADFs allows the complex *when* and *what* questions of predicate invention to be answered automatically.

1 Introduction

In machine learning, language biases applied to the hypothesis language restrict the number of expressible hypotheses. Whilst desirable from the point of view of efficiency and generalisation [16], such biases may hinder effectiveness as, in some cases, there will not exist a hypothesis consistent with the training set. For example, it is not possible to express relationships between attributes in the commonly-used attribute-value language. In such cases, the (biased) hypothesis language must be extended with appropriate constructs (e.g., attributes, functions). In fact, even when the hypothesis language need not be extended, it may still be advantageous to do so in order, for instance, to reduce the size of induced hypotheses.

The problem of deciding 1) *when* new constructs would be useful/necessary and 2) *what* the new constructs should consist of has become known as constructive induction or predicate invention [14, 29, 20]. The *when* question of predicate invention is undecidable in the general case [26]. The *what* question, although rather trivial in principle, begs the more difficult question of *utilisation*, i.e., which of many potential constructs are most appropriate [20]. Hence, progress has been slow within mainstream machine learning and has largely ignored related work on Automatically Defined Functions (ADFs) in the Genetic Programming (GP) community.

ADFs are an extension to GP involving the evolution of extra functions or modules that support code reuse. In the simplest case, the basic program tree of GP is enhanced to become a structure combining a main program and the subroutines, or ADFs, that may be used by that main program. The ADFs and the main program are evolved simultaneously. In some implementations, λ-abstractions are used to denote "anonymous" ADFs, that are evolved as arguments to higher-order functions in the main body of the program structure. There is a strong link between ADFs (and their extensions) and predicate invention since they address similar problems. Furthermore, the evolutionary nature of ADFs allows the *when* and *what* questions of predicate invention to be answered automatically. This review paper makes explicit, for the first time, the natural connection between predicate invention and ADFs.

2 Predicate Invention

We focus here on predicate invention within Inductive Logic Programming (ILP), as this setting subsumes the attribute-value one and is most closely related to the GP setting discussed later. In ILP, predicate invention is the process of introducing new predicates (i.e., functions into {0,1}) into the hypothesis language.

There are two motivations for predicate invention, namely, necessity and usefulness. They are formalised in the following definitions from [12]. An invented predicate is *necessary* if "... without inventing such new terms(s), the theory is not learnable according to the criterion of successful learning." An invented predicate is *useful* if "... its invention does not affect the learnability of the theory." These two types of invented predicates correspond to the two classes of predicate invention approaches identified in [26], namely, *demand-driven* approaches and *reformulation* approaches.

Reformulation approaches to predicate invention introduce new predicates to allow reformulation of an existing theory in order to express it in a more compact and concise way. The most commonly used technique is based on inverse resolution and implemented using the intra-construction operator

[1]This work was supported by EPSRC grant GR/L21884.

[17, 19], defined as follows: Given $h_1 \leftarrow \alpha\beta$ and $h_2 \leftarrow \alpha\gamma$, construct/invent h_3 such that $h_1 \leftarrow \alpha h_3$, $h_3 \leftarrow \beta$ and $h_3 \leftarrow \gamma$. The new predicate h_3 may be either accepted (and named adequately) by an oracle or rejected. One experiment based on Bacon's *Novum Organum* shows how the learning system CIGOL combines several existing laws about illumination and light sources to construct a new predicate for the concept of radiation [18], whilst experiments with the KPa7KR chess problem boast a 20% compaction using predicate invention [17]. Learning systems that carry out predicate invention using variants of inverse resolution include CIGOL [21], LFP2 [27], ITOU [23], RINCON [30], and Banerji's system [2]. Other systems, such as CIA [3] and FOCL [24], carry out reformulation using schemes, i.e., descriptions of useful literal combinations.

Demand-driven approaches to predicate invention attempt to detect situations where the hypothesis language is insufficient to produce a hypothesis that is consistent with the training examples, i.e., a consistent finite axiomatisation of the examples [12, 26]. For example, Ling reports on an experiment involving learning multiplication (i.e., `mul(x,y,z)` is true if `z = xy`) using only the successor function `s` and the constant `0` [12]. With the given language, `mul` is not finitely axiomatisable and the new predicate `fool(x,s(z1),s(z)) <- fool(x,z1,z)`, easily recognised as the definition of addition (i.e., `z = x + z1`), is invented. As determining that a hypothesis language is unable to produce a consistent finite axiomatisation of the examples is undecidable (unless the language and algorithm biases allow the full enumeration of the search space) [26], the decision about introducing new predicates must be driven by heuristics. Several learning systems use an over-general clause that can not be specialised with the existing hypothesis language as an indication that predicate invention is desirable. Of course, if over-generality is due to noise, inventing new predicates leads to overfitting. Learning systems that implement a demand-driven approach to predicate invention include INPP [13], DBC [8] and SIERES [28].

An interesting variant of predicate invention is found in the work on repeat learning [7], where predicates learned in a particular context are re-used in further learning tasks. In *intra* repeat learning, the learning task remains unchanged but is performed more than once. Hence, the predicates learned or invented on a given learning attempt are added to the hypothesis language before the next attempt. In *inter* repeat learning, the learning task changes and the predicates learned or invented while learning one task are added to the hypothesis language of another related learning task. These ideas are illustrated in [7] using the well-known chess end-game domain. It is interesting, in the light of this paper, to notice that both intra and inter repeat learning implement a form of code reuse.

3 Automatically Defined Functions

Automatically Defined Functions (ADFs) extend the traditional concept of sub-routines (or modules) to Genetic Programming (GP) [9]. The basic GP program tree is enhanced to become a structure combining a main program (i.e., one value returning branch) and the sub-routines or ADFs (i.e., one or more function defining branches) that are to be used by the main program. The bodies of the ADFs and of the main program are evolved simultaneously using suitably adapted genetic operators.

In practice, the use of ADFs typically leads to smaller, more general solutions to problems of a hierarchical nature, and tends to facilitate the production of solutions to problems challenging for ordinary GP (e.g., Even `N`-Parity for `N=11` [10]). In what follows, we give a brief overview of some of the most popular variants of ADFs.

Evolutionary Module Acquisition provides a general purpose technique for creating modules in any evolutionary based system [1]. Components of an evolving individual are chosen at random and collated into a module by a compress operation, which is isolated from further manipulation by the genetic operators. An expansion operation can be used to release components from a module in order that they may be manipulated by the genetic operators again. Modules survive based on fitness.

Automatically Defined Macros are evolved simultaneously with the main program [25]. A macro is an instruction that is replaced by a sequence of instructions prior to the compilation of the program. Macros are similar to subroutines since they allow programs to be modularised. However, they have the further advantage of allowing new control structures to be implemented through their influence over the evaluation of their arguments.

Adaptive Representation Learning is an alternative to ADFs that abstracts useful subtrees from program trees so that they may be used as subroutines in future populations [22]. A useful block of code is identified by looking at an individual's differential fitness compared to its least fit parent and block activation (i.e., number of times a particular block of code is activated during evaluation). Blocks of code with high activity in individuals with high positive differential fitness are generalised by replacing a random subset of terminals with variables, and placed as a new subroutine in the subroutine set. All subroutines are assigned a utility value according to the outcome of their use. Subroutines with low utility are deleted from the subroutine set to keep the size of the set below a specified threshold.

"Demand-driven" ADFs extend the standard use of ADFs by providing special operators that allow the individuals in the population to develop their own architecture (i.e., create/duplicate/delete ADFs and ADF arguments) [11]. The special operators are applied to a small percentage of the current population in conjunction with the standard genetic operators. Thus, subroutines and their arguments can be added or deleted on demand to fit the particular problem being solved.

λ-abstractions are used to denote anonymous functions where the name and type are not specified [31, 6]. λ-abstractions are evolved as arguments to higher-order functions in the main body of the program structure. They therefore evolve dynamically and, as with demand-driven ADFs, they are not a predetermined condition of evolution, but are evolved if helpful to the solution. The number and type of arguments of a λ-abstraction are implicitly decided by the type signature of the higher-order function of which they are a functional argument.

The above techniques allow sub-routines to be co-evolved with and used in a main program, thus producing solutions that are more compact and more general. For example, using λ-abstractions, PolyGP evolved a solution for the `N`-parity problem for any `N` [31]. In addition, the system invented the `XOR`, `EXOR` and `XAND` functions in the evolved solutions. Similarly, STEPS discovered useful properties in carcinogenic chemical compounds [5].

4 Predicate Invention and ADFs - The Missing Link

As seen in section 3, ADFs, and their variants, are used in conjunction with GP to reformulate the main evolving program. The co-evolved subroutines compress the amount of code required in the main program and can also facilitate the discovery of solutions to otherwise challenging problems.

Hence, ADFs clearly fulfill for GP a role similar to that of the reformulation approach to predicate invention in ILP. In fact, because of the iterative nature of the evolutionary process, a natural form of intra repeat learning takes place with ADFs, since the ADFs discovered in one generation become part of the alphabet in the next if they are preserved. Similarly, ADFs can be used for inter repeat learning, where the ADFs evolved for a particular problem are added to the alphabet of new related problems. Interestingly, well-known evolutionary techniques, such as seeding [15] and Species Adaption Genetic Algorithms (SAGA) [4], also implement both intra and inter repeat learning.

The correspondence between ADFs and the demand-driven approach to predicate invention is less trivial. ILP grounds its distinction between usefulness and necessity of predicate invention in logic theory, i.e., finite axiomatisation. GP makes no such distinction explicitly and introduces new functions/sub-routines heuristically based on fitness considerations. Hence, it is difficult to establish whether a sub-routine is created because it is strictly necessary or simply because it is useful. However, one can argue that the distinction between usefulness and necessity is of no real practical interest and of only limited theoretical value since it cannot be decided in general (see section 2). What seems most relevant for practitioners is that their system be capable of inventing predicates/functions whenever this proves valuable (be it useful or necessary) in finding a solution. In this light, it is clear that the ADF approach of GP is the same as the predicate invention approach of ILP. Both must rely on heuristics when it comes to answering the when question.

It is our contention that the GP approach to ADFs (and variants), which is driven solely by fitness considerations, offers a uniform way of treating both types of predicate inventions. Since there is no principled way of inventing predicates, the genetic paradigm, as in many similar circumstances, provides a valuable alternative. The recent approaches based on demand-driven ADFs and λ-abstractions seem most promising, as they leave the answers to the questions of what should be invented, and when it should be invented to be themselves learned thereby subsuming the implementations of both the reformulation and demand-driven approaches to predicate invention in ILP.

5 Conclusion

This paper reviews predicate invention and automatically defined functions and, for the first time, highlights the conceptual similarity between the two concepts. It further shows how the evolutionary nature of ADFs allows the complex *when* and *what* questions of predicate invention to be answered automatically. Experiments with benchmark problems are still needed to validate this claim. In particular, will the ADFs produced be predicates in a sense that

ILP researchers will recognise and be interested in? Preliminary results with PolyGP [31] and STEPS [5] are encouraging.

References

[1] P.J. Angeline and J.B. Pollack. Evolutionary module acquisition. In *Proc. of the Annual Conference on Evolutionary Programming*, pages 154–163, 1993.

[2] R.B. Banerji. Learning theoretical terms. In S. Muggleton, editor, *Inductive Logic Programming*, chapter 4. Academic Press, 1992.

[3] L. de Raedt and M. Bruynooghe. Interactive concept-learning and constructive induction by analogy. *Machine Learning*, 8(2):107–150, 1992.

[4] I. Harvey. Species adaptation genetic algorithms: a basis for a continuing SAGA. In *Proc. of the European Conference on Artificial Life*, pages 346–354, 1991.

[5] C.J. Kennedy. *Strongly Typed Evolutionary Programming.* PhD thesis, University of Bristol, 1999.

[6] C.J. Kennedy and C. Giraud-Carrier. An evolutionary approach to concept learning with structured data. In *Proc. of the Intl. Conference on Artificial Neural Networks and Genetic Algorithms*, pages 331–336, 1999.

[7] K. Khan, S. Muggleton, and R. Parsons. Repeat learning using predicate invention. In *Proc. of the Intl. Conference on Inductive Logic Programming*, volume LNAI 1446, pages 165–174. Springer-Verlag, 1998.

[8] B. Kijsirikul, M. Numao, and M. Shimura. Discrimination-based constructive induction of logic programs. In *Proc. of the Natl. Conference on Artificial Intelligence*, pages 44–49, 1992.

[9] J.R. Koza. *Genetic Programming: On the Programming of Computers by Means of Natural Selection.* MIT Press, 1992.

[10] J.R. Koza. *Genetic Programming II.* MIT Press, 1994.

[11] J.R. Koza, D. Andre, F. H Bennett III, and M. Keane. *Genetic Programming 3: Darwinian Invention and Problem Solving.* Morgan Kaufman, 1999.

[12] C. X. Ling. Inventing necessary theoretical terms. Technical Report Nr. 302, University of Western Ontario, Department of Computer Science, 1991.

[13] C. X. Ling. Introducing new predicates to model scientific revolution. *International Studies in the Philosophy of Science*, 9(1):19–36, 1995.

[14] C. Matheus. The need for constructive induction. In *Proc. of the Intl. Workshop on Machine Learning*, pages 173–177, 1991.

[15] Z. Michalewicz. *Genetic Algorithms + Data Structures = Evolution Programs.* Springer-Verlag, 1992.

[16] T.M. Mitchell. The need for biases in learning generalizations. Technical Report CBM-TR-117, Rutgers University, Department of Computer Science, 1980.

[17] S. Muggleton. Duce, an oracle based approach to constructive induction. In *Proc. of the Intl. Joint Conference on Artificial Intelligence*, pages 287–292, 1987.

[18] S. Muggleton. A strategy for constructing new predicates in first order logic. In *Proc. of the European Working Session on Learning*, pages 123–130, 1988.

[19] S. Muggleton. Inverting the resolution principle. *Machine Intelligence 12*, pages 93–104, 1991.

[20] S. Muggleton. Predicate invention and utilisation. *Journal of Experimental and Theoretical Artificial Intelligence*, 6(1):127–130, 1994.

[21] S. Muggleton and W. Buntine. Machine invention of first-order predicates by inverting resolution. In *Proc. of the Intl. Workshop on Machine Learning*, pages 339–352, 1988.

[22] J. P. Rosca and D. H. Ballard. Discovery of subroutines in genetic programming. In P. Angeline and K.E. Kinnear, editors, *Advances in Genetic Programming 2*, pages 177–202. MIT Press, 1996.

[23] C. Rouveirol. Extensions of inverse resolution applied to theory completion. In S. Muggleton, editor, *Inductive Logic Programming*, chapter 3. Academic Press, 1992.

[24] G. Silverstein and M.J. Pazzani. Learning relational clichés. In *Proc. of the IJCAI-93 Workshop on Inductive Logic Programming*, pages 71–82, 1993.

[25] L. Spector. Simultaneous evolution of programs and their control structures. In P. Angeline and K.E. Kinnear, editors, *Advances in Genetic Programming 2*, pages 137–154. MIT Press, 1996.

[26] I. Stahl. Predicate invention in ILP - an overview. In *Proc. of the European Conference on Machine Learning, LNAI 667*, pages 313–322, 1993.

[27] R. Wirth. Completing logic programs by inverse resolution. In *Proc. of the European Working Session on Learning*, pages 239–250, 1989.

[28] R. Wirth and P. O'Rorke. Constraints for predicate invention. In S. Muggleton, editor, *Inductive Logic Programming*, chapter 14. Academic Press, 1992.

[29] J. Wnek and R.S. Michalski. Hypothesis driven constructive induction in AQ17-HCI: A method and experiments. *Machine Learning*, 14:139–168, 1994.

[30] J. Wogulis and P. Langley. Improving efficiency by learning intermediate concepts. In *Proc. of the Intl. Joint Conference on Artificial Intelligence*, pages 657–662, 1989.

[31] T. Yu and C. Clack. Recursion, lambda abstractions and genetic programming. In *Proc. of the Annual Conference on Genetic Programming*, pages 422–431, 1998.

The Convergence Behavior of the PBIL Algorithm: A Preliminary Approach [1]

C. González, J.A. Lozano, P. Larrañaga*

*Department of Computer Science and Artificial Intelligence, University of the Basque Country, Spain, http://www.sc.ehu.es/isg, {cristina,lozano,pedro}@si.ehu.es

Abstract

In this paper the simplest version of Population Based Incremental Learning (PBIL) is used to minimize the *OneMax* function in two dimensions. After carrying out several experiments to reveal the limit behavior of the algorithm in this function we obtain that the convergence results depend on the initial vector $\mathbf{p}^{(0)}$, and on the α parameter value. This experienced behavior is guaranteed for mathematical proof. The probability that the algorithm converges to any point of the search space goes to 1 when $\mathbf{p}^{(0)}$ and α go to suitable values. Thus, even though the experimental results seem more stable when the α value is near to zero, we can not ensure that PBIL converges to the optimum.

1 Introduction

During the nineties many real combinatorial optimization problems have been solved successfully using Genetic Algorithms (GAs). But the existence of deceptive problems where the performance of GAs is very poor have motivated the search for new optimization algorithms. To overcome these difficulties a few group of researches have recently suggested a new family of algorithms called EDAs (Estimation of Distribution Algorithms) [1] and [2].

Like GAs, EDAs work with a population of individuals of the search space. The evolution of the population makes individuals to visit better space areas during the search. There are several differences between GAs and EDAs. The most important is that in EDAs the crossover and mutation operators are replaced by the estimation and simulation of a joint probability distribution. The point with this approach is the estimation of the joint probability distribution.

To overcome this problem, different algorithms have appeared that make different assumptions concerning the conditional independencies between the variables of the joint probability distribution. In the simplest case it is supposed that all the variables of the problem are independent, [3], [4], [5] (PBIL), [1] and [6]. In a second step second-order relations between the variables are considered, De Bonet et al. [7]. A step forward involves factorizing the joint probability distribution in a tree-like structure [8]. Recently, some works have appeared where the joint probability distribution is factorized by means of Bayesian networks. Mühlenbein et al. [9] have developed an algorithm for additive decomposable functions called FDA. The most general approaches have been developed by Etxeberria and Larrañaga [10] and Pelikan et al. [11].

In spite of this effort to obtain new algorithms, little attention has been given to the theoretical aspects of these.

In this work we study mathematically an instance of EDAs, the PBIL algorithm. We carry out an analysis of the behavior of this algorithm in a very simple function, the OneMax function in two dimensions. We prove that the limit behavior of the algorithm from a practical point of view depends on the value of two parameters.

The remainder of this work is organized as follows. Section 2 describes PBIL. Section 3 is dedicated to show some experimental results. A theoretical analysis is presented in Section 4, leaving Section 5 to draw conclusions.

2 An Introduction to PBIL

PBIL was introduced by Baluja [4] in 1994. This algorithm is based on the idea of substituting the individuals of the population by a set of statistics of them. In our case we suppose that the function to optimize is defined in the binary space $\Omega = \{0,1\}^l$. At each iteration of the algorithm a vector of probabilities $\mathbf{p} = (p_1, p_2, \dots, p_l)$ is maintained, being p_i the probability of obtaining 1 in the i-th component.

The algorithm works as follows. At each step, drawing the probability vector $\mathbf{p}$, λ individuals are obtained and the μ best of them ($\mu < \lambda$), $\mathbf{y}_{1:\lambda}, \mathbf{y}_{2:\lambda}, \dots, \mathbf{y}_{\mu:\lambda}$ are selected. These selected in-

[1] This work was supported by the University of the Basque Country under the grant 9/UPV/EHU/ 00140.226-12084/2000. C. González is also supported by UPV-EHU.

dividuals are used to modify the probability vector. A neural networks inspired rule is used to update the probability vector:

$$\mathbf{p}^{(t+1)} = (1-\alpha)\mathbf{p}^{(t)} + \alpha\frac{1}{\mu}\sum_{k=1}^{\mu}\mathbf{y}_{k:\lambda}^{(t)}$$

where $\mathbf{p}^{(t)}$ is the probability vector at the t-th step, and $\alpha \in (0,1)$ is an algorithm's parameter. Figure 1 shows a pseudo-code for PBIL.

```
Obtain an initial probability vector p^(0)
while no convergence do
  begin
  Using p^(t) obtain λ individuals y_1^(t), y_2^(t), ..., y_λ^(t)
  Evaluate y_1^(t), y_2^(t), ..., y_λ^(t)
  Select the μ best individuals y_{1:λ}^(t), y_{2:λ}^(t), ..., y_{μ:λ}^(t)
  p^(t+1) = (1 − α)p^(t) + α (1/μ) Σ_{k=1}^{μ} y_{k:λ}^(t)
  end
```

Figure 1: Pseudo-code for PBIL.

Three works are related with the theoretical analysis of PBIL. Höhfeld and Rudolph [13] analyse the behavior of the algorithm for linear pseudoboolean functions. The authors show that $\lim_{t\to\infty} \mathsf{E}[\mathbf{p}^{(t)}]$ exists and that it is equal to the optimal point. In the second work Berny [14] presents a statistical framework for combinatorial optimization over fixed-length binary strings and shows that PBIL can be derived from a gradient dynamical system acting on Bernoulli probability vectors. A different analysis based on dynamical systems can be consulted in [15].

3 Experimental Results

To make the behavior of the algorithm out we have carried out several experiments. In order to do that we have used the version of PBIL and the function that Höhfeld and Rudolph used in [13]. We consider that at each step two individuals ($\lambda = 2$) are generated and the best of them ($\mu = 1$) is selected to modify the probability vector $\mathbf{p}^{(t)}$. This is the simplest version of PBIL.

We minimize the function

$$OneMax(\mathbf{x}) = \sum_{i=1}^{2} x_i \ .$$

The optimum of $OneMax$ is, clearly, in the point $(0,0)$ and the worst value is in $(1,1)$.

We have carried out several experiments with different values of $\mathbf{p}^{(0)}$ and α. In particular we take four values for $\mathbf{p}^{(0)}$: $(.05,.05)$, $(.5,.5)$, $(.95,.05)$ and $(.95,.95)$, and three for α: .05, .5 and .95.

$\mathbf{p}^{(0)}$	α	$(0,0)$	$(0,1)$	$(1,0)$	$(1,1)$
(.05, .05)	.05	150,000	0	0	0
	.50	149,760	132	108	0
	.95	148,818	547	635	0
(.5, .5)	.05	150,000	0	0	0
	.50	107,220	20,335	20,405	2,040
	.95	69,213	37,195	35,376	8,216
(.95, .05)	.05	120,034	0	29,966	0
	.50	24,824	74	125,033	69
	.95	14,865	413	134,465	257
(.95, .95)	.05	88,178	28,055	27,948	5,819
	.50	2,942	21,501	21,501	104,056
	.95	897	14,445	13,922	120,736

Table I. Experimental results.

For each combination of these parameters we carried out 15×10^4 executions. Table I shows the results. Each entry of the table represents the number of executions in which PBIL with initial probability $\mathbf{p}^{(0)}$ and parameter α got a particular point of the search space. For instance, with $\mathbf{p}^{(0)} = (.95,.95)$ and $\alpha = .5$ the point $(1,1)$ was reached 104,056 times.

In view of the experimental results, we can extract the following deductions. PBIL does not converge to the global optimum. In every case the algorithm converges, but the limit point depends on the values of $\mathbf{p}^{(0)}$ and α. It seems that, when the value of α is near to zero, the algorithm converges to the optimum with high probability, but in other situations this does not happen. For instance, when $\mathbf{p}^{(0)}$ takes values near to $(1,1)$ and α takes values near to 1, the algorithm goes to $(1,1)$ with a very high probability. Similarly, when the algorithm starts with $\mathbf{p}^{(0)}$ near to $(1,0)$ and α near to 1 we obtain similar results for the point $(1,0)$. Next section proves mathematically some of the results suggested by the experiments.

4 Mathematical Modeling of PBIL

As we have seen empirically the behavior of PBIL depends on the initial vector $\mathbf{p}^{(0)}$ and on the value of the parameter α. Thus, the obtained experimental results suggest to show mathematically that the sequence $\{\mathbf{p}^{(t)}\}_{t=0,1,2,\ldots}$ (and hence the algorithm) can go to any point of the search space for suitable values of $\mathbf{p}^{(0)}$ and α.

The algorithm can be modeled using a Markov chain, because the probability vector $\mathbf{p}^{(t)}$ only depends on $\mathbf{p}^{(t-1)}$. For each $\mathbf{p}^{(0)}$ and α we have a different Markov chain. The state space of this chain is formed from the different values that the probability vector $\mathbf{p}^{(t)}$ can take:

$$\mathbf{p}^{(t)} = (1-\alpha)^t\mathbf{p}^{(0)} + \alpha\left(\sum_{k=0}^{t-1}(1-\alpha)^k\mathbf{y}_{1:2}^{(t-k-1)}\right),$$

where $\mathbf{y}_{1:2}^{(k)} \in \{0,1\}^2$. We can say that the state space of this chain is infinite numerable.

The analysis of this chain is very complex because every state is transient, however it is easier to study the probabilities of certain transitions in the chain. In this way it is possible to calculate the probability of a path chain driving the algorithm towards convergence to a particular point.

In particular, for each point of the search space we find a lower bound for the probability of a path chain driving the algorithm towards convergence to this point.

For example, using the point (1,1), we see how to calculate a lower bound of the probability of the chain path driving the algorithm convergence to this point. The arguments used for the point $(1,1)$ illustrate the procedure used in the others points. In this case the chain path is composed of the probability vectors that were generated if at each step of the algorithm the point $(1,1)$ would be obtained after the selection. It is important to realize that in this case the sequence $\{\mathbf{p}^{(t)}\}_{t=0,1,2,\ldots}$ converges to the point $(1,1)$.

Hence we want to calculate the probability that the Markov chain visits the states:

$$\mathbf{p}^{(0)},\ \mathbf{p}^{(1)},\ldots,\ \mathbf{p}^{(t)},\ldots$$

$$\text{where}\quad \mathbf{p}^{(t)} = (1-\alpha)\mathbf{p}^{(t-1)} + \alpha\mathbf{y}_{1:2}$$

being $\mathbf{y}_{1:2} = (1,1)$. It is similar to say that at each iteration of PBIL the point $(1,1)$ is obtained.

As seen in [13], the probability of obtaining the point $(1,1)$ in the $(t+1)$-th step of PBIL, starting from a probability vector $\mathbf{p}^{(t)} = (p_1^{(t)}, p_2^{(t)})$, can be written as:

$$(p_1^{(t)} p_2^{(t)})^2. \tag{1}$$

Therefore the probability of the chain path is expressed by an infinite product of type (1) probabilities. Based on calculating probabilities of paths we can establish the following theorem:

Theorem 1 *The sequence $\{\mathbf{p}^{(t)}\}_{t=0,1,2,\ldots}$ generated by PBIL in the OneMax function converges to $(a,b) \in \Omega = \{0,1\}^2$, with probability as near to 1 as we want when $\mathbf{p}^{(0)}$ and α go to (a,b) and 1 respectively.*

Proof 1 *We proof the case $(a,b) = (1,1)$, the others are similar.*
We want to calculate the probability that the chain visits the states:

$$\mathbf{p}^{(0)},\ \mathbf{p}^{(1)},\ldots,\ \mathbf{p}^{(t)},\ldots$$

$$\textit{where}\quad \mathbf{p}^{(t)} = (1-\alpha)\mathbf{p}^{(t-1)} + \alpha\mathbf{y}_{1:2} =$$

$$(1-\alpha)\mathbf{p}^{(t-1)} + \alpha(1,1) = (1-\alpha)\mathbf{p}^{(t-1)} + (\alpha,\alpha)\ .$$

We know that the probabilities of visiting $\mathbf{p}^{(t)}$ given $\mathbf{p}^{(t-1)}$ can be expressed as:

$$P\left(\mathbf{p}^{(t)} = (1-\alpha)\mathbf{p}^{(t-1)} + (\alpha,\alpha) \mid \mathbf{p}^{(t-1)}\right) =$$

$$P(\textit{to obtain vector } (1,1) \textit{ from } \mathbf{p}^{(t-1)}) = (p_1^{(t-1)} p_2^{(t-1)})^2\ .$$

Hence the probability we are looking for, can be expressed as the product of the probabilities of each transition. We suppose that $p_1^{(0)} = p_2^{(0)} = p$ to simplify notation. After several basic algebraic operations the probability can be written as:

$$\left[\prod_{t=0}^{\infty}\left[1+(1-\alpha)^t(p-1)\right]\right]^4. \tag{2}$$

We study the convergence of this infinite product, knowing that:

$$\prod_{n=0}^{\infty} x_n,\quad (x_n > 0)\ \textit{converges} \Leftrightarrow \sum_{n=0}^{\infty} \ln x_n\ \textit{converges.}$$

Therefore it is enough to see the behavior of the series

$$\sum_{t=0}^{\infty} \ln\left(1+(1-\alpha)^t(p-1)\right). \tag{3}$$

Using a classical criterion, we show that this series converges when α goes to 1. Instead of calculating the value of the product (2) we will give a lower bound for this product, bounding the series (3):

$$\sum_{t=0}^{\infty} \ln\left(1+(1-\alpha)^t(p-1)\right) =$$

$$\ln p + \sum_{t=1}^{\infty} \ln\left(1+(1-\alpha)^t(p-1)\right) >$$

$$\ln p + \sum_{t=1}^{\infty} \ln\left(1-(1-\alpha)^t\right) \overset{*}{=}$$

$$\ln p - \left[\sum_{t=1}^{\infty}\left(\sum_{k=1}^{\infty}\frac{(1-\alpha)^{tk}}{k}\right)\right] =$$

$$\ln p - \sum_{k=1}^{\infty}\left(\frac{1}{k}\,\frac{(1-\alpha)^k}{1-(1-\alpha)^k}\right) \geq$$

$$\ln p - \sum_{k=1}^{\infty}\left(\frac{(1-\alpha)^k}{1-(1-\alpha)^k}\right) \geq$$

$$\ln p - \sum_{k=1}^{\infty}\left(\frac{(1-\alpha)}{\alpha}\right)^k = \ln p - \left(\frac{1-\alpha}{2\alpha-1}\right).$$

Taking the limit of the last expression, when $\alpha, p \to 1$ *we obtain that*

$$\lim_{\alpha,p\to 1}\sum_{t=0}^{\infty}\ln\left(1+(1-\alpha)^t(p-1)\right) \geq$$

$$\lim_{\alpha,p\to 1}\left[\ln p - \left(\frac{1-\alpha}{2\alpha-1}\right)\right] = 0\ .$$

Now:

$$\ln\left[\prod_{t=0}^{\infty}\left(1+(1-\alpha)^t(p-1)\right)\right] = \ln x$$

$$\ln x \overset{\alpha,p\to 1}{\longrightarrow} 0 \Longrightarrow x = \prod_{t=0}^{\infty}\left[1+(1-\alpha)^t(p-1)\right] \longrightarrow 1$$

(*) We have used the power series expression of $\ln(1+x)$ *when* x *is near to zero.*

5 Conclusions and Future Work

In this paper we have shown that for PBIL with $\lambda = 2$ and μ=1 applied to the *OneMax* function :

$$P\left(\lim_{t\to\infty}\mathbf{p}^{(t)} = (a,b)\right) \longrightarrow 1$$

when $\alpha \to 1$, $\mathbf{p}^{(0)} \to (a,b)$, and $(a,b) \in \{0,1\}^2$. Although when the value of α is small the experimental results seem to be more stable, we can not conclude that PBIL converges to the optimum.

In our future research we will try, first, looking for the values of $\mathbf{p}^{(0)}$ and α that make PBIL to converge to the optimum of a linear function with high probability. In addition to this, we will try to model mathematically the behavior of other instances of EDAs.

References

[1] H. Mühlenbein and G. Paaß, "From Recombination of Genes to the Estimation of Distributions I. Binary Parameters", *Lecture Notes in Computer Science 1411: Parallel Problem Solving from Nature*, PPSN IV, pp. 178-187, 1996.

[2] P. Larrañaga and J.A. Lozano, *Estimation of Distribution Algorithms. A new tool for Evolutionary Computation*, Kluwer Academic Publishers, in press.

[3] G. Syswerda, "Simulated Crossover in Genetic Algorithms", *Foundations of Genetic Algorithms*, II, pp. 239-255, 1993.

[4] S. Baluja, "Population Based Incremental Learning: A Method for Integrating Genetic Search Based Function Optimization and Competitive Learning", *Carnegie Mellon Report*, CMU-CS-94-163, 1994.

[5] V. Kvasnicka, M. Pelikan, and J. Pospichal, "Hill Climbing with Learning (an Abstraction of Genetic Algorithms)," *Neural Networks World*, vol. 6, pp. 773-796, 1996

[6] H. Mühlenbein, "The equation for Response to Selection and its Use for Prediction", *Evolutionary Computation* vol. 5, pp. 303-346, 1998.

[7] J.S. De Bonet, C.L. Isbell and P. Viola, "MIMIC: Finding Optima by Estimating Probability Densities", *Advances in Neural Information Processing Systems*, vol. 9, 1997.

[8] S. Baluja and S. Davies, "Fast Probabilistic Modeling for Combinatorial Optimization", *AAAI-98*, 1998.

[9] H. Mühlenbein, T. Mahnig and A. Ochoa, "Schemata, Distributions and Graphical Models in Evolutionary Optimization", *Journal of Heuristics*, vol. 5, pp. 215-247, 1999.

[10] R. Etxeberria and P. Larrañaga, "Global optimization using Bayesian networks", *II Symposium on Artificial Intelligence, CIMAF99, Special Session on Distributions and Evolutionary Optimization*, pp. 332-339, 1999.

[11] M. Pelikan, D.E. Goldberg and E. Cantú-Paz, "BOA: Bayesian Optimization Algorithm", *GECCO'99:Proceedings of the Genetic and Evolutionary Computation Conference*, pp. 525-532, 1999.

[12] H. Mühlenbein and T. Mahnig, "The Factorized Distribution Algorithm for Additively Decomposed Functions", *Proceedings of the 1999 Congress on Evolutionary Optimization*, pp. 752-759, 1999.

[13] M. Höhfeld and G. Rudolph, "Towards a theory of population-based incremental learning", *4th IEEE Conference on Evolutionary Computation*, Piscataway, NJ: IEEE, Press, vol. 1, pp. 1-5, 1997.

[14] A. Berny, "Selection and Reinforcement Learning for Combinatorial Optimization," *Parallel Problem Solving from Nature*, PPSN-VI, pp. 601-610, 2000.

[15] C. González, J.A. Lozano and P. Larrañarraga, "Analyzing the PBIL algorithm by means of discrete dynamical systems", *Complex Systems*, accepted.

Information Dimension of a Population's Attractor in a Binary Genetic Algorithm

Paweł Kieś *

*Institute of Fundamental Technological Research, Polish Academy of Sciences, 21 Świętokrzyska Str., 00-049 Warsaw, Poland, phone: (+48)(22)8261281 ext. 144, fax: (+48)(22)8269815, e-mail: pkies@ippt.gov.pl

Abstract

The tools for a description of chaotic dynamics are applied to investigate the work of a binary genetic algorithm ($\mathcal{BGA}$). The method for determining strange attractors from $\mathcal{BGA}$'s populations is shown. Attractor's information dimension is taken as a measure of the state of $\mathcal{BGA}$'s activity. The equivalence between the information dimension of a population's attractor and the entropy of related bit positions for a given population is shown and confirmed experimentally.

1 Introduction

A binary genetic algorithm ($\mathcal{BGA}$) is a kind of an evolutionary algorithm. It usually is applied as an optimization method. The reader can find a good explanation of the basic ideas of a $\mathcal{BGA}$ in [8].

Since $\mathcal{BGA}$s are known, many researchers try to understand and explain how they work. One knows many papers that present mathematical models of simplified or modified GAs, for example [10], that analyses a behaviour of a GA with a continuous population. Also one can find excellent partial models of elements of a $\mathcal{BGA}$, for example [7] and [3], where models of the generalized crossover are proposed. Another well known model of dynamics of $\mathcal{BGA}$s uses Markov chains as a description of the job of genetic operators: crossover and mutation.

In this paper we propose a new approach taken from chaotic dynamics analysis. We hope that it can help to explain the role of genetic operators and a selection not only separately, but also the way how they cooperate and interact [1].

In Section 2 we explain why we can treat the $\mathcal{BGA}$ as a chaotic process. In Section 3 we show how to construct a population's attractor. Next, in Section 4 we define attractor's information dimension and in Section 5 we show how we can interpret it using notions known from the $\mathcal{BGA}$'s theory. In Sections 6 and 7 we describe realized experiments and draw our conclusions.

2 Genetic Algorithm as a Chaotic Process

A $\mathcal{BGA}$ is a probabilistic optimization algorithm, so it does not ensure that each its execution finds the optimum. The algorithm takes many random decisions during its run. Each such a decision can be treated as a realization of a certain random variable, and each particular execution is fully described by a sequence of values of these random variables. This causes that it is difficult to investigate the $\mathcal{BGA}$'s activity basing on its single execution, because it can vary in a very wide range. Also a simple averaging of many executions is not a good solution because we lose easily synchronization between related stages of executions. The possible solution is to identify somehow the current state of the $\mathcal{BGA}$ execution and to average measured parameters of related states for many executions.

First, let us make some assumptions about our $\mathcal{BGA}$. Its aim is to find such an element of the discrete search space X that maximizes the fitness function $f : X \to \mathbb{R}$. Each element of X is coded into an N-bit chromosome that is an element of the set $\mathcal{B} = \{0,1\}^N$, called the code space. The coding is realized by a bijection $\Psi : X \to \mathcal{B}$ called a coding function. The $\mathcal{BGA}$ operates on M-element populations that are vectors from the set $\mathcal{B}^M$.

We assume that the chromosomes from the current population describe the current state of the $\mathcal{BGA}$. So the state is given by the current information content of a population as defined below.

Definition 1 *Let $G \in \mathcal{B}^M$ be an M-element population. An* information content *is defined as a vector of probabilities $p(G) = (p_{\text{one}}(i|G))_{i=1,\ldots,N}$, where $p_{\text{one}}(i|G) = \frac{1}{M}\sum_{j=1}^{M} b_i^j$ and b_i^j is the i-th bit in j-th chromosome in G.* □

The information content is an extension of a so-called genetic pool. It describes the "genetic material" contained in a given population, i.e. the probability distribution of chromosomes that we can construct by combining randomly bits from related positions of chromosomes in a given population.

3 Determining a Population's Attractor

We can represent an information content in the domain of the search space X also. In order to do this we apply an idea from [7]. It defines the gen-

let $A_G := \emptyset$
for $w := 1$ to W do
    let $\underline{b}_1, \underline{b}_2$ be randomly chosen from G
    let u be randomly chosen from U
    let $(\underline{b}'_1, \underline{b}'_2) := c_u(\underline{b}_1, \underline{b}_2)$
    let $A_G := A_G \cup \Psi^{-1}(\underline{b}'_1) \cup \Psi^{-1}(\underline{b}'_2)$
end

Fig. 1. An algoritm that constructs an attractor A_G for a population G, where W – number of iterations, $\Psi^{-1} : \mathcal{B} \to X$ – decoding function, c_u – generalized crossover operator (see Def. 2).

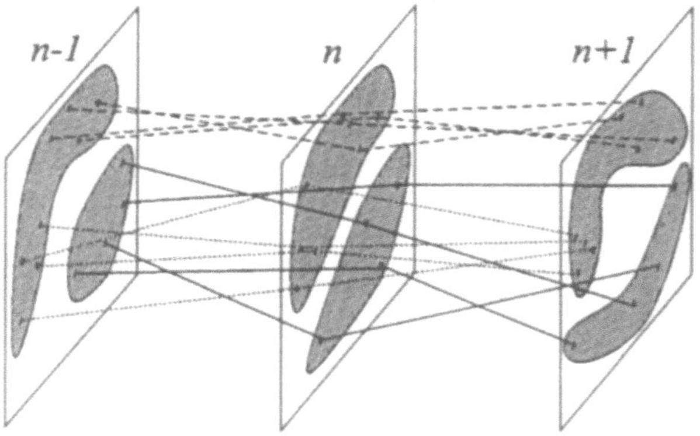

Fig. 2. Two population's attractors (lined areas) for three execution of a $\mathcal{BGA}$, where n – a population number, $M = 4$ – population size and $X \subset \mathbb{R}^2$ – the search space. The executions marked by dashed and dotted lines belong to the same attractor. The one marked by a solid line belongs to the other attractor.

eralized crossover operator as below and shows that it preserves the genetic material contained in parent chromosomes.

Definition 2 *We define the* generalized crossover operator *as a mapping* $c_u : \mathcal{B}^2 \to \mathcal{B}^2$, *where* $u \in U$ *is a particular crossover operation and* U *is the set of all possible crossovers.* □

For example in the one-point crossover we assume that $U = \{1, \ldots, N\}$ is the set of all possible crossing positions and the value of u is taken randomly from U.

We construct a simple algorithm based on this idea, shown at Fig. 1. Using it we find a subset $A_G \subset X$ related to the information content of a given population $G \in \mathcal{B}^M$. The received set has similar features to a set known form chaotic dynamics: a strange attractor.

An attractor is a basic notion in chaotic dynamics. Recently this branch of knowledge is more and more popular. The reader can find fundamental information on it in many works [2, 9]. We can think of a population's attractor as a Poincaré section of a $\mathcal{BGA}$'s trajectory, while we treat solutions coded by a given population as a set of at most M points in a phase space X and a sequence of such sets resulted from a single execution of the $\mathcal{BGA}$ – as a trajectory. One can see this idea at Fig. 2.

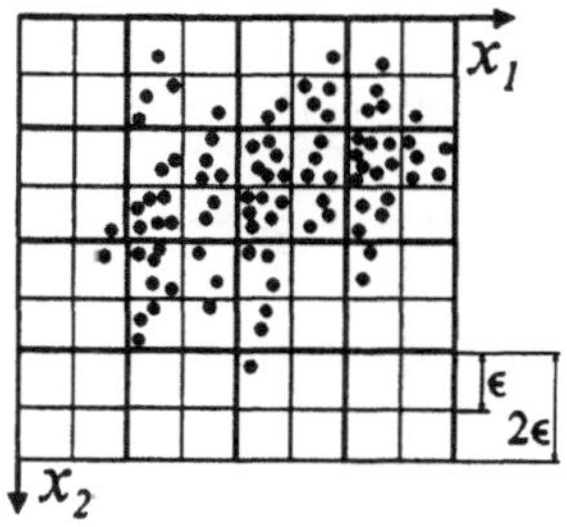

Fig. 3. Box covering of an example search space X for $K = 2$ for two sizes of a box: ϵ and 2ϵ.

4 Attractor's Information Dimension

The most important feature of an attractor is its dimension. In the case of an ordinary set the value of a dimension is a natural number. But the dimension of a strange attractor can be a real number.

In the literature one can find many kinds of a dimension of a set [2]. All of them provide similar information. We will be especially interested in the information dimension defined as follows [4].

Definition 3 *Let* $A \subset X$ *and* $X \subset \mathbb{R}^K$ *be a* K*-dimensional cube. Let us cover* X *by* C *boxes* $X_i \subset X$, *where* ϵ *is the length of an edge of each box. We define the* information dimension *of* A *as*

$$d_I(A) = -\lim_{\epsilon \to 0} \frac{H(A, \epsilon)}{\log_2 \epsilon}, \tag{1}$$

where

$$H(A, \epsilon) = -\sum_{i=1}^{C} p_i(A, \epsilon) \log_2 p_i(A, \epsilon) \tag{2}$$

is attractor's entropy for a given ϵ, $p_i(A, \epsilon)$ *is probability that a point* $x \in X_i \cap A$. □

The notion of a dimension of a set can give interesting information when applied in a special way to discrete sets [2]. We cannot use Def. 3 to determine the dimension of a population's attractor created by the algorithm presented at Fig. 1, because both sets A_G and X are discrete, thus the formula (1) gives 0. We need to derive a special formula for a discrete case that approximates the information dimension for a given ϵ.

First, we can assume that in Def. 3

$$p_i(A, \epsilon) = |X_i \cap A| / |A|, \tag{3}$$

where $|A| = WM$ and W is the number of iterations of the algorithm at Fig. 1.

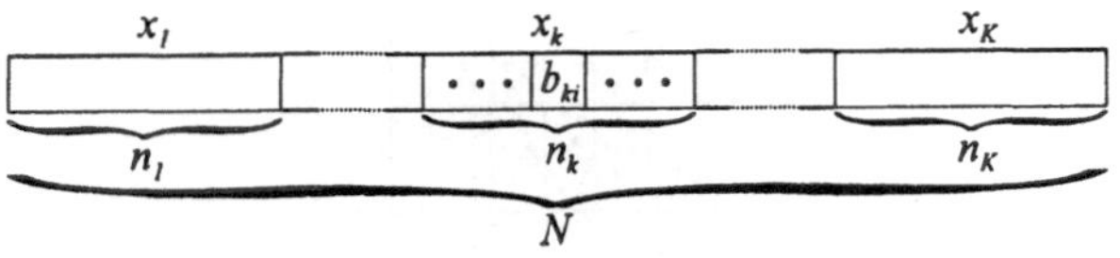

Fig. 4. Coding of solutions into chromosomes.

Let us notice that $C \sim \epsilon^{-d_I}$, so after calculations [6] we get

$$d_I(A,\epsilon) = H(A,\epsilon) - H(A,2\epsilon). \qquad (4)$$

The reader can find a detailed discussion on how the attractor's dimension changes during a $\mathcal{BGA}$'s execution in [5].

5 Meaning of Information Dimension in $\mathcal{BGA}$

Now we will show the meaning of the attractor's information dimension in the $\mathcal{BGA}$. First, let us look closer at the way how elements of the search space X are encoded into chromosomes.

We will assume that X is a K-dimensional cube, as in Def. 3, i.e.

$$X = \langle x_{1min}, x_{1max}\rangle \times \ldots \times \langle x_{Kmin}, x_{Kmax}\rangle \subset \mathbb{R}^K, \qquad (5)$$

where k-th component of X is encoded at n_k bits, so we have

$$\mathcal{B} = \{0,1\}^N = \{0,1\}^{n_1} \times \ldots \times \{0,1\}^{n_K}, \qquad (6)$$

and for convenience we introduce a new indexing of bits by a pair of numbers as shown at Fig. 4.

We cover X by boxes using the following rule. We use r most significant bits of the binary code of each component of X as the identificator of a box, thus we got totally $C = 2^{Kr}$ boxes. Next having that Ψ is a bijection we can write

$$\underline{\beta}_i = \Psi(X_i) \subset \mathcal{B} \qquad (7)$$

where $i = 1, \ldots, C$ is the number of a box, and the scheme [8] $\underline{\beta}_i$ represents the box $X_i \subset X$ in the code space $\mathcal{B}$. Similarly we define a set

$$B^A = \Psi(A) \subset \mathcal{B} \qquad (8)$$

representing the set $A \subset X$ in the code space $\mathcal{B}$.

We express (3) using (7) and (8) as

$$p_i(A,r) = \left|\underline{\beta}_i \cap B^A\right| / |B^A| \qquad (9)$$

where for convenience we use r as the size of a box, assuming the length of box edges equal to $\epsilon_k = \frac{x_{k\,max} - x_{k\,min}}{2^r}$ for $k = 1, \ldots, K$.

We apply (9) to a set $A_G \subset X$ being an attractor of the population $G \in \mathcal{B}^M$ generated by the algorithm at Fig. 1.

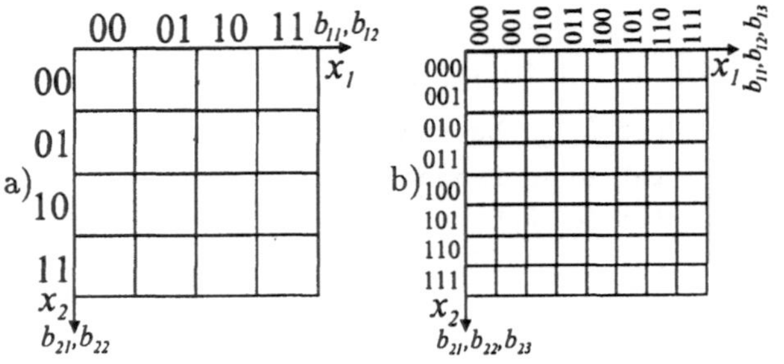

Fig. 5. An example of determining d_I: boxes for $r = 2$ (a) and 3 (b), number of X's components $K = 2$, and $n_1 = n_2 = 4$, so from (12) we get $d_I(A_G, 3) = H_p(1, 2|G) + H_p(2, 2|G)$.

Let us note that the events $x \in X_i \cap A$, mentioned in Def. 3, are cartesian products of Kr independent binary events related to fixed positions in schemes $\underline{\beta}_i$. So using Def. 1 and substituting ϵ by r, we can write (2) in the form

$$H(A_G, r) = \sum_{k=1}^{K}\sum_{j=1}^{r} H_p(k,j|G), \qquad (10)$$

where

$$H_p(k,j|G) = p_{\mathsf{one}}(k,j|G)\log_2 p_{\mathsf{one}}(k,j|G) + \\ + (1 - p_{\mathsf{one}}(k,j|G))\log_2(1 - p_{\mathsf{one}}(k,j|G)) \qquad (11)$$

is the entropy of j-th position of k-th component for a given population G.

Finishing we receive from (4) and (10) that

$$d_I(A_G, r) = H(A_G, r) - H(A_G, r-1) = \\ = \sum_{k=1}^{K} H_p(k, r|G). \qquad (12)$$

The above result means that the information dimension of a population's attractor for a given r equals to the information contained in r-th bits of each component code. One can see that this information is lost when we reduce the number of bits indexing boxes from r to $r-1$ (see Fig. 5).

6 The Experiments and Results

We investigated the $\mathcal{BGA}$ described in Section 2 with the elitism [11]. We assumed that the mutation probability $p_m = 0.01$, the crossover probability $p_c = 0.5$, the population size $M = 20$, and a constant number of generations GenNo = 200. The algorithm stopped when the error of the best chromosome in the current population was less than $\delta = 4 \cdot 10^{-9}$.

The search space was $X = \langle -2, 2\rangle^2$, and its elements were coded by 32-bit chromosomes. The fitness function was a very simple one:

$$f(x_1, x_2) = -\left[(x_1 - 1)^2 + (x_2 - 0.5)^2\right]$$

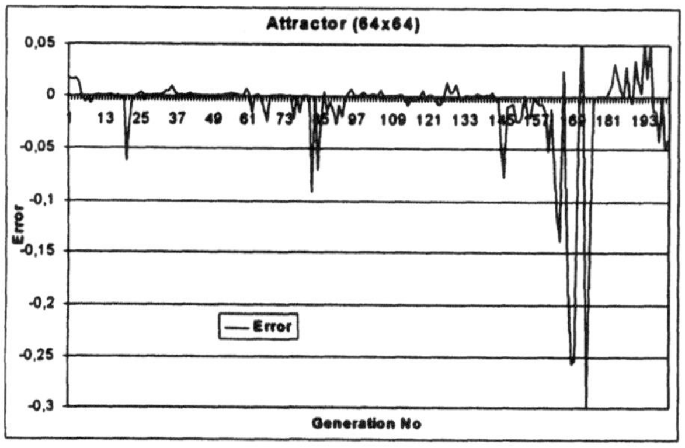

Fig. 6. A difference between the information dimension and the entropy of related bits for a 64 × 64 attractor.

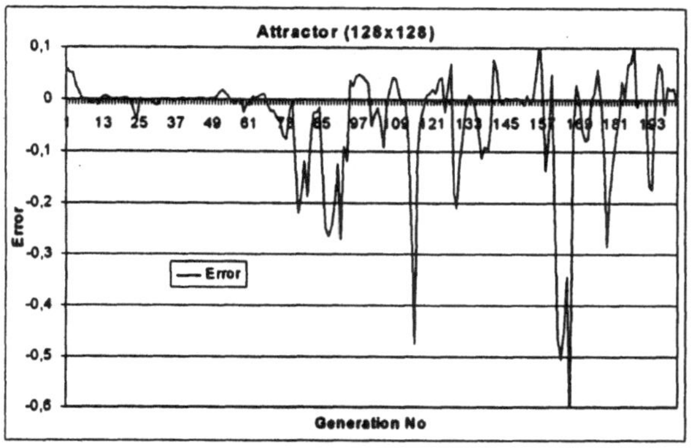

Fig. 7. A difference between information dimension and the entropy of related bits for a 128 × 128 attractor.

with the global maximum $f(1, 0.5) = 0$. In order to reconstruct attractors we applied the algorithm from Fig. 1 for $W = 10000$ for each population generated by the $\mathcal{BGA}$. The range of each of two components of the set X was discretized by 64 ($r = 6$) and 128 ($r = 7$) intervals. You can find examplary images with population's attractors in [5].

As a result we present at Fig. 6 and 7 graphs of the difference between attractor's information dimension and the entropy of related bits (6th and 7th) in a function of the generation number for a certain single $\mathcal{BGA}$'s execution.

7 Conclusions

The graphs presented at Fig. 6 and 7 confirm our theoretical reasoning realized in Section 5. We see that the difference between the two graphs is very small at the begining and increases in finishing populations. The cause of the increase is an higher error in reconstructing a population's attractor for later papulations. We suppose that the value of W should be higher when solutions are concentrated in few boxes only as well as when the number of boxes C is higher.

The received result is very useful because it facilitates an application of tools from the chaotic dynamics into $\mathcal{BGA}$ examinations. The algorithm at Fig. 1 has a high computational complexity because of intensive use of Ψ^{-1}, but now we can use much simpler computations in the domain of chromosomes instead. It seems that the presented approach can be useful in investigation of the $\mathcal{BGA}$'s dynamics, but it needs further research.

Acknowledgements

This material is based upon work supported by the grant 8 T11C 003 16 from the Polish State Committee for Scientific Research (KBN) in 1999-2000.

References

[1] Bäck T.: "The Interaction of Mutation Rate, Selection, and Self-Adaptation within a Genetic Algorithm", *Parallel Problem Solving from Nature*, Vol. 2, pp. 85-94, Amsterdam, North Holland, 1992.

[2] Baker G.L., Gollub J.P.: *Chaotic Dynamics: an Introduction*, Cambridge Univ. Press, Cambridge, 1996.

[3] De Jong K.A., Spears W.M.: "A Formal Analysis of the Role of Multi-point Crossover in Genetic Algorithms", *Annals of Math. and AI Journal*, Vol. 5, No. 1, pp. 1-26. J.C. Baltzer AG Scientific Publishing Co., 1992.

[4] Falconer K.J.: "Fractal geometry", *Math. Found. Appl.*, John Wiley, Chichester, pp.15-25, 1990.

[5] Kieś P.: "Dimension of attractors generated by a genetic algorithm", *Procs. of the IX-th Intern. Symposium on Intelligent Information Systems*, pp. 40-45, Bystra, Poland, June 2000.

[6] Kieś P.: "Meaning of the information dimension of a population's attractor in a binary GA", *IEEE Trans. on Evolutionary Computation*, 2001 (sent to print).

[7] Kolonko M.: "A Generalized Crossover Operation for Genetic Algorithms", *Complex Systems*, pp. 177-191, Vol. 9, 1995.

[8] Michalewicz Z.: *Genetic Algorithms + Data Structures = Evolution Programs*, Springer-Verlag, Berlin, 1996.

[9] Ott E.: *Chaos in Dynamical Systems*, Cambridge Univ. Press, Cambridge, 1996.

[10] Qi T., Palmieri F.: "Theoretical Analysis of Evolutionary Algorithms with an Infinite Population Size in Continuous Space, Part II: Analysis of the Diversification Role of Crossover", *IEEE Trans. of Neural Networks*, Vol. 5, No. 1, 1994.

[11] Rudolph G.: "Convergence analysis of canonical genetic algorithms", *IEEE Trans. on Neural Networks*, Special Issue on Evolutionary Computation, Vol. 5, No. 1, 1994.

An Evolutionary Approach to the Zero/One Knapsack Problem: Testing Ideas from Biology

Anabela Simões[*,†], Ernesto Costa[†]

[*] Instituto Superior de Engenharia de Coimbra, Quinta da Nora, 3030 Coimbra, Portugal [†] Centro de Informática e Sistemas da Universidade de Coimbra, Pinhal de Marrocos, 3030 Coimbra, Portugal, E-mail: abs@sun.isec.pt; ernesto@dei.uc.pt

Abstract

The transposition mechanism, widely studied in previous publications, showed that when used instead of the standard crossover operators, allows the genetic algorithm to achieve better solutions. Nevertheless, all the studies made concerning this mechanism always focused the domain of function optimization. In this paper, we present an empirical study that compares the performances of the transposition-based GA and the classical GA solving the 0/1 knapsack problem. The obtained results show that, just like in the domain of function optimization, transposition is always superior to crossover.

1 Introduction

Genetic Algorithms (GA) are a search paradigm that applies ideas from evolutionary biology (crossover, mutation, natural selection) in order to deal with intractable search spaces [5]. The basic iteration cycle of a GA proceeds on a population of individuals, each of which represents a search point in the space of potential solutions of a given optimization problem [2]. The standard GA starts with an initial population of individuals created at random. Then, this population evolves through time by a string manipulation process based in three genetic operators: reproduction, crossover and mutation. The power and success of GA is mostly achieved by the diversity of the individuals of a population. In the classical GA, diversity is maintained through the genetic operators crossover and mutation.

In nature, genetic diversity is caused and maintained by several mechanisms besides crossover and mutation. Some of those mechanisms are inversion, transduction, transformation, conjugation, transposition and translocation [3].

Several authors have already used some biologically inspired mechanisms besides crossover and mutation in evolutionary approaches. For instance, inversion [5], conjugation [4], [11], transduction [7], translocation [1] and transposition [8] were already used as the main genetic operators in the GA.

The goal of this paper is to enlarge the domain of application of the transposition mechanism, using it to solve different types of the 0/1 knapsack problem. The empirical study will focus the GA efficiency solving the 0/1 knapsack problem, first using the classical crossover operators and then, several variants of the transposition mechanism. The obtained results show that, in all the studied situations, transposition allows the GA to reach better solutions.

This paper is organized in the following manner. First, in section 2, we explain how transposition works in nature and we summarize previous work related to the transposition mechanism. In section 3, we introduce the 0/1 knapsack problem. Section 4 describes the characteristics of the experimental environment. In section 5, we present a summary of the obtained results using transposition and crossover. Finally, we present the relevant conclusions of the work.

2 Transposition

2.1 Biological transposition

Transposition is characterized by the presence of mobile genetic units inside the genome, moving themselves to new locations or duplicating and inserting themselves elsewhere. These mobile units are called transposons [3]. The movement of the transposons (also known as jumping genes) can take place in the same chromosome or to a different one. In order to a transposable element to transpose as a discrete entity, it is necessary for its ends to be recognized. Therefore, transposons within a chromosome are flanked by identical or inverse sequences, some of which are actually part of the transposon.

The point into which the transposon is inserted requires no homology with the point where the transposon was excised. This is in evident contrast to classical recombination, and consequently, transposition is sometimes referred to as illegitimate recombination.

2.2 Computational transposition

Previous publications studied three different variants of the transposition mechanism, all inspired in the biological process.

The first form of computational transposition was called simple transposition. After the selection of two parents for mating, the transposon is formed in one of them. The insertion point is searched in the second parent and the same amount of genetic material is exchanged between the two chromosomes [8].

In order to come closer to the biological mechanism, we proposed a new form of transposition: tournament-based transposition. The two selected parents become competitors in a tournament and the transposon will be searched in the winner chromosome. The insertion point will be located in the loser parent. Only this individual will be changed: the transposon is inserted, replacing the same number of bits after the insertion point [9].

In nature, the transposition mechanism can also occur in the same chromosome. Based on this principle, we introduced the asexual form of transposition, where all the process operates in the same individual [10].

3 The 0/1 Knapsack Problem

The well-known single-objective 0/1 knapsack problem (KP) is defined as follows: given a set of items (n), each with a weight $W[i]$ and a profit $P[i]$, with $i=1,..,n$. The goal is to determine the number of each item to include in the knapsack so that the total weight is less than some given limit (C) and the total profit is as large as possible.

4 Empirical Study

In our empirical analysis, we implemented the KP varying some parameters as suggested in [6]. We used several instances for the KP, taking in consideration several aspects such as, the algorithm used for evaluation of the individuals, the number of items, the correlation between the weights and the profits and the capacity of the knapsack. All those aspects will be detailed in the next sections. We will also refer to the parameters of the genetic algorithm used to solve the problem.

4.1 Algorithms used in the individual's evaluation

We used two types of algorithms in the evaluation of the individuals (**x**) of the population: algorithms based on penalty functions ($Ap[i]$) and algorithms based on repair methods ($Ar[i]$), where i is the index of a particular algorithm in each class [6].

Using algorithms based on penalty methods, a binary vector of length n represents a solution **x** for the problem each element of **x** can be 0 or 1. If **x**[i]=1 then the item i was selected for the knapsack. The fitness $f(x)$ for each binary string is determined as:

$$f(\mathbf{x}) = \sum_{i=1}^{n} \mathbf{x}[i]P[i] - Pen(\mathbf{x}) \qquad (1)$$

The penalty function $Pen(x)$ is zero to all feasible solutions and greater than zero otherwise.

There are many possibilities for assigning the penalty value. We consider three cases where the growth of the penalty function is logarithmic, linear and quadratic [6].

When using algorithms based on repair methods, the evaluation of the binary string **x** is determined as:

$$f(\mathbf{x}) = \sum_{i=1}^{n} \mathbf{x}'[i]P[i] \qquad (2)$$

where **x'**[i] is the repaired version of the original chromosome **x**. We used two different repair methods to obtain **x'**:

Ar[1] (random repair): the item to be removed from the knapsack is selected at random.

Ar[2] (greedy repair): all the items in the knapsack are ordered in the decreasing order of their profit to weight ratios. The deleted item is the last one.

Another aspect to consider in the repair-based algorithms is the percentage of repaired chromosomes to be replaced in the original population. Such replacement rate may vary between 0% and 100%. In our experiments, we used two replacement rates 100% (replace all) and 5%. We chose the 5% rate to test the 5% rule proposed by Orvosh et al. [6].

4.2 Parameters used in the evaluation algorithms

For each algorithm previously described, the empirical study analyzed the influence of the variation of several parameters such as the number of items in the knapsack, the relation between weights and profits and the capacity of the knapsack. In the following sections, we will detail how these parameters were changed in the experiments.

Concerning the number of items, we used three different values: n=100, n=250 and n=500.

Since the difficulty of the problem is affected by the correlation between profits and weights, we considered three different ways of randomly generate those vectors:

Uncorrelated: both vectors $W[i]$ and $P[i]$ are generated at random, using a uniform distribution. W[i]=(uniformly) *rand* ([1..v]); P[i]=(uniformly) *rand* ([1..v]).

Weakly correlated: vector $W[i]$ is created at random, but $P[i]$ is created with some correlation with $W[i]$. $W[i]=$ (uniformly) *rand* ([1..v]); $P[i]=W[i]+$ (uniformly) *rand* ([-r..r]).

Strongly correlated: $W[i]=$(uniformly) *rand* ([1..v]); $P[i]=W[i] + r$.

The values used for the parameters v and r were v=10 and r=5 (see [6]).

We used two types of capacity for the knapsack: restrictive knapsack capacity: $C_1 = 2.v$ and average knapsack capacity: $C_2 = 0.5\sum_{i=1}^{n} W[i]$.

4.3 Parameters of the genetic algorithm

We implemented the standard GA as described in [2]. The GA used, each time, a different genetic operator: 1-point crossover, 2-point crossover, uniform crossover, simple transposition, tournament-based transposition and asexual transposition. The population consisted in 100 binary strings. We used roulette-wheel as the selection method and an elitism rate of 10%. Mutation and transposition/crossover rates were fixed in 5% and 65%, respectively. The number of generations was 1000. Each experiment was repeated 25 times. When using transposition, the used flanking sequence length was always computed through the appropriated heuristic proposed in previous work [10].

5 The Results

Our empirical study was very extensive, but due to lack of space we will present the results obtained using the penalty algorithm *Ap[1]* and the repair method *Ar[1]*.
To measure the performance of the different algorithms we used the best solution found within the 1000 generations. The results synthesized in the following tables are mean values of the 25 experiments. The best solutions are marked in bold. The genetic operators used in the tables are identified by Cx1, Cx2, CxU, AT, ST and TT for 1-point, 2-point, uniform crossover, asexual, simple and tournament-based transposition, respectively.

5.1 Results obtained using the Ap[1] algorithm

Using the penalty-based algorithms, when the capacity of the knapsack was Restrictive (*C1*), no valid solutions were found by the GA. Using the Average capacity (*C2*), the GA using tournament-transposition achieved better results in all the situations. Concerning the remaining genetic operators, the other forms of transposition were always better than the crossover operators. The only exception was uniform crossover that had similar performances to asexual transposition in some particular cases. The main observation is that the results are influenced by the correlation of the data set. As we increase this correlation, the obtained results become more similar.
Table I shows the obtained results using the penalty algorithm *Ap[1]*. Using the penalty algorithms *Ap[2]* and *Ap[3]* the transposition-based GA had similar behavior.

Table I. Results Using the Penalty Algorithm Ap[1]

Correl.	Items	Capac.	Genetic Operator					
			Cx1	Cx2	CxU	AT	ST	TT
Uncorr	100	C2	478	500	518	518	534	**562**
	250	C2	1159	1258	1237	1356	1276	**1401**
	500	C2	2291	2332	2440	2460	2480	**2606**
Weak	100	C2	584	622	636	668	641	**681**
	250	C2	1481	1524	1539	1549	1555	**1587**
	500	C2	2819	2802	2817	2878	2822	**2887**
Strong	100	C2	1023	1028	1012	1051	1055	**1055**
	250	C2	2471	2491	2485	2523	2573	**2578**
	500	C2	4654	4654	4656	4659	4663	**4770**

5.2 Results obtained using Ar[1] algorithm

With this algorithm, we implemented two strategies (replace-all and replace 5%) but we will present the results achieved using the "replace-all" strategy. We choose to present these results, because they are inferior to those obtained with the replace 5% strategy (for the transposition mechanism). Using the *Ar[1]* algorithms, once again, transposition achieved better solutions. In particular, the tournament-based and simple transposition were the mechanisms that performed better. Table II synthesizes the achieved results.

Table II. Results Using the Repair Algorithm Ar[1] and the "replace-all" Strategy

Correl.	Itens	Capac.	Genetic Operator					
			Cx1	Cx2	CxU	AT	ST	TT
Uncorr	100	C1	516	524	524	524	**525**	524
		C2	341	372	377	385	**395**	**395**
	250	C1	1343	1347	1348	1350	**1353**	1352
		C2	860	931	896	943	**981**	**981**
	500	C1	2511	2518	2515	2539	2540	**2541**
		C2	1860	1873	1869	1882	1880	**1883**
Weak	100	C1	633	633	633	634	634	**635**
		C2	351	358	344	376	**392**	386
	250	C1	1557	1559	1560	1560	1563	**1564**
		C2	930	950	955	967	966	**973**
	500	C1	2878	2885	2884	2776	2892	**2907**
		C2	1899	1907	1879	1917	**1972**	1939
Strong	100	C1	1030	1033	1037	1039	1039	**1041**
		C2	549	548	558	570	573	**578**
	250	C1	2346	2527	2531	2538	2539	**2540**
		C2	1400	1407	1416	1417	1422	**1428**
	500	C1	4438	4835	4839	4860	4846	**4895**
		C2	2829	2843	2809	2843	2857	**2861**

5.3 Using smaller population with transposition

Previous work in the function optimization domain showed that, with 50 individuals, transposition always outperformed the standard crossover operators with 50, 100 and 200 individuals. In order to conclude about the influence of the population size solving the 0/1 KP, we executed some experiments using the three transposition mechanisms with only 50 individuals. These experiments focused the penalty algorithms with 100 items in the knapsack. Once again, transposition with smaller populations allowed the GA to achieve, in general, better solutions than when using crossover. Table III reports the obtained results.

Table III. Results using Transposition with 50 Individuals

Correl.	Penalty alg.	Nº Items	Genetic Operator					
			Cx1	Cx2	CxU	AT	ST	TT
			Pop = 100			Pop = 50		
Uncorr	Ap[1]	100	478	500	518	516	521	**536**
Weak	Ap[1]	100	584	622	636	636	637	**651**
Strong	Ap[1]	100	1023	1028	1012	1030	1029	**1046**
Uncorr	Ap[2]	100	349	358	359	366	360	**408**
Weak	Ap[2]	100	340	343	351	366	360	**378**
Strong	Ap[2]	100	556	557	558	558	**559**	**559**
Uncorr	Ap[3]	100	357	354	354	371	344	**379**
Weak	Ap[3]	100	346	346	347	347	348	**350**
Strong	Ap[3]	100	559	560	559	560	**562**	**562**

6 Conclusions

The goal of this paper was to use a biologically inspired genetic operator called transposition (already tested in the function optimization domain) in a different domain. The select problem was the 0/1 KP, several types of knapsacks were implemented and different genetic operators were used to solve it: the standard crossover operators (1-point, 2-point and uniform) and several variants of the transposition mechanism (asexual, simple and transposition-based).

The obtained results showed that, in all the implemented algorithms, the transposition mechanisms allowed the GA to achieve higher performances. We also reduced the size of the population for 50 individuals when using transposition and, even so, the GA obtained better results than when using crossover with 100 individuals. Those results reinforce our conviction that transposition is a powerful genetic operator alternative to crossover.

Acknowledgements

This work was partially financed by the Portuguese Ministry of Science and Technology under the Program Praxis XXI and by Coimbra Polytechnic.

References

[1] De Falco, I., Iazzetta, A., Tarantino, E., Della Cioppa, A.: On Biologically Inspired Mutations: The Translocation. In Late Breaking Papers at the 2000 Genetic and Evolutionary Computation Conference (GECCO'2000), pp. 70-77, Las Vegas, USA, 8-12 July 2000.

[2] Goldberg, D. E.: Genetic Algorithms in Search, Optimization and Machine Learning. Addison-Wesley Publishing Company, Inc, 1989.

[3] Gould, J. L., Keeton, W. T.: Biological Science. W. W. Norton & Company 1996.

[4] Harvey, I.: The Microbial Genetic Algorithm. Submitted as a Letter to Evolutionary Computation. MIT Press, 1996.

[5] Holland, J. H.: Adaptation in Natural and Artificial Systems: An Introductory Analysis with Applications to Biology, Control and Artificial Intelligence. 1st MIT Press edition, MIT Press 1992.

[6] Michalewicz, Z.: Genetic Algorithms + Data Structures = Evolution Programs. 3rd Edition Springer-Verlag 1999.

[7] Nawa, N., Furuhashi, T., Hashiyama, T., Uchikawa, Y.: A Study of the Discovery of Relevant Fuzzy Rules Using Pseudo-Bacterial Genetic Algorithm. IEEE Transactions on Industrial Electronics, 1999.

[8] Simões, A., Costa, E.: Transposition: A Biologically Inspired Mechanism to Use with Genetic Algorithms. In the Proceedings of the Fourth International Conference on Neural Networks and Genetic Algorithms (ICANNGA'99), pp. 612-619. Springer-Verlag 1999.

[9] Simões, A., Costa, E.: Transposition versus Crossover: An Empirical Study. Banzhaf, W., Daida, J., Eiben, A. E., Garzon, M. H., Honavar, V., Jakiela, M., and Smith, R. E. (eds.), Proceedings of the Genetic and Evolutionary Computation Conference (GECCO'99), pp. 612-619, Orlando, Florida USA, CA: Morgan Kaufmann 1999.

[10] Simões, A., Costa, E.: Using Genetic Algorithms with Asexual Transposition. In D. Whitley, D. Goldberg, E. Cantú-Paz, L. Spector, I. Parmee, H. Beyer. (eds.), Proceedings of the Genetic and Evolutionary Computation Conference (GECCO'2000), pp. 323-330, Las Vegas, USA, CA: Morgan Kaufmann 2000.

[11] Smith, P.: Conjugation - A Bacterially Inspired Form of Genetic Recombination. In Late Breaking Papers of the First International Conference on Genetic Programming. Stanford University, California, 1996.

An Evolutionary Approach to Identification of Nonlinear Dynamic Systems

Marcin Witczak*, Józef Korbicz†

*Institute of Control and Computation Engineering, Technical University of Zielona Góra, ul. Podgórna 50, 65-246 Zielona Góra, Poland, e-mail: *M.Witczak@irio.pz.zgora.pl* †Institute of Control and Computation Engineering, Technical University of Zielona Góra, ul. Podgórna 50, 65-246 Zielona Góra, Poland, e-mail: *J.Korbicz@irio.pz.zgora.pl*

Abstract

In this paper a nonlinear identification methodology founded upon NARX model description is presented. In particular, the model determination procedure is decomposed into the elementary model structures selection one. Those models are represented as fixed-depth trees and a genetic algorithm is used to obtain their appropriate form. To show the effectiveness of the proposed approach, the final part of the paper contains examples concerning modelling the juice temperature at the outlet of an evaporator at the Lublin sugar factory.

1 Introduction

The main objective of system identification is to obtain a mathematical description of a real system of interest. Unfortunately, the high complexity of a large majority of real systems makes it impossible to perform physical considerations underlying phenomenological models, which are built from physical laws governing the system studied. In such a situation, the *behavioural models*, which merely approximate the system's input-output behaviour, have to be employed. There are, of course, a lot of less or more sophisticated methods of building behavioural models, for a detailed description of them we refer the reader to [7, 10] and the references therein. In this work, the considerations are limited to the dynamic discrete-time nonlinear systems. Unfortunately, there is no general method applicable to a wide class of nonlinear systems. Thus, the nonlinear system identification problem is still open. The application of Artificial Intelligence (AI) methods to identification of nonlinear systems has been intensively studied for the last two decades. The most popular approach is to employ either neural networks [4] or fuzzy neural networks [9]. An alternative approach is to employ Genetic Programming (GP) [1, 6]. GP is an extension of genetic algorithms [8] which are a broad class of stochastic optimization algorithms inspired by some biological processes which allow populations of organisms to adapt to their surrounding environment. The main difference between these two approaches is that in GP the evolving individuals are parse trees instead of fixed-length binary strings. Such a methodology has proved its particular effectiveness in the context of system identification [3, 11, 12]. On the other hand, the trees underlying the genetic programming algorithm may become extremely sophisticated for complex multi-input systems. This is especially true for the models of dynamic systems, for which the number of variables increases according to the inputs and output lags. This work presents an alternative nonlinear system identification approach which decomposes the model determination procedure into the elementary model structures selection one. In particular, the model of the system is built from a group of elementary models represented by fixed-depth trees.

2 Nonlinear Identification Technique

2.1 Nonlinear system representation

The characterization of the set of possible candidate models $\mathbb{M}$ from which the system model will be obtained consists an important preliminary task in any system identification procedure. Knowing that the system exhibits a nonlinear characteristic, a choice of nonlinear model set must be made. In this work, an NARX (*AutoRegresive with eXogenous variable*) model representation was selected as the foundation for the identification methodology. Moreover, it is assumed that the system whose model has to be determined is an MISO (*Multi Input Single Output*) one.

The NARX model has the following form

$$\begin{aligned}\hat{y}_k = f(&\hat{y}_{k-1}, \ldots, \hat{y}_{k-n_y}, u_{1,k-1}, \ldots, u_{1,k-n_{1,u}},\\ &u_{1,k-1}, \ldots, u_{1,k-n_{1,u}}, \ldots,\\ &u_{m,k-1}, \ldots, u_{m,k-n_{m,u}}),\end{aligned} \tag{1}$$

where $\hat{y}_k \in \mathbb{R}$ is the output estimate, $\boldsymbol{u}_k \in \mathbb{R}^m$ is the input, $n_y, n_{1,u}, \ldots, n_{m,u}$ are the maximum lags in the output and inputs, $f(\cdot)$ is a nonlinear function. Thus, the system output is $y_k = \hat{y}_k + e_k$, where e_k is the output error which consists of structural deterministic errors, caused by the model-reality mismatch, and the measurement noise. The problem is to determine an unknown function $f(\cdot)$ and to estimate its parameters vector $\boldsymbol{p} \in \mathbb{R}^{n_p}$. To tackle this problem, the following criterion is introduced

$$\hat{f}(\cdot, \hat{\boldsymbol{p}}) = \arg \min_{f(\cdot) \in \mathbb{M}} \min_{\boldsymbol{p} \in \mathbb{R}^{n_p}} \jmath(f(\cdot, \boldsymbol{p})), \tag{2}$$

$$\jmath(f(\cdot, \boldsymbol{p})) = \frac{1}{n} \sum_{k=1}^{n} \frac{|e_k|}{|y_k|}, \tag{3}$$

where n denotes the number of data points, $\mathbb{M}$ is the set of possible model structures. As was already mentioned, the genetic programming technique can be successfully employed to determine $f(\cdot)$ and estimate its parameters $\boldsymbol{p}$ [11, 12] in such a way to satisfy (2). Indeed, each model structure can be represented as a tree, as shown in Fig. 1. In such a tree, two sets can be distinguished, namely the terminals $\mathbb{T}$ and functions $\mathbb{F}$ sets (e.g. $\mathbb{F} = \{+, *, /\}$, $\mathbb{T} = \{t_k, t_l\}$). Based on this sets, various trees representing various model structures can be created. Unfortunately, in the case of multi-input systems the trees become more and more complex thus making the parameter estimation problem, which is a nonlinear optimization one, extremely difficult. Since the parameter estimation process consists an important part of the fitness calculation procedure, inaccurate parameter estimates may lead to the rejection of the structures which are very likely to be well suited. To tackle this problem, it seems desirable to decompose the model structure selection procedure into the elementary model structures selection one. A particular way of searching the model structure through the determination of elementary model structures is the GMDH approach (*Group Method of Data Handling*) introduced by Ivakhnenko [5]. A thorough treatment of the GMDH approach is given in [2]. In this work, a fixed-depth tree, as shown in Fig. 1, is used to represent an elementary model structure.

2.2 Tree structure determination

Given a general form of the tree (Fig. 1), it is necessary to develop an algorithm which can be employed to obtain its structure and parameters in such a way to minimize the optimization criterion chosen. One way to settle this problem is to perform a full inspection of all possible structures and to estimate their parameters. Unfortunately, this is not a really practical approach because, even for such a simple form of the tree, the number of possible combinations is extremely large. Thus, the full inspection of all possible structures is rather not feasible. To overcome this problem, it seems desirable to use one of the evolutionary optimization methods and to search throughout the set of structures. Although it is still possible to use the genetic programming technique, it seems convenient, especially for the purpose of the implementation simplicity, to use a genetic algorithm. Since the maximum admissible depth of the tree is fixed, it can be easily expressed as a fixed length string. Thus, it is necessary to describe the chromosom which represents the structure of the tree. First, let us define the functions and terminals sets: $\mathbb{F} = \{+, *, /, \xi_1(\cdot), \ldots, \xi_l(\cdot)\}$, $\mathbb{T} = \{t_k, t_l\}$, where $\xi_i(\cdot)$ is a nonlinear univariate function. Given the above sets, the chromosom can be described as follows

l_1	l_2	l_3	l_4	l_5	l_6	l_7

where $l_i \in \mathbb{F} \cup \mathbb{T}$. As it can be seen, the length of the chromosom is determined by the maximum admissible number of non-terminal nodes, which, in the case of the tree in Fig. 1, is equal 7. Such a definition of the chromosom makes it possible to use the classical genetic algorithm to search for an appropriate structure of the tree. The only modification is that instead of the binary representation the values from the $\mathbb{F} \cup \mathbb{T}$ set are used. It should be also mentioned that the elements of the chromosom, which represent the nodes bellow the terminal nodes, have no effect on the structure of the tree and it does no matter what values they have. To perform the selection procedure, the fitness calculation process has to be detailed. Therefore, as a measure of fitness, expression (3) is adopted. As was already mentioned, to perform the fitness calculation process it is necessary to obtain the parameters vector $\boldsymbol{p}$. Based on numerous computer experiments, it was found that the extremely simple Adaptive Random Search (ARS) strategy [10] is especially suited for that pur-

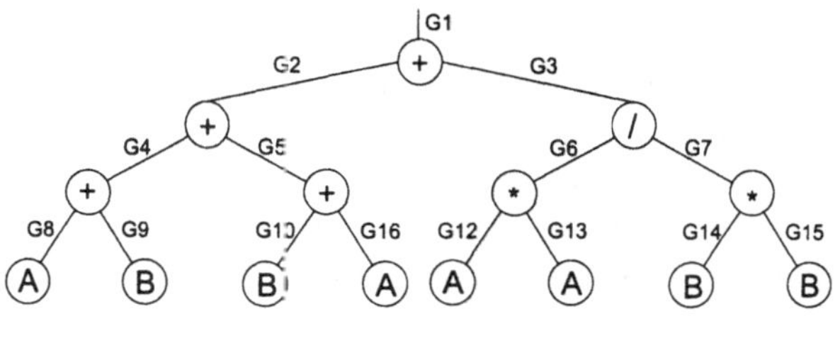

Fig. 1. An exemplary tree.

pose, although it is possible to use other nonlinear programming techniques. However, the maximum number of parameters to be estimated is not to great (i.e. is equal 15) it can be further reduced, this is because of the fact that most of them are not identifiable. For that purpose a few simple rules can be established:

$*, /$: A node of type either $*$ or $/$ has a parameter on the side of its parent. The node has no parameters if its child is of type $+$.

$+$: A node of type $+$ has parameters on the side of its children. If a parent of the node is of type $*$, its second child is of type $+$ and this child has parameters or its child is of type $+$, then the node has no parameters. If some sub-tree is connected with a sub-tree of the same structure through the node of type $+$ then only one of the sub-trees has a parameter on the side of its parent node.

ξ: A node of type ξ has all parameters.

To use these rules it is also necessary to assume that a parameter which is deleted for a certain node remains deleted for other nodes. The above rules are applied throughout the tree starting from each terminal (leaf). As an example, consider the tree shown in Fig. 1. Following the above rules, the resulting parameter vector has only three elements $\boldsymbol{p} = (p_8, p_9, p_3)$, and the elementary model is $\hat{y}_{k,l} = (1+p_8)t_k + (1+p_9)t_l + p_3 t_k^2/t_l^2$. It should be also pointed out that because of the application of crossover and mutation, a nonlinear univariate function $\xi_i(\cdot)$ can be introduced instead of one of the mathematical operators (i.e.: $+, *, /$). This means that one of the branches below the function has to be rejected. Although, various strategies can be employed for that purpose, the most straightforward way is to select the branch to be rejected randomly.

2.3 Structure identification of NARX models

Let us define an initial set of terminals

$$\mathbb{T}_0 = \{\hat{y}_{k-1}, \ldots, \hat{y}_{k-n_y}, u_{1,k-1}, \ldots, u_{1,k-n_{1,u}}, \ldots, u_{m,k-1}, \ldots, u_{m,k-n_{m,u}}\}, \tag{4}$$

The number of all admissible couples of terminals n_m is defined as, $n_u = \sum_{i=1}^{m} n_{i,u}$,

$$n_m = \binom{n_y + n_u}{2} - \binom{n_y}{2} = \frac{(n_u^2 - n_u)}{2} + n_u n_y. \tag{5}$$

The algorithm which can be employed to obtain the structure of the NARX model can be described as follows:

Step 0 : Select the functions set $\mathbb{F}$ (the functions should be chosen so as they be *a priori* useful in solving the problem) as well as inputs and output lags $n_{l,u}$, $l = 1, \ldots, m$ and n_y. Choose the values of control parameters in the genetic algorithm (e.g. crossover and mutation probabilities (P_{cross}, P_{mut})). Select the data sets $\mathbb{D}_e = \{u_k, y_k\}_{k=0}^{n_e}$ and $\mathbb{D}_v = \{u_k, y_k\}_{k=0}^{n_v}$. Set $i = 0$.

Step 1 : Using the algorithm of Section 2.2, the terminals set $\mathbb{T}_i$, and $\mathbb{D}_e$ data set determine the n_m elementary models.

Step 2 : Using the $\mathbb{D}_v$ data set select $n_u + n_y$ elementary models which are the most well fitted in terms of the criterion (2).

Step 3 : If the termination condition is reached (one of the models fits the data with a desired accuracy or the introduction of the new elementary models did not make a significant increase in the approximation abilities of the whole model) then STOP else set $i = i + 1$ and form a new terminals set $\mathbb{F}_i$ from the outputs of elementary models selected, and then go to the *Step 1*.

It should be also pointed out that the simulation programme must ensure robustness to unstable individuals. This can be easily attained when (3) is bounded by a certain maximum admissible value. This means that each individual which exceeds the above bound is penalized by stopping the calculation of its fitness and then fitness is set to a sufficiently large positive number.

3 Experimental Results

As was already mentioned, the problem is to model the juice temperature at the outlet of an evaporator in a sugar factory. The data used for the training and testing sets were collected, from two different shifts, in November 1998. A sampling rate of 10 s for both the sets was considered owing to small time constants of the evaporator process. The input vector $\boldsymbol{u}_k$ consists of juice flow F_s, steam flow F_p, steam temperature T_p, and juice temperature at the inlet of the evaporator T_{S1}: $\boldsymbol{u}_k = (F_s, F_p, T_p, T_{S1})$. The parameters exploited during the identification process are:

$\mathbb{F} = \{+, *, /\}$, $P_{\text{mut}} = 0.001$, $P_{\text{cross}} = 0.8$, respectively. The algorithm is stopped when the number of iterations exceeds 35. The population size is equal 500, and $n = 3000$. The tournament selection method is employed and hence the tournament population size is equal 10. Experimental results have shown that the input and output lags which provide the best approximation quality are equal to $n_y = n_{1,u} = n_{2,u} = n_{3,u} = n_{4,u} = 2$. Among 50 runs performed, the value of (3) (obtained for the test set), for the best model structure, is equal 0.132. The testing results are shown in Fig. 2, from which it can be seen that the obtained model possesses good approximation abilities.

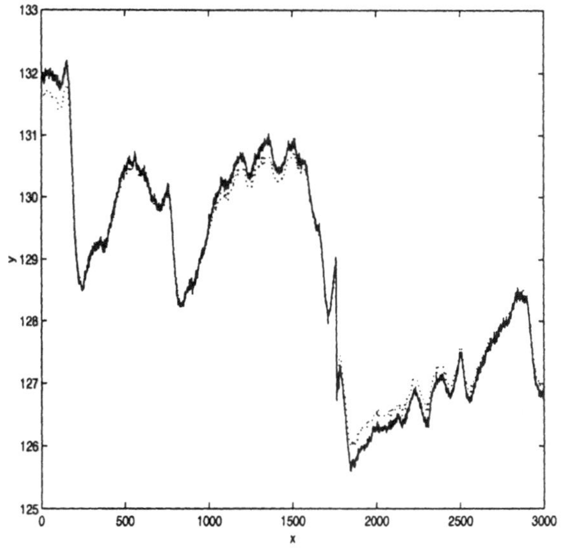

Fig. 2. The system output (solid) and its estimate (dotted).

4 Conclusions

The main objective of this work has been to propose a framework for the identification of nonlinear dynamic systems. In particular, the problem is to generate a suitable model set from which an appropriate model structure can be selected. This is usually a very difficult task which requires many trial and error iterations. To tackle this problem, a modification of the genetic programming technique has been proposed, which decomposes the model determination procedure into the elementary model structures selection one. This means that the model of the system is built from a group of elementary models represented by fixed-depth trees. Experimental results have shown that the proposed methodology provides a high modelling quality. The main drawback to this technique is a considerable amount of computation time which is necessary to identify the model structure. Thus, a decrease in the computational burden will be the subject of further research.

References

[1] Esparcia-Alcazar, A. I. (1998). *Genetic Programming for Adaptive Digital Signal Processing.* Ph. D. dissertation. Glasgow University.

[2] Farlow, S. J. (1984). *Self-organizing Methods in Modelling - GMDH Type Algorithms.* Marcel Dekker. New York.

[3] Gray, G. J, D. J. Murray-Smith,Y. Li, K. C. Sharman, T. Weinbrenner (1998). *Nonlinear model structure identification using genetic programming.*, Control Eng. Practice, Vol. 6. pp. 1341-1352.

[4] Hertz, J., R. Krogh and G. Palmer (1991). *Introduction to the Neural Computation.* Addison-Wesley Publishing Company, Inc.

[5] Ivakhnenko, A. G. T (1968). *The group method of data handling - A rival of stochastic approximation.* Sov. Autom. Contr. vol. 3, p. 43.

[6] Koza, J. R. (1992). *Genetic Programming: On the Programming of Computers by Means of Natural Selection.* The MIT Press. Cambridge.

[7] Ljung, L. (1987).*System Identification. Theory for the Users.* Prentice-Hall Inc. Englewood Cliffs. New Jersey.

[8] Michalewicz, Z. (1996). *Genetic Algorithms + Data Structures = Evolution Programs.* Springer. Berlin.

[9] Nuck, D. and, Klawonn F. and R. Krause (1997). *Foundations of Neuro-Fuzzy Systems.* John Wiley & Sons. Chichester.

[10] Walter, E. and L. Pronzato (1996). *Identification of Parametric Models from Experimental Data.* Springer. London.

[11] Witczak, M. and J. Korbicz (2000). *Genetic programming based observers for nonlinear systems.* Proc. of the IFAC Symp.: Fault Detection, Supervision and Safety of Technical Processes: SAFE-PROCESS'2000, June 14-16, Budapest, Hungary, pp. 967-972.

[12] Witczak, M. and J. Korbicz (2000). *Identification of nonlinear systems using genetic programming.* Proc. of the Sixth International Conference on *Methods and Models in Automation and Robotics,* August 28-31, Miedzyzdroje, Poland, pp. 795-800.

families of any hierarchical level and with multiple indices can be defined:

For $n = 1,2,...$ let there be given an index set $I(n) \neq \varnothing$, a set $\{R_{[i]}^{(n-1)} \mid i \in I(n-1)\}$ of non parameterized, non-empty relations of level n-1. $R_{[i]}^{(0)}$ is a set of "primitives" (families of level 0). We define $R^{(n)}$ on level n over the primitives by

$\varnothing \neq R^{(n)} \subseteq \prod_{i \in I(n)} R_i^{(n-1)}$, and the general relation

$\mathsf{R}[n] =_{def} \bigcup_{v=1,2,...n} R^{(v)}$.

2.2 Functions on relations

Let there be given a universe U of general, non parameterized relations R with families of form $(s_i)_{i \in I}$, $s_{[i]} \in \mathsf{S}$, S a fixed set, $I \subseteq \mathsf{I}$, I a fixed set. For any set $R \in \mathsf{U}$, $R \neq \varnothing$, a function $f: R \mapsto S \in \mathsf{U}$ with S being a singleton (one element set) is named a "concatenation", denoted by $\boldsymbol{K}$ (prefix-) or κ (infix-notation). If $S = \{s^{(n+1)} = (s_i^{(n)})_{i \in I},\}$ and if $s^{(n+1)}$ is a family of families $s_{[i]}^{(n)}$ of level n, then any function $\delta: S \mapsto \{s_i^{(n)} \mid i \in I\}$ is level lowering . Applied to above $\mathsf{R}[n]$, we define:

$\tilde{R}^{(n)} =_{def} R^{(n)}$,

for $v = n$-1, ..., 1 and for concatenations $\boldsymbol{K}$:

$\tilde{R}^{(v)} =_{def} R^{(v)} \cup \delta \boldsymbol{K} \tilde{R}^{(v+1)}$,

for $v = n, ..., 1$: $\tilde{\mathsf{R}}\,[v] =_{def} \tilde{\mathsf{R}}\,[v\text{-}1] \cup \tilde{R}^{(v)}$,

$\tilde{\mathsf{R}}\,[0] =_{def} \varnothing$.

We consider some examples:

(1) For a given $(s_i)_{i \in I}$ and an index set $J \subseteq \mathsf{I}$, the "projection" of $(s_i)_{i \in I}$ on J, $pr[J]\,(s_i)_{i \in I}$, is defined by $pr[J]\,(s_i)_{i \in I} = (s_i)_{i \in I \cap J}$, the "co-projection" with respect to J is defined by $cpr[J]\,(s_i)_{i \in I} = (s_i)_{i \in I \setminus J}$. For $pr[\{j\}]$ we write also $pr[j]$.

(2) $R = \{(s_{ji})_{i \in I[j]}) \mid j \in J\} \mapsto F = \{(s_{ji})_{ji \in M}\}$ with $M =_{def} \bigcup_{j \in J} I_j$ (i.e. $M \subseteq J \times I$) is a concatenation. In turn, a "cut" of F along j yields $F[j] = \{(s_{ji})_{i \in I[j]}\} = pr[V]\,R$ with $V = \{j\} \times I[j] \subseteq M$. Disregarding j, $pr[V]\,R$ and $\{(s_{[j]i})_{i \in I[j]}\}$ are isomorphic.

(3) If $\bigwedge j, j' \in J\,(j \neq j' \Rightarrow I[j] \cap I[j'] = \varnothing)$ and if we set $I = \bigcup_{j \in J} I[j]$, $s_{[j]i} = s_i$ for $i \in I[j]$, then $\{(s_{ji})_{i \in I[j]} \mid j \in J\} \mapsto \{(s_i)_{i \in I}\}$ is a concatenation.

(4) We assume: $R \subseteq \bigcup_{U \subseteq I} \prod_{i \in U} S_i$, $R = \{(s_{[j]i})_{i \in I[j]} \mid j \in J\}$, J finite, for each $S_{[i]}$ and each U, $\varnothing \neq U \subseteq J$, a (commutative) function $\#_i$: $S_{[i]}^{\,U} \rightarrow S_{[i]}$ is given with $\#_i((s)) = (s)$ for U being a singleton. Then R can be concatenated to $\boldsymbol{K}R = \{\#_i((s_{[ji]})_{j \in J[i]}) \mid i \in I\}$. If J is linearly ordered, the product is Cartesian and $\#_i$ is in general not commutative. For example, take a non-empty set D and $S =_{def} \mathrm{pow}\,D$ (the power set of D) with the set operations $\cap$, $\cup$. We consider the relation $R = \{(s_{ji})_{i \in I[j]} \mid j = 1, 2,..n\}$, with all $s_{[ji]} \in S$. The conjugate is $\{(s_{ji})_{j \in J[i]} \mid i \in I\}$ with $J[i] \subseteq \{1, 2,..n\}$, $I = \bigcup_{j=1,2,..n} I[j]$. R can be concatenated to $\{\#((s_{[ji]})_{j \in J[i]}) \mid i \in I\}$, where $\#$ represents $\cap$ or $\cup$.

We consider the parameterized relation $R = ((s_{ji})_{i \in I[j]})_{j \in J}$ and a partitioning of all $I[j]$ into two parts, i.e.: $I[j] = I[j]' \cup I[j]''$ with $I[j]' \cap I[j]'' = \varnothing$, with $I[j]' \neq \varnothing$. We set $x_j =_{def} (s_{ji})_{i \in I[j]'}$, $y_j =_{def} (s_{ji})_{i \in I[j]''}$, $X =_{def} \{x_{[j]} \mid j \in J\}$, $Y =_{def} \{y_{[j]} \mid j \in J\}$ and assume $\bigwedge j, j' \in J\,(y_{[j]} \neq y_{[j']} \Rightarrow x_{[j]} \neq x_{[j']})$. In this case we say the partition is functional and yields a function $f: X \xrightarrow{\text{onto}} Y$ with $x_{[j]} \underset{R}{\mapsto} y_{[j]}$ according R. In the following, $\underset{R}{\mapsto}$ is replaced by $\mapsto$ if the meaning is clear from context. We use the same notation f for the functional relation of all pairs $(x_{[j]}, y_{[j]})$, $f \subseteq X \times Y$.

From $x_{[j]} \underset{R}{\mapsto} y_{[j]}$ and $y_{[j]} = u_{[j]} \,\kappa\, v_{[j]}$, $u_{[j]}$ and $v_{[j]}$ being disjoint and not ε, follows $x_{[j]} \underset{\tilde{R}}{\mapsto} u_{[j]}$, $x_{[j]} \underset{\hat{R}}{\mapsto} v_{[j]}$ and $x_{[j]} \,\kappa\, u_{[j]} \underset{R}{\mapsto} v_{[j]}$. $\tilde{R}$ and $\hat{R}$ are restrictions of R. A partition $(x_{[j]}, y_{[j]})_{j \in J}$ is "coarser" than a partition $(x'_{[j]}, y'_{[j]})_{j \in J}$ if $x_{[j]} \subseteq x'_{[j]}$ for all $j \in J$. By this a partial ordering of the partitions of R is defined.

Given $x_{[j]} \mapsto y_{[j]}$ with respect to f, then the reciprocal image of $y_{[j]}$ is $\overset{-1}{f}(y_{[j]}) =_{def} \{x_{[j']} \mid j' \in J \wedge f(x_{[j']}) = y_{[j]}\}$, written $y_{[j]} \underset{\overset{-1}{f}}{\mapsto} \overset{-1}{f}(y_{[j]})$.

Let a set measure μ be given on $s_{[j]} = (s_{[j]i})_{i \in I[j]} = x_{[j]} \,\kappa\, y_{[j]}$. Then $\mu(x_{[j]}) \,/\, \mu(s_{[j]})$ can serve as a determinacy measure of $s_{[j]}$ being determined by $x_{[j]}$.

Example 1: $I = \{1, 2, 3, 4, 5, 6, 7\}$, $J = \{1, 2, 3, 4\}$, F is shown in Table 1. $M = \{(1, 1), (2, 2), (2, 3),....(3, 7)\}$. $R = \{s_1, s_2, s_3, s_4\}$, $R^c = \{s_1^c, ...s_7^c\}$. We assume $s_{[11]} = s_{[41]} = $ a, $s_{[22]} = s_{[32]} = s_{[42]} = $ b, $s_{[23]} = s_{[33]} = $ c, $s_{[34]} = $ d, $s_{[44]} = $ d′, $s_{[35]} = $ e, $s_{[26]} = $ f, $s_{[46]} = $ f′, $s_{[37]} = $ g. $S_{[1]} = \{$a$\}$, $S_{[2]} = \{$b$\}$, $S_{[3]} = \{$c$\}$, $S_{[4]} = \{$d, d′$\}$, $S_{[5]} = \{$e$\}$, $S_{[6]} = \{$f, f′$\}$, $S_{[7]} = \{$g$\}$. For two partitions marked by "]" and "}", $x_{[j]}$ is right, $y_{[j]}$ is left of] or }, respectively, for instance $x_{[1]} = (\mathrm{a}_{[1]1]}) \mapsto \mathrm{y}_{[1]} = \varepsilon$, $x_{[3]} = (\mathrm{d}_{[3]4}, \mathrm{e}_{[3]5}, \mathrm{g}_{[3]7}) \mapsto y_{[3]} = (\mathrm{b}_{[3]2}, \mathrm{c}_{[3]3})$ for partition "}". A determinacy measure of $s_{[3]}$ determined by $x_{[3]}$ is card $x_{[3]}$ / card $s_{[3]} = 3/5$. Partition "}" is coarser than partition "]". $(s_{[2]2}, s_{[2]3}) \overset{-1}{\underset{f}{\mapsto}} \{(s_{[2]6}), (s_{[3]4}, s_{[3]5}, s_{[3]7})\}$. According R^c, we have for example $(s_{1[1]}) \underset{R^c}{\mapsto} (s_{4[1]})$, $(s_{2[2]}) \underset{R^c}{\mapsto} (s_{3[2]}, s_{4[2]}),\dots (s_{3[7]}) \underset{R^c}{\mapsto} \varepsilon$.

Table 1

J	s_1^c	s_2^c	s_3^c	s_4^c	s_5^c	s_6^c	s_7^c	
4	s_{41}	s_{42}		s_{44}	]}	s_{46}		s_4
3		s_{32}	s_{33}	]}s_{34}	s_{35}		s_{37}	s_3
2		s_{22}	]s_{23}			}s_{26}		s_2
1	]}s_1 1							s_1
	1	2	3	4	5	6	7	I
	a_1	b_2	c_3	d'_4	e_5	f_6	g_7	$\boldsymbol{K}_{\#}R$

Let there be given $\#_1$ with (a, a) $\mapsto$ (a), $\#_2$ with (b, b, b) $\mapsto$ (b), $\#_3$ with (c, c) $\mapsto$ (c), $\#_4$ with (d, d′) $\mapsto$ (d′), $\#_5$, $\#_6$ with (f, f′) $\mapsto$ (f), $\#_7$. Then $\boldsymbol{K}_{\#}R = \{(a_1, b_2, c_3, d'_4, e_5, f_6, g_7)\}$.

2.3 Valuation of relations

In many applications the families $s^{(n+1)} = (s_i^{(n)})_{i\in I}$ of a hierarchical relation are "valuated", i.e. on level n a value $v_{[i]}^{(n)}$ taken of a domain $V^{(n)}$ is associated with $s_i^{(n)}$, written $(s_i^{(n)}, v_i^{(n)})$, and analogously $(s^{(n+1)}, v^{(n+1)})$. The valuations may be level dependent, but are i. g. independent of each other and usually assumed not to depend on the objects valuated. A particular case is the existence of a functional dependence $v^{(n+1)} = \varphi((v_i^{(n)})_{i\in I})$.

We explain this in detail: For levels $n = 0,1,2,..$ let there be given an index set $I^{(n+1)} \neq \varnothing$, for all $i \in I^{(n+1)}$ a set of relations $S_{[i]}^{(n)} \neq \varnothing$, and a set of valuations $V_{[i]}^{(n)}$. Any chosen $S^{(n+1)}$, $\varnothing \neq S^{(n+1)} \subseteq \prod_{i\in I^{(n+1)}} (S_i^{(n)} \times V_i^{(n)})$ is a relation on level n+1. We consider $S^{(n+1)}$ and assume, this relation is bijectively parameterized by the elements of a set J. Any pair (j,i) determines (s_{ji}) and v_{ji} uniquely, thus for given parameterizations, $(s_{ji}, v_{ji})_{i\in I[j]}$ can be represented by $(v_{ji})_{i\in I[j]}$.

Reducing the generality, we assume for all n and i that on level n all valuation types $V_{(i)}^{(n)}$ are equal to a type $V^{(n)}$. In many applications, the $V^{(n)}$ are complete lattices $(V^{(n)}, \leq_{(n)})$ with additional structures which we will study in the following:

For a complete lattice $(V, \leq)$ and an index set I, we consider $\tilde{V} =_{\text{def}} \bigcup_{\varnothing \subset I \subset \mathrm{I}} V^I$ with elements $v = (v_i)_{i\in I}$, $v_{[i]} \in V$. We want a reflexive and transitive extension "$\prec$" of "$\leq$" on V to $\tilde{V}$, i.e. $\prec = \leq$ on V. We define for $v = (v_i)_{i\in I}$, $v' = (v'_j)_{j\in J} \in \tilde{V} : v \prec v' \Leftrightarrow \bigwedge i \in I\, (\bigvee j \in J\, (v_{[i]} \leq v'_{[j]}))$. $\prec$ is in general no order relation, from $v \prec v'$ and $v' \prec v$ need not follow $v = v'$. A particular case is for example: if a bijection β: $I \to J$ exists and $\bigwedge i \in I\ (v_{[i]} \leq v_{[\beta(i)]})$, then $v \prec v'$. Given another complete lattice $(W, \leq)$, $W \cap V \neq \varnothing$ is admitted, then $\prec$ - homomorphisms λ: $\tilde{V} \to W$ (named "logic functions" in ref. 1) are considered. We have: $v \prec v' \Rightarrow \lambda v \leq \lambda v'$, $\bigsqcup v \leq \bigsqcup v' \Rightarrow \lambda\bigsqcup v \leq \lambda\bigsqcup v'$, $v \prec v'$ and $v' \prec v \Rightarrow \lambda v = \lambda v'$, $\bigsqcap v \prec v \prec v' \prec \bigsqcup v' \Rightarrow \lambda\bigsqcap v \leq \lambda v \leq \lambda v' \leq \lambda\bigsqcup v'$ and for $v = v'$: $\lambda\bigsqcap v \leq \lambda v \leq \lambda\bigsqcup v$. In case all components of v are equal to the element e, then $\lambda e = \lambda v$. Defining $v \prec' v' \Leftrightarrow \bigwedge i \in I\, (\bigvee j \in J\, (v_{[i]} \geq v'_{[j]})$, it follows $\bigsqcap v' \leq \bigsqcap v$. So far, for $w = \lambda((v_i)_{i\in I})$, λ depends on I. If only finite sets are admitted, λ may depend on card I only, i.e. $\lambda((v_i)_{i\in I}) = \lambda((v_{\beta(i)})_{i\in I})$ for a bijection β: $I \to J$ with $v_{[i]} = v_{\beta([i])}$.

Example 2: $V = W = \overline{\mathbf{R}}_+$ (the non-negative real numbers completed with $\nearrow$), $\mathrm{I} = \mathbf{N}_0$ (the natural numbers with 0), I finite, card $I = n$. Using suitable relations $\prec$, examples for λ are: $\lambda = \max$, $(\Sigma(...))/n$, $(*(...))^{1/n}$. Because of $((\min v)_i)_{i\in I} \prec v \prec ((\max v)_i)_{i\in I}$ it follows: $n \min(v) \leq \Sigma(v) \leq n \max(v)$, $(\min(v))^n \leq *(v) \leq (\max(v))^n$. Using $\prec'$, an example is $\lambda = \min$.

Applied to the above valuated hierarchical relations $S^{(n)}$, the $V^{(n)}$ can be complete lattices with given logic functions $\lambda^{(n)}$: $(V^{(n)})^{I^{(n+1)}} \to V^{(n+1)}$.

Example 3 (propositional calculus): Let be $\mathrm{I} = \mathbf{N}_0$, for all levels, $V^{(n)} = (\{f, t\}, f < t)$ (type Boolean), $S^{(0)} = \{a, b, c,..\}$ a set of propositions. λ-functions are $\wedge$, $\vee$ (homomorphisms), $\neg\wedge$, $\neg\vee$ (antimorphisms). For example
level 0: $\{a, b, c,...\}$,
level 1: $\{(a, t), (b, f), (c, t)\}$, card $I^{(1)} = 3$,
level 2: $\{(((a, t), (b, f)), \wedge(t, f) = f), (((a, t), (c, t)), \vee(t, t) = t)\}$, card $I^{(2)} = 2$. Indices are suppressed.

3 Variables and their control

We consider a non-empty family $(S_p)_{p\in P}$ of non-empty structures (general relations) S_p. For all p let be $S_{[p]} = \boldsymbol{K}_{[p]}(C, V_{[p]})$, $\boldsymbol{K}_{[p]}$ a concatenation of a p-independent component C and a p-dependent component $V_{[p]}$. C may be empty. To facilitate the representation of $\mathsf{S} = \{S_{[p]} \mid p\in P\}$ we define a variable var V on variability domain $\mathsf{V} = \{V_{[p]} \mid p \in P\}$ with respect to S, written var V : (V, S), and a variable $S(\text{var } V) \doteq \boldsymbol{K}(C, \text{var } V)$ on S. $\boldsymbol{K}$ depends on var V. We make the variables "controllable" by associating to S(var V) a function val with val: $P \times \{\text{var } V\} \times \{S(\text{var } V)\} \to \mathsf{V} \times \mathsf{S}$, with val: $(p, \text{var } V, S(\text{var } V)) \mapsto (V_{[p]}, S_{[p]})$, and to var V a function val: $P \times \{\text{var } V\} \to \mathsf{V}$ with val: $(p, \text{var } V) \mapsto V_{[p]}$. The val-functions are named "control-" or "assignment" functions, P is the set of control-/assignment parameters. For assignment according parameter p we write also val(S(var V)): $p \mapsto S_{[p]}$ and val(var V): $p \mapsto V_{[p]}$, or $S(\text{var } V) :=(p)\ S(V_{[p]})$, var $V :=(p)\ V_{[p]}$. $S(V_{[p]})$ and $V_{[p]}$ are "instantiations" of S(var V) and var V, respectively. Consecutive assignments by parameters p, q result in a substitution $(V_{[q]}, S_{[q]})$ for $(V_{[p]}, S_{[p]})$.

A structure can contain variables $(\text{var } w_j)_{j\in J}$ with var w_j : $W_j = \{w_{j[q]} \mid q \in Q_j\}$. The composite variable var w = $(\text{var } w_j)_{j\in J}$ is in general defined on $W \subseteq \prod_{j\in J} W_j$. For $W \subset \prod_{j\in J} W_j$ assignments to the var w_j are in general dependent on each other, in this case admissible assignment parameters $q = (q_j)_{j\in J}$ are from a set $Q \subset \prod_{j\in J} Q_j$. This has to be taken into account in case of partial assignments to var w, resulting in the partial variable $(u_j)_{j\in J}$ with $u_j = w_{j[q(j)]}$ for $j \in J_a$, u_j = var w_j for $j \in J \setminus J_a$.

The parameters $(q_j)_{j\in J}$ can themselves be the result of a computation $\varphi_{[r]} : (p_i)_{i\in I} \to (q_j)_{j\in J}$, and $\varphi_{[r]}$ can be an instantiation of a variable var φ with control function val(var φ). Continuing this way we have a hierarchy of control functions.

More than one variable can be defined on the same domain V and elements of V can be subject to certain relations $\mathsf{S}_{[q]}$, $q \in Q$. (V, $(\mathsf{S}_{[q]})_{q\in Q}$) is then also named "type" of the variables. For example, $\mathsf{S}_{[q]}$ can be a functional relation.

For any hierarchical level n = 1,2,... we admit variables on domains with variables of level n-1. On level 0 are constants only. Assignments are applied from higher to lower level and reduce the level.

Example 4: Revisiting *Example 1*. The relation can be described by (var x, var f, var y := var f(var x). var f: $\{f_{[p]} \mid p \in P = \{"]", ")"\}$ corresponding to the two partitions}, var x : $\{x_{[pq]} \mid p \in P \wedge q \in Q_p = Q = J\}$.
var $x_{["]"]}$: $\{x_{["]",j} \mid j \in Q\}$, var $y_{["]"]}$: $\{y_{["]",j} \mid j \in Q\}$,
$x_{["]",1]} = (s_{11})$, $\quad y_{["]",1]} = \varepsilon$,
$x_{["]",2]} = (s_{23}, s_{26})$, $\quad y_{["]",2]} = (s_{22})$,
$x_{["]",3]} = (s_{34}, s_{35}, s_{37})$, $\quad y_{["]",3]} = (s_{32}, s_{33})$,
$x_{["]",3]} = (s_{46})$, $\quad y_{["]",3]} = (s_{41}, s_{42}, s_{44})$,
similar for the $x_{["]",j]}$.

Introducing variables for the s_{ji} and for the parameters, we have the composite variable var $x_{[\text{var } p, \text{var } j]} = (\text{var } s_{[\text{var } p, \text{var } j]i})_{i\in \text{var } I}$ with (var p, var j) : $P \times J$, var $s_{[\text{var } p, \text{var } j]i}$: S_i independent of (var p, var j), var I : $\{I[j] \mid j \in J\} \subset$ pow I, depending on (var p, var j).

4 Topologized relations

To express neighborhood / similarity relations on the families of a relation $R \subseteq \bigcup_{U\in \text{pow } I} \prod_{i\in U} S_i$, we introduce topological structures by means of filter bases: Given a set S, then (pow S, $\subseteq$, $\cup$, $\cap$) is a complete Boolean lattice. Let $B_{[0]}$, $\varnothing \neq B_{[0]} \subseteq S$, be a set of points which by assumption are all equivalent with respect to the admitted accuracy. Further, let $\mathbf{B} = \{B_{[k]} \mid B_{[k]} \subseteq S$ and $k \in K\}$ be a set of upper sets of $B_{[0]}$ with the properties: from $B_{[k']} \in \mathbf{B}$ and $B_{[k'']} \in \mathbf{B}$ follows, it exists a $B_{[k]} \in \mathbf{B}$ with $B_{[k]} \subseteq B_{[k']}$ and $B_{[k]} \subseteq B_{[k'']}$. Then $\mathbf{B}$ is a "filter base". We assume: $B_{[0]} = \bigcap \mathbf{B}$ (denoted lim $\mathbf{B}$), $S \supseteq \bigcup \mathbf{B}$ (support of $\mathbf{B}$). We further assume, $\mathbf{B}$ is a complete lattice. Then for $s \in S$, $s^* \in$ lim $\mathbf{B}$, "generalized distances" $d_\cap(s^*, s) =_{\text{def}} \bigcap\{B \mid B\in\mathbf{B} \text{ and } s\in B\} \in \mathbf{B}$ and $d_\cup(s^*, s) =_{\text{def}} \bigcup\{B \mid B\in\mathbf{B} \text{ and } s\notin B\} \in \mathbf{B}$ can be defined. Let there be given a complete lattice $(V, \leq, \wedge, \vee)$. and a $\leq$-homomorphism ϕ: $\mathbf{B} \to V$, defining a valuation of $\mathbf{B}$. We use $v(s^*, s) =_{\text{def}} \phi(d_\cap(s^*, s))$ as a distance measure of s from s^*. "Fuzzy sets" in the engineering literature are examples of filter bases.

Let $\{\mathbf{B}_{[l]} = \{B_{[lk]} \mid k \in K\} \mid l \in L\}$ be a set of isomorphic filter bases, and for all k, let $B_{[lk]} \in \mathbf{B}_{[l]}$ be uniformly with respect to l valuated by a $\subseteq$-homomorphism ϕ into V, i.e. $\phi(B_{[lk]}) = \phi(B_{[l'k]})$. We further assume, for $l \neq l'$ holds lim $\mathbf{B}_{[l]} \cap$ lim $\mathbf{B}_{[l']} = \varnothing$. For more details we refer to ref. 1, 2, 3, 4.

We apply these topological concepts to the relation R: Let $R^* \subseteq R$ be a non-empty subset. We set $S = R$, $L = R^*$. To each $r^* \in R^*$ let be adjoined a neighborhood system in form of a filter base $\mathbf{B}_{[r^*]}$ as defined above, with $r^* \in$ lim $\mathbf{B}_{[r^*]}$, and with all sets lim $\mathbf{B}_{[r^*]}$ being disjoint. A trivial case is, $R = R^*$, lim $\mathbf{B}_{[r^*]} = \{r^*\}$.

For fixed r^*, $v(r^*, r) =_{\text{def}} \phi(d_\cap(r^*, r))$ defines an abstract distance of any $r \in \bigcup \mathbf{B}_{[r^*]}$ from any $r^* \in \lim \mathbf{B}_{[r^*]}$. Hence, to any $r^* \in \lim \mathbf{B}_{[r^*]}$ exists a neighborhood $\mathsf{N}(r^*)$ with elements $u \in R$ "nearer" to r^* than to any other comparable $\hat{r} \in R^* \setminus \lim \mathbf{B}_{[r^*]}$. If the relation R is hierarchical, then the topological reasoning can be applied to each hierarchical level.

Let us consider a family $(r_i, v_i)_{i \in N}$, N being a non-empty segment of integers, with $r_{[i]} \in R$, $v_{[i]}$ the abstract distance of $r_{[i]}$ to a given $r_{[i]}^* \in R^*$. Then, for a logic function λ, $\lambda((v_i)_{i \in N})$ can be defined as distance of $\mathbf{K}((r_i)_{i \in N})$ from $\mathbf{K}((r_i^*)_{i \in N})$, $\mathbf{K}$ denoting a concatenation. Examples are normed vector spaces with λ being a norm: Let $\mathbf{R}$ be the field of real numbers and let $\mathbf{R}^n$ be an n-dimensional normed vector space. For $r_{[i]}^*, r_{[i]} \in \mathbf{R}$ a distance is $\lambda(r_i^*, r_i) = |r_{[i]}^* - r_{[i]}|$, for $\mathbf{K}(((x_i))_{i=1,\dots n}) = (x_i)_{i=1,\dots n}$, $x_i \in \{r_i^*, r_i\}$, a distance from $r^* = (r^*_i)_{i=1,\dots n}$ to $r = (r_i)_{i=1,\dots n}$ is $\lambda(r^*, r) = \|r^* - r\|$. Well known examples for norms are $\max \{|r_{[i]}^* - r_{[i]}| \mid i = 1,\dots n\}$, $(\sum_{i=1}^{n} (r_{[i]}^* - r_{[i]})^2)^{1/2}$.

5 Applications

On a universe $\mathbf{R}$ of parameterized relations (structures) let there be given:

(1) A set of *structural* operations, e.g. definition/deletion of a particular object structure or index structure, composition and decomposition of relations, especially concatenations, parsings, cuts, projections, definition of variables, their domains and assignments, definition of topological structures.

(2) A set of *functional* operations on objects of relations, e.g. equality, being part of, search for and quantified selection by properties, comparison of objects, enumeration, sorting, deduction rules to derive implicitely given structures.

(3) A "*query language*" based on the structural and functional operations, to provide "answers" to formalized "questions".

Structural, functional operations and queries are subject to rules and constraints (grammars), which we are not going to formalize here. We name a relation $R \in \mathbf{R}$ a "knowledge module".

Example 5: We revisit *Example 1* with $s_{j.} =_{\text{def}} (s_{ji})_{i \in I[j]}$, $s_{.i} =_{\text{def}} (s_{ji})_{j \in J[i]}$, and with all $s_{[ji]}$ equal to the above defined constants. Examples of queries are given in Table 2:

Table 2

given:	find (one, all):	answer:
$U = \{(3, 1), (2, 3), (4, 6)\}$	$pr[U]R$	$\{c_{23}, f'_{46}\}$
$x = (a_1)$	j with $x \subseteq s_{j.}$	$j = 1, 4$
$x = (a_1)$	$s_{j.}$ with $x = s_{j.}$	s_1
$x = (a_1)$	$s_{j.}$ with $x \subseteq s_j$	s_1, s_4
b, d	j with b$\vee$c or b$\wedge$c $\in$ set of s_j	$j = 2, 3, 4$ $j = 2, 3$
$I^* = \{2, 4\}$	j with $I[j] = I^*$	empty
$i = 4$	cut of R along i	$s_i^c = (d_{34}, d'_{44})$
R	fctl. decompos. and dependencies	see *Example 4*
any x, y	$x \kappa y \in R$?	by trials of concatenations
any z	$x, y \in R$ with $z = x \kappa y$?	by trials of parsings

Examples of queries for knowledge modules with topologies are given in Table 3:

Table 3

given:	find (one, all):	answer:
$r^* \in R^* \subseteq R$, $\mathbf{B}_{[r^*]}$ $x \in R$	$d_\cap(r^*, x)$, $d_\cup(r^*, x)$	see *Example 6*
$x \in R$, all $\mathbf{B}_{[r^*]}$	nearest $r^* \in R^*$ to x	comparison of the $v(r^*, x)$ see *Example 6*
var $\mathbf{B}$ on R $S \subseteq R$, all elements of S valuated	assignment to var $\mathbf{B}$, fitting the valuations on S	interpolation of the valuations on S, see *Example 7*

Example 6 (use of topologies): Considering the real numbers $\mathbf{R}$, let be $r^* \in \mathbf{R}$, $\mathbf{B}_{[r^*]} =_{\text{def}} \{[r^* - 0.5*v, r^* + 0.5*v] \mid v \in V\}$ with $V = \mathbf{N}_0$ or $\mathbf{N}$. $\mathbf{B}(r^*)$ is a monotone filter base on pow $\mathbf{R}$ with $\lim \mathbf{B}_{[r^*]} = \{r^*\}$ or $\{[r^* - 0.5, r^* + 0.5]\}$, respectively. ϕ: $[r^* - 0.5*v, r^* + 0.5*v] \mapsto v$ is a valuation. E.g. for $V = \mathbf{N}$, $r^* = 3$, $r = 4$, $v(r^*, r) = \phi(d_\cap(r^*, r)) = 2$, $\phi(d_\cup(r^*, r)) = 1$. For $R^* =_{\text{def}} \{r^* = (3_1, 4_2, 5_3), r^{**} = (1_1, 2_2, 5_3)\} \subset \mathbf{R}^3$, $\lambda = \max$, $V = \mathbf{N}$, $x = (3.2, 3.9, 4.5) \in \mathbf{R}^3$, $\lambda(r^*, x) = 1$, $\lambda(r^{**}, x) = 4$.

Example 7 (adaptation and fitting): Let var $\mathbf{B}$ on $R \subseteq \mathbf{R}^3$ to be adapted to a sample $(s_n = (s_{nk})_{k=1,2,3})_{n=1,2,\dots N}$ of size N, the components $s_{[nk]}$ of s_n, $s_{[nk]} \in S$, $S \subseteq R$, assumed to be independently valuated by the relative frequencies $v_{[nk]}$ of their occurrences in the sample.

We choose an (approximate) interpolation of the frequency function on a closed domain C, $S \subseteq C \subseteq R$, which is then linearly scaled to a function φ with $\varphi(C) \subseteq [0, 1]$ and with $1 \in \varphi(C)$, i.e. φ can be used as generating ("membership"-) function to define a filter base (neighborhood system) $\mathbf{B} = \{B_\alpha = \{ \overset{-1}{\varphi}([\alpha, 1]) \mid 0 \leq \alpha \leq 1\}\}$ with $\lim \mathbf{B} = \overset{-1}{\varphi}(\{1\})$. The set function $\overset{-1}{\varphi}$ is the reciprocal to the point function φ. $\varphi(C)$ is the set of valuations $v(r)$ of elements r of C.

In the preceding example we used a statistical property for grouping elements to neighborhoods. As well other properties or explicitly described grouping could be used.

An interpretation of our general approach is: I is a set of "properties", J is a set of "objects", object j has all properties of $I[j]$, property $i \in I[j]$ is valuated by $v_{[ji]} \in V_{[j]}$, often all $V_{[j]}$ are identical and equal V. In particular, $V = \{1\}$, $V = \{0, 1\}$.

Our approach is a generalization of the so-called "relational data base" concept, obtained for $R \subseteq \prod_{i \in I} S_i$, finite I representing the set of "attributes", $S_{[i]}$ the set of attribute values to attribute i. Neighborhood sytems are not considered.

Frequently, for a given $\mathbf{B}$, the elements of $\bigcup \mathbf{B}$ are arguments to a function $f\colon \bigcup \mathbf{B} \to Z$. If $r \in \bigcup \mathbf{B}$, $f(r) = z$, and if r is valuated by $v(r)$, then $f(r)$ can be valuated by $v(r)$. If more than one r is mapped onto the same z, then, if existing, lub $v(\overset{-1}{f}(z))$ (lub means least upper bound) can serve as a valuation of z.

The generation and adaptation of neighborhood systems (filter bases) fitting certain input values in a "learning"/"training" phase is a basic feature of all adaptive classifying (pattern recognizing) devices.

To compare two families v^*, v of valuations with finite and distinct index sets I^*, I, $J =_{\text{def}} I^* \cap I \neq \varnothing$, and a distance $d = d(pr(J)v^*, pr(J)v) \in \overline{\mathbf{R}}_1$, i.e. $1 \leq d \leq \infty$, we can for example use a measure $\delta =_{\text{def}} \text{card}(I^* \cup I) / \text{card}(I^* \cap I)$, $1 \leq \delta < \infty$, to define $d\delta$ as distance of (v^*, v).

To incorporate the knowledge about neighborhood systems, in the definition $R \subseteq \bigcup_{U \in \text{pow}\, I} \prod_{i \in U} S_i$ the set of factors sets $S_{[i]}$ can be extended by a factor set $S^* = \{\mathbf{B}_{[r^*]} \mid r^* \in R^*\}$, the $\mathbf{B}_{[r^*]}$ represented explicitly or implicitly by generating functions.

Introducing a linearly or partially ordered model time $(T, \leq)$, processes of relations R can be considered. Examples are $(s_{tji})_{(tji) \in M \subseteq T \times J \times I}$ or $(((s_{tji})_{i \in I[tj]})_{j \in J[t]})_{t \in T}$. This way the definition of continuously adapting knowledge modules is possible.

References

[1] Albrecht R. F. (1998) "On mathematical systems theory". In: R. Albrecht (ed.): Systems: Theory and Practice, Springer, Vienna-New York, pp. 33-86

[2] Albrecht R. F (1999) "Topological Approach to Fuzzy Sets and Fuzzy Logic". In: A. Dobnikar, N. C. Steele, D. W. Pearson, R. F. Albrecht (eds.): Artificial Neural Nets and Genetic Algorithms, Springer, Vienna-New York, pp. 1-7

[3] Albrecht R. F. (1999) "Topological Theory of Fuzziness". In: B. Reusch (ed.): Computational Intelligence, Theory and Applications, LNCS vol. 1625, Springer, Heidelberg-New York, pp. 1-11

[4] Albrecht R. F. (2001) "Topological Concepts for Hierarchies of Variables, Types and Controls". In: G. Alefeld, J. Rohn, S. Rump, F. Yamamoto (eds.): Symbolic Algebraic Methods and Verification Methods, Springer Vienna-New York, pp. 3-10

[5] Albrecht R. F., Németh G. (1998) "A Generic Model for Knowledge Bases", Periodica Polytechnica, Electrical Engineering, TU Budapest, vol. 42, pp. 147-154

Handling categorical data in rule induction

Martin Burgess,[1] Gareth J. Janacek,[2] Vic J. Rayward-Smith[3]

[1 2 3]School of Information Systems, University of East Anglia, Norwich, UK

Abstract

In this paper we address problems arising from the use of categorical valued data in rule induction. By naively using categorical values in rule induction, we risk reducing the chances of finding a good rule in terms both of confidence (accuracy) and of support or coverage. In this paper we introduce a technique called *arcsin transformation* where categorical valued data is replaced with numeric values. Our results show that on relatively large databases, containing many unordered categorical attributes, larger databases incorporating both unordered and numeric data, and especially those databases that are small containing rare cases, this technique is highly effective when dealing with categorical valued data.

1 Introduction

We have a database, D, of records which for simplicity we assume is in the form of a flat file. Each record comprises m attribute values, one of which is a designated output, or target attribute. Attributes can either be numerical or categorical. Whereas numerical data has an ordering, categorical data is often unordered.

Rule induction seeks to find a rule of the form $\alpha \Rightarrow \beta$, where the postcondition, β, is a constraint on the output attribute and the precondition, α, is an expression involving the other attributes. For example,

$$\text{Salary} \leq 30,000 \wedge \text{Mortgage} > 100,000 \Rightarrow \text{Status} = \text{debtor} \quad (1)$$

is a rule where the precondition is a conjunction of constraints on two numeric attributes (Salary and Mortgage) and the postcondition is a constraint on a categorical attribute (Status). Rule induction algorithms aim to find rules which appear to hold in the database. The quality of the rule is usually defined in terms of confidence (i.e. the proportion of records satisfying the precondition for which the postcondition holds) and support (i.e. the proportion of the database for which both the precondition and the post condition hold). The term 'coverage' is sometimes used as a synonym for support; however will use it here to describe the proportion of the records for which both the precondition and postcondition hold. More formally, if

$$A = \{r|\alpha(r)\} \quad (2)$$

denotes the set of records in the database for which the precondition holds and similarly

$$B = \{r|\beta(r)\}, \qquad C = \{r|\alpha(r) \wedge \beta(r)\} \quad (3)$$

then

$$\text{Confidence (or accuracy)} = \frac{|C|}{|A|}, \quad (4)$$

$$\text{Support} = \frac{|C|}{|D|} \quad (5)$$

$$\text{and Coverage} = \frac{|C|}{|B|}. \quad (6)$$

where D is the database. Note that we assume $|A|$, $|D|$, $|B| \neq 0$.

In this paper, we consider the case where the precondition is a conjunction of simple tests on various non-output attributes and the postcondition is a predefined test on the output attribute. The software 'Data Lamp' [1] was utilised to generate the output discussed in this paper. It uses simulated annealing to find an optimal rule of this form with a specified β value which maximises the quality measure

$$\lambda|C| - |A|. \quad (7)$$

By adjusting the real value, λ, the rule can be more accurate (and have less coverage) or be less accurate (and have greater coverage). Note that

$$\lambda|C| - |A| \geq 1 \text{ iff confidence} = \frac{|C|}{|A|} \geq \frac{1}{\lambda}$$

assuming $\lambda > 0$.

Thus, if we find a rule with a fitness level $\geq$1 using λ=1.11 (say), this means we will have found a rule with a confidence of at least 90%.

If an attribute, A, is numerical then a simple test is of the form $A > x$ or $A \leq x$, where x is a value in the associated domain. Such tests can also be used on ordered categorical data. However, if the attribute is categorical and unordered then the only tests we can have are $A = x$

or $A \neq x$. If, moreover, such an attribute, A, has several values of which one is x, then $A \neq x$ is unlikely to result in an appropriately accurate rule. Conversely, $A = x$, is a very limiting constraint and any rule containing such a test is likely to have low levels of support. Thus, in either case, inclusion of constraints involving such categorical values may generate poor quality rules. Data Lamp does not allow conjunction of multiple intervals of the same attribute. Thus, if a set of categorical values is ordered $x_1, x_2, ..., x_k$, then it cannot achieve the test $A = x_i$, by $A > x_{i-1}$ and $A \leq x_i$.

2 Ordering of categorical data

The problem with categorical data is that, unlike numerical data, categorical values in most cases, do not have an inherent order. On some occasions a sensible ordering can be inferred. For example, consider the Adult database [2], the field *Education*, though categorical valued, has an inherent order to it. For example, a child will start at pre-school then progress through his/her educational learning, following an order to his/her learning in terms of schooling. Such an order would be termed *natural*. On the other hand, if we talk about colours-red, yellow, blue etc., there is no such ordering.

When looking at large data sets, we wish to extract rules which can then be used for predictive purposes. These rules are conjuncts of tests on the attributes which help to predict the target class. When looking for rules in a large data set, unless we impose some ordering on unordered categorical data, we are very much limiting our choice of finding the best rule in terms of confidence and support.

Replacing unordered categorical data by ordered data is generally not effective unless the number of distinct values is relatively large. In fact, if the number is less than four, it is never necessary. Say an attribute, A, has values $\{x_1, x_2, x_3\}$, then

$$
\begin{aligned}
A > x_1 &\text{ is equivalent to } A \neq x_1 \\
A \leq x_1 &\text{ is equivalent to } A = x_1 \\
A > x_2 &\text{ is equivalent to } A = x_3 \\
A \leq x_2 &\text{ is equivalent to } A \neq x_3 \\
A > x_3 &\text{ is invalid} \\
A \leq x_3 &\text{ is no constraint}
\end{aligned}
$$

Thus in this case the ordering has been detrimental since it has removed the option of tests of the form $A = x_2$ and $A \neq x_2$ unless we allow multiple tests on the attribute A.

2.1 Ordering categorical values

In this paper we consider two ways in which categorical field values may be ordered. We may wish to order according to $\hat{p}$ values or with u values (both are described later).

We claim that, ordering of unordered categorical data is better than no ordering at all. We illustrate this claim by experiments performed on three databases from the UCI repository [2], that is, *Mushroom*, *Auto Imports*, and *Adult*. With reference to section 3.1, with the *Mushroom* experiment, it is shown that with unordered categorical fields containing four or more values, ordering is advisable as shown by the results achieved. Where the database is small (see section 3.3–*Auto Imports* database) incorporating rare cases, not inferring some kind of ordering is unadvisable and can seriously hamper one's chances of discovering quality rules.

2.2 Proportion or p values

The technique of using $\hat{p}$ values is based on the proportion of records within the target class for a given categorical value. This technique generates $\hat{p}$ values without making any distributional assumptions about the populations. The justification of such an approach relies on asymptotic theory that is valid only if the sample sizes are reasonably large and well balanced across the populations. For small, sparse, skewed or heavily tied data, the asymptotic theory may not be valid. For some empirical results, and a more theoretical discussion see ref. [3], [4], [5]. The alternative approach based on u values addresses some of these concerns.

Let A be an unordered categorical attribute with k possible values, $case_1, case_2, ..., case_k$. Assume the output condition, β, is set. We first consider $p_i = P(\beta | case_i)$, where P is the probability. We define $\hat{p_i} = y_i/n_i$ where y_i is the number of cases where both $A = case_i$ and β hold, and n_i is the number of cases where $A = case_i$. We regard $\hat{p_i}$ as an estimate of the true value of p_i, where y_i is the outcome of a Binomial random variable with mean $n_i p_i$, and variance $n_i p_i(1 - p_i)$. The difficulty is that, as we only see $\hat{p_i}$, we have some uncertainty about the actual value and we need to take this into account. In order to stabilise that variance we propose that we consider the transformed estimate

$$Z_i = \arcsin \sqrt{\hat{p_i}}. \tag{8}$$

It is then possible to show that Z_i is asymptotically Normal with mean (approximately) $\arcsin \sqrt{p_i}$ and variance $1/4n_i$. Thus, for large n, $\arcsin \sqrt{p_i}$ has an approximate Normal distribution of mean $\arcsin \sqrt{p_i}$ and variance of $1/4n$, that is, a $\mathcal{N}(\arcsin \sqrt{p_i}, 1/4n)$ distribution if arcsin is measured in radians.

Using a result from ref. [6], we define

$$u_i = \arcsin(\sqrt{\hat{p}_i})\sqrt{4n_i} \quad (9)$$

which is standard normal. Dividing through by the standard deviation is to allow for scaling onto a standard scale. Whereas, with the $\hat{p}_i$ values, we have some uncertainty concerning the actual values, the u values have standardised this uncertainty. In [7], it is suggested that the following angular modification is used in order to stabilise the variance for values of $n_i \leq 50$. We use this an alternative to u_i when n_i is small.

$$u_i' = \frac{\arcsin\left[\sqrt{\frac{n}{n+1}\hat{p}_i} + \sqrt{\frac{n}{n+1}\hat{p}_i + \frac{1}{n+1}}\right]}{2\sqrt{(1/n + 1/2)}} \quad (10)$$

As before, we divide by standard deviation to allow for scaling onto a standard scale.

3 Multiple λ value experiment

3.1 Mushroom database

Our initial experiment used the Mushroom database [2]. The experiment was performed to illustrate the benefits of ordering categorical data and to investigate the differences between orderings achieved by $\hat{p}$ and u values.

Of the 8124 instances available, records containing unknown values were removed leaving 5644 records. The database contains 22 categorical fields with no continuous fields. The data was then randomly split into two parts, a training and a test set, with both sets containing 2822 records. The training set contains 1718 records of class *edible*, and 1104 records of class *poisonous*. The test set contains 1770 records of class edible, and 1052 of class poisonous.

Each field in our training data set is proportioned according to the target value.

Using the target class of *edible*, both $\hat{p}$ value and u value ordering produced the same fitness results producing the confidence and coverage levels of Table 3, which wre better than no ordering at all. For the target class of *poisonous*, both $\hat{p}$ value and u value ordering were again better than no ordering at all (refer to Table 4). In the following discussion we consider the more interesting case where the target class is poisonous.

If we refer to Table 1, we can see the breakdown of $\hat{p}$ and u values for the categorical values belonging to *cap_colour*. From the training data, from a total of 434 records there were for *white*, 160 records belonged to our target field of poisonous, and so the proportion of records is 160/434=0.369 (to three decimal places). For each categorical value in the field, *cap_colour*, the $\hat{p}$ value was calculated in a similar way. The u values were calculated using the formula from (9).

Table 1. Cap_colour from mushroom database

$\hat{p}$		u	
buff	1.000	gray	44.185
pink	0.904	yellow	42.306
yellow	0.627	white	27.187
gray	0.484	buff	25.523
white	0.369	brown	18.207
cinnamon	0.357	pink	18.108
brown	0.136	cinnamon	6.778
red	0.023	red	5.312

In Table 1, the values $\hat{p}$ and u are ordered according to their relevant values. The values are then replaced with integer values from 1=highest, ..., 8=lowest. The same technique was applied to the other 21 categorical valued fields. Similarly, the test set was set up using exactly the same encoding as the training set. The alternative u_i' technique (10) recommended when n_i is small results in a slightly different ordering as those achieved by the u values. The u_i' ordering results in *pink* and *brown* (Table 1) changing positions. These changes do not effect the overall results as shown in Table 2.

Table 2. Results from mushroom database

λ	$\hat{p}$	u	r
1.11	112(118)	110(115.3)	88(88)
1.25	252(265)	249(260)	198(199)
1.43	432(455)	426(445)	339(340)
1.67	671(707)	663(692)	527(529)
2	1007(1061)	994(1038)	790(794)
2.5	1511(1592)	1491(1557)	1185(1333)

Table 3. Confidence and coverage levels from mushroom test database, edible

	$\hat{p}$		u		r	
λ	Con.	Cov.	Con.	Cov.	Con.	Cov.
1.11	100	100	100	100	100	93.28
1.25	100	100	100	100	100	92.09
1.43	100	100	100	100	100	93.28
1.67	100	100	100	100	100	92.09
2	100	100	100	100	95.57	99.83
2.5	100	100	100	100	92.43	100

Table 4. Confidence and coverage levels from mushroom test database, poisonous

	$\hat{p}$		u		r	
λ	Con.	Cov.	Con.	Cov.	Con.	Cov.
1.11	100	95.72	100	94.5	100	75.10
1.25	100	95.72	100	94.5	100	75.10
1.43	100	95.72	100	94.5	100	75.10
1.67	100	95.72	100	94.5	100	75.10
2	100	95.72	100	94.5	100	75.10
2.5	100	95.72	100	94.5	100	75.10

For each λ value (see Table 2), there are a set of numbers which are called *fitness* values. These values describe how well each possible solution (rule) meets the criterion specified, in other words, we wish to maximise the quality measure given in (7). So for example, a λ value of 1.67 corresponds to a minimum confidence of 60%. The same applies to Tables 6 and 11 with the same λ values.

We first encoded all categorical data using integer values determined by the $\hat{p}$ values. We then ran the SA algorithm within Data lamp six times to find the best possible rule. For each value of λ, the fitness of this rule is reported in Table 2 on the column headed $\hat{p}$. The experiment was repeated but using the ordering defined by the u values, the corresponding fitness appearing in the column headed u. We compared our results with the results obtained using the raw data, that is the original unordered categorical data and the corresponding results are listed in the column headed r. The figures in brackets in Table 2, refer to results produced using the training database. The corresponding confidence and coverage levels for each value of λ (as shown in Table 2) are given in Table 4. In all cases, ordering the data proved effective.

3.2 Adult database

The second of our experiments used the Adult database [2]. This experiment compares the efficacy of ordering on $\hat{p}$ and u values with that of using a natural ordering. In the Adult database, the *educational-numbering* field is presented which provides a natural ordering, as an alternative to the field *Education* that was unordered. The Adult database is a database with both categorical and numerical attributes. However, some categorical fields have many values, for example, *occupation* has 14 values, and *education* has 16 values, and so, u values may still prove useful. The education field can be ordered in terms of academic progression although finding a linear ordering is not easy because we have both vocational and academic qualifications together; *education-numbering* assigns such a natural ordering. Using $\hat{p}$ values we get a slightly different ordering and u values give yet another quite different ordering that appears quite poor in places (for example, a doctorate is significantly lower than bachelor qualification). These orderings are compared in Table 5. In Table 6, experimental results are given where the field *education-numbering* is not used. Except for the smaller values of λ, the $\hat{p}$ ordering gives a much better result than the raw data. Since the database is large it is perhaps not surprising that $\hat{p}$ values prove better than u values.

Returning to Table 6, we repeat the experiment but use the *education-numbering* field instead of the *education* field. Comparing the raw values in Table 6, again we see that raw data but with the natural ordering of *education-numbering* gives much better results as the value of λ increases. However, in these cases the ordering of fields by $\hat{p}$ values is still the better option. The u ordering only gives good results as λ increases, that is, as the search moves towards widely applicable and less accurate rules.

Table 5. Ordering for educational field

Educational numbering	$\hat{p}$ ordering	u ordering
Doctorate	Prof-school	Bachelors
Prof-school	Doctorate	HS-grad
Masters	Masters	Some-college
Bachelors	Bachelors	Masters
Assoc-acdm	Assoc-voc	Prof-school
Assoc-voc	Assoc-acdm	Doctorate
Some-college	Some-college	Assoc-voc
HS-grad	HS-grad	Assoc-acdm
12th	12th	10th
11th	10th	11th
10th	7th-8th	7th-8th
9th	11th	12th
7th-8th	9th	9th
5th-6th	5th-6th	5th-6th
1st-4th	1st-4th	1st-4th
Preschool	Preschool	Preschool

The corresponding confidence and coverage levels for each value of λ (as shown in Table 6) are given in Tables 7 and 8.

Table 6. Results from Adult Database

No educational numbering				Educational numbering		
λ	$\hat{p}$	u	r	$\hat{p}$	u	r
1.11	62	63	63	62	63	63
1.25	149	149	150	149	149	149
1.43	260	260	260	260	260	260
1.67	492	408	413	502	407	422
2	1058	941	615	1012	919	1000
2.5	2152	1887	1404	2081	2149	2118

Table 7. Results from adult database with no educational numbering field

	$\hat{p}$		u		r	
λ	Con.	Cov.	Con.	Cov.	Con.	Cov.
1.11	99.18	16.41	99.34	16.22	99.19	16.59
1.25	98.89	16.81	99.04	16.73	99.04	16.81
1.43	98.89	16.81	98.89	16.81	98.89	16.81
1.67	75.39	39.08	98.89	16.81	94.34	18.46
2	70.66	48.81	68.63	46.89	98.73	16.84
2.5	65.07	60.41	55.98	71.38	56.43	52.16

Table 8. Results from adult database with educational numbering field included

	$\hat{p}$		u		r	
λ	Con.	Cov.	Con.	Cov.	Con.	Cov.
1.11	99.03	16.62	99.19	16.59	99.19	16.46
1.25	98.89	16.81	99.19	16.59	98.89	16.81
1.43	98.89	16.81	98.89	16.81	98.89	16.81
1.67	74.18	42.41	98.73	16.84	72.42	39.89
2	69.32	49.03	73.47	38.92	72.89	43.03
2.5	62.60	62.32	65.06	60.30	63.68	61.57

3.3 Auto imports database

Our final experiment used the Auto Imports database [2]. The experiment was set up to illustrate that, in small databases, where there are rare cases, the technique of u value ordering is very much preferred over $\hat{p}$ value ordering or no ordering at all.

With the Auto Imports experiments, we discretised the output field (*price*) into two clusters, *price* < 20970, and *price* ≥ 20970, with the aim of predicting membership of the first cluster. Because the database is so small, $\hat{p}$ values are often identical for particular attribute values. For the field *make*, all cars in the training database of *bmw, dodge, honda, mazda, mitsubishi, nissan, peugot, plymouth, saab, subaru, toyota, volkswagen* cost < 20970, and hence have $\hat{p}$=1. By chance, the cheaper *bmw* models have been selected in the training database. The total number of *bmw's* in the database is small and the $\hat{p}$ value has not been computed using a representative sample. In fact the sample is so small that even using u values is statistically unjustified. Whenever $n_i \leq 50$, the alternative u' values were used. These values introduced slight variations in ordering compared to that obtained using u values. In Table 10, position 5 is taken up with *volvo* with *plymouth* and *volkswagen* taking position 6. The u value ordering would have *plymouth* and *volkswagen* in position 5, and *volvo* in position 6. So the ordering in Table 10 is the ordering used in our experiments. Of the other categorical fields in the database–*body style*, *engine type*, and *fuel system*–the same ordering–*engine type*, and *fuel system* were ordered identically by u values and u' values. When both techniques were applied to the field *body style*, a different ordering is observed. The u' values generates the ordering *hatchback, sedan, wagon,* and *convertible*, whereas, the u values generates the ordering *sedan, wagon, hatchback,* and *convertible*.

Table 9. $\hat{p}$ value ordering from auto imports database

Position	Make
1	bmw,dodge,honda,mazda,mitsubishi,nissan, peugot,plymouth,saab,subaru,toyota, volkswagen
2	volvo

Table 10. Freeman value ordering from auto imports database

Make	Ordering
1	toyota
2	honda,subaru
3	mazda,nissan,peugot,saab
4	mitsubishi
5	volvo
6	plymouth,volkswagen
7	dodge
8	bmw

Looking at Table 9, the $\hat{p}$ values only impose a partial ordering and cluster 12 makes of cars together. We encoded all these cars using the value 1. The u' values are more discriminatory and deliver the partial ordering of Table 10. Using u' or the u value technique (as appropriate), we always produce rules which outperform

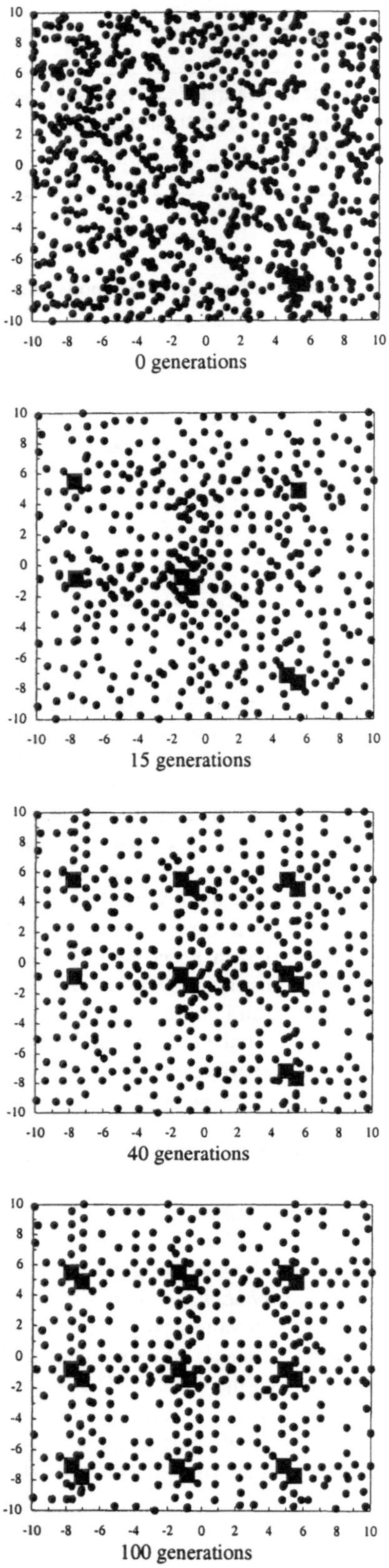

Fig. 2: Finding solutions for the Shubert function: the convergence of generations

current best individuals to be worth keeping. None of the techniques we have encountered in the literature addresses this problem.

We propose to solve the problems described above by attempting to partition the population into a set of dominated species and copying the dominating individual of each of these species into the new generation.

3.3 Conserving species

To be able to determine which individuals will be copied into the new generation we propose to partition the current population into a set of species centered around one of their dominating individuals. The dominating individuals defining this partitioning will be conserved in a way similar to copying the best individual in an elitist GA. Conserving the dominating individual for each of the species guarantees the survival of both good and diverse individuals.

This procedure has to be performed every generation. Thus it will introduce some overhead into the computations. The complexity of this additional computation can be characterized in terms of the number of times the distance between two individuals needs to be computed and it is of $O(n^2)$, where n is the population size. This is not higher than the complexity reported for other parallel population methods [9].

The outline of the GA implemeting our species conservation thechnique is given in Figure 1.

4 Some Experimental Results

To demonstrate the effectiveness of a GA using the species conservation technique we chose to present the results produced for the optimization of the Shubert function [10]:

$$f(x_1,x_2) = \sum_{i=1}^{5} i\cdot\cos[(i+1)x_1+1]\cdot\sum_{i=1}^{5} i\cdot\cos[(i+1)x_2+1]$$

where $-10 \le x_i \le 10$ for $i = 1,2$.

This is a very interesting function. The function has 760 local minima, 18 of which are global minima with -186.73. They are unevenly spaced. Despite the extensive search we have performed we could not find any report on results produced for this optimization problem. In a private communication Ken Price reported that he managed to obtain all the 18 global solutions by using the highly successful *Differential Evolution* (DE) [11] method. In order to obtain those solutions 50 trials were needed. Each of the trials required about 6,000 function evaluations, which means a total number of 300,000 function evaluations. This corresponds to an average number of approximately 16,667 function evaluations per solution.

For our experiments we used a SGA using species conservation with average crossover and uniform mutation. The parameters of the algorithm were set to the following values: population size 500, crossover probability 0.6, mutation probability 0.001 and species distance 0.8. The algorithm found *all* the solutions of the problem performing a total number of 28,025 function evaluations, that is an average number of 1,556 function evaluation per solution. Figure 2 shows a set of snapshots of generations converging onto the solutions.

To prove further test the effectiveness of our method we run another set of experiments for various instances of the *generalized Shubert*[1] function:

$$f(x_1,x_2,\ldots,x_l)=\prod_{i=1}^{l}\sum_{j=1}^{5} j\cdot\cos[(j+1)x_{i1}+1]$$

where $-10 \le x_{ii} \le 10$ for $i = 1,2,\ldots,l$.

Dim.	Sol.	Found	Evaluations	Average
2	18	18	7157	397
3	81	81	46607	798
4	324	324	469125	1448

Table I: Number of function evaluations for the generalized Shubert function

The l-dimensional Shubert function has $l\cdot 3^l$ unevenly spaced global solutions. Table I summarizes some results obtained by our algorithm (averaged over 10 runs for each of the problems).

5 Conclusions and Future Work

In this paper we have presented species conservation a new technique for evolving parallel subpopulation in a GA. We have demonstrated this technique by applying it to a family of hard multimodal optimization problems. The results produced clearly showed that a GA using species conservation can find several (possibly all) solutions of a multimodal optimization problem. We have conducted a large number of experiments with other hard multimodal optimization problems but the lack of space prevented us from presenting them here.

Our main objective for the future is to apply our technique to hard multimodal engineering design problems with the expectation to discover novel solutions. We will also perform further empirical studies on the performance of GAs using the species conservation method.

[1] We generalized the Shubert function by increasing the dimensionality of its domain

References

[1] D. Beasley, D.R. Bull and R.R. Martin, „A Sequential Niche Technique for Multimodal Function optimization“, *Evolutionary Computation*, No 2, Vol 1, pp. 101-125, 1993.

[2] Y. Davidor, "A Naturally Occurring Niche and Species Phenomenon: The Model and First results“, *Proceedings of the Fourth International Conference on Genetic Algorithms*, University of San Diego, pp. 257-263, 1991.

[3] K. Deb and D.E. Goldberg, „An Investigate of Niche and Species Formation in Genetic Function Optimization“, *Proceedings of the Third International Conference on Genetic Algorithms*, George Mason University, pp. 42-50, 1989.

[4] K.A. De Jong „An analysis of behavior of a class of genetic adaptive systems“, Doctoral *Dissertation, University of Michigan*, Dissertation Abstracts International 36(10), 5240B, 1975.

[5] F. Glover and M. Laguna, *Tabu Search*, Kluwer Academic Publishers, 1998.

[6] D.E. Goldberg and J. Richardson, „Genetic algorithms with sharing for multimodal function optimization“, *Proceedings of the second International Conference on Genetic Algorithm*, pp.41-49, 1987.

[7] D.E. Goldberg, Genetic Algorithms in search, optimization and machine learning, Reading, Addison-Wesley, 1989.

[8] J.H. Holland, *Adaptation in natural and artificial systems*, University of Michigan Press, 1975.

[9] O.J. Mengshoel and D.E. Goldberg, “Probability Crowding: Deterministic Crowding with Probabilistic Replacement”, *Proceedings of the Genetic and Evolutionary Computation Conference*, Orlando, Florida. pp. 409-416, 1999.

[10] Y. Michalewicz, *Genetic Algorithms + Data Structures = Evolutionary Programs*, Springer-Verlag Berlin Heidelberg, New York, 1996.

[11] R. Storn and K. Price, „Differential Evolution - a Simple and Efficient Heuristic for Global Optimization over Continuous Spaces“, *Journal of Global Optimization*, Kluwer Academic Publishers, Vol. 11, pp. 341-359, 1997.

Social Agents in Dynamic Equilibrium

Mark McCartney[1] and David Pearson[2]
[1]School of Computing and Mathematical Sciences, University of Ulster, Newtownabbey, County Antrim, Northern Ireland (m.mccartney@ulster.ac.uk) , [2]EURISE (Roanne Research Group), Jean Monnet University of Saint-Etienne, I.U.T. de Roanne, France (david.pearson@univ-st-etienne.fr)

Abstract

A simple model for social group interactions is introduced. The model is investigated for the minimal group size for the model N=3. Examples of model of model behaviour in terms of dynamic and static equilibrium are presented. Directions for further study of the model are considered.

1 Introduction

Sociodynamics is a relatively new field of interdisciplinary study which aims to provide quantitative models for how social groups evolve. It combines expertise form areas such as physics [1,2], mathematics [3], computer science [4] and sociology [5,6,7,8,9]. In this paper we present a model for sociodynamics based on fuzzy logic which builds upon earlier work [10,11,12].

Consider a social group of N agents who all interact with each other. Clearly, if the agents are people, the interactions between two agents j and k can occur in many different ways. At a psychological level we have, amongst others, verbal & visual interactions. A verbal interaction between j and k occurs via a conversation or written communication. A visual interaction between j and k may involve what the other person looks like. For example does j find k attractive (and vice versa)? Physical interaction could take on many forms. For example j might slap k in the face. Or in a panic situation individuals may find themselves, by their movement co-operating or competing to escape a disaster (see, for example, [2]).

Interactions within social groups are clearly dependant on a wide range of factors and are subtle. We however shall study a vastly simplified model of social interaction based on the *attitude* agents have to each other. Let $a_{jk}(t)$ be the time dependent function describing whether agent k *likes* agent j. The *like* function satisfies $a_{jk}(t) \in [0,1]$ where the larger the value of $a_{jk}(t)$ the more k likes j. Thus a value of 0 signifies that agent k definitely does not like agent j (j is an enemy) and a value of 1 signifies k definitely does like j (j is a close friend).

Obviously it is not necessary to have $a_{jk}(t) = a_{kj}(t)$ nor is it necessary for $a_{kj}(t)$ to have a value assigned to it even if $a_{jk}(t)$ does (you, for example, may have a view on Bob Dylan, but it is doubtful that Bob Dylan has a view on you). Thus if j does not know k we do not assign a value to $a_{kj}(t)$, but this does not mean that k does not know j and has a view on him/her. The term $a_{jj}(t)$ can be taken to signify j's self image, but we do not discuss that possibility here.

2 Evolving Attitudes

Initially let us assume that in the social group every agent has information about every other agent and has an opinion about how much they like them. A technique for generating this initial array of attitudes $[a_{ij}(0)]$ has been considered elsewhere [12].

If the group does not interact then we shall assume the attitudes do not change. (An alternative to this, which we do not consider in the current paper, would be factor into the model the effect that all the attitudes approach a (neutral) value of 0.5 with a characteristic 'half-life' or 'memory' in the absence of interaction.) Once the group does start to interact, attitudes begin to evolve. As noted above, in the real world our attitude to an individual will be influenced by a whole range of factors and interactions. In our simple model however we shall update how much k likes j by allowing k to ask a third party, m, their view of j. Then k updates their attitude to j to make it more, or less, like that of m's attitude to j, depending on k's attitude to m.

To formulate the model we first define the following step and ramp functions $H(x)$ and $R(x)$.

$$H(x)=\begin{cases}1 & x>0\\ 0 & x=0\\ -1 & x<-1\end{cases} \quad (1)$$

$$R(x)=\begin{cases}1 & x>1\\ x & 0\le x\le 1\\ 0 & x<0\end{cases} \quad (2)$$

We then define the change function

$$G(x,y,\Delta,\sigma)=\begin{cases}x+\Delta & y\ge\frac{1+\sigma}{2}\\ x & \frac{1-\sigma}{2}\le y<\frac{1+\sigma}{2}\\ x-\Delta & y<\frac{1-\sigma}{2}\end{cases} \quad (3)$$

where $\Delta,\sigma\in[0,1]$.

We can then construct an attitude updating rule of the form,

$$a_{jk}(t+1)= R\left[G(a_{jk}(t),a_{mk}(t),\Delta_k H(a_{jm}(t)-a_{jk}(t)),\sigma_k)\right] \quad (4)$$

The effect of $G\left(a_{jk}(t),a_{mk}(t),\Delta_k H\left(a_{jm}(t)-a_{jk}(t)\right),\sigma_k\right)$ is to increase the difference between a_{jk} and a_{jm} by an amount Δ_k if $a_{mk}<\frac{1}{2}(1-\sigma_k)$, decrease the difference between a_{jk} and a_{jm} by an amount Δ_k if $a_{mk}\ge\frac{1}{2}(1+\sigma_k)$, and to leave the value of a_{jk} unchanged otherwise.

We can say that $\Delta_k\in[0,1]$ is a measure of k's *volatility*. Thus if Δ_k is small then when k changes his attitude to another individual he does so in a 'cautious' way, making only small changes. Whereas if Δ_k is close to one when his k changes his attitude to another individual he does so in a volatile way, changing dramatically from extreme dislike to extreme fondness (or vice versa) in a single step.

Further, we define $\sigma_k\in[0,1]$ as a measure of k's stubbornness, or resistance to change, where the larger the value of σ_k the more stubborn k is.

Equation (4) thus gives a simple model for group social dynamics, with each agent, k, being defined in terms of two characteristics, volatility, Δ_k, and stubbornness σ_k. Given an initial array of attitudes $[a_{ij}(0)]$ and characteristic column vectors $[\Delta_k]$ and $[\sigma_k]$ equation (4) generates the evolution of attitudes $[a_{ij}(t)]$.

3 Results

In this paper we consider the minimal group of three people. This simplification means that when an individual k has decided to update his/her attitude towards another person j the third party, m, of whom they seek advise is uniquely defined. Examples of sociological meaningful methods for choosing m when the group is larger than three people are given elsewhere, as are methods for choosing the number of a_{ij} which are updated in a given time step [12].

Obviously the model, even though it is simple, is parameter rich and in this paper we consider only one set of initial attitude attitudes:

$$a_{21}=0.9,\ a_{31}=0.8,\ a_{12}=0.6,\ a_{32}=0.5,\ a_{13}=0.2,\ a_{23}=0.1 \quad (5)$$

Further we assume each member of the group has identical characteristics of stubbornness and volatility i.e.

$$\sigma_j=\sigma\ \forall j \qquad \Delta_j=\Delta\ \forall j \quad (6)$$

Given these conditions the system will either evolve to a state of static or dynamic equilibrium, as illustrated in figure 1. Note that for values of stubbornness, $\sigma>0.8$ the system always reaches static equilibrium. This is because if all members of the group are stubborn agents are unlikely to make any change to their attitudes i.e. $a_{jk}(t)=a_{jk}(0)\ \forall i,j$ and $t>0$. Apart from this trend,

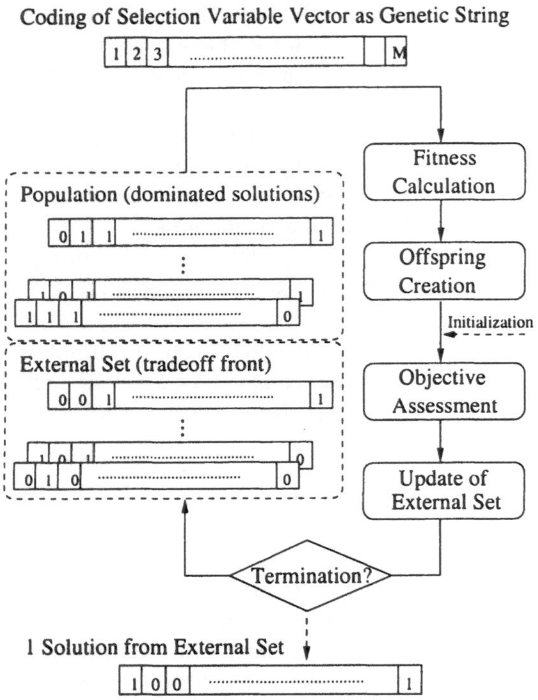

Fig. 2. Illustration of FSSPEA method.

turbine jet engine was monitored with regard to its operating range. The 24-dimensional data was generated from Fourier spectra obtained from the compressor air flow. Four operating regions were defined as classes. Five data sets with 375 samples each were drawn from a compressor set-up and filed as a Mech_1 to Mech_5 in a database including attribute values. The objective of the underlying research was the development of a *Stall-Margin-Indicator* for optimum jet engine operation. This medium sized database provides the first realistic application of the proposed method. Mech_1 and Mech_2 were chosen as training and test set, respectively. Table I shows the achieved results using q_{si} as cost function.

Table I. Selection results achieved for *Mechatronic.*

Method	Mech_1	Mech_2	Features
ORG	0.992500	0.981667	24
SFS	0.997500	0.928333	3
SBS	0.989333	0.981667	12
PM	0.997500	0.959167	6
SGA	0.997333	0.969167	12
FSSPEA	0.997500	0.928333	3

Figure 3 and Fig. 4 show the resulting feature space for the original 24-dimensional $Mech_1$ data and the optimum three-dimensional selection result, respectively. A dimensionality reducing projection, e.g., Sammon's mapping, is employed here to visualize the achievement of feature selection with regard to class region separability in an enhanced scatter plot [3]. The visual information complements the cost function results, e.g., q_{si}, and helps to assess the found solution for data analysis and recognition system design. In Table I the additional abreviations ORG, SFS, SBS, and PM denote the original data set, the selection by sequential forward selection, sequential backward selection, and perturbation method [3], respectively. Both the FSSPEA and the heuristic SFS found the minimum solution of three selected features for the training set.

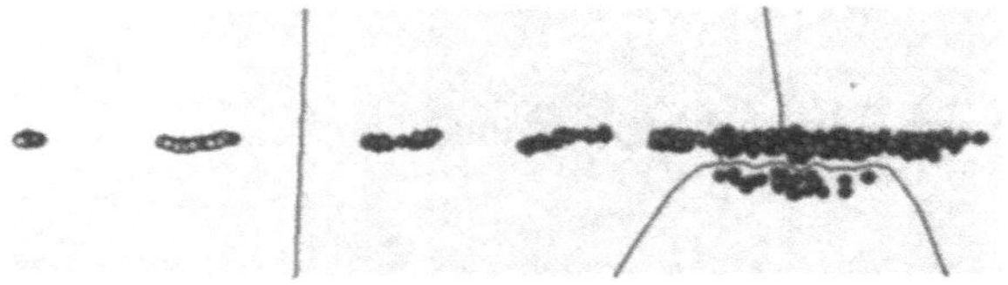

Fig. 3. Visualization of original $Mech_1$ data.

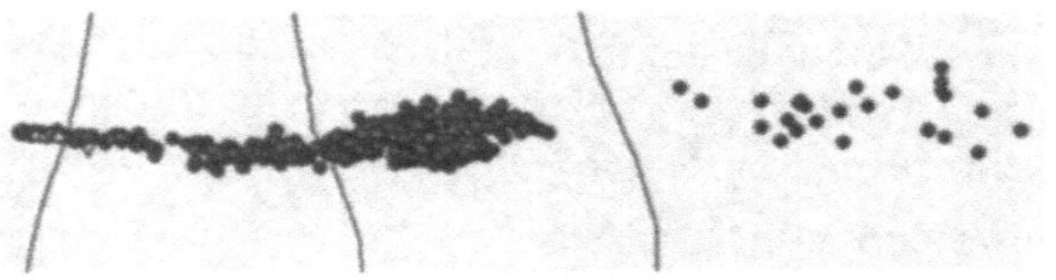

Fig. 4.. Visualization of FSSPEA for $Mech_1$ data.

However, for the test set and this minimum solution, a reduced generalization for the test set can be observed. This raises the issue of the stability of a selection solution in this optimization process, which requires the modification of the cost functions. This will be regarded in future work. The second example stems from an implementation effort of an integrated eye-tracking system [8]. The task of eye-classification is regarded here. It bases on a proven feature computation stage, employing a filter bank of gabor filter with a range of 12 different orientations and spatial frequencies. In a first step, the filter parameters have been manually optimized according to a training database of eye-shapes and non-eye-shapes. In the second step now, automatic feature selection is applied for further compaction and discriminance gain in the computation. Two data sets of 72 and 61 patterns were generated for this two-class problem. Table II shows the achieved selection results using q_{si} as cost function.

Also, Fig. 5 shows the feature space for the optimum selection result achieved for the eye-shape training data. As can be extracted from Table II, with regard to dimensionality reduction and discriminance gain for the training set FSSPEA with 50

Table II. Selection results achieved for eye-shapes.

Method	Train	Test	Features
ORG	0.858374	0.953647	12
SFS	0.947040	0.989362	6
SBS	0.938420	0.953647	6
PM	0.938420	0.896657	4
FSSPEA (50)	0.955670	0.896657	3
FSSPEA (10)	0.938420	0.896657	4

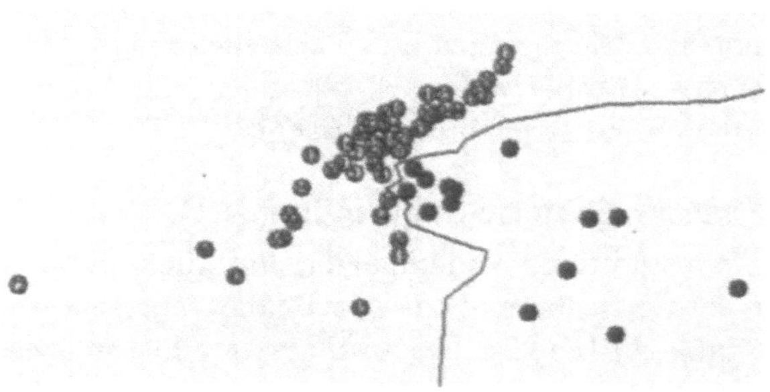

Fig. 5. Visualization of FSSPEA for eye-shape data.

generations found the optimum solution. With 10 generations the same result as for PM was found. Again, for the test set the generalization performance degraded with the achieved dimensionality reduction for all applied methods. Improved multicriterial optimization, e.g., using a validation set and a modified stopping criterion, will have to be developed for further improvement. A further improvement issue is, how appropriate automatic selection of one solution of the developped tradeoff or Pareto optimal front shall be executed.

5 Conclusions

The SPEA advanced optimization was applied to the problem of automatic feature selection, employing powerful dedicated nonparametric cost functions. Viability and competitiveness of the approach was demonstrated by applying it to practical problem data. The method was integrated in the framework of the QuickCog system and so is amenable for application in numerous classification or data analysis problems. For instance, the application to medical data base analysis, e.g., for type 2 diabetes prediction (s.e.g., [3]), is planned in the next step of work. Further, improvements of the cost function or fitness function with regard to the stability of the selection will be investigated. For instance, a validation set could be included in the selection process to provide a stopping criterion. A strong focus will be on the generalization of this FSSPEA approach to feature weighting as given in [9] or [6]. Evolutionary algorithms will in future work replace the currently applied gradient method [9]. The QuickCog system and the presented method of FS using GA will be made available from the QuickCog web page [10].

References

[1] K. Fukunaga, *Introduction to Statistical Pattern Recognition.* ACADEMIC PRESS, INC. Harcourt Brace Jovanovich, Publishers Boston San Diego New York London Sydney Tokyo Toronto, 1990.

[2] J. Kittler, *Feature Selection and Extraction.* ACADEMIC PRESS, INC. Tzai. Y. Young King Sun-Fu, Publishers Orlando San Diego New York Austin London Montreal Sydney Tokyo Toronto, 1986.

[3] A. König, "Dimensionality Reduction Techniques for Multivariate Data Classification, Interactive Visualization, and Analysis – Systematic Feature Selection vs. Extraction," in *Proc. of 4th Int. Conf. on Knowledge-Based Intelligent Engineering Systems & Allied Technologies KES'2000,* (University of Brighton, UK), pp. 44–56, August 2000.

[4] G. W. Gates, "The Reduced Nearest Neighbour Rule," in *IEEE Transactions on Information Theory, vol. IT-18,* pp. 431 – 433, 1972.

[5] E. Zitzler and L. Thiele, "Multiobjective evolutionary algorithms: A comparative case study and the strength pareto approach," *IEEE Transactions on Evolutionary Computation,* vol. 3, pp. 257–271, Nov. 1999.

[6] M. L. Raymer, W. F. Punch, E. D. Goodman, L. A. Kuhn, and A. K. Jain, "Dimensionality reduction using genetic algorithms," *IEEE Transactions on Evolutionary Computation,* vol. 4, pp. 164–171, July 2000.

[7] H. Wang, D. Hennecke, A. König, P. Windirsch, and M. Glesner, " Method for Estimating Various Operating States in a Single Stage Axial Compressor," in *AIAA Journal of Propulsion and Power 11(2),* pp. 385–387, 1995.

[8] A. König, A. Günther, A. Kröhnert, T. Grohmann, J. Döge, and M. Eberhardt, "Holistic Modelling and Parsimonious Design of Low-Power Integrated Vision and Cognition Systems," in *Proc. of 6th Int. Conf. on Soft Computing and Information/Intelligent Systems IIZUKA'2000,* (Iizuka, Fukuoka, Japan), pp. 710–717, October 2000.

[9] A. König, "A Novel Supervised Dimensionality Reduction Technique by Feature Weighting for Improved Neural Network Classifier Learning and Generalization," in *Proc. of 6th Int. Conf. on Soft Computing and Information/Intelligent Systems IIZUKA'2000,* (Iizuka, Fukuoka, Japan), pp. 746–753, October 2000.

[10] A. König, M. Eberhardt, and R. Wenzel, "QuickCog – HomePage," in *http://www.iee.et.tu-dresden.de/~koeniga/QuickCog.html,* 2000.

Feature Subset Selection Problems: A Variable-Length Chromosome Perspective

César M. Guerra-Salcedo*[1]

*Advanced Technology Group, Athene Software Inc. Boulder Colorado, USA

Abstract

Fixed-length subset problems occur where a solution to a problem is described by an unordered subset of a particular cardinality. Some fixed-length subset problems are known as "deceptive" problems. In a deceptive problem certain possible solutions tend to lead the search algorithm towards a locally optimal solution. In this research, deceptive problems are treated as feature selection problems, the goal is to find the right combination of features that solves the problem. We explore a family of genetic search methods known as messy genetic algorithms on artificially generated deceptive problems and real-world instance-based classification problems.

1 Introduction

One of the most challenging artificial feature subset problems is known as the fixed-length subset problem [2, 9]. Fixed-length subset problems occur where a solution to a problem is described by an unordered subset of particular cardinality. A particular class of fixed-length subset problems are known as "deceptive" [5]. In a deceptive problem certain possible solutions tend to lead the search algorithm towards a locally optimal solution.

In this research, deceptive problems are treated as feature selection problems, the main goal is to find the right combination of features that solves the problem. We explore a family of genetic search algorithms in which the chromosome is allowed to grow or to shrink during the execution of the search. This class of genetic search method is known as a messy genetic algorithm [5, 7]. We present applications of a variety of messy genetic algorithms to artificially generated feature selection problems and to real-world instance-based classification problems.

First, we introduce messy genetic algorithms and two variants of messy genetic algorithms known as Fast Messy Genetic Algorithms and Block Insertion Fast Messy Genetic Algorithms. Second, we present results of applying variable-length genetic algorithms as well as traditional genetic algorithms to artificially generated feature selection problems. Third, we present the results of experiments using real-world instance-based classification problems using a wrapper approach [8] and a search engine based on messy genetic algorithms.

2 Messy Genetic Algorithms (MGA)

In a traditional genetic algorithm, the problem to be solved is mapped to a particular representation suitable for the GA. For example, in a feature selection problem every feature is mapped as a bit in a chromosome. If the problem has l features, the chromosome has exactly l bits. The problem to be solved is neither underspecified nor overspecified in terms of the representation in the chromosome. Messy genetic algorithms allow variable-length chromosomes that may be underspecified or overspecified with respect to the problem being solved [5, 7].

A messy genetic algorithm typically has three phases: initialization, primordial phase and juxtapositional phase. During initialization, a population containing one copy of all substrings of length k is created ($k \ll l$ the problem length). During the primordial phase, selection based on the evaluation of the chromosome is performed to increase the number of "good" chromosomes.The evaluation is completed using a template to "fill in the blanks" the positions not present in the chromosome. During the juxtapositional phase, selection is used together with two operators: cut and splice. "Cut" cuts the chromosome at a random position. "Splice" attaches two chromosomes together.

The Fast Messy Genetic Algorithm (FMGA) was designed to cope with the problem of the large population size using the MGA [5, 7]. The initial chromosome length is set to l', $k < l' < l$ (e.g. l' is $l - k$). The probability of randomly selecting a gene combination of size k in a chromosome of length l' with l genes is given by Kargupta [7]

$$\binom{l-k}{l'-k} / \binom{l}{l'}.$$

From this probability, one can infer that in chromo-

[1] Part of this work was done while the author was a visiting researcher at Colorado State University.

somes of size l' created at random, one chromosome on average will have the desired gene combination of size k.

The advantage of the FMGA is the relatively small population size compared with the MGA. Nonetheless, for hard problems, the size of the initial population is on the order of thousands [7].

2.1 Fast messy genetic algorithm revisited: the block insertion fast messy GA (BIF-MGA)

We developed a modification on the final phase of the FMGA that introduces more variability to individual members of the population.

1. After the initial phase, the chromosome length is k. The chromosome length is increased (by applying juxtapositional phase several times) to a length which is about 80% of l.

2. The length of individuals is regulated by defining a new length l_{new} to be a fixed proportion of l (the problem size). For each individual i with length l_i, if $l_i > l_{new}$, l_i is reduced to l_{new} by randomly deleting genes. Otherwise no gene deletion is performed.

3. A procedure called *Block Insertion* redefines the chromosome by inserting new fixed-size messy gene blocks:

   (a) The block length is $l/3$.

   (b) The gene numbering is ordered and consecutive starting at $l/3 \times Random(0, 2)$ with randomly-generated allele values. $Random(0, 2)$ returns an integer in the range [0..2].

   (c) Let l_i be the individual length (fixed in 1.), if $l_i \geq l/3$ then the block is inserted at the beginning of the chromosome by changing the first $l/3$ messy genes and leaving $l_i - (l/3)$ messy genes without change. If $l_i < l/3$ then l_i messy genes taken from the block completely replaces the chromosome.

4. A juxtapositional phase of cut and splice is applied to increase the chromosome length.

These changes were made via empirical experimentation in an effort to reduce the population size required by the FMGA. The population size of the BIF-MGA was not larger than 100 individuals for all the experiments reported in this research.

Problem Type	Optimal Solution	Algorithm Employed	Best Found	Function Evaluations
Trap 100 5-bit	100	CHC	87	1500
		FMGA*	100	1005
		BIF-MGA	100	852
Subset 120-60-4	60	CHC	60	566.9
		FMGA	56	735
		BIF-MGA	60	314.9
Ugly 60-bit	600	CHC	598	5900
		FMGA	600	1250
		BIF-MGA	600	1000

Table I. Comparing CHC, FMGA and BIF-MGA for synthetic subset selection problems. Results labeled FMGA* are taken from Kargupta[7]. Function evaluations are in thousands.

3 Tests Results on Artificial Problems

We have compared the performance of CHC [4], FMGA and BIF-MGA using a wide variety of synthetic subset selection problems. CHC (Cross generational elitist selection, Heterogeneous recombination and Cataclysmic mutation) is a fixed-length-chromosome genetic algorithm that has been proved to be very effective for a wide variety of search problems including feature selection problems [4, 6]. The following test problems were used:

1. **Trap functions** as defined by Kargupta [7]. A trap function is a function of unitation that divides the search space into two basins in the Hamming space, one leading to a global optimum and the other leading to a local optimum. Experiments were conducted for a 100-bit trap problem with 5-bit subfunctions. Results are shown in Table I.

2. **Subset selection problems** as defined by Radcliffe and George [9]. Experiments were conducted for an epistatic 120-60-4 problem. Results are shown in Table I.

3. **Ugly-deceptive problems** as described by Goldberg [5]. For our experiments, the bits that jointly affect the fitness are randomly placed in the chromosome. We tested the algorithms with a 60-bit ugly-deceptive problem. Results are shown in Table I

All the results are the averages of 10 independent runs and are based on the total number of chromosomes evaluated to obtain the best fitness reported. We are presenting the optimal solution for each problem.

To assure that the block insertion process is not just a way of carrying out high mutation rates, we ran experiments using the FMGA while varying gene and allele mutation probabilities from 0 to 1 by 0.1 increments for each parameter. In all cases, the performance of BIF-MGA was better than FMGA.

In all the the problems, BIF-MGA was the algorithm with better performance being able to find the optimal solution every time. The population size for CHC was 50 and for BIF-MGA was 100 for all the experiments, FMGA used between 2600 and 5000 chromosomes (ugly 60-bit).

3.1 The sparse subset problem

We found the previously described problems to be relatively easy to solve. A new synthetic test problem, the *sparse subset problem*, was developed with the following characteristics. Two 60-bit blocks are composed of ten 6-bit sub-blocks. Each sub-block of six bits uses the following evaluation function based on the number of one bits in the sub-block.

Count of 1 bits	Contribution to Fitness
six	20
five	15
four	12
three	9
two	6
one (except 000001)	3
000001	12

For each of the two blocks of 60 bits, if the number of ones exceeds 13 there will be a penalty of $-2 \times ones(block)$ where $ones(block)$ returns the number of ones in a block of 60 bits. A maximum value of 240 is achieved when every sub-block of six bits has the pattern 000001. Results for this problem using ten random runs of each BIF-MGA, CHC, and FMGA are shown in Figure 1. We tested many different mutations levels, but FMGA never outperformed CHC. The BIF-MGA solved the problem every time, while the other algorithms never found an optimal solution.

Given the good results we obtained using messy genetic algorithms for artificially-generated feature selection problems, we decided to test them with real-world feature selection problems.

4 Application to Real-world Problems

We perform experiments using two real-world datasets. The first data set is the LandSat dataset from the UCI repository. The LandSat dataset consists of 4435 training cases and 2000 test cases. Each case represents a satellite image with 36 features. The data can be categorized into six different classes. The second data set is the Cloud[1] dataset for developing an automated system for cloud classification [1]. There are 1633 cases, each case representing a cloud with 204 features on a continuous-based range belonging to one of 10 different cloud types.

[1]The dataset was provided by Richard Bankert from the Naval Research Laboratory.

To evaluate the experiments, a wrapper approach [8] was used. The wrapper approach uses the evaluation of a particular subset of features found as a feedback that guides a feature search process. The classifier employed during the feature search is of the same type as the classifier that would be used for the final classification. The classifier employed for the wrapper was an EDT (Euclidean Decision Table) [6]. For CHC we use a population of 50 elements. For FMGA we used a population of 500 elements and for BIF-FMGA 50 elements. Unfortunately the results were not satisfactory. In all the problems the performance of the MGA family was much poorer than the performance of CHC. The results are summarized in Table II. The running time of FMGA and BIF-MGA were more than three times the running time of CHC. Even worse, for the Cloud dataset problem, the time it took the FMGA to generate a competitive solution was more than five days running on a SUN Sparc-Ultra station.

It seems that the MGA (and its variants) tend to find and exploit certain structure present in a partial solution. For example, in a 10-3-bit deceptive problem the MGA tends to find a partial solution of a sub-function (three bits only), and then constructs the global solution by 'splicing' several subsolutions together. In other words, FMGA is very good at solving sub-problems and using that information to construct the global solution. It seems that for at least some real-world classification problems, no such "repetitive" structure exists.

5 Summary

In this paper, we compared a traditional fixed-length genetic algorithm with variable-length genetic algorithms. The comparisons were performed for two different types of feature selection problems: Artificially generated feature selection problems and real-world feature selection problems. For artificially generated feature subset selection problems, MGAs proved to be more effective than CHC when the complexity of the problems increased. However, for real-world feature selection problems the performance of MGA and its variants was poor. We conjecture that this poor performance is due to the fact that many real-world feature selection problems cannot be decomposed in smaller well-defined sub-problems. By well-defined, we mean with a cer-

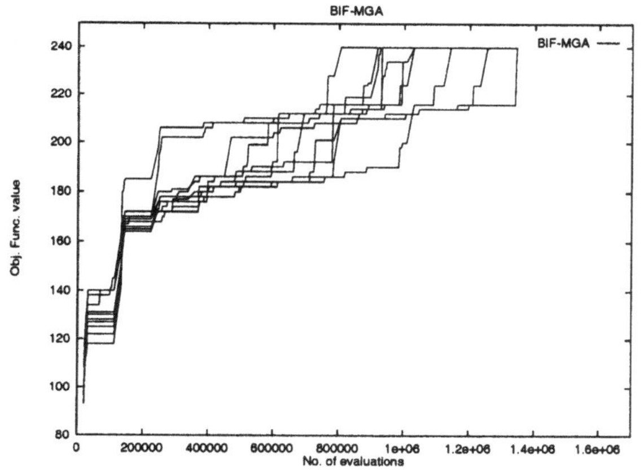

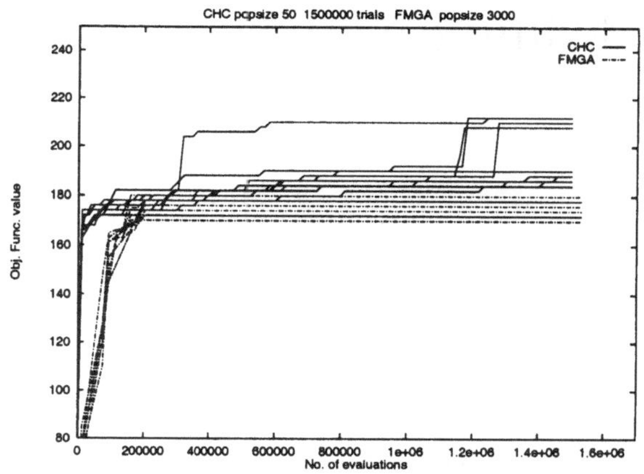

Fig. 1. Results for the 120-bit Sparse Subset Problem. On the left are results for BIF-MGA (solid lines), and on the right are the results for CHC (solid lines) and the standard FMGA (dashed lines). The optimal fitness is 240. Each graph represents ten runs for each algorithm.

Algorithm	Best Run	Worst Run	Average	Test Errors
CHC LandSat	55 errors 16 feat.	90 errors 18 feat.	66.3 errors 17.4 feat.	240
FMGA LandSat	65 errors 23 feat.	120 errors 28 feat.	75.8 errors 26.4 feat.	320
BIF-MGA LandSat	59 errors 20 feat.	110 errors 14 feat.	74.8 errors 21.5 feat.	269
CHC Cloud	74 err. 99 feat.	86 errors 93 feat.	67.2 errors 96.4 feat.	146.2
FMGA Cloud	245 errors 102 feat.	265 errors 145 feat.	248.3 err 132.3 feat.	523.2
BIF-MGA Cloud	212 errors 105 feat.	262 errors 128 feat	226.4 errors 111.3 feat	452.2

Table II. Comparing CHC, FMGA and BIF-MGA for real-world subset selection problems. Best Run column represents the best results reported for the GA on the data used for that run. Average column represents the average after 30 independent runs. The last column represents the average number of errors when the best set of features found is tested.

tain structure that can be exploited and replicated throughout the final solution (as was possible for the artificially-generated feature selection problems described in this research).

References

[1] Bankert, R. L., "Cloud Classification of avhrr Imagery in Maritime Regions Using a Probabilistic Neural Network", *Applied Metheorology*, vol. 33(8), pp. 909–918, 1994.

[2] Crawford, K. D., "The Role of Recombination in Genetic Algorithms for Fixed-length Subset Problems", *PhD thesis, The Graduate School of The University of Tulsa*, 1996.

[3] Deb, K. and Goldberg, D., "Analyzing Deception in Trap Functions", *In Whitley, L. D., editor, FOGA-2*, pp. 93–108, Morgan Kaufmann, 1993.

[4] Eshelman, L., "The CHC Adaptive Search Algorithm. How to Have Safe Search When Engaging in Nontraditional Genetic Recombination", *In Rawlins, G., editor, FOGA-1*, pp. 265–283, Morgan Kaufmann, 1991.

[5] Goldberg, D. E., Deb, K., Kargupta, H., and Harik, G., "Rapid, Accurate Optimization of Difficult Problems Using Fast Messy Genetic Algorithms", *In Forrest, S., editor, Proc. of the 5th Int'l. Conf. on GAs*, pp. 56–64, Morgan Kaufmann, 1993.

[6] Guerra-Salcedo, C. and Whitley, D., "Genetic Search for Feature Subset Selection: A Comparison Between CHC and GENESIS", In *Proceedings of the Third Annual Genetic Programming Conference*, pp. 504–509, Morgan Kaufmann, 1998.

[7] Kargupta, H., "SEARCH, Polynomial Complexity, And The Fast Messy Genetic Algorithm", *PhD thesis, Department of Computer Science University of Illinois at Urbana Champaign*, 1995.

[8] Kohavi, R., "Wrappers for Performance Enhancement and Oblivious Decision Graphs", *PhD thesis, Stanford University*, 1995.

[9] Radcliffe, N. J. and George, F. A. W., "A Study In Set Recombination", *In Forrest, S., editor, Proc. of the 5th Int'l. Conf. on GAs*, pp 23–30, Morgan Kaufmann, 1993.

Mining Numeric Association Rules with Genetic Algorithms

J. Mata *, J. L. Alvarez †, J.C. Riquelme ‡

*Dept. Ing. Electronica, Sist. Informaticos y Autom. Universidad de Huelva. Spain. email:mata@uhu.es
†Dept. Ing. Electronica, Sist. Informaticos y Autom. Universidad de Huelva. Spain. email:alvarez@uhu.es
‡Dept. Lenguajes y Sistemas Informaticos. Universidad de Sevilla. Spain. email:riquelme@lsi.us.es

Abstract

In this last decade, association rules are being, inside Data Mining techniques, one of the most used tools to find relationships among attributes of a database. Numerous scopes have found in these techniques an important source of qualitative information that can be analyzed by experts in order to improve some aspects in their environment.

Nowadays, there are different efficient algorithms to find these rules, but most of them are demanding of databases containing only discrete attributes. In this paper we present a tool, GENAR (GENetic Association Rules), that discover association rules in databases containing quantitative attributes. We use an evolutionary algorithm in order to find the different intervals. We also make use of the evolutionary methodology of iterative rule learning to not evolve always to the same rule. By means of this we get to discover the different association rules. In our approach we present a tool that obtain association rules with an undetermined number of numeric attributes in the antecedent of the rule.

1 Introduction

Association rules were introduced in [1] as a method to find relationships among attributes in a database. These techniques allow us to obtain a very interesting qualitative information with which we can take later decisions. In general terms, an association rule is a relationship between attributes in the way $C_1 \Rightarrow C_2$, where C_1 and C_2 are pair conjunctions (attribute, value) in the way $A_1 = v_1$ if it is a discrete attribute, or $A_1 \epsilon [x_1, y_1]$ if it is a continuous or numeric attribute.

Obviously, many associations of this kind can be found in a database but only the most interesting ones will be used for their later study. To define this notion of interesting two fundamental concepts were introduced [1]: **Support.** We will say that a rule $C_1 \Rightarrow C_2$ has a support value s, if a $s\%$ of the records contain C_1 and C_2. **Confidence.** We will say that a rule $C_1 \Rightarrow C_2$ has a confidence value c, if a $c\%$ of the records that contain C_1 also contain C_2.

For example, we suppose that we have a database with different measures taken from a river, from which we store among other information, temperature, pH, salinity, chlorophyll level, etc. The rule: $T \epsilon [12.1, 15.4] \wedge pH \epsilon [7.84, 7.96] \Rightarrow chlorophyll \epsilon [11.44, 16.41]$ has a *support* of 0.7 if the 70% of the measures realised have a temperature between 12.1 and 15.4, a pH between 7.84 and 7.96, and a chlorophyll level between 11.44 and 16.41. On the other hand, the rule has a *confidence* of 0.85 if the 85% of records where the temperature is between 12.1 and 15.4 and pH between 7.84 and 7.96, then the chlorophyll level is between 11.44 and 16.41.

The problem to solve consists in finding all the association rules that overcome some levels or minimum thresholds of support and confidence, defined by the user, called generally *minsup* and *minconf*. The usual algorithms of association rules work in two phases: firstly they find pair sets attribute-value which overcome *minsup* minimum threshold, secondly, departing from these sets, they discover the rules that overcome *minconf* minimum threshold. There are different algorithms to discover these association rules, that can be found in [2, 9].

The process to find frequent itemsets consists, basically, in building different combinations of attribute-value pairs and verifying that they are produced in a determined number of records. This process can be relatively easy and efficient when attributes are discrete and their domains span few values. For this reason the tools proposed in the previously referred articles, work with databases in which the domains of their attributes are formed by a finite set of values. But in the real world there are numerous databases where the information is numeric. In these databases, attributes have thousands of possibilities of taking one value, by this reason the process described above is unthinkable from a computational point of view.

There are some studies in which tools that handle with attributes with continuous domains are presented. In [6] the author propose to divide the quantitative attributes into a fixed number of intervals of the same size and to discover the rules departing from such intervals. One of the main problems is that rules are only discovered departing from such intervals. In [9] they go a step further allowing the union of consecutive intervals. In order that intervals do not cover all the domain of the attribute, the authors propose a new measure, *partial completeness*, that the intervals must not overcome. In [3], the concept of optimized association rule was introduced. Rastogi and Shim followed this line in [7] and [8], in which they permit rules that contain disjunctions over uninstantiated numeric attributes. The optimized association rules have the form: $U \wedge C_1 \Rightarrow C_2$, where U is a conjunction of one or two conditions over numeric attributes, and C_1 and C_2 are instantiated conditions.

Our contribution lies in the fact that these rules can have an undetermined amount of numeric attributes in the antecedent and a unique numeric attribute in the consequent. In this paper we present a technique to find association rules in numeric databases by using evolutionary algorithms (EA) [4].

2 Preliminaries

The tool developed in this approach is based on EA theory. In order to find the optimal rule, we depart from a population where the individuals are potential association rules. These individuals will be evolving by means of crossover and mutation operators, so that, the individual with the best fitness will correspond to the most significant rule in the last generation.

One of the problems we find when using EA theory is that, during the process, all the individuals tend to the same solution. In our study case, this means that all individuals evolve towards the same association rule, so that, rules composing the population of the last generation, provide, in practice, the same information. There are different techniques to solve this problem. Among them, the use of evolutionary algorithm with niches and the iterative rule learning [5]. In this tool we use iterative rule learning to find different association rules inside the database. The process consists in executing the genetic algorithm as many times as rules we want to obtain. In each iteration, we will mark the records covered by the obtained rule. This parameter affects the *fitness* function of the following generations, so that the algorithm will not search for rules that have been previously considered.

3 Practical Implementation

In the following subsections we will describe the general structure of the algorithm, the individuals representation and the meaning of the operators.

3.1 GENAR algorithm

In order to decide the completeness of the intervals that conform the rules, the algorithm only needs to know minimum and maximum values of each attributes domain. This value is needed for intervals not to grow up until spanning the total domain. **Definition 1.** We will define *amplitude* as the maximum size the interval of a determined attribute can get. We will obtain this value by 1.

$$amplitude(i) = \frac{M_i - m_i}{k} \tag{1}$$

Where M_i and m_i are maximum and minimum values of the domain of attribute i, and k is a value definable by the user, which we will call *AF (amplitude factor)*.

```
algorithm_GENAR
1.nRules = 0;
2.while ((nRules < NRULES) or
        (all records covered)) do
3.  nGen = 0;
4.  generate first population P(nGen);
5.  while (nGen < NGENERATIONS) do
6.    process P(nGen);
7.    P(nGen+1) = select individuals of P(nGen);
8.    complete population P(nGen+1) by crossover;
9.    make mutations in P(nGen+1);
10.   nGen++;
11. end_while
12. Rules[nRules] = choose the best of P(nGen);
13. penalize tuples covered by Rules[nRules];
14. nRules++;
15.end_while
end
```

As we can see, the algorithm is repeated until all rules (NRULES) have been obtained. This value is defined by the user depending on the number of rules he wants to obtain in the process. In step 4 the first rule population is generated. Intervals that overcome the *amplitude* defined in 1 are not allowed. The genetic algorithm is located among steps 5 to 11. In step 6, *process* carries out several functions: to calculate the *support*, *confidence* and *fitness* of each individual. In step 7, a percentage of individuals with the best fitness, according to 2 is selected. In 8 crossovers between selected individuals are made in order to complete the population.

Finally, in 9, mutations in individuals are carried out depending on a mutation factor.

In step 12 the best individual is chosen from the population formed in the last generation. This election will depend on the three factors pointed out previously: *support*, *confidence* and *fitness*. This individual is one of the rules that returns the algorithm. The operation made in step 13 is very important. In it records covered by the rule obtained in the previous step are penalized. Due to the fact that this factor is part of the adaptation function of the EA, we achieve that the next population does not repeat its search space, that is, it does not tend to generate the same association rule. To penalize the records we use a value named *PF (penalization factor)*, that will be defined by the user. This parameter takes its values from interval (0,1) and we use it to decrease the fitness of those individuals that are going to cover a record that has already been marked.

3.2 Genetic algorithm characteristics

GENAR algorithm uses as a search motor an EA with real codification for the individuals. During the evolutionary process a 15% of the individuals pass to the following generation according to the selection operator which will choose those with the best fitness. The rest of individuals to complete the population will be formed by the crossover operator. Besides, the individuals will be affected by a mutation operator depending on a mutation probability. At the end of the evolutionary process, that is, when all fixed generations have been completed, the best individual, depending on its *support*, *confidence* and *fitness* will be the association rule found.

3.3 Structure of individuals

In GENAR algorithm, each individual represents an association rule in which maximum and minimum values of the intervals of each numeric attribute are stored. The last interval is the rules consequent, while the rest of them conform the rules antecedent. Besides, certain additional information for each individual is stored, such as the number of attributes the rule has, *support*, *confidence* and *fitness*. In this paper we consider only those rules which involve all database attributes except the last one which acts as consequent.

In figure 1 we can see a graphic representation.

This individual represents the rule:

$A_1 \in [x_1,y_1] \wedge \ldots \wedge A_{n-1} \in [x_{n-1},y_{n-1}] \Rightarrow S \in [x_n,y_n]$

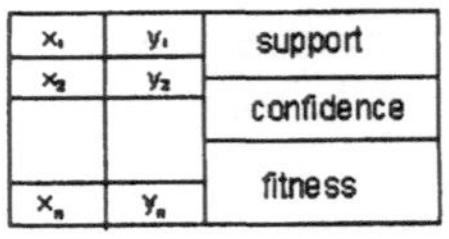

Fig. 1. Structure of an individual

3.4 Evaluation function

In order that in the different iterations the individuals tend to other search spaces, we have included a penalization factor in the evaluation function. By means of this we also achieve that association rules which our tool return form a non-hierarchical set, since we do not eliminate those cases covered by a rule, but we mark them in order that the same rule will not cover them again.

Evaluation function used appears in:

$$fitness = covered - (marked * PF) \qquad (2)$$

Where *covered* is the number of records which fit the rule and *marked* is a binary value which indicates if the record has already been covered by some previous rule.

3.5 Genetic operators

By means of selection operation the best individuals are chosen, that is, those with the best *fitness*, which are the ones that will go to the following generation. From each crossover between two individuals, two new ones appear whose intervals will be the first intervals of the first individual to cross and the following intervals of the second individual to cross and vice versa. The mutation operator consists in altering one of the intervals of the rule. For each bound of the chosen interval we can have two possibilities, to increase or to decrease its value. In this way we achieve four possible mutations: to shift the complete interval to the left or to the right, and to increase or to decrease its size. We have to be specially careful in not overcoming the fixed value of *amplitude*.

4 Results

To verify that the developed algorithm finds correctly numeric association rules, we have created two artificial databases in which certain rules, previously fixed, are fulfilled in an adequate number of records as to consider them interesting ones. In the first exemplary database, rules have no overlapping, that is, there are not any cases that can belong to two rules. Nevertheless, in order to fix results we have created a second database where there is overlapping among association rules. Due to the fact that the performance of the tool is based in a EA, we have carried out five times the proofs in both examples and the results fit in with the average values of such proofs.

4.1 Rules without overlapping

The first example database is formed by four numeric attributes. We have generated 1000 records distributed in 5 association rules. The rules will be formed then by a conjunction of the three first attributes in the antecedent and the fourth one in the consequent. The rules we pretend to find are the following ones:

1. [1,15] $\wedge$ [7,35] $\wedge$ [60,75] then [0,25]
2. [5,30] $\wedge$ [25,40] $\wedge$ [10,30] then [25,50]
3. [45,60] $\wedge$ [55,85] $\wedge$ [20,35] then [25,75]
4. [75,100] $\wedge$ [0,20] $\wedge$ [40,60] then [75,100]
5. [10,30] $\wedge$ [0,30] $\wedge$ [75,100] then [100,125]

rule	sup(%)	conf(%)	#r
1 [1,26] [6,35] [54,82] [0,31]	19.1	80.19	191
2 [1,32] [21,42] [10,36] [18,55]	19.9	100	199
3 [37,66] [56,84] [19,44] [23,64]	15.2	81.72	152
4 [70,99] [0,26] [34,61] [69,106]	19.1	100	191
5 [2,32] [1,30] [71,99] [92,124]	18.3	77.99	183

Table I. Obtained results without overlapping

The exact support of each of the rules artificially defined is 20%, since each of them cover 200 records. In table I we can see that the support of the obtained rules is very close to such value. Moreover, the values of the confidence are also close to 100%. In this case the number of covered records (#r) coincide with the support, since the rules have no overlapping.

4.2 Rules with overlapping

The second example database is formed by three numeric attributes. We have generated 600 records distributed in three association rules. These rules will be formed by a conjunction of the two first attributes in the antecedent and the third one in the consequent. The rules we pretend to find are the following ones:

1. [18,33] $\wedge$ [40,57] then [35,47]
2. [1,15] $\wedge$ [7,30] then [0,20]
3. [10,25] $\wedge$ [20,40] then [15,35]

In this case, the exact support of each of the rules artificially defined is 33,3%, since each of them cover 200 records. Again the obtained rules have a support very close to the expected value. Moreover,

rule	sup(%)	conf(%)	#r
1 [17,32] [35,56] [32,46]	28.5	89.1	166
2 [1,16] [7,30] [0,21]	32.33	82.91	195
3 [10,25] [17,38] [13,35]	31.55	86.25	180

Table II. Obtained results with overlapping

good values of confidence are obtained. In spite that the rules are with overlapping, they cover almost all the records assigned to every one of them.

References

[1] R. Agrawal, T. Imielinski and A. Swami, "Mining Association Rules Between Sets of Items in Large Databases", *Proc. of the ACM SIGMOD Conference on Management of Data*, pp. 207-216, Washington, D.C., 1993.

[2] R. Agrawal and R. Srikant, "Fast Algorithms for Mining Association Rules", *Proc. of the VLDB Conference*, pp. 487-489, Santiago (Chile), 1994.

[3] T. Fukuda, Y. Morimoto, S. Morishita and T. Tokuyama, "Mining Optimized Association Rules for Numeric Attributes", *Proc. of the ACM SIGACT-SIGMOD-SIGART Symposium on Principles of Databases Systems*, 1996.

[4] D.E. Goldberg, "Genetic Algorithms in Search, Optimization and Machine Learning", *Addison-Wesley*, New York.

[5] A. Gonzlez and F. Herrera, "Multi-stage Genetic Fuzzy System Based on the Iterative Rule Learning Approach", *Mathware & Soft Computing*, 4, 233-249.

[6] G. Piatestsky-Shapiro, "Discovery, Analysis and Presentation of Strong Rules", *Knowledge Discovery in Databases*, AAAI/MIT Press, 1991.

[7] R. Rastogi and K. Shim "Mining Optimized Association Rules for Categorical and Numeric Attributes", *Int'l Conference on Data Engineering*, Orlando, 1998.

[8] R. Rastogi and K. Shim "Mining Optimized Support Rules for Numeric Attributes", *Int'l Conference on Data Engineering*, Sydney, Australia, 1999.

[9] R. Srikant and R. Agrawal "Mining Quantitative Association Rules in Large Relational Tables", *Proc. of the ACM SIGMOD Conference on Management of Data*, 1996.

Takeover Times of Noisy Non-Generational Selection Rules that Undo Extinction[1]

Günter Rudolph*

*Department of Computer Science, University of Dortmund, Germany

Abstract

The takeover time of some selection method is the expected number of iterations of this selection method until the entire population consists of copies of the best individual under the assumption that the initial population consists of a single copy of the best individual. We consider a class of non-generational selection rules that run the risk of loosing all copies of the best individual with positive probability. Since the notion of a takeover time is meaningless in this case these selection rules are modified in that they undo the last selection operation if the best individual gets extinct from the population. We derive exact results or upper bounds for the takeover time for three commonly used selection rules via a random walk or Markov chain model. The takeover time for each of these three selection rules is $O(n \log n)$ with population size n.

1 Introduction

The notion of the *takeover time* of selection methods used in evolutionary algorithms was introduced by Goldberg and Deb [1]. Suppose that a finite population of size n consists of a single best individual and $n-1$ worse ones. The takeover time of some selection method is the expected number of iterations of the selection method until the entire population consists of copies of the best individual. Evidently, this definition of the takeover time becomes meaningless if all best individuals may get extinct with positive probability. Therefore we study a specific modification of those selection rules: If the all best individual have been erased by erroneous selection then these selection rules undo this extinction by reversing the last selection operation. Here, we concentrate on non-generational selection rules. For such rules Smith and Vavak [2] numerically determined the takeover time or takeover probability based on a Markovian model whereas Rudolph [3] offered a theoretical analysis via the same Markovian model. This work is an extension of [2, 3] as the modified selection rules introduced here have not been considered yet.

Section 2 introduces the particular random walk model, which reflects our assumptions regarding the selection rules, and our standard machinery for determining the takeover time or bounds thereof. Section 3 is of preparatory nature as it contains several auxiliary results required in section 4 in which our standard machinery is engaged to provide the takeover times for our modifications of random replacement selection, noisy binary tournament selection, and "kill tournament" selection. Finally, section 5 relates our findings to results previously obtained for other selection methods.

2 Model

Let N_t denote the number of copies of the best individual at step $t \geq 0$. The random sequence $(N_t)_{t\geq 0}$ with values in $S = \{1, 2, \ldots, n\}$ and $N_0 = 1$ is termed a Markov chain if

$$\mathsf{P}\{ N_{t+1} = j \mid N_t = i, N_{t-1} = i_{t-1}, \ldots, N_0 = i_0\} = \mathsf{P}\{ N_{t+1} = j \mid N_t = i\} = p_{ij}$$

for all $t \geq 0$ and for all pairs $(i,j) \in S \times S$. Since we are only interested in non-generational selection rules the associated Markov chains reduce to particular random walks that are amenable to a theoretical analysis. These random walks are characterized by the fact that $|N_t - N_{t+1}| \leq 1$ for all $t \geq 0$ as a non-generational selection rule chooses—somehow—an individual from the population and decides—somehow—which individual should be replaced by the previously chosen one.

Two special classes of random walks were considered in [3] in this context. Here, we need another class reflecting our assumption that the selection rules undo a potential extinction of the best individual by reversing the last selection operation. This leads to a random walk with one reflecting and one absorbing boundary which is a Markov chain with state space

[1]This work was supported by the Deutsche Forschungsgemeinschaft (DFG) as part of the *Collaborative Research Center "Computational Intelligence"*

$S = \{1, \ldots, n\}$ and transition matrix

$$P = \begin{pmatrix} r_1 & q_1 & 0 & \cdots & & & 0 \\ p_2 & r_2 & q_2 & 0 & \cdots & & 0 \\ 0 & p_3 & r_3 & q_3 & 0 & \cdots & 0 \\ \vdots & \ddots & \ddots & \ddots & \ddots & \ddots & \vdots \\ 0 & \cdots & 0 & p_{n-2} & r_{n-2} & q_{n-2} & 0 \\ 0 & \cdots & & 0 & p_{n-1} & r_{n-1} & q_{n-1} \\ 0 & \cdots & & & 0 & 0 & 1 \end{pmatrix}$$

with $p_i, q_i > 0$, $r_i \geq 0$, $p_i + r_i + q_i = 1$ for $i = 2, \ldots, n-1$ and $r_1 = 1 - q_1 \in (0,1)$. Notice that state n is the only absorbing state. The expected absorption time is $\mathsf{E}[T \mid N_0 = k]$ with $T = \min\{t \geq 0 : N_t = n\}$ and it can be determined as follows [4]. Let matrix Q result from matrix P by deleting its last row and column. If C is the inverse of matrix $A = I - Q$ with unit matrix I, then $\mathsf{E}[T \mid N_0 = k] = c_{k1} + c_{k2} + \cdots + c_{k,n-1}$ for $1 \leq k < n$. Since $N_0 = 1$ in the scenario considered here, we only need the first row of matrix $C = A^{-1}$ which may be obtained via the adjugate of matrix A. This avenue was followed in Rudolph [5] for a more general situation. Using the result obtained in [5] (by setting $p_1 = 0$) we immediately get

$$c_{1j} = \frac{\displaystyle\sum_{k=0}^{n-j-1} \left(\prod_{u=j+1}^{n-k-1} p_u \right) \left(\prod_{v=n-k}^{n-1} q_u \right)}{\displaystyle\prod_{k=j}^{n-1} q_k} \tag{1}$$

for $1 \leq j \leq n-1$. Thus, the plan is as follows: First, derive the transition probabilities for a non-generational selection rule that fulfills our assumptions. This is usually easy. Next, these expressions are fed into equation (1) yielding c_{1j}. The result may be a complicated formula; in this case it will be bounded in an appropriate manner. Finally, we determine the sum $\mathsf{E}[T \mid N_0 = 1] = \sum_{j=1}^{n-1} c_{1j}$ and we are done. For the sake of notational convenience we shall omit the conditioning $\{N_0 = 1\}$ and write simply $\mathsf{E}[T]$ for the expected takeover time.

3 Mathematical Prelude

In case of positive integers the Gamma function $\Gamma(\cdot)$ obeys the relationships $n\,\Gamma(n) = \Gamma(n+1) = n!$. For later purposes we need the following results:

Lemma 1 For $n \geq 1$,

$$\sum_{k=0}^{n-1} \frac{\Gamma(n+k+1)}{\Gamma(k+1)} = \frac{\Gamma(2n+1)}{(n+1)\,\Gamma(n)}.$$

Proof: See [3], p. 905. ∎

Lemma 2 Let $n \geq 2$ and $1 \leq j \leq n-1$. Then

$$S(n,j) = \frac{n^2\,\Gamma(n-j)\,\Gamma(n+j)}{\Gamma(j+1)\,\Gamma(2n-j+1)} \sum_{k=0}^{n-j-1} d_k \leq \frac{1}{2} + \frac{1}{4n}$$

where

$$d_k = \frac{\Gamma(n+k+1)\,\Gamma(n-k)}{\Gamma(2n-k)\,\Gamma(k+1)}.$$

Proof: Due to lack of space we only offer a sketch of the proof. First show that $S(n,0) \geq S(n,j)$ for $j = 1, \ldots, n-1$ and $n \geq 2$. Since the bound

$$2\,S(n,0) = n\,\frac{\Gamma(n)^2}{\Gamma(2n)} \sum_{k=0}^{n-1} d_k \leq 1 + \frac{1}{2n}$$

follows from [3], pp. 907-908, division by 2 yields the result desired. ∎

Moreover, the nth harmonic number H_n can be bracketed by $\log n < H_n = \sum_{i=1}^{n} \frac{1}{i} < \log n + 1$ for $n \geq 2$ and notice that

$$\sum_{i=0}^{n} a^{n-i}\,b^i = \frac{a^{n+1} - b^{n+1}}{a - b}$$

for $a \neq b$. Finally some notation: The set I_m^n denotes all integers between m and n (inclusive).

4 Analysis

4.1 Random replacement selection

Two individuals are drawn at random and the better one of the pair replaces a randomly chosen individual from the population. If the last best individual was erased by chance then the last selection operation is reversed. As a consequence, the transition probabilities of the associated Markov chain are $p_{nn} = 1$, $p_{11} = 1 - p_{12}$,

$$\forall i \in I_1^{n-1} : \; p_{i,i+1} = \frac{i}{n}\left(2 - \frac{i}{n}\right)\left(1 - \frac{i}{n}\right)$$

$$\forall i \in I_2^{n-1} : \; p_{i,i-1} = \left(1 - \frac{i}{n}\right)^2 \frac{i}{n}$$

and $p_{ii} = 1 - p_{i,i-1} - p_{i,i+1}$. Since $p_i = p_{i,i-1}$, $q_i = p_{i,i+1}$ and

$$\prod_{v=n-k}^{n-1} q_v = \frac{1}{n^{3k+1}} \frac{\Gamma(n+k+1)\,\Gamma(k+1)}{\Gamma(n-k)} \tag{2}$$

$$\prod_{u=j+1}^{n-k-1} p_u = \frac{1}{n^{3(n-k-j-1)}} \frac{\Gamma(n-j)^2}{\Gamma(j+1)} \frac{\Gamma(n-k)}{\Gamma(k+1)^2}$$

one obtains the numerator of eqn. (1) via

$$\frac{1}{n^{3(n-j-1)+1}}\,\frac{\Gamma(n-j)^2}{\Gamma(j+1)}\sum_{k=0}^{n-j-1}\frac{\Gamma(n+k+1)}{\Gamma(k+1)} =$$

$$\frac{1}{n^{3(n-j-1)+1}}\,\frac{\Gamma(n-j)}{\Gamma(j+1)}\,\frac{\Gamma(2n-j+1)}{n+1} \quad (3)$$

with the help of Lemma 1. Insertion of $k = n - j$ in equation (2) leads to

$$\prod_{v=j}^{n-1} q_v = \frac{\Gamma(2n-j+1)\,\Gamma(n-j+1)}{n^{3(n-j)+1}\,\Gamma(j)}. \quad (4)$$

After insertion of equations (3) and (4) in equation (1) we have

$$c_{1j} = \frac{n^3}{n+1}\cdot\frac{1}{j}\cdot\frac{1}{n-j} = \frac{n^2}{n+1}\left(\frac{1}{j}+\frac{1}{n-j}\right)$$

and finally $\mathsf{E}[\,T\,] = \sum_{j=1}^{n-1} c_{1j} = \frac{2n^2}{n+1} H_{n-1}$.

4.2 Noisy binary tournament selection

Two individuals are drawn at random and the best as well as worst member of this sample is identified. The worst member replaces the best one with some replacement error probability $\alpha \in (0, \frac{1}{2})$, whereas the worst one is replaced by the best one with probability $1-\alpha$. Again, if the last best copy has been discarded then the last selection operation is reversed. Therefore the transition probabilities are as follows: $p_{nn} = 1$, $p_{12} = s_1(1-\alpha)$, $p_{11} = 1 - p_{12}$ and $p_{i,i+1} = s_i(1-\alpha)$, $p_{i,i-1} = s_i\,\alpha$, $p_{ii} = 1 - s_i$ for $i = 2,\ldots,n-1$. Here, s_i denotes the probability that the sample of two individuals contains at least one best as well as one worse individual from a population with $i = 1,\ldots,n-1$ copies of the best individual, i.e.,

$$s_i = 1 - \left(\frac{i}{n}\right)^2 - \left(1-\frac{i}{n}\right)^2 = 2\,\frac{i}{n}\left(1-\frac{i}{n}\right).$$

According to equation (1) we need

$$\prod_{v=n-k}^{n-1} q_v = \left(\frac{2(1-\alpha)}{n^2}\right)^k \frac{\Gamma(n)\,\Gamma(k+1)}{\Gamma(n-k)}$$

$$\prod_{u=j+1}^{n-k-1} p_u = \left(\frac{2\alpha}{n^2}\right)^{2(n-j-1-k)} \frac{\Gamma(n-k)\Gamma(n-j)}{\Gamma(k+1)\Gamma(j+1)}$$

leading to the numerator of eqn. (1) via

$$\frac{2^{n-j-1}}{n^{2(n-j-1)}}\cdot\frac{\Gamma(n)\,\Gamma(n-j)}{\Gamma(j+1)}\sum_{k=0}^{n-j-1}\alpha^{n-j-1-k}(1-\alpha)^k =$$

$$\frac{2^{n-j-1}}{n^{2(n-j-1)}}\cdot\frac{\Gamma(n)\,\Gamma(n-j)}{\Gamma(j+1)}\cdot\frac{(1-\alpha)^{n-j}-\alpha^{n-j}}{1-2\alpha}. \quad (5)$$

Since

$$\prod_{v=j}^{n-1} q_v = \left(\frac{2(1-\alpha)}{n^2}\right)^{n-j}\cdot\frac{\Gamma(n)\,\Gamma(n-j+1)}{\Gamma(j)} \quad (6)$$

we get by inserting equations (5) and (6) into equation (1)

$$c_{1j} = \frac{n^2}{2}\,\frac{\Gamma(j)}{\Gamma(j+1)}\cdot\frac{\Gamma(n-j)}{\Gamma(n-j+1)}\cdot\frac{(1-\alpha)^{n-j}-\alpha^{n-j}}{(1-\alpha)^{n-j}(1-2\alpha)}$$

$$= \frac{n^2}{2}\cdot\frac{1}{j}\cdot\frac{1}{n-j}\cdot\frac{1}{1-2\alpha}\cdot(1-r^{n-j})$$

where $r = \alpha/(1-\alpha)$ and finally $\mathsf{E}[\,T\,] = \sum_{j=1}^{n-1} c_{1j} =$

$$\frac{n^2}{2}\cdot\frac{1}{1-2\alpha}\sum_{j=1}^{n-1}\frac{1}{j}\cdot\frac{1}{n-j}\cdot\left[1-\left(\frac{\alpha}{1-\alpha}\right)^{n-j}\right] =$$

$$\frac{n}{2(1-2\alpha)}\sum_{j=1}^{n-1}\left[\frac{1}{j}+\frac{1}{n-j}\right]\cdot\left[1-\left(\frac{\alpha}{1-\alpha}\right)^{n-j}\right] =$$

$$\frac{n}{1-2\alpha}\left[H_{n-1} - \frac{1}{2}\sum_{j=1}^{n-1}\frac{r^{n-j}}{j} - \frac{1}{2}\sum_{j=1}^{n-1}\frac{r^{n-j}}{n-j}\right] \le$$

$$\frac{n}{1-2\alpha}\left[H_{n-1} - \frac{1}{n}\sum_{j=1}^{n-1} r^j\right] =$$

$$\frac{n}{1-2\alpha}\left[H_{n-1} - \frac{1}{n}\cdot\frac{r-r^n}{1-r}\right] \le \frac{n\,H_{n-1}}{1-2\alpha}$$

where $r = \alpha/(1-\alpha) \in (0,1)$. The above bound is very accurate if α is not too close to $1/2$. For example, for $\alpha = \frac{1}{2} - \frac{1}{2n^k}$ this bound yields $\mathsf{E}[\,T\,] \le n^{k+1} H_{n-1}$ for $k \ge 0$ whereas the worst case ($\alpha = 1/2$) reveals[1] that $\mathsf{E}[\,T\,] \le n^2 H_{n-1}$ for $\alpha \in [0, 1/2]$. Moreover, notice that we get $\mathsf{E}[\,T\,] = n\,H_{n-1}$ in the best case ($\alpha = 0$) [3].

4.3 "Kill Tournament" selection

This selection method proposed in [2] is based on two binary tournaments: In the first tournament the best individual is identified. This individual replaces the worst individual identified in the second tournament (the "kill tournament"). If the last best

[1]If $\alpha = 1/2$ then the entire derivation collapses to simple expressions leading to $\mathsf{E}[\,T\,] = n^2 H_{n-1}$.

copy gets lost then the last selection operation is reversed. The transition probabilities are $p_{nn} = 1$, $p_{11} = 1 - p_{12}$,

$$\forall i \in I_1^{n-1} : p_{i,i+1} = \frac{i}{n}\left(2 - \frac{i}{n}\right)\left[1 - \left(\frac{i}{n}\right)^2\right]$$

$$\forall i \in I_2^{n-1} : p_{i,i-1} = \left(1 - \frac{i}{n}\right)^2 \left(\frac{i}{n}\right)^2$$

and $p_{ii} = 1 - p_{i,i-1} - p_{i,i+1}$. As in the previous cases we require the products

$$\prod_{v=n-k}^{n-1} q_v = \frac{\Gamma(n+k+1)\,\Gamma(k+1)\,\Gamma(2n)}{n^{4k+1}\Gamma(n-k)\,\Gamma(2n-k)}$$

$$\prod_{u=j+1}^{n-k-1} p_u = \frac{1}{n^{4(n-j-1-k)}}\,\frac{\Gamma(n-k)^2\,\Gamma(n-j)^2}{\Gamma(j+1)^2\,\Gamma(k+1)^2}$$

to evaluate the numerator of equation (1) via

$$\frac{\Gamma(n-j)^2\,\Gamma(2n)}{n^{4(n-j-1)+1}\,\Gamma(j+1)^2} \sum_{k=0}^{n-j-1} d_k \tag{7}$$

with

$$d_k = \frac{\Gamma(n+k+1)\,\Gamma(n-k)}{\Gamma(2n-k)\,\Gamma(k+1)}\,.$$

Since

$$\prod_{v=j}^{n-1} q_v = \frac{\Gamma(2n-j+1)\,\Gamma(n-j+1)\,\Gamma(2n)}{n^{4(n-j)+1}\,\Gamma(j)\,\Gamma(n+j)} \tag{8}$$

insertion of (7) and (8) in (1) yields

$$\begin{aligned} c_{1j} &= \frac{n^4}{j\,(n-j)}\,\frac{\Gamma(n-j)\,\Gamma(n+j)}{\Gamma(j+1)\,\Gamma(2n-j+1)} \sum_{k=0}^{n-j-1} d_k \\ &\le \frac{n^2}{j\,(n-j)} \cdot \left(\frac{1}{2} + \frac{1}{4n}\right) \qquad (9) \\ &= \left(\frac{1}{j} + \frac{1}{n-j}\right)\left(\frac{n}{2} + \frac{1}{4}\right) \end{aligned}$$

and finally

$$\mathsf{E}[T] = \sum_{j=1}^{n-1} c_{1j} \le \left(n + \frac{1}{2}\right) H_{n-1}\,.$$

Here, the inequality in (9) follows from Lemma 2.

5 Summary

Now we are in the position to compare the takeover times of the selection methods considered here with those examined in [3]. Table I offers an overview of the takeover times of replace worst selection (RW), quaternary (QT), ternary (TT) and binary (BT) tournament selection with $\alpha = 0$ [3], and kill tournament "with undoing" (KT_u), random replacement selection "with undoing" (RR_u) and noisy binary tournament selection "with undoing" and replacement error α ($BT_u(\alpha)$). For fixed $\alpha < 1/2$ the takeover times of all non-generational selection rules considered here and in [3] are of order $O(n \log n)$. Consequently, it does not matter which selection rule is used, provided that the takeover time is actually a key figure of the selection pressure.

selection method	takeover time
QT	$\le \frac{1}{2}\, n\, H_{n-1}$
RW	$\le \frac{1}{2}\, n\, H_{2n-1}$
TT	$\frac{2}{3}\, n\, H_{n-1}$
BT	$n\, H_{n-1}$
KT_u	$\le (n + \frac{1}{2})\, H_{n-1}$
$BT_u(\frac{1}{4})$	$\le 2\, n\, H_{n-1}$
RR_u	$\frac{2n^2}{n+1}\, H_{n-1}$
$BT_u(\frac{1}{2})$	$n^2\, H_{n-1}$

Table I. Survey of takeover times.

References

[1] D. E. Goldberg and K. Deb, "A comparative analysis of selection schemes used in genetic algorithms," in *Foundations of Genetic Algorithms* (G. J. E. Rawlins, ed.), pp. 69–93, San Mateo (CA): Morgan Kaufmann, 1991.

[2] J. Smith and F. Vavak, "Replacement strategies in steady state genetic algorithms: Static environments," in *Foundations of Genetic Algorithms 5* (W. Banzhaf and C. Reeves, eds.), pp. 219–233, San Francisco (CA): Morgan Kaufmann, 1999.

[3] G. Rudolph, "Takeover times and probabilities of non-generational selection rules," in *Proceedings of the Genetic and Evolutionary Computation Conference (GECCO 2000)* (D. Whitley *et al.*, eds.), pp. 903–910, San Fransisco (CA): Morgan Kaufmann, 2000.

[4] M. Iosifescu, *Finite Markov Processes and Their Applications.* Chichester: Wiley, 1980.

[5] G. Rudolph, "The fundamental matrix of the general random walk with absorbing boundaries," Technical Report CI-75 of the Collaborative Research Center "Computational Intelligence", University of Dortmund, October 1999.

Evolutionary Algorithms Aided by Sensitivity Information

Tadeusz Burczynski*† Piotr Orantek†

*Department for Strength of Materials and Computational Mechanics, Silesian University of Technology, Poland † Institute of Computer Modelling, Cracow University of Technology Poland, E-mail: burczyns@zeus.polsl.gliwice.pl and orantek@rmt4.kmt.polsl.gliwice.pl

Abstract

The paper deals with of coupling of the evolutionary and gradient algorithms. The detailed description of an approach based on an idea of a gradient mutation is presented. Three kinds of gradient mutation operators are proposed: (i) a steepest descent mutation operator, (ii) a conjugate gradient mutation operator and (iii) a variable metric mutation operator. Gradient mutation operators make use of information about sensitivity of the fitness function. An artificial neural network is proposed to control of working the gradient mutation. Several tests and practical examples of optimization and identification are examined. The proposed hybrid evolutionary algorithm is considerably more efficient than the evolutionary algorithm and its application makes then results more accurate.

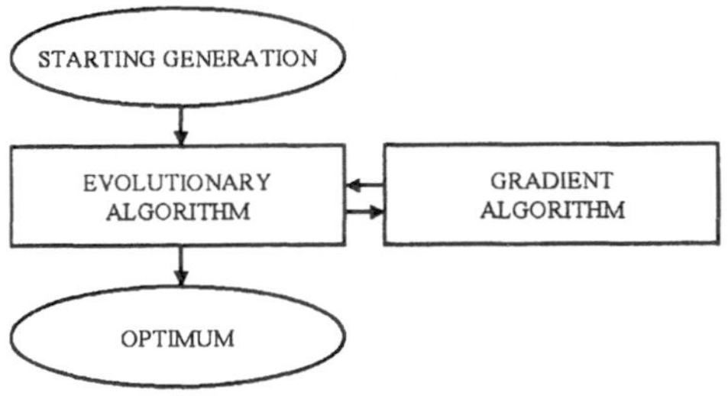

Fig. 1. Model of coupling of evolutionary and gradient algorithms

1 Introduction

The most common and widely used methods of optimization are classical gradient methods. The term - 'gradient methods' means here a family of optimization algorithms in which the direction of looking for optimum depends on an objective function gradient, e.g. a steepest descent method, a conjugate gradient method, a variable metric method. These methods are characterized by a very fast reaching the optimum, yet, they have a considerable disadvantage - converging the algorithm to the closest optimum. So you never know whether the result of the optimization is a local or a global optimum. Contrary to the classical methods the evolutionary algorithms (Goldberg, 1989; Michalewicz, 1996) cope with the global optimum searching quite well, but they are not as precise as the classical methods. Another disadvantage of the evolutionary algorithm is a big (sometimes very big) amount of computations. Until quite lately these two kinds of algorithms have been used separately, and if they were used competently the results were quite satisfactory. However, the more complex problems, the more evident disadvantages of both algorithms. An alternative is to coupling of evolutionary and gradient algorithms, what can give very good and interesting results provided their synthesis is competent. Such a algorithm is assumed to have advantages of both components and not to have their disadvantages. The present paper presents a suggestion of connecting the algorithms into one hybrid algorithm and contains the test results, which allow for comparison the evolutionary algorithm with the hybrid algorithm, which based a gradient mutation operator. Having the knowledge of advantages and disadvantages of the evolutionary and gradient algorithms a method which connects the above algorithms together with their advantages is created. The new hybrid evolutionary algorithm can be considered as a method, which belongs to family of methods of artificial intelligence.

2 Gradient Mutation

The approach (Fig.1) is based on the evolutionary algorithm, which has been enlarged with a new operator - a gradient mutation. The gradient mutation operator changes any chromosome on the ground of the gradient of a fitness function f. Consider a chromosome $\mathbf{C}h(k)$:

$$\mathbf{C}h(k) \equiv x(k) =< x_1(k), x_2(k), \ldots, x_n(k) > \quad (1)$$

where genes: $x_i, i = 1, 2, \ldots, n$, are real numbers. The chromosome after the gradient mutation takes the form:

$$\mathbf{C}h(k+1) =< x_1(k+1), \ldots, x_n(k+1) > \quad (2)$$

New genes are calculated as follows:

$$x_i(k+1) = \begin{cases} x_i(k) & \text{for } l{=}0 \\ x_i(k) + \Delta x_i(k) & \text{for } l{=}1 \end{cases} \quad (3)$$

where: $k = 0, 1, 2, \ldots, M$ - is the number of mutation, l - a random value which takes value $l = 0$ or $l = 1$. Correction values $\Delta x_i, i = 1, 2, \ldots, n$ create a vector Δx which takes form:

$$\Delta x(k) \equiv< \Delta x_1, \ldots \Delta x_n >= \alpha(k)\xi(x(k)) \quad (4)$$

where: α - a coefficient describing a step size towards ξ, ξ - a vector describing a direction of searching. The vector ξ can be constructed on the ground of our knowledge about a gradient and a hessian. In the hybrid algorithm three kinds of gradient mutation have been proposed. The formulas below describe how to determine the searching direction for different types of mutation:

- for the steepest descent gradient mutation:

  $$\xi(k) = -G(k) \quad (5)$$

  where: $G(k) = \nabla f(x(k))$- objective function gradient computed for the individual $\mathbf{C}h(k)$.

- for the conjugate gradient mutation:

  $$\xi(k) = -G(k) + \beta\xi(k-1) \quad (6)$$

  where: β - coefficient describing the influence of the previous gradient, $\xi(k)$, $\xi(k-1)$ - conjugate directions.

- for the variable metric gradient mutation:

  $$\xi(k) = -D(k)G(k) \quad (7)$$

  where: $D(k)$ - a hessian inverse matrix approximation computed on the base of the gradient:

  $$D(k) \approx \frac{1}{\eta}(\nabla^2 f)^{-1} \quad (8)$$

  where η- a scale ratio. The matrix $D(k)$ is updated according to the formula:

  $$D(k) = D(k-1) + \Delta D(k-1) \quad (9)$$

  where: $\Delta D(k-1)$- a correction matrix. For the particular case when we use the Davidon-Fletcher-Powell algorithm (D-F-P), $\eta = 1$, and:

  $$\Delta D(k-1) = \frac{c(k-1)c(k-1)^T}{c(k-1)^T d(k-1)} - \frac{D(k-1)d(k-1)d(k-1)^T D(k-1)^T}{d(k-1)^T D(k-1)^T d(k-1)} \quad (10)$$

  where:

  $$c(k-1) = x(k) - x(k-1) \quad (11)$$

  and:

  $$d(k-1) = G(k) - G(k-1) \quad (12)$$

Tests on such an algorithm gave very interesting and positive results. The algorithm quickly reaches the global optimum owing to the gradient operator. If on a certain stage of computations a point is generated near the global optimum and has a high value of the objective function, the gradient mutation will move the point towards the optimum in a few iterations. A turning point in this algorithm is generation of such a strong point. A traditional evolutionary algorithm needs about computation time for finding the optimum region, and after that it tunes in to the system. The hybrid algorithm can find the global optimum already at the beginning of the first stage. Moreover, the second stage is unnecessary. The above statements are confirmed by the test results presented in section 3. This type of the gradient mutation is one of the many possibilities. In numerical examples shown below all the genes were changed. It is possible that changing some of the genes better results could be obtained. Changing some of the genes is more similar to the classical mutation operator, but this approach maybe loose some information of the gradient. The difference between both approaches is a subject of new investigations. The next problem is how to control the number of the gradient mutations. At first the best chromosome was changed once or a few times per generation (Burczynski and Orantek, 1999). The number of changes was constant. A new approach which has been implemented controls the number of gradient mutations using fitness function value of the best chromosomes in a few last generations. This problem can be executed using different numerical techniques. An artificial neural network has been proposed to control the gradient mutation. The neural network (Fig. 2) has a three layers:

- the input layer - has ten neurons which have the identity active function,

- the hidden layer - has a four neurons which have the sigmoidal active function,
- the output layer - has one neuron which the output value is a result of acting the whole neural network.

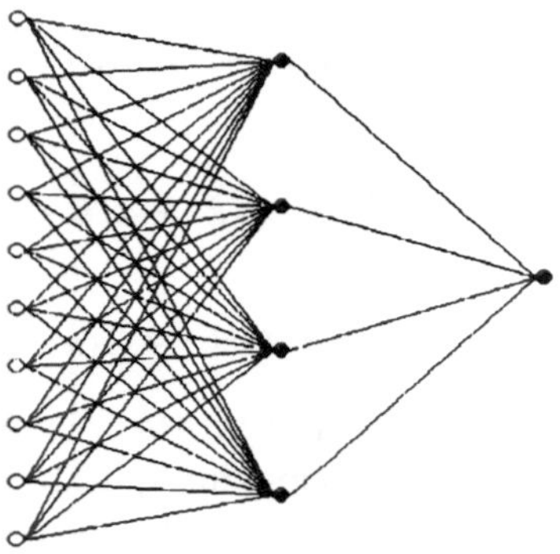

Fig. 2. Neural network which controls a gradient mutation

The neural network was trained by the 60 vectors and for the 60 vectors was verified. The procedure of training by the back propagation method with momentum for 2500 iterations was applied. The output result of this neural network is a real value from a interval [0,1]. The result was multiplied by a maximum number of mutations and rounded to an integral value. This value determines the number of gradient mutations. By using this type of the neural network the number of gradient mutations depends on variations of the fitness function. The number of gradient mutations was increased when the value of fitness function for a few generations was constant. When the value of the fitness functions for the best chromosome is constant during a few generations then the number of gradient mutation increases. When the value of the fitness function is increasing then the number of gradient mutation decreases.

3 Test

To test the hybrid algorithm efficiency the fitness function of the best individual from the successive populations in the whole computation process has been compared for the evolution algorithm and for the hybrid algorithm. The test below presents operation of three kinds of hybrid algorithm with the use of the following symbols:

- HEA/SDM - hybrid evolutionary algorithm using the Steepest Descent Mutation;
- HEA/CGM - hybrid evolutionary algorithm using the Conjugate Gradient Mutation;
- HEA/VMM - hybrid evolutionary algorithm using the Variable Metric Mutation.

The any parameters of the evolutionary algorithm (EA) and hybrid evolutionary algorithm (HEA) were constant for all tests. The algorithms made use of arithmetic, simple and heuristic crossover operators and boundary, uniform and non-uniform mutations. Probabilities for the all kind of the crossovers were $p_c = 0.25$, while for all kind of the mutations $p_m = 0.05$. The selection was done by a roulette wheel method. The test was performed to compare the evolution algorithm (EA) to every hybrid algorithm. The function whose maximum was looked for is described with the formula:

$$f(\mathbf{x}^*) = \sum_{i=1}^{n} \{[25 - (x_i - 5)^2] \cdot \cos[2 \cdot (x_i - 5)]\} \quad (13)$$

where: n - number of variables. 20 variables were assumed for the task. Maximum of the function was 500. Constraints for each variable were the following:

$$0 \leq x_i \leq 10 \qquad i := 1, 2, \ldots, 20; \quad (14)$$

The population consisted of 500 individuals and 200 generations were assumed. Three series of computations were performed. The evolution algorithm and every hybrid algorithm were computed in each series for the same starting population. Frequency of the gradient operator occurrence in the algorithms was 3; the operator only changed the best individual in the population. After that the results were averaged. The expression describing the gradient vector for any individual was calculated analytically. The gradient value was received by substitution of variable values kept in the chromosome into the formulas derived earlier. The evolution algorithm came closer to the optimum, but it did not find it with a satisfactory accuracy. In every hybrid algorithm the gradient operator application helped to find the maximum. The searching results and the generation number in which the best point was found are presented in Table I. Several other tests have been performed. All tests show that evolutionary algorithms aided by sensitivity information are considerably more efficient than the evolutionary algorithm.

4 The Conclusions

The hybrid evolutionary algorithm has been presented in the paper. In general the hybrid algorithm is considerably more efficient than the evolu-

Trial	Number of generation in which the algorithm found the optimum			
	EA	HEA/ SDM-3	HEA/ CGM-3	HEA/ VMM-3
1	199	48	52	52
2	198	46	85	50
3	199	47	57	46
4	199	40	59	52
5	197	50	47	76
6	198	50	43	60
7	199	50	47	89
8	200	47	50	85
9	197	49	46	50
10	199	62	80	53
mean	**199**	**49**	**57**	**61**

Table I. Searching results for the test 1

tionary algorithm and its application makes the result more accurate. A special type of mutation, so called a gradient mutation, has been applied for the evolutionary algorithm. This mutation is characterized by an evolutionary interference, which means a modification of genes making use of information about the fitness function gradient. Different kinds of the gradient mutation operators can be suggested. The present paper suggests three operators: (i) a steepest descent operator, (ii) a conjugate gradient operator and (iii) a variable metric operator. All the operators make use of the information coming from the computation of the fitness function gradient for a chosen chromosome. In many problems the computation of the fitness function gradient can be quite difficult. It can be made easier when applying the sensitivity analysis methods. In fact the gradient mutation operators, which modify a evolutionary structure of a chromosome make use of the information about sensitivity of the fitness function. An open question is how to work out a general and rational strategy of gradient mutation operators application to get a solution in the fewest generations form. In the numerical tests described in the paper the gradient operators were used separately for the best chromosome in each generation. Yet, other strategies taking bigger frequencies of these operators functioning into account may be considered, or this frequency can be made dependant on the generation number.

References

[1] D.E. Goldberg, "Genetic Algorithms in Search", *Optimization and Machine Learning*, 1989.

[2] Z. Michalewicz, "Genetic Algorithms + Data Structures = Evolution Programs", *Sprinter Verlag*, 1996.

[3] T. Burczynski, P. Orantek, "Coupling of the Genetic and Gradient Algorithms", *Proc. 3rd Conference on Evolutionary Algorithms and Global Optimization*, (in Polish), Zloty Potok 1999.

Ants and Graph Coloring

John Shawe-Taylor *, Janez Žerovnik †

*Department of Computer Science, Royal Holloway, University of London, Egham, Surrey TW20 0EX, UK, Email: jst@dcs.rhbnc.ac.uk †Faculty of Mechanical Engineering, University of Maribor, Smetanova 17, SI-2000 Maribor, Slovenia and Department of Theoretical Computer Science, IMFM, Jadranska 19, SI-1111 Ljubljana, Slovenia. Email: janez.zerovnik@imfm.uni-lj.si

Abstract

Ants algorithm is an evolutionary search method for solving combinatorial optimization problems. In this note we propose a version of the algorithm for coloring a graph with a fixed number of colors.

1 Introduction

Many combinatorial optimization problems are NP-hard. It is generally believed that NP-hard problems cannot be solved to optimality within times which are polynomially bounded functions of input size. Therefore, there is much interest in heuristic algorithms which can find near-optimal solutions within reasonable running times. Among heuristics there are several general approaches also called metaheuristics. Some popular metaheuristics are inspired by analogies to statistical mechanics as, for example, simulated annealing, and some others are inspired by biological systems as, for example, genetic algorithms or, more general, evolutionary algorithms.

One of the most studied NP-hard problems is the graph coloring problem [13]. Graph coloring has numerous applications in scheduling and other practical problems including register allocation, the design and operation of flexible manufacturing systems frequency assignment etc. It is well-known that both the problem of deciding whether a given graph can be colored by k colors for $k > 2$ and the problem of computing the chromatic number are NP–hard.

A version of the ants algorithm for estimating the chromatic number of a graph was recently proposed in [6]. 'Ants can color graphs' was the message of the paper, but unfortunately it was shown later [21] that the implementation of the ants paradigm in [6] did not even outperform simple memoryless restarts algorithm. Here we give an alternative ants algorithm for coloring a graph with a given fixed number of colors k. Our algorithm differs from algorithm of [6] both in the procedure performed by a single ant and in the way how the information experienced is memorized. Speedup over restarts and over long runs on a set of benchmark instances is shown experimentally.

There are many other approaches to graph coloring. A lot of both experimental and theoretical work has been done on various graph coloring methods, for example see [9, 12, 13, 14, 17], and many more. However, the aim of this paper is not to outperform the best graph coloring algorithms. The goal of our experiment is to show that the proposed way of storing the information may be useful for speeding up the computation. It remains open to see what are the exact conditions at which the speedup effect can be expected.

The present authors however strongly believe that no metaheuristic can produce competitive results on any problem if the procedures it uses are not competitive. For the ants algorithm, this includes the single ants procedure and in particular the method for combining and memorizing the information obtained by past single ants runs.

The rest of the paper is organized as follows. First, we briefly recall the graph coloring problem. Then we overwiev the ants metaheuristics and its two main ingredients, the local search type procedure which is performed by each ant and the method for memorizing the information experienced by the ants. Finally, we give results of the computational experiments and conclusions.

2 Graph Coloring and the Main Idea

If G is a graph, we write $V(G)$ for its vertex set and $E(G)$ for its edge set. $E(G)$ is a set of unordered pairs $xy = \{x, y\}$ of distinct vertices of G.

A *proper k-coloring* of vertices of a graph G is a function f from $V(G)$ onto a set (of colors) X, $|X| = k$, such that $uv \in E(G)$ implies $f(u) \neq f(v)$.

The decision problem of k–coloring is: given graph G and an integer k on input, answer the question "is G k–colorable?".

The problem is NP–hard for general graphs [10], but also for many special classes of graphs including planar graphs, regular graphs etc. It is also known that there is no polynomial approximation algorithm unless P=NP [15].

The ideas behind the randomized graph coloring algorithm reported in [17] have been extended in a series of papers [22, 23, 18, 2, 3], which have both analyzed and improved the heuristics. In [2] the energy landscape for Mean Field Annealing is altered by adapting weights on the graph edges. The paper [20] investigates an extension of this idea by introducing additional edges into the graph. This may seem a counter-intuitive strategy as it might appear to make the graph coloring problem more difficult. The idea is based on the fact that when restricted to the class of graphs with lowest vertex degree greater than αn for arbitrary $\alpha > 0$, the decision problem of 3-colouring is polynomial [7]. It is also known that the edge adding is error free with high probability provided some rather rough conditions are satisfied [20].

Another way of viewing the strategy is in terms of the way that humans often attempt to solve problems by further constraining them based on probabilistic reasoning from known constraints, hence reducing the size of the search space, while hopefully not excluding all feasible solutions. In the ants context, it seems likely that two vertices must be colored differently if many ants have proposed such near optimal colorings.

On the other hand, there is some theoretical evidence that repeated trials of various random heuristics (including simulated annealing and mean field annealing) on average perform better than one long run [8]. However, at each restart we loose all information obtained by previous unsuccessful trials, which may have obtained reasonably good near optimal colorings. Therefore, a possible way of accelerating the restarts algorithm is to somehow accumulate the information obtained by the near optimal colorings obtained so far. The idea of using the information of previous local searches is not new. In a different way it is explored by heuristics such as tabu search [11], reactive tabu search [1], and genetic algorithms [9].

3 Ants Algorithm for the Graph k-Coloring Problem

The algorithm is based on observations of an ant colony in search of a nearby feeding source. Entomological studies have shown that an ant colony converges towards an optimal solution whereas each single ant is unable to solve the problem by itself within a reasonable amount of time. Colorni et al [4] first developed Ant System for tackling so–called Assignment Type Problems (ATPs). The idea can be briefly explained as follows: The solution is produced by a colony of ants. Each ant builds a feasible solution at each iteration, i.e. it performs one run of a relatively fast algorithm which produces a near optimal solution. Initially, the ants do not have any knowledge of the problem to be solved. At the end of each iteration the experience of every ant is memorized. This information is then used in the next iteration.

algorithm ANTS($G, nants, ncycles, T, M, T_w$)
$G' := G$
for i:=1 to *ncycles* **do begin**
  for j:=1 to *nants* **do (in parallel)**
    PW(G', T, M)
  update(matrix $E(G')$, T_w)
  if k coloring of G found **then** exit
end

4 Single Ants Procedure, PW

The algorithm we use is a generalization of the algorithm of Petford and Welsh [17]. This algorithm and some of its generalizations have proved to be reasonably good heuristics for coloring various types of graphs including random k-colorable graphs, DIMACS challenge graphs [13], frequency assignment "realistic" graphs, and others [17, 23, 18].

The algorithm used here differs slightly from previous implementation because it uses the transformed edges set $E(G')$ for computing the new colors and the original edge set $E(G)$ for computing the cost function.

algorithm PW(G, G', T, M)
color vertices randomly using colors $1, 2, , \ldots, k$;
while coloring not proper and iterations $< M$ **do**
  select a bad vertex v (randomly)
  assign a new color to v
end while

In the first line of the while loop, a bad vertex is selected uniformly at random among the bad vertices, i.e. vertices which are endpoints of some edge which violates a constraint. In the version used in this paper, the bad vertices are defined with respect to edges of G (and not G').

A new color is assigned at random. The new color is taken from the set $\{1 \ldots k\}$ by sampling according to the distribution defined as follows: The probability of color i to be chosen as a new color of vertex v is proportional to

$$exp(-S_i/T)$$

where S_i is the number of edges of G' with one endpoint at v and the other colored by color i.

T is a parameter of the algorithm, which may be called temperature because of the analogy to temperature of the simulated annealing algorithm.

If no k-coloring is found before a time limit of M iterations is reached, the algorithm stops. (An iteration is a call of the function which computes a new color.) If the algorithm stops without finding a proper coloring, the best assignment can be regarded as an approximate solution, approximating the cost function which counts the number of bad edges, i.e. edges with both endpoints colored by the same color.

Remark: there are two decisions which are of crucial importance for the behavior of the algorithm: to define the number of iterations M and to define a good value for parameter T. We have no complete answer to these two questions. The choices we made work well on instances tested, but in general fine tuning is probably needed as is the usual case with other heuristics of similar type such as simulated annealing, tabu search, genetic algorithms, etc. The idea of adapting the temperature is addressed in [19]. A heuristic for tuning an analogous parameter is used with success in [16].

In this paper we do not aim to precisely tune the parameter T, therefore we either use $T = 0.7$ which corresponds to the value implicitly used in [17, 23], or use a presumably good temperature based on previous experiments (reported elsewhere [18, 19]). More tuning could only improve the performance of the algorithm.

5 Memorizing the Experience, Update

In the algorithm described below we will make use of the following quantities which are defined for the various colorings generated. Each coloring c is assigned a weight $w_G(c)$ given by

$$w_G(c) = \exp\{-b(c)/T_w\},$$

where $b(c)$ is the number of edges in G that connect vertices colored with the same color by c, and T_w is a parameter which defines the particular weighting function. Note that $w(c) \leq 1$, while $w(c) = 1$ for proper colorings.

The complement of a graph $G = (V, E)$ is the graph

$$\bar{G} = (V, (V \times V) \setminus (E \cup \{(i,i) | i \in V\})).$$

For a set of colorings C, we define the evidence

	ANTS $nants = 25$	STANDARD RESTARTS	ONE RUN
le450_5a: Average degree: 25.40	2.7(10)	6.2(7)	9.0(7)
le450_5b: Average degree: 25.48	3.8(10)	2.5(1)	11.1(3)
le450_5c: Average degree: 43.57	2.5(10)	6.1(10)	0.9(1)
le450_5d: Average degree: 43.36	2.7(10)	5.0(10)	5.7(2)

Table I. Results of the Algorithms on the 5-colourable Leighton Graphs. The average number of iterations ($\times 10^6$) over successfull runs are given. In parenthesis is the number of successfull runs (out of 10). Time limit was 22.5×10^6.

for an nonedge $(x, y) \in \bar{G}$ to be

$$E_{G,C}((x,y)) = \sum_{c \in C, c(x) \neq c(y)} w_G(c).$$

We also define the evidence of a set of colorings as

$$E_G(C) = \sum_{c \in C} w_G(c).$$

At a call of procedure update(E, T_w), the maximal evidence over nonedges is computed and all edges with maximal evidence are added.

6 Computational Experiments

The algorithms were first run on a series of four 'Leighton' graphs which are 5 colorable and which were proposed as part of the DIMACS Challenge in graph coloring [13] as examples of graphs that are hard to color. In the first set of experiments the parameters T, M, and T_w were set to 0.7, $200n$ and 3.0, respectively. As a comparison we also ran the Petford Welsh algorithm in two regimes. In the first instance we performed repeated short runs of the algorithm with the same number of iterations as used in the ants algorithm. In the second regime one long run was performed. Note that in both cases the same fixed temperature was used. The numbers in Table I indicate the total number of vertex updates performed whether or not the color of the vertex was changed.

For the second set of experiments we took the 15 colorable Leighton graphs. Here, the temperature $T = 0.7$ did not give good results in any regime. We therefore did another experiment with $T = 0.3$. In this case, the graph le450_15c was again never colored and the graph le450_15c was colored only

	ANTS	ANTS2	ANTS3	STAND. REST.	ONE RUN
le450_15a: Average deg.: 36.30	0.5(2)	1.0(3)	2.0(8)	8.2(7)	1.0(10)
le450_15b: Average deg.: 36.31	1.2(4)	0.9(4)	1.6(10)	5.2(10)	2.0(10)

Table II. Results of the Algorithms on the 15-colourable Leighton Graphs.

in 3 out of 10 runs of the ants algorithm while the restarts and long runs were never successful. Better results were observed for the other two graphs. Because of a relatively poor performance of the ants algorithm in this case we also tested two more combinations of the parameters of ants algorithm: in the first case (ANTS2) we increased *nants* to 50 and reduced *ncycles* to 5, in the second case (ANTS3) we doubled M and reduced *ncycles* to 5. The results are given in Table II.

7 Conclusions

In this note we show that on some instances the proposed ants algorithm outperforms both standard restarts (i.e. "ants not cooperating") and long runs (i.e. "a long single ant search"). The results on the second set of instances were less satisfactory. It seems clear that in the case when the single ants procedure is unable to produce good approximate solutions with few bad edges, the information accumulated is not reliable and thus can not be expected to be useful. For very difficult graphs, fine tuning of the parameters of the single ants procedure is neccessary. On the other hand, we note that the parameters *nants* and *ncycles* were guessed here, mainly based on the values used elsewhere [6]. Heuristics with some arguments for particular choices of these parameters may be worth studying which opens an avenue for both experimental and theoretical research.

References

[1] R.Battiti and G.Tecchiolli, The Reactive Tabu Search, *ORSA Journal of Computing*, 6 (1994) 126–140.

[2] P. Burge and J. Shawe-Taylor, Adapting the Energy Landscape for MFA, *Journal of Artificial Neural Networks* 2 (1995) 449–454.

[3] P. Burge and J. Shawe-Taylor, Bitstream Neurons for Graph Colouring, *Journal of Artificial Neural Networks* 2 (1995) 443–448.

[4] A. Colorni, M. Dorigo and V. Maniezzo, Distributed optimization by ant colonies, in *Proceedings of the first European Conference on Artificial Life* (F.J. Varela and P. Bourgine, Eds.), (1991) pp.134-142, MIT Press/Bradford Books, Cambridge, Massachusetts.

[5] D. Costa, A. Hertz and O. Dubuis, Embedding of a sequential algorithm within an evolutionary algorithm for coloring problems in graphs, *Journal of Heuristics*, **1** (1995) 105–128.

[6] D. Costa and A. Hertz, Ants can colour graphs, *Journal of the Operational Research Society*, **48** (1997) 295-305.

[7] K. Edwards, The complexity of colouring problems on dense graphs, *Theoretical Computer Science*, 43 (1986) 337–343.

[8] A.G. Ferreira and J. Žerovnik, Bounding the probability of success of stochastic methods for global optimization, *Journal of computers & mathematics with applications*, 25 (1993) 1–8.

[9] C. Fleurent and J.A. Ferland, Genetic and hybrid algorithms for graph coloring, *Annals of Operations Research* 63 (1996) 437–461.

[10] M.R. Garey, D.S. Johnson, *Computers and Intractability*, Freeman, San Francisco 1979.

[11] F. Glover and M. Laguna, Tabu Search, Kluwer, Dordrecht, 1998.

[12] D.S. Johnson, C.R. Aragon, L.A. McGeoch and C. Schevon, Optimization by Simulated Annealing: An Experimental Evaluation; Part II, Graph Coloring and Number Partitioning, *Operations Research*, 39:378–406, 1991.

[13] D.S. Johnson and M. Trick, (eds.), Cliques, Coloring and Satisfiability: Second DIMACS Implementation Challenge, American Mathematical Society 1996. DIMACS colouring benchmarks available via ftp at `rutgers.dimacs.edu`

[14] F. Leighton, A graph coloring algorithm for large scheduling problems, *J. Res. National Bureau of Standards*, **84** (1979) 489-505.

[15] C. Lund and M. Yanakakis, On the Hardness of Approximating Minimization Problems, *Journal of the Association for Computing Machinery*, 41 (1994) 960–981.

[16] C.Morgenstern, Distributed Coloration Neighborhood Search, pp. 335–357 of ref. [13].

[17] A. Petford and D. Welsh, A Randomised 3-colouring Algorithm, *Discrete Mathematics*, 74 (1989) 253–261.

[18] J. Shawe-Taylor and J. Žerovnik, Analysis of the Mean Field Annealing Algorithm for Graph Colouring, *Journal of Artificial Neural Networks* 2 (1995) 329–340.

[19] J. Shawe-Taylor and J. Žerovnik, Adapting temperature for some randomized local search algorithms, Preprint series Dept. Math., University of Ljubljana, 36 (1998) no.614.

[20] B.Rising, J. Shawe-Taylor and J. Žerovnik, Graph Colouring by Maximal Evidence Edge Adding, sumitted.

[21] A.Vesel and J. Žerovnik, How well can ants color graphs?, *Journal of computing and Information Technology - CIT* 8 (2000) 131–136.

[22] J. Žerovnik, A randomised heuristical algorithm for estimating the chromatic number of a graph, *Information Processing Letters*, 33 (1989) 213–219.

[23] J. Žerovnik, A Randomized Algorithm for k–colorability, *Discrete Mathematics*, 131 (1994) 379–393.

Evolutionary Optimization of a Wavelet Classifier for the Categorization of Beat-to-Beat Variability Signals

H.A. Kestler *†, M. Höher †, G. Palm*

*Neural Information Processing, University of Ulm, †Medicine II – Cardiology, University Hospital Ulm, Germany

Abstract

The beat-to-beat variation of the QRS and ST-T signal was assessed in healthy volunteers and in patients with malignant tachyarrhythmias using a novel wavelet based classifier designed by an evolutionary algorithm. High-resolution ECGs were recorded in 51 healthy volunteers and in 44 CHD patients with inducible sustained VT. QRS and ST-T variability was analyzed in 250 sinus beats. In each patient a variability signal was created from the standard deviation of corresponding data points of all beats. The complete variability signal was used. Analysis of the whole variability signal with the wavelet classifier results in an improved diagnostic ability of beat-to-beat variability analysis.

1 Background

High-resolution electrocardiography is used for the detection of fractionated micropotentials, which serve as a non-invasive marker for an arrhythmogenic substrate and for an increased risk for malignant ventricular tachyarrhythmias. Beat–to–beat variation of cardiac excitation and depolarization has been associated with electrical instability and an increased risk for arrhythmias [1]. Rosenbaum et al. [2] have shown that increased beat–to–beat microvariations of the T–wave, although visually inapparent, are associated with a decreased arrhythmia–free survival. Their method to quantify periodic electrical alternans of the T–wave amplitude has gained growing clinical acceptance as a non–invasive, electrocardiographic risk marker. Earlier high–resolution electrocardiographic studies already demonstrated periodic and non-periodic behaviour of ventricular late potentials at the terminal QRS [3–6]. Previous work of our group showed a significantly higher beat–to–beat variation of the duration of the filtered QRS [7] and an increased total beat–to–beat microvolt variation of both the QRS and the ST-T segment [8] among patients with an increased risk for ventricular tachycardias. The aim of this study was to utilize the complete intra–QRS beat–to–beat signal variation and ST-T signal variation in a novel classifier adapting its preprocessing stage.

2 Evolutionary Wavelet Classifier

The classifier consists of a preprocessing unit, extracting the features, and a nearest neighbour classifier for categorization. The objective of the optimization is to find a good set of features maximizing the classification accuracy. Features are extracted by evaluating the continuous wavelet transform at specific scales and locations of the input signal. The basic idea is to use the classification accuracy as a measure of fitness in the evolutionary design process. The nearest neighbour classifier is used as fast means of calculating the fitness. In each generation the classifier is trained by choosing a fixed number of data points randomly. A general outline of an evolutionary algorithm is given in Figure 1. Here,

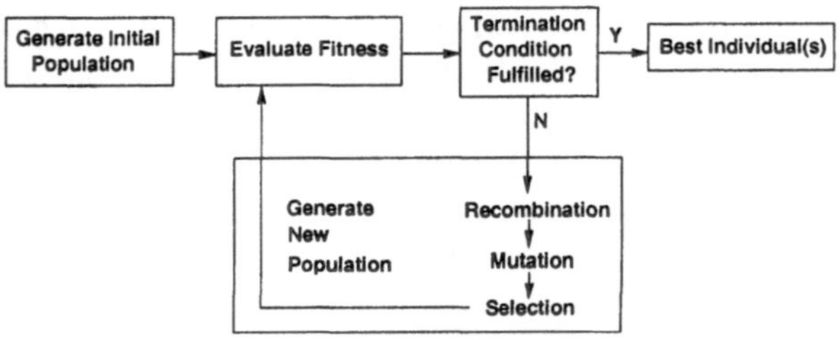

Fig. 1. Outline of an evolutionary algorithm.

we used only mutation as a probabilistic operator and restricted ourselves to the mexican hat wavelet.

We defined an individual I as a pointwise wavelet transformation:

$$W_{\psi,I}(f) = \begin{pmatrix} W_\psi(f)(a_1,b_1) \\ \vdots \\ W_\psi(f)(a_J,b_J) \end{pmatrix},$$

where a_i, b_i; $i = 1,\ldots,J$ denote the scaling and location parameters of the wavelet transform

$$W_\psi(f)(a_i,b_i) = \frac{1}{\sqrt{a_i}} \cdot \psi(\frac{t-b_i}{a_i}) * f,$$

where f is the input signal, $*$ denotes the convolution operation and $\psi(x) = (1 - x^2)e^{-x^2}$ is the mexican hat mother wavelet.

Classifier. There is a training set $F = \{(f^\mu, \omega^\mu) | \mu = 1, \ldots, N\}$ in which every input signal f^μ is labeled with its class membership $T(f^\mu) = \omega^\mu \in \{\omega_1, \ldots, \omega_l\}$, a classifier C which performs a mapping of a signal f to a class $C(f) \in \{\omega_1, \ldots, \omega_l\}$, and a finite subset $F^c \subset F$. Given an individual $I = ((a_1, b_1), \ldots, (a_J, b_J))$, the classification of a transformed input signal $W_{\psi,I}(f)$, denoted by $C(f)$, is determined by a nearest neighbour classification on the set $\{W_{\psi,I}(f^c) : f^c \in F^c\}$, so $C(f)$ is defined as the class of $f^* \in F^c$ where f^* is given by

$$f^* = \underset{f^c \in F^c}{\operatorname{argmin}} \{d(W_{\psi,I}(f), W_{\psi,I}(f^c))\}$$

here, $d(\cdot,\cdot)$ denotes the Euclidean distance.

Fitness. The fitness of an individual I is defined by the number of correct classified signals $f \in F \backslash F^c$:

$$\Phi(I) = \#\{f' \in F : T(f') = C(f')\}.$$

The relative fitness of an individual I_j of a population $P = \{I_j | j = 1, \ldots, K\}$ is

$$\phi(I_j) = K \frac{\Phi(I_j)}{\sum_{j'=1,\ldots,K} \Phi(I_{j'})}.$$

Selection and Reproduction. Offspring generation and selection was done by a semi–proportional assignment of the K individuals of the parent population to the K individuals of the offspring population. We defined $\phi_{int}(I) = \lfloor \phi(I) \rfloor$. Starting from the fittest individual to the one with the lowest fitness, every individual was identically reproduced $\phi_{int}(I) + 1$ times until the population size K was reached. If two or more individuals had the same fitness the one with the fewer number of wavelets J was assigned the higher rank.

Mutation. Apart from the object variables (here: a_i, b_i, and p as a size mutation parameter), the parameters of the optimization strategy (standard deviation of normally distributed mutations) were also incorporated into the search process [9]. For every individual there was one set of strategy parameters $\{\sigma_a, \sigma_b, \sigma_p\}$. Mutation was then as follows:

$$\begin{aligned} \sigma'_a &= \sigma_a e^{N(0,\tau_a)},\ \sigma'_b = \sigma_b e^{N(0,\tau_b)},\ \sigma'_p = \sigma_p e^{N(0,\tau_p)} \\ a'_i &= a_i + N(0,\sigma'_a), b'_i = b_i + N(0,\sigma'_b), \\ p' &= p + N(0,\sigma'_p). \end{aligned}$$

$N(0,\sigma)$ denotes the Gaussian distribution with mean 0 and standard deviation of σ. The size of the individual J is calculated as: $J = \lfloor p \rfloor$. If p exceeded the permitted range of wavelets it was set to predefined min or max values.

3 Subject Data

We compared a group of 51 healthy subjects (group A) with 44 cardiac patients at a high risk for malignant ventricular arrhythmias (group B, VT patients). All healthy volunteers (mean age 24.0±4.1 years) had a normal resting ECG and a normal echocardiogram, and no cardiac symptoms or coronary risk factors. The patients with a high–risk for malignant ventricular arrhythmias (mean age 61.2±8.9 years) were selected from our electrophysiologic database. Inclusion criteria were the presence of coronary artery disease, a previous myocardial infarction, a history of at least one symptomatic arrhythmia, and inducible sustained ventricular tachycardia (> 30 seconds) at electrophysiologic testing. Patients with bundle branch block or atrial fibrillation were excluded. All patients of group B underwent coronary angiography and programmed right ventricular stimulation due to clinical indications. Stimulation was done from the right apex and the right outflow tract. The stimulation protocol included up to 3 extrastimuli during sinus rhythm and at baseline pacing with a cycle length of 500 ms, and a maximum of 2 extrastimuli at baseline pacing with cycle lengths of 430 ms, 370 ms, and 330 ms. Group B consisted of 10 patients with single vessel disease, 17 patients with double vessel disease, and 17 patients with triple vessel coronary artery disease. Nineteen patients had a previous posterior infarction, 14 patients had a previous anterior infarction, and 11 patients had both a previous anterior and a previous posterior infarction. Mean left ventricular ejection fraction was 44.0%±14.9%. Forty-one patients had a documented episode of spontaneous, sustained ventricular tachycardia or ventricular fibrillation. Out of the remaining three patients, 1 patient had syncopes and non–sustained ventricular tachycardias on Holter monitoring, and 2 patients had syncopes of presumed cardiac origin.

4 ECG Recordings

High–resolution beat–to–beat electrocardiograms of 30 min duration were recorded during sinus rhythm from bipolar orthogonal X, Y, Z leads using the Predictor system (Corasonix Inc., Oklahoma, USA). Sampling rate was reduced to 1000 Hz. QRS triggering, reviewing of the ECG, and arrhythmia detection was done on a high–resolution ECG analysis platform developed by our group [10]. Before

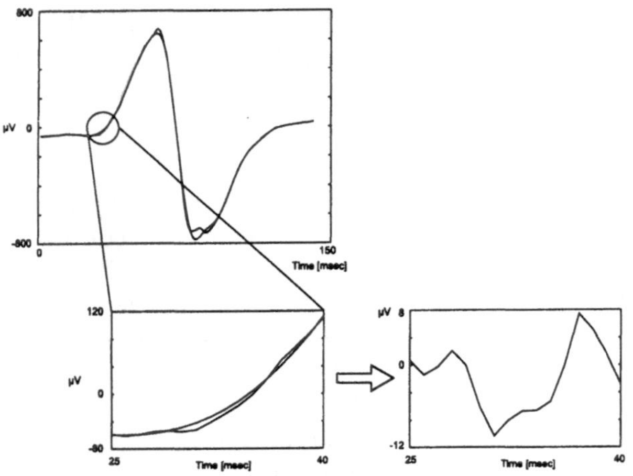

Fig. 2. Diagram of the spline–filtering procedure. The upper panel shows both signals, the QRS-complex (sum of the three leads) and the cubic spline. A zoom-in makes the differences more apparent (lower left panel). The resulting signal difference is shown on the lower right panel (note the different scaling of the Y-axis).

ECG recording antiarrhythmic drugs were stopped for at least four half-lives. The skin was carefully prepared and recordings were done with the subjects in reclining position in a Faraday cage. The three leads were summed into a signal $V = X + Y + Z$. From each recording 250 consecutive sinus beats preceded by another sinus beat were selected for subsequent beat-to-beat variability analysis.

In a first step the signals were aligned by maximizing the cross-correlation function [11] between the first and all following beats. Prior to the quantification of signal variability the beats were pre-processed to suppress the main ECG waveform, bringing the beat-to-beat micro-variations into clearer focus. To achieve this, the individual signal was subtracted from its cubic spline smoothed version (spline filtering, spline interpolation through every seventh sample using the not-a-knot end condition) [12], compare Figure 2. This method resembles a waveform adaptive, high-pass filtering without inducing phase-shift related artefacts. Next, for each individual beat the amplitude of the difference signal was normalized to zero mean and a standard deviation of 1 μV. Beat-to-beat variation of each point was measured as the standard deviation of the amplitude of corresponding points across all 250 beats. For the QRS we used a constant analysis window of 141 ms which covered all QRS complexes of this series [8]. The ST-T microvariation was measured the same way, but without any spline subtraction.

The resulting 141 dimensional variability vector

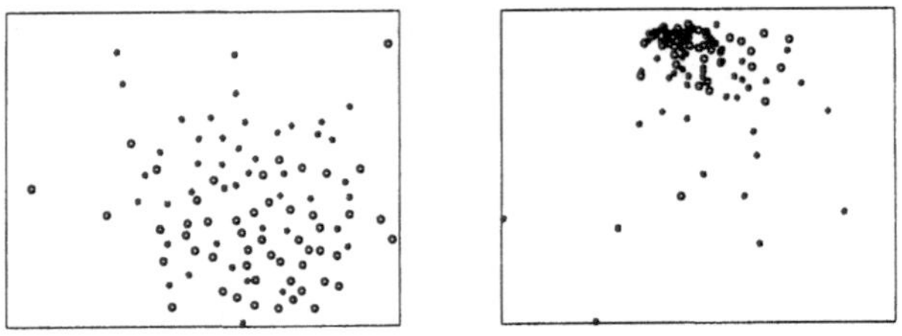

Fig. 3. Mapping (Sammon mapping) of the beat-to-beat QRS variability features (120D, left) and the ST-T variability data (400D, right) onto the 2 dimensional plane visualizing distance relations (using LVQ-Pak 3.1 Helsinki University of Technology). Samples from the healthy subject group are marked with ∘ those from the VT patient group with ∗.

was reduced to a 120 dimesional variability vector by discarding the equally spaced interpolation points. This signal was then used for classification into subject group A or B. The ST-T microvariation was measured the same way, but without any spline subtraction. For this, a constant window, starting at the QRS-offset, of 400ms was used. Figure 3 (right) shows a 2-dimensional visualization of the datasets [13].

5 Results

Two types of evolutionary algorithms were investigated: (a) self-adaptation of the strategy parameters $\{\sigma_a, \sigma_b\}$ was disabled they were both set to 0.15 and size mutation p was replaced by a constant probability of 0.0005 to delete a wavelet, and (b) self-adaptation with $\{\tau_a = 0.015, \tau_b = 0.015, \tau_p = 0.1\}$. All other parameters were set as follows: population size: 1000, wavelet number range: $[1, 50]$, number of data points used in the classifier: 20, and termination condition: 500 generations. Categorization performance is given in terms of re-validation, i.e. training and test with the complete data set, and as 10 fold cross-validation results, i.e. training on 90% of the data and test on 10% and repeating this ten times (test sets are disjoint).

The number of feature detectors, i.e. wavelets of the best individual of the last population, was in the range $[1, 22]$. The best re-validation result was attained with a feature detector consisting of 22 wavelets on the QRS data II. The variance of the number of wavelets in the best individual was generally lower with the self-adaptation strategy (coefficient of variation: (a) QRS: 1.5, ST-T: 1.2 and (b) QRS: 1.07, ST-T: 1,12). Although the results are still inconsistent (the significant difference between the re-validation and the cross-validation results indicates instability which is most probable a conse-

Table I. Classification results with no self-adaptation QRS var: 120D QRS variability data, ST-T var: 400D variability data of the ST-T segment, Acc: accuracy, Sensi: sensitivity, Speci: specificity, Re-val: training and test on the same data, Cross-val: test on independent data (see text)

		Acc	Sensi	Speci
QRS var.	Re-val	89.5%	88.6%	90.2%
	Cross-val	72.6%	68.2%	76.4%
ST-T var.	Re-val	87.4%	84.1%	90.2%
	Cross-val	75.9%	74.4%	83.3%

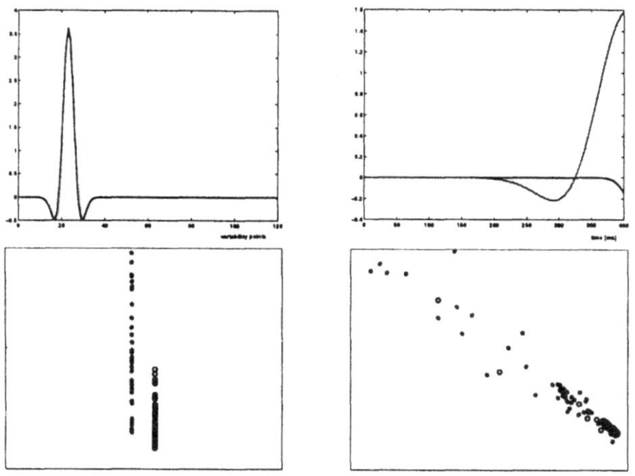

Fig. 4. Graphs of the feature detectors (upper panels) for the QRS (left) and ST-T (right) data corresponding to the re-validation results of Table I. Transforming each data set with these feature detectors yield the diagrams below. Samples from the healthy subject group are marked with ∘ those from the VT patient group with $*$.

quence of the small sample size) the re-validation results compare favourably to previous findings [8] using the sum of the variability signal as a feature together with a single cut-off value for categorization (QRS: Acc 73.7%, Sens 68.2%, Spec 78.4%; ST-T: Acc 77.9%, Sens 79.6%, Spec 76.5%).

6 Conclusion

Analysis of the QRS variability by means of a wavelet classifier improves the detection of patients with increased risk for malignant arrhythmias compared to the previously used numerical sum of intra-QRS variance with a single cut-off value. Future studies are necessary to further evaluate the feasibility of the method on variability data. Drawbacks are especially small sample sizes, which restrict the possible fitness values and long training runs (hours to days for a single run). Nevertheless, this strategy may give new insights into the emergence of invariant feature detectors.

Table II. Classification results with self-adaptation. Abreviations: see Table I

		Acc	Sensi	Speci
QRS var.	Re-val	90.5%	90.9%	90.2%
	Cross-val	74.7%	65.9%	82.4%
ST-T var.	Re-val	84.2%	88.6%	80.4%
	Cross-val	65.3%	61.4%	68.6%

References

[1] Smith J, Clancy E, Valeri C, Ruskin J, Cohen R. Electrical alternans and cardiac electrical instability. Circulation 1988;77(1):110–121.

[2] Rosenbaum D, Jackson L, Smith J, Garan H, Ruskin J, Cohen R. Electrical Alternans and Vulnerability to Ventricular Arrhythmias. N Engl J Med 1994;330(4):235–41.

[3] Hombach V, Kebbel U, Höpp HW, Winter U, Hirche H. Noninvasive beat-by-beat registration of ventricular late potentials using high resolution electrocardiography. Int J Cardiol 1984;6:167–183.

[4] Hombach V, Kochs M, Höpp HW, et al. Dynamic behavior of ventricular late potentials. In Hombach V, Hilger HH, Kennedy HL (eds.), Electrocardiography and cardiac drug therapy. Dordrecht, Netherlands: Kluwer Academic Publishers, 1989; 218–238.

[5] Sherif NE, Gomes J, Restivo M, Mehra R. Late potentials and arrhythmogenesis. Pacing Clin Electrophysiol 1985;8:440.

[6] Sherif NE, Gough W, Restivo M, Craelius W, Henkin R, Caref E. Electrophysiological basis of ventricular late potentials. Pacing Clin Electrophysiol 1990;13:2140–7.

[7] Höher M, Axmann J, Eggeling T, Kochs M, Weismüller P, Hombach V. Beat-to-beat variability of ventricular late potentials in the unaveraged high resolution electrocardiogram - effects of antiarrhythmic drugs. Eur Heart J 1993;14:E:33–39.

[8] Kestler HA, Wöhrle J, Höher M. Cardiac vulnerability assessment from electrical microvariability of the high-resolution electrocardiogram. Medical Biological Engineering Computing 2000;38:88–92.

[9] Bäck T. Evolutionary Algorithms in Theory and Practice. New York: Oxford University Press, 1996.

[10] Ritscher DE, Ernst E, Kammrath HG, Hombach V, Höher M. High–Resolution ECG Analysis Platform with Enhanced Resolution. Computers in Cardiology 1997;24:291–294.

[11] van Bemmel J, Musen M (eds.). Handbook of Medical Informatics. Heidelberg / New York: Springer Verlag, 1997

[12] de Boor C. A Practical Guide to Splines. Springer Verlag, 1978.

[13] Sammon J. A nonlinear mapping for data structure analysis. IEEE Transactions on Computers May 1969;C-18:401–409.

Genetic Implementation of a Classifier Based on Data Separation by Means of Hyperspheres

M. Jiřina, jr., J. Kubalík*, M. Jiřina†

*Department of Cybernetics, Faculty of Electrical Engineering, Czech Technical University Prague, Technická 2 166 27 Prague 6 - Dejvice, Czech Republic, {jirina,kubalik}@labe.felk.cvut.cz, † Institute of Computer Science, Academy of Sciences of the Czech Republic, Pod vodárenskou věží 2, 182 07 Prague 8, Libeň, Czech Republic, marcel@cs.cas.cz

Abstract:
This paper discusses a genetic implementation of the growing hyperspheres classifier (GHS) for high-dimensional data classification. The main idea of the GHS classifier consists in data separation by n-dimensional hyperspheres properly spread over the training data. First, the idea of training data representation is described. Then a brief description of a previous first representation by neural networks is reminded. The main part of this paper is focused on a precise description of the classifier implementation by genetic algorithms. Features of the new approach are discussed and compared. Finally, a task classifying data from a gamma telescope is presented to show the capabilities of the classifier.

1 Introduction

In contrast to other classifiers using complex hypersurfaces for separation of classes, the presented classifier uses hyperspheres. This principle gives a rather simple, understandable and quantifiable insight into classification. A really important feature is that the hyperspheres can easily cut out patterns that are enclosed inside the patterns of different classes and far from the main groups of patterns.

2 Classification using Separation by Means of Hyperspheres

The main idea of the suggested classifier is based on an extraction of learning patterns of the same class within a hypersphere. The centre of the sphere and its radius determine an area which contains learning patterns of the same class. All these information including the separated class are stored in a node. A linked list of nodes represents the structure of the GHS classifier. A disjunction of such spheres demarcates the area of one class that contains all patterns of this class and separates the learning patterns of different classes. The number of coordinates of the centre is the same as the feature dimension n. Therefore the classifier has n inputs. The radius is represented by only one real number and is independent on the feature dimension.

For a classification in k classes, k groups of hyperspheres should be used. However, for k classes only k-1 groups of hyperspheres are sufficient. The last, say the k-th, class sometimes called "don't know " class is just the one that does not belong to any other class.

Once we find the hyperspheres in the optimisation process we can simply classify new unknown patterns in individual classes. We run through all stored nodes and find one whose hypersphere contains this pattern. An appropriate class assigned to this hypersphere is just the class to which the pattern belongs. (Note that there can be several nodes whose hyperspheres contain the pattern but all represent the same class).

The most difficult problem is to find a suitable algorithm that finds these hyperspheres on the basis of a set of learning patterns. First suggested algorithm was precisely described in [1]. We briefly remind the main idea in the next section. A new approach to finding a minimal number of hyperspheres representing the classes describes this paper and is based on genetic background, see section 4.

3 Neural Approach to Finding Hyperspheres

The learning algorithm of the GHS neural network, see [1] and [2], works exactly this way. Let n be a feature dimension, k a number of classes, $N_1,\dots,N_k$ numbers of patterns in individual classes and $N=\sum_{i=1}^{k} N_i$ the total training set cardinality. The learning process will go over all k (or k-1, see previous section) classes and the same learning subprocedure will run for each class. Learning will finish after processing all classes. Note that the order in which we process the individual classes influences the total number of hyperspheres needed for the proper classification. Without a loss of generality we will suppose that we process the classes from class 1 to k.

A subprocedure works this way. First we randomly choose a pattern from a set containing patterns belonging to the first class. This pattern will represent centre $\vec{w}=(w_1,\dots,w_n)$ of a hypersphere now. The radius r of the hypersphere is found this way. First, we find the nearest pattern from any different class $\vec{h}=(h_1,\dots,h_n)$. The distance between the hypersphere centre $\vec{w}$ and the found pattern $\vec{h}$ is the primary radius

r^*. Further we reduce the primary radius to represent a boundary between some classes approximately in a half of the free space occurring between the classes. This is done this way. Using the just constructed hypersphere we find a pattern $\vec{g} = (g_1, \ldots, g_n)$ within the hypersphere that is nearest to the pattern $\vec{h}$. The final radius r is then recalculated according to formula

$$r = \sqrt{\sum_{i=1}^{n} \left(w_i - \frac{h_i + g_i}{2} \right)^2} \,. \quad (1)$$

The number of patterns within the hypersphere with centre $\vec{w}$ and radius r is counted and all these parameters are included in a new node.

It is probable that the hypersphere is not placed properly. Therefore a shift of the hypersphere to a new better position that would maximise the number of internal patterns is necessary. The shift is derived from hypersphere radius r. The new centre $\vec{w}^* = (w_1^*, \ldots, w_n^*)$ of the hypersphere is

$$w_i^* = w_i + d \cdot r \cdot s_i, \quad i = 1, \ldots, n \quad (2)$$

where d is a shift constant and $\vec{s} = (s_1, \ldots, s_n)$ a unit directional vector whose coordinates are calculated by formula

$$s_i = \frac{w_i - h_i}{\sqrt{\sum_{j=1}^{n} (w_j - h_j)^2}}, \quad i = 1, \ldots, n \,. \quad (3)$$

At this moment we have a new centre $\vec{w}^*$ of the hypersphere but we need to recalculate its radius. The calculation of the new radius runs in the same way as mentioned above for the initial hypersphere. Also the number of internal patterns is updated.

Now it is necessary to decide if the new shifted hypersphere is better than the original one. The quality of the hypersphere is measured by a number of internal patterns. If the new hypersphere has a greater number of patterns, we continue with its expansion. The more sophisticated algorithm of a hypersphere expansion is described in [1].

4 Genetic Approach to Finding the Configuration of GHS

Genetic algorithms (GAs) are probabilistic search and optimisation techniques, which operate on a population of chromosomes, representing potential solutions of the given problem [3]. Each chromosome is assigned a fitness value expressing its quality reflecting the given objective function. In the main loop of GA, chromosomes are reproduced and recombined to generate a new population of chromosomes, i.e. new sample points from hopefully more promising parts of the search space. This is repeatedly performed until some given termination-condition is fulfilled. The best chromosome encountered so far is then considered as the found solution.

The implementation of the genetic algorithm used in this work for learning the classifier will be described in the following paragraphs.

4.1 Representation & fitness function

The output of the learning process is the optimal configuration of the classifier in terms of the number, positions and radii of used hyperspheres. So we chose quite simple and natural form of the chromosome for this problem as a string of real-valued vectors; each vector expressing centre coordinates of one used hypersphere. The hyperspheres must not cover points of any class but the separated class. For the given hypersphere centre the radius is determined according to the formula 1.

Chromosomes are of a variable length L up to the pre-defined upper bound L_{max}. Classifiers with more hyperspheres than L_{max} are not allowed. The purpose is to find the smallest possible set of hyperspheres that would provide a satisfactory classification of the training set samples.

In order to assess particular individuals (classifiers) in the population we used a compound fitness function that reflects following two requirements imposed upon the classifier:

1. Minimise the number of used hyperspheres. Thus the set of as few hyperspheres as necessary to fully separate the given class is looked for.
2. Maximise the performance of the classifier, i.e. maximise a number of the points of the separated class that are covered by the set of hyperspheres.

The fitness is calculated according to the formula

$$fitness = \frac{N_{class} - N_{remain}}{N_{class}} + \frac{L_{max} - L}{L_{max}}, \quad (4)$$

where N_{class} is the cardinality of the separated class and N_{remain} is the cardinality of that class that are not covered by any out of L hyperspheres. The first part of the formula expresses the performance of the classifier and the second part expresses the robustness of the classifier in terms of the number of hyperspheres needed to get such a performance. In other words we are looking for the simplest classifier that would provide the best classification and that will the best generalise over the training data.

4.2 Initialization

The role of initialisation is to properly compose the population from which the evolution will start. It is clear that the content of this population affects further calculation and the quality of obtained results. So it is strongly recommended to use any background knowledge about the solved problem in order to generate the most appropriate initial population.

When generating a new chromosome (a new classifier) the primary goal is to cover as much of the patterns of the separated class as possible. It means that those hyperspheres that cover many patterns can be very useful. On the other hand it may not hold that every pattern falls into the region controlled by some of such large hyperspheres. So we can not focus only on the hyperspheres with a large number of patterns. Instead we should take into account also such hyperspheres that identify isolated patterns or small group of patterns of the separated class within the training set. This knowledge was incorporated into the initial strategy used.

Generally, each time a new hypersphere should be added to the classifier one of the uncovered patterns is chosen as its centre and the radius is calculated according to the formula 1. This is a simplest way to ensure that the new hypersphere will definitely reduce a number of uncovered patterns at least by 1. No other operation for optimising the hypersphere position and its radius is performed.

The initialisation strategy works in two steps:

1. First, go through the training set and find at most $L_{max}/2$ hyperspheres that cover isolated patterns. It means take into account such hyperspheres that does not cover more than $N_{isolated}$ patterns.
2. Then until $N_{remain}>0$ and $L< L_{max}$ go through the training set and use any of the uncovered patterns as a centre around which a new hypersphere is constructed. In this phase the hyperspheres are used regardless of the number of patterns they cover.

In this way the population is filled with promising hyperspheres that cover a big number of patterns as well as with the hyperspheres that cover small groups of patterns which are very likely to stay uncovered otherwise.

4.3 Crossover

The crossover operator used in this application is designed so that in the first phase it tries to use as many hyperspheres from the parental chromosomes as possible. It works in following steps:

1. Alternately choose among the first and the second parent's classifier.
2. Find an arbitrary unused hypersphere of this classifier.
3. Apply either mutation1 or mutation2 (mutation operators are described in the next paragraph) on the centre of the hypersphere. Take the mutated centre's coordinates if any the following conditions is not true, otherwise take the original centre:
   - The mutation has shifted the centre so that any usable hypersphere can not be found for the new centre[1].
   - The hypersphere expanded from the new centre covers less patterns than the original hypersphere.
4. Add the centre's coordinates to the generated child chromosome (classifier) iff the corresponding hypersphere decreases N_{remain}.
5. Mark the hypersphere as used no matter it was added to the new chromosome or not.
6. If ($N_{remain}>0$) and still some unused hyperspheres left then go to the step1).

If after this phase the set of uncovered patterns is not empty ($N_{remain}>0$) and the number of generated hyperspheres is still less than L_{max} than the second phase is carried out in order to decrease N_{remain} as much as possible. To achieve this a strategy similar to that implemented in the initialisation procedure is used.

4.4 Mutation

We used two mutation operators in our GA – *blind* and *informed*. Both operators have been used with the same probability 0.5. The mutation operates on the centre's coordinates. Blind mutation changes one of its coordinates by a value of up to ¼ of the radius. Informed mutation shifts the centre's coordinates in the direction away from the closest extraneous pattern. Again the rate of imposed changes is up to ¼ of the original hypersphere radius.

4.5 Merge

The last operator used to improve the search capabilities of the GA is operator *merge*. This operator can be applied only on a chromosome which is shorter than L_{max}. The idea is to replace two hyperspheres that are very close each to other within the classifier by the hypersphere with the centre just in the middle between the two hyperspheres. It works as follows:

1. Choose an arbitrary hypersphere in the classifier.
2. Find the hypersphere that is closest to the first one.

[1] This happens if the pattern that is closest to the centre does not belong to the separated class.

3. Calculate coordinates of the new hypersphere so it lies in the middle between the two hyperspheres.
4. Iff a usable hypersphere can be constructed around the new centre than add the centre to the classifier.

This operator was applied only on the best chromosome in the population.

5 Experiments and Results

The GHS classifier was tested on the task of particle physics data classification obtained from CERN [4]. They describe a physical phenomena observed in a gamma telescope. The data is in fact a mix of patterns of events (phenomena) looked for (signal, class 0) and patterns from other different but similar events (background, class 1). The data has been split into the learning and testing sets, see Table I.

The task is to classify the data into two classes. The quality of classification is expressed by kinds of errors. First, the ratio of properly recognised patterns of class 0 to the cardinality of class 0 is *signal efficiency*. Second, the ratio of the number of patterns of class 1, which are considered as patterns of class 0 to the cardinality of class 1, is *background error*.

There are two groups of experiments separated by double line in Table I. The first group of experiments shows results when the signal was separated from the background and the second group shows results when the background was separated from the signal.

6 Conclusions

The use of a hypersphere as a separation surface is very natural because real data usually forms a non-angular data cluster in the pattern space. Moreover, the data with common features (and thus belonging to one class) is often concentrated near one another around a centre and form a nearly hypersphere's shape.

The results in the Table I show that in all performed experiments the genetic algorithm was better in finding the classifier configuration (measured by the number of used hyperspheres) as well as in the *signal efficiency* ratio obtained. On the other hand the *background error* was slightly worse with the GA than with the neural method. Both the aspects follow from that the GA better generalised over patterns of the training set.

It seems interesting to explore the increase of the number of hyperspheres with feature dimension. This problem is currently studied. Up-to-date results have shown a relative decrease of the need of number of hyperspheres up to the 50th feature dimension on a benchmark example.

Acknowledgement

The research has been carried out partly under the support of the Czech Ministry of Education grants No. VS 96047 and partly under the support of the Czech Ministry of Industry and Commerce in the project No. RP-4210/69/97 "Cooperation of the Czech Republic with CERN".

References

[1] Jiřina, M., jr., Jiřina, M.: Neural Network Classifier Based on Growing Hyperspheres, Neural Network World Proceeding 2000,

[2] Jiřina, M., jr., Jiřina, M.: Neural Classifier Using Hyperspheres, ISCI 2000,

[3] Goldberg, D. E.: Genetic Algorithms in Search, Optimization and Machine Learning, Addison-Wesley, Reading, MA 1989,

[4] Bock, R. K.: Personal Communications, CERN 1999.

Table I. A comparison between genetic and neural implementation of the GHS classifier

Name of Experiment	Learning set size		Testing set size		Genetic implementation of GHS			Neural implementation of GHS		
	Signal	Backgr.	Signal	Backgr.	#hyper.	bcgr. err.	sign. eff.	#hyper.	bcgr. err.	sign. eff.
A18a1	2991	15009	3009	14992	234	0.0271	0.8996	347	0.0232	0.8943
All18a2	1500	15009	3009	14992	169	0.0199	0.8568	266	0.0181	0.7956
All18a4	750	15009	3009	14992	121	0.0155	0.7787	202	0.0117	0.7092
All18aA	300	15009	3009	14992	79	0.0110	0.6304	128	0.0065	0.4855
All18aA2	300	15009	3009	14992	79	0.0107	0.6371	130	0.0057	0.4988
All1aX	823	177	25007	4994	16	0.1324	0.9746	32	0.0637	0.9685
All2aX	1659	341	25007	4994	32	0.1428	0.9787	52	0.0623	0.9708
All3aX	2478	522	25007	4994	44	0.1296	0.9774	82	0.1039	0.9640
All6aX	4994	1006	25007	4994	80	0.1149	0.9813	133	0.0713	0.9727
All18aX	15009	2991	14992	3009	223	0.1040	0.9813	356	0.0837	0.9707

Evolving Order Statistics Filters for Image Enhancement

Cristian Munteanu*, Agostinho Rosa†

*LaSEEB, ISR - IST, Av. Rovisco Pais, Torre Norte 6.21, cmunteanu@pop.isr.ist.utl.pt †LaSEEB, ISR - IST, Av. Rovisco Pais, Torre Norte 6.20, acrosa@isr.ist.utl.pt

Abstract

This paper describes an effective method for performing both image denoising and contrast/brightness enhancement to images corrupted with a wide category of noise. The method employs a Real-Coded Genetic Algorithm with subjective fitness and a novel crossover operator called Gaussian Uniform Crossover. The algorithm evolves the structure of a general Order Statistics Filter (OSF). Results are presented that indicate the efficiency of the method proposed as compared to classical filtering methods.

1 Introduction

Evolutionary Computation is a modern field of optimization and search based on natural paradigms of evolution and natural selection. These paradigms were found to be efficient in solving difficult optimization tasks like the ones required for several applications in image processing and signal processing. This paper employs a Real Coded Genetic Algorithm (GA) with subjective fitness to the problem of finding the best Order Statistics Filter to perform image denoising and enhancement.

Image enhancement [3] is the task of improving the visual quality of a given image by increasing its contrast, brightness, and detail. A particular image enhancement task is the removal of noise whenever images are corrupted [7]. Given the result of image enhancement being evaluated by a human interpreter, the criteria to judge the effectiveness of a method are hard to establish, especially when it comes to contrast/brightness improvement [3]. It is also difficult to find a filtering method that removes a large category of noise simultaneously, and moreover it is difficult to find efficient methods that do both brightness/contrast/detail enhancement and noise removal.

The paper introduces a novel approach for doing both contrast/brightness enhancement and denoising, by employing an evolutionary computation model capable of *adapting* the filter to the human interpreter's demands.

2 The General Order Statistics Filter

OSF have been proven to be quite efficient in removing different types of noise from images, a well known OSF being the median filter [6]. The General Order Statistics Filter (GOSF), also known as the L Filter [6] can be defined as follows:

$$y_i = \frac{\sum_{j=1}^{n} a_j x_{(j)}}{\sum_{j=1}^{n} a_j} \tag{1}$$

where $x_{(j)}$ are the inputs arranged in increasing order of magnitude, y_i are the filter outputs, and a_j are the filter coefficients. In image processing applications, the inputs are the ordered pixel intensity values in a sliding window centered on the pixel for which the output is calculated. The moving average, median, rank ordered, α-trimmed mean, and midpoint filters are special cases of (1) if the filter coefficients are chosen appropriately. OSF are based on robust statistics principles, but despite their clear mathematical background, design methods to build the filters do not exist or are difficult to implement [6]. There are also difficulties in finding a method for adaptation of the filter coefficients to an image that has been corrupted with more complex combinations of noise, or corrupted with unknown noise. The strategy we propose in this paper comes to fill the gap left by not having a direct method to find the filter's coefficients given any corrupted image.

3 Real-Coded GA for Evolving the GOSF Coefficients

We employ a Real-Coded GA [4] to find the GOSF coefficients for any given image. Real-Coded GA have proven to be quite effective in solving various applications due to their very natural coding scheme: each real-coded gene represents a parameter of the problem to be optimized. The present application employs a variant of GA with subjective fitness: the population is composed of chromosomes representing the string of GOSF coefficients a_j. When evaluated for selection, the respective filter is applied to the input image, the output filtered image being evaluated by the human interpreter who gives a score between 0 (worst) and 1 (best). The score represents the fitness of the chromosome, which is a subjective fitness rather than an objective one (e.g. as given by the output of a function). Subjective fitness allocation and selection are not new, they have been applied successfully to a series of applications (see for example [1]). The user has a better control on the process of enhancement. His particular demands regarding performance, visual quality of the output image, as well as several other subjective factors of image interpretation can be embedded in the GOSF evolved in a direct and quite transparent fashion.

The GA runs with a small population and a short length chromosome ($l = 9$ genes corresponding to a 3×3 sliding window). Premature convergence is a phenomenon that has to be combated in this case, therefore we have chosen a binary tournament selection with 1-elitism, and a highly explorative crossover called Gaussian Uniform Crossover (GUX). Also a mutation that chooses randomly an allele from the allowed allele domain, has been employed. GUX was introduced by the authors in [5] and it was shown empirically that it attains higher levels of genetic diversity in the population. The present paper further develops on GUX by giving a brief theoretical analysis in the following. GUX can be defined starting with a bitmask of length l: $mask = \{mask(i)\}_{i=\overline{1\cdots l}}$ with:

$$mask(i) = \begin{cases} 1 & \text{if } r_i \leq 0.5 \\ 0 & \text{otherwise} \end{cases} \quad (2)$$

with r_i a sample from a uniform distributed random variable $\Re \sim U(0,1)$. Second, the distance between each corresponding gene in the two parents x_1 and x_2 is computed as: $d_j = |x_{1j} - x_{2j}|, \forall j = \overline{1\cdots l}$. Finally, the offspring genes in x_1^o and x_2^o are calculated from the parental genes as:

$$x^o_{\{1,2\},j} = \begin{cases} x_{\{1,2\},j} + \frac{d_j}{3}\chi_j & \text{if } mask(j) = 0 \\ x_{\{2,1\},j} + \frac{d_j}{3}\chi_j & \text{if } mask(j) = 1 \end{cases} \quad (3)$$

where χ_j is a sample from a standard normal distribution $N(0, \sigma_\chi^2)$. From [2] we know that a good crossover operator should keep the mean of the parents equal to the mean of the offspring (preserves the mean), and only should vary the covariance (or variance). We check these features on GUX, by first assuming that the genes in the two parents are i.i.d. random variables with mean $\overline{x}$ and variance σ_x^2. Thus, we can study the chromosomes statistics at each locus. We can rewrite (3) as: $x^o_{\{1,2\},j} = x_{\{1,2\},j}h(r_j - 0.5) + x_{\{2,1\},j}h(0.5 - r_j) + \frac{d_j}{3}\chi_j$, with $h(\cdot)$ the Heaviside function. Tacking the mean of this expression we get:

$$\overline{x^o_{\{1,2\}}} = \prod_{j=1}^{l} \overline{x^o_{\{1,2\},j}} = \prod_{j=1}^{l} \overline{x_j} = (\overline{x})^l = \overline{x_{\{1,2\}}} \quad (4)$$

where we have used the fact that $\overline{\chi_j} = 0$ and $\overline{h(r_j - 0.5) + h(0.5 - r_j)} = 1$, as well as the i.i.d. assumption. For the derivation of the offspring genes' variance we can rewrite (3) as: $x^o_{\{1,2\},j} = x_{\{1,2\},j} + \frac{|x_{\{1,2\},j} - x_{\{2,1\},j}|}{3}\chi_j$ where we are still referring to genes at locus j, and we have used the i.i.d. assumption in ignoring the order of the subscripts $\{1,2\}$, and the independence of χ_j and χ_j^2 from the parental genes. From this expression we get after straightforward derivation that:

$$\sigma^2_{x^o_{\{1,2\},j}} = \sigma^2_{x_{\{1,2\},j}} \left(1 + \frac{2}{9}\sigma_\chi^2\right) \quad (5)$$

From (4) we can see that GUX preserves the mean of the parents, while from (5) we can see that the variance of the genes at the same locus in the offspring is increased or left unchanged (when $\sigma_\chi^2 = 0$) compared to the variance in the parental genes at the same locus. This proves that GUX has a bias towards increasing exploration, which has been the expected behaviour.

4 Experimental Results

The performance of the Real-Coded GA has been tested on several images, only results for one image will be presented in this paper, due to lack of space [1]. The GA-based enhancement method has been compared to classical filtering techniques like average, median and adaptive (Wiener) filters. Parameters used for the GA based method are given in Table I. The test image (i.e. airplane image) has been corrupted with several types of noise as follows: gaussian (additive) noise with mean -0.35 and variance 0.01, speckle (multiplicative) noise with noise

Allele domain	l	N	T_{max}	P_c	P_m	σ_χ^2
[0, 1]	9	10	10	0.8	0.2	1

Table I. GA parameters: chromosome length: l, population size: N, maximum number of generations: T_{max}, crossover rate: P_c, mutation rate: P_m, variance for GUX: σ_χ^2

[1]Detailed result images can be found at our home page http://laseeb.ist.utl.pt/cristi

density 0.02, salt and pepper (impulse) noise with density 0.1, as well as combinations between these types of noise: gaussian plus speckle, gaussian plus salt and pepper, and salt and pepper plus speckle.

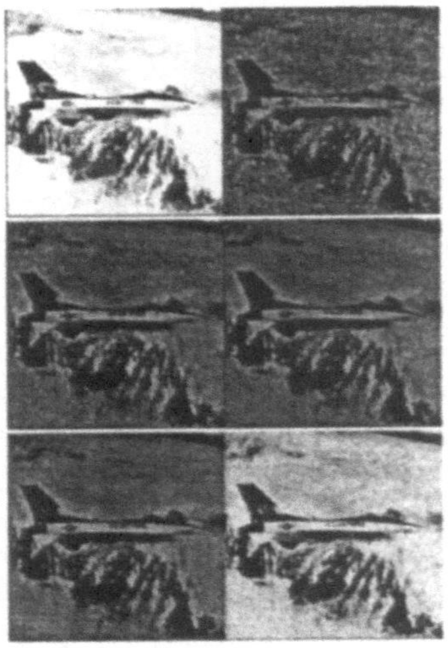

Fig. 1. Results for gaussian noise: original (upper-left), corrupted (upper-right), median (middle-left), average(middle-right), adaptive(lower-left), GA-OSF(lower-right)

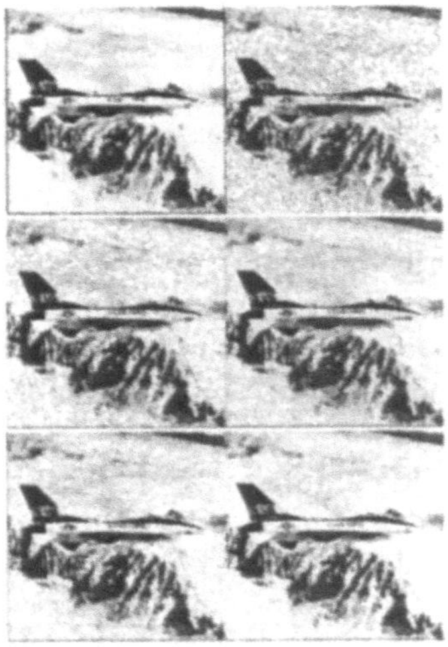

Fig. 2. Results for speckle noise: original (upper-left), corrupted (upper-right), median (middle-left), average(middle-right), adaptive(lower-left), GA-OSF(lower-right)

From Figures 1 to 6 one can note that the GA-OSF method performs good for the image denoising task, compared to the other methods, but it also achieves contrast/brightness enhancement simultaneously, a feature that the classical methods do not possess. The contrast/brightness enhancement is apparent when negative mean gaussian noise is present. The efficiency of enhancement can be seen also from the results obtained on images corrupted with combinations of noise, images for which the classical methods give poor results.

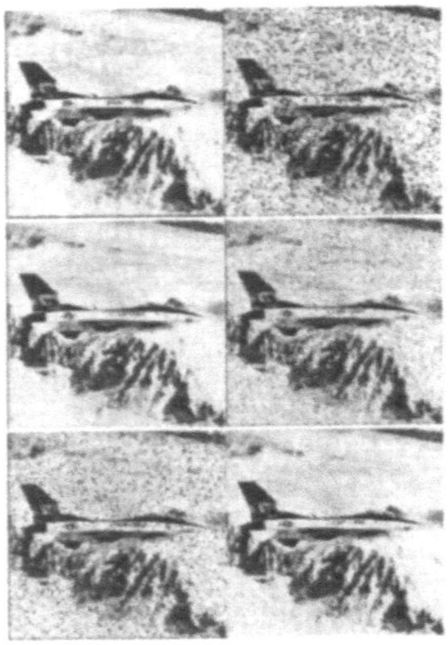

Fig. 3. Results for salt and pepper noise: original (upper-left), corrupted (upper-right), median (middle-left), average(middle-right), adaptive(lower-left), GA-OSF(lower-right)

It is also instructive to look at the OSF coefficients the method evolves. For example, in case of the salt and pepper noise, the GA-OSF evolves a filter having the coefficients: $a_1 = 0.022, a_2 = a_3 = a_6 = a_7 = a_8 = 0, a_4 = 0.123, a_5 = 0.862, a_9 = 0.088$ which is very close to the median filter known to be optimal in removing the salt and pepper noise, the median filter having coefficients $a_5 = 1$, and the remaining $a_{1\div 9} = 0$. For the gaussian noise, the OSF evolved is not the average filter, but a filter that better preserves the edges, being well known that the average filter although optimal in removing gaussian additive noise, doesn't preserve the edges. The PSNR in dB are given in Table II, for each method and each image.

PSNR	Average	Median	Adaptive	GA
g	8.37	8.29	8.39	17.53
s p	21.95	31.04	17.70	28.70
sp	26.22	23.93	26.48	23.20
g s p	8.57	8.27	8.33	14.58
g sp	8.36	8.15	8.36	13.55
s p sp	21.11	23.11	17.59	23.13

Table II. PSNR for test images

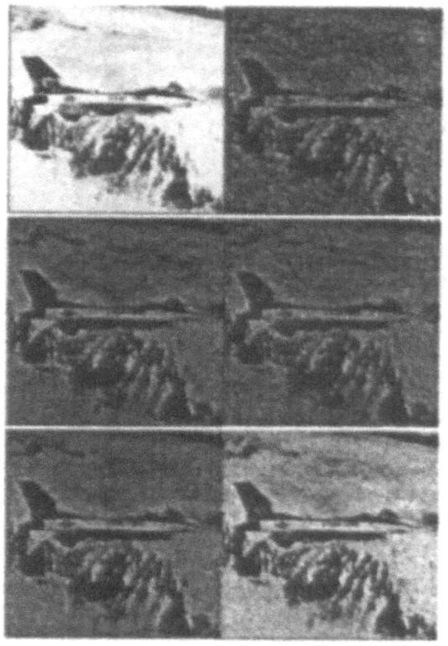

Fig. 4. Results for gaussian plus speckle noise: original (upper-left), corrupted (upper-right), median (middle-left), average(middle-right), adaptive(lower-left), GA-OSF(lower-right)

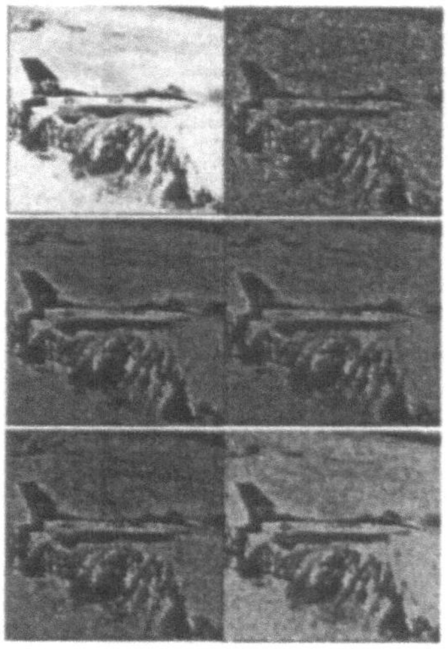

Fig. 5. Results for gaussian and salt and pepper noise: original (upper-left), corrupted (upper-right), median (middle-left), average(middle-right), adaptive(lower-left), GA-OSF(lower-right)

5 Conclusions

The paper describes an effective method of finding the best OSF. The user guides the OSF construction process through the evolution of a real-coded GA, by evaluating the filtered image and giving a score to the result. The results obtained are compared to classical filtering methods, the conclusion being that the GA-OSF method is capable of satisfactory filtering the image without significantly blurring the edges, for a large category of noise. The most interesting feature of GA-OSF is that it is capable of enhancing the contrast/brightness, which translates for the gaussian noise in removing additive gaussian noise with non-zero mean. This latter feature clearly makes the difference between the GA-OSF strategy and the classical filtering methods.

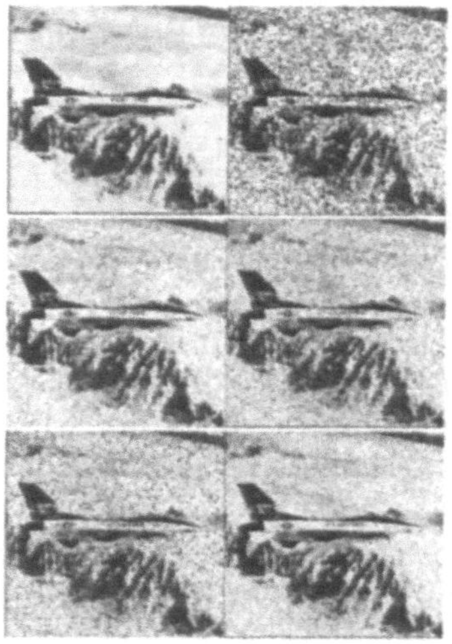

Fig. 6. Results for speckle and salt and pepper noise: original (upper-left), corrupted (upper-right), median (middle-left), average(middle-right), adaptive(lower-left), GA-OSF(lower-right)

References

[1] Herdy, M.: Evolutionary Optimization based on Subjective Selection. Proc. of EUFIT'97 *1*, 640 (1997).

[2] Kita, H., Ono, I., Kobayashi, S.: Theoretical Analysis of the UNDX for Real-Coded GA. Proc. of IEEE ICEC'98 *1998*, 529.

[3] Jain, A. K.: Fundamentals of Digital Image Processing, Englewood Cliffs, NJ: Prentice Hall 1989.

[4] Michalewicz, Z.: Genetic Algorithms + Data Structures = Evolution Programs, 3rd ed. Berlin: Springer 1996.

[5] Munteanu, C., Lazarescu, V.: Evolutionary Contrast Stretching and Detail Enhancement. Proc. of MENDEL'99 *1*, 94 (1999).

[6] Pitas, I., Venetsanopoulos, A. N.: Order Statistics in Digital Image Processing. Proc. of the IEEE *80*, 1893 (1992).

[7] Ramponi, G., Strobel, N.: Nonlinear Unsharp Masking Methods for Image Contrast Enhancement. Journal of Electronic Imaging *20*, 353 (1996).

GA-Based Non-Linear Harmonic Estimation

Maamar Bettayeb[*], Uvais Qidwai[†]

[*]Electrical/Electronics and Computer Engineering Department, University of Sharjah, P. O. Box 27272, Sharjah, United Arab Emirates, maamar@sharjah.ac.ae, [†]Electrical and Computer Engineering Dept., University of Massachusetts, Dartmouth, North Dartmouth, MA 02747, USA, g_uqidwai@umassd.edu

Abstract

Knowing the frequency contents present in a network enables the operation and design engineers to predict the functioning of the system with respect to certain equipment and can also be used to identify the location and extent of use of filters and other protective devices. The usual frequency estimation techniques are dependent on some linear approximation for the nonlinear distortion present in the system. This paper presents the use of Genetic Algorithms (GA) with nonlinear system structure to estimate the amplitudes and phases of various frequency components present in the system. In this paper, the general nonlinear formulation problem of estimation of harmonics is solved using Genetic Algorithms (GAs).

1 Basics of Genetic Algorithms (GAs)

GAs are stochastic global search algorithms [1,2]. They operates on the population of current approximations -individuals- initially drawn at random, from which improvement is sought. Individuals are encoded as strings -the chromosomes- constructed over some particular character set, so that the chromosome values, the genotypes, are exactly mapped into the decision variable (phenotype) domain. Once the phenotype of the current population is calculated, individual performance is assessed according to the objective function which characterizes the problem to be solved. The objective function establishes the basis of selection. At the reproduction stage, a fitness value is derived from the raw individual performance measure as given by the objective function, and is used to bias the selection process. Highly fit individuals will have a higher probability to go to the next stage. The selected individuals are then modified using genetic operators. After this, individual chromosomes are decoded, evaluated, and selected according to their fitness, and the process continues for different generations [2-5].

GAs do not rely on Newton-like gradient descent methods, and this makes them less likely to be trapped in local minima. By using parallel search strategies, contrary to point search as in gradient methods, the GAs can effectively explore many regions of search space simultaneously, rather than a single region. Consequently, GAs do not require a complete understanding of the system or the model. The only problem-specific requirement is the ability to evaluate the trial solutions on their relative fitness, i.e., the objective function.

2 GA in Harmonic Estimation

For a given distorted signal, composed of a fundamental frequency along with n harmonics buried in random noise, the problem is to estimate the amplitude and phase of the desired frequency components in the signal.

Unlike the usual estimation techniques, GA does not require any calculation of the gradient and is not susceptible to local minimum problems occurring in the multi-modal systems, specially, when parameters are non-linear [6-8]. In order to make use of GA for harmonic estimation some kind of Fitness function or Objective function is needed. This is the only system-related information that is required for the algorithm.

Signals in power system can be represented by expanding them into Fourier series representation as follows:

$$x(t) = \sum_{n=1}^{N} A_n \sin(2n\pi f_o t + \phi_n) + \eta(t) \tag{1}$$

where $x(t)$ represents the distorted signal with additive noise $\eta(t)$. The original frequency of the signal, the fundamental frequency is f_o. A_n and ϕ_n are the respective amplitude and phase for the nth harmonic. The estimated signal corresponds to the following series:

$$\hat{x}(t) = \sum_{m=1}^{M} \hat{A}_m \sin\left(2m\pi f_o t + \hat{\phi}_m\right) \tag{2}$$

Where $\hat{x}$, $\hat{A}_m$ and $\hat{\phi}_m$ represents the estimates for the distorted signal without noise, and the amplitude and phase of the mth harmonic, respectively. M may or may not be equal to N. The performance measure for the system is taken as the Euclidean norm for the estimates for a time window of T seconds, and is given by:

$$J = \int_0^T \left[x(t) - \hat{x}(t)\right]^2 dt \qquad (3)$$

The estimation algorithm used in this paper utilizes the original GA discussed earlier with modified objective function. A flow diagram for this algorithm is given in Figure 1.

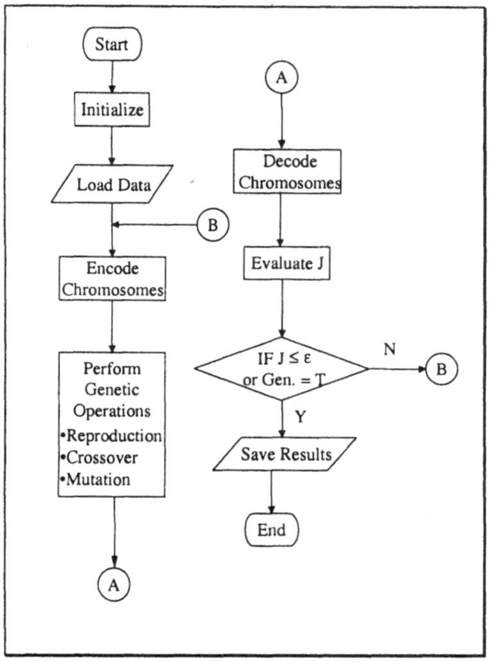

Figure 1. Flow diagram for Estimation algorithm.

3 Simulation Results

Figure 2 shows the sample system which comprises of a two bus three phase system with a full wave six pulse bridge rectifier at the load bus. The test signal is a distorted voltage waveform at the terminals of the load bus for the system of Figure 2. Table 1 shows the harmonics present in the system, and Figure 3 shows the resulting signal with its frequency spectrum.

Table I. Frequency content of the test signal, Voltage at the load bus.

Harmonic Order	Amplitude (P.U)	Phase (Degrees)
Fundamental (60Hz)	0.95	-2.02
5th (300 Hz)	0.09	82.1
7th (420 Hz)	0.043	7.9
11th (660 Hz)	0.03	-147.1
13th (780 Hz)	0.033	162.6

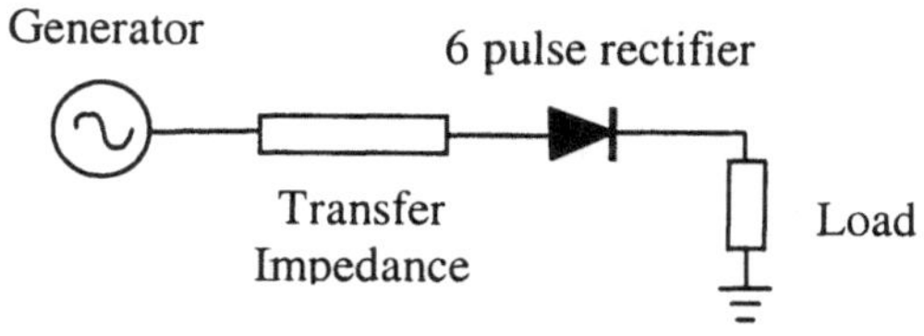

Figure 2. Sample system, a two bus architecture with six pulse full wave bridge rectifier supplying the load.

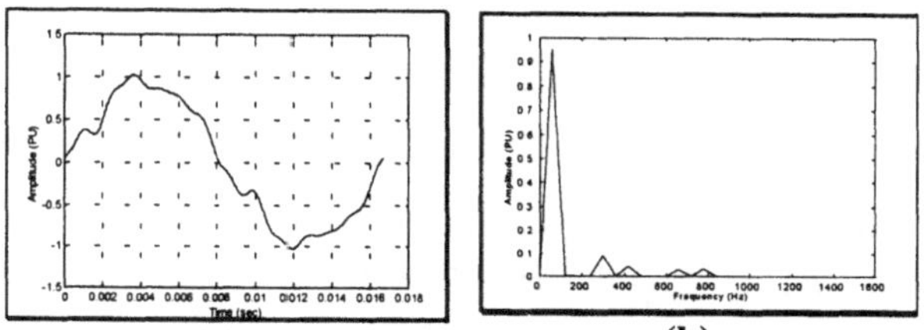

(a) (b)

Figure 3. Test signal (a) and its spectrum (b).

Figure 4 shows the sample signals and their spectra for different set of Gaussian noise, with three different Signal to Noise Ratios (SNRs) in dB, 0, 10, 20. The no noise case is the one in Figure 3. The test signal was generated in MATLAB with a sampling frequency of 1620 Hz. The chromosome size was selected as 19 bits based upon the range of ±3 with precision divisions of 0.00001.

With 100 chromosomes in the genetic pool selected randomly, the GA was ran for a maximum of 100 generations. However, it was observed that the algorithm converged to the final solution in approximately 40 generations. Keeping the mutation probability fixed at 1%, the algorithm was tested with various crossover probabilities. Using the best-obtained results, the signal was reconstituted and then compared with the actual fundamental frequency component present in the system. The results of these graphical comparisons are given in Figure 5, where the actual and estimated waveforms are essentially identical.

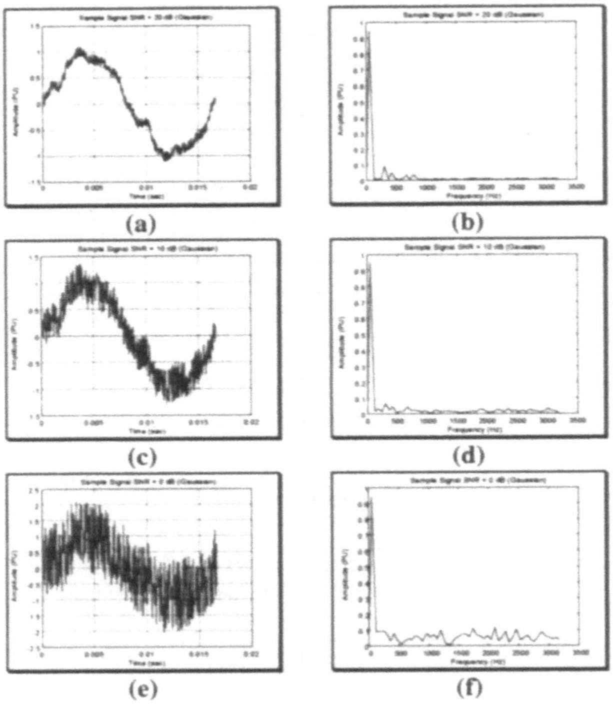

Figure 4. Sample signal and its spectrum in Gaussian Noise, (a) No noise, (b) SNR = 20, (c), (d) SNR = 10, and (e), (f) SNR = 0 dB.

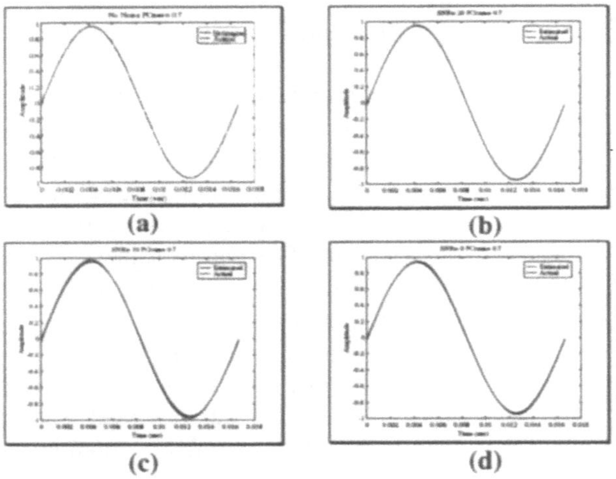

Figure 5. Actual and Estimated waveforms for fundamental harmonic, for (a) No noise, (b) SNR=20 dB, (c) SNR=10 dB, (d) SNR=0 dB, additive Gaussian noise.

The procedure for estimating the fundamental harmonic can be extended to multiple frequency estimation. This requires the incorporation of all the desired frequency sinusoids in the performance index.

The number of parameters to be estimated in this case is 10, five amplitudes and five phases for the assumed 1st, 5th, 7th, 11th, and 13th harmonic components. Using the estimated values, the waveform was reconstructed and compared with the original wave (without noise). These graphical comparisons are shown in Figure 6. Figure 7 represents the convergence behavior of the algorithm performance index.

Similar results were obtained for the simulated data in uniform noise. These results are shown in Figures 8-10.

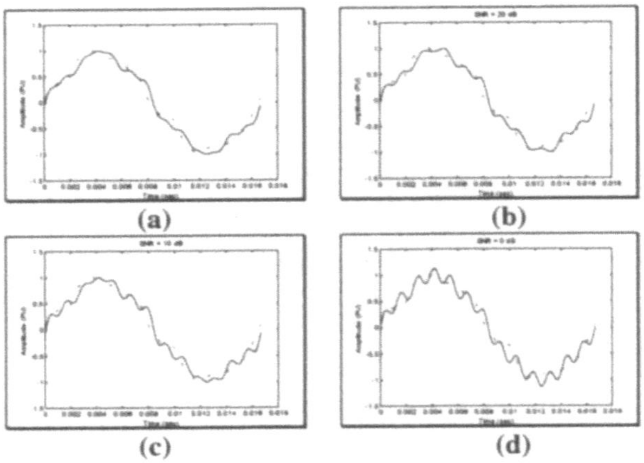

Figure 6. Actual and Estimated waveforms, for (a) No noise, (b) SNR=20 dB, (c) SNR=10 dB, (d) SNR=0 dB, additive Gaussian noise.

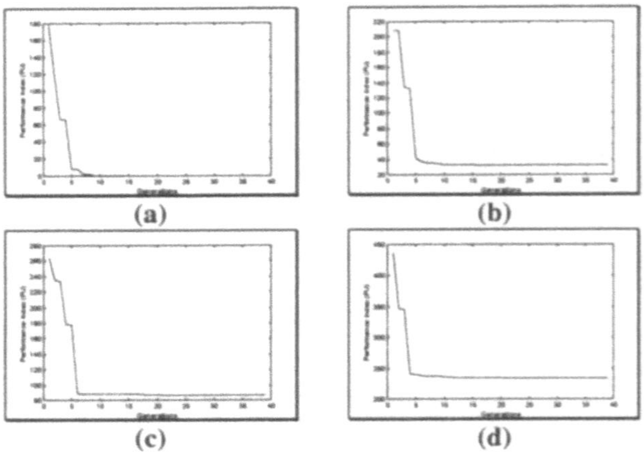

Figure 7. Performance Index, for (a) No noise, (b) SNR=20 dB, (c) SNR=10 dB, (d) SNR=0 dB, additive Gaussian noise.

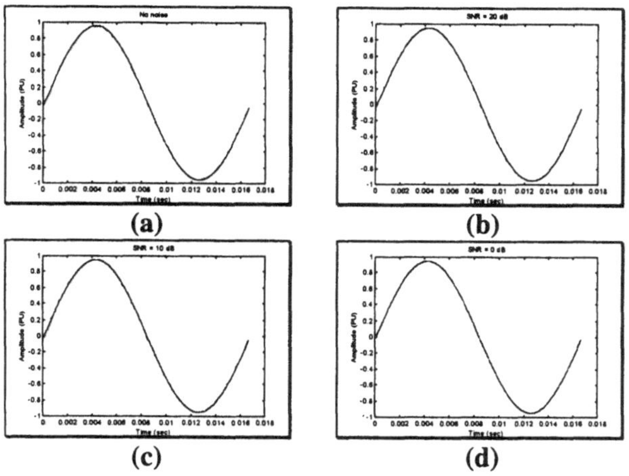

Figure 8. Actual and Estimated waveforms for fundamental harmonic, for (a) No noise, (b) SNR=20 dB, (c) SNR=10 dB, (d) SNR=0 dB, additive Uniform Noise.

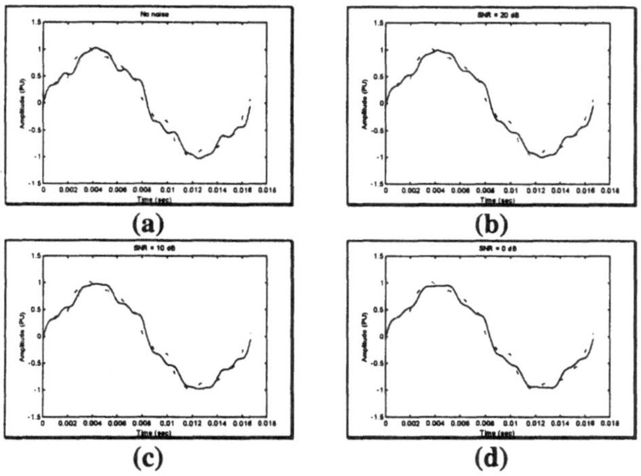

Figure 9. Actual and Estimated waveforms, for (a) No noise, (b) SNR=20 dB, (c) SNR=10 dB, (d) SNR=0 dB, additive Uniform Noise.

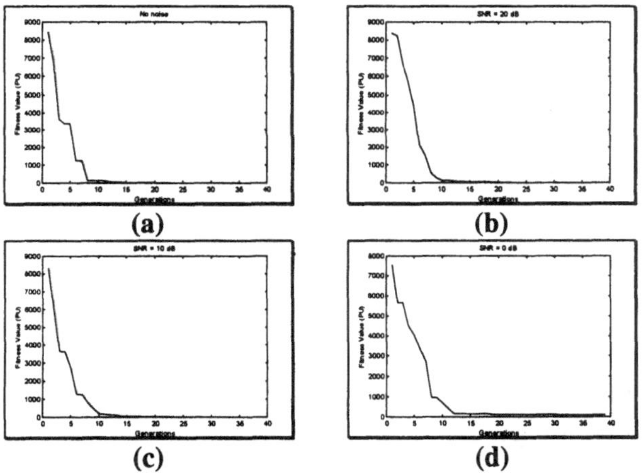

Figure 10. Performance Index, for (a) No noise, (b) SNR=20 dB, (c) SNR=10 dB, (d) SNR=0 dB, additive Uniform noise.

4 Conclusion

This paper has presented the application of Genetic Algorithms for harmonic estimation in power signals. The same algorithm can also be applied to other type of signals from communication channels, telephones, and other encrypted signals. The use of the nonlinear problem formulation removes the problems of approximations and other limitations posed by the linear ones. The need for minimal information about the actual system is emphasized here which is an inherent characteristics for the nonlinear formulation and fitness function. For higher noise cases, the convergence takes place using more generations and requires more time compared to the low or no noise cases. As it is, the algorithm can be used to estimate the system harmonic contents before hand as a design step for system augmentation or commissioning the necessary protective instruments in the system.

References

[1] Frenzel, J.F.: Genetic Algorithms, A new breed of Optimization, IEEE Potentials *1993*, 21.

[2] Fleming, P.J. and Fonseca, C.M.: Genetic Algorithms in Control Systems Engineering: A brief introduction, IEE Colloquium on Genetic Algorithm for Control System Engineering *1993*.

[3] Goldberg, D.E.: Genetic Algorithms in search, Optimization and Machine Learning, Addison-Wesley Publishing Company Inc., 1989.

[4] Davis, L.: Handbook of Genetic Algorithms, Van Nostrand Reinhold, 1991.

[5] Buckles, B.P. and Petry, F.E.: Genetic Algorithms, IEEE Computer Society Press, 1992.

[6] Kristinsson, K. and Dumont, G. A.: System Identification and Control using Genetic Algorithms, IEEE Transactions on Systems, Man, and Cybernetics *1992*, 1033.

[7] Yao, L. and Sethares, W., Nonlinear Parameter Estimation via Genetic Algorithm, IEEE Transactions on Signal Processing *1994*, 927.

[8] Hunt, K.J.: System Identification with Genetic Algorithms, IEE Colloquium on Genetic Algorithm for Control System Engineering, 1993.

Scheduling Multiprocessor Tasks with Correlated Failures Using Population Learning Algorithm

Piotr Jędrzejowicz, Ewa Ratajczak * [1]

*Chair of Information Systems, Gdynia Maritime Academy, Gdynia, Poland, email:{pj,ewra}@wsm.gdynia.pl

Abstract

The paper considers a problem of scheduling multiprocessor tasks with correlated failures to maximize schedule reliability under time constraints. Since the problem is NP-hard the approximation algorithms based on a population learning algorithm are proposed.

1 Introduction

As it was pointed out by [1] there exist scheduling problems where tasks are processed on more than one processor at a time. During the execution of these multiprocessor (m-p) tasks communication among processors executing the same task is implicitly hidden in a "black box"denoting an assignment of this task to a subset of processors during some time interval. The paper deals with scheduling multiple variant software on multiple processors to maximize schedule reliability. Multiple variant software is modeled by m-p tasks. The proposed approach addresses one of the frequently encountered conflicts involving high dependability and safety standards required versus system performance and cost.

It should be noted that the m-p task concept could be used to model variety of the fault-tolerant structures, since all require processing of redundant variants, internal communication and execution of an adjudication algorithm. In such a model, the size of a task would correspond to the number of its redundant variants.

Scheduling m-p tasks is understood as assigning processing elements to tasks and, at the same time, deciding on task size and structure, in such a way that all constraints are satisfied and some overall performance goal is optimized. Scheduling m-p tasks differs from traditional, single-variant task scheduling problems, by the extended solution space. It includes not only an assignment of tasks to processors but also a decision as to which combination of redundant variants, from a set of the available ones, should be used to construct each task (at least one variant of each task has to be included within a schedule). Algorithms for scheduling multiprocessor tasks constructed from multiple program variants have been proposed in [3], [2]. The discussed algorithms are based on the assumption that failures of multiple program variants are statistically independent. The present approach differs by allowing correlation of failures within a task. Hence, the earlier assumption is not needed any longer.

In the paper global optimization criterion is schedule reliability understood as a probability that all scheduled tasks will be executed without failures. Schedule reliability is to be maximized under the requirement that all tasks meet their respective deadlines. Unfortunately the discussed scheduling problem belongs to the NP-hard class. In view of the above it has been decided to investigate a possibility of solving the discussed problem using a population learning algorithm which is a new approach in the class of population based methods.

2 Problem Formulation

A set N of n of multiprocessor tasks to be processed under time constraint is considered. It is assumed that the following information with respect to each task in N is available:

- Ready time - a_j, $j = 1, \ldots, n$
- Deadline - d_j, $j = 1, \ldots, n$
- Number of variants - NV_j, $j = 1, \ldots, n$
- Variant processing times - p_{ji}, $j = 1, \ldots, n$, $i = 1, \ldots, NV_j$;
- Variant reliabilities - r_{ji}, $j = 1, \ldots n$, $i = 1, \ldots, NV_j$
- Level of failure correlation between variants - θ_j, $j = 1, \ldots, n$

The considered problem of scheduling multiprocessor tasks, denoted using Graham's notation as $P|a_j, m-p|R$, is characterized by a set of multiple, identical processors P, and a set of multiple-variant tasks N. Each task has the maximum size NV_j, and the minimum size equal to 1. Tasks are independent and non-preemptable with ready times and deadlines differing per task. Tasks have processing

[1]The work was supported by the KBN grant NR 8T11F02019

times, which may differ per combination of variants chosen to be run. Task j of the size $|Gj| = m$ requires m parallel processors for processing. Optimization criterion is schedule reliability calculated as $R = \prod R_{ji}$. Task reliabilities are calculated using reliability model proposed in [5]. It allows that reliabilities of the redundant variants may vary and assumes that failures of variants within a task can be statistically correlated. Decision variables include assignment of tasks to processors and combination of variants to be run for each task. Tasks can not be delayed.

3 Population Learning Algorithm

Population learning algorithm (PLA) proposed in [4] has been inspired by analogies to a social phenomenon rather than to natural processes. Whereas evolutionary algorithms emulate basic features of natural evolution including natural selection, hereditary variations, the survival of the fittest and production of far more offspring than are necessary to replace current generation, population learning algorithms take advantage of features that are common to education systems.

The following features characterize contemporary education systems and learning processes therein:

- a massive number of individuals enter the system;
- individuals learn through organized tuition, interaction, self-study and self-improvement;
- learning process is inherently parallel (with different schools, curricula, teachers, etc.);
- learning process is divided into stages;
- more advanced stages are entered by a diminishing number of individuals from an initial population;
- at higher stages more advanced education techniques are used and more hard work, knowledge, skills, talent and abilities are required;
- the final stages can be reached by only a few of the best.

Effectiveness of the social learning process is the main factor behind an attempt of using some of its principles to solving difficult optimization problems. Population learning algorithms, as all other p-b methods, handle population of individuals. An individual could be a solution of the considered problem or it could be part of a solution or, in fact, it could be any other object that can be somehow transformed into a solution.

Initially, a massive population of individuals is generated. The number of individuals in the initial population should be sufficient to represent adequately the whole space of feasible solutions. What precisely "sufficient" and "adequately" in this context mean can not be easily defined in general terms. Sufficient number of individuals relates to the need of covering the neighborhood of all of the local optima. Adequate representation of these neighborhood is related to the need of assuring that the improvement processes, originated at the initial stage, should be effective enough to carry at least some individuals to the highest stages of learning.

Generating the initial population could be, simply, based on some random mechanism assuring the required representation of the whole feasible solution space. Once the initial population has been generated its individuals enter the first learning stage. It involves applying some, possibly basic and elementary, improvement schemes or conducting learning sessions. The improved individuals are than evaluated and better ones pass to the subsequent stage. A strategy of selecting better or more promising individuals must be defined and applied. At the following stages the whole cycle is repeated. Individuals are subject to improvement and learning, either individually or through information exchange, and the selected ones are again promoted to a higher stage and the remaining are dropped from the process. At the final stage problem solution is selected. Learning process at early stages can be run in parallel. At certain level the best from all groups join together to form higher level groups where improvement and learning processes are still carried in parallel. At some stage selected individuals are brought together to complete education. At different stages of the process different improvement schemes and learning procedures are applied. These gradually become more and more sophisticated and time consuming as there are less and less individuals to be taught.

4 PLA for Scheduling Multiprocessor Tasks

Three versions of the PLA solving $P|a_j, m-p|R$ problem have been designed: simple, parallel and modified parallel. All three versions use 2-optimal heuristics and local search methods as learning tools in the process of educating population of solutions.

The 2-optimal heuristics is based on finding an optimal solution for the subset of tasks (exactly two of them) by the exhaustive search. Search is continued for different pairs of tasks until a schedule can be constructed. Strategy for choosing pairs of tasks is based on the tabu-search technique. The local search, in turn, involves checking consecutively the neighborhood of each task accepting, eventually, changes that bring improvement to the value of the fitness factor of the individual in question.

Simple PLA proposed in the paper involves three learning stages and randomly generated initial population of individuals representing feasible solutions to the problem at hand. At the first stage (primary school) 2-optimal heuristics is applied for 10 iterations. To the secondary school only individuals which pass selection criterion are admitted. Several random crossovers are performed to extend population of individuals. In the process five best individuals are involved. In the secondary school 2-optimal heuristics is applied for 10 iterations. At the third stage - "university one" only the selected individuals are improved using a local search algorithm. After learning phases have been completed the best individual is considered as a final solution.

The simple-PLA algorithm involves executing steps shown in the following pseudo-code:

Begin

Generate primary population.

Improve tardy individuals by the Primary_School procedure to make them admissible.

Generate new solutions based on the best from the population.

Remove individuals which reliability is lower then average reliability in the population.

Improve individuals by the Secondary_School procedure based on optimum 2 with tabu search technique algorithm.

Remove individuals which reliability is lower then average reliability in the population.

Improve individuals by the University procedure based on tabu search algorithm.

end;

A parallel PLA is an extension of the simple case. There are two parallel primary schools, secondary schools and universities. Learning algorithms and selection criteria at parallel stages differ but are based on variations of 2-optimal heuristics and local search. Here, best individuals are also allowed to exchange information.

Finally the modified parallel PLA involves four primary schools, two secondary ones and one university. At intermediary stage individuals exchange information to create offspring. The approach is still under development.

5 Numerical Example and Computational Experiment Results

To evaluate the proposed approach a computational experiment has been carried out. It has involved 20 randomly generated data sets. The number of tasks varies between 10 to 29, each task has 2 to 5 variants. Number of processors varies between 2 and 8. Table I presents first 4 from 11 tasks from one of the data sets.

task	NV_j	a_j	p_{ji}	d_j	r_{ji}	θ_j
1	5	0	4	9	0.9250	0.1705
			4		0.9301	
			6		0.9365	
			6		0.9873	
			7		0.9998	
2	5	0	5	16	0.9450	0.1533
			5		0.9569	
			5		0.9600	
			6		0.9723	
			6		0.9825	
3	4	1	6	21	0.9029	0.1579
			6		0.9563	
			7		0.9762	
			8		0.9835	
4	4	3	9	25	0.9441	0.1289
			10		0.9619	
			11		0.9558	
			11		0.9608	
...	...	...	...	...	...	...

Table I. A part of the example data set.

Results obtained by s-PLA algorithm for this set are shown in the pictures. The best solution from the primary population has reliability of 0.49004253. In figure 1 the best individual from the population graduated from primary school is presented (reliability 0.90635751). This stage of the computation contributes to the biggest improvement. Next, the best individual educated from secondary school has reliability 0.91290253. The last improvement procedure called university, changes the result to the 0.91718554 (Figure 2).

Results for described data sets are presented in table II. The following measures have been used to evaluate the algorithms:

- mean relative error related to the best known solution – MRE;
- minimum relative error encountered – MinRE;
- maximum relative error encountered – MaxRE.

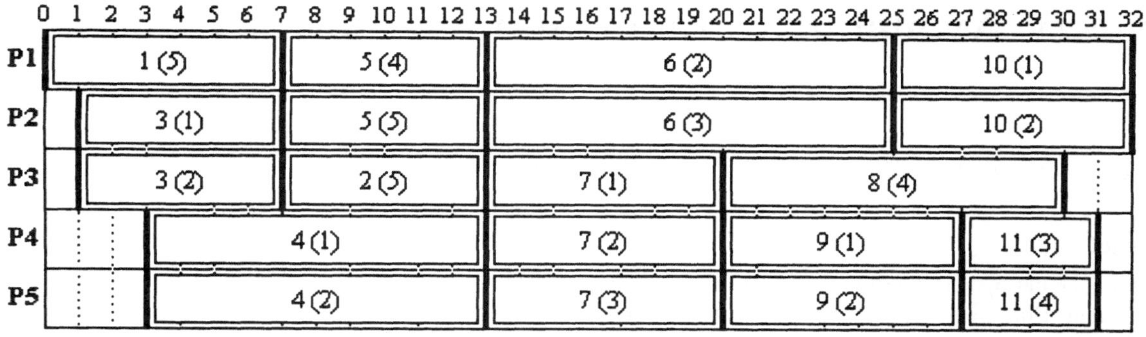

Fig. 1. Best solution after primary school stage, reliability 0.90635751

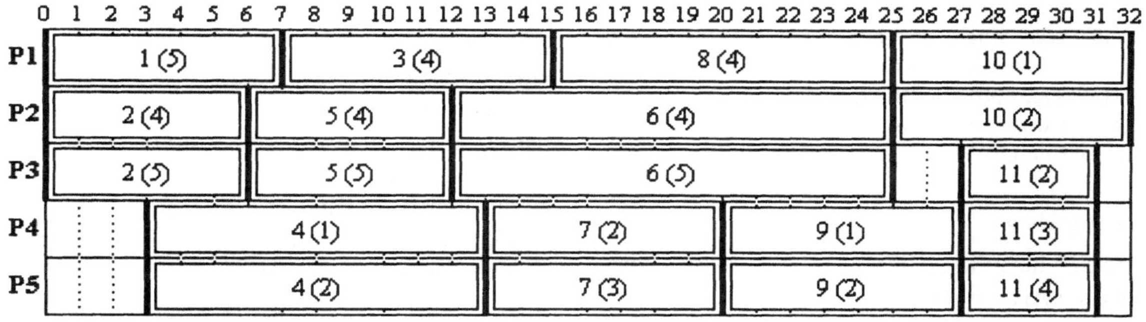

Fig. 2. Best solution after university stage, reliability 0.91718554

	simple PLA	parallel PLA	modified parallel PLA
MRE	0.71%	0.42%	0.40%
MinRE	0.00%	0.00%	0.00%
MaxRE	4.78%	3.20%	1.59%

Table II. Computational experiment results.

All presented algorithms are effective with regard to a quality of the results. The worst solution obtained for the described data set by simple-PLA has an error 4.78%. Mean relative errors for all data sets are below 1%. The MRE for parallel algorithms is about 0.3% lower then for the simple one.

6 Conclusions

It can be concluded that the proposed approach is promising and worth further research. Quality of solutions obtained by applying PLA is high. Critical resource, from the point of view of the effectiveness of the approach, is the computational time. Parallel versions of the PLA algorithm can be of help in reducing time needed to obtain good quality results. Further research should concentrate on investigating different learning tools and improvement algorithms as well as using parallel nature of the approach to reduce computational time requirements.

References

[1] J. Błażewicz, K.H. Ecker, E. Pesch, G. Schmidt, J. Węglarz, Scheduling Computer and Manufacturing Processes, Springer, Berlin 1996.

[2] I. Czarnowski, P. Jędrzejowicz, E. Ratajczak, Scheduling Fault-Tolerant Programs on Multiple Processors to Maximize Schedule Reliability, Lecture Notes in Computer Science *1698*, Springer (1999), pp. 385–395.

[3] P. Jędrzejowicz, I. Czarnowski, H. Szreder, A. Skakowski, Evolution-Based Scheduling of Fault-Tolerant Programs on Multiple Processors, Parallel and Distributed Processing, Lecture Notes in Computer Science *1586*, Springer (1999), pp. 210–219.

[4] P. Jędrzejowicz, Social Learning Algorithm as a Tool for Solving Some Difficult Scheduling Problems, Foundation of Computing and Decision Sciences *24*, 2 (1999), pp. 51–66.

[5] I. Czarnowski, W.J. Gutjahr, P. Jędrzejowicz, E. Ratajczak, A. Skakowski, I. Wierzbowska, Scheduling multiprocessor tasks in presence of the correlated failures, Procedings of the 2nd Intl. Workshop on Soft Computing Applied to Software Engineering, Rotterdam 2001, (Accepted paper).

Fuzzy Availability Agent in a Virtual Environment

J.R. King, M. Razaz*

*School of Information Systems, University of East Anglia, Norwich, England, {king,mr}@sys.uea.ac.uk

Abstract

This paper details the development of a fuzzy availability agent that operates within the Forum, a collaborative virtual environment (CVE) developed at the British Telecomms (BT) Laboratories. The agent (APIA) uses information collected by the Transponder from each participant's computer to determine presence and availability. Each user has a unique Fuzzy User Model (FUM) that contains a fuzzy representation of his work patterns and provides the knowledge based used by APIA. A real-world trial was conducted and the results are presented here.

1 Introduction

The use of fuzzy logic in multimedia applications and collaborative virtual environments (CVE) in particular has received very little attention in the literature [1]. The data presented in a typical multimedia application can be in the form of sound, images and texts or a combination thereof. Many features of such information, visual or otherwise, are not precisely specified, and therefore can benefit from the use of fuzzy logic.

The application of fuzzy logic to various aspects of the Forum, a CVE developed at BT Labs is being studied. In particular, a software system known as APIA (Availability of Person, Intelligent Agent), concerned with determining the availability and presence of participants in the Forum is being developed. A typical application for this system would be to aid in communication between people at remote sites where direct contact is not available. A software module of APIA is the Transponder system that runs on client machines, collecting the appropriate data. The creation method of unique fuzzy user models for each user as well as results from the trials and their analysis. Finally, we will give a summary and review of the information presented.

2 Collaborative Virtual Environments

A collaborative virtual environment (CVE) is a multi-user virtual reality system that explicitly supports co-operative work. The central concept of CVE is the support of mutual awareness between users [2]. This involves not only the recognition of presence amongst users, but also some form of identifying the user's role and activities within the environment. Once users are aware of one another it becomes critical to provide rich communication and collaboration techniques. This generally indicates the presence and support of audio, text, video, gesture, etc.

Researchers in this field are attempting to provide rich communication and collaboration between users in disparate locations. One focus of this research is in the provision of appropriate embodiments, or avatars to users. Embodiments are the virtual representation of users in the CVE. As put by [3], "the inhabitants of collaborative virtual environments (and other kinds of collaborative system [sic]) ought to be directly visible to themselves and to others through a process of direct and sufficiently rich embodiment".

3 The Forum

The lessons learned from the Virtuosi project [4, 5, 6] has been used to build the Forum, a system developed at BT Labs that focuses on facilitating contact and communication between participants. Two environments have been constructed that provide these facilities, the Contact and Meeting Spaces. The Contact Space promotes interaction between participants by placing users in zones arranged within 'action layers' that allow users to meet others who are performing similar actions, in addition to those with similar interests. The Forum Meeting Space provides an area in which participants can come together in a controlled virtual environment, generally for scheduled meetings. Razaz and King first introduced the application of fuzzy logic to various aspects of Virtuosi [1]. Early results from trials of the Forum have been reported in [7].

Neither of these two systems currently indicate the presence and availability of participants. This can result in awkward situations where a user is attempting to maintain a dialogue with another who is not present. The Meeting Space is considered a synchronous system whereby each user is present fully, but it is possible for a user to be called away from a meeting leaving behind a motionless avatar. Users

Idle time
Idle churn
New messages since last checked email
Last email time
Do Not Disturb flag
Scheduled now
Time since last appointment finished
Time until next appointment start

Table I. Available input variables for FUM

are encouraged to explore the different zones and layers within the Contact Space with the explicit goal of making contact with other users. Participants in this space will come into contact with a large number of users which will greatly increase the likelihood of interacting with an empty avatar. By providing feedback in the form of presence and availability, many more than two states are created.

4 APIA and the Transponder

Availability of Person - Intelligent Agent (APIA) is a software system within the Forum that operates by monitoring a variety of equipment used in the working environment. APIA is designed to both collect and provide information about users to other parts of the Forum with the purpose of aiding in the collaboration between users. APIA requires data to be collected from the host computer, a task performed by the Transponder. The Transponder was designed and developed to monitor the user's computer and collect usage data. There are six main categories of data collected: User, Email, Idle, Calendar, Applications, and Web Browser.

5 Fuzzy User Model

A Fuzzy User Model (FUM) based on fuzzy logic is created to provide a unique model for each user. This model allows our fuzzy software to make deductions about availability and presence for each user based on their specific situation. Another way to view the FUM is as a unique knowledge base which contains information on the working patterns of each user. A typical list of inputs available in a FUM is given in Table I. The Do Not Disturb and Scheduled inputs are Boolean variables while Idle churn is a measure of the volatility of the user's idle time.

The construction of FUMs utilises expert knowledge. An expert can determine rules from a visual investigation of the supplied data. The goal from the expert's point of view is to generate a fuzzy knowledge base for each user which makes accurate determinations of presence and availability. This task requires the expert to have a firm understanding of the various input variables as well as knowledge of the particular habits of individual users. The FUM is designed to work with our fuzzy logic controller (FLC) [8] and an example is given in Table II.

Each user may have patterns to their work that are based on the day of the week. For instance, a user may have a day set aside for meetings. Therefore, five days of collected data were used to create a FUM for each user. The remaining data were used to test the resulting FUM.

6 FUM Validation

In order to both create and determine the accuracy of a given FUM, additional validation data is required. Availability feedback data is collected and used to construct or train a fuzzy model for that user. A software program asks each user the following question every 30 minutes,

"How available are you if someone wanted to contact you now?"

The user is given five options ranging from 'I am NOT available' (1) to 'I am available' (5) and asked to select one. The intermediate choices of (2), (3), and (4) are evenly spaced answers between the two extremes, indicating different levels of availability . The question is a reasonable one, as the final goal of the software is to determine the user's availability. This data provides a valuable feedback mechanism that will allow a more accurate model to be determined for that user.

A direct measurement of presence is required for constructing the FUM. This is accomplished during the trial using a digital camera placed in a location that gives a clear view of each users' workspace. Each users' computer and desk space is visible within the image obtained. Due to the unique configuration of the work area, the common space used by all five users is also visible. This allowed the user's status in each image to be marked for four different cases,

1. The user is not visible in the image
2. The user is very near his computer
3. The user is within their work area, but not close to his computer
4. The status cannot be accurately determined.

The images are taken once every minute and allow for the direct determination of presence.

7 The Trial

Test data was collected in a two week trial at BT Labs. The purpose of the trial was to collect real-world data to test and validate our new fuzzy

if (IdleTime is Low)	then	(Availability is High)	and	(Presence is High)
if (IdleTime is Medium)	then	(Availability is High)	and	(Presence is MH)
if (IdleTime is High)	then	(Availability is High)	and	(Presence is Low)
if (LastEmail is Short)	then	(Availability is High)	and	(Presence is High)
if (IdleChurn is Medium)	then	(Presence is MH)		
if (IdleChurn is High)	then	(Presence is High)		
if (LastEmail is Medium)	and	(NewEmail is Normal)	then	(Availability is Medium)
if (LastEmail is Medium)	and	(NewEmail is High)	then	(Availability is Low)
if (LastEmail is Long)	and	(NewEmail is Normal)	then	(Availability is Low)
if (LastEmail is Long)	and	(NewEmail is High)	then	(Availability is Low)
If (LastAppt is Short)	or	(NextAppt is Short)	then	(Availability is Low)

Table II. FUM example

method for determining availability and presence. Five users were tracked for a period of ten working days. Each user's computer had the Transponder installed and was initially setup to collect the maximum amount of information. Each user was free to change the collection settings, and some took advantage of this.

In most cases the collected data contains gaps during periods when the Transponder was not running. These gaps occur when the user's computer is turned off (very rare), logged out (uncommon), or rebooted (common). Each of the five users is a software developer who uses his computer extensively during the working day. Breaks taken can range from just a few minutes when the user is getting a cup of coffee, to two or three hours for a meeting. Lunch times vary from the quick sandwich while working to two hour long breaks. Typically once or twice per day the users stop working and socialise with one another, although they may socialise while working as well. This generally takes place in the workplace and can come at any point during the day.

8 Results

It is not possible to provide the full results of the trial in this short paper. Therefore, a small subset of representative results is presented here. The availability and presence results for User Three on day two are shown in Figures 1 and 2 respectively.

Circles and crosses in Figure 1 indicate the value obtained from availability feedback collection. A circle is placed at a value selected from {-5,25,50,75,105} based on the availability selection of the user {1,2,3,4,5}. A cross is placed at -10 to indicate that the user did not select a response before the dialog box timed-out. The image and availability feedback is plotted along the left-hand y-axis.

The data is plotted off the 0 and 100 lines due to the large amount of input and output data plotted there. This aids in making the graphs clearer.

The availability results are shown in Figure 1. There are two periods where the user is scheduled, resulting in the two periods of 0 availability. At one point (around 15:00) during the appointment from 12:00 to 18:00, the user selects a 100% availability level. The user has marked himself as scheduled, but is in fact at his computer and available at that time. The fuzzy system detected two other times when the user selects 100% availability. The periods where the user selects lower availability are less well determined. The two periods where the user selects 0% are easily identified as the user is scheduled during those times. Of the ten times when the user does not provide feedback, or indicates 0% availability, the system detects eight. This would suggest an 80% success rate for those periods. Of the ten times the user indicates any availability, the system detects six within roughly 20%, indicating a 60% success rate. If success is determined to indicate a decision within 25% of the user supplied value, 80% of the determinations are successful.

In summary, the fuzzy system returns values for availability that are reasonable, 14 out of the 18 times when feedback is given, or roughly 75%. This analysis only takes into account those times when the user provides feedback, which occurs once every 30 minutes. As the system makes a determination for availability every minute, feedback is only available one in every 30 times (3.33%). Therefore the 75% success rate only applies to these 3.33% of values. A visual inspection of the input and output graphs shows that the fuzzy system gives reasonable results roughly 75%.

The numerical analysis of availability is subjective as definitive comparisons of expected and actual results are only possible for 3.33% of the actual results. A quantitative analysis would therefore be lacking and a qualitative approach is used instead. This is also in keeping with the fuzzy approach used in solving this problem.

Notice that when the user is visible the presence value in Figure2 is typically between 80 and 100 percent. The results for presence track exceptionally well the given user visibility. As the images obtained

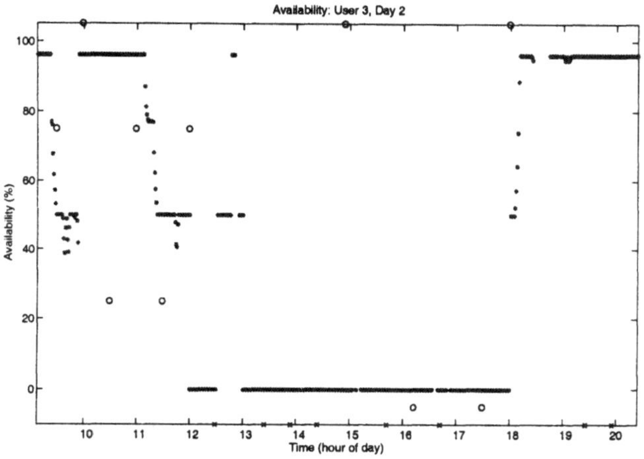

Fig. 1. Availability Results

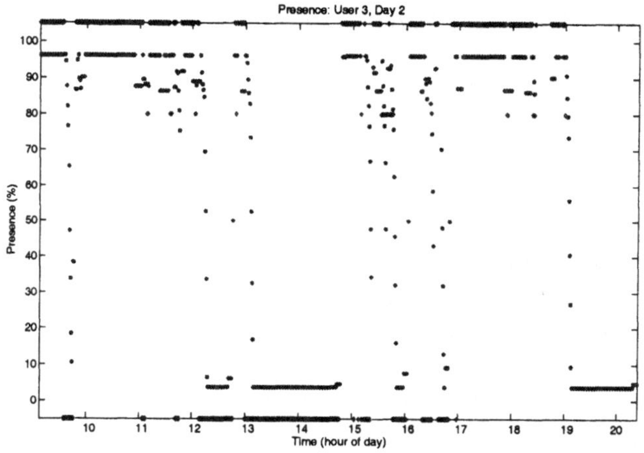

Fig. 2. Presence Results

during the trial were taken at one minute intervals, it is likely that the user is moving about the office but this is not obvious from the images. The idle churn data provides clues as to when this occurs. When the user is absent for periods of more than a few minutes, the presence value quickly reduces to zero. In periods where the user is moving about the office greatly, such as between 15:00 and 15:20, the presence value remains very high. In-between situations where the user is moving about but is only away for periods less than a minute are detected using the idle churn measure. This creates a characteristic pattern on the presence results where the values are between 80 and 100. This result is reasonable, as the user is actually present but is being highly mobile. A visual inspection of the input and output graphs suggests that the presence determinations are very reasonable.

It is not possible to calculate a success rate for each user, as shown above. Calculating that the fuzzy system is capable of a 75% success rate using 3.33% of the availability data does not translate to a 75% success rate for the entire data set. The success rate is a subjective quantity. Presence determinations must also be analysed using a subjective approach. A numerical analysis will fail due to the quantisations used on the camera data. While data are available for comparison for each data point, the quantised nature of the camera data makes such a comparison meaningless.

9 Conclusions

A new application area of fuzzy logic to multimedia has been explored. In particular, fuzzy logic for the first time was applied to determining availability of a person in a collaborative virtual environment. The determination of presence and availability has clearly been shown to be enhanced with the use of fuzzy logic and a range of input variables. It is possible that improved results can be obtained using better fuzzy user models and more extensive data collection. In either case, the fuzzy software developed is more flexible and powerful than its non-fuzzy counterpart.

References

[1] M. Razaz and J. King, "Application of fuzzy logic to multimedia technology," in Fuzzy Logic and Applications ISFL '97 (N. Steele, ed.), (Canada / Switzerland), pp. 333–337, ICSC Academic Press, February 1997.

[2] A. Steed and J. Tromp, "Experiences with the evaluation of cve applications," 1997.

[3] S. Benford, J. Bowers, L. Fahlen, C. Greenhalgh, and D. Snowdon, "User embodiment in collab orative virtual environments," in Proceedings of ACM Conference on Human Factors in Comput ing Systems (CHI'95), pp. 242–249, ACM Press, Addison Wesley, 1995.

[4] S. Benford and D. Snowdon, "Integrating com munication embodiment and information," tech. rep., BT Research Report, July 1995.

[5] A. Rogers, "Virtuosi – virtual support for group working," BT Technical Journal, vol. 12, no. 3, pp. 81–89, 1994.

[6] A. Duke, "Virtuosi: Availability," BT Research Report, December 1995.

[7] A. McGrath, "The forum," in Siggroup Bulletin, vol. 19, pp. 21–25, New York: ACM Press, 1998.

[8] M. Razaz and R. May, "Design of a fuzzy logic controller," University of East Anglia, Research Report, pp. 1–36, June 1995.

The Flowering of Fuzzy CoCo: Evolving Fuzzy Iris Classifiers

Carlos Andrés Peña-Reyes, Moshe Sipper*

*Logic Systems Laboratory, Swiss Federal Institute of Technology in Lausanne, CH-1015 Lausanne, Switzerland, E-mails: Carlos.Pena@epfl.ch, Moshe.Sipper@epfl.ch

Abstract

Combining the search power of coevolutionary computation with the expressive power of fuzzy systems, we present *Fuzzy CoCo:* Fuzzy Cooperative Coevolution. We demonstrate the efficacy of our algorithm by applying it to a hard problem—flower classification—obtaining the best classification performance to date, coupled with high human-interpretability.

1 Introduction

Fuzzy logic is a computational paradigm that provides a mathematical tool for representing and manipulating information in a way that resembles human communication and reasoning processes [1]. *Fuzzy modeling* is the task of identifying the parameters of a fuzzy inference system so that a desired behavior is attained [1]. This task becomes difficult when the available knowledge is incomplete or when the problem space is very large, thus motivating the use of automatic approaches to fuzzy modeling—such as evolutionary algorithms. In this paper we apply coevolutionary fuzzy modeling to a well-known benchmark classification problem: Fisher's iris data.

2 Fuzzy CoCo: A Cooperative Coevolutionary Approach to Fuzzy Modeling

Fuzzy CoCo is a Cooperative Coevolutionary approach to fuzzy modeling, wherein two coevolving species are defined: database (membership functions) and rule base. This approach is based primarily on the framework defined by Potter [2].

In Fuzzy CoCo, the fuzzy modeling problem is solved by two coevolving cooperative species. Individuals of the first species encode values which define completely all the membership functions for all the variables of the system. Individuals of the second species define a set of rules of the form:

if (v_1 **is** A_1) **and** ... (v_n **is** A_n) **then** (*output* **is** C),

where the term A_v indicates which one of the linguistic labels of fuzzy variable v is used by the rule. The two evolutionary algorithms used to control the evolution of the two populations are instances of a simple genetic algorithm. The genetic algorithms apply fitness-proportionate selection to choose the mating pool, and apply an elitist strategy with an elitism rate Er to allow a given proportion of the best individuals to survive into the next generation. Standard crossover and mutation operators are applied with probabilities P_c and P_m, respectively.

An individual undergoing fitness evaluation establishes cooperations with one or more representatives of the other species, i.e., it is combined with individuals from the other species to construct fuzzy systems. The fitness value assigned to the individual depends on the performance of the fuzzy systems it participated in. Representatives, or *cooperators*, are selected both fitness-proportionally and randomly from the last generation since they have already been assigned a fitness value. In Fuzzy CoCo, N_{cf} cooperators are probabilistically selected according to their fitness, usually the fittest individuals, thus favoring the exploitation of known good solutions. The other N_{cr} cooperators are selected randomly from the population to represent the diversity of the species, maintaining in this way exploration of the search space. For a more detailed exposition of Fuzzy CoCo see [3].

3 Applying Fuzzy CoCo to Fisher's Iris Data

Fisher's iris data, wherein iris flowers are classified according to external features, has been widely used to test classification and modeling algorithms, recently including fuzzy models [4, 5, 6, 7, 8]. We propose herein two types of fuzzy logic-based systems to solve the iris data classification problem: (1) fuzzy controller-type (as used by Shi *et al.* [4] and Russo [5]), and (2) fuzzy classifier-type (as used by Hong and Chen [6], Wu and Chen [7], and Hung and Lin [8]). Both types consist of a fuzzy inference subsystem whose output is fed to a selection unit.

In the fuzzy controller the fuzzy subsystem computes a single continuous value estimating the class to which the input vector belongs. Note that each class is assigned a numeric value: based on the iris data distribution, we assigned values 1, 2, and 3 to the classes ***setosa***, ***versicolor***, and ***virginica***, respectively (such an assignment makes sense only under

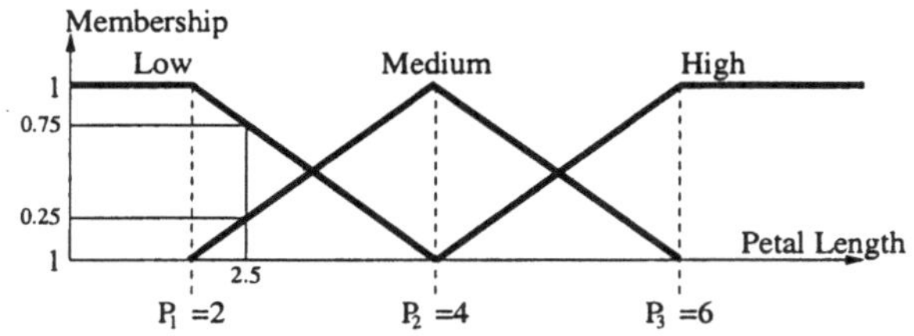

Fig. 1. Fuzzy variable *Petal Length.*

the assumption that *versicolor* is an intermediate species in between *setosa* and *virginica*). The selection unit approximates this value to the nearest class value using a stair function.

In the fuzzy classifier the fuzzy inference subsystem computes a continuous membership value for each of the three output classes. The selection unit chooses the most active class, provided that its membership value exceeds a given threshold (which we set to 0.5).

The two fuzzy subsystems thus differ in the number of output variables: a single output (with values {1,2,3}) for the controller-type and three outputs (with values {0,1}) for the classifier-type. In general, controller-type systems take advantage of data distribution while classifier-type systems offer higher interpretability because the output classes are independent; these latter systems are harder to design.

Fuzzy CoCo searches for four parameters: input membership-function values, relevant input variables, and antecedents and consequents of rules. The genomes of the two species are constructed as follows:

- Species 1: Membership functions. There are four input variables (*SL*, *SW*, *PL*, and *PW*), each with three parameters P_1, P_2, and P_3, defining the membership-function edges (Figure 1).
- Species 2: Rules (Controller-type systems). The *i*-th rule has the form:
  if (*SL* **is** A^i_{SL}) **and** ... **and** (*PW* **is** A^i_{PW}) **then** (*output* **is** C^i),
  A^i_j can take on the values: 1 (*Low*), 2 (*Medium*), 3 (*High*), or 0 (*Other*). C^i can take on the values: 1 (*setosa*), 2 (*versicolor*), or 3 (*virginica*).
- Species 2: Rules (Classifier-type systems). The *i*-th rule has the form:
  if (*SL* **is** A^i_{SL}) **and** ... **and** (*PW* **is** A^i_{PW}) **then** {(*setosa* **is** C^i_{set}), (*versicolor* **is** C^i_{ver}), (*virginica* **is** C^i_{vir})},
  A^i_j can take on the values: 1 (*Low*), 2 (*Medium*), 3 (*High*), or 0 (*Other*). C^i_j can take on the values: 0 (*No*), or 1 (*Yes*).

Table I delineates the parameter encoding for both species' genomes, which together describe an entire fuzzy system. Table II delineates values and ranges of values of the evolutionary parameters.

Table I. Genome encoding.

Species 1: Membership functions

Parameter	Values	Bits	Qty	Total bits
P_i	$[V_{mn} - V_{mx}]$	5	3×4	60

Species 2: Rules (Controller-type)

Parameter	Values	Bits	Qty	Total bits
A	{0,1,2,3}	2	$4 \times N_r$	$8 \times N_r$
C	{1,2,3}	2	$N_r + 1$	$2\times(N_r+1)$
	Total Genome Length			$10\times N_r+2$

Species 2: Rules (Classifier-type)

Parameter	Values	Bits	Qty	Total bits
A	{0,1,2,3}	2	$4 \times N_r$	$8 \times N_r$
C	{0,1}	1	$3\times(N_r+1)$	$3\times(N_r+1)$
	Total Genome Length			$11\times N_r+3$

Table II. Fuzzy CoCo set-up.

Parameter	Values
Population size N_p	{60,70}
Maximum generations G_{max}	$500 + 100 \times N_r$
Crossover probability P_c	1
Mutation probability P_m	{0.02,0.05,0.1}
Elitism rate E_r	{0.1,0.2}
"Fit" cooperators N_{cf}	1
Random cooperators N_{cr}	{1,2}

Our fitness function combines three criteria: (1) F_c: classification performance, computed as the percentage of cases correctly classified; (2) F_{mse}: a value dependent on the mean square error (*mse*), measured between the continuous values of the outputs and the correct classification given by the iris data set ($F_{mse} = 1 - mse$); and (3) F_v: a rule-length dependent fitness with value 0 when the average number of variables per active rule is maximal and equal to 1 in the hypothetical case of zero-variable rules. The fitness function combines these three measures:

$$F = \begin{cases} F_c \times F^{\beta}_{mse} & \text{if } F_c < 1 \\ (F_c - \alpha F_v) \times F^{\beta}_{mse} & \text{if } F_c = 1, \end{cases}$$

where $\alpha = 1/150$ and $\beta = 0.3$.

4 Results

In this section we present the fuzzy systems evolved using Fuzzy CoCo for the two setups described above. We compare our systems with those presented in recently published articles, and detail two high-performance systems obtained.

Table III. Comparison of results. Parentheses show average number of variables per rule.

Rules per system	Shi *et al.* [4] best	FuGeNeSys [5] best	Fuzzy CoCo average	Fuzzy CoCo best
2	--	--	98.71% (1.9)	99.33% (2)
3	--	--	99.10% (1.3)	100% (1.7)
4	98.00% (2.6)	--	99.12% (1.3)	100% (2.5)
5	--	100% (3.3)	--	--

Database

	SL	*SW*	*PL*	*PW*
P_1	5.68	3.16	1.19	1.55
P_2	6.45	3.16	1.77	1.65
P_3	7.10	3.45	6.03	1.74

Rule base

Rule 1	**if** (*PL* **is** *High*) **then** (*output* **is** *virginica*)
Rule 2	**if** (*SW* **is** *Low*) **and** (*PW* **is** *Low*) **then** (*output* **is** *virginica*)
Rule 3	**if** (*SL* **is** *Medium*) **and** (*PW* **is** *Medium*) **then** (*output* **is** *setosa*)
Default	**else** (*output* **is** *setosa*)

Fig. 2. The best evolved, controller-type system with three rules. It exhibits a classification rate of 100%, and an average of 1.7 variables per rule.

4.1 Controller-type systems

We performed a total of 145 evolutionary runs, searching for controller-type systems with 2, 3, and 4 rules, all runs of which found systems whose classification performance exceeds 97.33% (i.e., the worst system misclassifies only 4 cases). The average classification performance of these runs was 98.98%, corresponding to 1.5 misclassifications. 121 runs led to a fuzzy system misclassifying 2 or less cases, and of these, 4 runs found perfect classifiers.

Table III compares our best controller-type systems with the top systems obtained by two other evolutionary fuzzy modeling approaches. Shi *et al.* [4] used a simple genetic algorithm with adaptive crossover and adaptive mutation operators. Russo's FuGeNeSys method [5] combines evolutionary algorithms and neural networks to produce fuzzy systems. The main drawback of these two methods is the low interpretability of the generated systems. As they do not define constraints on the input membership-function shapes, almost none of the semantic criteria favoring interpretability are respected [9]. As evident in Table III, the evolved fuzzy systems described in this section surpass those obtained by the two other approaches in terms of performance, while maintaining high interpretability. Our approach not only produces systems exhibiting high performance, but also ones with less rules and less antecedents per rule (which systems are thus more interpretable).

Fuzzy CoCo found controller-type systems with 3 and 4 rules exhibiting perfect performance (no misclassifications). Among these, we consider as best the system with fewest rules and variables. Figure 2 presents one such three-rule system, with an average of 1.7 variables per rule.

4.2 Classifier-type systems

We performed a total of 144 evolutionary runs, searching for controller-type systems with 2, 3, and 4 rules, all runs of which found systems whose classification performance exceeds 95.33% (i.e., the worst system misclassifies 7 cases). The average classification performance of these runs was 97.40%, corresponding to 3.9 misclassifications. 104 runs led to a fuzzy system misclassifying 5 or less cases, and of these, 13 runs found systems with a single misclassification.

Table IV compares our best classifier-type systems with the top systems obtained by three other fuzzy modeling approaches. Hong and Chen [6] and Wu and Chen [7] proposed sequential learning methods to progressively construct fuzzy systems. These two approaches are able to find systems with either a few [7] or simple rules [6]. They do not, however, constrain the input membership functions, thus rendering the obtained systems less interpretable. Hung and Lin [8] proposed a neuro-fuzzy hybrid approach to learn classifier-type systems. As their learning strategy hinges mainly on the adaptation of the connection weights, their systems exhibit low interpretability. The evolved fuzzy systems described herein surpass those obtained by these three approaches in terms of both performance and interpretability. As evident in Table IV, our approach not only produces systems exhibiting higher performance, but also ones with less rules and less antecedents per rule (which are thus more interpretable).

Fuzzy CoCo found classifier-type systems with 3 and 4 rules exhibiting the highest classification performance to date (i.e., 99.33%, corresponding to 1 misclassification). We consider as most interesting the system with the smallest number of conditions (i.e., the total number of variables in the rules). Figure 3 presents one such three-rule system with an average of 2.3 variables per rule, corresponding to a total of 7 conditions.

Table IV. Comparison of results. Parentheses show average number of variables per rule.

Rules per system	Hong and Chen [6]	Wu and Chen [7]	Hung and Lin [8]	Fuzzy CoCo	
	best	average	average	average	best
2	--	--	--	96.47% (2.1)	98.00% (1.5)
3	--	96.21% (4)	--	97.51% (2.4)	**99.33% (2.3)**
4	--	--	97.40% (4)	98.21% (2.3)	99.33% (2)
8	97.33% (2)	--	--	--	--

Database

	SL	*SW*	*PL*	*PW*
P_1	4.65	2.68	4.68	0.39
P_2	4.65	3.74	5.26	1.16
P_3	5.81	4.61	6.03	2.03

Rule base

Rule 1	**if** (*PW* **is** *Low*) **then** {(*setosa* **is** *Yes*), (*versicolor* **is** *No*), (*virginica* **is** *No*) }
Rule 2	**if** (*PL* **is** *Low*) **and** (*PW* **is** *Medium*) **then**{(*setosa* **is** *No*), (*versicolor* **is** *Yes*), (*virginica* **is** *No*)}
Rule 3	**if** (*SL* **is** *High*) **and** (*SW* **is** *Medium*) **and** (*PL* **is** *Low*) **and** (*PW* **is** *High*) **then**{(*setosa* **is** *No*), (*versicolor* **is** *Yes*), (*virginica* **is** *No*)}
Default	**else**{(*setosa* **is** *No*), (*versicolor* **is** *No*), (*virginica* **is** *Yes*)}

Fig. 3. The best evolved, classifier-type system with three rules. It exhibits a classification rate of 99.33%, and an average of 2.3 variables per rule.

5 Concluding Remarks

We presented Fuzzy CoCo, a cooperative coevolutionary approach to fuzzy modeling, and applied it to Fisher's iris data problem. Comparing our results with other fuzzy-modeling approaches, we conclude that our coevolved systems attain higher classification performance and better interpretability. These promising results have incited us to engage in further investigation, specifically: (1) application of Fuzzy CoCo to more complex problems, and (2) improving and expanding upon the methodology presented herein. Our underlying goal is to provide an approach for automatically producing high-performance, interpretable fuzzy systems for real-world problems.

References

[1] R. R. Yager and D. P. Filev, *Essentials of Fuzzy Modeling and Control.* New York: John Wiley & Sons., 1994.

[2] M. A. Potter and K. A. DeJong, "Cooperative coevolution: An architecture for evolving coadapted subcomponents," *Evolutionary Computation*, vol. 8, pp. 1–29, spring 2000.

[3] C. A. Peña-Reyes and M. Sipper, "Applying Fuzzy CoCo to breast cancer diagnosis," in *Proceedings of the 2000 Congress on Evolutionary Computation (CEC00)*, vol. 2, pp. 1168–1175, IEEE Press, Piscataway, NJ, USA, 2000.

[4] Y. Shi, R. Eberhart, and Y. Chen, "Implementation of evolutionary fuzzy systems," *IEEE Transactions on Fuzzy Systems*, vol. 7, pp. 109–119, April 1999.

[5] M. Russo, "FuGeNeSys - A fuzzy genetic neural system for fuzzy modeling," *IEEE Transactions on Fuzzy Systems*, vol. 6, pp. 373–388, August 1998.

[6] T. P. Hong and J. B. Chen, "Processing individual fuzzy attributes for fuzzy rule induction," *Fuzzy Sets and Systems*, vol. 112, pp. 127–140, May 2000.

[7] T.-P. Wu and S.-M. Chen, "A new method for constructing membership functions and fuzzy rules from training examples," *IEEE Transactions on Systems, Man and Cybernetics, Part B: Cybernetics*, vol. 29, pp. 25–40, February 1999.

[8] C. A. Hung and S. F. Lin, "An incremental learning neural network for pattern classification," *International Journal of Pattern Recognition and Artificial Intelligence*, vol. 13, pp. 913–928, September 1999.

[9] C. A. Peña-Reyes and M. Sipper, "A fuzzy-genetic approach to breast cancer diagnosis," *Artificial Intelligence in Medicine*, vol. 17, pp. 131–155, October 1999.

Genetic Algorithm Based Parameters Identification for Power Transformer Thermal Overload Protection

V. Galdi, L. Ippolito, A. Piccolo, A. Vaccaro *

*Department of Electronic & Electrical Engineering, University of Salerno, Fisciano (SA), I-84084, Italy

Abstract

Recent studies by various authors have shown as the IEEE Transformer Loading Guide model and the more recent modified equations, proposed by the Working Group K3 of the IEEE "Power System Relaying Committee", are lacking in accuracy in prediction the winding hottest spot temperature of a power transformer in presence of overload conditions. This is mainly due to the deviation of the parameters of the thermal model of the power transformer in presence of overload conditions. In the paper a novel technique to identify the thermal parameters to be used for the estimation of the hot spot temperature is presented. The proposed method is based on a Genetic Algorithm (GA) which, working on the load current and on the measured hot spot temperature pattern, permits to identify a corrected set of parameters for the thermal model of the power transformer. Thanks to data obtained from experimental tests, the GA based method is tested to evaluate the performance of the proposed method in terms of accuracy.

1 Introduction

Monitoring and protection of mineral-oil-filled power transformers are of critical importance in power systems, since they can cause widespread power outages of the distribution power systems.

Today, protection of power transformer is of critical importance considering that the utilities, in order to increase system operation margins in presence of overload conditions, are compelled to adopt a new approach aimed to load power transformers beyond nameplate ratings, for a short or a long time.

This approach, although it allows the exploitation of the full capacity of the transformer, requires an accurate monitoring of the transformer thermal state and in particular of the evolution of the hot spot temperature of the windings [1,2,3].

Undoubtedly a direct measurement of the windings hottest spot temperature can guarantee a very accurate transformer thermal monitoring, but it needs of a optical fibre based measurement station, which could be expensive for medium power electrical transformers.

An alternative technique frequently adopted for the thermal transformer protection is based on the estimation of the windings hottest spot temperature by analytical models descriptive of the heating equations driving the internal heat flow exchanges.

In particular, amongst the possible approaches, the procedure reported in the IEEE load guide [4], with the improvement proposed in [5], seems to be handy and therefore particularly suitable for the implementation on a programmable unit.

As recent studies have shown [3,6,7,8], the main limitation of this approach is that the model accuracy decays drastically in presence of overload conditions due to the deviation of the characteristic parameters from their nameplate values. Therefore in order to guarantee a good level of accuracy in the hot spot estimation also in presence of different overload conditions it is necessary to manage this parameter fluctuations evaluating proper correction factors.

Starting from this considerations in the paper it is presented a new identification method for the correction of the thermal parameters of a power transformer in presence of overload conditions. The proposed approach, starting from measured data acquired directly from a laboratory prototype mineral-oil-immersed power transformer, identifies by a genetic based procedure a set of corrected transformer thermal parameters that minimise, on a typical daily load pattern, the global error between the measured and the estimated hot spot temperature.

2 Mathematical Model of the Transformer Heating

The instantaneous evolution of the winding hot spot temperature in the top or in the center of the high or low voltage winding of a power transformer could be identified solving the transient heating equation describing the dynamic of the source and the sink of transformer heating. The mathematical formalisation of such a problem however leads to the definition of a set of non linear coupled differential equations whose resolution is hardware and time consuming.

In order to overcome this limits alternative approaches, based on simplified version of the transient heating equations, could be employed.

For this purpose it is possible to adopt the analytical model for the estimation of the winding hot spot temperature described in the IEEE Transformer

Loading Guide [4] with the improvements proposed in [5]. In accordance to this methodology the hottest-spot temperature could be calculated as the sum of two components, the top oil temperature Θ_{TO} and the hot spot rise above top oil temperature $\Delta\Theta_H$ as expressed in the following equation:

$$\Theta_H = \Theta_{TO} + \Delta\Theta_H \tag{1}$$

As detailed reported in [5] an estimation of the evolution of such variables could be identified starting from the transformer characteristic parameters according to the follows equations:

$$\begin{cases} \tau_{TO}\dfrac{d\Theta_{TO}}{dt} = \left[\Delta\Theta_{TO,U} + \Theta_A\right] - \Theta_{TO} \\ \tau_H \dfrac{d\Delta\Theta_H}{dt} = \Delta\Theta_{H,U} - \Delta\Theta_H \\ \Delta\Theta_{TO,U} = \Delta\Theta_{TO,R}\left[\dfrac{I_L^2 R + 1}{R+1}\right]^q \\ \Delta\Theta_{H,U} = \Delta\Theta_{H,R} I_L^{2m} \end{cases} \tag{2}$$

Usually, the values to be used for the model parameters are indicated in the test reports provided by the manufacturer as nameplate data [5].

As recent studies have shown, the main limitation of this kind of approach is that the accuracy of the thermal model based on the nameplate data decay drastically in correspondence of overload conditions [8].

Therefore in order to achieve a good level of accuracy in the hot spot estimation also in presence of different overload conditions it is necessary to manage this parameters fluctuation evaluating proper correction factors.

In order to do this in the next section a novel technique, based on a genetic identification procedure, for the estimation of the optimal transformer thermal parameters in presence of different load conditions, is developed.

3 Parameter Identification Method

The main objective of the parameters identification procedure is to identify the set of transformer thermal parameters Γ belonging to the solution space Ω that minimise, on a typical daily load pattern, the global error between the measured $\Theta_{H,i}^m$ and the estimated $\Theta_{H,i}^e(\Gamma)$ hot spot temperature samples.

In particular the overall problem can be regarded as an unconstraint minimisation of the objective function defined as the weighted sum of the estimated error:

$$\begin{cases} \min\limits_{\Gamma\in\Omega} f(\Gamma) \\ f(\Gamma) = \sum\limits_i w_i \left[\Theta_{H,i}^m - \Theta_{H,i}^e(\Gamma)\right]^2 \end{cases} \tag{3}$$

Observing the expression of the goal function, that obviously can assume different formulation depending on the case of interest, it is important to emphasise the function of the weighting factors w_i that represents a set of degree of freedoms, which can be used to take into account the relevance of the estimated error in dependence of the operating conditions.

2.1 Proposed solution

To solve the aforementioned problem an iterative optimisation method is proposed. The method, starting from the nameplate values of the thermal parameters, fixes at first the boundary of the design variables and then identifies, using a minimum search technique, the corrected value of the parameters thermal model that minimise the objective function (eqn. 3).

For this optimisation problem Genetic Algorithms (GAs) seem to be particular suitable. If compared with conventional optimisation methods, such as calculus-based and enumerative strategies, they are robust, global and may be generally applied without recourse to domain-specific heuristics [9,10].

The starting point in applying a genetic search technique to the scheduled problem is to choose a chromosome representation to describe each individual in the population of interest and a fitness function to evaluate its status.

Starting from the previous argumentation it is worth to assume as chromosome a vector of seven elements representing a possible set of transformer thermal parameters:

$$(\Delta\Theta_{TO,R}, \Delta\Theta_{H,R}, \tau_{TO}, \tau_H, R, m, q) \tag{4}$$

Each gene is therefore a floating point variable that can vary in a proper neighbour of the respective nameplate value. The form of the selected fitness function is $k/f(\Gamma)$, where k is a very large number used to amplify the value of the function $(1/f(\Gamma))$ so that the fitness value of the chromosomes will be in a wide range.

Once the chromosome representation and fitness function has been defined it needs to fix the other features of the genetic algorithm in terms of selection mechanism and genetic operators.

In particular the selection function adopted is based on a tournament selection while as genetic operators the heuristic crossover, which takes two parents and performs an extrapolation along the line formed by the two parents outward in the direction of the better parent, and the non-uniform mutation, which changes one of the parameter of the parent based on a non-uniform probability distribution, are used [11].

4 Experimental Application

In order to prove the validity of the proposed approach the results of an experimental activity on a laboratory prototype mineral-oil-immersed power transformer are presented.

4.1 Data acquisition

The measurement station was equipped with an array of optical fibre based temperature sensors for direct measurement of duct oil, bulk oil and conductor temperatures [6,8].

With the above mentioned measurement station, the test program has executed simulating various realistic daily transformer loading current. The gathered data are then organised into two different sets.

4.2 Simulation results

The proposed parameter identification method, using the experimental data set giving the load current profile and the hot spot temperature reported in figs. 1 and 2, identifies the corrected parameters.

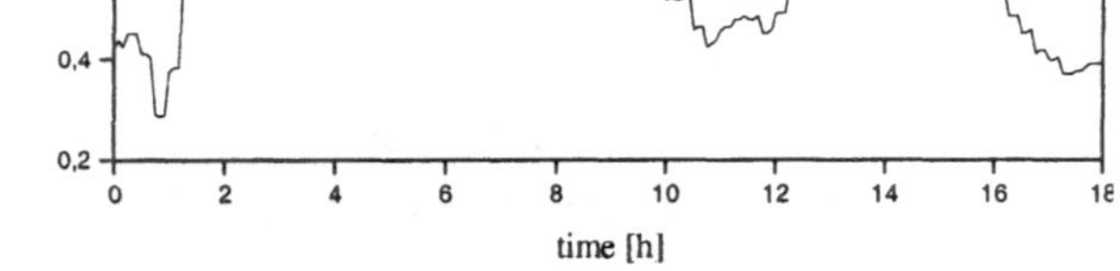

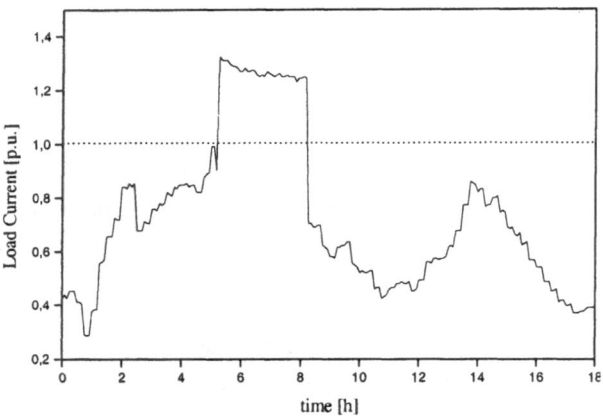

Fig. 1: Measured load current against time in presence of a overload about 3 hours long

As it can be noted the load pattern is relative to a 18-h observation period with $N_C = 220$ sample points. The adopted weighting factors w_i are assumed linearly correlated to the measured load current samples $I_{L,i}^m$:

$$w_i = \frac{I_{L,i}^m}{I_{L,rated}} \tag{5}$$

Starting from an initial population randomly generated and after 300 generations the optimisation procedure identifies the corrected parameters set reported in tab. I, where there are also indicated the nameplate data furnished by the transformer manufacturer.

Table I: Summary of the obtained results

Parameter	Nameplate value	Identified value
τ_H	0.1	0.15
$\Delta\Theta_{H,R}$	4	5.45
m	1.6	1.58
τ_{TO}	3	2.52
$\Delta\Theta_{TO,R}$	40	37.97
R	4	4.94
x	0.8	0.99

The adoption of such parameters in the resolution of the model (eqn. 2) led to the results shown in fig. 2, where the estimated and measured hot spot temperature curves are depicted. In the same diagram it is reported also the hot spot winding temperature profile estimated solving the eqn. 2 using the nameplate data.

The comparison shows that while the accuracy of the model (eqn. 2) solved with the nameplate data decays rapidly as the overload conditions become severe, the hot-spot estimation processed using the identified parameters set insure a better accuracy especially in correspondence of the overload. In particular the comparison of the mean squared error values, reported in tab. II, for the estimated winding hot-spot temperature with that one obtained solving the thermal model evidences a strong reduction of the deviation from the measured value.

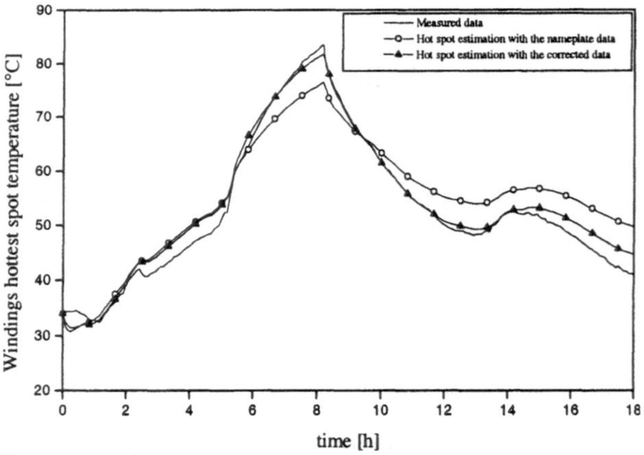

Fig. 2: Winding hot-spot temperature profiles in presence of a overload about 3 hours long

At the end in order to verify the generalised capability of the model another set of experimental data characterised of the load current profile shown in fig. 3 is employed for a further hot-spot estimation.

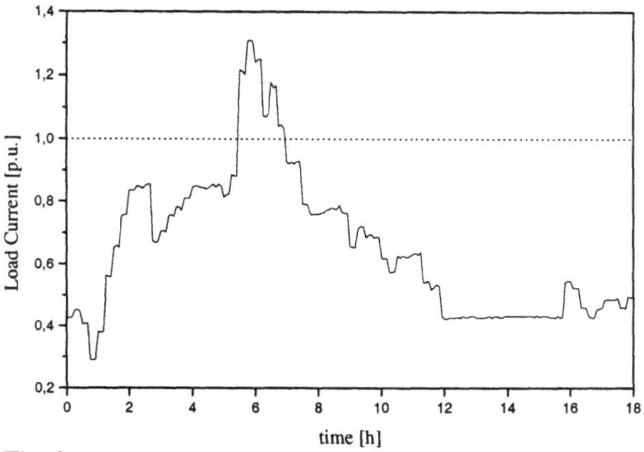

Fig. 3: Measured load current with a overload condition about 2 hours long

Table II: Comparison of the results

	MSE	
	with nameplate parameters	with identified parameters
3h of overload	21.25	3.45
2h of overload	13.38	3.07

The analysis of the hot-spot curves depicted in fig. 4 and value of the mean squared error (see tab. II) for the estimated temperature shows an improvement of the estimation accuracy also for this data set that describes an operating condition extremely different from that considered in the identification procedure.

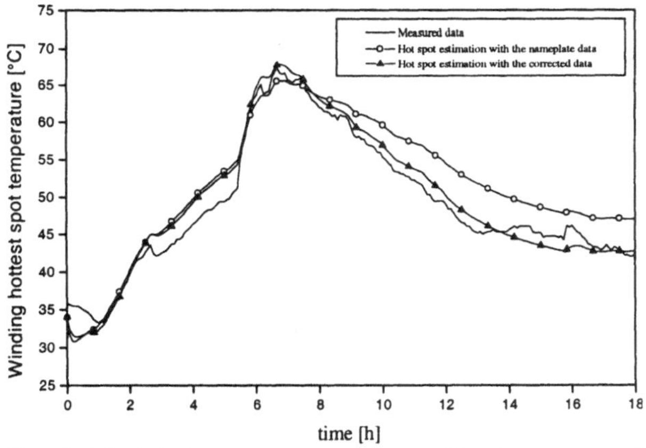

Fig. 4: Winding hot-spot temperature profiles in presence of a overload about 2 hours long

5 Conclusion

In this paper, a GA based parameters identification for power transformer thermal overload protection has been developed. The performance of the proposed method in the computational experiments was quite satisfactory.

The better accuracy compared with the results obtained using the nameplate data, as suggested by the Working Group K3 of the IEEE Power System Relay Committee, allows to increase power transformer loadability and, consequently, the power system operation margins in presence of overload conditions.

References

[1] B. D. Lahoti, D. E. Flowers: Evaluation of transformer loading above nameplate rating. IEEE Trans. on Power Apparatus and Systems PAS-100 (4), 1989 (1981).

[2] CIGRE WG 12.09 (Thermal Aspect of Transformers): Survey of power transformer overload field practices. CIGRE 1995, 147.

[3] CIGRE WG 12.09 (Thermal Aspect of Transformers): A survey of facts and opinion on the maximum safe operating temperatures under emergency conditions. CIGRE 1995, 25.

[4] IEEE: IEEE Guide for Loading Mineral -Oil-Immersed Transformers. Std. C57.91 1995.

[5] IEEE WG K3 (Transformer Thermal Overload Protection): Adaptive transformer thermal overload protection. IEEE 1999.

[6] P. Daponte, D. Grimaldi, A. Piccolo, D. Villacci: A neural diagnostic system for the monitoring of transformer heating. Measurement, Elsevier 18(1), 35 (1996).

[7] F. Gagliardi, L. Ippolito, D. Villacci: Il controllo e la protezione termica dei trasformatori di potenza mediante impiego di reti neurali. L'Energia Elettrica 74(2), 135 (1997).

[8] V. Galdi, L. Ippolito, A. Piccolo, A. Vaccaro: Neural diagnostic system for thermal overload protection. IEE Proc. Electric Power Applications 147(5), 415 (2000).

[9] D. Goldberg: Genetic algorithms in search optimisation and machine learning, Addison Wesley 1989.

[10] J. S. R. Jang, C. T. Sun, E. Mizutani: Neuro-Fuzzy and Soft Computing, 1th ed. Upper Saddle River: Prentice-Hall Inc. 1997.

[11] K. F. Man, K. S. Tang, S. Kwong: Genetic algorithms: concepts and applications. IEEE Trans. on Industrial Electronics 43(5), 1996.

Optimal Municipal Bus Routing Using a Genetic Algorithm

Jay M. Suiter, Donald H. Cooley *

*Computer Science Department, Utah State University, Logan, Utah, USA, cooley@don.cs.usu.edu

Abstract

Efficient utilization of mass transit is a cost effective and environmentally sound means to meet urban transportation needs. One of the most common forms of mass transit is the municipal bus. The process of bus route assignment is extremely complex when issues such as efficiency, ridership, access, etc. are considered. This paper describes a genetic algorithmic (GA) approach for the efficient design of municipal bus routes. For complex problems, the GA-based algorithm is more efficient and flexible than manual or other computer-based processes. The genetic algorithm in this study finds an optimal or near optimal route based on the features of a given bus route service area. For this study, four features were considered: ridership, the significance of visited sites, distance, and impediments. Other attributes such as fuel efficiency, time of travel, etc. can be added to the system, and the user can change the relative importance of attributes to allow for a "what if" type of analysis.

1 Introduction

Efforts to optimize municipal bus routes (MBR's) have generally been in the areas of bus and driver scheduling. For general MBR optimization, little work has been done. This is because such optimizations are complex and computationally intensive [3]. This paper presents a new and effective approach to the problem of MBR optimization. The approach uses a genetic algorithm (GA) to find optimum routes in the kind of large complex search spaces associated with a MBR.

2 Previous Work and Goals

There has been considerable research on school bus and delivery route optimization [1], [3], [7]. In school bus routing, the goal is to find the shortest overall distance while visiting all specified bus-stop sites. The delivery route goal is to find the shortest route to pick-up and deliver packages at all sites. Both are basically traveling salesperson problems. The constraints and goals of such problems differ from those of the MBR problem. An optimal MBR is one in which several diverse community attributes are utilized in the optimization. Ceder [2] describes the bus route problem as a network of sub-problems. Route generation is accomplished by first defining an origin and a destination. Possible routes from the origin to the destination are then almost randomly selected, and the routes are graded based on cost and customer service criteria. The route scoring the highest becomes the "optimal" route. With a large service area, the probability of randomly picking an optimum route is extremely low. In van Ness [10], a model was created that emphasized network design rather than route design. A set of individual routes with different origin-destination pairs was input to and a complete network with these routes and fitness values was output. The individual routes that are input to the model are determined manually, and thus optimality is unlikely. In Pattnaik [8], a GA was used to select from a set of candidate routes. The individual routes input to the GA were determined using Dijakstra's shortest path algorithm. The shortest path algorithm does not sufficiently represent the bus route problem. Several route attributes in addition to distance should be considered. Some of these will be positive and some negative in terms of the optimum. Furthermore, the weight values for these attributes will change depending on issues such as multiple traversal of a street, etc.In Ramirez and Seneviratne [9] a shortest path algorithm was also used with the weights of the paths between nodes based on two conflicting attributes (ridership and distance).

As a minimum, in order for a bus route optimum to be found, several attributes need to be considered (optimized):

- The travel route should serve the most riders.
- The route should be adjacent to popular or significant places.
- Travel distance should be minimized.
- Impediments such as four-way stops, construction, etc. should be minimized.

The goal of this research was to show that a genetic algorithm (GA), could effectively determine an optimum route or transit based on such attributes.

3 Problem Space

A service area can be represented as a graph with nodes and edges. The nodes represent intersections, and the edges represent roads. To accommodate one-way streets, there are either one or two edges between nodes. Four weights were assigned to each edge (road) denoting ridership, distance, significance, and impedance.

The impedance of an edge is an indication of the overall difficulty a bus will encounter while traveling that road. The significance of an edge indicates the presence of significant or popular facilities such as, hospitals, schools, handicapped rider locations, etc. The potential number of riders for each edge is the ridership. An accurate technique for ridership determination is difficult to arrive at. For study purposes, ridership values were randomly assigned to each edge. Time of travel, fuel consumed, emissions generated, etc. are also route attribute to minimize. Four route attributes provided sufficient complexity to test the GA, and hence other attributes were not considered. Furthermore, the distance traveled attribute is related to many such attributes. Ridership and significance are route positive route attributes, while impedance and distance are negative route attributes.

The fitness of a chromosome is a measure of the quality of the route it represents. In this work, fitness was the accumulation of these four attributes. In the calculations, each attribute's fitness was normalized by its population maximum. To reduce the number of twice-traveled edges, ridership was only counted the first time an edge was visited. The inherent penalty for visiting an edge twice was that distance and impedance were accumulated. To further reduce the number of twice-traveled edges, the impedance penalty was doubled each time an edge was re-visited. To reduce the likelihood that both directions of an edge would be traveled, the significance of both edges between two nodes was included in the fitness calculation. To complicate the problem, there are conflicts among attributes. For example, a small distance will usually produce a small ridership value and a small significance value. In order to produce effective bus routes, all features whether they conflict or not must be considered in calculating the fitness of a bus route. Included with each fitness feature such as distance was a weight. Changing and attribute's weight changes its importance in the overvall fitness expression.

The total fitness for a route for chromosome i is given as:

$$TF_i = WD*\Sigma NDF_i + WI*\Sigma NIF_i + WR*\Sigma NRF_i + WS*\Sigma NSF_i \quad (1)$$

Where:

- WD, WI, WR, and WS are user set weights
- $N\text{-}F_i$ represents the normalized distance (D), impediance (I), ridership (R), and significance (S) values for chromosome i

4 Chromosome Encoding

The most important element of a GA is the encoding scheme. In this work, each chromosome was encoded by listing the node number visited in order visited. Since the number of nodes visited was not fixed, the length of a chromosome varied.

4.1 Population initialization

A GA begins by randomly generating an initial population of chromosomes or in this case, routes. To reduce the size of the search space, a "guided" random technique for population generation was developed. This technique insured that all initial chromosomes represented legal transits. i.e. a path of finite length, with a correct start and end node. In the technique developed, relatively efficient routes with no loops or U-turns were generated To create such a chromosome, two strings were alternatively generated. One string represented the start node and subsequent connected nodes, and the other represented the end node and subsequent connected nodes. Alternating between each string, the algorithm randomly selects the next adjacent node to the current node. This process continues until one of the strings contains a node that is in the other string. Their paths thus have crossed and can therefore be joined. To prevent loops and U-turns, a "visited bit" was maintained for each node. Neither string was allowed to contain a duplicate node. If one string became stuck in a "dead-end", i.e. all adjacent nodes already visited, it was restarted. Each chromosome thus generated represents a valid bus route.

4.2 Reproduction

In a GA, reproduction is the process by which the GA searches the solution space. Reproduction consists of first selecting mating pairs and then performing crossover and mutation to produce a child that has a portion of each parent's genetic material.

The mate selection process was implemented using recombinative hill climbing (RHC) [6]. In RHC, a pseudo random number is generated as an offset to determine mating pairs. Once mating pairs are selected, single point crossover is used to generate an offspring [5].

To ensure that the offspring produce valid routes, crossover must occur at one of the common nodes between parents. If there were no common nodes, no child was produced. When there were common nodes, crossover was implemented by randomly selecting a crossover point from a list of common nodes. As is done in RHC, only one child was produced for each mating pair. This child contained the portion of the first parent's chromosome preceding the random common node, and the second parent's portion beyond this common node.

4.3 Mutation and replacement

Mutation is the means by which genetic diversity is introduced into a population, and thus, it is a means to move from one area of a search space to another. Furthermore, it is applied only infrequently so that the search process does not become simply a random search process. After the crossover and mutation operations have been performed, if the fitness of the child exceeds the parent's fitness, the child replaces the parent. Thus, the hill climbing designation of RHC.

When applied, mutation must produce a valid transit. In this process, a start and an end node were randomly selected in a child. These two random nodes were then input to the chromosome initialization function to create a new mutated route.

Both mating and mutation can cause a node to appear more than once in a route. If an edge is traveled more than once, there is a penalty. The degree of the penalty is proportional to how recently the duplicated node was visited. There is an additional natural penalty in that no additional ridership or significance is accumulated, but extra distance and impedance are accumulated. If, through mating and/or mutation the resultant chromosome (child) contained more nodes than the number of nodes in the map, the child was discarded.

5 Results

5.1 Verification

A map of only 20 nodes with no circuits and each node connected to exactly 4 other nodes has ~1,000,000 different transits. Thus verification is difficult. To demonstrate that the GA worked correctly, a street network was set up with the optimal route known. The feature data were contrived so that there were four separate optimal routes, i.e. one each, for distance, significance, impedance, and ridership. For each such case, the program found the optimal route.

5.2 Identical initial populations

One of the features of a GA is that it is stochastic. This means that multiple runs with the same initial population may not give the same result, especially when there is more than one optimum. In this work, two runs were made with the same initial populations. While the routes produced were different, their overall or total fitness values were approximately the same. To an extent, the choice of the "best" route is subjective, and very much influenced by the weights given to the attributes.

5.3 Optimum route

The actual optimum route produced by the GA was dependent on the range of the data used for the fitness features, and the weight factors used. Changing weights altered the results. This adds a significant tuning component to the GA. Tuning of the weight factors is a necessary part of the process. In our case, the weight factors were tuned so that most (70% or greater) of the significant sites were visited, and more than 70% of the total ridership was acquired. In other instances, one might want to visit all of the significant sites or avoid all impediments. The weight factors can be adjusted to suit a wide range of preferences.

One method of "fine tuning" the weight factors, is to first execute the program with all weight factors equal. Examine the route produced and find a small area of the route where the impedance, significance and/or ridership are present. In the small area of the route, determine if the route visits enough of the significant sites, avoids the higher impedance roads, and travels roads where bus riders are located. Adjust the weight factors and try again. From experimentation it was revealed that if the route met expectations in a small area, expectations were usually met along the entire route.

In actuality, tuning is an important feature of the algorithm. It allows the specialist to perform a "what-if" analysis on the service area.

6 Conclusions

The GA as defined and developed in this study is an applicable problem solving technique for the municipal bus route problem. Implicit in the use of this technique is the requirement that all features in a bus route scenario be accurately included as attributes. There are likely some subtleties of certain streets in actual scenarios that would not be reflected in the impedance or the significance values. Also, the ridership value is more complicated than a single integer for a given edge. One way to address these issues is to interview bus drivers familiar with the area to assist with the assignment of parameter values.

It is also recognized that feature values for edges change during the day. While this will make it difficult to arrive at a single day-long fitness value for a transit, the speed of the GA (typically under 3 minutes for 60 nodes on a Sun Sparcstation) makes it feasible to perform route analyses for different times of the day.

Obviously the closer the four feature values and weights reflect the actual bus route conditions, the more effective the GA will be at designing municipal bus routes. Most of the work using the GA in an actual scenario would be in gathering, analyzing, scaling, and placing the fitness feature data and weights in a file. Much of this feature data is readily available from a GIS and other sources.

7 Future research

While the results of this work are very encouraging, additional research and development work is needed to build a system that can easily and most effectively be used by traffic engineers. Among the needs and issues are:

1. Consider more complex route impediments, for example, those associated with road intersections where it is more difficult to turn left than right.
2. Include in the optimization goal bounds on attribute values such as ridership, i.e. only so many people can ride a bus at a given time.
3. Consider solving for several routes simultaneously in a network, and thus allow the GA to determine the route start and end points.

References

[1] Bertsimas, Dimitris. "Computational Approaches to Stochastic Vehicle Routing Problems." Trans. Sci., 29, 342-352 (1995).

[2] Ceder, Avishai, and Wilson, Nigel H. M., "Bus Network Design." Trans. Res.-B, 20B, 4, 331-344 (1986).

[3] Duhamel, Christophe, "A Tabu Search Heuristic for the Vehicle Routing Problem with Backhauls and Time Windows." Trans. Sci., 31, 49-59 (1997).

[4] Goldberg, David E. 1989 "Genetic Algorithms in Search, Optimization, and Machine Learning." Reading, MA: Addison-Wesley Publishing.

[5] Hooper, Dale C., Flann, Nicholas S., and Fuller, Stephanie R, "Recombinative Hill-Climbing: {A} Stronger Search Method for Genetic Programming." Genetic Programming 1997: Proceedings of the Second Annual Conference, 174-179 (1997).

[6] Kuo, J., "Optimal Bus Route Search Using Genetic Algorithms." A report submitted in partial fulfillment of the requirements for the degree of a Master of Science in Computer Science, Utah State University. Logan, Utah. (1997).

[7] Malandraki, Chryssi. "Time Dependent Vehicle Routing Problems: Formulations, Properties and Heuristic Algorithms." Trans. Sci., 26 185-200 (1992).

[8] Pattnaik, S. B.., Mohan, S., and Tom, V.M., "Urban Bus Transit Route Network Design Using Genetic Algorithm." Trans. Eng., 124, 4, 368-375. (1998).

[9] Ramirez, Ana I., and Seneviratne, Prianka N., "Transit Route Design Applications Using GIS." Trans. Res. Rec. No. 1557, 10-14. (1996).

[10] van Nes R., Hamerslag, R., and Immers, B. H., "Design of Public Transport Networks." Trans. Res. Rec. No. 1202, 74-83. (1986).

Rostering with a Hybrid Genetic Algorithm

Matthias Gröbner*, Peter Wilke†

*Lehrstuhl für Programmiersprachen, Universität Erlangen-Nürnberg, Martensstrasse 3, D-91058 Erlangen, e-mail: Groebner@informatik.uni-erlangen.de † Centre for Intelligent Information Processing Systems, The University of Western Australia, Nedlands, 6907, Australia, e-mail: wilke@ee.uwa.edu.au

Abstract

Human workforce is an expensive resource and therefore should be used as efficient as possible. This optimization task is quite difficult especially in situations where staff members are different in their skills, qualifications or the details of their employment contracts, i.e. these constraints make the optimization of rosters a challenging and difficult task. In this paper we show how Genetic Algorithms combined with problem specific knowledge can be successfully applied to solve such scheduling and planning problems.

1 Introduction

One of the most common reasons for the existence of a company is their effort to make profit, if this is not *the* reason. Therefore management has to keep an eye on cost management. One of the most expensive resources is human workforce and therefore a demand for computer aided scheduling tools exists.

The classic approach is scheduling by hand, i.e. the supervisor has to create a roster for a certain period, typically one month. In certain cases specially developed planning software can assist in finding a schedule. Software available today is specially designed for these cases, i.e. tailored to match the constraints of the specific circumstances of the case. This results in high expenses for the software or the adaption of off the shelf programs, e.g. office tools.

Until now off the shelf scheduling software has not been developed since no concept to handle the variety of constraints and differences between the applications and their environment exists. So our goal was to develop a method that facilitates companies in a number of economical sectors to create rosters automatically. The key idea is to establish a standard toolkit which covers a broad variety of scheduling tasks and can easily be adopted to the specific circumstances. With this approach we accept the fact that this standard toolkit will only be suitable for most of the cases while it isn't appropriate for some.

Based on the fact that Genetic Algorithms have been successfully applied to some other similar timetabling and scheduling problems [1] [2] [6] our toolkit uses a Genetic Algorithm which is described in the following sections.

2 The Data

Fortunately we could use data from real world scheduling tasks. Figure 1 shows a part of a scheduled roster using this data. To make comparison easier these data sets have been standardized as follows:

- All datasets describe work around the clock divided in shifts.
- In each shift different positions are to be filled, e.g. supervisor, engineer, driver etc. Each position requires one or more specific functions to be performed and therefore can only be filled by a worker qualified for these functions.
- For each position within a shift there are requirements regarding the number of staff members on duty able to fill this position (staffing levels). These requirements are noted according to the following scheme: minimal requirement / target requirement / maximal requirement.
- There is a data base entry for each staff member indicating if he meets the qualifications required by the positions to be filled.
- The employees may have different number of weekly working hours.

3 Constraints

When a roster is about to be scheduled different constraints have to be considered. Commonly constraints are divided in two sets: hard constraints which have to be fulfilled under all circumstances and soft constrains which should be fulfilled if possible.

Examples for hard constraints are:

January	1	2	3	4	5	6	7	8	9	10	11	12	13	14	15
Weekday	Th	Fr	Sa	Su	Mo	Tu	We	Th	Fr	Sa	Su	Mo	Tu	We	Th
Shift Early 1 (E1)															
SURGERY1	3	2	1	1	2	1	3	2	1	1	1	1	2	2	3
SURGERY2	1	2	1	1	2	1	2	2	2	1	1	2	1	1	2
ASSISTANT	1	1			1		1	1	1			1	1	1	1
CHIEF	1	1			1		1	1	1			1	1	1	1
Shift Early 2 (E2)															
SURGERY1	2	2	1	1	1	1	2	2	1	1	1	1	1	2	2
SURGERY2	1	1	0	0	0	1	0	1	0	0	1	1	1	0	1
ASSISTANT	1	1			0		0	1	1			1	1	1	1
CHIEF															
Late Shift (LS)															
SURGERY1	2	2	1	1	2	1	2	2	2	1	1	2	2	2	2
SURGERY2	2	2	1	1	2	1	2	2	2	1	1	2	2	2	2
ASSISTANT	1	1			1		1	1	1			1	1	1	1
CHIEF	1	1			1		1	1	1			1	1	1	1
... (further shifts)															

January	1	2	3	4	5	6	7	8	9	10	11	12	13	14	15
Weekday	Th	Fr	Sa	Su	Mo	Tu	We	Th	Fr	Sa	Su	Mo	Tu	We	Th
Adam,T.	LS	LS	LS	LS			E2	E2	E2					E1	E1
Alexander,P.	NS	NS	NS	NS			E2	E2			LS	LS	LS	LS	LS
Huber,E.	E1	E1			LS	LS	LS	NS	NS	NS			E2	E2	E2
Beisser,H.	E1	E1			LS		E1	E1	E1			LS	LS	LS	E1
Cäsar,J.	E1	E1	E1	E1	E1			LS	LS			NS	NS	NS	LS
Escher,H.	E1	E1			NS	NS	NS	NS	NS			LS	LS	LS	E1
Huber,G.		LS	NS	NS	NS	NS	NS				E2	E2			LS
... (further employees)															

Fig. 1. Example roster. The upper part shows the number of employees assigned to the positions of each shift. The lower part shows the shifts the employees have been assigned to.

- By law no employee is allowed to works in more than one shift per day.
- By law breaks between two successive shifts have a minimal length.
- By law the monthly working hours (within a tolerance) limit must not be exceeded.
- The assigned number of staff members should be within the lower and upper limit of the staffing level for all separate positions within a shift.

Soft constraints are:

- The number of assigned staff members should be close to the target staffing level.
- The working and holiday blocks of the employees should be as compact as possible, i.e. eemployees should work in sequence as many days as possible respectively have as many days off as possible. Singular working days or holidays should be avoided.
- Similar shift patterns within each employee's working block are preferred, e.g. not always alternate between night shift and early shift.
- The shifts, especially the night and weekend shifts, should be shared fair - in most cases this will be a uniform distribution - among the employees.

4 The Algorithm

The core of our algorithm is a standard Genetic Algorithm [3] with tournament selection and 2-point crossover with three individuals. Furthermore we use an Elitist strategy by substituting only the worst individual of the population if the fitness value of the new individual is better.

4.1 Coding

Direct coding has been chosen to represent the monthly roster in a chromosome, i.e. the genes represent the sequence of workplaces of the functions of each shift. Thus we have a gene for each possible assignment of an employee to a function (see figure 2). The consequence of this coding method is a quite long gene string.

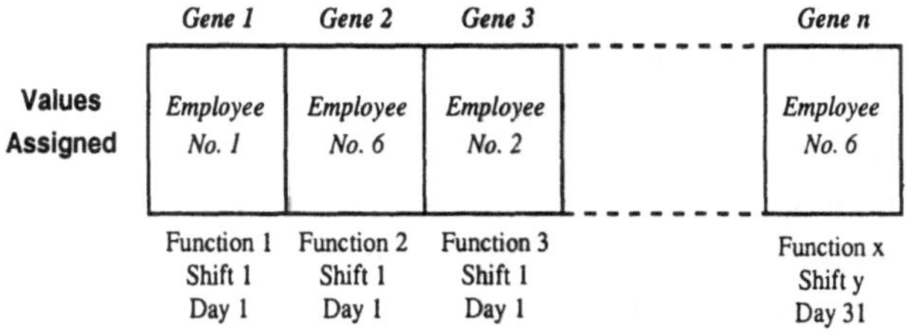

Fig. 2. Coding of the roster

The alleles represent the employees respectively the "empty" employee. Assignments of employees are allowed only if the employees are able to fulfill the required function of the position.

Using the above encoding scheme yields assignments of the employees to the functions of the shifts which are directly apparent and no time consuming decoding algorithms are needed. The drawback is that several hard constraints are not enforced, i.e. the genotype represents invalid solutions, e.g. multiple assignments of one employee to one shift. However, a significant larger part of the search space is searched by this method as if certain assignments would be excluded from search by the coding scheme. For this reason direct representation has been given preference to implicit representation [1]. Of course the resulting violations have to be sorted out in a later stage of the algorithm.

4.2 Mutation

The chosen coding scheme induces random mutations to the assigned employee. The initialy defined mutation rate is reduced linearly down to a lower limit of two changes. This is to ensure that the algorithm converges towards an optimum [5].

4.3 Fitness function

The individual fitness is computed by assigning penalty costs for the violation of constraints. As mentioned above the violation of most hard constraints is allowed by the chosen coding scheme. To sort things out violations result in lower fitness scores and therefore increase the evolutionary pressure on these individuals.

Penalty costs are assigned to each occurrence of a constraint violation. The penalty is calculated depending on the severity of the violation. E.g. for the penalization of the violation of fairness constraints like "uniform distribution of weekend shifts among the employees" the average number of weekend shifts has to be calculated first and the penalty costs are assigned representing deviation of each employee's number of weekend shifts from the average value.

5 Repair Operators

The Genetic Algorithm described so far produced relatively good rosters, but still an unacceptable high number of hard constraints remain violated. To improve the quality and speed of the algorithm it was necessary to introduce problem specific knowledge to accelerate convergence.

The idea is to apply repair operators which outperform the simple penalization of constraint violations. These new operators are applied to the individuals after selection, recombination and mutation, but before fitness calculation.

The disadvantage is that the search behavior becomes more locally. So the parameters of the algorithm have to be chosen carefully to ensure that the reachable part of the search space isn't reduced too fast,i.e. that the search becomes locally too early.. So the developed repair operators are applied best if their impact is gradually increased. This impact is to be fine tuned during the experiment.

The repair operators carry out the following modifications:

Canceling of assignments When an employee has been assigned more than once to one day all assignments can be canceled except one.

Selected assignment An employee can be assigned to days not previously been assigned to if his current monthly working hours are less than his target working hours.

Selected canceling of an assignment and reassignment of the employee. Sometimes it is more favorable with respect to the staffing levels of the shifts to cancel the assignment of an employee and to reassign him to another position within a shift of the same day.

Swapping of assignments of two employees to produce a more homogenous sequence of shifts for the employees.

The improvements of the roster found by the repair operators are subsequently re-decoded to the genotype. This establishes a kind of Lamarckian evolution.

6 Results

The Genetic Algorithm presented here has been tested on several real world databases and produced descent solutions for the problems in nearly all cases i.e. a roster without violations of hard constraints. An adapted version of the algorithm has been integrated in the employee planning software SP-EXPERT© [1] and is successfully used by several customers since then. The algorithm calculates a valid month roster with approx. 25 employees in less than 10 minutes on a 600 MHz Pentium computer. Solutions found after this period typically do not have a significant higher solution quality. The presented technique shows that Genetic Algorithms are able to create rosters automatically. But it seems to be necessary to use problem specific knowledge — in our example repair operators — to produce acceptable rosters. Figure 3 shows the effect of the repair operators and proves that no feasible solutions can be found even after a large number of generations when renouncing the repair operators.

Furthermore the presented Genetic Algorithm verifies that Lamarckian strategies can speed up the search process despite the danger to converge in local optima [6]. We tried to avoid this by introducing the repair operators smoothly during the run of the evolution process.

It hasn't escaped our attention that the chosen fitness calculation is not completely without problems because there is a large variety of incompatible constraints. The question is how the algorithm

[1]Contact: Astrum GmbH, Am Wolfsmantel 2, D-91058 Erlangen, Germany or www.astrum.de

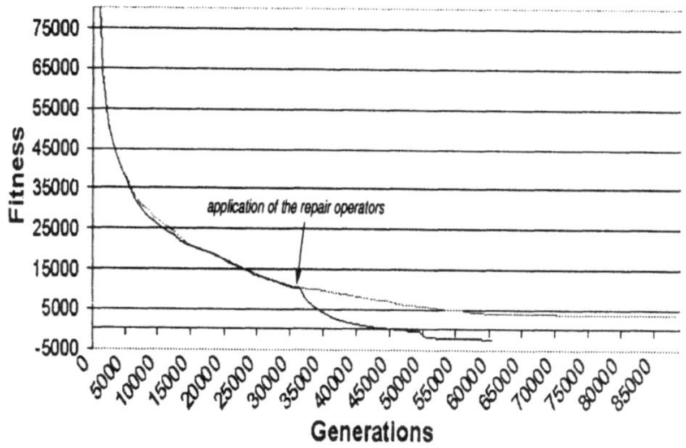

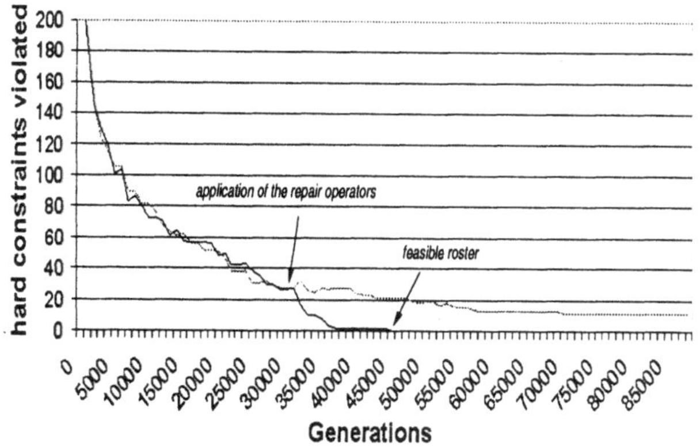

Fig. 3. Comparison of two test runs with and without repair operators. The upper figure shows the different trends of the fitness values (penalty costs), the figure below shows the number of remaining violations of hard constraints.

"knows" that a solution with a better fitness value really violated a lower number of hard constraints. We solved this problem in our test databases by appropriate parameter values for the penalties used by the fitness function.

7 Conclusion and Outlook

Our tests showed that the presented Genetic Algorithm is flexible with respect to the integration of future constraints. The only thing to do is to extend the fitness function to calculate the appropriate penalty costs. Normally the convergence is slowed by this and sometimes some soft constraints are not satisfied in the final solution. In some cases the repair operators have to be adapted to take the new constraints in account. Otherwise the repairs would not be able to improve the solution found so far. It looks promising to develop a more general version of this problem specific method to provide the possibility to integrate new constraints easier and to be able to apply this method to similar combinatorial optimisation problems without big effort.

Two problems are connected with the repair operators. The first problem is the choice of the penalty costs for the fitness function. Often it is rather difficult for users to express the different constraint violations in penalty costs. A rating in linguistic terms ("very important", "doesn't matter", "not so important") would make sense. Thus a combination of our the presented algorithm and fuzzy logic would be interesting.

Another problem is the control of the impact of the repairs operators. If these operators are introduced too early the search space turns locally, if they are introduced to late the repair may not be complete to yield valid solutions. In our tests the optimal point in time to begin with the repair operators has been determined heuristically. Further research will be directed towards finding an automated mechanisms to control the repair operator's impact on the search space.

References

[1] D. Corne and P. Ross and H.-L. Fang, "Evolutionary Timetabling: Practice, Prospects and Work in Progress", in *Proceedings of the UK Planning and Scheduling SIG Workshop*, University of Strathclyde, 1994

[2] C. Fernandes and J. P. Caldeira and F. Melicio and A. Rosa, "High School Weekly Timetabling by Evolutionary Algorithms", in *Proceedings of 14th Annual Acm Symposium On Applied Computing*, San Antonio, Texas, 1999

[3] D. E. Goldberg, "Genetic Algorithms in Search, Optimization and Machine Learning", Addison-Wesley, 1989

[4] M. Gröbner, "Optimierung der Einsatzplanung für Personal im Schichtdienst", *Master Thesis*, Universität Erlangen-Nürnberg, October 1998.

[5] I. Rechenberg, "Evolutionsstrategie: Optimierung technischer Systeme nach Prinzipien der biologischen Evolution", Fromann-Holzboog, 1973

[6] R. Weare and E. Burke and D. Elliman, "A Hybrid Genetic Algorithm for Highly Constrained Timetabling Problems", in *Proceedings of the Sixth International Conference on Genetic Algorithms*, pp. 605-610, 1995.

Design of Discrete Non-Linear Two-Degrees-of-Freedom PID Controllers Using Genetic Algorithms

P. B. de Moura Oliveira*

*Departamento de Engenharias, Universidade de Trás-os-Montes e Alto Douro, 5000 Vila Real, Portugal,
E-mail: oliveira@utad.pt

Abstract

Genetic algorithms are proposed to design two-degrees-of-freedom non-linear PID controllers for single input-single output systems. The evolutionary scheme proposed is able to design simultaneously a feedforward compensator and a non-linear picewise PID controller. A time-domain cost function subjected to a performance constraint is deployed in order to obtain a good compromise between the set-point tracking design and the disturbance rejection design. This evolutionary approach is illustrated by a simulation example and compared with the corresponding linear configuration.

1 Introduction

The importance and role of PID controllers in the past of feedback control systems is well reported by Bennett (2000). Despite major developments of advanced control techniques over the last thirty years, the most popular controller used in industrial processes is of the PID type. This is because of it's simple structure and reliability in a wide range of operating conditions. Some important considerations about the possible future of PID controllers are stated by Ĺström and Hägglund (2000).

It is well known that non-linear PID controllers have a superior performance than linear ones. Many design methods and tuning rules have been proposed for linear PID controllers, with some of the more established ones developed by Ziegler and Nichols (1942), Lopez *et al.* (1969), etc. However, this is not the case for non-linear PID controllers as it can be verified by the small number of research papers published about this topic (Lelic and Gajic (2000)).

One of the limitations associated with PID controllers is the poor performance obtained when the dynamics of the output disturbance are significantly slower that the ones of the plant to be controlled (Sung and Lee, 1996). Indeed, most of the PID designs are based on optimum set-point tracking and have a bad performance in rejecting slow output disturbances. In some cases it is impossible to obtain both satisfactory set-point tracking and disturbance rejection responses due to the conflicting control objectives. To overcome this difficulty two degrees-of-freedom control configurations were proposed by Horowitz (1963), to decouple the Disturbance Rejection (*DR*) design from the Set-Point Tracking (*SPT*) design. The design of a two degrees-of-freedom control system should be carried out in one single step, while the classical approach is to perform first the disturbance rejection design, and then perform the set-point tracking. The reason for this approach is that the number of parameters involved in the control optimisation makes the problem difficult to tackle with conventional optimisation techniques.

Evolutionary techniques such as *Genetic Algorithms (GAs)* (Holland (1975)) and *Population Based Incremental Learning (PBIL)* (Baluja (1994)) have proved to be an excellent optimisation tool for the design and tuning of two-degrees-of-freedom linear PID controllers Oliveira and Jones (1998) and non-linear PID controllers (Jones (1995), Oliveira and Pires (1999)).

In this paper a two-degrees-of-freedom control configuration using a non-linear PID controller is proposed. The optimisation of the control system is accomplished in the digital domain by using GAs. A simulation example that compares the performance of the two-degrees-of-freedom control system with the proposed picewise non-linear profile is compared with the corresponding linear configuration.

2 Problem Statement

Horowitz (1963) has proposed several two-degrees-of-freedom (2DOF) control configurations. One of these configurations is illustrated in Figure 1, in which $G_f(z)$ is the transfer function of the pre-filter and $Gc(z)$ is the transfer function of a PID controller. The classical approach to the design of 2DOF control structure is to design first the PID controller for DR and then used the fixed PID controller to design the pre-filter. However, when the feedback control loop is designed for optimum *DR*, without considering the effect in the design of the pre-filter, may restrict and compromise the possible performance in terms of the optimum *SPT*. The simultaneous design of both prefilter and PID controller in the continuous time was reported by Oliveira and Jones (1998), in which a *Multi-Objective GA* (MOGA) was used with respect to the objectives of *DR* and *SPT*. This resulted in a non-dominated Pareto front from which the end-user could select an optimal compromise between the two design objectives.

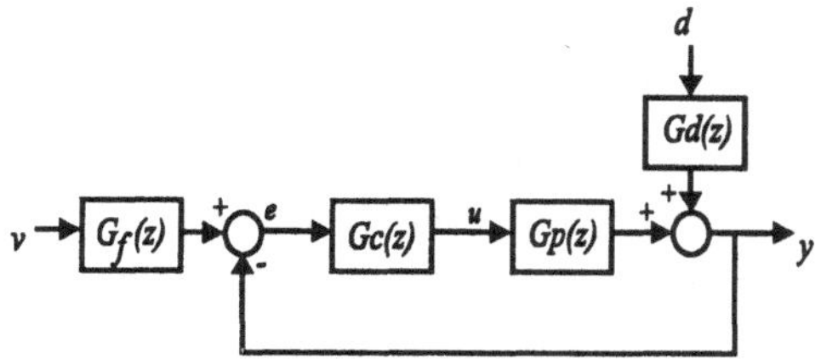

Figure 1: 2DOF digital configuration I.

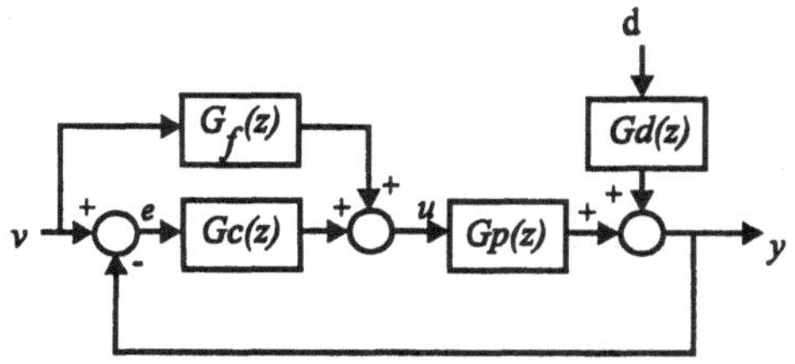

Figure 2: 2DOF control configuration II.

Another *2DOF* configuration well known and widely used (Taguchi and Araki (2000)) is illustrated in Figure 2 and it is used in this research. In Figure 2 $G_f(z)$ represents the discrete transfer function of a reference feedforward controller. The problem to be addressed in the next section is how to design simultaneously the feedforward controller and a non-linear PID controller using GAs.

3 Genetic Design of Non-Linear Two-Degrees of Freedom PID Controllers

A usual choice for the prefilter of Figure 1 and the feedforward controller of Figure 2 is a lead-lag network (Skogestad and Postlethwaite (1996)) that can be expressed in the discrete time set $\{0,T,2T,\ldots,kT,\ldots\}$ by :

$$G_f(z)=\frac{\left(\frac{T_{lead}}{T_{lag}}\right)+\left(1-\frac{T_{lead}}{T_{lag}}-e^{-\frac{T}{T_{lag}}}\right)z^{-1}}{1-e^{-\frac{T}{T_{lag}}}z^{-1}} \qquad (1)$$

in which $T\in R^{+}$ is the sampling period. It is well known that digital PID controllers can be governed on the discrete time set by control law equations of the *incremental* form:

$$\Delta u(kT)=K_p\left[\Delta e(kT)+K_p e(kT)+K_i\Delta^2 e(kT)\right] \quad (2)$$

in which K_p, K_i and K_d are the proportional, integral and derivative gains respectively, Δu is the controller output increment and:

$$e(kT)=v(kT)-y(kT) \quad (3)$$

$$\Delta e(kT)=e(kT)-e((k-1)T) \quad (4)$$

$$\Delta^2 e(kT)=e(kT)-2e((k-1)T)+e((k-2)T) \quad (5)$$

are the error, first and second backward error differences, respectively.

The non-linear profiles used for the PID gains are picewise linear as the one illustrated in Figure 3 for the proportional gain.

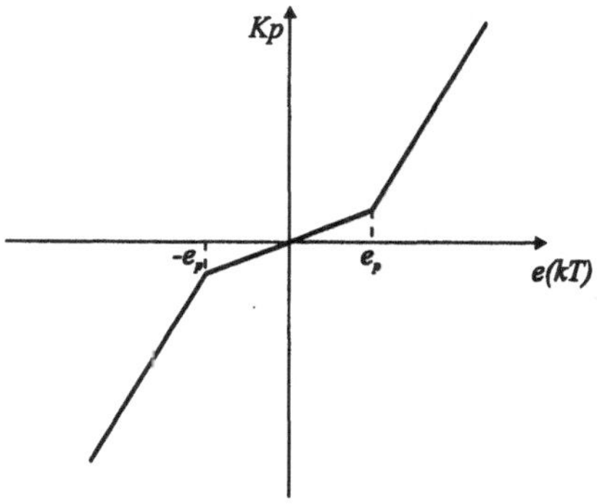

Figure 3: Picewise linear proportional gain.

The proportional, integral and derivative gains are defined by expressions (6), (7) and (8) respectively.

$$K_p=\begin{cases}k_{p1}\Leftarrow -e_p\le e(kT)\le e_p\\ k_{p2}\Leftarrow \left(e(kT)>e_p\right)\vee\left(e(kT)<-e_p\right)\end{cases} \quad (6)$$

$$K_i=\begin{cases}k_{i1}\Leftarrow -e_i\le e(kT)\le e_i\\ k_{i2}\Leftarrow \left(e(kT)>e_i\right)\vee\left(e(kT)<-e_i\right)\end{cases} \quad (7)$$

$$K_d=\begin{cases}k_{d1}\Leftarrow -e_d\le e(kT)\le e_d\\ k_{d2}\Leftarrow \left(e(kT)>e_d\right)\vee\left(e(kT)<-e_d\right)\end{cases} \quad (8)$$

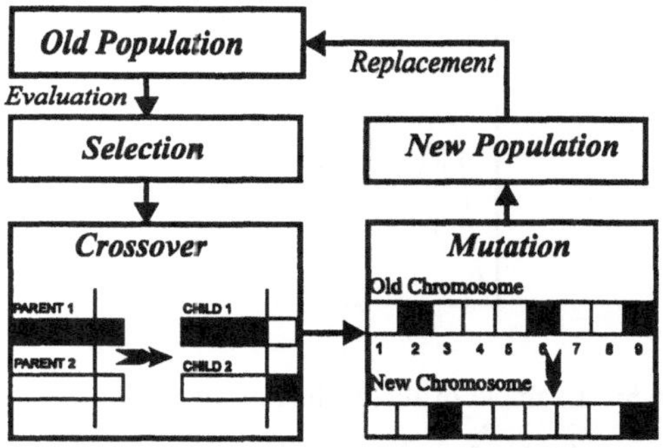

Figure 4: Basic genetic algorithm cycle.

The design of the *2DOF* control system illustrated by Figure 2, requires the appropriate selection of two parameters for the lead-lag feedforward compensator, and nine parameters for the proposed non-linear PID controller.

$$\{T_{lead}, T_{lag}, k_{p1}, k_{p2}, e_p, k_{i1}, k_{i2}, e_i, k_{d1}, k_{d2}, e_d,\} \quad (9)$$

In order to use *GAs* to design this *2DOF* control system it is necessary to encode the controllers parameters involved in the design equations in accordance with a system of concatenated, multiparameter, mapped, fixed-point coding.. Thus, each set of tuning gains, represented by expression (9) is represented by a string of binary digits.

The first population is initialised randomly by using an interviewing stage in order to guaranty a minimum fitness level of the potential solution candidates. Then entire generations of strings can be readily processed in accordance with the basic genetic operations of selection, crossover, and mutation (Figure 4). In particular, the selection process ensures that successive generations of model parameters produced by GAs exhibit progressively improving behaviour in respect of a fitness measure. In this case, the cost function to be minimised by the GA is the discrete *Integral of Square Error (ISE)* represented by expression (10):

$$ISE = \alpha \ ISE_{SPT} + \beta \ ISE_{DR} \quad (10)$$

in which α and β are weighting factors, ISE_{SPT} is the *ISE* obtained when a unit step is applied only to the reference input and ISE_{DR} is the *ISE* obtained when a step disturbance is applied only at the output disturbance *d*. To avoid the selection of the weights prior to optimisation, biasing the search direction a *MOGA* can be apllied in order to obtain a multiobjective non-dominated family of solutions instead of a single solution (Oliveira and Jones (1998)). The same technique is not used in this research as the overall goal is to compare the non-linear 2DOF configuration performance to the corresponding linear one.

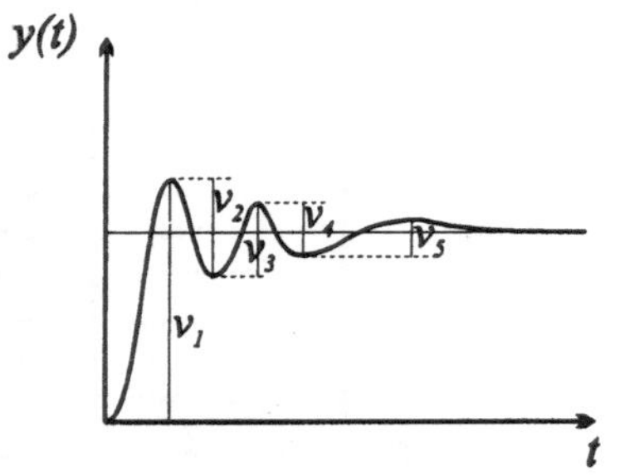

Figure 5: Example of the evaluation of the Total variation: $TV = \sum_i v_i$ for the SPT response.

As it is easier to identify a good *SPT* response than a *DR* response, the minimisation of expression (10) is subjected to a minimum *Total Variation (TV)* constraint, as defined in Skogestad and Postlethwaite (1996) (Figure 5).

4 Illustrative Example

The example that it is presented here has the following process and disturbance discrete transfer functions, for a sampling period of *T=0.05* seconds:

$$G_p(z) = \frac{0.00499 z^{-31}}{1 - 0.995 z^{-1}} \quad (11)$$

$$G_d(z) = \frac{0.001667 z^{-1}}{1 - 0.9983 z^{-1}} \quad (12)$$

The results obtained by designing a discrete *2DOF* control system with a lead-lag feedforward controller and a linear *PID* controller, by optimising the parameters expressed by:

$$\{T_{lead}, T_{lag}, K_p, K_i, K_d\} \quad (13)$$

are shown in Figure 6 for the unit *SPT* response *(y_spt)*, the unit output *DR* response *(y_dr)* and the output of the digital feedforward controller *(uf)*.

A *GA* with a population of size of *n=100*, a crossover probability *pc=0.75*, and a mutation probability of *pm=0.001* was used. Both in the *2DOF* linear and non-linear design the selection of weighting factors in expression (10) results in the following cost function:

$$ISE = 0.3 \sum_{k=0}^{999} e_{SPT}(kT) + 0.7 \sum_{k=0}^{999} e_{DR}(kT) \quad (14)$$

for a simulation time of 50 seconds. In this expression the disturbance rejection design is given a higher weighting value than the corresponding set-point tracking design because of the *Total Variation* constraint used that contributes to the optimisation of the latest. In this case a *TV* constraint of *1.2* was used. The controller parameters obtained are: *{5.25,3.0,5.0,6.19,4.5}* with $ISE_{SPT}=44.3$, $ISE_{DR}=58.8$ and *TV=1.19*.

The results obtained by designing a discrete *2DOF* control system with a lead-lag feedforward controller and a non-linear PID controller, by optimising the parameters expressed by (9) are shown in Figure 7 for the unit *SPT* response *(y_spt)*, the unit output *DR* response *(y_dr)* and the output of the digital feedforward controller *(uf)*. In this case a *TV* constraint of *1.1* was used. The feedforward controller parameters obtained are *{0.047,4.97}*. The non-linear PID piecewise gains are governed by expressions (14) (15) and (16) for the proportional, integral and derivative gains with $ISE_{SPT}=50.0$, $ISE_{DR}=23.3$ and *TV=1.07*.

$$K_p = \begin{cases} 6.00 \Leftarrow -0.89 \le e(kT) \le 0.89 \\ 2.97 \Leftarrow (e(kT) > 0.89) \vee (e(kT) < -0.89) \end{cases} \quad (14)$$

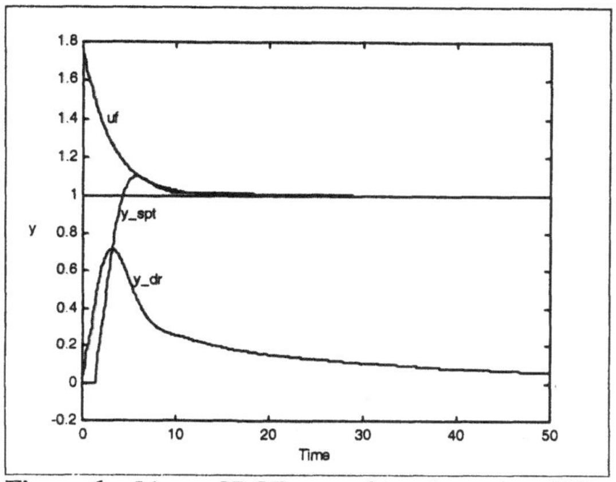

Figure 6: Linear 2DOF control configuration SPT (y_spt), DR (y_dr) and feedforward controller (uf) responses.

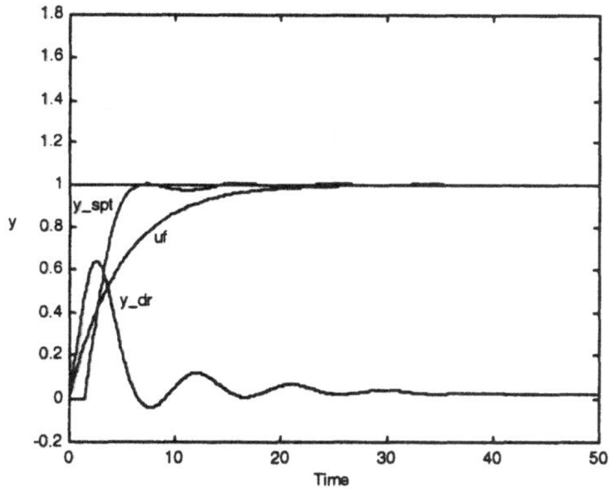

Figure 7: Non-linear 2DOF control configuration SPT (y_spt), DR (y_dr) and feedforward controller (uf) responses.

$$K_i = \begin{cases} 0.48 \Leftarrow -0.43 \leq e(kT) \leq 0.43 \\ 0.25 \Leftarrow \left(e(kT) > 0.43\right) \vee \left(e(kT) < -0.43\right) \end{cases} \quad (15)$$

$$K_d = \begin{cases} 5.92 \Leftarrow -0.2 \leq e(kT) \leq 0.2 \\ 2.96 \Leftarrow \left(e(kT) > 0.2\right) \vee \left(e(kT) < -0.2\right) \end{cases} \quad (16)$$

As it can be observed by comparing the results in Figure 6 and 7 and the corresponding ISEs, while the *SPT* response for the non-linear *2DOF* configuration is slightly worst in terms of *ISE* compared with the linear configuration the *DR* response is significantly improved with the non-linear PID design.

5 Conclusion

Genetic algorithms were proposed to design two-degrees-of-freedom non-linear PID controllers for single input-single output systems. The evolutionary scheme proposed is able to design simultaneously a feedforward compensator and a non-linear picewise PID controller. A time-domain cost function subjected to a performance constraint was deployed in order to obtain a good compromise between the set-point tracking design and the disturbance rejection design. This evolutionary approach was illustrated by a simulation example and compared with the corresponding linear configuration.

References

[1] Åström K. J. and Hägglund T. (2000), "The Future of PID Control", IFAC Workshop on Digital Control: Past, present and future, Spain, Terrassa, pp. 19-30.

[2] Baluja S., (1994), " Population Based Incremental Learning: A Method for Integrating Genetic search Based Function Optimization and Competitive Learning", Technical report CMU-CS-95-163, School of Computer Science, Carnegie Melon University, USA.

[3] Bennet S., (2000), " The Past of PID Controllers", IFAC Workshop on Digital Control: Past, present and future, Terrassa, Spain, pp. 3-13.

[4] De Moura Oliveira and Jones A. H.,(1998), " Genetic Design of Two Degrees-of-Freedom PID Controllers for Set-Point Tracking and Disturbance Rejection, 3rd Portuguese Conference on Automatic Control, Coimbra, Portugal, pp 111-116.

[5] De Moura Oliveira P. B and Solteiro Pires E J (1999), "Projecto de Controladores PID Discretos Nao-Lineares Usando PBIL", XX Jornadas de Automatica, Salamanca, Spain, pp.229-234.

[6] Holland J H, (1975). Adaptation in Natural and Artificial Systems, 1st MIT Press ed., ISBN: 0-262-58111-6.

[7] Horowitz I M, (1963). Synthesis of Feedback Systems, pp.246-249 Academic Press, ISBN: 63-12033.

[8] Jones A H, (1995), "Genetic Tuning of Neural Non-Linear PID Controllers", Artificial Neural Nets and Genetic Algorithms", Pearson *et al.* Editors, Springer Verlag, pp. 412-415.

[9] Lelic M. and Gajic Z. (2000), "A Reference Guide to PID Controllers in the Nineties", IFAC Workshop on Digital Control: Past, present and future, Terrassa, Spain, pp. 73-82.

[10] Lopez A, Murril P and Smith C, (1969). An Advanced Tuning Method, British Chemical Eng., Vol.14 (11), pp. 1553.

[11] Skogestad S and Postlethwaite I, (1996). Multivariable Feedback Control, Wiley, ISBN: 471942774, pp. 51-52.

[12] Sung S W and Lee I-B, (1996). Limitations and Countermeasures of PID Controllers, Industrial Engineering Chemical Research, Vol. 35, pp. 2596-2610.

[13] Taguchi H and Araki M, (2000), "Two-Degree-of-Freedom PID controllers- Their Functions and Optimal Tuning", IFAC Workshop on Digital Control: Past, present and future, Terrassa, Spain, pp. 95-100.

[14] Ziegler J G and Nichols N B (1942). Optimum Settings for Automatic Controllers, Trans. ASME, No. 64, pp. 759-768.

Modification of the Particle Swarm Optimizer for Locating All the Global Minima

K.E. Parsopoulos, M.N. Vrahatis*

*Department of Mathematics, University of Patras Artificial Intelligence Research Center (UPAIRC), University of Patras, GR-261.10 Patras, Greece, e–mail: {kostasp, vrahatis}@math.upatras.gr

Abstract

In many optimization applications, escaping from the local minima as well as computing all the global minima of an objective function is of vital importance. In this paper the Particle Swarm Optimization method is modified in order to locate and evaluate all the global minima of an objective function. The new approach separates the swarm properly when a candidate minimizer is detected. This technique can also be used for escaping from the local minima which is very important in neural network training.

1 Introduction

Many recent advances in science, economics and engineering rely on numerical techniques for computing globally optimal solutions to corresponding optimization problems. These problems are extremely diverse and include economic modeling, neural networks training, image processing and engineering design and control [3]. Due to the existence of multiple local and global optima all these problems cannot be solved by classical nonlinear programming techniques.

During the past three decades, however, many new algorithms have been developed and new approaches have been implemented, resulting to powerful optimization algorithms such as the Evolutionary Algorithms [6]. In contrast to other adaptive algorithms, evolutionary techniques work on a set of potential solutions, which is called *population*, and find the optimal solution through cooperation and competition among the potential solutions. These techniques can often find optima in complicated optimization problems faster than traditional optimization methods. The most commonly used population–based evolutionary computation techniques, such as Genetic Algorithms and Artificial Life methods, are motivated from the evolution of nature and the social behavior.

It is worth noting that, in general, Global Optimization (GO) strategies possess strong theoretical convergence properties, and, at least in principle, are straightforward to implement and apply. Issues related to their numerical efficiency are considered by equipping GO algorithms with a "traditional" local optimization phase. Global convergence, however, needs to be guaranteed by the global–scope algorithm component which, theoretically, should be used in a complete, "exhaustive" fashion. These remarks indicate the inherent computational demand of the GO algorithms, which increases non–polynomially, as a function of problem–size, even in the simplest cases.

In practical applications, most of the aforementioned methods can detect just *sub–optimal solutions* of the objective function. In many cases these sub–optimal solutions are acceptable but there are applications where an optimal solution is not only desirable but also indispensable. Moreover, in many applications there are many global minima that have to be computed quickly and reliably. Therefore, the development of robust and efficient GO methods is a subject of considerable ongoing research.

Recently, Eberhart and Kennedy (1995) proposed the *Particle Swarm Optimization* (PSO) algorithm: a new, simple evolutionary algorithm, which differs from other evolution–motivated evolutionary computation techniques in that it is motivated from the simulation of social behavior [2, 4]. Although, in general, PSO results in global solutions even in high–dimensional spaces, there are some problems whenever the objective function has many global and few (or not at all) local minima.

In this paper we propose a strategy that finds all global minima (or some of them if their number is infinite) of an objective function using a modification of the PSO technique and show, through simulation experiments, that this strategy is efficient and effective.

The paper is organized as follows: the background of the PSO is presented in Section 2. The proposed strategy is derived in Section 3. In Section 4 some results are presented and discussed, and finally conclusions are drawn in Section 5.

2 The Particle Swarm Optimizer

As it is already mentioned, PSO is different from other evolutionary algorithms. Indeed, in PSO the population dynamics simulates a "bird flock's" behavior where social sharing of information takes place and individuals can profit from the discoveries and previous experience of all other companions during the search for food. Thus, each companion, called *particle*, in the population, which is now called *swarm*, is assumed to "fly" over the search space in order to find promising regions of the landscape. For example, in the minimization case, such regions possess lower functional values than other visited previously. In this context, each particle is treated as a point in a D–dimensional space which adjusts its own "flying" according to its flying experience as well as the flying experience of other particles (companions). There are many variants of the PSO proposed so far. In our experiments we used a new version of this algorithm, which is derived by adding a new inertia weight to the original PSO dynamics [1]. This version is described in the following paragraphs.

First, let us define the notation adopted in this paper: the i-th particle of the swarm is represented by the D–dimensional vector $X_i = (x_{i1}, x_{i2}, \dots, x_{iD})$ and the best particle in the swarm, i.e. the particle with the smallest function value, is denoted by the index g. The best previous position (the position giving the best function value) of the i-th particle is recorded and represented as $P_i = (p_{i1}, p_{i2}, \dots, p_{iD})$, and the position change (velocity) of the i-th particle is $V_i = (v_{i1}, v_{i2}, \dots, v_{iD})$.

The particles evolve according to the equations

$$\begin{aligned} v_{id} &= w\, v_{id} + c_1\, r_1\, (p_{id} - x_{id}) + \\ &\quad + c_2\, r_2\, (p_{gd} - x_{id}), \qquad (1) \\ x_{id} &= x_{id} + v_{id}, \qquad (2) \end{aligned}$$

where $d = 1, 2, \dots, D$; $i = 1, 2, \dots, N$, and N is the size of population; w is the inertia weight; c_1 and c_2 are two positive constants; r_1 and r_2 are two random values in the range $[0, 1]$.

The first equation is used to calculate i-th particle's new velocity by taking into consideration three terms: the particle's previous velocity, the distance between the particle's best previous and current position, and, finally, the distance between swarm's best experience (the position of the best particle in the swarm) and i-th particle's current position. Then, following the second equation, the i-th particle flies toward a new position. In general, the performance of each particle is measured according to a predefined fitness function, which is problem–dependent.

The role of the inertia weight w is considered very important in PSO convergence behavior. The inertia weight is employed to control the impact of the previous history of velocities on the current velocity. In this way, the parameter w regulates the trade–off between the global (wide–ranging) and local (nearby) exploration abilities of the swarm. A large inertia weight facilitates global exploration (searching new areas), while a small one tends to facilitate local exploration, i.e. fine–tuning the current search area. A suitable value for the inertia weight w usually provides balance between global and local exploration abilities and, consequently, a reduction on the number of iterations required to locate the optimal solution. A general rule of thumb suggests that it is better to initially set the inertia to a large value, in order to make better global exploration of the search space, and gradually decrease it to get more refined solutions, thus a time decreasing inertia weight value is used. The initial population can be generated either randomly or by using a Sobol sequence generator [8] which ensures that the D-dimensional vectors will be uniformly distributed into the search space.

From the above discussion it is obvious that PSO, to some extent, resembles evolutionary programming. However, in PSO, instead of using genetic operators, each individual (particle) updates its own position based on its own search experience and other individuals' (companions) experience and discoveries. Adding the velocity term to the current position, in order to generate the next position, resembles the mutation operation in evolutionary programming. Note that in PSO, however, the "mutation" operator is guided by particle's own "flying" experience and benefits by the swarm's "flying" experience. In another words, PSO is considered as performing mutation with a "conscience", as pointed out by Eberhart and Shi [1].

3 Locating All the Global Minima of an Objective Function Using the PSO Method

Let $f : \mathcal{B} \to \mathbb{R}$ be an objective function that has many global minima inside a hypercube $\mathcal{B}$. If we use the plain PSO algorithm to compute just one global minimizer, i.e. a point $\bar{x} \in \mathcal{B}$ such that $f(\bar{x}) \leq f(x)$, for all $x \in \mathcal{B}$, there are two things that might happen: either the PSO will find one global minimum (but we don't foreknow which one) or the swarm will ramble over the search space failing to decide where to land. This last behavior is due to the equal "good" information that each global minimizer has. Each particle moves toward a global minimizer and influences the swarm in order to move toward that direction, but it is also affected by the rest of the particles in order to move toward the other global minimizer that they target. The result of this interaction between particles is a cyclic movement over the search space and disability to detect a minimum. A strategy to overcome these problems and find all global minimizers of f is described in the rest of this section.

In many applications, such as neural networks training, the goal is to find a global minimizer of a nonnegative function. The global minimum value is a priori known and is equal to zero, but there is a finite (or infinite in neural networks case) number of global minimizers. In order to avoid the problem mentioned in the previous paragraph, we can do as follows: we determine a not–so–small threshold $\epsilon > 0$ (e.g. if the desired accuracy is 10^{-5}, a threshold around 0.01 or 0.001 will work) and whenever a particle has a functional value that is smaller than ϵ, we pull this particle away from the population and isolate it. Simultaneously, we apply deflation or *stretching* [7] to the original objective function f at that point, in order to repel the rest of the swarm from moving toward it and add a new particle (randomly generated) in the swarm.

"Stretching" is a new technique that provides a way of escape from the local minima when PSO's convergence stalls. It consists of a two–stage transformation to the form of the original function f and can be applied soon after a local minimum $\bar{x}$ of the function f has been detected:

$$G(x) = f(x) + \gamma_1 \frac{\|x - \bar{x}\|(\text{sign}(f(x) - f(\bar{x})) + 1)}{2}, \quad (3)$$

$$H(x) = G(x) + \gamma_2 \frac{\text{sign}\,(f(x) - f(\bar{x})) + 1}{2\,\tanh\,(\mu(G(x) - G(\bar{x})))}, \quad (4)$$

where γ_1, γ_2 and μ are arbitrary chosen positive constants, and sign($\cdot$) defines the well known three–valued sign function [7]:

$$\text{sign}(x) = \begin{cases} +1, & x > 0, \\ 0, & x = 0, \\ -1, & x < 0. \end{cases} \quad (5)$$

Thus, after isolating a particle, we check its functional value. If the functional value is far from the desired accuracy, we can generate a small population of particles around it and constrain this small swarm in the isolated neighborhood of f to perform a finer search while the big swarm continues searching the rest of the search space for other minimizers. If we set the threshold to a slightly higher value, then the isolated particle is probably a local minimizer and during the local search, a global minimum will not be detected but we have already helped PSO to avoid it by deflating or stretching it. If we know how many global minimizers of f exist in $\mathcal{B}$ then, after some cycles, we will find all of them. In case we do not know the number of global minimizers, we can ask for a specific number of them or let the PSO run until it reaches the maximum allowable number of iterations in a cycle. This will imply that no other minimizers can be detected by PSO. The whole algorithm can be parallelized and run the two procedures (for the big and the small swarm) simultaneously, saving this way a lot of time.

In the next section we will discuss a simulation of a simple yet difficult problem that can be solved using the presented algorithm.

4 Some Experimental Results

Let f be the 2-dimensional function

$$f(x_1, x_2) = \cos(x_1)^2 + \sin(x_2)^2, \quad (6)$$

where $(x_1, x_2) \in \mathbb{R}^2$. This function has infinite number of minima in $\mathbb{R}^2$, at the points $(\kappa\frac{\pi}{2}, \lambda\pi)$, where $\kappa = \pm1, \pm3, \pm5, \ldots$ and $\lambda = 0, \pm1, \pm2, \pm3, \ldots$. We assume that we are interested only in the subset $[-5, 5]^2$ of $\mathbb{R}^2$. Into this hypercube, the function f has 12 global (equal to zero) minima.

If we try to find a single minimizer of f then we find out that the swarm moves back and forth as described in the previous section, until the maximum number of iterations is reached, failing to detect the minimizer. As already mentioned, this happens due to the same information (i.e. functional value) that each minimizer has. Thus, we could say that the swarm is so excited that it cannot decide where to land. Applying the algorithm given above, after 12 cycles of the method, we found all global minimizers with accuracy 10^{-5} and there even was no need for further local search.

In a second experiment we tried to find all minima of a notorious two dimensional test function, called the *Levy No. 5*:

$$\begin{aligned} f(x) = & \sum_{i=1}^{5} i\cos[(i+1)x_1 + i] \times \\ & \times \sum_{j=1}^{5} j\cos[(j+1)x_2 + j] + \\ & + (x_1 + 1.42)^2 + (x_2 + 0.80)^2, \end{aligned} \quad (7)$$

where $-10 \leq x_i \leq 10, i = 1, 2$. There are about 760 local minima and one global minimum with function value $f(x^*) = -176.1375$ located at $x^* = (-1.3068, -1.4248)^\top$. The large number of local optimizers makes it extremely difficult for any method to locate the global minimizer. Using the presented algorithm, we are able to compute all minima (global and local) of this function in cpu time that does not surpass 760 times the mean cpu time needed to compute each minimizer separately.

In another experiment a neural network has been trained using the PSO to learn the XOR Boolean classification problem. The XOR function maps two binary inputs to a single binary output and the network that was trained to solve the problem had two linear input nodes, two hidden nodes with logistic activations and one linear output node. Training the network corresponds to the minimization of a 9-dimensional objective function [5, 9]. It is well known from the neural networks literature that successful training in this case, i.e. reaching a global minimizer, strongly depends on the initial weight values and that the error function of the network presents a multitude of local minima. To solve this problem, we use the new algorithm as follows: we set a threshold of 0.1 and start the algorithm as above but if the standard deviation of the population in an iteration is too close to zero without having functional value close to the threshold (e.g. if the error value of the network is around 0.5, where there is a well known local minimum of the function), we pull the best particle of the population away and isolate it. This particle is probably a local minimizer (or near one) and thus we provide to it some new particles (in our simulation the size of the population was 40 thus we were adding 10 particles to the isolated one) and perform a local search in the vicinity of it (we took an area of radius 0.01 around it) while the rest of the big swarm continues searching the rest of the space. If the local search yields a global minimizer, we add it to our list of found global minima, otherwise it is a local minimizer and we have already avoided it. In this way, we are able to detect an arbitrarily large number of global minimizers while simultaneously avoiding local ones.

5 Conclusions

A new strategy for locating efficiently and effectively all the global minimizers of a function with many global and local minima has been introduced. Experimental results indicate that the proposed modification of the PSO method is able to detect effectively all the global minimizers instead of rambling over the search space or attracting by the local minima.

The algorithm provides stable and robust convergence and thus a better probability of success for the PSO. Also, it can be straightforwardly parallelized.

Extensive testing on higher–dimensional and more complicate real–life optimization hard tasks is necessary to fully investigate the properties of the proposed algorithm, as well as give some hints of modifications that will probably improve its performance.

References

[1] R.C. Eberhart and Y.H. Shi, "Evolving Artificial Neural Networks", *Proc. Int. Conf. on N.N. and Brain*, Beijing, P.R. China, 1998.

[2] R.C. Eberhart, P.K. Simpson and R.W. Dobbins, "Computational Intelligence PC Tools", Academic Press Professional, Boston, 1996.

[3] R. Horst, P.M. Pardalos and N.V. Thoai, "Introduction to Global Optimization", Kluwer Academic Publishers, 1995.

[4] J. Kennedy and R.C. Eberhart, "Particle Swarm Optimization", *Proc. IEEE Int. Conf. on N.N.*, Piscataway, NJ, pp. 1942–1948, 1995.

[5] G.D. Magoulas, M.N. Vrahatis and G.S. Androulakis, "Effective back–propagation with variable stepsize", *Neural Networks*, vol. 10, pp. 69–82, 1997.

[6] Z. Michalewicz, "Genetic Algorithms + Data Structures = Evolution Programs", Springer, New York, 1996.

[7] K.P. Parsopoulos, V.P. Plagianakos, G.D. Magoulas and M.N. Vrahatis, "Objective function "stretching" to alleviate convergence to local minima", *Nonlinear Analysis, T.M.A.*, 2001, to appear.

[8] W.H. Press, W.T. Vetterling, S.A. Teukolsky and B.P. Flannery, "Numerical Recipes in Fortran 77", Cambridge University Press, 1992.

[9] M.N. Vrahatis, G.S. Androulakis, J.N. Lambrinos and G.D. Magoulas, "A class of gradient unconstrained minimization algorithms with adaptive stepsize", *J. of Comp. and App. Math.*, vol. 114, pp. 367–386, 2000.

Computational Issues in Model Predictive Control

Luigi Chisci, Giovanni Zappa *

*DSI, Università di Firenze, Firenze, Italy. e-mail: chisci,zappa@dsi.unifi.it

Abstract

Model Predictive Control (MPC) is an effective but computationally demanding control design methodology. The paper surveys different techniques used to alleviate the on-line computational burden in MPC.

1 Introduction

Model Predictive Control (MPC) [1] has gained widespread acceptance among industrial practictioners thanks to its capability of taking into account both performance criteria and hard constraints, e.g. actuator limits, directly in the control design procedure. A drawback of MPC is its high computational cost which, in practice, has limited its applicability to slow, e.g. chemical and petro-chemical, processes and ruled out potential applications to faster, e.g. electromechanical, systems. Several research efforts have been devoted to alleviating the computational burden of on-line optimization for MPC. To this end, two different approaches have been, either independently or jointly, undertaken: **(1)** to speedup on-line optimization by fast algorithms tailored to the structure of the specific optimal control problem; **(2)** to shift as much as possible the computational load off-line.

Within the first approach, several fast algorithms for *Linearly Constrained Linear Quadratic* (LCLQ) optimal control have been developed [3]-[8]. They can be classified into three different methods for constrained optimization, namely: *Active Set* (AS), *Interior Point* (IP) and *Mixed Weights* (MW) methods.

Within the second approach are, for instance, the explicit solution of LCLQ [9] and the approximation of MPC via neural networks [10].

The aim of this paper is to critically and comparatively survey existing computational techniques for MPC design.

2 MPC Background

Let us consider the LTI discrete-time model, quadratic cost and, respectively, linear constraints

$$x_{k+1} = Ax_k + Bu_k, \tag{1}$$

$$J(x_0, u) = \sum_{k=0}^{\infty} x_k^T Q x_k + u_k^T R u_k \tag{2}$$

$$\|z_k\|_\infty \le 1, \; z_k = Gx_k + Eu_k + d, \quad \forall k \ge 0. \tag{3}$$

In (1)-(3) u denotes the manipulated variable and x a possibly augmented state vector including all the relevant information on which u may depend. An MPC feedback is defined from (1)-(3), according to the RHC strategy, by **(a)** setting, at time t, $x_0 = x(t)$; **(b)** minimizing w.r.t. $u = \{u_0, u_1, \ldots\}$ the cost $J(x_0, u)$ in (2) subject to (3); and **(c)** applying $u(t) = u_0$ to the plant. From LQ theory it is well known that the unconstrained solution of (1)-(3), i.e. the solution of (1)-(2), is an LTI feedback $u_k = u_k^* = Fx_k$, $\forall k \ge 0$, for an appropriate gain F computed from the solution P of a Riccati equation. Notice that the problem (1)-(3) has an infinite number of degrees of freedom $u_0, u_1, \ldots$ and also an infinite number of constraints. However, it has been shown [11] that for any x_0 there exists a sufficiently large N, depending on x_0, such that the constrained optimum $\hat{u}_k$ of (1)-(3) coincides with the unconstrained optimum $u_k^* = Fx_k$ for $k \ge N$. In words, the solution of the infinite-horizon problem (1)-(3) coincides with the solution of the finite-horizon problem

$$\min_{u_0,u_1,\ldots,u_{N-1}} \left\{ x_N^T P x_N + \sum_{k=0}^{N-1} x_k^T Q x_k + u_k^T R u_k \right\}$$
$$\text{s.t.} \begin{cases} x_{k+1} = Ax_k + Bu_k \\ z_k = Gx_k + Eu_k + d \\ \|z_k\| \le 1, \quad k = 0, 1, \ldots, N-1 \end{cases} \tag{4}$$

with only N degrees of freedom and N constraints. In addition, for a given compact set X_0 of initial conditions, a suitably large N can be chosen [11] such that for any $x_0 \in X_0$ the solution of (1)-(3) coincides with the solution of (4). Introducing the deviation $c_k = u_k - u_k^* = u_k - Fx_k$ from the unconstrained optimum, the problem (4) is more conveniently re-expressed as

$$\min_c J(x_0, c), \; J(x_0, c) = \sum_{k=0}^{N-1} c_k^T \overline{R} c_k, \; \overline{R} \triangleq R + B^T P B$$
$$\text{s.t.} \begin{cases} x_{k+1} = (A + BF)x_k + Bc_k \\ z_k = (G + EF)x_k + Ec_k + d \\ \|z_k\| \le 1, \quad k = 0, 1, \ldots, N-1 \end{cases} \tag{5}$$

Notice that with the new reparametrization, the unconstrained optimum becomes simply $c_k \equiv 0$. The dynamic optimization (5) can clearly be reformulated as a static QP problem of the form

$$\min_{\mathbf{c}} \mathbf{c}^T \Psi \mathbf{c} \quad \text{s.t.} \quad M\mathbf{c} \le v + Lx_0 \tag{6}$$

where $\mathbf{c} = [c_0^T, c_1^T, \ldots, c_{N-1}^T]^T$ and M, L, v are precomputable from the matrices A, B, E, G, d, Q, R.

3 Fast Algorithms for On-Line Design of MPC

3.1 Fast AS algorithms

The AS method finds the solution of a QP problem by iteratively searching the active set, i.e. the set of active constraints in the optimum. Starting from an empty active set and from a feasible initial point c^0, this search is carried out as follows.
For $i = 1, 2, \ldots$:
1) the best point $\mathbf{c}^*$ on the current active set is found by solution of an equality-constrained QP problem;
2) if $\mathbf{c}^*$ is feasible and satisfies the KKT (Karush-Kuhn-Tucker) optimality conditions, then $\hat{\mathbf{c}} = \mathbf{c}^*$ is the desired optimum and the procedure is stopped;
3) if $\mathbf{c}^*$ is feasible but suboptimal, i.e. does not satisfy the KKT conditions, a constraint (the one with largest negative Lagrange multiplier) is removed from the active set and $\mathbf{c}^i$ is set equal to $\mathbf{c}^*$ before proceeding to step $i+1$;
4) if $\mathbf{c}^*$ is infeasible, a new feasible point $\mathbf{c}^i$ is found along the line segment joining $\mathbf{c}^{i-1}$ and $\mathbf{c}^*$, as close as possible to $\mathbf{c}^*$, and a constraint (the one which holds with equality in $\mathbf{c}^i$) is added to the active set before proceeding to step $i+1$.
This procedure guarantees that the cost $J(\mathbf{c}^i)$ decreases at each step and, hence, that an active set is visited at most once. Thus, the AS method guarantees convergence in a finite number of steps as the number of possible active sets is finite, even if exponentially growing with the problem size. In practice, the AS method requires on average few iterations. The most demanding part of an AS algorithm is the solution of a sequence of equality-constrained QP subproblems in which each subproblem differs from the previous one either by the inclusion or by the exclusion of a single constraint. In [3]-[5] it has been shown how to efficiently update the optimal solution of an LCLQ problem after the activation/deactivation of a single constraint. Let us consider, for simplicity, z_k scalar. In this case, possible active constraints are $z_k = 1$ or $z_k = -1$ and the active set can be described by a vector $a = [a_0, a_1, \ldots, a_{N-1}]^T$ where

$$a_k = \begin{cases} 0, & \text{if no constraint on } z_k \text{ is active} \\ 1, & \text{if } z_k = 1 \text{ is active} \\ -1, & \text{if } z_k = -1 \text{ is active.} \end{cases}$$

For each a, the optimal unconstrained solution on the corresponding active set, is of the affine feedback form

$$c_k = F_k(a)x_k + g_k(a), \qquad k = 0, 1, \ldots, N-1 \tag{7}$$

for appropriate gains $F_k(a), g_k(a)$ depending on a. In particular, for $a = 0$ (empty active set), $F_k(a) = 0$ and $g_k(a) = 0$ for all k. For the AS method it is therefore required to update the gains F_k and g_k after a change $a \to a'$ such that

$$\begin{aligned} a'_q &\neq a_q, \quad q \in \{0, 1, \ldots, N-1\} \\ a'_i &= a_i, \quad i \neq q \end{aligned}$$

In particular, two cases need to be distinguished

$$\begin{array}{lll} \textit{constraint activation:} & a'_q = \pm 1, & a_q = 0 \\ \textit{constraint deactivation:} & a'_q = 0, & a_q = \pm 1 \end{array}$$

The transformations $F_k(a), g_k(a) \to F_k(a'), g_k(a')$ for the above two cases are derived in [4]. The details are outside the scope of this paper; however it is important to highlight that both activation and deactivation of the constraint q modify F_k, g_k only for $k \leq q$, i.e.

$$F_k(a') = F_k(a), \; g_k(a') = g_k(a) \qquad \forall k > q.$$

The computational cost of the activation/deactivation of constraint q is either $O(qn^2)$ for a generic state-space representation or $O(qn)$ for a canonical one, n being the model order. This compares favourably with standard (general purpose) AS solvers for QP which require $O(N^3)$ operations. An extension of the fast activation/deactivation procedure to linear constraints of general form has been presented in [5].

3.2 Fast IP algorithms

Interior Point (IP) methods solve an optimization problem starting from an initial point and iteratively updating it so as to enforce satisfaction of the optimality conditions. There is a large variety of IP methods which differ from each other for **(1)** the choice of the initial solution (feasible or infeasible); **(2)** the calculation of the update direction; **(3)** the calculation of the step-size. Let us consider the optimal control problem (4) of sect. 2 and re-express it conveniently as follows

$$\begin{aligned} &\min_{u_k, x_k} \; x_N^T P x_N + \sum_{k=0}^{\infty} x_k^T Q x_k + u_k^T R u_k \\ &\text{s.t.} \; \begin{cases} x_{k+1} = A x_k + B u_k \\ H x_k + D u_k \leq h. \end{cases} \end{aligned} \tag{8}$$

Notice that, unlike the standard formulation, in (8) also the states x_k, besides the inputs u_k, are regarded as optimization variables and the model dynamics are retained as equality constraints. This seemingly unmotivated increase of variables and constraints is the key factor for the derivation of a fast algorithm. In fact, it makes the relevant matrices banded with a bandwidth independent of N so that, by exploiting *dynamic programming*, the cost per iteration becomes linear rather than cubic in N. To be more specific IP methods, applied to (8), aim at finding the solution vector $S^T = [u_0^T, p_0^T, t_0^T, \lambda_0^T, x_1^T, u_1^T, \ldots, \lambda_1^T, \ldots, x_{N-1}^T, \ldots, \lambda_{N-1}^T, x_N^T]$. In S, besides the states x_k and inputs u_k, also the Lagrange multipliers p_k and λ_k for the equality constraints $x_{k+1} = Ax_k + Bu_k$ and, respectively, inequality constraints $Hx_k + Du_k \leq h$ as well as the slack variables t_k for the latter inequality constraints, have been included. The optimum $\widehat{S}$ is found at convergence by an iterative scheme

$$S^{j+1} = S^j + \alpha^j \delta S \tag{9}$$

which is stopped whenever S^j satisfies the KKT optimality conditions within some prespecified accuracy. The main computational burden of the iteration (9) is due to the computation of the update direction δS. This requires solution of a large, but highly sparse, set of linear equations with a banded matrix having bandwidth which depends on the model order n and the number of inputs m, but not on N. The solution can be obtained by an efficient backward elimination procedure which involves iteration of a suitable Riccati equation. The details are presented in [6]. The resulting computational complexity of this fast IP (FIP) algorithm is $O(N(n+m)^3)$ per iteration and compares to $O(N^3)$ of standard IP solvers. An important feature of IP methods which make them attractive for large-scale optimization problems (i.e. problems with many variables and/or many constraints) is the weak dependence of the number of iterations on the problem size.

3.3 Fast MW algorithms

A further approach is referred to as *Mixed Weights* (MW). According to MW, suitably weighted quadratic penalties of constraint violations are added to the cost, thus giving an unconstrained least-squares problem. At each iteration, the weights of the constraint penalties are updated so as to make the algorithm converge to the constrained optimum. With reference to (5), the MW method yields the following sequence of unconstrained least-squares problems

$$\mathbf{c}^i = \arg\min_{\mathbf{c}} \left\{ w^i \sum_{k=0}^{N-1} c_k^T \bar{R} c_k + \sum_{k=0}^{N-1} z_k^T W_k^i z_k \right\} \quad (10)$$

where w^i is a positive scalar weight and $W_k^i = diag\{w_{kj}^i\}$ a diagonal positive weight matrix. The weights can be updated as follows

$$w^{i+1} = \frac{w^i}{w}, \quad w_{kj}^{i+1} = \frac{w_{kj}^i |z_{kj}^i|}{w}, \quad w = \sum_k \sum_j w_{kj}^i |z_{kj}^i|$$

which guarantees convergence of $\mathbf{c}^i$ to the constrained optimum. The solution of (10) can be obtained efficiently by exploting dynamic programming. The details are reported in [7, 8]. The proposed FMW (Fast MW) algorithm requires $O(Nn^3)$ operations per iteration compared to $O(N^3)$ of the standard MW approach.

The MW method does not require a feasible starting point and clearly does not guarantee a feasible solution at each iterate but only at convergence. However, convergence speed may be quite slow and also very sensitive to weight initialization as well as to the choice of the stopping criterion.

3.4 Comparisons

The aim of this section is to compare several QP solvers based on AS, IP and MW methods. Specifically, the following algorithms will be considered: AS, FAS, IP, FIP, FIIP, MW, FMW where the prefix "F" refers to the "fast" version of respectively AS, IP, IIP, MW algorithms and IP, IIP stand for feasible, respectively, infeasible IP. Table 1 reports the computational cost (number of flops) of various QP solvers (implemented in MATLAB) for several values of N. The value 100% represents the cost of AS (QP routine of MATLAB). All values in the table refer to the same test problem, relative to a 2nd order system with bound constraints on the input, and have been obtained by averaging 10 different optimizations relative to randomly generated initial conditions for which the problem is feasible. The examination of Table 1 reveals clearly that the fast algorithms, tailored to the specific LCLQ problem, outperform their general-purpose counterparts. Fig. 1 plots the computational cost of fast QP solvers versus N. It is evident that FAS is preferable for low-medium values of N while FIP is better for large N. This could be expected since the cost per iteration of FIP is by far larger than for FAS while it is well known that the number of operations required by IP methods is weakly dependent on N (almost constant).

Another type of test has been performed in order to evaluate the control performance of the various methods when there is a limitation on the computing time. In this context, only methods which provide a feasible solution at each iteration, can be considered. In fact, in such a case it is possible to stop the computation when the available time has elapsed and use the suboptimal feasible solution obtained at the last iteration. In this respect, the MW is ruled out, while AS and feasible IP methods can be used. Hence, we compared the performance of FAS and FIP assuming a limitation on the computing time to a given percentage of the time required to get the optimal solution. Simulations have been performed on a 3rd order system with bound constraints on the input. Below, we compare the cost J_{opt}, obtained without limitations on the computing time (100%), and the costs J_{FAS} and J_{FIP} obtained with FAS and, respectively, FIP with a limitation of 50%:

$$J_{opt} = 8.3344, \quad J_{FAS} = 8.3995, \quad J_{FIP} = 20.8860.$$

Note that. although only 50% of the required time is given to the QP solver, the performance obtained with FAS is very close to optimal. Conversely the performance of FIP is significantly worsened. This discrepancy is emphasized for stronger limitations (e.g. 10%, 20%). In general, it has been found that FAS gives superior performance in presence of computing time limitations.

4 Off-line Design of MPC

MPC provides a nonlinear feedback implicitly defined via RH implementation of an optimal control trajectory. In fact, given the state x, the MPC defines a feedback mapping $u(x)$ as follows: **(1)** set $x_0 = x$; **(2)** solve an optimal control problem like e.g. (4) w.r.t.

$u_0, u_1, \ldots, u_{N-1}$ and **(3)** set $u(x) = u_0$. An alternative approach to the numerical on-line determination of $u(x)$ given the value of x, is the off-line determination of the feedback mapping $u(x)$ as a function of x. This would clearly shift the computational burden off-line, thus alleviating the on-line implementation task. Hereafter, two different methodologies for the off-line design of $u(x)$ will be briefly described.

4.1 Explicit solution of LCLQ

Consider the MPC defined by RH implementation of the LCLQ problem (4). Then the associated feedback mapping $u(x)$ is PWA (Piece-Wise Affine) in x i.e. it is possible to partition the state space X into convex polyhedral regions $\mathcal{X}_i$ such that $u(x) = F_i x + g_i, \quad \forall x \in \mathcal{X}_i$. This follows immediately from (7). In fact, let $a(x)$ be the vector characterizing the active set corresponding to the state x. Then clearly $u(x) = [F + F_0(a(x))]x + g_0(a(x))$. is an uniquely defined affine function of x. [9] provides an algorithmic procedure to find a partition of the state space into polyhedral regions $\mathcal{X}_i$ and the corresponding gains F_i, g_i. The number of regions can be, in principle, combinatorially growing with the number of constraints. In practice, however, there are problems of reasonably low size (i.e. model order and control horizon) for which such a number is manageable. In this case, it is possible to store all the matrix descriptions of the regions $\mathcal{X}_i = \{x : M_i x \leq v_i\}$ and a lookup table containing the gains F_i, g_i. Hence MPC implementation only requires to find the region $\mathcal{X}_i$ to which the current state x belongs and to compute the corresponding $u(x) = F_i x + g_i$.

4.2 Neural approximation of a RH controller

Another interesting approach is to approximate the mapping $u(x)$ via a *multilayer feedforward Neural Network* (NN) [10]. The NN provides an output $\hat{u}(x, w)$ which depends on the input x and on an adjustable weight vector w. The latter is updated off-line so as to minimize the approximation error between the RH controller and its neural approximation, by feeding them in parallel with a randomly generated sequence of state vectors. The choice of the number of layers ℓ and of neurons ν per layer is a crucial issue. The number of neurons ν to get a given accuracy may grow exponentially with the model order n and give rise to the curse of dimensionality. Notice that this approach, unlike the previous one, encompasses nonlinear models and/or nonquadratic criteria and/or nonlinear constraints in a natural way.

5 Conclusion

Research efforts towards implementation of MPC have been essentially in two directions: (1) to develop fast optimization solvers tailored to the needs of MPC and/or (2) to move the MPC computations off-line. In particular, the second approach can be ultimately pushed towards the explicit or approximate off-line determination of the MPC feedback control law; this can be indeed a viable and efficient solution for small problems where the complexity of the MPC feedback is manageable, but may easily suffer from the curse of dimensionality for problems of larger size, for which fast solvers provide therefore a valuable tool. In the authors' opinion, a suitable combination between on-line computation, using fast solvers, and off-line computation should give the best option in many circumstances.

N	FAS	IP	FIP	MW	FMW
10	6.0%	408%	109.0%	1965%	392.0%
20	3.2%	276%	17.6%	689%	37.0%
30	2.2%	310%	5.1%	321%	7.2%
40	1.6%	344%	2.9%	270%	3.6%
50	1.1%	324%	1.6%	196%	1.7%
60	0.9%	288%	1.1%	220%	1.4%
70	0.8%	251%	0.7%	182%	0.9%
80	0.6%	216%	0.5%	152%	0.6%
90	0.5%	193%	0.4%	129%	0.4%
100	0.4%	168%	0.3%	110%	0.3%

Table I. Computational cost of QP solvers

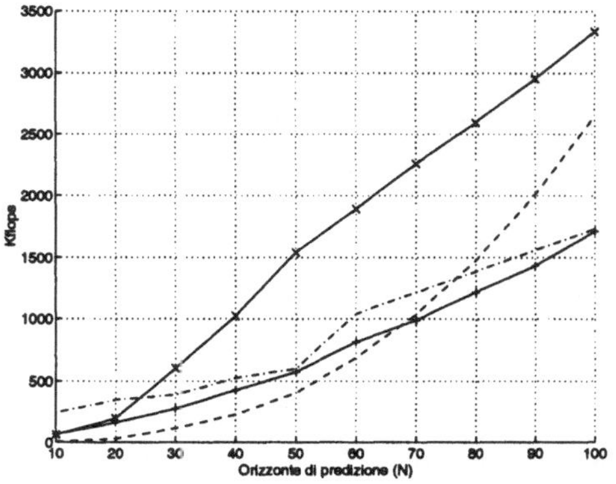

Fig. 1. Average number of kflops vs. N. FMW: dash-dotted; FAS: dashed; FIP: +; FIIP: ×.

References

[1] J.B. Rawlings, "Tutorial overview of model predictive control", *IEEE Control Systems Magazine*, vol. 20, pp. 38–52, 2000.

[2] E.F. Camacho and C. Bordons, *Model predictive control in the process industry*, Springer, London, 1995.

[3] L. Chisci and G. Zappa, "Fast algorithm for a constrained infinite-horizon LQ problem", *Int. Journal of Control*, vol. 72, pp. 1020–1026, 1999.

[4] L. Chisci, J.A. Rossiter and G. Zappa, "Iterative active set method for efficient on-line MPC design", in *Nonlinear model predictive control*, pp. 336–345; F. Allgower and A. Zheng (Eds.), Birkhauser, Basel, 2000.

[5] L. Chisci and G. Zappa, "Fast QP algorithms for predictive control", *Proc. 38th IEEE CDC*, pp. 4589–4595, Phoenix, U.S.A., 1999.

[6] C.V. Rao, S.J. Wright and J.B. Rawlings, "On the application of interior point methods to model predictive control", *JOTA*, vol. 99, pp. 723–757, 1998.

[7] L. Chisci and J.A. Rossiter, "Efficient mixed weights least-squares algorithm for constrained predictive control", *Systems Science*, vol. 23, pp. 5–17, 1997.

[8] L. Chisci and J.A. Rossiter, "Iterative weighted least-squares approach to constrained LQ predictive control", *Proc. IEEE Conf. on Control Applications*, pp. 1145–1149, Trieste, 1998.

[9] A. Bemporad, M. Morari, V. Dua and E.N. Pistikopoulos, "The explicit solution of model predictive control via multiparametric quadratic programming", *Proc. ACC 2000*, Chicago, U.S.A., 2000.

[10] T. Parisini and R. Zoppoli, "A receding-horizon regulator for nonlinear systems and a neural approximation", *Automatica*, vol. 31, pp. 1443-1451, 2000.

[11] D. Chmielewski and V. Manousiouthakis, "On constrained infinite-time linear quadratic optimal control", *Systems and Control Letters*, vol. 29, pp. 121–129, 1996.

Motor Control using Adaptive Time-Delay Learning

Claudia Ungerer*

*Department of Computing Science, Munich University of Technology, D-80290 Munich, Germany, e-mail: ungerer@in.tum.de

Abstract

This paper presents the application of time-delay learning to motor control. An extended version of the Adaptive Time-Delay Neural Network is used guaranteeing a more precise delay learning. This approach is applied to motor control by supporting a traditional PI-controller when controlling a combustion engine as a neural co-controller showing that it outperforms the traditional Adaptive Time-Delay Neural Network.

1 Introduction

This paper addresses the time-delay problem in nonlinear control systems. A system happens to be a time-delay system if it has at least one component, whose input affects the output not immediately but after a certain time delay. Such time-delay components often appear in control systems but have a negative influence on control, especially if the time delays are changing over time.

This paper is devided into two major parts. The first part describes the architecture and the extended version of the learning algorithm for the Adaptive Time-Delay Neural Network (ATNN). The ATNN, introduced in [1] and [2], is a feedforward neural network with internal, adaptable time delays. Traditional feedforward networks immediately generate a specific output to a particular input. There is no memory incorporated for information from past time steps. Within the ATNN however, the internal delay lines explicitly store recent information from the input signal. Adaptation occurs online with continuous time-varying signals using an extended form of the backpropagation algorithm. Due to the incorporated time delays this ATNN learning algorithm requires a so-called aging system in order to keep it causal. In [3] two useful extension to the ATNN are presented: the use of linear interpolation for the retrieval of past values in the network and a dynamic aging algorithm allowing a better delay learning. Now, these two extension are combined in a network that will be called LV-ATNN.

The second part of the paper shows the application of both ATNN and LV-ATNN to motor control, where time delays play an important role. In [4] it has already been reported how neural networks can successfully support a traditional PI-controller when controlling a combustion engine by acting as a neural co-controller. A testbed simulation presented in [4] will be used in order to show, that the introduced LV-ATNN leads to a better result than the ATNN when applied as a neural co-controller.

2 Adaptive Time-Delay Neural Network

2.1 Architecture

The Adaptive Time-Delay Neural Network is a multilayer feedforward neural network. Beside the weight there is a time delay defined on each connection. Each layer is fully connected and there may be more than one connection between two neurons. Consequently, different time delays can be defined between two neurons and therefore information from different time steps can be processed together. Figure 1 shows two neurons i and j of an ATNN having n connections between each other. The total input to neuron i at time t is defined as

$$x_i(t) = \sum_j \sum_k w_{ijk}(t)\, y_j(t - \tau_{ijk}(t)) \qquad (1)$$

where $w_{ijk}(t)$ is the weight of the k-th connection from neuron j to neuron i at time t, $\tau_{ijk}(t)$ is the time delay on that connection at time t and y_j is either the output of neuron j, $y_j(t) = f_j(x_j(t))$, or a network input. The activation function f is sigmoidal. The connections of the network can be seen as the memory of the network. They operate as buffers holding the neurons outputs. The time delay values define which past output gets processed to the next neuron.

2.2 Learning algorithm

The learning algorithm is an extended form of the backpropagation algorithm for traditional multilayer feedforward neural networks. Therefore, to minimize the error function, the weights and time delays must adapt with changes proportional to the

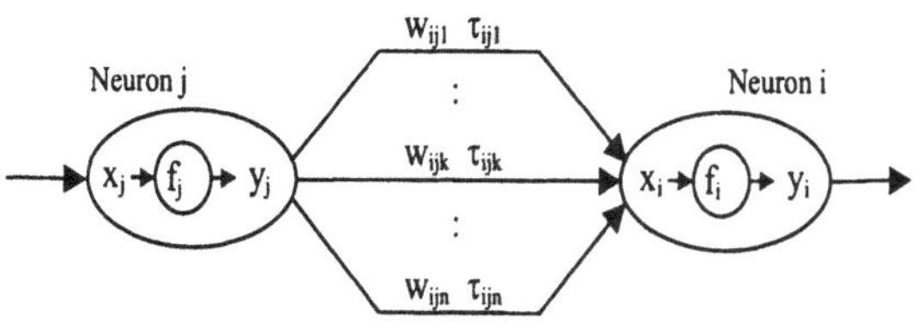

Fig. 1. Section of an Adaptive Time-Delay Neural Network showing two neurons i and j having n connections between each other. Each connection holds a weight w_{ijk} and a time delay τ_{ijk}

negative gradients

$$\Delta w_{ijk}(t) = -\mu \frac{\partial E(t)}{\partial w_{ijk}(t)} \quad (2)$$

$$\Delta \tau_{ijk}(t) = -\eta \frac{\partial E(t)}{\partial \tau_{ijk}(t)} \quad (3)$$

where μ and η are learning rates.

In [5] the learning algorithm was newly defined by means of the Jacobian matrix. The derivation of this algorithm is described in detail there. Here, the update rules for the weights and time delays are summerized as

$$\Delta \tau_{ijk}(t) = \eta \; opt_a \; jac\_\delta_i^a(t - a_i) \; \tau\_term \quad (4)$$

$$\Delta w_{ijk}(t) = -\mu \; opt_a \; jac\_\delta_i^a(t - a_i) \; w\_term (5)$$

where $\tau\_term = w_{ijk}(t - a_i)\; y'(t - \tau_{ijk}(t - a_i) - a_i)$ and $w\_term = y(t - \tau_{ijk}(t - a_i) - a_i)$ and opt_a is the Jacobian matrix of the error function E calculated by

$$opt_a = \frac{\partial E(t)}{\partial y_a(t)} \quad (6)$$

where $y_a(t)$ is the output of the a-th output neuron. $jac\_\delta_i^a(t - a_i)$ is the error signal associated with each neuron i and is defined as

$$jac\_\delta_i^a(t - a_i) = \frac{\partial y_a(t)}{\partial x_i(t - a_i)} \quad (7)$$

The parameters a_i are the aging parameters of the network associated to each layer due to causality constraints. Each layer i holds an age a_i, where $a_i > a_{i+1}$ for two adjacent layers i and $i + 1$. Note that the age of the output layer is equal to zero and all ages have to be integer numbers.

Appropriate age parameters have to be chosen, surely depending on the requirements of a particular application. For a given set of ages, each τ_{ijk} must be kept smaller than $a_j - a_i$ by imposing a limiting procedure during learning. As one can see, the choice of the ages crucially impacts the learning performance. It is desirable to make each a_i as small as possible in order to minimize storage requirements, maintain stability, and ensure that adaption most closely approximates the true gradient descent. In [1] a simple solution is suggested, that is to assign an age of zero to the output layer, and increment the age by a fixed amount Δa for each preceding layer, so that $a_i = a_{i+1} + \Delta a$. In a network with four layers and $\Delta a = 5$, the ages would be as follows: $a_{input\_layer} = 15$, $a_{first\_hidden\_layer} = 10$, $a_{2nd\_hidden\_layer} = 5$ and $a_{output\_layer} = 0$. The age parameter set stays fixed throughout the learning. As stated above, the ages impose an upper border for the adaptable time delays. If the ages are badly chosen in the beginning, two possible problems may arise. The ages might have been chosen too big, so that the calculations are not close to the true gradient descent or the ages might have been chosen too small, so that the proper time delays cannot be adapted. Note, that in many applications, no appropriate upper bounds for the time delays can be given as they are unknown. To avoid these problems, a dynamic aging system was suggested in [3]. The idea of the dynamic aging is to impose an algorithm to the network putting the ages as close as possible to the actual time delays while keeping the learning causal. Starting with the simple solution (assign an age of zero to the output layer and increment the age by an arbitrary, but fixed amount Δa for each preceding layer) every n learning cycles [1], Algorithm 1 is applied: each layer in the network is checked, starting with the output layer and then going backwards through the network till the input layer is reached. The current layer is called *target_layer* (shortened: *tl*) and its preceding layer is called *source_layer* (shortened: *sl*) with respect to the connections between them. Then, the age difference a_{diff} between these two layers is determined (note, that in the beginning this difference equals to Δa) as well as the maximal time delay τ_{max} of all connections connecting these layers. The maximal time delay is rounded up to the next integer value and used as age increment to the next layer. This guarantees a learning very close to the true gradient. In case, that the maximal time delay got close to its upper bound ($a_{diff} - \tau_{max} < 0.5$), the age increment is further increased by one in order to overcome the upper bound. Otherwise, it would only be possible to decrease the ages. Now, this approach is combined with the use of linear interpolation while retrieving the past values in the network. Otherwise, only integer times of the sampling rate could

[1] n is supposed to be chosen equal 1, but higher values can be sufficient too, e.g. $n = 10$

Algorithm 1 New aging algorithm given in pseudocode

```
tl = Output Layer;
new_a_tl = 0;
repeat
  sl = Preceding Layer of tl;
  a_diff = a_sl − a_tl;
  Determine the maximal time delay τ_max from
  sl to tl;
  new_a_sl = new_a_tl + ⌈τ_max⌉;
  if (a_diff − τ_max < 0.5) then
    new_a_sl = new_a_sl + 1;
  end if
  tl = sl;
until (sl == Input Layer)
```

be modelled as time delays, a restriction present in the traditional ATNN.
Assume that x lies between x_0 and $x_1 = x_0 + h$ with $y_0 = f(x_0)$ and $y_1 = f(x_1) = y_0 + \Delta$, then the functional value of x can be calculated as $f(x) = f(x_0) + \frac{x - x_0}{h} \cdot \Delta$.
Here, x corresponds to the time-delay values and the function f maps the time delay to the correpondig past output value. As $h = 1$, the linearly interpolated current output value is retrieved by $f(x) = (1 - (x - x_0)) \cdot f(x_0) + (x - x_0) \cdot f(x_1)$.
Therefore, the total input to neuron i at time t (1) is rewritten to

$$x_i(t) = \sum_j \sum_k w_{ijk}(t)\, y_j^{lip}(t - \tau_{ijk}(t)) \qquad (8)$$

where y_j^{lip} denotes the linearly interpolated output value of neuron j or a network input.

3 Application to Motor Control

This section describes the application of the introduced LV-ATNN to motor control showing that it outperforms the traditional Adaptive Time-Delay Network. The used simulation model will be described as well as the simulation results presented.

3.1 The simulation

Developing combustion engines or aggregates of cars like gearboxes, catalytic converters and control boxes means testing them on a testbed. The ability to follow a prescribed speed table with a combustion engine on a testbed with high precision, is one of the main characteristic features of a controller for a combustion engine.
Presently, throttle angle based control of combustion engines is done using traditional controllers like PI-controller and inverse engine maps. This has the major drawback of high measurement effort resulting in high stress for the combustion engine and also the fact, that the inverse engine maps are static.
A Matlab/Simulink model of a testbed environment shown in Figure 2 is used. It was provided by one of our industry partners and reflects a state-of-the-art testbed. The Simulink model consists of a three phase AC induction machine, a spark ignition engine described by a mean value method and a car model, all described in detail in [4].

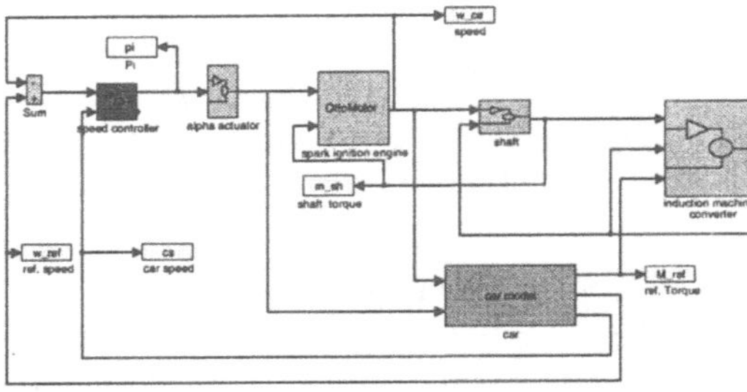

Fig. 2. The Simulink Simulation of the testbed and the combustion engine

A neural network, that is able to identify nonlinear dynamic systems can be trained to learn the inverse plant dynamics. Given the plants current output as input, the neural network is trained to compute the controller signal generated by a traditional controller. Once trained, the network should be capable of predicting the appropriate controller signal from the desired plant output.
Generally it can not be proven, that a neural network predicts the correct output given any possible inputs. For this reason the trained network is not used as a stand alone controller, but as an assistance to the traditional controller. Whenever the network produces the correct output, the controller signal is driven to zero. If the network is unable to compute the appropriate controller signal, the traditional controller minimizes the resulting control error.
To incorporate the trained neural network into a controller of PI type, the network output is added to the proportional and integral part as shown in Figure 3. As one can see, the anti-windup and limiter mechanism is located *behind* this point to take the network input into account as well. If the network is unable to produce a suitable output, the controller nevertheless minimizes the resulting control error by using the PI part. This architecture automatically provides a fallback mechanism if the network fails.

The control task is to follow a prescribed speed schedule as good as possible. The aim is to yield

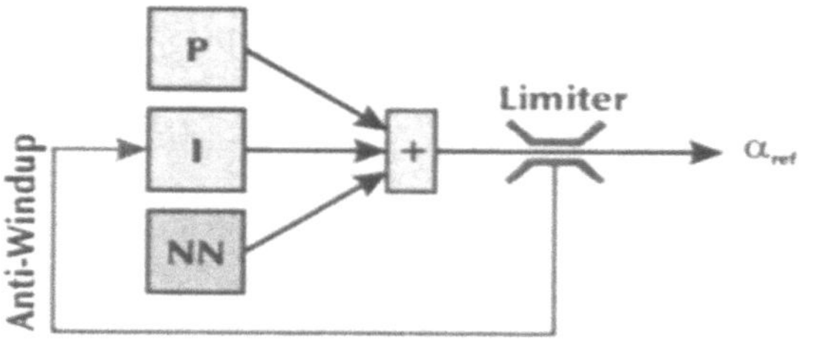

Fig. 3. A traditional PI controller assisted by a neural co-controller

higher precision with the assistance of a neural co-controller. The Euklidean distance between the Matlab vectors v_{ref} and v_{car} describing the driving schedule and the speed curve resulting from the controller actions is a measure of the control performance. When controlling the plant with a PI controller without assistance by a neural co-controller the Euklidean distance $||v_{ref} - v_{car}||_2$ equals to 92.4613.

The main design decision is the choice of adequate neural network models. Looking at the structure of the control system under concern yields to choose the ATNN for this task. The system contains some time delay components, i.e., components, whose input affects the output not immediately but after a certain time delay (e.g. the delay between intake and power production). In [4] it has already been shown, that the ATNN performs well in this task. Here, it is shown, that its extension, the presented LV-ATNN outperforms its traditional relative.

3.2 Simulation results

As described above, the ATNN and LV-ATNN are used to identify the inverse plant dynamics of the combustion engine. The training data set was produced by running the simulation with a predefined driving schedule and consists of the torque and the speed of the combustion engine as network inputs and the measured controller signal as target. The measurements were taken over a time interval of 50 seconds, one sample every 0.01 second. Both inputs as well as the target were scaled to the interval [0; 1]. The used ATNN is a fully connected network with k=1 for each connection having 2 input neurons, 7 hidden neurons in one hidden layer, and one output neuron. The age of the hidden layer was set to 5. Out of ten random network initializations the best one was chosen and trained as neural co-controller. The same initialization was used for the LV-ATNN now allowing the ages to vary and using linear interpolation. The new aging algorithm was applied every 10th learning cycle.

The simulation results show that the LV-ATNN outperforms the ATNN. The Euklidean distance was $||v_{ref} - v_{car}||_2 = 88.3105$ for the ATNN network and $||v_{ref} - v_{car}||_2 = 86.8197$ when using the LV-ATNN network. This equals to an improvement of 4,5% resp. 6,1% compared to traditional PI-Control.

4 Conclusion

This paper presented the application of time-delay learning to motor control. A testbed simulation in Matlab/Simulink is used, where the task is to control a combustion engine. Traditionally, this is done by using a PI-controller and a static inverse engine map. The approach in this paper is to support the traditional PI-controller using a neural network as co-controller. The neural network has to be able to suitably model the inverse system behaviour. For this task the LV-ATNN is defined in this paper, an extended version of the Adaptive Time-Delay Neural Network. Two extension are combined in this model: dynamic aging and linear interpolation. It is shown, that the LV-ATNN leads to better results when applied as neural co-controller than its normal relative. An improvement of over 5% compared to the traditional PI-control was rated as a good result by our industry partners.

References

[1] S. P. Day, M. R. Davenport, "Continous-time temporal back-propagation with adaptable time delays", *IEEE Transactions on Neural Networks*, vol. 4, no. 2, pp. 348–354, 1993.

[2] D.-T. Lin, J. Dayhoff, P. Ligomedes, "A learning algorithm for adaptive time-delays in a temporal neural network", tech. rep., University of Maryland, 1992.

[3] C. Ungerer, A. Radan, "Identifying time delays using neural networks", in *Proc. of the Int. Conf. on Engineering Applic. of Neural Networks* (D. Tsaptsinos, ed.), pp. 237–244, 2000.

[4] C. Ungerer, D. Stübener, C. Kirchmair, M. Sturm, "Supporting traditional controllers of combustion engines by means of neural networks", in *Proceedings of the 6th Fuzzy Days* (B. Reusch, ed.), pp. 132–141, Springer, 1999.

[5] C. Ungerer, "Identifying time-delays in nonlinear control systems: An application of the adaptive time-delay neural network", in *CIMCA 99 - Neural Networks & Andvanced Control Strategies* (M. Mohammadian, ed.), pp. 99–104, IOS Press, 1999.

Computer Aided Prototyping of Repetitive Distributed Control

Jarosław Stańczyk*

*Technical University of Zielona Góra, ul. Podgórna 50, 65-246 Zielona Góra, Poland, e-mail: J.Stanczyk@irio.pz.zgora.pl

Abstract

This paper addresses the problem of designing the steady-state behavior of a system composed of a set of repetitive processes competing for access to shared resources. Its objective is to provide a tool allowing to prototype a robust distributed control policy as a function of the characteristics of the component processes and dispatching rules involved. The proposed logistics-oriented methodology integrates production routes design, buffers capacity assignment, and allocation of the dispatching rules that control the work-flow as well as allows one to evaluate system's behavior including its steady state and transient state.

1 Introduction

The integration of resource allocation over time (production planning) and inventory/transportation policy making (logistic planning) is receiving increasing attention. The aim is to develop a scheduling strategy that can simultaneously reduce an impact of both logistic and manufacturing costs, while maintaining high levels of customer service. So, the question is how to achieve a well-synchronized behaviour of the dynamically interacting components, where the right quantity of the right material is provided in the right place, and at the right time. In order to meet this objective many different assumptions regarding, for instance, the capacity of available AGVs and/or local buffers, allocation of working cells and warehouses, production routings and guidepaths, etc. have to be taken into account [2].

The optimal solution (if any) can be obtained, however, for a special combination of those constraints. Therefore, the solution can be obtained either through searching of a space of all possible variants (each variant has to be examined either it is feasible one or not) or through a synthesis of an admissible, e.g., guaranteeing the presumed qualitative and quantitative system behaviour.

Some conditions for designing the control procedures that allow achieving the required performance of the system composed of a set of repetitive processes competing for access to common resources, while preserving some of its qualitative constraints have been already presented [4]. A dispatching strategy that results from a set of dispatching rules allocated to the common resources (it is assumed that each dispatching rule specifies an order the competing process can access a shared resource) is possessing a self-synchronized property, i.e. encompasses a work-flow robust control. Such a self-synchronized system is capable of returning to a unique steady state from any state it was forced into as a result of an accidental disturbance, i.e., each time a disturbance (e.g., caused by operation time change, and delivery time delay) disappear a system results in the same, nominal, cyclic steady state behaviour. Application of such design method [4] supporting evaluation of system time behaviour is limited, however, to an algebraic structure (i.e. determining a coupling operation) which in turn depends on dispatching policy actually employed. In order to overcome this disadvantage an event driven simulator implemented in Java language has been developed [5]. Besides a capability to evaluate a system performance in a transient state it also provides a workbench for evaluation of different variants of work-flow structure, buffers capacity, dispatching rules, and initial states (i.e. initial processes allocation to a system resources).

This paper summarizes our previous work [3, 4] and presents a modeling framework that provides a logistics-oriented methodology for rapid prototyping of a robust control policy. The methodology proposed integrate production routes design, buffers capacity assignment, and allocation of the dispatching rules that control the work-flow as well as allows one to evaluate system's behavior including its steady state and transient state.

2 Software Package

The idea was to combine the modern and attractive capabilities of Java language for platform-independent and distributed work-flow control with an analytical approach to synthesis of distributed discrete dynamic systems that incorporate concurrently interacting repetitive processes [1], [4]. Thus, both Java and algebraic structure supporting evaluation of system time behaviour would complement each other, compensating each others disadvantages: a Java based software would incorporate mathematically proven formulas and algebraic structure based designing method would increase its flexibility. Properly compound they would act as a mean for rapid prototyping of a distributed control policy. Consequently, similar complement would not only allow to proof-test the work-flows control policy, but also would present an inexpensive mean for investigation, illustration and even advertisement of cyclic scheduling strategies.

The challenge of designing of a Java based event driven simulator was to test whether the use of event driven models would allow to simulate the dynamic interaction of concurrent work-flows under different distributed control policies. Run of simulation should enable to trace the qualitative (e.g. a rate of system resources utilization) as well as quantitative (e.g. deadlock-free and starvation-free processes execution) parameters of system's behaviour. Of course, it is important to allow not only slowed-down, but also real-time and speeded-up simulation, so that each viewer could concentrate on those aspects of the process, which are most important for him.

Another challenge was to exploit the portability and network-suitability of Java based simulator for offering platform-independent and distributed or place-independent (in sense of models, model data and processing programs located on different hosts) verification of work-flow distributed control over local or wide-side processing and mixed processing of the information.

Besides of the above mentioned challenges, the development of conditions sufficient for a deadlock-free work-flows interaction was also considered. Such conditions (expressing relations among an initial processes allocation, dispatching rules, and buffers capacities) result then in control rules, providing required (i.e. deadlock, and starvation-free) work-flows interaction.

The main objective of this work is to demonstrate ability of the software package for discovering the relationship existing between the behaviour of a system composed of a set of cyclic processes and the structural as well as functional parameters of its components. In this context, the problem is to find, for a given class of systems, the conditions that will guarantee the required quality of concurrently executed work-flows.

3 Implementation

Consider the Distributed Control Rapid Prototyping System (CRP for short) which is Java written decision support tool [5] aimed at modelling and performance evaluation of concurrently interacting processes. Its design (see Fig. 1) enables one to specify (including operation times, capacity constraints and dispatching rules), and then to simulate repetitive processes interacting via a set of shared resources.

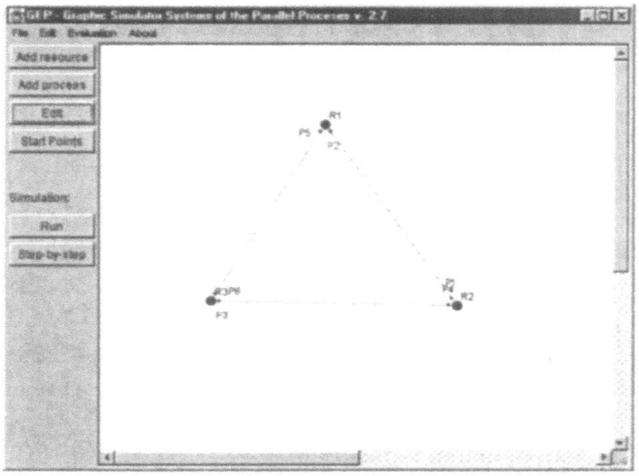

Fig. 1. CRP user interface.

3.1 Repetitive processes

In order to illustrate its application, let us examine a repetitive processes system. As shown in Fig 2, the system consists of the three resources and three processes. Access to resources is realised in mutual exclusion mode. To ensure proper system's functionality (cyclic realisation), at least one resource has to have capacity equal two.

In graphic specification of system from Fig. 2 (see Fig. 3) the following notation is used: instead of arrows the small dots are used (so in the case of the process P_1 instead of the circle composed of two arcs (see Fig. 2) the edge, ended both sides with dotes, is used (see Fig. 3)). The arcs occurring in Fig. 2 are labelled in CRP specification (see Fig. 3).

Let $DR_i = (P_{i1}, P_{i2}, ..., P_{in})$ denotes the dispatching rule determining the order the process $P_{i1}, P_{i2}, ..., P_{in}$ can access to the the i-th resource. It assures that after the n-th access (i.e., of the

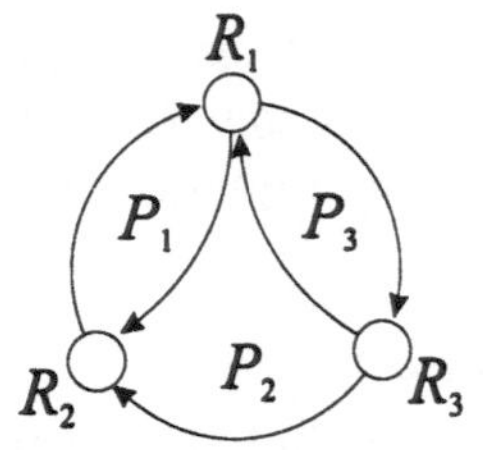

Fig. 2. Graphic representation of repetitive processes system.

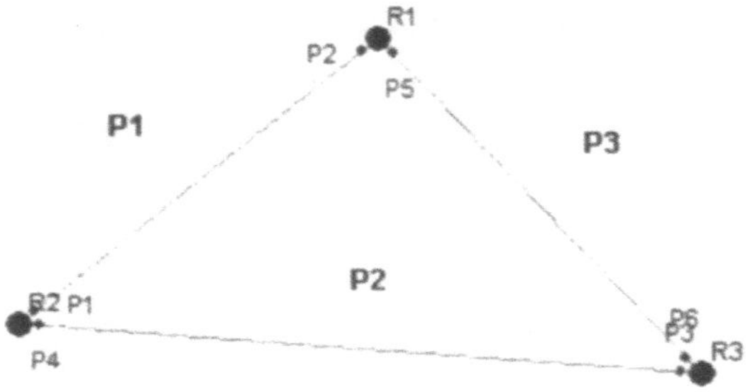

Fig. 3. Graphic representation of repetitive processes system in CRP specification.

P_{in} process) the next one is P_{i1}, and then P_{i2}, and so on. Assuming the following dispatching rules: $DR_2 = (P_1, P_1, P_2)$, $DR_3 = (P_2, P_3)$, $DR_1 = (P_3|\{P_1, P_3\}, P_1)$, and the following initial processes allocation: the process P_1 is allocated to the resource R_2, the process P_2 to R_3, and P_3 to R_1.

The task at hand is to determine the shortest cycle time T and the performance index η of the system considered determined by the following formula:

$$\eta = \frac{nT - \sum_{i=1}^{n} t_{is}}{nT} \tag{1}$$

where:
n — the number of all the component processes,
T — the cycle time of the system,
t_{is} — the time of the i-th process is suspended.

3.2 Computer experiment

In order to illustrate the application package to evaluation of the qualitative functionality properties of a set of processes, let us considered the system composed of three processes, as shown in Fig. 2.

Let the operation times for the processes considered are as follows: for the process P_1: $t_{R_1} = 1$, $t_{R_2} = 1$, and for P_2: $t_{R_2} = 1$, $t_{R_3} = 2$, and for P_3: $t_{R_1} = 3$, $t_{R_3} = 1$.

Example 1. The three possible additional capacity allocations are considered, assuming that the given dispatching rules and initial state are constant. The initial allocation of processes: the process P_1 is allocated to the resource R_2, the process P_2 to R_3, and P_3 to R_1. The dispatching rules allocated to resources are: $DR_1 = (P_3|\{P_1, P_3\}, P_1)$, $DR_2 = (P_1|\{P_1, P_2\}, P_1|\{P_1, P_2\}, P_2|\{P_1, P_2\}|P_1)$, $DR_3 = (P_2, P_3)$. In variant I resources capacity are: R_1 - 2, R_2 - 1, and R_3 - 1. In variant II additional capacity is allocated on resource R_2, it means, that capacity R_1 is 1, capacity R_2 is 2, and capacity R_3 is 1. In variant III additional capacity is allocated on resource R_3: that is, capacity R_1 is 1, capacity R_2 is 1, and capacity R_3 is 2.

Table I. Variants of the additional capacity allocations.

	Additional capacity on		
	R_1	R_2	R_3
Variant	I	II	III
T	4	4	6
η	0.917	0.75	0.75

Example 2. An exemplary application of various distributed control procedures represented by the local dispatching rules. In particular cases, the following initial state and capacity allocation are assumed. Initial state: the process P_1 is allocated to the resource R_2, P_2 to R_3, and P_3 to R_1. Capacity for R_1 is 2, for R_2 and R_3 is 1.

Table II. Variants of the dispatching rules allocation.

Dispatching rules	T	η
$DR_1 = (P_3\|\{P_1, P_3\}, P_1)$, $DR_2 = (P_1, P_1, P_2)$, $DR_3 = (P_2, P_3)$	4	0.917
$DR_1 = (P_3\|\{P_1, P_3\})$, $DR_2 = (P_1, P_2, P_2)$, $DR_3 = (P_2, P_3)$	8	0.667
$DR_1 = (P_3, P_1, P_1)$, $DR_2 = (P_1, P_2, P_2)$, $DR_3 = (P_2, P_3)$	1	deadlock

The examples presented show that for a given topology of processes it is possible to determine a pair: an initial state (processes allocation) and a set of dispatching rules that guarantee deadlock-free and starvation-free processes execution. A still open problem is how to determine a relevant pair guaranteeing systems optimality, e.g., in the sense of η performance index. Another remaining problem is how to refine operators that could support an algebraic approach to a system synthesis from a given set of presumed components (processes).

3.3 Using the CRP

The definition of the investigated process should be started from resources description. First, it is necessary to give their capacities and, in the case of shared resources, access modes (mutual exclusion or randezvous) as well as access rules which are necessary while solving conflicts, e.g. user defined, FIFO, random. The second step, is to define particular processes which are realized in the system. First, it is necessary to give elements which form the process, e.g. sub-processes and operation times.

If the system is fully defined, the simulation process can be started. The system operation can be analyzed "step by step" by tracing particular stages of processes' realization (these information are available on the screen). It is also possible to start the so-called continues simulation. To do so, it is necessary to establish the maximum number iterations. This opportunity is especially useful for the purpose of the fast result analysis, e.g. performance indexes. Before the simulation, the user has to define the system initial condition.

The presented package provides:

- Gantt's chart exploration (Fig. 4);
- calculation of the basic performance indexes (e.g. system period, rate of i-th resource utilization, rate of resources utilization, efficiency index);
- save the simulation data to file, to later processing.

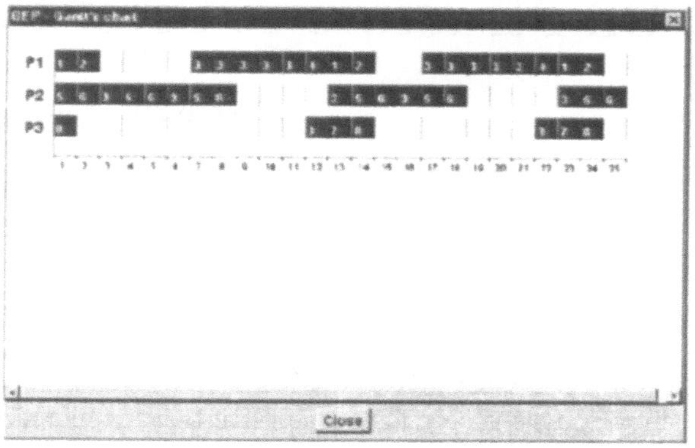

Fig. 4. Exemplary Gantt's chart.

4 Conclusion

This paper presents a new approach to the problem of designing of a steady-state behaviour of a system composed of a set of repetitive processes that share a set of reusable resources.

The approach presented is relevant to and can be of significant interest for evaluation and synthesis of a variety of distributed discrete dynamic systems that incorporate concurrently interacting repetitive processes, and whose evaluation in time depends on complex interactions of various parameters and constraints. Besides of the manufacturing systems the computer/communication networks are the most representatives ones.

References

[1] F.G. Baccelli, et al. *Synchronization and Linearity, An Algebra for Discrete Event Systems*, John Wiley & Sons, Chichester, 1992.

[2] Z.A. Banaszak, M.B. Zaremba and L.S. Yampolski, Logistics Approach to Repetitive Workflows Modelling. In: J. Jedrzejewski (ed.), Innovative and Integrated Manufacturing, Wrocaw University of Technology, 1999, pp. 45–51.

[3] M. Madry, J. Stańczyk and Z.A. Banaszak, Computer Aided Prototyping of Workflow Control, *Scientific Papers of the Institute of Production Engineering and Automation of Wrocaw University of Technology*, Vol. 76, 2000, pp. 176–184.

[4] M.B. Zaremba, K.J. Jedrzejek and Z.A. Banaszak, Design of Steady-State Behaviour of Concurrent Repetitive Processes: An Algebraic Approach. *IEEE Trans. on Systems, Man, and Cybernetics*, Vol. 28, **2**, 1998, pp. 199–212.

[5] The WWW presentation of the Distributed Control Rapid Prototyping System, `http://www.iiz.pz.zgora.pl/ppt/iiz/crp/`, 1999.

On Structure of Local Models for Hybrid Controllers

Tatiana Valentine Guy, Miroslav Kárný*[1]

* Department of Adaptive Systems, Institute of Information Theory and Automation, Academy of Sciences of the Czech Republic, P.O.Box 18, 182 08, Prague, Czech Republic, e-mail: guy@utia.cas.cz

Abstract

A set of models suitable for high-rate filtering of involved signals is specified. Its description is tailored to Bayesian structure estimation having a great potential in the areas relying on multiple local modelling.

1 Introduction

Autoregressive (AR) model with external (X) inputs (ARX) is extensively used for modelling of stochastic dynamic systems, especially in adaptive control [1]. The use of a long regression vector is often required that makes the whole controller complex and non-robust. Through that, relatively long sampling periods are used in discrete controllers. This yields low order models but may lead to undesirable loss of information between sampling points. Two contradictory requirements are thus encountered: (i) use as short sampling period as possible to improve quality of modelling; (ii) prevent extensive increase of model complexity and induced sensitivity to noise, including numerical one.

The paper proposes a way for search of an optimal compromise between these requirements by design a suitable continuous-time model that can be simply related to sampled-data space. Specifically, combination of convolution description of the continuous time controlled system together with wavelet signals decomposition defines the set of model candidates for each considered sampling rate. Then, Bayesian structure estimation [2] provides a justified, data-based compromise between (i) and (ii).

The following assumptions are adopted:

(A1) the controlled system is linear (linearisable) and time-invariant or slowly time-varying;

(A2) convolution operators relate the system output to the system input and process noise;

(A3) the past history of the controlled process has only a limited-time effect on the future outputs;

(A4) the system has scalar input and output.

[1] The work was partially supported by grants EC IST-99-12058, GAČR 102/00/P045, GAČR 102/99/1292

2 Preliminaries

2.1 Convolution description of system

The affine, time-invariant continuous stochastic system relates output $y(\cdot)$, input $u(\cdot)$, external measurable disturbance $v(\cdot)$ and noise $\bar{e}(\cdot)$. Unknown linear time-invariant, causal operators $\bar{\mathbf{A}}, \bar{\mathbf{B}}, \bar{\mathbf{D}}$ and $\mathbf{C}$ specify the relation between signals:

$$\bar{\mathbf{A}}\, y(\cdot) + \bar{\mathbf{B}}\, u(\cdot) + \bar{\mathbf{D}}\, v(\cdot) + \mathbf{C}\, \bar{e}(\cdot) = 0. \qquad (1)$$

Suppose, there is such an operator $\mathbf{C}^*$ that the transformed noise signal $e(\cdot) = \mathbf{C}^*\, \mathbf{C}\, \bar{e}(\cdot)$ becomes white discrete process after sampling with the shortest technically feasible period:

$$\mathbf{A}\, y(\cdot) + \mathbf{B}\, u(\cdot) + \mathbf{D}\, v(\cdot) + e(\cdot) = 0. \qquad (2)$$

The transformed operators $\mathbf{A} = \mathbf{C}^*\, \bar{\mathbf{A}}$, $\mathbf{B} = \mathbf{C}^*\, \bar{\mathbf{B}}$ and $\mathbf{D} = \mathbf{C}^*\, \bar{\mathbf{D}}$ are time invariant casual linear operators. The model (2), which is equivalent to (1), represents a *continuous-time non-parametric description* of the system and is used further on as a basic model.

Applying the convolution theorem to the components in (2), the basic model can be written in the integral form as follows:

$$\int_0^t k_A(\tau) y(t-\tau) d\tau + \int_0^t k_B(\tau) u(t-\tau) d\tau + \int_0^t k_D(\tau) v(t-\tau) d\tau + e(\cdot) + O_t = 0, \qquad (3)$$

where the kernel $k_\bullet(\tau)$ is considered to be a casual ($k_\bullet(\tau) = 0$ for $\tau \le 0$, $\bullet = \{A, B, D\}$) smooth integrable function of a finite length. $O_t = O_t^A + O_t^B + O_t^D$ denotes the total offset reflecting the initial conditions that tends to zero for the considered stable system.

2.2 Filter banks

To calculate the wavelet transform coefficients, a multirate filter bank [3], producing sequences of coefficients at different rates is employed.

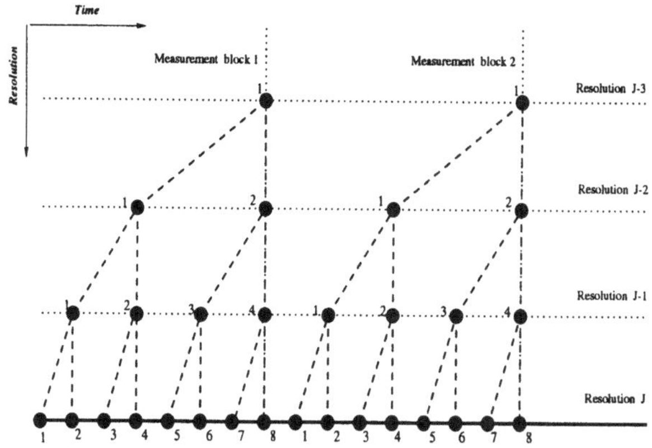

Fig. 1. Example of multiresolution measurements tree for two subsequent measurement blocks and four levels of resolution.

As any physical measuring device measures a real signal at a *finite* resolution and with a *finite* number of samples, there is an upper scale $j = J$, above which the signals details are negligible small. Starting with the high resolution description of a signal, the analysis tree calculates the discrete wavelet transform (DWT) down to as low a resolution ($j = \mathrm{j}$) as desired by having ($J - \mathrm{j}$) stages. Fig.1 illustrates the analysis tree for three levels.

Therefore, for any signal $f(t) \in V_J$ a finite scale approximation in terms of dilated scaling functions and scaling-function coefficients $c_J(k)$ can be written as follows (ϕ denotes scaling function):

$$f(t) = \sum_k c_J(k)\phi_{J,k}(t). \quad (4)$$

Taking into account the multiresolution analysis conception [4], the sum in (4) can be further decomposed in such a way (ψ denotes wavelet function)

$$f(t) = \sum_k c_{\mathrm{j}_0}(k)\phi_{\mathrm{j},k}(t) + \sum_k \sum_{j=\mathrm{j}}^{J-1} d_j(k)\psi_{j,k}(t), \quad (5)$$

where $d_j(k)$ are scaling-function coefficients. The value j sets the coarsest scale whose space is spanned by scaling functions $\phi_{\mathrm{j},k}(t)$. The rest of $L^2(\mathbb{R})$ is spanned by wavelets $\psi_{j,k}(t)$, providing the high resolution details of the signal. Hence, the first sum gives a coarse approximation of $f(t)$, while second term adds details (with increasing index j a higher resolution function is added).

Using low pass filter $h_L(n)$, derived from scaling function, together with high pass filter $h_H(n)$, gained from corresponding wavelet function, the wavelet decomposition is fully represented by sequence of scaling-function and wavelet coefficients

$$c_j(k) = \sum_n h_L(2n-k)c_{j+1}(k), \quad (6)$$

$$d_j(k) = \sum_n h_H(2n-k)d_{j+1}(k). \quad (7)$$

Scaling function coefficients (6) represent a lower rate resolution signal sequence, obtained by sampling of the low pass filter output and wavelet coefficients (7) provide additional details.

In practice, there is no need to deal with the scaling functions or wavelets directly. The only coefficients $c_{\mathrm{j}}(k)$, $d_j(k)$ in (5) and scaling function filters h_L and h_H in (6), (7) are to be considered.

To perform a filtering, let us suppose a measurement block measured at the resolution level J (original data sequence) with the length of data block M[1] in the following form:

$$\mathcal{F}_k^J = [f(J, k-M+1), \ldots, f(J,k)]'. \quad (8)$$

The lower-resolution signal can be obtained by low-pass filtering of $\mathcal{F}_k^J$ with subsequent downsampling of the output from low-pass filter (6).

The filter bank decomposition of the signal $f(t)$ into low-passed $\mathcal{F}_{k\mathbf{L}}^{J-1}$ and high-passed $\mathcal{F}_{k\mathbf{H}}^{J-1}$ sequences written in the operator form reads

$$\mathcal{F}_{k\mathbf{L}}^{J-1} = \mathbf{H_L}^{J-1}\ \mathcal{F}_k^J, \quad (9)$$

$$\mathcal{F}_{k\mathbf{H}}^{J-1} = \mathbf{H_H}^{J-1}\ \mathcal{F}_k^J, \quad (10)$$

where $\mathbf{H_L}$ and $\mathbf{H_L}$ are composed of low and high pass filters' responses. In order to filter noise corrupting the signals, outputs from low pass filter $\mathcal{F}_{k\mathbf{L}}^{J-1}$ (9) are only used. The transformation matrix making low-pass filtering up to the desired level of resolution j is given by:

$$\mathbf{T_L^{\mathrm{j}|J}} = \prod_{l=1}^{J-\mathrm{j}} diag\{\underbrace{\mathbf{H_L}, \ldots, \mathbf{H_L}}_{2^{J-\mathrm{j}-l}}\}. \quad (11)$$

2.3 Bayesian estimation: basic formulae

In statistical set-up, the system is modelled by the conditional probability density function (c.p.d.f.)

$$p(y_t|D^{t-1}, \Theta, S). \quad (12)$$

It relates the probability of the system output y_t to a known function of previous data D^{t-1}. It is usually formed by a vector of delayed outputs and inputs or their filtered values. Incomplete knowledge

[1]The data block length should satisfy the requirement $M = 2^{J-\mathrm{j}}$, where j denotes the desired level of resolution.

of the relationship is expressed by parameterising (12) with respect to the particular structure S and related finite dimensional unknown parameter Θ.

Using the measured data, the probabilistic description of uncertainty of Θ for the particular structure S is fully given by the functional recursion [2]

$$p(\Theta_t|D^t,S) = \frac{p(y_t|D^{t-1},\Theta_t,S)p(\Theta_t|D^{t-1},S)}{\int p(y_t|D^{t-1},\Theta_t,S)p(\Theta_t|D^{t-1},S)\,d\Theta_t} = \frac{p(y_t|D^{t-1},\Theta_t,S)p(\Theta_t|D^{t-1},S)}{p(y_t|D^{t-1},S)},$$

$$p(\Theta_{t+1}|D^t,S) = \int_\Theta p(\Theta_{t+1}|D^t,\Theta_t,S)p(\Theta_t|D^t,S)d\Theta$$

provided the prior p.d.f. $p(\Theta,S) = p(\Theta|S)p(S)$ and evolution model $p(\Theta_{t+1}|D^t,\Theta_t,S)$ are specified. Thus, the prior p.d.f., set on a basis of available prior information, is updated using the measured data and evolution model. The predictive p.d.f. $p(y_t|D^{t-1},S)$, obtained as by-product of the estimation and used in Bayes rule, updates the prior belief $p(S)$ to competitive structures $\{S\}$ as follows

$$p(S|D^t) = \frac{p(y_t|D^{t-1},S)p(S|D^{t-1})}{\sum_S p(y_t|D^{t-1},S)p(S|D^{t-1})}.$$

The last formula, combined with tailored search in (usually) large set $\{S\}$, gives all information required for the choice of the optimal structure [2].

3 Approximate Resolution Dependent Model

This section focuses on the building of the approximate system model using the multiresolution approximations of the input-output signals. The real time implementation of the multiresolution approximation algorithm is performed by filter banks. It employs the measurement blocks containing a predefined finite number of measurements (see footnote on the previous page). The measurement block (8) within the range given by control period is chosen. It can be done as the considered control period is greater than the measurement period.

Substituting input/output signals by their multiresolution approximations (4), (3) converts into:

$$\int_0^t k_A(t-\tau)\sum_k c^y_J(k)\phi^y_{J,k}(\tau)d\tau + \int_0^t k_B(t-\tau)\sum_k c^u_J(k)\phi^u_{J,k}(\tau)d\tau + \int_0^t k_D(t-\tau)\sum_k c^v_J(k)\phi^v_{J,k}(\tau)d\tau + e(t) = 0. \tag{13}$$

Starting with a high resolution description of a signal, the DWT is calculated down to a desired resolution j (see (5)). By omitting the finer second term in (5), noise is filtered and (13) reads

$$\sum_k \left\{c^y_j(k)\int_0^t k_A(t-\tau)\phi^y_{j,k}(\tau)d\tau + c^u_j(k)\int_0^t k_B(t-\tau)\phi^u_{j,k}(\tau)d\tau + c^v_j(k)\int_0^t k_D(t-\tau)\phi^v_{j,k}(\tau)d\tau\right\} + e(t) = 0 \tag{14}$$

with $c^\bullet_j(k)$ $(\bullet = \{y,u,v\})$ are scaling-function coefficients at the desired level of resolution j.

The values $c^\bullet_j(k)$ are outputs of low-pass filter in filter bank, cf. (9). Hence, (14) can be written

$$\sum_k \left\{y^j_{\mathbf{L}}(k)\int_0^t k_A(t-\tau)\phi^y_{j,k}(\tau)d\tau + u^j_{\mathbf{L}}(k)\int_0^t k_B(t-\tau)\phi^u_{j,k}(\tau)d\tau + v^j_{\mathbf{L}}(k)\int_0^t k_D(t-\tau)\phi^v_{j,k}(\tau)d\tau\right\} + e(t) = 0, \tag{15}$$

where $y^j_{\mathbf{L}}(k)$, are *discrete approximations of the output signal at the resolution level* j obtained by

$$y^j_{\mathbf{L}}(k) = \mathbf{T}^{J-j}_{\mathbf{L}}\ \mathcal{Y}^j_k. \tag{16}$$

$\mathbf{T}^{J-j}_{\mathbf{L}}$ stands for the low-pass part of transformation matrix (see (11)) for the filter bank decomposition and $\mathcal{Y}^j_k$ denotes a kth measurement block for the output signal $y(t)$.

The approximations of the input $u^j_{\mathbf{L}}(k)$ and measurable disturbance $v^j_{\mathbf{L}}(k)$ are made similarly.

Let us define the following coefficients for $k \in \mathbf{Z}$

$$
\begin{aligned}
{}^j a_k(t) &= \int_0^{min\{t,L_A\}} k_A(t-\tau)\phi^y_{j,k}(\tau)d\tau\\
{}^j b_k(t) &= \int_0^{min\{t,L_B\}} k_B(t-\tau)\phi^u_{j,k}(\tau)d\tau\\
{}^j d_k(t) &= \int_0^{min\{t,L_D\}} k_D(t-\tau)\phi^v_{j,k}(\tau)d\tau.
\end{aligned} \tag{17}
$$

The majority of scaling functions have finite support or they are at least fast decaying. This fact together with the assumption (A3) ensures that the number of non-zero coefficients, defined by (17) for a fixed t, is finite. This number is determined by lengths L_A, L_B and L_D of kernel supports and by width of the used scaling function. Consequently, the only finite summation over k in (15) is required. The model

(15) under the notations (17) can be rewritten as follows

$$\sum_{k=0}^{{}^{j}m_A-1} {}^{j}a_{{}^{j}m_A-k}(t)\, y_{\mathrm{L}}^{j}(k) + \sum_{k=0}^{{}^{j}m_B-1} {}^{j}b_{{}^{j}m_B-k}(t)\, u_{\mathrm{L}}^{j}(k)$$
$$+ \sum_{k=0}^{{}^{j}m_D-1} {}^{j}d_{{}^{j}m_D-k}(t)\, v_{\mathrm{L}}^{j}(k) + e(t) = 0, \qquad (18)$$

where ${}^{j}m_A$, ${}^{j}m_B$, ${}^{j}m_D$ denote finite numbers of non-zero coefficients ${}^{j}a_k(t)$, ${}^{j}b_k(t)$ and ${}^{j}d_k(t)$.

The sampling moments t_i are considered such that make the coefficients (17) time invariant and ensure the sampled noise $\{e_i\}_{i=0}^{\infty}$ is white discrete-time process with $e_i = e(t_i)$.

Let us order the filtered data into the data vector:

$$\begin{aligned} {}^{j}z_k = [&y_{\mathrm{L}}^{j}(k), \ldots, y_{\mathrm{L}}^{j}(k - m_A), \\ &u_{\mathrm{L}}^{j}(k), \ldots, u_{\mathrm{L}}^{j}(k - m_B), v_{\mathrm{L}}^{j}(k), \ldots]', \end{aligned} \qquad (19)$$

and the corresponding unknown coefficients into the vector of unknown parameters:

$${}^{j}\Theta' = [a_0, \ldots, a_{m_A}, b_0, \ldots, b_{m_B}, d_0, \ldots, d_{m_D}]\big|_{j}. \qquad (20)$$

Then, (18) reads

$${}^{j}\Theta'\, {}^{j}z_k + e_k = 0, \quad k = 1, 2, \ldots \qquad (21)$$

which is the linear regression type model with regression vector (19) composed of *filtered* values of involved signals – their multiresolution approximations. Formula (21) represents a family of model considerations depending on the desired level of resolution $j = J, J-1, J-2, \ldots$.

4 Algorithmic Description

The proposed solution is based upon a Bayes decision algorithm. The optimal resolution level j represents a model structure S. Set of model candidates is given by (21). Then, accordingly to the Bayes approach, the posterior probability $p(j, |D^t)$ contains the information needed for model selection.

The key steps of the algorithm are the following:

- Define a set of all hypothesis about the approximate system model. Each hypothesis is represented by the model gained at each resolution level. The upper resolution J is determined by technical restrictions.
- Assign to each model some prior p.d.f. $p({}^{j}\Theta_1|j)p(j)$ using the best available knowledge.
- Compute the posterior probabilities $p(j|D^t)$ of the predefined hypothesises using the available data D^t, time evolution models $p({}^{j}\Theta_{t+1}|{}^{j}\Theta_t, D^t)$ and Bayes rules.

5 Concluding Remarks

The paper proposes an algorithm for the choosing of an appropriate structure model among the previously defined set of model candidates. The algorithm relies on the hybrid modelling [5] of continuous time system for discrete adaptive control [1]. The model combines a convolution description of the system with wavelet-based approximation of involved noisy signals. Under realistic assumptions it leads to the ARX model with regression vector formed by *multiresolution approximations* of input/output signals. The multiresolution approximations are calculated from real noisy measurements at a particular resolutions by employing filter banks [3] which suppresses corrupting details of processed signals. The probabilistic interpretation of the model allows to use a well elaborated Bayesian approach to structure selection.

In summary, the resulting algorithm offers:

- filtration of measurements for generating controller outputs;
- essential decrease of regression vectors length and, consequently, computational complexity of the whole algorithm;
- the Bayesian methodology for the choice of the best model structure, which possesses a good signal filtering and preserves information about the signal.

The described algorithm origins in adaptive prediction and control area. It is, however, applicable in a much wider context as it provides guidelines how to decide on structure of multiple local models.

References

[1] M. Kárný, A. Halousková, J. Böhm, R. Kulhavý, P. Nedoma, "Design of linear quadratic adaptive control: Theory and algorithms for practice", *Kybernetika*, vol. 21, 1985.

[2] V.Peterka, "Bayesian system identification", in *Trends and Progress in System Identification*, P.Eykhoff,Ed., Oxford, Pergamon Press, 1981.

[3] M. Vetterli, C. Herley, "Wavelets and filter banks: theory and design", *IEEE Trans. on Acoustics, Speech and Signal Processing*, vol. 40, no. 9, pp. 2207–2232, 1992.

[4] S.G. Mallat, "A theory of multiresolution decomposition: The wavelet representation", *IEEE Trans. Pattern Analysis and Machine Intelligence*, vol. 11, no. 7, pp. 674–693, 1989.

[5] T.V. Guy, M. Kárný, "Possible way of improving the quality of modelling for adaptive control", in *8th IFAC Symposium Computer Aided Control Systems Design*, Salford, UK, 2000.

Cerebellar Climbing Fibers Anticipate Error in Motor Performance

Y. Burnod, M. Dufossé *, A.A. Frolov †, A. Kaladjian*, S. Řízek ‡1

* INSERM U483, Univ. P.M. Curie, Paris, France †Institute of Higher Nervous Activity and Neurophysiology, Russian Academy of Sciences, Butlerova 5a, Moscow, Russia, E-mail: aafrolov@mail.ru ‡Institute of Computer Science, Academy of Sciences of the Czech Republic, Pod Vodárenskou věží 2, 182 07 Prague 8, Czech Republic, E-mail: rizek@cs.cas.cz

Abstract

The proposed formulation lies upon four main preliminaries, the Marr-Albus-Ito theory of cerebellar learning, the equilibrium point theory of motor control, the columnar organization of the cerebral cortex and the thoery of the differential neurocontroller. It is shown that: 1) any linear combination of cerebral motor commands which generates olivary signals is able to drive the cerebellar learning processes; 2) climbing fiber activities which supervise the cerebellar learning may originate from the generation of the cerebral commands, before any error of performance occurs, and thus early enough to improve these commands.

1 Introduction

Cerebellar learning was hypothesized to be located at the parallel fiber contacts with the Purkinje cells [14], [1]. Experiments in alert animals [9] [7] and in anaesthetized preparation [11] confirmed the hypothesis. The repeated conjunction of climbing and parallel fiber activities on a Purkinje cell produces a long term decrease in the synaptic strength of the parallel fiber synapse. This process of long-term depression (LTD) [6] can be interpreted as an error learning rule, in the sense that the more a climbing fiber discharges, the less the Purkinje cell will discharge, for the same level of its parallel fiber activities. This definition, made at the cellular level, refers to a 'local error'. By contrast, the term of error was often referred to a functional motor error, such as a mismatch between the final position and the target during reaching.

Two questions about the generation of olivary source of climbing fiber error signals are here addressed. The first question is related to the temporal domain. In this domain, an error in performance occurs far too late for being used as a supervising signal for movement learning. The second question is related to the spatial (somatotopic) domain. In this domain, Purkinje cells or cerebellar microzones must receive error signals related to the synergies they produce. Unhappily, it is unlikely that the inferior olivary nucleus computes any error signal the Purkinje cells need.

As another prerequisite of the model, the equilibrium point (EP) theory assumes that the trajectory results from a simple planned trajectory of an equilibrium position or skeletal configuration. Due to the skeleto-muscular inertial and viscoeleastic parameters, muscle forces result from the difference between the actual and virtual trajectories, and the control system must only solve the static inverse problem to execute a movement.

The third precondition of the model is the theory of cerebral column performance developed in [4] and [5] which emphasizes different properties of information processing at different column levels.

The fourth main idea, on which the model is based, is the theory of the differential neurocontroller developed in [17]. It suggests that the visual-motor transformation determines only displacement but not the absolute position along the movement trajectory. This assumption makes it possible to reduce the problem to the linear consideration.

Three sites of plasticity are known: the cortical long-term potentiation and depression (LTP) and (LTD), the cerebellar LTD at the site of parallel fiber synapses to Purkinje cells and LTP and LTD in the cerebello-thalamo-cortical pathway. First and third learning rules are here considered as being unsupervised Hebbian rules. By contrast, the cerebellar learning rule is considered as a supervised by the inferior olive.

[1]The paper was supported by grant GA ČR 201/01/1192

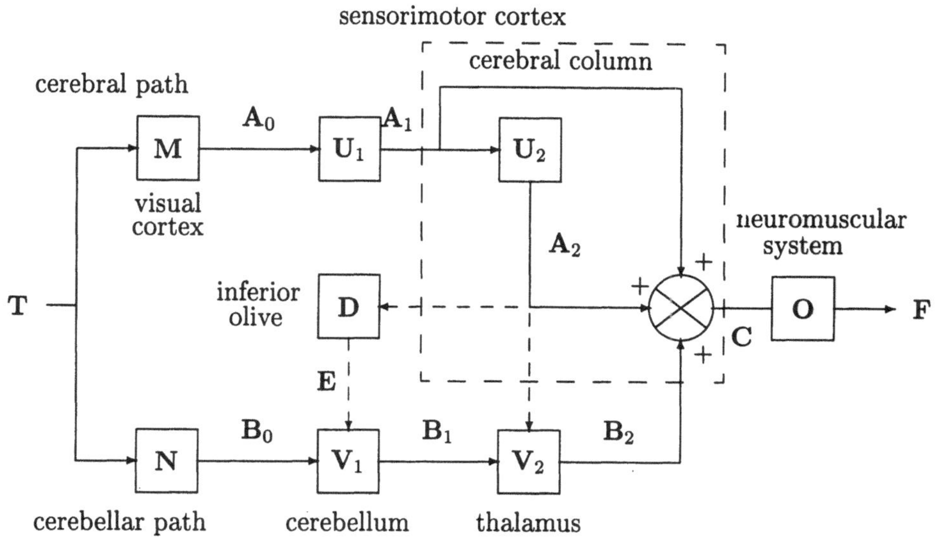

Fig. 1. Linear model of the sensorimotor transformation. Two pathways, cerebral (top) and cerebellar (bottom) link a visual target **T** (left) to a final arm position **F** (right). Vectors $\mathbf{A}_0$, $\mathbf{A}_1$, $\mathbf{A}_2$, $\mathbf{B}_0$, $\mathbf{B}_1$, $\mathbf{B}_2$, **C** and **E** show the cell activities. Boxes **M**, **N**, $\mathbf{U}_1$, $\mathbf{U}_2$, $\mathbf{V}_1$, $\mathbf{V}_2$, **O**, and **D** are matrices of synaptic weights, four of them having adaptive values: $\mathbf{U}_1$ and $\mathbf{U}_2$ for the cerebral cortico-cortical plastic synapses, $\mathbf{V}_1$ for the parallel fibers/Pukinje cell plastic synapses and $\mathbf{V}_2$ for the cerebello-thalamo-cortical plastic synapses.

2 The Model Description

It is generally accepted that the brain controls the working point (WP), such as the finger tip in a reaching movement, moving it from an initial to a final position, with the aim of reaching a target position. The desired WP displacement is internally represented in the visual cortex by the set of cell activities which code the projections of this displacement to cells in preferred directions. Then it is represented in the sensorimotor cortex by the set of cell activities which code the contribution of cells in motor execution. Activity of each cell corresponds to the projection of WP displacement to the 'direction of action' of this cell. Direction of action of the cell is defined as elementary motor action produced due to its activation. To produce the movement, the central nervous system must perform visual-motor transformation, i.e transform desired WP displacement from the visual to sensorimotor presentation.

Two brain pathways allow the brain to perform this transformation. In a cerebral pathway (Fig. 1, top), a first matrix **M**, with fixed random coefficients, performs projections of the vector target (**T**) on preferred directions (vectors corresponding to rows of matrix **M**), uniformly distributed in space. This matrix provides an internal visual representation of the target **T** in the form $\mathbf{A}_0 = \mathbf{MT}$. The second matrix $\mathbf{U}_1$, which coefficients are adaptive, performs the visual-motor internal transformation of the target which provides an internal motor target representation in the form $\mathbf{A}_1 = \mathbf{U}_1\mathbf{A}_0$. The third matrix $\mathbf{U}_2$ which coefficients are also adaptive performs the correction of internal motor target presentation and computation of the error signal used in the learning of cerebellum. The result of the third transformation is vector of neurons activity $\mathbf{A}_2 = \mathbf{U}_2\mathbf{A}_1$.

In the cerebellar side-pathway (Fig. 1, bottom), matrix of projection **N** provides another internal visual target representation in terms of the parallel fiber activities given by the vector $\mathbf{B}_0 = \mathbf{NT}$. The first adaptive matrix $\mathbf{V}_1$ performs the parallel fiber/Pukinje cell transformation, which provides the vector of Purkinje cell activities $\mathbf{B}_1 = \mathbf{V}_1\mathbf{B}_0$. The second adaptive matrix $\mathbf{V}_2$ performs the cerebello-thalamo-cortical transformation, which provides another internal motor representation of the target in terms of the vector of cerebral activities $\mathbf{B}_2 = \mathbf{V}_2\mathbf{B}_1$.

Vectors $\mathbf{A}_1$, $\mathbf{A}_2$ and $\mathbf{B}_2$ are summed into the control signal **C**. Components of vectors $\mathbf{A}_1$, $\mathbf{A}_2$ and $\mathbf{B}_2$ with the same indices represent the activities of the same cerebral column (Fig. 2) but at its different parts. The sum of these activities is the output of the column. This output defines 'Direction of Action', i.e. the vectorial contribution of this column to the vector **F** of the final position in the 3D external space. The transformation of the control signal **C** into the final arm position **F** is performed by its transformation from sensorimotor to motor cortex and successive its transformation into the muscle

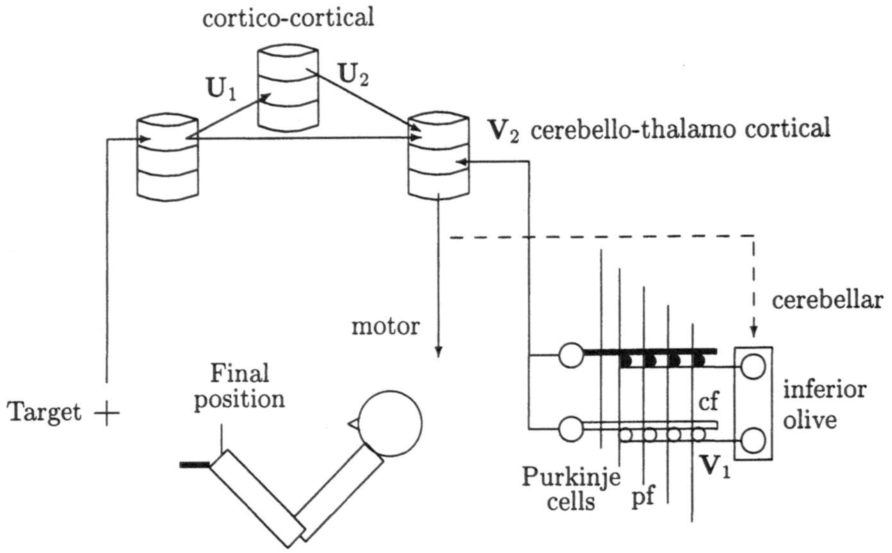

Fig. 2. Sites of learning. Four sites are shown: $\mathbf{U}_1$ is the matrix of synaptic connections which provides transformation of target presentation in visual cortex into its rough presentation in sensorimotor cortex; $\mathbf{U}_2$ is the matrix of the cortico-cortical connections in sensorimotor cortex which produces the correction of the target presentation in the sensorimotor cortex and estimation of the error signal (climbing fibers (cf) activity) used to modify the cerebellar path of sensorimotor transformation; $\mathbf{V}_1$ is the matrix of connection between parallel fibers (pf) and Purkinje cells in cerebellum; $\mathbf{V}_2$ is the matrix of conections in the cerebello-thalamo-cortical route.

forces by the neuromuscular apparatus of the executive system. Into the simplest linear case under consideration this transformation is determined by the output matrix $\mathbf{O}$. Then $\mathbf{F} = \mathbf{OC}$. Each column of matrix $\mathbf{O}$ represents the action which is produced in the external space due to activation of one cortical column. It is assumed that directions of actions are uniformly distributed in the external space.

The motor command $\mathbf{C}$ determines the change of the hand equilibrium position in the external space. The components of $\mathbf{C}$ are the activities of columns in the cerebral sensorimotor area. Each column contributes to the total change of WP equilibrium position by WP displacement into specific direction (direction of action), and the movement is the sum of the activities at the lower layers of the column multiplied by these directions of action [8] (Fig. 2).

During voluntary goal directed arm movements, error signals arise in sensorimotor areas of the cerebral cortex, cerebellar learning being slower than the cerebro-cerebral learning and plasticity of the cerebello-thalamo-cortical pathway occurring still later [16].

Four types of learning are present in the model to take into account these neurophysiological findings.

1. **Learning of block $\mathbf{U}_1$.** It is assumed that this block learns in advance before learning of all other blocks which learn simultaneously. This block learns by Hebbian rule. It is assumed that during the learning of block $\mathbf{U}_1$ some random vectors of columnar activity $\mathbf{A}_1$ are generated in the sensorimotor cortex. As the result of this activity the hand displaces in the external space. This displacement is observed by the visual system and produces the activity $\mathbf{A}_0$ in the visual cortex which is associated with the activity $\mathbf{A}_1$ in the sensorimotor cortex producing the observed movement. Thus

$$\Delta \mathbf{U}_1 = \epsilon_1 \mathbf{A}_1 {\mathbf{A}_0}^T = \epsilon_1 \mathbf{A}_1 {\mathbf{A}_1}^T \mathbf{O}^T \mathbf{M}^T$$

where ϵ gives the learning rate and $\mathbf{A}^T$ is a transposed $\mathbf{A}$. Since vectors $\mathbf{A}_1$ are assumed to be uniformly distributed in the internal motor space then $< \mathbf{A}_1 {\mathbf{A}_1}^T >$ tends to the matrix $K_1 \mathbf{I}_U$ where $< \cdot >$ denotes time averaging and $\mathbf{I}_U$ denotes the unit matrix of the dimension of columns in the sensorimotor area. Then as a result of learning $\mathbf{U}_1 \simeq K_1 \mathbf{O}^T \mathbf{M}^T$.

2. **Learning of block $\mathbf{U}_2$.** This block also learns by Hebbian rule. It is assumed that the activity in the sensorimotor cortex $\mathbf{A}_1$ is associated with the activity provoked in the same cortex by the error in movement performance. Then

$$\Delta \mathbf{U}_2 = \epsilon_2 \mathbf{U}_1 \mathbf{M}(\mathbf{T} - \mathbf{F})(\mathbf{U}_1 \mathbf{M} \mathbf{T})^T.$$

3. **Learning of block $\mathbf{V}_1$.** This block learns by the error estimation produced by the block $\mathbf{U}_2$. Error estimation $\mathbf{E}$ is the result of transformation of the motor cortex activity $\mathbf{A}_2$ into activity of climbing fibers with the use of random fixed matrix $\mathbf{D}$. Then

$$\Delta \mathbf{V}_1 = \epsilon_3 \mathbf{D} \mathbf{A}_2 \mathbf{B}_0^T.$$

4. **Learning of block $\mathbf{V}_2$.** This block also learns by error estimation produced by the block $\mathbf{U}_2$. Since outputs of this block project to sensorimotor cortex, signal $\mathbf{A}_2$ can be directly used for its learning. Then

$$\Delta \mathbf{V}_2 = \epsilon_4 \mathbf{A}_2 \mathbf{B}_1^T.$$

3 Discussion

So, two brain structures, the cerebral cortex (cortico-cortical connections $\mathbf{U}_2$) and the cerebellar cortex (parallel fibers/Purkinje cells connections $\mathbf{V}_1$) learn in parallel. In order to explain how these two learning processes may interact, we stressed on the cerebello-thalomo-cortical pathway $\mathbf{V}_2$ whose plasticity has been described [2], [3], [16]. Our model explains also how the cerebral cortex can control the cerebellar climbing signals. The cortical origin of the climbing signals was strictly established in [13], [12]. In addition, the cerebral cortical activities which are directed to the inferior olive, originate from lower layer of the cerebral cortex [18]. In our model the cerebral-origin olivary/climbing error-like signals are not computed explicetely and, despite the low velocity of the olivo-cerebellar pathway and the low-frequency climbing fiber discharge, their timing is appropriate for optimal learning.

The model eplains also the time coarse of climbing fibers activity experimentally observed in motor learning tasks. During goal directed arm movements performed by awake monkeys, Gilbert and Thach [9] have shown that the learning of a new task protocol is accompanied by a transient increase of climbing fiber activities related to a long lasting decrease of simple spike activities. The transient increase of this activity corresponds to the transient increase of the matrix $\mathbf{U}_2$ efficiency observed in the model. The efficiency of this matrix increases from zero to some maximum at the first stage of learning and then decreases to zero due to increase of the cerebellar contribution to the movement performence.

The proposed formulation shows that: 1) any linear combination of cerebral motor commands generate olivary signals able to drive the cerebellar learning processes, 2) climbing fiber activities which supervises the cerebellar learning may originate before any error of performance can occur, from the generation of the cerebral commands, arising early enough to improve (or even to replace) these commands.

References

[1] Albus J.A.: *Math. Biosci.* 10, 1971, 25-61.

[2] Asanuma H., Keller A.: *Concepts in Neurosci.* 2(1), 1991, 1-30.

[3] Baranyi A., Feher O.: *Exp. Brain Res.* 33, 1978, 283-298.

[4] Burnod Y.: An adaptive neural networks: the cerebral cortex, Masson, Paris.

[5] Burnod Y., Dufosse M.: In: *"Brain and Space"*, J. Paillard (Ed), Oxford Univ., pp 446-458.

[6] Daniel H., Levenes C., Crepel F.: *Trends in Neurosci.* 21(9), 1998, 401-407.

[7] Dufossé M., Ito M., Jastreboff P. J., Miyashita Y.: *Brain Res.* 150, 1978, 611-616.

[8] Georgopoulos A.P., Caminiti R., Kalaska J.F.: *Brain Res.* 54, 1984, 447-454.

[9] Gilbert P.F.C., Thach W.T.: *Brain Res.* 128, 1977, 309-328.

[10] Ito M.: The cerebellum and neural control. Raven Press, New York, 1984.

[11] Ito M., Sakurai M., Tongroach P.: *J. Physiol.* (Lond.) 324, 1982, 113-134.

[12] Kitazawa S., Kimura T., Yin P.B.: *Nature* 392, 1998, 494-497.

[13] Mano N.L., Kanazawa I., Yamamoto K.I.: *Neurophysiol* 56, 1986, 137-158

[14] Marr D.: A theory of cerebellar cortex. *J. Physiol.* (Lond.) 202, 1979, 437-470.

[15] Nakano E., Imamizu H., Osu R., Gomi H., Yoshioka T., Kawato M.J.: *Neurophysiol.* 81, 1999, 2140-2155.

[16] Rispal-Padel L., Pananceau M., Meftah E.M.: *J. Physiol* (Paris) 90, 1996, 373-379.

[17] Řízek S., Frolov A.A.: Differential control by neural networks. *Neural Networks World*, 4, 1994, 494-508.

[18] Saint-Cyr J.A., Courville J.: In: *The inferior olivary nucleus*, Raven, 1980, 97-124.

Model of Neurocontrol of Anthropomorphic Systems

A. Frolov*, S. Řízek† [1], M. Dufossé‡

*Inst. of Higher Nervous Activity and Neurophys., Russian Acad. of Sci., Butlerova 5a, Moscow, Russia
†Inst. of Computer Science, Acad. of Sci. of the Czech Republic, Pod Vodárenskou věží 2, 182 07 Prague 8, Czech Republic, E-mail: rizek@cs.cas.cz ‡INSERM U483, Univ. P. M. Curie, Paris, France

Abstract

The proposed neural model of control of multijoint anthropomorphic systems imitates the visual-motor transformations performed in living creatures. It involves three subtasks: development of the model of the human arm biomechanics; modelling of the neuromuscular apparatus of living creatures; design of the central neurocontroller. The Equilibrium Point theory simplifies the task (reaching movement) performed by the central neurocontroller to the inverse static problem. The contribution of various nonlinear effects of muscle force generation on the accuracy of linear approximation have been tested and qualified. It has been found that the presence of the time delay is substantial. The proposed complex model may provide a scientific base for the design of anthropomorphic robots and manipulators.

1 Introduction

Development of systems for robotic motor control based on neural network models of the cerebellum is one of the most interesting and perspective applications of the neural network theory. This problem, however, still cannot be classified as practical application. Even though only illustrative demonstrations of these systems have been performed, the practical experiments are very impressive [4].

It is generally accepted that movement in living systems is planned as a motion in extracorporal space. For movement execution, the nervous system must transform this kinematic plan into muscle forces. The Equilibrium Point (EP) control hypothesis derives this transformation from the viscoelastic properties of the neuromuscular apparatus [1]. The planned hand trajectory represents the shift of the arm equilibrium position, and the control system must solve only the *static* inverse problem to execute movement.

It is believed that the transformations from the external visual space into the internal motor space, and then into the muscle forces space, are performed in the cerebellum, while corrections of transformations are performed in the motor areas of the neurocortex. Thus, unfamiliar movements, for which the transformations have not yet been formed precisely, are first of all controlled by the neurocortex. On the other hand, the skilled movements are mainly controlled by the cerebellum and the motor area of the neurocortex is made free from current movement corrections.

This paper presents the neural model of control of the human arm reaching movement based on the neurophysiological findings in the supraspinal as well as spinal parts of the central neural system.

2 Model of Control of the Arm Neuromuscular Apparatus

2.1 Equilibrium point hypothesis

The human arm is generally a complex nonlinear system. Control of arm reaching movement should take into account all its nonlinear dynamic properties. It means that the control system must be able to solve the inverse *dynamic* problem to perform the task. The attractive idea which simplifies the control problem is the Equilibrium Point (EP) hypothesis. It suggests that the transformation of control signals into the muscle forces might result from the viscoelastic properties of the neuromuscular apparatus [1], [4]. Then, the planned (or virtual) hand trajectory corresponds to the shift of the arm equilibrium position, and the muscle forces result from the difference between the actual and virtual trajectories. Hence, the system only needs to compute the control signals required to stabilize the arm in each virtual position, and so it must solve the inverse *static* problem.

[1]The paper was supported by grant GA AV ČR No.2030801

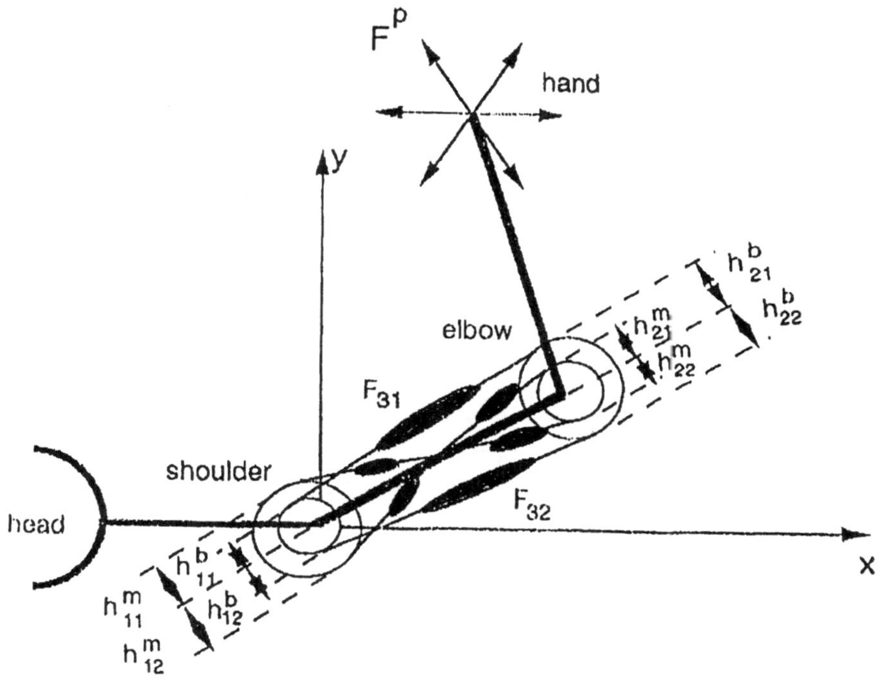

Fig. 1. Human arm model

It was shown [3] that the control signal calculated by the spring – like model of the human arm with the time delay $\tau = 40$ms is in a good agreement with that of the full nonlinear model, while absence of the time delay $\tau = 0$ gives rather inaccurate estimations. So, we can assume in our neural network model of the reaching movement control that it solves the inverse static problem instead of the inverse dynamic problem.

2.2 Human arm neuromuscular apparatus

We have used the model of the human arm according to Fig. 1 taken form [3], [6]. It consists of two links, two joints (shoulder, elbow) and six muscles. The arm movement is described by the motion equation [4]

$$\mathbf{I}(\boldsymbol{\Theta})\ddot{\boldsymbol{\Theta}} + \dot{\mathbf{C}}(\boldsymbol{\Theta}, \dot{\boldsymbol{\Theta}})\dot{\boldsymbol{\Theta}} = \mathbf{T} \qquad (1)$$

where $\boldsymbol{\Theta}$ is the vector of the joint angles (θ_1 for the shoulder, θ_2 for the elbow), $\mathbf{I}$ is the matrix of inertia, $\mathbf{C}$ is the matrix of the centrifugal and Coriolis forces, and $\mathbf{T}$ is the vector of the joint muscle torques calculated from a non-linear model of the neuromuscular apparatus.

The model of muscle force generation was taken from [3]. It consists of a pool of alpha-motoneurons, extrafusal muscle fibres and intrafusal fibres. The pool of alpha-motoneurons receives inputs from the supraspinal level signals and from the stretch reflex loop. Extrafusal muscle fibres produce the muscle forces and intrafusal fibres contain sensors that produce signals for the stretch reflex loop, depending on the current muscle length and contraction velocity.

The activity of alpha-motoneuron pool for each muscle is essentially a non-linear function of its input control signal. The resulting forces involve several nonlinear operations, as nonlinear generation of static forces, nonlinearity in the stretch reflex loop, calcium kinetics, e.t.c. Nevertheless, the whole control system can be approximated to a certain extent by linear functions [3].

The supraspinal control signal is presented in the *lambda* version of the EP theory for every muscle by the control signal λ, which equals to the threshold muscle length of the stretch reflex activation under static conditions. The six components λ (associated with the six muscles) constitute the vector control signal $\boldsymbol{\Lambda}$, which uniquely determines the arm equilibrium position and equilibrium joint angles [1].

It can be assumed that the arm reaching movement is performed by the shift of the control signal $\boldsymbol{\Lambda}$ from its initial to final position with a constant rate. We have shown that the nonlinear model of neuromuscular apparatus can be rather accurately

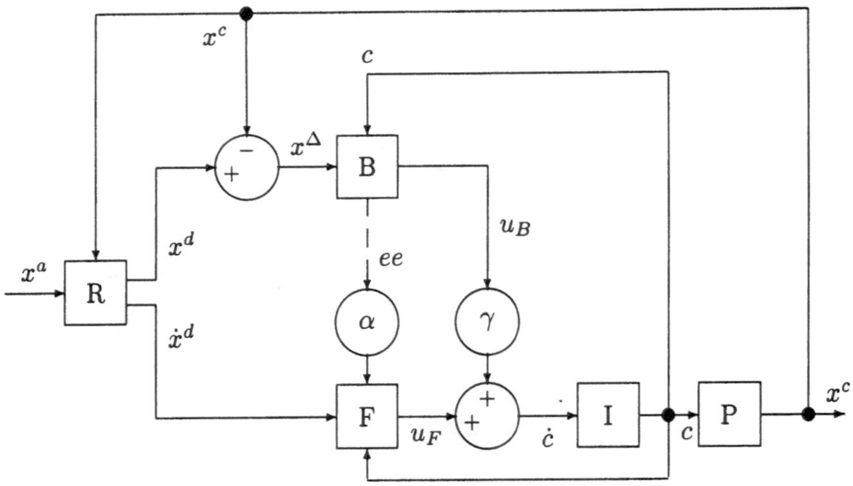

Fig. 2. Scheme of the differential neurocontroller

approximated by the linear spring-like model with the time delay τ:

$$\mathbf{T}(t) = \mathbf{S}(\mathbf{\Theta}^{eq}(t) - \mathbf{\Theta}(t-\tau)) - \mathbf{V}\dot{\mathbf{\Theta}}(t-\tau) \quad (2)$$

where $\mathbf{S}$ and $\mathbf{V}$ are the matrices of stiffness and viscosity. These matrices can be estimated by a linear approximation giving the best fit between $\mathbf{T}$, $\mathbf{\Theta}$, $\dot{\mathbf{\Theta}}$ and τ or directly from the experimental data according to the procedure [3]. Thus the linear shift in the control signal Λ causes the linear transition of the equilibrium hand position along the straight line between the initial and final static arm position.

2.3 Model of the neurocontroller

The proposed scheme of the differential neurocontroller is shown in Fig 2. It involves: F - feedforward control block, B - feedback control block, R - movement planner, I - integrator, P - plant. These blocks process the following vector signals: $x^a, x^c(t)$ and $x^d(t)$ are the target, current and desired outputs of the plant; $x^\Delta(t) = x^d(t) - x^c(t)$ is the discrepancy between the desired and current plant output; $\dot{x}^d(t)$ is the time derivative of the desired output, i.e. the desired output velocity; $c(t)$ is the control vector determined by the joint angles and $\dot{c}(t)$ is its time derivative; $u_B(t)$ and $u_F(t)$ are the partial control vectors generated by the control blocks B and F; ee is the estimated control error; t is the time parameter. The learning coefficient α and the feedback coefficient γ are the global free parameters of the control scheme; they influence the learning process and must be normalized in some way. It should be mentioned that the control signal c defines uniquely the state of the plant and determines its output coordinates. On the other hand the scheme is redundant, as many control vectors c may cause the same output vector x.

The neurocontroller is described in detail in [2]. It should be stressed that the plant P is essentially nonlinear, so that it cannot be described by a linear model within the whole working space. The control blocks B and F should also be nonlinear, but they can be linearized with respect to local changes of their input signals x^Δ and $\dot{x}^d$. The proposed scheme with the differential approach performs it by using layered neural networks with multiplicative neurons with the hidden layers of rather high dimensionality. The basic blocks perform the functions $x^c = P(c)$, $u_B = B(c) \cdot x^\Delta$, $u_F = F(c) \cdot \dot{x}^d$.

The differential neurocontroller is biologically motivated. It imitates basic control functions of special brain areas in reaching movement. However, it exhibits also some cybernetical approach that may not be biologically plausible, epecially simple summation of different classes of signals.

The simple scheme in Fig. 2 is convenient for solving the static inverse problem, when time constants of all transient processes in the plant are much less than the time constants of the planned (desired) trajectory. This condition is met in the case of EP control. The system is described by the equation

$$\dot{x}^c + \gamma \cdot J(c) \cdot B(c) \cdot x^c = \gamma \cdot J(c) \cdot B(c) \cdot x^d + J(c) \cdot F(c) \cdot \dot{x}^d \quad (3)$$

where $J(c)$ is the Jacobian of the plant. Its solution

is stable if the product $J(c) \cdot B(c)$ is a positive definite matrix. It can be reached by the proper design of the block B according to the relation

$$B(c) = \beta \cdot \sum \Delta c \cdot \Delta x^T \quad (4)$$

where β is the gain normalizing constant. Then $B(c) \sim J^T(c)$.

The feedback block B learns by Hebbian rule and resembles cortical structures of living creatures. It ensures that the plant output will reach the target position x^a.

The feedforward block F learns by error back propagation procedure (BP) and imitates the function of cerebellum. The BP algorithm according to the steepest descent method leads to the relation

$$\dot{F} = \alpha \cdot ee \cdot (\dot{x}^d)^T \quad (5)$$

where α is the normalized learning coefficient and ee is the estimated error generated by block B. In the linear case $ee \sim \dot{u}_B$ [2]. The exact control error is not available in the scheme. In the learning process, targets are chosen in the working output space of the plant, and the control problem is solved through time for sequences of random trials.

The output error is integrated for training trials to produce the total error of the learning process. The dependence of learning rate on global parameters α and γ was studied in detail in [5]. The increasing feedback coefficient γ lowers the converegnce rate but stabilizes the complex system. The growing learning coefficient α accelerates the convergence rate till its saturation which can be theoretically forecast and may cause numerical problems.

The feedforward block F ensures that the plant output will follow the prescribed trajectory with the prescribed time derivative. Both blocks B and F are implemented in three-layer neural networks with multiplicative neurons [2].

3 Conclusions

The differential neurocontroller makes it possible to locally linearize the control structure with respect to input signals x^Δ and $\dot{x}^d$, which results in good learning convergence. Characteristics of the scheme depend on the global parameters α and γ and on the complexity of the control blocks B and F. The main advantages of the scheme are its simplicity, learning, good convergence, generalization and possibility of parallel computing.

Computer simulations demonstrate that the linear spring-like model of joint torque generation provides the fairly accurate approximation of the neuromuscular apparatus, if the time delay between current joint angular variables and torques is included in the model. It provides the possibility for the estimation of the time course of the hand equilibrium trajectory, which can be associated, according to the equilibrium point hypothesis, with the time course of the control signals. The proposed model can give a theoretical basis for using the simple linear regression technique to analyse the dependence of the torques on the joint angles and angular velocities in experiments on human reaching movements.

References

[1] Feldman A.G.: Once More on the Equilibrium-Point Hypothesis (λ Model) for Motor Control. J. Motor. Behav., 18, 1986, 17-54.

[2] Frolov A., Řízek S.: Differential Neurocontrol of Multidimensional Systems. In: Dealing with Complexity: A Neural Network Approach. (ed. Kárný M., Warwick K., Kůrková V.), Springer, London, 1998, 238-251.

[3] Frolov A., Dufossé M., Řízek S., Kaladjian A.: On the Possibility of Linear Modelling the Arm Neuromuscular Apparatus. Biological Cybernetics, 82, 2000, 449-515.

[4] Gomi H., Kawato M.: Human Arm Stiffness and Equilibrium-Point Trajectory during Multi-Joint Movement. Biol. Cyber., 76, 1997, 163-171.

[5] Řízek S., Frolov A.A.: Influence of feedback upon learning of the differential neurocontroller. Neural Network World, 3, 1996, 347-353.

[6] Gribble P.L., Ostry D.J., Sanguineti V., Laboissière R.: Are Complex Control Signals Required for Human Arm Movement? J. Neurophysiol. 79, 1998, 1409-1424.

A Neural Model for Animats Brain

Guillaume Beslon, Hédi Soula, Joël Favrel*

*PRISMa Lab., INSA de Lyon 69621 Villeurbanne, France, e-mail: gbeslon@prisma.insa-lyon.fr

Abstract

We propose a model of neural controller, the NeuroReactive controller, which is designed to exhibit both the learning abilities of artificial neural networks and the modular structure of reactive control. This model is based on Asynchronous Spikes Propagation (ASP) in a rank-based neural network. The asynchronous propagation of activity interacts with the internal/external loops in which the animat is involved, leading behavioral modules to emerge in the network, in the form of functional clusters.

1 Introduction

Althought modular organization appears as one of the fundamental features in animats architectures (see, for instance, *subsumption* architecture [2]), the *NeuroControl* approach of animat design mainly remains on centralized architectures. Since neural networks exhibit a monolithic structure, the adaptive process has to learn the control policy as a whole. Thus, animat designers generally have to choose between learning and modularity, i.e. between *NeuroControl* and *Reactive Control*.

In order to provide Behavior-Based controllers with the learning abilities of neural networks, we attempt to develop a dual architecture. However, merging NeuroControl and Behavior-Based Control involves the introduction of an intermediate represention level between the cellular level (neurons) and the network level. Namely, we want *Neural Assemblies* to emerge from our monolithic network in order to create a *NeuroReactive* Controller.

Some recent advances have been made in the field of modular neural networks [4]. In some aspects, modules can be viewed as an artificial implementation of biological neural assemblies. However, as S. Nolfi emphasized it, modular systems are generally based on a "distal" decomposition of the global policy while effective control must be based on a "proximal" approach [6]. This means that the intermediate level cannot be "hard-coded" as usual neural modules are: It has to be "soft-coded" inside the global neural network. Far from being independant modules communicating through specific links, neural assemblies are to be functional groups of neurons [9] that are able to act and react together in order to perform *different* behaviors in a *single* network.

2 The NeuroReactive Architecture

The main objective of the NeuroReactive approach is to develop an intermediate model which exhibits both the learning abilities of neural networks and the modular organization of Behavior-Based control. This means that the neural controller has to branch out *during the learning process* into several functional groups of neurons (neural assemblies) encoding different elementary behaviors.

From a biological point of view, neural assemblies are characterized by a time-coherent activity [3]. Besides, synchronization appears to be a fundamental mechanism of neural assemblies growth [8]. Moreover, Behavior-Based approach emphasizes the asynchrony of the elementary behaviors. That's why our architecture is based on an asynchronous network allowing temporaly organized knowledge processing.

Temporal coding of information in neural networks has received a great interest since it appears as a basic mechanism for sensory feature binding in the cortex [3]. Moreover, Asynchronous Spikes Propagation[1] has been proposed as possible mechanism for information processing in the sensory cortex [5][7]. However, relatively few works have been devoted to the temporal coding for motor patterns generation let alone for sensory-motor association.

Yet in a situated animat, time is no longer an abstract external concept: In a neural animat, the neural network is involved in a sensory-motor loop. Consequently, any action induced by a sensorial occurence may lead to modify the animat relation with the environment. Thus, the initial sensorial occurence itself will be modified. In classical neural networks, this fundamental process has no consequence at all since the update rule (i.e. internal neural time) is independant of the animat time and the sensory inputs are considered as independant "patterns". On the opposite, in ASP networks, the sensorial modification may occur *before* the spikes have all been generated. Consequently, only a subset of neurons is involved in the generation of a particular behavior. Repeated over and over during the animat's life, this simple scheme leads to neural assemblies emergence in the network. Moreover, an animat is not only involved in a sensori-motor loop, but also in multiple internal loops (somato-sensorial system, neural recurrences, ...). The ASP mechanism also interacts with all these internal loops thus enhancing its influence on neural assemblies emergence.

In this approach, the time is not only considered as a

[1] i.e. taking into account the *order* in which the neurons fire instead of the spike frequency.

new coding scheme for neural network. It is a fundamental factor that, while combined with internal or external loops, leads neural assemblies to emerge in a monolithic network.

Given the above-mentioned principles, the implementation is quite simple. It relies on a specific update rule in artificial neural networks: The scheduled mode.

We define the neural imbalance d_i as the difference between the state of a neuron i and its incoming stimulation (1). The imbalance characterizes the network's influence on the neuron: The higher it is, the most the received stimuli leads the neuron into changing its state. Then, at each time step, only the most imbalanced neuron changes its state (2).

$$d_i(t) = \begin{cases} \sum_{k=1}^{n} W_{ik}x_k(t) - \theta_i & \text{if } x_i(t) = 0, \\ -\left[\sum_{k=1}^{n} W_{ik}x_k(t) - \theta_i\right] & \text{if } x_i(t) = 1. \end{cases} \quad (1)$$

where $x_i(t)$ is the state of the neuron i ($i \in [1..n]$) at a given time t ($x_i(t) \in \{0,1\}$, $1 \leq i \leq n$), θ_i its bias and W_{ik} the synaptic weight between neuron k and neuron i.

$$x_i(t+1) = \begin{cases} \mathrm{H}\left(\sum_{k=1}^{n} W_{ik}x_k(t) - \theta_i\right) \\ \qquad \text{if } d_i(t) = \mathrm{MAX}_{j=1}^{n}(d_j(t)) \\ \qquad \text{and } d_i(t) > 0, \\ x_i(t) \\ \qquad \text{otherwise.} \end{cases} \quad (2)$$

where H() is the Heavyside function.

The main difference between scheduled mode and classical update modes in neural networks (synchronous/asynchronous modes) is that the cell's state is not periodically checked. Thus, the neural activity quickly propagates into the network and an answer can be computed without taking into account lower imbalances. Then the animat acts. Acting modifies its relation to the environment and, consequently, its perceptions.

Although the scheduled mode can be used whatever the network architecture is, we use a recurrent network. Then, as argued previously, the network is not only involved in a large sensori-motor loop but also in local/internal loops that will interact with the neural activity propagation. Mutual influence of rank-*scheduled*-coding and multiple retroactions is the key that enables neural assemblies to emerge in the network.

According to (1) and (2), the scheduled mode appears as a simplified implementation of temporal-coded neural networks: The neural signal propagates into the network but its *relative* speed depends on the strength of the synaptic connexions.

3 Application: Learning the Conveying Task

To validate our model, we have designed a NeuroReactive Conveying Animat which task is to carry pieces in a manufacturing workshop (see [1] for a detailed description of the environment).

3.1 The conveying task and the conveying animat

The animat has to learn to perform a transportation task in a workshop composed of three machines, an input desk and an output desk (figure 1). Both machines and desks are fitted out with robots that are able to load/offload pieces when the animat comes close enough. All the loading robot are identified by a beacon but the identification signal are shadowed by any obstacle in the workshop[2].

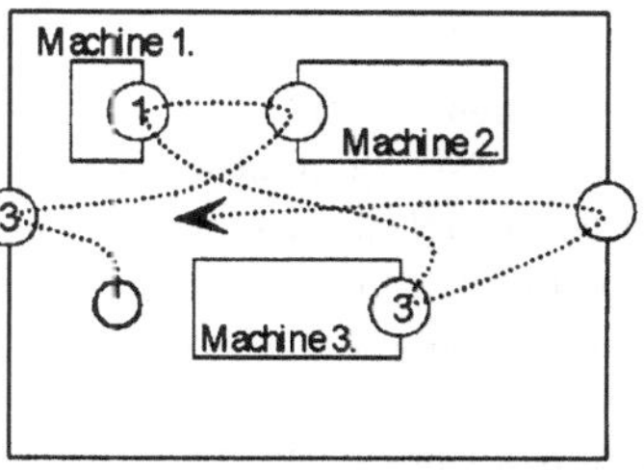

Figure 1. The simulated workshop layout.

Obviously, the transportation task is difficult to perform with a monolithic network exhibiting a single complex policy because the animat has to learn the global task incrementaly according to the piece routing. On the opposite, a structured network which can learn and perform "independant" behaviors thanks to independant substructures (i.e. neural assemblies) would be more efficient.

The conveying animat is round shaped with six perception sectors. Each of them contains a binary proximity sensor and a beacon detector that indicates which beacon(s) face(s) this sector (figure 2.A). Five binary inputs indicate to the animat which beacon it should reach. Moreover, in order to study the neural network ability to develop modular behavior, we provide the animat with the direction of its destination. Of course, the animat doesn't receive any information about its destination if the corresponding beacon is out of sight.

The animat has two propulsion wheels (figure 2.B). Each of them is controled by a boolean value. Consequently, at each time, the animat can choose one action from four elementary ones (move ahead, move back, turn left, turn right).

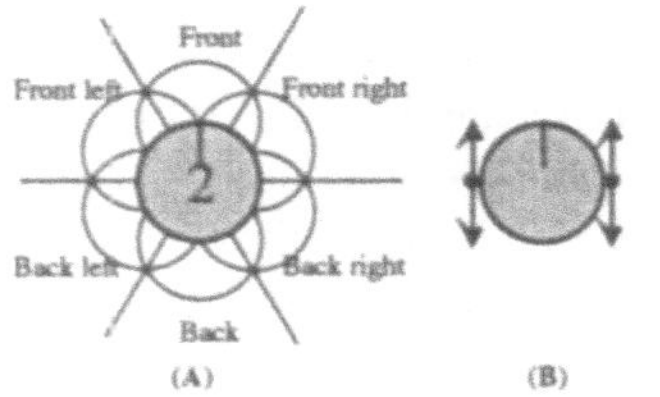

Figure 2. Agent's perceptions (A) and actions (B).

The animat is driven by mean of a neural controller. The neural network has 47 input units and 2 output

[2]The workshop layout is voluntarily sub-optimal and the conveying task is difficult because large shadow-areas prevent the animat from seeing its destination's beacon.

units. The hidden layer is a fully interconnected network made of 40 neurons. At each time step of the simulation, the simulator computes the animat's perceptions and the 47 input neurons are updated. Then the most imbalanced neuron is updated and the animat acts according to the networks output. Note that no output unit may have been updated if the most imbalanced neuron is an hidden one. Consequently, as we emphasized it in section 2, environment time, animat time and neural time are strongly linked.

3.2 Adaptation process

To evolve the neural controller, we use a simple genetic algorithm. Each population is composed of 100 genotypes. Each genotype is composed of 28160 binary genes that encode the synaptic weights of the neural controller. The initial genotypes are randomly generated. Then, at each generation, all the animats are evaluated in the workshop during 52500 time steps. However, in order to quicken the computations, the complete run is split into 21 sessions. If the number of pieces an animat has carried happens to be lower than the number of sessions, then the animat dies. Moreover, when an animat got stucked in front of an obstacle, it dies immediately.

All the animats in a generation are evaluated according to a fitness function which depends on the number of pieces carried and on how closest the animat got from its next destination (if it is on sight).

When all the individuals have been evaluated, the best 20 are selected for reproduction. While reproducing, the genotypes are randomly altered thank to mutation and crossover. Then, the evaluation-selection-reproduction process has been repeated until the stopping criterion has been reached. In our experiments the simulation stopped when 40 pieces have been delivered to the machines.

3.3 Results

We ran five simulations with this architecture. Moreover, in order to evaluate the performance of the ASP model, we used the genetic algorithm with another model of animat in which the neural architecture is exactly the same as the one described above except that it uses a classical synchronous update rule.

Figure 3 shows the fitness values obtained for the best animat at each generation for both architectures (according to the fitness function, if N_p is the number of pieces delivered during the animats life, and f the fitness value, then $N_p \approx f/10$). Each point is the mean value of five simulations (when a simulation has reached the stopping criterion, the later fitness values are artificially set to 400 in the mean computation). It shows that our animat efficiently learns the conveying task twice faster than "synchronous animat".

The great performance of the animat at the end of the adaptive process is confirmed by visual observation (figure 5) and by comparison with the results obtained with

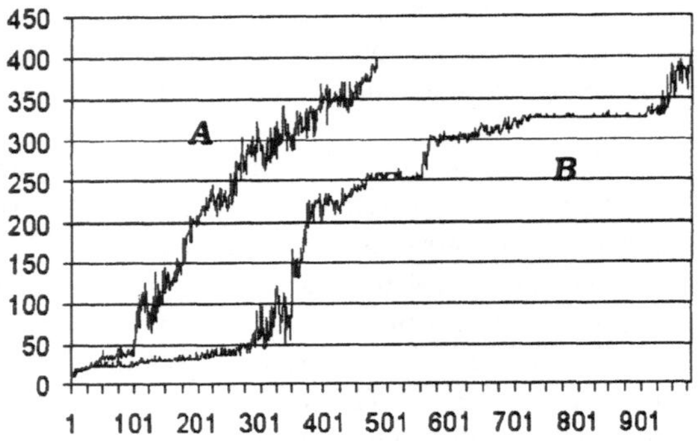

Figure 3. Fitness evolution for the ASP model (A) and for classical synchronous model (B).

hand-coded conveying agents. In particular, our animat is at least as efficient as "Compass Agents" which solve the path-planning problem using hand-coded rules [1].

These results show that our adaptive animat is able to acquire complex behaviors. Moreover, due to the organization of learning, the behaviors have to be learnt incrementally since the animat discovers the beacons one after the other. When looking more precisely to the results of an individual animat (figure 4), one can notice that it slowly acquires the main behaviors during the first learning period; Then, it stabilizes these behaviors and makes them secure; Finally, it is able to reach safely all the machines and its performance grows regularly while it enhances its behavior. The simulation presented on figure 4 has been continued over the convergence criterion (up to generation 700). Figure 5 shows the animat behavior at the end of this simulation. It illustrates the great animat performance (which grows up to 95% of the maximum value regarding its speed and the workshop dimension).

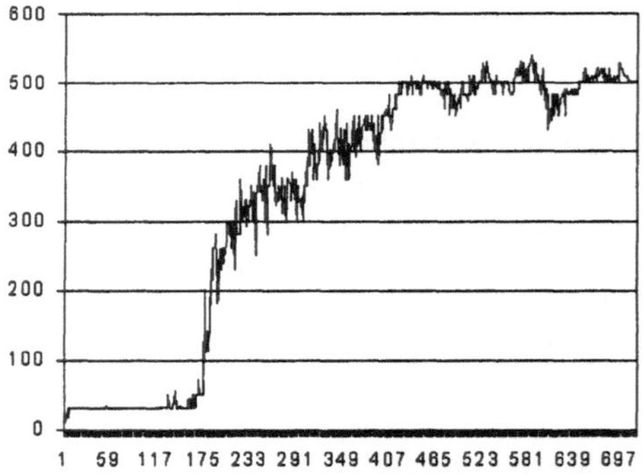

Figure 4. Evolution of the fitness of the best animat during 700 generations (simulation 5).

4 Discussion and Futur Work

We have presented a neurocontrol architecture, named NeuroReactive architecture, in which a neural network is able to acquire incrementally some complex behaviors. Thanks to a specific neural update rule, based on Asynchronous Spikes Propagation, our model is able to split a complex task into simpler subtasks im-

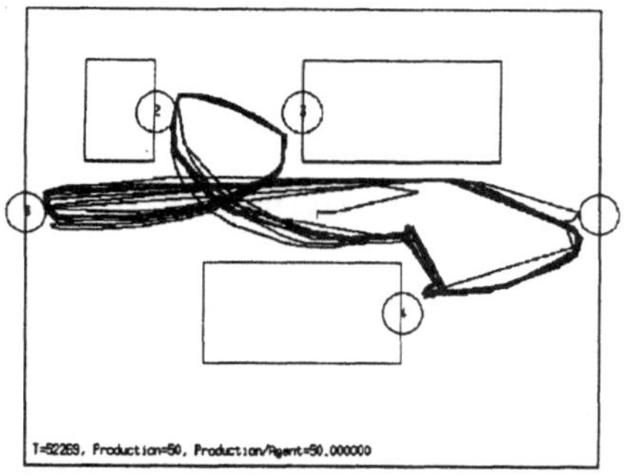

Figure 5. Example of animat behavior at the end of the learning period (generation 700).

plemented as neural assemblies emerging in the initial neural net. Then, a NeuroReactive controller, whose weights were genetically determined, has been used to drive an animat that performed a conveying task in a simulated workshop. This experiment have shown the high performance of the model when combined with a genetic algorithm.

We now plan to use an on-line reinforcement learning algorithm instead of the genetic algorithm. Indeed, we argue that, although genetic algorithms are useful to validate the static performance of a modular architecture (by finding an appropriate set of weights), they are useless if we want to prove the network ability to break down a complex behavior into simpler ones *during an on-line learning process.*

Thus, the development of a learning algorithm is one of the most important stage in the NeuroReactive way to intelligent control.

References

[1] G. Beslon, F. Biennier, and J. Favrel. A flexible conveying system based on autonomous vehicles. In *proc. of CARs and FoF'95, Colombia*, pages 115–120, 1995.

[2] R.A. Brooks. Intelligence without reason. In *Proc. of IJCAI'91*, pages 569–595, 1991.

[3] E. Domany, J.L.V. Van Hemmen, and K. Schulten (Eds). *Model of Neural Networks II; Temporal Aspects of Coding and Information Processing in Biological Systems.* Springer-Verlag, 1994.

[4] A.J.C. Sharkey (Eds). *Combining Artificial Neural Nets; Ensemble and Modular Multi-Net Systems.* Springer-Verlag, 1999.

[5] J.J. Hopfield. Pattern recognition computation using action potential timing for stimulus representation. *nature*, 376(6535):33–36, jul 1995.

[6] S. Nolfi. Using emergent modularity to develop control systems for mobile robots. *Adaptive Behavior*, 5(3/4):343–363, 1997.

[7] M. Samuelides, S. Thorpe, and E. Veneau. Implementing hebbian learning in a rank-based neural network. In *Proc. of ICANN'97, Lausanne*, pages 145–150, 1997.

[8] W. Singer. Synchronization of cortical activity and its putative role in information processing and learning. *Annual Review of Physiology*, (55):349–374, 1993.

[9] T. Ziemke. On parts and wholes of adaptive behavior: Functional modularity and diachronic structure in reccurrent neural robot controllers. In *Proc. of Int. Conf. on Simulation of Adaptive Behavior, Paris*, pages 115–124, 2000.

A Fuzzy Logic System Applied in Lightning Models

André Nunes de Souza, Ivan Nunes da Silva , José Alfredo Covolan Ulson *

*University of São Paulo – UNESP, Department of Electrical Engineering, CP 473, CEP 17033-360, Bauru-SP, Brazil, andrejau@bauru.unesp.br

Abstract

This paper describes a novel approach for mapping lightning processes using fuzzy logic. The estimation process is carried out using a fuzzy system based on Sugeno's architecture. Simulation results confirm that proposed approach can be efficiently used in these types of problem.

1 Introduction

Among on most electric energy systems, lightning is the main cause of unscheduled supply interruptions and several experimental tests and theoretical investigations have been carried out to obtain characteristics and parameters associated with the lightning processes [1,2].

Therefore the lightning phenomenon can be explained as a sudden transfer of electric charge in micro seconds duration due to the failure of insulating properties of atmospheric air at a particular location. In this process, thousands of Volts in electric potential are acquired by the clouds, which are neutralised by the ground potential through the charge injection of luminous leader breakdown channels.

Lightning consists basically on a high current, i.e., transient atmospheric electric discharge with path length around several kilometres. It is due to a great amount of electric charge accumulated in thunderclouds and it occurs when the electric field exceeds locally the air electric insulation. The most common producer of lightning is the thundercloud (cumulonimbus).

During the last years a great improvement on lightning protection methodology has been made. In fact the major step forward in this field is relevant to the evaluation and identification of risk of damage due to lightning related to the protection of systems and components [3,4].

The effects of lightning can have serious consequences on human activities. Its cost for human lives and the devastation it might produce, over huge areas of land through the years are few examples of this power. Lightning can also disrupt industrial processes by causing recurrent power outages and significant equipment failures.

According to the current literature the main core regarding lightning process is to identify and to model those uncertain information on mathematical principles. I mean, the lightning process involves several non-linear features (electrical field, pressure, temperature, humidity, polarity, wave form, etc), that our current mathematical tools would not be able to model.

Another remarkable aspect that will be addressed on this paper is related both to length of time with the electrical current amplitude of the phenomenon.

The theory of fuzzy sets and the various mathematical representations and measurements of uncertainty and information have a virtually unrestricted applicability. The possibilities for application include any field in which the complexity of the necessary knowledge requires some form of simplification.

In this paper, we provide a formal framework to model lightning process using fuzzy logic systems. This system provides a mechanism for modeling and making inferences from imprecise functional relationships. From data obtained in laboratory, it is developed a fuzzy system to represent the process under imprecision and uncertainty environments.

Thus, the Fuzzy Logic System might be used as an additional tool to evaluate the lightning performance. Simulations results are presented and compared with other ones.

2 Fuzzy Systems

Fuzzy logic systems, introduced by Zadeh [5], provide a powerful framework for manipulating uncertain information. In fuzzy logic, a system can represent imprecise concepts, not only by linguistic variables (such as fast, low and small), but also through mathematical principles; moreover it can use these concepts to make deductions about the system. In a rule-based fuzzy system, input values are normalized and converted to fuzzy representations. The rule bases are executed in parallel, consequently producing a fuzzy region for each variable. After this is done, the regions are converted into crisp values to determine the expected value for each solution variable.

In the theory of fuzzy systems, a fuzzy set F, is characterized by a membership function $\mu_F(.)$ in a given universe of discourse U, where u is a generic element in [0,1]. The membership grade of u in F is given by $\mu_F(u)$. Considering two fuzzy sets A and B in U, the three basic operations (union, intersection and complement) can be defined as follows:

$$\text{Union}: \mu_{A \cup B} = \max\{\ \mu_A(u)\ ,\ \mu_B(u)\ \}$$

$$\text{Intersection}: \mu_{A \cap B} = \min\{\ \mu_A(u)\ ,\ \mu_B(u)\ \}$$

$$\text{Complement}: \mu_{\bar{A}}(u) = 1 - \mu_A(u)$$

A fuzzy decision-making system can be formed by a set of production rules using the fuzzy implications described by the previous operations. A typical fuzzy rule "IF (x is A_i and y is B_i) then (z is C_i)" is implemented by a fuzzy implication (relation) R_i with membership function μ_{R_i} defined as follows:

$$\mu_{R_i} = \mu_{(A_i \text{ and } B_i \rightarrow C_i)}(u,v,w)$$
$$= [\ \mu_{A_i}(u) \text{ and } \mu_{B_i}(v)\ ] \rightarrow \mu_{C_i}(w)$$

where A_i and B_i is a fuzzy set $A_i \times B_i$ in $U \times V$; $R_i = (A_i \text{ and } B_i) \rightarrow C_i$ is a fuzzy relation in $U \times V \times W$; and the operator ($\rightarrow$) denotes the fuzzy implication function. Considering rules of the type "IF (u is A_i) then (v is B_i)", the membership function μ_{R_i} is defined as follows:

$$\mu_{R_i} = \min\{\mu_A(u)\ ,\ \mu_B(v)\};\ u \in U\ ,\ v \in V$$

Using Zadeh's compositional rule of inference [5], given the relation R, U to V and a fuzzy set of U denoted by A', the fuzzy set B' of V, inferred from A', has the following membership function:

$$\mu_{B'}(v) = \max\{\min(\mu_{A'}(u)\ ,\ \mu_R(u,v))\}$$

where $u \in U$ and $v \in V$.

3 Process of Fuzzy Inference

The developed fuzzy approach consists of using a fuzzy inference system of the type Sugeno. The choice of this type is especially due to the fact of presenting a good computation efficiency and a more compact representation of the knowledge when compared with other fuzzy architectures, besides being possible the inclusion of optimization techniques in its algorithm.

Other positive point of Sugeno's architecture is the integrity guaranty of the system outputs since the membership functions belonging to output variables are either linear or constant. A more detailed description of this architecture can be found in [6,7].

Besides a larger integrity in relation to the values of its output variables, the use of this Sugeno's fuzzy system type allows that, through adaptive techniques, the fuzzy system find an appropriate relationship between the input and output variables. An efficient algorithm for this purpose is the Adaptive Neuro-Fuzzy Inference System (ANFIS), which is one mainly applied in Sugeno's fuzzy system type.

The ANFIS method uses an algorithm of hybrid learning that combines least-mean-square algorithm with the back-propagation learning algorithm used in training processes of artificial neural networks, in this way, the parameters of the input membership functions are adjusted until the precision is reached. A more detailed description of this method can be found in [7-12].

The fuzzification of the input fuzzy variables consists of converting crisp input values to membership degrees associated with each fuzzy predicate represented by the linguistic variables. The input fuzzy variable of the fuzzy system is the time (s), and the output fuzzy variable is the current (A).

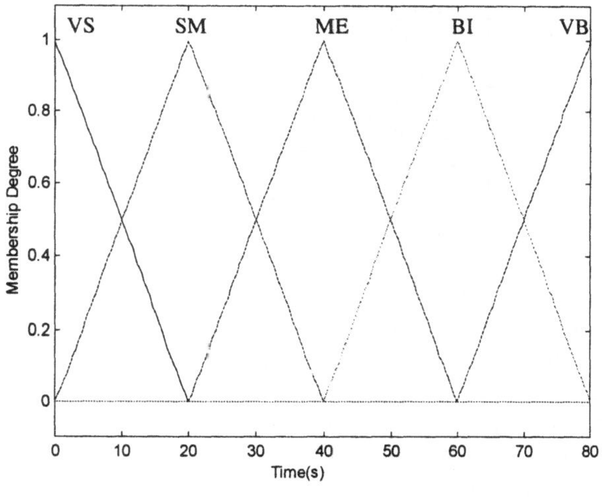

Figure 1. Membership functions

The input linguistic variable is expressed by the linguistic values with membership functions having the shapes shown in Figure 1. The linguistic terms have the following meanings:

- VS – Very Small
- SM – Small
- ME – Medium
- BI – Big
- VB – Very Big

Figure 2 illustrates the variation of the electrical current intensity (kA) with the time (µs).

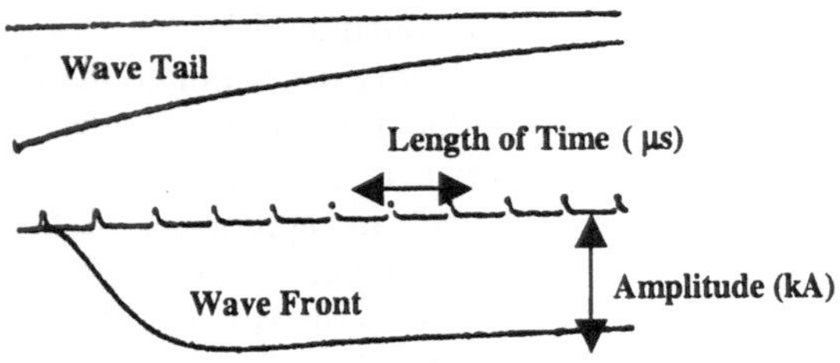

Figure 2. Electrical current intensity

A series of simulations using the developed method was conducted to validate the proposed approach. The simulation results obtained by the fuzzy method are presented in Table I.

In this table, the 'Relative Error' column provides the relative error between the values estimated by the fuzzy method and those obtained by measurements.

Table I. Simulation Results

method / time (µs)	Current (A) Fuzzy Method	Current (A) Experimental Method	Relative Error (%)
20	752.0272	778.9500	3.46
25	719.4879	721.0500	0.22
35	624.1188	621.0500	0.49
40	576.1491	578.6500	0.43
45	537.4015	536.8400	0.10
55	463.9046	463.1600	0.16
60	430.5621	431.5800	0.24
65	400.2774	400.0000	0.07
75	347.9311	347.3700	0.16
80	325.8691	326.3200	0.14

These results show the efficiency of the fuzzy approach used for estimation of parameters related to lightning process. The length of time and the electrical current amplitude were taken into account to simulate the phenomenon.

From analysis of the results presented in Table 1, it is verified that the relative error between values provided by the fuzzy method and those obtained by experimental measurements is very small. In this case, the greatest relative error is 3.46 %.

4 Conclusions

This paper has presented a novel methodology for estimation of parameters related to lightning process fuzzy systems techniques. The estimation process is carried out using a fuzzy system based on Sugeno's architecture. Simulation results confirm that proposed approach can be efficiently used in these types of problem.

Those results obtained are preliminary. In sake of a better understanding, the process has still been developed towards to an accurate identification of the whole process. Another variables such as, resistivity of soil, frequency, etc, will be considered to validate the fuzzy approach.

References

[1] Uman, M. A. *The Lightning Discharge*. Academic Press, New York, 1987.

[2] Diendorfer, G. "An improved return stroke model with specified channel-base current", *J. Geophys. Res.*, vol. 95, pp. 617-630, 1990

[3] IEEE WG Report. "A simplified method for estimating lightning performance of transmission lines". *IEEE Trans. on Power Apparatus and Systems*, vol. PAS-104, no. 04, pp. 919-932, April 1985.

[4] Lin, Y. Y. "Lightning return stroke models". *J. Geophis. Res.*, vol. 85, pp. 1571-1583, 1980.

[5] Zadeh, L. A. "Outline of a new approach to the analysis of complex systems and decision process", *IEEE Trans. Systems, Man and Cybernetics*, Vol. 3, pp. 28-44, January 1973.

[6] Sugeno, M. "Fuzzy measures and fuzzy integrals: a survey," (M.M. Gupta, G. N. Saridis, and B.R. Gaines, editors) *Fuzzy Automata and Decision Processes*, pp. 89-102, North-Holland, New York, 1977.

[7] Sugeno, M. *Industrial applications of fuzzy control*, Elsevier Science Pub. Co., 1985.

[8] Jang, J.-S. R. "Fuzzy modeling using generalized neural networks and kalman filter algorithm," *Proc. of the Ninth National Conf. on Artificial Intelligence (AAAI-91)*, pp. 762-767, July 1991.

[9] Jang, J.-S. R. "ANFIS: Adaptive-network-based fuzzy inference systems," *IEEE Transactions on Systems, Man, and Cybernetics*, Vol. 23, No. 3, pp. 665-685, May 1993.

[10] Jang, J.-S. R. and N. Gulley, "Gain scheduling based fuzzy controller design," *Proc. of the International Joint Conference of the North American Fuzzy Information Processing Society Biannual Conference, the Industrial Fuzzy Control and Intelligent Systems Conference, and the NASA Join Technology Workshop on Neural Networks and Fuzzy Logic*, San Antonio, Texas, Dec. 1994.

[11] Jang, J.-S. R. and C.-T. Sun, "Neuro-fuzzy modeling and control," *Proceedings of the IEEE*, March 1995.

[12] Jang, J.-S. R. and C.-T. Sun, Neuro-Fuzzy and Soft Computing: *A Computational Approach to Learning and Machine Intelligence*, Prentice Hall, 1997.

Multi-Agent Environment for Hybrid AI Models

Roman Neruda[1], Pavel Krušina, Zuzana Petrová*

*Institute of Computer Science, Academy of Sciences of the Czech Republic, P.O. Box 5, 18207 Prague, Czech Republic, roman@cs.cas.cz.

Abstract

We describe a system which represents hybrid computational models as communities of cooperating autonomous software agents. It supports easy creation of combinations of modern artificial intelligence methods, namely neural networks, genetic algorithms and fuzzy logic controllers, and their distributed deployment over a cluster of workstations. The adaptive agents paradigm allows for semiautomated model generation, or even evolution of hybrid schemes.

1 Introduction

Since the practical use of artificial intelligence methods, such as neural networks, genetic algorithms as well as their simple combinations seem to be widely explored [2], we have turned our effort to more complex combinations, which have not been studied much yet, probably also because of the lack of a unified software platform that would allow for experiments with hybrid models.

Design of our system called Bang2 pursues two goals. The first is to build a library of various artificial intelligence methods. Moreover the unified interface allows to switch easily e.g. between several learning methods, and to choose the best combination for application design. Parallel processing is the useful advance here as well as a rapid and easy design. Second goal of Bang2 design involves creation of more complex models, semi-automated model generation and even the evolution of hybrid models.

For distributed and relatively complex system as Bang2 it is favorable to make it modular and to prefer the local decision making against global intelligence, and therefore to take advantage of agent technology. Employing software agents also simplifies the implementation of new AI components and even their dynamic changes.

In our previous work we have tested some of these ideas on the previous implementation of the Bang system with encouraging results on several benchmark test [5].

2 Overview of Bang2

Bang2 consists of a population of agents living in the environment, which provides support for creation of agents, their communication, distribution of processes (parallelism, load balancing). Each agent provides and requires services (e.g. statistic agent provides statistic preprocessing of data and requires data to process). Agents communicate via special communication language encoded in XML. There are several special agents necessary for Bang2's run (like the Yellow Pages agent maintains information about all living agents and about services they provide). Most of the agents realize various computational methods ranging from simple statistics to advanced evolutionary algorithm.

For introduction to software agents see [6]. Generally software agent is a computer program, which is autonomous, reacts to its environment (e.g. to user's commands or messages from other agents) in pursue of its own agenda. It can be adaptive and intelligent in a sense that it is able to obtain information it needs by asking somebody (other agent, a human, a server). Moreover it is usually mobile and persistent. We do not consider other types of agents that for example try to simulate human character, etc.

3 Architecture

Bang2 environment is a living space for all the agents. It supplies resources and services the agents need and serves as a communication layer. One example of such an abstraction is a location transparency in communication between agents — the goal is to make the communication simple for the agent programmer and identical for local and remote case while still exploiting all the advantages of the local one. There should be no difference from the agent point of view between communication to local and remote agent. On the other hand, we want to provide an easy way how to select synchronous, asynchronous or deferred synchronous mode of op-

[1]This work has been partially supported by the Grant Agency of the Czech Republic under grants no. 201/00/1489 and 201/99/P057.

Medium	XML strings	CData*	function parameters
Call	Sync	BinSync	UFastNX
Generality	High	Run-time	Hardwired
Speed	Normal	Fast	The fastest

Table I. Communication functions properties: Sync is a blocking call of the given agent returning its answer, Async is non-blocking call discarding answer and Dsync is non-blocking call storing answer at negotiated place. BinSync and BinDsync are same as Sync and Dsync but the exchange binary data instead of XML strings. UFastNX is a common name for set of functions with number of different parameters of basic types usually used for proprietary interfaces.

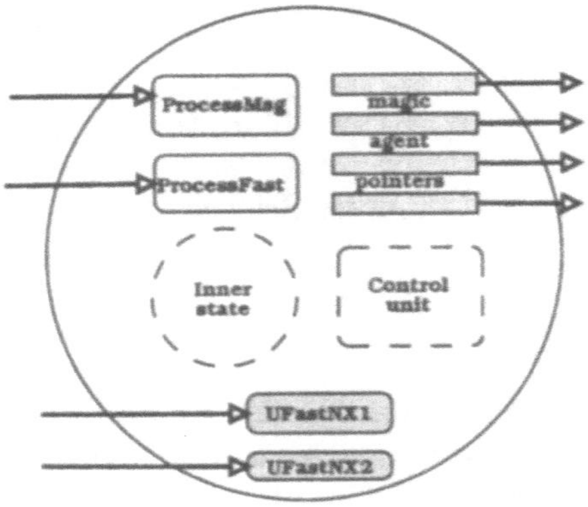

Fig. 1. Agent structure

eration for any single communication act. The communication should be efficient both for passing XML strings and binary data.

As the best abstraction for the agent programmer we have chosen the model of object method invocation. This approach has several advantages in contrast to the most common model of message passing. Among them let us mention the fact that programmers are more familiar with concept of function calling then message sending and that the model of object method invocation simplifies the trivial but most common cases while keeping the way to the model of message passing open and easy.

3.1 Agents

All agents in Bang2 are regular C++ classes derived from base class Agent which provide common services and connection to environment (Fig. 1). Agent behavior is mainly determined by its ProcessMsg function which serves as the main message handler. The ProcessMsg function parses the given message, runs user defined triggers via RunTriggers function and finally, if none is found, the DefaultBehavior function. The last mentioned function provides standard processing of common messages. Agent programmer can either override the ProcessMsg function on his own or (preferably) write trigger functions for messages he wants to process. Triggers are functions with specified XML tags and attributes. RunTriggers function calls matching trigger functions for a received XML message and fills up the variables corresponding to specified XML attributes with the values and composes the return statement from the triggers return values (see 4).

There are several helper classes and functions prepared for the programmers. Magic agent pointer, which is one of them, is an association of a regular pointer to Agent object with its name which is usable as a regular pointer to an agent class but has the advantage of being automaticly updated, when the targeted agent moves around.

The agent inner state is a general name for values of relevant member variables determining the mode of agent operation and its current knowledge. The control unit is its counterpart — program code manipulation with the inner state and performing agent behavior, it can be placed in all ProcessMsg functions or triggers.

4 Communication Language

Agents need a communication language for various negotiations and for data transfer between them. The language should be able to describe basic types of communication, such as requests, acceptance, denial, queries. Also, the language should be able to describe quite wide range of data formats, such as the inner state of a neural network. The language should also be human readable, to some extent, although there might be other tools that can provide better means of communication with the user. Last but not least, we expect reliable protocol for message delivery (TCP/IP).

Several existing agent communication languages for agents already try to solve these problems. ACL ([3]) and KQML ([7] — widely used, de facto standard) are lisp-based languages for definition of message headers. KIF (KQML group — [4]), ACL-Lisp (ACL group — [3]) are languages for data transfer. They both came out of predicate logic and both are lisp-based, enriched with keywords for predicates, cycles etc. XSIL [8] and PMML [1] are XML-based languages designed for transfer of complex data structures through the simple byte stream.

Messages in Bang2 have adopted XML syntax. Headers are not necessary, because of the inner en-

```
<broadcast><halt/></broadcast>
<inform>
<created myid="!000000000001"
  name="Lucy"
  type="Neural Net.MLP"/>
</inform>

<ok>Agent Lucy, id=!000000000001,
type=Neural Net.MLP created</ok>

<request><ping/></request>
```

Fig. 2. Example of Bang2 language for agent negotiation

vironment representation of messages — method invocation— the sender and receiver are known. First XML tag defines the type of the message (similar to message types defined in an ACL header). Available message types are:

- *request* (used when an agent require another agent to do something),
- *inform* (information providing),
- *query* (information gathering),
- *ok* (reply, no error),
- *ugh* (reply, an error occurs).

Content (everything between outermost tags) contains commands (type request), information provisions, etc. Some of them are understandable to all agents (ping, kill, move, ...), others are specific to one agent or a group of agents. Nevertheless, agent is able to indicate whether he understands a particular message.

There are two ways how to transfer data: as a XML string, or as a binary stream. The former is human readable, but lacks performance. This is not fatal in agents' negotiation stage (as above), but is a great disadvantage during extensive data transfers. The latter way is much faster, but the receiver has to be able to decode it. Generally in Bang2, the XML way of data transfer is implicit and the binary way is possible after the agents make an agreement about format of transferred data.

5 Conclusion

For now, the design and implementation of the environment is complete. We have started to create a set of agents of different purpose and behavior to be able to start designing and experimenting with adding more sophisticated agent oriented features to

```
<query><vector row="45"/></query>
<query><vector/></query>
<ok><data separator=",">
Here are binary data
</data></ok>
<query><bin><query>
<vector/>
</query></bin></query>
<ok session="5" funcnum="1"/>
```

Fig. 3. Example of Bang2 language for data transfer

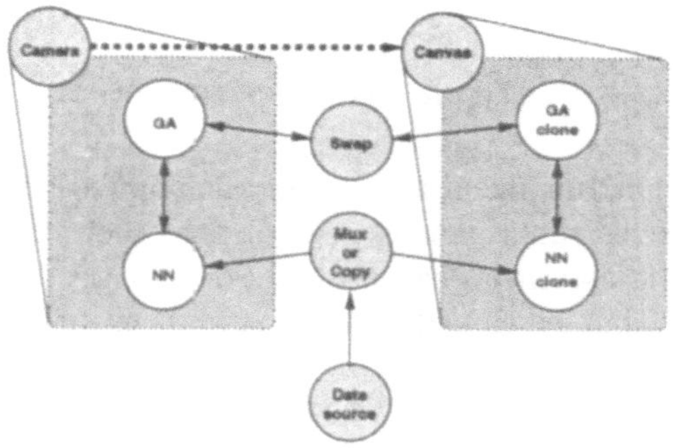

Fig. 4. Task parallelization

the system. There are GA and RBF network agents currently, more is coming soon.

For experimenting with agent schemes, we need agents of various types. We want to try mirrors, parallel execution, automatic scheme generating and evolving. Also the concept of an agent working as the other agent's brain by means of delegating the decision seems to be promising. Another thing is the design of load balancing agent able to adapt to changing load of host computers and to changing communication/computing ratio. To make interaction with human more comfortable we work on a user-friendly graphical user interface. Preferably in a way allowing easy swap to non-graphical representation. And finally we think about some form of inter Bang2-sites communication.

In the following we discuss some of these directions in more detail.

5.1 Task parallelization

There are two ways of parallelization: by adding ability to parallelize its work to a computation agent or by creating generic parallelization agent able to manage non-parallel agent schemes. Both have their good and weak sides, but there is no reason not to implement both and let the user or agent programmer to choose. Consider an example of a genetic algorithm. It can explicitly parallelize by cloning fit-

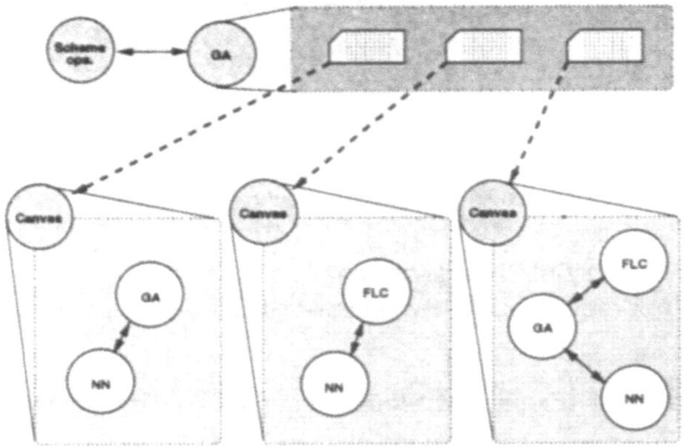

Fig. 5. Scheme evolving

ness function agent and letting the population being fitness-ed simultaneously. Or on the other hand, the genetic algorithm can use only one fitness function agent, but be cloned together with it and share the best genoms with its siblings via a special purpose genetic operator. We can see this in figure 4, where agents of Camera and Canvas are used to automatize the sub scheme-cloning. Camera looks over the scheme we want to replicate and produces its description. Canvas receives such description and creates the scheme from new agents. One can imagine cases where each of the above approaches is better then the other, so it make sense to defer this decision till the real task is considered.

5.2 Agents scheme evolving

When thinking about implementing the task parallelization, we found it very useful to have a way of encoding scheme descriptions in a way which is understandable by regular agents. Namely, we think of some kind o XML description. This leads to idea of agents not only creating and reading such a description, but also manipulating it. All we need to be able to evolve agent schemes by generic genetic algorithm, is to create a suitable genetic operator package. As a fitness, one can employ the part of generic task parallelization infrastructure (namely the Canvas, see fig. 5). For genetic evolving of schemes we can use the Canvas for testing newly modified schemes. In fact, the only thing that needs to be added is the actual scheme genetic operator package, which we see as a nice nice proof of reusability and sound design.

5.3 Agent as a brain of other agent

As it is now, the agent has some autonomous — or intelligent — behavior encoded in standard

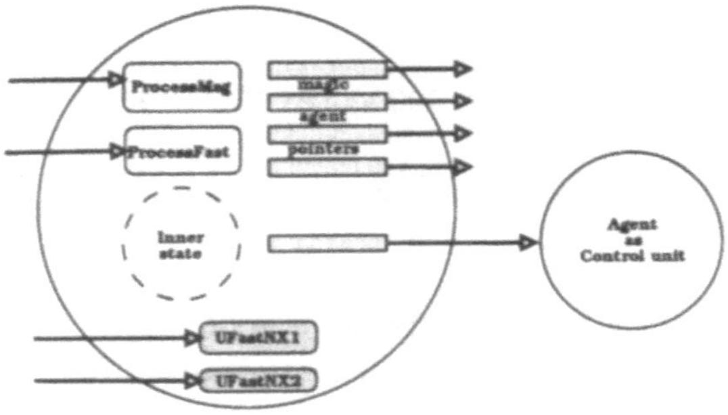

Fig. 6. Agents as brains

responses for certain situations and messages. A higher degree of intelligence can be achieved by hard-coding some consciousness mechanisms into an agent. One can think of creating a planning agents, Brooks subsumption architecture agents, layered agents, or Franklin "conscious" agents. We plan to create a universal mechanism via which a standard agent can delegate some or all of its control to a specialized agent that serves as its external brain. This brain can independently seek for supplementary information, create its own internal models, etc, and finally advise the original agent what to do.

References

[1] Pmml v1.1 predictive model markup language specification. Technical report, Data Mining Group, 2000.

[2] Pietro P. Bonissone. Soft computing: the convergence of emerging reasoning technologies. *Soft Computing*, 1:6–18, 1997.

[3] Foundation for Intelligent Physical Agents. *Agent Communication Language*, October 1998.

[4] Richard Fikes et. al. Michael Genesereth. Knowledge interchange format, v3.0 reference manual. Technical report, Computer Science Department, Stanford University, March 1995.

[5] Pavel Krušina Roman Neruda. Creating hybrid AI models with Bang. *Signal Processing, Communications and Computer Science*, I:228–233, 2000.

[6] Art Graesser Stan Franklin. Is it an agent, or just a program?: A taxonomy for autonomous agents. In *Intelligent Agents III*, pages 21–35. Springer-Verlag, 1997.

[7] James Mayfield Tim Finnin, Yannis Labrou. Kqml as an agent communication language. *Software Agents*, 1997.

[8] Roy Williams. Java/xml for scientific data. Technical report, California Institute of Technology, 2000.

Using Neural Networks and Genetic Algorithms as Building Blocks for Artificial Life Simulations

Gerd Beuster*

*AI Research Group, University Koblenz-Landau, Germany, gb@uni-koblenz.de

Abstract

Artificial neural networks and genetic algorithms are used very often in artificial life simulations. In this paper we describe Artificial Life Environment (ALE). ALE uses neural networks and genetic algorithms, among other parts, as building blocks to set up artificial life simulations.

1 Creating Simulations from Building Blocks

Writing software for artificial life simulations is a complex, time-consuming and error-prone task. With ALE (Artificial Life Environment) we are developing a tool to make the creation of simulations easier. The basic idea of ALE is to set up a simulation from building blocks. Since many artificial life simulations share common characteristics, these common characteristics should be identified and encapsulated in interchangeable building blocks with common interfaces. The advantages of this approach include:

- The process of writing simulations is sped up, because with a common grounding, the researcher can focus on his or her simulation, and has not to bother with user interfaces design and other elements not central to the simulation, because these can be provided by the simulation framework.
- The software contains less bugs if it is used and debugged by a larger group of people.
- Software written for a specific problem is usually highly specialized and can only be used by the people who programmed it. With a common system, sharing simulation components and results between groups of researchers becomes easier.

2 Artificial Life Environment

ALE consists of a two parts. The main part is a C++-class-hierarchy which provides the building blocks for artificial life simulations. Custom building blocks are constructed by inheritance from the base classes. The second part of ALE is a graphical user interface which allows easy access to the simulations created with ALE.

ALE focuses on (though it is not limited to) simulations with populations of autonomous entities who interact in a spatial environment. For this, ALE provides four kinds of basic building blocks:

- `Body`

  Artificial life simulations are usually driven by autonomous entities. ALE has a building block for these, called `Body`. In each step of the simulation, all `Bodies` are successively given a chance to act. When an entity gets activated, it examines its environment, and decides on how to act. How the `Body` perceives the environment, how it decides about its actions, and which actions it can perform, depends on how the researcher implemented this building block.

  Two other building blocks are provided to help in the definition of the internal structure of the `Body`: `Chromosome` and `NeuralNetwork`.

- `Chromosome`

  The `Chromosome` contains the genetic information of the entity. This is usually a string of integer or float point values, together with crossover and mutation operators. How this genotype information is reflected in the phenotype, for example who it influences the strength or agility of the entity, depends on how the `Body` building block is implemented.

- `NeuralNetwork`

  The neural network is the "brain" of the entity. The entity uses it to decide how to act. It is the task of the `Body` to feed the information about the current state of the entity and its environment into the neural network, and to interpret the result of the network's calculation as an action. For example, the `Body` might have a perception of the surrounding telling it

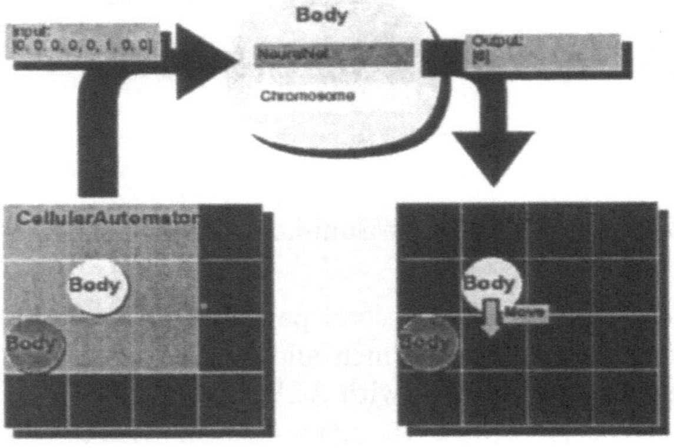

Fig. 1. The steps an entity does when it gets activated: Examining its environment, calculating the next activity, acting.

that some food-source is close by. This information is converted into a format that can be fed into the neural network. The neural network calculates an output value. The output value is interpreted as an action, for example to move closer to the food source.

By subclassing from the base `NeuralNetwork` class, the researcher can create variations of this building block with different neural network architectures and parameters.

- `CellularAutomaton`

  The fourth and last building block for ALE simulations provides the environment of the entities. For this, we use a `CellularAutomaton`. Our cellular automata have — in difference to the commonly used models — asynchronous update functions instead of synchronous ones, and support inhomogenous cells. Again, if the researchers wants a different type of environment, he or she can reimplement this class and use her class instead as a building block for the simulation.

The interaction of these building blocks is shown in figure 1. First, the entity examines its environment. (In this example its Moore neighborhood.) This information is fed into the neural network. The result of the neural network's calculation is interpretated as an activity. (In this case, to move in direction South.)

3 An Example Simulation

Simulations are set up by combining these four classes or subclasses of it. ALE has mostly been used

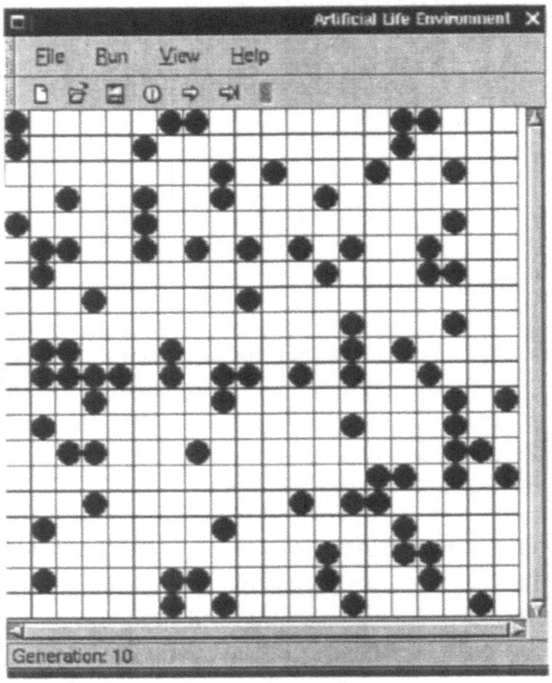

Fig. 2. A screenshot from the predator-prey-simulation. The lighter spots are the prey entities, the darker spots are the predators.

to simulate the co-evolution of populations of predators and preys. In these simulations, the predators and preys are special subclasses of `Body`. They interpret their `Chromosomes` as instructions on how to construct their `NeuralNetworks`. The entities have a certain energy level which can raise or fall, depending on their actions. When the energy level of an entity drops below zero, it dies. When the energy raises above a threshold, the entity reproduces. In the reproduction process, the `Chromosome` of the child is copied from the parents `Chromosome` with some mutations.

There are four parameters affecting the energy household of an entity:

1. The energy level at birth
2. The energy threshold for reproduction
3. The energy usage per move
4. The energy gain for eating other entities

An entity eats another entity when it moves on the cell occupied by the other entity. The only difference between the `Predator` and the `Prey` subclass of `Body` is that preys can not eat other entities, i.e. they are not able to move on cells occupied by other entities. Therefore, in order to survive, the prey-entities have

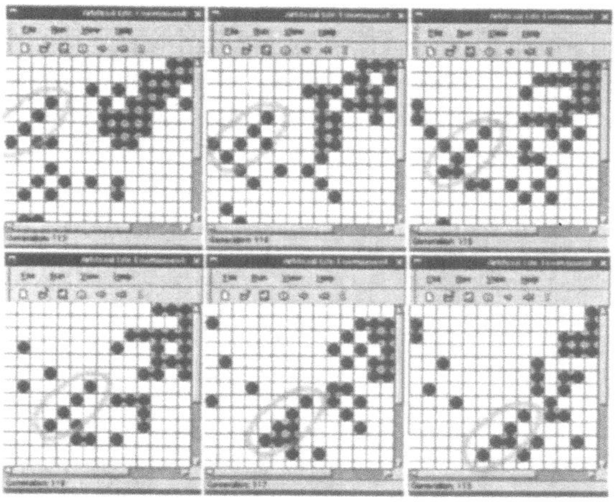

Fig. 3. From upper left to lower right: Six screenshots of predators "fishing" for preys. There is a circle drawn around the predators who are using the fishing-technique.

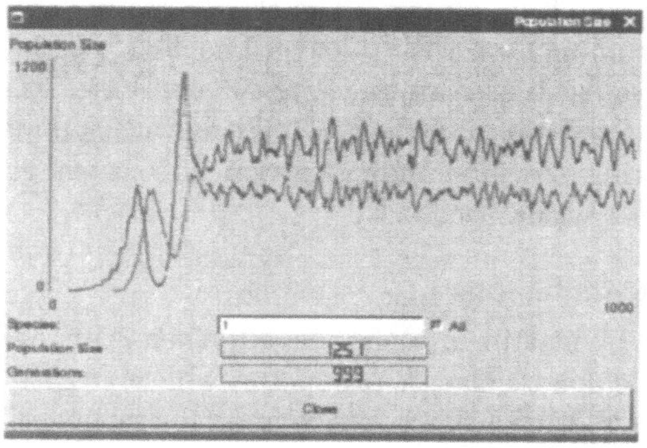

Fig. 4. Development of the population sizes of the predator- (upper line) and the prey-populations (lower line) over 1000 generations.

to have a negative energy consumption. With negative energy consumption, an entity gains energy on every move. We can think about this as the process of taking energy from the environment, like plants are faciliating sunlight. With this setup, it must be the implicit goal of the predators to move onto cells occupied by other entities, and the goal of the preys to move away from the predators. Figure 2 shows a screenshot of the simulation.

The `NeuralNet` of the entities is defined by the rest of the `Chromosome`. We used a very simple scheme to encode the neural network. The structure of the networks is fixed. We use fully connected feed-forward networks without shortcuts. The single genes in the `Chromosme` define the weights of the neural network's connections. By using this very simple scheme, we can not expect the evolutionary process to generate very sophisticated behaviors. We can observe some improvement in the behavior of the entities, though. One interesting phenomenon to observe is a a rudimentary form of cooperation. The predators start to "fish" for preys by moving diagonally in rows. This form of cooperation is quite efficient, because the predators are systematically wandering over the *Cellular Automaton*, fishing for preys. In evolutionary terms, it is also very easy to develop this kind of behavior. When an entity reproduces, the new entity is placed on a cell adjacent to the parent entity's cell. The only behavior the entities have to show is to always move diagonally in the same direction. Figure 3 shows a sequence of updates in which predators use the fishing-technique to hunt preys.

A second, more interesting phenomenon was observed. When creating eco-system simulations, there are many free variables. In our example simulation, we need sensible values for the original size of the predator- and prey-population, and for the energy-household of the entities. If these parameters are not well chosen, the eco-system breaks down very quickly. Either, the preys vanish, followed by the predators who find no more food, or the predators die-out, and the preys take over the whole `CellularAutomaton`. There are several approaches to solve this problem. One simple yet unsatifsfying solution is to inject new entities whenever one of the populations is about to die out. This keeps the simulation running, but is not a very natural approach to handel the problem. Another approach is to search for good parameter combinations by apropriate search methods, for example by a genetic algorithm.

Surprisingly, there is a very easy and natural solution to this problem. When we make the energy household of the entitites subject to evolution, the eco-system gets stable very quickly. This is done by including the four parameters who govern the energy household of the entities into the *Chromosome*. Figure 4 shows who quickly the population sizes stabilizes when these parameters are subject to evolution: After about 300 generations, the population sizes become quite stable. We see also a phenomenon that can also be observed in natural eco-systems: The size of the prey-population is somewhat larger than the size of the predator-population.

4 Teaching

ALE has a number of features which make it very well suited as a tool to teach general principles of artificial life simulations. With ALE, a beginner in the field of artificial life does not have to write simulations from scratch. The clear structure of ALE, in combination with a set of predefined classes and a graphical user interface, allow him or her to get a quick start into the field by playing with and manipulating existing building blocks. The system gives a direct, visual feedback.

5 Conclusions

ALE is still in alpha stage. When ALE has become a more mature product, it will serve as a valuable tool both for research and for teaching in the field of artificial life. For researches, it provides a set of building block which allow to develop a project without spending time on the programming of low level features. The building block concept also makes it easy to exchange parts of the simulations and to compare simulation results. Additonally, the developer gets tool for the analysis of simulations.

In the realm of teaching artificial life, the graphical user interface and the already existing building blocks allow the beginner in the field of artificial life to get a direct experience for simulations and how they are affected by different parameters, without having to program a whole system.

The current version of ALE is available for download at http://www.uni-koblenz.de/~gb/ale/. This version is not ready for productive use. It is only of interest for programmers who might want to participiate in the development of the system.

References

[1] Gerd Beuster, *Artificial Life Environment*, Master's thesis, University Koblenz-Landau, Koblenz, 1999, http://www.uni-koblenz.de/~gb/ale/studienarbeit_ale.ps.gz

[2] A. K. Dewdney, *Simulated evolution: wherein bugs learn to hunt bacteria*, Scientific American, May 1989.

A Hybrid Intelligent System for Image Matching, Used as Preprocessing for Signature Verification

József Valyon, Gábor Horváth *

*Technical University of Budapest, Budapest, Hungary, e-mail: valyon@mit.bme.hu, horvath@mit.bme.hu

Abstract

A complex, hybrid intelligent system for two-dimensional image matching is described, in the context of off-line signature verification. The proposed method can be used as the preprocessing step of a verification process, or it may be employed to determine the measure of similarity for two signatures. The main idea is to apply a nonlinear transformation -commonly used in remote sensing- to the images, in order to reduce their differences, and permit a more exact and reliable comparison.

1 Introduction

Signature is widely used and accepted for proving the authority and validity of documents. Signature verification can also be applied for user identification, where the use of biometric measures has several advantages compared to traditional practices, which involve the use of PIN numbers, passwords or access cards. Biometric measures are hard to duplicate, copy and cannot be stolen or lost. Because of these advantages they provide more security, whilst requiring the least amount of user effort [1].

A signature verification system can be either *off-line* (static)[7-14] or *on-line* (dynamic)[1-6]. These systems differ both in their field of use and the method of validation. In the first case only the signature images are available, while in an on-line system information is obtained also about the dynamics of the signing process, such as, pressure, speed, and acceleration, utilizing a kind of special input device, e.g., a digitizing board. A verification process can be divided into three stages: *preprocessing, feature extraction* and *classification*. The main difference between proposed systems is in the definition of the feature vector, which is used to represent the signature. Experiments show, that the vectors consisting both global and local features are more successful in separating the valid and forged instances, than the ones made up of only one of these feature types. Global parameters are describing the signature as a whole, while local features are calculated to small segments of the signatures. The computation and comparison of local parameters require the matching of two different sets of signature data. The discriminating power of these features depends on the accuracy of this match, therefore we focus on establishing a good, precise point-to-point correspondence between the two datasets. To achieve this we are proposing the use of a transformation, which is calculated with the help of landmarks. The fitting of two signature patterns allows us to compare them precisely, either by determining the closeness of the match or through the use of features. In the first case the feature extraction step isn't necessary, while in the second case the transformation is only a preprocessing step. To minimize the error rates of the verification one has to use the feature vector, which provides the largest discriminating power between valid and forged samples. By using the proposed transformation the local parameters included in this vector can be more precise, and of significant contribution in enhancing the performance of the system. The goal of the feature extraction process is to provide a vector space, where the valid patterns make a small, dense cluster, which is far from all forgeries, in the means of a predefined distance measure. The classification step has to separate these two sets.

2 Overview

This paper describes an algorithm to determine a point-to-point match between the tested and the reference signature. This is done, by applying a two-dimensional transformation -commonly used in remote sensing- on the images [17,18]. The transformation step is described in Section 3. The parameters representing the nonlinear transformation, are calculated from a number of reference points. First these landmarks must be determined. This is done by selecting special points on the reference signature, and then searching for the counterparts of these on the tested image. This method is described in the next section (Section 4.). A genetic algorithm is applied to the resulting point pairs to filter out mismatches and to determine the usable landmarks (Section 5). To test the algorithm, a very

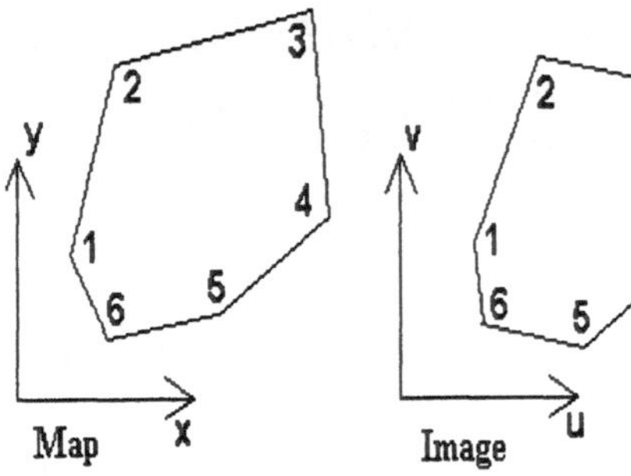

Fig. 1. The coupling of the reference images *(x,y)* points and the tested pictures *(u,v)* points.

simple decision making was implemented, based on the difference between the images (Section 6).

3 Transformation

The transformation that defines the relationship between the reference and the test image uses two functions: $u=f(x,y)$ and $v=g(x,y)$, where (x,y) and (u,v) denote corresponding pixels on the reference and on the test image respectively. The task is to find f and g. For the sake of simplicity we assume that they are polynomials. The following polynomials may be used for the fitting of two signature images:

$$u = a_0 + a_1x + a_2y + a_3xy + a_4x^2 + a_5y^2$$

$$v = b_0 + b_1x + b_2y + b_3xy + b_4x^2 + b_5y^2$$

For the calculation of the coefficients we can utilize some known point pairs called landmarks. The minimal number of needed known pairs depends on the power of the polynomials used (in our case it is six). Experiments show, that the results provided by quadratic polynomials are good enough for our goals. By using polynomials of higher order, we can achieve smaller errors on the fix points, but in this case the equations allow a larger freedom in the areas between them, which results in much larger overall differences [17].

4 Coupling Landmarks

In order to be able to parameterize the transforming equations (in our case polynomials) some point pairs, defining the relationship between the images, must first determined. It is done in two steps. First some reference points (landmarks) are selected on one of the signatures, then a search is done for their matches on the other, as shown on Figure 1. The minimal number of the necessary points is determined by the order of the transforming polynomial. The calculation of a precise match requires much

Fig. 2. a.) Selecting a point to search for. b.) The window containing the searched neighborhood. c.) The searched area of the other image. d.) The point showing maximal resemblance.

more points, than this, therefore the coefficients are calculated by minimizing the squared errors.

Because of the distortions between signature images, the automatic search for the matching points is quite difficult. In the present work two possible solutions are proposed for seeking the corresponding points, referred to as maximal correlation and fuzzy system.

Maximal correlation.

This method creates a small image of predefined size and shape containing the neighborhood of the searched *(x,y)* point. Than this image is swapped over the other signature, whilst the correlational coefficient between this small image and the covered area is calculated. This similarity measure is calculated for every point in a larger area where we expect the matching point to be found. This means searching that half or quarter etc. of the other signature that corresponds to the part where the point was selected from. The point having the maximal coefficient is accepted as the *(u,v)* match [18]. The second best match is also taken into consideration, because in some cases there are more segments that look the same, which can be misleading. The correct match is determined later, with the help of a genetic algorithm, that also takes the fuzzy answers into consideration. Figure 2. shows the main steps of the above described algorithm.

Fuzzy system.

The main idea is to create a similarity measure that is based on the overall location of the point relative to the other parts of the signature. From a selected point the distance of another signature point is calculated in eight directions. The pixel distance of the next line in a given direction is then fuzzified. The membership functions are trapezoids representing the following intervals: 'right next to', 'within a letters distance', 'one letter away', 'one word away', and 'very far away'. This distance calculation is done for every *(u,v)* point of the searched signature, and for the *(x,y)* points we are looking for on the other signature. The systems output is a similarity measure, based on the above described fuzzy system. The match of a point is the pixel with the mostly resembling distance results.

5 Genetic Filtering

The two search methods introduced in the previous section provide more than one possible matching point pair for every *(x,y)* point selected on the reference signature. Experiments show that it is enough to use the best fuzzy and the closest two correlational match in order to find good, precisely matching landmarks. In this case we have three *(u,v)* delegates for every *(x,y)* point that we intended to find a pair for. A genetic algorithm is created to select the best pair from these three choices. For every landmark the algorithm has four options: one of three possible matches and no match at all. In the last case all the three possible pairs are considered to be mismatches. In this case this landmark is discarded. The genetic algorithm operates with chromosomes represented as vectors. Each element of this vector stand for a selected landmark, and consists one choice out of the four possibilities (one of the three matches, or none of these). The objective function is the "precision of the match", therefore the fitness function is defined as the squared error calculated for the landmarks in the transformation. This is the sum of the squared distances between the calculated *(u',y')* and the real *(u,v)* pixels, where $u'=f(x,y)$ and $v'=g(x,y)$. The genetic optimization minimizes this error. This error is zero if only the minimal number of required points are available for the transformation (in this case $u'=u$ and $v'=v$ for all the landmarks), therefore we have to define the minimum number of elements that must be used for the transformation, otherwise the genetic optimization results in a vector that consists to many elements meaning "none of the found matches"! The point pair configuration is determined by the fittest chromosome after a fix number of generations. The genetic algorithm uses the elitist strategy, therefore this specimen is the fittest of all considered throughout the process.

Fig. 3. The reference a.) and the tested b.) signature.

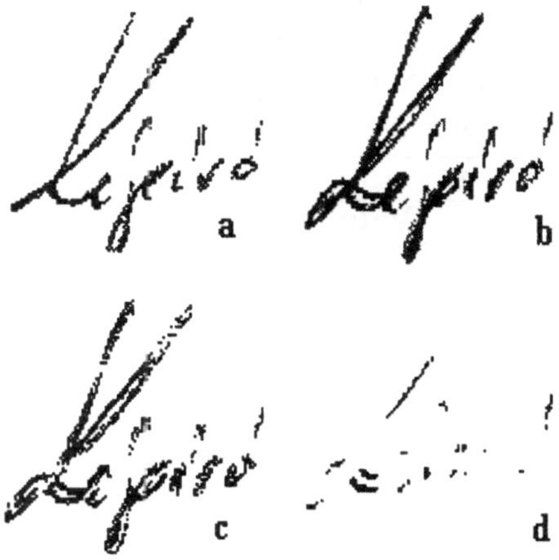

Fig. 4. a.) The transformed signature, b.) the two signatures together, c.) the not overlapping pixels, d.) pixels outside a 4 pixel window.

6 Decision Making

In this case no feature extraction is utilized, the fitted images provide the input of the decision making process. The precision of the fit can be measured by the number of overlapping pixels. Similarly, the difference between the input images can be represented by the distances of the signature lines. To count these distances we can move a window over the signatures, whilst clearing all the pixels inside. The remaining pixels are further away than the window size. A full transformation and decision making process is shown on Figure 3-4.

7 Results

The main idea of this paper, is the use of a transformation as a preprocessing step. This nonlinear transformation is determined, with the use of some landmarks. In section 4 and 5 a hybrid intelligent solution was shown to determine these point pairs. There are several other possible ways to determine these landmarks [15,16], therefore in this section the goal to us is to concentrate on the usefulness of the transformation step alone, without considering the additional errors caused by the uncertainties of the point pair matching. This is especially important when forged signatures are concerned, because in this case, the sometimes large differences result in very bad point pair matches, and transformation. It is not a problem when we are validating the signatures, because these false transformations result in good classification (large errors for forged signatures). To characterize the transformation, the point pairs were selected by hand, so the above mentioned additional errors do not disturb the results. In the next table the error numbers calculated for both valid and forged signatures are enumerated. The transformation was done, by utilizing 21 preselected landmark points:

no.	a	b	c	d	e
Valid signatures					
1	126.537028	9630	2382	468	49
2	142.942031	8440	1235	348	120
3	171.604656	11872	4514	1127	307
4	172.472311	9824	2896	473	142
5	122.273076	9030	2196	271	50
Forged signatures					
1	344.358339	13328	10356	8211	6845
2	249.917618	11278	5845	3176	2058
3	467.377721	12221	6987	3973	2358
4	442.096978	12084	7019	4875	3484
5	236.559947	12248	7776	5123	3285

Table I. a: Squared error. b: Non overlapping pixels. c: Pixels outside a 4x4 window. d: Pixels outside a 8x8 window. e: Pixels outside a 12x12 window.

References

[1] R. Martens and L. Claesen, "Incorporating Local Consistency Information into Online Signature Verification Process", *International Journal on Document Analysis and Recognition*, vol. 1, pp. 110-115, 1998.

[2] R. Kashi, J. Hu, W.L. Nelson and W. Turin, "A Hidden Markov Model Approach to Online Handwritten Signature Verification", *International Journal on Document Analysis and Recognition*, vol. 1, pp. 102-109, 1998.

[3] Quen-Zong Wu, Suh-Yin Lee and I-Chang Jou, "On-line Signature Verification Based on Logarithmic Spectrum", *Pattern Recognition*, vol. 31, no. 12, pp. 1865-1871, 1998.

[4] Rjean Plamondon, "The Design Of An On-Line Signature Verification System: From Theory To Practice", *International Journal of Pattern Recognition and Artificial Intelligence*, vol. 8, no. 3, pp. 795-811, 1994.

[5] L. Yang, B. K. Widjaja and R. Prasad, "Application of Hidden Markov Models for Signature Verification", *Pattern Recognition*, vol. 28, no. 2, pp. 161-170, 1995.

[6] Chan F. Lam and David Kamins, "Signature Recognition Through Spectral Analysis", *Pattern Recognition*, vol. 22, no. 1, pp. 39-44, 1989.

[7] J.-P. Drouhard, R. Sabourin and M. Godbout, "Neural Network Approach to Off-line Signature Verification using Directional PDF", *Pattern Recognition*, vol. 29, no. 3, pp. 415-424, 1996.

[8] Reena Bajaj and Santanu Chadhury, "Signature Verification Using Multiple Neural Classifiers", *Pattern Recognition*, vol. 30, no. 1, pp. 1-7, 1997.

[9] Kai Huang and Hong Yan, "Off-line Signature Verification Based on Geometric Feature Extraction and Neural Network Classification", *Pattern Recognition*, vol. 30, no. 1, pp. 9-17, 1997.

[10] V. E. Ramesh, M. Narasimha Murty, "Off-line Signature Verification using Genetically Optimized Weighted Features", *Pattern Recognition*, vol. 32, pp. 217-233, 1999.

[11] R. Sabourin, Ginette Genest, and Francoise J. Prteux, "Off-Line Signature Verification by Local Granulometric Size Distributions", *IEEE Transactions on Pattern Analysis And Machine Intelligence*, vol. 19, no. 9, September 1997.

[12] Yingyong Qi and Bobby R. Hunt, "A Multiresolution Approach to Computer Verification of Handwritten Signatures", *IEEE Transactions on Image Processing*, vol. 4. no. 6. December 1995.

[13] Andrew W. Senior, and Anthony J. Robinson, "An Off-Line Coursive Hanwriting Recognition System", *IEEE Transactions on Pattern Analysis And Machine Intelligence*, vol. 20, no. 3, pp. 309-321 March 1998.

[14] Fathallah Nouboud, "Handwriten Signature Verification: A Global Approach", *Fundamentals in Handwriting Recognition*,Springer-Verlag, Berlin, Series F:, *Computer and Systems Science*, 124, pp. 455-459, 1991.

[15] Shih-Hsu Chang, Fang-Hsuan Cheng, Wen-Hsing Hsu and Guo-Zua Wu, "Fast Algorithm for Point Pattern Matching: Invariant to Translations, Rotations and scale Changes", *Pattern Recognition*, vol. 30, no. 2., pp. 311-320, 1997.

[16] K. Zhang, I. Pratikakis, J. Cornelis and E. Nyssen, "Using Landmarks to Establish a Point-to-Point Correspondence between Signatures", *Pattern Analysis And Applications*, vol. 3, pp. 69-75, 2000.

[17] John A. Richards, "Remote Sensing Digital Image Analysis An Introduction", Springer-Verlag Berlin Hedelberg New York London Paris Tokyo 1986.

[18] Wayne Niblack, "An Introduction to Digital Image Processing", Prentice-Hall Int., Inc.1986.

A GA-ANN for the Eulerian Cycle Problem

T.Tambouratzis*

*Institute of Nuclear Technology – Radiation Protection, NCSR "Demokritos", Aghia Paraskevi 153 10, Athens, Greece, email tatiana@ipta.demokritos.gr

Abstract

A novel approach for solving the Eulerian cycle problem is proposed. The approach constitutes a combination of genetic algorithms and artificial neural networks and accomplishes the consistent production of optimal solutions: on one hand, the existence of a Eulerian cycle of a given graph is determined; on the other hand, either a Eulerian cycle (if it exists) or a path encompassing the greatest possible number of edges is constructed.

1 Introduction

The Eulerian cycle problem [7] is a well-known problem of graph theory, which - though apparently recreational in nature (e.g. edge-tracing graphs, diagram-tracing puzzles, maze-escaping) - is of particular historical and theoretical importance. Furthermore, the Eulerian cycle problem and its variants have been applied to a number of practical tasks (e.g. snow-clearing route-finding, the rotating drum problem, the Chinese postman problem). Assuming a graph $G=\{V,E\}$, where

- $V=\{v_1,v_2,\dots,v_{VS}\}$ the set of vertices (of size VS),
- $E=\{e_1,e_2,\dots,e_{ES}\}$ the set of edges (of size ES) with $e_i=(v_{i1},v_{i2})$, $i=1,2,\dots,ES$ and $v_{i1},v_{i2}\in V$,
- $EX=\{e_1,e_2,\dots,e_{ES},e_{ES+1},e_{ES+2},\dots,e_{2.ES}\}$ the expanded set of oriented edges (of size $2.ES$) with $e_i=(v_{i1},v_{i2})$ and $e_{ES+i}=(v_{i2},v_{i1})$ edges of opposite direction,

the Eulerian cycle problem consists of finding a closed path of edges (i.e. a path that terminates at the originating vertex) such that each edge of E is traversed exactly once; the vertices of V may be visited more than once. Euler supplied a necessary and sufficient condition for the existence of a Eulerian cycle of G: the degree (number of incident edges) of every vertex of V must be even. However, this condition is not constructive in the sense that it does not supply a means of producing a Eulerian cycle. Following Fleury's algorithm [4], linear [1-2,5] as well as parallel graph [5] algorithms have been put forward. A novel approach is proposed here for simultaneously:

- determining whether a Eulerian cycle of G exists,
- constructing either a Eulerian cycle (if it exists) or a path encompassing the greatest possible number of edges of E.

The Eulerian cycle problem is treated as a constraint optimization problem. The proposed approach constitutes a combination of genetic algorithms (GAs) [3] and harmony theory artificial neural networks (HT ANNs) [6], where the harmony consensus function of the HT ANN is employed as the fitness function Ff of the GA. Optimal solutions (either Eulerian cycles or paths of maximum length where each edge is traversed no more than once) are consistently constructed. Furthermore, the existence of a Eulerian cycle is determined directly by the highest fitness value Ff_{max} of the chromosomes in the population: $Ff_{max}=ES$ if and only if a Eulerian cycle is produced, while $Ff_{max}<ES$ if at least one edge of E is not included in the path of greatest length or if the path is open.

2 Genetic Algorithms

Chromosomes constitute the structure-encoding devices of living organisms. Natural evolution operates on the chromosomes - and via them on the living organisms - of each generation in the following ways:

- recombination; material from more than one chromosome is used in order to create the chromosomes of the next generation,
- mutation; random variations are introduced in the chromosomes of each generation,
- natural selection; fit (according to some fitness criterion) chromosomes dominate over and reproduce more often than less fit chromosomes of each generation.

As a result, natural evolution creates increasingly fit living organisms at each generation. Similar to natural evolution, GAs are general-purpose stochastic optimization search algorithms. GAs are initialized with a population of chromosomes C_k ($k=1,2,\dots,POP$, where POP the population size). Each chromosome constitutes a set of binary elements BE^k_j ($j=1,2,\dots,L$, where L the chromosome length) and corresponds to a candidate solution of the problem. A fitness function Ff is used to measure the quality (fitness) Ff^k of chromosome C_k as a solution of the problem. The initial population evolves through successive iterations (generations) in the same manner as that described above concerning living organisms. After a number of generations, highly fit chromosomes, which are analogous to good (optimal or near-optimal) solutions of the problem, emerge. GAs are determined by the following components:

- encoding the potential solutions into chromosomes,
- constructing the fitness function of the chromosomes,
- creating the initial population,
- employing genetic operators for the creation of the next generation-population, and
- selecting control parameters (e.g. population size, probability of applying a genetic operator, termination criterion).

3 The Eulerian Cycle GA-ANN

The aforementioned components are detailed for the Eulerian cycle problem.

Chromosome encoding. In order for all the optimal solutions to be obtainable, it must be possible for the proposed cycle/path to begin from any vertex of V and for both directions of edge traversal to be allowed; this is especially important when a Eulerian cycle does not exist, whereby the a priori selection of an initial vertex/edge may produce sub-optimal solutions.

The notion of time step $\boldsymbol{TS}$ has been employed for specifying the time at which a particular edge is traversed. Since a Eulerian cycle contains exactly $\boldsymbol{ES}$ edges, at most $\boldsymbol{ES}$ time steps are allowed in the search for a cycle or a path of maximum length of G (exactly $\boldsymbol{ES}$ time steps if a Eulerian cycle exists or if traversal of all the edges results in an open path; less than $\boldsymbol{ES}$ time steps if at least one edge is not included in the path). The potential solutions constitute combinations of oriented edges and time steps during which the oriented edges are traversed. Each chromosome comprises $L=2.ES^2$ binary elements $BE^k_j=(e_q,TS_r)$ ($q=1,2,\ldots,2.ES$ and $r=1,2,\ldots,ES$) representing the exhaustive set of combinations of oriented edges of $\boldsymbol{EX}$ and time steps. A +1 value of BE^k_j (active element) denotes that oriented edge e_q is traversed at time step TS_r, while a value of 0 (inactive element) that e_q is not traversed at TS_r. The proposed paths are represented by the active elements of the chromosomes, i.e. combinations of oriented edges and time steps during which they are traversed.

Fitness function construction. This has been derived from the harmony consensus function of the HT ANN and exposes the constraints that apply between the active elements of the same chromosome. In fact, the harmony consensus function quantifies the compatibility between traversal of the various oriented edges at specific time steps, as these are given by the

Table I. The relation between active elements $BE^k_j=(e_q=(v_{q1},v_{q2}),TS_r)$ and $BE^k_l=(e_s=(v_{s1},v_{s2}),TS_t)$ of chromosome C_k. As shown, incompatibility may be due to reasons such as (a) traversal of the same oriented edge at another time step, (b) traversal of the opposite oriented edge at any time step, (c) traversal of any other edge at the same time step, (d) non-continuity of traversal. The cyclic nature of TS (if $TS_r=ES$ then $TS_r+1=1$; if $TS_r=1$ then $TS_r-1=ES$) promotes the production of closed rather than open paths.

BE^k_l v_{s1}	v_{s2}	TS_t	Relation of BE^k_j to BE^k_l
v_{q1}	v_{q2}	TS_r	compatible
v_{q1}	v_{q2}	TS_r -1	incompatible
v_{q1}	v_{q2}	TS_r +1	incompatible
v_{q1}	v_{q2}	other TS	incompatible
v_{q1}	other v	TS_r	incompatible
v_{q1}	other v	TS_r -1	incompatible
v_{q1}	other v	TS_r +1	incompatible
v_{q1}	other v	other TS	irrelevant
v_{q2}	v_{q1}	TS_r	incompatible
v_{q2}	v_{q1}	TS_r -1	incompatible
v_{q2}	v_{q1}	TS_r +1	incompatible
v_{q2}	v_{q1}	other TS	incompatible
v_{q2}	other v	TS_r	incompatible
v_{q2}	other v	TS_r -1	incompatible
v_{q2}	other v	TS_r +1	irrelevant
v_{q2}	other v	other TS	irrelevant
other v	v_{q1}	TS_r	incompatible
other v	v_{q1}	TS_r -1	irrelevant
other v	v_{q1}	TS_r +1	incompatible
other v	v_{q1}	other TS	irrelevant
other v	v_{q2}	TS_r	incompatible
other v	v_{q2}	TS_r -1	incompatible
other v	v_{q2}	TS_r +1	incompatible
other v	v_{q2}	Other TS	irrelevant
other v	other v	TS_r	incompatible
other v	other v	TS_r -1	incompatible
other v	other v	TS_r +1	incompatible
other v	other v	other TS	irrelevant

active elements of the chromosome. Each active element $BE^k_j = (e_q, TS_r)$ of C_k contributes a value of +1 to Ff^k if and only if it is compatible with (or irrelevant to) all the active elements of C_k; otherwise, its contribution is zero. Consequently, active elements that combine into acceptable traversal of edges contribute to Ff^k, while elements causing invalid traversal of edges do not. $Ff^k=ES$ denotes that a Eulerian cycle is produced, i.e. that all the edges of E are traversed once each, in a continuous manner, and with coincident terminating and originating vertex. Conversely, $Ff^k<ES$ denotes that only some edges of E are involved in the production of a closed path, that all/some edges of E are involved in the creation of an open path, or that there exists (at least) one time step during which traversal of more than one oriented edge occurs (invalid path). The relation of two active elements of the same chromosome is given in Table I.

Initial population creation. Each of the POP chromosomes of the initial population is constructed with at most ES randomly selected active elements. This has been performed since a Eulerian cycle has exactly ES active elements, while any other path has ES or less. Chromosomes with more than ES active elements necessary correspond to invalid solutions (i.e. edges traversed more than once).

Genetic operator application. The following operators have been employed for the creation of the chromosomes of the next generation-population:

- The crossover operator. Two (parent) chromosomes are selected from the population, the same random crossover point is chosen at both parents, and two new (offspring) chromosomes are created by exchanging the elements of the two parents up to the crossover point. $POP/2$ crossovers are performed in each generation. The parents are selected at random, whereby some may be used more than once in the crossovers of a generation while others may not be used at all. Random selection of the parents (instead of roulette-wheel or rank selection) promotes diversity of the next generation-population.
- The mutation operator. Each element of every offspring is subjected to inversion (from +1 to 0 and vice versa) given a fixed probability p. Mutation furthers the diversity of the offsprings.
- The selection operator. Both the chromosomes of the current generation and the offsprings resulting from crossover and mutation are subjected to selection of the POP chromosomes to be entered in the next generation-population. Four constraints have been applied:
  (a) Replace invalid. Chromosomes with more than ES active elements (representing invalid paths) are immediately discarded and not considered for insertion in the next generation-population.
  (b) Replace worst (elitistic selection). Chromosomes are ranked by fitness and the POP fittest are placed in the next generation-population.
  (c) Replace most similar. Chromosomes are not inserted in the next generation-population if their active elements constitute a subset of the active elements of an already entered chromosome (of higher fitness). Such chromosomes are redundant since the corresponding solutions constitute sub-paths of the path represented by the entered chromosome. This constraint also eliminates the appearance of identical chromosomes in the next generation-population.
  (d) Insert random. If less than POP chromosomes appear in the next generation-population, new chromosomes are created exactly as for the initial population creation.

  While constraints (a)-(b) restrict the search space and drive the GA towards convergence at a good solution, constraints (c)-(d) promote diversity of the next generation-population and counteract premature convergence at a non-optimal solution.

Control parameter selection.

- Three values of the mutation probability have been tried, namely p=0.05, 0.10 and 0.15.
- The size of the population has been kept at a similar order of magnitude as L and varies as a function of ES: $POP=ES^{3/2}, ES^2, ES^{5/2}$.
- The maximum number of generations I_{max} has been limited to 10000. An additional termination criterion has been derived from Euler's necessary and sufficient condition for the existence of a Eulerian cycle:
  (a) If all the vertices of V are of even degree, the GA is terminated once $Ff_{max}=ES$, whereby the chromosome(s) of fitness ES correspond to Eulerian cycles. If $Ff_{max}<ES$ after I_{max} generations, the GA has converged to a sub-optimal solution.
  (b) If at least one vertex of V is of odd degree, the GA is terminated once $Ff_{max}=ES$-1, whereby the chromosome(s) of fitness ES-1 correspond to paths involving the greatest possible number of edges of E. If $Ff_{max}<ES$-1 after I_{max} generations, either the optimal solution involves less than ES-1 edges or the GA has converged to a sub-optimal solution.

4 Results

Complete and incomplete graphs with $ES \in [5,45]$ have been tested; for the incomplete graphs, average connectivity ranges in [3,10]. Pairs of graphs with similar characteristics have been used, where Eulerian cycle(s) exist only for one graph of the pair. Each graph has been subjected to a total of 450 trials, with 50 trials performed for each combination of ***POP*** and ***p***. Performance has been measured in terms of:

(a) Accuracy, i.e. the frequency of reaching an optimal solution in at most I_{max} generations. In the case of the non-existence of a Eulerian cycle and if $Ff_{max}<ES$-1, the greatest fitness reached in the 450 trials has been considered as the actual Ff_{max} (this assumption has been made since the 450 trials cover a significant number of tests and a considerable range of GA characteristics, for which the proposed approach has been found able to reach Ff_{max} on the tested graphs with $Ff_{max} \geq ES$-1).

(b) Computational efficiency, measured by the total number of chromosomes (the product of ***POP*** and the number I_c of generations until convergence); only successful trials have been considered.

No significant difference in the performance of the GA-ANN approach has been observed between graphs that have Eulerian cycles and those that do not. Trials employing large populations ($POP=ES^{5/2}$, especially for $ES \geq 20$) are accurate but not efficient for any tested p; raising p produces slower convergence, i.e. further reduces efficiency. Conversely, trials employing small populations ($POP=ES^{3/2}$) are efficient but not accurate; although raising p causes an increase in accuracy, a 100% accuracy is not always attained for larger graphs.

Fig. 1. Average number of generations until convergence (I_c average) as a function of ES; $POP=ES^2$ and p=0.10.

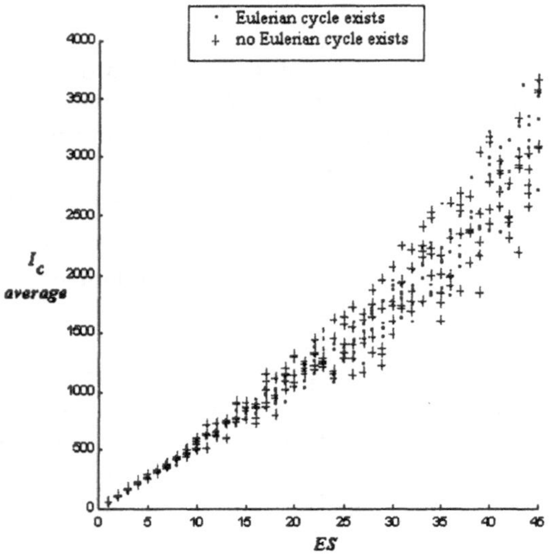

The combination of $POP=ES^2$ and p=0.10 optimally combines 100% accuracy with efficiency. For these control parameter values, each trial terminates with between one and four optimal solutions; subsequently, the optimal solutions obtained for each tested graph cover a variety of starting vertices and edges, an assortment of paths as well as edge traversal in either direction. Fig. 1 illustrates the average number of generations until convergence as a function of ***ES***, where dots (crosses) denote graphs with (without) Eulerian cycles. It can be seen that I_c is related to ***ES*** in a similar fashion for both kinds of graphs, although greater variation is observed for graphs without Eulerian cycles. For both kinds of graphs, I_c is dependent on average connectivity, with graphs of larger average connectivity requiring higher I_c than graphs of the same size but of lower connectivity.

5 Conclusions

A combination of genetic algorithms and artificial neural networks has been proposed for solving the Eulerian cycle problem. The Eulerian cycle problem is treated as a combinatorial optimization problem: paths of maximum length - with each edge traversed no more than once - (i.e. optimal solutions) are consistently constructed. Furthermore, owing to the redundant chromosome representation, all the optimal solutions are obtainable. The existence of a Eulerian cycle is determined directly by the highest fitness value of the chromosomes in the population after convergence: the highest fitness equals the number of edges in the graph if and only if a Eulerian cycle is produced.

It is of interest to further investigate the potential of the proposed GA-ANN approach to other combinatorial optimization problems as well as to NP-hard problems.

References

[1] Ebert J.: Computing Eulerian trails. Information Processing Letters *28*, 93 (1988).

[2] Even S.: Graph Algorithms. Rockville, MD: Computer Science Press 1979.

[3] Holland J.H.: Adaptation in Natural and Artificial Systems. Ann Arbor, Michigan: University of Michigan Press 1975.

[4] Lucas E.: Récréations Mathématiques. Paris: Gauthier-Villares 1891.

[5] Manber U.: Introduction to Algorithms. Reading MA: Addison-Wesley 1989.

[6] Smolensky P.: Information processing in dynamical systems: foundations of harmony theory. Parallel Distributed Processing: Foundations (ed Rumelhart D.E., McClelland J.L.). Cambridge MA: MIT Press 1986.

[7] R.J. Wilson R.J., Watkins J.J.: Graphs: An Introductory Approach. New York: John Wiley & Sons 1990.

Design of RBF Networks by Cooperative/Competitive Evolution of Units

A.J. Rivera*; J. Ortega, A. Prieto†

*Departamento de Informática, Universidad de Jaén, †Departamento de Arquitectura y Tecnología de Computadores, Universidad de Granada

Abstract

This paper presents a new evolutionary procedure to build optimal networks of Radial Basis Functions (RBFs) that addresses the problem of assigning credits to the population of RBFs. Credit assignment is done after considering three main factors: the weight of each neuron in the RBF Network, the overlapping between neurons, and the distances from neurons to the points where the approximation is worst. The procedure applies transformations that add, delete and move neurons in a distributed way.

1 Introduction

To design a neural network it is necessary to determine the number of elements that constitute the network, the topology of the connections between these elements, and the values of their parameters, including the weights. This paper deals with the design of neural networks, more specifically networks of Radial Basis Functions [2,3,4], by using evolutionary algorithms. A Radial Basis Function Network (RBFN) implements the mapping $f(x)$ from R^n onto R:

$$f(x) \approx w_0 + \sum_{i=1}^{m} w_i \phi_i(x)$$

where $x \in R^n$, and the m RBF functions ϕ_i have the form $\phi_i = \phi(\|x - c_i\| / d_i)$ with a scaling factor $d_i \in R$ and a centre $c_i \in R^n$. Of the several possible choices of ϕ, in this paper we consider gaussian RBFs, $\phi(r)=exp(-r^2/2)$, where r is the scaled radius, $\|x - c\| / d$, and $\|x - c\|$ is a euclidean norm on R^n. Thus, the problem is to determine the simpler RBFN structure, and the centres, widths and weights of the RBFs, to achieve the best approximation and generalization performance [2,3,4]. The optimization of a RBFN can be considered as a multicriteria optimization problem [13] in which the two criteria are the complexity of the network measured as the number of parameters, and its performance evaluated as the approximation error. The research work done in the field of genetic and evolutionary algorithms, can be classified according to the following major trends:

- *Competition among neural networks.* The population is a set of whole neural networks that compete by using a fitness function corresponding to the performance of each network. Alternatives within this general procedure include: (a) to evolve only the structural specification of the untrained networks, and use conventional nonevolutionary learning algorithms to train the networks [7], and (b) to use evolutionary algorithms to determine both the weights and the structure of the network [6]. Other methods use a compromise between the two previous approaches [14].

- *Incremental/Decremental algorithms.* These are based on adding [8] and/or deleting neurons [9] in the hidden layer one by one. These methods, by optimizing one unit at a time, can become trapped in local optima.

- *Evolving cooperating and competing units.* Each individual in the population is a RBF of the network, and the neurons reproduce or die depending on their performance [5]. Two problems must be solved in order to reach this goal. First, the credit apportionment problem [11], which involves determining a performance measure for each individual in the population; this is the measure that has to be optimized, and which can be evaluated. The second problem is that of niching problem [15], which implies to maintain a population of neurons evolving to different parts of the overall task. Thus, the individuals in the population have to both cooperate and compete. In [5], the credit assigned to each RBF is based on the contribution of this RBF to the overall performance of the network, and niche creation is implicitly obtained [12] by changing the intensity of competition between neurons according to the degree of overlap in their activations.

The procedure proposed here can be included in this latter group of methods. It uses a population of RBFs that cooperate and compete to reach a set of RBFs distributed across the whole input domain of the application. It adds and deletes neurons, as an incremental/decremental algorithm, but these transformations can be applied concurrently, and several neurons may appear/disappear simultaneously. The procedure also differs from previously proposed algorithms in the set of transformations and rules used to accomplish evolution within the set of RBFs.

The next section gives a description of the procedure. Section 3 describes the experiments implemented and the experimental results obtained, and Section 4 gives the conclusions of the paper.

2 Description of the Algorithm

The steps of the procedure are listed in Figure 1 and explained below.

1. Initialize the RBFN.
2. Evaluate fitness of RBFs
3. Select the worst RBFs
4. Apply operators to RBFs
5. Train the RBFN.
6. If Not End goto 2 else terminate

Fig 1. RBF network optimization algorithm

Step 1. Firstly, the algorithm builds an initial RBFN. This is done by starting from a population of r_{init} RBFs with centres allocated to randomly selected points and widths set to a given initial value. After defining the population of RBFs, a given number of LMS training iterations is applied to adapt the weights of the RBFN:

$$w_{k+1} = w_k + \alpha \frac{e_k \cdot x_k}{|x_k|^2}$$

α is modified during the training iterations by decreasing it if the error is reducing.

Step 2. The fitness of each RBF is evaluated by considering the weight of the corresponding RBF in the present network of RBFs, its distance to badly predicted points, and the distances to other RBFs in the network. The effect of the value of the weight associated with a RBF, ϕ_i, is taken into account by using a function, e_i, of its weight, w_i, as follows:

$$e_i = K(1 - (1/e^{|w_i|}))$$

The influence on the fitness of the closeness of neuron i to the p worst approximated points of the training set is quantified by the parameter m_i , defined as:

$$m_i = 1 + \sum_j^p m_{ij}$$

where each m_{ij} is positive and related to the closeness of ϕ_i and point j (pt_j), by the expression:

$$m_{ij} = \begin{cases} (1 - (D(\phi_i, pt_j)/d_i))/\gamma & \text{if } D(\phi_i, pt_j) < d_i \\ 0 & \text{otherwise} \end{cases}$$

while γ, is a parameter used to fit the shape of m_{ij}.

The overlapping of RBFs is quantified by a function s_i which is assigned to the RBF i and defined as:

$$s_i = 1 + \sum_j^m s_{ij}$$

where m is the number of RBFs and s_{ij} measures the overlapping between the RBFs i and j, (ϕ_i and ϕ_j):

$$s_{ij} = \begin{cases} (1 - (D(\phi_i, \phi_j)/d'))/\delta & \text{if } D(\phi_i, \phi_j) < d' \\ 0 & \text{otherwise} \end{cases}$$

in which the parameter δ controls the shape of the functions s_{ij}. The parameter d' is related to the width d_j of each RBF and is set to $d' > d_j$.

The fitness associated to each RBF ϕ_i is:

$$fitness\ (\phi_i) = e_i/(s_i * m_i)$$

Step 3. After assigning the corresponding fitness to each RBF, the set of RBFs to be modified is determined. This contains the r RBFs with the worst fitness values, where r is taken as a small value to retain a parsimonious evolution of the RBF network and maintain the main characteristics of the network behaviour. In this step, the p worst approximated points are also determined.

Step 4. A set of operators is applied to the population of RBFs selected in Step 3, in order to improve the performance of the network. These operators are:

- *OP1 (RBF creation)*: A new RBF is added in a zone where there are badly approximated points. First, the centre of the new RBF is set to the coordinates of the selected point and is modified by applying a randomly selected vector with module less than *dc*. The width of the RBF is determined by taking into account the closeness of the remaining RBFs and the parameter *er*. Thus, the actual radius for the new neuron is obtained by dividing *er* (which sets the higher value for the radius in a new neuron) by a coefficient related to the nearness of other neurons. The value of *er* is modified in each iteration of the algorithm by multiplying it by a constant CE, lower than 1.0.
- *OP2 (RBF elimination)*: A RBF is pruned.
- *OP3 (RBF small movement)*: The centre and the width of a selected RBF are changed by a small randomly selected percentage of the parameter *pc*.
- *OP4 (RBF movement)*: The centre of a given neuron moves towards one of the badly approximated points. The width of the RBF may also be changed. This is applied only if one of the worst predicted points is at a distance less than $d_i/1.5$ (in our present implementation) from the initial RBF centre.

A set of rules that depend on the values s_i and m_i of a given RBF ϕ_i are used to decide the operators to apply to this RBF. Upper and lower thresholds for s_i, (called *lss* and *lis*, respectively) are also defined. In our experiments *lss* is set to 30% above the average of the values of s_i, and *lis* is set to 30% below the same average. In the same way, the thresholds *lsm* and *lim* are defined from the values of m_i.

For the given RBF ϕ_i, the rules use the situation of m_i with respect to their thresholds. Thus, it is determined if there are badly approximated points near ϕ_i. Then, the value of s_i is also situated with respect to *lss* and *lis* to consider whether there exist any other RBFs near ϕ_i.

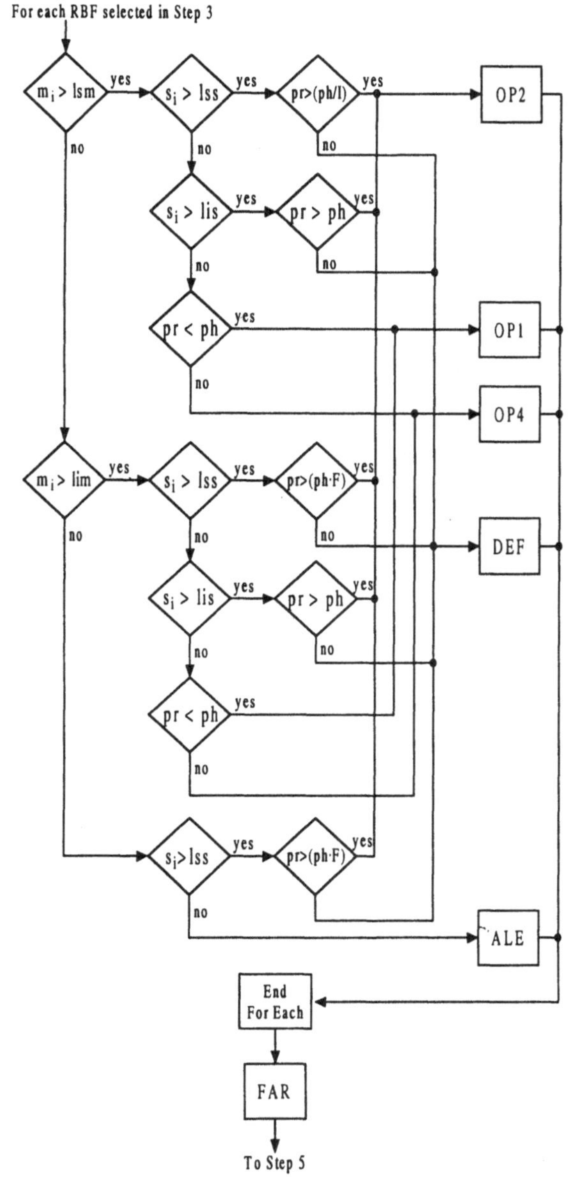

Fig. 2. Rules to apply the Operators

The rules to determine the operators to be applied are described in Figure 2. The procedures DEF, ALE, and FAR in Figure 2 are the following:

DEF: if (pr1<=A) OP4;
if (A<pr1<=B) OP3;
if (B<pr1<=C) OP1;
if (C<pr1) OP2;
ALE: if (pr1<=E) OP3;
if (E<pr1<=G) OP1;
if (G<pr1) OP2;
FAR: if (pr<ph) and ($D(\phi_i, pt_j) > d_j/2$) OP1;

In Figure 2 and in the DEF, ALE, and FAR procedures, *pr* and *pr1* are random numbers in [0,1], F is set to 1.5, and I to 2. In the rules described in Figure 2, there are some parameters, *ph*, *er* and *dn*, whose values are dynamically modified during the execution of the algorithm according to the number of generations or RBFs. The parameter *ph* (in the range [0,1]) controls the generation of new RBFs through a probability that decreases linearly as the number of existing neurons grows. Finally, the parameter *dn* determines the number of RBFs selected to be modified by the different operators. It is also modified with the iterations.

Step 5. The weights of the new network of RBFs are adjusted again by implementing some iterations of the LMS algorithm, as at the end of Step 1.

Step 6. The performance of the solution obtained is compared with the desired levels of approximation error, number of neurons, etc. If these levels are not attained, the procedure returns to Step 2.

3 Experimental Results

As an example of application, we consider a time series prediction problem. The time series used is the frequently used Mackey-Glass series [1]. The inputs to the network of RBFs consist of four past data points, x(t), x(t-6), x(t-12), and x(t-18) and the output is x(t+85). The training data were extracted from points 4000 to 4500, and the following 500 data points were used as testing data. The normalized root-mean-square error (RMS) was used to evaluate the perfomance of the algorithm. The algorithm was executed several (50) times with initial population sizes of 15, 20 and 25 RBFs. Table I summarizes the results of these executions.

Table I. Parameters and results

RBFs	*Best Fitness*	*Av. Fitness*	*Std Desv*
10	0.410	0.465	0.043
12	0.375	0.446	0.038
15	0.280	0.376	0.043
18	0.256	0.331	0.024
20	0.262	0.312	0.024
22	0.261	0.296	0.020
25	0.252	0.285	0.025
28	0.251	0.268	0.017
30	0.259	0.266	0.009

Figure 3 represents the number of RBFs and the corresponding RMS obtained by our algorithm from initial solutions with 15, 20, and 25 RBFs. As the number of RBFs is increased, an improvement in the performance of the network is obtained. The values obtained define a curve of non dominated solutions for the multiobjective optimization problem that takes into account the performance and the complexity of the network.

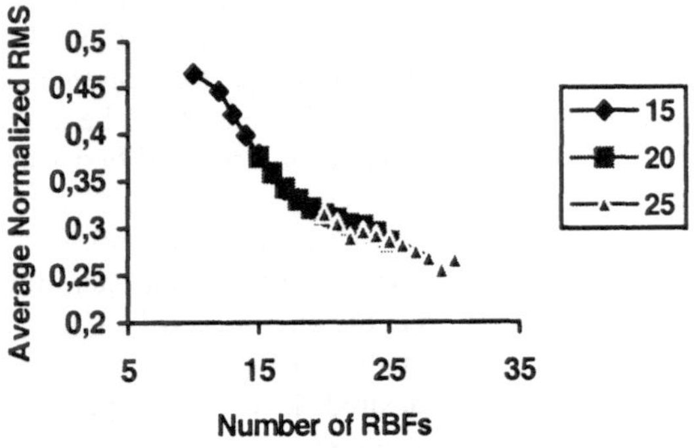

Fig. 3. Solutions obtained from different initial ones

These results were compared with other time series prediction procedures using the same benchmark [5]. The results for the normalized RMS provided by two of these algorithms are given in Table II. As can be seen, the procedure proposed here improves on the performance of both methods.

Table II. Results for other algorithms

Algorithm	*RBFs*	*Normalized RMS*
Fitness sharing [5]	25	0.29
K-means clustering [5]	25	0.53

4 Conclusions

This paper presents a new evolutionary procedure to design networks of radial basis functions. The procedure describes a new method to calculate the fitness of a given RBF according to its contribution to the final performance of the neural network, thus proposing a solution to the credit assignment problem. Other characteristics and elements of the procedure are a set of rules which control the application of some transformations to the RBFs, the dynamic-adaptative character of the parameters, and the parsimonious evolution of the behaviour of the network as the procedure advances.

Good experimental results were obtained, and the method performs better than others, more mature methods, previously proposed. Future research aims are the relations between the major parameters of this algorithm, i.e. the parameters which calculate the fitness and the design of the set of rules, and to include a learning procedure to determine the optimal set of rules and to adapt the values of the parameters that define the rules of Figure 2. Moreover, as the algorithm has a distributed character, we consider that a parallel version could significantly improve the speed of the procedure.

Acknowledgements. This paper has been supported by project TIC2000 -1348 of the Spanish *Ministerio de Ciencia y Tecnología*.

References

[1] M. C. Mackey and L. Glass, "Oscillation and chaos in physiological control system" Sci vol.197,pp. 287-289, 1.977.
[2] J. Platt, "A resource-allocating network for function interpolation", Neural Computation 3, 213-225. 1.991
[3] L. Yingwei, N. Sundararajan, P. Saratchandran, "A sequential learning scheme for function approximation using minimal radial basis function neural networks." Neural Computation 9, 461-478. 1.997
[4] J. Moody and C. Darken, "Fast learning networks of locally-tuned processing units," Neural Computation, vol. 3 n. 4 pp.579-588, 1.991
[5] Whitehead, B.A.; Choate, T.D.:"Cooperative-competitive genetic evolution of Radial Basis Function centers and widths for time series prediction". IEEE Trans. on Neural Networks, Vol.7, No.4, pp.869-880. July, 1996.
[6] Angeline, P.J.; Saunders, G.M.; Pollack, J.B.:"An evolutionary algorithm that constructs recurrent neural networks". IEEE Trans. on Neural Networks, Vol.5, No.1, pp.54-65. January, 1994.
[7] Maniezzo, V.:"Genetic evolution of the topology and weight distribution of neural networks". IEEE Trans. on Neural Networks, Vol.5, No.1, pp.39-53. January, 1994.
[8] Fahlman, S.E.; Lebiere, C.:"The cascade-correlation learning architecture"..*Advances in Neural Information Processing Systems, 2,* Lippmann, P.; Moody, J.E.; and Touretzky, D.S. , Morgan Kaufmann, pp.524-532, 1991.
[9] Hwang, J.-N.; Lay, S.-R.; Maechler, M.; Martin, R.D.; Schiemert, J.:"Regression modeling in backpropagation and projection pursuit learning". IEEE Trans. on Neural Networks, Vol.5, No.3, pp.342-353. March, 1994.
[10] Hintz-Madsen, M.; Hansen, L.K.; Larsen, J.; Pedersen, M.W.; Larsen, M.:"Neural classifier construction using regularization, pruning and test error estimation". Neural Networks, 11, pp.1659-1670, 1998.
[11] Smalz, R.; Conrad, M.:"Combining evolution with credit apportionment: a new learning algorithm for neural nets". Neural Netwroks, Vol.7, No.2, pp.341-351, 1994.
[12] Horn, J.; Goldberg, D.E.; Deb, K.:"Implicit niching in learning classifier system: Nature's way". Evolutionary Computation, Vol.2, No.1, pp.37-66, 1994.
[13] Coello, C.C.:"An Updated Survey of Evolutionary Multiobjective Optimization Techniques: State of the art and future trends". CEC'99, pp.3-13, 1999.
[14] Whitehead, B.A.; Choate, T.D.:"Evolving space-filling curves to distribute radial basis functions over an input space". IEEE Trans. on Neural Networks, Vol.5, No.1, pp.15-23. January, 1994.
[15] Sareni, B.; Krähenbühl, L.:"Fitness sharing and niching methods revisited". IEEE Trans. on Evolutionary Computation, Vol.2, No.3, pp.97-106. September, 1998.

Avoiding Local Minima in ANN by Genetic Evolution

R. Mendes*, P. Vale*, J.M. Sousa*[1], J.A. Roubos†

*Technical University of Lisbon, Instituto Superior Técnico, Dept. of Mechanical Engineering/GCAR, Av. Rovisco Pais, 1049-001 Lisbon, Portugal, E-mail: j.sousa@dem.ist.utl.pt †Control Laboratory, Faculty of Information, Technology and Systems, Delft University of Technology, P.O. Box 5031, 2600 GA Delft, The Netherlands

Abstract

A novel approach is proposed to avoid the problem of local minima in neural networks learning when using the backpropagation algorithm. This problem is solved by applying a genetic evolution. The unification of both powerful technics can be viewed as an example of a mix between Darwinian and Lamarckian learning. The wine data, a high-dimensional classification problem, is given as an example.

1 Introduction

Artificial neural networks (ANN) have been known to provide an extremely robust approach to the pattern recognition problem [1]. Problems such as learning to interpret complex real-world sensor data have proven ANN to be among the most effective learning methods currently known [2]. The results provided by the use of an ANN to approximate a certain target function depend on its topology and initialization. A common approach to overcome this subjectiveness is to use a pool of different ANN and an appropriate collective decision strategy to obtain the better and more robust results. The existence of ANN with different accuracy levels causes this kind of selection to be less appropriate when searching for an optimal pattern recognition algorithm. One of the most commonly used training algorithms for ANN is the backpropagation algorithm. When using this training algorithm in multilayer feedforward ANNs, the problem of local minima may be a great barrier towards the optimal fitness. Thus, the backpropagation algorithm, despite being one of the best training algorithms, may not be the most appropriate when this kind of problems occurs. Recently, several attempts were made to reinforce the power of ANN by combining them with other soft computing tools such as genetic algorithms (GAs) [3, 4]. A population of different ANN is created and tuned by using the power of natural selection to promote its evolution. These combined methods have obtained interesting results. The use of the naturally powerful selection mechanism provided by GAs provides an approach to handle several pending problems such as the existence of local minima.

This paper is focused on using GAs with a population of ANN to solve the problem of local minima. Section 2 presents a brief introduction of ANN and the inherent training algorithms. Section 3 introduces GAs and their use in combination with the backpropagation algorithm. Section 4 presents the wine data, a high-dimensional classification problem, and the classification results obtained by using the proposed technique. Finally, Section 5 concludes the paper.

2 Artificial Neural Networks

Biological learning systems are a good example of the nature's richness. Human learning is based on very complex webs of interconnected neurons like the brain. In a similar way, ANN are based on the interconnection between a set of simple processing units (neurons). Each of these neurons contains a linear or nonlinear transformation function. The connections between the neurons have an associated weight which must be trained in order to adjust the performance of the network to the purpose of its use.

Feedforward ANN allow the mapping between an input set and an output set. This paper focuses on the application of ANN to classification tasks. Since the classification problem translates into the problem of mapping the feature space into the set of possible output classes, an ANN can be considered as a classifier. This paper deals with the three-layer ANN, which is the most common structure. Additional hidden layers increase the complexity and learning capabilities of ANN. However, if the number of hidden layers is exaggerated the risk of overfitting the training data is very high. One can consider adding additional layers when the learning results are poor. A three-layer ANN has F neurons in the input layer (F is the number of features), H neurons in the hidden layer (H is an appropriately selected number) and C neurons in the output layer (C can be the number of classes), in which adjacent layers are fully connected. Given an input $X = \{x_1, x_2, \ldots, x_F\}$ and the class set $\Omega = \{\omega_1, \omega_2, \ldots, \omega_C\}$, each output neuron produces y_i of belonging to a class by

[1]This work was supported by Sapiens, ref. 34058/99, FCT, Ministério da Ciência e Tecnologia, Portugal.

$$P(\omega_i \mid X) \simeq y_i f\left\{\sum_{k=1}^{H} w_{ik} f\left(\sum_{j=1}^{T} w_{kj} x_j\right)\right\} \quad (1)$$

where w_{kj} is a weight between the jth input neuron and the kth hidden neuron, w_{ik} is a weight from the kth hidden neuron to the ith class output and f is a sigmoid function of the type $f(x) = \frac{1}{1+e^{-x}}$.

In order to train the ANN, the *backpropagation algorithm* [1] is used, which is commonly recognized as a powerful, robust training algorithm. This algorithm aims at learning the weights of a multilayer ANN with a fixed number of units and interconnections, by using gradient descent in order to minimize the squared error between the network outputs and the correspondent target values.

Given an ANN with F inputs, H neurons in the hidden layer, C neurons in the output layer and a vector $X = \{x_1, x_2, \ldots, x_F\}$ with the desired output $T = \{t_1, t_2, \ldots, t_C\}$, the backpropagation algorithm is as follows:

1. Compute the output o_u of every neuron u in the network given X.
2. For each network output k, compute its error δ_k:
$$\delta_k = o_k(1 - o_k)(t_k - o_k) \quad (2)$$
3. For each hidden unit h calculate its error δ_h:
$$\delta_h = o_h(1 - o_h)\sum_{k=1}^{C} w_{kh}\delta_k \quad (3)$$
4. Update each network weight w_{ji}
$$w_{ji} = w_{ji} + \Delta w_{ji} \quad (4)$$
where $\Delta w_{ji} = \eta \delta_j x_{ji}$ and η is the learning rate.

The backpropagation algorithm can be viewed as a searching algorithm whose state space corresponds to all possible weights for the network units. The algorithm uses the squared error between the network outputs and the target values as a fitness function to guide the gradient descend. The correspondent error surface can thus be considered multidimensional and may have multiple local minima. This characteristic guarantees only that the gradient descent will converge to a local minima and not necessarily to the global minimum error.

3 Genetic Algorithms

As ANN, GAs have also been developed as a reflection of nature's teachings. Inspired by the biological process of Darwinian evolution, GAs perform selection, crossover and mutation over a population, in order to achieve a global optimum. At each iteration of this genetic process, an evolution is obtained by replacing elements of the population by offspring of the most fit elements of that same population. In this way, the best fit hypotheses have a higher possibility of having their offspring (that represent variations of itself) included in the next generation. GAs provide the capability of searching through very complex spaces. Considering that ANN are elements of the population, it can be also considered that each network's unit is a part of it with an unprecise (and difficult to model) relation to the network's fitness. This approach would represent a consolidated view of both the Lamarckian (ANN) and the Darwinian (genetic algorithm) evolution models, in which the best hypotheses is obtained by transforming and combining other hypotheses that have been refined by using ANN. This combined learning method should allow to overcome the limitations associated with the training of ANN (especially the local minima problem) by introducing a random factor in the process. One of the difficulties rising from this combination is to adjust all the inherent parameters in a correct way. In order to fully adapt the parameters of the domain in study, a long period of tests must occur and many combinations must be tested, since both the ANNs training algorithm and the GA must be parameterized. In the following sections the main aspects of the proposed combined learning algorithm are discussed.

3.1 ANN representation in the GA

The GA described in this paper is based on the real-coded genetic algorithm proposed in [5]. The fundamental part of integrating the use of ANN with GAs is, of course, the representation of the neural networks as elements of the population (the chromosomes). As explained above, each ANN is characterized by the number of its layers, the units that compose each one of these layers, the interconnections between the units composing each layer and the ones that compose the following layer and all the associated weights. Since we are considering that all the networks that compose the population have similar topologies (the same number of layers, units and interconnections), we can obtain a straightforward representation of each network by encoding all its weights in a single sequence that composes the chromosome. By fixing a clear correspondence between each position in the chromosome sequence and a particular weight in the correspondent ANN, a simple process is created which permits the exchange of information between the ANN information format (necessary to perform the Lamarckian learning - the network training process) and the GA's chromosomatic representation (to allow the Darwinian evolution of the population).

3.2 GA details

The evolutionary process presented in this paper is supported by a roulette wheel elitist selection method.

This means that the chromosomes which yield a better fitness have a higher chance to survive and generate offspring and that the best fit chromosome in a certain generation always survives and evolves to the following generation. To promote the evolution of the population towards a better fitness in the concerning domain, two major kinds of genetic operators are used with three subclasses each [5]: the crossover (simple arithmetic crossover, whole arithmetic crossover and heuristic crossover) and the mutation (uniform, multiple uniform and gaussian). When a chromosome is selected for operation, the chance of its manipulation by a crossover operator is set at 80% and the probability of a mutation to occur is set at 20%. When a chromosome is selected for a mutation or a crossover, each of the subclass-operators has an equal chance of being applied.

The complete algorithm contains the following steps:

1. Create and simulate the initial population:
   (a) Create a population of ANNs according to the selected topology;
   (b) Train the ANNs of the initial population using the backpropagation algorithm for a specified number of training epochs;
   (c) Calculate the fitness of the individual ANNs in the population.
2. Repeat the genetic optimization cycle:
   (a) Select chromosomes (ANNs) for operation and deletion;
   (b) Create a new generation: operate on the chromosomes selected for operation and substitute the chromosomes selected for deletion by the resulting offspring;
   (c) Train the ANNs of the new generation using the backpropagation algorithm.
   (d) Evaluate the population.
3. Select the best chromosome (solution) from the final generation.

4 Experiments and Results

In order to test the proposed method, a wine data classification problem was used [6]. The wine data is the result of a chemical analysis of wines grown in the same region in Italy but derived from three different cultivars. The analysis determined the quantities of 13 constituents in each of the three types of wines. All attributes are continuous and there are 59 class 1 instances, 71 class 2 instances and 48 class 3 instances. The data set was subdivided in three subsets with randomly chosen data-pairs:

1. **Train set** - Used to train the ANN before and during the GA.
2. **Test set** - Used to calculate each ANN fitness function. The GA search is driven by the ANN performance in this set.
3. **Validation set** - Used to check the ANN performance for not previously seen examples. While the test set is normally used for this task, it was decided to use yet another set since the performances of the ANN in the test set are used as the GAs fitness function.

The ANN proposed for the wine classification has a hidden layer with three neurons. In order to perform classification, the output of the ANN was used with the following classification rule:

$$c_k = \begin{cases} 1 & \text{if } y < 0.25, \\ 2 & \text{if } 0.25 \leq y < 0.75, \\ 3 & \text{if } y \geq 0.75. \end{cases} \tag{5}$$

The GA's fitness function combined the mean squared error (MSE) of the ANN prediction with the number of classification errors (in the test set):

$$J_W = \frac{1}{N}\left(\sum_{k=1}^{N}(y_k - \hat{y}_k)^2 + \sum_{k=1}^{N}(c_k \neq \hat{c}_k)\right) \tag{6}$$

The MSE is needed to differentiate between solutions with the same number of classification errors. Moreover, it was found to enhance the optimization process.

Several tests were performed with the purpose of evaluate the relation between the following three parameters and the obtained results:

- The number of *a priori* training cycles;
- The number of operations involved in the generation of new offsprings;
- The number of training epochs for each ANN in each step of the GA optimization cycle.

A correct setting of these parameters is important for obtaining the best behavior from the algorithm. The obtained results are summarized in Table I. In all the tests, the number of chromosomes was 20 and the crossover and mutation probability was 0.8 and 0.2, respectively. The maximum number of iterations was 300.

First, the number of *a priori* training epochs is varied. As the number of *a priori* training epochs increases, so does the probability of obtaining training/testing data sets overfitness. Secondly, number of operations in each evolutionary step is varied. In order to get the right amount of influence from the genetic operations in the learning process, a sufficient number of operations involved in each evolutionary step must be available. These results confirm that the possibility of having a higher number of modifications on the weight structure of the ANNs helps to overcome the risk of overfitness.

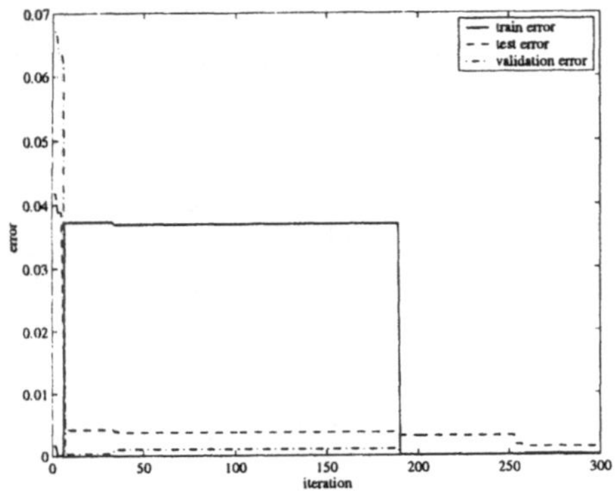

Fig. 1. Best result obtained: 15 operations, 0 *a priori* training epochs and 5 training epochs in each generation

Thirdly, the number of training epochs is changed. After the creation of each new generation, there is a learning period in which each network is trained. This training is necessary to soften the abrupt variations in the network's weights introduced by the genetic operators. Although it may seem inappropriate (because it will modify the achievements of the genetic crossovers and mutations), this training is very useful since the network's weights are modified by involving a random factor, which increases the possibility of overcoming any existing local minima. However, if a high number of training epochs is used, the overfitness problem may arise (as in the *a priori* training).

O	P	T	LE	TE	VE	M
15	0	0	0.486	0.361	0.27	4
15	0	5	0	0	0.067	1
15	0	10	0	0	0.129	2
15	10	0	0.001	0.040	0.069	1
15	10	5	0	0	0.206	3
15	20	5	0.020	0.013	0.136	2
15	50	5	0	0.012	0.140	2
6	0	5	0.021	0.012	0.208	3
6	10	5	0	0	0.268	4

Table I. Results for different O: operations, P: *a priori* training, T: training epochs, LE: learning error, TE; test error, VE: validation error, M: wrongly classified.

The correct adjustment of all these parameters should allow this algorithm to overcome any local minima and obtain nearly perfect results for the used domain. Figure 1 exposes the best of 10 tests using 15 operations, 0 *a priori* training epochs and 5 training epochs (Table II).

As can be seen from the table, using 15 operations, 0 *a priori* training epochs and 5 training epochs in each generation, it is possible to obtain the following general (and very satisfying) results: best classification rate 94.4%, average classification rate 89.98% and worst classification rate 77.8%. In order to gain a better insight into the obtained results, comparisons were made with ANN learning only. An ANN with one hidden layer with three neurons was used. The domain data was divided in a training set with 120 elements and a test set with 58 elements. Ten ANN were trained for 500 epochs. The results were: best classification rate 98.3%, average classification rate 90.3%, and worst classification rate 77.6%. The results are also very good and quite similar.

LE	TE	VE	M	CC
0	0	0.069	1	94.4
0	0	0.069	1	94.4
0.024	0.022	0.204	3	83.3
0.001	0.003	0.070	1	94.4
0	0	0.136	2	88.9
0	0	0.277	4	77.8
0	0	0.129	2	88.9
0	0	0.131	2	88.9
0	0	0.067	1	94.4
0.053	0.031	0.071	1	94.4

Table II. Results for 10 trials. LE: learning error, TE: test error, VE: validation error, M: misclassifications, CC: correct classification in validation set.

5 Conclusions and Future Work

This paper proposes the introduction of GAs in ANN learning. The obtained results provide some evidence supporting the idea of mixing Darwinian and Lamarckian learning. The proposed method is a powerful tool to overcome one of the most important problems introduced by the backpropagation algorithm: *local minima*. In the future, the genetic operators must be generalized in order to modify and crossover the structure of ANN. This approach may prove to be more powerful, since it would provide the possibility of finding the ideal structure for the ANN, using however a more complex implementation.

References

[1] S. Haykin. *Neural Networks - A comprehensive foundation.* Macmillan College Publishing Company, 1994.

[2] T. Mitchell. *Machine Learning*. McGraw-Hill, 1997.

[3] S.-B. Cho. "Pattern recognition with neural networks combined by genetic algorithms", *Fuzzy Sets and Systems*, 103:339–347, 1999.

[4] A. Blanco, M. Delgado, M.C. Pegalajar. "A genetic algorithm to obtain the optimal recurrent neural network", *Int. J. Approx. Reasoning*, 23:67–83, 2000.

[5] M. Setnes, J.A. Roubos. "GA-Fuzzy modeling and classification: complexity and performance", *IEEE Trans. on Fuzzy Systems*, 8(5):509–522, Oct. 2000.

[6] *UCI Repository of machine learning databases*, http://www.ics.uci.edu/~mlearn/, 1998.

A Genetic Designed Beta Basis Function Neural Network for Approximating Multi-Variables Functions

Chaouki Aouiti[*], Adel M. Alimi[†], Aref Maalej[*]

[*]LASEM: Laboratory of Electromechanical Systems University of Sfax ENIS, Department of Mechanical Egineering, BP W - 3038, Sfax, Tunisia. e-mail: Chaouki.Aouiti@isetsf.rnu.tn ; aref.maalej@enis.rnu.tn, [†]REGIM: Research Group on Intelligent Machines, University of Sfax, ENIS, Department of Electrical Engineering, BP W - 3038, Sfax, Tunisia. e-mail: *adel.alimi@ieee.org*

Abstract

We propose in this paper a new genetic algorithm for Beta basis function neural networks (BBFNN). The proprieties of this genetic algorithm are the representation used and the ability to obtain the optimal structure of the BBFNN for approximating a multi-variable function.

Each network is coded as a matrix for which the number of rows is equal to the number of parameters in the function. The genetic algorithm operators change the number of neurons in the hidden layer. Some applications to functions with one and two variables are considered to demonstrate the performance of the BBFNN and of their genetic algorithm based design.

1 Introduction

Gas are useful for dealing with large complex problems with many local optima. So, GAs are used to perform various tasks in the case of designing a neural network, such as connection weight training, architecture design, etc.

The first step in the implementation of GAs is the choice of the representation. If the goal is the evolution of connection weights, two methods are used. In the first representation used in canonical Gas, binary strings are used to encode alternative solutions [4]. The second method is the real-number representation, i.e., one real number per connection weight [5] and [6].

GA are used for the evolution of architectures for many reason, like the surface where each point represents an architecture is infinitely large, the surface is non-differentiable since changes in the number of node or connections are discrete, the surface is deceptive since similar architectures may have quit different performance and the surface is multimodal since different architectures may have similar performance [7].

When we use GAs for the evolution of architectures, we decide how much information about architectures should be encoded in the chromosome. In the first method the chromosome can specify the entire connexion weights and nodes of architecture. This method is the direct encoding. For the second method, which is called indirect encoding, only the most important parameters of architecture are encoded.

For the BBFNN we have used in [3] a real method for the representation. In this paper, we will present a new method and we will prove that we can use GA to approximate a function with many variables.

The rest of the paper is organized as follows: section 2 is devoted to describe the BBFNN; in section 3, we present our GA for the design of BBFNN; the experimental results and discussion are in section 4.

2 Beta Basis Function Neural Network

The first idea of using Beta functions for the design of Beta Basis Function Neural Network (BBFNN) that are extended versions of RBFNN was introduced by Alimi in 1997 [2]. The Beta function is used as kernel functions for many reasons, such as their great flexibility and their universal approximation characteristics. The Beta function is defined by:

$$\beta(x)=\beta(x,x_0,x_1,p,q)=\begin{cases}\left(\frac{x-x_0}{x_C-x_0}\right)^p\left(\frac{x_1-x}{x_1-x_C}\right)^q & if\ x\in[x_0,x_1]\\ 0 & elsewhere\end{cases} \quad (1)$$

where p > 0, q > 0, x_0 and x_1 are real parameters, and:

$$x_C=\frac{px_1+qx_0}{p+q} \quad (2)$$

is the center of the Beta function.

The Beta function has the following properties:

$$\beta(x_0)=\beta(x_1)=0\ ;\beta(x_c)=1 \quad (3)$$

Let's define the width of the Beta function by:

$$D=x_1-x_0 \quad (4)$$

In the multi-dimensional case and if

$$X=\left(x^1,\ldots,x^N\right),\ X_0=\left(x_0^1,\ldots,x_0^N\right),\ X_1=\left(x_1^1,\ldots,x_1^N\right)$$

$$P=\left(p^1,p^2\ldots,\ p^N\right)\text{and } Q=\left(q^1,\ldots,q^N\right)\text{ Then}$$

$$\beta\left(X,X_0,X_1,P,Q\right)=\prod_{i=1}^{N}\beta\left(x^i,x_0^i,x_1^i,p^i,q^i\right) \quad (5)$$

It has been shown in [1] and [2] that if we have a given continuous real function and for any arbitrary precision, there exists a Beta fuzzy basis function expansion that approximates it.

3 The Genetic Algorithm

It has bee proven that genetic algorithms are able to find near-optimal solutions to complex problems. The first step in this method is the choice of the representation of a possible solution. In this paper we will introduce a new method to represent a possible solution in genetic algorithms. Each chromosome that represents a network is a matrix. The number of lines in this matrix is equal to the number of variables in the function to be approximated. The number of columns is variable since each chromosome represents a network and the hidden layer has a variable number of neurons, each sequence of four genes codes a beta and the whole chromosome codes the set of the parameters of the hidden layer (see figure 1).

X_c^1	D_x^1	P_x^1	Q_x^1			X_c^N	D_x^N	P_x^N	Q_x^N
Y_c^1	D_y^1	P_y^1	Q_y^1			Y_c^N	D_y^N	P_y^N	Q_y^{3N}

Fig. 1. A chromosome that represents a BBFNN with *N* hidden neurons (the function has two variables)

Since repeated usage of the same neuron in a neural network didn't improve the approximation capability of the network, each chromosome is formed by distinct sequences.

After the choice of the representation, we will choose the objective function. Let f be the function that we will approximate and X the variable, so the output of the network is:

$$f^*(X)=\sum_{j=1}^{n_c} w\beta_j(X) \quad (6)$$

where n_c is the number of neurons in the hidden layer and W_j is the jth weight. Given a set of data (x_i,y_j) (i= 1,2,…,N, j=1,2,…,N) we can obtain the connection weights, the parameters of each neuron and the number of neurons by minimizing the following objective function:

$$J_1(n_c,w,x_0,x_1,p,q)=\sum_{i=1}^{N}\left(y_i-f^*(x_i)\right)\left(y_i-f^*(x_i)\right)^T \quad (7)$$

If we use this function as an objective function, the best structure that minimizes it will have *N* nodes in the hidden layer. To provide a trade-off between the network performance and the network structure the objective function to be minimized should be

$$J_2(n_c,w,x_0,x_1,p,q)=J_1(n_c,w,x_0,x_1,p,q)+alpha*n_c \quad (8)$$

alpha is a parameter of the problem.

In the initial population the number of neurons in the hidden layer is fixed between *Min_neu* and *Max_neu*, so each chromosome has between 4**Min_neu* and 4* *Max_neu* columns. Now that we have described the chromosome representation and the objective function, the genetic operators used in our algorithms will be described in detail.

3.1 The R^n-crossover operator

The most important operator in a Genetic Algorithm is the crossover operator. This operator recombines the genetic material from two parents into their children.

In our algorithm we have used two-crossover operators. One of the operators changes the number of columns of each chromosome so it changes the number of neurons in the hidden layer. The second operator didn't change the number of neurons in the hidden layer.

12	4	10	2	34	12	1	54	11	32	19	20
14	47	98	20	1	9	23	61	7	28	17	3

17	9	32	11	3	27	8	35	22	13	14	30
23	11	8	12	16	27	2	5	15	31	6	14

↓

12	4	10	2	34			35	22	13	4	30
14	47	98	7	1			5	15	31	6	14

17	9	32	11	11	32	19	20
23	11	8	12	7	28	17	3

Fig. 2. The first R^n- crossover operator

For the first operator, after the choice of the two chromosomes for which we will apply the crossover operator we choose an arbitrary position *a* in the first chromosome and a position *b* in the second chromosome according to *a*. Then, we exchange the second parts of the two chromosomes. If one child has more than *Max_neu* or less than *Min_neu* neurons in the hidden layer, we apply the second crossover operator (see figure 2).

For the second crossover operator, we let *Min_point*=Min(a,b), then we change the values of a and b to *Min_point*. In this case, we have necessarily a first child having the same length as the second chromosome and the second child having the same length as the first chromosome (see figure 3).

12	4	10	2	34	12	1	54	11	32	19	20
14	47	98	7	1	9	3	61	7	28	17	3

17	9	32	11	3	27	8	35	22	13	4	30
23	11	8	12	6	11	2	5	15	31	6	14

12	4	10	2	34	12	1	54	22	13	14	30
14	47	98	20	1	9	23	61	15	31	6	14

17	9	32	11	3	27	8	35	11	32	19	20
23	11	8	12	16	27	2	5	7	28	17	3

Fig. 3. The second R^n- crossover operator

3.2 The R^n-mating operator

Before applying the crossover operator we began by applying a mating operator (see figure 4). If we change the position of a neuron in the hidden layer, the output of the network didn't change. For this reason we can change randomly the position of the neurons in the hidden layer. This will transform the crossover operator to multi-point type and overcome the limitations of the one-point crossover operators.

7	11	26	13	36	1	7	32	12	4	2	5
1	29	7	33	19	2	44	21	87	11	15	71

12	4	2	5	7	11	26	13	36	1	7	32
87	11	15	71	1	29	7	33	19	2	44	21

Fig. 4. The Rn- Mating operator

3.3 The R^n- mutation operator

The goal of applying the mutation operator is to inject some new information into the population because, generally, the initial population does not have all the information that is essential to the solution.

We begin by associating to each gene a real value between 0 and 1. If this real is less than the probability of mutation P_m then the mutation operator is applied to the gene by changing its value of the gene.

3.4 The R^n- addition and the R^n- elimination operators

Classic training algorithms for Beta Basis Function Neural Networks (BBFNN) start with a predetermined network structure and the quality of the response of the BBFNN depends strongly on its structure. Using two genetic operators R^n- Addition and R^n- Elimination will solve this problem. The first operator adds a neuron in the hidden layer (see figure 5) and the second operator eliminates a neuron (see figure 6).

22	15	7	95	3	36	10	2
18	11	3	54	12	37	44	19

22	15	7	95	3	36	10	2	10	27	12	9
18	11	3	54	12	37	44	19	13	41	13	11

Fig.5. The R^n- Addition operator

12	4	10	2	34	12	1	54	11	32	19	20
14	47	98	20	1	9	23	61	7	28	17	3

12	4	10	2	34	12	1	54
14	47	98	20	1	9	23	61

Fig.6. The R^n- Elimination operator

3.5 The compete genetic algorithm

The algorithm is structured as follows:

1. We choose randomly the initial population. Each chromosome defines a BBFNN that has between the minimum and the maximum number of neurons in the hidden layer (these numbers are chosen randomly).
2. Decode each chromosome in the population.
3. Compute the connection (The weights are the solutions in the least squares sense of the system of equations [*Beta*][*w*]=[*d*]. [*Beta*] is the matrix of the values of the Beta functions in the hidden layer on the training points).
4. Find the fitness of each chromosome and apply the proportional selection.
5. Apply genetic operators to the population of chromosomes.
6. If (the number of generation is equal to an integer N_{Max}) or (relative error<ε) then stop else return to step 2.

4 Experimental Results

We will now evaluate the performance of the proposed genetic algorithm to produce a BBFNN able to approximate some functions.

4.1 Example 1

In the first example we will approximate the function:

$$f_1(x)=10*Arctg\left(\frac{(x-0.2)(x-0.7)(x+0.8)}{(x+1.4)}\right)$$

$$x\in[-1,1] \tag{9}$$

The training points are 101 points chosen from $[-1,1]$.

The initial population has 50 chromosomes chosen randomly. Each chromosome represents a neural network that has between 4 and 15 neurons in the hidden layer. The centers are chosen from the training points. The width is chosen from 100 points in [0,2] and P and Q are chosen from 500 points in [0,5]. The probability of the crossover operator is equal to 0.7, the probability of the mutation operator is equal to 0.05 and both the addition operators and the elimination operator has a probability equal to 0.1.

Our goal isn't to find a minimal approximation but to show that with genetic algorithms we can optimize both the neural network architecture and the relative error. The best neural network has 11 neurons in the hidden layer and the relative error between the desired function and the obtained function is equal to 0.8%. In figure 7 we plot the output of the best neural network and the desired output.

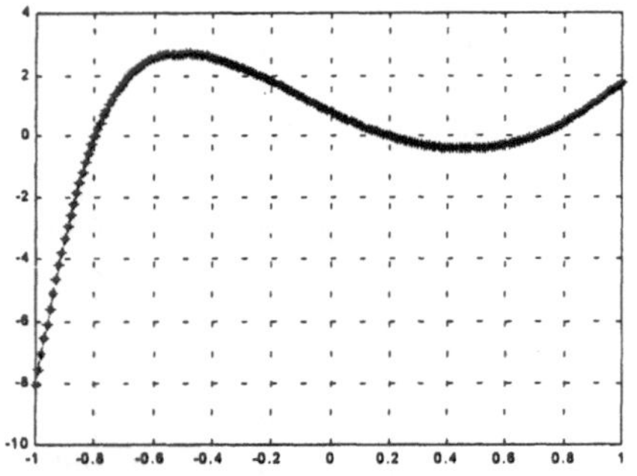

Fig.7. The output of the BBFNN and the desired function

4.2 Example 2

Our second example is to find the best network that approximates the function g_1:

$$g_1(x,y)=3x(x-1)(x-1.9)(x-0.7)(x+1.8)\sin(y)$$

$$-2\leq x,y\leq 2\,. \tag{10}$$

For this example, the initial population has 50 chromosomes and each chromosome represents a network that has between 40 and 60 neurons in the hidden layer. The probability of the crossover operator is equal to 0.8, the probability of the mutation operator is equal to 0.2. For the addition and elimination operator the probability is equal to 0.2. The centers are chosen from the training points. The widths are chosen from 100 points in [0,2] and P and Q are chosen from 10000 points in [0,10]. For this example, *alpha* is equal to 0.05. The best network has 47 neurons in the hidden layer and the relative error is equal to 4%.

5 Conclusion

In this paper we presented a new GA to train Beta Basis Function Neural Networks. The problem was to find the optimal network structure and the different parameters of the network.

The originality of this work resides in the method used for encoding each Beta function and the approximation of functions with many variables. Some examples were presented to show that the proposed algorithm can automatically determine the appropriate neural network structure as well as its parameters (weights, centers, etc).

In summary, the numerical results showed that:

- the BBFNN can approximate functions with many variables.
- the new method for representing each Beta function is successfully validated.
- genetic algorithms were able to solve this problem.

References

[1] Alimi, M.A.: The Beta Fuzzy System: Approximation of Standard Membership Functions. Proc. 17ème Journées Tunisiennes d'Électrotechnique et d'Automatique: JTEA'97, Nabeul, Tunisia, Nov., *1*, *1997*, 108-112.

[2] Alimi, M.A.: Beta Fuzzy Basis Functions for the Design of Universal Robust Neuro-Fuzzy Controllers. Proc. Séminaire sur la Commande Robuste & ses Applications: SCRA'97, Nabeul, Tunisia, Feb., *1997*, C1-C5.

[3] Aouiti, C., Alimi, M.A. and Maalej, A.: Genetic Algorithms to Construct Beta Neuro-Fuzzy Systems. Proc. Int. Conf. Artificial& Computational Intelligence for Decision, Control & Automation: ACIDCA'2000, Monastir, Tunisia, March, *2000*, 88-93.

[4] Janson, D. J. and Frenzel, J. F.: Application of genetic algorithms to the training of higher order neural networks. J. Syst. Eng., *2* .272-276 (1992).

[5] Montana, D. and Davis, L.: Training feedforward neural network neural networks using genetic algorithms. in Proc. 11th Int. Joint Conf. Artificial Intelligence. San Mateo, CA: Morgan Kaufmann, *1989*,. 762-767.

[6] Porto, V. W., Fogel, D. B., and Fogel, L. J.: Alternative neural network training methods", IEEE Expert, *10*, 16-22, (1995).

[7] Yao, X.: Evolving Artificial Neural Networks. IEEE Proc. Computational Intelligence,*87*, No. 9, 1423-1447 (1999).

Optimization with Implicitly Known Objective Functions Using RBF Networks and Genetic Algorithms

Hirotaka Nakayama *, Masao Arakawa †, Rie Sasaki *

*Department of Applied Mathematics, Konan University, e-mail: nakayama@konan-u.ac.jp †Department of Reliability-based Information System Engineering, Kagawa University, e-mail: arakawa@eng.kagawa-u.ac.jp

Abstract

In many practical engineering design problems, the form of objective function is not given explicitly in terms of design variables. Given the value of design variables, under this circumstance, the value of objective function is obtained by some analysis such as structural analysis, fluidmechanic analysis, thermodynamic analysis, and so on. Usually, these analyses are considerably time consuming to obtain a value of objective function. In order to make the number of analyses as few as possible, we suggest a method by which optimization is performed in parallel with predicting the form of objective function. In this paper, radial basis function networks (RBFN) are employed in predicting the form of objective function, and genetic algorithms (GA) in searching the optimal value of the predicted objective function. The effectiveness of the suggested method will be shown through some numerical examples.

1 Introduction

Our aim in this paper is to optimize objective functions whose forms are not explicitly known in terms of design variables. In many engineering design problems under this circumstance, values of objective functions are obtained by some complicated analysis of large scale with multi-peaked characters that are often seen in elastic-plastic analysis and so on. These analyses usually require considerably expensive computational time. Therefore, if these functions are optimized by existing methods, it takes an unrealistic order of time to obtain a solution. For this situation, the number of necessary analyses should be as few as possible. To this end, we suggest a method consisting of two stages. The first stage is to predict the form of objective function by RBFN(Radial Basis Function Networks). The second stage is to optimize the predicted objective function by GA(Genetic Algorithms). A major problem in this method is how to get a good approximation of the objective function based on as few sample data as possible. To this end, the form of objective function is revised by relearning on the basis of additional data step by step. Our discussion will be focused on this additional learning and how to select additional data in the following sections.

2 Additional Learning in RBFN

Since the number of sample points for predicting objective functios should be as few as possible, we adopt some additional learning techniques which predict objective functions by adding learning data step by step. RBFN are effective to this end. The output of RBFN is given by

$$f(\boldsymbol{x}) = \sum_{j=1}^{m} w_j h_j(\boldsymbol{x}).$$

where $h_j (j = 1, \ldots, m)$ are radial basis functions, e.g.,

$$h_j(\boldsymbol{x}) = e^{-\|x-\mu_j\|^2/r_j}.$$

Given the training data $(\boldsymbol{x}_i, \hat{y}_i)$ $(i = 1, \cdots, p)$, the learning of RBFN is usually made by solving

$$E = \sum_{i=1}^{p} (\hat{y}_i - f(\boldsymbol{x}_i))^2 \; + \; \sum_{j=1}^{m} \lambda_j w_j^2 \;\to\; \text{Min}$$

where the second term is introduced for the purpose of regularization.

Letting $A = (H_p^T H_p + \Lambda)$, we have as a necessary condition for the above minimization

$$A\hat{\boldsymbol{w}} = H_p^T \hat{\boldsymbol{y}}.$$

Here

$$H_p^T = [\boldsymbol{h}_1 \; \cdots \; \boldsymbol{h}_p]$$

where $\boldsymbol{h}_j^T = [h_1(\boldsymbol{x}_j), \ldots, h_m(\boldsymbol{x}_j)]$, and Λ is a diagonal matrix whose diagonal components are $\lambda_1 \cdots \lambda_m$.

Therefore, the learning in RBFN is reduced to finding

$$A^{-1} = (H_p^T H_p + \Lambda)^{-1}$$

The additional learninig in RBFN can be made by adding new data and/or a basis function, if necesary. Since the learning in RBFN is equivalent to the matrix inversion A^{-1}, the additional learning here is reduced to the incremental calculation of matrix inversion. The following algorithm can be seen in [3]:

(i) Adding a New Training Data

Adding a new data $\boldsymbol{x}_{p+1}$, the additional learning in RBFN can be made by the following simple update formula: Let

$$H_{p+1} = \begin{bmatrix} H_p \\ \boldsymbol{h}_{p+1}^T \end{bmatrix}$$

where $\boldsymbol{h}_{p+1}^T = [h_1(\boldsymbol{x}_{p+1}), \ldots, h_m(\boldsymbol{x}_{p+1})]$.

Then

$$A_{p+1}^{-1} = A_p^{-1} - \frac{A_p^{-1}\boldsymbol{h}_{p+1}\boldsymbol{h}_{p+1}^T A_p^{-1}}{1 + \boldsymbol{h}_{p+1}^T A_p^{-1}\boldsymbol{h}_{p+1}}$$

(ii) Adding a New Basis Function

In cases in which we need a new basis function to improve the learning for a new data, we have the following update formula for the matrix inversion: Let

$$H_{m+1} = \begin{bmatrix} H_m & \boldsymbol{h}_{m+1} \end{bmatrix}$$

where $\boldsymbol{h}_{m+1}^T = [h_{m+1}(\boldsymbol{x}_1), \ldots, h_{m+1}(\boldsymbol{x}_p)]$.

Then

$$A_{m+1}^{-1} = \begin{bmatrix} A_m^{-1} & \boldsymbol{0} \\ \boldsymbol{0}^T & 0 \end{bmatrix} +$$

$$\frac{1}{\lambda_{m+1} + \boldsymbol{h}_{m+1}^T (I_p - H_m A_m^{-1} H_m^T)\boldsymbol{h}_{m+1}}$$

$$\times \begin{bmatrix} A_m^{-1} H_m^T \boldsymbol{h}_{m+1} \\ -1 \end{bmatrix} \begin{bmatrix} A_m^{-1} H_m^T \boldsymbol{h}_{m+1} \\ -1 \end{bmatrix}^T$$

Remark It is important to decide parameters μ_j and r_j of radial basis functions moderately. We assign a basis function at each learning data $(\boldsymbol{x}_i, \hat{y}_i)$ $(i = 1, \cdots, p)$ so that $\mu_i = \boldsymbol{x}_i$ $(i = 1, \cdots, p)$. Modifying the formula given by [2] slightly, the value of r_j is determined by

$$r = d_{max} / \sqrt[n]{nm}$$

where d_{max} is the maximal distance among the data, n is the dimension of data, m is the number of basis function.

3 How to Select Additional Data

If some stop condition is not satisfied, we add some data in order to improve the approximation of objective function. Now, how to select such additional data becomes our issue.

If the current optimal point is taken as such additional data, the estimated optimal point tends to converge to a local maximum (or minimum) point. This is due to lack of global information in predicting the objective function.

On the other hand, if addtional data are taken to be far from the existing data, it is difficult to obtain more detailed information near the optimal point. Therefore, it is hard to obtain a solution with a high precision. This is because of insufficient information near the optimal point.

We suggest a method which gives global information for predicting the objective function and local information near the optimal point at the same time. To this end, we take two kinds of additional data for relearning the form of objective function. One of them is selected from a neighborhood of the current optimal point in order to add local information near the (estimated) optimal point. The size of this neighborhood is controlled during the convergence process. The other one is selected to be far from the currenct optimal value in order to give a better prediction of the form of objective function. The first additional data gives more detailed information near the current optimal point. The second data prevents from converging to local maximum (or minimum) point.

In this paper, the neighborhood of the current optimal point is given by a square S, whose center is the current optimal point, with the length of a side l. Let S_0 be a square, whose center is the current optimal point, with the fixed length of a side l_0. The square S is shrinked according to the number $C_{\#}$ of optimal points appeared continuously in S_0 in the past. Namely,

$$l = l_0 \times \frac{1}{C_{\#} + 1} \tag{1}$$

The first additional data is selected inside the square S at random. The second additional data is selected in a part, in which the existing learning data are sparse, outside the square S. A part with sparse existing data may be found as follows: First, a certain number of data are generated randomly outside the square S. Denote d_{ij} the distance from this random data p_i $(i = 1, \ldots, N_{rand})$ to the existing learning data q_j $(j = 1, \ldots, N)$. Select the shortest k distance $\tilde{d}_{ij}$ $(j = 1, \ldots, k)$ for each p_i, and sum up these k distances, i.e., $D_i = \sum_{j=1}^{k} \tilde{d}_{ij}$. Take p_t which maximizes $\{D_i\}_{(i=1,\dots,N_{rand})}$ as an addtional data outside S.

The algorithm is sumarized as follows:

Step.1 Predict the form of objective function by RBFN on the basis of given training data.

Step.2 Estimate an optimal point for the predicted objective function by GA.

Step.3 Count the number of optimal points appeared continuously in the past in S_0. This number is represented by $C_{\#}$.

Step.4 If $C_{\#}$ is larger than or equal to the given C_0 a priori, terminate the iteration. Otherwise calculate l by (1), and go to the next step.

Step.5 Select an additional data near the current optimal value, i.e., inside S.

Step.6 Select another additional data outside S in a place in which the density of training data is low as stated above. Go to Step.1.

4 GA for Optimizing Predicted Objective Functions

We use GA for optimizing predicted objective functions in this paper. Another method, of course, can be applied to this end. The reason why we use GA is that GA is simple and easy to use, and can provide an approximate solution to the global optimum.

Among GAs, we applied BLX-α method which can treat in a simple way continuous design varibles [4]. The method uses continuous values of design valriables as codes of individuals as they are. In crossover, children are generated randomly inside a hyperbox including their parents. It has been observed that BLX-α method without mutation is effective in problems with continuous design variables. Although there have been developed several sophisticated methods which can treat continuous variables (e.g., Arakawa *et al.*[1]), we apply BLX-α method due to the simplicity of its algorithm in this paper.

5 A Numerical Example

Consider an example given by

$$f(x_1, x_2) = 10\exp(-0.01(x_1 - 10)^2 - 0.01(x_2 - 15)^2)\sin x_1 \qquad (2)$$

$$(0 \leq x_1 \leq 15, \quad 0 \leq x_2 \leq 20)$$

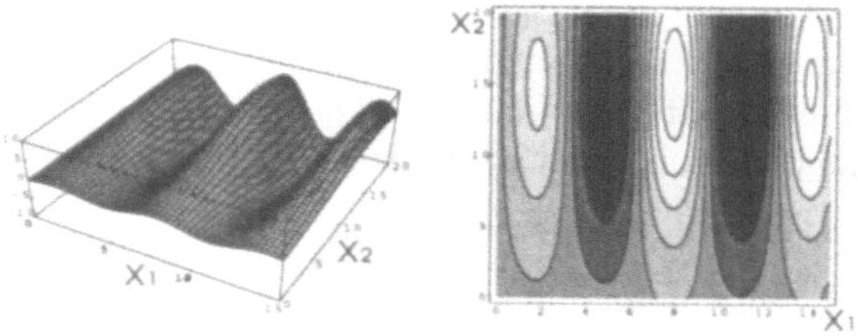

$x_1 = 7.8960, x_2 = 15.0000, f = 9.5585$

Fig. 1. Example

This function has a maximum value 9.5585 at $x_1 = 7.8960$ and $x_2 = 15.0000$ as shown in Figure 1.

Set λ in RBFN: 0.01, population in GA: 10, generation in GA: 50, l_0: 4.0, C_0: 10, N_{rand}: 50, and k: 2. At the begining, five data ($(x_1, x_2) = (0,0), (15,0), (0,20), (7.5,10), (15,20)$) were taken for the initial learning. The contour of the forecasted objective function by RBFN is shown in Figure 2. In figures, here after, "⊙" shows the training data, "×" the current optimal point, and "□" the correct optimal point to (2).

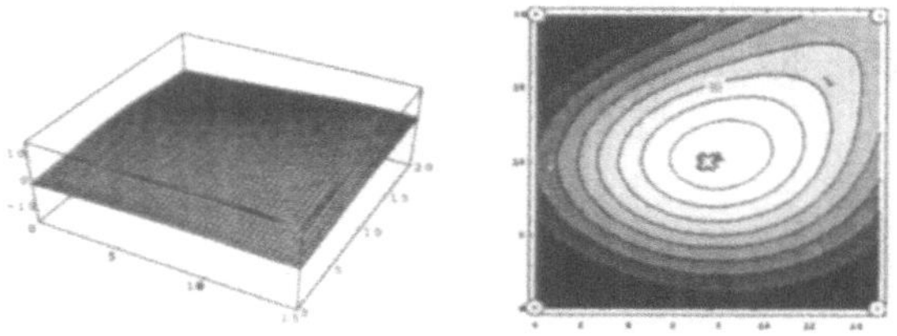

$x_1 = 7.9947, x_2 = 10.5039, f = 6.8400$

Fig. 2. Result at the first iteration(5 training data)

Figure 3 shows that two additional data ("•") are selected according to the method stated in Section 3. In this figure, there is no past optimal points in the square S_0. Therefore, the squares S and S_0 are identical at this stage. An additional data is selected randomly inside the square S, and another additional data is outside S in a part in which the density of training data is low.

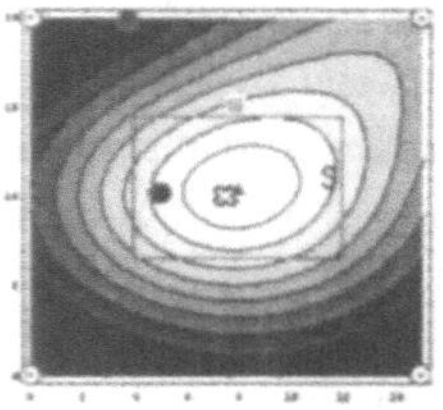

Fig. 3. Additional data denoted by •

After 69 iterations, we have a good estimatte as shown in Figure 5. Figure 4 shows the result at an intermediate 31 iterations. The numerical result of simulation is given by Table I.

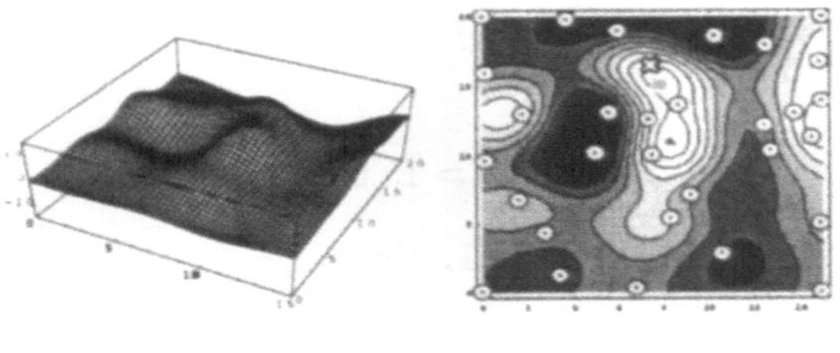

$x_1 = 7.4645$, $x_2 = 16.4721$, $f = 8.4892$

Fig. 4. Result of simulation (31 training data)

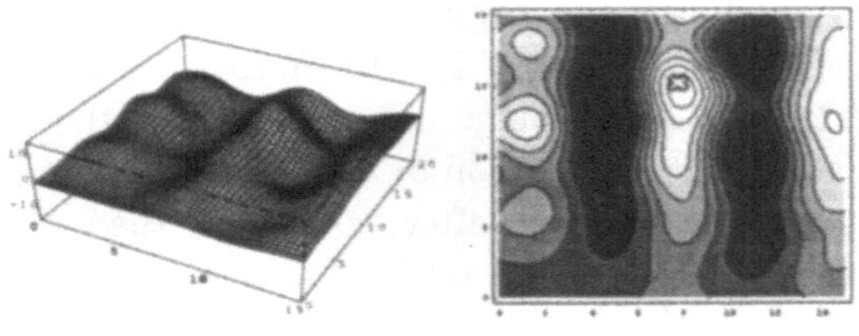

$x_1 = 7.8725$, $x_2 = 15.2023$, $f = 9.5519$

Fig. 5. Result of simulation (69 training data)

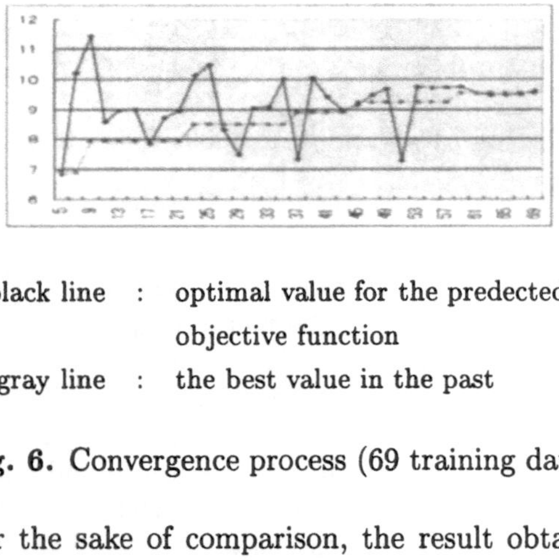

black line : optimal value for the predected objective function

gray line : the best value in the past

Fig. 6. Convergence process (69 training data)

For the sake of comparison, the result obtained by only GA is given in Table II. The population is 10 as well as in the proposed method. In optimizing by only GA, when the top 10 individuals with high ranks among parents and children converge within a given fixed range, the calculation is stopped. Table II shows the total number of generated individuals (represented by #analysis) until the calculation is stopped. "*" shows converging to a local maximum point different from the exact solution. In this table, one may see that the average number of evaluation of objective function in GA only is almost twice as the suggested method in order to get similar precision.

Table I. The result of simulation

	x_1	x_2	f
Theoretical value	7.8960	15.0000	9.5585
Suggested method	7.8725	15.2023	9.5519

Table II. The result of simulations with only GA

—	# analysis	x_1	x_2	f
1	100	8.008	14.678	9.4869
2	60	7.980	13.653	9.3532
3	100*	14.093	13.851	8.3383
4	90	8.000	12.650	8.9946
5	80	7.902	14.937	9.5580
6	780	6.912	14.634	5.3383
7	70	7.876	14.845	9.5542
8	70*	1.762	14.745	4.9773
9	60*	7.675	6.310	4.3812
10	130	8.066	13.864	9.2966
Average	154	—	—	—

6 Concluding Remarks

In optimizing objective functions which are not given explicitly in terms of design variables, RSM(Response Surface Method) is well known. However, RSM uses polynomial functions (usually, quadratic functions for the simplicity in implementation) for predicting objective functions. This causes a difficulty in highly nonlinear cases. The authors already have applied the method to an actual problem in designing the shape of turbin blade. The details will be mentioned at the conference.

References

[1] M. Arakawa and I. Hagiwara, "Nonlinear Integer, Discrete and Continuous Optimization Using Adaptive Range Genetic Algorithms", Proc. of ASME Design Technical Conferences (in CD-ROM), 1997

[2] S. Haykin, "*Neural Networks: A Comprehensive Foundation*", Macmillan College Publishing Company, 1994

[3] M.J.L.Orr, "Introduction to Radial Basis Function Networks", http://www.cns.ed.ac.uk/people/mark.html, Apr. 1996

[4] N.J.Radcliffe, "Forma Analysis and Random Respectful Recombination", Proceedings of the Forth International Conference on Genetic Algorithms, pp222-229, 1991

Hybrid Model of Cooling Tower Based on First Principles and Neural Networks

Nenad Milosavljevic*, Henrik Saxén†

*Valmet Corporation, Air Systems, Pansio, FIN-20240 Turku, Finland, †Heat Eng. Lab., Åbo Akademi University, Biskopsg. 8, FIN-20500 Turku, Finland

Abstract

Mathematical modelling in paper industry is to a large extent based on physical models that usually contain empirical correlations and assumptions. Milosavljevic and Heikkilä [1] derived a mathematical model for a counter-flow wet cooling tower based on one-dimensional heat and mass balance equations using an expression for the volumetric heat transfer coefficient. In the present study, volumetric heat transfer coefficient values determined by the above-mentioned model on the basis of measurements from a pilot cooling tower were approximated with feed-forward neural networks. It was found that a considerably more accurate approximation of the coefficient was obtained with the neural network than by conventional non-linear regression. The implementation of the expression in the first-principles model can lead to more accurate dimensioning of cooling towers, which results in lower equipment costs and better energy utilization.

1 Introduction

Almost every industrial process has some requirement for temperature control. For this reason, cooling towers are nowadays part of many plant installations. Their operation is based on a principle where energy is removed from warm water in direct contact with a relatively cool and dry air. In a counter-flow cooling tower the process consists of a gas phase (air) flowing upwards, a liquid phase (water film) flowing downwards, and a large interface between the two phases. Convective heat and mass transfer take place between the air and water. The evaporative cooling process occurs at the interface between the water film and the air stream. It is very difficult to determine the exact value of the available area for heat and mass transfer, because it is formed by a falling water film and also includes the surface area of separate water droplets. Therefore, a volumetric heat transfer coefficient is usually applied in mathematical modelling, but its value strongly depends on the type of packing used in the tower, and on the flow rates of water and air.

Milosavljevic and Heikkilä [1] derived a mathematical model for a counter-flow wet cooling tower, which is based on one-dimensional heat and mass balance equations using a given heat transfer coefficient. The balance equations [2], including differential and algebraic equations, were solved numerically to predict the temperature change of air and water, as well as the humidity as a function of the cooling tower height from known initial conditions. The volumetric heat transfer coefficient, α^*, for the given type of filling material was correlated in the form

$$\alpha^* = c_1 \cdot \left(1 + \left(\frac{m_w}{A}\right)^{C_2}\right) \cdot \left(\frac{m_a}{A}\right)^{C_3} \tag{1}$$

where c_1, c_2 and c_3 of the equation were determined using nonlinear regression analysis on data obtained from a cooling tower test rig.

In this work an attempt was made to improve the accuracy of the model by using an artificial neural network for the prediction of α^*. The results of the approximation are compared with the regression-based model. Since the neural model turned out to be clearly superior, it will be used as part of the simulation model in future calculations.

2 Neural Network Modeling

The study is based on standard multilayer perceptron networks using one layer of hidden neurons, where the input variables are the mass flow rate of supply air, $\dot{m}_a$, and water, $\dot{m}_w$, and the height of the filling material, h. The networks were trained to make the signal from the output node match the reported values of the volumetric heat transfer coefficient. The data set consisted of 158 vectors based on the data obtained from the experimental measurements at the pilot cooling tower. The set was divided into two parts, one for training (110 vectors) and one for testing (48 vectors). It is important that the models be applied as predictive tools only within the range of the process variables used in the parameter estimation (training), since interpolation may produce acceptable results whereas extrapolation usually leads to loss of accuracy. The two data sets were therefore chosen so as to have common characteristics in terms of the input vectors. The ranges of the input variables used in this study are reported in Table I.

Table I Ranges of process variables for the pilot cooling tower experiments.

Process variable	Range
Mass flow rate of air, kg/(m^{2}s)	1.74 -5.57
Mass flow rate of water, kg/(m^{2}s)	1.94 -6.00
Filling material height, m	0.60 -1.50

The networks were trained using the Levenberg-Marquardt method [3,4] implemented in the MATLAB Neural Network Toolbox 2.0, using the error square sum as the criterion of the fit. The number of units in the hidden layer was varied in order to find a proper network size. The networks will be denoted by *(i, j, k)*, where *i* (=3) represents the number of input units, *j* the number of units in the hidden layer, and *k* (=1) the number of output units.

3 Results and Discussion

On the basis of the normalized sum of squares of errors, SSE, given in Table II, it was found that the (3, 5, 1) network gave sufficient accuracy in approximating the volumetric heat transfer coefficient. It also showed a low error on the test set. Using a smaller number of hidden nodes led to worse performance, while a larger number resulted in overfitting, observed in increasing test set errors. By comparison with the performance reported on the last row of the table, it can be concluded that the selected network performs considerably better that the nonlinear regression formula of eq. (1).

The results of the (3,5,1) network and the regression model for the test set are illustrated in Figures 1 and 2. The average deviation (and maximum deviation) between the predicted and the experimental volumetric heat transfer coefficient was 178 W/(m^{3}K) (and 826 W/(m^{3}K)), while the corresponding value for the nonlinear regression was 455 W/(m^{3}K) (and 1925 W/(m^{3}K)).

Table II Performance of different model configurations studied

Configuration of model	Training set		Test set	
	SSE	R^2	SSE	R^2
(3, 2, 1)	0.1761	0.9578	0.1542	0.9230
(3, 3, 1)	0.1256	0.9118	0.1870	0.9031
(3, 4, 1)	0.0959	0.9397	0.1552	0.9242
(3, 5, 1)	0.0767	0.9571	0.1124	0.9589
(3, 6, 1)	0.0633	0.9713	0.1651	0.9377
(3, 7, 1)	0.0575	0.9751	0.1499	0.9548
Regression analysis	0.2303	0.7977	0.205	0.8107

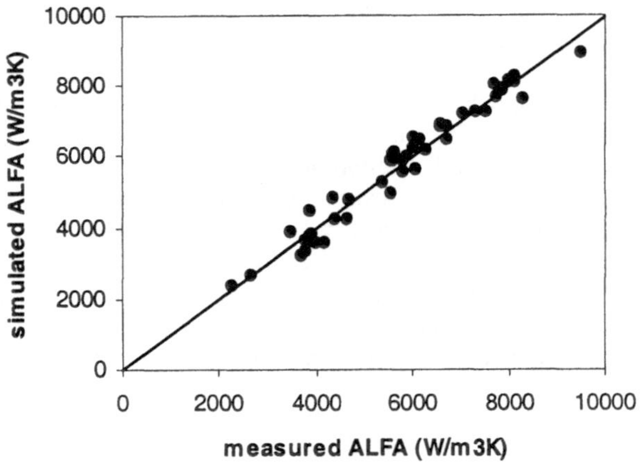

Fig. 1. Comparison between the volumetric heat transfer coefficient calculated from the measurements and the results from the (3,5,1) neural network model.

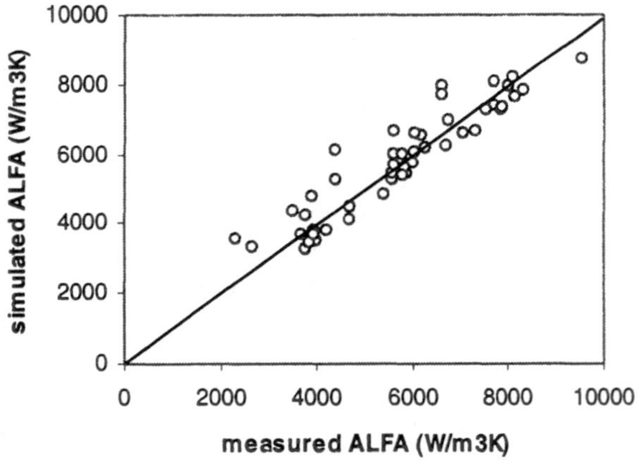

Fig. 2. Comparison between the volumetric heat transfer coefficient calculated from the measurements and results from the nonlinear regression approximation model for the test set.

4 Conclusions

Experimental results from a pilot cooling tower have been used in combination with a mathematical model to yield estimates of the volumetric heat transfer coefficient under different operating conditions of the tower. The dependence between these values and the process inputs was modeled by non-linear regression and neural networks. It was found that the networks were clearly more accurate in describing the correlations required in the mathematical model. By incorporating the network in the model, the hybrid concept is expected to bring about improved accuracy in dimensioning of cooling towers, and therefore to savings in material and utility costs.

References

[1] Milosavljevic, N. and Heikkilä, P.: A Comprehensive Approach to Cooling Tower Design. To appear in *Applied Thermal Engineering,* (2001).

[2] Fredman, T. and Saxén H.: Modelling and Simulation of a Cooling Tower. Proceedings of European Simulation Multiconference, 66-70, (1995).

[3] Levenberg, K.: A Method for Solution of Certain Non-linear Problems in Least-Squares. Quart. Appl. Math. *2*, 164-168, (1944).

[4] Marquardt, D.W.: An algorithm for least-squares estimation of nonlinear parameters, J. SIAM. *11*, 431-441, (1963).

Dynamic Handwriting Recognition Based on an Evolutionary Neural Classifier

Stéphane Gentric, Lionel Prevost, Maurice Milgram*

*LISIF / PARC, Université Pierre & Marie Curie, Boîte 164, 4, Place Jussieu, 75252 Paris cedex 05, France,
E-mail : Lionel.Prevost@lis.jussieu.fr / gentric/maum@ccr.jussieu.fr

Abstract

At the present time, most of the classification problems have to deal with heterogeneous data presenting a strong variability, even within a class. It seems therefore relevant to substitute to the notion of class, the notion of sub-class, the latter regrouping a relatively homogeneous subset of examples. In order to generate these sub-classes (and models that are associated them) automatically, we developed an evolutionary neural classifier . At the beginning, it is made of as many networks as the number of classes of the problem. During the training, the number of networks evolves in order to modelize to the best the different sub-classes and to decrease the overall confusion rate between classes. An application of this classifier is the recognition of unconstrained dynamic handwriting: the multiplication of character models (called allographs) makes essential the automatic sub-class generation. Results, tested on some 25000 letters of the Unipen database are very encouraging.

1 Introduction

1.1 Dynamic handwriting recognition

Handwriting recognition covers two investigation areas, mainly differentiated by the data acquisition mode. Off-line (or static) recognition works on images of the data to be classified, images obtained through an optical scanner. On the contrary, on-line (also called dynamic) recognition, manages signals coming from a graphic tablet. Static data (represented by a BitMap matrix) are therefore two-dimensional while dynamics (representing the continuation of the stroke coordinates) are mono-dimensional. Fields of application in static recognition vary from the automatic recognition of check amounts and zip-codes to that of the handwriting in order to store data. Similarly, dynamic recognition leads to many applications: signature authentication, handwriting recognition in order to create pen-based computers without keyboard.

The notion of allograph is specific to dynamic handwriting. It includes characters with the same static representation (i.e the same image) but presenting very variable dynamics in term of the number of strokes composing the character, of their senses and directions. On the other hand, the various models used for writing a given character: cursive, script ... (Fig. 1) are also allographs. In the writer-independent framework, the multiplication of allographs affects the different classes very heterogeneously: some will be represented by a single allograph (like "o"), others will need many.

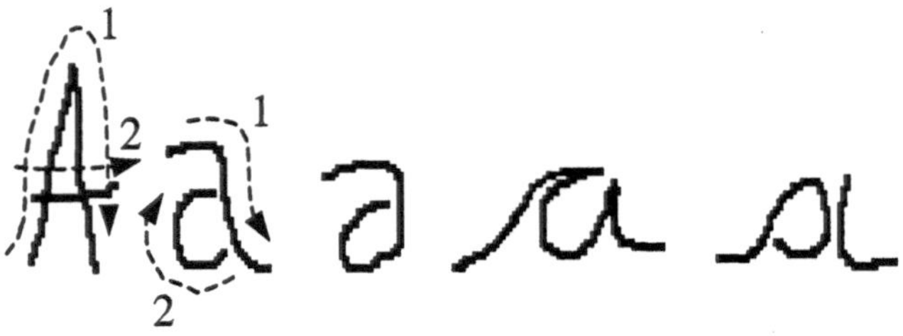

Fig. 1. Allographic variability.

1.2 State of art

There are two sets of classifiers in dynamic handwriting recognition.

Discriminating classifiers tempt to estimate frontiers between classes. They are generally connectionist ones: TDNN'like or derived from it. Their architecture is very complex and they reach high recognition rates on isolated characters recognition (Guyon 1991), (Manke 1994) (Bengio 1995) and word recognition (Schenkel 1995) (Manke 1995). Unfortunately their implementation in an extensive alphabet recognition system seems excessively delicate.

On the contrary, generating classifiers make cooperate several experts (each dedicated to a single class), and an expert takes the final decision. This solution proves more flexible than the precedent (it is easy to activate or to inhibit a part of the experts during the classification phase) and incremental (the addition of new classes pass by the only determination of the corresponding models and does not need a complete re-training). These last years, many generating classifiers appeared, using the most varied algorithmics and having in common the ultimate goal of modelizing optimally the different allographs of each character: hybrid neuro-markovian models (Garcia 1996), diabolos networks (Schwenk 1996) and radial basis function networks (Gentric 1993), self–organizing maps (Morasso 1996) k-NN classification using prototypes stemmed from pruning (Prevost 2000) or genetic algorithm (Bontempi 1994)...

All these classifiers undergo a training stage and face generally two problems mainly linked to the presence of heterogeneous classes. The first focuses on the model consistency : a single model will not be able to represent correctly a class with a strong allographic variability. From then on, if several models are used for one very class, the necessity appears to distribute examples in distinct corpus so as to train the different models. Once this partition realized, the problem of the model training appears : the marginal allographs, under–represented, generate a small corpus generally insufficient to train the model.

We analyze in section 2 the interest of an evolutionnary neural classifier to overcome these problems. Then, we formalize the evolution process. The next section is devoted to experimental results on the selected application: dynamic handwriting recognition. They show that it is possible to realize a very robust modelization by using only a small number of networks, even on this very heterogeneous problem. Finally, we expose our conclusions in section 4.

2 The Classifier

2.1 From a fixed to an evolutionary classifier

The classifier is built from associative MLP with three layers. Their name is auto-associative or diabolo networks owing to their conical shape. The input and output layers have the same number of cells while the hidden one has a smaller size. Used initially for data compression, they gave excellent results on classification problems because each one modelizes quite well a specific class and does a large error on others (Schwenk 1996). Each network is dedicated to one very class and only trained with examples of this latter. Classification is done by choosing the best fitting model i.e the network with the smallest reconstruction error. Transformation invariance is achieved by using constraint tangent distance (Simard 1993) as objective function during training as well as generalization instead of the simple euclidean distance. The initial classifier has N networks (one per class: fig. 2.a). It achieved poor results because of the class heterogeneity.

Then, it seemed interesting to use several networks (instead of one) for each class. The new classifier is build around sub-class networks (so called because each of them is dedicated to one particular sub-class) and then optimize their number to decrease confusions (fig. 2.b). At the end of the training, each sub-class-network modelize as well as possible one allographic sub-class. The associated training set has been found in a totally unsupervised way.

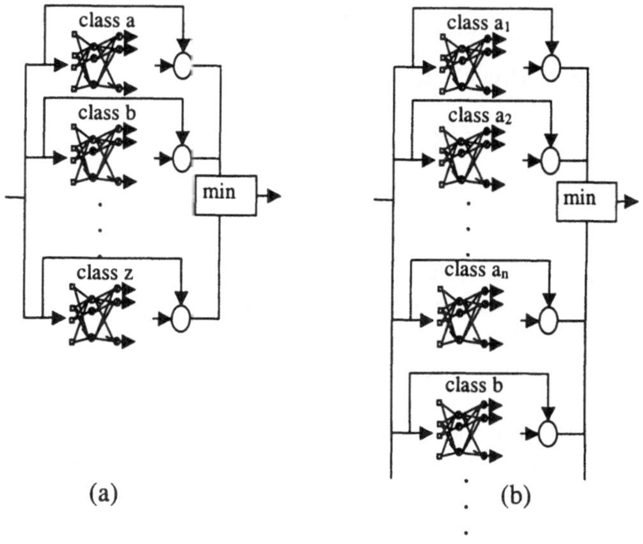

Fig. 2 : (a) orginal & (b) incremental classifiers.

2.2 The birth and death strategy

As the final goal is to increase the recognition rate, let's tackle the problem from the confusion starting-point. Sub-classes will be generated according to confusions produced by the classifier. In other words, if an example of the class C_1 is classified as a C_2, the class C_1 requires a new sub-class to take account of this example. Each sub-class is then represented by a specific diabolo network and the corresponding set of examples used to train this network.

This strategy leads to the emergence of new sub-classes while sub-class networks get more and more specialized. From time to time during the training, the number of examples associated to each sub-class allows to fix birth and death sub-class criteria as follows:

- **$N > N_B$ (birth)** : the example number is sufficient to build (and to train) a new sub-class network.
- **$N < N_D$ (death)** : not enough examples to train this sub-class network (deletion of the network).

a) Initialization:

Training of the initial classifier (26 class-networks R_i^1 $i \subset [1,26]$ using back-propagation algorithm.

Test on the training data. Each training set is separated in two sub-sets: N_i^1 examples well-classified by their own class-network and N_i^* mis-classified examples (better score on another class).

Decision : for each of classes, if $N_i^* > N_B$, a new sub-class network R_i^2 is build, to be trained on examples N_i^* (figure 2.a).

b) Iteration:

Training on all sub-class-networks. Each associated corpus is split up into training and validation sets (respectively 4/5 and 1/5 of the whole corpus). The associated network is trained (with training stopped on the validation set).

Test: on the training data. For every class, each example is re-assigned to the corpus of an existing sub-class if it got its better score on a network of its class, to the one of the mis-classified if it got its better score on a network belonging to another class.

Decision:

- **Birth**: For each class, if $N_i^* > N_B$, a new subclass-network R_i^{Nsc+1} is build (fig 3.b), N_{sc} is the number of sub-class-network of the class i at the current iteration.

- **Death** : let N_i^k be the number of examples associated to the subclass-network R_i^k (k = 1,2... $N_{sc}(i)$). If $N_i^k < N_D$, R_i^k is deleted (fig 3.c). For each of these examples, one computes its score on all the networks of its class (each one representing a given sub-class) and affects it to the network producing the best score.

c) go to stage b) or **stop** if :

- Training sets no further change ;

- Error rate is minimum on the validation set

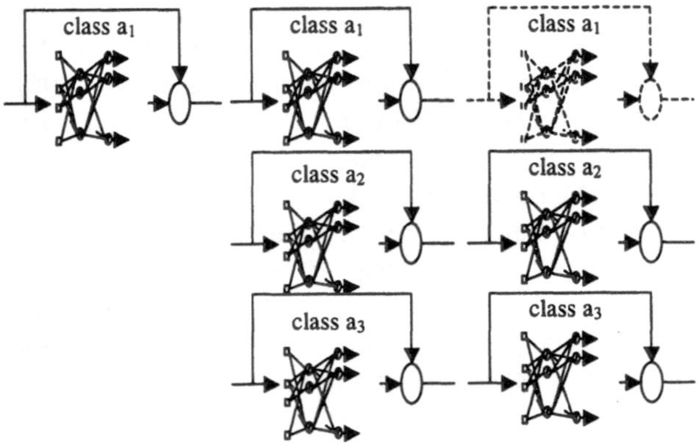

(a) initialization (b) birth (c) death.

Fig. 3. Learning process.

3 Dynamic Handwriting Recognition

3.1 The database

This survey has been achieved on isolated letters (upper and lower cases) of the Unipen corpus Train R01-V07 (Guyon 94), regrouping in one same class upper and lower cases. Therefore our basis is constituted of 26 classes distributed as follows (Table I) in training, validation and test sets. We applied a simple pre-processing: characters are resampled isometrically with 21 points (inducing 42 cells on the network input layer) and then centred and size normalised in [-1,1].

	Example number
Training set	36748
Validation set	8970
Test set	17798

Table I : Database sizes.

3.2 The results

We propose here the final test done on the Unipen basis of isolated letters described previously. The final classifier (243 sub-class networks 42-7-42: 10 network per class) reaches the following error rates: 4.05% on the training set and 7.53% on the test set.

These results have been achieved with a life criterion fixed to 4% and a death criterion of 2%. One will notice the error rate on the training set: 4.05%. In fact this latter depends directly on the life criterion. To the extreme, if this criterion is nil, many sub-classes should be created, in order to make disappear all confusions. But the associated networks should specialized on very few examples (notice that for less than 8 examples, our networks overfit their examples).

Fig. 4 shows distribution in 9 sub-classes of examples of the j " class " presented according to their frequency. We can note the visual homogeneity inside a sub-class is not fixed by the system. It is just a consequence of the confusion reduction process.

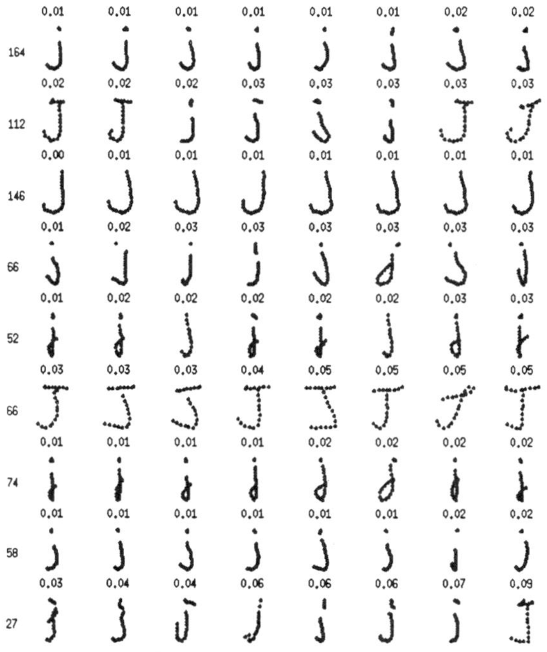

Fig. 4. The 9 sub-classes of the class "j".

4 Conclusions

We proposed a new system for dynamic handwriting recognition in the writer-independent framework. The classification uses neural networks called diabolos networks . These networks are able to specialize on a letter during training and to identify it during recognition. The sub-class generation process allows to take in account the variability of character classes. A letter is then modelized by several networks. These are generated automatically during training in order to decrease confusions between classes. The error rate on isolated letters, 7.53% is encouraging. On the same basis, a training by MDCA clustering (Prevost 2000) followed-up by a *1*-ppv classification achieved 5.05%. But this engine, due to its nature, is not too speed (10 characters per second on Pentium 200) while the neural one run at 60 characters per second.

The incorporation of the classifier in a complete system of cursive word recognition gives excellent results (Gentric 98).

Many research axes are yet opened, particularly the improvement of the automatic sub-class generation process. It seems that the system creates too many networks and a survey on their elimination or fusion would probably reduce their number as well as classification time.

References

[1] Bengio Y., LeCun Y., Nohl C. & Burges C., LeRec : A NN/HMM Hybrid for On-Line Handwriting Recognition, *Neural Computation, 7(6), 1289-1303, 1995.*

[2] Bontempi B. & Marcelli A., A genetic learning system for on-line character recognition, *ICPR'94, 83-87, 1994.*

[3] Garcia-Salicetti S., Dorizzi B., Gallinari P. & Wimmer Z., Adaptive Discrimination in an HMM-based neural predictive system for on-line word recognition, *ICPR'96, (D), 515-519, 1996.*

[4] Gentric P., Handwritten character recognition using neural networks, *ICOHD'93, pp 13-15, 1993.*

[5] Gentric S. & Milgram M, Reconnaissance de mots cursifs et génération non supervisée de sous-classes, *CIFED'98, pp 384-393, 1998.*

[6] Guyon I., Albrecht P., LeCun Y., Denker J. & Hubbard W., Design of a Neural Network Character Recognizer for a Touch Terminal, *Pattern Recognition, 24(2), 105-119, 1991.*

[7] Guyon I., Schomaker L., Plamondon R., Liberman M. & Janet S., UNIPEN project of on-line data exchange and recognizer benchmarks, *ICPR'94, pp. 29-33, 1994.*

[8] Manke S., Finke M. & Waibel A., Combining Bitmaps with Dynamic Writing Information for On-Line Handwriting Recognition, *ICPR'94, 596-598, 1994.*

[9] Manke S., Finke M. & Waibel A., Npen++ : a writer-independent large vocabulary on-line handwriting recognition system, *ICDAR'95, pp 403-408, 1995.*

[10] Morasso P., Barberis L., Pagliano S. & Vergano D., Recognition experiments of cursive dynamic handwriting with self-organising networks, *Pattern Recognition, 26(3), 451-460, 1996.*

[11] Prevost L. & Milgram M., Modelizing character allographs in omni-scriptor frame : a new non-supervised algorithm, Pattern Recognition Letters, 21(4), pp 295-302, 2000.

[12] Schwenk H. & Milgram M., Constraint tangent distance for on-line character recognition, ICPR'96, (D), 520-524, 1996.

[13] Schenkel M., Guyon I. & Henderson D., On-line Cursive Script Recognition using Time-Delay Neural Networks and Hidden Markov Models, *Machine Vision and Applications, (8), pp 215-223, 1995.*

[14] Simard P., Le Cun Y. & Denker J., Efficient pattern recognition using a new transformation distance, NIPS (5), pp 50-58, 1993.

Initial Description of Multi-Modal Dynamic Models

Miroslav Kárný, Petr Nedoma, Ivan Nagy, Markéta Valečková * [1]

*ÚTIA, AV ČR, Pod vodárenskou věž 4, Prague 8, Czech Republic e-mail: school@utia.cas.cz.

Abstract

Multiple models, neural networks, cluster analysis and probabilistic mixtures are prominent examples of situations when complex multi-modal models [1] are built using vast amount of data. Complexity and non-unicity of modelled situation imply that resulting description depends heavily on the initial phase of search. The safest repetitive purely random search is mostly inhibited by computational complexity of the addressed task. For this reasons, various techniques have been designed. None of them, to our best knowledge, suits to cases when *dynamic* models are constructed. The paper describes a novel technique that fills this gap in a promising way. Essentially, the trial description is gradually split whenever there is possibility that a unimodal sub-model hides more modes.

1 Introduction

Modelling of relationships in sequences of multivariate records is addressed in many ways and has a wide range of applications covering majority of human activities. As a rule, the surface in the modelled data space is multi-modal and random. Thus, probabilistic mixtures are its "natural" description. Their estimation is a computationally hard task that has to be approximated using heuristic techniques. Many of them have been developed and successfully used. Predominantly, they assume that processed data records are independent, that the modelled system is static. At the same time, dynamics, i.e. dependence of future on past, is an important feature of majority of practically important cases. The available techniques are able to cope with them at least approximately but the decisive task – the search for initial guess of the model – is almost unsupported. This led us to a development of a novel technique that suits to this case. The paper specialises it to normal mixtures. The proposed technique is also expected to be useful out of probabilistic context and to serve to radial-basis neural networks, multiple regression, cluster analysis etc.

The assumed preliminary knowledge of probabilistic language can be well reduced to the basic algebra of probability density functions (pdf), i.e. their non-negativity, normalisation to unity, chain rule, rule for marginalisation and Bayes rule. A broader engineering insight into the used tools gives [2].

<u>**Notation:**</u> The following symbols are useded:
x^* means the set of all values of x,
$\mathring{x}$ denotes cardinality of x^*,
x_t is x at discrete time $t \in t^* \equiv \{1, \ldots, \mathring{t}\}$,
$x(t) \equiv x_1, \ldots, x_t$, $x_{i;t}$ is i-th entry of x_t,
$\Theta \in \Theta^*$ denotes unknown model parameters,
$f(\cdot|\cdot)$ is a common symbol for conditional pdfs: versions are distinguished by identifiers in arguments,
$\propto$ means equality up to a normalising factor.

<u>**Basic scenario:**</u> A sequence $d(\mathring{t})$ of data records d_t is observed. Their relationships are modelled by joint pdf conditioned on unknown parameters Θ

$$f(d(\mathring{t})|\Theta) = \prod_{t \in t^*} f(d_t|d(t-1), \Theta).$$

The considered *mixture model* has the form

$$f(d_t|d(t-1), \Theta) = \sum_{c \in c^*} \alpha_c f(d_t|d(t-1), \Theta_c, c). \quad (1)$$

The individual pdfs $f(d_t|d(t-1), \Theta_c, c)$ are called *components*. The unknown parameter Θ is formed by probabilistic weights of components $\alpha \equiv (\alpha_1, \ldots, \alpha_{\mathring{c}}) \in \alpha^* \equiv \{\alpha_c \geq 0, \sum_{c \in c^*} \alpha_c = 1\}$ and by individual parameters Θ_c, $c \in c^*$, of components. The components are decomposed by the chain rule

$$f(d_t|d(t-1), \Theta_c) = \prod_{i=1}^{\mathring{d}} f(d_{i;t}|\psi_{ic;t}, \Theta_{ic}, i, c) \quad (2)$$

where $f(d_{i;t}|\psi_{ic;t}, \Theta_{ic}, i, c)$ are called *parameterised factors*. They predict scalar entries $d_{i;t}$ of d_t. They are assumed to depend on regression vectors $\psi_{ic;t}$ that consist of values of known finite-dimensional mappings $d^*(t-1), d^*_{1;t}, \ldots, d^*_{i-1;t} \rightarrow \psi^*_{ic;t}$. The factorisation (2) allows us to deal with parsimonious description of components and to combine entries of logical and continuous nature.

The adopted Bayesian estimation modifies a chosen prior pdf $f(\Theta)$ and applies Bayes rule [2] in order

[1] This research was partially supported by EC, grant IST-99-12058, e-mail: school@utia.cas.cz

get posterior pdf $f(\Theta|d(\mathring{t}))$

$$f(\Theta|d(\mathring{t})) \propto f(d(\mathring{t})|\Theta)f(\Theta). \qquad (3)$$

This pdf is the most general result of Bayesian estimation and may serve for obtaining point estimates of the unknown Θ, provides information about precision of these estimates and can be directly used for computing predictors that relate arbitrary predicted data item $d_{i;t}$ with an arbitrary regression vector ψ_{ic} within c-th component

$$f(d_i|\psi_{ic}, d(\mathring{t})) = \int_{\theta^*_{ic}} f(d_i|\psi_{ic}, \Theta_{ic}, i, c) f(\Theta_{ic}|d(\mathring{t}))\, d\Theta_{ic}.$$

The posterior pdfs $f(\Theta_{ic}|d(\mathring{t}))$ describing parameterised factors are called *factors.*

2 Addressed Problem and Idea of Solution

Application of the Bayes rule requires evaluation of the likelihood function, i.e. the pdf $f(d(\mathring{t})|\Theta)$ taken as a function of parameters. For mixtures, it is computationally hard as it consists of an exponentially blowing number of terms depending on (high-dimensional) Θ. The evaluation can be done approximately, see e.g. [3]. The quality of the approximation depends heavily on the choice of the mixture structure and on the choice of the prior pdf $f(\Theta)$. This choice is addressed here.

The solution is based on a simple iterative search. First, the non-iterative Bayesian evaluation of the posterior pdf is converted into an iterative application of Bayes rule followed by an appropriate flattening. Flattening prevents the iterations to over-fit data. Second, (approximate) Bayes rule is applied for a given number of components. In the component describing majority of data, a most uncertain factor (pdf on Θ^*_{ic}!) is selected as being suspicious for hiding more modes. This factor is appropriately split in pair of factors so that a new richer mixture arises. The splitting is performed till predictive abilities of the mixture are increasing.

3 Solution Details

During evaluations, we need to measure distances between various pdfs involved. Kullback-Leibler (KL) distance [4] is a prominent tool for this. For a pair of pdfs $f, \bar{f}$ posttive on Θ^*, it is defined

$$\mathcal{D}(f||\bar{f}) \equiv \int_{\Theta^*} f \ln\left(\frac{f}{\bar{f}}\right) d\Theta. \qquad (4)$$

The KL distance is convex in f, non-negative and equals to zero iff $f = \bar{f}$ almost everywhere.

Iterative Bayes learning: Let $f_n(\Theta)$ be n-th guess of the prior pdf and $\mathcal{L}_n(\Theta) \approx f(d(\mathring{t})|\Theta)$ the corresponding approximation of the likelihood function. Then the posterior pdf $f_n(\Theta|d(\mathring{t})) \propto \mathcal{L}_n(\Theta)f_n(\Theta)$ gives a better clue about appropriate parameters and is a candidate for a new guess of the prior pdf. It is, however, over-concentrated. Thus, we select a new guess of the prior pdf $f_{n+1}(\Theta)$ as a compromise between $f_n(\Theta|d(\mathring{t}))$ and a flat prior pdf $\bar{f}(\Theta)$. A construction of such a compromise is described by the proposition below. It also hints how to choose the knob that controls it. The omitted proof relies on properties of $\mathcal{D}(\cdot||\cdot)$.

Proposition 3.1 (Iterative Bayes learning)
1. Let pdfs $f, \bar{f}$ have a common support Θ^*. Then the pdf $\hat{f}$ minimising $\mathcal{D}(\hat{f}||f) + q\mathcal{D}(\hat{f}||\bar{f})$, $q > 0$ is

$$\hat{f} \propto f^{\Lambda}\bar{f}^{1-\Lambda}, \ \Lambda = \frac{1}{1+q} \in (0,1). \qquad (5)$$

2. Let us generate sequence of pdfs $f_n(\Theta)$ by flattening the result of the Bayesian estimation $f \propto \mathcal{L}_n(\Theta)f_n(\Theta)$ to the flat prior pdf $\bar{f}$, i.e.

$$f_{n+1}(\Theta) \propto [\mathcal{L}_n(\Theta)f_n(\Theta)]^{\Lambda}\left[\bar{f}(\Theta)\right]^{1-\Lambda}, \ f_0 \equiv \bar{f}. \qquad (6)$$

If the likelihood function $\mathcal{L}_n(\Theta)$ is evaluated exactly, then $f_{n+1}(\Theta)$ converges for $n \to \infty$ to the exact posterior pdf iff $\Lambda = 0.5$.

Factor shifting: As outlined in Section 2, a suitable factor is to be split. The created factors are required to have different modes than the split one. Expected value a suitable function $g(\Theta)$ is used as a mode characterisation. Uncertainty of the factor f is measured by its KL distance $\mathcal{D}(f||\bar{f})$ to a flat factor $\bar{f}$. KL distance $\mathcal{D}(\hat{f}||f)$ measures similarity of the new factors $\hat{f}$ to the split f.

Proposition 3.2 (Shifted factor) *Let pdf f and a function g be defined on Θ^*. Then the pdf $\hat{f}$*
- *minimising the KL distance $\mathcal{D}(\hat{f}||f)$,*
- *having a fixed norm $\omega > 0$ of a shift in expectation*

$$\left\langle \int \Theta(\hat{f}(\Theta) - f(\Theta))\, d\Theta, \int g(\Theta)(\hat{f}(\Theta) - f(\Theta))\, d\Theta \right\rangle$$

has the form $\hat{f} \propto f\exp(- < \mu, g(\Theta) >)$. (7)

The solution is "indexed" by an array μ that is compatible with a scalar product $< \cdot, \cdot >$. This converts infinite-dimensional optimisation into finite-dimensional search for μ indexing solutions (7).

Proof: The Lagrangian of this convex optimisation on the convex set $\hat{f}^* \equiv \{\hat{f} \geq 0, \int \int f = 1\ \&(7)\}$ is

$\mathcal{J} \equiv \mathcal{D}(\hat{f}||f) + \nu \int \hat{f}(\Theta)\, d\Theta +$
$\rho \left\langle \int g(\Theta)(\hat{f}(\Theta) - f(\Theta))\, d\Theta, \int g(\Theta)(\hat{f}(\Theta) - f(\Theta))\, d\Theta \right\rangle.$
It is parameterised by scalars ν, ρ that have to be chosen so that restrictions are met. Its variation $\delta\mathcal{J}$, that should equal to zero for the optimal $\hat{f}$, reads $\int \hat{\delta}[\ln(\hat{f}/f) + \nu + 2\rho < \tilde{\mu}, g(\Theta) >]\, d\Theta$, where

- $\hat{\delta} \equiv \hat{\delta}(\Theta)$ is an arbitrary variation of $\hat{f}(\Theta)$
- $\tilde{\mu} \equiv \int g(\Theta)(\hat{f}(\Theta) - f(\Theta))\, d\Theta$.

This proves the claimed results with $\mu = \rho\tilde{\mu}$. □

4 Application to Normal Factors

Normal factor and conjugate prior: The considered normal factors have the form

$$f(d|\psi,\Theta) = \mathcal{N}_d(\theta'\psi, r) = (2\pi r)^{-0.5} \exp\left\{-\frac{1}{2r}\left([-1,\theta']\Psi\right)^2\right\} \text{ where} \quad (8)$$

$'$ denotes transposition,
$\Theta = [\theta, r] = $ [regression coefficients, noise variance],
$\Psi = [d, \psi']' = $ [regressand, regression vector].

The conjugate prior pdf $f(\Theta)$ that preserves its functional form during Bayes estimation of the model (8) is Gauss-inverse-Wishart (GiW) pdf [2]

$$f(\Theta) = GiW_{[\theta,r]}(L, D, \kappa) \propto r^{-\frac{\kappa}{2}} \exp\left\{-\frac{1}{2r}[-1,\theta']L'DL[-1,\theta']'\right\} \text{ where} \quad (9)$$

$\kappa > 0$ is an effective counter $f(\Theta)$,
$L'DL$ is an extended information matrix in numerically advantageous $L'DL$ decomposition in which L is lower triangular matrix with a unit diagonal, D is diagonal matrix with positive entries. The split version of $L'DL$ decomposition

$$L \equiv \begin{bmatrix} 1 & 0 \\ L_{d\psi} & L_{\psi} \end{bmatrix}, D = \text{diag}[D_d, D_{\psi}] \quad (10)$$

can be related to well known least squares (LS) quantities $\hat{\theta} = L_{\psi}^{-1}L_{d\psi}$ is LS estimate of θ, D_d/κ is LS estimate of r, $D_d/\kappa L_{\psi}^{-1}D_{\psi}^{-1}(L'_{\psi})'$ is covariance matrix of the LS estimate of θ.

Shifting of *GiW* factors: We use the KL distance of a pair of GiW pdfs. Proof is omitted.

Proposition 4.1 (KL distance of *GiW* pdfs) *Let $\hat{f} \equiv GiW_{\Theta}(\hat{L}, \hat{D}, \hat{\kappa})$, $f \equiv GiW_{\Theta}(L, D, \kappa)$ be a pair of GiW pdfs of parameters $\Theta \equiv (\theta, r)$. Their KL distance is given by the formula*

$$\mathcal{D}(\hat{f}||f) = \ln\left(\frac{\Gamma(0.5\kappa)}{\Gamma(0.5\hat{\kappa})}\right) - 0.5\ln\left(\prod_{i=2}^{\mathring{\Psi}} \frac{D_{ii}}{\hat{D}_{ii}}\right) + \quad (11)$$
$$+ \quad 0.5\kappa \ln\left(\frac{\hat{D}_d}{D_d}\right) + 0.5(\hat{\kappa} - \kappa)\frac{\partial}{\partial(0.5\hat{\kappa})}\ln(\Gamma(0.5\hat{\kappa}))$$
$$- \quad 0.5\mathring{\psi} + 0.5\text{tr}[\hat{L}_{\psi}^{-1}\hat{D}_{\psi}^{-1}(\hat{L}'_{\psi})^{-1}L'_{\psi}D_{\psi}L_{\psi}] +$$
$$+ \quad 0.5\frac{\hat{\kappa}}{\hat{D}_d}[(\hat{\hat{\theta}} - \hat{\theta})'\hat{L}'_{\psi}\hat{D}_{\psi}\hat{L}_{\psi}(\hat{\hat{\theta}} - \hat{\theta}) - \hat{D}_d + D_d]$$

where Γ denotes Euler gamma function.

Proposition 4.2 (Shifted GiW factor) *Let $f = GiW_{\theta,r}(L, D, \kappa)$ and $g(\Theta)' = \theta'/r$. Then the pdf $\hat{f}$ minimising the KL distance $\mathcal{D}(\hat{f}||f)$ and having a fixed norm $\omega > 0$ of the shift*

$$\left\langle \int g(\Theta)(\hat{f}(\Theta) - f(\Theta))\, d\Theta, \int g(\Theta)(\hat{f}(\Theta) - f(\Theta))\, d\Theta \right\rangle$$

has the form $\hat{f} = GiW_{\theta,r}(\hat{L}, \hat{D}, \hat{\kappa})$ with

$$\hat{L} \equiv \begin{bmatrix} 1 & 0 \\ L_{d\psi} + \rho L_{\psi}x & L_{\psi} \end{bmatrix} \Leftrightarrow \hat{\hat{\theta}} = \hat{\theta} + \rho x$$
$$\hat{D} = D, \ \hat{\kappa} = \kappa.. \quad (12)$$

The vector x in (12) is a unit eigenvector of the matrix $D_d L_{\psi}^{-1} D_{\psi}^{-1}(L'_{\psi})^{-1}$ corresponding to the maximum eigenvalue. The choice of the optional scalar $\rho \neq 0$ replaces the choice of ω.

Proof: Proposition 3.2 implies functional form (12) of $\hat{f}$. The choice of the function $g(\Theta)$ preserves in GiW class. The form (12) allows no changes in κ, L_{ψ} and D. Thus, $\hat{L}_{d\psi}$ is the only optional quantity influencing just $(\hat{\hat{\theta}} - \hat{\theta})'L'_{\psi}D_{\psi}L_{\psi}(\hat{\hat{\theta}} - \hat{\theta})$.

The restriction (12) with $\hat{D} = D$ enforces $(\hat{\hat{\theta}} - \hat{\theta})'(\hat{\hat{\theta}} - \hat{\theta}) > \hat{r}\omega > 0$. Obviously, $\hat{\hat{\theta}} - \hat{\theta} = \gamma x$ with x being eigenvector corresponding to the smallest eigenvalue of $D_d^{-1}L'_{\psi}D_{\psi}L_{\psi}$ (to the largest one of its inversion) is the minimiser searched for. The normalising factor ρ has to meet the considered restriction. □

Overall algorithm: The particular results in the outlined algorithm give *initiation of normal mixtures*. Only GiW pdfs are involved used. Their statistics L, D, κ are distinguished by indices $(icn;t)$ pointing to (factor,component,iteration;time). This allows us to use simplified notation

$$GiW(icn;t) \equiv GiW_{\theta_{ic},r_{ic}}(L_{icn;t}, D_{icn;t}, \kappa_{icn;n})$$

* Set iteration counter $n = 0$ and $\mathring{c} = 1$.
* Select a guess of prior pdf $\hat{f}_0(\Theta) \equiv \prod_{i=1}^{\mathring{d}} GiW(i1n;0)$ corresponding to the trivial mixture with single component.

• Evaluate approximately $f_n(\Theta|d(\mathring{t})) = \prod_{c\in c^*}\prod_{i=1}^{\mathring{d}} GiW(icn;t)$, (quasi-Bayes estimation [3] reduces to weighted recursive LS).
$*$ Evaluate the value of the corresponding predictive pdf $f(d(\mathring{t}))$ (a by-product of the step •).
$*$ Select the component $\bar{c}$ with the largest sum of $\kappa_{icn;t}$ (explaining the majority of data).
$*$ Set flags to all factors within the component $\bar{c}$ to the value "admissible".
•• Find among all estimated factors with the flag "admissible" the most uncertain factor $GiW(\bar{i}\bar{c}n;\mathring{t})$, i.e. $\mathcal{D}(GiW(\bar{i}\bar{c}n;\mathring{t})||\bar{f}) \leq \mathcal{D}(GiW(i\bar{c}n;\mathring{t})||\bar{f})$ where $\bar{f}$ is a flat (uniform) pdf.
$*$ Generate new from $GiW(\bar{i}\bar{c}n;\mathring{t})$ shifted factors, say $GiW(\bar{i}^{+}\bar{c}n;\mathring{t})$, $GiW(\bar{i}^{-}\bar{c}n;\mathring{t})$, Proposition 4.2.
$*$ Create a new trial estimate of the mixture $f_{n+1|n}(\Theta|d(\mathring{t}))$ by replacing the splitted factor $GiW(\bar{i}\bar{c}n;\mathring{t})$ by the pair $GiW(\bar{i}^{+}\bar{c}n;\mathring{t})$, $GiW(\bar{i}^{-}\bar{c}n;\mathring{t})$ of new ones in the selected component with index $\bar{c}$. It splits the component $\bar{c}$ in two and increases $\mathring{c} = \mathring{c} + 1$.
$*$ Apply the flattening operation to $f_{n+1|n}$, see Proposition 3.1, to get a trial prior pdf $f_{n+1|n}(\Theta) = \prod_{c\in c^*}\prod_{i=1}^{\mathring{d}} GiW(ic(n+1|n);0)$.
$*$ Make the trial estimation with this prior and evaluate the corresponding value of predictive pdf.
$*$ Set the flag of the split factor to the value not admissible, take all intermediate evaluations as trial ones and go to the step •• if no prediction improvement is recorded and there are admissible factors. Stop if no admissible factor remains.
$*$ Set $f_{n+1} = f_{n+1|n}(\Theta)$ if an increase of of predictive pdf is recorded, set $n = n + 1$, go to •.

5 Illustrative Example

The algorithm has been thoroughly tested and found satisfactory. The example just illustrates its behaviour. In it, $\mathring{t} = 1000$ records of two-dimensional data from a static normal mixture were simulated. Mean values of components together with iso-probability ellipsoids encircling probability 0.75 are shown on left-hand side of the 1st figure. Data samples are on its right hand side. A bit denser leftist and rightist clusters correspond to the choice $\alpha = [0.3, 0.2, 0.2, 0.3]$. The subsequent figures show centres and the same ellipsoids found during three iterations of the proposed algorithm.

6 Conclusions

The proposed novel algorithm of initiating mixture estimation was found satisfactory. Its applicability to dynamic mixtures is its exceptional feature. Naturally, it is not able to cope with all cases and should be combined with other techniques containing random search steps. Its applicability to non-normal components is worth noticing. The adopted factorised description of components opens a way for describing mixed discrete-continuous mixtures. Also, the algorithm generates potentially mixtures with factors that are common to several components. This feature is important both for physical modelling and parsimonity of the final description.

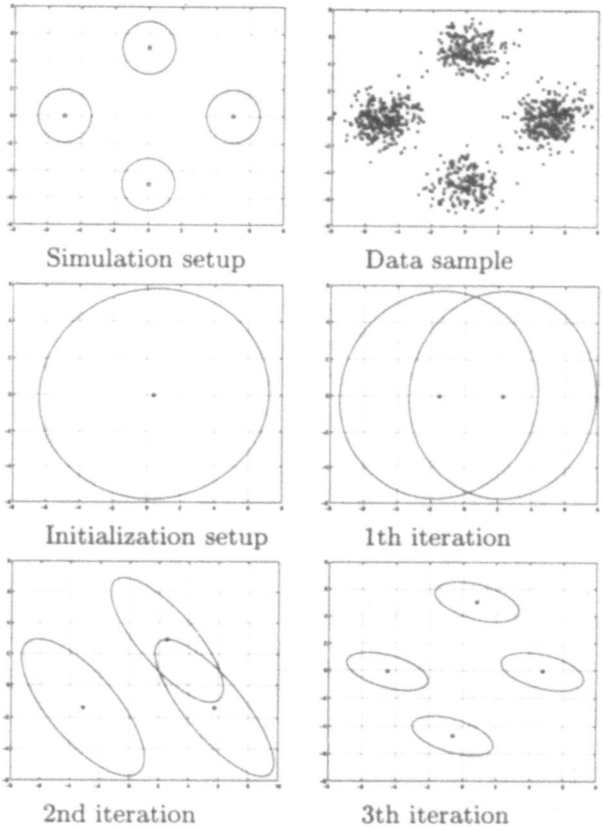

References

[1] R. Murray-Smith and T. Johansen, *Multiple Model Approaches to Modelling and Control.* London: Taylor & Francis, 1997.

[2] V. Peterka, "Bayesian system identification," in *Trends and Progress in System Identification* (P. Eykhoff), Oxford: Pergamon Press, 1981.

[3] M. Kárný, et al, "Quasi-Bayes estimation applied to normal mixture," in *3rd IEEE Workshop CMP* (J. Rojíček et al), ÚTIA, 1998.

[4] S. Kullback and R. Leibler, "On information and sufficiency," *Annals of Mathematical Statistics*, vol. 22, 1951.

Factorized EM Algorithm for Mixture Estimation

Ivan Nagy*, Petr Nedoma and Miroslav Kárný † 1

*FD ČVUT, Konviktská 20, Prague 1, Czech Republic e-mail: nagy@utia.cas.cz †ÚTIA, AV ČR, Pod vodárenskou věž 4, Prague 8, Czech Republic e-mail: nedoma@utia.cas.cz, school@utia.cas.cz.

Abstract

A classical version of the EM algorithm is considered in the paper. Its numerical properties are improved using factorized algorithms for maximization in M step of the algorithm. The results are illustrated on simulated examples.

1 Introduction

Expectation and Maximization (EM) algorithm is well known and widely spread procedure for parameter estimation in cases when some of the modeled data can and some cannot be measured [1]. The parameter estimates are computed iteratively with monotonic convergence guaranteed [2].

Estimation of mixture models [3] is a well known theoretical application of the EM algorithm. Here, the measured data are represented by the output of the system. The internal pointers, indicating the active components of the mixture at respective time moments t, are not known. They correspond to the unmeasured part of the data.

Mixture models have a wide practical use as they have universal approximation property: they can be interpreted as approximators of pdfs or approximators based on radial basis functions. In practical applications, they may serve for describing rather large systems with many output variables and rather high number of components. In such cases, however, numerical instabilities occur while manipulating estimates of covariances that are extensively used in each iterative estimation step. These estimates of covariance matrices can easily loose their theoretical positive definiteness. Then, the subsequent estimation produces unbounded values.

The aim of this paper is to utilize numerical procedures adopted from factorized on-line estimation [4, 5] and to combine them with EM algorithm to improve its numerical properties. Indeed, both the single tasks are known and successfully used for a long time. It is astonishing that they were not coupled, yet.

The mostly used way, how to reduce the instabilities mentioned, is to add a small diagonal to the estimated covariances. This solution helps, but it adds an exogenous element to the computations, even if negligible. The solution suggested adds nothing and so it is purer and is hoped to give generally better results.

2 EM Algorithm

Let us consider a system generating at time t data $d_t = [y_t, c_t]$. The part y_t of the data is measured and thus known, the rest c_t cannot be measured and thus it is unknown to us. We suppose to have a model of the system, but this model depends also on the unknown part of the data c_t. Consequently, it cannot be directly used in estimation for constructing likelihood function. Instead, the expectation of the logarithmic likelihood taken with respect to unknown data c_t is used for parameter estimation.

2.1 General task

The model of the system is supposed in the form of conditional probability density function (pdf)

$$f(d_t|d_1,\ldots,d_{t-1},\Theta) = f(d_t|\Theta) = f(y_t,c_t|\Theta), \quad \text{where}$$

d_t is the complete data pair $d_t = [y_t, c_t]'$,
y_t are measured data,
c_t are unmeasured data and
Θ is a vector of all unknown parameters of the pdf.

The data y_t are measured at discrete time instances $t = 1, 2, ..., \mathring{t}$. By Y, we denote a vector of all collected measured data $Y = [y_1, y_2, ..., y_{\mathring{t}}]$. Similarly we denote a vector of unmeasured part of data c_t by C and complete data by D. The likelihood is

$$L(\Theta, Y, C) = \prod_{t=1}^{\mathring{t}} f(y_t, c_t|\Theta).$$

As the likelihood contains unknown variables C, it is a random variable. The main trick of the EM algorithm is that it minimizes expectation of its logarithm. The expected value of the log-likelihood,

[1]The work has been partially supported by EC IST-99-12058

with respect to C, given data Y and current estimate of the parameters Θ, denoted by $\hat{\Theta}$, is

$$\begin{aligned}\mathcal{L}(\Theta,Y) &= E[\log L(\Theta,Y,C)|Y,\hat{\Theta}] = \\ &= \sum_{C\in C^*} \log L(\Theta,Y,C) f(C|Y,\hat{\Theta}),\end{aligned}$$

where the summation is performed over all possible sequences of unmeasured data $C = [c_1, c_2, ..., c_{\mathring{t}}] \in C^*$; C^* denotes the sample space of the random vector C. The estimation of parameters Θ is based on maximization of this expectation of log-likelihood. The maximum is searched iteratively. In each iteration step, the expectation is computed on the basis of old parameter estimates (E step). Then the resulting function of Θ is maximized giving new parameter estimates (M step).

2.2 Normal mixture model

Let us specify the system under consideration in the following way. It consists of several subsystems, on which the set of data y_t, $t = 1, 2, ..., \mathring{t}$ is measured. Each subsystem is modeled by Gaussian pdf

$$f_c(y_t|\theta_c) \sim \mathcal{N}_{y_t}(\mu_c, R_c), \quad c \in c^*, \text{ where}$$

$c \in c^* = \{1, 2, ..., \mathring{c}\}$ is a label of a particular submodel (mixture component),
$\theta_c = [\mu_c, R_c]'$ is a sub-model parameter with μ_c and R_c being the corresponding mean and covariance of y_t, respectively.

We suppose that at each time instant, only one of the models is valid and we take the sequence $c_t \in c^*$ for $t = 1, 2, ..., \mathring{t}$ as pointers to the active subsystem at each time instant t. As we do not know which model is currently active, the individual terms of the sequence are random. They correspond to the unmeasured system output from the general EM algorithm formulation. We suppose them to be independent of $y_\tau, \tau \neq t$ at time t and of the component parameters $\theta_c, c \in c^*$. Their probabilities are

$$f(c_t|\Theta) = \alpha_{c_t}, \quad c_t \in c^*, \quad t = 1, 2, ..., \mathring{t}, \text{ where}$$

$\Theta = [\theta', \alpha']'$,
$\theta = [\theta_1', \dots, \theta_{\mathring{c}}']'$ are parameters of the components,
$\alpha = [\alpha_1, \dots, \alpha_{\mathring{c}}]$ are component probabilities.

With this assumption, the system model is equivalent to a mixture with $\mathring{c}$ static Gaussian components, each component having the probability α_c, $c \in c^*$, to be active at time $t = 1, 2, ..., \mathring{t}$. The complete description of such mixture model at time t is given by the joint pdf

$$\begin{aligned}&f(y_t, c_t|d_1, \dots, d_{t-1}, c_1, \dots, c_{t-1}, \Theta) = \\ &= f(y_t|c_t, \Theta) f(c_t|\Theta) = f_{c_t}(y_t|\theta)\, \alpha_{c_t}.\end{aligned}$$

The pdf $f_{c_t}(y_t|\theta)$ models the c_t-th component of the mixture. We have used its assumed independence on α and independence of pdfs $f(c_t|\Theta)$ on θ.

3 EM Algorithm for Normal Mixtures

Our task is to identify the parameters of the mixture μ_c, R_c, α_c for $c \in c^*$. For the solution of the task, we will follow the standard EM algorithm with the only exception that for the numerically sensitive parts of the data covariance estimation we will use factorized algorithms, worked up in connection with the Bayesian approach to identification [5].

3.1 Distribution of unmeasured data

To be able to take the expectation of the log-likelihood, we need the probability $f(C|Y, \Theta)$ of all unknown pointers conditioned not only on parameters Θ but also on measured data y_t, $t = 1, 2, ...\mathring{t}$. This joint probability is the product

$$f(C|Y, \Theta) = \prod_{t=1}^{\mathring{t}} f(c_t|y_t, \Theta)$$

in which we have used the independency of the pointer c_t of other pointers c_τ and data y_τ for $t \neq \tau$. The particular probability $f(c_t|y_t, \Theta)$ can be viewed as a prediction of the activity of a component at time t based on the measured data. This probability is determined by Bayes rule

$$f(c_t|y_t, \Theta) \propto f(y_t|c_t, \Theta) f(c_t|\Theta) = f_{c_t}(y_t|\theta)\alpha_{c_t}.$$

It uses the models of components and their probabilities. In accordance with EM algorithm, the expectation defined by this pdf is evaluated with the current parameter estimates substituted for the unknown parameter values. Using the measured data Y we can compute values of the needed pdf $f(c_t|y_t, \Theta = \hat{\Theta})$ for all t and at each time t for all possible values $c \in c^*$ of random variable c_t. To stress that the results are just numbers - probabilistic weights for expectation and consequently for weighted data accumulation, we denote them

$$w_{c,t} = f(c_t = c|y_t, \hat{\Theta}),$$

which is the probability, that the component labeled by c is active at time t.

3.2 E step

The expectation of log-likelihood $\mathcal{L}(\Theta, Y) =$

$$= \sum_{C\in C^*} \left[\sum_{i=1}^{\mathring{t}} \log[\alpha_{c_i} f_{c_i}(y_i|\theta)] \prod_{j=1}^{\mathring{t}} f(c_j|y_j, \hat{\Theta}) \right] =$$

$$= \sum_{c=1}^{\mathring{c}} \sum_{t=1}^{\mathring{t}} \log[\alpha_c f_c(y_t|\theta)] w_{c,t} =$$

$$= \underbrace{\sum_{c=1}^{\mathring{c}} \sum_{t=1}^{\mathring{t}} \log[\alpha_c] w_{c,t}}_{\mathcal{L}_1} + \underbrace{\sum_{c=1}^{\mathring{c}} \sum_{t=1}^{\mathring{t}} \log[f_c(y_t|\theta)] w_{c,t}}_{\mathcal{L}_2} .$$

The first equality is implied by the fact, that probabilities sum to unity.

3.3 M step

This step consists in maximization of the expectation of log-likelihood $\mathcal{L}(\Theta, Y)$ constructed in the previous step. The maximization can be performed independently in two steps. The first term $\mathcal{L}_1$ gives the estimate of α. For explicit maximization of the second term, we suppose the component models in the form of static and normally distributed pdfs

$$f_c(y_t|\theta) \propto |R_c|^{-\frac{1}{2}} \exp\left\{-\frac{1}{2}(y_t - \mu_c)' R^{-1} (y_t - \mu_c)\right\}$$

for $c = 1, 2, ...\mathring{c}$, where
μ_c is the expectation of the cth component,
R_c is the covariance matrix of the cth component.

Then the second term $\mathcal{L}_2$ gives estimates of μ_c and R_c for $c = 1, 2, ..., \mathring{c}$. The classical results can be derived by the method of Lagrange multipliers. In their simplified description, the symbols without the sign ˆ mean estimation results obtained at previous iteration step. For $c = 1, 2, ..., \mathring{c}$, the formulae are

$$w_{c,t} = \frac{|R_c|^{-0.5} \exp\left\{-\frac{1}{2}(y_t - \mu_c)' R_c^{-1} (y_t - \mu_c)\right\}}{\sum_{\tilde{c}} |R_{\tilde{c}}|^{-0.5} \exp\left\{-\frac{1}{2}(y_t - \mu_{\tilde{c}})' R_{\tilde{c}}^{-1} (y_t - \mu_{\tilde{c}})\right\}}$$

$$\hat{\alpha}_c = \frac{1}{\mathring{t}} \sum_{t=1}^{\mathring{t}} w_{c,t}, \qquad \hat{\mu}_c = \frac{1}{\mathring{t}\hat{\alpha}_c} \sum_{t=1}^{\mathring{t}} y_t w_{c,t},$$

$$\hat{R}_c = \frac{1}{\mathring{t}\hat{\alpha}_c} \sum_{t=1}^{\mathring{t}} (y_t - \mu_c)(y_t - \mu_c)' w_{c,t}.$$

These formulas are simple and clearly interpretable. The explicit form of the weights $w_{c,t}$ shows the source of numerical troubles. If an iteration fails to give numerically a positive definite $\hat{R}_c$ then the next iteration breaks down. It reflects that the maximized function $\mathcal{L}_2$ is not concave any more and its maximizer has infinite-valued entries.

4 Factorized Estimation

The evaluation of $\hat{R}_c$ whose inversion and determinant are used in weighting is bottle neck of the classical formulas. To avoid difficulties mentioned, we will update not the whole matrix but its LD factorization. The formula for computing weights shows, that the inverse of the matrix $\hat{R}_c$ is needed. That is why we will compute directly the LD factorization of the inverse of $\hat{R}_c$.

The unnormalized version of the estimate

$$\tilde{R}_c = \sum_{t=1}^{\mathring{t}} (y_t - \mu_c)(y_t - \mu_c)' w_{c,t}$$

can be simply updated recursively

$$\tilde{R}_{c,t} = \tilde{R}_{c,t-1} + (y_t - \mu_c)(y_t - \mu_c)' w_{c,t} = \tilde{R}_{c,t-1} + z_{c,t} z'_{c,t}$$

where $z_{c,t} = (y_t - \mu_c)\sqrt{w_{c,t}}$, $\tilde{R}_{c,0} = 0$, $\hat{R}_c = \frac{\tilde{R}_{c,\mathring{t}}}{\mathring{t}\hat{\alpha}_c}$.

As we have stated, we need to update LD factorization of the inverse of $\tilde{R}_c$. We drop the fixed subscript $_c$ and introduce factorizations of inversions of $\tilde{R}_{t-1}$ and $\tilde{R}_t$

$$\tilde{R}_{t-1}^{-1} = L\,D\,L', \; \tilde{R}_t^{-1} = \bar{L}\bar{D}\bar{L}', \text{ where}$$

$L, \bar{L}$ are lower triangular matrices with unit diagonals and
$D, \bar{D}$ are diagonal matrices with positive diagonals.

All involved square matrices have dimension n.

With these notations, the recursion for $\tilde{R}$ becomes

$$\begin{aligned} \bar{L}\bar{D}\bar{L}' &= \left[(L\,D\,L')^{-1} + z_t z_t'\right]^{-1} = \\ &= L - \left[D\frac{Dff'D}{1 + f'Df}\right] L', \text{ with } f = L'z_t. \end{aligned}$$

The definition of LD decomposition implies that the expression in brackets [ ] can be decomposed analytically into the LD form, say $H\bar{D}H'$, see [4]. The following algorithm provides it

for $i = n, n-1, ..., 1$

$$\sigma^2 = 1 + \sum_{k=i}^{n} D_k f_k^2$$

$$\bar{D}_{i,i} = D_{i,i} \frac{\sigma_{i+1}^2}{\sigma_i^2}$$

$$H_{i,i} = 1$$

for $j = i+1, i+2, ..., n$

$$H_{j,i} = -\frac{f_i f_j}{\sigma_{i+1}^2} D_{j,j}$$

end for j

end for i

We attach to this algorithm the multiplications $f = L'z_t$ and $\bar{L} = LH$ and organize carefully manipulations with individual entries. It gives the overall very compact algorithm that replaces L, D by $\bar{L}, \bar{D}$ after including new data vector z_t:

```
s = 1
for j = N to 1
  h = z_j
  for i = j + 1 to N
    h = h + L_ij
  end j
  a = h/s
  v_j = D_j * h
  r = s + v_j * h
  D_j = s/r * D_j
  s = r
  for i = j + 1 to N
    p = L_ij - v_i * a
    v_i = v_i + L_ij * v_j
    L_ij = p
  end i
end j
```

5 Illustrative Example

The EM algorithm and its factorized version have been thoroughly tested. The following example illustrates advantages of the proposed factorized modification. A simulated data sample of the length of 1000 data items was generated from a static normal mixture with 10 components. The mean values of the components together with iso-probability ellipsoids encircling probability 0.75 are shown on the left-hand side of the 1st figure. The data sample is plotted on its right hand side. The weights of the components grow linearly from the first one to the last as it can be seen from the plot. The estimation setup is on left-hand side of the 2nd figure. The centers of the initial ellipsoid differ from the simulation ones by a random value. Uniform initial estimates of component weights were chosen. The standard EM algorithm broke down after the 1st iteration step when at least one estimate of R_c lost its definiteness. When running the square root version of EM algorithm, no problems were encountered. The estimated mixture after 20 iterations is shown in the right-hand side of the figure 2.

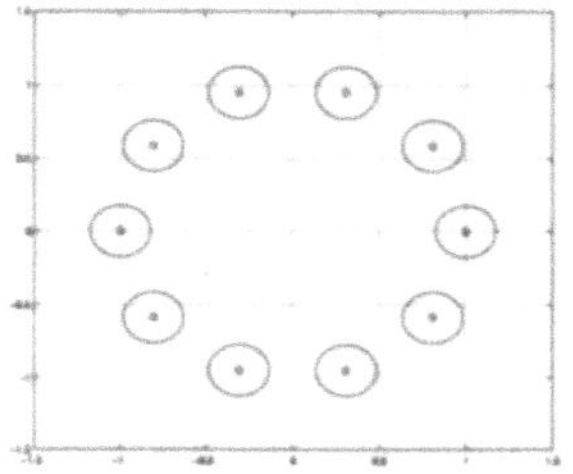

Simulaton setup

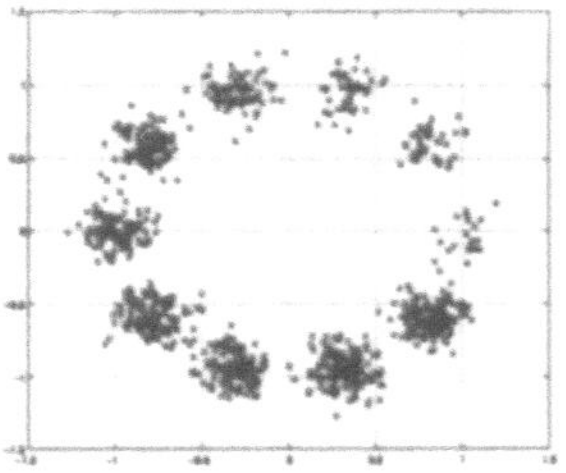

Data sample

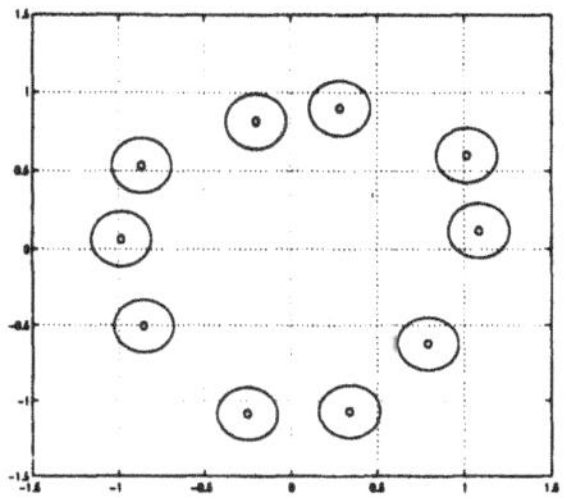

Estimation setup

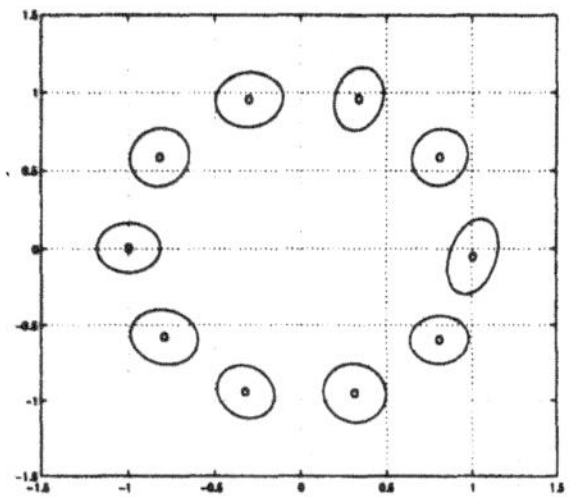

Estimation result

6 Conclusions

A new numerical solution to the estimation of normal mixture by EM algorithm has been proposed. The algorithm accepts original estimation of component probabilities α and means θ. Instead of the estimate of the noise covariances R, it updates LD decomposition of its inversion. It simplifies evaluation of weights and, first of all, it prevents these estimates from loosing their positive definiteness. The numerical stability of the proposed algorithm in comparison with the "rough" solution has been demonstrated on simulated examples.

The approach can be extended to the case of dynamic components when the expected value of the predicted data is a linear combination of estimated regression coefficients and known regression vector. The overall description is made in probabilistic terms. It opens a way of including a nontrivial prior probability on estimated parameters giving maximum a posterior probability estimates. The interpretation can be also suppressed and the algorithm used for estimating normal radial basis neural network. These options will be elaborated in detail in foreseen future.

References

[1] A.P.Dempster, N.M.Lair, and D.B.Rubin, "Maximum-likelihood from incomplete data via the em algorithm", *J.Royal Statist. Soc. Ser. B.*, vol. 39, 1977.

[2] C.F.J.Wu, "On the convergence properties of the em algorithm", *The Annals of Statistics*, vol. 11, pp. 95–103, 1983.

[3] R.Redner and H.Walker, "Mixture densities, maximum likelihood and the em algorithm", *SIAM Review*, vol. 26, 1984.

[4] G.J. Bierman, *Factorization Methods for Discrete Sequential Estimation*, Academic Press, New York, 1977.

[5] V. Peterka, "Bayesian approach to system identification", in *Trends and Progress in System Identification*, P. Eykhoff, Ed., pp. 239–304. Pergamon Press, Oxford, 1981.

Number of Components and Initialization in Gaussian Mixture Model for Pattern Recognition

Pavel Paclík *, Jana Novovičová †

*Pattern Recognition Group, TU Delft, P.O. Box 5046, 2600 GA, Delft, The Netherlands, pavel@ph.tn.tudelft.nl, Faculty of Transportation Sciences, Czech Technical University, Na Florenci 25, 110 00, Prague, Czech Republic † Institute of Information Theory and Automation, Academy of Sciences of the Czech Republic, Pod vodarenskou vezi 4, 182 08, Prague, Czech Republic, novovic@utia.cas.cz

Abstract

Number of components and initial parameter estimates are of crucial importance for successful mixture estimation using Expectation-Maximization (EM) algorithm. In the paper a method for the complete mixture initialization based on a product kernel estimate of probability density function is proposed. The mixture components are assumed here to correspond to local maxima of optimally smoothed kernel density estimate. The gradient method is used for local extrema finding. Then, local extrema are grouped together to form component candidates and these are merged by the 4hierarchical clustering method. Finally, the initial mixture parameters are estimated. A comparison to scale-space approaches for finding of the number of components is given on examples.

1 Introduction

In Pattern Recognition finite mixture models have been used to model distributions where the observations arise from separate groups, but individual membership is unknown [10]. The mixture model problem involves the estimation of unknown parameters from a given set of observations. This problem is normally solved by using the EM algorithm [3], which is based on maximizing the likelihood of the observations. Unfortunately, the EM algorithm is an iterative procedure that depends heavily on the selected initial parameter estimates as only the local maxima convergence is guaranteed. Moreover, the number of component densities is assumed to be known or given, in the traditional EM algorithm. There have been already presented several different approaches for the EM initialization.

Ueda and Nakano [11], for example, used deterministic annealing method (DAEM) to overcome the initialization problem. They have introduced posterior parametrized by a parameter β which could be interpreted as an annealing temperature. The algorithm is started at β which will smooth the energetic function. EM algorithm is then executed repeatedly in so-called annealing loop with β slowly increasing (decreasing temperature). This helps to overcome local maxima problem and also makes the EM algorithm robust to randomly chosen initial conditions. Anyway, the number of components must be supplied as an algorithm parameter.

Scale-space theory [7] has been widely used in the cluster analysis community for estimating of the number of clusters in the data [2, 6]. It is based on idea that at lowest scale each data point forms a cluster while at the largest level the only one cluster will be observed in the data.

Kehtarnavaz and Nakamura [5] have published a solution based on the scale-space approach estimating both component count and initial parameters in the EM preprocessing step. Their method is based on the idea of potential field function computed from the data and parametrized by scale σ. The number of components corresponds to the local minima count of the field function at the particular scale. True number of components in the data is estimated as the count persisting for the longest time (inter-cluster *lifetime* criterion). The initial component parameters are then obtained by finding the optimal scale according to the intra-cluster criterion called *drift speed* and by the nonlinear numerical optimization method.

There have also been published other approaches for the estimation of the number of components arising from the mixture model. For example Celeux and Soromenho [1] published method based on normalized entropy criterion (NEC) and Tenmoto *et al.* [9] suggested to use Minimum Description Length (MDL) criterion for the same purpose. Mixture initialization is random in the first case and based on fuzzy c-means method in the second one.

Grim *et al.* [4] proposed a method starting with optimally smoothed kernel estimate which is subsequently optimally weighted. Local maxima, found on weighted estimate, are then taken as component centers. The algorithm gives both component count and initial parameters of mixture components.

2 Mixture Model

A finite mixture model is a probabilistic model of the form

$$p(\mathbf{x}|\Theta) = \sum_{k=1}^{K} w_k p(\mathbf{x}|\theta_k), \quad \mathbf{x} \in R^D \tag{1}$$

where K is the number of mixture components, w_k are mixing proportions ($w_k \geq 0, \sum_{k=1}^{K} w_k = 1$), and $p(\mathbf{x}|\theta_k)$ are component densities specified by a parameter vector θ_k. Θ denotes the set of parameters $\{w_1, ..., w_K, \theta_1, ..., \theta_K\}$.

3 Algorithm Description

Algorithm 1 Estimation of number of components and their parameters for initialization of EM algorithm for mixtures

1: **input:** D-dimensional data $\mathbf{x}_i$, $i = 1, ..., N$
2: **smoothing:** find h_j, $j = 1, ..., D$ maximizing cross-validated log-likelihood function (by EM algorithm)
3: **local maxima searching:** find local maxima $\mathbf{m}_i$, $i = 1, ..., N$ corresponding to each data point $\mathbf{x}_i$ (by BFGS gradient method)
4: **grouping:** group together near local maxima $\mathbf{m}_i$, $i = 1, ..., N$ and return the list of component candidates $\mathbf{g}_k$, $k = 1, ..., N_g$ together with corresponding weight estimates $\hat{w}_k$, $N_g \leq N$
5: **postprocessing:** merge component candidates $\mathbf{g}_k$ and return merged list of components $\mathbf{c}_l$, $l = 1, ..., N_c$, $N_c \leq N_g$
6: **parameter estimation:** use mapping $\mathbf{x}_i \to \mathbf{c}_l$, $i = 1, ..., N$, $l = 1, ..., N_c$ to estimate component parameters $\hat{\mu}_l$ and $\hat{\Sigma}_l$ from the data
7: **output:** component count N_c and component parameters $\hat{w}_l$, $\hat{\theta}_l = \{\hat{\mu}_l, \hat{\Sigma}_l\}$, $l = 1, ..., N_c$

Given a set of N observations $\{\mathbf{x}_1, \mathbf{x}_2, ..., \mathbf{x}_N\}$, the optimally smoothed product Gaussian kernel density estimate is computed in the initialization step. More detailed description of the product kernel density estimate and its smoothing by the EM-based maximization of the cross-validated log-likelihood is given in [8].

Then, a hill-climbing procedure is run from each data point finding the corresponding local maximum $\mathbf{m}_i$ of the probability density estimate. Quasi-Newton method has been used - namely its *Broyden-Fletcher-Goldfarb-Shanno* mutation (BFGS).

In the next step, the obtained local maxima are grouped together by thresholding the distance $||\mathbf{H}^{-1}(\mathbf{m}_i - \mathbf{m}_j)||$, $i \neq j$, between the local maxima. The symbol $\mathbf{H}$ stands for the diagonal matrix of smoothing parameters. The threshold value t_g is set according to the analysis of distribution of local maxima distances. The histogram of local maxima distances contains a distinct peak for very small distance values. It is caused by small differences between numerical solutions of the hill climbing procedure. The aim is to identify these small distances by thresholding t_d and grouping the corresponding points together. Generally, the value t_g should not exceed the smallest inter-component distance. The actual value of the t_g threshold does not seem to influence the result much due to the following postprocessing step. We have used the same setting $t_g = 1.0$ in all presented experiments. The grouping reveals the information about the number of points which have ended-up in the particular local maximum. This information allows us to estimate the weight $\hat{w}_k$ of each *component candidate*.

The following step is the *postprocessing* based on the agglomerative hierarchical clustering algorithm. The aim is to merge close component candidates together. Mutual distances are computed for all candidates and the closest pair $\mathbf{c}_i$,$\mathbf{c}_j$ is replaced by the new component $\mathbf{c}_{new}$ reflecting the weights of its both "parents".

$$\mathbf{c}_{new} = \frac{\hat{w}_i \mathbf{c}_i + \hat{w}_j \mathbf{c}_j}{\hat{w}_i + \hat{w}_j} \tag{2}$$

The postprocessing is stopped as soon as

$$\min(\hat{w}_i, \hat{w}_j) > t_m \quad \forall\, i, j \quad \text{such that} \quad i \neq j. \tag{3}$$

Thus, all the components have weights higher or equal t_m, after the merging process. The threshold t_m prevents us from mixtures with many negligible components and is set-up according to the user's needs. We have used $t_m = 0.05$ in our experiments.

As it is known for each data point into which component it belongs we use this information for estimation of component weights, means and covariance matrices. The estimated values may be then used for initialization of the EM algorithm for mixture estimation.

4 Experimental Results

4.1 Experiment 1

The first experiment uses artificial dataset of 600 data points generated from three-component Gaussian mixture as published in [5]. If the EM algorithm for mixture estimation is run using initial parameters defined in [5] it will find a good solution, but the component count must be supplied by user. In figure 1 we display the results of presented initialization algorithm - estimated number of components and their initial parameters. *Dashed ellipses* denote the Gaussian components used for data generation. *Thick ellipses* reflect the results of presented initialization algorithm. It has found three components and estimated their means and covariances. *Thin lines* then

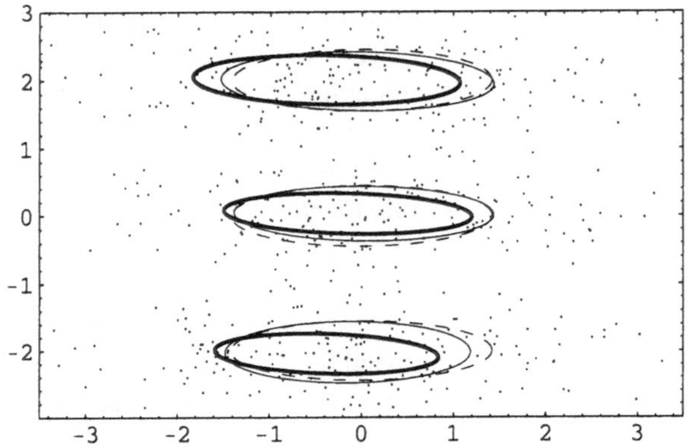

Fig. 1. Estimation of number of components and their initial parameters for 2D three-component mixture

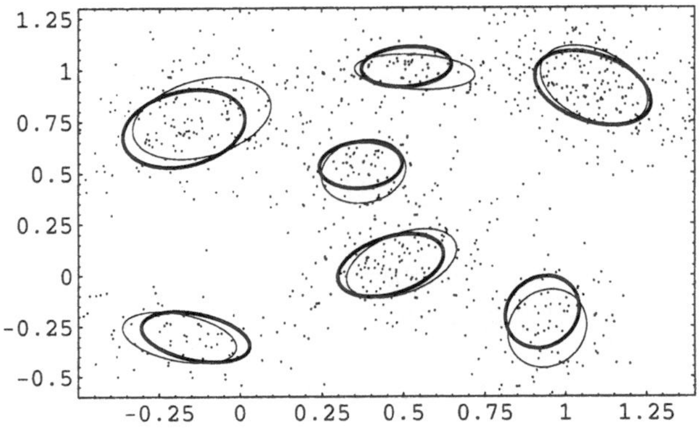

Fig. 2. Results of EM mixture estimation initialized by the algorithm on data from [4]

show the result of EM algorithm for mixture estimation initialized by our algorithm.

4.2 Experiment 2

In the second experiment we work with mixture used by Grim *et al.* in [4]. It contains seven components in 2D. We have generated 1000 data points from the mixture and ran initialization algorithm followed by EM algorithm for mixture estimation. The results may be seen on figure 2. *Thin lines* are again the final results obtained by EM algorithm for mixture estimation initialized by number of components and parameters (*thick ellipses*) estimated by the initialization algorithm.

Let us briefly discuss the difference between method of Grim *et al.* and our approach. Grim *et al.* firstly optimally smooth and then optimally weight the kernel estimate. Local extremes are then found on the result. The main idea is that optimal weighting reveals the most important kernels (component candidates). On the other hand, our approach uses optimally smoothed estimate for local extremes search directly. Therefore, the key difference is in the kind of components both methods return. The method of Grim *et al.* finds more components (namely 28 on this mixture) reflecting also local anomalies in the data. Our method returns the most distinct components showing the very basic structure of the data.

4.3 Comparison to scale-space approach

In this section we would like to show the relationship between presented algorithm and scale-space approaches to component count estimation. The scale-space approach is based on empirical fact that the salient structures are those persisting on broad range of different scales [7]. In the case of component count estimation we observe the number of clusters in the data on many different scales and propose *the true one* as that with the longest *lifetime*. It's natural and robust but also extremely computationally intensive method.

We think that the optimal scale may be chosen *apriori* by some data-driven criterion. In the case of presented algorithm the smoothing parameter may be viewed as the measure we use for the data observation. It is selected to maximize the cross-validated log-likelihood function.

An example of this approach for 1D data generated from two-component mixture is given at figure 3. The two-component solution persists for the longest interval of scales - it would be chosen by the scale-space approach. The maximum likelihood estimate of smoothing is drawn as $\mathbf{h}_{MLE}$ point.

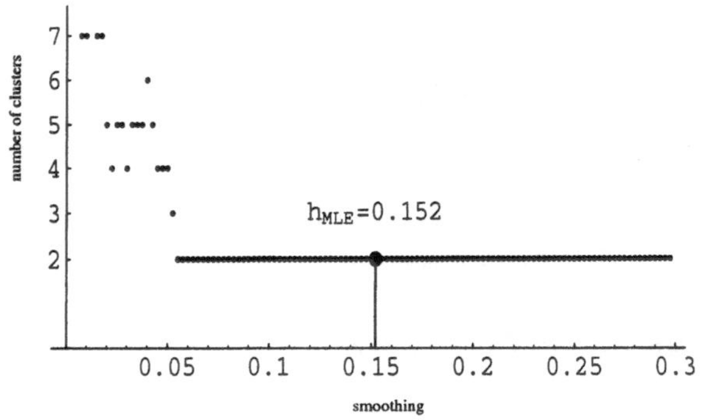

Fig. 3. Estimated component count for a range of scales

In the scale-based algorithms the only measure is used even for multivariate data. In our approach each smoothing parameter $h_j, j = 1, ..., D$ of the product kernel may be viewed as the individual measure for the corresponding dimension.

Let us show the scale-space analogy for the 2D data (three-component mixture) from experiment 1 (section 4.1). Estimated number of components is shown for several smoothing vectors $\mathbf{h}$ on figure 4. The black point denotes the maximum likelihood estimate $\mathbf{h}_{MLE} = \{0.489, 0.045\}$ taken again as the apriori chosen scale. Finally, let us discuss the scale selection for

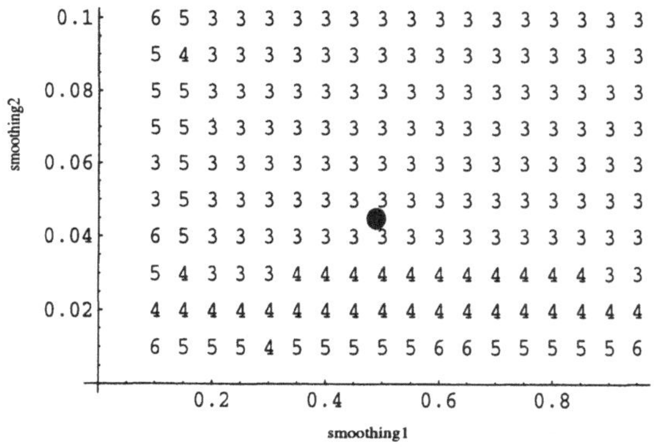

Fig. 4. Scale-space and ML estimate of smoothing for the data from experiment 1

seven-component mixture (experiment 2, section 4.2). The figure 5 shows again the scale-space and maximum likelihood estimate of smoothing parameters. The maximum likelihood estimate of smoothing $\mathbf{h}_{MLE} = \{0.0067, 0.0057\}$ is denoted by the black point. It again fell into the region with the longest lifetime.

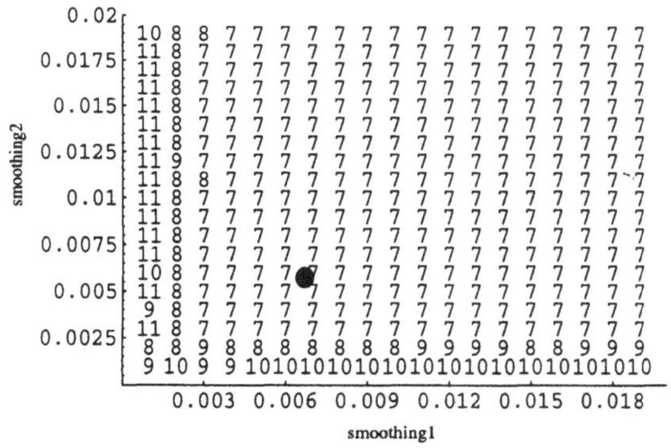

Fig. 5. Scale space and ML estimate of smoothing for the data from experiment 2

4.4 Speed consideration

The algorithm has been implemented in the C language and tested on 450 MHz Pentium III processor. The time spent on experiments follows in the table. Initialization algorithm (init.alg.) column covers local extrema search, grouping, postprocessing and parameter estimation. It can be seen that the most time demanding procedure is the EM algorithm for smoothing of the kernel estimate.

data	samples	smoothing	init.alg.	sum
exp 1	600	47 sec / 118 iter.	23 sec	70 sec
exp 2	1000	110 sec / 103 iter.	13 sec	123 sec

5 Conclusion and Discussion

In the paper the algorithm for estimation of both component count and initial mixture parameters has been presented. The algorithm results serve for the initialization of EM algorithm for the mixture estimation. The main idea is that components coincide with local maxima of optimally smoothed kernel estimate. The algorithm investigates a general structure of the data and tries to roughly estimate the component parameters. For large datasets, a randomly chosen subset of the original data may offer an effective solution.

For the product kernel estimate the optimal smoothing corresponds to the measure in scale-based approaches. It has been shown on examples, that the maximum likelihood estimate of smoothing may be taken as apriori chosen scale.

Acknowledgements

This work has been partially supported by the Grant of the Ministry of Education No. VS 96063 of the Czech Republic and the Complex research project of the Academy of Sciences of the Czech Republic No. K1075601.

References

[1] G. Celeux and G. Soromenho. An entropy criterion for assessing the number of clusters in a mixture model. *Journal of Classification*, 13:195–212, 1996.

[2] S. V. Chakravarthy and J. Ghosh. Scale-Based Clustering Using the Radial Basis Function Network. *IEEE Trans.on Neural Networks*, 7(5):1250–1261, 1996.

[3] A. Dempster, N. Laird, and D. Rubin. Maximum likelihood from incomplete data via EM algorithm. *J. Royal Stat.Soc.* vol.39, pp.1-38, 1977.

[4] J. Grim, J. Novovičová, P. Pudil, P. Somol, and F. Ferri. Initialization normal mixtures of densities. In *Proceedings of the 14th ICPR*, pages 886–890, Australia, 1998.

[5] N. Kehtarnavaz and E. Nakamura. Generalization of the EM algorithm for mixture density estimation. *Pattern Recognition Letters*, 19:133–140, 1998.

[6] R. Kothari and D. Pitts. On finding the number of clusters. *Pattern Recognition Letters*, 20:405–416, 1999.

[7] T. Lindeberg. Scale-space theory: A basic tool for analysing structures at different scales. *Journal of Applied Statistics*, 21(2):225–270, 1994.

[8] P. Paclík, J. Novovičová, P. Pudil, and P. Somol. Road Sign Classification using Laplace Kernel Classifier. *Pattern Recognition Letters*, 21(13-14):1165–1173, 2000.

[9] H. Tenmoto, M. Kudo, and M. Shimbo. MDL-Based Selection of the Number of Components in Mixture Models for Pattern Recognition. In *Lecture Notes in Computer Science 1451: Advances in Pattern Recognition*, pages 831–836, 1998.

[10] D. Titterington, A. Smith, and U. Makov. *Statistical analysis of finite mixture distributions*. John Wiley& Sons: Chichecter, Singapore, New York, 1985.

[11] N. Ueda and R. Nakano. Deterministic annealing EM algorithm. *Neural Networks*, (11):271–282, 1998.

Parallel Factorised Algorithms for Mixture Estimation

Milan Tichý, Bohumil Kovář * [1]

*UTIA, AV ČR, Pod vodárenskou věží 4, Prague 8, Czech Republic

Abstract

This paper describes software aspects of an advisory system based on finite-mixture estimation. Factorised algorithms have been designed. Parallelism is used as a principal approach to acceleration of learning and processing phases. A coarse grain granularity parallelism is used in the first phase of the work. The Parallel Virtual Machine (PVM) is used for parallel implementation.

1 Introduction

This work has been accomplished as a part of the ProDaCTool (Decision Support Tool for Complex Industrial Processes Based on Probabilistic Data Clustering) project. The aim of the project is to develop a probabilistic advisory system. The algorithms should be robust and fast and should be amenable to many applications, such as industry, transportation, glass manufacture, and medicine.

This paper deals with software aspects of the system. Matlab is used as the development environment but the applications must not rely on it. ANSI C code independent of Matlab has therefore been developed. Thus an ANSI C library emulating Matlab Application Program Interface (API) must be developed and appropriate data conversions between Matlab and ANSI C must be provided. We will discuss problems encountered during coarse grain granularity parallelization in this paper. We decided to use the Parallel Virtual Machine (PVM) for this purpose. The PVM is the system that allows us to use a heterogeneous collection of computers and to view them as a single parallel computer.

A description of the appropriate mixture structure and algorithms for finite-mixture estimation [2] will be provided also. We use Bayesian fitting of a finite mixture of probability density functions (pdfs) to a sample of multivariate data. Individual pdfs forming the mixture are called components. Using the chain rule for pdfs, each component can be written as a product of single-variate conditional pdfs. These are called factors. The channel that they predict is referred to as the output channel. The factorised algorithms have been developed in the Matlab environment [4].

2 Factorised Algorithm

A brief review of the factorised algorithm for finite-mixture estimation is now given. Some conventions are necessary:

f *denotes probability (density) function* $(p(d)f)$

x^* *denotes the range of the quantity* x

$\mathring{x}$ *denotes cardinality of the set* x^* *(possibly infinite)*

x_t *is the quantity* x *at discrete time* $t \in t^* \equiv \{1, \ldots, \mathring{t}\}$

$x_{i;t}$ *is the* i*th entry of array* x *at time* t

$x(t)$ *denotes the sequence* $x_1, \ldots, x_t$

2.1 Mixture structure

The pdf $f(d(\mathring{t}))$ can be factorised according to the chain rule $f(d(\mathring{t})) = \prod_{t\in t^*} f(d_t|d(t-1))$. The system is at least Markovian of finite order. Thus,

$$f(d(\mathring{t})) = \prod_{t\in t^*} f(d_t|\phi_{t-1}), \tag{1}$$

where $\phi_t \in \phi^*$, $t \in t^*$ are known finite-dimensional vectors called (observable) states. It is assumed that the ϕ_t can be evaluated in a recursive manner; i.e. $\phi_t = \Phi(\phi_{t-1}, d_t)$ where Φ is a known function.

We assume that the outer model $\{f(d_t|\phi_{t-1})\}_{t\in t^*}$ is time invariant. The multiple mode probabilistic models can rather often be approximated by a finite mixture of unimodal models, called components [2]. Hence, we describe the system by a time-invariant Markov parameterised finite-mixture model:

$$f(d(\mathring{t})|\Theta) \equiv \prod_{t\in t^*} f(d_t|\phi_{t-1}, \Theta) \tag{2}$$

with

$$f(d_t|\phi_{t-1}, \Theta) = \sum_{c\in c^*} \alpha_c f(d_t|\phi_{c;t-1}, \Theta_c, c) \tag{3}$$

$$\phi_{c;t} = \Phi_c(\phi_{c;t-1}, d_t). \tag{4}$$

[1]This paper was done with support of EU project IST ProDaCTool No. IST-1999-12058, GAČR 102/99/1564, and OK 317/2000

Using the chain rule, the individual components can be written as the product of pdfs predicting individual entries of d_t. Entries of d_t can be permuted before applying the chain rule. Let π_c denote the permutation used in c-th component:

$$\pi_c(d) \equiv \pi_c([d_1, \ldots, d_{\mathring{d}}]) \equiv d_c \equiv [d_{1c}, \ldots, d_{\mathring{d}c}],$$

with $d_{ic} = d_{j_i}$. Here, j_i is the i-th entry of the permuted indices $[1, \ldots, \mathring{d}]$. Thus

$$f(d_t|\phi_{c;t-1}, \Theta_c, c) = \prod_{i=1}^{\mathring{d}} f(d_{ic;t}|\psi_{ic;t}, \Theta_{ic}, i, c) \quad (5)$$

with $\psi_{ic;t} = d_{1c;t}, \ldots, d_{(i-1)c;t}, \phi_{c;t-1}$. The additional subscript i in Θ_{ic} indicates that some entries of Θ_c need not enter the i-th pdf. Similarly, the vector $\psi_{ic;t}$ is a sub-vector of the vector $[d_{1c;t}, \ldots, d_{(i-1)c;t}, \phi'_{c;t-1}]'$.

2.2 Parameter estimation

For the design and description of the parameter estimation problem we introduce discrete random pointers $c(\mathring{t}) = (c_1, \ldots, c_{\mathring{t}})$, $c_t \in c^*$, to particular components. These are assumed to be mutually independent with time-invariant probabilities α_c. Given these pointers, the parameterized model (3) can be interpreted as the marginal pdf $f(d_t|d(t-1), \Theta)$ of the following joint pdf:

$$f(d_t, c_t|d(t-1), \Theta) = \prod_{c \in c^*} [f(d_t|\phi_{c;t-1}, \Theta_c, c)\alpha_c]^{\delta_{cc_t}}$$

$$\underbrace{=}_{(5)} \prod_{c \in c^*} \left[\alpha_c \prod_{i=1}^{\mathring{d}} f(d_{ic;t}|\psi_{ic;t}, \Theta_{ic}, i, c)\right]^{\delta_{cc_t}}, \quad (6)$$

where δ_{cc_t} is the *Kronecker symbol* defined by the formula

$$\delta_{c\tilde{c}} = \begin{cases} 1 & \text{if } c = \tilde{c} \\ 0 & \text{otherwise} \end{cases}. \quad (7)$$

The marginal pdf $f(d_t|d(t-1), \Theta)$ of the pdf $f(d_t, c_t|d(t-1), \Theta)$ is found by summing (6) over all $c_t \in c^*$. It is the mixture (3).

Let the pdf $f(\Theta|d(t))$ be of the product form:

$$f(\Theta|d(t)) \propto \prod_{c \in c^*} \prod_{i=1}^{\mathring{d}} f(\Theta_{ic}|d(t))\alpha_c^{\kappa_{c;t}-1}. \quad (8)$$

$f(\Theta_{ic}|d(t))$ are posterior pdfs of individual factors and $\kappa_{c;t} > 0$ are known values of scalar functions of $d(t)$. Together (8), (6) and Bayes rule give

$$f(\Theta, c_{t+1}|d(t+1)) \propto \prod_{c \in c^*} \prod_{i=1}^{\mathring{d}} \mathcal{F}_{ic}^{\delta_{cc_{t+1}}} \mathcal{P}_{ic}\alpha_c^{\beta_c}, \quad (9)$$

with

$$\mathcal{F}_{ic} = f(d_{ic;t+1}|\psi_{ic;t+1}, \Theta_{ic}, i, c),$$
$$\mathcal{P}_{ic} = f(\Theta_{ic}|d(t)),$$
$$\beta_c = \kappa_{c;t} + \delta_{cc_{t+1}} - 1.$$

In order to obtain an approximation of the desired pdf $f(\Theta|d(t+1))$, we have to eliminate c_{t+1} from (9). The correct marginalisation with respect to c_{t+1} destroys the tractable product form (8). The latter is, however, preserved if $\delta_{cc_{t+1}}$ is replaced by its estimate. We approximate $\delta_{cc_{t+1}}$ by its conditional expectation

$$\delta_{cc_{t+1}} \approx \mathcal{E}[\delta_{cc_{t+1}}|d(t+1)]$$
$$= f(c_{t+1} = c|d(t+1)). \quad (10)$$

By integrating (9) over the parameters Θ, it can be shown that this probability has the form

$$w_{c;t+1} \equiv f(c_{t+1} = c|d(t+1))$$

$$= \frac{\prod_{i=1}^{\mathring{d}} f(d_{ic;t+1}|d(t), i, c)\kappa_{c;t}}{\sum_{\tilde{c} \in c^*} \prod_{i=1}^{\mathring{d}} f(d_{ic;t+1}|d(t), i, \tilde{c})\kappa_{\tilde{c};t}}. \quad (11)$$

Here, $F_i = f(d_{ic\,t+1}|d(t), i, c)$ is the Bayesian prediction for a single factor identified by indices i, c. The formula (11) can be interpreted as Bayes rule applied to the discrete unknown random variable $c_{t+1} \in c^*$ with prior probability $w_{c;t} \propto \kappa_{c;t}$.

By inserting the approximation (10) and (11) into (9), the approximate updated posterior pdf has the same functional product form as in (8), with

$$\kappa_{c\,t+1} = \kappa_{c;t} + w_{c;t+1} \quad (12)$$

and

$$f(\Theta_{ic}|d(t+1)) \propto \mathcal{F}_{ic}^{w_{ic;t+1}} \mathcal{P}_{ic}. \quad (13)$$

2.3 The algorithm

We can now specify the estimation algorithm. Initialization requires selection of prior pdfs $f(\Theta_{ic})$, of the individual factors and, initial values $\kappa_{c;0} > 0$. Then the following algorithm is followed:

1. Acquire the data record d_{t+1}.
2. Compute values of the pdfs F_i for each individual factor $i \in i^*$ for all components $c \in c^*$ and the measured data record d_{t+1}.
3. Compute values of the pdfs $f(d_{t+1}|d(t), c)$ for measured data d_{t+1} and each individual component $c \in c^*$.
4. Compute the probabilities $w_{c;t+1}$, using (11).

5. Update scalars $\kappa_{c;t+1}$ according to (12).

6. Update Bayesian parameter estimates of different factors $f(\Theta_{ic}|d(t+1))$, using (13), with

$$w_{ic;t+1} = \sum_{\tilde{c} \in c_i^*} w_{\tilde{c};t+1}, \tag{14}$$

where $c_i^* \equiv$ the set of pointers to components that contain the i-th factor.

3 Parallel Approach

We have described a sequential algorithm in the previous section (2.3). The issue now is to determine which parts of the sequential algorithm can be effectively parallelized. This problem will be discussed in Section 3.2. The general properties of the PVM will be given first.

3.1 Parallel virtual machine

The Parallel Virtual Machine [1] is designed to link computing resources and provide users with a parallel platform for running their computer applications, irrespective of the number of different computers they use and where the computers are located.

The PVM system uses the message-passing model to allow programmers to exploit distributed computing across a wide variety of computer types. A key concept in PVM is that it makes a collection of computers appear as one large virtual machine.

The computers available on a network may be made by different vendors or have different compilers. When we wish to exploit a collection of networked computers, we may have to contend with several different types of heterogeneity: architecture, data format, computational speed, machine load, and network load.

The PVM computing model is based on the notion that an application consists of several tasks. Each task is responsible for a part of the application's computational workload of the application.

3.2 Computation analysis

Forming the mixture, we consider units or tens of components and tens or hundreds of factors. We now attempt to discuss the algorithm in Section 2.3 in relation to these values.

Looking at that algorithm, we find that the critical steps of the algorithm—from the computational viewpoint—are steps 2 and 6.

The processing data in step 2—where predictive pdfs, F_i, are computed—is close to the recursive least squares (RLS) algorithm [3]. The same is true of step 6. The update of the parameter estimates, $f(\Theta_{ic}|d(t+1))$, is—in the case of individual factors—close to the weighted RLS algorithm [3]. All these computations are repeated for each factor.

If we consider the sequential algorithm, with tens of factors, and the fact that the RLS algorithm involves $\mathcal{O}(N^2)$ operations for each factor, the computational load is very heavy. This problem is, however, very amenable to parallelization.

The other steps of the algorithm are much less complex to compute. Furthermore, unlike the previously discussed computations, they only operate probabilities for units (maximal tens) of components. The significant difference between computation of factors and components is that the values of the probabilities of components are highly dependent on the parameters of the factors. Thus, prohibitive levels of communication between computational nodes is necessary. The parallel implementation of these parts of the algorithm has not been attempted.

3.3 Data decomposition

The master-slave model of virtual parallel machine architecture has been chosen. A separate "control" program termed the master is responsible for process spawning, initialization, collection of results, and timing of functions. The slave programs perform the necessary computations, and they are allocated their workloads by the master. Thus, the master is responsible for load balancing.

The complete graph interconnection network is used. During computation, individual slaves have to communicate both with the master and with each other. Because of the facilities of the PVM, it is relatively straightforward to implement this interconnection network.

Before the slaves' computations are performed, the master has to do work that can be termed "workload allocation". The master establishes connections between individual nodes and performs data decomposition or partitioning. This is done statically; i.e. each process knows *a priori* its share of workload.

During the time when the slaves are active, the master performs timing and synchronization. Then the master collects and evaluates results. It can follow the sequential algorithm or send the results to other separated modules of the system.

Suppose we use P slaves to compute N values of probability. Then, each slave will perform operations on $\frac{N}{P}$ factors; i.e. the p-th slave computes values for factors F_i with $i \in i^* = \{\frac{N}{P}(p-1)+1,\ldots,\frac{N}{P}p\}$ and $p \in p^* = \{1,\ldots,P\}$. We assume that $\frac{N}{P} \in \mathcal{N}^+$.

When computing the values of the predictive pdfs F_i for each factor (see step 2 of the sequential algorithm), primitive parallelism can be used. This is because individual factors are independent of each other. Thus, minimal communication is needed between individual slaves.

If we consider that the computational complexity of step 2 is $T(n) = \mathcal{O}(N^3)$, using the parallelization, we can theoretically achieve linear speedup, $T(n,p) = \mathcal{O}(\frac{N^3}{P})$, in solving this part of the problem. The practical speedup is rather less due to data distribution during initialization and due to collection of results.

The situation is more complex in the case of updating Bayesian parameter estimates of factors: $f(\Theta_{ic}|d(t+1))$. As implied by (13), the slaves have to compute $w_{ic;t+1}$ for individual factors but information about $w_{\tilde{c};t+1}$ is contained in the initial data. Thus, no communication between slaves is necessary and the computational complexity and speedup are the same as discussed in the previous paragraph.

There is minimal overhead caused by the slaves' communication. But, we have to realize that we have not attempted to parallelize all steps of the algorithm. Thus, we cannot achieve linear speedup of the algorithm as a whole. As implied by Amdahl's law [5], speedup $S(n,p) \leq \frac{1}{f_s}$, where f_s is relative sequential part of the algorithms.

4 Other Software Aspects

Factorised algorithms in Matlab, as well as MEX source files, have been developed. Consequently, we have designed a library of functions that make similar processing as a set of Matlab API so that it is possible to compile the MEX source files using the standard ANSI C compiler without rewriting them. There are two reasons for this: first, run-time library, independent of Matlab should be available; second, the PVM system supports C, C++, and Fortran only.

Matlab works with only a single object type, the Matlab array. All Matlab variables, including scalars, vectors, matrices, strings, cell arrays, structures, and objects are stored as Matlab arrays. In ANSI C, the Matlab array is declared to be of type `mxArray`. We have designed our own specific (Matlab compatible) structure `mxArray`.

Another problem was data conversion between Matlab and ANSI C on different computer platforms. Four routines—two in Matlab and two in ANSI C—have been prepared for this purpose.

The current version of the ANSI C support requires Matlab version 5.x (Win32 or Unix) and the corresponding ANSI C compiler.

5 Future Work

There remains a lot of work in order to complete the parallelization. The parallelization may play significance in other, more complex, tasks in mixture processing. We have to attend to the problem of parallelization of those tasks. Furthermore, a detailed theoretical complexity analysis of the parallel algorithms that are already implemented is required. We then need to compare measured and theoretical performance values.

The next part of the work will focus on implementation of dynamic scheduling; i.e. the case where individual process workloads vary as the computation progresses. We also want to consider the problem of optimization of the parallel algorithms and fine grain granularity parallelization.

6 Conclusion

We have discussed algorithms for finite-mixture estimation in this paper. We have described the structure of a mixture and the sequential factorised algorithm for finite-mixture estimation.

Finally, we have considered the problem of parallelization of associated factorised algorithms and the development of the necessary library emulating Matlab API.

References

[1] A. Geist et al, "PVM: Parallel Virtual Machine", *A User's Guide and Tutorial for Networked Parallel Computing*, The MIT Press, Cambridge, Massachusetts, 1994.

[2] M. Kárný et al, "ProDaCTool — theory, algorithms, and software", *EU Project IST ProDaCTool No. IST-1999-12058*, Research Report, 2000.

[3] M. Moonen, "Introduction to Adaptive Signal Processing", K.U. Leuven, Leuven, Belgium, 1999.

[4] P. Nedoma et al, "MixTools, MATLAB Toolbox for Mixtures", *EU Project IST ProDaCTool No. IST-1999-12058*, Research Report, 2000.

[5] P. Tvrdík, "Parallel Systems and Algorithms", Lecture Notes, Czech Technical University, August 1996.

Method for Artefact Detection and Suppression Using Alpha-Stable Distributions

Ludvík Tesař, Anthony Quinn * [1]

*University of Dublin, Trinity College, Dublin 2, Ireland

Abstract

This paper describes a method for artefact detection and suppression based on α-Stable distributions. The reason for choosing the α-stable distribution is, that it is heavy-tailed distribution ideal for modeling of data polluted by outliers. A method for on-line data processing is emphasized. The artefact suppression is based on the idea that data are modeled by a Symmetric α-Stable distribution, parameters of which are estimated. Then the data are regenerated from the Gaussian distribution with parameters, that correspond to the original parameters of the α-Stable distribution. The new data is free of any outliers.

1 Introduction

Methods, that use the raw data, might be vulnerable to high amplitude noise coming from measurement faults or other sources. This includes some Neural Networks or mixture modeling methods. But unfortunately, most data measured by sensors suffer from these artefacts. This is why we are investigating new methods for data pre-processing, that would remove this impulsive noise and provide clean data for processing. The work on data pre-processing techniques was performed as the part of the work for the ProDaCTool project, which aims to create an advisory system for operator-controlled processes based on clustering with a mixture model consisting of normal clusters.

This paper describes methods for artefact detection based on modeling of the underlying process by the Symmetric α-Stable distribution (SαS) [1, 2]. The data are assumed to be independent and parameters are assumed to be slowly varying. We are re synthesizing data via nearest equivalent Gaussian. This enable later processing of data by methods relying on their Gaussian distribution.

At first the parameters of the SαS distribution are estimated, than the new data are generated based on the assumption, that data are distributed by the Gaussian distribution with the same mean (positional parameter) and dispersion as the SαS distribution.

In Section 2 we define the α-Stable distribution. In Section 3 we describe in detail the outlier detection and removal. In Section 4, we describe the one-shot estimation of parameters of SαS, if all data are given and in Section 5, we introduce the algorithm for on-line estimation of data. Since in Section 5 we need to calculate the weighted median, the Section 6 describes the algorithm to calculate it.

2 α-Stable Distribution

The Gaussian distribution is very good for modeling processes that occur in nature. However, in many processes, there occur occasional large-amplitude bursts and in such cases, the use of the Gaussian distribution is not reasonable. Such cases are usually denoted as impulsive noise and underlying distributions are denoted as heavy-tailed [3]. One of the classical examples of heavy-tailed distributions is the Cauchy distribution. However, the Cauchy distribution is not sufficient to describe all cases of heavy-tailed distributions.

A very general model for the heavy-tailed distribution is the α-Stable distribution [3]. It can be understood as a generalization of Gaussian and Cauchy distributions. The α-Stable distribution is parameterized by the parameter α, what describes the measure of heaviness of the tails. In two special cases of values of parameter α the distribution is Gaussian and Cauchy (it is Gaussian for $\alpha = 2$ and Cauchy for $\alpha = 1$)[1]. It is much more general than Gaussian and Cauchy distributions, as it can also be skewed to one side. The symmetric (non-skewed) case is called the Symmetric α-Stable distribution (SαS). For our applications, only this case is considered. An interesting property of α-Stable distributions is that after a generalization of the Central Limit Theorem,

[1]Supported by the European project ProDaCTool, IST-1999-12058

[1]There is also the second parameter called β (defined later in this Section) and this parameter does not affect the shape of distribution for $\alpha = 2$ and for $\alpha = 1$ it must be equal to zero in order to represent the Cauchy distribution.

the α-Stable distribution replaces the Gaussian distribution in the generalized Central Limit Theorem.

The α-Stable distribution is defined by its characteristic function as follows, if $\alpha \neq 1$ then:

$$\phi(\omega) = \exp\left(-|\gamma\omega|^{\alpha}\left[1 - i\text{sign}(\omega)\beta\tan\frac{\pi\alpha}{2}\right] + i\delta\omega\right) \tag{1}$$

If $\alpha = 1$ then:

$$\phi(\omega) = \exp\left(-|\gamma\omega|\left[1 - i\text{sign}(\omega)\beta\frac{2}{\pi}\log|\omega|\right] + i\delta\omega\right) \tag{2}$$

where i is the complex unit and where:

- α, the characteristic exponent $\alpha \in (0, 2]$, determines the thickness of tails;
- $\beta \in [-1, 1]$ is the skewness parameter; if $\beta = 0$ then the distribution is symmetric (SαS);
- $\gamma > 0$ is a scale (or dispersion) parameter;
- δ is the location parameter.

Two special cases are as follows:

- If $\alpha = 2$, $\forall\beta$, it is Gaussian distribution (parameters are $\mu = \delta$ and $\sigma^2 = 2\gamma$);
- If $\alpha = 1$ and $\beta = 0$, then it is the Cauchy distribution;
- If $\alpha = 1/2$ and $\beta = -1$, then it is the Pearson distribution.

It is not possible to give the algebraic expression for the probability density function for any other than the special cases of the Gaussian, Cauchy and Pearson distributions, so it is defined as the inverse Fourier transform of its characteristic function:

$$f(\alpha, \beta, \gamma, \delta; x) = \tag{3}$$

$$= \frac{1}{2\pi}\int_{-\infty}^{\infty} e^{-|\gamma\omega|^{\alpha}[1 - i\text{sign}(\omega)\beta w(\omega,\alpha)] + i\delta\omega} e^{-i\omega x} d\omega$$

where:

$$w(\omega, \alpha) = \tan\frac{\pi\alpha}{2} \tag{4}$$

for $\alpha \neq 1$, and:

$$w(\omega, \alpha) = \frac{2}{\pi}\log|\omega| \tag{5}$$

for $\alpha = 1$.

$S(\alpha, \beta, \gamma, \delta)$ denotes the α-Stable distribution with parameters α, β, γ and δ.

Figure 1 shows the SαS distribution $S(1.5, 0, 0.5, 0)$ (dashed) compared with $\mathcal{N}(0, 1)$ (solid line). $\mathcal{N}(0, 1)$ is the same as $S(2, 0, 0.5, 0)$.

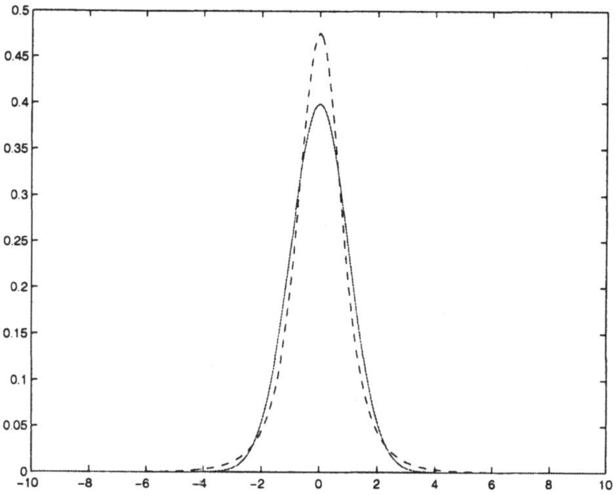

Fig. 1. Dashed: $S(1.5, 0, 0.5, 0)$, solid line: $\mathcal{N}(0, 1) = S(2, 0, 0.5, 0)$

3 Outline of the Method for Outlier Removal Based on the α-Stable Distribution

Outlier removal using the α-Stable distribution is based on the idea that we estimating on-line all the parameters of the α-Stable distribution. Then we regenerate the data with the non heavy-tailed Gaussian distribution with parameters $\mu = \delta$ and dispersion $\sigma^2 = 2\gamma$. Hence, the parameter α is forced to be 2. Newly generated data are non heavy-tailed, i.e. not polluted by outliers.

As the first step, we have to estimate of parameters of the α-Stable distribution assuming that given measured data are generated from this distribution. Section 4 describes one shot methods for estimating parameters of the Symmetric α-Stable (SαS) distribution.

4 Estimation of Parameters of the Symmetric α-Stable Distribution

The assumption is that the data $x_1, x_2, \ldots, x_t$ are independently generated from the Symmetric α-Stable distribution $S(\alpha, 0, \gamma, \delta)$.

For estimation of location parameter δ of the SαS distribution, the median estimator can be used (assuming that $\alpha < 2$):

$$\hat{\delta} = \text{median}\{x_1, x_2, \ldots, x_t\} \tag{6}$$

The median from the data sample forms the maximum-likelihood estimate for a two-sided exponential distribution and for the SαS case is very close to maximum-likelihood estimate See [4]. Also it is considered as very robust estimate of parameter δ.

For estimation of α and γ, we use the so called fractional lower order moments (FLOM), $\mathbf{E}(|x - \delta|^p)$.

FLOM are known to be finite for $p \in (-1, \alpha)$. We will use statistics [5]:

$$A_p = \frac{1}{T}\sum_{k=1}^{T} |x_k - \hat{\delta}|^p \qquad (7)$$

Parameter α can be estimated via the so-called SINC estimator by solving:

$$\mathrm{sinc}(\frac{p}{\hat{\alpha}}) = \frac{sin(\frac{p\pi}{\hat{\alpha}})}{\frac{p\pi}{\hat{\alpha}}} = \frac{\tan q}{qA_pA_{-p}} \qquad (8)$$

where $q = \frac{p\pi}{2}$. Once we have an estimate $\hat{\alpha}$ of α, we can estimate the parameter γ:

$$\hat{\gamma} = \left(\frac{A_p}{C(p,\alpha)}\right)^{\frac{\alpha}{p}} \qquad (9)$$

where:

$$C(p,\alpha) = \frac{2^p\Gamma(\frac{p+1}{2})\Gamma(1-\frac{p}{\alpha})}{\sqrt{\pi}\Gamma(1-\frac{p}{2})} \qquad (10)$$

For estimation of α and γ the variable p must be selected in the interval $0 < p < \alpha/2$. This gives the lower bound for estimates of α.

5 On-line Estimation of Parameters of the SαS Distribution

Our aim is to remove outliers from the data in an on-line manner, so that we would be able to process data in real time. In Section 4, we described methods that can be used only for one-shot estimation of parameters, given all data.

Now we will assume that parameters α, γ, δ are changing slowly in time. For slowly changing parameters, there exists a heuristic but rationally based extension of Bayesian theory, that is called exponential forgetting [7]. This extension is based on idea, that parameter estimates are calculated from collected statistics, which are computed cumulatively from data. The influence of older datum to collected statistics and parameter estimates decreases with time. Because the cumulated information is decreased in each step by given multiplicative constant smaller than 1, the proportional information of every collected datum is decreasing exponentially with adding new data. It is the reason, why this type of forgetting is called exponential.

Based on the exponential forgetting theory we introduce weight $q_i = \ell^{t-i}$ for every single datum x_i (we are still assuming, that we have data $x_1, x_2, \ldots, x_t$, we are using i as the time index). The constant ℓ is called the forgetting factor. After normalization we will get weights:

$$w_i = \frac{q_i}{\sum_{k=1}^{t} q_k} = \ell^{t-i}\left(\frac{1-\ell}{1-\ell^t}\right) \qquad (11)$$

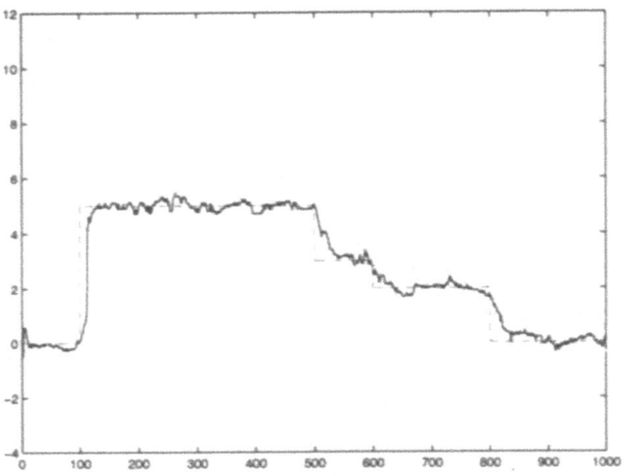

Fig. 2. Estimate of $\hat{\delta}$ calculated from equation (12). data - dots, signal - dashed, estimated $\hat{\delta}$ is solid line

The exponential forgetting theory can be understood as calculation of parameters by using w_i as weights for data x_i. We will use these weights to compute parameters α, β and δ of the SαS distribution.

First we calculate the estimate $\hat{\delta}$ of parameter δ. Instead of computing the median, we will calculate the weighted median of data x_1 to x_t.:

$$\hat{\delta} \doteq \mathrm{wmedian}(x_1, \ldots, x_t; w_1, \ldots, w_t) \qquad (12)$$

Explanation of how to calculate the weighted median is more complicated, what is why it is given in the Section 6.

For estimates of α and γ we modify statistics A_p and A_{-p} from equation (7), to take account of the weights w_i:

$$A_p = \sum_{k=1}^{T} w_k |x_k - \hat{\delta}|^p \qquad (13)$$

In this way, we get estimates, $\hat{\alpha}$ and $\hat{\gamma}$, by equations (9) and (8) respectively.

Figure 2 shows the estimate $\hat{\delta}$ computed via the median with forgetting (12). Data are represented by dots, the underlying signal without noise is represented by the dashed line and the estimated $\hat{\delta}$ is represented by the solid line. The underlying signal is jumping between a small number of values, i.e. it is a piecewise constant signal. It is apparent that the estimate is following this signal.

6 Calculation of the Weighted Median

In Section 5 we calculate the weighted median for given data. In contrast to the usual definition of mean, where all data have the same weight, here we have the data x_1 to x_t with weights w_1 to w_t, where $\sum_{i=1}^{t} w_i = 1$. The motivation for this is to make an analogy to the weighted mean, which is calculated as $\sum_{i=1}^{t} w_i x_i$.

The following algorithm describes the idea of the weighted median:

1. Sort the data x_i, and denote the sorted data by $x_{<i>}$. Reorder w_i appropriately, so that if $x_{<j>} = x_i$ then $w_{<j>} = w_i$.

2. Find j such that $\sum_{k=1}^{j} w_{<k>} \leq 0.5$ and $\sum_{k=1}^{j+1} w_{<k>} > 0.5$. It is obvious that there must be exactly one such j.

3. If $\sum_{k=1}^{j} w_{<k>} < 0.5$, then let the median be $m = x_{<j+1>}$; otherwise let the median be $m = \frac{x_{<j>}+x_{<j+1>}}{2}$.

Algorithm that explains the weighted median above might look too complicated, but the idea is very simple. In fact, if weights were rational number w_i, we would be able to explain the idea of weighted mean in much simpler way. For w_i rational, we can put weights into the same denominator d, then numbers $n_i = w_i \times d$ are natural. Now, it is possible to create a new data set, so that data x_i repeat n_i times and calculate the median from the data set created like this. The algorithm for real w_i above, is a direct generalization of this intuitive idea.

7 On-Line Method for Outlier Removal Based on the α-Stable Distribution Assumption

The outlier removal method is based on the idea that the data distributed by the α-stable distribution are replaced by data distributed by the Gaussian distribution. Methods from Section 6 are used for on-line estimation of parameters. Let us denote these estimates $\hat{\alpha}_t$, $\hat{\gamma}_t$ and $\hat{\delta}_t$. The new data z_t are randomly generated from the Gaussian distribution, so that $z_t \sim \mathcal{N}(\delta_t, 2\gamma_t)$. This Gaussian distribution is equivalent to the α-stable distribution with the same positional and scaling parameter, but not heavy-tailed, i.e. with α forced to be 2 (see Figure 1).

For practical reasons it is feasible not to remember all the history of data x_i, but only data for which the weight would be above a given threshold. Therefore, the algorithm may work only with data newer than $t - t_0$, where $\ell^{t_0} < h$, and h is some threshold. If h is sufficiently small, then this decreasing of the data memory has hardly any influence on the algorithm.

8 Conclusion

This article describes an approach to removal of outliers from data by modeling them as realizations from a Symmetric α-Stable (SαS) distribution. The method is based on estimation of parameters of the SαS distribution and and re-generation of the data from the Gaussian distribution with appropriate parameters. A new contribution of the paper is also the algorithm for on-line estimation of the parameters of the SαS distribution with an exponential forgetting feature.

References

[1] E. E. Kuruoğlu, W. J. Fitzgerald, and P. J. Rayner, "Near optimal detection of signals in impulsive noise modeled with a symmetric alpha-stable distribution," *IEEE Comm. Letters*, vol. 2, pp. 1–14, October 1998.

[2] E. E. Kuruoğlu, P. J. Rayner, and W. J. Fitzgerald, "Least ℓ_p-norm impulsive noise cancellation with polynomial filters," *Signal Processing*, vol. 69, pp. 1–14, 1998.

[3] C. L. Nikias and M. Shao, *Signal processing with alpha-stable distributions and applications*. Adaptive and learning systems for signal processing, communications, and control, Chicester: Wiley, 1995.

[4] G. A. Tsihrintzis and C. L. Nikias, "Fast estimation of the parameters of alpha-stable impulsive interference," *IEEE Transactions on Signal Processing*, vol. 44, pp. 1492–1503, June 1996.

[5] X. Ma and C. L. Nikias, "Parameter estimation and blind channel identification in impulsive signal environments," *IEEE Transactions on Signal Processing*, vol. 43, pp. 2884–2897, December 1995.

[6] T. Bayes, "An essay towards solving a problem on the doctrine of chances," *Philosophical Transactions of the Royal Society of London*, vol. 53, pp. 370–418, 1763.

[7] V. Peterka, "Bayesian approach to system identification," in *Trends and Progress in System Identification* (P. Eykhoff, ed.), ch. 8, pp. 239–304, Oxford: Pergamon Press, 1981.

Model Selection with Small Samples[1]

Masashi Sugiyama*, Hidemitsu Ogawa*

*Department of Computer Science, Tokyo Institute of Technology, Tokyo, Japan, E-mail: sugi@og.cs.titech.ac.jp, URL: http://ogawa-www.cs.titech.ac.jp/~sugi

Abstract

Recently, a new model selection criterion called the subspace information criterion (SIC) was proposed. SIC gives an unbiased estimate of the generalization error with finite samples. In this paper, we theoretically and experimentally evaluate the effectiveness of SIC in comparison with existing model selection techniques. Theoretical evaluation includes the comparison of the generalization measure, approximation method, and restriction on model candidates and learning methods. The simulations show that SIC outperforms existing techniques especially when the number of training examples is small and the noise variance is large.

1 Introduction

Supervised learning is estimating unknown input-output dependency from available input-output examples. Once the dependency has been accurately estimated, it can be used for predicting output values corresponding to novel input points. This ability is called the *generalization capability.*

The level of the generalization capability depends heavily on the choice of the *model*, which indicates, for example, the number and type of basis functions used for learning. The problem of choosing the model that provides the optimal generalization capability is called *model selection.* Model selection has been extensively studied from various standpoints: information statistics [1][5][3][4], Bayesian statistics [?], stochastic complexity [6], and structural risk minimization principle [7][8].

Recently, a new model selection criterion called the *subspace information criterion* (SIC) was proposed by the authors [9]. SIC gives an unbiased estimate of the generalization error with finite samples. In this paper, we evaluate the effectiveness of SIC in comparison with existing model selection techniques.

[1]A detailed version of this article is to appear in *Machine Learning, Special Issue on New Methods for Model Selection and Model Combination*, 2001.

2 Subspace Information Criterion (SIC) for Subset Regression

Let us consider the regression problem of obtaining, from a set of M training examples, an approximation to a target function $f(x)$ of L variables defined on $\mathcal{D} \subset \mathbf{R}^L$. The training examples are made up of *sample points* $x_m \in \mathcal{D}$ and corresponding *sample values* $y_m \in \mathbf{R}$. We suppose that y_m is degraded by additive noise ϵ_m, i.e., $y_m = f(x_m) + \epsilon_m$.

Let θ be a set of factors which determine learning result functions, for example, the type and number of basis functions. We call θ a *model.* Let $\hat{f}_\theta(x)$ be a learning result function obtained with a model θ. We measure the generalization error of $\hat{f}_\theta(x)$ by

$$J_G[\theta] = \mathrm{E}_\epsilon \int (\hat{f}_\theta(u) - f(u))^2 p(u)du, \qquad (1)$$

where E_ϵ denotes the ensemble average over the noise and $p(\cdot)$ is the probability density function of future (test) input points u. Then the problem of model selection considered in this paper is to select, from a set $\mathcal{M}$ of model candidates, the best model $\hat{\theta}$ that minimizes the generalization error J_G.

The model selection criterion called the *subspace information criterion* (SIC) [9] gives an unbiased estimate of the generalization error J_G. Here, we briefly review SIC for subset regression.

The following conditions are assumed:

(a) The learning target function $f(x)$ is a linear combination of a given set $\{\varphi_p(x)\}_{p=1}^{\mu}$ of μ linearly independent functions.

(b) The $M \times \mu$-dimensional *design matrix* A with (m,p)-th element being $\varphi_p(x_m)$ has the rank μ.

(c) The number M of training examples is larger than the number μ of basis functions.

(d) The mean noise is zero and the noise covariance matrix is given as $\sigma^2 I_M$ where $\sigma^2 > 0$ and I_M is the M-dimensional identity matrix.

(e) The μ-dimensional covariance matrix U with (p,p')-element being $\int \varphi_{p'}(u)\varphi_p(u)p(u)du$ is known.

(f) A model θ indicates a subset of indices $\{1,2,\ldots,\mu\}$, and the learning result function $\hat{f}_\theta(x)$ is defined as a minimizer of the training error $\frac{1}{M}\sum_{m=1}^{M}(\hat{f}(x_m) - y_m)^2$ in a subspace spanned by

$\{\varphi_p(x)\}_{p\in\theta}$. In this case, $\hat{f}_\theta(x)$ is given as

$$\hat{f}_\theta(x) = \sum_{p\in\theta}[A_\theta^\dagger y]_p \varphi_p(x), \qquad (2)$$

where A_θ is an $M \times \mu$ matrix with (m,p)-th element being $\varphi_p(x_m)$ if $p \in \theta$ otherwise 0. $A_\theta^\dagger$ denotes the *Moore-Penrose generalized inverse* of A_θ and $[\cdot]_p$ denotes the p-th element of a vector.

Under the above assumptions, SIC is given as

$$\begin{aligned}\mathrm{SIC}[\theta] = &\langle U(A_\theta^\dagger - A^\dagger)y, (A_\theta^\dagger - A^\dagger)y\rangle \\ &- \hat{\sigma}^2 \mathrm{tr}(U(A_\theta^\dagger - A^\dagger)(A_\theta^\dagger - A^\dagger)^\top) \\ &+ \hat{\sigma}^2 \mathrm{tr}(UA_\theta^\dagger (A_\theta^\dagger)^\top), \end{aligned} \qquad (3)$$

where $\hat{\sigma}^2 = \langle y - AA^\dagger y, y\rangle/(M-\mu)$ and $\top$ denotes the transpose of a matrix. It is shown that SIC is an unbiased estimate of J_G [9]:

$$\mathrm{E}_\epsilon \ \mathrm{SIC}[\theta] = J_G[\theta]. \qquad (4)$$

3 Theoretical Evaluation of SIC

In this section, SIC is compared with the traditional leave-one-out cross-validation (CV), Mallows's C_P [10], Akaike's information criterion (AIC) [1], Sugiura's corrected AIC (cAIC) [5], Schwarz's Bayesian information criterion (BIC) [?], Rissanen's minimum description length criterion (MDL) [6], and Vapnik's measure (VM) [8].

3.1 Generalization measure

SIC can adopt any generalization measure expressed as $\mathrm{E}_\epsilon\|\hat{f}_\theta - f\|^2$ as long as it is computable (e.g. the covariance matrix U is known). $\|\cdot\|$ denotes the norm in the functional Hilbert space spanned by $\{\varphi_p(x)\}_{p=1}^{\mu}$. The derivatives of the functions $\hat{f}_\theta(x)$ and $f(x)$ can also be included in the generalization measure (with the Sobolev norm).

C_P adopts the predictive training error $\frac{1}{M}\mathrm{E}_\epsilon\sum_{m=1}^{M}(\hat{f}_\theta(x_m) - f(x_m))^2$ as the error measure, which is equivalent to Eq.(1) with $p(u)$ being replaced by the empirical distribution. Note that the predictive training error does not evaluate the error at future sample points u.

CV adopts the so-called leave-one-out error $\frac{1}{M}\sum_{m=1}^{M}(\hat{f}_\theta^{(m)}(x_m) - y_m)^2$ as the error measure, where $\hat{f}_\theta^{(m)}$ denotes the learning result function obtained with the training examples without (x_m, y_m). The leave-one-out error also does not directly evaluate the error at future sample points u. The relation between the leave-one-out error and Eq.(1) is not well recognized yet.

AIC and cAIC adopt the expected Kullback-Leibler information over all possible training sets $\{(x_m, y_m)\}_{m=1}^{M}$ as the generalization measure, which is conceptually similar to the expectation of Eq.(1) over training sample points $\{x_m\}_{m=1}^{M}$ [3]. Although $p(\cdot)$ can be unknown in AIC and cAIC, instead training sample points $\{x_m\}_{m=1}^{M}$ and future sample points u are assumed to be independently subject to the same probability density function $p(\cdot)$ and the generalization measure is further averaged over training sample points. If one adopts the generalization measure averaged over training sample points, the purpose of model selection is to obtain the model that gives good learning result functions on average. In contrast, if one adopts the generalization measure which is *not* averaged over training sample points, the purpose of model selection is to obtain the model that gives the optimal learning result function from a given, particular training set. This implies that the latter standpoint is suitable for acquiring the best prediction performance from given training examples.

BIC gives an estimate of the posterior probability of parameters, and MDL gives an estimate of the description length of the model and data. The relation between the posterior probability, description length of the model and data, and generalization error is not clear.

The generalization measure of VM is a probabilistic upper bound of the risk functional $\int(\hat{f}_\theta(u) - f(u))^2 p(u)du$, where $p(\cdot)$ can be unknown but training sample points $\{x_m\}_{m=1}^{M}$ and future sample points u are assumed to be independently subject to the same probability density function $p(\cdot)$ instead.

3.2 Approximation methods

C_P, AIC, cAIC, BIC, MDL, and VM are expressed with the training error $\frac{1}{M}\sum_{m=1}^{M}(\hat{f}_\theta(x_m) - y_m)^2$. In contrast, CV and SIC directly evaluate the error measures.

C_P is an unbiased estimate of the predictive training error with finite samples. Since the predictive training error asymptotically agrees with the generalization error Eq.(1) if training sample points $\{x_m\}_{m=1}^{M}$ are subject to $p(\cdot)$, it can be regarded as an approximation of Eq.(1). Although asymptotic optimality of C_P is shown, its effectiveness with small samples is not theoretically sure.

In CV, the leave-one-out error can be regarded as an approximation of the generalization error (i.e., the error at future sample points u) since it is shown that the model selection by CV is asymptotically equivalent to that by AIC. Although it is known that CV practically works well, its mechanism in small sample cases is not well recognized yet.

Although AIC directly evaluates the generalization error, it is assumed in the derivation that the number of training examples is very large. This means that when the number of training examples is small, the approximation is no longer valid. BIC and MDL also use asymptotic approximation so they have the same drawback.

cAIC, VM, and SIC do not assume the availability of a large number of training examples for evaluating the generalization error. Therefore, they will work well with small samples. cAIC is a modified AIC with consideration of small sample effect for faithful models (i.e., models which include the learning target function). However, its performance for unfaithful models is not sure. VM gives a probabilistic upper bound of the risk functional based on the VC theory [7]. Although VM is derived under general setting, some heuristics are used in its derivation and the tightness of the upper bound is not evaluated yet.

SIC utilizes only the noise characteristics in its derivation, and it gives an unbiased estimate of the generalization error J_G with finite samples. However, its variance is not theoretically investigated yet. In order to calculate SIC, rather restrictive conditions should be assumed (see Sec. 2). However, these conditions do not have to be rigorously satisfied in practice. For example, when basis functions $\{\varphi_p(x)\}_{p=1}^{\mu}$ which include the learning target function $f(x)$ are unknown (see Assumption (a) in Sec .2), basis functions $\{\varphi_p(x)'\}_{p=1}^{\mu'}$ with the following properties are practically adopted:

(i) $\{\varphi_p(x)'\}_{p=1}^{\mu'}$ approximately include the learning target function $f(x)$.

(ii) The number μ' of basis functions is less than the number M of training examples.

When the covariance matrix U (see Assumption (e) in Sec. 2) is unknown, it can be estimated by using unlabeled sample points $\{x'_m\}_{m=1}^{M'}$ (i.e., sample points without sample values $\{y'_m\}_{m=1}^{M'}$) as $[\hat{U}]_{p,p'} = \frac{1}{M'}\sum_{m=1}^{M'}\varphi_{p'}(x'_m)\varphi_p(x'_m)$. If the training sample points $\{x_m\}_{m=1}^{M}$ are used instead of unlabeled sample points, then SIC agrees with Mallows's C_P. For this reason, SIC can be regarded as an extension of C_P (see also [9]).

3.3 Restriction on model candidates

AIC and cAIC are valid only when model candidates in the set $\mathcal{M}$ are nested [11][3], the fact is known to those who work on AIC, but it is still not well known to those who apply AIC in practice. In contrast, SIC imposes no restriction on models.

3.4 Restriction on learning methods

AIC, cAIC, BIC, and MDL are specialized for maximum likelihood estimation. A generalized AIC [3][4] relaxed the restriction of maximum likelihood estimation. C_P is specialized for the training error minimization learning with linear regression models. An extension of C_P called C_L [10], VM, and SIC are applicable to various learning methods expressed by linear mapping ($A_\theta^\dagger$ in Eq.(2)), including regularization learning with quadratic regularizers (ridge regression). Note that in VM, the VC-dimension [7] of models should be explicitly calculated.

4 Experimental Evaluation of SIC

In this section, SIC is experimentally compared with existing model selection techniques through computer simulations.

Let the learning target function $f(x)$ be $f(x) = \frac{1}{10}\sum_{p=1}^{50}(\sin px + \cos px)$ defined on $[-\pi, \pi]$. Let us consider a set of 201 basis functions $\{1, \sin px, \cos px\}_{p=1}^{100}$ which includes $f(x)$. Let the set $\mathcal{M}$ of model candidates be $\mathcal{M} = \{\theta_0, \theta_{10}, \theta_{20}, \ldots, \theta_{100}\}$, where θ_n indicates a regression model with $\{1, \sin px, \cos px\}_{p=1}^{n}$. Let us assume that the training sample points $\{x_m\}_{m=1}^{M}$ and future sample points u are independently subject to the same uniform distribution on $[-\pi, \pi]$. Let the noise ϵ_m be independently subject to the same normal distribution with mean 0 and variance σ^2. We compare SIC, CV, C_P, AIC, cAIC, BIC (which is the same as MDL), and VM. Note that the covariance matrix U (see Assumption (e) in Sec. 2) is the identity matrix in the above setting. We shall measure the error of a learning result function $\hat{f}_{\theta_n}(x)$ by $\frac{1}{2\pi}\int_{-\pi}^{\pi}(\hat{f}_{\theta_n}(x) - f(x))^2 dx$.

The simulation is performed 100 times with changing the noise $\{\epsilon_m\}_{m=1}^{M}$ in each trial. Fig. 1 show the distributions of the selected order n of models (upper) and error obtained by the selected model (lower) by 100 trials. 'OPT' indicates the optimal model that minimizes the error. When $(M, \sigma^2) = (500, 0.2)$, all model selection criteria work well. When $(M, \sigma^2) = (250, 0.2)$, AIC tends to select larger models and BIC (MDL) is inclined to select smaller models, so they provide large errors. This may be caused since AIC and BIC (MDL) are derived under the assumption that the number M of training examples is very large. When $(M, \sigma^2) = (500, 0.6)$, BIC (MDL) and VM show a tendency to select smaller models and they result in large errors. This implies that BIC (MDL) and VM are not robust against the noise. Finally, when

$(M, \sigma^2) = (500, 0.2)$ $(M, \sigma^2) = (250, 0.2)$ $(M, \sigma^2) = (500, 0.6)$ $(M, \sigma^2) = (250, 0.6)$

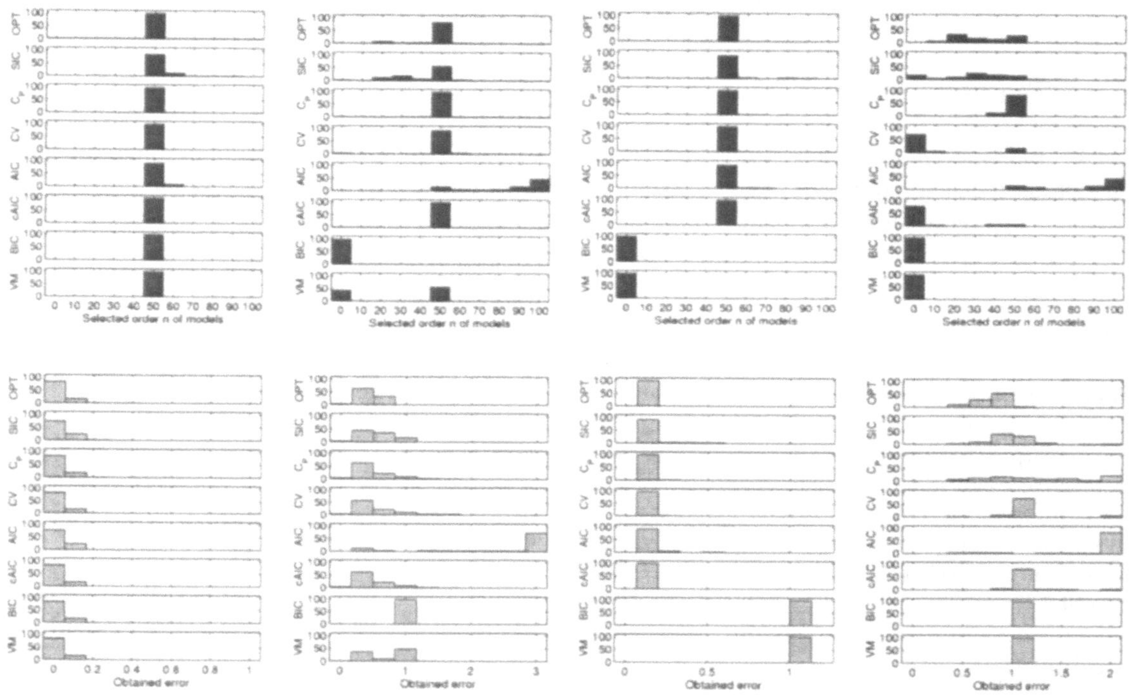

Fig. 1. Distributions of the selected order and error by 100 trials.

$(M, \sigma^2) = (250, 0.6)$, SIC works better than other criteria. In this case, C_P almost always selects θ_{50}, AIC tends to select larger models, and other criteria tend to select smaller models. As a result, they give large errors.

The simulation results show that SIC outperforms other model selection criteria especially when the number M of training examples is small and the noise variance σ^2 is large. It should be noted that C_P almost always selects the true model θ_{50} in any cases. This implies that C_P is more suitable for finding the true model than finding the model with minimum generalization error.

References

[1] Akaike, H.: A new look at the statistical model identification. IEEE Transactions on Automatic Control, vol. AC-19, no. 6, pp. 716723, (1974).

[2] Sugiura, N.: Further analysis of the data by Akaikes information criterion and the finite corrections. Communications in Statistics. Theory and Methods, vol. 7, no. 1, pp. 1326, (1978).

[3] Murata, N., Yoshizawa, S. and Amari, S.: Net-work information criteriondetermining the num-ber of hidden units for an artificial neural network model. IEEE Transactions on Neural Networks, vol. 5, no. 6, pp. 865872, (1994).

[4] Konishi, S. and Kitagawa, G.: Generalized information criterion in model selection. Biometrika, vol. 83, pp. 875890, (1996).

[5] Schwarz, G.: Estimating the dimension of a model. Annals of Statistics, vol. 6, pp. 461464, (1978).

[6] Rissanen, J.: Modeling by shortest data description. Automatica, vol. 14, pp. 465471, (1978).

[7] Vapnik, V.N.: The Nature of Statistical Learning Theory. Berlin: Springer-Verlag, (1995).

[8] Cherkassky, V., Shao, X., Mulier, F.M. and Vapnik, V.N.: Model complexity control for regression using VC generalization bounds. IEEE Transac-tions on Neural Networks, vol. 10, no. 5, pp. 10751089, (1999).

[9] Sugiyama, M. and Ogawa, H.: Subspace information criterion for model selection. Neural Computation, (2001), (to appear).

[10] Mallows, C.L.: Some comments on CP. Technometrics, vol. 15, no. 4, pp. 661675, (1973).

[11] Takeuchi, K.: On the selection of statistical models by AIC. Journal of the Society of Instrument and Control Engineering, vol. 22, no. 5, pp. 445453, (1983), (in Japanese).

A Data-Reusing Stochastic Approximation Algorithm for Neural Adaptive Filters

Danilo P. Mandic*, Igor R. Krcmar†, Warren Sherliker ‡, George Smith §

*School of Information Systems, University of East Anglia, Norwich, UK. e-mail:d.mandic@uea.ac.uk
†Faculty of Electrical Engineering, University of Banjaluka, Banjaluka, BH. e-mail:ikrcmar@etf-bl.rstel.net
‡Department of Electrical Engineering, Imperial College, London, UK. e-mail:w.sherliker@ic.ac.uk §School of Information Systems, University of East Anglia, Norwich, UK. e-mail:gds@uea.ac.uk

Abstract

A data-reusing stochastic approximation algorithm for adaptation of a neural adaptive filter is derived. The proposed algorithm is of the gradient-descent (GD) type and incorporates the data-reusing technique and the learning-rate annealing schedule. The convergence analysis is undertaken upon contraction mapping, and bounds on the learning rate parameter η are provided. This algorithm outperforms the linear LMS and NLMS for prediction of speech.

1 Introduction

A single-neuron neural adaptive filter, trained by gradient-descent (GD) can be described as

$$e(k) = d(k) - \Phi(\mathbf{x}^T(k)\mathbf{w}(k)) \tag{1}$$

$$\mathbf{w}(k+1) = \mathbf{w}(k) - \eta\nabla_{\mathbf{w}}E(e(k)) \tag{2}$$

where k denotes a discrete time instant, $e(k)$ is the instantaneous error at the output neuron, $E(\cdot)$ is the filter cost function, $d(k)$ is some training (desired) signal, $\mathbf{x}(k) = [x_1(k), ..., x_N(k)]^T$ is the input vector, $\mathbf{w}(k) = [w_1(k), ..., w_N(k)]^T$ is the weight vector, $\Phi(\cdot)$ represents a nonlinear activation function of a neuron, η denotes the learning rate parameter and $(\cdot)^T$ denotes the vector transpose. The most common choice for the cost function $E(\cdot)$ is

$$E(e(k)) = \frac{1}{2}e^2(k) \tag{3}$$

The algorithm for on-line adaptation of such an adaptive filter, described by the equations (1), (2), and (3), is referred to as nonlinear gradient descent algorithm (NGD). Due to its inherent simplicity and nonlinearity, this filter might be considered as a suitable solution for applications in nonlinear and/or nonstationary prediction.

The NGD algorithm, however, might suffer from slow convergence and local minima. A constant learning rate parameter η can be considered as one of the factors contributing to these problems. Further, a recent result [1] indicates an inherent relationship between the learning rate parameter η and slope of the nonlinear activation function of an output neuron β, which has a negative impact on the convergence properties of the NGD algorithm with fixed η. In order to achieve fast convergence in the beginning of operation of the NGD algorithm, it is advisable to adopt a large learning rate parameter. On the other hand, to ensure convergence to an optimal weight vector $\mathbf{w}^*$, when close to the optimal value the learning rate parameter has to be small [2]. Also, large constant η may result in algorithms' instability. The cost function given by equation (3) represents an instantaneous estimate of the ensemble average $\langle e^2(k)\rangle$, thus introducing a gradient noise in the operation of the algorithm. This noise will help algorithm to escape from poor local minima, but will reduce the convergence rate of the algorithm [3].

A posteriori techniques have been considered in the area of linear adaptive filtering [4]. It has been shown that instantaneous *a posteriori* output error $\bar{e}(k)$ is smaller in magnitude than corresponding *a priori* error $e(k)$, thus resulting in improved convergence and accuracy. In the limit, *a posteriori* techniques result in an adaptive learning rate algorithm [5, 4]. In the case of neural adaptive filters, results that relate the instantaneous *a posteriori* output error and *a priori* output error, learning rate parameter and slope of the nonlinear activation function of a neuron have been presented in [5].

The learning-rate annealing schedules that behave in the "search-then-converge" manner have been presented in [6]. They ensure fast convergence and small mismatch between actual and optimal weight vector.

Here we propose a novel data-reusing stochastic approximation algorithm, that is built upon the NGD algorithm and incorporates data-reusing techniques and the learning-rate annealing schedule. The convergence analysis of the proposed algorithm is carried out based upon the notion of a contraction mapping. Simulations are carried out on speech, nonlinear and nonstationary signals, in order to verify our analysis.

2 Derivation of the Algorithm

The proposed algorithm is of the GD type. It is derived for a single-neuron forward feed neural adaptive filter. An output neuron nonlinear activation function of such a filter is of sigmoid type, e.g. logistic function.

The algorithm operates in two steps at every time instant k. The first step is a standard annealing step modified for the NGD algorithm, whereas the second step is the data-reusing step.

In the first phase, an update of the learning rate parameter is performed, according to the so-called "search-then-converge" learning rate annealing schedule [6]

$$\eta(k) = \frac{\eta_0}{1+\frac{k}{\tau}} \tag{4}$$

where τ denotes the search time constant. In the early stages of adaptation when time step k is small compared to the search constant τ, the learning rate parameter is approximately equal to η_0, and the algorithm operates essentially as with the constant learning rate parameter $\eta(k) = \eta_0$. For time step k large compared to τ, the learning rate parameter behaves as c/k, where $c = \tau\eta_0$.

The data-reusing step consists of L *a posteriori* iterations. During the data-reusing phase learning rate parameter $\eta(k)$ is kept constant. The proposed algorithm hence increases computational burden L times with two additional divisions and one addition per iteration.

In order to ensure proper behaviour of the algorithm in data-reusing phase, our aim is, as in [5], to preserve the following relationship among the successive *a posteriori* output errors

$$|e_{i+1}(k)| \leq \gamma|e_i(k)|, 0<\gamma<1 \tag{5}$$

where $i, i = 1, ..., L$ denotes the ith *a posteriori* iteration. If we adopt the cost function given by the equation (3), the data-reusing phase can be described as follows

$$\mathbf{w}_{i+1}(k) = \mathbf{w}_i(k) + \eta(k)\Phi'(\mathbf{x}^T(k)\mathbf{w}_i(k))e_i(k)\mathbf{x}(k) \tag{6}$$

$$e_i(k) = d(k) - \Phi(\mathbf{x}^T(k)\mathbf{w}_i(k)) \tag{7}$$

subject to

$$|e_{i+1}(k)| \leq \gamma|e_i(k)|, 0<\gamma<1, i=1,...,L \tag{8}$$

where $\mathbf{w}_i(k)$ denotes the weight vector at the ith *a posteriori* iteration and $e_i(k)$ is the ith *a posteriori* prediction error. Clearly the proposed algorithm incorporates the linear case as a special case. In the case of the linear activation function of an output neuron, this algorithm leads to the modification of the standard least mean square (LMS) algorithm.

3 Analysis of the Data-Reusing Stochastic Approximation Updates

First, we shall introduce the following notation

$$\mathbf{w}(k) = \mathbf{w}_1(k)$$
$$\mathbf{w}(k+1) = \mathbf{w}_{L+1}(k) \tag{9}$$

After iterating the equation (6) L times we have

$$\begin{aligned} \mathbf{w}(k+1) &= \mathbf{w}_{L+1}(k) \\ &= \mathbf{w}_L(k) + \eta(k)e_L(k)\Phi'(\mathbf{x}^T(k)\mathbf{w}_L(k))\mathbf{x}^T(k) \\ &= \mathbf{w}(k) + \sum_{i=1}^{L}\eta(k)e_i(k)\Phi'(\mathbf{x}^T(k)\mathbf{w}_i(k))\mathbf{x}(k) \end{aligned} \tag{10}$$

Recall that for the case of a conctractive activation function of an output neuron the following relationship holds

$$\Phi(a+b) \leq \Phi(a) + \Phi(b) \tag{11}$$

Following the results from [5], in the case of a contractive activation function of an output neuron, the following holds

$$e_{i+1}(k) > [1 - \eta(k)\Phi'(k)\|\mathbf{x}\|_2^2]e_i(k) \tag{12}$$

Here, we assumed that $\Phi'(\mathbf{x}^T(k)\mathbf{w}_i(k)), i = 1, ..., L$ does not change significantly during the iteration, and $\Phi'(\mathbf{x}^T(k)\mathbf{w}_i(k)) = \Phi'(k), i = 1, ..., L$. Also, that $e_i(k), i = 1, ..., L$ and $e(k)$ have the same sign. By iterating the relationship (12) we have [5]

$$e(k+1) > [1 - \eta(k)\Phi'(k)\|x(k)\|_2^2]^L e(k) \tag{13}$$

For the algorithm given by the equations (5), (6) and the constraint (7) to be feasible the term $[1 - \eta(k)\Phi'(k)\|\mathbf{x}(k)\|_2^2]$ must have the L_2 norm less than unity. In that case the whole procedure is the fixed point iteration. This leads us to the following constraint on the learning rate parameter

$$0 < \eta(k) < \frac{1}{\Phi'(k)\|\mathbf{x}(k)\|_2^2} \tag{14}$$

Combining equations (10) and (13) yields

$$\mathbf{w}(k+1) < \mathbf{w}(k) + \frac{1 - [1 - \eta(k)\Phi'(k)||\mathbf{x}(k)||_2^2]^L}{||\mathbf{x}(k)||_2^2} e(k)\mathbf{x}(k) \tag{15}$$

Thus, for the case of a contractive activation function of the neuron, the upper bound of the weight vector update of the proposed algorithm, after L *a posteriori* iterations at the time instant k is given by

$$\Delta\mathbf{w}(k) < \frac{1 - [1 - \eta(k)\Phi'(k)||\mathbf{x}(k)||_2^2]^L}{||\mathbf{x}(k)||_2^2} e(k)\mathbf{x}(k) \tag{16}$$

Recall that in the case of an expansive activation function of the neuron, the following relationship holds

$$\Phi(a+b) \geq \Phi(a) + \Phi(b)$$
$$\Phi'(\zeta) > \zeta, \forall \zeta \in \mathbf{R} \tag{17}$$

Following the analysis carried out in the fashion similar to the one above, in the case of an expansive activation function of the neuron we have

$$e_{i+1}(k) < [1 - \eta(k)\Phi'(k)||\mathbf{x}(k)||_2^2] e_i(k) \tag{18}$$

Thus, in the case of an expansive activation function of an output neuron and having in mind the constraint (7), the lower bound of the weight vector update of the proposed algorithm, after L *a posteriori* iterations at the time instant k is given by

$$\Delta\mathbf{w}(k) > \frac{1 - [1 - \eta(k)\Phi'(k)||\mathbf{x}(k)||_2^2]^L}{||\mathbf{x}(k)||_2^2} e(k)\mathbf{x}(k) \tag{19}$$

This is why in practice we try to avoid an expansive activation function of an output neuron.

4 Convergence Analysis of the Proposed Algorithm

Even though the upper bound for the data-reuse error has been established, in the case of an expansive activation function of an output neuron, according to the contraction mapping theorem (CMT), the a priori output error $e(k)$ can grow without a limit. Hence, in this section we shall constrain our analysis to the case of the contractive activation function of the neuron.

Let us introduce the error weight vector $\mathbf{v}$ as $\mathbf{v}(k) = \mathbf{w}(k) - \mathbf{w}^*(k)$, where $\mathbf{w}^*(k)$ are some optimal weights, and

$$e(k) = e^*(k) - \Phi(\mathbf{x}^T(k)\mathbf{v}(k)) \tag{20}$$

From the equations (15) and (20) we have

$$\mathbf{v}(k+1) < \mathbf{v}(k) + \frac{1 - [1 - \eta(k)\Phi'(k)||\mathbf{x}(k)||_2^2]^L}{||\mathbf{x}(k)||_2^2} e^*(k)\mathbf{x}(k) - \frac{1 - [1 - \eta(k)\Phi'(k)||\mathbf{x}(k)||_2^2]^L}{||\mathbf{x}(k)||_2^2} \mathbf{x}^T(k) \times \Phi(\mathbf{x}^T(k)\mathbf{v}(k)) \tag{21}$$

With the assumption of contractivity of Φ, the homogeneous part of (21) becomes

$$\mathbf{v}(k+1) < \Big[I - \frac{1 - [1 - \eta(k)\Phi'(k)||\mathbf{x}(k)||_2^2]^L}{||\mathbf{x}(k)||_2^2} \mathbf{x}(k)\mathbf{x}^T(k)\Big]\mathbf{v}(k) \tag{22}$$

In order to ensure convergence of (22), the term in the square parenthesis of (22) has to be a contraction mapping operator, which has already been shown. Now, we shall consider the influence of the learning-rate annealing schedule, given by the equation (4), on the convergence of the proposed algorithm. From the equation (4) it is obvious that $0 < \eta(k) \leq \eta_0, \forall k \geq 1$. Thus, condition (22) will hold, for suitable chosen values of η_0 and τ.

5 Experimental Results

In order to verify our analysis we performed several experiments on the speech, which is a nonlinear and nonstationary signal, this is shown in Figure 1. The order of the filter in the performed experiments was $N = 10$, and the logistic function was chosen as the nonlinear activation function of neuron. The logistic function performs contraction mapping for $\beta < 4$. In the first experiment we compared the performances of the proposed algorithm and the NGD algorithm. The quantitative performance measure was the standard prediction gain, a logarithmic ratio between the expected signal and error variances

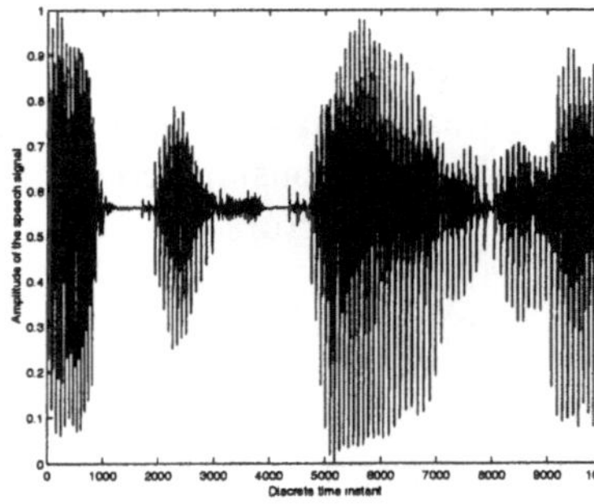

Fig. 1. Speech signal

$R_p = 10\log(\hat{\sigma}_s^2/\hat{\sigma}_e^2)$. The slope of the nonlinear activation function of an output neuron β was set to be $\beta = 1$. For the NGD algorithm the learning rate parameter was set to be $\eta = 0.3$. In the case of the proposed, data-reusing stochastic approximation algorithm, the search constant τ was set to be $\tau = 500$, parameter η_0 was set to be $\eta_0 = 0.3$ and the number of *a posteriori* iterations L was set to be $L = 10$. The prediction gain for the standard NGD algorithm was $R_p = 1.99dB$, compared to $R_p = 5.62dB$ for the data-reusing stochastic approximation algorithm. It is obvious that the proposed algorithm outperforms the standard NGD algorithm.

In the second experiment we have compared the performance of the LMS algorithm, normalised least mean squares (NLMS) algorithm and the proposed, data-reusing stochastic approximation algorithm. Also, we have modified LMS algorithm weight adaptation according to the proposed algorithm and tested that modified LMS algorithm. Simulations were carried out on the same speech signal as in the first experiment. The slope of the nonlinear activation function of the neuron β was set to be $\beta = 4$, since this value makes Φ close to the linear function in the vicinity of the origin. For the standard LMS algorithm learning rate parameter η was set to be $\eta = 0.3$ and in that case prediction gain was $R_p = 7.71dB$. Further, in the case of the LMS algorithm with learning rate annealing schedule (4), for different values of search time constant τ, i.e. $\tau = 100, 500, 1000, 5000$, the prediction gain R_p was $5.89, 6.57, 6.99$, and $7.69dB$ respectively. In the case of the NLMS algorithm for the L set to be $L = 1$ (standard NLMS algorithm) and the prediction gain was $R_p = 8.72dB$. After increasing the value of L, i.e. $L = 2$ the prediction gain was $R_p = 32.30dB$. Further increase of the value of L, i.e. $L = 3, 4, ...$ gave no improvement. For the modified LMS algorithm and the proposed algorithm parameter η_0 was set to be $\eta_0 = 0.3$. Results for the modified LMS algorithm and the proposed algorithm, for the different values of the number of *a posteriori* iterations L, are summerized in Tables I and II. The proposed algorithm outperforms the standard LMS and NLMS algorithms and exhibits improved convergence. Also, it has been observed that modified LMS algorithm approaches NLMS algorithm as the number of *a posteriori* iterations L increases, which is in accordance with the results in [4, 5].

Number of the *a posteriori* iterations $L = 10$	
τ	Pred. gain [dB] (mod.LMS/proposed)
100	9.76/9.27
500	14.40/15.22
1000	18.18/19.01
5000	26.68/28.18

Table I. Performance of the modified LMS algorithm and the proposed algorithm, $L = 10$

Number of the *a posteriori* iterations $L = 100$	
τ	Pred. gain [dB](mod.LMS/proposed)
100	20.61/20.29
500	32.13/31.43
1000	32.20/33.20
5000	32.20/34.11

Table II. Performance of the modified LMS algorithm and the proposed algorithm, $L = 100$

6 Summary

A data-reusing stochastic approximation algorithm for neural adaptive filters has been proposed. The proposed algorithm operates with a time-varying learning rate and a data-reusing gradient modification. The convergence analysis of the proposed algorithm based on the notion of contraction mapping has been undertaken. Experiments on the speech show that the proposed algorithm outperforms the LMS, NLMS and modified LMS algorithms.

References

[1] D. P. Mandic and J. A. Chambers, "Relationship between the slope of the activation function and the learning rate for the RNN", *Neural Computation*, vol. 11, no. 5, pp. 1069–1077, 2000.

[2] V. J. Mathews and Z. Xie, "A stochastic gradient adaptive filter with gradient adaptive step size", *IEEE Transactions on Signal Processing*, vol. 41, no. 6, pp. 2075–2087, 1993.

[3] C. Darken, J. Chang, and J. Moody, "Learning schedules for faster stochastic gradient search", *IEEE Workshop on NNSP '92*, pp. 3–10, 1992.

[4] J. R. Treichler, C. R. Johnson, and M. G. Larimore, *Theory and design of adaptive filters*, John Wiley & Sons, 1987.

[5] D. P. Mandic and J. A. Chambers, "Relationships between the a priori and a posteriori errors in nonlinear adaptive neural filters", *Neural Computation*, vol. 12, pp. 1285–1292, 2000.

[6] C. Darken and J. Moody, "Towards faster stochastic gradient search", *Neural Information Processing Systems* (J. E. Moody, S. J. Hanson, and R. P. Lippman, eds.), vol. 4, pp. 1009–1016, Morgan Kaufman, 1992.

Robust On-Line Statistical Learning

Enrico Capobianco* [1]

*CNR - Consiglio Nazionale delle Ricerche, Italy

Abstract

We describe possible ways of endowing neural networks with statistically robust properties. We especially look at learning schemes resistant to outliers by defining error criteria able to handle deviations from convenient probability distribution assumptions. It comes out to be convenient to cast neural nets in state space representations and apply both Kalman Filter and Stochastic Approximation procedures in order to suggest statistically robustified solutions for on-line learning.

1 Introduction: Stochastic Approximation

We consider a nonlinear function $f(X_t, \theta)$, where $f: R^k \times \Theta \to R$, X_t is a $k \times 1$ random input vector and $\theta \in \Theta \subset R^p$ represents the vector of unknown parameters, and use this set-up for forecasting the random variable y_t through a single hidden layer feedforward network structure:

$$f(X,\theta) = \alpha_0 + \sum_{j=1}^{q} \alpha_j F(X'\beta_j) \qquad (1)$$

where $\theta = (\alpha', \beta')'$ represents the weight vector and $F: R \to R$ is a bounded and continuosly differentiable activation function localized at the hidden layers. At the output level we set an identity matrix. One can clearly see that $f(X,\theta)$ is an approximation of the objective function $g(X_t) = E(y_t/X_t)$. Thus, in this *nonlinear least square* frame we seek a solution θ^* to the problem $min_\theta[E([y_t - f(X_t,\theta)]^2)]$, or, equivalently, to the *first order conditions* equation $E(\nabla_\theta f(X_t,\theta)[y_t - f(X_t,\theta)]) = 0$, with ∇_θ representing the gradient $k \times 1$ vector calculated w.r.t. θ. The Robbins-Monro (RM) *Stochastic Approximation* (SA) algorithm stores the current approximation value X_t and approximates the objective function locally and linearly via its gradient. It can also be adapted for a nonlinear regression:

$$\hat{\theta}_{t+1} = \hat{\theta}_t + \delta_t \nabla_\theta f(X_t, \hat{\theta}_t)[y_t - f(X_t, \hat{\theta}_t)] \qquad (2)$$

Since this recursion is equivalent to that of a *Stochastic Gradient* method, it generalizes the well-known *Backpropagation* (BP) algorithm [7], popular in neural network learning theory, by allowing for a time varying learning rate.

2 Artificial Learning

In batch identification algorithms the prediction error is the building block for the chosen optimization criterion, i.e., $Loss_N(\theta) = \frac{1}{N}\sum_{t=1}^{N} \epsilon'_{t\theta}\epsilon_{t\theta}$, where $\epsilon_{t\theta} = y_t - \hat{y}_{t\theta}$. Learning schemes seek to minimize the loss function via iterations according to the negative gradient direction, i.e., the steepest descent algorithm, or a Gauss-Newton search direction. Alternatively, we could find the prediction error ϵ_t at each step, in a recursive fashion. The approximation to $E(y_t/X_t)$ is only locally optimal, but it's nevertheless important to relax the usually retained i.i.d. (independently and identically distributed) assumption about the stochastic process generating the data. Thus, both batch and recursive algorithms converge to a local minimum with probability one, i.e., $\hat{\theta}_t \to \theta^*$ local minimum of $lim_t E[y_t - f(x_t,\theta)]^2$; so does the BP algorithm, therefore.

Artificial Neural Networks can be cast in a state space representation and, as in the case of SA algorithms, this is a way to allow for a generalized BP to learn how to approximate nonlinear functions in a time series set-up. The Kalman filter algorithm thus relates to the BP procedure expressed in this re-formatted neural net set-up; we can apply the *Extended Kalman Filter (EKF)* and the *Iterated Kalman Filter (IKF)* procedures as in [1], and this last algorithm is known to be an application of the Gauss-Newton (GN) method. It's common in statistics and econometrics to work with estimators that aim at minimizing sums of squared residuals like $S(\theta) = \sum_t \epsilon_t^2$, whose Gradient is $v(\theta) = \frac{\partial S(\theta)}{\partial \theta} = 2\sum \frac{\partial \epsilon_t}{\partial \theta}\epsilon_t$ and whose Hessian is $V(\theta) = \frac{\partial^2 S(\theta)}{\partial\theta\partial\theta'} = 2\sum[\frac{\partial\epsilon_t}{\partial\theta}\frac{\partial\epsilon_t}{\partial\theta'} - \frac{\partial^2\epsilon_t}{\partial\theta\partial\theta'}\epsilon_t]$. Several schemes are able of iteratively finding a solution to the initial minimization problem. The most general one is the *Newton-Raphson* (NR) method, which is given by:

$$\theta^* = \hat{\theta} - [\sum(\frac{\partial\epsilon_t}{\partial\theta}\frac{\partial\epsilon_t}{\partial\theta'} - \frac{\partial^2\epsilon_t}{\partial\theta\partial\theta'}\epsilon_t)]^{-1}\sum\frac{\partial\epsilon_t}{\partial\theta}\epsilon_t \qquad (3)$$

Since the term involving second derivatives is usually small when compared to the first derivatives

[1]The work preparation was supported by a grant at the time of an author's visit to the Technical University of Denmark, Department of Mathematical Modelling, CONNECT group, Lyngby (DK).

product term, the GN scheme approximates the above iterative solution and presents a formula that is identical to NR, apart from the term with second derivatives.

3 Likelihood-Based Inference

Once a neural network architecture is cast in a state space representation, the most important aspect is that from the Kalman Filter algorithm and its variants we straightforwardly obtain the likelihood function through the *prediction error decomposition* (PED). The likelihood function for the whole time series, when temporally dependent observations are considered, is obtained by the joint conditional probability density funtion, i.e. $L(y,\theta) = \Pi_{t=1}^{N} p(y_t/Y_{t-1})$ (by considering Y_{t-1} the set of observations up to and including y_{t-1}). Since the innovation or prediction error computed by the filter is $\eta_t = y_t - E(y_t/Y_{t-1})$ and $var(\eta_t) = D_t$, when the observations are normally distributed the multivariate likelihood function can be expressed as:

$$\log L = -c - \frac{1}{2}\sum_{t=1}^{N} \log \mid D_t \mid - \frac{1}{2}\sum_{t=1}^{N} \eta_t' D_t^{-1} \eta_t \quad (4)$$

with $c = \frac{KN}{2}\log 2\pi$, η_t a $k \times 1$ vector and N observations. Note that under Gaussianity the filter delivers an optimal *Minimum Mean Squared* solution for the estimation problem; under hypotheses different from the Gaussian, the filter gives only a *Minimum Mean Square Linear* solution and the values which are computed are *Quasi Maximum Likelihood* estimates, less efficient but consistent (and therefore useful to start a recursive or multi-step procedure).

For the case we study here, the solution is therefore sub-optimal and close to the optimal one according to the accuracy of the approximation involved. From the likelihood function as given by (4) we can calculate its derivatives either analytically or numerically, and in both cases we can leave the filter to compute these quantities together with the other ones representing the core of the algorithm. The ith element of the Score vector $\frac{\partial \log L}{\partial \theta_i}$ is given by:

$$-\frac{1}{2}\sum_t [tr[(D_t^{-1}\frac{\partial D_t}{\partial \theta_i})(I - D_t^{-1}\eta_t\eta_t')] - \frac{\partial \eta_t'}{\partial \theta_i} D_t^{-1}\eta_t] \quad (5)$$

therefore requiring the evaluation of the $k \times k$ matrices of derivatives $\frac{\partial D_t}{\partial \theta_i}$ and the $k \times 1$ vector of derivatives $\frac{\partial \eta_t}{\partial \theta_i}$, for $i = 1, \ldots, p$ and $t = 1, \ldots, N$.

4 Computational Asymptotic Aspects

From equation (4) $log\ L = \sum_{t=1}^{N} l_t$, where for univariate time series $l_t = -\frac{1}{2}log\ 2\pi - \frac{1}{2}log \mid D_t \mid -\frac{1}{2}\eta_t' D_t^{-1}\eta_t$; we differentiate l_t with respect to the jth element of the parameter vector θ[1] and obtain $\frac{\partial l_t}{\partial \theta_j} - \frac{1}{2}tr[D_t^{-1}\frac{\partial D_t}{\partial \theta_j}] - \frac{1}{2}[\frac{\partial \eta_t'}{\partial \theta_j}D_t^{-1}\eta_t - \eta_t' D_t^{-1}\frac{\partial D_t}{\partial \theta_j}D_t^{-1}\eta_t + \eta_t' D_t^{-1}\frac{\partial \eta_t}{\partial \theta_j}]$. Re-write the expression as $\frac{\partial l_t}{\partial \theta_j} = -\frac{1}{2}tr[(D_t^{-1}\frac{\partial D_t}{\partial \theta_j})(I - D_t^{-1}\eta_t\eta_t')] - (\frac{\partial \eta_t}{\partial \theta_j})' D_t^{-1}\eta_t$ and differentiate w.r.t. the rth element of θ, obtaining the $(jr)^{th}$ element of the IM,

$$\begin{aligned}\frac{\partial^2 l_t}{\partial \theta_j \partial \theta_r} = & -\frac{1}{2}tr[\frac{\partial (D_t^{-1}\frac{\partial D_t}{\partial \theta_j})}{\partial \theta_r}][I - D_t^{-1}\eta_t\eta_t'] - \\ & \frac{1}{2}tr[D_t^{-1}\frac{\partial D_t}{\partial \theta_j}D_t^{-1}\frac{\partial D_t}{\partial \theta_r}D_t^{-1}\eta_t\eta_t'] + \\ & \frac{1}{2}tr[D_t^{-1}\frac{\partial D_t}{\partial \theta_j}D_t^{-1}(\frac{\partial \eta_t}{\partial \theta_r}\eta_t' + \eta_t\frac{\partial \eta_t'}{\partial \theta_r})] - \\ & \frac{\partial^2 \eta_t'}{\partial \theta_j \partial \theta_r}D_t^{-1}\eta_t - \frac{\partial \eta_t'}{\partial \theta_j}\frac{\partial D_t^{-1}}{\partial \theta_r}\eta_t - \frac{\partial \eta_t'}{\partial \theta_j}D_t^{-1}\frac{\partial \eta_t}{\partial \theta_r}\end{aligned} \quad (6)$$

By applying the *law of iterated expectations* we take expectations in the above equations conditional on the information available up to time t-1; thus, with innovations involved, many of these terms result 0 by definition, and we can find a much simpler IM expression. We have that $E_z[E_{y/z}(y/z)] = E_y(y)$, or in other terms the random variable $E(y/z)$ has the same expectation as the y one; thus, when z represent previous observations, time series too are accounted for. In our set-up this means that since $\eta_t = y_t - \hat{y}_t = y_t - E_{t-1}(y_t)$, we can express the innovation derivatives as functions of the last expression, i.e., dependent on the $E_{t-1}(y_t)$ factor: $\frac{\partial \eta_t}{\partial \theta_j} = -\frac{\partial}{\partial \theta_j}E_{t-1}(y_t)$.

Thus: $E_{t-1}(\frac{\partial \eta_t'}{\partial \theta_j}\eta_t) = \frac{\partial \eta_t'}{\partial \theta_j}E_{t-1}(\eta_t) = 0$, given that $E_{t-1}(\eta_t) = 0$; it is the case that $E_{t-1}(\frac{\partial^2 \eta_t'}{\partial \theta_j \partial \theta_r}D_t^{-1}\eta_t) = \frac{\partial^2 \eta_t'}{\partial \theta_j \partial \theta_r}D_t^{-1}E_{t-1}(\eta_t) = 0$, for the same reason of the previous case. The first term in the first row of the master expression too disappears, since $E_{t-1}(\eta_t\eta_t') = D_t$ and thus the last squared parenthesis simplifies to 0. Then, we are left with a simplified second term plus the very last one in the expression; all the other terms drop out.

The final expression for the $(jr)^{th}$ element of the IM is given by:

$$\begin{aligned}IM_{(j,r)}(\theta) \quad = \quad & \frac{1}{2}\sum_t tr[D_t^{-1}\frac{\partial D_t}{\partial \theta_j}D_t^{-1}\frac{\partial D_t}{\partial \theta_r}] + \\ & E[\sum_t (\frac{\partial \eta_t}{\partial \theta_j})' D_t^{-1}\frac{\partial \eta_t}{\partial \theta_r}]\end{aligned} \quad (7)$$

[1] We use two rules for symmetric matrices: $\frac{\partial |X|}{\partial t} = \mid X \mid tr\ [X^{-1}\frac{\partial X}{\partial t}]$, $\frac{\partial X^{-1}}{\partial t} = -X^{-1}\frac{\partial X}{\partial t}X^{-1}$.

where the $\sum_t$ now appear by the definition $-E[\frac{\partial^2 log\ L}{\partial\theta_j \partial\theta_r}] = -E[\sum_t \frac{\partial^2 l_t}{\partial\theta_j \partial\theta_r}]$ and the E left in the last term has, asymptotically, a negligible impact on the argument's order of magnitude.

5 Adapted Artificial Learning

The above derivatives may be computed through p additional passes of the KF; if we consider a new run of the filter with $\theta = [\theta_1 \ldots\ldots \theta_i + \delta_i \ldots\ldots \theta_p]$ we obtain a new set of innovations $\eta_t^{(i)}$ and variances $D_t^{(i)}$ and the numerical approximations of the derivatives are $\delta_i^{-1}[\eta_t^{(i)} - \eta_t]$ and $\delta_i^{-1}[D_t^{(i)} - D_t]$. However, for large complex networks computations become heavy. If we denote with $\hat{\theta}$ our ML estimate for θ and $IM(\theta)$ represents the *Fisher Information Matrix* $\int(\frac{\partial log\ L}{\partial\theta})^2 dF$ such that $lim \frac{1}{N} IM(\theta) = I^*(\theta)$, then statistical theory says that under regularity conditions the $\sqrt{N}(\hat{\theta} - \theta) \sim Gaussian(0, [I^*(\theta)]^{-1})$ law goes through asymptotically. Therefore, IM is a crucial quantity because it gives an estimate of the asymptotic covariance matrix of the ML estimator.

In the State Space Filter set-up the PED likelihood function yields an IM which depends on first derivatives only. With the Kalman Filter we can calculate these first derivatives, numerically or analytically, but always through parallel computations with respect to the main algorithm, and then the IM too, quite simply, given the structure of it once it's derived through the prediction errors or innovations. Kalman filter numerical estimates, in particular, are more efficient because obtained as a byproduct of the filter runs and, moreover, their computation comes from the PED likelihood or quasi-likelihhod function and related derivatives, which benefits of the inherited asymptotic properties of such estimates.

When accurate estimates are seeked, then the Berndt, Hall, Hall and Hausman (*BHHH*) algorithm, which suggests the approximation of the Hessian by the well-known outer-product formula, i.e. $\sum_{t=1}^{N} \frac{\partial \log L_t}{\partial\theta} \frac{\partial \log L_t}{\partial\theta'}$, is the most indicated choice we have in order to improve, at least asymptotically, the efficiency of the initial GN or equivalently IKF estimator $\hat{\theta}$ according to the recursion:

$$\theta^* = \hat{\theta} + \lambda \left[\sum_{t=1}^{N} \frac{\partial \log L_t}{\partial\theta}' \frac{\partial \log L_t}{\partial\theta'}\right]^{-1} \sum_{t=1}^{N} \frac{\partial \log L_t}{\partial\theta} \quad (8)$$

where λ is a variable step length chosen so that the likelihood function is maximized in a given direction. In the likelihood framework, where one maximizes $log\ L$ or equivalently minimizes $-log\ L$, the NR iteration in (3) is effective; if instead of working with the matrix of second derivatives one uses its expectation, the *Scoring* method is employed and thus the information matrix now appears in the update step by simply multiplying the expectation by minus one,

$$\theta^* = \hat{\theta} + IM^{-1}(\hat{\theta})\ \partial\ log\ L(\hat{\theta}) \quad (9)$$

For $\hat{\theta}$ consistently estimated in the above formulations, with just one iteration an estimator with the same asymptotic distribution as the ML estimator can be found; therefore, asymptotically efficient estimates are achieved.

6 M-Estimation Approach

Following [2, 4], consider the choice of the following error criterion function, $E_n(\hat{\theta}) = \sum_{i=1}^{n} \rho(y_i - \hat{y}_i)$, where $\hat{y}_i$ is computed by the net as a superposition of functions of the input x and the weights θ, as explained before. The ρ function is a statistically robust function if satisfies certain properties. Basically, it has either to trim out the abnormal data points or allow for a smoothly descending-to-0 influence function, and still preserve efficiency in estimating the bulk of the observations distribution[2].

Thus, $\psi(.) = \rho'(.)$ is the *Influence Function* IF to be analyzed and we require it to be bounded and continuous; well-known examples are the classic Huber's minimax, Hampel's skipped means or Lorentzian error functions. Instead of minimizing the ρ criterion function, we could equivalently reason in terms of *first order conditions* (FOC), the equations expressed before through the function ψ. And given that the system of equations that one obtains is nonlinear we must use the iterative techniques shown before to get solutions. The one-step M-estimators are definitely useful in these circumstances since they improve the efficiency of an initial consistent estimator, which in learning problems could be the backpropagation algorithm that runs for some epochs and then is stopped based on a cross-validation method.

By considering the FOC equation for the local solution θ^* we saw that we must find solution for an equation like $V(\theta) = 0$, where $V(\theta) = \sum_i v(e_i(\hat{\theta})) = \sum_i(\nabla_\theta f(x_i, \hat{\theta})[y_i - f(x_i, \hat{\theta}))$, with $e_i(.)$ the residual computed at pattern i. The importance of the proposed methods depends on at least two strategies:

1. *It would intuitively help, for robustness purposes, to put a leverage on the influence func-*

[2]It means that it must be fully informative about the middle region of the data distribution, i.e., efficient at the gaussian model, but also able to account for the possible thick tails of the distribution.

tion in the described inference context; this step would allow us to merge the M-estimation approach directly in the learning paradigm [5].

The Procedure: by simply using the chain rule one can decompose the error criterion function first derivative such that:

$$\frac{\partial Loss(e_i(\theta))}{\partial\theta} = \frac{\partial Loss(e_i(\theta))}{\partial e_i}\frac{\partial e_i}{\partial\theta} \quad (10)$$

and thus write the original expression equivalently as a function of the influence function; here the IF is playing the role of the weight or influence of the individual residuals on their own first derivatives, i.e.,

$$\frac{\partial Loss(e_i(\theta))}{\partial\theta} = \psi_i\frac{\partial e_i(\theta)}{\partial\theta} \quad (11)$$

2. *It is possible to extend the first strategy to a recursive design, by applying the same principles in the KF set-up.* We can indeed exploit the fact that the KF delivers a set of prediction errors, one at each run, and from them we can set a likelihood function equation as in (4). This formula basically corresponds to a functional which we can express in the following compact way: $\Gamma(\hat{\theta}_k) = E[L(\eta,\theta_k)]$, where η is the vector of prediction errors, and thus we can come up with an empirical functional like $\frac{1}{n}\sum_{i=1}^{n} L(\eta_i,\theta_k)$, after the KF's runs[3].

The Procedure: in order to exploit the IF in such a recursive framework, we note that the same functional Γ numerically optimized in the SSF framework can be treated in a recursive way when a SA algorithm is adapted to include it in its updating step, i.e., $\hat{\theta}_{n+1} = \hat{\theta}_n + a_n W_n(\eta_n)$, where the W_n term is going to modulate (or equivalently measure the influence of) the impact of the likelihood prediction errors, like for the gain factor in the KF update equation. By just allowing for the W_n matrix to be decomposed as in equation (11), we have the IF entering the update step.

The W_n is thus defined as the factorization between the ratio of the first derivatives of the error (or loss) function w.r.t the prediction errors derivatives and the ratio of the prediction errors derivatives calculated w.r.t. the weights derivatives. We are thus considering the influence of possible outliers in the data set onto the prediction errors derivatives.

[3]In the engineering literature this is known as ***prediction error identification*** criterion, where the model can also deal with misspecification but still retain the asymptotic properties, in the same spirit of the QML criteria [6].

When instead W_n combines a gradient function with a weighting matrix, we are in the case of the IKF. In learning we interpret this step as a generalized BP algorithm where the W_n can again be defined as the factorization described before.

7 Conclusions

From the statistical inference perspective it is important to have the possibility of adopting estimators that use likelihood information in order to reach better asymptotic properties for the final estimates. Equivalently, for neural networks learning processes, it would be ideal to be able to characterize the estimates not only in terms of consistency but also of efficiency. Depending on the error function we may use the information coming from the influence function, by letting it to enter the recursive steps characterizing the Kalman filter or Stochastic Approximation algorithms, and thus the generalized Backpropagation, so to modulate the impact on estimates of deviations from the assumed model or by using it so to correct for efficiency an initial robust estimate via the one-step Newton-Raphson.

References

[1] E.Capobianco: "A Unifying view of Stochastic Approximation, Kalman Filter and Backpropagation". *NNSP-IEEE Proceedings*, vol. V, pp 87-94, 1995.

[2] F.R.Hampel, E.M.Ronchetti, P.J.Rousseeuw and W.A.Stahel: *Robust Statistics: the approach based on Influence Function*. New York, US: Wiley, 1986.

[3] A.C. Harvey: *Forecasting, Structural Time Series Models and the Kalman Filter*. Cambridge, UK: Cambridge University Press, 1989.

[4] P.J. Huber: *Robust Statistics*. New York, US: Wiley, 1981.

[5] K. Liano: "Robust Error Measure for Supervised Neural Network Learning with Outliers". *IEEE Transactions on Neural Networks*, vol. 7, pp 246-250, 1996.

[6] L.Ljiung, G.Pflug and H.Walk: *Stochastic Approximation and Optimization of Random Systems*. Basel, CH: Birkhauser Verlag, 1992.

[7] D.E. Rumelhart, G.E. Hinton, R.J. Williams: "Learning internal representations by error propagation", in *Parallel Distributed Processing: Explorations in the Microstructure of Cognition*, D.E. Rumelhart and J.L. McClelland, Eds. Cambridge, US: MIT Press, vol 1, pp 318-362, 1986.

Efficient Sequential Minimal Optimisation of Support Vector Classifiers

Gavin C. Cawley*[1]

*School of Information Systems, University of East Anglia, Norwich, Norfolk, U.K. NR4 7TJ. E-mail: gcc@sys.uea.ac.uk

Abstract

This paper describes a simple modification to the sequential minimal optimisation (SMO) training algorithm for support vector machine (SVM) classifiers, reducing training time at the expense of a small increase in memory used proportional to the number of training patterns. Results obtained on real-world pattern recognition tasks indicate that the proposed modification can more than halve the average training time.

1 Introduction

The Support Vector Machine (SVM) classification method has demonstrated impressive results in several difficult, real-world pattern recognition tasks. The Sequential Minimal Optimisation (SMO) algorithm has proved one of the most popular training algorithms for support vector machines, due to its simplicity and competitive training times possible. Platt reports that the run-time of the sequential minimal optimisation algorithm is dominated by evaluation of the kernel function. This paper describes a simple modification that reduces the number of kernel evaluations involving patterns associated with Lagrange multipliers at the upper bound C. This modification can be implemented without a significant increase in the complexity of the code, and is demonstrated to more than halve training time at the expense of a modest increase in storage requirements.

The remainder of this paper is structured as follows: Section 2 provides an overview of the support vector pattern recognition method, introducing the notation adopted in this paper, and section 3 an overview of the sequential minimal optimisation algorithm. Section 4 describes a simple modification improving the efficiency of the sequential minimal optimisation algorithm. Results obtained on real-world benchmark tasks are presented in section 5. The key results are summarised in section 6.

[1]This work was supported by the European Commission, grant number IST-99-11764, as part of its Framework V IST programme.

2 Support Vector Classification

The support vector machine [1, 2], given labelled training data

$$\{(\boldsymbol{x}_i, y_i)\}_{i=1}^{\ell}, \qquad \boldsymbol{x}_i \in \boldsymbol{X} \subset \mathbb{R}^d, \quad y_i \in \{-1, +1\},$$

constructs a maximal margin linear classifier in a high dimensional feature space, $\Phi(\boldsymbol{x})$, defined by a positive definite kernel function, $k(\boldsymbol{x}, \boldsymbol{x}')$, specifying an inner product in the feature space,

$$\Phi(\boldsymbol{x}) \cdot \Phi(\boldsymbol{x}') = k(\boldsymbol{x}, \boldsymbol{x}').$$

A common kernel is the Gaussian radial basis function (RBF),

$$k(\boldsymbol{x}, \boldsymbol{x}') = e^{-||\boldsymbol{x} - \boldsymbol{x}'||^2/2\sigma^2}.$$

The function implemented by a support vector machine is given by

$$f(\boldsymbol{x}) = \left\{ \sum_{i=1}^{\ell} \alpha_i y_i k(\boldsymbol{x}_i, \boldsymbol{x}) \right\} - b. \tag{1}$$

To find the optimal coefficients, $\boldsymbol{\alpha}$, of this expansion it is sufficient to maximise the functional,

$$W(\boldsymbol{\alpha}) = \sum_{i=1}^{\ell} \alpha_i - \frac{1}{2} \sum_{i,j=1}^{\ell} y_i y_j \alpha_i \alpha_j k(\boldsymbol{x}_i, \boldsymbol{x}_j), \tag{2}$$

in the non-negative quadrant,

$$0 \leq \alpha_i \leq C, \qquad i = 1, \ldots, \ell, \tag{3}$$

subject to the constraint,

$$\sum_{i=1}^{\ell} \alpha_i y_i = 0. \tag{4}$$

C is a regularisation parameter, controlling a compromise between maximising the margin and minimising the number of training set errors. The

Karush-Kuhn-Tucker (KKT) conditions can be stated as follows:

$$\alpha_i = 0 \quad \Rightarrow \quad y_i f(\boldsymbol{x}_i) \geq 1, \tag{5}$$
$$0 < \alpha_i < C \quad \Rightarrow \quad y_i f(\boldsymbol{x}_i) = 1, \tag{6}$$
$$\alpha_i = C \quad \Rightarrow \quad y_i f(\boldsymbol{x}_i) \leq 1. \tag{7}$$

These conditions are satisfied for the set of feasible Lagrange multipliers, $\boldsymbol{\alpha}^0 = \{\alpha_1^0, \alpha_2^0, \ldots, \alpha_\ell^0\}$, maximising the objective function given by equation 2. The bias parameter, b, is selected to ensure that the second KKT condition is satisfied for all input patterns corresponding to non-bound Lagrange multipliers. Note that in general only a limited number of Lagrange multipliers, $\boldsymbol{\alpha}$, will have non-zero values; the corresponding input patterns are known as support vectors. Let $\mathcal{I}$ be the set of indices of patterns corresponding to non-bound Lagrange multipliers,

$$\mathcal{I} = \{i \quad : \quad 0 < \alpha_i^0 < C\},$$

and similarly $\mathcal{J}$ be the set of indices of patterns with Lagrange multipliers at the upper bound C,

$$\mathcal{J} = \{i \quad : \quad \alpha_i^0 = C\}.$$

Equation 1 can then be written as an expansion over support vectors,

$$f(\boldsymbol{x}) = \left\{ \sum_{i \in \{\mathcal{I}, \mathcal{J}\}} \alpha_i^0 y_i k(\boldsymbol{x}_i, \boldsymbol{x}) \right\} - b. \tag{8}$$

For a full exposition of the support vector method, see the excellent books by Vapnik [3, 4] or Cristianini and Shawe-Taylor [5].

3 Sequential Minimal Optimisation

The sequential minimal optimisation (SMO) algorithm (Platt [6]) implements an extreme form of the decomposition method of Osuna *et al.* [7]. SMO iteratively solves the constrained quadratic optimisation problem, given in equations 2–4, via a series of smaller optimisation problems, each involving a single pair of Lagrange multipliers, which may be solved analytically. The popularity of SMO stems from the competitive training times that can be obtained without the need for a complex numerical optimisation library. The functional given in equation 2 can be written in terms of a pair of Lagrange multipliers, say α_1 and α_2, as

$$\begin{aligned} W(\alpha_1, \alpha_2) &= \alpha_1 + \alpha_2 - \alpha_1 y_1 v_1 - \alpha_2 y_2 v_2 \\ &- \frac{1}{2} K_{11} \alpha_1^2 - \frac{1}{2} K_{22} \alpha_2^2 - s K_{12} \alpha_1 \alpha_2 \\ &+ W_{\text{constant}}, \end{aligned} \tag{9}$$

where

$$v_i = \sum_{j \neq \{1,2\}} \alpha_j y_j k(\boldsymbol{x}_j, \boldsymbol{x}_i),$$

and

$$K_{ij} = k(\boldsymbol{x}_i, \boldsymbol{x}_j).$$

The linear equality constraint can be expressed in terms of α_1 and α_2 alone as

$$\alpha_1 + s\alpha_2 = \gamma. \tag{10}$$

Each iteration of the sequential minimal optimisation algorithm begins with the selection of a pair of Lagrange multipliers to jointly optimise. The second multiplier, α_2, is then updated to minimise the functional (9), within the feasible region defined by the linear equality and box constraints (10, 3). The first multiplier, α_1, is then updated, substituting the the new value for the second multiplier into the linear equality constraint (10). All that remains is to update the bias term, b, preserving the second Karush-Kuhn-Tucker condition (6).

The first Lagrange multiplier is selected by an outer loop, alternating between single iterations over the entire training set and multiple passes over patterns corresponding to non-bound Lagrange multipliers. A series of passes over non-bound multipliers ends if a pass is made without updating a single Lagrange multiplier. The algorithm terminates following a pass over all training patterns not resulting in a reduction in the objective function. The second multiplier is chosen according to a sequence of heuristics: The first heuristic selects a non-bound multiplier corresponding to the error maximising $|E_1 - E_2|$, where E_i is the error for the ith training pattern, as this is likely to increase the magnitude of the update to the second multiplier. If no progress is made using the first choice, the algorithm iterates over the remaining non-bound multipliers, starting from a random location. If this also fails the algorithm iterates over all bound multipliers, again starting from a random location.

4 Method

Platt reports that the run-time of the sequential minimal optimisation algorithm is dominated by evaluation of the kernel function. Evaluation of the kernel function, involving a training pattern corresponding to a Lagrange multiplier at the upper bound ($\alpha_i = C, i \in \mathcal{J}$), arises in two situations; firstly during joint optimisation of the Lagrange multipliers associated with such patterns, but also whenever the output of the support vector machine is evaluated. The former situation is unavoidable, and is most likely to occur only during the intermittent passes through the entire training set made by the outer loop of the SMO algorithm. The second situation, however is likely to result in a significant proportion of kernel function evaluations, simply because each joint optimisation involving a Lagrange multiplier at the upper or lower bound will result in at least one evaluation of the output function. The support vector expansion given by (8) can be written as

$$f(\boldsymbol{x}) = -b + \sum_{i \in \mathcal{I}} \alpha_i^0 y_i k(\boldsymbol{x}_i, \boldsymbol{x}) + C \sum_{i \in \mathcal{J}} y_i k(\boldsymbol{x}_i, \boldsymbol{x}).$$

Note that the value of the second summation is likely to remain constant for much of the time during the training procedure. We therefore cache the value of this term for each training pattern, updated only when a Lagrange multiplier reaches or leaves the upper bound, C, during a joint optimisation (note this involves evaluation of only a single column of the kernel matrix). This is an example of a time-space optimisation, improving speed at the expense of additional storage requirements, however the costs are low (in the author's implementation, an additional double precision floating point variable for each training pattern).

5 Results

In this section we present results obtained using the modified and standard sequential minimal optimisation algorithms on a publically available bench mark data set. Both the standard sequential minimal optimisation algorithm (Platt [6]) and the enhanced SMO algorithm, described in the previous section, were implemented in the Java programming language (Arnold *et al.* [8]). Figure 1 shows a graph of training time as a function of the number of training patterns for support vector classification networks trained on the Adult benchmark problem. A Gaussian radial basis kernel was used in each case, with a variance of 10, and with the regularisation parameter, C, set at unity, conforming to the experimental conditions used in (Platt [6]). Ten networks were trained for each of the eight partitions of the data, the mean training times for each partition are shown in Figure 1.

It can easily be seen that the enhanced sequential minimal optimisation algorithm is significantly faster than the standard algorithm for all partitions of the data, the improvement in performance increasing slightly as the number of training patters increases. The variability in training time using the enhanced algorithm also appears to be reduced. It

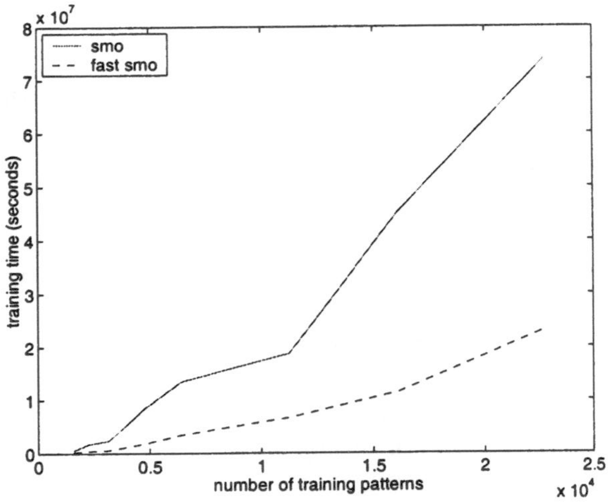

Fig. 1. Mean training time for 10 support vector machines trained on the Adult bench mark data set, as a function of the number of training patterns used.

appears that the sequential minimal optimisation algorithm sometimes experiences difficulty converging to the optimal solution, resulting in a large number of passes through the entire data set. The proposed modification is most useful in such circumstances as it reduces the cost of evaluations of the support vector expansion.

6 Summary

This paper describes a simple method improving the efficiency of the sequential minimal optimisation algorithm for support vector classification. This technique is likely to be most useful for difficult pattern recognition tasks, with a high degree of overlap in the distribution of patterns belonging to each class and therefore a large number of bound support vectors. Naturally the improvement in training time will be greatest for computationally expensive kernel functions, and smallest (i.e. nil) in the case of linear support vector machines, where in principle the output function can be computed without evaluating the kernel function [6].

Acknowledgements

The author would like to thank Rob Foxall for his helpful comments on previous drafts of this manuscript.

References

[1] B. Boser, I. Guyon, and V. N. Vapnik, "A training algorithm for optimal margin classifiers," in *Proceedings of the fifth annual workshop on computational learning theory*, (Pittsburgh), pp. 144–152, ACM, 1992.

[2] C. Cortes and V. Vapnik, "Support vector networks," *Machine Learning*, vol. 20, pp. 1–25, 1995.

[3] V. N. Vapnik, *The Nature of Statistical Learning Theory*. New York: Springer-Verlag, 1995.

[4] V. N. Vapnik, *Statistical Learning Theory*. Wiley Series on Adaptive and Learning Systems for Signal Processing, Communications and Control, New York: Wiley, 1998.

[5] N. Cristianini and J. Shawe-Taylor, *An Introduction to Support Vector Machines (and other kernel-based learning methods)*. Cambridge, U.K.: Cambridge University Press, 2000.

[6] J. C. Platt, "Fast training of support vector machines using sequential minimal optimization," in *Advances in Kernel Methods - Support Vector Learning* (B. Schölkopf, C. J. C. Burges, and A. J. Smola, eds.), ch. 12, pp. 185–208, Cambridge, Massachusetts: MIT Press, 1999.

[7] E. Osuna, R. Freund, and F. Girosi, "An improved training algorithm for support vector machines," in *Neural Networks for Signal Processing VII - Proceedings of the 1997 IEEE Workshop* (J. Principe, L. Gile, N. Morgan, and E. Wilson, eds.), (New York), pp. 276–285, IEEE, 1997.

[8] K. Arnold, J. Gosling, and D. Holmes, *The Java Programming Language*. Addison-Wesley, third ed., 2000.

Model Selection for Support Vector Machines via Adaptive Step-Size Tabu Search

Gavin C. Cawley*[1]

*School of Information Systems, University of East Anglia, Norwich, Norfolk, U.K. NR4 7TJ. E-mail: gcc@sys.uea.ac.uk

Abstract

The generalisation properties of a support vector classification network are typically governed by a regularisation parameter, C, and a small number of parameters specifying the kernel function. The process by which the optimal values of these parameters are obtained is known as model selection. This paper describes an automated model selection procedure based on minimisation of an upper bound on the leave-one-out cross-validation error, via a simple tabu search strategy with adaptive step size adjustment.

1 Introduction

Support vector machines have demonstrated impressive performance in a wide range of notable real world classification problems. The major parameters of the support vector classification network are given by the solution of a linearly constrained quadratic optimisation problem, for which efficient algorithms are available (e.g. Platt [1]). However the optimal choice of kernel function and the values of a small number of hyper-parameters, consisting of the kernel parameters and the regularisation parameter C, must also be determined. This task, known as model selection, is most often performed by training a number of classifiers with different permutations from a range of kernel functions and hyper-parameters, and retaining the configuration resulting in optimal performance on an independent set of validation patterns. In this paper we present a simple and efficient tabu search method, with a robust adaptive step size adjustment heuristic, that can be used to find the value of these hyper-parameters, via minimisation of a recent upper bound on the leave-one-out cross-validation error rate. Model selection is then fully automated and allows all of the available data to be used during training.

The remainder of this paper is structured as follows: Section 2 briefly describes the support vector classifier and introduces the notation used. Section 3 describes a suitable model selection criteria based on a recent upper bound on the leave-one-out cross-validation error. Section 4 details a model selection procedure, based on Tabu search for the minimiser of this criteria. Initial results obtained on a small, but real-world classification task are presented in section 5, and the work summarised in section 6.

[1]This work was supported by the European Commission, grant number IST-99-11764, as part of its Framework V IST programme.

2 Support Vector Classification

The support vector machine [2, 3], given labelled training data

$$\{(\boldsymbol{x}_i, y_i)\}_{i=1}^{\ell}, \quad \boldsymbol{x}_i \in \boldsymbol{X} \subset \mathbb{R}^d, \quad y_i \in \{-1, +1\},$$

generates a maximal margin linear decision rule of the form $h(\boldsymbol{x}) = \mathrm{sign}(\boldsymbol{w} \cdot \boldsymbol{x} - b)$. The weight vector, $\boldsymbol{w}$, is given by the solution of the primal optimisation problem: minimise

$$V(\boldsymbol{w}, \boldsymbol{\xi}) = \boldsymbol{w} \cdot \boldsymbol{w} + C \sum_{i=1}^{\ell} \xi_i \tag{1}$$

subject to

$$y_i[\boldsymbol{w} \cdot \boldsymbol{x}_i - b] \geq 1 - \xi_i, \qquad i = 1, 2, \ldots, \ell, \tag{2}$$

and

$$\xi_i \geq 0, \qquad i = 1, 2, \ldots, \ell. \tag{3}$$

The slack parameters, ξ_i, allow training patterns to be misclassified in the case of linearly non-separable problems. The parameter C sets the penalty applied to margin-errors, and therefore can be viewed as a regularisation parameter, controlling the trade-off between the width of the margin and training set error. A non-linear decision rule can be constructed using a maximal margin linear classifier in a high dimensional feature space, $\Phi(\boldsymbol{x})$, defined by a positive definite kernel function, $k(\boldsymbol{x}, \boldsymbol{x}')$, specifying an inner product in the feature space,

$$\Phi(\boldsymbol{x}) \cdot \Phi(\boldsymbol{x}') = k(\boldsymbol{x}, \boldsymbol{x}').$$

A common kernel is the Gaussian radial basis function (RBF),

$$k(\boldsymbol{x}, \boldsymbol{x}') = e^{-\gamma\|\boldsymbol{x}-\boldsymbol{x}'\|^2}.$$

The function implemented by a support vector machine is given by

$$f(\boldsymbol{x}) = \left\{ \sum_{i=1}^{\ell} \alpha_i y_i k(\boldsymbol{x}_i, \boldsymbol{x}) \right\} - b. \qquad (4)$$

The optimal coefficients, $\boldsymbol{\alpha}$, of this expansion are given by the solution of the Wolfe dual of the primal optimisation problem: maximise

$$W(\boldsymbol{\alpha}) = \sum_{i=1}^{\ell} \alpha_i - \frac{1}{2} \sum_{i,j=1}^{\ell} y_i y_j \alpha_i \alpha_j k(\boldsymbol{x}_i, \boldsymbol{x}_j), \qquad (5)$$

in the non-negative quadrant,

$$0 \le \alpha_i \le C, \qquad i = 1, 2, \ldots, \ell, \qquad (6)$$

subject to the constraint,

$$\sum_{i=1}^{\ell} \alpha_i y_i = 0. \qquad (7)$$

The Karush-Kuhn-Tucker (KKT) conditions can be stated as follows:

$$\alpha_i = 0 \;\Rightarrow\; y_i f(\boldsymbol{x}_i) \ge 1, \qquad (8)$$

$$0 < \alpha_i < C \;\Rightarrow\; y_i f(\boldsymbol{x}_i) = 1, \qquad (9)$$

$$\alpha_i = C \;\Rightarrow\; y_i f(\boldsymbol{x}_i) \le 1. \qquad (10)$$

These conditions are satisfied for the set of feasible Lagrange multipliers, $\boldsymbol{\alpha}^0 = \{\alpha_1^0, \alpha_2^0, \ldots, \alpha_\ell^0\}$, maximising the objective function given by equation 5. The bias parameter, b, is selected to ensure that the second KKT condition is satisfied for all input patterns corresponding to non-bound Lagrange multipliers. Note that in general only a limited number of Lagrange multipliers, $\boldsymbol{\alpha}$, will have non-zero values; the corresponding input patterns are known as support vectors. Let $\mathcal{I}$ be the set of indices corresponding to non-bound Lagrange multipliers,

$$\mathcal{I} = \{i \;:\; 0 < \alpha_i^0 < C\},$$

and similarly $\mathcal{J}$ be the set of indices corresponding to Lagrange multipliers at the upper bound C,

$$\mathcal{J} = \{i \;:\; \alpha_i^0 = C\}.$$

Equation 4 can then be written as an expansion over support vectors,

$$f(\boldsymbol{x}) = \left\{ \sum_{i \in \{\mathcal{I}, \mathcal{J}\}} \alpha_i^0 y_i k(\boldsymbol{x}_i, \boldsymbol{x}) \right\} - b. \qquad (11)$$

For a full exposition of the support vector method, see the excellent books by Vapnik [4] or Cristianini and Shawe-Taylor [5].

3 Model Selection Criteria

Model selection is performed for most classifiers on the basis of validation set error. An alternative approach is possible in the case of support vector classifiers as theoretical bounds on generalisation performance are available. In this paper we adopt the latter approach. Joachims [6] demonstrates that the leave-one-out cross-validation error of a stable soft-margin support vector classifier is bounded by,

$$\mathrm{Err}_{\xi\alpha}^{\ell} = \frac{d}{\ell}, \; d = |\{i \;:\; (\rho\alpha_i^0 R_\Delta^2 + \xi_i) \ge 1\}|, \qquad (12)$$

where ρ equals 2, and R_Δ^2 is an upper bound on $k(\boldsymbol{x}, \boldsymbol{x}) - k(\boldsymbol{x}, \boldsymbol{x}')$, $\forall \boldsymbol{x}, \boldsymbol{x}'$. The inequality $\rho\alpha_i^0 R_\Delta^2 + \xi_i \ge 1$ holds for any training pattern corresponding to an error in the leave-one-out procedure, equation 12 therefore provides an upper bound on the leave-one-out error that can be efficiently computed from the solution of the primal and dual optimisation problems given by equations 1-3 and 5-7 respectively. Unlike other bounds on the generalisation ability of support vector classifiers, this bound is directly applicable to soft-margin support vector machines incorporating a bias parameter.

The upper bound on the leave-one-out cross-validation error (12) is discrete, and therefore is less than ideal for model selection based on most heuristic search methods, as a small change in the hyper-parameters in general will not produce a change in the value of the bound. Instead we minimise the following continuous model selection criteria:

$$E = \sum_{i \in \mathcal{I}, J} e_i,$$

where

$$e_i = \begin{cases} \rho\alpha_i^0 R_\Delta^2 + \xi_i - 1, & \rho\alpha_i^0 R_\Delta^2 + \xi_i \ge 1 \\ 0, & \rho\alpha_i^0 R_\Delta^2 + \xi_i < 1 \end{cases}.$$

This criteria also penalises patterns that may correspond to leave-one-out errors, but the penalty is linear in the deviation from the boundary between definite correctness and possible error in the leave-one-out cross-validation procedure.

4 Model Selection via Tabu Search

The tabu search procedure (e.g. Glover and Laguna [7]) used to minimise the cost function described in the previous section is based on a simple iterative local search heuristic. At each step the cost function is evaluated at the set of points given by positive and negative perturbations of each parameter around the current solution. The point minimising the cost function then forms the starting point for the subsequent iteration. Tabu search heuristics disallow moves likely to recover recently encountered configurations. In this case, the simplest tabu produces good results; the direction of a legal step for the current iteration must not oppose the most recently accepted step. For kernels with more than one parameter a more substantial tabu is likely to further reduce the computational expense of model selection.

4.1 Adaptive Step Size Adjustment

A separate step size parameter is associated with each hyper-parameter optimised during the model selection process. Each step size parameter is adjusted adaptively at the end of every epoch according to a procedure based on that used in the RPROP algorithm [8]. Each time the value of the cost function is reduced the step size is multiplied by a factor greater than unity (in this case 1.1), if the cost increases the step size is multiplied by a factor less than unity (in this case 0.1). The step size is only updated if the corresponding hyper-parameter was modified during that epoch. It is assumed that the kernel parameters, like the regularisation parameter C, are strictly positive, and so the step size parameters are moderated to ensure that a hyper-parameter cannot be reduced below a fixed fraction (in this case 0.5) of its current value during the subsequent epoch.

5 Results

This section presents initial results obtained using the model selection procedure outlined in the previous section on a small, but real-world pattern recognition task. The well known Iris data set (Fisher [9]) consists of 150 records describing the lengths and widths of the sepals and petals of three varieties of Iris (Setosa, Versicolour and Virginica). In this work we aim to find the optimal classifier separating examples of Versicolour from Setosa and Virginica varieties, using only the petal length and width attributes. Figure 1 shows the decision boundary formed by the support vector classifier used as the initial estimate in the model selection procedure (Gaussian radial basis kernel, $C = 100, \gamma = 0.5$). This classifier achieves a training set error of 4% and a $\xi\alpha$ bound on the leave-one-out error of 8.7%.

Model selection procedures based on local and tabu search heuristics, both using adaptive step size selection, were then performed. Table I summarises the number of times the cost function is evaluated, the number of iterations performed and the total elapsed time for each search method. Note that tabu search, as might be expected, out-performs local search in every respect. The tabu heuristic helps to prevent evaluation of the cost function for configurations unlikely to lead to a reduction in cost.

Search	Epochs	Evals	Time
Local	36	144	18.9913 sec
Tabu	23	70	8.7215 sec

Table I. Summary of performance statistics for model selection methods based on local and tabu search heuristics.

Figure 2 shows the true leave-one-out cross validation error and $\xi\alpha$-bound during model selection using tabu search. Note it appears that the bound is not always tight, and that a decrease in the cost function does not always result in a decrease in the bound on the leave-one-out error. It should be noted that the Iris data set is fairly small, with only a small number of patterns close to the decision boundary. Only a small number of patterns are then likely to correspond to leave-one-out errors for any sensible classifier. The resulting highly quantised nature of the leave-one-out cross-validation error may in part explain the variation in the quality of the $\xi\alpha$ bound.

Figure 3 shows the decision boundary formed by the support vector classifier resulting from the tabu search model selection procedure ($C = 0.04462, \gamma = 0.3345$). Note that this classifier is far more heavily regularised, having

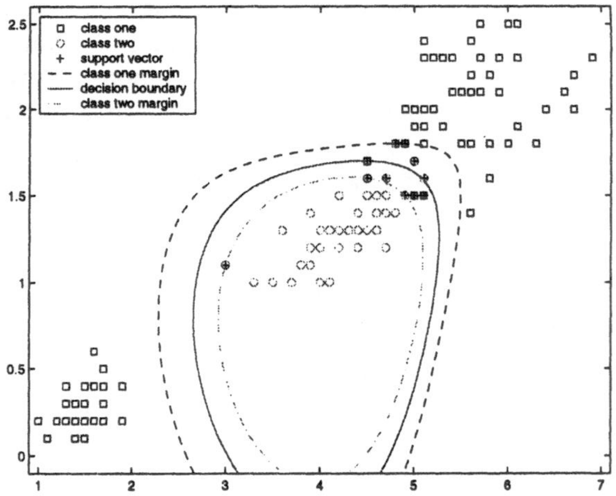

Fig. 1. Decision surface for the support vector classifier used as the starting point for the model selection process.

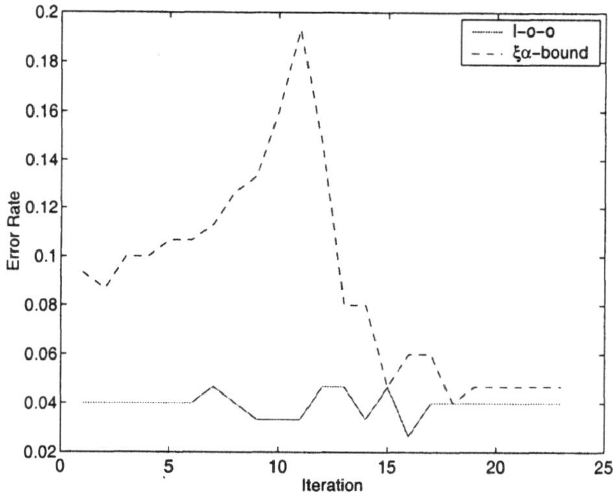

Fig. 2. True leave-one-out cross-validation error and $\xi\alpha$-bound during model selection via tabu search.

a broader margin, but also having a much larger number of (bound) support vectors. This classifier achieves a training set error and $\xi\alpha$ bound on the leave-one-out error of 4%. Although the true leave-one-out error has not decreased, minimising the $\xi\alpha$-bound has produced a subjectively better solution as Occam's Razor tells us that it is sensible to prefer heavily regularised decision rules.

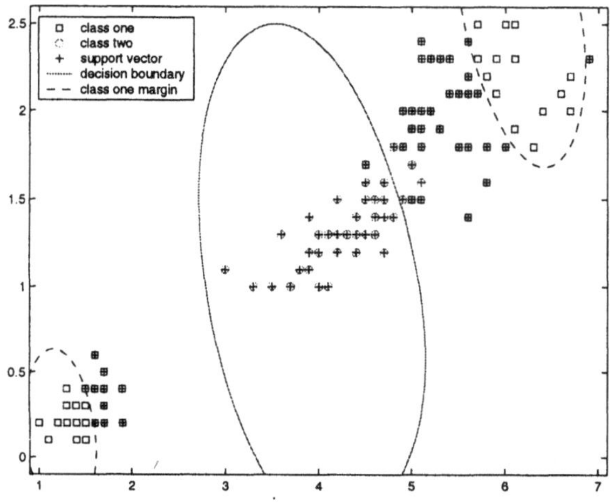

Fig. 3. Decision surface for the support vector classifier resulting from the tabu search model selection process.

6 Summary

This paper describes a practical automatic model selection procedure for support vector classifiers based on tabu search with adaptive step size selection. This heuristic seems to behave efficiently and robustly for a range of initial conditions and search parameters. Further work is needed to further refine the cost function and to evaluate performance on large-scale benchmark pattern recognition tasks.

Acknowledgements

The author would like to thank Rob Foxall for his helpful comments on previous drafts of this manuscript.

References

[1] J. C. Platt, "Sequential minimal optimization: A fast algorithm for training support vector machines," Tech. Rep. MSR-TR-98-14, Microsoft Research, 1998.

[2] B. Boser, I. Guyon, and V. N. Vapnik, "A training algorithm for optimal margin classifiers," in *Proceedings of the fifth annual workshop on computational learning theory*, (Pittsburgh), pp. 144–152, ACM, 1992.

[3] C. Cortes and V. Vapnik, "Support vector networks," *Machine Learning*, vol. 20, pp. 1–25, 1995.

[4] V. N. Vapnik, *The Nature of Statistical Learning Theory.* New York: Springer-Verlag, 1995.

[5] N. Cristianini and J. Shawe-Taylor, *An Introduction to Support Vector Machines (and other kernel-based learning methods).* Cambridge, U.K.: Cambridge University Press, 2000.

[6] T. Joachims, "Estimating the generalization performance of a SVM efficiently," Tech. Rep. LS-8 number 25, Univerität Dortmund, Fachbereich Informatik, 1999.

[7] F. Glover and M. Laguna, "Tabu search," in *Modern Heuristic Techniques for Combinatorial Problems* (C. R. Reeves, ed.), Advanced Topics in Computer Science, ch. 3, pp. 70–150, Mc Graw Hill, 1995.

[8] M. Riedmiller and H. Braun, "A direct adaptive method for faster backpropagation learning: The RPROP algorithm," in *Proceedings of the IEEE International Conference on Neural Networks* (H. Ruspini, ed.), (San Francisco, CA), pp. 586–591, 1993.

[9] R. A. Fisher, "The use of multiple measurements in taxonomic problems," *Annual Eugenics*, vol. 7, no. II, pp. 179–188, 1936.

Data Clustering Based on a New Objective Function

Zhi-Qiang Liu*†, Ya-Jun Zhang†

*School of Creative Media, City University of HongKong, HongKong, China †Department of Computer Science and Software Engineering, The University of Melbourne, Australia

Abstract

Clustering is an unsupervised learning process that is usually based on the mean-squared distortion as the objective function. Thus, it is a minimization process to partition data into different clusters with the least distortion. In this paper we propose a new clustering approach which is more robust than the classical LMS-type learning algorithms. It is based on a new objective function which takes three clustering properties into account. Experimental results show that the new clustering algorithm has the capability to find higher quality solutions especially in detecting overlapped clusters.

1 Introduction

Clustering is a process that partitions data into different classes with the least distortion. In the neural network literature, it is always achieved by competitive learning which can be viewed as a process of minimization. The performance of classical competitive learning (eg. [1, 3]) is usually measured by the average squared-error distortion. The statistics of the source are represented by a finite training set of patterns. The objective of the design procedure is to minimize this distortion [4]. The learning is said to be optimal if no other prototype set has a smaller average distortion, for the same source, vector dimension, and the number of prototypes. There are two well-known necessary conditions to achieve the optimal learning: *Nearest Neighbor Condition* and *Centroid Condition* [2].

The remainder of this paper is organized as follows. In Section 2 we propose a data clustering algorithm in the competitive learning literature based on a new objective function, and in Section 3 we apply it to some experiments in comparison to the soft competitive learning algorithm. Finally Section 4 gives the summary and conclusions.

2 The Algorithm

By incorporating the gradient scheme into the deterministic SOFM, we introduce a new on-line clustering algorithm, which can be viewed as an adaptive version of the unsupervised competitive learning algorithm. The first task in this process is to find a good objective function which can well reflect the clustering quality. Generally, a good partition on data should have the following properties:

1. minimal volume of each cluster;
2. maximal number of data within a cluster;
3. large intercluster seperations.

Let us define these three properties quantitatively. Suppose that there are N training vectors $\{\vec{X}_1, \ldots, \vec{X}_N\}$ and K cluster prototypes $\{\vec{P}_1, \ldots, \vec{P}_K\}$. Assuming crisp partitioning and Euclidean geometry, first we define the average squared error of cluster k,

$$v_k = \frac{\sum_{i=1}^{N} w_{ik} v_{ik}}{N_k^2}; \qquad N_k = \sum_{i=1}^{N} w_{ik}. \tag{1}$$

where

$$v_{ik} = \|\vec{X}_i - \vec{P}_k\|^2; \quad w_{ik} = \begin{cases} 1 & \text{if } v_{ik} = \min_l v_{il}, \\ 0 & \text{otherwise;} \end{cases} \tag{2}$$

and a vector $\vec{V}$ for all clusters,

$$\vec{V} = (v_1, \ldots, v_K), \tag{3}$$

which is an indicator of the first two properties. For the third property, let us define a separation vector,

$$\vec{S} = (s_1, \ldots, s_K), \tag{4}$$

with

$$s_k = \sum_{j=1(\neq k)}^{K} \frac{N_j N_k}{\|\vec{P}_j - \vec{P}_k\|^2}. \tag{5}$$

s_k is the sum of inverse intercluster distance.

Based on the definitions of the three clustering properties, we define the objective function as the inner product of $\vec{V}$ and $\vec{S}$:

$$\begin{aligned} E &= \vec{V}\vec{S}^T = \sum_{k=1}^{K} v_k s_k \\ &= \sum_{i=1}^{N}\sum_{k=1}^{K}\sum_{j=1(\neq k)}^{K} \frac{w_{ik}\|\vec{X}_i - \vec{P}_k\|^2 N_j}{\|\vec{P}_j - \vec{P}_k\|^2 N_k} \\ &= \sum_{i=1}^{N} E_i. \end{aligned} \tag{6}$$

E is expected to be minimized. To minimize E, we must minimize E_i for each sample vector. Let us update the prototypes using the gradient descent method:

$$\vec{P}_k(n+1) - \vec{P}_k(n) = -\eta \frac{\partial E_i}{\partial \vec{P}_k(n)}. \quad (7)$$

where η is the learning rate. In our case it is a constant with the value $\frac{1}{2}$.

For a small update step, $\partial \vec{P}_k(n)$ actually takes the value $\vec{P}_k(n+1) - \vec{P}_k(n)$, thus, according to Eq.(7), the differential of E_i satisfies $\partial E_i \leq 0$. That is, the individual distortion upon the update scheme of prototypes defined in Eq.(7) is always decaying. If the distortion reaches minimum, $\partial E = 0$. As a result, the prototype set converges and it may be said to be optimal.

According to Eq.(6) and (7), we get

$$\begin{aligned} -\frac{1}{2}\frac{\partial E_i}{\partial \vec{P}_k(n)} &= \frac{[\vec{X}_i - \vec{P}_k(n)]}{N_k} \sum_{j=1(\neq k)}^{K} \frac{N_j}{\|\vec{P}_j - \vec{P}_k\|^2} \quad (8) \\ &- \frac{\|\vec{X}_i - \vec{P}_k(n)\|^2}{N_k} \sum_{j=1(\neq k)}^{K} \frac{[\vec{P}_j(n) - \vec{P}_k(n)]N_j}{\|\vec{P}_j - \vec{P}_k\|^4}, \\ -\frac{1}{2}\frac{\partial E_i}{\partial \vec{P}_j(n)} &= [\vec{P}_j(n) - \vec{P}_k(n)]\frac{\|\vec{X}_i - \vec{P}_k(n)\|^2 N_j}{\|\vec{P}_j - \vec{P}_k\|^4 N_k}, \end{aligned}$$

where $\vec{P}_k$ is the Frequency-Sensitive Square Distance (FSSD) winner. The FSSD is defined as

$$FSSD(\vec{P}_i, \vec{X}) = \gamma_i \|\vec{X} - \vec{P}_i\|^2, \quad (9)$$

$$\gamma_i = \frac{n_i}{\sum_{j=1}^{K} n_j}, \quad (10)$$

where n_j is the winning counter for the jth prototype. Since the denominator is always equal to the number of learning cycle C, hence it can be rewritten simply as,

$$\gamma_i = \frac{n_i}{C}. \quad (11)$$

It is understood that C has the same value for each prototype, therefore, in the practice, γ_i can be simply replaced by n_i. Using FSSD to determine the winner will stop "dead nodes" from happening.

Alternatively, we can easily modify this algorithm into the Winner-Take-Most (WTM) strategy, in which for each input vector every prototypes is a winner to a certain degree. In this paradigm, the Eculidean Distance winner is updated using the first formula in Eq.(8) while the others using the second.

The algorithm has the following steps:

1. Initialize all sample vectors as active.

	$\bar{D}_1$	$\bar{D}_2$	$\bar{D}_3$	$\bar{D}_4$	Global $\bar{D}$
Obtained by SCL	7.6	7.8	7.3	6.9	7.4
Obtained by GCL	5.2	5.1	5.8	5.6	5.4

Table I. Distortions obtained in Fig.1.

2. Take an arbitrary active sample vector $\vec{X}$ from training set.
3. Find the winning prototype $\vec{P}_k$, giving the minimum of

$$FSSD(\vec{P}_k, \vec{X}) = \min_l FSSD(\vec{P}_l, \vec{X}). \quad (12)$$

4. Increment the winning counter for the winner and mark it as inactive.
5. Adjust the weight of winning prototype according to the Eq.(7) and (8).
6. If no active samples found, goto (1) to continue learning or quit loop as a specific loop number is reached.

3 Experimental Results

In this section we apply our new gradient algorithm to clustering analysis. In the experiments, we compared the performance of our algorithm with that of *Soft Competitive Learning* (SCL) [4] which is an unsupervised learning algorithm. The sample space is from -100 to 100 on both X and Y coordinates. In Fig.1, we arbitrarily initialized four 100-pixel clusters with centers at (-20.8, 60.2), (-77.9, 7.6), (44.6, 7.6), and (-20.8, -57.8), respectively. To detect these four clusters, four prototypes were randomly initialized in the feature space. Learning by SCL algorithm, the four prototypes moved into the four clusters and labeled them with a global average distortion 7.4 and individual distortions as 7.6, 7.8, 7.3, 6.9, respectively (Table I). The learning trajectory and final place of each prototype is shown in Fig.1(a) and Tabel II. To make a fair comparison, the same clusters and the same prototypes were initialized in the feature space for testing our algorithm. It can be observed from Fig.1(b) that the four prototypes moved into the four clusters very smoothly without osillation, and the global average distortion reduced sharply to 5.4 (the last column in Table I). This result is remarkablely better than the performance of SCL algorithm. Since in SCL all the prototypes are treated as winners to a certain degree, some of them will be detracted from their cooresponding clusters. To make it simpler, in the following parts we use GCL to represent the new gradient competitive learning algorithm.

	Cluster1	*Cluster2*	*Cluster3*	*Cluster4*
Expected Centroid	(-20.8, 60.2)	(-77.9, 7.6)	(44.6, 7.6)	(-20.8, -57.8)
Obtained by SCL	(-21.2, 64.3)	(-79.2, 8.1)	(42.3, 6.9)	(-22.2, -55.4)
Obtained by GCL	(-20.6, 60.7)	(-77.3, 7.4)	(43.9, 7.7)	(-21.1, -57.3)

Table II. Learning by SCL and GCL in Fig.1.

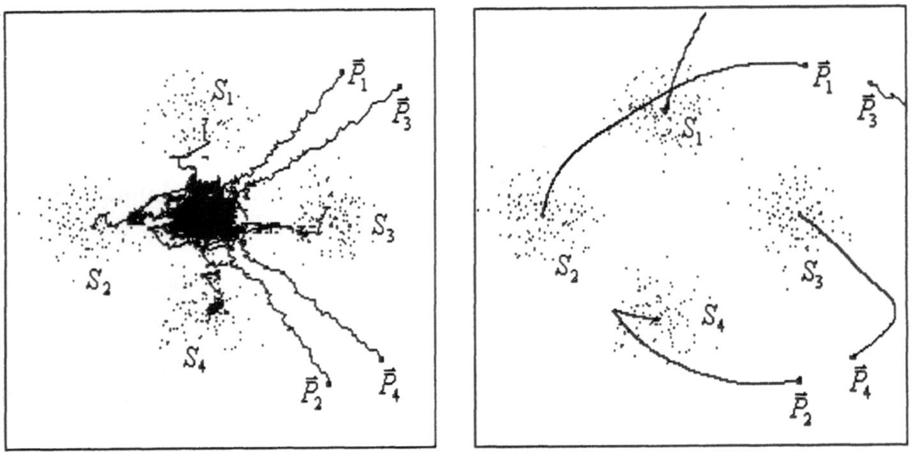

Fig. 1. Apply the algorithm to clustering: (a) the trajectory of learning process by SCL; (b) the trajectory of learning process by the new gradient CL algorithm.

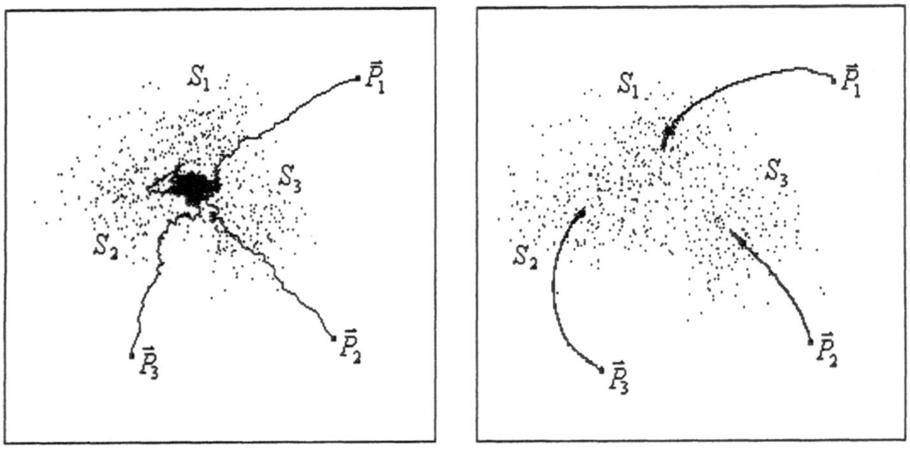

Fig. 2. Clustering on three overlapped clusters: (a) the trajectory of learning process by SCL; (b) the trajectory of learning process by GCL.

	Cluster1	*Cluster2*	*Cluster3*
Expected Centroid	(-4.2, 52.4)	(-37.5, 3.4)	(33.2, -2.8)
Obtained by SCL	(-10.2, 11.3)	(-5.4, 15.6)	(4.3, 14.8)
Obtained by GCL	(-4.6, 54.4)	(-37.9, 3.4)	(33.8, -2.7)

Table III. Learning by SCL and GCL in Fig.2.

	$\bar{D}_1$	$\bar{D}_2$	$\bar{D}_3$	Global $\bar{D}$
Obtained by SCL	17.6	14.8	16.2	16.2
Obtained by GCL	7.2	7.1	6.8	7.0

Table IV. Distortions obtained in Fig.2.

	Obtained by SCL		*Obtained by GCL*	
	Quantity	Confusion	Quantity	Confusion
$\vec{P}_1$	102	49%	205	2.5%
$\vec{P}_2$	238	19%	203	1.5%
$\vec{P}_3$	260	30%	192	4.0%

Table V. The confusion matrix in Fig.2.

Fig.2 shows another example using the GCL in clustering. In this example, three overlapped 200-pixel clusters were initialized in the feature space with Gaussian variance 6.0. Their expected cluster centers are (-4.2, 52.4), (-37.5, 3.4) and (33.2, -2.8), respectively. As shown in Fig.2(a), learning by SCL, the three prototypes were trapped into the global centroid, thus it failed in classifying the overlapped clusters. Using the new gradient algorithm, however, we were able to discover the three clusters and obtained very low distortion and confusion rates. This example demonstrates the remarkable capability of our algorithm in solving overlapped clusters. Tabel III shows that after learning by GCL, each prototype discovered a cluster and its place was very close to the center of its associated cluster; whereas after learning by SCL, the result was far from being satisfactory. Tabel 3 gives the comparison of the distortions obtained by SCL and GCL. From this table, we find that the distortions obtained by SCL are very large simply because the prototypes were trapped into a local minimum. It is interesting to note that learning by GCL, three prototypes made initial detours to avoid going directly to the global centroid as shown in Fig.2(b). As a result, the global average distortion obtained by GCL reduced sharply in comparison with that of SCL (Tabel 3). In Tabel V we list the confusion matrix obtained by SCL and GCL. Each prototype was expected to win 100 sample points. Learning by SCL, there were 102 sample points associated with $\vec{P}_1$, 238 sample points and 260 sample points with $\vec{P}_2$ and $\vec{P}_3$, respectively. However, learning by GCL, three clusters were discovered with low confusion rates at 2.5%, 1.5% and 4.0%, respectively. From this table we may conclude that GCL is successful indeed in detecting overlapped clusters with small confusion rates.

Moreover, GCL is consistant and robust compared to SCL and other unsupervised clustering algorithms. We have investigated the sensitivity of several other competitive learning algorithms to random sampling by running each ten times. We observed that GCL yielded consistant results whereas others varied results.

4 Conclusion

We have presented a new competitive clustering algorithm in this paper. It is based on a new objective function and a gradient descent method. The experimental results are remarkably better than a more recent clustering algorithm. One additional benefit of our approach is that it is robust and consistant in discovering clusters although random sampling is applied. Such an ability is prerequisite for any system to achieve high performance in data clustering and vector quantization applications.

References

[1] S. C. Ahalt, A. K. Krishnamurty, P. Chen, and D. E. Melton, "Competitive learning algorithms for vector quantization," *Neural Networks*, vol. 3, no. 3, pp. 277-291, 1990.

[2] Z.-Q. Liu, M. Glickman, and Y.-J. Zhang, "Soft-Competitive Learning Paradigms," in *Soft Computing and Human-Centered Machines*, Z.-Q. Liu and S. Miyamoto (eds), pp. 131-161, Tokyo: Springer-Verlag, 2000.

[3] L. Xu, A. Krzyzak, and E. Oja, "Rival penalized competitive learning for clustering analysis, RBF Net, and curve detection," *IEEE Trans. Neural Networks*, vol. 4, pp. 636-649, July 1993.

[4] E. Yair, K. Zeger, and A. Gersho, "Competitive learning and soft competition for vector quantizer design," *IEEE Trans. Signal Processing.*, vol. 40, no. 2, Feb., pp. 294-309, 1992.

Dynamical Cluster Analysis for the Detection of Microglia Activation

A. Baune* †, A. Wichert*, G. Glatting†, F. T. Sommer*

*Neural Information Processing Department, University of Ulm, D-89069 Ulm, Germany, email: friedrich.sommer@informatik.uni-ulm.de †Department of Nuclear Medicine, Univeristy of Ulm, D-89069 Ulm, Germany

Abstract

Dynamical cluster analysis (DCA) was used to extract sets of representative time courses to detect brain lesions using positron emmision tomography (PET) data. DCA is an adaptive hard-clustering algorithm where the number of clusters k is not initially fixed but is dynamically changed by generation and fusion of clusters during runtime. We analyzed PET data sets of 9 patients applying DCA repeatedly. We compared the results that vary in the number of clusters even on the same data set. As validation measure we used the mean square quantization error (MSQE). We found that the MSQE was strictly correlated with k only on 4 of the 9 data sets. We propose DCA for extracting the reference time course required in reference tissue modeling [7]. In the case of one patient, we checked the ability of DCA to characterize directly the three most interesting regions, reference tissue, the veins and the lesion and how this ability relates to high validation scores. The characterisation of all three regions was not reproducible in all of the runs, however, runs rated high in validity by the MSQE were able to reproduce all the three regions.

1 Introduction

Explorative analysis methods play an important role in many areas of functional brain imaging. A (preprocessed) functional data set of a subject $\mathcal{D} := \{x^i \in \mathbf{R}^T : i = 1, ..., N\}$ contains for every voxel i within the field of view a signal course x^i measured at T sample values. Various explorative analysis methods have been suggested, like principal component analysis [3], independent component analysis [9], self-organizing maps and clustering algorithms. Various alternatives to the standard k-means cluster analysis (k-CA) have been proposed, as for example mixture models [1] fuzzy-k clustering [3, 8], hierarchical aglomerative hard clustering [5, 6], and a neural network dynamical hard clustering [4]. Cluster analysis extract a reduced set of cluster centers $\mathcal{C} := \{x^c \in \mathbf{R}^T : c = 1, ..., k\}$ (with $k << N$) that retain characteristic properties of $\mathcal{D}$. With hard clustering the membership to cluster centers are a disjunct partition of the data $\mathcal{P}_k(\mathcal{D}) := \{\mathcal{D}_c \subset \mathcal{D} : c = 1, ..., k\}$, mixture models and fuzzy cluster methods generate a soft assignment through a $k \times N$ membership matrix.

In the following we will apply DCA to a the domain of functional imaging, i.e., positron emission tomography using the ligand $[^{11}C]PK11195$, which can be used to localize different brain lesions quantitatively [2]. We analyzed data sets of nine patients suffering from cerebral insult (CI), Kennedy syndrom (KS), amyotropic lateral sclerosis (ALS), or myotonic dystrophy (MD), see table I.

symbol	N	T	disease
○	260857	18	ALS
×	232809	21	ALS
+	231323	21	CI
∗	209056	20	KS
□	225036	20	MD
◇	221354	21	MD
▷	321039	20	MD
▽	214675	21	MD
△	223498	6	MD

Table I. PET data sets analyzed, with symbol assignment used in Fig.2

The motivation behind our analysis approach is: i) Hard clustering is computationally less time consuming than fuzzy clustering where the membership matrix are $N \times k$ additional free parameters that have to be aligned. A hard clustering result, however, provides similar information about memberships since for every data point the distances to the cluster centers can be computed. ii) DCA was tested on EPI-fMRI data [4] and yielded robust results, in particular, the reproducibility increased compared with results of the classical k-means method.

2 Dynamical Cluster Analysis

Here we can only briefly revisit dynamical cluster analysis (DCA). For more detailed description of adaptive clustering see [10], for its application in functional imaging, see [4]. DCA extents the classical k-means method where the data set $\mathcal{D}$ is approximated by fitting a fixed number k of cluster centers. Fitting is provided by a gradient descent on the objective function

$$E^k = \frac{1}{N} \sum_{c=1}^{k} \sum_{i \in \mathcal{D}_c} (x^c - x^i)^2 \tag{1}$$

DCA approximates the data $\mathcal{D}$ not only by fitting cluster centers but also by cluster generation. Additional cluster fusion processes avoid too close neighborhood of cluster centers. The initial parameters θ_g and θ_f in DCA govern the occurence of generation and fusion processes: If a data point is farther away than θ_g from the closest cluster center, it is added as a new cluster center. If any pair of cluster centers comes closer than θ_f, a fusion into one cluster center takes place. There is no objective function describing the process of moving, generation and fusion of cluster centers: After a declining phase k reaches a plateau, see Fig. 1.

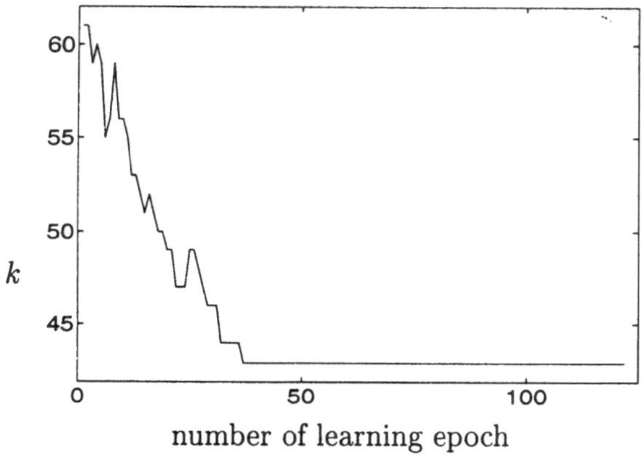

Fig. 1. Typical example for the development of the number of clusters k during a run of DCA.

However, not always a stable clustering partition is reached – sometimes intermittent generation and fusion can persist. In DCA the thresholds are successively changed so that in the k plateau phase generation and fusion becomes more and more unlikely. Finally the ordinary k-means fitting is reached where the convergence to a stable local minimum of (1) is ensured. Instead of choosing a fixed number of clusters DCA requires the presetting of the thresholds θ_g and θ_f. To compare different data sets it is proposed to set the thresholds for each data set in a way that the corresponding ball centered at the grand mean separates the data with the same ratio. The comparison between DCA and k-means CA has shown that the initial constraints imposed with DCA allow a better adjustment of the solution to a particular data set increasing the reproducibility of solutions in repeated runs [4].

3 Results

To study the robustness of DCA on the PET data we repeated ten analysis runs on data sets of the nine patients.

First we checked criteria of validity assessment for the DCA results. Formula (1) is a common validation measure computing the mean squared quantization error (MSQE) of a cluster result. The MSQE value depends on the position of centers but, of course, also on their number k. One expects MSQE and k to be anticorrelated. In general, MSQE $= 0$ can only be achieved for $k = N$ and the centers positioned at the data points. Low values of the MSQE can therefore only be expected for high k-values.

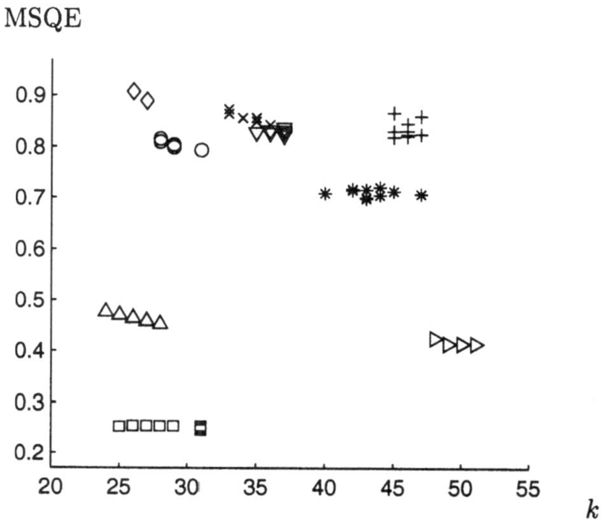

Fig. 2. MSQE of cluster results on the data sets of 9 patients. On each data set the analysis was repeated ten times. Analysis results of the same data set are labeled by the same symbol.

Fig. 2 displays the MSQEs of the different cluster results over the number of clusters. First, it can be seen that the number of extracted clusters vary in repeated DCA runs. Second, in the variation range of k given by the different runs on the same data, a high k corresponds with a low MSQE only in 4 of the 9 the data sets.

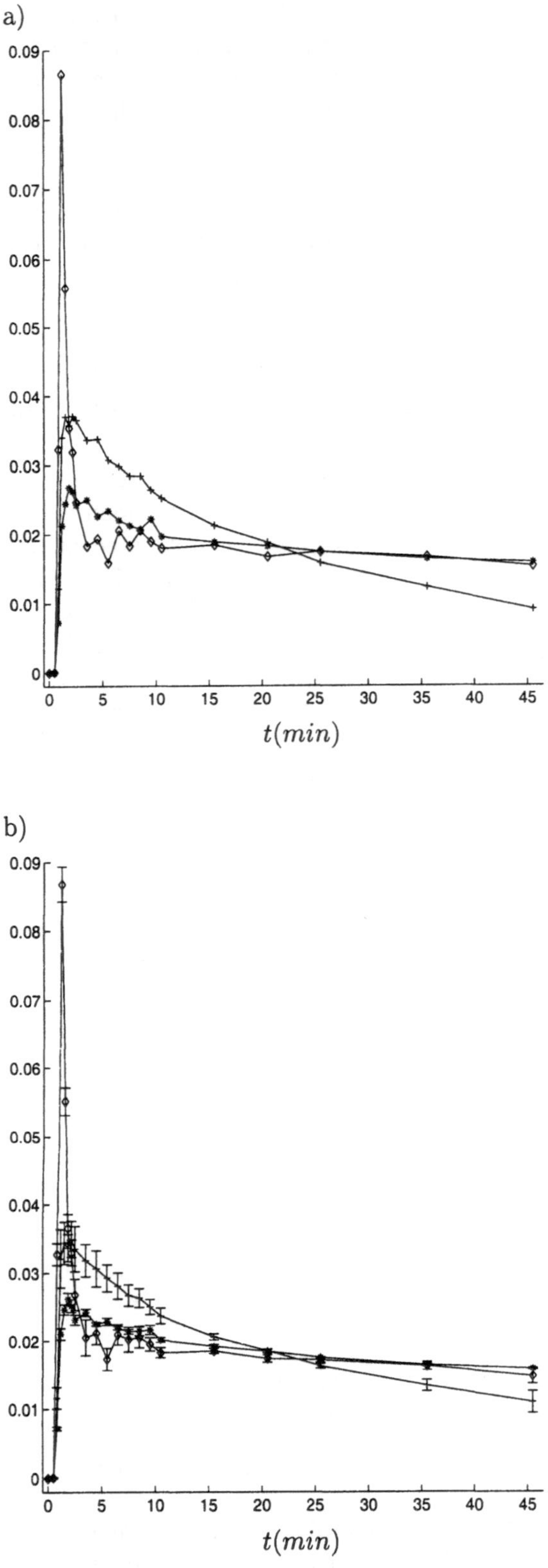

Fig. 3. Time courses of clusters representing the insult (∗), reference tissue (+) and veins (◇) for one data set. a) mean timecurses and deviation, b) result of one run (members: ∗: 66011, +: 32515, ◇: 14328).

Tissue with high microglia activation is currently quantified by determining the spatial distribution of the binding potential. One of the main problems with this is the elimination of the nonspecific binding of the ligand. For this purpose the reference tissue model (RTM) had been proposed [7], that critically depends on how well one can extract the time course of a reference tissue with no specific binding. So far, by visual inspection of a physician a region of reference tissue is selected by hand. This determination of the reference time course is feasible only for localized lesions (for instance, insults), though, with low reproducibility. For wide spread lesions like in Alzheimer desease, no good way is known so far to determine the reference time course. The idea is therefore to use the DCA cluster analysis result.

In the following we show for one patient, (+ in Fig. 2), how well DCA directly discerns the three most interesting regions, the reference tissue R, the veins V, and the lesioned tissue L (For this patient the three regions could also be characterized in the MR image). The three regions could be well separated in 5 of the 10 runs. In the remaining runs in three cases R and L was not discerned, in two cases none of the regions was separated from the others. Fig 3 a) shows for the three regions the mean time courses and standard deviations averaged over the runs where the region has been separated. The MSQE was a well suited validation measure with regard to this question: The 5 runs separating all the regions had low MSQE values. Fig 3 b) shows the different time courses for one of the runs with low MSQE. The difference in decay between the curves (+) and (◇) reflects the difference between specific and nonspecific ligand binding.

Fig. 4 shows the spatial distribution of the clusters displayed in Fig 3 b). These corresponded to the lesion caused by the insult, reference tissue and veins.

4 Discussion

We have presented a method for analyzing PET measurements in order to quantify brain lesions using adaptive cluster analysis DCA. We discussed how to validate clustering results from repeated analysis runs that with DCA differ in the number of clusters. Only on 4 of the 9 data sets the MSQE was strongly anticorrelated with the number of clusters. We showed that low MSQE runs of DCA were able to detect directly the three most interesting regions, the reference tissue, the lesion, and veins. The extracted time course of the reference tissue can be used in reference tissue modelling. RTM is a method

a)

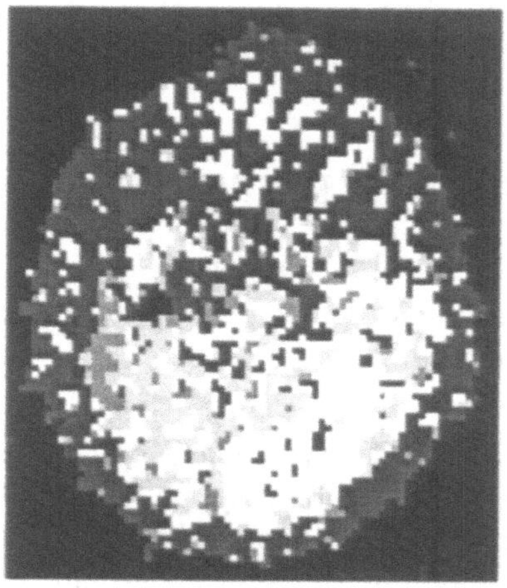

b)

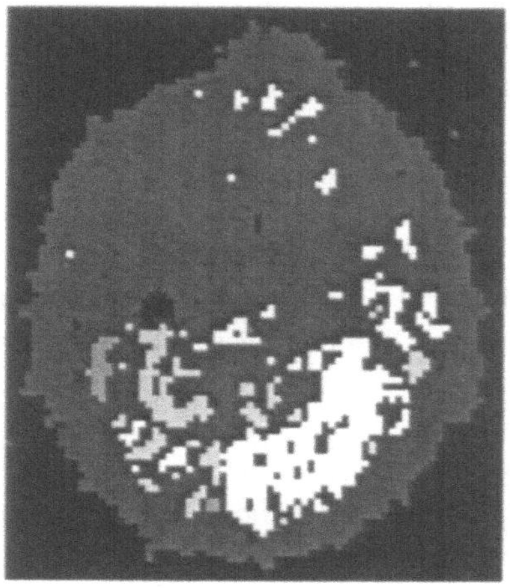

Fig. 4. Spatial pattern of cluster representing the insult (white), reference tissue (lightgray) and veins (gray) in one slice of data volume (+). a) all members b) voxels with minimal 15 neighbours of the same cluster.

to characterize the lesion tissue more quantitatively. A detailed assessment of the diagnostic power of the method will be given in a subsequent publication.

References

[1] Ashburner, J., Haslam, J., Taylor, C., Cunningham, V. J., and Jones, T. (1996). A cluster analysis approach for the characterization of dynamic PET data. In Myers, R., Cunningham, V., and Bailey, D., editors, *Quantificction of Brain Function Using PET*, pages 301–306. Academic Press, New York.

[2] Banati, R. B., Goerres, G. W., Myers, R., Gunn, R. N., Turkheimer, F. E., Kreutzberg, G. W., Brooks, D. J., Jones, T., and Duncan, J. S. (1999). $[^{11}C]$(R)-PK11195 positron emission tomography imaging of activated microglia. *Neurology*, 53(9):2199 - 2203.

[3] Baumgartner, R., Ryner, L., Richter, W., Summers, R., Jarmasz, M., and Somorjai, R. (2000). Comparison of two exploratory data analysis methods for fMRI: fuzzy clustering vs. principal component analysis. *Magnetic Resononance Imaging*, 18(1):89 - 94.

[4] Baune, A., Sommer, F. T., Erb, M., Wildgruber, D., Kardatzky, B., and Grodd, W. (1999). Dynamical cluster analysis of cortical fMRI activation. *NeuroImage*, 6(5):477 - 489.

[5] Filzmoser, P., Baumgartner, R., and Moser, E. (1999). A hierarchical clustering method for analyzing functional MR images. *Magnetic Resonance Imaging*, 17(6):817 - 826.

[6] Goutte, C., Toft, P., Rostrup, E., Nielsen, F. A., and Hansen, L. K. (1999). On clustering fMRI time series. *NeuroImage*, 9:298–310.

[7] Gunn, R. N., Lammertsma, A. A., Hume, S. P., and Cunningham, V. J. (1997). Parametric imaging of ligand-receptor binding in PET using a simplified reference region model. *NeuroImage*, 6(4):279–287.

[8] Masulli, F., Schenone, A., and Massone, A. M. (2000). Fuzzy clustering methods for the segmentation of multi-modal medical images. In Szczepaniak, P. S., Lisboa, P. J. G., and Tsumoto, S., editors, *Fuzzy Systems in Medicine*, Studies in Fuzziness and Soft Computing, pages 335–350. Springer-Verlag.

[9] McKeown;, M. J. and Sejnowski, T. J. (1998). Independent component analysis of fMRI data: Examining the assumptions. *Human Brain Mapping*, 6:368–372.

[10] Schalkoff, R. J. (1989). *Digital Image Processing and Computer Vision*. Wiley, New York.

Interpretation of Event-Related fMRI Using Cluster Analysis

A. Wichert, A. Baune, J. Grothe, G. Grön, H. Walter, F. T. Sommer *

*Neural Information Processing Department, University of Ulm, D-89069 Ulm, Germany, email:{andreas.wichert; axel.baune; friedrich.sommer}@informatik.uni-ulm.de, {jo.grothe; georg.groen; henrik.walter} @medizin.uni-ulm.de

Abstract

Event-related fMRI can access more complex experimental paradigms than traditional block designs. A new problem, however, becomes the data analysis, i.e., the generation of appropriate statistical models. For parametric paradigms where a set of parameters describing task and stimulus can be varied, the number of different conditions becomes large. Even if the combinatorics of contrasts between conditions is reduced by assumptions, the number of contrasts that potentially may contribute to the functional interpretation becomes huge.

We propose a new explorative method of data interpretation for paradigms with many different conditions. The method is based on a cluster analysis (dynamical clustering [2]) that extracts a set of represenatative time courses from the original data. We select interesting clusters using cluster size and a compactness criterion. On this highly reduced data set it is possible to study systematically the temporal patterns of activation and derive a functional interpretation of the data.

1 Introduction

The event-related technique in functional magnetic resonance imaging (efMRI) provides a timing of the data aquisition synchronized with events in the experimental paradigm [4]. Therefore, efMRI can temporarily resolve brain activity in experimental designs that are more structured in time than the simple block design with just two different conditions, the on- and the off-phase. However, with many different experimental conditions the functional interpretation of the data becomes a much harder problem. With block design one can apply an inferential data evaluation, for instance using the general linear model [3]. The statistical model to be tested is quite obvious, the on/off box-car function. (folded with a canonical hemodynamic response function). But with many experimental conditions, many contrasts possible by the combinatorics between them are candidates that may potentially contribute to the functional interpretation. Even in simple cases the number of contrasts becomes so high that a systematic exploration by inferential evaluations becomes impossible, in practice. Furthermore, when analyzing multiple contrasts, the analysis theoretically would have to be corrected for multiple comparisons, which, however rarely is done. Here we will present a new method to extract good candidates for statistical models from the data. We use an explorative method, cluster analysis to determine representative time courses from the data set. To select the time courses suited for a functional interpretation in a subsequent inferential analysis we search for high correlations in the whole set of possible contrasts. Since cluster analysis yields the time courses of activation the determination of contrasts with high correlations is possible without an exhaustive search.

We will demonstrate the method on a working memory experiment. Working memory is the cognitive function of continuously updating and actively maintaining information "on-line" for the guidance of ensuing behavior [1]. The neural network supporting working memory has been studied widely with PET and in the last years with fMRI [6]. We chose a working memory paradigm (delayed match-to-sample: stimulus set presentation, delay without stimulus, target presentation with decision if the target was included in the stimulus set, 2×3 factorial design with 3 load conditions and 2 delay conditions) because we know already enough about the underlying neural network so that we could interpret our results intelligibly. The new method of efMRI for the first time allows to asses delay related activity noninvasively in humans. However, the analysis of such data is known to be particularly complex and is still a subject of current discussion [5].

2 Data Evaluation for Parametric Paradigms

One large class of imaging experiments accessible with efMRI are those with parametric designs where properties of the task and the stimulus are gradually changed resulting in a set of conditions with different parameter settings. The usual way of data interpretation is to study contrasts in a voxel-based analysis between different phases/conditions in the paradigm. If in an experimental design l parameters can be varied and each parameter i is sampled on d_i discrete values, there are $r = \prod_l d_l$ different experimental conditions. If we assume a monotonic dependency between parameter setting and brain activity the interesting contrasts are between high and low settings of parameters. For a parameter with d_i sample points the number of high-low partitions (with a "no-care" gap with size between 0 and $d_i - 2$ sample points) is $P_i = d_i(d_i - 1)/2$. Over all the parameters the number of interesting contrasts is then given by

$$G(d_1, ..., d_l) = \prod_{i=1}^{l} P_i + \sum_{j=1}^{l} \prod_{i=1, i \neq j}^{l} P_i$$

$$+ \sum_{j=1}^{l} \sum_{j'=1}^{l} \prod_{i \neq j, i \neq j'} P_i + ... + \sum_{i=1}^{l} P_i$$

On the RHS of equation the first term counts all the contrasts between single conditions, the second term counts contrasts where one variable is averaged out, and the last term counts contrasts where all but one variable are averaged out. If during a condition $j = 1, ..., r$ the experiment is devided into f_j functionally distinct phases, the number of contrasts between different phases is $F(f_1, ..., f_r) = \prod_{j=1}^{r} f_j$ and the total number of contrasts is $H = F(f_1, ..., f_r)G(d_1, ..., d_l)$. For the case $d_i = d \; \forall i$ and $f_j = f \; \forall j$ we obtain

$$H(d, l, f) = \left[\left(1 + \frac{d(d-1)}{2}\right)^l - 1 \right] f^{d^l}$$

For the working memory (WM) paradigm even with only three sample points for delay time and memory load, and with three phases, stimulus presentation, delay and retrieval, the number of potentially interesting contrasts is $H(3, 2, 3) = 295245$, and the experimenter has to preselect according to his/her expectations what can be examined on a voxel-by-voxel basis. The idea is in the following to use first an explorative analysis to extract clusters, i.e., voxel sets that show similarity in activation. Using these clusters and their mean time courses contrasts with high correlations can be found without exhaustive search and the functional role of the corresponding cortical regions can be explored more effectively.

The cluster analysis uses minimal assumptions, consisting only in the choice of data preprocessing/selection and the used similarity measure. An interpretation of a cluster result requires further assumptions:

Cluster selection: We assume that the functional entities detectable by fMRI strech over a set of spatially adjacent voxels. Thus, interesting clusters were selected by two criteria: a) A limit on the mimimal cluster size. b) A spatial compactness criterion calling for a large enough subset of spatially clustered voxels allowing a localisation. Since the data where not spatially smoothed at all, a compact distribution of cluster voxels expresses spatial homogenity of the hemodynamic activation.

Labeling of the measurements: For examination of the mean signal courses in selected clusters we assumed a phase shift between a neuronal activation and the hemodynamic response of 6 seconds. We then labeled the measurement time points according to the ongoing experimental condition. Based on this labeling one can define step functions with values high, low, and zero corresponding to contrasts between different conditions. The correlations between contrast functions and mean signal courses allows to study the temporal pattern of activations with respect to conditions and experimental phases.

3 Results

We examined the data sets of two subjects (A and B) performing a delayed match-to-sample-task with delay times of 2 or 6 seconds and memory loads of 1, 3 or 6 items. (fMRI acquisition: Siemens 1.5 Tesla, TR=2.6 seconds, 21 slices). The error score was low in both cases: 3.4 % for A, and 2.5 % for B. Two cluster analyses were performed on each realigned and slice timed data set: (I) for the whole experiment, (II) for the data points acquired during the delay period, an interesting period to reveal regions involved in working memory. For the selection of measurements corresponding to the delay period we assumed – similar as in section 4 – a phase shift of 6 seconds for the hemodynamic response.

To compare the quality measures we performed on each data set repetitions of the cluster analysis. The cluster selection process yielded a limited set of clusters satifsfying our criteria which differed depending on the period analyzed: A: 4 clusters in I, 2 clusters in II, B: 5 in I and 4 in II. Clusters were located in superior and inferior frontal regions, as well as in temporal, parietal and orbitofrontal regions. A comparison between the two subjects revealed one clear difference: In frontal areas subject A showed a cluster only in the left prefrontal cortex, see Fig. 1, whereas subject B showed a cluster in the left as well in the right prefrontal cortex. (An analysis for the main WM effect with SPM99 [Wellcome Department of Cognitive Neurology, London] showed a left prefrontal cluster in subject A and a right prefrontal cluster in Subject B but no left frontal cluster in Subject B).

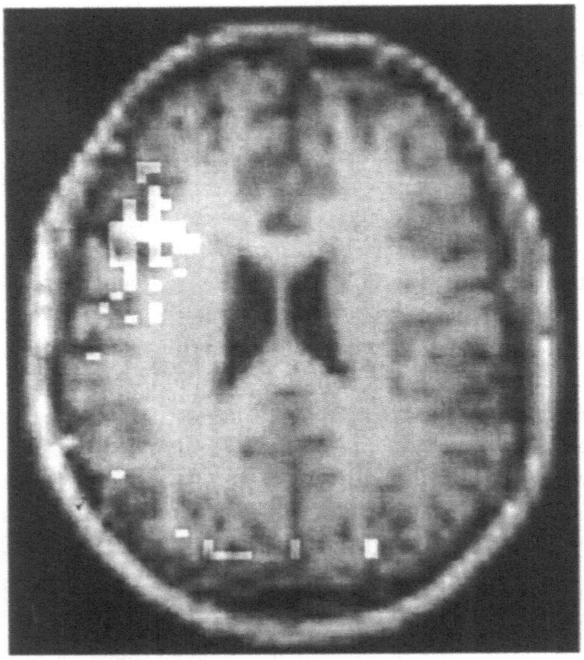

Fig. 1. Spatial pattern of a selected cluster in data set AII: prefrontal region

To analyze the functional roles of the clusters further we compared the correlations of the cluster centers obtained with cluster analysis with 16 contrasts (1 for the main effect, 15 for the delay period). In subject B the right prefrontal cluster was active when contrasting the whole task with the no task period (main WM effect). It also was active when contrasting high load versus low load conditions during the delay period. However, when contrasting the last third with the first third of the long delay we found a negative correlation in all load conditions. The left prefrontal cluster showed a different pattern of correlations. It was only active when contrasting late and early delay for the 6 seconds delay, was deactivated during the short delay for high versus low load and showed no other positive or negative correlations for different contrasts during the delay. In subject A the cluster in left prefrontal cortex showed a pattern which was very similar to the pattern of the right prefrontal cluster in B.

4 Discussion and Conclusion

We have presented a method for analyzing efMRI data from a complex experimental design. The approach is as far as possible data driven, and not hypothesis constrained. Cluster analysis yields regions with a common – yet not preselected – signal time course during the experiment. The method employs a hard clustering algorithm. For interpretation of the cluster results our method exploits the assumption of spatial homogenity of cortical activation accessible by fMRI, and the experimentally estimated time shift of six seconds for the hemodynamic response. By this it is possible to identify activated regions and the temporal relation of their activation time courses with the experimental design for further analysis. Specific questions can be asked about the functional role of the spatial and temporal patterns by correlating it with experimental design parameters in a fairly conventional way. For instance, because the two prefrontal cluster in subject B where separated they must have different functional roles. We interpret the above described correlation patterns in such a way that the right prefrontal cortex in subject B is active load dependently during the delay period. However, when the delay is six seconds long the right prefrontal cortex gets less active towards the end of the delay whereas the left prefrontal cortex takes over as shown by its increase in activity.

References

[1] Baddeley, A. (1986). *Working Memory.* Oxford University Press, Oxford, England.

[2] Baune, A., Sommer, F. T., Erb, M., Wildgruber, D., Kardatzky, B., and Grodd, W. (1999). Dynamical cluster analysis of cortical fMRI activation. *NeuroImage*, 6(5):477 – 489.

[3] Friston, K. J., Holmes, A. P., Worsley, K. J., Poline, J. B., Frith, C. D., and R.S., R. S. F. (1995). Statistical parametric maps in functional imaging: A general linear approach. *Human Brain Mapping*, 2:189–210.

[4] Josephs, O. and Henson, N. A. (1999). Event-related funcitonal magnetic resonance imaging: modelling, inference and optimization. *Philosophical Transactions of the Royal Society of London B Biological Sciences*, 354:1215–1228.

[5] Postle, B. R., Zarahn, E., and D'Esposito, M. (2000). Using event-related fmri to asses delay-period activity during performance of spatial and nonspatial working memory tasks. *Brain Research Protocolls*, 5:57–66.

[6] Smith, E. E., Jonides, J. J., Marshuetz, C., and Koeppe, R. A. (1997). Components of verbal working memory: evidence from neuroimaging. *Proceedings of the National Academy of Sciences USA*, 95:875–882.

A Special-Purpose Neural Network Recogniser to Detect Non-Random Pattern on Control Charts

A. Anglani, M. Pacella*, Q. Semeraro†

*Dip. di Ing. Innovazione – Università degli Studi di Lecce, Via per Arnesano, Lecce, 73100 Italy, e-mail: massimo.pacella@unile.it, †Dip. di Meccanica – Politecnico di Milano, Via Bonardi 9, Milano, 20132 Italy

Abstract

With the growing employment of automatic data-collection methods and the enhancements on computerised plotting on control charts, a demand exists to automate the analysis of process data. Computerised recognition techniques can provide an actual alternative to conventional methods for analysing control charts with little or no human intervention. In this paper, a neural network approach is discussed and applied to trend-pattern recognition on control charts. In the proposed approach the neural network is trained to recognise both "natural" and "unnatural" distribution of points. Experimental results are compared to a combined Shewhart-CUSUM approach in terms of Average Run Length (ARL).

1 Introduction

The main objective of process quality control consists in maintaining a constant and acceptable level of some process characteristics. Usually in a process, a certain amount of variability may occur; the sources of this variability may be referred as *unassignable* and *assignable* causes. The variations due to unassignable causes are intrinsic to the process and cannot be removed or it is not convenient to remove them. When only unassignable causes are in effect, a process is considered to be in a *natural* state (in-control). The variations due to assignable causes, preclude the process being it *stable* and has to be eliminated. In such a situation, the process is said to be in an *unnatural* state (out-of-control).

Conventional Shewhart control charts have been the most popular charts providing the capacity to discover unnatural processes. A chart can show when to start investigating for source of instability. One approach to enhance the operation of control charts is to adopt a control method based on identifying unnatural pattern behaviour [1]. Typically, *pattern-recognition* has been based on the visual opinion of the operator and it requires considerable experience.

Many different neural network architectures and learning algorithms have been used to develop control chart pattern recognisers. Hwarng and Hubele [2] proposed a multilayer perceptron (MLP) trained with back-propagation algorithm to detect six unnatural control chart patterns: trend, cycle, stratification, systematic, mixture and sudden shift. In order to improve the performance of MLP architecture, a modular neural network (MNN) was proposed by Cheng [3]. Hwarng and Chong [4] used an ART-recogniser (Adaptive Resonance Theory) in order to improve the model adaptation ability in a variable environment. Smith [5] described a combined X-bar and R chart feed-forward back-propagation neural network to analyse both mean and variance shifts. Chang and Aw [6] used a neural fuzzy approach to both recognise abnormal patterns and estimate essential parameters of the specific pattern involved.

While the previous approaches prove a fine discriminatory ability in identifying different kind of patterns (outperforming typical approaches as a combined Shewhart-CUSUM scheme), a common difficulty encountered is to improve discrimination between *natural* and *unnatural* patterns especially for small and moderate change of the process mean. In the mentioned approaches, neural networks are trained to recognise simultaneously several typical kinds of unnatural patterns. Moreover, all the above studies are based on the assumption that the neural network is trained only on the unnatural patterns across a wide spectrum of possible pattern parameters. As a consequence, the majority of the wrong classifications are incorrect detection of a pattern when no assignable cause exists. That is, the network signals too many unnatural-pattern recognition false alarms when the process is actually in control.

In this paper, a different neural network approach is proposed for process control. In detail, a neural network is solely trained for discriminating "natural" by a specific "unnatural" distribution of points. The objective is to show that this kind of "*special-purpose pattern recogniser*" can achieve a better performance in specific pattern analysis than that one of a general-purpose pattern recogniser which can identify various pattern classes.

A *supervised* neural network has been chosen in this work. The utilisation of a supervised neural network requires that user has a set of known unnatural pattern behaviours that are to be detected and identified. Commonly the implementation prerequisite of a special-purpose pattern recogniser is that a specific pattern is known to occur more frequently than other one. In order to compare the proposed approach to those reported in the literature, a linear upward and downward trend (Fig. 1) has been adopted to model an unnatural pattern. Obviously, another kind of unnatural pattern can be chosen (shift, mixture, cycle, and et.). However, the goal of the proposed approach is to detect only the structural change in the process under hypothesis of a prior knowledge of the unnatural pattern shape and it is not to identify unnatural patterns per se. The proposed work represents a preliminary study that aims to verify the potential of utilising a pattern recognition that operates only to recognise a natural by an unnatural process behaviour. Since a supervised neural network has been chosen for this work a prior knowledge of both natural and unnatural patterns must be assumed.

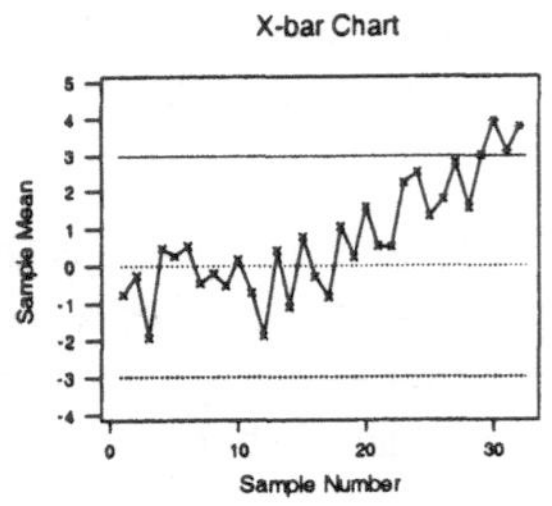

Fig. 1. Linear Trend Pattern

Many traditional analysis utilities are available for pattern recognition. For example a linear regression can be used to recognise a linear trend. However this kind of tool can be applied only if a determinist mathematical formulation can be used to model the unnatural pattern class. As consequence, these tools are not able to recognise non-deterministic patters as mixtures. Neural networks offer significant advantages over traditional classification tools. Neural networks do not need explicit rules to be derived for the recognition process, they simply "learn" to distinguish among pattern of different classes.

The rest of the paper is organised in the following manner. Section 2 gives an overview of the neural network architecture. Section 3 explains the training data set generation. Section 4 describes the training approach. Section 5 discusses the performance of the artificial neural network. Finally, conclusions are given.

2 The Neural Network Architecture

There are different supervised neural network architectures that can be used for pattern recognition. In this study, a 3-layer (input-hidden-output) fully connected *Feed–Forward* network trained with *Back-Propagation* (FFBP) has been adopted [7]. The FFBP neural network has been used successfully in many control chart pattern applications in literature. The Back-Propagation training-algorithm tends to be relatively slow; nevertheless the recall process is very fast. Since the network training can be implemented off-line, this feature facilities the use of the proposed model in an on-line real time mode.

Recurrent neural networks (e.g. Elman networks) can also be used. However, Elman networks are not as reliable as Feed–Forward networks since training occur using an approximation of the error gradient. Moreover, our preliminary research showed that, unlike a recurrent network, the FFBP neural network generalises well in regions of the input space where there are no training data. As a consequence, the Feed-Forward Back Propagation network is appeared to be the best choice for this study.

To ensure efficient convergence of the network training and the desired performance of the trained network, several parameters must be determined. At present, an established method is not available to determine the optimal configuration and the parameters of a neural network for control chart pattern-recognition. Therefore, most of the design parameters are application-dependent and must be empirically determined. In this study the parameter values reported in table I have been adopted. They have been decided after preliminary simulations, in which it has been observed their effects on the pattern-recogniser performance in terms of both in control and out-of-control alarm rates.

Parameters	Value/Type
momentum coefficient	0.4
learning coefficient	0.15
number of hidden nodes	24
window size (number of input nodes)	32
hidden transformation function	hyperbolic tangent

Table I. Parameters of the proposed neural network

In particular, the definition of the number of nodes in the input layer is critical. This number is referred to the recognition *window size* in on-line SPC applications. In the proposed neural network, the input layer contains 32 nodes used as input data for 32 consecutive points in the control chart. The output layer consists of one node only that is used for identifying the pattern addressed in this study. The hidden layer contains 24 nodes. As the number of hidden nodes is increased, the learning results are generally improved too. However, our experience shows that if the number of nodes exceeds 24, it does not improve learning and, instead, increases the total training time. In the proposed model, hyperbolic tangent function is used as the transformation function for the hidden layer and no transformation function is used for the output layer. The neural network in this study is coded using the NNET toolbox of MATLAB (by MATHWORKS).

3 The Training Data Set

This study is based on the assumption that a natural pattern can be modelled by a normal distribution of known mean and standard deviation, and the unnatural pattern can be modelled as linear upward or downward pattern trend (Fig. 1). This assumption allows generating data by mathematical model in order to mimic the true outcome of the process, and thus, can be used to train and testing a *supervised* neural network system. The correct choice of training data is critical for the performance of a neural network. Since large numbers of example pattern are necessary, in this study a Monte Carlo simulation approach is used to generate the required sets of control chart for the training phase. The trend pattern is specified to consist of an in-control mean value combined with two noise components: $\varepsilon(n)$ represents common-cause variation and $d(n)$ the special disturbance. In detail, the expressions used for pattern generation is:

$y(n) = \mu + \varepsilon(n) + d(n);$

where:

n is the number of sample;

$y(n)$ is the nth sample value;

μ is the process mean when the process is in-control;

$\varepsilon(n)$ is the common cause variation at n^{th} sample following a normal distribution with zero mean and standard variation σ;

$d(n)$ is the special disturbance at the nth sample modelled by:

$d(n) = n \cdot s \cdot \sigma;$

where s is the slope of the trend in terms of σ.

Input data are then coded in order to reduce the effect of a common cause variation (noise). The coding scheme filters the small random variations while maintaining the main features in the data. It allows network convergence to occur more readily. The variable range is divided into 64 zones with a width of 0.25σ, each returning an integer code. More exactly, data are coded as described by the following expression.

$$z(n) = \begin{cases} +8; & \text{if } y(n) > \mu + 8\sigma \\ -8; & \text{if } y(n) < \mu - 8\sigma \\ 0; & \text{if } y(n) = \mu \\ w\left[\left|\dfrac{y(n)-\mu}{w\sigma}\right|\right]\dfrac{y(n)-\mu}{|y(n)-\mu|}; & \text{otherwise} \end{cases}$$

Where w (in this study set to 0.25) is the zone width expressed in units of standard deviation, and [.] is the floor operator that returns the largest integer less than its argument.

4 Network Training

During training, input and output streams are presented in a random fashion to the neural network. The input values are received by the input-layer and the desired output (i.e. the known process status) is associated with the output-layer. In particular, the required output is set to 1 for a positive change of the mean (upward trend), –1 for a negative change in the mean (downward trend) and 0 for no change in the mean (natural distribution). As a consequence, an output value close to 1 (to –1) would indicate an upward linear trend (a downward linear trend). An output value close to 0 would indicate that no process change has been detected.

The total number of training streams is 8000 including 4000 natural 2000 upward trends and 2000 downward trends. Each stream is a window of 32 points. For training purposes the linear trend slope s has been chosen as follow:

$-0.5 \le s \le -0.025$ for downward trend

$0.025 \le s \le 0.5$ for upward trend

The step increment has been fixed to 0.025 therefore 20 values of negative slope and 20 values of positive slope are generated for neural network training. For each value of s 100 streams are generated. It will be noticed that the neural network is trained only with data streams presenting a small change of the mean. In such a way, the simulated patterns occur mostly within the range of ±3σ around the process mean. The training session has been stopped after 1000 learning iterations (the root-mean square has been reduced to 0.0036).

5 Testing Results

Due to the large number of patterns needed for assessing the performance of the pattern recognition model based on the proposed policy, extensive Monte Carlo simulations are been made. The performance has been measured using the Average Run Length (ARL) and has been compared to those reported in Cheng [8]. The ARL is calculated based on 1,000 simulation runs both for natural and unnatural streams. Each stream consisted in 1,032 data from which moving windows of 32 observations is obtained with step of a single sample. All the unnatural patterns began at point 33 in the simulated process data stream. The first 32 data points are generated from a normal distribution. It is assumed that unnatural pattern do not exists in the data for the first recognition window.

A threshold or cut-off value (θ) is applied to neural network output. Any value above θ is considered to signal the presence of an unnatural pattern. With actually no unnatural pattern presents, a Type I error occurred when the output value is equal to or greater than θ. The ARL performances of the proposed model are compared to those of the previous approach described in the published literature. The combined Shewhart-CUSUM scheme [1] is a special control chart scheme which aims to detect large as well as small process mean changes simultaneously. The one-sided Shewhart-CUSUM schema is designed to detect changes in one direction only (i.e. upward or downward).

Table II compares (for each pattern parameter setting) the ARL's of the one-sided Shewhart-CUSUM scheme K=0.25, H=8 and the neural network results. The ARL of the combined one-sided Shewhart-CUSUM scheme in the presence of linear trend are obtained from simulation results reported in [8]. Column "ν" contains the ratios of the ARL values obtained by the neural network presented in this study to the corresponding ARL values of the Shewhart-CUSUM ([8]). Column "ρ" reports the analogous ratio obtained from the neural network recogniser reported in [8] for linear trend detection. If the process is in out-of-control state, a ν ratio less than one indicates better performances of the proposed neural network than those ones of the combined Shewhart-CUSUM scheme reported in [8]. Moreover, the performance of the proposed approach can be compared to those proposed by Cheng [8] relating ν to ρ. For the comparison between CUSUM and proposed approach to be unbiased, it has been chosen a cut-off value in order to maintain the in-control ARL ($s=0.0\sigma$) of both methods approximately the same ($\theta=0.7$).

s	Cusum*	ARL	St.d. RL	ν	ρ†
0	721	727.8	49.8	1.01	1.00
0.0001	471	299.6	19.5	0.64	0.83
0.0005	260	195.6	15.0	0.75	0.80
0.001	185	119.1	11.4	0.64	0.81
0.002	129	96.7	10.1	0.75	0.81
0.005	76.8	66.8	9.6	0.87	0.84
0.01	51.9	41.2	6.8	0.79	0.84
0.02	35.2	29.6	5.6	0.84	0.84
0.05	21.5	18.7	2.6	0.87	0.82
0.1	14.9	12.4	2.2	0.83	0.82
0.5	6.88	5.7	1.1	0.83	0.82
1	4.93	4.0	0.8	0.81	0.86
3	2.86	2.6	0.7	0.91	0.97

Table II. Comparison of ARL between neural network and one-sided Shewhart-CUSUM (K=0.25, H=8, θ=0.7)

It can be noticed that the proposed model has better performance than that one with the combined Shewhart-CUSUM scheme when detecting upward and downward trends for small slope (0.0001, 0.0005, 0.001 and 0.002). This implies that the proposed model possesses better detection capability against small trends than the combined Shewhart–CUSUM scheme, maintaining, at the same time, an equivalent ARL value for the in-control process. Therefore, the proposed model improves the discriminatory capability to detect small change of the means that a traditional control chart cannot address outperforming, moreover, the combined Shewhart-CUSUM schema.

6 Concluding remarks

The system discussed in this paper is intended for an automated pattern-recognition procedure in which the operator of a manufacturing process is not required to monitor the control chart for patterns. This paper extends the above research by considering an artificial neural network trained to recognise both an in-control and out-of-control observation distribution. Simulation has been used to evaluate the performance of the proposed approach. Controlled conditions has been adopted in the experiments (i.e. no real data have been used) in order to compare the obtained results to those ones reported in the literature. The numerical results show that the proposed model outperforms both neural network recognition systems and combined Shewhart-CUSUM reported in the published literature for detecting moderate slope linear trend. Finally, a few potential improvements are outlined as follows.

1) A systematic study must be conducted to analyse the relationship between window size, number of hidden nodes and cut-off value in order to evaluate their influence on the ARL (both in-control and out-of-control) for the proposed neural network.

2) An interesting alternative approach is to use artificial intelligence not to identify unnatural patterns per se, rather to detect only the structural change in the process. This way, the special-purpose recogniser will be specialised to recognise only the natural structure and to signal any type of change in the process mean. In this case an *unsupervised* architecture (as ART neural network) can be used for detecting unnatural process behaviour.

Acknowledgement

This work has been partially funded by Ministero dell'Università e della Ricerca Scientifica e Tecnologica MURST and the National Research Council of Italy CNR.

References

[1] Montgomery, D. C.: Introduction to Statistical Quality Control, 3rd ed. New York: Wiley, 1996.

[2] Hwarng, H. B. and Hubele, N. F.: Back-propagation pattern recognizers for X-bar control charts: methodology and performance. Comp. & Ind. Eng., *24*, 219-235 (1993).

[3] Cheng, C. S.: A neural network approach for the analysis of control chart patterns. Int. J. of Production Research, *35*, 667-697 (1997).

[4] Hwarng, H. B. and Chong, C. W.: Detecting process non-randomness through a fast and cumulative learning ART-based pattern recognizer. Int. J. of Production Research, *33*, 1817-1833 (1995).

[5] Smith, A. E.: X-bar and R control chart interpretation using neural computing. Int. J. of Production Research, *32*, 309-320 (1994).

[6] Chang SI, Aw CA.: A neural fuzzy control chart for detecting and classifying process mean shifts. Int. J. of Production Research, *34*, 2265-2278 (1996).

[7] Haykin, S.: Neural Networks: a comprehensive foundation, 2nd ed, Prentice-Hall, 1999.

[8] Cheng, C. S.: A multi-layer neural network model for detecting changes in the process mean. Comp. & Ind. Eng., *28*, 51-61 (1995).

* Simulation results reported in Cheng [8], Table 7.

† From Table 7 of Cheng [8].

Data Mining and Automation of Experts Decision Process Applied to Machine Design for Furniture Production

G. Klene, A. Grauel*, H. J. Convey, A. J. Hartley†

*University of Paderborn, Soest Campus, Unit 16, Department Mathematics, Steingraben 21, D-59494 Soest, Germany, e-mail: klene@ibm16.uni-paderborn.de, †Bolton Institute of Higher Education, Faculty of Technology, Technology Development Unit, Deane Campus, Bolton, BL3 5AB, England, e-mail: hc1@bolton.ac

Abstract

A software concept based on data mining and knowledge discovery for a multi-spindle drilling gear configuration and optimisation applied to a machine used in furniture production process is proposed. The objective is to find the minimum number of supports and the optimised configuration of the multi-spindle drilling gears. Intelligent analysis of input data and an automated system covering the human design procedure are applied to configure multi-drilling gears. The input data presented as digitalised customer engineering drawings and furthermore technology data describing the basic constraints of the machine construction are presented. Moreover the transfer of acquired manual design experience from a human expert to a software strategy to solve the multi-criteria optimisation problem is shown.

1 Introduction

The target machine for the research work is a large flexible machine consisting of up to eight drill supports and each drill support has one or two drilling gears each having up to 40 individual drill locations. Each machine manufactured has to be specifically designed with regard to the minimum number of drill supports and the optimised configuration of the drilling gears. Configuration means the determination of location and properties of each drill – properties like diameter and type of drill – on one drilling gear. The gears usually have different configurations to optimise the furniture production process with regard to minimise production time. In other words there are two minimisation goals: the optimisation of the machine design and the optimisation of the furniture production.

The design of multi-spindle drilling machines itself requires an extensive knowledge of the machine construction in order to configure it with respect to the end customer's manufactured board specifications. Each board specification is given by a structural component engineering drawing. Usually more than 500 drawings have to be considered. When this data is coupled with the particularities of the machine [1], analytical processes help to design the machine and in more detail to define the number of supports, the number of drilling gears as well as the configuration of each gear. This design is done in respect to minimise the number of supports, drilling gears and drilling tools and furthermore to minimise the production time for each board. The production time is minimised by minimising the number of drilling cycles during production. One drilling cycle is one machine step during which a part of holes or all holes are drilled. In one step the gears are positioned and the holes are drilled by moving up the spindles for selected drills.

The design process [2] normally takes three man-months to be completed and is done by a human expert who has many years of experience. The aim of the project is to automate the design of such multi-drilling gear machines. Depending on the complexity of the problem it is not sufficient to use traditional optimisation methods only. An automated multi-spindle gear configuration using database concepts combined with data mining techniques and knowledge processing of the human expert design procedure is introduced to solve this multi-criteria optimisation problem.

2 Automated Configuration of Multi-Drilling Gears

The automated configuration of multi-drilling gears is based on database concepts combined with data mining techniques and an intelligent covering of the human design procedure [3]. The customer engineering drawings are stored in a database and techniques for primitive detection – primitives are arrangements of holes in a neighbourhood area of one drill – are used to generate project-individual database tables, which serve for further design steps using data-mining and fuzzy data processing.

Following the human expert the problem of automated configuration of the drilling gears can be divided in two major tasks [4]. The first step is to find a generalised pre-placement of drills and the second step is an iterative process which processes each board by defining the placement of the board in the area of work of the machine, by finding optimised positions for each support and gear and by achieving the possibility to produce each board by defining cycles and suitable drills. The drill pre-placement is the result of analysing all customer engineering drawings regarding their similarities. A sequence for the consideration of boards is determined using the board complexity (see 2.2) before the iterative configuration process starts. During this configuration process restrictions related to the parameters of the machine have to be observed and each structural component has to be checked concerning the feasibility of production.

2.1 Generalised pre-placement

The generalised pre-placement is based on analysing all boards in respect of characteristic primitives on different boards and on different locations on one and the same board. Primitives appearing on different boards can be generalised and accessory drills can be pre-placed on the drilling-gears. The program works analog to the human experts action which is to classify the boards into structural parts like cupboard units like side- and middle walls, bottom boards and doors. This is done by identifying characteristic primitives e.g. sides- and middle walls contain X-rows of holes, mounting-plate holes, dowel holes or metal-fitting holes. To give an example for a primitive X-rows or X-sets of holes are at least three holes in combination which fulfil the following conditions:

- placed on the same work piece,
- equal diameter,
- equal mode,
- equal Y-coordinate, and
- distance between holes = grid OR 2 times grid.

The most important primitives are

- X-rows with number of holes $l \geq 3$ within a specified grid,
- Y-rows with number of holes $l \geq 2$ within a specified grid and,
- metal-fitting holes l with diameter $\geq$ 12mm and near the edge of the board.The holes for metal fittings can be divided into drill through holes and fitting holes, which can be classified in more detail to single-holes, and holes having further holes, e.g. for mounting near the main hole.

X-row processing

To reach the aim of primitives identification and generalisation easily database techniques are applied. A database in which all X-rows are stored is created during program execution. The table definitions are dependent on the project and in more detail on the diameter-mode combinations. For important diameter-mode combinations a table for the X-rows is generated and a reference table to save related information like name of workpiece. The amount of columns in each table depends on the largest board (board with maximal X-direction) holding holes of the regarded diameter-mode combination. The amount of X-rows is calculated by:

$$NumXRows = \mathbf{ROUND}\left[\frac{X_l}{grid}\right]_+ \quad \text{with } X_l \text{ the length}$$

of the board with maximal X-direction, grid the standard grid of the project, and $ROUND_+$ an operator giving the next integer value if remainder greater than zero.

The holes of a identified primitive are stored in a fuzzy way to the generated tables. Similarity analysis reduces the complexity of data to a more global view on all boards to be produced (the whole project). Looking at the work-place of the machine a lower and upper work area can be specified (Fig. 1). This information is used for generalizing the primitives. The X-rows are generalized to a upper row and a lower row. To define the length of the sum-rows, the longest row of each board is taken and all this rows are shifted to a predefined grid. After this the sum row is calculated by OR-connection of each single row. The generalized primitives are used for the pre-placement of drills on the gears. Similar to the human expert the drills are placed on the upper and lower gears. The following rules show how the experts rules are used to find an optimised pre-location for the drills. For upper gears do:

a) Search on all boards and find that X-row, which has the maximum distance from the stopper-side.

b) Find for that X-row the spindle-row on the upper gears which has the maximal distance from the stopper-side and fulfils the machine restrictions maximal stopper position and maximal distance of feed-in piston.

c) Save this spindle-row as maximal spindle-row.

For lower gears do:

d) Search on all boards and find the X-row which has the minimum distance from the stopper-side.

e) Find for that drill-row on the specified board the spindle-row on the lower gears, which has the minimal distance from the stopper-side and fulfils the machine restrictions maximal stopper position and maximal distance of feed-in piston.

f) Save this spindle-row as minimal spindle-row.

To find the optimised spindle places for the drills count the free spindle rows between minimal and maximal spindle-row, split the free rows and shift the rows for pre-placing of the drills so far, that the distance between the spindle-places on upper and lower gears are smaller than the minimal distance between two X-rows on one the board with the smallest X-row distance.

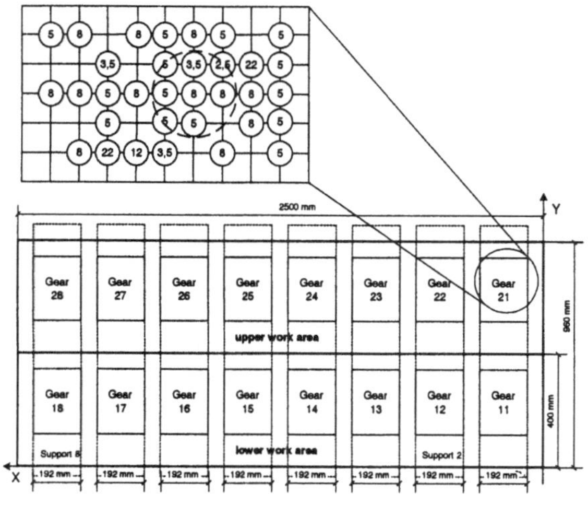

Fig. 1. Schematic of multi-drilling gear and machine.

Further primitives processing

In similar way like described for the X-rows (see X-row processing) generalized primitives can be found for Y-rows of holes and metal-fitting holes. Processing of this primitives results as well in generalised primitives used for spindle pre-placment.

Following the experience of the human expert, a fuzzy decision system [5-7] seems to be applicable to evaluate the number of multi-drilling gears required. At this stage the information from the human expert to specify the number and size of gears is used. This value is the input to the configuration program for the multi-drilling gears.

2.2 Computation of board complexity

During the drilling gear configuration two knowledge areas have to be taken into account. The generalised view about all boards which defines long drill rows or often seen primitives for metal fittings, and each board by itself with his specific holes. Considering each board it is important to start with the most complex board because the degree of freedom on the gears decreases with every board that the drills are positioned for. The complexity of each board can be characterised by:

- *Amount of primitives for metal fittings* per board; the consideration of metal fittings is hard because they usually have holes that appear not in a specified grid and therefore two cycles or gears holding drills out of the grid have to be planned.

- *Number of X-rows;* usually the board with X-rows has two of them. Sometimes there are three rows e.g. for bottom-clips. As usual two resulting long drill rows are placed on the gears. In the case of three hole-rows a second cycle or a third drill hole has to be considered.

- *Distance between primitives out of grid and X-rows;* the drills for X- rows are placed as result from the pre-processing before each board is processed board by board. If the distance between X-row holes and primitive holes out of grid are close together, it is hard to consider them in one cycle. If the distance is big enough there is the chance to uses separate gears.

- *Distance between the edge of the board and the holes;* this is important for construction holes usually lying in y-rows. The supports that hold the board are mounted above the gears. It is necessary to check that the supports are adequate to hold the board.

- *Size of the board;* for small boards in the X-direction the supports have to be checked to ensure that the board can not drop down, for large boards in y-direction the position of vertical feed-in piston and vertical stopper lead to problems. The complexity of each board can be calculated by OR-connection of all mentioned items.

2.3 Iterative Configuration Process for multi-drilling gears

The design of the multi-drilling gears related to each board is an iterative process. After placing the drills for the most important generalised primitives (see 2.1) the system has to check the feasibility regarding each board to be manufactured. Due to the fact that only the important rows have been considered initially, some drills will be missing. These have to be placed in an appropriate position. To find an optimised layout of the drilling gears, intelligent and rule-based search- and positioning algorithms are required. These algorithms cover the human design procedure.

During optimisation different constraints of the machine and the gears have to be checked continuously which is described as follows.

The most important constraints of the gears are:
- maximal number of spindle per vertical gear,
- maximal number of spindles in X-direction and
- maximal number of spindles in y-direction.

The constraints to be checked regarding the supports and gears are:
- minimal distance between 2 supports,
- minimal distance between y = 0 and middle of lower gears,
- maximal distance between y = 0 and centre of upper gears.

The constraints to be checked regarding the placement of boards:
- maximal position of vertical feed-in piston and
- maximal position of vertical stopper.

The boards can be placed in each position of Y-direction as long as the constraints for placement are fulfilled. With regard to the design on the drilling gears it is important to find an optimised position for each board. The most important rule taken into account for placement of the board is: Drills of upper gears (Fig. 1) work from Y > 400 mm. Due to the fact that the work pieces have different dimension in Y-direction they are placed in that way, that the middle of the work piece is located to Y=400 mm if stopper and feed-in piston allow this position. The stopper can be pulled out to a length of 500 mm and the feed-in piston – which presses the board to the vertical stopper – has a maximal length of 1060 mm. The second rule considered is that if there are two X-rows on the board the lower row has to be drilled by row-drillers of lower gears.

After fixing the position for the board optimised positions for the gears and the supports have to be found. Regarding the positioning of the gears the same rule used for board placement is considered. If X-rows where identified on the actual board then the position of the gears have to size that the rows can be drilled by upper and lower gears in one cycle. In principle that position for gears is the best where a maximum number of holes can be drilled with already equipped drills and with suitable free spindle places.

The optimised position of supports is iteratively found by starting with an initial position value. The start value results from a symmetric distribution of all supports. Applying an iterative alignment process the optimised support positions can be found and required drills can be placed on the gears. This process stops if the work piece can be produced with the computed design.

Conclusion

After briefly introducing the target machine important design goals for multi-spindle drilling gear configuration where shown. Solution strategies and a technical concept for automated multi-drilling gear configuration dependent on a generalised pre-placement and a iterative configuration process where proposed. The generalised pre-placement of the automated concept is based on characteristic primitives while the iterative process is based on the complexity of each board.

References

[1] Klene, G., Grauel, A., Convey, H. J., Hartley, A. J.: Optimisation under Constraints for Multi-Drilling Head Techniques. Computational Intelligence im industriellen Einsatz, VDI Bericht 1526 (ISBN 3-18-091526-9), VDI Verlag, Düsseldorf, pp. 353–358, 2000.

[2] Klene, G., Grauel, A., Convey, H. J., Hartley, A. J.: Intelligent Multi-Drilling Gear Optimisation for Industrial Automation, Proc. World Conference on Intelligent Systems for Industrial Automation (ISBN 3-933609-07-0), WCIS 2000, b-Quadrat Verlag, Kaufering, pp. 14–19, 2000.

[3] Klene, G., Grauel, A., Convey, H. J., Hartley, A. J.: Intelligent Data Analysis for Design of Multi-Drilling Gear Machines, Proc. European Symposium on Intelligent Techniques (ISBN 3-89653-797-0), ESIT2000, Fotodruck Mainz GmbH, Aachen, pp. 257–262, 2000.

[4] Klene, G., Grauel, A., Convey, H. J., Hartley, A. J.: Automated Multi-Drilling Gear Design in the Framework of Computational Intelligence, Proc. 4th Portuguese Conference on Automatic Control (ISBN 972-98603-0-0), Controlo´2000, Igreja de Oliveira, Guimaraes-Portugal, pp. 482–486, 2000.

[5] Grauel, A., Fuzzy-Logik: Einführung in die Grundlagen mit Anwendungen, B.I.-Wissenschaftsverlag, Mannheim, 1995.

[6] Grauel, A., Klene, G., Ludwig, L. A. ECG Diagnostics by Fuzzy Decision Making, Int. Journal of Uncertainty, Fuzziness and Knowledge-Based Systems, **6**, 2, pp. 201-210, 1998.

[7] Grauel, A., "Fuzzy-Logik", chapter 5.8 in Taschenbuch der Mathematik, 4th ed., eds. I. N. Bronstein, K. A. Semendjajew, G. Musiol and H. Mühlig, pp. 719-739. Verlag Harri Deutsch, Frankfurt, 1999.

Estimating Hourly Solar Radiation on Tilted Surface via ANNs

M. Gazela*, T. Tambouratzis†

*Solar and Other Energy Systems Laboratory, NCSR 'Demokritos', †Institute of Nuclear Technology-Radiation Protection, NCSR 'Demokritos', Aghia Paraskevi 153 10, Athens, Greece, email:tatiana@ipta.demokritos.gr

Abstract

A novel approach is proposed for performing on-line estimation of hourly global radiation on tilted surface from hourly global horizontal radiation at the same location and vice versa. The approach, which is based on back-propagation artificial neural networks, is computationally simple, accurate and compares favourably with theoretical, parametric and non-parametric models. Additionally, it demonstrates completion power, filtering potential, robustness to noise and data generation capability.

1 Introduction

The values of solar radiation on tilted surface (I_T) and solar horizontal radiation (I) determine the performance of weather-driven energy systems, i.e. solar collectors, solar hot-water systems and solar cells. A novel approach is presented here for estimating I_T directly from I and vice versa. The proposed approach is based on back propagation (BP) [1] artificial neural networks (ANNs); ANNs have recently been successfully implemented for the estimation of a number of meteorological parameters [2,3,4].

The proposed approach is of practical interest since the high cost of pyranometers usually precludes the installation of two pyranometers at the same meteorological station, whereby only either I_T or I is measured (usually I, which is of greater importance in meteorology). Furthermore, the proposed approach is appealing, especially when compared with existing theoretical models, in that:

a) I_T estimation is performed directly from I (and vice versa); this effectuates a significant reduction in computational effort, while the errors encountered in theoretical models that are owing to the utilisation or calculation of other meteorological parameters (i.e. diffuse solar radiation and ground reflectance) are not involved.

b) On-line operation is accomplished, while completion power, filtering potential and robustness to noise are demonstrated.

c) Long-term estimation of I_T (or I) can be performed, provided that the respective measurements of I (or I_T) are available.

The dataset employed here comprises 4185 hourly measurements of $I_T(t)$ and $I(t)$ - where t is the local apparent time (LAT) - collected from Julian Day (JD) 1 to JD 365 of the year 1998. The measurements have been accumulated at the meteorological station of the Laboratory of Solar and Other Energy Systems, Institute of Nuclear Technology – Radiation Protection, which is permanently stationed at NCSR 'Demokritos'. Fig. 1 illustrates the values of I_T versus I for the 1998 dataset.

Fig. 1. I_T versus I for the 1998 dataset

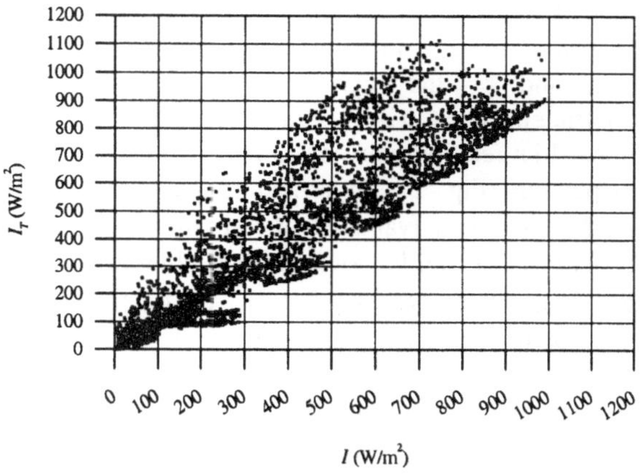

Only the results concerning the estimation of I_T from I are reported here (the results concerning the estimation of I from I_T are essentially identical). The estimation of I_T from I constitutes a problem of greater practical interest, as far as energy systems are concerned: the collecting surfaces of solar collectors and solar cells are tilted by an amount directly related to the longitude of the location where they are installed.

2 The Estimation of I_T Using ANNs

The main aim is to find the most appropriate back-propagation (BP) ANN structure – or a set thereof – in order to accomplish optimal estimation of hourly values of I_T from I. Apart from the selection of the parameters concerning BP ANN structure (e.g. number and nature of inputs, number of hidden nodes) and training (e.g. batch or on-line, η and α parameter

Table I. Characteristics and *SSE*s [W/m^2] of the of BP ANNs employed for the estimation of I_T

BP ANN	η	a	epochs	Training patterns	*SSE*	Validation patterns	*SSE*	Test patterns	*SSE*
ANN$_w$	0.6	0.96	25000	796	43.07	398	47.48	398	44.50
ANN$_s$	0.6	0.97	25000	1118	20.11	559	21.58	559	23.04
ANN	-	-	-	-	29.66	-	32.35	-	31.96

values), the quality of estimation depends on the partitioning of the 1998 dataset, with one BP ANN assigned to each partition; partitioning is especially beneficiary in cases where the different dataset partitions demonstrate distinct problem characteristics.

2.1 Selection of the appropriate BP ANN structure

The following criteria concerning BP ANN construction have been derived from the observation of the characteristics as well as the relationship of the two radiations:

- The selected BP ANN structure must be simple (small BP ANN with a single hidden layer and a few hidden nodes). This is supported by the similar shapes of the I_T and I curves during any day of the year and by the high value of the correlation coefficient (ρ_c=0.9146), which together suggest a linear-like relation between I_T and I with considerable dispersion (Fig. 1). It is mentioned that small BP ANNs have the advantages of fast convergence, good generalisation and few overfitting problems [5].
- Previous values ($I_T(t$-1), $I_T(t$-2), ...) together with t must be used as BP ANN inputs for the estimation of $I_T(t)$.This is supported by the bell-shaped I_T curve during any sunshine day which implies a dependence of $I_T(t)$ on previous values $I_T(t$-1), $I_T(t$-2), ..., as well as on the LAT t.
- The current value of $I(t)$ must be included in the BP ANN inputs. This is supported by the simultaneous disruption of the bell-shaped I_T and I curves during any intermittently cloudy day.

Partitioning of the 1998 dataset into two partitions according to the I_T/I ratio has been deemed expedient for the optimal prediction of I_T from I. The resulting partitions are:

- The "winter" partition (from JD 1 to 80 - Vernal Equinox - and from JD 267 to 365) featuring generally larger than 1.3 and of significant variance I_T/I values.
- The "summer" partition (from JD 81 to 266 - Autumnal Equinox) featuring practically constant and smaller than 1.3 I_T/I values.

From the different BP ANN structures and training parameters employed, the best estimations of $I_T(t)$ at LAT t, have been achieved with two 4-4-1 BP ANNs, namely ANN$_w$ and ANN$_s$ constructed and trained on the "winter" and "summer" partition, respectively. The training characteristics and parameters of both ANN$_w$ and ANN$_s$ are summarised in Table I and involve the sigmoid activation function and the batch mode of training. The ANN inputs correspond to hourly global horizontal radiation values $I(t)$ and $I(t$-1) at LAT t and t-1, respectively, the value of hourly radiation on tilted surface $I_T(t$-1) at LAT t-1 and LAT t. The inputs are scaled in the interval [-1.4,1.4], the output in the interval [0.2,0.8] and weight initialisation occurs in the interval [-0.01,0.01]. The training of ANN$_w$ was terminated when e_{av} of the validation set started increasing, while that of ANN$_s$ was terminated when e_{av} of the training set stopped decreasing. The *SEE* (standard error of estimates) of I_T for each partition has been used as a criterion of the quality of the corresponding BP ANN estimation. It is measured in W/m^2 and is given by

$$SSE = \sqrt{\frac{\sum_{d=1}^{TN} (I_{T\,d} - I_{Tann\,d})^2}{TN - p}} \tag{1}$$

where $I_{T\,d}$ is the actual value of the dth measurement of I_T, $I_{Tann\,d}$ is the BP ANN estimate (output) of $I_{T\,d}$, TN is the number of measurements of I_T and p is the number of model parameters; ANNs constitute non-parametric models, whereas p = 0. The *SSEs* of ANN$_w$ and ANN$_s$ are shown in Table I, separately for training, validation and test sets. Additionally, the total *SSE* of the ANN approach is given by

$$SSE = \frac{TN_w \cdot SSE_w + TN_s \cdot SSE_s}{TN_w + TN_s} \tag{2}$$

where TN_w and TN_w are the total numbers of measurements for the winter and summer partition, respectively, and SSE_w and SSE_s are the standard errors of estimates of the two partitions.

Table II. Comparative evaluation of the I_T estimators

Approach	*SSEs*		
	ANN$_w$	ANN$_s$	ANN
Linear	92.08	70.33	79.38
10th order	88.74	67.70	76.45
Perez	73.13	30.10	47.99
ANN	44.50	23.04	31.96

Fig. 2. I_T estimation during an intermittently cloudy day

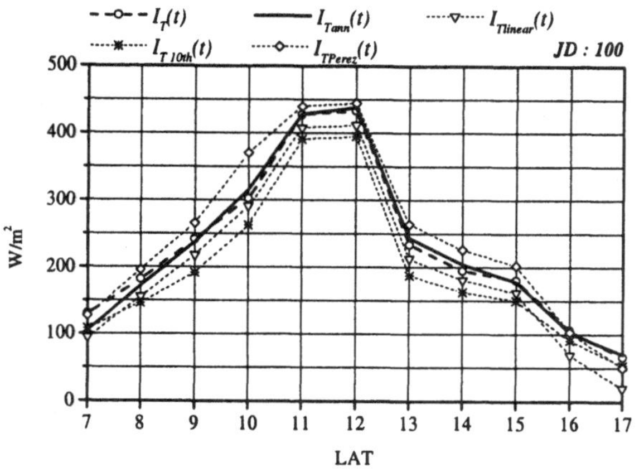

2.2 Comparative evaluation of estimation

A number of distinct approaches have been used to evaluate the quality of the BP ANN estimation:

- Parametric approaches, namely linear regression models of order 1 (linear least squares fitting [6]) up to 20. Superior results have been obtained with the 1st and 10th order linear regression models.
- A non-linear model (non-linear least squares fitting [6]). This is not discussed in the following as its implementation has shown that it is not appropriate for the estimation of I_T.
- Theoretical models. The most accurate of these, namely the model of Perez [7] with the diffuse radiation on horizontal surface calculated according to Orgill and Hollands [8], is shown next.

The superiority of the ANN approach over the selected parametric models and the theoretical model of Perez is illustrated in Fig. 2 as well as in Table II, which tabulates the *SSE*s for the two ANN test sets (separately for ANN_w and ANN_s, as well as collectively).

2.3 Further analysis of I_T estimation via ANNs

The ANN approach has been subjected to further analysis concerning its completion power, filtering potential, robustness to noise, and data generation capability.

Completion power. A problem frequently encountered with the available meteorological datasets is the existence of gaps in the time series. Customary approaches for filling in missing values of $I_T(t)$ are linear interpolation between previous- and next-day values of the same data series, linear interpolation between $I_T(t)$ of other nearby meteorological stations, and theoretical models.

The main drawbacks of these techniques are that:

Table III. *SSE*s [W/m²] for various levels of white noise

SNR / Approach	no noise	50	15	10	5
Linear	79.38	79.63	81.95	84.98	99.34
10th order	76.45	76.69	79.00	82.06	97.32
Perez	47.99	48.72	54.47	60.17	86.45
ANN	31.96	33.02	42.90	53.66	91.98

(i) The I_T measurements used in the interpolation may be significantly different from the missing values, especially in winter where the alterations between sunshine and intermittently cloudy days are often.

(ii) Meteorological datasets from nearby stations as well as long-term data are not always available.

(iii) The use of theoretical models demands the calculation of many parameters, which is neither convenient nor practical.

Employing the proposed approach for completing an isolated missing value of I_T (e.g. $I_T(t)$) is performed by using $I_T(t\text{-}1)$ as BP ANN input. For consecutive missing values of I_T (e.g. $I_T(t)$, $I_T(t+1)$, ..., $I_T(t+\tau)$) the ANN approach begins from the known $I_T(t\text{-}1)$ measurement, estimates (fills in) $I_T(t)$, uses the estimated $I_T(t)$ as BP ANN input for estimating (filling in) $I_T(t+1)$ and so on until $I_T(t+\tau)$ is filled in, i.e. until the entire gap is completed. Hence, the completion of missing values via the proposed approach is performed on-line (minimal computational effort and direct response), estimation is not in any way related to the I_T measurements of previous or next days, while there is no need for long-term data or data from other stations.

Filtering Potential. Owing to its accuracy in estimating I_T and filling in missing measurements, the approach can also be used for filtering existing I_T measurements. Discrepancies between I_T and I_{Tann}, especially when the shapes of I and I_T deviate significantly from each other, can provide evidence of erroneous measurements of I_T. Filtering constitutes a matter of further research as the allowable amount of discrepancy between I_T and I_{Tann} is a function of LAT t, season of the year, magnitude of the I value, as well as confidence in the measured I.

Robustness to noise. In order to examine the soundness of the proposed approach, white uncorrelated noise has been added to the 1998 dataset measurements. The results of the application of noise to the ANN approach are compared with those of the linear, the 10th order and the Perez models. Table III illustrates the *SSE*s of each approach with no added noise as well as with noise levels of signal to noise ratio (SNR) equalling 50, 15, 10 and 5. It is seen that SNRs up to 15 have

Table IV. *SSE*s [W/m^2] of each methodology for long-term I_T estimation via ANNs

Trials	*SSE* "winter"	*SSE* "summer"	Total
I	100.34	36.92	63.30
II	70.84	60.89	65.03
III	77.08	50.46	61.53
IV	79.03	59.85	65.33
V	80.01	50.93	63.03

practically no effect on the estimation of I_T for all approaches. For smaller SNRs and owing to the denoising and generalisation properties of BP ANNs, the estimation of I_T using the ANNs is more accurate than that of the other approaches; finally, for SNRs less than 10 all of the approaches are disrupted.

Data generation capability. In the case of complete absence of I_T measurements, the proposed approach can be employed for estimating long-term I_T data. Owing to the fact that the approach requires $I_T(t\text{-}1)$ for the estimation of $I_T(t)$, an approximation of $I_T(t\text{-}1)$ is necessary. A simple way of accomplishing this is:

I. $I_T(t\text{-}1) = R_b(t\text{-}1)\cdot I(t\text{-}1)$ for every LAT t, where R_b is the ratio of hourly beam radiation on tilted surface to hourly beam horizontal radiation.

More efficient alternatives exploit the completion capability of the ANN approach, whereby only the initial value of I_T ($I_T(t_{sr})$) is approximated at the sunrise hour of each day (t_{sr}) and the BP ANNs progress in the manner described above for the estimation (completion) of the subsequent I_T values of each day. A number of methodologies have been tested for the evaluation of $I_T(t_{sr})$:

II. $I_T(t_{sr}) = I(t_{sr})$; the latter is known.

III. $I_T(t_{sr})$ given by the model of Perez.

IV. $I_T(t_{sr}) = r\cdot I(t_{sr})$, with r given by

$$r = \frac{\sum_{sr=1}^{T_t} I_T(t_{sr})/I(t_{sr})}{T_t} \tag{3}$$

where T_t is the total number of days considered for data generation; in other words, r constitutes the average of the ratio $I_T(t_{sr})/I(t_{sr})$ of every JD. Two r values are calculated here, one for the "winter" and the other for the "summer" partition.

V. Same as (IV), but r is calculated independently for each month.

Table IV illustrates the *SSE*s of methodologies I to V for the generation of long-term I_T values; greater *SSE*s are observed in the "winter" partition because of the high variation in the relationship between I_T and I. Methodology II is the most appropriate for data generation during the "winter" partition (70.84 W/m^2, see Table IV), while methodology I is the most appropriate for the "summer" partition (36.92 W/m^2, see Table IV). Their combination results in a data-generation *SSE* of 51.03 W/m^2 for the ANN approach. A comparison with the model of Perez shows a slight superiority of the latter (47.99 W/m^2, see Table II). However, the proposed implementation for calculating long-term I_T data is simpler and more efficient, while it is more accurate than the Perez model in the "winter" partition (70.84 versus 73.13 W/m^2 from Tables IV and II, respectively).

3 Conclusions

A novel approach for performing on-line estimation of hourly global radiation on tilted surface directly from hourly global horizontal radiation at the same location (and vice versa) via back-propagation artificial neural networks has been proposed. The approach is simple as well as accurate and compares favourably with a number of conventionally used models (theoretical, parametric and non-parametric). Apart from its computational efficiency, the proposed approach combines the advantages of on-line operation, completion power, filtering potential and robustness to noise. Finally, the proposed approach can be employed for the long-term estimation of hourly global radiation.

References

[1] Rumelhart D.E., Hinton G.E. and Williams R.J.: Learning representations by back-propagating errors. Nature *323*, 533(1986)

[2] Shyi-Ming Chen and Jeng-Ren Hwang: Temperature prediction using fuzzy time series. IEEE Trans. on systems, man and cybernetics-Part B: Cybernetics *30*, 263(2000)

[3] M. Mohandes, A. Balghonaim, M. Kassas, S. Rehman and T.O. Halawani: Use of radial basis functions for estimating monthly mean daily solar radiation. Solar Energy *68*, 161(2000)

[4] A. Sfetsos and A.H. Coonick: Univariate and multivariate forecasting of hourly solar radiation with artificial intelligence techniques. Solar Energy *68*, 169(2000)

[5] S. Haykin: Neural Networks - A comprehensive foundation, 138, U.S.A.: Prentice-Hall INC 1994.

[6] A. Papoulis: Probability, random variables, and stochastic processes, 175, 3rd ed., Singapore: Mc Graw-Hill International Editions 1991.

[7] R. Perez, R. Seals, P. Ineichen, R. Stewart and D. Menicucci: A new simplified version of the Perez diffuse irradiation model for tilted surfaces. Solar Energy *39*, 221(1987)

[8] J.F. Orgill and K.G.T. Hollands: Correlation equation for hourly diffuse radiation on a horizontal surface. Solar Energy *19*, 357(1977)

On the Typology of Daily Courses of Tropospheric Ozone Concentrations

Emil Pelikán*, Kryštof Eben*, Jiří Vondráček*, Michal Dostál*, Pavel Krejčíř*, Josef Keder†

*Institute of Computer Sciences, Academy of Sciences of the Czech Republic, Prague, Czech Republic †Czech Hydrometeorological Institute, Prague, Czech Republic

1 Introduction

During last two decades extensive efforts have been made in modelling and prediction of the tropospheric ozone concentrations. Many statistical methods work on daily characteristics as maximal, average or other values. In the framework of the European union project APPETISE (Air Pollution Episodes: Modelling Tools for Improved Smog Management) a database has arisen holding hourly measurements of ozone concentrations on stations located in five European countries. In this contribution we try to classify typical daily courses of these concentrations. The stations are compared with respect to the occurrence of different types throughout the two-years period 1998-1999. We employed the Kohonen self-organizing maps and multivariate statistical analysis to visualize similarities and dissimilarities between involved stations.

2 Classification of Daily Courses of Ozone Concentrations

Tropospheric ozone arises as a result of a photochemical reaction of ozone precursors with oxygen during warm days with high solar radiation. Therefore we have confined ourselves to the period between April and October.

The data were preprocessed by the methods described below. This first analysis discovered many days with unrealistic or erratic behaviour of daily measurements, suggesting a problem with the measuring device. These outlying observations were excluded from the following analysis.

The preprocessed data constisted of 6046 24-dimensional vectors of hourly ozone concentrations measured on 21 stations. We will call these vectors "profiles" henceforth. Obviously neighbouring measurements are highly correlated and the correlation decreases with time interval between them. We formed the sample covariance matrix of the profiles regardless of date and station. The profiles were characterized by first six principal components of the covariance matrix. They represent 95 percent of the total variance. In our case, they have a natural interpretation: the first component (60.8 % variance) can be reflected as the weighted mean of the concentrations, the second component (18.8 % variance) represents quadratic term and the third one (7.5 % variance) the linear term in the daily course of ozone concentration. Remaining three components (7 % variance) seem to represent higher order terms.

The six-dimensional vectors arisen in the described manner were clustered with the help of procedure FASTCLUS (SAS System [5]) into ten clusters. This clustering procedure (method of k-means) creates in the multidimensional space clusters of locally minimal within-class variation. The number of clusters was selected so as to avoid clusters with low numbers of observations. Fig. 1 illustrates the average courses of daily ozone concentrations in selected clusters. These average profiles provide a typology of daily courses under different meteorological conditions and geographical location. The clusters with highest daily maxima correspond to an episodic situation.

Obviously there is no sense in seeking a single "typical" daily course of a particular station. Therefore, the statistical behaviour of the stations during the analyzed period was characterized by a vector of frequencies of occurrence of the above profiles.

Let p_{ij}, $i = 1, \ldots, 21$, $j = 1, \ldots, 10$ be relative frequencies of ozone profiles belonging to the j-th cluster on the i-th station. It is $\sum_j p_{ij} = 1$. The dissimilarity of the stations $i, k = 1, \ldots, 21$ was quantified with the help of Matusita's metric [4]

$$d_{ik} = \left[\sum_{j=1}^{10} \left(\sqrt{p_{ij}} - \sqrt{p_{kj}} \right)^2 \right]^{1/2} .$$

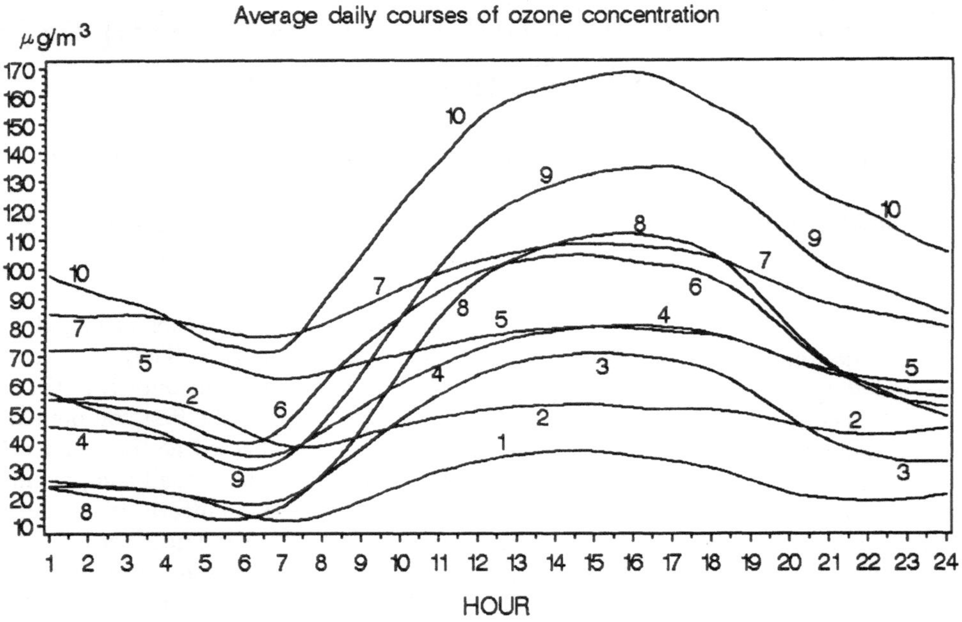

Fig. 1. Ten average profiles representing the typology of daily courses of ozone concentration

Group of stations	Typical profile									
	1	2	3	4	5	6	7	8	9	10
1	27.2	35.4	13.6	6.6	11.6	0.9	1.2	2.9	0.6	.
2	6.0	10.8	14.6	17.3	9.7	12.4	4.1	16.2	8.0	0.9
3	5.1	13.1	17.3	18.8	16.7	14.5	7.2	2.8	3.3	1.2
4	1.6	14.1	5.5	18.5	24.7	5.8	20.5	1.8	4.0	3.6
5	98.8	0.6	0.6	.	.	.	.	.	.	.

Table I. Frequencies of occurrence of typical profiles in groups of similar stations

To visualize similarities in statistical behaviour of the stations we employed the Kohonen [3] self-organizing maps and hierarchical clustering procedures [1], [2]. The result of complete linkage procedure is shown in Fig. 2, other clustering methods (single linkage, average, Ward) gave very similar results. All of the clustering method suggest forming of the same five groups of similar stations.

For the Kohonen network, input cases were ten-dimensional vectors of relative frequencies as above. A Kohonen network with 10×10 neurons in the Kohonen layer was trained, using STATISTICA Neural Networks software [6]. We assigned labels to winning units using the the names of measuring stations in the training set. The Kohonen network relates similar classes to each other in two-dimensional space of the Kohonen map. The result is presented in Fig. 3, where we can distinguish similar groups of stations as was shown in Fig. 2. Generally there is a very good correspondence between the two visualisation techniques.

In the table I we present the percentage of occurrence of the 10 typical profiles in the groups of stations (1-5). Table I. together with Fig. 1 could serve for interpretation of similarities in terms of geographic location and type of station (urban, rural, etc.), e.g. group 1 corresponds to urban stations with lower occurence of episodes (northern, close to the see) whereas group 2 consist of Berlin and Prague, middle European cities with higher occurence of episodes (continental, industrial region). Nevertheless, each of the stations has its own specificities, e.g. Yarner Wood lies in a nature reserve in Devon which is a popular recreation area with lots of traffic during summer, causing high levels of ozone. It turns out that the daily courses of ozone concentration differ both in shape and frequency of occurrence according to the site, region, etc.

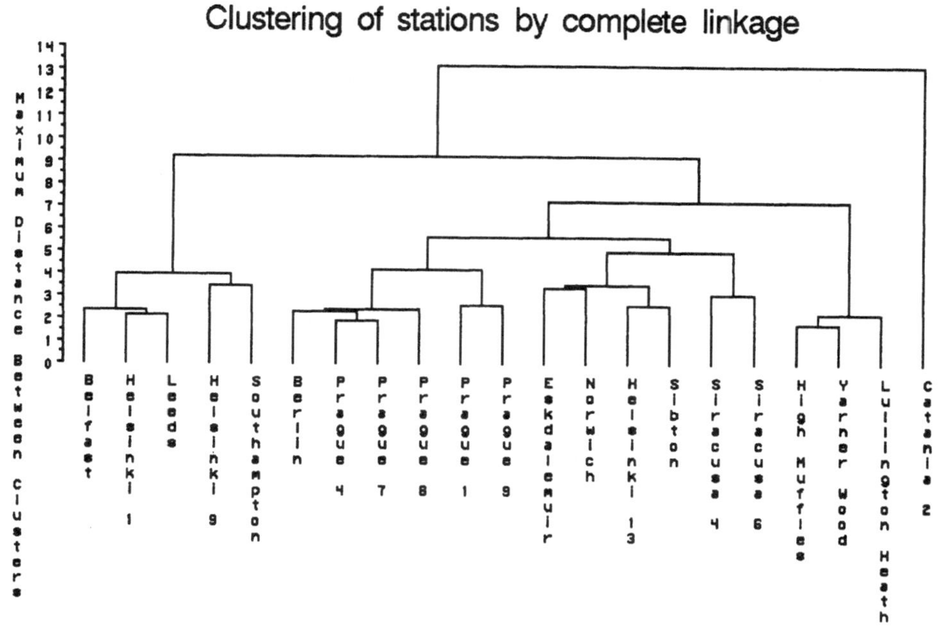

Fig. 2. Clustering of the stations by complete linkage

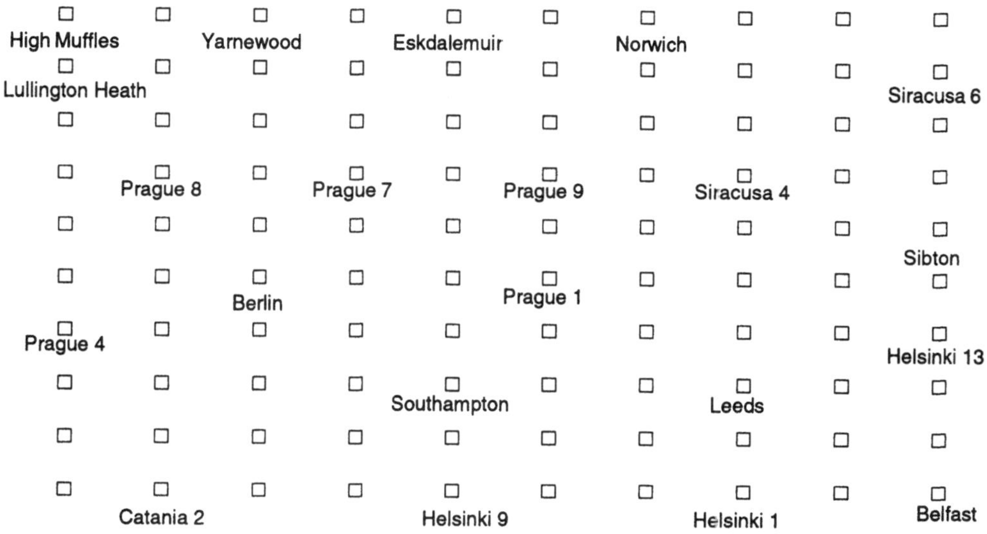

Fig. 3. Clustering of the stations by the Kohonen maps with 10 × 10 neurons

3 Conclusions

We presented an approach for dimension reduction, summarization, visualisation and other preprocessing for complex datasets arising in monitoring of tropospheric ozone. In particular these methods have lead to quick detection of erratic observations in the database.

Obviously the presented approach cannot directly lead to a design of a tool for succesful modelling and predicting ozone concentrations, since these are dependent on complicated chemical and meteorological processes. However, the suggested methods can be used for getting insight into multidimensional dependencies.

It will be interesting to compare the performance of ozone episode detectors on sites similar/dissimilar in the above sense. This can contribute to the comparison of different statistical modelling methods, which is the one of the goals of the APPETISE project.

Aknowledgement

This work was supported by the EU project APPETISE No. IST-99-11764 and by the grant project of Czech Ministry of Environment No. VaV/740/2/00.

References

[1] Anderberg, M.R.: Cluster Analysis for Applications. Academic Press, New York, 1973.

[2] Hartigan, J.A.: Clustering Aloritms, John Wiley, New York, 1975.

[3] Kohonen,T.: Self-Organizing Maps. Second edition. Springer, Berlin, 1997

[4] Matusita, K.: Distance and decision rules. Ann. Inst. Stat. Math. 16, The twentieth Anniversary Volume, Part II, 1964, 305-315.

[5] SAS/STAT User's Guide, Version 6, Fourth Edition. Volume 1, Volume 2. Cary, NC:SAS Institute Inc., 1989.

[6] STATISTICA Neural Networks, StatSoft, Inc., Tulsa, OK, USA, 1998.

GUHA Analysis of Air Pollution Data

David Coufal*[1]

*Institute of Computer Science, Academy of Sciences of the Czech Republic, Pod Vodárenskou věží 2, 182 07 Praha 8, Czech Republic, e-mail: coufal@cs.cas.cz

Abstract

This paper presents an exploratory analysis of air pollution data employing GUHA method. The aim is to examine if there are any significant relations between quantity of ozone (O_3) air pollution and quantity of other types of pollution (NO, NO_2). Influence of physical conditions (humidity, temperature) is examined too. The analysis is performed on base of data collected at several measuring stations localized across the Czech Republic.

1 Introduction

An *exploratory analysis* in GUHA (General Unary Hypotheses Automaton) sense is a process of systematic generation of hypotheses, theirs testing and selection of "interesting" ones. This process differs from standard *confirmatory analysis*, commonly used in statistics, where tested hypothesis needs to be formulated in advance usually on base of some extra information. The advantage of an exploratory analysis over a confirmatory one becomes clear when we have no or only little idea about relations within analyzed data (therefore it is hard to formulate some relevant hypotheses to confirmatory testing in advance) then exploratory analysis enables us systematically formulate and test a broad class of hypotheses. Hypotheses supported by data are then revealed and give us basic orientation within analyzed data. These "interesting" hypotheses can consequently serve as inputs to further classical methods of statistical analysis.

From data analysis point of view GUHA method can be seen as a method of data mining and in fact as one of the oldest method - first papers appears already in mid-sixties and it still develops. Within this paper GUHA method is applied on analysis of air pollution data with intention to get basic insight to interactions among particular pollutants (O_3, NO, NO_2) together with examination of influence of physical conditions (humidity, temperature) on these interactions. The analysis is driven by an effort for improving current methods of ozone concentration prediction and it is performed as a part of APPETISE international project [1].

The rest of the paper is organized as follows. The second section gives basic review of GUHA method. The third section describes structure and preprocessing of air pollution data used for analysis. The fourth section refers to results of analysis and presents revealed hypotheses. The paper concludes by the fifth section.

[1]This work was supported by APPETISE project grant EU No. IST-99-11764 and by GA CR 201/00/1489.

2 Basics of GUHA Method

In previous section main idea of exploratory analysis was outlined. In this section this idea is concretized for the case of GUHA method. Our review is a very basis one so reader interested in an deeper insight into the method is referred to [2, 3].

2.1 Source database and data matrix

Let's start with a notion of *source database*. Source database is formally a matrix with N rows and M columns, where rows correspond to objects and columns to variables defined on these objects. Each variable V^j, $j = 1, \ldots, M$ has its value $V^j(o_i)$ for each object o_i, $i = 1, \ldots, N$. Value $V^j(o_i)$ then represents the value in ith row and jth column of source database. A source database is in fact a set of data we are given to process by GUHA method.

Variables of source database can be *nominal* or *ordinal*. For a nominal variable set of its possible values (range of variable) is finite, usually not ordered. The elements of this set are called classes. For an ordinal variable set of its possible values is a linearly ordered set, possibly not finite. For purpose of this paper only subsets of real numbers are considered.

A basic object the GUHA works on is a dichotomized source database which is a matrix of zeros and ones. This matrix, called *data matrix*, is created on base of source database by process of variables *categorization*. During this process each variable is associated with a set of its categories. For a nominal variable a category is given by some of its classes. For an ordinal variable a category is given by an interval defined within the set of its possible values.

Rows of data matrix represent objects of source database and the columns represent categories. Which categories are defined to create data matrix is an user depend task and in fact it is driven by the overall purpose of GUHA analysis. Denote by C^j some category derived on base of jth variable V^j. For each object o_i and each category C^j of data matrix there is 1 in ith row and category's respective column if and only if $V^j(o_i) \in C^j$. If $V^j(o_i) \notin C^j$, then there is 0.

2.2 Hypotheses formation and testing

A general form of hypothesis in GUHA is $A \approx B$. The first part A is called *antecedent*, the second part S *succedent*. To form the whole hypothesis, antecedent with succedent are bind by a *quantifier* $\approx$. Antecedent and succedent (together called as cedents, when they need not be distinguished) are actually boolean propositions formed on base of categories employing standard *and* & and *negation* $\neg$ boolean connectives. Denote as *literal* L term C or $\neg C$, where C is a category, then cedent has a general form

$$L_1 \& L_2 \& \ldots L_{n-1} \& L_n, \qquad (1)$$

where n is denoted as *length of cedent*. Since cedents are boolean propositions we are able to determine for particular object o_i and particular cedent if a boolean proposition given by this cedent has the value 1 for the object o_i (it is satisfied by o_i) or it has value 0 (it is not satisfied by o_i). For given antecedent A and succedent S we can perform this test for all objects of source database. The information about cedents behavior then can be expressed in a form of so called four fold table (ff-table)

	S	$\neg S$	
A	a	b	r
$\neg A$	c	d	s
	k	l	m

where a is the number of objects both satisfying A and S; b is the number of objects satisfying A and not satisfying S; c is the number of object not satisfying A and satisfying S; and d is the number of objects both not satisfying A and not satisfying S. Values r, s, k, l are marginal sums, $r = a + b$, $k = a + c$, etc. Note that $m = a + b + c + d = N$.

The semantics of a quantifier is given by its associated function. An associated function is a boolean function operating on ff-tables (on values a, b, c, d). Consider some data matrix, then for each hypothesis $H : A \approx S$ we can determine its ff-table generated by this hypothesis antecedent and succedent. On base of values of this ff-table, value of associated function of quantifier $\approx$ can be computed. If this value is 1, then we say that hypothesis H is valid in data matrix, if value is 0 then we say that the hypothesis H is not valid. We mention here only FISHER quantifier because we will employ it in our analysis but many others quantifiers are defined in GUHA method.

Fisher quantifier – *FISHER*: $\approx_{\alpha,base}$

FISHER quantifier's associated function has two optional parameters - *base* $\in \mathcal{N}$ and $\alpha \in \mathcal{R}$ restricted to values *base* ≥ 1 and $\alpha \in (0, 0.5]$. Computation of associated function is given by formula (2) and in fact it represents fisher test known from statistics.

$$\approx_{\alpha,base}(a,b,c,d) = \begin{cases} 1 & ad > bc,\ a \geq base \text{ and} \\ & fsh(\cdot) \leq \alpha \\ 0 & \text{otherways} \end{cases} \qquad (2)$$

where $fsh(\cdot) = fsh(a,b,c,d)$ is given by

$$fsh(a,b,c,d) = \sum_{i=a}^{\min(r,k)} \frac{\binom{k}{i}\binom{m-k}{r-i}}{\binom{m}{r}}. \qquad (3)$$

If some hypothesis $H : A \approx_{\alpha,base} S$ is valid according to FISHER quantifier then it means that we have accepted on significance level α (statistical) hypothesis $P(A|S) \geq P(A|\neg S)$ on conditional probabilities. In words, a valid hypothesis H can be represented as "presence of succedent increases presence of antecedent", i.e., there is some association between them.

Now we can approach to process of hypotheses generation and testing. This process is based on so called *task definition*. During a task definition a rough template of generated hypotheses is specified. That is, specification of maximal lengths of antecedent and succedent - n_A^{max}, n_S^{max} is given together with two sets of categories Cat_A, Cat_S. On base of these specifications broad set of *relevant* antecedents and succedents is given. A relevant antecedent is a cedent (1) of a length $n_A \leq n_A^{max}$, where literals are from set Cat^A. Similarly for relevant succedent with n_S^{max} and Cat^S. The last step of task definition is a choice of certain quantifier together with the values of its parameters. Relevant hypothesis is then a hypothesis formed from some relevant antecedent and some relevant succedent bind together with the chosen quantifier. Each possible relevant hypothesis is then generated and tested for its validity. When hypothesis is valid then it is revealed. Obviously, GUHA method can be used only if it is implemented in a form of computer program. Currently there is an implementation denoted as GUHA +– available at http://www.cs.cas.cz/ics/software.html web page.

3 Air Pollution Data

Air pollution data grounded GUHA analysis were collected from 37 measuring stations localized across the Czech Republic within the period of 1.4.1994 - 31.8.1998 (not all days were covered). Theoretically, values of variables given in Table I. were measured at each station each half hour during a day. Practically, there were missing data for some variables in case of some stations. Data are available at APPETISE project web site [1] to public use.

Code	Variable
WV	Wind velocity [m·s^{-1}]
WD	Wind direction [grades $\in [0, 360)$]
NO2	NO_2 concentration [μg·m^{-3}]
NO	NO concentration [μg·m^{-3}]
H	Humidity [%]
O3	Ozone concentration [μg·m^{-3}]
T2M	Temperature in 2 meters [°C or K]

Table I.

Station	WV	NO2	NO	H	O3	T2M	No.Rec.
Libus		⋆	⋆		⋆		748
Prostejov	⋆	⋆	⋆	⋆	⋆	⋆	732
Sokolov	⋆	⋆	⋆	⋆	⋆	⋆	877
Liberec		⋆	⋆		⋆		691
Karvina	⋆	⋆	⋆	⋆	⋆	⋆	882
Ces.Bud.		⋆	⋆		⋆		718
Teplice	⋆	⋆	⋆	⋆	⋆	⋆	759
Serlich	⋆			⋆	⋆	⋆	177

Table II.

Since number of 37 stations and the fact that several stations are localized in near distance from each other, e.g. there are 5 stations in Prague, we have selected 8 representative stations. The selection was done on base of fuzzy c-means cluster analysis [4] of stations considered as points in a plane with values of coordinates given by geographic coordinates of longitude and latitude. Selected stations were the ones in the nearest distance from found clusters centers. In Table II. there are presented these stations together with variables available for them $(\star)$[1].

[1] Variable WD was excluded from other considerations because it was available only for one station.

3.1 Data preprocessing

The first step of raw data preprocessing was averaging of available records per day. This was done for each day and station. Number of reduced records obtained is given in the last column of Table II.

To process averaged data by GUHA method data need to be transformed into a form of a source database. Since we are interested in interactions among variables from prediction point of view, source database for each station was created as it is demonstrated in the figures Fig.1 and Fig.2. In the Fig.1 there are few first rows of averaged data for Libus station. These data are transformed into source database in such way to be in each row values of variables for given day together with averaged values for three past days. This is done for all variables, see Fig.2.

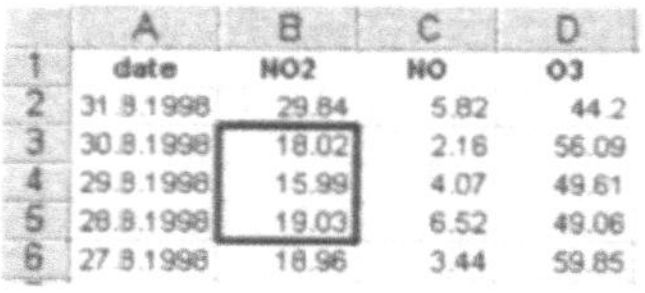

	A	B	C	D
1	date	NO2	NO	O3
2	31.8.1998	29.84	5.82	44.2
3	30.8.1998	18.02	2.16	56.09
4	29.8.1998	15.99	4.07	49.61
5	28.8.1998	19.03	6.52	49.06
6	27.8.1998	18.96	3.44	59.85

Fig. 1. Per day averaged raw data.

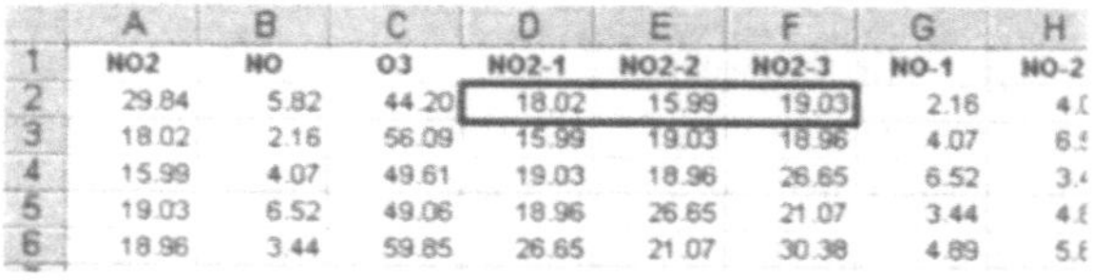

	A	B	C	D	E	F	G	H
1	NO2	NO	O3	NO2-1	NO2-2	NO2-3	NO-1	NO-2
2	29.84	5.82	44.20	18.02	15.99	19.03	2.16	4.
3	18.02	2.16	56.09	15.99	19.03	18.96	4.07	6.
4	15.99	4.07	49.61	19.03	18.96	26.65	6.52	3.
5	19.03	6.52	49.06	18.96	26.65	21.07	3.44	4.
6	18.96	3.44	59.85	26.65	21.07	30.38	4.89	5.

Fig. 2. Source database.

By this procedure, source databases for all 8 stations were created.

4 GUHA Analysis of Air Pollution Data

GUHA analysis of above specified source databases was performed utilizing GUHA+– software. Employing this software respective data matrices were determined by particular variables categorization. All variables were considered as ordinal type. Variables NO2, NO, O3 were each categorized into three categories represented by intervals. Limit points of intervals were set individually for each source database, however, each three intervals divided range of respective variable into three categories which could be characterized as *low*, *middle* and *high*. Variables WV, H and T2M were categorized into two categories represented by intervals interpretable as *low* and *high*. The same kind of categorization was used for variables NO2-x, NO-x, O3-x, WV-x, H-x and T2M-x where $x = \{1, 2, 3\}$ i.e., for variables representing values of past days.

Regarding task definition we have aimed on specification of ozone concentration values, that is, set Cat^S forming succedent's allowed literals was given by categories of O3 variable. Set Cat^A then was given by all categories of all other (available) variables NO2-x, NO-x, O3-x, WV-x, H-x, T2M-x, $x = \{1,2,3\}$. The length of antecedent was set to $n_A^{max} = 2$ and length of succedent to $n_S^{max} = 1$.

GUHA analysis of above specified data matrices was performed from FISHER quantifiers point of view. The value of *base* parameter was set to be equal to 10% of number of objects of source database (after rounding to nearest integer). Value of α was set to standard value $\alpha = 0.05$.

4.1 Results

In the following table there are summarized numbers of valid hypotheses found for each station.

Libus	Prostejov	Sokolov	Liberec
30	34	108	65
Karvina	Ces.Bud.	Teplice	Serlich
59	23	83	113

Table III.

Found valid hypotheses were analyzed by interpretation software which is a part of GUHA +- package. This software enables sort and filter found hypotheses. On base of this processing we have concluded these relations between ozone concentration and past values of other variables.

1) Trivial results: These are hypotheses

O3-x:high $\approx$ O3:high
O3-x:low $\approx$ O3:low

i.e., if in some of past three days value of ozone concentration was high/low then it increases probability of high/low value of ozone concentration in the following day(s).

2) Interaction with NO and NO_2:

NO-x:low $\approx$ O3:high
NO-x:high $\approx$ O3:low
NO2-x:low $\approx$ O3:high
NO2-x:high $\approx$ O3:low

These hypotheses are surprising at first glance. They tell us that future O_3 concentration is associated with past low NO concentration and vice versa. Similarly, but in less strength, for NO_2. This observation has probably background in chemical reactions of O_3 with NO (very fast) and NO_2 of the forms

O_3+NO $\longrightarrow$ O_2+NO_2, O_3+2NO_2 $\longrightarrow$ O_2+N_2O_5

3) Relation with WV:

WV-x:low $\approx$ O3:high
WV-x:high $\approx$ O3:low

These hypotheses reflect dispersion of ozone on base of wind velocity.

4) Relation with H:

H-x:low $\approx$ O3:high
H-x:high $\approx$ O3:low

The reason of this relation between humidity and ozone concentration is probably given by influence of humidity on progress of some chemistry reaction.

5) Relation with T2M:

T2M-x:high $\approx$ O3:high
T2M-x:low $\approx$ O3:low

These hypotheses point out on positive correlation between ozone concentration and increasing temperature.

After listing found and selected hypotheses add that other hypotheses found (Table III.) were given by hypotheses with antecedents of length $n = 2$. Antecedents of theses hypotheses were given by combination of single antecedents of above listed hypotheses with the same succedents therefore these hypotheses did not bring any new information and they were not referred. In the end note that presented hypotheses were revealed for all source databases (with respect to available variables), therefore it seems that they have global validity across the area of the Czech Republic.

5 Conclusion

In this paper, GUHA analysis of air pollution data collected across the area of the Czech Republic was referred. There were observed several relations among ozone concentration and other variables. These relations can be employed in design of an ozone concentration predictor. To complete the research these hypotheses should be tested by standard statistical methods, too.

References

[1] http://www.uea.ac.uk/env/appetise/

[2] Hájek P., Havel, Chytil M. (1966) The GUHA method of automatic hypotheses determination. Computing **1**, 293–308.

[3] Hájek P., Havránek T. (1977) The GUHA method - its aims and techniques, Int. J. Man-Machine Studies **10**, 3-22.

[4] Höppner F. et al. (1999) Fuzzy cluster analysis. John Wiley & Sons

Application of the Surrogate Test to Detect Dynamic Non-Linearity in Ground-Level Ozone Time-Series from Berlin

Uwe Schlink, Peggy Haase [*]

[*]Department of Human Exposure Research and Epidemiology at UFZ-Centre for Environmental Research Leipzig-Halle, Permoserstrasse 15, 04318 Leipzig, Germany

Abstract

Recent applications of non-parametric methods to forecast ground level ozone concentrations are based on dynamic non-linearity of the data series. We explain the surrogate method to test this assumption, illustrate the method with non-linear data generated by the Lorenz system, and discuss our test results for Berlin ozone time-series. We find that the null-hypothesis of linearity is clearly rejected for 12- and 24-step-ahead predictions of hourly ozone concentrations.

1 Introduction

During last decades non-linear time-series analysis (NTSA) became a valuable tool to describe irregular phenomena which are not of stochastic nature, but may be explained by deterministic non-linearity (or chaos). Non-linear phenomena occur in many scientific fields and in particular in atmospheric science.

A very early example is the system of coupled non-linear differential equations that were developed by Lorenz [5] to describe convective motion of the Rayleigh-Bénard type. Here the variables represent the velocity of the fluid, the temperature difference between ascending and descending fluid, and the deviation of the temperature profile from linearity.

In a number of papers the assumption of non-linear dynamics was used to apply forecasting procedures that are based on NTSA [4,6,1,3]. Unfortunately, preceding tests for non-linearity were not applied in these papers.

The aim of the present paper is to explain a particular test for non-linearity, the surrogate test, and to apply this test to ozone data measured in the suburbs of Berlin. For comparison we also describe the results of the test for data that were generated by the Lorenz equations.

2 Phase Space Embedding

One of the crucial points in NTSA is the embedding theorem. Assume the existence of coupled and non-linear equations of motion for a system described by variables, like x, y, and z. The solution is a trajectory in a hyperspace (the so-called attractor) in the state space that is spanned by x, y, and z. Assume further, a time-series was observed for only one component, say $(x_t)_{t=1}^{N}$. We then may compose m-dimensional vectors $\underline{x}_t = (x_{t-(m-1)\tau}, \ldots, x_{t-\tau}, x_t)$, t=(m-1)τ+1,...,N of delayed x_t values with suitably chosen delay time τ. The vectors $\underline{x}_t$ represent an embedding of the time-series in the m-dimensional phase space and describe an attractor as well.

Takens [10] was able to show that for sufficiently large m the attractor embedded in phase space is topologically equivalent to the system's attractor in state space. That means, delay embedding is capable of reconstructing the dynamics of the whole non-linear system from only *one* observed component.

The embedded attractor may be used to make predictions. For that purpose the measured time-series is divided into two parts. A training set is used to reconstruct the attractor and a test set is used to evaluate the quality of forecasts. Successively the points of the test set are taken as query points for which predictions are to be determined. This is done by help of such points on the attractor that are next to the query point. This way a Δt-step-ahead forecast is

$$\hat{x}_{i+\Delta t} = \frac{1}{\left|U_\varepsilon(\underline{x}_i)\right|} \sum_{\underline{x}_j \in U_\varepsilon(\underline{x}_i)} x_{j+\Delta t} \qquad (1)$$

where $U_\varepsilon(\underline{x}_i)$ denotes the neighbourhood of $\underline{x}_i$ with points that have a distance to $\underline{x}_i$ less than ε and $\left|U_\varepsilon(\underline{x}_i)\right|$ is the number of nearest neighbours within this neighbourhood.

This non-parametric prediction method is able to handle non-linear dynamics if ε is selected suitable. A very large ε smoothes out the structure of the attractor and non-linearity is not maintained in the forecasting process. On the other hand, a very small ε causes very noisy forecasts since the density of points in phase space is limited and the existence of an attractor point in an extreme small neighbourhood of the query point may occur just by coincidence. In case we do not find any neighbours within the radius ε we might just increase ε. In the literature there are a lot of similar algorithms (e.g. [2]).

The accuracy of non-parametric forecasts may be measured by the correlation coefficient between predicted and observed values. The maximum in the plot of prediction accuracy vs. embedding dimension m gives the optimal embedding dimension.

The plot of prediction accuracy vs. prediction time Δt enables to distinguish between stochastic noise and deterministic chaos. In the latter case the accuracy is large for short Δt and decreases for larger Δt while prediction accuracy is generally poor for time-series of stochastic noise [11].

3 Surrogate Test

The main idea of the method is to compare the quality of non-parametric predictions of the original time-series with the prediction accuracy obtained for artificially generated time-series. Such so-called surrogate time-series have the same linear properties like the original one, i.e., they have the same autocorrelation function and, consequently, the same spectrum.

We so may generate surrogates by identifying an AR model for the original data and then utilising this model for simulations. The other way is to conserve the spectrum and to destroy non-linear relationships in the original time-series. We used the latter method, applied discrete Fourier transformation into the complex domain, randomised phase angles, and transformed the data back to time domain.

The null hypothesis for the test is that the data results from a Gaussian linear stochastic process. If we allow for a 5% chance ($\alpha = 0.05$) that the null hypothesis may be rejected although it is in fact true, then the test is said to be valid at a 95% significance level. That means, more than one wrong result out of 20 is usually not considered acceptable.

Suppose the data follows the null hypothesis. When there are $1/\alpha - 1$ surrogates and one data set, we have a total of $1/\alpha$ sets following the null hypothesis. Consequently, each of the sets with probability α by chance yields the smallest value of prediction accuracy. Assume that we want to test whether the measured prediction accuracy is larger than the expected value for data with respect to the null hypothesis. A natural strategy will then be to reject the null hypothesis whenever the data gives a prediction accuracy which is larger than all the values obtained for the surrogates.

4 Results and Discussion

4.1 Simulated Lorenz data

We chose the Lorenz system because it was originally developed in an atmospheric context and may therefore be relevant for comparison with our ozone data. In particular it shows (if properly transformed) properties that are very similar to the ozone data, like quasi-periodicity and strong autocorrelation.

The length of the generated time series is 1000. The algorithm used is based on an Adams Pece scheme with local extrapolation [12]. The generated data were squared to make them comparable to ozone time-series (cf. Fig. 1).

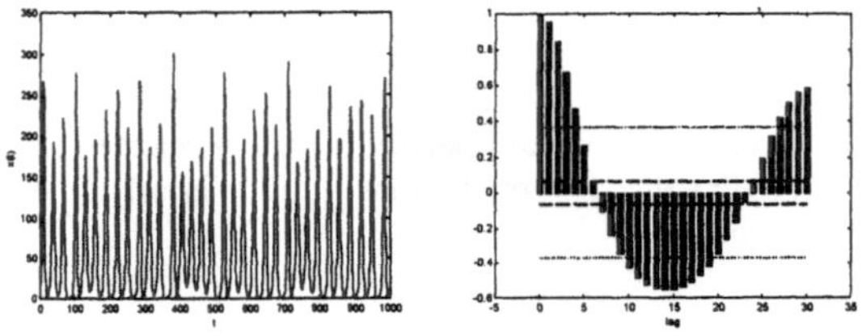

Fig. 1. Time-series of squared Lorenz data of length 1000 (integration step size 0.025) and ACF with confidence limits (dashed line) and limits ± 1/e (dotted line)

There is a period of about 30. For τ=5 the value of the autocorrelation function first decays to 1/e. We chose a test set of 500 values.

The result of the surrogate test is shown in Fig. 2. The test clearly identifies non-linear structure in the Lorenz time series since prediction accuracy of original data is larger than prediction accuracy of all surrogates for all embedding dimensions m and prediction times Δt.

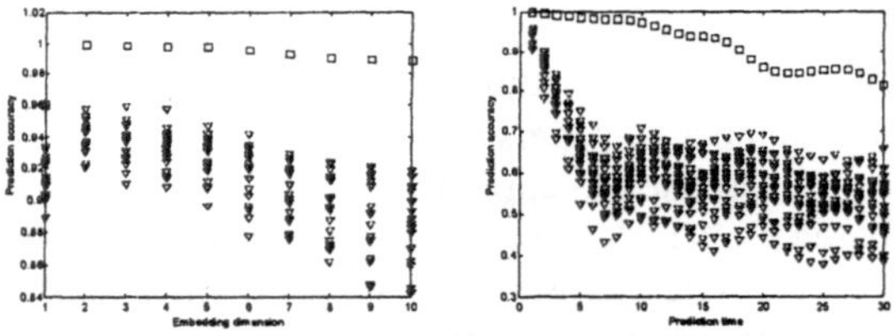

Fig. 2. Surrogate test for the Lorenz time series. Prediction accuracy is given vs. embedding dimension (Δt =1) (left) and vs. prediction time (m=3) (right hand side). The remaining parameters were ε=0.00005, τ=5, L=500. Surrogate data are represented by triangles, original data by squares.

4.2 Berlin ozone data

Ozone concentrations were observed by a local monitoring station in Berlin-Marienfelde from January 1997 to December 1999. The station is situated in the southern suburbs of Berlin and is characterised by flat terrain and low buildings. About 3% of data were missing. We replaced missing values by an interpolation method, that covers trends, day oscillations and 12 hour rhythms (cf. [8]). The highest ozone concentration 203 μg/m³ occurred on 22nd August 1997. The threshold of 180 μg/m³ was broken just in 17 cases during August 1997 and July/September 1999, in the year 1998 not at all. We confined ourselves to the period April to September 1999. The surrogate test result is given in Figs. 3-5.

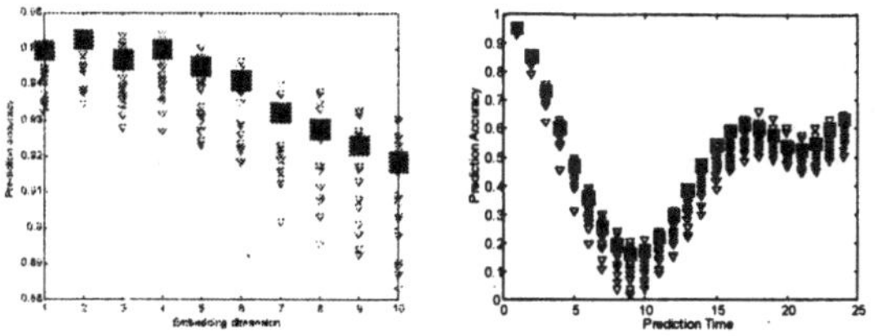

Fig. 3. Surrogate test for Berlin ozone data. Prediction accuracy is given vs. embedding dimension (left, for $\Delta t = 1$) and vs. prediction time (right hand side, for m=2). The remaining parameters were L=3000, τ=6, ε=0.5. Surrogate data are represented by triangles, original data by squares.

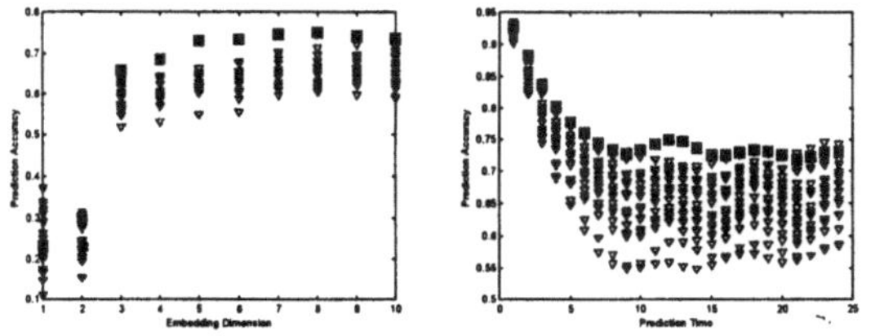

Fig. 4. Same as Fig. 3 with $\Delta t = 12$ (left) and m=8 (right hand side).

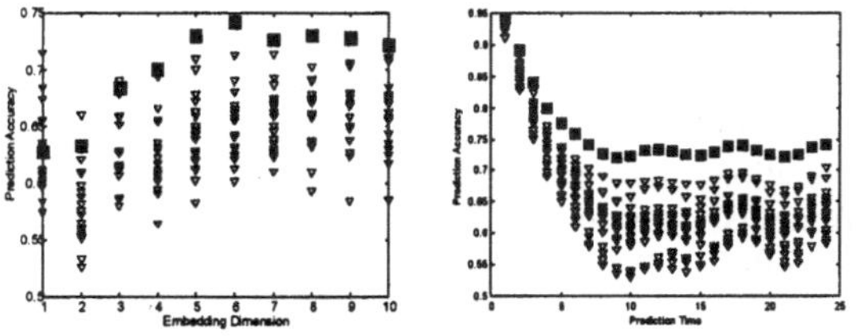

Fig. 5. Same as Fig. 3 with $\Delta t = 24$ (left) and m=6 (right hand side).

For short prediction time (e.g. $\Delta t = 1$) the test seems to reject linearity for m=1 and m=2 (Fig. 3, left). However, the decay of prediction accuracy with prediction time (Fig. 3, right) clearly demonstrates that this is only true for $\Delta t = 1$. Based on this figure we also may suppose that linearity is rejected for predictions across 12 and 24 hours. These prediction times are studied in Figs. 4 and 5, and we find optimal embedding dimensions of 8 and 6, respectively.

4.3 Trend of Berlin ozone data

The dynamic of ozone data is controlled by a number of factors. For example, the diurnal cycle is fixed astronomically and, therefore, may not be expected to behave non-linearly. Another factor is the passing of weather fronts that may produce ozone changes of lower frequency. Such low-frequency components, the so-called trend, may be calculated by low-pass filtering of ozone time-series [7]. The irregular movement of cyclones and anticyclones may show non-linearity and so may the low-frequency component of ozone. Since the Kalman-Filter gives the smoothed trend as well as its derivative we can again use a phase space. Instead of time delayed coordinates we now use the derivative. Figure 6 shows such phase space portrait for the trend function of Berlin ozone data of the year 1999. The cut off frequency of the low-pass is chosen that way, that the trend component does not contain a daily period any more. For comparison we also give a delay embedding of the original data (with τ=1) and the trend component (τ=8). Obviously the trend and its derivative are not independent stochastic processes. Such behaviour of trajectories is only possible if they are functional dependent.

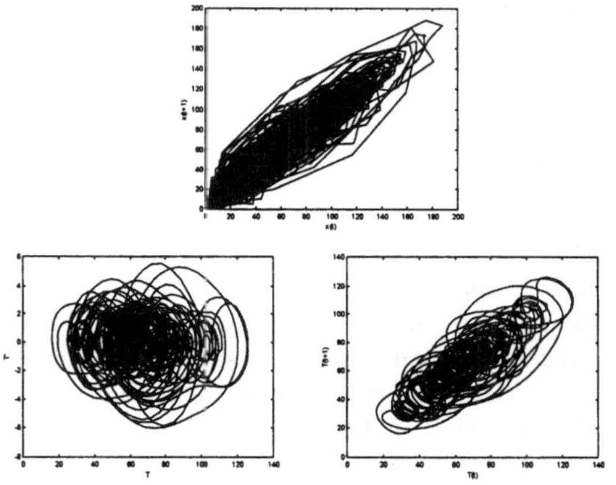

Fig. 6. Phase space reconstruction of the original time series by delay time embedding (τ=1, top), phase space of the trend and its derivative (bottom left) and delay embedding of the trend (τ=8) (bottom right hand side).

Visually Fig. 6 may suggest the existence of an attractor. The surrogate test, however, does not reject the null hypothesis of linearity (Fig. 7).

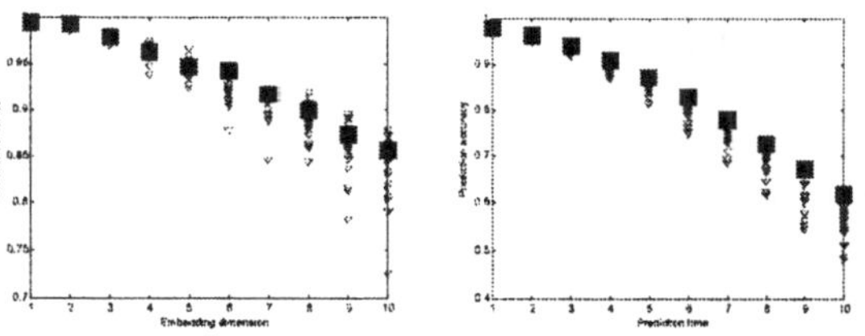

Fig. 7. Surrogate test for ozone trend-component. Prediction accuracy is given vs. embedding dimension ($\Delta t = 1$) (left) and vs. prediction time (m=3). Remaining parameters were L=3000, ε=0.00005, τ=34. Results for surrogate data are represented by triangles, for original data by squares.

4.4 Daily maximums of ozone concentration

Air quality criteria often work with threshold values that may not be broken during the day. Therefore we finally consider daily maximums of the Berlin ozone time-series. For the surrogate test we used 183 daily maximums observed in 1999. Similar to our findings for the trend component, for daily maximums the test does not reject the null hypothesis of linearity (see Fig. 8).

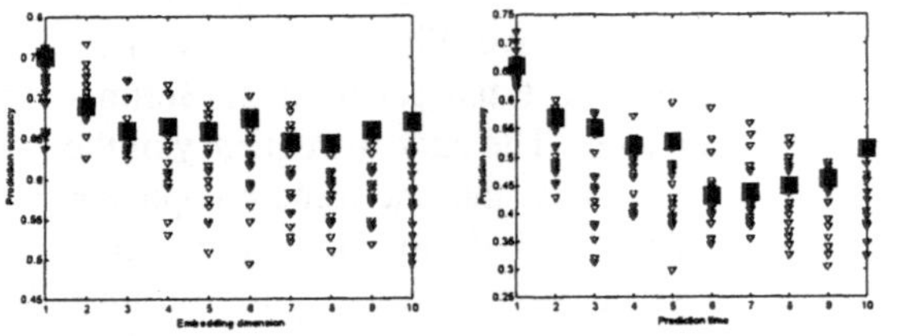

Fig. 8. Surrogate test for daily maximums of Berlin ozone data. Prediction accuracy is given vs. embedding dimension (Δt =1) (left) and vs. prediction time (m=3) (right hand side). The remaining parameters were L=3000, ε=0.00005, τ=34. Surrogate data are represented by triangles, original data by squares.

5 Conclusions

The methods of phase space embedding and surrogate testing described above are able to identify dynamic non-linearity. This was demonstrated for data series that were generated by the Lorenz equations which is known to have non-linear solutions. The methods were also applied to three different forms of Berlin ozone data.

Hourly mean values of ozone concentration seem to behave non-linearly exceptionally for *very* short prediction times. Surprisingly the surrogate test rejected the null hypothesis of linearity for prediction times of 12 and 24 hours ahead and identified embedding dimensions of 8 and 6. These embedding dimensions are in accordance to the values given in the literature [1].

The finding of a possible existence of non-linearity across 12 and 24 hours supports our discussion about meteorological factors that might cause non-linear dynamics. However, there is no rejection of the null-hypothesis for ozone trend and the daily ozone maximums. Considering this and the result for hourly average values one might suspect that the detection of non-linearity might be an influence of the diurnal cycle, which can cause spurious non-linearity due to flaws in surrogates. There are indeed slight differences between the ACF of the original data and a set of surrogates. They amount to about zero to nine percent. However, our initial trials based on an improved methodology of the surrogate test [9] led to almost the same result. Identifying the reasons for this behaviour will be a subject of future work.

Acknowledgements

The results have been achieved within the APPETISE project that is funded by the European Commission (IST-1999-11764). Data were supplied by "Senatsverwaltung für Stadtentwicklung, Umweltschutz und Technologie", Berlin who participates in the project HEAVEN. We are grateful to Christian Merkwirth, University of Goettingen for providing the C routines [12].

References

[1] Chen, J.L., Islam, S., Biswas, P.: Nonlinear Dynamics of Hourly Ozone Concentrations: Nonparametric Short Term Prediction. Atmospheric Environment 32(11), 1839-1848 (1998).

[2] Farmer, J.D., Sidorowich, J.J.: Predicting Chaotic Time Series. Physical Review Letters 59(8), 845-848 (1987).

[3] Kocak, K., Saylan, L., Sen, O.: Nonlinear Time Series Prediction of O_3 Concentration in Istanbul. Atmospheric Environment 34, 1267-1271 (2000).

[4] Li, I.F., Biswas, P., Islam, S.: Estimation of the Dominant Degrees of Freedom for Air Pollutant Concentration Data: Applications to Ozone Measurements. Atmospheric Environment 28(9), 1707-1714 (1994).

[5] Lorenz, E.N.: Deterministic Nonperiodic Flow. Journal of Atmospheric Sciences 20, 130-141 (1963).

[6] Raga, G.B., Moyne, L. Le: On the Nature of Air Pollution Dynamics in Mexico City – I. Nonlinear Analysis. Atmospheric Environment 30(23), 3987-3993 (1996).

[7] Schlink, U., Herbarth, O., Tetzlaff, G.: A component time-series model for SO2 data: forecasting, interpretation and modification. Atmospheric Environment 31(9), 1285-1295 (1997).

[8] Schlink, U., Herbarth, O., Richter, M.: Ozone data analysis to develop an immediate warning system. Proceedings of the Air Pollution '96 conference 1996.

[9] Schreiber, T., Schmitz, A.: Surrogate time series. Physica D 142, 346 (2000).

[10] Takens, F.: Detecting Strange Attractors in Turbulence. In Rand, D.A., Young, L.S. (Eds.): Dynamical Systems and Turbulence. Lecture Notes in Mathematics 898, 366-381, Berlin: Springer 1981.

[11] Kantz, H., Schreiber, T.: Nonlinear Time Series Analysis. Cambridge: Cambridge University Press 2000.

[12] Merkwirth, C.: TSTOOL Documentation, http://www.dpi.physik.uni-goettingen.de/tstool, (2000).

Nonlinearity and Prediction of Air Pollution

Milan Paluš, Emil Pelikán, Kryštof Eben, Pavel Krejčíř and Pavel Juruš * [1]

*Institute of Computer Science, Academy of Sciences of the Czech Republic, Pod vodárenskou věží 2, 182 07 Prague 8, Czech Republic

Abstract

A presence of nonlinearity in time series of concentrations of air pollutants and in their relations to time series of meteorological variables is tested using information-theoretic functionals and the surrogate data approach. The results are discussed in relation to predictability of the pollutant concentrations aimed to alert smog episodes.

1 Introduction

The most used types of air quality models are either deterministic models or models given by simple regression-based statistics. Their success, however, is limited either by their failure to capture the nonlinear behaviour of air pollutants, or our incomplete understanding of the physical and chemical processes involved. The APPETISE project (Air Pollution ePisodes: modElling Tools for Improved Smog managEment, see `http://www.uea.ac.uk/env/appetise/`) aims to develop and test the suitability of novel nonlinear statistical methods to improve our ability to accurately forecast variations in air quality. The work is being carried out over a period of 2 years by a consortium from 9 institutions from 5 European countries and is founded under the European Union Fifth Framework Programme. The project will work towards the construction of a prototype air quality prediction and warning system and is concentrated on 4 key pollutants: nitrogen oxides, particulates, sulphur dioxide and ground level ozone. The latter is the main research topic for the group of investigators represented by the authors of this paper.

Before trying to enhance the existing linear models by nonlinear ones it is suitable to test a presence of nonlinearity in the dynamics of time series of the ground level ozone (GLO) concentration as well as in relations of these data to time series of the most influential meteorological variables and to concentrations of other pollutants.

[1] This study is supported within the European Union Fifth Framework Programme project APPETISE (IST–99–11764).

2 Testing Nonlinearity

In this section we briefly review a method for detection and characterization of nonlinear relations in multivariate as well as in univariate time series. The method employs the technique of uni- and multivariate surrogate data and information-theoretic functionals called redundancies. The test for nonlinearity based on the redundancy – linear redundancy approach, combined with the surrogate data is described in detail in Ref. [3], its multivariate version in Ref. [4]. The surrogate data have been introduced in Ref. [6], and their multivariate version in Ref. [5]. More details about the information-theoretic functionals can be found in Ref. [1].

Consider n discrete random variables $X_1, \ldots, X_n$ with sets of values $\Xi_1, \ldots, \Xi_n$, and probability distribution functions (PDF) $p(x_1), \ldots, p(x_n)$, respectively, and the joint PDF $p(x_1, \ldots, x_n)$. The redundancy $R(X_1; \ldots; X_n)$, in the case of two variables also known as mutual information (MI) $I(X_1; X_2)$, quantifies average amount of common information, contained in the n variables $X_1, \ldots, X_n$:

$$R(X_1; \ldots; X_n) = \sum_{x_1 \in \Xi_1} \cdots \sum_{x_n \in \Xi_n} p(x_1, \ldots, x_n) \log \frac{p(x_1, \ldots, x_n)}{p(x_1) \ldots p(x_n)}. \quad (1)$$

Now, let the n variables $X_1, \ldots, X_n$ have zero means, unit variances and correlation matrix $\mathbf{C}$. Then, we define the *linear redundancy* $L(X_1; \ldots; X_n)$ of $X_1, X_2, \ldots, X_n$ as

$$L(X_1; \ldots; X_n) = -\frac{1}{2} \sum_{i=1}^{n} \log(\sigma_i), \quad (2)$$

where σ_i are the eigenvalues of the $n \times n$ correlation matrix $\mathbf{C}$.

If $X_1, \ldots, X_n$ have an n-dimensional Gaussian distribution, then $L(X_1; \ldots; X_n)$ and $R(X_1; \ldots; X_n)$ are theoretically equivalent (see [3] and references therein). The general redundancies R detect all dependences in data under study, while the linear redundancies L are sensitive only to linear structures [3].

The basic idea in the surrogate-data based nonlinearity test is to compute a *nonlinear* statistic from data under study and from an ensemble of realizations of a linear stochastic process, which mimics "linear properties" of the studied data. If the computed statistic for the original data is significantly different from the values obtained for the surrogate set, one can infer that the data were not generated by a linear process; otherwise the null hypothesis, that a linear model fully explains the data, is accepted. For the purpose of such test the surrogate data must preserve the spectrum and consequently, the autocorrelation function of the series under study [6]. In the multivariate case also cross-correlations between all pairs of variables must be preserved [5].

Like in [3] we define the test statistic as the difference between the redundancy obtained for the original data and the mean redundancy of a set of surrogates, in the number of standard deviations (SD's) of the latter. The result is considered significant if the difference is clearly larger than 2 SD. In this study only 2-variable mutual information $I(X;Y)$ was applied: the univariate version $I(X(t);X(t+\tau))$ when dynamical properties and nonlinearity of individual series (variables) were studied, and the bivariate version $I(X(t);Y(t+\tau))$ when dynamical relations between two variables were investigated. The mutual information $I(X;Y)[o]$ from the scrutinized data and the mean mutual information $I(X;Y)[s]$ from the surrogates, as well as the test statistics, defined above, were plotted as functions of lag τ. Significant differences found between $I(X;Y)[o]$ and $I(X;Y)[s]$ were used to infer nonlinearity in dynamics of a variable (in univariate case), or in a relation between two variables (in bivariate case). The values of $I(X;Y)[o]$ indicate a "coherence" or predictability of a variable, i.e., the dependence between $x(t)$ and $x(t+\tau)$ (in univariate case), or a strength of the link between two variables (in bivariate case), both as a function of the lag τ.

3 Data

Time series of GLO, NO_2 and NO_x concentrations, as well as air temperature, wind speed and relative humidity were selected from a database of 37 Czech stations and several stations from UK, Germany, Finland and Italy. The sampling time is 30 min. in the data from the Czech stations and 1 hour otherwise. The lengths of processed data segments were 2048 and 4096 samples, according to data availability. Since all the data are dominated by the diurnal cycle, in addition to raw data also the following two slow components were extracted from the data and processed: a) *trends,* (Fig. 1, top panel) obtained from the raw data by simple moving average (MA, window length equal to 49 samples), b) *instantaneous amplitude* (Fig. 1, bottom panel) of the diurnal cycle. The latter has been obtained from the MA filtered data by using the analytic signal concept of Gabor [2]. For any signal $s(t)$, the analytic signal $\psi(t)$ is a complex function of time defined as

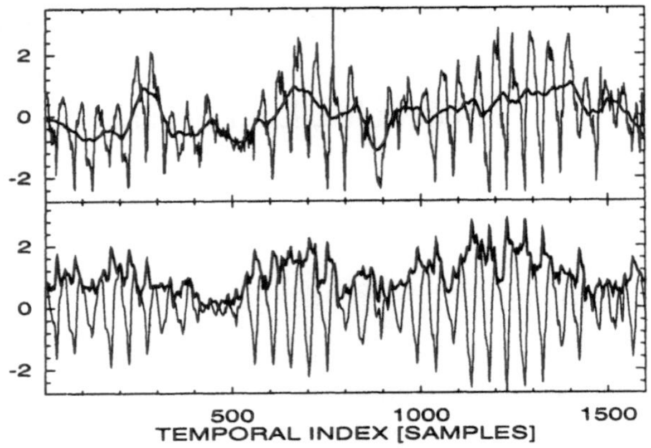

Fig. 1. Top panel: A segment of the ground level ozone concentration time series – the raw data (thin line) and the MA trend (thick line). Bottom panel: The MA filtered diurnal oscillations of the above GLO data (thin line) and their instantaneous amplitude (thick line).

$$\psi(t) = s(t) + j\hat{s}(t) = A(t)e^{j\phi(t)}, \qquad (3)$$

where the function $\hat{s}(t)$ is the Hilbert transform of $s(t)$

$$\hat{s}(t) = \frac{1}{\pi}\ \text{P.V.} \int_{-\infty}^{\infty} \frac{s(\tau)}{t-\tau} d\tau. \qquad (4)$$

(P.V. means that the integral is taken in the sense of the Cauchy principal value.) The instantaneous amplitude is then

$$A(t) = \sqrt{s(t)^2 + \hat{s}(t)^2}. \qquad (5)$$

4 Results

The results from the above described nonlinearity test obtained from the raw GLO concentration data (the Czech Station Tušimice, 4096 half-hour samples from the 1997 season) are presented in Fig. 2. Results from the linear MI (Fig. 2a,c) show no significant differences between the data and the surrogates, i.e., the surrogates correctly reflect the linear properties of the studied data and the significant

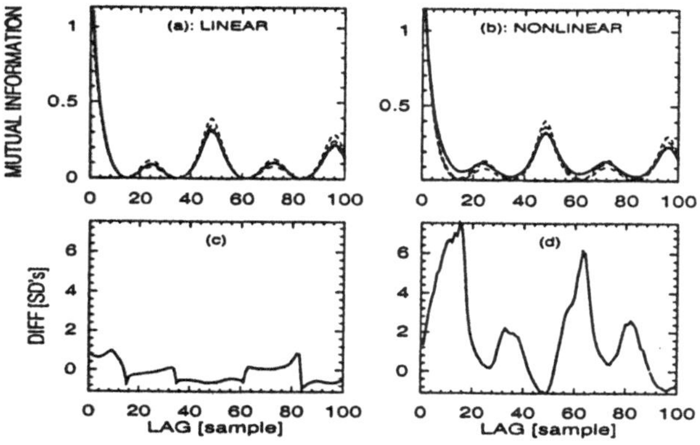

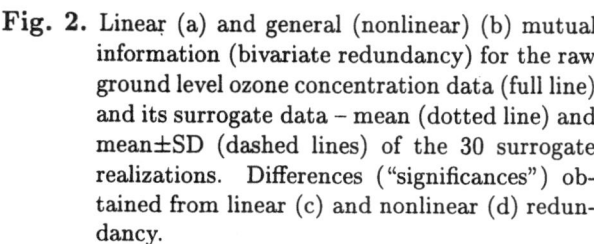

Fig. 2. Linear (a) and general (nonlinear) (b) mutual information (bivariate redundancy) for the raw ground level ozone concentration data (full line) and its surrogate data – mean (dotted line) and mean±SD (dashed lines) of the 30 surrogate realizations. Differences ("significances") obtained from linear (c) and nonlinear (d) redundancy.

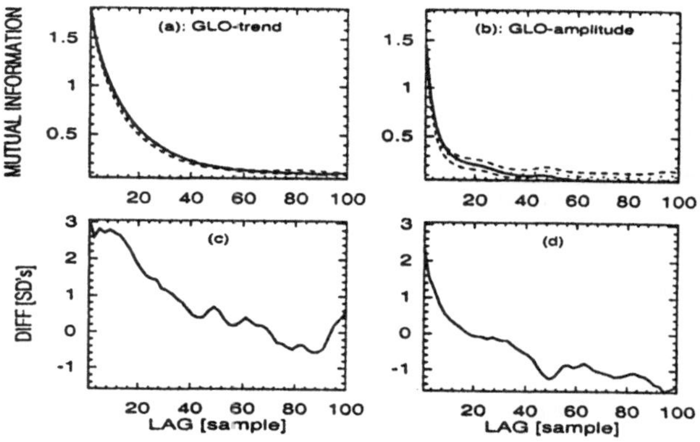

Fig. 3. The (nonlinear) mutual information (a,b) and the related difference statistics (c,d) for the trend (a,c) and the instantaneous amplitude (b,d) obtained from the ground level ozone concentration data (full line) and its surrogate data (in a,b: mean (dotted line) and mean±SD (dashed lines) of the 30 surrogate realizations).

differences, detected by the nonlinear MI (Fig. 2b,d) should not be due to flawed surrogates. On the other hand, the formal rejection of the null hypothesis of a linear stochastic process is not a conclusive evidence for nonlinear character of the process underlying the GLO concentration time series. It is clear that the diurnal cycle is externally driven and could bring a formal statistical long-term dependence on linear or nonlinear level which, however, does not represent any causal connection between $x(t)$ and $x(t+\tau)$. For the purpose of predicting the GLO concentration we should study the series of the trends and amplitudes, obtained from the raw data as described above. In the following tests the surrogate data were constructed from the raw data, and their trends and amplitudes were obtained from the surrogates in the same way as from the raw data. The results for the trend and amplitude of the above GLO concentration data (the Czech Station Tušimice, 4096 half-hour samples from the 1997 season) are presented in Fig. 3. Without the diurnal cycle the serial dependence (and predictability) of this series falls quickly, esp. in the case of amplitudes where it lasts less then 20 hours (Fig. 3b), while the serial dependence of the trend spreads to lags of approx. 30 hours (Fig. 3a). This dependence is predominantly linear, a nonlinear dependence can be detected for short lags up to 10 hours for the trend (Fig. 3c), while for the amplitude it is practically negligible (Fig. 3d).

In the following we analyse pairs of simultaneously recorded time series using the bivariate surrogate data. The relation between the air temperature and the GLO concentration (in terms of trends and amplitudes) is analyzed in Fig. 4 using 4096 half-hour sample data from the Czech station Teplice, 1997 season. There is a slowly decreasing, long term, entirely linear dependence between the temperature and GLO trends (Fig. 4a,c), while the dependence between the temperature and GLO amplitudes is decreasing quickly, however, for the short lags (up to 10 hours) it is stronger than the dependence of the trends (Fig. 4b) and is also nonlinear (Fig. 4d). The linear dependence between the trends of the relative air humidity and the GLO concentration has a similar long-term slowly decreasing character as the air temperature - GLO trends dependence, however, unlike the latter case there is a strong nonlinear relation between the humidity and GLO trends, though confined to short time lags (up to approx. 20 hours, however, with a high significance again only to 10 hours, see Fig. 5a,c). The relation between the amplitudes is again linear, long-term slowly decreasing one (Fig. 5b,d). The above results were obtained from 4096 half-hour samples, recorded at the Czech Station Tušimice, in the 1997 season. Comparable results have been obtained from other Czech stations as well as from some UK and German stations.

In further analyses, the short-lag nonlinear dependence has also been found in the relations of the trends of the wind velocity and the GLO concentration in the data from several Czech stations, however, has not been confirmed in the German and UK data. Only a weak and short linear dependence

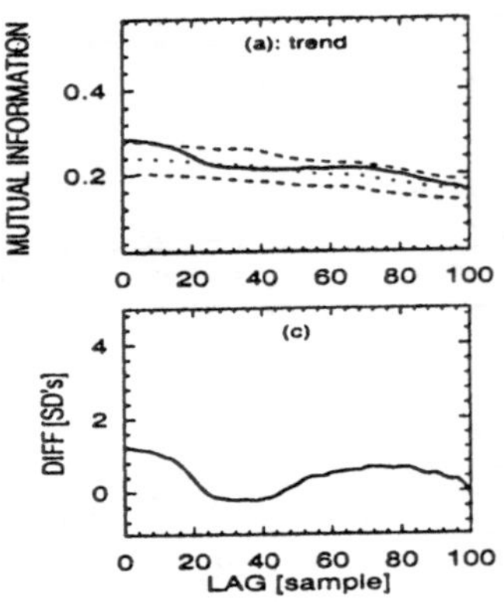

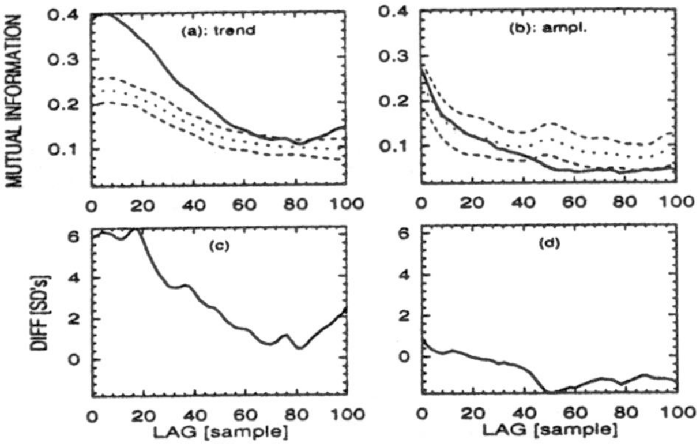

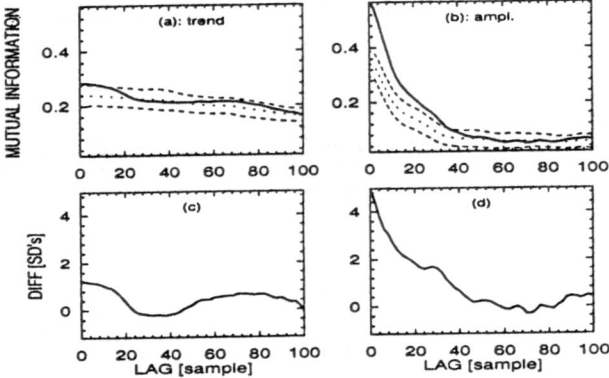

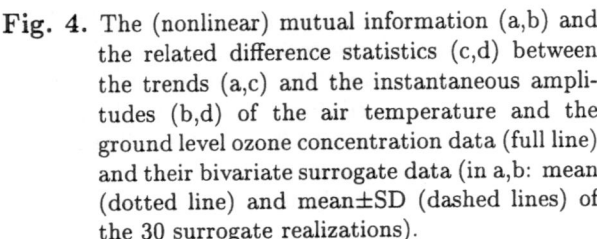

Fig. 4. The (nonlinear) mutual information (a,b) and the related difference statistics (c,d) between the trends (a,c) and the instantaneous amplitudes (b,d) of the air temperature and the ground level ozone concentration data (full line) and their bivariate surrogate data (in a,b: mean (dotted line) and mean±SD (dashed lines) of the 30 surrogate realizations).

Fig. 5. The (nonlinear) mutual information (a,b) and the related difference statistics (c,d) between the trends (a,c) and the instantaneous amplitudes (b,d) of the air relative humidity and the ground level ozone concentration data (full line) and their bivariate surrogate data.

has been found in the relations of the amplitudes of wind velocity and GLO data.

The analyses of GLO relations to other pollutants are in their introductory state, so only as a preliminary result we can state the short-lag nonlinearity in the GLO – NO_2 amplitudes relation, while the GLO – NO_2 trends and both the trends and amplitudes of the GLO – NO relations appear to be limited to a weak linear dependence.

5 Discussion

The presence of nonlinearity in the dynamics of time series of the ground level ozone (GLO) concentration as well as in relation of this data to time series of the most influential meteorological variables and to concentrations of other pollutants has been investigated by the test for nonlinearity employing the mutual information and the surrogate data method. The analysis of the raw data indicated long-term dependence and some formally proven nonlinearity caused by the diurnal cycle which dominates all the studied data sets. Since the externally driven diurnal cycle has practically no implications for predictability of the scrutinized time series, for further analyses the data had been preprocessed in order to obtain the long term trends and the instantaneous amplitude of the diurnal cycle. The GLO concentration has been found related to the influential meteorological variables (air temperature, relative humidity and wind velocity) by the slowly decreasing long-term linear dependence in some cases enhanced by a short-lag (up to 10 hours) nonlinearity. Thus nonlinear time series models, such as neural networks, can improve only the short term (several hours) GLO concentration forecasts. For predictions with day or longer horizons the statistical models should be combined with deterministic models and forecasted meteorological variables.

References

[1] T.M. Cover and J.A. Thomas, *Elements of Information Theory* (J. Wiley & Sons, New York, 1991).

[2] Gabor D., "Theory of Communication," *J. IEE London* vol. **93** pp. 429–457, 1946.

[3] M. Paluš, "Testing for nonlinearity using redundancies: Quantitative and qualitative aspects," *Physica D* vol. **80** pp. 186–205, 1995.

[4] M. Paluš, "Detecting nonlinearity in multivariate time series," *Phys. Lett. A* vol. **213** pp. 138–147, 1996.

[5] D. Prichard and J. Theiler, "Generating surrogate data for time series with several simultaneously measured variables," *Phys. Rev. Lett.* vol. **73** pp. 951–954, 1994.

[6] J. Theiler, S. Eubank, A. Longtin, B. Galdrikian and J.D. Farmer, "Testing for nonlinearity in time series: the method of surrogate data," *Physica D* vol. **58** pp. 77–94, 1992.

On Nonlinear Processing of Air Pollution Data

Rob Foxall*[1], Igor Krcmar†, Gavin Cawley*, Stephen Dorling‡, Danilo P. Mandic*

*School of Information Sytems, University of East Anglia, Norwich, UK, e-mail: {rjf, gcc, mandic}@sys.uea.ac.uk †Faculty of Electrical Engineering, University of Banjaluka, Banjaluka, e-mail: ikrcmar@etf-bl.rstel.net ‡School of Environmental Sciences, University of East Anglia, Norwich, UK, e-mail: s.dorling@uea.ac.uk

Abstract

Three methods - DVS plots, attractor reconstruction, and variance analysis of delay vectors - for detecting nonlinearities in time series are compared on an air pollution dataset. For rigour each method is also used on a surrogate dataset, based on a high-order linear fit to the original data. Finally, a comparison of a standard linear analysis to a neural network model analysis of the air pollution dataset is provided.

1 Introduction

Air pollutants such as surface Ozone (O_3), Nitrogen Oxides (NO_X), Sulphur Dioxide (SO_2) and Particulates have significant health effects associated with them at high concentrations. A rigorous analysis of pollutant data requires consideration of a number of meteorological variables (e.g. wind speed) and non-meteorological variables (e.g. traffic density). To obtain an insight into the underlying structure, however, it is worthwhile to look initially at each pollutant time series individually with the standard linear methods, and to do nonlinearity analysis only if it appears that a linear model is inadequate. A NO_2 time series of hourly measurements taken over a four-year period from the Leeds meteo station is used throughout this paper.

2 Linear Analysis of Time Series

A standard model of linear time series, the ARIMA(p, d, q) model popularised by Box and Jenkins [1], assumes that the time series $x(j)$ is generated by a succession of "random shocks" $\epsilon(j)$, drawn from a distribution with zero mean and variance σ_ϵ^2. If $x(j)$ is non-stationary, then successive differencing of $x(j)$ via the differencing operator, $\nabla x(j) = x(j) - x(j-1)$ can provide a stationary process. A stationary process $z(j) = \nabla^d x(j)$ can be modelled as an autoregressive moving average

$$z(j) = \sum_{i=1}^{p} a_i z(j-i) + \sum_{i=1}^{q} b_i \epsilon(j-i) + \epsilon(j) \quad (1)$$

Of particular interest are pure autoregressive (AR) models, which have an easily understood relationship to the nonlinearity detection technique of DVS (Deterministic Versus Stochastic) plots. The order of the AR model can be chosen by the point where the *autocorrelation function* (ACF) essentially vanishes for all subsequent lags, other methods, such as AIC or BIC, can also be used. Figure 1 shows

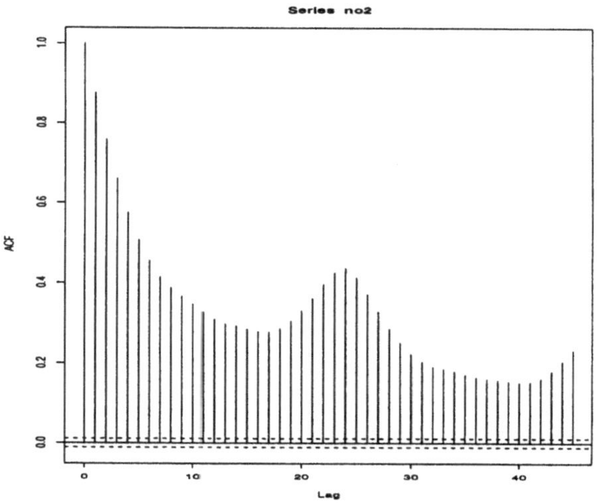

Fig. 1. ACF plot of NO_2 series.

the autocorrelation function for 40 lags for the NO_2 dataset; the ACF does not vanish and a high-order AR model is necessary.

3 The Method of Surrogate Data

Following the approach from [2], to gauge efficacy of the techniques for detecting nonlinearity, a surrogate dataset is simulated from a high order autoregressive model fit to the original series. The coefficients a_i from an $AR(45)$ model were used to generate the surrogate series, with surrogate residuals

[1]This work was supported by the European Commission, grant number IST-99-11764, as part of its Framework V IST programme.

$\epsilon(j)$ taken as a random permutation of the residuals from the original series. Evidence of nonlinearity from any method of detection is negated if the method gives a similar result when applied to the surrogate series, which is known to be linear [2].

4 Attractor Reconstruction

Existence and/or discovery of an attractor in phase space demonstrates whether the system is deterministic, purely stochastic, or somewhere in between. To reconstruct the attractor examine plots in m-dimensional space of $[x(j), x(j-\tau), \ldots, x(j-(m-1)\tau)]^T$. It is critically important for the dimension of the space, m, in which the attractor is to viewed, to be large enough to "untangle" the attractor. This is known as the *embedding dimension.* The value of τ, the *lag time* or *lag spacing*, is also important, particularly with noise present. The first inflection point on the autocorrelation function is a possible starting value for τ [3]. Alternatively, if the series is known to be sampled coarsely, the value of τ can be taken as unity [4].

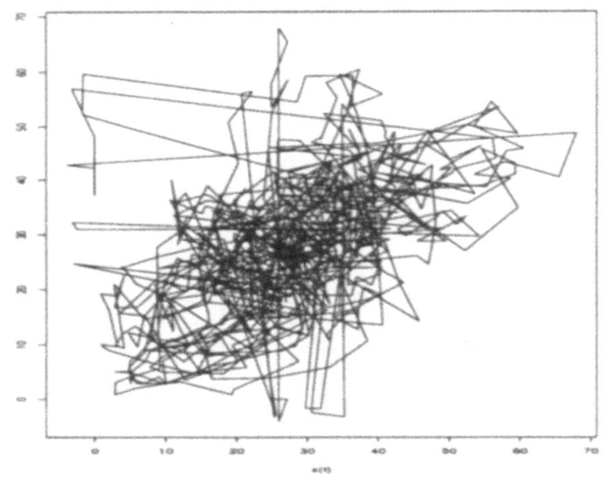

Fig. 2. NO_2 time series embedded into 2-dimensional phase-space with delay time $\tau = 22$.

Figure 2 shows the two-dimensional attractor reconstruction for the NO_2 time series after it has been passed through a linear filter to remove some of the noise present. Although the graph shows some regularity, if an attractor exists it is in a higher dimensional space.

5 DVS Plots

DVS plots [4] display the (robust) prediction error $E(k)$ for local linear models against the number of nearest neighbours, k, used to fit the model, for a range of embedding dimensions m. The last 500 values of the series are set aside for prediction purposes, these values are known as the ***test*** set. For each element in the test set $x(j)$, construct the delay vector $\mathbf{x}(j) = [x(j), x(j-\tau), \ldots, x(j-(m-1)\tau)]^T$. The k nearest neighbours are defined to be the k vectors $\mathbf{x}(j')$ from the series which have the shortest Euclidean distance to $\mathbf{x}(j)$, these k nearest neighbours are used to fit the local linear model.

If the optimal k, taken to be the value of k giving the lowest prediction error $E(k)$, is at or close to the maximum possible k, then globally linear models perform best and there is no indication of nonlinearity. In this case the model is equivalent to an AR model of order m when $\tau = 1$. Small optimal k suggests local linear models perform best, indicating nonlinearity and/or chaotic behaviour.

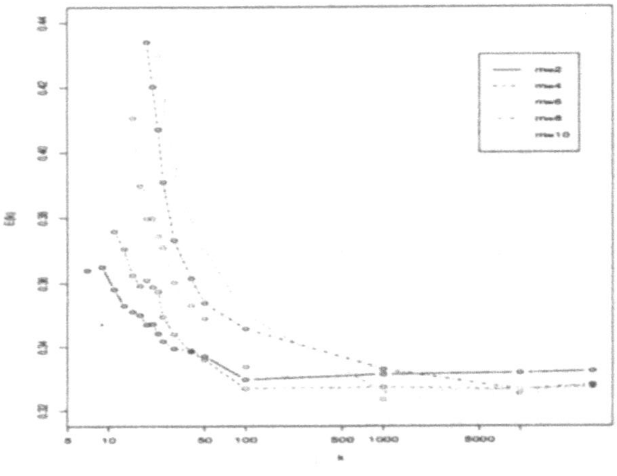

Fig. 3. DVS plot for the NO_2 time series with lead time $T = 1$ and lag delay $\tau = 22$.

In Figure 3 the DVS plot for $m = \{2, 4, 6, 8, 10\}$ is shown. For each value of m, the optimal k is less than the maximum, but the difference in the prediction error is minimal.

Figure 4 displays the equivalent DVS plot for a surrogate dataset simulated from the $AR(45)$ model fit to the series. The behaviour for the surrogate data is similar to the original data, suggesting that the underlying structure of the series has only a small nonlinear component.

6 Variance Analysis of Delay Vectors

Closely related to DVS plots is the nonlinearity technique introduced in [5]. For each observation $x(i), i \geq m+1$ construct the group, Ω_i, of nearest neighbours by

$$\Omega_i = \{\mathbf{x}(j) : j \neq i \quad \& \quad d_{ij} \leq \alpha A_x\}$$

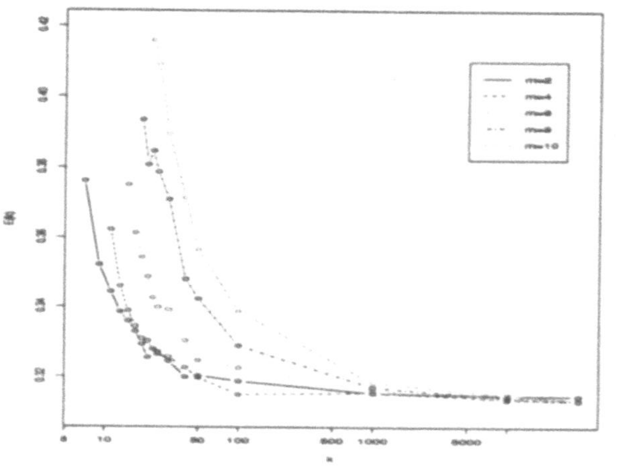

Fig. 4. DVS plot for the simulated NO_2 time series with lead time $T = 1$ and lag delay $\tau = 22$.

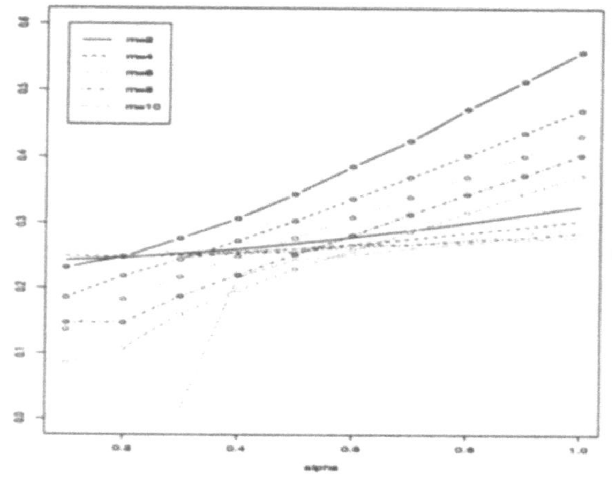

Fig. 5. Normalised variance plot. The lines with points correspond to the actual NO_2 series; the lines without points to the simulated linear series.

where $\mathbf{x}(j) = \{x(j-1), x(j-2), \dots, x(j-(m-1))\}$, $d_{ij} = \|\mathbf{x}(j) - \mathbf{x}(i)\|$ is the Euclidean norm, $0 < \alpha \leq 1$, $A_x = \frac{1}{N} \sum_{i=m+1}^{N} |x(i)|$ and N is the length of the time series. If the series is linear, then the similar patterns $\mathbf{x}(j)$ belonging to a group Ω_i will map onto similar $x(j)$s. For nonlinear series, the patterns $\mathbf{x}(j)$ will not map onto similar $x(j)$s. This is measured by the variance σ^2 of each group Ω_i

$$\sigma_i^2 = \frac{1}{|\Omega_i|} \sum_i (x(j) - \mu_i)^2, \qquad \mathbf{x}(j) \in \Omega_i.$$

The measure of nonlinearity is taken to be the mean of σ_i^2 over all the Ω_i, denoted $\bar{\sigma}_N^2$, normalised by dividing through by σ_x^2, the variance of the entire time series $\bar{\sigma}^2 = \frac{\bar{\sigma}_N^2}{\sigma_x^2}$. The larger the value of $\bar{\sigma}^2$ the greater the suggestion of nonlinearity [5].

The results are shown in Figure 5. Apart from a few exceptions for $\alpha < 0.5$, the normalised variance of similar delay vectors for the simulated series is much lower than for the real series, an indication that the series is nonlinear.

7 Neural Adaptive Filters in the Air Pollution Time Series Prediction

As stated above prediction of air pollution time series is a difficult task due to the complex and cyclic nature of the underlying process that generates atmospheric pollutants. In addition, some results, i.e. DVS plots and attractor reconstruction, indicate inherent nonlinearity of the air pollution time series. Thus, in order to obtain good prediction of the future value of the time series, based on the past measurements, an efficient algorithm should be employed, an algorithm that is inherently nonlinear and/or adaptive. Gradient-descent (GD) based neural adaptive filters, due to inherent simplicity and nonlinearity, are adequate choices for the prediction of time series that represents atmospheric pollution data. Furthermore, the structure of neural adaptive filters could be chosen to reflect the nature of the underlying process, i.e. it could be feedforward or recurrent.

The adaptation of a GD based neural adaptive filter can be described by the following set of equations

$$\begin{aligned} v(k) &= \mathbf{w}^T(k)\mathbf{u}(k) && (2)\\ e(k) &= d(k) - \Phi(v(k)) && (3)\\ \mathbf{w}(k+1) &= \mathbf{w}(k) - \eta \nabla_{\mathbf{w}} E(e(k)) && (4) \end{aligned}$$

where $d(k)$ is some training (desired) signal, $e(k)$ is the instantaneous error at the output neuron, $E(\cdot)$ is the filter cost function, η denotes the learning rate parameter, $\mathbf{w}(k) = [w_1(k), ..., w_N(k)]^T$ is the weight vector, and $\Phi(\cdot)$ represents a nonlinear activation function of a neuron. Definition of the vector $\mathbf{u}(k)$ depends on the structure of a neural adaptive filter. In the case of the feedforward filter $\mathbf{u}(k)$ contains only samples of the input signal $x(k)$, and it is defined as $\mathbf{u}(k) = [x(k-1), ..., x(k-N)]^T$. The most common choice for the cost function $E(\cdot)$ is

$$E(e(k)) = \frac{1}{2} e^2(k). \qquad (5)$$

Computation of the gradient of the cost function, denoted by $\nabla_{\mathbf{w}} E(e(k))$, depends on the structure of a neural adaptive filter. For the feedforward type of a filter, this gradient is given by

$$\nabla_{\mathbf{w}} E(e(k)) = e(k)[x(k-1), ..., x(k-N)]^T. \quad (6)$$

The algorithm described by equations (3) – (6) is usually referred to as the nonlinear gradient-descent (NGD) algorithm. The gradient of the cost function for a nonlinear ARMA(p, q) recurrent perceptron is defined as

$$\nabla_{\mathbf{w}} E(e(k)) = e(k)\mathbf{\Pi}(k) \quad (7)$$

where $\mathbf{\Pi}(k) = \left[\frac{\partial y(k)}{\partial w_1(k)}, \ldots, \frac{\partial y(k)}{\partial w_N(k)}\right]$ represents the gradient at the output of the neuron. The normalized nonlinear gradient-descent (NNGD) algorithm exhibits optimal behaviour in the sense that it minimizes instantaneous prediction error, thus providing an adaptive learning rate η [6]. In the case of the linear activation function of an output neuron, the NNGD algorithm reduces to the normalized least mean squares (NLMS) algorithm.

8 Experimental Results

Air pollution data represent hourly measurements of the concentration of nitrogen dioxide (NO_2), in the period 1994 – 1997, provided by the Leeds meteo station. In the performed experiments the logistic function was chosen as the nonlinear activation function of an output neuron. The logistic function performs contraction mapping for the slope β set to be $0 \leq \beta \leq 4$ [7]. The quantitative performance measure was the standard prediction gain, a logarithmic ratio between the expected signal and error variances $R_p = 10\log(\hat{\sigma}_s^2/\hat{\sigma}_e^2)$. The slope of the nonlinear activation function of the neuron β was set to be $\beta = 4$, since this value makes Φ close to the linear function in the vicinity of the origin. The learning rate parameter η in the NGD algorithm, was set to be $\eta = 0.3$, and the constant C in the NNGD algorithm, was set to be $C = 0.1$. The order of the feedforward filter N was set to be $N = 10$ [8]. The order of the MA part q and the AR part p, of the nonlinear ARMA recurrent perceptron, were set to be $q = 3$ and $p = 1$. Due to saturation type of logistic nonlinearity, input data was prescaled to fit the range of an output neuron activation function. The summary of the performed experiments is given in Table I.

It is obvious that nonlinear algorithms for adaptation of a neural adaptive filter have better performance comparing to the best linear adaptive algorithm (NLMS).

	NGD	NNGD	Rec.Perc.	NLMS
Pred. gain [dB]	5.78	5.81	6.04	4.75

Table I. Performance of the algorithms employed in the prediction of the NO_2 time series

9 Conclusions

An insight into the dynamical properties of an air pollutant dataset has been provided. Nonlinear adaptive algorithms have been compared with the linear algorithms on the air pollution series and have provided better results. The time series is nonlinear and cyclic and therefore the recurrent perceptron has exhibited the best performance, corroborating the results given by the measures of nonlinearity.

References

[1] G. E. P. Box and G. M. Jenkins, *Time Series Analysis: Forecasting and Control.* San Fransisco: Holden-Day, Inc., 1970.

[2] J. Theiler, P. S. Linsay, and D. M. Rubin, *Time Series Prediction : Forecasting the Future and Understanding the Past*, ch. Detecting Nonlinearity in Data with Long Coherence Times, pp. 429–455. Addison Wesley, 1993.

[3] D. Beule, H. Herzel, E. Uhlmann, J. Kruger, and F. Becker, "Detecting nonlinearities in time series of machining processes," *Proceedings of the American Control Conference*, pp. 694–698, 1999.

[4] M. C. Casdagli and A. S. Weigend, *Time Series Prediction : Forecasting the Future and Understanding the Past*, ch. Exploring the Continuum Between Deterministic and Stochastic Modeling, pp. 347–366. Addison Wesley, 1993.

[5] A. A. M. Khalaf and K. Nakayama, "A hybrid nonlinear predictor: Analysis of learning process and predictability for noisy time series," *IEICE Trans. Fundamentals*, vol. E82-A, no. 8, pp. 1420–1427, 1999.

[6] D. P. Mandic, "NNGD algorithm for neural adaptive filters," *Electronic Letters*, vol. 36, no. 9, pp. 845–846, 2000.

[7] D. P. Mandic and J. Chambers, "Relationship between the slope of the activation function and the learning rate for the rnn," *Neural Computation*, vol. 11, no. 5, pp. 1069–1077, 2000.

[8] I. R. Krcmar, D. P. Mandic, and R. J. Foxall, "On predictability of atmospheric pollution time series," in *To appear in the Proceedings of the 5th ICANNGA*, 2001.

On Predictability of Atmospheric Pollution Time Series

Igor R. Krcmar*, Danilo P. Mandic†, Robert J. Foxall‡

*Faculty of Electrical Engineering, University of Banjaluka, Banjaluka, BH. e-mail:ikrcmar@etf-bl.rstel.net
†School of Information Systems, University of East Anglia, Norwich, UK. e-mail: d.mandic@sys.uea.ac.uk
‡School of Information Systems, University of East Anglia, Norwich, UK. e-mail: rjf@sys.uea.ac.uk

Abstract

Atmospheric pollution is a health hazard. Thus, an accurate prediction of atmospheric pollution time series is almost a necessity nowdays. The existence of missing data further complicates this challenging problem. The cubic spline interpolation method is applied on the hourly measurements of nitrogen oxide (NO), nitrogen dioxide (NO_2), ozone (O_3), and dust particles ($PM10$). In order to asses predictability of an air pollution time series, a class of gradient-descent based neural adaptive filters is employed. Results indicate that, yet simple, this class of neural adaptive filters is a suitable solution.

1 Introduction

Atmospheric pollution affects human life in a short, as well as, in a long term. The risk of disease, such as lung cancer, hard metal poisoning, and pregnancy complications, just to mention a few, caused by an air pollution, makes us very keen to provide an accurate prediction of atmospheric pollution data. It is a difficult task, due to the nonlinear, cyclic and very complex nature of the process of atmospheric pollutants production.In addition, processes of production and disintegration of different air pollutants and gases contained in the Earth's atmosphere are interconnected (coupled).

The time series prediction is a twofold task [1]: discovering the properties of a time series at hand and inferring the model from observed data. These tasks, even though complementary, are strongly interconnected. In order to obtain a good characterization of a time series at hand, we have to ask ourselves some difficult questions: is the time series linear or nonlinear, is it maybe chaotic, is it deterministic or stochastic? These questions gave rise to the development of wide range of techniques, such as: deterministic vs stochastic (DVS) maps [2], attractor reconstruction [3], Lyapunov exponents [4, 3], Poincare maps [4], etc. Search for an appropriate model in the area of nonlinear systems might be a problem, due to the impossibility to decompose an output of the system into an input signal and independent transfer function. Furthermore, complex systems, with a structure hard to comprehend, provide us with a plenty of observed data. These facts form a firm ground for development of, so called, week models such as neural networks. A week model combines broad generality with insight how to manage its complexity [1]. It is important to note that, from the time series prediction point of view, process of learning, in the neural networks framework, without generalization is pointless. Measurements in time series are usually corrupted by noise, thus process of learning might provide us with a good fit to the observed data, and on the other hand with a poor performance on a novel data.

Due to temporary sensor failure or data transmission errors, measurements may be incomplete, i.e. some measurements might be missing . In order to interpolate or estimate missing data in nonlinear time series or to perform nonlinear system identification different techniques have been developed. Solution for the problem of missing and noisy data in nonlinear time series prediciton from a probabilistic point of view is given in [5]. In [6], an application of stochastic neural networks in identification of nonlinear systems with missing data is presented. Also, there are methods with simple graphical interpretation such as linear, quadratic, spline and least squares (LS) polynomial. They perform a fit of different geometrical shapes into the observed data, with respect to possible constraints .

A class of gradient-descent (GD) based neural adaptive filters, due to inherent simplicity and nonlinearity, might be considered as an adequate choice for the prediction of time series that represents atmospheric pollution data. GD is one of the most important ideas in the theory of learning. It attempts to provide a simple guiding principle for the overall organization of weight changes within the network [7]. The results [8] establish conditions for the stability of a class of GD based neural adaptive filters

upon the notion of the contraction mapping, thus making it even more attractive for application in a nonlinear time series prediction. An algorithm for adaptation of the GD based, single-neuron neural adaptive filter, with a fixed learning rate parameter η is referred to as the nonlinear gradient-descent (NGD) algorithm. The NGD algorithm, however, might suffer from slow convergence and local minima. A constant learning rate parameter η can be considered as one of the factors contributing to these problems [9]. Further, a recent result[10] indicates an inherent relationship between the learning rate parameter η and slope of the nonlinear activation function of an output neuron β, which has a negative impact on the convergence properties of the NGD algorithm with fixed η.
A successful attempt of design of an algorithm for adaptation of GD based, single-neuron neural adaptive filter is given in [11]. Normalized nonlinear gradient-descent (NNGD) algorithm exhibits optimal behaviour in the sense that minimizes instantaneous prediction error, thus providing adaptive learning rate η.

2 Interpolation of Missing Data

We shall follow the method presented in [12], in the formal treatment of the missing data interpolation. Synthesis of missing data can be defined as the process of estimating a function over a set of open intervals from the known values of the function over a set of closed intervals. Let $f(t)$ be a single valued function defined over the closed interval domain

$$I = \{t : t_0 \leq t \leq t_{2N-1}\} \tag{1}$$

Further, we partition the domain I into a set of N closed and $N-1$ open intervals of the form

$$\begin{aligned} I_{2n-1} &= \{t : t_{2n-2} \leq t \leq t_{2n-1}\}; n = 1, \ldots, N \\ I_{2n} &= \{t : t_{2n-1} < t < t_{2n}\}; n = 1, \ldots, N-1 \end{aligned} \tag{2}$$

respectively, where

$$t_0 \leq t_1 \leq t_2 \leq \ldots \leq t_{2N-1} \tag{3}$$

and

$$I = I_1 \bigcup I_2 \bigcup \ldots \bigcup I_{2N-1} \tag{4}$$

Now, the missing data interpolation, or the gap filling, problem is to estimate $f(t)$ over the open set of intervals

$$O = \{I_n : n = 2, 4, 6, \ldots, 2N-2\} \tag{5}$$

from knowledge of $f(t)$ over the closed set of intervals

$$C = \{I_n : n = 1, 3, 5, \ldots 2N-1\} \tag{6}$$

We focus our attention on the problem where the known region consists of areas of complete knowledge of $f(t)$ interspersed with regions with no knowledge of $f(t)$.
Spline interpolation has been considered for the the missing data interpolation in the air pollution time series due to its inherent nonlinearity, and clear graphical interpretation. It performs the best fit of the cubic polynomials to the measurement data, with the possible continuity constraints.

3 A Class of GD Based Neural Adaptive Filters

The equations that describe dynamics of a single-neuron neural adaptive filter, trained by gradient-descent (GD) are

$$e(k) = d(k) - \Phi(\mathbf{x}^T(k)\mathbf{w}(k)) \tag{7}$$

$$\mathbf{w}(k+1) = \mathbf{w}(k) - \eta \nabla_{\mathbf{w}} E(e(k)) \tag{8}$$

where k denotes a discrete time instant, $e(k)$ is the instantaneous error at the output neuron, $E(\cdot)$ is the filter cost function, $d(k)$ is some training (desired) signal, $\mathbf{x}(k) = [x_1(k), ..., x_N(k)]^T$ is the input vector, $\mathbf{w}(k) = [w_1(k), ..., w_N(k)]^T$ is the weight vector, $\Phi(\cdot)$ represents a nonlinear activation function of a neuron, η denotes the learning rate parameter and $(\cdot)^T$ denotes the vector transpose. The most common choice for the cost function $E(\cdot)$ is

$$E(e(k)) = \frac{1}{2}e^2(k) \tag{9}$$

As stated above, the algorithm for on-line adaptation of such an adaptive filter, described by the equations (1), (2), and (3), is referred to as the NGD algorithm . Due to its inherent simplicity and nonlinearity, this filter might be considered as a suitable solution for applications in nonlinear and/or nonstationary prediction.
In order to overcome drawbacks of the NGD algorithm, coming from the fixed learning rate parameter η, modifications of the basic algorithm that provide an adaptive learning rate were developed, e.g. NNGD. Furthermore, for nonlinear systems, learning algorithms with an adaptive learning rate are most desirable [13].
Recall that normalized algorithms are designed to minimize the *a posteriori* output error (11), i.e.

$$\begin{aligned} &\text{minimize} && \| \mathbf{w}(k+1) - \mathbf{w}(k) \|_p && (10) \\ &\text{subject to} && d(k) - \Phi\left(\mathbf{w}^T(k+1)\mathbf{u}(k)\right) = 0 && (11) \end{aligned}$$

where $\| \cdot \|_p$ denotes the $\mathcal{L}_p$ norm [14]. Thus, the NNGD algorithm belongs to a class of *a posteriori* (or data reusing) algorithms. It minimizes the *a posteriori* output error (11) via an adaptive learning rate [11]

$$\eta_{OPT}(k) = \frac{1}{C + [\Phi'(\mathbf{x}^T(k)\mathbf{w}(k))]^2 \|\mathbf{x}(k)\|_2^2} \quad (12)$$

It is obvious, that in the case of linear activation function of an output neuron, the NNGD algorithm reduces to its linear counterpart, the normalized least mean squares (NLMS) algorithm.

4 Experimental Results

In order to support our analysis and to investigate predictability of an air pollution time series, we performed several experiments. Air pollution data represent hourly measurements of nitrogen oxide (NO), nitrogen dioxide (NO_2), ozone (O_3), and dust particles ($PM10$), in the period 1994 - 1997. This data was provided to us, courtesy of the Leeds meteo station. There are a lot of missing data in the structure of the raw time series. It is important to note that air pollution time series are nonstationary which implies that interpolation has to be performed on the time window of certain length. Note that concept of the time window complies with the rigourous formulation of the missing data interpolation. In the performed experiments the spline interpolation method has been applied, and the length of the time window was set to be 12, meaning that 12 measurements, prior to and after the gap of missing data were used, in order to perform the gap filling. In the performed experiments the logistic function was chosen as a nonlinear activation function of an output neuron. The logistic function performs contraction mapping for the slope β set to be $\beta < 4$. The quantitative performance measure was the standard prediction gain, a logarithmic ratio between the expected signal and error variances $R_p = 10\log(\hat{\sigma}_s^2/\hat{\sigma}_e^2)$. The slope of the nonlinear activation function of the neuron β was set to be $\beta = 4$, since this value makes Φ close to the linear function in the vicinity of the origin. The learning rate parameter η in the NGD algorithms, was set to be $\eta = 0.3$, and the constant C in the NNGD algorithm, was set to be $C = 0.1$. A neural adaptive filter, employed in the prediction of air pollution time series, is of the FIR type. Due to saturation type of logistic nonlinearity, input data was prescaled to fit the range of an output neuron activation function.
In order to investigate the predictability of the time series at hand, order of a neural adaptive filter, denoted by L, was varied from 1 to 25. Thus, the results are obtained in the form of diagrams Prediction gain vs Tap length (L), for every time series and different algorithms.
The results of the prediction of the NO and the NO_2 time series, by the NLMS, NGD, and NNGD algorithms are shown in Figures 1 and 2, respectively.

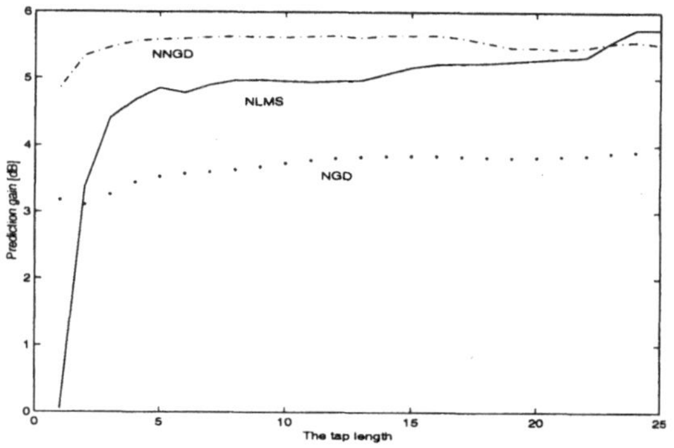

Fig. 1. Performance of the NGD, NNGD, and NLMS algorithms in the prediction of NO time series

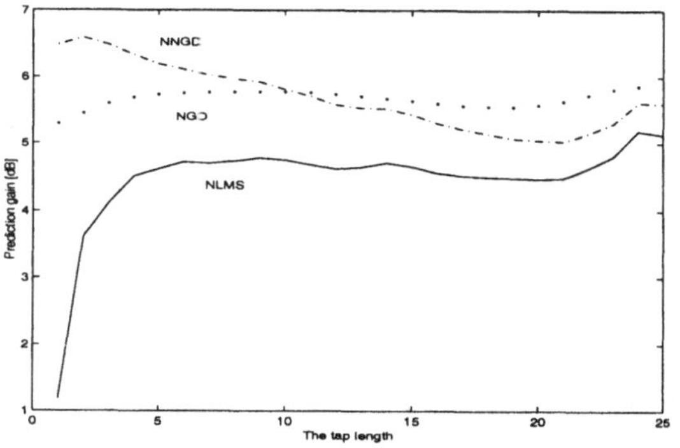

Fig. 2. Performance of the NGD, NNGD, and NLMS algorithms in the prediction of NO_2 time series

The results for the prediction of the O_3 and the $PM10$ time series, by the NLMS, NGD, and NNGD algorithms are shown in Figures 3 and 4, respectively.

It is obvious that algorithms for adaptation of a neural adaptive filter with an adaptive learning rate η have better performance comparing to the algorithms with the fixed η. Finally, we want to add that in the companion paper [15] on nonlinear processing of air pollution data there is the analysis of recurrent perceptron applied to prediction of air pollution time-series. In our opinion, it would have as much merit as including multilayer perceptron analysis.

5 Conclusions

Analysis of the predictability of atmospheric pollution time series is performed. Presented results

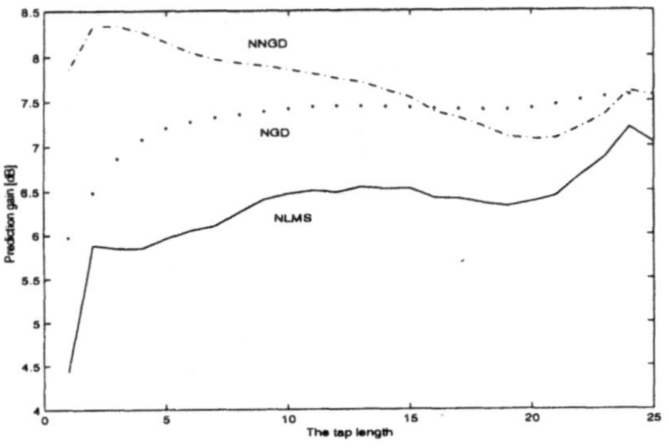

Fig. 3. Performance of the NGD, NNGD, and NLMS algorithms in the prediction of O_3 time series

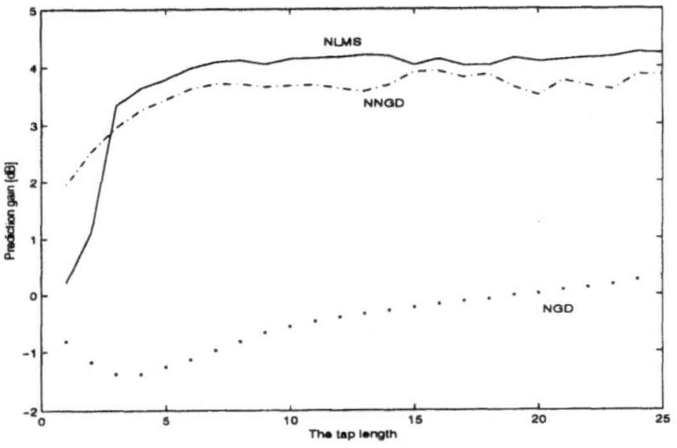

Fig. 4. Performance of the NGD, NNGD, and NLMS algorithms in the prediction of $PM10$ time series

indicate inherently nonlinear, cyclic and very complex nature of the analyzed time series. GD based algorithms with an adaptive learning rate, such as NNGD and NLMS, have better overall performance comparing to the algorithms with the fixed learning rate parameter η, i.e. the NGD algorithm. In addition, NNGD algorithm, as inherently nonlinear, outperforms NLMS algorithm.

References

[1] N. A. Gershenfeld and A. S. Weigend, "The future of time series: learning and understanding", *Proc. of the NATO Advanced Research Workshop on Comparative Time Series Analysis*, pp. 1–70, 1992.

[2] M. Casgaali and A. S. Weigend, "Exploring the continuum between deterministic and stochastic modeling", *Proc. of the NATO Advanced Research Workshop on Comparative Time Series Analysis*, pp. 347–366, 1992.

[3] S. Haykin, "Making sense of a complex world", *IEEE Signal Procssing Magazine*, vol. 8, pp. 66–81, 1998.

[4] H. Kantz, "Noise reduction by local reconstruction of the dynamics", *Proc. of the NATO Advanced Research Workshop on Comparative Time Series Analysis*, pp. 475–490, 1992.

[5] V. Tresp and R. Hofmann, "Nonlinear time-series prediction with missing and noisy data", *Neural Computation*, vol. 10, pp. 731–747, 1998.

[6] M. Tanaka, "Identification of nonlinear systems with missing data using stochastic neural network", *Proceedings of the 35th Conference on Decision and Control*, pp. 933–934, 1996.

[7] P. Baldi, "Gradient descent learning algorithm overview: A general dynamical systems perspective", *IEEE Transactions on Neural Networks*, vol. 6, no. 1, pp.182–195, 1995.

[8] D. P. Mandic and J. A. Chambers, "Global asymptotic stability of nonlinear relaxation equations realised through a recurrent perceptron", *Proceedings of the International Conference on Acoustics, Speech and Signal Processing (ICASSP-99)*, vol. 2, pp. 1037–1040, 1999.

[9] S. Haykin, *Neural networks - A comprehensive foundation*, Prentice Hall, 1994.

[10] D. P. Mandic and J. A. Chambers, "Relationship between the slope of the activation function and the learning rate for the RNN", *Neural Computation*, vol. 11, no. 5, pp. 1069–1077, 2000.

[11] D. P. Mandic, "NNGD algorithm for neural adaptive filters", *Electronic Letters*, vol. 36, no. 9, pp. 845–846 2000.

[12] D. S. Mazel and W. N. Sirgany, "Synthesizing missing data points with iterated function systems", *IEEE Workshop on NNSP '92*, pp. 583–586, 1992.

[13] S. C. Douglas and A. Cichocki, "On-line step-size selection for training of adaptive systems", *IEEE Signal Processing Magazine*, vol.14, no. 6. pp. 45–46, 1997.

[14] S. C. Douglas, "A family of normalized LMS algorithms", *IEEE Signal Processing Letters*, vol. 1, no. 3, pp. 49–51, 1994.

[15] R. J. Foxall, I. R. Krcmar, and D. P. Mandic, "On nonlinear processing of air pollution data", *Proceedings of the International Conference on Artificial Neural Networks and Genetic Algorithms (ICANNGA-2001)*, 2001.

Estimating the Costs Associated with Worthwhile Predictions of Poor Air Quality

Gavin C. Cawley*[1], Stephen R. Dorling†, Robert J. Foxall*, Danilo P. Mandic*

*School of Information Systems, University of East Anglia, Norwich, Norfolk, U.K. NR4 7TJ. E-mail: {gcc,rjf,mandic}@sys.uea.ac.uk †School of Environmental Sciences, University of East Anglia, Norwich, Norfolk, U.K. NR4 7TJ. E-mail: s.dorling@uea.ac.uk

Abstract

In this study we investigate the effect of varying the ratio of false-positive and false-negative misclassification costs on the sensitivity and selectivity of binary predictions of exceedences of atmospheric pollutants. This allows us to determine a window of values for this ratio for which it is worthwhile making definite rather than probabilistic predictions. The support vector machine provides a suitable statistical pattern recognition method for this work.

1 Introduction

It is rarely the case in real world classification tasks that the penalties associated with false-negative and false-positive misclassifications are exactly equal, although this is frequently an implicit assumption of practical statistical pattern recognition algorithms. For instance in diagnosis of a medical disorder a false-positive result is likely to be cause for some concern for the patient and may incur financial costs due to the conduct of further unnecessary tests, however a false-negative result may lead to a potentially serious disorder developing undetected, with far more severe consequences. Likewise misclassifications in the prediction of episodes of poor air quality also incur asymmetric social, healthcare and financial penalties. These costs are complex, difficult to adequately quantify and vary for different end users. As the prior probability of an exceedence of a given pollutant is relatively low, for a prediction of poor air quality to be possible, either there must be little overlap in the distributions of patterns representing good and poor air quality, or the penalty associated with false-negative misclassifications must be sufficiently higher than that associated with false-positive errors. In this work, we aim to estimate the range of values the ratio of misclassification costs can take for which worthwhile predictions of poor air quality can still be made.

[1]This work was supported by the European Commission, grant number IST-99-11764, as part of its Framework V IST programme.

2 Support Vector Classification

The support vector machine [1, 2], given labelled training data

$$\{(\boldsymbol{x}_i, y_i)\}_{i=1}^{\ell}, \quad \boldsymbol{x}_i \in \boldsymbol{X} \subset \mathbb{R}^d, \quad y_i \in \{-1, +1\},$$

generates a maximal margin linear decision rule of the form $h(\boldsymbol{x}) = \mathrm{sign}(\boldsymbol{w} \cdot \boldsymbol{x} - b)$. The weight vector, $\boldsymbol{w}$, is given by the solution of the primal optimisation problem: minimise

$$V(\boldsymbol{w}, \boldsymbol{\xi}) = \boldsymbol{w} \cdot \boldsymbol{w} + C \sum_{i=1}^{\ell} \xi_i \tag{1}$$

subject to

$$y_i[\boldsymbol{w} \cdot \boldsymbol{x}_i - b] \geq 1 - \xi_i, \qquad i = 1, 2, \ldots, \ell, \tag{2}$$

and

$$\xi_i \geq 0, \qquad i = 1, 2, \ldots, \ell. \tag{3}$$

The slack parameters, ξ_i, allow training patterns to be misclassified in the case of linearly non-separable problems. The parameter C sets the penalty applied to margin-errors, and therefore can be viewed as a regularisation parameter, controlling the trade-off between the width of the margin and training set error. A non-linear decision rule can be constructed using a maximal margin linear classifier in a high dimensional feature space, $\Phi(\boldsymbol{x})$, defined by a positive definite kernel function, $k(\boldsymbol{x}, \boldsymbol{x}')$, specifying an inner product in the feature space,

$$\Phi(\boldsymbol{x}) \cdot \Phi(\boldsymbol{x}') = k(\boldsymbol{x}, \boldsymbol{x}').$$

A common kernel is the Gaussian radial basis function (RBF),

$$k(\boldsymbol{x}, \boldsymbol{x}') = e^{-\gamma \|\boldsymbol{x} - \boldsymbol{x}'\|^2}.$$

The function implemented by a support vector machine is given by

$$f(\boldsymbol{x}) = \left\{ \sum_{i=1}^{\ell} \alpha_i y_i k(\boldsymbol{x}_i, \boldsymbol{x}) \right\} - b. \tag{4}$$

To find the optimal coefficients, $\boldsymbol{\alpha}$, of this expansion it is sufficient to maximise the functional,

$$W(\boldsymbol{\alpha}) = \sum_{i=1}^{\ell} \alpha_i - \frac{1}{2} \sum_{i,j=1}^{\ell} y_i y_j \alpha_i \alpha_j k(\boldsymbol{x}_i, \boldsymbol{x}_j), \quad (5)$$

in the non-negative quadrant,

$$0 \leq \alpha_i \leq C, \qquad i = 1, \ldots, \ell, \quad (6)$$

subject to the constraint,

$$\sum_{i=1}^{\ell} \alpha_i y_i = 0. \quad (7)$$

The Karush-Kuhn-Tucker (KKT) conditions can be stated as follows:

$$\alpha_i = 0 \;\Rightarrow\; y_i f(\boldsymbol{x}_i) \geq 1, \quad (8)$$
$$0 < \alpha_i < C \;\Rightarrow\; y_i f(\boldsymbol{x}_i) = 1, \quad (9)$$
$$\alpha_i = C \;\Rightarrow\; y_i f(\boldsymbol{x}_i) \leq 1. \quad (10)$$

These conditions are satisfied for the set of feasible Lagrange multipliers, $\boldsymbol{\alpha}^0 = \{\alpha_1^0, \alpha_2^0, \ldots, \alpha_\ell^0\}$, maximising the objective function given by equation 5. The bias parameter, b, is selected to ensure that the second KKT condition is satisfied for all input patterns corresponding to non-bound Lagrange multipliers. Note that in general only a limited number of Lagrange multipliers, $\boldsymbol{\alpha}^0$, will have non-zero values; the corresponding input patterns are known as support vectors. Equation 4 can then be written as an expansion over support vectors,

$$f(\boldsymbol{x}) = \left\{ \sum_{\text{support vectors}} \alpha_i^0 y_i k(\boldsymbol{x}_i, \boldsymbol{x}) \right\} - b. \quad (11)$$

For a full exposition of the support vector method, see Vapnik [3].

3 Support Vector Machines and Asymmetric Misclassification Costs

In the case of binary classification, for any risk functional that is a linear combination of penalties for each observation, the imposition of asymmetric false-positive and false-negative misclassification costs is equivalent to an unequal replication of positive and negative training examples. Consider a generalised empirical risk functional,

$$R^*_{\text{Emp}} = \frac{1}{\ell} \sum_{i=1}^{\ell} C_i \theta(y_i, f(\boldsymbol{x}_i, \boldsymbol{\alpha})), \quad (12)$$

where C_i is the cost associated with the error for pattern i. For binary pattern recognition, where $y_i, f \in \{-1, +1\}$, typically

$$\theta(y, f(\boldsymbol{x}, \boldsymbol{\alpha})) = \begin{cases} 0 & y = f(\boldsymbol{x}, \boldsymbol{\alpha}) \\ 1 & y \neq f(\boldsymbol{x}, \boldsymbol{\alpha}) \end{cases}.$$

To implement asymmetric misclassification costs for positive and negative examples,

$$C_i = \begin{cases} C^+ & y_i = +1 \\ C^- & y_i = -1 \end{cases},$$

where C^+ is the cost associated with false-negative and C^- the cost associated with false-positive misclassifications. Clearly the generalised risk functional given by equation 12 is equivalent to the standard empirical risk,

$$R_{\text{Emp}} = \frac{1}{\ell'} \sum_{i=1}^{\ell'} \theta(y_i, f(\boldsymbol{x}_i, \boldsymbol{\alpha})),$$

evaluated over a second dataset consisting of C^+ replicates of each positive training example and C^- replicates of each negative example.

For the support vector machine, the symmetry of the optimisation problem given by equations 5-7 suggests that identical training patterns can safely be assigned identical Lagrange multipliers. A notional resampling of training patterns is then implemented by the solution of a modified optimisation problem, maximise

$$W(\boldsymbol{\alpha}^*) = \sum_{i=1}^{\ell'} \zeta_i \alpha_i^* - \frac{1}{2} \sum_{i,j=1}^{\ell'} y_i y_j \zeta_i \zeta_j \alpha_i^* \alpha_j^* k(\boldsymbol{x}_i, \boldsymbol{x}_j),$$

in the non-negative quadrant,

$$0 \leq \alpha_i^* \leq C, \qquad i = 1, \ldots, \ell,$$

subject to the constraint,

$$\sum_{i=1}^{\ell} y_i \zeta_i \alpha_i^* = 0,$$

where ζ_i is the replication factor for pattern i. A change of variables, such that $\alpha_i = \zeta_i \alpha_i^*$, reveals that the solution of the modified optimisation problem is identical to that of the original problem subject to the modified box constraint,

$$0 \leq \alpha_i \leq \zeta_i C, \qquad i = 1, \ldots, \ell.$$

Unequal misclassification costs can therefore be accommodated using the modified box constraint,

$$\begin{cases} 0 \leq \alpha_i \leq C^+ C & y_i = +1 \\ 0 \leq \alpha_i \leq C^- C & y_i = -1 \end{cases}$$

(c.f. Lin *et al.* [4], Veropoulos *et al.* [5]).

4 Method

For the majority of air quality time series, exceedences of a given pollutant are likely to be relatively rare. As a result it may be the case that it is only worthwhile predicting exceedences if the cost of false-negative predictions outweighs that of false positives. In order to investigate the effect of asymmetric misclassification costs on prediction of exceedences, radial basis function support vector machines can be trained using a range of misclassification costs, in this case for the task of predicting SO_2 exceedences in Belfast. The input vector for the support vector network consists of variables representing todays mean SO_2 concentration, sin and cosine components representing the day of the week and the Julian day and also meteorological variables representing tomorrows' weather, namely mean temperature, sea level pressure, wind speed and wind direction. The network is trained to predict the existence of an exceedence twenty-four hours in advance. The support vector machines were trained using a freely available MATLAB toolbox [6]. The value of the overall regularisation parameter, C, and kernel parameter, γ, are chosen in accordance with the model selection procedure [7], which attempts to minimise an upper bound on the leave-one-out cross-validation error [8]. The $\xi\alpha$ bound on the leave-one-out error of a support vector machine is given by,

$$\mathrm{Err}^{\ell}_{\xi\alpha} = \frac{d}{\ell}, \; d = |\{i \; : \; (\rho\alpha_i^0 R_\Delta^2 + \xi_i) \geq 1\}|, \qquad (13)$$

where ρ equals 2, and R_Δ^2 is an upper bound on $k(\boldsymbol{x}, \boldsymbol{x}) - k(\boldsymbol{x}, \boldsymbol{x}')$, $\forall \boldsymbol{x}, \boldsymbol{x}'$. The inequality $\rho\alpha_i^0 R_\Delta^2 + \xi_i \geq 1$ holds for any training pattern corresponding to an error in the leave-one-out procedure, equation 13 therefore provides an upper bound on the leave-one-out error that can be efficiently computed from the solution of the primal and dual optimisation problems (equations 1-3 and 5-7). For support vector machines with a Gaussian radial basis kernel, $R_\Delta^2 = 1$.

The performance of classifiers will be reported in terms of three statistics,

$$\begin{aligned} \text{recall} &= 1 - \frac{e_+}{n_+}, \\ \text{precision} &= \frac{n_+ - e_+}{n_+ - e_+ + e_-}, \\ \text{accuracy} &= 1 - \frac{e_+ + e_-}{n_+ + n_-}, \end{aligned}$$

where n_+ and n_- represent the number of positive and negative examples respectively, and e_+ and e_- represent the number of false-negative and false-positive leave-one-out errors respectively,

$$\begin{aligned} e_+ &= |\{i : y_i = +1 \wedge (\rho\alpha_i R_\Delta^2 + \xi_i) \geq 1\}|, \\ e_- &= |\{i : y_i = -1 \wedge (\rho\alpha_i R_\Delta^2 + \xi_i) \geq 1\}|, \\ n_+ &= |\{i : y_i = +1\}|, \\ n_- &= |\{i : y_i = -1\}|. \end{aligned}$$

5 Results

Figure 1 shows a graph of recall against the ratio of misclassification costs, over the training set, using the $\xi\alpha$ bound on the leave-one-out cross-validation error and the true leave-one-out error. It can be seen that for symmetric misclassification costs, it is only marginally worthwhile to make any positive classifications. For a ratio of costs of just over 6:1 or more, all exceedences are reliably predicted.

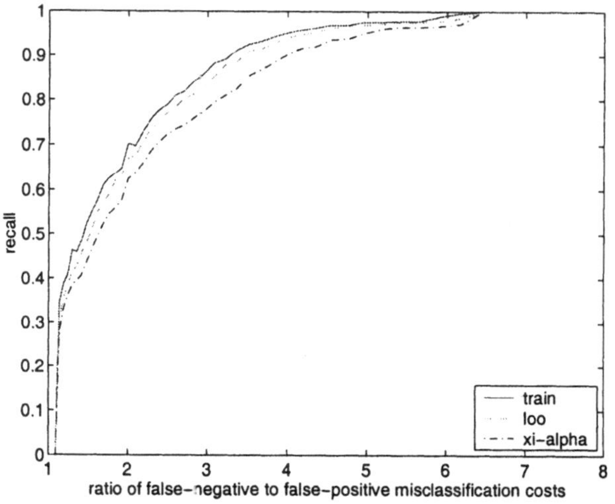

Fig. 1. Graph of recall against ratio of misclassification costs.

Figure 2 shows a graph of precision against the ratio of misclassification costs. Naturally the number of false positive errors increases with an increasing ratio of misclassification costs. If a ratio of more than approximately 6:1 is used, all patterns are predicted as exceedences.

Figure 3 shows a graph of accuracy against the ratio of misclassification costs. Note the large discontinuities at either extreme of the graph. It appears that the model selection criterion used favours the simplest possible classifier, with a constant output, rather too strongly for marginally worthwhile predictions. Better classifiers for the extremes may result from a different model selection criterion.

6 Summary

In this paper we have demonstrated that the performance of a classifier strongly depends on the costs

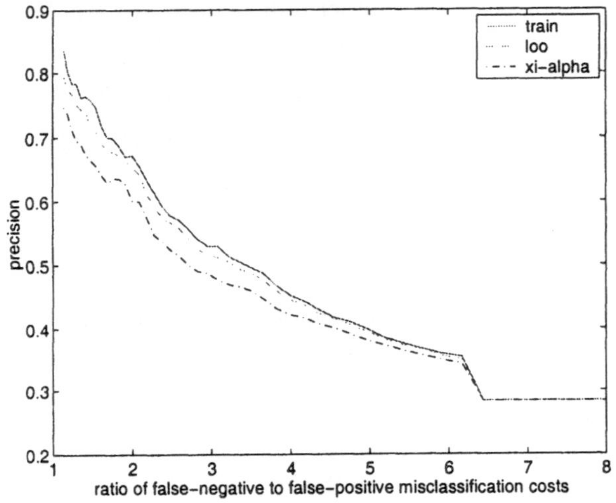

Fig. 2. Graph of precision against ratio of misclassification costs.

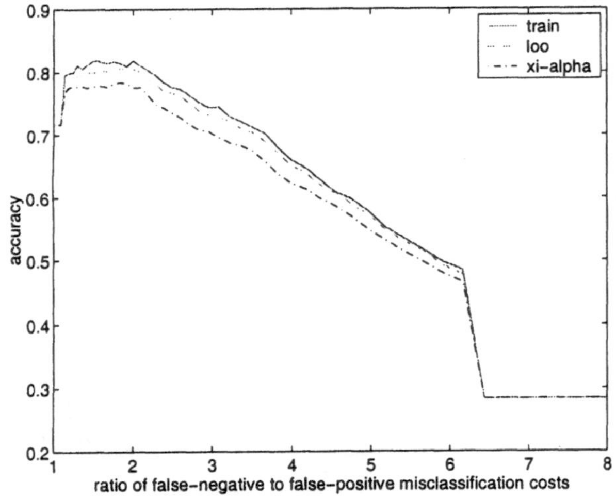

Fig. 3. Graph of accuracy against ratio of misclassification costs.

associated with false-positive and false-negative misclassification errors. We have also experimentally determined the range of misclassification costs for which it is worthwhile actively making predictions of poor air quality due to SO_2 in Belfast.

References

[1] B. Boser, I. Guyon, and V. N. Vapnik, "A training algorithm for optimal margin classifiers," in *Proceedings of the fifth annual workshop on computational learning theory*, (Pittsburgh), pp. 144–152, ACM, 1992.

[2] C. Cortes and V. Vapnik, "Support vector networks," *Machine Learning*, vol. 20, pp. 1–25, 1995.

[3] V. N. Vapnik, *Statistical Learning Theory.* Wiley Series on Adaptive and Learning Systems for Signal Processing, Communications and Control, New York: Wiley, 1998.

[4] Y. Lin, Y. Lee, and G. Wahba, "Support vector machines for classification in nonstandard situations," Tech. Rep. 1016, Department of Statistics, University of Wisconsin, 1210 West Drayton St., Madison, WI 53706, March 2000.

[5] K. Veropoulos, C. Campbell, and N. Cristianini, "Controlling the sensitivity of support vector machines," in *Proceedings of the International Joint Conference on Artificial Intelligence (Workshop ML3)*, (Stockholm, Sweeden), pp. 17–21, 1999.

[6] G. C. Cawley, "MATLAB SVM toolbox (v0.50) [`http://theoval.sys.uea.ac.uk/~gcc/svm/toolbox`]." University of East Anglia, School of Information Systems, Norwich, Norfolk, U.K. NR4 7TJ, 2000.

[7] G. C. Cawley, "Model selection for support vector machines via adaptive step-size tabu search," in *Proceedings of the International Conference on Artificial Neural Networks and Genetic Algorithms (*accepted for publication*)*, (Prague), April 2001.

[8] T. Joachims, "Estimating the generalization performance of a SVM efficiently," Tech. Rep. LS-8 number 25, Univerität Dortmund, Fachbereich Informatik, 1999.

Modelling Air Pollution Time-Series by Using Wavelet Functions and Genetic Algorithms

G. Nunnari, L. Bertucco *

*Dipartimento Elettrico Elettronico e Sistemistico, Università degli Studi di Catania, Viale A. Doria, 6, 95125 Catania (Italy),

Abstract

The peculiarity of the proposed approach is that of combining the use of wavelets and genetic algorithms for searching the best wavelets parameters in order to model a given pollution time series. The results are compared with a neural approach.

1 Introduction

The literature provides several approaches for modelling air pollution time series such as the traditional parametric identification methods based on stochastic models of ARMAX type or more sophisticated approaches based on the use of artificial neural networks or neuro-fuzzy approaches [1]-[3]. This paper deals with the possibility of using wavelet function for modelling air pollution time series. The use of wavelets for identification has been introduced in literature by Benveniste et al. [4]. The main advantage of modelling with wavelets consists in the fact that is possible to represent in more efficiènt way the transitory characteristics of the time series. This advantage derives from the fact that the wavelets are limited duration functions; moreover the shape of the wavelets used for modelling can be chosen time to time depending on the peculiarities of the time series to be modelled. The peculiarity of the proposed approach is that of combining the use of wavelets and genetic algorithms for searching the best wavelet parameters in order to model a given pollution time series. An example is reported and the results are compared with a neural approach.

2 Some Mathematical Background

The modelling of a time series $y(t)$ starts with an analysis of the series and other available correlated series to determine what kind of variables (i.e. the exogenous model inputs) the time series mainly depends on and to what extent (number of regressions).

$$\begin{aligned} y(t+s) = f(&y(t), y(t-1), \dots y(t-n_y+1), u_1(t), \\ &u_1(t-1), \dots, u_1(t-n_1+1), \dots, \\ &\dots, u_q(t), u_q(t-1), \dots, u_q(t-n_q+1)) \end{aligned} \quad (1)$$

Since models of pollutant time series are usually non-linear, they can be represented as NARX (Non-linear Auto-Regressive with eXogenus inputs) models of the form (1) where f is an unknown non-linear function, $u_1, \dots, u_q$ are the exogenous model inputs, s represents the number of steps ahead for the prediction model, and $n_y, n_1, \dots n_q$ are integer numbers related to the model order. NARX models can be considered a generalisation of the well-known ARX models. The decision to use exogenous inputs, which can always be done using a *trial and error* procedure, should be made not only on the availability of such data but also on the basis of their degree of correlation with the time series being modelled. Formally, the modelling problem is that of finding a suitable approximation of the unknown function f.

In this paper we propose to approximate y as shown in (2):

$$y(t+s) = \sum_{i=1}^{k} h_i \cdot \varphi_i(X) \quad (2)$$

where $X=(y(t), y(t-1)\dots y(t-n_y+1), u_1(t),\dots u_1(t-n_1+1),\dots u_q(t),\dots u_q(t-n_1+1))$, $h \in \Re$, and $\varphi(X): \Re^d \to \Re$ with $d = dim(X)$ are wavelet functions of vectorial arguments such as the following (3)-(5):

$$\varphi(X) = \alpha_1 X_1 X_1^{'} e^{-\alpha_2 X_1 X_1^{'}} \quad (3)$$

$$\varphi(X) = \alpha_1 2^{-1/4} e^{-\pi X_1 X_1^{'}} \cos\left(\alpha_2 X_1 X_1^{'} - \alpha_3\right) \quad (4)$$

$$\varphi(X) = \alpha_1 2^{-1/4} e^{-\pi X_1 X_1^{'}} - \alpha_2 2^{-1/4} e^{-\pi X_2 X_2^{'}} \quad (5)$$

where $X_1 = A \otimes (X - T)$, $X_2 = A \otimes (-X - T)$ and X, A, $T \in \Re^d$, $\alpha_1, \alpha_2, \alpha_3 \in \Re$ and $\otimes$ is the component to component product.

One of the problems that immediately arise by using a similar approach is that of finding the parameters contained these expressions, i.e. the coefficients α_i and the components of the vectors A and T. To solve this problem in a optimal way we propose to use the Genetic Algorithm (GA) approach, as discussed in the following section.

3 Finding the Parameters

Let us define an cost function to be minimised such as:

$$\begin{aligned} &J = \|y - y^*\|, \qquad or\ J = \max\|y - y^*\|, \\ &or\ J = \|y - y^*\| + b * \max\|y - y^*\|, \\ &or\ J = \|(y - y^*)^2\| \end{aligned} \tag{6}$$

or similar, being y and y^* the true and simulated time series respectively.

The proposed identification algorithm can be synthesised as follows:

1) let us assign $i = 1$ and $R_i = y$;
2) find the best parameters of the chosen wavelet function φ_i by using GA, after defining the object function (or cost function) as indicated above,
3) compute the residual function as: $R_{i+1} = R_i - \varphi_i$
4) increment i by 1 and repeat steps 1) through 3) until $i \leq k$

4 Some Numerical Results

The proposed procedure was applied to model air pollution data recorded in the industrial area of Syracuse (Italy). The monitoring network covers a wide area comprising several towns in the Syracuse hinterland. The atmospheric pollution monitored in the area is essentially caused by industrial emissions, as there are large petrochemical plants nearby. In particular in this paper SO_2 time series are considered.

The procedure described above was considered for modelling an SO_2 time series. For this pollutant during 1995-1999 a considerable number of overshoots of the attention level threshold (125 $\mu g/m^3$, daily mean value) were observed. The results reported below refer to the recording station named Melilli. Since in this station meteorological data are not recorded, the station named Priolo, which is about 8 km far from Melilli (Southeast direction), supplies meteorological information.

In Fig. 1 (a) and (b) the SO_2 daily mean values recorded at the Melilli station during 1997 and 1999 respectively are shown.

Analysis of SO_2 time series in the considered area allowed to point out that the level of pollution is strongly influenced by local marine breezes. Hence Wind Speed (WS) and Wind Direction (WD) were considered as exogenous inputs of SO_2 prediction models. For more details on this see also the paper [5] published in this volume.

In the paper [5] it is shown that NARX prediction models of SO_2 time series can be suitably implemented by using neural networks. The main aim of this paper is

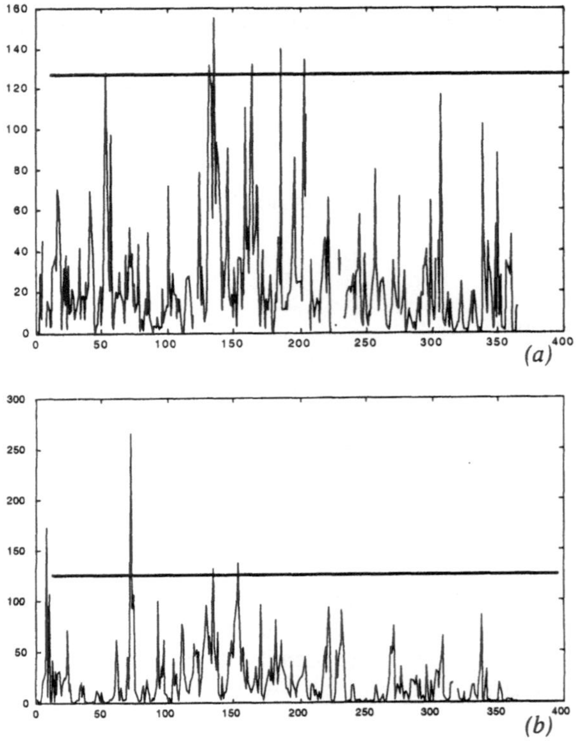

Fig. 1. Melilli (Syracuse, Italy), SO_2 concentration, daily mean values recorded in (a) 1997 and (b) 1999. The attention level (125 $\mu g/m^3$) was indicated with a horizontal line.

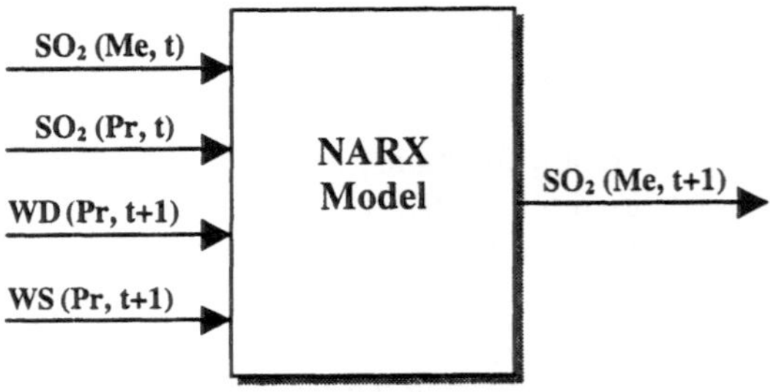

Fig. 2. Structure of the NARX model: Me=Melilli, Pr=Priolo, WD=Wind Direction, WS=Wind Speed.

to show that using a wavelet function approach can solve the same problem. The structure of the NARX prediction model considered to compare the two approaches is shown in Fig. 2.

In more detail the prediction model considers the following exogenous inputs:

Exogenous n. 1: SO_2 Priolo: daily averages computed from 8 a.m. to 6 p.m. with 1 regression and delay of 1 with respect to target.

Exogenous n. 2: WD Priolo: daily averages computed from 8 a.m. to 6 p.m. with 1 regression and delay of 0 with respect to the target (meteorological forecast).

Exogenous n. 3: WS Priolo: daily averages computed from 8 a.m. to 6 p.m. with 1 regression and delay of 0 with respect to the target (meteorological forecast).

From Fig. 2 it is possible to recognise that the model considers as regressed input the daily average concentration of SO_2 at Melilli at the day t. The output of the prediction model is the average SO_2 concentration at the day $t+1$ at the Melilli station.

As stated above the identification of the model was carried out by using both a Multi-Layer Perceptron (MLP) neural network and a wavelet approach WGA (WGA stands for Wavelet with Genetic algorithms optimisation Approach) below.

The same data set was considered for the identification of both MLP and WGA models in order to compare the results. In particular, data recorded at Melilli during the period 1995-1998 were used for model identification while data recorded during 1999 was considered for the testing phase.

The topology of the MLP prediction model was obtained by a trial and error approach; the best results were obtained by using a 4-40-1 network, i.e. 4 neurones in the input layer, 40 in the unique hidden layer and 1 in the output layer. The training of the MLP neural network was performed by using a standard back-propagation algorithm. The wavelet function considered for the identification of the WGA model is that given by expression (5) since, based on the trials performed, it gives the best results among the others considered in the present paper (see expressions (3) and (4)). The number k of wavelet functions considered (see expression (2)) was limited to 3 since it has been observed that a greater number does not significantly improves the results.

The main parameters considered for the genetic optimisation algorithm were the following: number of individuals = 20, number of generation = 10, gap between generations = 0.8. Two different WGA models were identified referred as WGA-1 and WGA-2 below. The model WGA-1 was identified by using a standard objective function of quadratic type (see the first expression in (6)) while WGA-2 was identified by using the following objective function (7):

$$J = \sum_{i=1}^{N} \sqrt{(y_i - \bar{y}_i)^2 a}$$

$$\text{with} \begin{cases} a = a^* > 1 & \text{if} \quad y_i > Th \\ a = 1 & \text{if} \quad y_i \le Th \end{cases} \qquad (7)$$

In particular in the considered case of study it was assumed: $a^* = 1.5$, $Th = 125\ \mu g/m^3$.

The performance of each prediction model was tested by computing a set of indexes defined according with the recommendation of the European Topic Centre on Air Quality [6].

Thr	70	80	100	125	MaxT	234
Index					MinT	0.3
M	22	16	6	3	MaxC	131
F	23	17	11	4	MinC	-12
A	10	8	3	2	MeaE	-9.6
SP (%)	45	50	50	67	MSDE	24.9
SR (%)	43	47	27	50	NEF	0.77
FA (%)	57	53	73	50	TCC	0.53
IS (%)	41	47	48	66	Skill	-8.4

Table I. SO_2 ($\mu g/m^3$) Performance indexes computed for MLP model (testing data set, topology of the net 4-40-1), learning pattern: 100, testing pattern: 327.

The meaning of the symbols in the tables (I-IV) is the following: **Thr**: Prefixed threshold, **M**: Observed overshoots, **F**: Predicted overshoots, **A**: Number of overshoots correctly predicted, **SP**: % of overshoots correctly predicted, **SR**: % of predicted overshoots actually occurring, **FA**: % of false alarms, **IS:** Index of successes, **MaxT, MinT**: Max and Min values of the true series, **MaxC, MinC**: Max and Min values of the computed model, **MeaE**: Mean of errors, **MSDE**: Mean square deviation of errors, **NEF**: Unexplained fraction, **TCC**: True/computed correlation, **Skill**: Skill-Score index.

Thr	70	80	100	125	MaxT	234
Index					MinT	0.3
M	22	16	6	3	MaxC	138
F	31	17	10	3	MinC	-22
A	11	8	4	0	MeaE	-6.9
SP (%)	50	50	67	0	MSDE	24.3
SR (%)	35	47	40	0	NEF	0.73
FA (%)	65	53	60	100	TCC	0.6
IS (%)	43	47	65	0	Skill	2.8

Table II. SO_2 ($\mu g/m^3$) Performance indexes computed for WGA-1 model (testing data set) - wavelet (5), learning pattern: 100, testing pattern: 327.

Thr	70	80	100	125	MaxT	234
Index					MinT	0.3
M	22	16	6	3	MaxC	238
F	17	14	9	8	MinC	0
A	8	8	3	2	MeaE	1.1
SP (%)	36	50	50	67	MSDE	24
SR (%)	47	57	33	25	NEF	0.73
FA (%)	53	43	67	75	TCC	0.7
IS (%)	33	48	48	65	Skill	10.9

Table III. SO_2 ($\mu g/m^3$) Performance indexes computed for WGA-2 model (testing data set) - wavelet (5), learning pattern: 100, testing pattern: 327.

The results obtained by using the neural prediction model are summarised in Table I while the results obtained by using the WGA-1 and WGA-2 are reported in Table II and Table III respectively.

From these tables it is possible to observe that the performances of the MLP and WGA-2 models are quite similar and are superior both to WGA-1and to the so-called persistent model (see Table IV) which often is used as reference model. However some statistical indexes such as MeaE, MSDE, NEF, TCC and Skill point out a superiority of WGA-2 with respect MLP.

Thr	70	80	100	125	MaxT	234
Index					MinT	0.3
M	22	16	6	3	MaxC	234
F	22	16	6	3	MinC	0.3
A	9	5	1	0	MeaE	0
SP (%)	41	31	17	0	MSDE	25.7
SR (%)	41	31	17	0	NEF	0.8
FA (%)	59	69	83	100	TCC	0.6
IS (%)	37	28	15	0	Skill	0

Table IV. SO_2 ($\mu g/m^3$) Performance indexes computed for persistent model (testing data set)

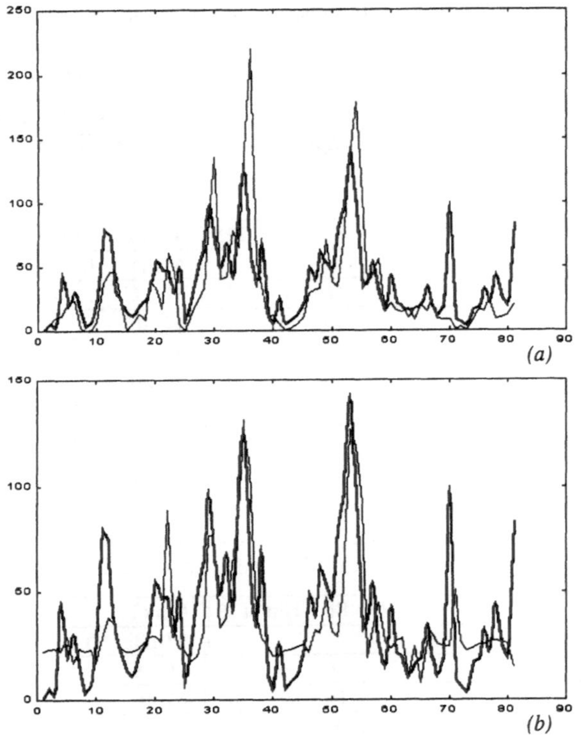

Fig. 3. True (thick black) and 1 step prevision model (thin black), SO_2 concentration in $\mu g/m^3$ - pattern number (a) WGA-2 model (b) MPL model.

The output of the WGA-2 and MLP models are compared with the true time series in Fig. 3 (a) and Fig. 3 (b) respectively. It is possible to observe that the WGA-2 model tends to overestimate the true time series while the MLP seems to underestimate.

5 Conclusions

In this paper a novel wavelet function based approach to identify air pollution prediction models is proposed. One of the peculiarities of the approach is represented by the use of GA to search the best model parameters. The results obtained so-far show that the proposed approach performs as good as the MLP neural approach, which at the present is considered one of the best techniques available for the identification of NARX models. However it is the author's opinion that the proposed approach is more flexible and can be easily adapted to the features of the time series to be modelled. Work is in progress to further validate the technique also considering different types of air pollutants (i.e. Ozone, NO_2 etc).

Acknowledgement

The paper has been financially supported by the EU under the framework of the APPETISE project (Contract N. IST-99-11764).

The authors are also grateful to the Municipal and Provincial authorities in Syracuse and Municipal authorities in Catania (Italy) for providing pollution and meteorological data considered in the paper.

References

[1] Boznar M., Lesjak M., Mlakar P.: A Neural Network-Based Method for Short-Term Predictions of Ambient SO_2 Concentrations in Highly Polluted Industrial Areas of Complex Terrain. Atmospheric Environment, 27B, 2, pp. 221-230 (1993).

[2] Arena P, Baglio S., Castorina C., Fortuna L., Nunnari G.: A Neural Architecture to Predict Pollution in Industrial Areas. ICNN, 4, pp. 2107-2112, Washington, (1996).

[3] Nunnari G., Nucifora A., Randieri C.: The Application of Neural Techniques to the Modelling of Time Series of Atmospheric Pollution Data. Ecological Modelling, 111, pp. 187-205 (1998).

[4] Benveniste A., Juditsky A., Delion B., Zhang Q., Glorennec P. Y.: Wavelets in identification. Proc. SYSIS '94, 2, pp. 27-48 (1994).

[5] Nunnari G., Bertucco L., Milio D.: Predicting Daily Average SO_2 Concentrations in the Industrial Area of Syracuse (Italy) ICANNGA 2001, Prague, (2001).

[6] Van Aalst R. M., De Leeuw F. A. A. M. (editors), National Ozone Forecasting System and International Data Exchange in Northwest Europe, European Topic Centre on Air Quality, 1997.

Predicting Daily Average SO_2 Concentrations in the Industrial Area of Syracuse (Italy)

G. Nunnari, L. Bertucco, D. Milio *

*Dipartimento Elettrico Elettronico e Sistemistico, Università degli Studi di Catania, Viale A. Doria, 6, 95125 Catania, Italy

Abstract

In this paper artificial neural networks are used to build 1-day-ahead SO_2 prediction models. The structure of the model was obtained following appropriate statistical analysis of the time series.

1 Introduction

Sulphur emissions, mainly SO_2, have represented the world's greatest and most common air pollution problem. Although a considerable decrease in SO_2 emissions in the atmosphere has been observed in the last decade [1], thanks to improvements in the quality of fuels with lower sulphur content [2], SO_2 is still among the most harmful pollutants in highly industrialised areas in both developing countries and several European states [3]. The concentration of SO_2 in a given area depends on a number of parameters of a meteorological and topographical nature. In this context the availability of prediction models is extremely useful to avoid the occurrence of critical situations.

Modelling atmospheric pollution phenomena is a complex task on account of the highly non-linear influence of several external variables and the presence of non-stationary phenomena. In order to handle the difficulties which arise in identifying non-linear systems with traditional techniques, in this paper neural networks are used to build 1-day-ahead prediction models. The structure of the model was obtained following appropriate statistical analysis of the time series.

2 The Case Study Considered

The area studied is the province of Syracuse, in the South-east of Sicily (Italy). In the post-war period one of the largest concentrations of petrochemical industries in Europe developed here and it is considered to be an area of high environmental risk.

The air quality in this area is monitored by an interconnected network, run by both public authorities at different levels (Provinces and Municipalities) and private organisations, such as the CIPA Consortium (Industrial Consortium for Environmental Protection) and ENEL (the Italian National Agency for Electrical Energy). One of the most important parameters for evaluating the degree of pollution due to SO_2 is the 24h mean value, i.e. the average concentration computed during a day. The attention and alarm levels imposed by national law are indicated below.

Pollutant	Computing Mode	Attention Level	Alarm Level
SO_2	Daily Mean	125 $\mu g/m^3$	250 $\mu g/m^3$

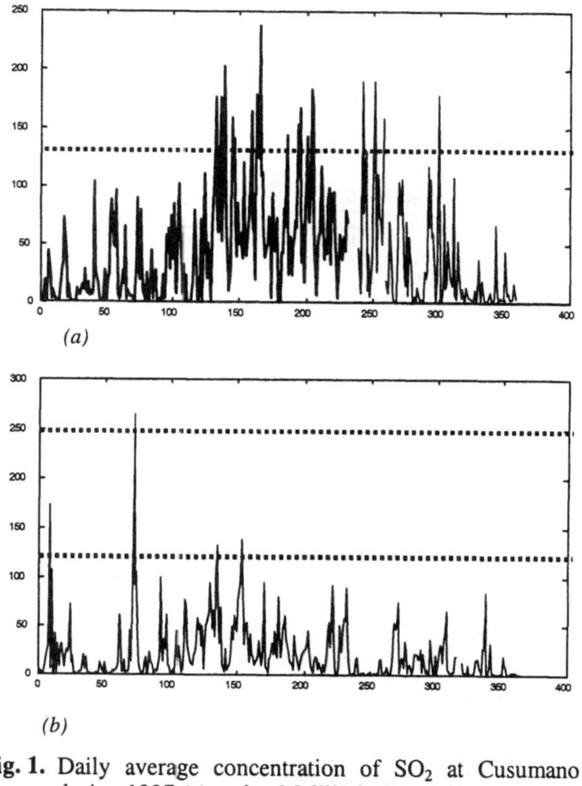

Fig. 1. Daily average concentration of SO_2 at Cusumano during 1997 (a) and at Melilli during 1999 (b)

The 24h average concentration recorded at the Cusumano and Melilli stations during 1997 and 1999 are shown in Figs. 1 (a) and 1 (b) below. As can be seen from these figures, the SO_2 attention level is exceeded several times. The main aim of this paper is to identify a 1-day-ahead SO_2 prediction model for the daily mean concentration of SO_2. This is not a easy task and it can be shown that traditional ARX linear models give poor results when used to predict the level of air pollution [4]. For this reason we use NARX (Non linear ARX) neural structures.

Neural networks for six hours-ahead predictions in the area being considered were previously proposed in [5].

Data concerning the SO_2 emissions is not available, so the prediction can only be performed on the basis of the past series of SO_2 and the related time series (meteorological data and other related pollutants).

3 The Structure of the Prediction Model

In order to define the prediction model structure preliminary data analysis was carried out.

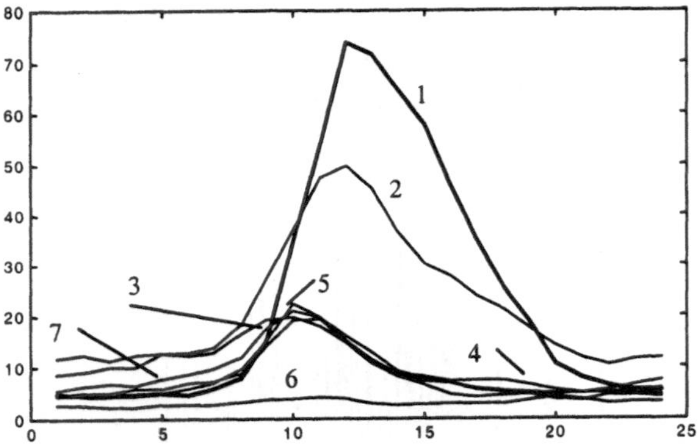

Fig. 2. Typical daily SO_2 distribution at seven stations during 1999, station names: 1 Cusumano 2 Melilli 3 Belvedere 4 Priolo 5 Ciapi 6 Augusta 7 Scala Greca

$$\hat{SO_2}(i) = \frac{1}{365}\sum_{j=1}^{365} SO_{2i,j} \quad i = 1 \ldots 24 \tag{1}$$

Initial results gave the typical daily SO_2 concentration in Fig. 2, which shows data recorded by seven stations during 1999.

By typical daily concentration we mean the curve obtained by averaging the level of SO_2 recorded at the same time of day throughout the year. In other terms, if $SO_{2i,j}$ indicates the level of SO_2 at the *i-th* hour *[i=1...24]* of the *j-th* day *[j=1...365]*, the typical daily concentration of SO_2 is a curve whose samples are computed by expression (1). The results are surprising, since it seems unrealistic to assume that peaks in the rate of SO_2 emissions between 10 a.m. and 12 a.m. local time are due to particular production strategies on the part of the companies responsible for the emissions. It seems more reasonable to attribute the peaks to the local atmospheric conditions. It should also be noted, in fact, that the SO_2 peaks are correlated with wind direction (WD) and wind speed (WS) as shown in Figs. 3 (a) and (b), respectively for the Cusumano station. Moreover, it was observed that higher SO_2 levels are mostly reached during the hot season (see for instance Fig. 1 (a).

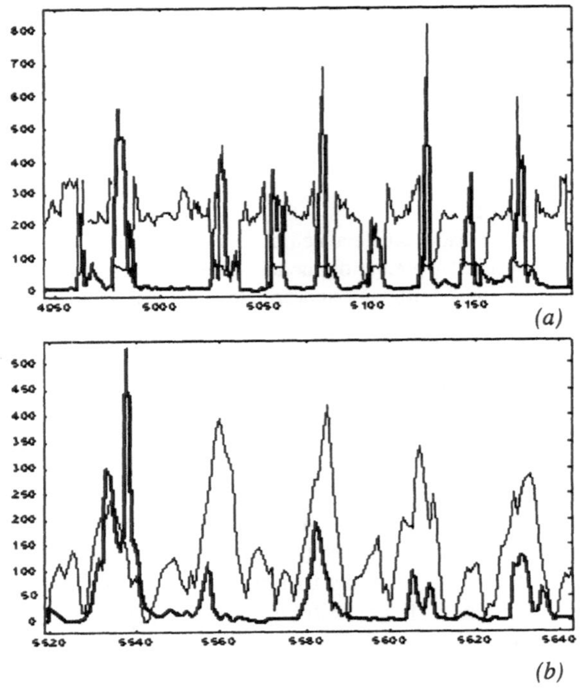

Fig. 3. (a) Cusumano, 1998, SO_2 (thick black, $\mu g/m^3$)-WD (gray, Deg), average per hour (b) SO_2 (thin black, $\mu g/m^3$)-WS (gray, m/s x50).

A possible explanation of what has been observed might be given in terms of local sea breeze circulation, the onset of which is strongly linked to variations in solar radiation, as argued by some meteorologists (S. Dorling and J. Keder).

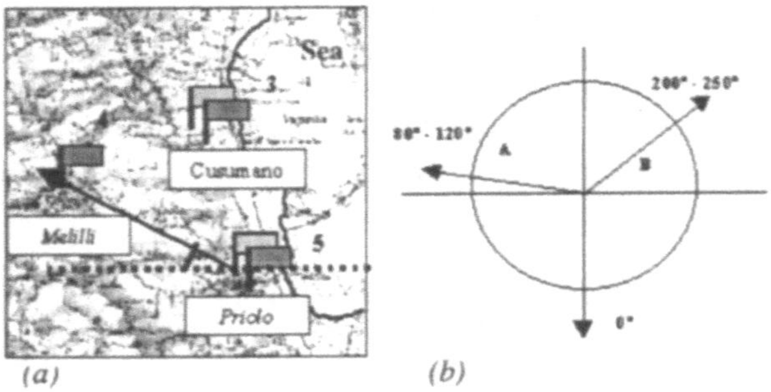

Fig. 4. (a) Area monitored by Melilli -Priolo - Cusumano stations; dark grey flags refer to stations recording pollutant data, clearly grey to stations recording meteorological data (b) Priolo, wind direction: origin and main directions. The wind vector rotates from position A to B during the night, time as it is typical of marine breezes.

The topography of the area, which features a chain of mountains to the west and the sea to the east of the three stations in the network monitoring SO_2 levels, helps to interpret the daily distribution of SO_2 levels previously shown. Peaks in the SO_2 concentration are recorded during the daytime when the wind blows from the sea towards the land where probably SO_2 accumulates due to the presence of mountains on the west side. Lower values of SO_2 concentration are recorded when the wind blows from the land to the sea (see the vector B in Fig. 4 (b)). It is to be observed that the rotation of the wind vector from position A to B during the night time is the typical behaviour of marine breezes.

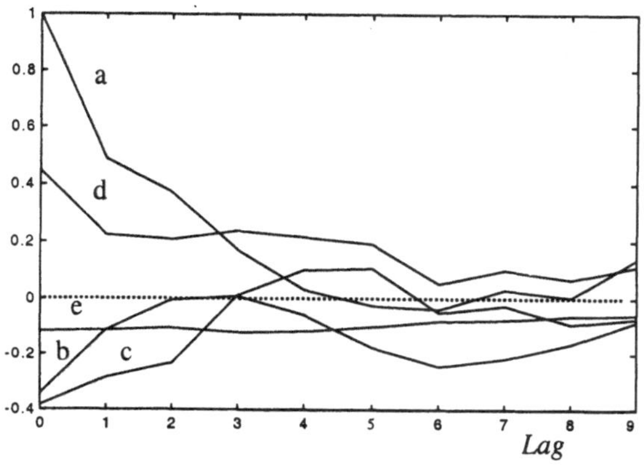

Fig. 5. Correlation computed on time series of daily average values recorded at Cusumano during 1998. (a) SO_2 autocorrelation (b) correlation between SO_2 and WD (c) SO_2 and WS (d) SO_2 and solar radiation (e) SO_2 and relative humidity

On the basis of these results it was decided to consider the following variables as exogenous inputs for the prediction model: the daily average values of wind speed, wind direction and solar radiation.

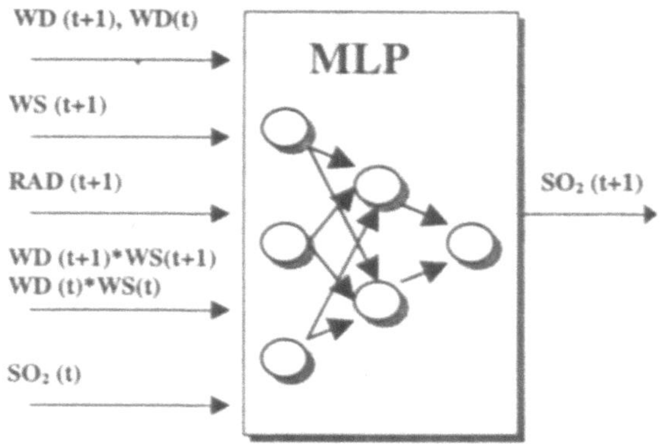

Fig. 6. Structure of a NARX Neural Prediction Model

Temperature was not considered as an exogenous variable in order to simplify the model and also because attempts made demonstrated that using temperature as an additional input variable does not lead to any significant improvement in the performance of the model.

Th	80	100	125	150	200	250	MaxT	238
Index							MinT	0.6
M	63	44	27	17	4	*	MaxC	203
F	72	61	29	12	1	*	MinC	-24.9
A	58	42	19	11	1	*	MeaE	-1.8
SP (%)	92	95	70	65	25	*	MSDE	19.8
SR (%)	81	69	66	92	100	*	NEF	0.12
FA (%)	19	31	34	8.3	0	*	TCC	0.94
IS (%)	78	79	63	64	25	*	Skill	77%

Table I. Performance indexes computed using the 162 learning patterns available

Th	80	100	125	150	200	250	MaxT	209
Index							MinT	0.2
M	31	18	7	3	1	*	MaxC	151
F	40	15	4	1	0	*	MinC	-22.8
A	25	9	2	1	0	*	MeaE	-3.6
SP (%)	81	50	29	33	0	*	MSDE	21.9
SR (%)	63	60	50	100	0	*	NEF	0.4
FA (%)	38	40	50	0	*	*	TCC	0.8
IS (%)	74	47	28	33	0	*	Skill	64%

Table II. Performance indexes computed using the 250 testing patterns available

Th	80	100	125	150	200	250	MaxT	209
Index							MinT	0.2
M	31	18	7	3	1	*	MaxC	209
F	31	18	7	3	1	*	MinC	0.2
A	8	5	1	0	0	*	MeaE	-0.02
SP (%)	26	28	14	0	0	*	MSDE	36.9
SR (%)	26	28	14	0	0	*	NEF	1.0
FA (%)	74	72	86	100	100	*	TCC	0.5
IS (%)	15	22	12	0	0	*	Skill	0%

Table III. Persistent Model, 250 Patterns

Th: Prefixed threshold, **M**: Observed overshoots, **F**: Predicted overshoots, **A**: Number of overshoots correctly predicted, **SP**: % of overshoots correctly predicted, **SR**: % of predicted overshoots actually occurring, **FA**: % of false alarms, **IS**: Index of successes, **MaxT, MinT**: Max and Min values of the true series, **MaxC, MinC**: Max and Min values of the computed model, **MeaE**: Mean of errors, **MSDE**: Mean square deviation of errors, **NEF**: Unexplained fraction, **TCC**: True/computed correlation, **Skill**: Skill-Score index.

The structure of the NARX neural model used is shown in Fig. 6. The inputs to the neural network are as follows: the average SO_2 level and the average wind direction on day t, and the predicted average wind speed, wind direction and solar radiation values on day (t+1). In addition, the product of the predicted average wind speed and direction values on days (t+1) and t was used as reinforced learning input. It should also be observed that attempts made during the learning process suggested using different weights for the neural inputs. A weight of 2 was used for all the inputs except for the WD*WS product and SO_2, which were weighted by 1. The model output is the average level of SO_2 on day (t+1).

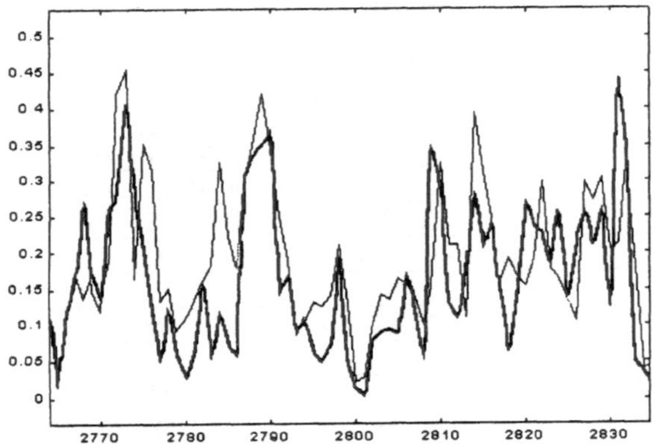

Fig. 7. True (thick black) and 1 step of prevision model (thin black) (concentration in $\mu g/m^3$ - Pattern Number)

The learning pattern was constructed using SO_2 data recorded at the Melilli and Cusumano stations during 1995, 1996 and 1997, while the data used for testing was recorded at Cusumano during 1998.

The meteorological data used was recorded at the Cusumano station (except for WD in 1997, which was not available for Cusumano: in this case the WD recorded at Priolo, which is the station closest to Cusumano, was used).

The topology of the Multilayer Perceptron neural network considered to identify the model was established by a trial and error procedure and the best results were obtained using 10 neurons in a single hidden layer. Hence the topology of the neural network is 11-10-1, i.e. it has 11 neurons in the input layer, 10 in the hidden layer and 1 in the output layer. The performance indexes of the prediction model, as suggested by the European Topic Centre on Air Quality [6], computed on the learning and test data, are shown in Tab. 1 and Tab. II respectively, while the true and simulated data of the daily average SO_2 value at Cusumano during 1998 is shown in Fig. 7. Tab. III shows the performance of the persistent model computed on the data used for the test.

The results obtained clearly show that the proposed neural model performs much better than the persistent model but its performance degrades with a higher threshold level. This not surprising behaviour might be accounted for by the consideration that the number of critical episodes is limited and for this reason the neural network cannot produce a satisfactory learning phase. Work is still in progress to overcome this drawback. Further improvements are also expected by exploring other configurations of the prediction model.

Acknowledgement

The authors are grateful to Dr S. Dorling and Dr. J. Keder for the useful suggestions concerning the interpretation of the daily SO_2 profiles. The paper has been financially supported by the EU under the framework of the APPETISE project (Contract N. IST-99-11764).

The authors are also grateful to the Municipal and Provincial authorities in Siracusa and Municipal authorities in Catania for providing pollution and meteorological data considered in this paper.

References

[1] Holland D. M., Principe P. P., Sickles J. E.: Trends in atmospheric sulfur and nitrogen species in the eastern United States for 1989-1995. Atmospheric Environment 33, pp. 37-49 (1999).

[2] Zannetti P.: Air Pollution Modeling, Theories, Computational Methods and Available Software Ed. Van Nostrand Reinhold, New York, 1990.

[3] Boznar M, Lesjak M., Mlakar P.: A Neural Network-Based Method for Short-Term Predictions of Ambient SO_2 Concentrations in Highly Polluted Industrial Areas of Complex Terrain. Atmospheric Environment, 27B, 2, pp. 221-230 (1993).

[4] Nunnari G., Nucifora A., Randieri C.: The Application of Neural Techniques to the Modelling of Time Series of Atmospheric Pollution Data. Ecological Modelling, 111, pp. 187-205 (1998).

[5] Arena P., Baglio S., Castorina C., Fortuna L., Nunnari G.: A Neural Architecture to Predict Pollution in Industrial Areas. Proceedings of ICNN, 4, pp. 2107-2112, Washington (1996).

[6] Van Aalst R. M., De Leeuw F. A. A. M. (editors), National Ozone Forecasting System and International Data Exchange in Northwest Europe, European Topic Centre on Air Quality, 1997.

Forecasting of Air Pollution at Unmonitored Sites

Sofia Lopes, Mahesan Niranjan, Jeremy Oakley *

*Department of Computer Science, The University of Sheffield, Sheffield, England,
S.Lopes/M.Niranjan/J.Oakley@dcs.shef.ac.uk

Abstract

We address the problem of forecasting air-pollution at a site where there is no monitoring station by constructing data-driven models. We assume synchronous measurements of pollution are available at other sites in the vicinity, and that the spatial correlation carries information relevant for prediction. A Gaussian Process type spatial model is assumed, for interpolating pollution, and the time variation of the hyperparameters of the GP model is considered. An illustration of the method on synthetic data is presented.

1 Introduction

The distribution of air pollution, both in space and time, is of interest to a number of organisations and individuals. The monitoring and prediction of the formation of so called pollution hot-spots is important in planning the layouts of cities, control of the amout of traffic and other sources of pollution as well as early warning of potential health risks. The convenional approach to modelling the distribution, and hence forecasting of pollution episodes, is via a model based approach, sometimes called deterministic modelling. The idea in deterministic modelling is to start from a description of the physics of the process by which pollution can propagate in the atmosphere. Differential equations arising from such a description can be solved by finite-element type techniques, with a whole host of assumptions on the boundary conditions involved. These include estimates and characterisations of the sources of pollution as well as assumptions to do with wind conditions approximated as laminar flows and so on.

An alternate approach we take in this paper, to dealing with characterisation and prediction of pollution episodes, is a data driven one. Forecasting at a particular site using time series prediction techniques has been addressed by a number of authors. Linear models of the Box-Jenkins type as well as nonlinear models such as neural networks have been introduced to achieve site specific forecasts in a data driven way. Such datadriven approaches usually either consider forecasting at a specific site by time series analysis, or use spatial interpolation techniques-such as kriging to interpolate pollution concentration at an unmonitored site. In this paper we report on our efforts as part of the EC funded APPETISE consortium to study the joint characterisation in space and time of air pollution. Specifically we assume that the spatial variation of pollution can be characterised by a model belonging to the family of Gaussian processes; the interpolation technique of Radial Basis Functions is similar, but the Gaussian process formulation permits a Bayesian treatment of the problem in which parameter and model uncertainties can be handled in a principled manner. In a simple form of the approach, parallel predictions are made using a dynamical systems approach at a number of sites at which pollution concentration is-made. Superposed on this is the Gaussian Process model of spatial variation which enables the interpolated prediction at an unmonitored site. A more elaborate formulation of our approach treats the underlying hyperparameters of the GP model as time varying dynamical system. We then apply the technique of particle filtering to predict from the model.

We first describe the formulation of the model and its performance on synthetic data. Further work on real data, including monitoring the concentration of CO in a narrow street canyon (Silver Street, Cambridge, UK) will be reported in a future publication.

2 Time Series Prediction and Spatial Interpolation

At a particular monitoring site it is possible to forecast future pollution concentrations by well known techniques of time series prediction. Such prediction may be based on past values of the pollution concentration as well as other related factors such as meteorological measurements and variables relating to the causes of pollution, *e.g.* the flow of traffic. The mathematical formulation of a time series predictor may either be linear (such as in ARX models) or nonlinear, the most general form of which is known as NARMAX models. Usually one estimates parameters of such models from a temporal window of data and applies the model to previously unseen data in the future.

The learning, or parameter estimation, problem in the prediction of a univariate time series is cast as an interpolation problem. The data is of the form:

$$\{ \mathbf{x}_n, \, y_n \}_{n=1}^{N} ,$$

where vector $\mathbf{x}_n$ contains values of past observations as well as other variables relevant to the prediction. We impose a parametric functional form $f(\mathbf{x}; \theta)$ as predictor and set its parameters to minimise the error:

$$E = \sum_{n=1}^{N} \{y_n - f(\mathbf{x}_n; \theta)\}^2 + g(\theta),$$

where the second term on the right hand side is a regularisation term, included to avoid powerful classes of function approximators 'overfitting' the training data. If the environment generating the time series in nonstationary, it may be more appropriate to apply sequential estimation and prediction, the linear case of which is the Kalman filter.

Simple spatial interpolation may be performed by a parametric interpolation function defined in space. An example of such an interpolation function may be the Radian Basis Function class of methods, interpolating the concentration at spatial location $\mathbf{x}$

$$g(\mathbf{x}) = \sum_{j=1}^{N} \lambda_j \, \phi(\mathbf{x} - \mathbf{x}_j)$$

where $\mathbf{x}_j$ are the locations at which measurements y_j of pollution concentration are aviailable. Learning the interpolation function is then a problem of estimating the λ_j's by minimising a regularised error function similar to the estimation of the single time series seen earlier.

There are two possible ways of combining the time series and spatial models to addressing the problem of forecasting at unmonitored sites. One possibility is to do one step ahead time series prediction at each of the monitoring stations, and then apply spatial interpolation on the predictions. The other is to perform spatial interpolation at each point in time to obtain a time series of interpolated pollution values at the unmonitored site. One could then apply some NARMAX type time series prediction on this interpolated series. Our objective in this work is to develop a method that integrates both these techniques. We believe the temporal variation of pollution (at all sites taken together) happens at a lower dimensional space than there are monitoring stations. We thus resort to the approach of Gaussian Processes, which will also give us a systematic way of quantifying and propagating the uncertainties involved, and model the time variations as dynamics in the hyper-parameters. The following sections describe early attmpts along this and presents some results on simulated data.

3 Modeling Spatio-Temporal Data

The Gaussian process approach to modeling spatio-temporal data involves describing the joint distribution of

$$\{Z(x_1, t_1), \ldots, Z(x_N, t_N)\}$$

for any set of locations in space and points in time $\{(x_1, t_1), \ldots, (x_N, t_N)\}$. The covariance between $Z(x,t)$ and $Z(x',t')$ is assumed to be given by some function $c(x, x', t, t', \theta)$ for some set of parameters θ. Examples of this are [1] and [2]. In[2], the objective was to interpolate observations in space and time over the period in which the data was collected. It is not clear that this approach is suitable for forecasting. This is because the pollutant is generally modeled as a *function* of space x and time t, and the data is used to learn about this function. If our goal is to forecast future values of the pollutant, then this involves estimating the function in a region of the input space where we have no data. We think it is preferable to first forecast $Z(x,t)$ at each monitoring site, since we are able to model $Z(x,t)$ as a function of past values $Z(x, t-1), Z(x, t-2), \ldots$ (and possibly past values of other correlated variables that have been measured) directly. We can then interpolate these forecasted values to a new location to give us a forecast at an unmonitored site.

4 Sequential Interpolation of the Forecasted Values

Given a sample from the joint distribution of $\{Z(x_1, t+1), \ldots, Z(x_n, t+1)\}$, we now have to estimate the value of $Z(x', t+1)$. It will also be important to quantify the uncertainty we have about $Z(x', t+1)$. To achieve this we suppose that the joint distribution of $\{Z(x_1, t), \ldots, Z(x_N, t)\}$ for any set of locations $x_1, \ldots, x_N$ at any time t is multivariate normal, i.e. we model $Z(., t)$ as a Gaussian process. A priori we state that

$$E\{Z(x,t)|\beta\} = \beta,$$

and

$$Cov\{Z(x,t), Z(x',t)||\sigma^2, \delta^2, b, \beta\} = \sigma^2 c(||x - x'||) + \delta^2 I_{\{x=x'\}}, \quad (1)$$

for unknown parameters σ^2, δ^2, b and β. We denote σ^2, δ^2 and b collectively by ϕ. The function $c(d)$

decreases monotonically with d and $I_{x=x'}$ is the indicator function. The parameter δ^2 represents measurement and other noise that may be independent at each location. Note that in (1) we have assumed that the covariance is stationary and isotropic, i.e. it is a function of the distance between two sites only. Though this is common practise in spatial modeling, this has been criticised in [3]. Non stationary spatial modeling is considered in [3], [4] and [5]. Implementing this in this context is an area for future research.

In (4) we have assumed that the expected value of the pollutant is constant at all locations. If we have prior knowledge about how the pollutant may vary with location, than we define the prior mean to be some function of x. In some cases, there may be a deterministic model that is considered to be a good approximation of the spatial dispersion, and we will be able to use this model to improve our interpolation. Inference using both observed data and deterministic models is described in detail in [6].

Given data $D = \{Z(x_1, t+1), \ldots, Z(x_n, t+1)\}^T$ we can now update the distribution of $Z(x', t+1)$. If we assume a flat noninformative prior for β then using properties of multivariate normal distributions (see for example [7]) it can be shown after a little algebra that

$$E\{Z(x,t+1)|D,\phi\} = \hat{\beta} + r(x)'V^{-1}(D - H\hat{\beta})$$

and

$$\begin{aligned} &Cov\{Z(x,t), Z(x',t)|D,\phi\} = \sigma^2 c(||x-x'||) \\ &+\delta^2 I_{\{x=x'\}} - r(x)'V^{-1}r(x')' \\ &+(1 - r(x)'V^{-1}H)^T(H^TV^{-1}H)^{-1} \\ &\times(1 - r(x')'V^{-1}H), \end{aligned} \quad (2)$$

where

$$\begin{aligned} r(x) &= Cov(Z(x,t+1), D|\phi), && (3)\\ V &= Var(D|\phi), && (4)\\ H &= 1, 1, \ldots, 1^T && (5)\\ \hat{\beta} &= (H^TV^{-1}H)^{-1}H^TV^{-1}D. && (6) \end{aligned}$$

Note the similarity between (4) and the estimator derived in Kriging [8]. The model we have given here is commonly used for interpolating spatial data. See for example [1] and [2].

Finally, we need to remove the conditioning on ϕ. One possibility is to use Markov Chain Monte Carlo techniques to sample from the posterior distribution of ϕ, an in [9]. However, in this application, rather than performing a one-off interpolation, we will want to do a new interpolation at every time point. Furthermore, the hyperparameters ϕ may vary over time. This can be established by taking successive batches of measurements and plotting sample correlations against distance. Consequently, we suppose that ϕ evolves over time, and we write

$$\phi_t = \phi_{t-1} + \epsilon_t, \quad (7)$$

where the ϵ_t for $t = 1, 2, \ldots$ are drawn independently from some common distribution. Markov Chain Monte Carlo methods become impractical as time increases due to the ever increasing parameter space. In [10] we consider using a sequential Monte Carlo approach based on particle filters, where the prior distribution of ϕ_t at each time t is represented by a discrete set of weighted samples. This set of samples and weights is then updated at each time point as new information arrives. References giving further details of sequential Monte Carlo methods see [11], [12] and [13].

5 Example with Synthetic Data

We first generate at random eleven locations in two dimensional space. The output data is generated from a time varying sum of radial basis functions. We then suppose that the output of the function is known at each time point at ten locations, and we forecast the output at the eleventh location one step ahead in time. The locations are plotted in figure 1

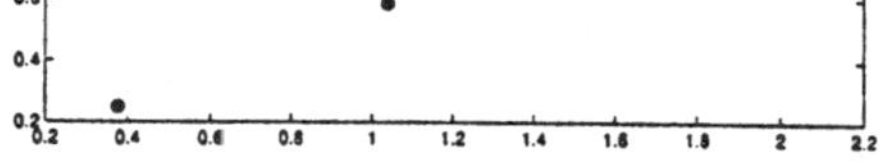

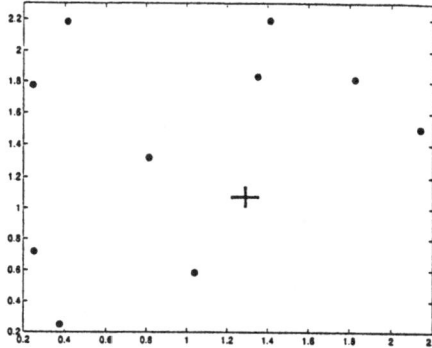

Fig. 1. Locations of the synthetic monitoring sites (•) and unmonitored site (+)

We denote the ten monitoring sites by $x_1, \ldots, x_n$ and the unmonitored site by x^*. At each time t we use the Kalman filter to make draws from the predictive distribution of $Z(x_1, t+1), \ldots, Z(x_10, t+1)$. We denote a single draw by $D_{i,t+1} = \{Z_i(x_1, t+1), \ldots, Z_i(x_10, t+1)\}$. Then we obtain a sample from the distribution of $\phi_{t+1}|D_{i,t+1}$ which we denote by $\phi_{i,t+1}$. Using $\phi_{i,t+1}$ and $D_{i,t+1}$ we sample a value of $Z(x^*, t+1)$ using the Gaussian process model. Repeating this for $i = 1, \ldots, N$, we can then give the median forecast of $Z(x^*, t+1)$ and various percentiles, so that we can give a description of the uncertainty about $Z(x^*, t+1)$. We define the forecast distribution to be the distribution of $Z(x^* t+1)$ given observations at the ten monitoring sites at time t. The results for fifty time steps are plotted in figure 2

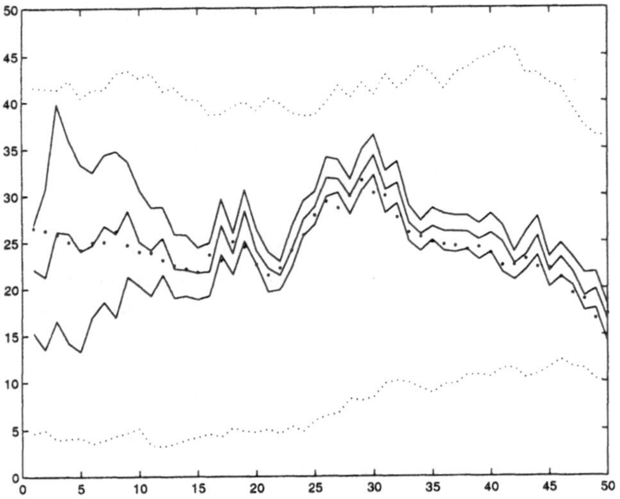

Fig. 2. True output (.), 10th, 50th and 90th percentiles from the forecast distribution (solid lines), and maximum and minimum observed values (dotted lines)

Note that the uncertainty decreases over time. This is to be expected, as we have little information about the spatial correlation initially. The Kalman filter was chosen for convenience, but the sequential interpolation method can be coupled with any time series forecasting method that can provide a sample of forecasted values at each time point.

Acknowledgements

This work was performed under the European Community funded project APPETISE IST-99-11764; `http://www.uea.ac.uk/env/appetise`

References

[1] K. V. Mardia and C. R. Goodall, "Spatio-temporal analysis of multivariate environmental monitoring data," in *Multivariate Environmental Statistics* (G. P. Patil and C. R. Rao, eds.), North-Holland, 1993.

[2] T. Follestad and G. Høst, "Spatial interpolation of ozone exposure in Norway from space-time data," in *Geostatistics for Environmental Applications* (J. Gomez-Hernandez, A. Soares, and R. Froideveaux, eds.), Kluwer Academic Publishers, 1999.

[3] P. D. Sampson and P. Guttorp, "Nonparametric estimation of nonstationary spatial covariance structure," *J. Am. Statist. Assoc*, vol. 87, pp. 108–119, 1992.

[4] D. Higdon, J. Swall, and J. Kern, "Non-stationary spatial modelling," in *Bayesian Statistics 6* (J. M. Bernardo, J. O. Berger, A. P. Dawid, and A. F. M. Smith, eds.), Oxford: University Press, 1999.

[5] A. M. Schmidt and A. O'Hagan, "Bayesian inference for nonstationary spatial covariance structure," Tech. Rep. 498/00, Department of Probability and Statistics, University of Sheffield, 2000. Submitted to *J. Am. Statist. Assoc.*

[6] M. C. Kennedy and A. O'Hagan, "Bayesian callibration of complex computer models," tech. rep., Department of Probability and Statistics, University of Sheffield, 2000. To appear in *J. Roy. Statist. Soc. Ser. B.*

[7] W. J. Krzanowski, *Principles of Multivariate Analysis, a User's Perspective.* Oxford: University Press, 1988.

[8] G. Matheron, "Principles of geostatistics," *Economic Geol.*, vol. 58, pp. 1246–1266, 1963.

[9] R. Neal, "Regression and classification using gaussian process priors," in *Bayesian Statistics 6* (J. M. Bernardo, J. O. Berger, A. P. Dawid, and A. F. M. Smith, eds.), pp. 69–95, Oxford: University Press, 1999.

[10] J. Oakley and M. Niranjan, "Sequential interpolation of pollution monitoring data with particle filters," tech. rep., Department of Computer Science, University of Sheffield, 2000.

[11] N. J. Gordon, D. J. Salmond, and A. F. M. Smith, "Novel approach to nonlinear/nostationary Gaussian Bayesian state estimation," *IEE-Proceedings-F*, vol. 140, pp. 107–113, 1993.

[12] J. S. Liu and R. Chen, "Sequential Monte Carlo methods for dynamic systems," *J. Am. Statist. Assoc.*, vol. 93, pp. 1032–1044, 1998.

[13] A. Doucet, S. J. Godsill, and C. Andrieu, "On sequential Monte Carlo sampling methods for Bayesian filtering," *Statist. Comp.*, vol. 10, pp. 197–208, 2000.

Applications of Neural Networks in Processing and Regionalization of Radiation Data

Ladislav Metelka, Stanislava Kliegrova [*]

[*]Czech Hydrometeorological Institute, Branch Hradec Králové, Dvorská 410, 503 11 Hradec Králové-Svobodné Dvory, Czech Republic, e-mail: metelka@chmi.cz,stanislava.kliegrova@chmi.cz

1 Introduction

Regionalization of the fields of climatological characteristics may be regarded as one of the most important tasks in data preparation and processing for TRY (Test Reference Years) determination. Broad spectrum of measured climatological characteristics may lead in extremely large number of their possible mutual combinations. But only some characteristics and their combinations are important for technical usage. Moreover, the optimal combination of climatological characteristics is strongly user-dependent.

Another problem is that the coverage of the area of interest by direct climatological measurements may be found as insufficient (low density of stations which measured the characteristics of interest, short period of measurement etc.). In this case estimated values of climatological characteristics are needed to cope with the problem of the lack of data.

This contribution may serve as test of application of relatively new statistical method - neural networks - both in estimating of missing values and for objective regionalization of climatological characteristics.

2 Methods

Neural networks is relatively new statistical method which is based on artificial intelligence and originally was inspired by the signal processing in biological neural systems. Artificial neural networks are built from the elements - artificial neurons - which are interconnected by synapses. Neural network may be regarded as the parallel distributed system of elementary processors (neurons) which exchange information by means of synapses so that the system works as a whole. The ability of learning and generalization may be regarded as the most important features of intelligence, artificial intelligence including. The ability of artificial neural networks to learn is ensured by presence of feedback between network error and free parameters of the network. In the iterative learning process free parameters of network are slightly changed in each step (dependently on the overall network error), to improve network performance. The ability of generalization manifests itself as the ability of reproducing general features of the relation between dependent and independent data, while the rare or ambiguous features of relation are suppressed. Neural networks are able to solve the tasks of both regression and classification (or even clustering) nature. In comparison with "classical" regression approaches (non-linear multiregression) neural networks do not need any a priori analytical description of the relation of interest, they are able to derive this relation directly from data. Moreover, neural networks are more robust in processing of ambiguous, erroneous or noisy data (which is usually the case of climatological data). On the other hand, the interpretation of signal processing inside the neural network is more difficult than the interpretation of the analytical relation obtained by methods of "classical" regression (neural network works as a "black box"). Despite this fact, neural networks are very powerful tool for climatological data processing.

The aim of this study was to regionalize the Czech Republic territory with respect to long-term characteristics of global radiation. Due to the lack of directly measured data, it was impossible to do it with the help of directly measured data only. It was necessary to evaluate a method of calculation of estimates of global radiation with the help of sunshine duration data (STEP 1). Consecutively it was necessary to simulate the geographic (on latitude and longitude) and orographic (on altitude) dependency of the estimated global radiation to get uniform coverage of the Czech Republic territory by the data (STEP 2). Finally, these data were objectively clustered to get the regions with similar global radiation characterstics (STEP 3). Neural networks were used in all 3 steps of processing.

3 Results

3.1 Step 1 – Estimating of global radiation from sunshine duration data at 19 stations with hourly sunshine duration data available

Simultaneous measurements of hourly global radiation and hourly sunshine duration were available from these stations and periods of time: Hradec Kralove (50°11'N, 15°50'E, 1964-1996), Svratouch (49°44'N, 16°02'E, 1984-1996), Kucharovice (48°53'N, 16°05'E, 1984-1996) and Usti nad Labem (50°41'N, 14°02'E, 1984-1996).

Regression task was solved with latitude (LAT), longitude (LON), altitude (ALT), hourly sunshine duration (HSD) and hourly extraterrestrial radiation (HEXT) as independent variables and hourly sum of global radiation (HGLOB) as dependent variable. Multilayer perceptron neural network was built with 5 neurons in the input layer (corresponding to LAT, LON, ALT, HSD and HEXT), 3 neurons in the hidden layer and 1 neuron in the output layer (HGLOB). This network simulates relation of HGLOB on LAT, LON, ALT, HEXT and HSD. Data from the period 1984-1996 were used for network training while data from Hradec Králové 1964-1983 were used as the independent data set for testing of network performance. Moreover, data from the period 1984-1996 were randomly divided into two equaly numerous data sets (training and verification sets) and training process was cross-validated even during network training. BEP (Back Error Propagation) training algorithm was used for network training. 12 neural networks were trained in this way, one for each month, to cope with possible annual course of the dependency of HGLOB on the individual independent variables. Then values of LAT, LON, ALT, HSD and HEXT for 19 stations were introduced into trained networks, hour by hour (19 stations with measured HSD were available). Trained networks transformed the input data into output values that can be regarded as the estimates of HGLOB.

Cross-validation and the independent testing showed that models work very well in summer period while model performance could be slightly worse in winter months. It may be caused by the fact that some influences may exist which are not included into models but may affect the global radiation especially in the winter period. Main portion of differences between measured and calculated values of HGLOB is caused by the fact that the values of HSD are not continuous but discrete (they can reach 11 discrete values only, from 0% to 100% with the step of 10%). This is the reason why these models simulate well the mean value of HGLOB but are not able to reproduce well the high-frequency variability of HGLOB. Low-frequency variability should not be affected (summarizing of these hourly values over sufficiently long time interval will suppress the inaccuracies). Possible way to cope with this problem is to add some (possibly white) noise to the network output data (estimates of HGLOB). Proper parameters of the noise (spectral properties, variance, possible autocorrelation structure) need to be specified by further investigation.

Based on this fact, the decision was made not to continue with the processing of hourly values of global radiation but with the long-term monthly averages. They were calculated for all 19 stations of interest.

3.2 Step 2 - Modelling of geographic and orographic dependency of long-term monthly averages of global radiation within the Czech Republic

Estimates of long-term monthly averages of global radiation at 19 stations were available from the step 1. Neural networks were then build, month by month again, that should reproduce the relation of these long-term monthly averages of global radiation (MGLOB) on LAT, LON and ALT (multilayer perceptron neural networks with 3 input neurons, 3 neurons in the hidden layer, 1 neuron in output layer and BEP training algorithm). Trained networks were then fed by the data from Navy Worldbath "ETOPO5" topography with grid 5 by 5 minutes in geographic coordinates. In this way the estimates of monthly long-term averages of global radiation for 1476 ETOPO5 grid points at the Czech Republic territory were calculated. Monthly averages were then summarized into seasonal and annual long-term averages of global radiation.

The long-term averages of global radiation are geographically dependent especially in autumn, winter and spring (astronomical influences in the Nort-South direction combined with increasing continentality eastward). This geographical dependency is modified by the dependency on altitude, with exception of autumn. In the summer the dependency on altitude strongly prevails. The relation between global radiation and altitude may be both positive (increasing MGLOB with increasing ALT) and negative (decreasing MGLOB with increasing ALT). Positive relation is present from October to March (winter type), negative from April to September (summer type). The winter type is affected by more frequent temperature inversions in the Bohemian basin which can reach the levels of about

600-800 MSL (meters above the sea level). Summer type is then affected by enhanced cloudiness in mountain regions.

Negligible dependency of MGLOB on ALT is indicated in autumn as a whole but it is not the case of individual months within this season (September with negative relation, October and November with positive relation – not shown). Negligible relation in autumn as a whole is caused by summarizing of the fields with opposite signs of the relation MGLOB vs. ALT.

Annual field shows weak negative dependency MGLOB vs. ALT combined with geographical dependency (see fig.1)

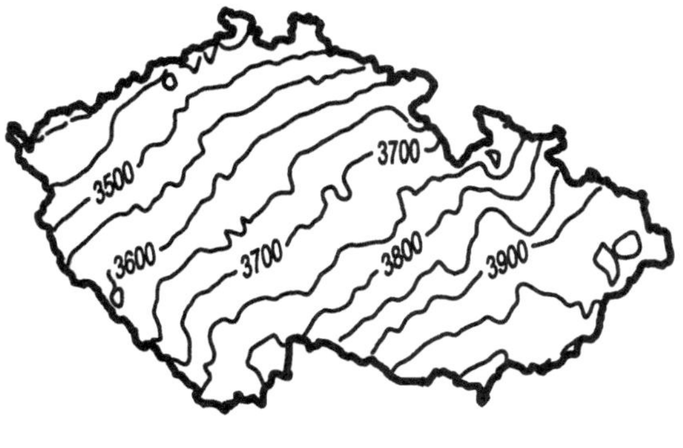

Fig.1. Annual sums of global radiation at the Czech Republic territory [MJ/m^2]

Results indicate that both relation ALT vs. MGLOB and the "strength" of geographic dependency strongly vary during the year. For this reason the annual average of global radiation by itself is not a good characteristic for regionalization. Even the localities (stations) with equal annual averages of global radiation may differ at monthly time scale and in the shape of annual course. Conclusion could be made here that the shape of annual course should be better characteristic for regionalization than the annual average of global radiation by itself. For this reason the vectors of 12 long-term monthly averages of global radiation were used for regionalization instead of one annual average only. In this case the regionalization should take into account not only the overall average value but also the shape of annual course.

3.3 Step 3 - Objective regionalization with the help of Kohonen neural network

Kohonen neural network was used in this step. Kohonen network is neural network with special architecture and training algorithm which is used for data clustering. It may be regarded as neural counterpart of cluster analysis. It groups the input data vectors into similar clusters so that the vector which belongs to some cluster is closer to the center ("centroid") of "its" cluster then to the centroids of all other clusters.

Vectors of monthly long-term averages of global radiation were clustered with the help of Kohonen network into 12 clusters. Consecutively the average vector of monthly long-term global radiation values was calculated for each cluster ("centroid"). Then the interregional variability (differences between centroids of different clusters) was compared with intraregional variability (variability of the vectors within the individual cluster). It was found that intraregional variability is significantly lower than interregional variability which indicates that clusters are well separated. Map of the regions is shown in fig.2.

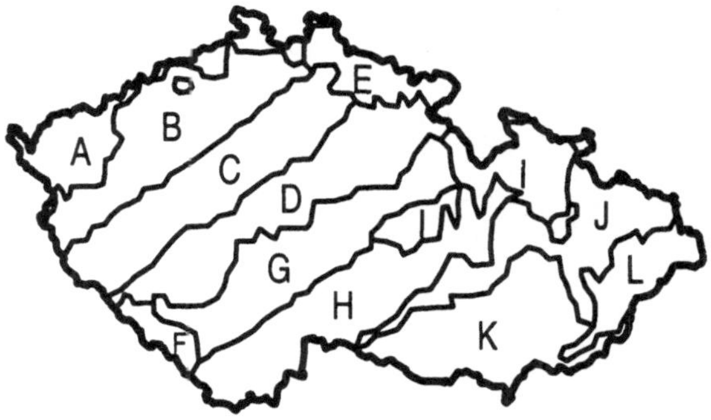

Fig.2. Regionalization of the Czech Republic territory with respect to annual course of global radiation

Strong geographic influence prevails in Bohemian lowlands but it is combined with the influence of surrounding mountains. On the other hand the geographic influence is weaker in Moravia.

4 Conclusions

Monthly and seasonal fields of global radiation reproduce well both its geographic dependency and the dependency on altitude. 12 regions were defined with respect to mean annual course of global radiation. Individual regions are separated well (intraregional variability is significantly lower then interregional variability).

Clustering to different number of clusters is possible but the number of clusters needed is strongly user-dependent. It should be stressed that the higher number of clusters, the better separation of individual regions (measured by interregional/intraregional variability relation) but the more complicated and less lucid results.

Centroids of individual clusters (regions) may be regarded as typical characteristics (values) within that cluster.

Neural networks could be used for regionalization of other climatological characteristics or for "blended" regionalization (several characteristics together). User has to decide about the weights of the individual characteristics (which are more and which are less important). This may be both user-dependent and problem-dependent.

References

StatSoft, Inc. (1999). STATISTICA for Windows [Computer program manual]. Tulsa, OK: StatSoft, Inc., 2300 East 14th Street, Tulsa, OK 74104, phone: (918) 749-1119, fax: (918) 749-2217, email: info@statsoft.com, WEB: http://www.statsoft.com

This study was supported by the Grant Agency of the Czech Republic within the frame of the research project No. GA 103/97/1199 „Reference Climatic Years for Simulation and Evaluation of Energy Demands of Buildings in the Czech Republic"

Springer-Verlag
and the Environment

WE AT SPRINGER-VERLAG FIRMLY BELIEVE THAT AN international science publisher has a special obligation to the environment, and our corporate policies consistently reflect this conviction.

WE ALSO EXPECT OUR BUSINESS PARTNERS – PRINTERS, paper mills, packaging manufacturers, etc. – to commit themselves to using environmentally friendly materials and production processes.

THE PAPER IN THIS BOOK IS MADE FROM NO-CHLORINE pulp and is acid free, in conformance with international standards for paper permanency.